Applied Statistics

in Business and Economics

2024 Release

The McGraw Hill Series in Operations and Decision Sciences

Applied Statistics

in Business and Economics

2024 Release

David P. Doane
Oakland University

Lori E. Seward
University of Colorado

APPLIED STATISTICS IN BUSINESS AND ECONOMICS, 2024 RELEASE

Published by McGraw Hill LLC, 1325 Avenue of the Americas, New York, NY 10019. Copyright ©2024 by McGraw Hill LLC. All rights reserved. Printed in the United States of America. Previous editions ©2022, 2019, and 2016. No part of this publication may be reproduced or distributed in any form or by any means, or stored in a database or retrieval system, without the prior written consent of McGraw Hill LLC, including, but not limited to, in any network or other electronic storage or transmission, or broadcast for distance learning.

Some ancillaries, including electronic and print components, may not be available to customers outside the United States.

This book is printed on acid-free paper.

1 2 3 4 5 6 7 8 9 LWI 29 28 27 26 25 24

ISBN 978-1-265-03210-4 (bound)
MHID 1-265-03210-6 (bound)
ISBN 978-1-266-27182-3 (loose-leaf)
MHID 1-266-27182-1 (loose-leaf)

Portfolio Manager: *Rebecca Olson*
Product Developer: *Christina Sanders*
Marketing Manager: *Kristen Salinas*
Content Project Managers: *Pat Frederickson and George Theofanopolous*
Manufacturing Project Manager: *Laura Fuller*
Content Licensing Specialist: *Sarah Flynn*
Cover Image: *Xuanyu Han/Moment/Getty Images*
Compositor: *Straive*

All credits appearing on page or at the end of the book are considered to be an extension of the copyright page.

Library of Congress Cataloging-in-Publication Data

Names: Doane, David P., author. | Seward, Lori Welte, 1962- author.
Title: Applied statistics in business and economics / David P. Doane,
  Oakland University, Lori E. Seward, University of Colorado.
Description: 2024 Release. | New York, NY : McGraw Hill LLC, 2024. |
  Includes index. | Audience: Ages 18+
Identifiers: LCCN 2023048503 (print) | LCCN 2023048504 (ebook) |
  ISBN 9781265032104 (paperback) | ISBN 9781266276897 (epub)
Subjects: LCSH: Commercial statistics. | Management–Statistical methods. |
  Economics–Statistical methods. | Statistics.
Classification: LCC HF1017 .D55 2024 (print) | LCC HF1017 (ebook) |
  DDC 519.5–dc23/eng/20231016
LC record available at https://lccn.loc.gov/2023048503
LC ebook record available at https://lccn.loc.gov/2023048504

mheducation.com/highered

About the Authors

Courtesy of David Doane

David P. Doane

David P. Doane is accredited by the American Statistical Association as a Professional Statistician (PStat®). He is professor emeritus in Oakland University's Department of Decision and Information Sciences. He earned his Bachelor of Arts degree in mathematics and economics at the University of Kansas and his PhD from Purdue University's Krannert Graduate School. His research and teaching interests include applied statistics, forecasting, and statistical education. He is co-recipient of three National Science Foundation grants to develop software to teach statistics and to create a computer classroom. He is a longtime member of the American Statistical Association, serving in 2002 as president of the Detroit ASA. He has consulted with governments, health care organizations, and local firms. He has published articles in many academic journals. He currently belongs to ASA chapters in San Diego and Orange County/Long Beach.

Courtesy of Lori Seward

Lori E. Seward

Lori E. Seward is a teaching professor in The Leeds School of Business at the University of Colorado in Boulder. She earned her Bachelor of Science and Master of Science degrees in Industrial Engineering at Virginia Tech. After several years working as a reliability and quality engineer in the paper and automotive industries, she earned her PhD from Virginia Tech and joined the faculty at The Leeds School in 1998. Professor Seward has served as the faculty director of Leeds' MBA programs since 2017. She currently teaches as well as coordinates the core statistics course for the Leeds full-time, Professional, and Executive MBA programs. She served as the chair of the INFORMS Teachers' Workshop for the annual 2004 meeting. Her teaching interests focus on developing pedagogy used for Leeds' various hybrid MBA programs. Leeds' recent hybrid course designs received the 2022 Brandon Hall Award for Best Use of Video for Learning. Her most recent article, co-authored with David Doane and Kevin Murphy, was published in the *Journal of Aging Research* (2022).

Dedication

To Robert Hamilton Doane-Solomon

David

To all my students who challenged me to make statistics relevant to their lives.

Lori

From the Authors

"How often have you heard people/students say about a particular subject, 'I'll never use this in the real world'? I thought statistics was a bit on the 'math-geeky' side at first. Imagine my horror when I saw α, R^2, and correlations on several financial reports at my current job (an intern position at a financial services company). I realized then that I had better try to understand some of this stuff."

—*Jill Odette (an introductory statistics student)*

As recently as a decade ago our students used to ask us, "**How** do I use statistics?" Today we more often hear, "**Why** should I use statistics?" *Applied Statistics in Business and Economics* has attempted to provide real meaning to the use of statistics in our world by using real business situations and real data and appealing to your need to know *why* rather than just *how*.

With over 50 years of teaching statistics between the two of us, we feel we have something to offer. Seeing how students have changed over the last few decades has required us to adapt and seek out better ways of instruction. So we wrote *Applied Statistics in Business and Economics* to meet four distinct objectives.

Objective 1: Communicate the Meaning of Variation in a Business Context Variation exists everywhere in the world around us. Successful businesses know how to measure variation. They also know how to tell when variation should be responded to and when it should be left alone. We'll show how businesses do this.

Objective 2: Use Real Data and Real Business Applications Examples, case studies, and problems are taken from published research or real applications whenever possible. Hypothetical data are used when it seems the best way to illustrate a concept.

Objective 3: Incorporate Current Statistical Practices and Offer Practical Advice With the increased reliance on computers, statistics practitioners have changed the way they use statistical tools. We'll show the current practices and explain why they are used the way they are. We also will tell you when each technique should *not* be used.

Objective 4: Provide More In-Depth Explanation of the Why and Let the Software Take Care of the How It is critical to understand the importance of communicating with data. Today's computer capabilities make it much easier to summarize and display data than ever before. We demonstrate easily mastered software techniques using the common software available. We also spend a great deal of time on the idea that there are risks in decision making and those risks should be quantified and directly considered in every business decision.

Our experience tells us that students want to be given credit for the experience they bring to the college classroom. We have tried to honor this by choosing examples and exercises set in situations that will draw on students' already vast knowledge of the world and knowledge gained from other classes. Emphasis is on thinking about data, choosing appropriate analytic tools, using computers effectively, and recognizing limitations of statistics.

What's New in This 2024 Release?

In this release, we have listened to you and have made many changes that you asked for. We sought advice from students and faculty who are currently using the textbook and reviewers at a variety of colleges and universities. At the end of this preface is a detailed list of chapter-by-chapter improvements, but here are just a few of them:

- Improved design and Connect connectivity for e-format users.
- Continued strong focus on Excel and business applications, including new Integrated Excel exercises and concept videos in Connect.
- New *Analytics in Action* briefings (e.g., Dark Data, Data Privacy, Sharpe Ratio, Will Big Data Kill Statistics).
- Improved Appendix J with side-by-side comparison of statistics functions in Excel and R.
- Improved Appendix K with an easy walk-through to get started with R and RStudio.

- New and updated test bank questions matched with topics and learning objectives.
- New and updated Mini Cases for economics and business.
- New and updated exercises, data sets, web links, *Big Data Sets* (e.g., NFL free agent contracts), and *Related Reading.*
- New guided examples on Connect to aid learning.
- New *LearningStats* demonstrations on using R for key topics in each chapter.

Software

Excel is used throughout this book because it is available everywhere. Some calculations are illustrated using MegaStat and Minitab because they offer more capability than Excel's Data Analysis Tools. In recognition of growing interest in analytics training beyond Excel, our textbook provides an optional introduction to R with illustrations of topics in each chapter. Our support for R is further enhanced with *LearningStats* modules, tables of R functions, and R-compatible Excel data sets. To further assist students we provide Connect tutorials or demonstrations on using Excel, Minitab, MegaStat, and R. At the end of each chapter is a list of *LearningStats* demonstrations that illustrate the concepts from the chapter.

Math Level

The assumed level of mathematics is college algebra, though there are rare references to calculus where it might help the better-trained reader. All but the simplest proofs and derivations are omitted, though key assumptions are stated clearly. The learner is advised what to do when these assumptions are not fulfilled. Worked examples are included for basic calculations, but the textbook does assume that computers will do the calculations after the statistics class is over, so *interpretation* is paramount. End-of-chapter references and suggested websites are given so that interested readers can deepen their understanding.

Exercises

Simple practice exercises are placed within each section. End-of-chapter exercises tend to be more integrative or to be embedded in more realistic contexts. Attention has been given to revising exercises so that they have clear-cut answers that are matched to specific learning objectives. A few exercises invite short answers rather than just quoting a formula. Answers to odd-numbered exercises are in the back of the book (all of the answers are in the instructor's manual).

LearningStats

Connect users can access *LearningStats,* a collection of Excel spreadsheets, Word documents, and PowerPoints for each chapter. It is intended to let students explore data and concepts at their own pace, ignoring material they already know and focusing on things that interest them. *LearningStats* includes deeper explanations on topics such as how to write effective reports, how to perform calculations, or how to make effective charts. It also includes topics that did not appear prominently in the textbook (e.g., partial *F* test, Durbin–Watson test, sign test, bootstrap simulation, and logistic regression). Instructors can use *LearningStats* PowerPoint presentations in the classroom, but Connect users also can use them for self-instruction. No instructor can "cover everything," but students can be encouraged to explore *LearningStats* data sets and/or demonstrations, perhaps with an instructor's guidance.

David P. Doane
Lori E. Seward

How Are Chapters Organized to Promote Student Learning?

Chapter Contents

Each chapter begins with a short list of section topics that are covered in the chapter.

Chapter Learning Objectives

Each chapter includes a list of learning objectives students should be able to attain upon reading and studying the chapter material. Learning objectives give students an overview of what is expected and identify the goals for learning. Learning objectives also appear next to chapter topics in the margins.

CHAPTER LEARNING OBJECTIVES

When you finish this chapter, you should be able to

LO 1-1 Define statistics and explain some of its uses.

LO 1-2 List reasons for a business student to study statistics.

LO 1-3 Explain the uses of statistics in business.

LO 1-4 State the common challenges facing business professionals using statistics.

LO 1-5 List and explain common statistical pitfalls.

Section Exercises

Multiple section exercises are found throughout the chapter so that students can focus on material just learned.

Section Exercises

connect

3.1 (a) Make a stem-and-leaf plot for these 24 observations on the number of customers who used a downtown Citibank ATM during the noon hour on 24 consecutive workdays. (b) Make a dot plot of the ATM data. (c) Describe these two displays. (*Hint:* Refer to center, variability, and shape.) Citibank

39	32	21	26	19	27	32	25
18	26	34	18	31	35	21	33
33	9	16	32	35	42	15	24

3.2 (a) Make a stem-and-leaf plot for the number of defects per 100 vehicles for these 32 brands. (b) Make a dot plot of the defects data. (c) Describe these two displays. (*Hint:* Refer to center, variability, and shape.)

Mini Cases

Every chapter includes two or three mini cases, which are solved applications. They show and illustrate the analytical application of specific statistical concepts at a deeper level than the examples.

Mini Case 4.8

What Is the DJIA? DJIA

The Dow Jones Industrial Average (commonly called the DJIA) is the oldest U.S. stock market price index, based on the prices of 30 large, widely held, and actively traded "blue chip" public companies in the United States (e.g., Coca-Cola, Microsoft, Walmart, Walt Disney). Actually, only a few of its 30 component companies are "industrial." The DJIA is not a simple average but rather a *weighted average* based on the prices of its component stocks. Originally a simple mean of stock prices, the DJIA now is calculated as the sum of the 30 stock prices divided by a "divisor" that compensates for stock splits and other changes over time. The divisor is revised as often as necessary (see *The Wall Street Journal* for the latest divisor value). Because high-priced stocks comprise a larger proportion of the sum, the DJIA is more strongly affected by changes in high-priced stocks. That is, a 10 percent price increase in a $10 stock would have less effect than a 10 percent price increase in a $50 stock, even if both companies have the same total market capitalization (the total number of shares times the price per share; often referred to as "market cap"). Broad-based market price indexes (e.g., NSDQ, AMEX, NYSE, S&P 500, Russ 2K) are widely used by fund managers, but the venerable "Dow" is still the one you see first on CNN or MSNBC.

Analytics in Action

These NEW features bring in real-world examples to illustrate data analytics in action.

Analytics in Action

Walmart, Big Data, and Retail Analytics

Walmart processes over a million customer transactions each hour, which translates into two to three petabytes of data each hour. (A petabyte is a million gigabytes!) What to do with all these data? Walmart is practicing *data democratization*. This term means making large amounts of data available to everyone in the organization so that employees can quickly react to changes in their customers' behaviors.

Walmart operates a Data Café that can be accessed by everyone in the company. When store managers notice changes in sales for particular products, they can go to the Data Café and look at data across all their stores to figure out why the changes

Figures and Tables

Throughout the text, there are hundreds of charts, graphs, tables, and spreadsheets to illustrate statistical concepts being applied. These visuals help stimulate student interest and clarify the text explanations.

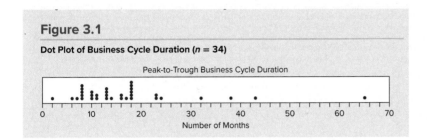

Figure 3.1

Dot Plot of Business Cycle Duration ($n = 34$)

Peak-to-Trough Business Cycle Duration

Number of Months

Table 2.6

Some Ways to Get 10 Random Integers between 1 and 875

Excel	Enter =RANDBETWEEN(1,875) into 10 spreadsheet cells. Press F9 to get a new sample.
R	Enter sample(875,10,1)
Internet	The website www.random.org will give you many kinds of random numbers (integers, decimals, etc.).
Pocket Calculator	Press the RAND key to get a random number in the interval [0, 1], multiply by 875, and then round up to the next integer.

Examples

Examples of interest to students are taken from published research or real applications to illustrate the statistics concept. For the most part, examples are focused on business, but there are also some that are more general and don't require any prerequisite knowledge. And there are some that are based on student projects.

Example 3.3

Birth Rates and Life Expectancy

Source: *The World Factbook 2003.* Central Intelligence Agency, 2003. www.cia.gov.

Figure 3.11 shows a scatter plot with life expectancy on the *X*-axis and birth rates on the *Y*-axis. In this illustration, there seems to be an association between *X* and *Y*. That is, nations with higher birth rates tend to have lower life expectancy (and vice versa). No cause-and-effect relationship is implied because, in this example, both variables could be influenced by a third variable that is not mentioned (e.g., GDP per capita).

Figure 3.11

Scatter Plot of Birth Rates and Life Expectancy ($n = 153$ nations) BirthLife

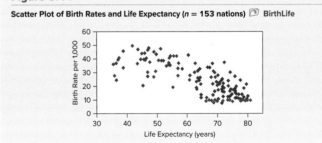

Data Set Icon

A data set icon is used throughout the text to identify data sets used in the figures, examples, and exercises that are included in Connect for the text.

 USTrade

How Does This Text Reinforce Student Learning?

Chapter Summary

Chapter summaries provide an overview of the material covered in the chapter.

Chapter Summary

The **mean** and **median** describe a sample's **center** and also indicate **skewness**. The **mode** is useful for discrete data with a small range. The **trimmed mean** eliminates extreme values. The **geometric mean** mitigates high extremes but cannot be used when zeros or negative values are present. The **midrange** is easy to calculate but is sensitive to extremes. Variability is typically measured by the **standard deviation**, while relative dispersion is given by the **coefficient of variation** for nonnegative data. **Standardized data** reveal **outliers** or unusual data values, and the **Empirical Rule** offers a comparison with a normal distribution. In measuring dispersion, the **mean absolute deviation** or **MAD** is easy to understand but lacks nice mathematical properties. **Quartiles** are meaningful even for fairly small data sets, while **percentiles** are used only for large data sets. **Box plots** show the quartiles and data range. The **correlation coefficient** measures the degree of linearity between two variables. The **covariance** measures the degree to which two variables move together. We can estimate many common descriptive statistics from **grouped data**. Sample coefficients of **skewness** and **kurtosis** allow more precise inferences about the **shape** of the population being sampled instead of relying on histograms.

Key Terms

Key terms are highlighted and defined within the text. They are also listed at the ends of chapters to aid in reviewing.

Key Terms

Center	Variability	Shape	Other
geometric mean	Chebyshev's Theorem	bimodal distribution	box plot
mean	coefficient of variation	kurtosis	covariance
median	Empirical Rule	kurtosis coefficient	five-number summary
midhinge	mean absolute deviation	leptokurtic	interquartile range
midrange	outliers	mesokurtic	method of medians
mode	population variance	multimodal distribution	quartiles
trimmed mean	range	negatively skewed	sample correlation coefficient
weighted mean	sample variance	Pearson 2 skewness coefficient	Sharpe's ratio
	standard deviation	platykurtic	
	standardized data	positively skewed	
	z-score	Schield's Rule	
		skewed left	
		skewed right	
		skewness	
		skewness coefficient	
		symmetric data	

Commonly Used Formulas

Some chapters provide a listing of commonly used formulas for the topic under discussion.

Commonly Used Formulas in Descriptive Statistics

Sample mean: $\bar{x} = \dfrac{1}{n}\sum_{i=1}^{n} x_i$

Geometric mean: $G = \sqrt[n]{x_1 x_2 \ldots x_n}$

Growth rate: $GR = \sqrt[n-1]{\dfrac{x_n}{x_1}} - 1$

Range: $\text{Range} = x_{max} - x_{min}$

Midrange: $\text{Midrange} = \dfrac{x_{max} + x_{min}}{2}$

Sample standard deviation: $s = \sqrt{\dfrac{\sum_{i=1}^{n}(x_i - \bar{x})^2}{n-1}}$

Chapter Review

Each chapter has a list of questions for student self-review or for discussion.

Chapter Review connect

1. What are descriptive statistics? How do they differ from visual displays of data?

2. Explain each concept: (a) center, (b) variability, and (c) shape.

3. (a) Why is sorting usually the first step in data analysis? (b) Why is it useful to begin a data analysis by thinking about how the data were collected?

4. List the strengths and weaknesses of each measure of center and write its Excel function: (a) mean, (b) median, and (c) mode.

5. (a) Why must the deviations around the mean sum to zero? (b) What is the position of the median in the data array when n is even? When n is odd? (c) Why is the mode of little use in continuous data? (d) For what type of data is the mode most useful?

6. (a) What is a bimodal distribution? (b) Explain two ways to detect skewness.

7. List strengths and weaknesses of each measure of center and give its Excel function (if any): (a) midrange, (b) geometric mean, and (c) 10 percent trimmed mean.

8. (a) What is variability? (b) Name five measures of variability. List the main characteristics (strengths, weaknesses) of each measure.

9. (a) Which standard deviation formula (population, sample) is used most often? Why? (b) When is the coefficient of variation useful?

10. (a) To what kind of data does Chebyshev's Theorem apply? (b) To what kind of data does the Empirical Rule apply? (c) What is an outlier? An unusual data value?

11. (a) In a normal distribution, approximately what percent of observations are within 1, 2, and 3 standard deviations of the mean? (b) In a sample of 10,000 observations, about how many observations would you expect beyond 3 standard deviations of the mean?

Chapter Exercises

Exercises give students an opportunity to test their understanding of the chapter material. Exercises are included at the ends of sections and at the ends of chapters. Some exercises contain data sets, identified by data set icons. Data sets can be accessed through Connect and used to solve problems in the text.

Chapter Exercises ■connect

4.54 (a) For each data set, calculate the mean, median, and mode. (b) Which, if any, of these three measures is the weakest indicator of a "typical" data value? Why?

   a. Number of e-mail accounts (12 students): 1, 1, 1, 1, 2, 2, 2, 3, 3, 3, 3, 3

   b. Number of siblings (5 students): 0, 1, 2, 2, 10

   c. Asset turnover ratio (8 retail firms): 1.85, 1.87, 2.02, 2.05, 2.11, 2.18, 2.29, 3.01

4.55 If the mean asset turnover for retail firms is 2.02 with a standard deviation of 0.22, without assuming a normal distribution, within what range will at least 75% of retail firms' asset turnover fall?

4.56 For each data set: (a) Find the mean, median, and mode. (b) Which, if any, of these three measures is the weakest indicator of a "typical" data value? Why?

   a. 100 m dash times ($n = 6$ top runners): 9.87, 9.98, 10.02, 10.15, 10.36, 10.36

   b. Number of children ($n = 13$ families): 0, 1, 1, 2, 2, 2, 2, 2, 2, 2, 2, 2, 6

(a) Find Bob's standardized z-score. (b) By the Empirical Rule, is Bob's SAT score unusual?

4.61 Find the data value that corresponds to each of the following z-scores.

   a. Final exam scores: Allison's z-score = 2.30, $\mu = 74$, $\sigma = 7$

   b. Weekly grocery bill: James' z-score = -1.45, $\mu = \$53$, $\sigma = \$12$

   c. Daily video game play time: Eric's z-score = -0.79, $\mu = 4.00$ hours, $\sigma = 1.15$ hours

4.62 The average time a Boulder High varsity lacrosse player plays in a game is 30 minutes with a standard deviation of 7 minutes. Nolan's playing time in last week's game against Fairview was 48 minutes. (a) Calculate the z-score for Nolan's playing time against Fairview. (b) By the Empirical Rule, was Nolan's playing time *unusual* when compared to the typical playing time?

4.63 The number of blueberries in a blueberry muffin baked by EarthHarvest Bakeries can range from 18 to 30 blueberries. (a) Use the Empirical Rule to estimate the standard deviation of the number of blueberries in a muffin. (b) What

More Learning Resources

LearningStats provides a means for Connect users to explore data and concepts at their own pace. Applications that relate to the material in the chapter are identified by topic at the end of each chapter.

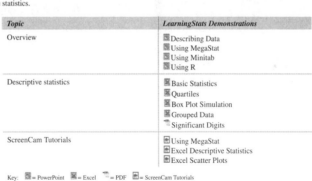

CHAPTER 4 More Learning Resources

You can access these *LearningStats* demonstrations through Connect to help you understand descriptive statistics.

Topic	LearningStats Demonstrations
Overview	Describing Data Using MegaStat Using Minitab Using R
Descriptive statistics	Basic Statistics Quartiles Box Plot Simulation Grouped Data Significant Digits
ScreenCam Tutorials	Using MegaStat Excel Descriptive Statistics Excel Scatter Plots

Key: ▣ = PowerPoint ▣ = Excel ▣ = PDF ▣ = ScreenCam Tutorials

Exam Review Questions

At the end of a group of chapters, students can review the material they covered in those chapters. This provides them with an opportunity to test themselves on their grasp of the material.

Exam Review Questions for Chapters 1–4

1. Which type of statistic (descriptive, inferential) is each of the following?

   a. Estimating the default rate on all U.S. mortgages from a random sample of 500 loans.

   b. Reporting the percentage of students in your statistics class who use Verizon.

   c. Using a sample of 50 iPhones to predict the average battery life in typical usage.

2. Which is *not* an ethical obligation of a statistician? Explain.

   a. To know and follow accepted procedures.

   b. To ensure data integrity and accurate calculations.

   c. To support the client's wishes in concluding from the data.

3. "Driving without a seat belt is not risky. I've done it for 25 years without an accident." This *best* illustrates which fallacy?

   a. Unconscious bias.

   b. Conclusion from a small sample.

4. Which data type (categorical, numerical) is each of the following?

   a. Your current credit card balance.

   b. Your college major.

   c. Your car's odometer mileage reading today.

5. Give the type of measurement (nominal, ordinal, interval, ratio) for each variable.

   a. Length of time required for a randomly chosen vehicle to cross a toll bridge.

   b. Student's ranking of five cell phone service providers.

   c. The type of charge card used by a customer (Visa, Mastercard, AmEx, Other).

6. Tell if each variable is continuous or discrete.

   a. Tonnage carried by an oil tanker at sea.

   b. Wind velocity at 7 o'clock this morning.

   c. Number of text messages you received yesterday.

7. To choose a sample of 12 students from a statistics class of 36 students, which type of sample (simple random, systematic,

A complete course platform

Connect enables you to build deeper connections with your students through cohesive digital content and tools, creating engaging learning experiences. We are committed to providing you with the right resources and tools to support all your students along their personal learning journeys.

65%
Less Time Grading

Laptop: Getty Images; Woman/dog: George Doyle/Getty Images

Every learner is unique

In Connect, instructors can assign an adaptive reading experience with SmartBook® 2.0. Rooted in advanced learning science principles, SmartBook 2.0 delivers each student a personalized experience, focusing students on their learning gaps, ensuring that the time they spend studying is time well spent.
mheducation.com/highered/connect/smartbook

Study anytime, anywhere

Encourage your students to download the free ReadAnywhere® app so they can access their online eBook, SmartBook® 2.0, or Adaptive Learning Assignments when it's convenient, even when they're offline. And since the app automatically syncs with their Connect account, all of their work is available every time they open it. Find out more at **mheducation.com/readanywhere**

"I really liked this app—it made it easy to study when you don't have your textbook in front of you."

Jordan Cunningham, a student at *Eastern Washington University*

Effective tools for efficient studying

Connect is designed to help students be more productive with simple, flexible, intuitive tools that maximize study time and meet students' individual learning needs. Get learning that works for everyone with Connect.

Education for all

McGraw Hill works directly with Accessibility Services departments and faculty to meet the learning needs of all students. Please contact your Accessibility Services Office, and ask them to email **accessibility@mheducation.com**, or visit **mheducation.com/about/accessibility** for more information.

Affordable solutions, added value

Make technology work for you with LMS integration for single sign-on access, mobile access to the digital textbook, and reports to quickly show you how each of your students is doing. And with our Inclusive Access program, you can provide all these tools at the lowest available market price to your students. Ask your McGraw Hill representative for more information.

Solutions for your challenges

A product isn't a solution. Real solutions are affordable, reliable, and come with training and ongoing support when you need it and how you want it. Visit **supportateverystep.com** for videos and resources both you and your students can use throughout the term.

Updated and relevant content

Our new Evergreen delivery model provides the most current and relevant content for your course, hassle-free. Content, tools, and technology updates are delivered directly to your existing McGraw Hill Connect® course. Engage students and freshen up assignments with up-to-date coverage of select topics and assessments, all without having to switch editions or build a new course.

Additional Connect Features

Excel Data Sets A convenient feature is the inclusion of an Excel data file link in many problems using data files in their calculation. The link allows students to easily launch into Excel, work the problem, and return to Connect to key in the answer.

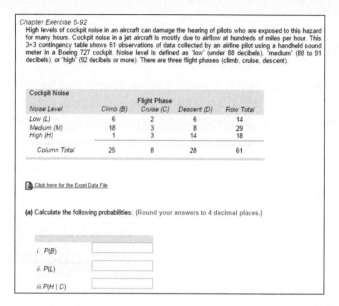

Guided Examples These narrated video walkthroughs provide students with step-by-step guidelines for solving selected exercises similar to those contained in the text. The student is given personalized instruction on how to solve a problem by applying the concepts presented in the chapter. The narrated voiceover shows the steps to take to work through an exercise. Students can go through each example multiple times if needed.

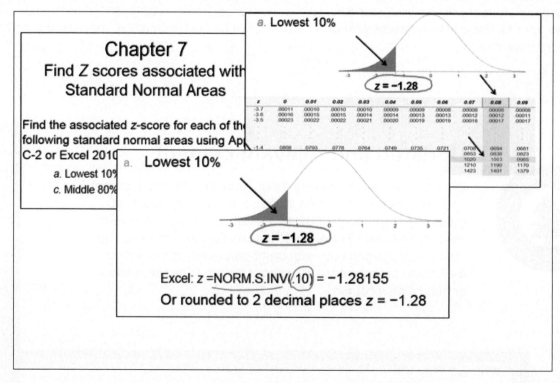

What Resources Are Available for Students?

The following software tools are available to assist students in understanding concepts and solving problems.

LearningStats

LearningStats allows students to explore data and concepts at their own pace. It includes demonstrations, simulations, and tutorials that can be downloaded from Connect.

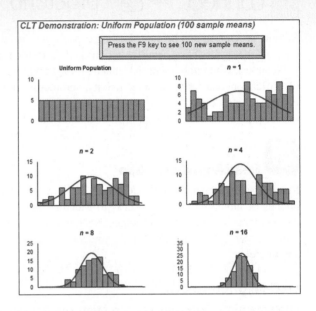

MegaStat® for Excel®

Access Card (ISBN: 0077425995) or online purchase at www.mhhe.com/megastat.

MegaStat is a full-featured Excel add-in that is available with this text. It performs statistical analyses within an Excel workbook. It does basic functions such as descriptive statistics, frequency distributions, and probability calculations as well as hypothesis testing, ANOVA, and regression.

MegaStat output is carefully formatted, and ease-of-use features include Auto Expand for quick data selection and Auto Label detect. Because *MegaStat* is easy to use, students can focus on learning statistics without being distracted by the software. *MegaStat* is always available from Excel's main menu. Selecting a menu item pops up a dialog box. *MegaStat* is updated continuously to work with the latest versions of Excel for Windows and Macintosh users.

Minitab® and Minitab Express®

Free trials and academic versions are available from Minitab at **http://www.minitab.com/en-us/academic/.**

R and RStudio

R and RStudio provide a sophisticated programming language for statistical computing and graphics plus an integrated development environment. This textbook offers detailed instructions for downloading, installing, and using free versions of R (**https://www.r-project.org/**) and RStudio (**https://rstudio.com/**).

Proctorio

Remote Proctoring & Browser-Locking Capabilities

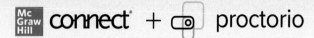

Remote proctoring and browser-locking capabilities, hosted by Proctorio within Connect, provide control of the assessment environment by enabling security options and verifying the identity of the student.

Seamlessly integrated within Connect, these services allow instructors to control the assessment experience by verifying identification, restricting browser activity, and monitoring student actions.

Instant and detailed reporting gives instructors an at-a-glance view of potential academic integrity concerns, thereby avoiding personal bias and supporting evidence-based claims.

 **ReadAnywhere® App**

Read or study when it's convenient with McGraw Hill's free ReadAnywhere® app. Available for iOS and Android smartphones or tablets, give users access to McGraw Hill tools including the eBook and SmartBook® or Adaptive Learning Assignments in McGraw Hill Connect®. Students can take notes, highlight, and complete assignments offline–all their work will sync when connected to Wi-Fi. Students log in with their Connect username and password to start learning–anytime, anywhere!

OLC-Aligned Courses

Implementing High-Quality Instruction and Assessment through Preconfigured Courseware

In consultation with the Online Learning Consortium (OLC) and our certified Faculty Consultants, McGraw Hill has created preconfigured courseware using OLC's quality scorecard to align with best practices in online course delivery. This turnkey courseware contains a combination of formative assessments, summative assessments, homework, and application activities, and can easily be customized to meet an individual instructor's needs and desired course outcomes. For more information, visit https://www.mheducation.com/highered/olc.

Test Builder in Connect

Available within McGraw Hill Connect, Test Builder is a cloud-based tool that enables instructors to format tests that can be printed, administered within a Learning Management System, or exported as a Word document. Test Builder offers a modern, streamlined interface for easy content configuration that matches course needs, without requiring a download.

Test Builder allows you to:

- access all test bank content from a particular title.
- easily pinpoint the most relevant content through robust filtering options.
- manipulate the order of questions or scramble questions and/or answers.
- pin questions to a specific location within a test.
- determine your preferred treatment of algorithmic questions.
- choose the layout and spacing.
- add instructions and configure default settings.

Test Builder provides a secure interface for better protection of content and allows for just-in-time updates to flow directly into assessments.

Writing Assignment

Available within McGraw Hill Connect, the Writing Assignment tool delivers a learning experience to help students improve written communication skills and conceptual understanding. Assign, monitor, grade, and provide feedback on writing more efficiently and effectively.

Polling

Every learner has unique needs. Uncover where and when you're needed with the new Polling tool in McGraw Hill Connect! Polling allows you to discover where students are in real time. Engage students and help them create connections with your course content while gaining valuable insight during lectures. Leverage polling data to deliver personalized instruction when and where it is needed most.

Evergreen

Content and technology are ever-changing, and it is important that you can keep your course up to date with the latest information and assessments. That's why we want to deliver the most current and relevant content for your course, hassle-free.

Doane and Seward *Applied Statistics in Business and Economics* is moving to an Evergreen delivery model, which means it has content, tools, and technology that is updated and relevant, with updates delivered directly to your existing McGraw Hill Connect course. Engage students and freshen up assignments with up-to-date coverage of select topics and assessments, all without having to switch editions or build a new course.

Create

Your Book, Your Way

McGraw Hill's Content Collections Powered by Create® is a self-service website that enables instructors to create custom course materials—print and eBooks—by drawing upon McGraw Hill's comprehensive, cross-disciplinary content. Choose what you want from our high-quality textbooks, digital products, articles, cases, and more. Combine it with your own content quickly and easily, and tap into other rights-secured, third-party content such as cases, articles, readings, cartoons, and labs. Content can be arranged in a way that makes the most sense for your course, and you can select your own cover and include the course name and school information as well. Choose the best format for your course: color print, black-and-white print, or eBook. The eBook can be included in your Connect course and is available on the free ReadAnywhere app for smartphone or tablet access as well. When you are finished customizing, you will receive a free digital copy to review in just minutes! Visit McGraw Hill Create—www.mcgrawhillcreate.com—today and begin building!

Reflecting the Diverse World Around Us

McGraw Hill believes in unlocking the potential of every learner at every stage of life. To accomplish that, we are dedicated to creating products that reflect, and are accessible to, all the diverse, global customers we serve. Within McGraw Hill, we foster a culture of belonging, and we work with partners who share our commitment to equity, inclusion, and diversity in all forms. In McGraw Hill Higher Education, this includes, but is not limited to, the following:

- Refreshing and implementing inclusive content guidelines around topics including generalizations and stereotypes, gender, abilities/disabilities, race/ethnicity, sexual orientation, diversity of names, and age.
- Enhancing best practices in assessment creation to eliminate cultural, cognitive, and affective bias.
- Maintaining and continually updating a robust photo library of diverse images that reflect our student populations.
- Including more diverse voices in the development and review of our content.
- Strengthening art guidelines to improve accessibility by ensuring meaningful text and images are distinguishable and perceivable by users with limited color vision and moderately low vision.

Integrated Excel NEW!

Integrated Excel assignments pair the power of Microsoft Excel with the power of Connect. A seamless integration of Excel within Connect, Integrated Excel questions allow students to work in live, auto-graded Excel spreadsheets—no additional logins, no need to upload or download files. Instructors can choose to grade by formula or solution value, and students receive instant cell-level feedback via integrated Check My Work functionality.

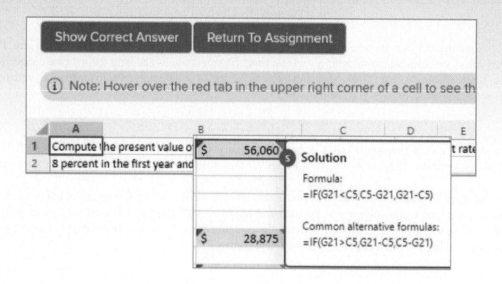

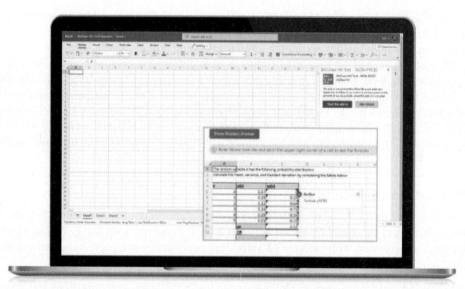

Microsoft Excel

Acknowledgments

The authors would like to acknowledge some of the many people who have helped with this book. T. J. Wharton, Mark Isken, Ron Tracy, and Robert Kushler gave generously of their time as expert statistical consultants. Jonathan G. Koomey of E.O. Lawrence Berkeley National Laboratory offered valuable suggestions on visual data presentation.

We are grateful to Farrukh Abbas and Issariya Sirichakwal for their careful scrutiny of the text and for offering ideas on improving the text and exercises. Mark Isken has reliably provided Excel expertise and has suggested health care applications for examples and case studies. J.D. Power and Associates generously provided permission to use vehicle quality data. The Public Interest Research Group in Michigan (PIRGIM) has generously shared data from its field survey of prescription drug prices.

Phil Rogers has offered numerous suggestions for improvement in both the textbook exercises and Connect. Milo A. Schield shared his research on "quick rules" for measuring skewness from summarized data. We owe special thanks to Aaron Kennedy and Dave Boennighausen of Noodles & Company; to Mark Gasta, Anja Wallace, and Clifton Pacaro of Vail Resorts; to Jim Curtin and Gordon Backman of Ball Corporation; and to Santosh Lakhan from The Verdeo Group for providing suggestions and access to data for Mini Cases and examples. For reviewing the material on quality, we wish to thank Kay Beauregard, administrative director at William Beaumont Hospital, and Ellen Barnes and Karry Roberts of Ford Motor Company. Amy Sheikh provided a new Facebook Friends data set, along with other excellent suggestions and reports from the "front lines" of her classes. Issariya Sirichakwal updated the test bank and added new test bank questions for each chapter, as well as the accompanying chapter PowerPoints.

A special debt of gratitude is due to Eric Weber, Kristen Salinas, Pat Frederickson, and Ryan McAndrews for their direction and support and George Theofanopolous for managing the Connect pieces of the project. Thanks to the many reviewers who provided such valuable feedback including criticism that made the book better, some of whom reviewed several previous editions of the text. Any remaining errors or omissions are the authors' responsibility. Thanks too, to the participants in our focus groups and symposia on teaching business statistics, who have provided teaching ideas and insights from their experiences with students in diverse contexts. We hope you will be able to see in our book and the teaching package consideration of those ideas and insights.

Farrukh Abbas, *National University of Modern Languages (NUML), Islamabad, Pakistan*

Heather Adams, *University of Colorado—Boulder*

Sung Ahn, *Washington State University*

Mostafa Aminzadeh, *Towson University*

Scott Bailey, *Troy University*

Hope Baker, *Kennesaw State University*

Saad Taha Bakir, *Alabama State University*

Steven Bednar, *Elon University*

Adam Bohr, *University of Colorado—Boulder*

Katherine Broneck, *Pima Community College—Downtown*

Mary Beth Camp, *Indiana University—Bloomington*

Alan Cannon, *University of Texas—Arlington*

Deborah Carter, *Coahoma Community College*

Kevin Caskey, *SUNY—New Paltz*

Michael Cervetti, *University of Memphis*

Paven Chennamaneni, *University of Wisconsin—Whitewater*

Alan Chesen, *Wright State University*

Wen-Chyuan Chiang, *University of Tulsa*

Chia-Shin Chung, *Cleveland State University*

Joseph Coleman, *Wright State University—Dayton*

Robert Cutshall, *Texas A&M University—Corpus Christi*

Terry Dalton, *University of Denver*

Douglas Dotterweich, *East Tennessee State University*

Jerry Dunn, *Southwestern Oklahoma State University*

Michael Easley, *University of New Orleans*

Jerry Engeholm, *University of South Carolina—Aiken*

Mark Farber, *University of Miami*

Soheila Kahkashan Fardanesh, *Towson University*

Mark Ferris, *St. Louis University*

Stergios Fotopoulos, *Washington State University*

Vickie Fry, *Westmoreland County Community College*

Joseph Fuhr, *Widener University*

Bob Gillette, *University of Kentucky*

Malcolm Gold, *Avila University*

Don Gren, *Salt Lake City Community College*

Karina Hauser, *University of Colorado—Boulder*

Clifford Hawley, *West Virginia University*

Yijun He, *Washington State University*

Natalie Hegwood, *Sam Houston State University*

Eric Hernandez, *Miami Dade College*

Allen Humbolt, *University of Tulsa*

Patricia Igo, *Northeastern University*

Alam M. Imam, *University of Northern Iowa*

Marc Isaacson, *Augsburg College*

Kishen Iyengar, *University of Colorado—Boulder*

Christopher Johnson, *University of North Florida*

Jennifer Johnson, *San Jose State University*

Ronald Johnson, *Central Alabama Community College*

Linda Jones, *Maryville University*

Jerzy Kamburowski, *University of Toledo*

Mohammad Kazemi, *University of North Carolina—Charlotte*

Bob Kitahara, *Troy University*

Drew Koch, *James Madison University*

Agnieszka Kwapisz, *Montana State University*

Kenneth Lawrence, *New Jersey Institute of Technology*

Bob Lynch, *University of Northern Colorado*

Bradley McDonald, *Northern Illinois University*

Richard McGowan, *Boston College*

Kelly McKillop, *University of Massachusetts*

Larry McRae, *Appalachian State University*

Robert Mee, *University of Tennessee—Knoxville*

John Miller, *Sam Houston State University*

Shelly Moore, *College of Western Idaho*

James E. Moran Jr., *Oregon State University*

Geraldine Moultine, *Northwood University*

Gourab Mukherjee, *University of Southern California*

Adam Munson, *University of Florida*

Joshua Naranjo, *Western Michigan University*

Anthony Narsing, *Macon State College*

Robert Nauss, *University of Missouri–St. Louis*

Pin Ng, *Northern Arizona University*

Thomas Obremski, *University of Denver*

Don Oest, *University of Colorado*

Grace Onodipe, *Georgia Gwinnett College*

Mohammad Reza Oskoorouchi, *California State University—San Marcos*

Ceyhun Ozgur, *Valparaiso University*

Ed Pappanastos, *Troy University*

Nitin Paranjpe, *Oakland University*

Mahour Mellat Parast, *University of North Carolina—Pembroke*

Eddy Patuwo, *Kent State University*

John Pickett, *University of Arkansas—Little Rock*

James Pokorski, *Virginia Polytechnic Institute & State University*

Stephan Pollard, *California State University—Los Angeles*

Claudia Pragman, *Minnesota State University*

Tammy Prater, *Alabama State University*

Michael Racer, *University of Memphis*

Azar Raiszadeh, *Chattanooga State Community College*

Phil Rogers, *University of Southern California*

Milo A. Schield, *Augsburg College*

Sue Schou, *Idaho State University*

Elizabeth Scofidio, *Colorado State University*

Sankara N. Sethuraman, *Augusta State University*

Don Sexton, *Columbia University*

Thomas R. Sexton, *Stony Brook University*

Murali Shanker, *Kent State University*

Issariya Sirichakwal, *University of Washington*

Gary W. Smith, *Florida State University*

Courtenay Stone, *Ball State University*

Paul Swanson, *Illinois Central College*

Bedassa Tadesse, *University of Minnesota—Duluth*

Rahmat Ola Tavallali, *Walsh University*

Deborah Tesch, *Xavier University*

Dharma S. Thiruvaiyaru, *Augusta State University*

Michael Urizzo, *New Jersey City University*

Jesus M. Valencia, *Slippery Rock University*

Bhavneet Walia, *Western Illinois University*

Rachel Webb, *Portland State University*

Simone A. Wegge, *City University of New York*

Chao Wen, *Eastern Illinois University*

Alan Wheeler, *University of Missouri—St. Louis*

Blake Whitten, *University of Iowa*

Charles Wilf, *Duquesne University*

Anne Williams, *Gateway Community College*

Janet Wolcutt, *Wichita State University*

Frank Tian Xie, *University of South Carolina—Aiken*

Mustafa R. Yilmaz, *Northeastern University*

Ye Zhang, *Indiana University–Purdue University Indianapolis*

Enhancements for Doane/Seward ASBE 2024 Release

Changes were motivated by advice from reviewers and users of the textbook. Besides hundreds of small edits and improved topic organization, these changes were common to most chapters:

- Expanded and updated test bank questions.
- *LearningStats* walk-throughs for R coding for key topics beyond existing end-of-chapter *Software R Supplements*.
- Updated *Appendix J* and *K* comparing R and Excel functions.
- Updated exercises, *Big Data Sets* (e.g., NFL draft contracts).

- Updated *Related Readings* and *Web Sources* for students who want to "dive deeper."
- New and enhanced *LearningStats* demonstrations to illustrate concepts beyond what is possible in a textbook.
- New *Guided Examples* with step-by-step explanations.

Chapter 1—Overview of Statistics
New *Analytics in Action* (Dark Data).

Updated section on Critical Thinking.

New Apple financial performance example.

Updated exercises and *Related Readings*.

Chapter 2—Data Collection
New *Analytics in Action* (Data Privacy).

Leaner and updated discussion of sampling concepts, sampling methods, surveys, question scaling, and data sources.

Updated exercises and *Related Readings*.

Chapter 3—Describing Data Visually
Reorganized presentation of chart types based on actual instructor usage.

Updated examples (e.g., U.S. business cycles, U.S. recessions).

Updated Excel histogram and scatter plot presentations.

Leaner discussion of line charts, pie charts, and deceptive charts.

Updated *Related Readings* for recent views.

Replaced, deleted, or updated 12 exercises.

Chapter 4—Descriptive Statistics
Updated J.D. Power data on defect rates by vehicle brand to illustrate key topics.

Leaner treatment of some topics (e.g., dot plots, boxplots, fences) based on instructor feedback.

New or updated examples and Mini Cases (Presidents' ages, duration of bear markets, vehicle defect rates over time, stock prices for energy and info sectors).

New *Analytics in Action* (Sharpe's ratio).

Updated *Related Readings*.

Chapter 5—Probability
Twelve new or revised exercises.

Updated example (videos on demand).

Two new *LearningStats* demonstrations on probability using Excel and R.

Chapter 6—Discrete Probability Distributions
Several revised exercises.

Improved discussion of Poisson distribution and variable transformations.

Two new *LearningStats* demonstrations: discrete distributions using Excel and R.

Chapter 7—Continuous Probability Distributions
New *Analytics in Action* (Distributions and Simulation).

Two new *LearningStats* demonstrations: continuous distributions using Excel and R.

Chapter 8—Sampling Distributions and Estimation
Reorganized and renumbered sections to explain sampling distributions before showing their role in interval estimation.

Improved discussion of estimators, sampling error, confidence intervals, and the finite population correction.

More examples using Excel for *t*-tests and confidence intervals.

Six updated exercises and one example (OpenTable).

Two new *LearningStats* demonstrations: estimation using Excel and R.

Chapter 9—One-Sample Hypothesis Tests
More on hypothesis formulation and testing.

New *Analytics in Action* (Will Big Data Kill Statistics).

New proportion example (airline bag loss).

Six new or revised exercises.

One new *LearningStats* demonstration: using R for one-sample tests.

Updated *Related Readings* ("Moving to a World Beyond $p < .05$").

Chapter 10—Two-Sample Hypothesis Tests

New Mini Case (Price-Earnings Ratios).

Five new or updated exercises.

Two new *LearningStats* demonstrations: using Excel and R for two-sample tests.

Chapter 11—Analysis of Variance

Revised Tukey test statistic table using R.

One new *LearningStats* demonstration: using R for ANOVA tests.

Chapter 12—Simple Regression

Revised examples (golf shoe sales, median home prices, stock prices)

Two new *LearningStats* demonstrations: simple regression using Excel and R.

Chapter 13—Multiple Regression

New examples (insulin monitoring, NFL free agent salaries).

More explanation of how to interpret estimated regression coefficients (home prices).

New Data Set L (NFL free agent contracts).

One new *LearningStats* demonstration: using R for multiple regression.

Chapter 14—Time-Series Analysis

Deleted tables of R^2 calculations and forecasts that are usually done in Excel.

Changed notation for trend models to match regression chapters and Excel more closely.

Updated several examples and outputs.

Updated exercises (presidential voting, law enforcement officer deaths, lightning deaths, snowboarder visits).

Updated *Related Readings*.

One new *LearningStats* demonstration: using R for time-series trend analysis.

Chapter 15—Chi-Square Tests

Updated examples (private pilot accidents).

Leaner treatment of Poisson GOF test.

New or updated exercises (running shoe purchases, U.S. smartphone sales, retail discounts, Boston Bruins hockey, Kentucky Derby, President ages).

One new *LearningStats* demonstration: using R for chi-square independence tests.

Chapter 16—Nonparametric Tests

Comparison of R functions and Excel for common non-parametric tests.

One new *LearningStats* demonstration: using R for non-parametric tests.

Chapter 17—Quality Management

No significant changes.

Chapter 18—Simulation

No significant changes.

Brief Contents

Contents

1 Overview of Statistics

CHAPTER CONTENTS

CHAPTER LEARNING OBJECTIVES

When you finish this chapter, you should be able to

LO 1-1 Define statistics and explain some of its uses.

LO 1-2 List reasons for a business student to study statistics.

LO 1-3 Explain the uses of statistics in business.

LO 1-4 State the common challenges facing business professionals using statistics.

LO 1-5 List and explain common statistical pitfalls.

hen managers are well informed about a company's internal operations (e.g., sales, production, inventory levels, time to market, warranty claims) and competitive position (e.g., market share, customer satisfaction, repeat sales), they can take appropriate actions to improve their business. Managers need reliable, timely information so they can analyze market trends and adjust to changing market conditions. Better data also can help a company decide which types of strategic information it should share with trusted business partners to improve its supply chain. *Statistics* and *statistical analysis* permit *data-based decision making* and reduce managers' need to rely on guesswork.

Statistics is a key component of the field of *business intelligence*, which encompasses all the technologies for collecting, storing, accessing, and analyzing data on the company's operations in order to make better business decisions. Statistics helps convert unstructured "raw" data (e.g., point-of-sale data, customer spending patterns) into *useful information* through online analytical processing (OLAP) and data mining, terms that you may have encountered in your other business classes. Statistical analysis focuses attention on key problems and guides discussion toward issues, not personalities or territorial struggles. While powerful database software and query systems are the key to managing a firm's data warehouse, relatively small Excel spreadsheets are often the focus of discussion among managers when it comes to "bottom line" decisions. That is why Excel is featured prominently in this textbook.

In short, companies increasingly are using **business analytics** to support decision making, to recognize anomalies that require tactical action, or to gain strategic insight to align business processes with business objectives. Answers to questions such as "How likely is this event?" or "What if this trend continues?" will lead to appropriate actions. Businesses that combine managerial judgment with statistical analysis are more successful.

1.1 WHAT IS STATISTICS?

LO 1-1

Define statistics and explain some of its uses.

Statistics is the science of collecting, organizing, analyzing, interpreting, and presenting data. Some experts prefer to call statistics **data science**, a trilogy of tasks involving data modeling, analysis, and decision making. A **statistic** is a single measure, reported as a number, used to summarize a sample data set. Statistics may be thought of as a collection of methodologies to summarize, draw valid conclusions, and make predictions from empirical measurements. Statistics helps us organize and present information and extract meaning from raw data. Although it is often associated with the sciences and medicine, statistics is now used in every academic field and every area of business.

Plural or Singular?

Statistics	The science of collecting, organizing, analyzing, interpreting, and presenting data.
Statistic	A single measure, reported as a number, used to summarize a sample data set.

Many different measures can be used to summarize data sets. You will learn throughout this textbook that there can be different measures for different sets of data and different measures for different types of questions about the same data set. Consider, for example, a sample data set that consists of heights of students in a university. There could be many different uses for this data set. Perhaps the manufacturer of graduation gowns wants to know how long to make the gowns; the best *statistic* for this would be the *average* height of the students. But an architect designing a classroom building would want to know how high the doorways should be and would base measurements on the *maximum* height of the students. Both the average and the maximum are examples of a *statistic*.

You may not have a trained statistician in your organization, but any college graduate is expected to know something about statistics, and anyone who creates graphs or interprets data is "doing statistics" without an official title.

There are two primary kinds of statistics:

- **Descriptive statistics** refers to the collection, organization, presentation, and summary of data (either using charts and graphs or using a numerical summary).

- **Inferential statistics** refers to generalizing from a sample to a population, estimating unknown population parameters, drawing conclusions, and making decisions.

Figure 1.1 identifies the tasks and the text chapters for each.

Figure 1.1

Overview of Statistics

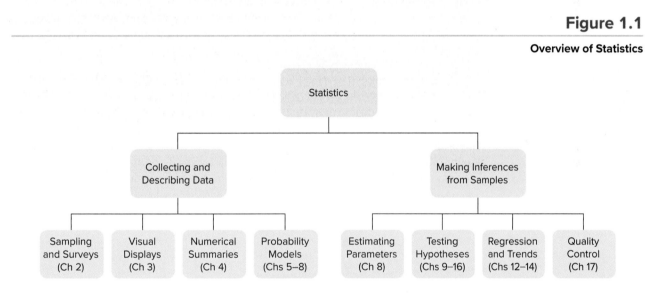

What Is Business Analytics?

Analytics is a broad field that uses statistics, mathematics, and computational tools to extract information from data. Analytics tools fall into three categories: *descriptive, predictive,* and *prescriptive.* This terminology derived from the questions we are trying to answer.

What happened? Businesses use *descriptive analytics* tools to analyze historical data and help them identify trends and patterns.

What is likely to happen next? Businesses use *predictive analytics* tools to predict probabilities of future events and help them forecast consumer behavior.

What actions do we take to achieve our goals? Businesses use *prescriptive analytics* tools to help them make decisions on how to achieve objectives within real-world constraints.

Analytics in Action

Using Analytics to Improve Business

Descriptive Analytics: Google Analytics provides a wide range of metrics for companies that want to better understand how their websites operate. This type of *descriptive analytics* allows companies to see, for example, loading time for their pages in different browsers and on different devices; how much time a viewer spends on the page; and the number of visits per hour, day, or week. In addition, companies can see performance measures such as bounce rate or click-through rate. Bounce rate shows the percentage of viewers who leave the site after viewing only the homepage. Click-through rate is the percentage of viewers who click on the ad when they view it on a webpage. Both of these metrics are important for evaluating the effectiveness of a web design.

Predictive Analytics: In 2015, Oracle acquired Datalogix (now called Oracle Advertising), a company that analyzes consumer purchasing behavior to help its customers design the most effective digital marketing campaigns. The acquisition of Datalogix strengthened the tools Oracle provides as part of its data cloud offerings. *Predictive analytics* has been the key to Datalogix's success. Dr Pepper is one of its happy customers. When Dr Pepper wanted to increase sales, it turned to Datalogix for help in designing a successful Facebook advertising campaign. Through the use of analytics, Datalogix showed that by targeting Facebook users who purchased carbonated beverages and those who purchased Dr Pepper products in particular, Dr Pepper was able to increase its sales by 1.5 percent. And for every $1 spent on advertising, Dr Pepper saw a $3 increase in revenue.

Prescriptive Analytics: Hotel rooms and airplane seats are perishable, just like apples and lettuce. When a room or seat is not booked, the company has lost the ability to earn revenue for that night or flight. Setting the right price for the right customer on the right day is complicated. Hotels and airlines feed large data sets into mathematical algorithms that choose the optimal pricing scheme to maximize the revenue for that room or seat. This practice is called revenue management and is an example of *prescriptive analytics*. When large data sets are analyzed along with system constraints, companies can quickly simulate many different outcomes related to a wide variety of decisions. This allows the decision makers to choose a set of actions that will result in the best outcome for the company and its customers.

1.2 WHY STUDY STATISTICS?

LO 1-2

List reasons for a business student to study statistics.

A *Bloomberg Businessweek* article called statistics and probability "core skills for business people" in order to know when others are dissembling, to build financial models, or to develop a marketing plan. The 2019 Job Outlook survey conducted by the National Association of Colleges and Employers (NACE) found that 72 percent of employers ranked "analytical and quantitative skills" as one of the most important attributes for a new hire. When asked to rank the attributes most sought after, employers said that the top seven attributes were (1) communication skills (written), (2) problem-solving skills, (3) ability to work on a team, (4) initiative, (5) analytical/quantitative skills, (6) strong work ethic, and (7) communication skills (verbal). Will these skills earn you a higher salary? According to the Bureau of Labor Statistics, the median salary for a statistician in 2021 was $96,280 with job growth projected to be 30 percent over the next 10 years. (See www.bls.gov/ooh/math/mathematicians-and-statisticians.htm.)

Data Skills Count

"We look to recruit and groom leaders in our organization who possess strong quantitative skills in addition to a passion for what we do—delivering exceptional experiences at our extraordinary resorts every day. Knowing how to use and interpret data when making important business decisions is one of the keys to our Company's success."

Rob Katz, chairman and chief executive officer of Vail Resorts

Knowing statistics will make you a better consumer of other people's data analyses. You should know enough to handle everyday data problems, to feel confident that others cannot deceive you with spurious arguments, and to know when you've reached the limits of your expertise. Statistical knowledge gives your company a competitive advantage versus those that cannot understand their internal or external market data. Mastery of basic statistics gives you, the individual manager, a competitive advantage as you work your way through the promotion process or when you move to a new employer. For specialized training, many universities now offer master's degrees in business analytics. But here are some reasons for anyone to study statistics.

Communication   The language of statistics is widely used in science, social science, education, health care, engineering, and even the humanities. In all areas of business (accounting, finance, human resources, marketing, information systems, operations management), workers use statistical jargon to facilitate communication. In fact, statistical terminology has reached the highest corporate strategic levels (e.g., "Six Sigma" at GE and Motorola). And in the multinational environment, the specialized vocabulary of statistics permeates language barriers to improve problem solving across national boundaries.

Computer Skills   Whatever your computer skill level, it can be improved. Every time you create a spreadsheet for data analysis, write a report, or make an oral presentation, you bring together skills you already have and learn new ones. Specialists with advanced training design the databases and decision support systems, but you must handle daily data problems *without* experts. Besides, you can't always find an "expert," and, if you do, the "expert" may not understand your application very well. You need to be able to analyze data, use software with confidence, prepare your own charts, write your own reports, and make electronic presentations on technical topics.

Information Management   Statistics can help you handle either too little or too much information. When insufficient data are available, statistical surveys and samples can be used to obtain the necessary market information. But most large organizations are closer to drowning in data than starving for them. Statistics can help summarize large amounts of data and reveal underlying relationships. You've heard of data mining? Statistics is the pick and shovel that you take to the data mine.

Technical Literacy   Many of the best career opportunities are in growth industries propelled by advanced technology. Marketing staff may work with engineers, scientists, and manufacturing experts as new products and services are developed. Sales representatives must understand and explain technical products like pharmaceuticals, medical equipment, and industrial tools to potential customers. Purchasing managers must evaluate suppliers' claims about the quality of raw materials, components, software, or parts.

Process Improvement   Large manufacturing firms like Boeing or Toyota have formal systems for continuous quality improvement. The same is true of insurance companies, financial service firms like Vanguard or Fidelity, and the federal government. Statistics helps firms oversee their suppliers, monitor their internal operations, and identify problems. Quality improvement goes far beyond statistics, but every college graduate is expected to know enough statistics to understand its role in quality improvement.

Analytics in Action

Can Big Data Predict Airfares?

When you book an airline ticket online, does it annoy you when the next day you find a cheaper fare on exactly the same flight? Or do you congratulate yourself when you get a "good" fare followed by a price rise? Ticket price variation can be predicted, even though there is some uncertainty in the actual price you end up paying. KAYAK, a subsidiary of Priceline.com, is a travel planning website and mobile app used by millions of people every day to help forecast prices on airfares as well as hotels and rental cars. How does KAYAK predict airfare variation? It uses big data. Each day, KAYAK's analysts process over a billion online queries to forecast whether an airfare will go up or down. The mathematical models they use also provide a statistical confidence in their prediction so you can make decisions on when to purchase your airline ticket. If you travel a lot and take the time to use services such as KAYAK, you could save money. (See www.kayak.com/price-trend-explanation.)

LO 1-3

Explain the uses of statistics in business.

1.3 APPLYING STATISTICS IN BUSINESS

You've seen why statistics is important. Now let's look at some of the ways statistics is used in business.

Auditing   A large firm pays over 12,000 invoices to suppliers every month. The firm has learned that some invoices are being paid incorrectly, but it doesn't know how widespread the problem is. The auditors lack the resources to check all the invoices, so they decide to take a sample to estimate the proportion of incorrectly paid invoices. How large should the sample be for the auditors to be confident that the estimate is close enough to the true proportion?

Marketing   Many companies use Customer Relationship Management (CRM) to analyze customer data from multiple sources. With statistical and analytics tools such as correlation and data mining, they identify specific needs of different customer groups, and this helps them market their products and services more effectively.

Health Care   Health care is a major business (20 percent of U.S. GDP). Hospitals, clinics, and their suppliers can save money by finding better ways to manage patient appointments, schedule procedures, or rotate their staff. For example, an outpatient cognitive retraining clinic for victims of closed-head injuries or stroke evaluates incoming patients using a 42-item physical and mental assessment questionnaire. Each patient is evaluated independently by two experienced therapists. Are there significant differences between the two therapists' evaluations of functional status? Are some assessment questions redundant? Do the initial assessment scores accurately predict the patients' lengths of stay in the program?

Quality Improvement   A manufacturer of rolled copper tubing for radiators wishes to improve its product quality. It initiates a triple inspection program, sets penalties for workers who produce poor-quality output, and posts a slogan calling for "zero defects." The approach fails. Why?

Purchasing   A food producer purchases plastic containers for packaging its product. Inspection of the most recent shipment of 500 containers found that 3 of the containers were defective. The supplier's historical defect rate is .005. Has the defect rate really risen or is this simply a "bad" batch?

Medicine   An experimental drug to treat asthma is given to 75 patients, of whom 24 get better. A placebo is given to a control group of 75 volunteers, of whom 12 get better. Is the new drug better than the placebo, or is the difference within the realm of chance?

Operations Management   The Home Depot carries 50,000 different products. To manage this vast inventory, it needs a weekly order forecasting system that can respond to developing patterns in consumer demand. Is there a way to predict weekly demand and place orders from suppliers for every item without an unreasonable commitment of staff time?

Product Warranty   A major automaker wants to know the average dollar cost of engine warranty claims on a new hybrid vehicle. It has collected warranty cost data on 4,300 warranty claims during the first six months after the engines are introduced. Using these warranty claims as an estimate of future costs, what is the margin of error associated with this estimate?

Mini Case 1.1

How Do You Sell Noodles with Statistics?

"The best answer starts with a thorough and thoughtful analysis of the data," says Aaron Kennedy, founder of Noodles & Company.

(Visit www.noodles.com to find a Noodles & Company restaurant near you.)
Kristoffer Tripplaar/Alamy Stock Photo

Noodles & Company introduced the *quick casual* restaurant concept, redefining the standard for modern casual dining in the United States in the 21st century. Noodles & Company first opened in Colorado in 1995 and has not stopped growing since. It has over 400 restaurants all across the United States from Portland and San Diego to Alexandria and Silver Spring with stops in cities such as Omaha and Naperville.

Noodles & Company has achieved this success with a customer-driven business model and fact-based decision making. Its widespread popularity and high growth rate have been supported by careful consideration of data and thorough statistical analysis. Noodles uses analytics to help it make decisions about issues such as

- Should we offer continuity/loyalty cards for our customers?
- How can we increase the use of our extra capacity during the dinner hours?
- Which new city should we open in?
- Which location should we choose for the new restaurant?

- How do we determine the effectiveness of a marketing campaign?
- Which meal maximizes the chance that a new customer will return?
- Are Rice Krispies related to higher sales?
- Does reducing service time increase sales?

Kennedy says that "using data is the strongest way to inform good decisions. By assessing our internal and external environments on a continuous basis, our Noodles management team has been able to plan and execute our vision."

"I had no idea as a business student that I'd be using statistical analysis as extensively as I do now," says Dave Boennighausen, chief executive officer at Noodles & Company. In the coming chapters, as you learn about the statistical tools businesses use today, look for the Noodles logo next to examples and exercises that show how Noodles uses data and statistical methods in its business functions.

New Frontiers

Statistics is only one of the key tools of modern data analytics. You should be familiar with the terminology of rapidly evolving cross-disciplinary approaches to exploring massive data sets in real time, making predictions, and even implementing decisions with minimal human intervention.

- **Machine learning** (ML) refers to using observed data and algorithms to train computers to classify events and predict outcomes in a useful way without task-specific rules. Statistics is a core component of ML to code and clean large input data sets, perform data transformations, train and test algorithms, and evaluate predictions. Programming skills are also essential to access data warehouses, collect real-time information, and implement ML technology. Business subject knowledge (e.g., accounting, finance, marketing) is essential to yield applicable results.

- **Artificial intelligence** (AI) refers to an area of computer science that seeks to create intelligent machines that can think and behave like humans to solve problems and act autonomously. While it relies on machine learning, AI is a broader field of study involving high-order emulations of human capabilities using image recognition and natural language processing (speech recognition, text translation, content extraction). AI seeks to equal and improve outcomes in tasks that humans already perform. While AI might replace some human jobs (especially in tedious work such as scanning text or images), new jobs are being created for individuals who design, test, and implement AI algorithms. That's why business graduates need strong skills in programming and statistics.

- **Artificial neural networks** (ANN) or simply *neural nets* are a key component of ML and AI. These are "black boxes" whose internal connections mimic the human brain, learning to process raw inputs and produce outputs or conclusions based on examples that are provided. Their node structures assign weights randomly at first and then learn to modify these weights based on training data. However, the weights used by the ANN are not easily interpreted and may even be unknown. Such systems can keep learning as long as they are supplied new data but in the future may also self-learn using feedback about the accuracy of their own predictions. Statistics will play a major role in oversight and management of such systems.

Section Exercises

1.1 Give an example of how statistics might be useful to the person in the scenario.

    a. An auditor is looking for inflated broker commissions in stock transactions.

    b. An industrial marketer is representing her firm's compact, new low-power OLED screens to the military.

    c. A plant manager is studying absenteeism at vehicle assembly plants in three states.

1.2 Give an example of how statistics might be useful to the person in the scenario.

    a. A personnel executive is examining job turnover in different restaurants in a fast-food chain.

    b. An intranet manager is studying e-mail usage rates by employees in different job classifications.

    c. A retirement planner is studying mutual fund performance for six different types of asset portfolios.

1.3 (a) Should the average business school graduate expect to use computers to manipulate data, or is this a job better left to specialists? (b) What problems arise when an employee is weak in quantitative skills? Based on your experience, is that common?

1.4 "Many college graduates will not use very much statistics during their 40-year careers, so why study it?" (a) List several arguments for and against this statement. Which position do you find more convincing? (b) Replace the word "statistics" with "accounting" or "foreign language" and repeat this exercise.

1.5 (a) How much statistics does a student need in *your* chosen field of study? Why not more? Why not less? (b) How can you tell when the point has been reached where you should call for an expert statistician? List some costs and some benefits that would govern this decision.

Analytics in Action

Dark Data?

Data that you don't have (or don't even know about) are "dark data," a term coined by statistician David J. Hand. You may even know of their existence and possible importance but have no way to obtain the data. Many illnesses during the COVID-19 epidemic were not reported or were recorded incorrectly. This made it difficult for clinicians to obtain reliable measures of the effectiveness of various treatments. The COVID-19 underreporting problem continues now that home testing is common. David Hand cites similar examples in studies of diabetes and autism. A dramatic example of missing data is the *Challenger* space shuttle disaster. Data on successful launches were not included in studies of leaks in O-ring seals versus ambient temperature. Had complete data been considered, the doomed launch would have been scrubbed.

Common problems in business data include nonresponse in surveys (see Chapter 2), unclear definition of variables, excessive rounding, and aggregating of data in ways that may conceal subtle but important effects. Sometimes we can use logic or prior information to anticipate "known unknowns." In 2018 only 0.6 percent of World Anti-Doping Agency blood and urine tests showed violations, while experts privately believe the figure is nearer 10 percent. Poorly performing stocks or failed companies may drop out of the calculation of average performance measures, creating an upward bias.

Sources: See David J. Hand, *Dark Data: Why What You Don't Know Matters* (Princeton, NJ: Princeton University Press, 2020); David J. Hand, "Dark Data," *Significance* 17, no. 3 (June 2020), pp. 42–44. See also "Sport Is Still Rife with Doping," *The Economist* 440, no. 9254 (July 14, 2021) (www.economist.com/science-and-technology/2021/07/14/sport-is-still-rife-with-doping).

1.4 STATISTICAL CHALLENGES

Business professionals who use statistics are not mere number crunchers who are "good at math." As Jon Kettenring succinctly said, "Industry needs holistic statisticians who are nimble problem solvers" (www.amstat.org). The ideal data analyst

- Is technically current (e.g., software-wise).
- Communicates well.
- Is proactive.

LO 1-4

State the common challenges facing business professionals using statistics.

- Has a broad outlook.
- Is flexible.
- Focuses on the main problem.
- Meets deadlines.
- Knows their limitations and is willing to ask for help.
- Can deal with imperfect information.
- Has professional integrity.

Role of Communication Skills

"Leaders differentiate themselves by *how* they get jobs done. The how is largely about communication. By communication, I mean both written and oral skills, and both listening and presentation skills. . . . Leaders effectively engage and listen to others, ultimately gaining buy-in and a comprehensive solution. These tasks are dependent upon excellent communication skills—a core competency for leaders at all levels."

Comments on leadership skills by Mark Gasta, former senior vice president and chief human resources officer, Vail Resorts Management Company

Imperfect Data and Practical Constraints

In mathematics, exact answers are expected. But statistics deals with the messy interface between theory and reality. For instance, suppose a new airbag design is being tested. Is the new airbag design safer for children? Test data indicate the design may be safer in some crash situations, but the old design appears safer in others. The crash tests are expensive and time-consuming, so the sample size is limited. A few observations are missing due to sensor failures in the crash dummies. There may be random measurement errors. If you are the data analyst, what can you do? Well, you can know and use generally accepted statistical methods, clearly state any assumptions you are forced to make, and honestly point out the limitations of your analysis. You can use statistical tests to detect unusual data points or to deal with missing data. You can give a range of answers under varying assumptions. Occasionally, you need the courage to say, "No useful answer can emerge from these data."

You will face constraints on the type and quantity of data you can collect. Automobile crash tests can't use human subjects (*too risky*). Telephone surveys can't ask a female respondent whether she has had an abortion (*sensitive question*). We can't test everyone for COVID (*the world is not a laboratory*). Survey respondents may not tell the truth or may not answer all the questions (*human behavior is unpredictable*). Every analyst faces constraints of time and money (*research is not free*).

Business Ethics

In your business ethics class, you learned (or will learn) the broad ethical responsibilities of business, such as treating customers in a fair and honest manner, complying with laws that prohibit discrimination, ensuring that products and services meet safety regulations, standing behind warranties, and advertising in a factual and informative manner. You learned that organizations should encourage employees to ask questions and voice concerns about the company's business practices and give employees access to alternative channels of communication if they fear reprisal. But as an individual employee, *you* are responsible for accurately reporting information to management, including potential sources of error, material inaccuracies, and degrees of uncertainty. A data analyst faces a more specific set of ethical requirements.

Surveys of corporate recruiters show that ethics and personal integrity rank high on their list of hiring criteria. The respected analyst is an honest broker of data who uses statistics to find out the truth, not to represent a popular point of view. Scrutinize your own motives carefully. If you manipulate numbers or downplay inconvenient data, you may succeed in fooling your competitors (or yourself) for a while. But what is the point? Sooner or later the facts will reveal themselves, and you (or your company) will be the loser. Quantitative analyses in business can quantify the risks of alternative courses of action and events. Statistics can help managers set realistic expectations on sales volume, revenues, and costs. An inflated sales forecast or an understated cost estimate may propel a colleague's favorite product program from the planning board to an actual capital investment. But a poor analysis may cost both of you your jobs.

Headline scandals such as Bernard L. Madoff's financial pyramid that cost investors as much as $65 billion or tests of pain relievers financed by drug companies whose results turned out to be based on falsified data (see *New Scientist,* March 21, 2009, p. 4) are easily recognizable as willful lying or criminal acts. You might say, "I would never do things like that." Yet in day-to-day handling of data, you may not *know* whether the data are accurate. You may not *know* the uses to which the data will be put. You may not *know* of potential conflicts of interest. You and other employees (including top management) will need training to recognize the boundaries of what is or is not ethical within the context of your organization and the decision at hand.

Find out whether your organization has a code of ethics. If not, initiate efforts to create such a code. Fortunately, ideas and help are available (e.g., www.ethicsweb.ca/codes/). Because every organization is different, the issues will depend on your company's business environment. Creating or improving a code of ethics will generally require employee involvement to identify likely conflicts of interest, to look for sources of data inaccuracy, and to update company policies on disclosure and confidentiality. Everyone must understand the code and know the rules for follow-up when ethics violations are suspected.

Analytics in Action

Ethical Issues in Artificial Intelligence

Predictions from machine learning (ML) and artificial intelligence (AI) face major concerns about reproducibility, potential bias, transparency, and causality. For example, if a mortgage loan is denied because the ANN predicts a high probability of default, how can the bank explain its decision to a customer? If a medicine is predicted to be effective in treating a disease but the causal mechanism is not explicit, should the medicine be prescribed? Was the training data somehow biased (e.g., by race or gender)? Deep ethical problems arise when AI is used to make decisions that affect people. People are demanding that AI explain its methods more clearly.

- Will Google's AI search engines (e.g., BERT) ingest biases from web data sources that it scans without human direction? Online digitized information is likely to reflect common gender stereotypes, racial profiles, or posts by hate groups and bots.

- Other search engines (e.g., Siri, Alexa) face the same issues. This is a challenge, but also a source of business opportunity that creates demand for statistical expertise in improving these search engines and the reliability of data sources.

- AI decisions may require ethical judgments. For example, in self-driving vehicles, which of several bad options is best when a crash is unavoidable? Faced with a choice of sacrificing one life to save several, what should AI do?

Sources: *See* David J. Hand, "Dark Data," *Significance* 17, no. 3 (June 2020), pp. 34–37; S. U. Noble, *Algorithms of Oppression: How Search Engines Reinforce Racism* (New York: NYU Press, 2018).

Upholding Ethical Standards

Ethical requirements apply to anyone who analyzes data and writes reports for management. You need to know the rules to protect your professional integrity and to minimize the chance of inadvertent ethical breaches. Ask questions, think about hidden agendas, and dig deeper into how data were collected. If unacceptable client pressure persists, you can withhold professional services or express no opinion. Here are some basic rules for the data analyst:

- Know and follow accepted procedures.
- Maintain data integrity.
- Carry out accurate calculations.
- Report procedures faithfully.
- Protect confidential information.
- Cite sources.
- Acknowledge sources of financial support.

Because legal and ethical issues are intertwined, there are specific ethical guidelines for statisticians concerning treatment of human and animal subjects, privacy protection, obtaining informed consent, and guarding against inappropriate uses of data. For further information about ethics, see the American Statistical Association's ethical guidelines (www.amstat .org), which have been extensively reviewed by the statistics profession.

Ethical dilemmas for a nonstatistician are likely to involve conflicts of interest or competing interpretations of the validity of a study and/or its implications. For example, suppose a market research firm is hired to investigate a new corporate logo. The CEO lets you know that she strongly favors a new logo, and it's a big project that could earn you a promotion. Yet the market data have a high error margin and could support either conclusion. As a manager, you will face such situations. Statistical practices and statistical data can clarify your choices.

A *perceived* ethical problem may be just that—*perceived*. For example, it may appear that a company promotes more men than women into management roles while, in reality, the promotion rates for men and women are the same. The perceived inequity could be a result of fewer female employees to begin with. In this situation, organizations might work hard to hire more women, thus increasing the pool of women who are promotable. Statistics plays a role in sorting out ethical business dilemmas by using data to uncover *real* versus *perceived* differences and identify root causes of problems.

Working in Teams

Business activity is often handled in teams. Forget the stereotype of a lonely data analyst tapping computer keys or poring over tables of numbers. Today, technical experts spend much of their time in meetings, discussing project milestones, preparing team reports, and arguing about methodology. All teams face similar issues. What is the best way to organize the team? What communication channels will the team use? How can the team's resources best be utilized? What is the role of the team leader? Do we need a subcommittee? Your organizational behavior classes can offer advice on such matters. And, hopefully, your statistics class will include team projects so you can practice for the real world.

Using Consultants

Students often comment on the first day of their statistics class that they don't need to learn statistics because businesses rely on consultants to do the data analyses. This is a misconception. Today's successful companies expect their employees to be able to perform statistical analyses, from simple descriptive analyses to more complex inferential analyses. They also expect employees to be able to interpret the results of a statistical analysis, even if it was completed by an outside consultant. Organizations have been asking business schools across the nation to increase the level of quantitative instruction that students receive and, when hiring, are increasingly giving priority to candidates with strong quantitative skills.

Jobs for Data Scientists

Many companies hire data scientists. Examples include social networks (Facebook, Twitter, LinkedIn), web service providers (Google, Microsoft, Apple, Yahoo!), financial institutions (Bank of America, Citi, JPMorgan Chase), credit card companies (Visa, American Express), online retailers (Amazon, eBay), computer software and database developers (IBM, SAP, Oracle, Tableau, SAS, Hewlett-Packard, Cisco Systems), accounting firms (KPMG, Deloitte, Ernst & Young), and retailers (Target, Sears). Governments, defense contractors, and engineering firms also hire data scientists, sometimes calling them operations research analysts.

Here are examples of some recent job postings: Risk Analytics Manager, Data Management Specialist, Analytics Associate, Digital Strategy Consultant, Quantitative Research Analyst, Big Data Enterprise Architect, Software Engineer, Predictive Modeling Consultant, Security Data Analyst, Mobile App Developer, and Cloud Professional. To find out more about analytics jobs and the required skills, see www.datasciencecentral.com.

When an organization is faced with a decision that has serious public policy implications or high cost consequences, an hour with an expert at the *beginning* of a project could be the smartest move a manager can make. When should a consultant be hired? When your team lacks certain critical skills or when an unbiased or informed view cannot be found inside your organization. Expert consultants can handle domineering or indecisive team members, personality clashes, fears about adverse findings, and local politics. Large and medium-sized companies may have in-house statisticians, but smaller firms only hire them as needed. Obtain agreement at the outset on the consultant's rates and duties. Fees should be by the hour or job (never contingent on the findings). Make better use of the consultant's time by learning how consultants work. Read books about statistical consulting. If your company employs a statistician, take them to lunch!

Keep It Simple

"When we communicate statistics, there are two things I want to make sure don't happen. One is showing off and using too much statistical jargon to our clients. The second thing is adding too many details . . . I prefer a two-sentence explanation or overview in language that would be clear to our clients."

From an interview with Mary Batcher, executive director of Ernst & Young's Quantitative Economics and Statistics Group, American Statistical Association.

Communicating with Numbers

Numbers have meaning only when communicated in the context of a certain situation. Strong *domain knowledge* (such as accounting, finance, marketing, or engineering) will make you a more effective data analyst because you know what the numbers represent. Busy managers rarely have time to read and digest detailed explanations of numbers. **Appendix I** contains advice on technical report writing and oral presentations. You probably already know that attractive graphs will enhance a technical report and help other managers quickly understand the information they need to make a good decision. For example, compare Table 1.1 and Figure 1.2. Which is more helpful in understanding Apple's recent financial performance? Chapter 3 will give detailed directions for effective tables and graphs using Excel.

Table 1.1

Apple Revenue and Profit, 2015–2021 (Billions)

Year	2015	2016	2017	2018	2019	2020	2021
Revenue	233.7	215.6	229.2	265.6	260.2	274.5	365.8
Profit	93.6	84.3	88.2	101.8	98.4	105.0	152.8

Source: www.macrotrends.net.

Figure 1.2

Apple Revenue and Profit, 2015–2021 (Billions)
🖫 **Apple**

Source: Apple

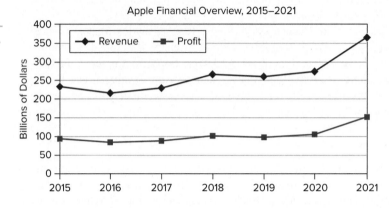

Section Exercises

1.6 The U.S. Public Interest Research Group Education Fund (USPIRG) recently published a report titled *The Campus Credit Card Trap: A Survey of College Students about Credit Card Marketing*. You can find this report and more information about campus credit card marketing at The Campus Credit Card Trap. Read this report and then answer the following questions about how statistics plays a role in resolving ethical dilemmas.

    a. What is the perceived ethical issue highlighted in this report?
    b. How did USPIRG conduct its study to collect information and data?
    c. What broad categories did its survey address?
    d. Did the survey data verify that the issue was a real, instead of a perceived, ethical problem?
    e. Do you agree with the study's assessment of the issue? Why or why not?
    f. Based on the results of the survey, is the issue widespread? Explain.
    g. Describe the report's suggested solutions to confront unethical business practices.

1.7 Using your favorite web browser, enter the search string "business code of ethics." List five examples of features that a business ethics code should have.

Mini Case 1.2

Lessons from GM's Ignition Switch Debacle

The late President Lyndon Baines Johnson observed, "A President's hardest task is not to *do* what is right, but to *know* what is right." What's missing is wisdom, not courage. Given incomplete or contradictory data, people have trouble making decisions (remember *Hamlet?*). Sometimes the correct choice is obvious in retrospect, as in General Motors' ignition switch defect that caused vehicle crashes, fatalities, and injuries over the course of 11 years. This defect led to the recall of 2.6 million vehicles. The defective ignition switches could accidentally turn to the "off" position if jiggled, which would then shut down the car's electrical system, including the power steering, and disable the airbags. Without the ability to steer the car, a crash could result in injury or death if the airbags failed to deploy. This did happen in many documented cases, and 13 deaths have been attributed to a faulty ignition switch.

The defect was first identified in 2001 when it was discovered that the ignition switch producer was not meeting specifications. GM engineers did recognize that this defect could cause the vehicle to stall. In fact, GM sent notices to dealers alerting them to the potential problem. However, the engineers did not correctly assess the fatal consequences of the problem. Because the consequences were not deemed serious, the National Highway Traffic Safety Administration (NHTSA) was not notified of the safety issue.

This is a *statistical* issue because the ignition switches were estimated to be defective in only 1 out of 2,000 vehicles. Because not *all* defective switches would result in a vehicle crash, the chance of a crash occurring due to a faulty switch was extremely low. However, this is also an *ethical* issue. To what extent was GM responsible for safety of product components from its suppliers? Did organizational inertia, pressure to launch new vehicles, and pressure to keep costs low prevent GM from reporting the defect and redesigning the ignition switch?

In May 2014 GM was required to pay a $34 million penalty for delaying the report of the faulty switch to the NHTSA. These failures remind us that decisions involving data and statistics are embedded in organizational culture. Our government has oversight of products that have the potential to harm users, but if information is withheld, then the regulatory entities cannot do their job. While 13 fatalities over the course of 11 years seems small compared to the number of vehicle fatalities that occur each day (there were on average 90 deaths due to vehicle crashes each day in 2014), these deaths might have been prevented if GM had responded earlier. (See https//en.wikipedia.org/wiki/General_Motors_ignition_switch_recalls.)

1.5 CRITICAL THINKING

LO 1-5

List and explain common statistical pitfalls.

Statistics is an essential part of **critical thinking** because it allows us to test an idea against empirical evidence. Random occurrences and chance variation inevitably lead to occasional outcomes that could support one viewpoint or another. But the science of statistics tells us whether the sample evidence is convincing. In this book, you will learn how to use statistics correctly in accordance with professional standards to make the best decision.

> "Critical thinking means being able to evaluate evidence, to tell fact from opinion, to see holes in an argument, to tell whether cause and effect has been established, and to spot illogic."
> Source: Sharon Begley, "Science Journal: Critical Thinking / Part Skill, Part Mindset," *The Wall Street Journal,* October 22, 2006.

We use statistical tools to compare our prior ideas with **empirical data** (data collected through observations and experiments). If the data do not support our theory, we can reject or revise our theory. In *The Wall Street Journal,* in *Money* magazine, and on CNN, you see stock market experts with theories to "explain" the current market (bull, bear, or pause). But each year brings new experts and new theories, and the old ones vanish. Logical pitfalls abound in both the data collection process and the reasoning process. Let's look at some.

Pitfall 1: Conclusions from Small Samples   "My aunt Harriet smoked all her life and lived to 90." Good for her. But does one case prove anything? If 10 patients try a new asthma medication and one gets a rash, can we conclude that the medication caused the rash? How large a sample is needed to make reliable conclusions? Until you learn the rules for sample size in Chapter 8, it's OK to raise your pennant hopes when your favorite baseball team wins five games in a row.

Pitfall 2: Conclusions from Non-Random Samples   "Rock stars die young. Look at Jimi Hendrix, Janis Joplin, Jim Morrison, and Kurt Cobain." But what about those who are alive and well or who lived long lives? Almost 25 percent of the notable performers in Wikipedia's "List of Deaths in Rock and Roll" lived to past age 70. Similarly, studies of vaccine-related complications must also include those who have no complications. In Chapter 2, you will learn proper sampling methods to make reliable inferences.

Pitfall 3: Conclusions from Rare Events

Unlikely events happen if we take a large enough sample. Your friend Bill may predict every Notre Dame football win this season. But millions of sports fans make predictions. A few of them will call every game correctly. Likewise, a few lucky stock analysts may enjoy a winning streak and a few pollsters may correctly call all 50 state presidential votes in the next election. In Chapter 5, you will learn about the law of large numbers, which explains unlikely events such as these.

Pitfall 4: Poor Survey Methods

Did your instructor ever ask a question like "How many of you remember the simplex method from your math class?" One or two timid hands (or maybe none) are raised, even if they do recall the topic. It's difficult for students to respond to such a question in public, for they assume (often correctly) that if they raise a hand, the instructor is going to ask them to explain it, or that their peers will think they are showing off. An anonymous survey or a quiz on simplex would provide better insight. In Chapter 2, you will learn rules for survey design and response scaling.

Pitfall 5: Assuming a Causal Link

In your economics class, you may have learned about the **post hoc fallacy** (the mistaken conclusion that if *A* precedes *B,* then *A* is the *cause* of *B*). For example, the divorce rate in Mississippi fell in 2005 after Hurricane Katrina. Did the hurricane cause couples to stay together? A little research reveals that the divorce rate had been falling for the previous two years, so Hurricane Katrina could hardly be credited.

The *post hoc fallacy* is a special case of the general fallacy of *assuming causation* anytime there is a *statistical association* between events. For example, there is the "curse of the ballfield," which says that teams that play in named ballparks (e.g., Citi Field for the New York Mets) tend to lose more games than they win. Perhaps in a statistical sense this may be true. But it is actually the players and managers who determine whether a team wins. Association does not prove causation. You've probably heard that. But many people draw unwarranted conclusions when no cause-and-effect link exists or "cherry pick" the data to find odd patterns [see Gary Smith, "Full Moons and Forking Paths," *Significance* 19, no. 4 (August 2022), pp. 32–35.]

On the other hand, association may warrant further study when common sense suggests a potential causal link. For example, does vaping lead to cancer or emphysema? In Chapter 12, you will learn tests to decide whether a correlation is within the realm of chance.

Pitfall 6: Generalization to Individuals

"Men are taller than women." Yes, but only in a statistical sense. Men are taller *on average,* yet many women are taller than many men. We should avoid reading too much into **statistical generalizations**. Instead, ask how much *overlap* is in the populations that are being considered. Often, the similarities transcend the differences. In Chapter 10, you will learn precise tests to compare two groups.

Pitfall 7: Unconscious Bias

Without obvious fraud (tampering with data), researchers can unconsciously or subtly allow bias to color their handling of data. For example, for many years it was assumed that heart attacks were more likely to occur in men than women. But symptoms of heart disease are usually more obvious in men than women, so doctors tend to catch heart disease earlier in men. Studies actually show that heart disease is the number one cause of death for American women (www.americanheart.org). In Chapter 2, you will learn about sources of bias and error in surveys.

Pitfall 8: Significance versus Importance

Statistically significant effects may lack practical importance. Cost-conscious businesses know that some product improvements will not justify an increase in price. Consumers may not perceive small improvements in durability, speed, taste, and comfort if the products already are "good enough." For example, Seagate's Cheetah 600GB disk drive already has a mean time between failure (MTBF) rating of 1.4 million hours (about 160 years in continuous use). Would a 10 percent improvement in MTBF matter to anyone?

Analytics in Action

Algorithms: Friend or Foe?

In movies, artificial intelligence (AI) sometimes is portrayed as malevolent (e.g., self-aware terminators that seek to destroy humanity). But AI can also produce happy outcomes such as creating new ice cream flavors or teaching a robot to sing. On a more practical level, business applications use statistics and AI to strengthen profits and improve customer experiences.

- Spam filters use AI to examine sender location and keywords to screen unwanted e-mails. AI must continually evolve to keep up with tireless cyber-vandals.

- Pandora employs humans to categorize songs based on hundreds of attributes (e.g., vocal style, rhythm) but uses AI to present songs in batches that make sense.

- Uber uses AI algorithms to choose routes, set fares, and estimate arrival times. Lyft recently hired its first director of ML and AI, and more jobs in this area are being created.

- Self-driving cars by Google, GM, and Tesla rely on AI to allow vehicles to act autonomously and safely.

- Cogito (www.cogitocorp.com) uses AI to improve customer interactions by analyzing sales conversations and advising salespeople on ways to improve.

- Altexsoft.com (https://www.altexsoft.com/ai-artificial-intelligence-in-travel/) uses AI to customize travel plans for specific customer needs, just as a travel agent would.

- Computer-managed funds represent over half of all stock trading activity, basing their decisions on statistical indicators such as volatility.

Section Exercises

1.8   Recently, the same five winning numbers (4, 21, 23, 34, 39) came up both on Monday and on Wednesday in the North Carolina Lottery. "That's so unlikely that it must be rigged," said Mary. Which fallacy, if any, do you see in Mary's reasoning?

1.9   A National Health Interview Survey conducted by the U.S. Centers for Disease Control and Prevention reported that using a cell phone instead of a landline appeared to double a person's chances of binge drinking. "I guess I'd better give up my cell phone," said Bob. Which fallacy, if any, do you see in Bob's reasoning?

1.10   A study found that radar detector users have lower accident rates, wear their seat belts more, and even vote more than nonusers. (a) Assuming that the study is accurate, do you think there is cause-and-effect? (b) If everyone used radar detectors, would voting rates and seat-belt usage rise?

1.11   A lottery winner told how he picked his six-digit winning number (5-6-8-10-22-75): number of people in his family, birth date of his wife, school grade of his 13-year-old daughter, sum of his birth date and his wife's, number of years of marriage, and year of his birth. He said, "I try to pick numbers that mean something to me." Based on your understanding of how a lottery works, would someone who picks 1-2-3-4-5-6 because "it is easy to remember" have a lower chance of winning?

1.12   "Smokers are much more likely to speed, run red lights, and get involved in car accidents than nonsmokers." (a) Can you think of reasons why this statement might be misleading? *Hint:* Make a list of six factors that you think would cause car accidents. Is smoking on your list? (b) Can you suggest a causal link between smoking and car accidents?

1.13   An ad for a cell phone service claims that its percent of "dropped calls" was significantly lower than that of its main competitor. In the fine print, the percents were given as 1.2 percent versus 1.4 percent. Is this reduction likely to be *important* to customers (as opposed to being *significant*)?

1.14 What logical or ethical problems do you see in these hypothetical scenarios?

    a. Dolon Privacy Consultants concludes that its employees are not loyal because a few samples of employee e-mails contained comments critical of the company's management.

    b. Calchas Financial Advisors issues a glowing report of its new stock market forecasting system, based on testimonials of five happy customers.

    c. A consumer group rates a new personal watercraft from Thetis Aquatic Conveyances as "Unacceptable" because two Ohio teens lost control and crashed into a dock.

1.15 A recent study of 231,164 New Jersey heart attack patients showed that those admitted on a weekday had a 12.0 percent death rate in the next three years, compared with 12.9 percent for those admitted on a weekend. This difference was statistically significant. "That's too small to have any practical importance," said Sarah. Do you agree with Sarah's conclusion? Explain.

1.16 When Pennsylvania repealed a law that required motorcycle riders to wear helmets, a news headline reported, "Deaths Soar after Repeal of Helmet Law." After reading the story, Bill said, "But it's just a correlation, not causation." Do you agree with Bill's conclusion? Explain.

Chapter Summary

Statistics (or **data science**) is the science of collecting, organizing, analyzing, interpreting, and presenting data. A **statistician** is an expert with a degree in mathematics or statistics, while a **data analyst** is anyone who works with data. **Descriptive statistics** is the collection, organization, presentation, and summary of data with charts or numerical summaries. **Inferential statistics** refers to generalizing from a sample to a population, estimating unknown parameters, drawing conclusions, and making decisions. Statistics is used in all branches of business. **Statistical challenges** include imperfect data, practical constraints, and ethical dilemmas. Statistical tools are used to test theories against empirical data. Pitfalls include non-random samples, incorrect sample size, and lack of causal links. The field of statistics is relatively new and continues to grow as new applications arise and technical frontiers expand.

Key Terms

artificial intelligence	data science	machine learning	statistical generalization
artificial neural networks	descriptive statistics	post hoc fallacy	statistics
business analytics	empirical data	statistic	
critical thinking	inferential statistics		

Chapter Review

1. Define (a) statistic and (b) statistics.

2. List three reasons to study statistics.

3. List three applications of statistics.

4. List four skills needed by statisticians. Why are these skills important?

5. List three *practical* challenges faced by statisticians.

6. List three *ethical* challenges faced by statisticians.

7. List five pitfalls or logical errors that may ensnare the unwary statistician.

Chapter Exercises

1.17 A survey of beginning students showed that a majority strongly agreed with the statement "I am afraid of statistics." (a) Why might this attitude exist among students who have not yet taken a statistics class? (b) Would a similar attitude exist toward an ethics class? Explain your reasoning.

1.18 U.S. Food and Drug Administration (FDA) standards for food contaminants allow up to 15 fruit fly eggs per 100 grams of tomato sauce. How could statistical sampling be used to see that this standard is being met by producers? (Source: www.fda.gov.)

1.19 A statistical consultant was retained by a linen supplier to analyze a survey of hospital purchasing managers. After looking at the data, she realized that the survey had missed several key geographic areas and included some that were outside the target region. Some survey questions were ambiguous. Some respondents failed to answer all the questions or gave silly replies (one manager said he worked 40 hours a day). Of the 1,000 surveys mailed, only 80 were returned. What alternatives are available to the statistician?

1.20 Ergonomics is the science of making sure that human surroundings are adapted to human needs. How could statistics play a role in the following:

a. Designing an office chair so that 95 percent of the employees (male and female) will feel it is the "right height" for their legs to reach the floor comfortably.

b. Defining a doorway width so that a "typical" wheelchair can pass through without coming closer than 6 inches from either side.

c. Setting the width of a parking space to accommodate 95 percent of all vehicles at your local Walmart.

d. Choosing a font size so that a highway sign can be read in daylight at 100 meters by 95 percent of all drivers.

1.21 Analysis of 1,064 deaths of famous popular musicians showed that 31 percent were related to alcohol or drug abuse. "But that is just a sample. It proves nothing," said Hannah. Do you agree with Hannah's conclusion? Explain.

1.22 A recent study showed that women who lived near a freeway had an unusually high rate of rheumatoid arthritis. Sarah said, "They should move away from freeways." Is there a fallacy in Sarah's reasoning? If so, explain which one.

1.23 "Lacrosse helmets are not needed," said Tom. "None of the guys on my team have ever had head injuries." Is there a fallacy in Tom's reasoning? If so, explain which one.

1.24 A European study of thousands of men found that the PSA screening for prostate cancer reduced the risk of a man's dying from prostate cancer from 3.0 percent to 2.4 percent. "But it's already a small risk. I don't think a difference of less than 1 percent would be of practical importance," said Ed. Do you agree with Ed's conclusion? Explain.

1.25 The 2019 national Youth Risk Behavior Survey showed that twice as many "D" students used tobacco products compared with "A" students. (a) List in rank order six factors that you think affect grades. Is tobacco use on your list? (b) If tobacco use is not a likely cause of poor grades, can you suggest reasons why these results were observed? (c) Assuming these statistics are correct, would "D" students who give up tobacco use improve their grades? Why or why not?

1.26 A research study showed that adolescents who watched more than 4 hours of TV per day were more likely to start smoking than those who watched less than 2 hours a day. The researchers speculate that TV actors' portrayals of smoking as personally and socially rewarding were an effective indirect method of tobacco promotion. List in rank order six factors that you think cause adolescents to start smoking. Did TV portrayals of attractive smokers appear on your list?

1.27 The Graduate Management Admission Test (GMAT) is used by many graduate schools of business as one of their admission criteria. Using your own reasoning and concepts in this chapter, criticize each of the following conclusions.

a. "Last year, 29,688 engineering majors took the GMAT, compared with only 3,589 English majors. Clearly, more students major in engineering than in English."

b. "Last year, physics majors averaged 100 points higher on the GMAT than marketing majors. If marketing students majored in physics, they would score better on the GMAT."

c. "On average, physics majors score higher on the GMAT than accounting majors. Therefore, physics majors would make the best managers."

1.28 (a) Which of these two displays (table or graph) is more helpful in visualizing the relationship between restaurant size and interior seating for 74 Noodles restaurants? Explain your reasoning. (b) Do you see anything unusual in the data? (Source: Noodles & Company.)  **NoodlesSqFt**

Number of Restaurants in Each Category (*n* = 74 restaurants)

Interior Seats	Square Feet Inside Restaurant				Row Total
	1,000 < 1,750	1,750 < 2,500	2,500 < 3,250	3,250 < 4,000	
105 < 130	0	0	0	3	3
80 < 105	0	4	17	0	21
55 < 80	0	21	24	0	45
30 < 55	1	4	0	0	5
Col Total	1	29	41	3	74

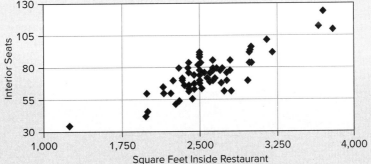

Restaurant Size and Number of Interior Seats

1.29 (a) Which of these two displays (table or graph) is more helpful in describing the salad sales by Noodles & Company? Why? (b) Write a one-sentence summary of the data. (Source: Noodles & Company.) 🖫 **NoodlesSalad**

Average Daily Salads Sold by Month, Noodles & Company

Month	Salads	Month	Salads
Jan	2,847	Jul	2,554
Feb	2,735	Aug	2,370
Mar	2,914	Sep	2,131
Apr	3,092	Oct	1,990
May	3,195	Nov	1,979
Jun	3,123	Dec	1,914

1.30 Choose *three* of the following statisticians and use the web to find out a few basic facts about them (e.g., list some of their contributions to statistics, when they did their work, whether they are still living, etc.).

Florence Nightingale	Kimberly Sellers	Judith M. Tanur
David Cox	John Wilder Tukey	Edward Tufte
Gertrude Cox	William Cochran	Genichi Taguchi
Sir Francis Galton	Simeon Poisson	Helen Walker
W. Edwards Deming	Jake Porway	George Box
The Bernoulli family	S. S. Stevens	Sam Wilks
William H. Kruskal	Sandra Stinnett	Carl F. Gauss
Frederick Mosteller	George Snedecor	William S. Gosset
Susan Murphy	Karl Pearson	Thomas Bayes
Jerzy Neyman	C. R. Rao	Bradley Efron
Egon Pearson	Abraham De Moivre	Nate Silver

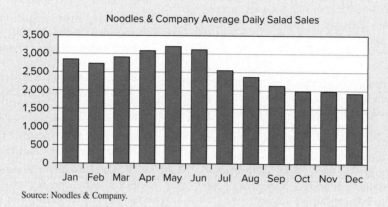

Noodles & Company Average Daily Salad Sales

Source: Noodles & Company.

Related Reading

Practical Guides

Best, Joel. *Stat-Spotting: A Field Guide to Dubious Data.* University of California Press, 2013.

Davenport, Thomas H., and Jeanne G. Harris. *Competing on Analytics: The New Science of Winning.* Harvard Business School Press, 2007.

Dodge, Yadolah. *The Concise Encyclopedia of Statistics.* Springer, 2010.

Everitt, B. S. *The Cambridge Dictionary of Statistics.* 4th ed. Cambridge University Press, 2013.

Fung, Kaiser. *Numbers Rule Your World: The Hidden Influence of Probabilities and Statistics on Everything You Do.* McGraw Hill, 2010.

Levitin, Daniel J. *A Field Guide to Lies: Critical Thinking in the Information Age.* Penguin, 2016.

Newton, Rae R. *Your Statistical Consultant.* 2nd ed. Sage Publications, 2013.

Seife, Charles. *Proofiness: The Dark Arts of Mathematical Deception.* Viking, 2010.

Utts, Jessica. "What Educated Citizens Should Know about Statistics and Probability." *The American Statistician* 57, no. 2 (May 2003), pp. 74–79.

Ethics

Hartman, Laura P., Joseph DesJardins, and Chris MacDonald. *Business Ethics: Decision Making for Personal Integrity & Social Responsibility.* 5th ed. McGraw Hill, 2021.

Scharding, Tobey. *This Is Business Ethics: An Introduction.* Wiley-Blackwell, 2018.

Shaw, William H. *Business Ethics.* 9th ed. Cengage, 2017.

Vardeman, Stephen B., and Max D. Morris. "Statistics and Ethics: Some Advice for Young Statisticians." *The American Statistician* 57 (February 2003), pp. 21–26.

Velasquez, Manuel G. *Business Ethics: Concepts and Cases.* 8th ed. Pearson, 2018.

CHAPTER 1   More Learning Resources

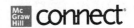

You can access these *LearningStats* demonstrations through Connect to give you an overview of statistics.

Topic	LearningStats Demonstrations
Microsoft® Office	Excel Tips Excel for PC Excel for Mac
Excel	Using R and R Studio Excel and R Functions
Math Review	Math Review Significant Digits
Web Stuff	Web Resources Statistics Software

Key: = PowerPoint = PDF

Software Supplement

R and RStudio

R is a free software environment for statistical computing and graphics. R was created by the R Foundation (a not-for-profit organization) that runs on UNIX platforms, Windows, and MacOS. RStudio is an integrated development environment (IDE) for R. It includes an editor for direct code execution and tools for data analysis, plotting, history, debugging, and data management. RStudio is available in open source and commercial editions. R and RStudio together provide a versatile programming language and an integrated suite of well-documented (www.r-project.org/other-docs.html) functions and procedures for data manipulation, calculation, and graphical display.

Installing and Using R and RStudio

If your instructor is utilizing R, then **Appendix K** would be imperative. It explains how to download, install, and get started using R and RStudio. Chapters in this textbook will illustrate key R functions and tools, along with Excel (also Minitab and MegaStat). As your statistics training progresses, **Appendix J** will be a useful reference to R functions and procedures and comparing them to corresponding Excel features. Notation and symbols in **Appendix J** are chosen to be consistent with this textbook.

Excel or R?

Excel is widely used in business, government, health care, and research because it is widely available and offers powerful features for data organization, analysis, and visualization. It is easy to learn and intuitive to use, yet it can be customized for specific needs. Excel training is required in almost all business programs and is used in many core and elective classes (e.g., accounting, economics, finance, marketing, human resources). Although not designed for statistical analysis, Excel offers many of the most common procedures needed in data analytics. Although R and RStudio are not new, there is growing interest in introducing them into business curricula. In part this is because they are free, but they also offer entry into sophisticated and versatile tools of data analysis. Unlike Excel, they offer core skill training in programming as well as advanced analytical procedures. This matches the expectation for college graduates to be capable in quantitative reasoning and data analytics.

2 Data Collection

CHAPTER CONTENTS

CHAPTER LEARNING OBJECTIVES

When you finish this chapter, you should be able to

LO 2-1 Use basic terminology for describing data and samples.

LO 2-2 Explain the difference between numerical and categorical data.

LO 2-3 Explain the difference between time series and cross-sectional data.

LO 2-4 Recognize levels of measurement in data and ways of coding data.

LO 2-5 Recognize a Likert scale and know how to use it.

LO 2-6 Use the correct terminology for samples and populations.

LO 2-7 Explain the common sampling methods and how to implement them.

LO 2-8 Find everyday print or electronic data sources.

LO 2-9 Describe basic elements of survey types, survey designs, and response scales.

I n scientific research, data arise from experiments whose results are recorded systematically. In business, data usually arise from accounting transactions or management processes (e.g., inventory, sales, payroll). Much of the data that statisticians analyze were observed and recorded without explicit consideration of their statistical uses, yet important decisions may depend on the data. How many pints of type A blood will be required at Mt. Sinai Hospital next Thursday? How many dollars must State Farm keep in its cash account to cover automotive accident claims next November? How many yellow three-quarter-sleeve women's sweaters will Lands' End sell this month? To answer such questions, we usually look at historical, or empirical, data.[1]

Andersen Ross Photography Inc/Getty Images

2.1 VARIABLES AND DATA

Data Terminology

An **observation** is a single member of a collection of items that we want to study, such as a person, firm, or region. An example of an observation is an employee or an invoice mailed last month. A **variable** is a characteristic of the subject or individual, such as an employee's income or an invoice amount. The **data set** consists of all the values of all of the variables for all of the observations we have chosen to observe. In this book, we will use **data** as a plural and *data set* to refer to a collection of observations taken as a whole. Data usually are entered into a spreadsheet or database as an $n \times m$ matrix. Specifically, each column is a variable (m columns), and each row is an observation (n rows). Table 2.1 shows a small data set with eight observations (8 rows) and five variables (5 columns).

LO 2-1

Use basic terminology for describing data and samples.

Obs	Name	Age	Salary	Position	Veteran?	Education
1	Frieda	45	87,100	Personnel director	No	Master's
2	Stefan	32	76,500	Operations analyst	No	Doctorate
3	Barbara	55	108,200	Marketing VP	Yes	Master's
4	Donna	27	79,000	Statistician	No	Bachelor's
5	Larry	46	46,000	Security guard	No	High School
6	Alicia	52	98,500	Comptroller	Yes	Master's
7	Alec	65	125,200	Chief executive	No	Master's
8	Jaime	50	71,200	Public relations	No	Bachelor's

Table 2.1

Multivariate Data (5 Variables, 8 Observations)
SmallData

A data set may consist of many variables. The questions that can be explored and the analytical techniques that can be used will depend upon the data type and the number of variables. This textbook starts with **univariate data sets** (one variable) and then moves to **bivariate data sets** (two variables) and **multivariate data sets** (more than two variables), as illustrated in Table 2.2.

---

[1]Is *data* singular or plural? *Data* is the plural of the Latin *datum* (a "given" fact). But in the popular press you will often see "data" used synonymously with "information" and hence as a singular ("Employee data is stored in the cloud.").

Table 2.2

Number of Variables and Typical Tasks

Data Set	Variables	Example	Typical Tasks
Univariate	One	Income	Histograms, basic statistics
Bivariate	Two	Income, Age	Scatter plots, correlation
Multivariate	More than two	Income, Age, Gender	Regression modeling

LO 2-2

Explain the difference between numerical and categorical data.

Categorical and Numerical Data

A data set may contain a mixture of *data types,* as shown in Figure 2.1. **Categorical data** (also called *qualitative* data) have values that are described by words rather than numbers. For example, structural lumber can be classified by the lumber type (e.g., fir, hemlock, pine), automobile styles can be classified by size (e.g., full-size, mid-size, compact, subcompact), and movies can be categorized using common movie classifications (e.g., action and adventure, children and family, classics, comedy, documentary).

Figure 2.1

Data Types and Examples

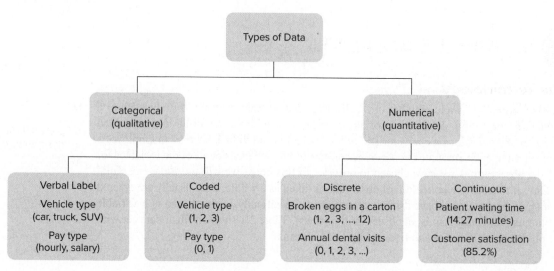

Because categorical variables have nonnumerical values, it might seem that categorical data would be of limited statistical use. In fact, there are many statistical methods that can handle categorical data, which we will introduce in later chapters. On occasion the values of the categorical variable might be represented using numbers. This is called **coding**. For example, a database might code payment methods using numbers:

1 = cash   2 = check   3 = credit/debit card   4 = gift card

Coding a category as a number does *not* make the data numerical, and the numbers do not typically imply a rank. But on occasion a ranking does exist. For example, a database might code education degrees using numbers:

1 = Bachelor's   2 = Master's   3 = Doctorate

Some categorical variables have only two values. We call these **binary variables**. Examples include employment status (e.g., employed or unemployed), mutual fund type (e.g., load or no-load), and marital status (e.g., currently married or not currently married). Binary variables are often coded using a 1 or 0. For a binary variable, the 0-1 coding is arbitrary, so the choice is equivalent. For example, a variable such as pay type could be coded as

1 = hourly, 0 = salary   *or*   0 = hourly, 1 = salary

Numerical data (also called *quantitative* data) arise from counting, measuring something, or some kind of mathematical operation. For example, we could count the number of auto insurance claims filed in March (e.g., 114 claims) or sales for last quarter (e.g., $4,920). Most accounting data, economic indicators, and financial ratios are quantitative, as are physical measurements.

A numerical variable with a countable number of distinct values is **discrete**. Often, such data are integers. You can recognize integer data because their description begins with "number of"—for example, the number of Medicaid patients in a hospital waiting room (e.g., 2) or the number of takeoffs at Chicago O'Hare International Airport in an hour (e.g., 37). These are integer variables because we cannot observe a fractional number of patients or takeoffs.

A numerical variable that can have any value within an interval is **continuous**. This would include things like physical measurements (e.g., distance, weight, time, speed), financial variables (e.g., sales, assets, price/earnings ratios, inventory turns), or the weight of a package of Sun-Maid raisins (e.g., 427.31 grams). These are continuous variables because any interval such as [425, 429] grams can contain infinitely many possible values. Sometimes we round a continuous measurement to an integer (e.g., 427 grams), but that does not make the data discrete.

Ambiguity between discrete and continuous is introduced when we round continuous data to whole numbers. Consider a package of Sun-Maid raisins that is labeled 425 grams. The underlying measurement scale is continuous (e.g., on an accurate scale, its weight might be 425.31 grams). Precision depends on the instrument we use to measure the continuous variable. Shoes are sized in discrete half steps (e.g., 6½) even though the length of your foot is a continuous variable. Conversely, we sometimes treat discrete data as continuous when the range is very large (e.g., SAT scores) and when small differences (e.g., 604 or 605) aren't of much importance. We generally treat financial data (dollars, euros, pesos) as continuous even though retail prices go in discrete steps of .01 (i.e., we go from $1.25 to $1.26). If in doubt, just think about how X was measured and whether its values are countable.

Time Series Data and Cross-Sectional Data

If each observation in the sample represents a different equally spaced point in time (years, months, days), we have **time series data**. The *periodicity* is the time between observations (annual, quarterly, monthly, weekly, daily, hourly, and so on). Examples of *macroeconomic* time series data would include national income (GDP, consumption, investment), economic indicators (Consumer Price Index, unemployment rate, Standard & Poor's 500 Index), and monetary data (M1, M2, M3, prime rate, T-bill rate, consumer borrowing). Examples of *microeconomic* time series data would include a firm's sales, market share, debt/equity ratio, employee absenteeism, and inventory turnover. For time series, we are interested in *trends and patterns over time* (e.g., personal bankruptcies, as shown in Figure 2.2).

If each observation represents a different individual unit (e.g., a person, firm, geographic area) at the same point in time, we have **cross-sectional data**. Thus, traffic fatalities in the 50 U.S. states for a given year, debt/equity ratios for the *Fortune* 500 firms in the last quarter of a certain year, last month's Visa balances for a bank's new mortgage applicants, or GPAs of students in a statistics class would be cross-sectional data. For cross-sectional data, we are interested in *variation among observations* (e.g., accounts receivable in 20 Subway franchises) or in *relationships* (e.g., whether accounts receivable are related to sales volume in 20 Subway franchises).

LO 2-3

Explain the difference between time series and cross-sectional data.

Figure 2.2

Examples of Time Series versus Cross-Sectional Data

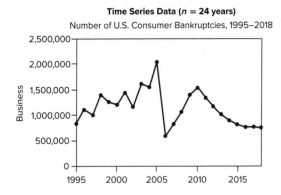

Time Series Data (*n* = 24 years)
Number of U.S. Consumer Bankruptcies, 1995–2018

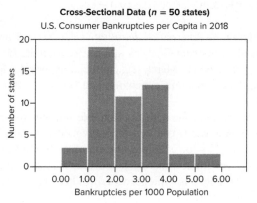

Cross-Sectional Data (*n* = 50 states)
U.S. Consumer Bankruptcies per Capita in 2018

Source: American Bankruptcy Institute (www.abiworld.org).

Some variables (such as unemployment rates) could be either time series (monthly data over each of 60 months) or cross-sectional (January's unemployment rate in 50 different cities). We can combine the two (e.g., monthly unemployment rates for the 13 Canadian provinces or territories for the last 60 months) to obtain *pooled cross-sectional and time series data.*

Section Exercises

2.1   What type of data (categorical, discrete numerical, or continuous numerical) is each of the following variables? If there is any ambiguity about the data type, explain why the answer is unclear.

   a. The manufacturer of your car.
   b. Your college major.
   c. The number of college credits you are taking.

2.2   What type of data (categorical, discrete numerical, or continuous numerical) is each of the following variables? If there is any ambiguity, explain why the answer is unclear.

   a. Length of a TV commercial.
   b. Number of peanuts in a can of Planter's Mixed Nuts.
   c. Occupation of a mortgage applicant.
   d. Flight time from London Heathrow to Chicago O'Hare.

2.3   What type of data (categorical, discrete numerical, or continuous numerical) is each of the following variables? If there is any ambiguity about the data type, explain why the answer is unclear.

   a. The miles shown on your car's odometer.
   b. The fat grams you ate for lunch yesterday.
   c. The name of the airline with the cheapest fare from New York to London.
   d. The brand of cell phone you own.

2.4   (a) Give three original examples of discrete data. (b) Give three original examples of continuous data. In each case, explain and identify any ambiguities that might exist. *Hint:* Do not restrict yourself to published data. Consider data describing your own life (e.g., your sports performance, financial data, or academic data). You need *not* list all the data; merely describe them and show a few typical data values.

2.5   Which type of data (cross-sectional or time series) is each variable?

   a. Scores of 50 students on a midterm accounting exam last semester.
   b. Bob's scores on 10 weekly accounting quizzes last semester.
   c. Average score by all takers of the state's CPA exam for each of the last 10 years.
   d. Number of years of accounting work experience for each of the 15 partners in a CPA firm.

2.6 Which type of data (cross-sectional or time series) is each variable?

a. Value of Standard & Poor's 500 stock price index at the close of each trading day last year.
b. Closing price of each of the 500 stocks in the S&P 500 index on the last trading day this week.
c. Dividends per share paid by General Electric common stock for the last 20 quarters.
d. Latest price/earnings ratios of 10 stocks in Bob's retirement portfolio.

2.7 Which type of data (cross-sectional or time series) is each variable?

a. Mexico's GDP for each of the last 10 quarters.
b. Unemployment rates in each of the 31 states in Mexico at the end of last year.
c. Unemployment rate in Mexico at the end of each of the last 10 years.
d. Average home value in each of the 10 largest Mexican cities today.

2.8 Give an original example of a time series variable and a cross-sectional variable. Use your own experience (e.g., your sports activities, finances, education).

2.2 LEVEL OF MEASUREMENT

LO 2-4

Recognize levels of measurement in data and ways of coding data.

Data types can be further classified by their measurement level. Statisticians typically refer to four levels of measurement for data: nominal, ordinal, interval, and ratio. This typology was proposed over 60 years ago by psychologist S. S. Stevens. The allowable mathematical operations, statistical summary measures, and statistical tests depend on the measurement level. The criteria are summarized in Figure 2.3.

Nominal Measurement

Nominal measurement is the weakest level of measurement and the easiest to recognize. **Nominal data** (from Latin *nomen,* meaning "name") merely identify a *category.* "Nominal" data are the same as "qualitative," "categorical," or "classification" data. To be sure that the categories are collectively exhaustive, it is common to use **Other** as the last item on the list. For example, the following survey questions yield nominal data:

Did you file an insurance claim last month?

1. Yes 2. No

Which cell phone service provider do you use?

1. AT&T 2. Sprint 3. T-Mobile 4. Verizon 5. Other

Figure 2.3

Determining the Measurement Level

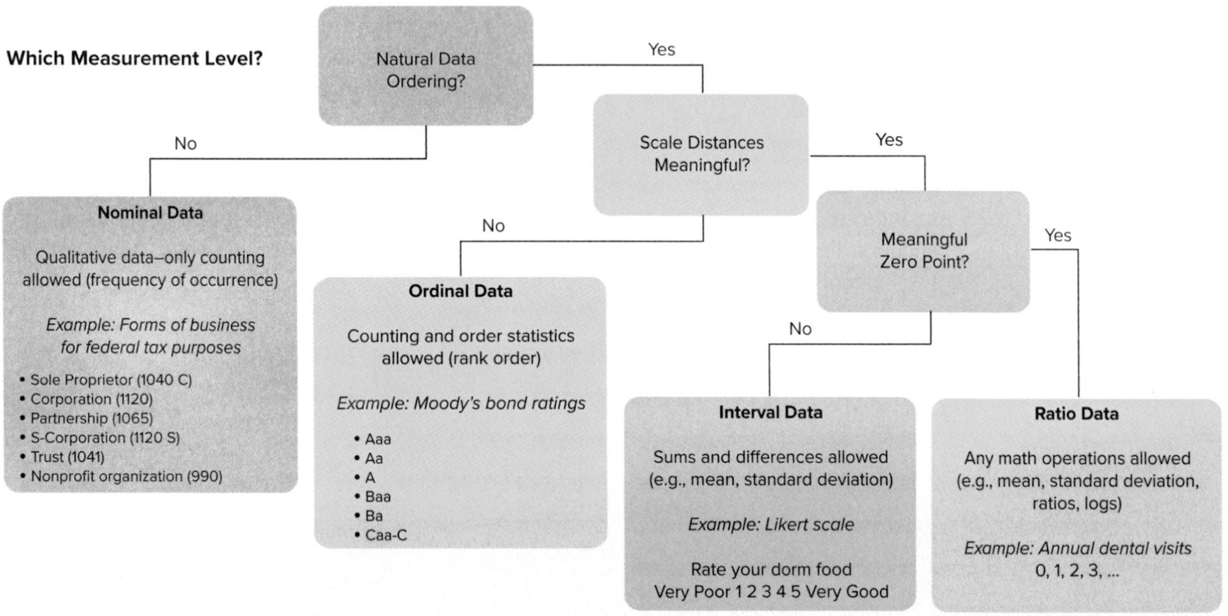
Which Measurement Level?
Natural Data Ordering?
Yes
No
Scale Distances Meaningful?
Yes
No
Meaningful Zero Point?
Yes
No
Nominal Data
Qualitative data–only counting allowed (frequency of occurrence)
Example: Forms of business for federal tax purposes
• Sole Proprietor (1040 C)
• Corporation (1120)
• Partnership (1065)
• S-Corporation (1120 S)
• Trust (1041)
• Nonprofit organization (990)
Ordinal Data
Counting and order statistics allowed (rank order)
Example: Moody's bond ratings
• Aaa
• Aa
• A
• Baa
• Ba
• Caa-C
Interval Data
Sums and differences allowed (e.g., mean, standard deviation)
Example: Likert scale
Rate your dorm food
Very Poor 1 2 3 4 5 Very Good
Ratio Data
Any math operations allowed (e.g., mean, standard deviation, ratios, logs)
Example: Annual dental visits
0, 1, 2, 3, ...

We usually code nominal data numerically. However, the codes are arbitrary placeholders with no numerical meaning, so it is improper to perform mathematical analysis on them. For example, we would not calculate an average using the cell phone service data (1 through 5). This may seem obvious, yet people have been known to do it. Once the data are in the computer, it's easy to forget that the "numbers" are only categories. With nominal data, the only permissible mathematical operations are counting (e.g., frequencies) and a few simple statistics such as the mode.

Ordinal Measurement

Ordinal data codes connote a *ranking* of data values. For example:

What size automobile do you usually drive?

1. Full-size   2. Compact   3. Subcompact

How often do you use Microsoft Access?

1. Frequently   2. Sometimes   3. Rarely   4. Never

Thus, a 2 (Compact) implies a larger car than a 3 (Subcompact). Like nominal data, these ordinal numerical codes lack the properties that are required to compute many statistics, such as the average. Specifically, there is no clear meaning to the *distance* between 1 and 2, or between 2 and 3, or between 3 and 4 (what would be the distance between "Rarely" and "Never"?). Other examples of ordinal scales can be found in a recruiter's rating of job candidates (outstanding, good, adequate, weak, unsatisfactory), S&P credit ratings (AAA, AA+, AA, AA−, A+, A, A−, B+, B, B−, etc.), or job titles (president, group vice president, plant manager, department head, clerk). Ordinal data can be treated as nominal, but not vice versa. Ordinal data are especially common in social sciences, marketing, and human resources research. There are many useful statistical tests for ordinal data.

Interval Measurement

LO 2-5

Recognize a Likert scale and know how to use it.

The next step up the measurement scale is **interval data**. Interval data are used frequently and are important in business. Interval data often arise from surveys where customers are asked to rate their satisfaction with a service or product on a numerical scale. While these scale points are expressed as numbers (e.g., 1–10), the numbers are not a count or a physical measure, and the value "0" has no meaning. However, we often say that the distances between scale points have meaning. The difference between a rating of 4 and 6 is treated the same as the difference between 7 and 9. If intervals between numbers represent distances, we can do mathematical operations such as taking an average. If the zero point of the scale is arbitrary, we can't say that a customer who rates our service an 8 is twice as satisfied as a customer who rates our service a 4. Nor can we say that 60°F is twice as warm as 30°F. That is, *ratios* are not meaningful for interval data. The absence of a meaningful zero is a key characteristic of interval data.

The **Likert scale** is a special case that is frequently used in survey research. You have undoubtedly seen such scales. Typically, a statement is made and the respondent is asked to indicate their agreement/disagreement on a five-point or seven-point scale using verbal anchors. The *coarseness* of a Likert scale refers to the number of scale points (typically 5 or 7). For example:

"College-bound high school students should be required to study a foreign language." (check one)

❏   ❏   ❏   ❏   ❏

Strongly   Somewhat   Neither Agree   Somewhat   Strongly
Agree   Agree   nor Disagree   Disagree   Disagree

A neutral midpoint ("Neither Agree nor Disagree") is allowed if we use an *odd* number of scale points (usually 5 or 7). Occasionally, surveys may omit the neutral midpoint to force the respondent to "lean" one way or the other. Likert data are coded numerically (e.g., 1 to 5), but any equally spaced values will work, as shown in Table 2.3.

Likert Coding: 1 to 5 scale	Likert Coding: −2 to +2 scale
5 = Will help a lot	+2 = Will help a lot
4 = Will help a little	+1 = Will help a little
3 = No effect on investment climate	0 = No effect on investment climate
2 = Will hurt a little	−1 = Will hurt a little
1 = Will hurt a lot	−2 = Will hurt a lot

Table 2.3

Examples of Likert-Scale Coding: "How Will Inflation Affect the Investment Climate?"

But do Likert data qualify as interval measurements? By choosing the verbal anchors carefully, many researchers believe that the *intervals* are the same (e.g., the distance from 1 to 2 is "the same" as the *interval,* say, from 3 to 4). However, ratios are not meaningful (i.e., here 4 is not twice 2). The assumption that Likert scales produce interval data justifies a wide range of statistical calculations, including averages, correlations, and so on. Researchers use many Likert-scale variants.

"How would you rate your Internet service provider?" (check one)

❏ Terrible   ❏ Poor   ❏ Adequate   ❏ Good   ❏ Excellent

Instead of labeling every response category, marketing surveys might put verbal anchors only on the endpoints. This avoids intermediate scale labels and permits any number of scale points. Likert data usually are discrete, but some web surveys use a continuous response scale that allows the respondent to position a "slider" anywhere along the scale to produce continuous data (actually, the number of positions is finite but very large). For example:

Likert (using discrete scale points)

Very Poor 1 2 3 4 5 6 7 Very Good

Likert (using a slider)

Very Poor _____▼_____ Very Good

Vendors offer professional templates for sliders and other survey scales. For example:

Source: www.qualtrics.com.

Ratio Measurement

Ratio measurement is the strongest level of measurement. **Ratio data** have all the properties of the other three data types but in addition possess a *meaningful zero* that represents the absence of the quantity being measured. Because of the zero point, ratios of data values are meaningful (e.g., $20 million in profit is twice as much as $10 million). Balance sheet data, income statement data, financial ratios, physical counts, scientific measurements, and most engineering measurements are ratio data because zero has meaning (e.g., a company with zero sales sold nothing). Having a zero point does *not* restrict us to positive data. For example, profit is a ratio variable (e.g., $4 million is twice $2 million), yet firms can have negative profit (i.e., a loss).

Zero does *not* have to be observable in the data. Newborn babies, for example, cannot have zero weight, yet baby weight clearly is ratio data (i.e., an 8-pound baby is 33 percent heavier than a 6-pound baby). What matters is that the zero is an absolute reference point. The Kelvin temperature scale is a ratio measurement because its absolute zero represents the absence of molecular vibration, while zero on the Celsius scale is merely a convenience (note that 30°C is not "twice as much temperature" as 15°C).

Lack of a true zero is often the quickest test to defrock variables masquerading as ratio data. For example, a Likert scale (+2, +1, 0, −1, −2) is *not* ratio data despite the presence of zero because the zero (neutral) point does not connote the absence of anything. As an acid test, ask yourself whether 2 (strongly agree) is twice as much "agreement" as 1 (slightly agree). Some classifications are debatable. For example, college GPA has a zero, but does it represent the absence of learning? Does 4.00 represent "twice as much" learning as 2.00? Is there an underlying reality ranging from 0 to 4 that we are measuring? Most people seem to think so, although the conservative procedure would be to limit ourselves to statistical tests that assume only ordinal data.

Although beginning statistics textbooks usually emphasize interval or ratio data, there are textbooks that emphasize other kinds of data, notably in behavioral research (e.g., psychology, sociology, marketing, human resources).

We can recode ratio measurements *downward* into ordinal or nominal measurements (but not conversely). For example, doctors may classify systolic blood pressure as "normal" (under 130), "elevated" (130 to 140), or "high" (140 or over). The recoded data are ordinal because the ranking is preserved. Intervals may be unequal. For example, U.S. air traffic controllers classify planes as "small" (under 41,000 pounds), "large" (41,001 to 254,999 pounds), and "heavy" (255,000 pounds or more). Such recoding is done to simplify the data when the exact data magnitude is of little interest; however, we discard information if we map stronger measurements into weaker ones.

Section Exercises

2.9   Which measurement level (nominal, ordinal, interval, ratio) is each of the following variables? Explain.

   a. Number of hits in Game 1 of the next World Series.
   b. Field position of a randomly chosen baseball player (catcher, pitcher, etc.).
   c. Temperature on opening day (Celsius).
   d. Salary of a randomly chosen American League pitcher.
   e. Freeway traffic on opening day (light, medium, heavy).

2.10   Which measurement level (nominal, ordinal, interval, ratio) is each of the following variables? Explain.

   a. Number of employees in the Walmart store in Hutchinson, Kansas.
   b. Number of merchandise returns on a randomly chosen Monday at a Walmart store.
   c. Name of the cashier at register 3 in a Walmart store.
   d. Manager's rating of the cashier at register 3 in a Walmart store.
   e. Social Security number of the cashier at register 3 in a Walmart store.

2.11   Which measurement level (nominal, ordinal, interval, ratio) is each of the following variables? Explain.

   a. Number of passengers on Delta Flight 833.
   b. Waiting time (minutes) after gate pushback before Delta Flight 833 takes off.
   c. Brand of cell phone owned by a cabin attendant on Delta Flight 833.
   d. Ticket class (first, business, or economy) of a randomly chosen passenger on Delta Flight 833.
   e. Passenger rating (on a 5-point Likert scale) of Delta's in-flight food choices.

2.12 Which measurement level (nominal, ordinal, interval, ratio) is the response to each question? If you think that the level of measurement is ambiguous, explain why.

a. How would you describe your level of skill in using Excel? (check one)
   ❏ Low   ❏ Medium   ❏ High

b. How often do you use Excel? (check one)
   ❏ Rarely   ❏ Often   ❏ Very Often

c. Which version of Excel for Windows do you use? (check one)
   ❏ 2007   ❏ 2010   ❏ 2013   ❏ 2016

d. I spend _____ hours a day using Excel.

2.13 Here is a question from a ski resort guest satisfaction survey that uses a five-point scale. (a) Would the measurement level for the data collected from this question be nominal, ordinal, interval, or ratio? (b) Would it be appropriate to calculate an average rating for the various items? Explain. (c) Would a 10-point scale be better? Explain.

"Rate your satisfaction level on numerous aspects of *today's* experience, where 1 = Extremely Dissatisfied and 5 = Extremely Satisfied."

1. Value for Price Paid:	1	2	3	4	5
2. Ticket Office Line Wait (if went to ticket window):	1	2	3	4	5
3. Friendliness/Helpfulness of Lift Operators:	1	2	3	4	5
4. Lift Line Waits:	1	2	3	4	5
5. Ski Patrol Visibility:	1	2	3	4	5

2.14 (a) Would the measurement level for the data collected from this Microsoft® survey question be nominal, ordinal, interval, or ratio? (b) Would a "6" response be considered twice as good as a "3" response? Why or why not? (c) Would a 1–5 scale be adequate? Explain.

Microsoft **Quality of Support Survey**

Please rate the overall quality of support you received from Microsoft on this particular issue, using a 9-point scale where 9 is Excellent and 1 is Very Poor.

Excellent								Very Poor	Don't Know
9	8	7	6	5	4	3	2	1	
○	○	○	◉	○	○	○	○	○	○

Source: Microsoft.

2.3 SAMPLING CONCEPTS

There are almost 2 million retail businesses in the United States. It is unrealistic for market researchers to study them all in a timely way. But since 2001, a firm called ShopperTrak RCT (www.sensormatic.com/shoppertrak-retail-traffic-insights) has been measuring purchases at a sample of 45,000 mall-based stores and using this information to advise clients quickly of changes in shopping trends. This application of sampling is part of the relatively new field of *retail intelligence*. In this section, you will learn the differences between a **sample** and a **population** and why sometimes a sample is necessary or desirable.

LO 2-6

Use the correct terminology for samples and populations.

Population or Sample?

Population	All of the items that we are interested in. May be either finite (e.g., all of the passengers on a particular plane) or effectively infinite (e.g., all of the Cokes produced in an ongoing bottling process).
Sample	A subset of the population that we will actually analyze.

Sample or Census?

A *sample* involves looking only at some items selected from the population, while a **census** is an examination of all items in a defined population. The accuracy of a census can be illusory. For example, the 2000 U.S. decennial census is believed to have overcounted by 1.3 million people, while the 2010 census count is thought to have overestimated the U.S. population by only 36,000. Reasons include the extreme mobility of the U.S. population and the fact that some people do not want to be found (e.g., illegal immigrants) or do not reply to the mailed census form. Further, budget constraints make it difficult to train enough census field workers, install data safeguards, and track down incomplete responses or nonresponses. For these reasons, U.S. censuses have long used sampling in certain situations.

When the quantity being measured is volatile, there cannot be a census. For example, The Arbitron Company tracks American radio listening habits using over 2.6 million "Radio Diary Packages." For each "listening occasion," participants note start and stop times for each station. Panelists also report their age, sex, and other demographic information. Table 2.4 outlines some situations where a sample rather than a census would be preferred, and vice versa.

Table 2.4

Sample or Census?

When a Sample May Be Preferred	When a Census May Be Preferred
Infinite Population	**Small Population**
Population is of indefinite size (an assembly line can keep producing cars).	Population is small or the effort of data collection is small.
Destructive Testing	**Large Sample Size**
The act of measurement may destroy or devalue the item (battery life, vehicle crash tests).	If the required sample size approaches the population size, we might as well go ahead and take a census.
Timely Results	**Database Exists**
Sampling may yield more timely results (checking vegetables for salmonella contamination).	If data are in a database, we can examine all the cases.
Accuracy	**Legal Requirements**
Instead of spreading resources thinly to attempt a census, budget might be better spent to improve sampling.	Banks must count *all* the cash in bank teller drawers at the end of each business day.
Cost	
Even if a census is feasible, the cost, in either time or money, may exceed our budget.	
Sensitive Information	
Sexual harassment might best be studied by selected in-depth employee interviews.	

Parameters and Statistics

From a sample of n items, chosen from a population, we compute **statistics** that can be used as estimates of **parameters** found in the population. To avoid confusion, we use different symbols for each parameter and its corresponding statistic. Thus, the population mean is denoted μ (the lowercase Greek letter mu), while the sample mean is $\bar{x}$. The population proportion is denoted π (the lowercase Greek letter pi), while the sample proportion is p. Figure 2.4 illustrates this idea.

For example, suppose we want to know the mean (average) repair cost for auto air-conditioning warranty claims or the proportion (percent) of 25-year-old concertgoers who have permanent hearing loss. Because a census is impossible, these parameters would be estimated using a sample. For the sample statistics to provide good estimates of the population parameters, the population must be carefully specified and the sample must be drawn scientifically so the sample items are representative of the population.

Figure 2.4

Population versus Sample

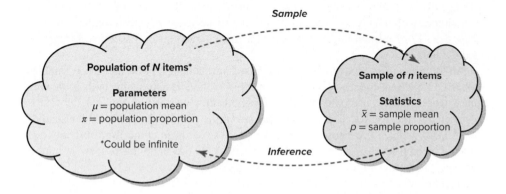

Parameter or Statistic?

Parameter   A measurement or characteristic of the population (e.g., a mean or proportion). Usually unknown because we can rarely observe the entire population. Usually (but not always) represented by a Greek letter (e.g., μ or π).

Statistic   A numerical value calculated from a sample (e.g., a mean or proportion). Usually (but not always) represented by a Roman letter (e.g., $\bar{x}$ or p).

Target Population

The **target population** contains all the individuals in which we are interested. Suppose we wish to estimate the proportion of potential consumers who would purchase a $20 Harley-Davidson desk calendar. Is the target population all drivers? Only male drivers over age 16? Only drivers with incomes over $25,000? Only motorcycle owners? By answering questions such as these, we not only identify the target population but also are forced to define our business goals more clearly. The **sampling frame** is the group from which we take the sample. If the frame differs from the target population, then our estimates might not be accurate. Examples of frames are phone directories, voter registration lists, alumni association mailing lists, or marketing databases.

> The sample for the U.S. Energy Information Administration's survey of gasoline prices is drawn from a frame of approximately 115,000 retail gasoline outlets, constructed from purchased private commercial sources and EIA sources, combined with zip codes from private lists. Individual frames are mapped to the county level by using zip codes, and outlets are assigned to standard metropolitan statistical areas from Census Bureau definitions. (For details, see www.eia.doe.gov.)

Example 2.1

Gasoline Price Survey

Mini Case 2.1

How Many People Watch the Super Bowl Each Year?

The Super Bowl, the annual championship football game between the NFC and AFC football conferences, is the most-watched television program every year. The game is broadcast on national TV, and viewership has been steadily growing since the game was first played on January 15, 1967. In 2018 (Philadelphia Eagles versus New

England Patriots), Nielsen reported that the Super Bowl had a 43.1 rating and an average 103.4 million viewers. Nielsen estimated that 68 percent of the 116 million U.S. homes with television were tuned in to the broadcast at any given minute and that there were 170.7 million Super Bowl social media interactions on Facebook, Instagram, and Twitter.

All of these measures are estimates based on samples. How are these samples collected? Nielsen collects data from a sample of households using both a viewing diary, where viewers manually track the programs they've watched, and a meter that is connected to the TV set. The sample sizes are in the magnitude of 25,000 households and 50,000 people. The people who make up these samples come from different age groups, income levels, and geography. The quality of the sampling process is important because much of these data are used to guide decisions about advertising. And by some estimates, advertising is a $70 billion industry! The Nielsen sampling process is overseen by the Media Ratings Council and is audited each year by the public accounting firm Ernst & Young. More information on its sampling techniques can be found at www.nielsen.com.

Section Exercises

2.15 Would you use a sample or a census to measure each of the following? Why?

a. The model years of the cars driven by each of your five closest friends.
b. The model years of the cars driven by each student in your statistics class.
c. The model years of the cars driven by each student in your university.

2.16 Would you use a sample or a census to measure each of the following? Why? If you are uncertain, explain the issues.

a. The mean time battery life of your laptop computer in continuous use.
b. The number of students in your statistics class who brought laptop computers to class today.
c. The average price paid for a laptop computer by students at your university.

2.17 The target population is all stocks in the S&P 500 index. Is each of the following a parameter or a statistic?

a. The average price/earnings ratio for all 500 stocks in the S&P index.
b. The proportion of all stocks in the S&P 500 index that had negative earnings last year.
c. The proportion of energy-related stocks in a random sample of 50 stocks.

LO 2-7

Explain the common sampling methods and how to implement them.

2.4 SAMPLING METHODS

There are two main categories of sampling methods. In **random sampling** items are chosen by randomization or a chance procedure. The idea of random sampling is to produce a sample that is representative of the population. **Non-random sampling** is less scientific but is sometimes used for expediency.

Random Sampling Methods

We will first discuss the four random sampling techniques shown in Table 2.5 and then describe three commonly used non-random sampling techniques, summarized in Table 2.8.

Table 2.5

Random Sampling Methods

Simple Random Sample	Use random numbers to select items from a list.
Systematic Sample	Select every *k*th item from a list or sequence.
Stratified Sample	Select randomly within defined strata (e.g., by age).
Cluster Sample	Select random geographical regions (e.g., zip codes).

We denote the population size by N and the sample size by n. In a **simple random sample**, every item in the population of N items has the same chance of being chosen in the sample of n items. A physical experiment to accomplish this would be to write each of the N data values on a poker chip and then to draw n chips from a bowl after stirring it thoroughly. But we can accomplish the same thing if the N population items appear on a numbered list, by choosing n integers between 1 and N that we match up against the numbered list of the population items.

For example, suppose we want to select one student at random from a list of 15 students (see Figure 2.5). If you were asked to "use your judgment," you would probably pick a name in the middle, thereby biasing the draw against those individuals at either end of the list. Instead we rely on a **random number** to "pick" the name using Excel's function =RANDBETWEEN(1,15) to pick a random integer between 1 and 15. There is no bias because all values from 1 to 15 are *equiprobable* (i.e., equally likely to occur). An equivalent R function for choosing a single random integer between 1 and 15 is sample(1:15,1,1).

Random person **12**

1	Adam	6	Haitham	11	Moira
2	Addie	7	Jackie	12	**Stephanie**
3	Don	8	Judy	13	Stephen
4	Floyd	9	Lindsay	14	Tara
5	Gadis	10	Majda	15	Xander

Figure 2.5

Picking on Stephanie

Sampling without replacement means that once an item has been selected to be included in the sample, it cannot be considered for the sample again. The Excel function =RANDBETWEEN(a,b) uses **sampling with replacement**. This means that the same random number could show up more than once. Using the bowl analogy, if we throw each chip back in the bowl and stir the contents before the next draw, an item can be chosen again. Instinctively most people believe that sampling without replacement is preferred over sampling with replacement because allowing duplicates in our sample seems odd. In reality, sampling without replacement can be a problem when our sample size n is close to our population size N. At some point in the sampling process, the remaining items in the population will no longer have the same probability of being selected as the items we chose at the beginning of the sampling process. This could lead to a bias (a tendency to overestimate or underestimate the parameter we are trying to measure) in our sample results. In a list of items to be sampled (a vector x), the R function sample(x, n, 1) will choose a random sample of n items with replacement, or use sample(x, n, 0) to sample without replacement.

When should we worry about sampling without replacement? Only when the population is finite and the sample size is close to the population size. Consider the Russell 3000® Index, which has 3,000 stocks. If you sample 100 stocks, without replacement, you have "used" only about 3 percent of the population. The sample size $n = 100$ is considered small relative to the population size $N = 3,000$. A common criterion is that a finite population is *effectively infinite* if the sample is less than 5 percent of the population (i.e., if $n/N < .05$). In Chapter 8, you will learn how to adjust for the effect of population size when you make a sample estimate. For now, you only need to recognize that such adjustments are of little consequence when the population is large.

Infinite Population?

When the sample is less than 5 percent of the population (i.e., when $n/N < .05$), then the population is effectively infinite. An equivalent statement is that a population is effectively infinite when it is at least 20 times as large as the sample (i.e., when $N/n \geq 20$).

Table 2.6 shows a few alternative ways to choose 10 integers between 1 and 875. All are based on a software algorithm that creates uniform decimal numbers between 0 and 1. Excel's function =RAND() does this, and many pocket calculators have a similar function. The R function runif(*n*) will generate *n* uniform decimal numbers between 0 and 1. We call these *pseudorandom* generators because even the best algorithms eventually repeat themselves. Thus, a software-based random data encryption scheme could conceivably be broken (see *The Economist,* Dec. 4, 2021, pp. 79–80). To enhance data security, Intel and other firms are examining hardware-based methods (e.g., based on thermal noise or radioactive decay) to prevent patterns or repetition. Fortunately, most applications don't require that degree of randomness. For example, Apple's music "shuffle" is not strictly random because its random numbers are generated by an algorithm that eventually repeats. However, the repeat period is so great that a user would never notice. Excel's random numbers are good enough for most purposes.

Table 2.6

Some Ways to Get 10 Random Integers between 1 and 875

Excel	Enter =RANDBETWEEN(1,875) into 10 spreadsheet cells. Press F9 to get a new sample.
R	Enter sample(875,10,1)
Internet	The website www.random.org will give you many kinds of random numbers (integers, decimals, etc.).
Pocket Calculator	Press the RAND key to get a random number in the interval [0, 1], multiply by 875, and then round up to the next integer.

Randomizing a List

To randomize a list (assuming it is in a spreadsheet), we can insert the Excel function =RAND() beside each row. This creates a column of random decimal numbers between 0 and 1. Copy the random numbers and paste them in the same column using Paste Special > Values to "fix" them (otherwise they will keep changing). Then sort all the columns by the random number column, and *voilà*—the list is now random! The first *n* items on the randomized list can now be used as a random sample. This method is especially useful when the list is very long (perhaps millions of lines). The first *n* items are a random sample of the entire list, for they are as likely as any others.

Figure 2.6

Systematic Sampling

Another method of random sampling is to choose every *k*th item from a sequence or list, starting from a randomly chosen entry among the first *k* items on the list. This is called **systematic sampling**. Figure 2.6 shows how to sample every fourth item, starting from item 2, resulting in a sample of *n* = 20 items.

An attraction of systematic sampling is that it can be used with unlistable or infinite populations, such as production processes (e.g., testing every 5,000th light bulb) or political polling (e.g., surveying every 10th voter who emerges from the polling place). Systematic sampling is also well suited to linearly organized physical populations (e.g., pulling every 10th patient folder from alphabetized filing drawers in a veterinary clinic).

A systematic sample of n items from a population of N items requires that periodicity k be approximately N/n. For example, suppose we would like to estimate the median CEO compensation for the largest 500 companies in the United States. It would be tedious to research the published compensation of 500 executives. Instead, we could choose a *systematic sample* of every 20th company in a list of firms, alphabetized by ticker symbol, starting (randomly) with the 13th company. This would yield a sample of 25 companies ($k = 500/20 = 25$). Then we could look up the compensation of the CEO in each sampled company. This reduces our research time yet provides a representative cross-section of CEOs.

Analytics in Action

Too Much Randomness?

Spotify recently altered its "shuffle" feature because it was perceived by some to be "too random." Its research suggested that customers might respond better when songs by favorite artists were stretched more evenly along a playlist. Do you agree?

There are other reasons to depart from pure random sampling. For example, you would not expect the same song to be played twice in a row. But should it be allowed to play again with only one intervening song? Two? Three? How much can we "tweak" a random algorithm without defeating its purpose? Research shows that humans try to perceive patterns even in random sequences. Might "tweaking" give rise to a different type of customer complaint?

Systematic sampling should yield acceptable results unless patterns in the population happen to recur at periodicity k. For example, weekly pay cycles ($k = 7$) would make it illogical to sample bank check cashing volume every Friday. A less obvious example would be a machine that stamps a defective part every 12th cycle due to a bad tooth in a 12-tooth gear, which would make it misleading to rely on a sample of every 12th part ($k = 12$). But periodicity coincident with k is not typical or expected in most situations.

Stratified sampling is sometimes used to improve our sample efficiency by utilizing prior information about the population. This method is applicable when the population can be divided into relatively homogeneous subgroups of known size (called *strata*). Within each *stratum,* a simple random sample of the desired size could be taken. Alternatively, a random sample of the whole population could be taken, and then individual strata estimates could be combined using appropriate weights. This procedure can reduce cost per observation and narrow the error bounds. For a population with L strata, the population size N is the sum of the stratum sizes: $N = N_1 + N_2 + \ldots + N_L$. The weight assigned to stratum j is $w_j = N_j/N$ (i.e., each stratum is weighted by its known proportion of the population).

To illustrate, suppose we want to estimate measles-mumps-rubella (MMR) vaccination rates among employees in state government, and we know that our target population (those individuals we are trying to study) is 55 percent male and 45 percent female. Suppose our budget only allows a sample of size 200. To ensure the correct gender balance, we could sample 110 males and 90 females. Alternatively, we could just take a random sample of 200 employees. Although our random sample probably will not contain *exactly* 110 males and 90 females, we can get an overall estimate of vaccination rates by *weighting* the male and female sample vaccination rates using $w_M = 0.55$ and $w_F = 0.45$ to reflect the known strata sizes.

Mini Case 2.2

Sampling for Safety

To help automakers and other researchers study the causes of injuries and fatalities in vehicle accidents, the U.S. Department of Transportation developed the National Accident Sampling System (NASS) Crashworthiness Data System (CDS). Because it is impractical to investigate every accident (there were 6,159,000 police-reported accidents in 2005), detailed data are collected in a common format from 24 primary sampling units, chosen to represent all serious police-reported motor vehicle accidents in the United States during the year. Selection of sample accidents is done in three stages: (1) The country is divided into 1,195 geographic areas called Primary Sampling Units (PSUs) grouped into 12 strata based on geographic region. Two PSUs are selected from each stratum using weights roughly proportional to the number of accidents in each stratum. (2) In each sampled PSU, a second stage of sampling is performed using a sample of Police Jurisdictions (PJs) based on the number, severity, and type of accidents in the PJ. (3) The final stage of sampling is the selection of accidents within the sampled PJs. Each reported accident is classified into a stratum based on type of vehicle, most severe injury, disposition of the injured, tow status of the vehicles, and model year of the vehicles. Each team is assigned a fixed number of accidents to investigate each week, governed by the number of researchers on a team. Weights for the strata are assigned to favor a larger percentage of higher-severity accidents while ensuring that accidents in the same stratum have the same probability of being selected, regardless of the PSU. The NASS CDS database is administered by the National Center for Statistics and Analysis (NCSA) of the National Highway Traffic Safety Administration (NHTSA). These data are currently helping to improve the government's "5 Star" crashworthiness rating system for vehicles.

See www-nrd.nhtsa.dot.gov/Pubs/NASS94.pdf.

Cluster samples are taken from strata consisting of geographical regions. We divide a region (say, a city) into subregions (say, blocks, subdivisions, or school districts). In one-stage cluster sampling, our sample consists of all elements in each of k randomly chosen subregions (or clusters). In two-stage cluster sampling, we first randomly select k subregions (clusters) and then choose a random sample of elements within each cluster. Figure 2.7 illustrates how four elements could be sampled from each of three randomly chosen clusters using two-stage cluster sampling. Because elements within a cluster are proximate, travel time and interviewer expenses are kept low. Cluster sampling is useful when

- Population frame and stratum characteristics are not readily available.
- It is too expensive to obtain a simple or stratified sample.
- The cost of obtaining data increases sharply with distance.
- Some loss of reliability is acceptable.

Although cluster sampling is cheap and quick, it is often reasonably accurate because people in the same neighborhood tend to be similar in income, ethnicity, educational background, and so on. Cluster sampling is useful in political polling, surveys of gasoline pump prices, studies of crime victimization, vaccination surveys, or determination of lead contamination in soil. A hospital may contain clusters (floors) of similar patients. A warehouse may have clusters (pallets) of inventory parts. Forest sections may be viewed as clusters to be sampled for disease or timber growth rates.

Cluster sampling is also widely used in marketing and economic surveys. The Bureau of Labor Statistics relies on multistage cluster sampling for economic indicators such as the Consumer Price Index (CPI) and employment rates. The CPI is estimated from a two-stage cluster sampling process. The sampling process begins with 87 urban areas in the United States. Within these urban areas, prices on over 200 categories are gathered from approximately 50,000 housing units and 23,000 retail establishments.

Figure 2.7

Two-Stage Cluster Sampling: Randomly Choose Three Clusters, then Randomly Choose Four Items in Each Cluster

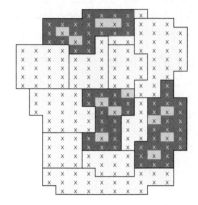

Non-Random Sampling Methods

Table 2.7 describes three commonly used non-random sampling techniques. Businesses often rely on these techniques to quickly gather data that might be used to guide informal decisions or as preliminary data to help design formal studies that use random samples.

Table 2.7

Non-Random Samples

Judgment Sample	Use expert knowledge to choose "typical" items (e.g., which employees to interview).
Convenience Sample	Use a sample that happens to be available (e.g., ask co-workers' opinions at lunch).
Focus Groups	In-depth dialogue with a representative panel of individuals (e.g., Spotify users).

Judgment sampling is a non-random sampling method that relies on the expertise of the sampler to choose items that are representative of the population. For example, to estimate the corporate spending on research and development (R&D) in the medical equipment industry, we might ask an industry expert to select several "typical" firms. Unfortunately, subconscious biases can affect experts, too. In this context, "bias" does not mean prejudice but rather *non-randomness* in the choice. Judgment samples may be the best alternative in some cases, but we can't be sure whether the sample was random. *Quota sampling* is a special kind of judgment sampling in which the interviewer chooses a certain number of people in each category (e.g., men/women).

The sole virtue of **convenience sampling** is that it is quick. The idea is to grab whatever sample is handy. An accounting professor who wants to know how many MBA students would take a summer elective in international accounting can just survey the class the professor is currently teaching. The students polled may not be representative of all MBA students, but an answer (although imperfect) will be available immediately. A newspaper reporter doing a story on perceived airport security might interview co-workers who travel frequently.

You might think that convenience sampling is rarely used or, when it is, that the results are used with caution. However, this does not appear to be the case. Because convenience samples often sound the first alarm on a timely issue, their results have a way of attracting attention and have probably influenced quite a few business decisions. The mathematical properties of convenience samples are unknowable, but they do serve a purpose, and their influence cannot be ignored.

A **focus group** is a panel of individuals chosen to be representative of a wider population, formed for open-ended discussion and idea gathering about an issue (e.g., a proposed new product or marketing strategy). Typically five to 10 people are selected, and the interactive discussion lasts one to two hours. Participants are usually individuals who do not know each other but who are prescreened to be broadly compatible yet diverse. A trained moderator guides the focus group's discussion and keeps it on track. Although not a random sampling method, focus groups are widely used, both in business and in social science research, for the insights they can yield beyond "just numbers."

Other Data Collection Methods

Businesses now have many other methods for collecting data with the use of technology. Point-of-sale (POS) systems can collect real-time data on purchases at retail or convenience stores, restaurants, and gas stations. Your customer loyalty accounts (using your phone number or e-mail) offer better deals but also build a profile of your purchasing behavior. Businesses send out e-mail surveys to loyal customers on a regular basis to get feedback on their products and services. Google also tracks Internet searches and provides these data through its Google Analytics services.

Mini Case 2.3

Pricing Accuracy

Bar code price scanning using the Universal Product Code (UPC) became the standard in most retail businesses following the rapid improvement in scanning technology during the 1970s. Since that time, federal and state agencies have monitored businesses to regulate pricing accuracy at their checkouts. Because a census is impossible for checking price accuracy, sampling is an essential tool in enforcing consumer protection laws. The National Institute for Standards and Technology (NIST) has developed a handbook for inspection agencies that provides guidance on how to conduct a pricing sampling inspection.

Arizona's Department of Weights and Measures (DWM) has set up a UPC scanner pricing sampling inspection process for the retail establishments in that state. A UPC inspection will be based on either a stratified sample (e.g., a cosmetics department) or a simple random sample from throughout the store. The inspector will select between 25 and 50 items based on the sample size recommendation from NIST. The items will be taken to the register for scanning, and the inspector will count the number of items that show a difference between the display and scanned price. Arizona requires that the retail store have 98 percent accuracy.

Between 2001 and 2006, in the state of Arizona, Walmart failed 526 price accuracy inspections. The Arizona attorney general filed a lawsuit against Walmart in 2006. The lawsuit was settled when Walmart agreed to a financial settlement of $1 million and modifications of its pricing practices.

See www.azag.gov/press-release/terry-goddard-announces-1-million-pricing-settlement-wal-mart.

Sample Size

The necessary sample size for a reliable estimate of a parameter depends on the inherent variability of the quantity being measured and the desired precision of the estimate. For example, the caffeine content of Red Bull is fairly consistent because each can is filled at the factory, so a small sample size would suffice to estimate the mean. In contrast, the amount of caffeine in an individually brewed cup of Bigelow Raspberry Royale tea varies widely because people let it steep for varying lengths of time. The purposes of the investigation, the costs of sampling, the budget, and time constraints also are taken into account in deciding on sample size. Chapter 8 explains how to choose a proper sample size.

Sources of Error or Bias

In sampling, the word *bias* does not refer to prejudice. Rather, it refers to a systematic tendency to over- or underestimate a population parameter of interest. However, the words *bias* and *error* are often used interchangeably. The word *error* generally refers to problems in sample methodology. No matter how careful you are when conducting a survey, you will encounter potential sources of error. Table 2.8 lists a few of these.

Source of Error	Characteristics
Nonresponse bias	Respondents differ from nonrespondents
Selection bias	Self-selected respondents are atypical
Response error	Respondents give false information
Coverage error	Incorrect specification of frame or population
Measurement error	Unclear survey instrument wording
Interviewer error	Responses influenced by interviewer
Sampling error	Random and unavoidable

Table 2.8

Potential Sources of Survey Error

Nonresponse bias occurs when those who respond have characteristics different from those who don't respond. For example, people with caller ID, answering machines, blocked or unlisted numbers, or cell phones are likely to be missed in telephone surveys. Because these are generally more affluent individuals, their socioeconomic class may be underrepresented in the poll. A special case is **selection bias**, a self-selected sample. For example, a talk show host who invites viewers to take a web survey about their sex lives will attract plenty of respondents. But those who are willing to reveal details of their personal lives (and who have time to complete the survey) are likely to differ substantially from those who dislike nosy surveys or are too busy (and probably weren't watching the show anyway).

Response error occurs when respondents deliberately give false information to mimic socially acceptable answers, to avoid embarrassment, or to protect personal information. Silly or hoax replies are common in anonymous online surveys.

Coverage error occurs when some important segment of the target population is systematically missed—for example, a survey of Notre Dame University alumni will fail to represent noncollege graduates or those who attended public universities. **Measurement error** results when the survey questions do not accurately reveal the construct being assessed. When the interviewer's facial expressions, tone of voice, or appearance influences the responses, data are subject to **interviewer error**.

Sampling error is uncontrollable random error that is inherent in any random sample. Even when using a random sampling method, it is possible that the sample will contain unusual responses. This cannot be prevented and is generally undetectable. It is *not* an error on your part.

2.18 The target population is all students in your university. You wish to estimate the average current Visa balance for each student. How large would the university student population have to be in order to be regarded as effectively infinite in each of the following samples?

   a. A sample of 10 students.
   b. A sample of 50 students.
   c. A sample of 100 students.

2.19 Suppose you want to know the ages of moviegoers who attend a theater's last showing on a Friday night. What kind of sample is it if you (a) survey the first 20 persons to emerge from the theater, (b) survey every 10th person to emerge from the theater, and (c) survey everyone who looks under age 12?

Section Exercises

connect

2.20 Suppose you want to study the number of e-mail accounts owned by students in your statistics class. What kind of sample is it if you (a) survey each student who has a student ID number ending in an odd number, (b) survey all the students sitting in the front row, and (c) survey every fifth student who arrives at the classroom?

2.21 Below is a 6 × 8 array containing the ages of late-night moviegoers (see file 🎬 **LateShow**). Treat this as a population. Select a random sample of 10 moviegoers' ages by using (a) simple random sampling with a random number table, (b) simple random sampling with Excel's =RANDBETWEEN() function, (c) systematic sampling, (d) judgment sampling, and (e) convenience sampling. Explain your methods.

32	34	33	12	57	13	58	16
23	23	62	65	35	15	17	20
14	11	51	33	31	13	11	58
23	10	63	34	12	15	62	13
40	11	18	62	64	30	42	20
21	56	11	51	38	49	15	21

2.22 (a) In the previous population, what was the proportion of all 48 moviegoers who were under age 30? (b) For each of the samples of size $n = 10$ that you took, what was the proportion of moviegoers under age 30? (c) Was each sample proportion close to the population proportion?

2.23 In Excel, type a list containing names for 10 of your friends into cells B1:B10. Choose three names at random by randomizing this list. To do this, enter =RAND() into cells A1:A10, copy the random column and paste it using Paste Special > Values to fix the random numbers, and then sort the list by the random column. The first three names are the random sample.

Analytics in Action

Data Privacy?

You probably know that you can be identified from just a few of your online transactions or social media postings, so you take steps to protect yourself. What about the 2020 U.S. Census that you were legally required to complete? It includes data on age, sex, race, and other characteristics of 331 million individuals. Besides informing public policy, such data are useful to businesses and organizations. To allow access yet to prevent identification of individuals, released data can be sampled, aggregated (e.g., by zip code), or otherwise anonymized. The U.S. Census Bureau requires aggregation of categorical variables with fewer than 10,000 individuals.

The trade-off is that these privacy-protecting measures diminish the research utility of the data. It is tempting to think that we could publish individual data but remove identifying data (e.g., social security numbers) to protect identity. We could remove unusual cases that might be identifiable or "jitter" the data slightly while preserving the essential relationships (e.g., adding a small amount of "noise" to income). Alas, powerful machine learning tools can probe data in ways that were not imaginable a few decades ago.

One novel method to thwart individual identification is to publish only a sample that is small enough to allow "plausible deniability" of individual identity. Another method is to publish synthetic data "like" the original in terms of essential relationships. But this presumes that the synthesizing model is appropriate. Statistics will be at the forefront of the quest to improve data utility while protecting privacy.

Source: See Claire McKay Bower, "The Art of Data Privacy," *Significance* 19, no. 1 (February 2022), pp. 14–19.

2.5 DATA SOURCES

LO 2-8

Find everyday print or electronic data sources.

One goal of a statistics course is to help you learn where to find data that might be needed. Fortunately, many excellent sources are widely available, either in libraries or through private purchase. Table 2.9 summarizes a few of them.

Table 2.9

Useful Data Sources

Type of Data	Examples
U.S. job-related data	U.S. Bureau of Labor Statistics
U.S. economic data	*Economic Report of the President*
Periodicals	*The Economist, Bloomberg Businessweek, Fortune, Forbes*
Indexes	*The New York Times, The Wall Street Journal*
Databases	Compustat, Citibase, U.S. Census
World data	*CIA World Factbook*
Web	Wikipedia, Google, Yahoo!, MSN

The U.S. Census Bureau and the U.S. Bureau of Labor Statistics are rich sources of data on many different aspects of life in the United States. The publications library supported by the Census Bureau can be found at www.census.gov. The monthly, quarterly, and annual reports published by the Bureau of Labor Statistics can be found at www.bls.gov. You can create nice charts using FRED (https://fred.stlouisfed.org/).

For annual and monthly time series economic data, try the *Economic Report of the President* (*ERP*), which is published every February. The tables in the *ERP* can be downloaded for free in Excel format. Data on cities, counties, and states can be found in the *State and Metropolitan Area Data Book,* published every few years by the Bureau of the Census and available on CD-ROM in many libraries.

Annual surveys of major companies, markets, and topics of business or personal finance are found in magazines such as *Bloomberg Businessweek, Consumer Reports, Forbes, Fortune,* and *Money.* Indexes such as the *Business Periodical Index, The New York Times Index,* and *The Wall Street Journal Index* are useful for locating topics. Financial data about publicly traded companies can be found at www.morningstar.com. Libraries have web search engines that can access many of these periodicals in abstract or full-text form.

Specialized computer databases (e.g., CRSP, Compustat, Citibase, U.S. Census) are available (at a price) for research on stocks, companies, financial statistics, and census data. The web allows us to use search engines (e.g., Google, Yahoo!, MSN) to find information. Sometimes you may get lucky, but web information is often undocumented, unreliable, or unverifiable. Better information is available through private companies or trade associations, though often at a steep price.

Often-overlooked sources of help are your university librarians. University librarians understand how to find databases and how to navigate databases quickly and accurately. Librarians can help you distinguish between valid and invalid Internet sources and then help you put the source citation in the proper format when writing reports.

Web Data Sources

Source	Website
Bureau of Economic Analysis	www.bea.gov
Bureau of Justice Statistics	www.bjs.gov
Bureau of Labor Statistics	www.bls.gov

Source	Website
Central Intelligence Agency	www.cia.gov
Economic Report of the President	www.gpo.gov/erp
Environmental Protection Agency	www.epa.gov
Federal Reserve System	www.federalreserve.gov
Food and Drug Administration	www.fda.gov
National Center for Education Statistics	nces.ed.gov
National Center for Health Statistics	www.cdc.gov/nchs
U.S. Census Bureau	www.census.gov
U.S. Federal Statistics	www.usa.gov/statistics
World Bank	www.worldbank.org
World Demographics	www.demographia.com
World Health Organization	www.who.int/en

LO 2-9

Describe basic elements of survey types, survey designs, and response scales.

2.6 SURVEYS

Survey design and administration require thoughtful preparation to capture the right information and to ensure a high response rate. What do you really need to know from the survey? What in-house staff expertise is available? What skills are best hired externally? What precision is required? To ensure a good response and get useful data, you may need to use commercial software and hire a consultant. Most survey research follows the same basic steps. These steps may overlap in time:

- Step 1: State the goals of the research.
- Step 2: Develop the budget (time, money, staff).
- Step 3: Create a research design (target population, frame, sample size).
- Step 4: Choose a survey type and method of administration.
- Step 5: Design a data collection instrument (questionnaire).
- Step 6: Pretest the survey instrument and revise as needed.
- Step 7: Administer the survey (follow up if needed).
- Step 8: Code the data and analyze it.

Designing and administering a survey is much easier than it used to be. Software can automate much of the process, allowing you to choose different question formats and visualize the layout. Because most surveys are now administered online, survey software can include features that allow the respondent to remain anonymous or prevent respondents from taking the survey twice. One free application is SurveyMonkey (www.surveymonkey.com), or you can choose a commercial vendor such as Qualtrics (www.qualtrics.com).

Survey Types

Surveys differ in cost, response rate, data quality, time required, and survey staff training requirements. **Mail surveys** require a current mailing list (people move a lot) and yield low response rates and nonresponse bias. **E-mail surveys** of customers can be well-targeted (except customers who never registered) but still face nonresponse bias. **Web surveys** work best for a well-defined target group (e.g., vehicle owners) but still have non-response bias (too busy, distrust of scams). **Telephone surveys** work best if the customer has a stake in the transaction ("Would you be willing to take a brief survey after this call to tell us how we did?").

Consider the cost per *valid* response. For example, a telephone survey might seem cheap, but over half the households in some metropolitan areas have unlisted phones, and many have answering machines or call screening to send unknown callers directly to voicemail. The sample you get may not be very useful in terms of reaching the target population.

Questionnaire Design

You should consider hiring a consultant, at least in the early stages, to help you get your survey off the ground successfully. Resources are available on the web to help you plan a survey. The American Statistical Association (www.amstat.org) offers brochures *What Is a Survey* and *How to Plan a Survey*. Additional materials are available from the Pew Research Center (www.pewresearch.org) and the Insights Association (www.insightsassociation.org). Entire books have been written to help you design and administer your own survey (see Related Reading). You can even find free survey forms (e.g., Google Surveys).

The layout must not be crowded (use lots of white space). Begin with very short, clear instructions, stating the purpose, assuring anonymity, and explaining how to submit the completed survey. Divide the survey into sections if the topics fall naturally into distinct areas. Let respondents bypass sections that aren't relevant to them. Include an "escape option" where it seems appropriate (e.g., "Don't know" or "Does not apply"). Use wording and response scales that match the reading ability and knowledge level of the intended respondents. Pretest and revise. Keep the questionnaire as short as possible. Table 2.10 lists a few common question formats and response scales.

Table 2.10

Question Format and Response Scale

Type of Question	Example
Open-ended	Briefly describe your job goals.
Fill-in-the-blank	How many times did you attend formal religious services during the last year? _____ times
	What is your most common method of communication? ❑ Cell Phone Call ❑ Text Message ❑ E-mail ❑ Facebook ❑ Other
Ranked choices	Please evaluate your dining experience:

Ranked choices:

	Excellent	Good	Fair	Poor
Food	❑	❑	❑	❑
Service	❑	❑	❑	❑
Ambiance	❑	❑	❑	❑

Pictograms: What do you think of the president's economic policies? (circle one)
☺ ☺ ☺ ☹ ☹

Likert scale: Statistics is a difficult subject.

Strongly Agree	Slightly Agree	Neither Agree nor Disagree	Slightly Disagree	Strongly Disagree
❑	❑	❑	❑	❑

Survey Validity and Reliability

Surveys are often called *instruments* because they are thought of as a measurement tool. As a measurement tool, researchers want the survey to be both **valid** and **reliable**. A valid survey is one that measures what the researcher wants to measure. Are the questions worded in such a way that the responses provide information about what the researcher wants to know? A reliable survey is one that is consistent. In other words, over time will the responses from similar respondents stay the same?

Question Wording

The way a question is asked has a profound influence on the response. For example:

Version 1: I would be disappointed if Congress cut its funding for public television.

Version 2: Cuts in funding for public television are justified to reduce federal spending.

In a poll of 1,031 people, Version 1 showed 40 percent in favor of cuts, while version 2 showed 52 percent in favor of cuts. To "rig" the poll, emotional overlays or "loaded" mental images can be attached to the question. In fact, it is often difficult to ask a neutral question without any context. For example:

Version 1:   Shall state taxes be cut?

Version 2:   Shall state taxes be cut, if it means reducing highway maintenance?

Version 3:   Shall state taxes be cut, if it means firing teachers and police?

An unconstrained choice (version 1) makes tax cuts appear to be a "free lunch," while versions 2 and 3 require the respondent to envision the consequences of a tax cut. An alternative is to use version 1 but then ask the respondent to list the state services that should be cut to balance the budget after the tax cut.

Another problem in wording is to make sure you have covered all the possibilities. For example, how could a widowed independent voter answer questions like these?

Are you married?   What is your party preference?

❑ Yes   ❑ Democrat

❑ No   ❑ Republican

Avoid overlapping classes or unclear categories. What if the father is deceased?

How old is your father?

❑ 35–45   ❑ 45–55   ❑ 55–65   ❑ 65 or older

Survey responses usually are coded numerically (e.g., $1 =$ male, $2 =$ female), although software packages also can use text variables (nominal data) in certain kinds of statistical tests. We can denote missing values by a special character (e.g., blank, period, or asterisk). If too many entries on a given respondent's questionnaire are flawed or missing, you may decide to discard the entire response. Other data screening issues include multiple responses (i.e., the respondent chose two responses where one was expected) or inconsistent replies (e.g., a 25-year-old respondent who claims to receive Medicare benefits). Sometimes a follow-up is possible, but in anonymous surveys you must make the best decisions you can about how to handle anomalous data. Document your data-coding decisions—not only for the benefit of others but also in case you are asked to explain how you did it (it is easy to forget after you have moved on to other projects).

Survey Software

Designing and creating a survey is much easier than it used to be. Software is available that automates much of the process, allowing you to use different question formats, skip questions and move to a new section, easily visualize the layout, and other features. Because most surveys are now administered online, survey software also includes features that allow the respondent to remain anonymous if warranted and prevent respondents from taking the survey twice. One of the most commonly used free applications is SurveyMonkey (www.surveymonkey.com). Qualtrics (www.qualtrics.com) offers a 14-day free trial but is used in larger enterprises such as universities. Many other software applications exist that offer similar features to SurveyMonkey or Qualtrics. It is important to remember that survey design, creation, and administration require thoughtful preparation and planning in order to capture the right information and to ensure a high response rate.

Section Exercises

Mc Graw Hill connect

2.24 What sources of error might you encounter if you want to know (a) about the dating habits of college men, so you go to a dorm meeting and ask students how many dates they have had in the last year; (b) how often people attend religious services, so you stand outside a particular church on Sunday and ask entering individuals how often they attend; (c) how often people eat at McDonald's, so you stand outside a particular McDonald's and ask entering customers how often they eat at McDonald's?

2.25 What kind of survey (mail, telephone, interview, web, direct observation) would you recommend for each of the following purposes, and why? What problems might be encountered?

    a. To estimate the proportion of students at your university who would prefer a web-based statistics class to a regular lecture.

    b. To estimate the proportion of students at your university who carry backpacks to class.

    c. To estimate the proportion of students at your university who would be interested in taking a two-month summer class in international business with tours of European factories.

2.26 What kind of survey (mail, telephone, interview, web, direct observation) would you recommend that a small laundry and dry cleaning business use for each of the following purposes, and why?

    a. To estimate the proportion of customers preferring opening hours at 7 a.m. instead of 8 a.m.

    b. To estimate the proportion of customers who have only laundry and no dry cleaning.

    c. To estimate the proportion of residents in the same zip code who spend more than $20 a month on dry cleaning.

Mini Case 2.4

Roles of Colleges

A survey of public opinion on the role of colleges was conducted by *The Chronicle of Higher Education.* Results of the survey showed that 77 percent of respondents agreed it was highly important that colleges prepare their undergraduate students for a career. The percentage of respondents who agreed it was highly important for colleges to prepare students to be responsible citizens was slightly lower, at 67 percent. The survey utilized 1,000 telephone interviews of 20 minutes each, using a random selection of men and women aged 25 through 65. It was conducted February 25, 2004. The survey was administered by TMR Inc. of Broomall, Pennsylvania. Data were collected and analyzed by GDA Integrated Services, a market research firm in Old Saybrook, Connecticut.

The Likert-type scale labels are weighted toward the positive, which is common when the survey items (roles for colleges in this case) are assumed to be potentially important and there is little likelihood of a strong negative response. Respondents also were asked for demographic information. Fifty-eight percent were women and 42 percent were men, coming from all states except Alaska and Hawaii. Eleven percent were African American (similar to the national average), but only 6 percent were Hispanic (about 8 percent below the national average). The underrepresentation of Hispanics was due to language barriers, illustrating one difficulty faced by surveys. However, the respondents' incomes, religious affiliations, and political views were similar to the general U.S. population. The random selection method was not specified. Note that firms that specialize in survey sampling generally have access to commercial lists and use their own proprietary methods.

2.27 What would be the difference in student responses to the two questions shown?

    Version 1:   I would prefer that tuition be reduced.

    Version 2:   Cuts in tuition are a good idea even if some classes are canceled.

2.28 What problems are evident in the wording of these two questions?

    What is your race?   What is your religious preference?

    ❏ White   ❏ Christian

    ❏ Black   ❏ Jewish

Chapter Summary

A **data set** consists of all the values of all the variables we have chosen to observe. It often is an array with *n* rows and *m* columns. Data sets may be **univariate** (one variable), **bivariate** (two variables), or **multivariate** (three or more variables). There are two basic data types: **categorical data** (categories that are described by labels) or **numerical** (meaningful numbers). Numerical data are **discrete** if the values are integers or can be counted or **continuous** if any interval can contain more data values. **Nominal** measurements are names, **ordinal** measurements are ranks, **interval** measurements have meaningful distances between data values, and **ratio** measurements have meaningful ratios and a zero reference point. **Time series** data are observations measured at *n* different points in time or over sequential time intervals, while **cross-sectional** data are observations among *n* entities such as individuals, firms, or geographic regions. Among **random samples, simple random** samples pick items from a list using random numbers, **systematic** samples take every *k*th item, **cluster** samples select geographic regions, and **stratified** samples take into account known population proportions. **Non-random** samples include convenience or judgment samples, gaining time but sacrificing randomness. **Focus groups** give in-depth information. **Survey design** requires attention to question **wording** and **scale definitions. Survey techniques** (mail, telephone, interview, web, direct observation) depend on time, budget, and the nature of the questions and are subject to various sources of error.

Key Terms

binary variable	discrete data	observation	sampling with replacement
bivariate data sets	focus group	ordinal data	sampling without replacement
categorical data	interval data	parameter	selection bias
census	interviewer error	population	simple random sample
cluster sample	judgment sampling	random numbers	statistics
coding	Likert scale	random sampling	stratified sampling
continuous data	measurement error	ratio data	systematic sampling
convenience sampling	multivariate data sets	reliability	target population
coverage error	nominal data	response error	time series data
cross-sectional data	non-random sampling	sample	univariate data sets
data	nonresponse bias	sampling error	validity
data set	numerical data	sampling frame	variable

Chapter Review

1. Define (a) data, (b) data set, (c) observation, and (d) variable.

2. How do business data differ from scientific experimental data?

3. Distinguish (a) univariate, bivariate, and multivariate data; (b) discrete and continuous data; (c) numerical and categorical data.

4. Define the four measurement levels and give an example of each.

5. Explain the difference between cross-sectional data and time series data.

6. (a) List three reasons why a census might be preferred to a sample. (b) List three reasons why a sample might be preferred to a census.

7. (a) What is the difference between a parameter and a statistic? (b) What is a target population?

8. (a) List four methods of random sampling. (b) List two methods of non-random sampling. (c) Why would we ever use non-random sampling? (d) Why is sampling usually done without replacement?

9. List five (a) steps in a survey, (b) issues in survey design, (c) survey types, (d) question scale types, and (e) sources of error in surveys.

10. List advantages and disadvantages of different types of surveys.

Chapter Exercises

DATA TYPES

2.29 Which type of data (categorical, discrete numerical, continuous numerical) is each of the following variables?

  a. Age of a randomly chosen tennis player in the Wimbledon tennis tournament.

  b. Nationality of a randomly chosen tennis player in the Wimbledon tennis tournament.

  c. Number of double faults in a randomly chosen tennis game at Wimbledon.

2.30 Which type of data (categorical, discrete numerical, continuous numerical) is each of the following variables?

  a. Number of spectators at a randomly chosen Wimbledon tennis match.

  b. Water consumption (liters) by a randomly chosen Wimbledon player during a match.

  c. Gender of a randomly chosen tennis player in the Wimbledon tennis tournament.

2.31 Which measurement level (nominal, ordinal, interval, ratio) is each of the following variables?

a. A customer's ranking of five new hybrid vehicles.

b. Noise level 100 meters from the Dan Ryan Expressway at a randomly chosen moment.

c. Number of occupants in a randomly chosen commuter vehicle on the San Diego Freeway.

2.32 Which measurement level (nominal, ordinal, interval, ratio) is each of the following variables?

a. Number of annual office visits by a Medicare subscriber.

b. Daily caffeine consumption by a six-year-old child.

c. Type of vehicle driven by a college student.

2.33 Below are five questions from a survey of MBA students. Answers were written in the blank at the left of each question. For each question, state the data type (categorical, discrete numerical, or continuous numerical) and measurement level (nominal, ordinal, interval, ratio). Explain your reasoning. If there is doubt, discuss the alternatives.

_____ Q1 Can you conduct simple transactions in a language other than English? (0 = No, 1 = Yes)

_____ Q2 What is your approximate undergraduate college GPA? (1.0 to 4.0)

_____ Q3 About how many hours per week do you expect to work at an outside job this semester?

_____ Q4 What do you think is the ideal number of children for a married couple?

_____ Q5 On a 1 to 5 scale, which best describes your parents? 1 = Mother clearly dominant ↔ 5 = Father clearly dominant

2.34 Below are five questions from a survey of MBA students. Answers were written in the blank at the left of each question. For each question, state the data type (categorical, discrete numerical, or continuous numerical) and measurement level (nominal, ordinal, interval, ratio). Explain your reasoning. If there is doubt, discuss the alternatives.

_____ Q6 On a 1 to 5 scale, assess the current job market for your undergraduate major. 1 = Very bad ↔ 5 = Very good

_____ Q7 During the last month, how many times has your schedule been disrupted by car trouble?

_____ Q8 About how many years of college does the more-educated one of your parents have? (years)

_____ Q9 During the last year, how many traffic tickets (excluding parking) have you received?

_____ Q10 Which political orientation most nearly fits you? (1 = Liberal, 2 = Middle-of-Road, 3 = Conservative)

2.35 Write the required Excel function.

a. Choose a random integer between 1 and 100.

b. Choose a random decimal between 0 and 1.

c. Choose a random decimal between 0 and 100.

2.36 Identify the following data as either time series or cross-sectional.

a. The 2023 CEO compensation of the 500 largest U.S. companies.

b. The annual compensation for the CEO of Coca-Cola Enterprises from 2018 to 2023.

c. The weekly revenue for a Noodles & Company restaurant for the 52 weeks in 2023.

d. The number of skiers on the mountain on Christmas Day 2023 at each of the ski mountains owned by Vail Resorts.

2.37 Identify the following data as either time series or cross-sectional.

a. The number of rooms booked each night for the month of January 2023 at a Vail Resorts hotel.

b. The amount spent on books at the start of this semester by each student in your statistics class.

c. The number of Caesar salads sold for the first week of April this year at each Noodles & Company restaurant.

d. The stock price of Coca-Cola Enterprises on the last trading day of the year for each of the last 10 years.

SAMPLING METHODS

2.38 Would you use a sample or a census to measure each of the following? Why? If you are uncertain, explain the issues.

a. The number of cans of Campbell's soup on your local supermarket's shelf today at 6:00 p.m.

b. The proportion of soup sales last week in Boston that was sold under the Campbell's brand.

c. The proportion of Campbell's brand soup cans in your family's pantry.

2.39 Would you use a sample or census to measure each of the following?

a. The number of workers currently employed by Campbell Soup Company.

b. The average price of a can of Campbell's Cream of Mushroom soup.

c. The total earnings of workers employed by Campbell Soup Company last year.

2.40 Is each of the following a parameter or a statistic? If you are uncertain, explain the issues.

a. The number of cans of Campbell's soup sold last week at your local supermarket.

b. The proportion of all soup in the United States that was sold under the Campbell's brand last year.

c. The proportion of Campbell's brand soup cans in the family pantries of 10 students.

2.41 Is each of the following a parameter or statistic?

a. The number of visits to a pediatrician's office last week.

b. The number of copies of John Grisham's most recent novel sold to date.

c. The total revenue realized from sales of John Grisham's most recent novel.

2.42 Recently, researchers estimated that 76.8 percent of global e-mail traffic was spam. Could a census be used to update this estimate? Why or why not?

2.43 A certain health maintenance organization (HMO) is studying its daily office routine. It collects information on three variables: the number of patients who visit during a day, the patient's complaint, and the waiting time until each patient sees a doctor. (a) Which variable is categorical? (b) Identify the two quantitative variables and state whether they are discrete or continuous.

2.44 There are 327 official ports of entry in the United States. The Department of Homeland Security selects 15 ports of entry at random to be audited for compliance with screening procedures of incoming travelers through the primary and secondary vehicle and pedestrian lanes. What kind of sample is this (simple random, systematic, stratified, cluster)?

2.45 The IRS estimates that the average taxpayer spent 3.7 hours preparing Form 1040 to file a tax return. Could a census be used to update this estimate for the most recent tax year? Why or why not?

2.46 The General Accounting Office conducted random testing of retail gasoline pumps in Michigan, Missouri, Oregon, and Tennessee. The study concluded that 49 percent of gasoline pumps nationwide are mislabeled by more than one-half of an octane point. What kind of sampling technique was most likely to have been used in this study?

2.47 Arsenic (a naturally occurring, poisonous metal) in home water wells is a common threat. (a) What sampling method would you use to estimate the arsenic levels in wells in a rural county to see whether the samples violate the EPA limit of 10 parts per billion (ppb)? (b) Is a census possible?

2.48 Would you expect Starbucks to use a sample or census to measure each of the following? Explain.

   a. The percentage of repeat customers at a certain Starbucks on Saturday mornings.
   b. The number of chai tea latte orders last Saturday at a certain Starbucks.
   c. The average temperature of Starbucks coffee served on Saturday mornings.
   d. The revenue from coffee sales as a percentage of Starbucks' total revenue last year.

2.49 Would you expect Noodles & Company to use a sample or census to measure each of the following? Explain.

   a. The annual average weekly revenue of each Noodles restaurant.
   b. The average number of weekly lunch visits by customers.
   c. The customer satisfaction rating of a new dessert.

   d. The number of weeks in a year that a restaurant sells more bottled beverages than fountain drinks.

2.50 A financial magazine publishes an annual list of major stock funds. Last year, the list contained 1,699 funds. What method would you recommend to obtain a sample of 20 stock funds?

2.51 Examine each of the following statistics. Which sampling method was most likely to have been used (simple random, systematic, stratified, cluster)?

   a. A survey showed that 30 percent of U.S. businesses have fired an employee for inappropriate web surfing, such as gambling, watching porn, or shopping.
   b. Surveyed doctors report that 59 percent of patients do not follow their prescribed treatment.
   c. The Internal Revenue Service reports that, based on a sample of individual taxpayers, 80 percent of those who failed to pay what they owed did so through honest errors or misinterpretation of the tax code.
   d. In Spain, per capita consumption of cigarettes is 1,265 compared with 1,083 in the United States.

2.52 The Department of Health and Human Services performed a detailed audit of adverse medical events on a random sample of 780 drawn at random without replacement by assigning a random number to each patient on a list of 999,645 Medicare patients who were discharged from acute care hospitals and then choosing random integers between 1 and 999,645. (a) What kind of sample is this (random, systematic, stratified, cluster)? (b) Is this population effectively infinite?

2.53 Prior to starting a recycling program, a city decides to measure the quantity of garbage produced by single-family homes in various neighborhoods. This experiment will require weighing garbage on the day it is set out. (a) What sampling method would you recommend, and why? (b) What would be a potential source of sampling error?

2.54 A university wanted to survey alumni about their interest in lifelong learning classes. They mailed questionnaires to a random sample of 600 alumni from their database of over 30,000 recent graduates. Would you consider this population to be effectively infinite?

2.55 The U.S. Fisheries and Wildlife Service requires that scallops harvested from the ocean must weigh at least 1/36 pound. The harbormaster at a Massachusetts port randomly selected 18 bags of scallops from 11,000 bags on an arriving vessel. The average scallop weight from the 18 bags was 1/39 pound. (a) Would the population of 11,000 bags be considered effectively infinite in this case? (b) Which value represents a sample statistic: 1/36 or 1/39? (See *Interfaces* 25, no. 2 [March–April 1995], p. 18.)

2.56 A marketing research group wanted to collect information from existing and potential customers on the appeal of a new product. They sent out surveys to a random sample of 1,200 people from their database of over 25,000 current and potential customers. Would you consider this population to be effectively infinite?

2.57 Households can sign up for a telemarketing "no-call list." How might households who sign up differ from those who don't? What biases might this create for telemarketers promoting (a) financial planning services, (b) carpet cleaning services, and (c) vacation travel packages?

SURVEYS AND SCALES

2.58 Suggest response check boxes for these questions. In each case, what difficulties do you encounter as you try to think of appropriate check boxes?

   a. Where are you employed?
   b. What is the biggest issue facing the next U.S. president?
   c. Are you happy?

2.59 Suggest both a Likert scale question and a response scale to measure the following:

   a. A student's rating of a particular statistics professor.
   b. A voter's satisfaction with the president's economic policy.
   c. A patient's perception of waiting time to see a doctor.

2.60 What level of measurement (nominal, ordinal, interval, ratio) is appropriate for the movie rating system that you see in *TV Guide* (☆, ☆☆, ☆☆☆, ☆☆☆☆)? Explain your reasoning.

2.61 Insurance companies are rated by several rating agencies. The Fitch 20-point scale is AAA, AA+, AA, AA−, A+, A, A−, BBB+, BBB, BBB−, BB+, BB, BB−, B+, B, B−, CCC+, CCC, CCC−, DD. (a) What level of measurement does this scale use? (b) To assume that the scale uses interval measurements, what assumption is required?

2.62 A tabletop survey by a restaurant asked the question shown below. (a) What kind of response scale is this? (b) Suggest an alternative response scale that would be more sensitive to differences in opinion. (c) Suggest possible sources of bias in this type of survey.

Were the food and beverage presentations appealing?
      ❑ Yes   ❑ No

MINI-PROJECTS

2.63 Below are 64 names of employees at NilCo. Colors denote different departments (finance, marketing, purchasing, engineering). Sample eight names from the display shown by using (a) simple random sampling, (b) systematic sampling, and (c) cluster sampling. Try to ensure that every name has an equal chance of being picked. Which sampling method seems most appropriate? 📁 **PickEight**

Finance			
Floyd	Anne	Erik	Peter
Nathan	Sid	Joel	Jackie
Lou	Ginnie	LaDonna	Moira
Loretta	Tim	Mario	
Marketing			
Don	Gadis	Balaji	Al
Graham	Scott	Lorin	Vince
Bernie	Karen	Ed	Liz
Purchasing			
Tom	Dmitri	Erika	Dan
Claudia	Mike	Nicholas	Brian
Jean	Molly	Kevin	Margie
Juanita	Ted	Jeremy	
Marnie	Takisha	Carl	
Mabel	Jody	Greg	
Engineering			
Bonnie	Duane	Deepak	Amanda
Blythe	Ryan	Dave	Sam
Chad	Tania	Doug	Ralph
Buck	Gene	Janet	Pam

2.64 From the display below, pick five cards (without replacement) by using random numbers. Explain your method. Why would the other sampling methods not work well in this case?

A ♠	A ♥	A ♣	A ♦
K ♠	K ♥	K ♣	K ♦
Q ♠	Q ♥	Q ♣	Q ♦
J ♠	J ♥	J ♣	J ♦
10 ♠	10 ♥	10 ♣	10 ♦
9 ♠	9 ♥	9 ♣	9 ♦
8 ♠	8 ♥	8 ♣	8 ♦
7 ♠	7 ♥	7 ♣	7 ♦
6 ♠	6 ♥	6 ♣	6 ♦
5 ♠	5 ♥	5 ♣	5 ♦
4 ♠	4 ♥	4 ♣	4 ♦
3 ♠	3 ♥	3 ♣	3 ♦
2 ♠	2 ♥	2 ♣	2 ♦

2.65 Photocopy the exhibit below (omit these instructions) and show it to a friend or classmate. Ask that individual to choose a number at random and write it on a piece of paper. Collect the paper. Repeat for *at least* 20 friends/classmates. Tabulate the results. Were all the numbers chosen equally often? If not, which were favored or avoided? Why? 📁 **PickOne**

0	11	17	22
8	36	14	18
19	28	6	41
12	3	5	0

2.66 Ask each of 20 friends or classmates to choose a whole number between 1 and 5. Tabulate the results. Do the results seem random? If not, can you think of any reasons?

2.67 You can test Excel's algorithm for selecting random integers with a simple experiment. Enter =RANDBETWEEN(1,2) into cell A1 and then copy it to cells A1:E20. This creates a data block of 100 cells containing either a one or a two. In cell G1 type =COUNTIF(A1:E20,"=1") and in cell G2 type =COUNTIF(A1:E20,"=2"). Highlight cells G1 and G2 and create a column chart. Click on the vertical axis scale and set the lower limit to 0 and upper limit to 100. Then hold down the F9 key and observe the chart. Are you convinced that, on average, you are getting about 50 ones and 50 twos? *Ambitious Students:* Generalize this experiment to integers 1 through 5. 📁 **RandBetween**

Related Reading

Guides to Data Sources

Clayton, Gary E., and Martin Giesbrecht. *A Guide to Everyday Economic Statistics.* 8th ed. McGraw Hill, 2019.

Sampling and Surveys

Best, Joel. *Damned Lies and Statistics: Untangling Numbers from the Media, Politicians, and Activists.* University of California Press, 2012.

Fowler, Floyd J. *Survey Research Methods.* 5th ed. Sage, 2019.

Lohr, Sharon L. *Sampling: Design & Analysis.* 3rd ed. Chapman and Hall/CRC, 2021.

Mathieson, Kieran, and David P. Doane. "Using Fine-Grained Likert Scales in Web Surveys." *Alliance Journal of Business Research* 1, no. 1 (2006), pp. 27–34.

Scheaffer, Richard L., William Mendenhall, and R. Lyman Ott. *Elementary Survey Sampling.* 7th ed. Brooks/Cole, 2012.

Schindler, Pamela. *Business Research Methods.* 14th ed. McGraw Hill, 2022.

Thompson, Steven K. *Sampling.* 3rd ed. Wiley, 2012.

CHAPTER 2 More Learning Resources

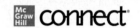

You can access these *LearningStats* demonstrations through Connect to help you understand random sampling.

Topic	LearningStats Demonstrations
Sampling	⊠ Sampling Methods ⊠ Who Gets Picked? ⊠ Randomizing a List ⊠ Pick a Card ⊠ Excel's RANDBETWEEN Function
Data sources	⊡ Web Stats Resources ⊡ Survey Tips ⊡ Sampling Plans ⊡ Sampling Using R

Key: ⊠ = Excel ⊡ = PowerPoint

CHAPTER

3 Describing Data Visually

CHAPTER CONTENTS

CHAPTER LEARNING OBJECTIVES

When you finish this chapter, you should be able to

LO 3-1   Make a stem-and-leaf or dot plot.

LO 3-2   Create a frequency distribution for a data set.

LO 3-3   Make a histogram with appropriate bins.

LO 3-4   Identify skewness, modal classes, and outliers in a histogram.

LO 3-5   Make an effective line chart.

LO 3-6   Make and interpret a scatter plot.

LO 3-7   Make an effective column chart or bar chart.

LO 3-8   Make an effective pie chart.

LO 3-9   Make simple tables and pivot tables.

LO 3-10   Recognize deceptive graphing techniques.

3.1 STEM-AND-LEAF DISPLAYS AND DOT PLOTS

Managers need information that can help them identify trends and adjust to changing conditions. But it is hard to assimilate piles of raw data. How can a business analyst convert raw data into useful information? Statistics offers methods that can help organize, explore, and summarize data in a succinct way. The methods may be *visual* (charts and graphs) or *numerical* (statistics or tables). In this chapter, you will see how visual displays can provide insight into the characteristics of a data set *without* using mathematics. The type of graph you use to display your data is dependent on the type of data you have. Some charts are better suited for quantitative data, while others are better for displaying categorical data. This chapter explains several basic types of charts, offers guidelines on when to use them, advises you how to make them effective, and warns of ways that charts can be deceptive.

We begin with a set of n observations $x_1, x_2, \ldots, x_n$ on one variable (univariate data). Such data can be discussed in terms of three characteristics: **center**, **variability**, and **shape**. Table 3.1 summarizes these characteristics as *questions* that we will be asking about the data.

LO 3-1

Make a stem-and-leaf or dot plot.

Table 3.1

Characteristics of Univariate Data

Characteristic	Interpretation
Measurement	What are the units of measurement (e.g., dollars)? Are the data values integers or continuous? Any missing observations? Any concerns with accuracy or sampling methods?
Center	Where are the data values concentrated? What seem to be typical or middle data values?
Variability	How much dispersion is there in the data? How spread out are the data values? Are there unusual values?
Shape	Are the data values distributed symmetrically? Skewed? Sharply peaked? Flat? Bimodal?

Preliminary Assessment

Before calculating any statistics or drawing any graphs, it is a good idea to *look at the data* and try to visualize how they were collected. Because the companies in the S&P 500 index are publicly traded, they are required to publish verified financial information, so the accuracy of the data is not an issue. Because the intent of the analysis is to study the S&P 500 companies

Example 3.1

Price/Earnings Ratios

Price/earnings (P/E) ratios—current stock price divided by earnings per share in the last 12 months—show how much an investor is willing to pay for a stock based on the stock's earnings. P/E ratios are also used to determine how optimistic the market is for a stock's growth potential. Investors may be willing to pay more for a lower-earning stock than a higher-earning stock if they see potential for growth. Table 3.2 shows P/E ratios for a random sample of companies ($n = 44$) from Standard & Poor's 500 index. We are interested in learning how the P/E ratios of the companies in the S&P 500 compare to each other and what the distribution of P/E ratios looks like. Visual displays can help us describe and summarize the main characteristics of this sample.

Table 3.2

P/E Ratios for 44 Companies 📁 **PERatios**

Company	P/E Ratio	Company	P/E Ratio	Company	P/E Ratio
American Tower Corp. A	59	FMC Corporation	20	NetApp	37
Analog Devices Inc.	16	Gap (The)	12	Occidental Petroleum	19
Applied Materials Inc.	20	The Hartford	11	O'Reilly Automotive	22
Best Buy Co. Inc.	10	Hess Corporation	7	PepsiCo Inc.	16
Big Lots Inc.	11	Hospira Inc.	24	PG&E Corp.	16
Carefusion Corporation	38	Intel Corp.	11	PPL Corp.	14
Coventry Health Care Inc.	10	Invesco Ltd.	27	Reynolds American Inc.	19
Cummins Inc.	23	Johnson Controls	17	Roper Industries	26
Dell Inc.	13	King Pharmaceuticals	42	Starbucks Corp.	26
Dentsply International	18	Kroger Co.	13	Sunoco Inc.	28
Donnelley (R.R.) & Sons	31	Macy's Inc.	17	Titanium Metals Corp.	50
Eastman Chemical	16	Mattel Inc.	14	United Health Group Inc.	9
Entergy Corp.	10	Medco Health Solutions Inc.	21	Ventas Inc.	37
Exelon Corp.	10	MetroPCS Communications Inc.	21	Wal-Mart Stores	13
Fiserv Inc.	18	Murphy Oil	15		

Source: www.finance.yahoo.com, accessed 12-30-2010. Each S&P 500 company on an Alphabetical list was assigned a random number using Excel's =RAND() function. Companies were then sorted on the =RAND() column. The first 44 companies on the sorted list were chosen as a random sample.

at a *point in time,* these are *cross-sectional* data. (Financial analysts also study time series data on P/E ratios, which vary daily as stock prices change.) Although rounded to integers, the measurements are continuous. For example, a stock price of $43.22 divided by earnings per share of $2.17 gives a P/E ratio of 43.22/2.17 = 19.92, which would be rounded to 20 for convenience. Because there is a true zero (refer back to Figure 2.3), we can speak meaningfully of ratios and can perform any standard mathematical operations. Because the analysis is based on a sample (not a census), we must allow for *sampling error,* that is, the possibility that our sample is not representative of the population of all 500 S&P 500 firms due to the nature of random sampling.

As a first step, it is helpful to sort the data. This is a visual display, although a very simple one. From the sorted data, we can see the range, the frequency of occurrence for each data value, and the data values that lie near the middle and ends.

44 Sorted Price/Earnings Ratios

7	9	10	10	10	10	11	11	11	12	13
13	13	14	14	15	16	16	16	16	17	17
18	18	19	19	20	20	21	21	22	23	24
26	26	27	28	31	37	37	38	42	50	59

When the number of observations is large, a sorted list of data values is difficult to analyze. Further, a list of numbers may not reveal very much about center, variability, and shape. To see broader patterns in the data, analysts often prefer a *visual display* of the data.

Stem-and-Leaf Display

One simple way to visualize small data sets is a **stem-and-leaf plot**. The stem-and-leaf plot is a tool of *exploratory data analysis* (EDA) that seeks to reveal essential data features in an intuitive way. A stem-and-leaf plot is basically a frequency tally, except that we use digits instead of tally marks. For two-digit or three-digit integer data, the *stem* is the tens digit of the data, and the *leaf* is the ones digit. Excel does not make stem-and-leaf plots, but you can get them from Minitab or MegaStat. To plot the 44 P/E ratios in R, we could create a vector named something like PERatios and then use the command stem(PERatios) to get this:

```
0 |  79
1 |  00001112333445666778899
2 |  00112346678
3 |  1778
4 |  2
5 |  09
```

For example, the data values in the fourth stem are 31, 37, 37, 38. We always use equally spaced stems (even if some stems are empty). The stem-and-leaf can reveal *central tendency* (24 of the 44 P/E ratios were in the 10–19 stem) as well as *dispersion* (the range is from 7 to 59). The stem-and-leaf has the advantage that we can retrieve the raw data by concatenating a *stem digit* with each of its *leaf digits*. For example, the last stem has data values 50 and 59.

A stem-and-leaf plot works well for small samples of integer data with a limited range but becomes awkward when you have decimal data (e.g., $60.39) or multidigit data (e.g., $3,857). In such cases, it is necessary to round the data to make the display "work." Although the stem-and-leaf plot is rarely seen in presentations of business data, it is a useful tool for a quick tabulation of small data sets.

Dot Plots

A **dot plot** is another simple graphical display of n individual values of numerical data. The basic steps in making a dot plot are to (1) make a scale that covers the data range, (2) mark axis demarcations and label them, and (3) plot each data value as a dot above the scale at its approximate location. If more than one data value lies at approximately the same X-axis location, the dots are piled up vertically as shown here for the 44 P/E ratios.

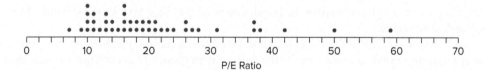

Dot plots are an attractive tool for data exploration because they are easy to understand. A dot plot shows *variability* by displaying the range of the data. It shows the *center* by revealing where the data values tend to cluster and where the midpoint lies. A dot plot also can reveal the *shape* of the distribution if the sample is large enough.

While they are easy to understand, dot plots have limitations. They don't reveal very much information about the data set's shape when the sample is small, and they become awkward when the sample is large (what if you have 100 dots at the same point?) or when you have decimal data. The next section of this chapter explains how to use Excel to create histograms and other visual displays that work for any sample size. You can make a dot plot yourself (if the sample is small) using a straightedge and a pencil. Excel doesn't offer dot plots, but you can get them from most statistical software packages.

You may download data files indicated by the symbol 🗁 followed by a file name (e.g., 🗁 **Housing**) from the problems in Connect or through your instructor. At the end of each chapter, there are additional learning resources that expand on topics in the textbook (e.g., Excel demonstrations).

Section Exercises

Mc Graw Hill **connect**

3.1 (a) Make a stem-and-leaf plot for these 24 observations on the number of customers who used a downtown Citibank ATM during the noon hour on 24 consecutive workdays. (b) Make a dot plot of the ATM data. (c) Describe these two displays. (*Hint:* Refer to center, variability, and shape.) 🗁 **Citibank**

39	32	21	26	19	27	32	25
18	26	34	18	31	35	21	33
33	9	16	32	35	42	15	24

3.2 (a) Make a stem-and-leaf plot for the number of defects per 100 vehicles for these 32 brands. (b) Make a dot plot of the defects data. (c) Describe these two displays. (*Hint:* Refer to center, variability, and shape.)

Defects per 100 Vehicles (alphabetical by brand) 🗁 JDPower					
Brand	**Defects**	**Brand**	**Defects**	**Brand**	**Defects**
Acura	86	Hyundai	102	MINI	133
Audi	111	Infiniti	107	Mitsubishi	146
BMW	113	Jaguar	130	Nissan	111
Buick	114	Jeep	129	Porsche	83
Cadillac	111	Kia	126	Ram	110
Chevrolet	111	Land Rover	170	Scion	114
Chrysler	122	Lexus	88	Subaru	121
Dodge	130	Lincoln	106	Toyota	117
Ford	93	Mazda	114	Volkswagen	135
GMC	126	Mercedes-Benz	87	Volvo	109
Honda	95	Mercury	113		

Source: J.D. Power and Associates 2010 Initial Quality Study™.

Mini Case 3.1

U.S. Business Cycles

Businesses need to anticipate the probable length of a recession to form strategies for debt management and future product releases. Fortunately, good historical data are available from the National Bureau of Economic Research, which keeps track of business cycles. The length of a contraction is measured from the peak of the previous expansion to the beginning of the next expansion based on the real gross domestic product (GDP). Table 3.3 shows the durations, in months, of 33 U.S. recessions.

From the dot plot in Figure 3.1, we see that the 65-month contraction (1873–1879) was quite unusual, although four recessions did exceed 30 months. Most recessions have lasted less than 20 months. Only 7 of 33 lasted less than 10 months. The 2-month 2022 recession during the COVID-19 epidemic was the shortest.

Table 3.3

U.S. Business Contractions, 1857–2021 (n = 34) 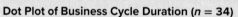 **Recessions**

Peak	Trough	Months	Peak	Trough	Months
June 1857 (II)	December 1858 (IV)	18	May 1923 (II)	July 1924 (III)	14
October 1860 (III)	June 1861 (III)	8	October 1926 (III)	November 1927 (IV)	13
April 1865 (I)	December 1867 (I)	32	August 1929 (III)	March 1933 (I)	43
June 1869 (II)	December 1870 (IV)	18	May 1937 (II)	June 1938 (II)	13
October 1873 (III)	March 1879 (I)	65	February 1945 (I)	October 1945 (IV)	8
March 1882 (I)	May 1885 (II)	38	November 1948 (IV)	October 1949 (IV)	11
March 1887 (II)	April 1888 (I)	13	July 1953 (II)	May 1954 (II)	10
July 1890 (III)	May 1891 (II)	10	August 1957 (III)	April 1958 (II)	8
January 1893 (I)	June 1894 (II)	17	April 1960 (II)	February 1961 (I)	10
December 1895 (IV)	June 1897 (II)	18	December 1969 (IV)	November 1970 (IV)	11
June 1899 (III)	December 1900 (IV)	18	November 1973 (IV)	March 1975 (I)	16
September 1902 (IV)	August 1904 (III)	23	January 1980 (I)	July 1980 (III)	6
May 1907 (II)	June 1908 (II)	13	July 1981 (III)	November 1982 (IV)	16
January 1910 (I)	January 1912 (IV)	24	July 1990 (III)	March 1991 (I)	8
January 1913 (I)	December 1914 (IV)	23	March 2001 (I)	November 2001 (IV)	8
August 1918 (III)	March 1919 (I)	7	December 2007 (IV)	March 2009 (I)	18
January 1920 (I)	July 1921 (III)	18	February 2020 (I)	April 2020 (II)	2

Source: U.S. Business Contractions found at www.nber.org.

The table supplies information that the dot plot cannot. For example, during the 1930s, there were actually *two* major contractions (43 months from 1929 to 1933, 13 months from 1937 to 1938), which is one reason why that period seemed so terrible to those who lived through it. The Great Depression of the 1930s was so named because it lasted a long time and the economic decline was deeper than in most recessions.

Figure 3.1

Dot Plot of Business Cycle Duration (n = 34)

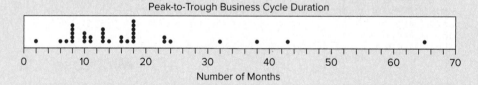

Peak-to-Trough Business Cycle Duration

Number of Months

FREQUENCY DISTRIBUTIONS AND HISTOGRAMS

LO 3-2

Create a frequency distribution for a data set.

Frequency Distributions

A **frequency distribution** is a table formed by classifying n numerical data values into k classes called *bins*. The table shows the *frequency* of data values that fall within each bin. Frequencies also can be expressed as *relative frequencies* or *percentages* of the total number of observations.

Frequency Distribution

A tabulation of n data values into k classes called *bins,* based on values of the data. The *bin limits* are cutoff points that define each bin. Bins generally have equal interval widths, and their limits cannot overlap.

The basic steps for constructing a frequency distribution are to (1) sort the data in ascending order, (2) choose the number of bins, (3) set the bin limits, (4) put the data values in the appropriate bin, and (5) create the table. Let's walk through these steps.

Step 1: Find Smallest and Largest Data Values

Sorted Price/Earnings Ratios										
7	9	10	10	10	10	11	11	11	12	13
13	13	14	14	15	16	16	16	16	17	17
18	18	19	19	20	20	21	21	22	23	24
26	26	27	28	31	37	37	38	42	50	59

For the P/E data, we get $x_{min} = 7$ and $x_{max} = 59$ (highlighted). You might be able to find x_{min} and x_{max} without sorting the entire data set, but it is easier to experiment with bin choices if you have already sorted the data.

Step 2: Choose Number of Bins   Because a frequency distribution seeks to condense many data points into a small table, we expect the number of bins k to be much smaller than the sample size n. When you use *too many* bins, some bins are likely to be sparsely populated or even empty. With *too few* bins, dissimilar data values are lumped together. Left to their own devices, people tend to choose similar bin limits for a given data set. Generally, larger samples justify more bins. According to **Sturges' Rule**, a guideline proposed by statistician Herbert Sturges, every time we double the sample size, we should add one bin, as shown in Table 3.4.

Sample Size (*n*)	Suggested Number of Bins (*k*)
16	5
32	6
64	7
128	8
256	9
512	10
1,024	11

Table 3.4

Sturges' Rule

For the sample sizes you are likely to encounter, Table 3.4 says that you would expect to use from $k = 5$ to $k = 11$ bins. Sturges' Rule can be expressed as a formula:

(3.1)   Sturges' Rule: $k = 1 + 3.3 \log(n)$

For the P/E data ($n = 44$), Sturges' Rule suggests

$$k = 1 + 3.3 \log(n) = 1 + 3.3 \log(44) = 1 + 3.3(1.6435) = 6.42 \text{ bins}$$

Sturges' formula suggests using 6 or 7 bins for the P/E data. However, to get "nice" bin limits, you may choose more or fewer bins. Picking attractive bin limits is often an overriding consideration (not Sturges' Rule). When the data are skewed by unusually large or small data values, we may need more classes than Sturges' Rule suggests.

Step 3: Set Bin Limits Just as choosing the number of bins requires judgment, setting the bin limits also requires judgment. For guidance, find the approximate width of each bin by dividing the data range by the number of bins:

(3.2)
$$\text{Bin width} \approx \frac{x_{max} - x_{min}}{k}$$

Experimentation may be required to get aesthetically pleasing bins that cover the data range. For example, for this data set, the smallest P/E ratio was 7 and the largest P/E ratio was 59, so if we want to use $k = 6$ bins, we calculate the approximate bin width as

$$\text{Bin width} \approx \frac{59 - 7}{6} = \frac{52}{6} = 8.67$$

To obtain "nice" limits, we could round the bin width up to 10 and choose bin limits of 0, 10, 20, 30, 40, 50, 60. Make sure that your bins don't overlap and that they will include all the data values.

Step 4: Prepare a Table of Frequencies Count the data values in each bin. You can choose to show only the absolute frequencies, or counts, for each bin, but we often also include the relative frequencies and the cumulative frequencies, as shown in Table 3.5. Relative frequencies are calculated as the frequency for a bin divided by the total number of data values.

Table 3.5

Frequency Distribution of P/E Ratios Using Six Bins **PERatios**

Note: The < symbol is used in this table to indicate that the upper bin limit is not included in the interval.

Bin Limits			Frequency (f)	Relative Frequency (f/n)	Percent	Cumulative Frequency	Percent
0	<	10	2	2/44 = .0455	4.55	2	4.55
10	<	20	24	24/44 = .5455	54.55	26	59.09
20	<	30	11	11/44 = .2500	25.00	37	84.09
30	<	40	4	4/44 = .0909	9.09	41	93.18
40	<	50	1	1/44 = .0227	2.27	42	95.45
50	<	60	2	2/44 = .0455	4.55	44	100.00
			44		100.00		

Histograms

LO 3-3

Make a histogram with appropriate bins.

A **histogram** is a graphical representation of a frequency distribution. A histogram is a column chart whose Y-axis shows the number of data values (or a percentage) within each bin of a frequency distribution and whose X-axis ticks show the end points of each bin. There should be no gaps between bars (except when there are no data in a particular bin). The appearance of a histogram is identical, regardless of whether the vertical axis displays *frequency, relative frequency,* or *percent.*

Choosing the number of bins and bin limits requires judgment. Creating a histogram is often a trial-and-error process. Our first choice of bins and limits may not be our final choice for presentation. Figure 3.2 shows histograms for the P/E ratio sample using three different bin definitions.

Our perception of the shape of the distribution depends on how the bins are chosen. For example, the skewed shape of the P/E distribution becomes more obvious when we use more than six bins. In this example, we might wish to depart from Sturges' Rule to show additional detail. You can use your own judgment to determine which histogram you would ultimately include in a report.

Figure 3.2

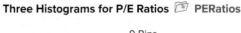

Three Histograms for P/E Ratios PERatios

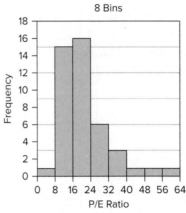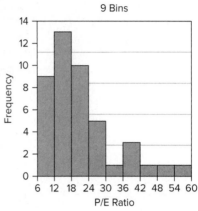

Making an Excel Histogram

The simplest way to make an Excel histogram is to highlight the data and select Insert > Statistics Chart > Histogram. Each histogram bar is labeled with its bin limits. You can edit the colors and spacing, but the bins are chosen automatically.

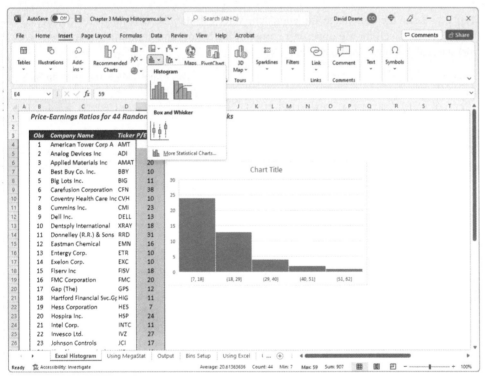

Source: Microsoft Excel.

The bin limits may not be what you would prefer. To choose your own bins, follow these steps:

Step 1: Key in the upper bin limits that will cover the data range. Each bin upper limit will be *included* in the bin. Only the *upper* bin limit is used by Excel (both are shown here for clarity). You may wish to experiment with different bin limits, as illustrated in the example shown here. Click on the Data ribbon, select the Data Analysis icon, choose the Histogram option, and click OK. If you don't see Data Analysis on the Data ribbon, click File (upper left corner), click the Excel Options button at the bottom of the screen, click Add-Ins, select Analysis Tool Pak, and click OK.

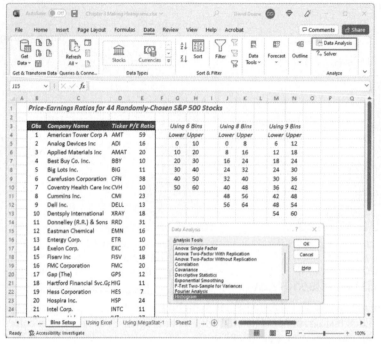

Source: Microsoft Excel.

Step 2:   Enter the cell range for your data values. Enter the cell range for the upper bin limits. Specify a cell for the upper left corner of the histogram output range (or choose a new worksheet ply). Check Chart Output and click OK.

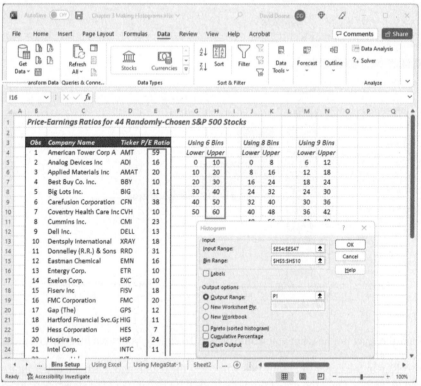

Source: Microsoft Excel.

Step 3:   Excel's default histogram is quite basic, so you will most likely want to customize it. To do this, click on the chart and use the *Chart Elements* to edit a specific chart feature (e.g., *Chart Title, Axis Titles, Gridlines*) or right-click on any feature of the chart and choose the edit options you want. For example, to improve Excel's skinny histogram bars, right-click on the histogram bars, choose Format Data Series, and reduce the gap width.

Default

Customized

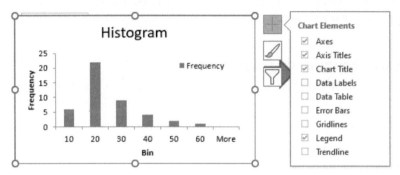

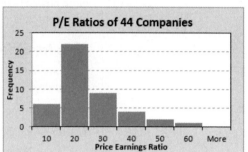

Shape

A histogram suggests the *shape* of the population we are sampling, but unless the sample is large, we must be cautious about making inferences. Our perception is also influenced by the number of bins and the way the bin limits are chosen. The following terminology is helpful in discussing shape.

A **modal class** is a histogram bar that is higher than those on either side. A histogram with a single modal class is *unimodal,* one with two modal classes is *bimodal,* and one with more than two modes is *multimodal.* However, modal classes may be artifacts of the way the bin limits are chosen. It is wise to experiment with various ways of binning and to make cautious inferences about modality unless the modes are strong and invariant to binning. For example, the Excel histogram in Step 3 shows a single modal class for P/E ratios between 10 and 20.

A histogram's *skewness* is indicated by the direction of its longer tail. If neither tail is longer, the histogram is **symmetric**. A **right-skewed** (or positively skewed) histogram has a longer right tail, with most data values clustered on the left side. A **left-skewed** (or negatively skewed) histogram has a longer left tail, with most data values clustered on the right side. Few histograms are exactly symmetric. Business data tend to be right-skewed because they are often bounded by zero on the left but are unbounded on the right (e.g., number of employees). You may find it helpful to refer to the templates shown in Figure 3.3.

LO 3-4

Identify skewness, modal classes, and outliers in a histogram.

Figure 3.3

Prototype Distribution Shapes

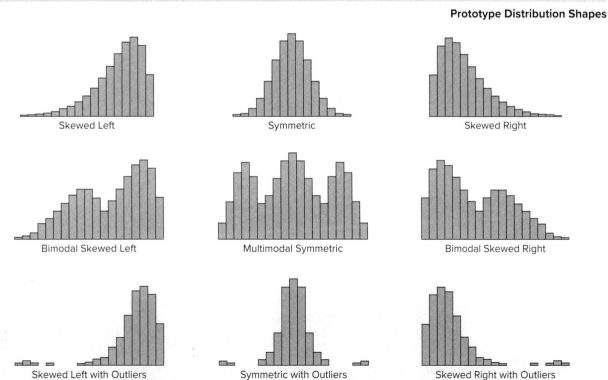

An **outlier** is an extreme value that is far enough from the majority of the data that it probably arose from a different cause or is due to measurement error. We will define outliers more precisely in the next chapter. For now, think of outliers as unusual points located in the histogram tails.

Tips for Effective Frequency Distributions

Here are some general tips to keep in mind when making frequency distributions and histograms.

1. Check Sturges' Rule first, but only as a suggestion for the number of bins.
2. Choose an appropriate bin width.
3. Choose bin limits that are multiples of the bin width.
4. Make sure that the range is covered, and add bins if necessary.
5. Skewed data may require more bins to reveal sufficient detail.

Frequency Polygon and Ogive

Figure 3.4 shows two more frequency plots based on bin frequencies. A **frequency polygon** is a line graph that connects the midpoints of the histogram bin intervals plus extra intervals at the beginning and end so that the line will touch the *X*-axis. It serves the same purpose as a histogram but is attractive when you need to compare two data sets (because more than one frequency polygon can be plotted on the same scale). An **ogive** (pronounced "oh-jive") is a line graph of the cumulative frequencies. It is useful for finding percentiles or in comparing the shape of the sample with a known benchmark such as the normal distribution (that you will be seeing in the next chapter).

Figure 3.4

Frequency Polygon and Ogive   PERatios

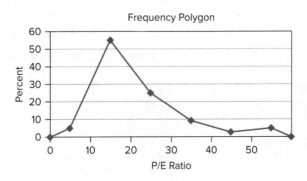
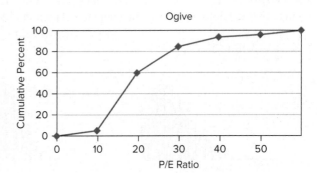

Mini Case 3.2

Duration of U.S. Recessions

Table 3.6 shows two "nice" ways to bin the data on the duration of 33 U.S. recessions (for details, see **Mini Case 3.1**). Sturges would recommend using six bins, that is, $k = 1 + 3.3 \log(n)$ $= 1 + 3.3 \log(34) = 6.05$. Using six bins works out well, with a bin size of 12 months. However, we also can create a nice histogram using seven bins of width 10 months. You can surely think of other valid possibilities.

Table 3.6

Some Ways to Tabulate 34 Business Contractions 🗂 **Recessions**

Using *k* = 6 bins				Using *k* = 7 bins		
From	**To (not included)**	***f***		**From**	**To (not included)**	***f***
0	12	13		0	10	8
12	24	16		10	20	19
24	36	2		20	30	3
36	48	2		30	40	2
48	60	0		40	50	1
60	72	1		50	60	0
				60	70	1
	Total	34			Total	34

Both histograms in Figure 3.5 suggest right-skewness (long right tail, most values cluster to the left). Each histogram has a single modal class (e.g., the *k* = 7 bin histogram says that a recession most often lasts between 10 and 20 months). The long recession of 1873–1879 (65 months) can be seen as a possible outlier in the right tail of both histograms.

Figure 3.5

Histograms for 6 and 7 Bins

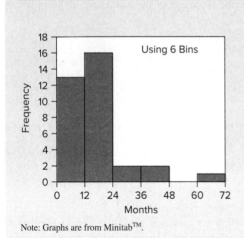
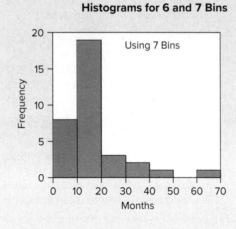

Note: Graphs are from Minitab™.

3.3 (a) Make a frequency distribution and histogram of the lengths of Sarah's last 48 cell phone calls (in minutes) using bin limits of width 10 starting at 0. (b) Describe the distribution. (c) Does your histogram follow Sturges' Rule?

6	9	4	1	31	5	1	6	5	22	38	28	41	14	5	2
29	29	23	2	23	35	45	24	33	1	3	2	3	17	1	2
33	8	27	23	22	5	3	4	3	26	64	6	21	39	12	27

3.4 (a) The table shows the number of days on the market for the 36 recent home sales in the city of Sonando Hills. Construct a frequency distribution and histogram, using nice (round) bin limits. (b) Describe the distribution and note any unusual features. 🗂 **Homes**

18	70	52	17	86	121	86	3	66
96	41	50	176	26	28	6	55	21
43	20	56	71	57	16	20	30	31
44	44	92	179	80	98	44	66	15

Section Exercises

🔗 connect

3.5 (a) The table shows raw scores on a state civil service exam taken by 24 applicants for positions in law enforcement. Construct a frequency distribution and histogram, using nice (round) bin limits. (b) Describe the distribution and note any unusual features. 📁 **Civil**

83	93	74	98	85	82	79	78
82	68	67	82	78	83	70	99
18	96	93	62	64	93	27	58

3.6 (a) Make a frequency distribution and histogram (using appropriate bins) for these 28 observations on the amount spent for dinner for four in downtown Chicago on Friday night. (b) Repeat the exercise, using a different number of bins. Which is preferred? Why? 📁 **Dinner**

95	103	109	170	114	113	107
124	105	80	104	84	176	115
69	95	134	108	61	160	128
68	95	61	150	52	87	136

3.7 (a) Make a frequency distribution and histogram for the monthly off-campus rent paid by 30 students. (b) Repeat the exercise, using a different number of bins. Which is preferred? Why? 📁 **Rents**

730	730	730	930	700	570
690	1,030	740	620	720	670
560	740	650	660	850	930
600	620	760	690	710	500
730	800	820	840	720	700

3.8 (a) Make a frequency distribution and histogram for the annual compensation of 40 randomly chosen CEOs (millions of dollars). (b) Describe the shape of the histogram. (c) Identify any unusual values. 📁 **CEOComp40**

5.33	18.3	24.55	9.08	12.22	5.52	2.01	3.81
192.92	17.83	23.77	8.7	11.15	4.87	1.72	3.72
66.08	15.41	22.59	6.75	9.97	4.83	1.29	3.72
28.09	12.32	19.55	5.55	9.19	3.83	0.79	2.79
34.91	13.95	20.77	6.47	9.63	4.47	1.01	3.07

3.9 For each frequency distribution, suggest "nice" bins. Did your choice agree with Sturges' Rule? If not, explain.

    a. Last week's MPG for 35 student vehicles ($x_{min} = 9.4$, $x_{max} = 38.7$).
    b. Ages of 50 airplane passengers ($x_{min} = 12$, $x_{max} = 85$).
    c. GPAs of 250 first-semester college students ($x_{min} = 2.25$, $x_{max} = 3.71$).
    d. Annual rates of return on 150 mutual funds ($x_{min} = .023$, $x_{max} = .097$).

3.10 Below are sorted data showing average spending per customer (in dollars) at 74 Noodles & Company restaurants. (a) Construct a frequency distribution. Explain how you chose the number of bins and the bin limits. (b) Make a histogram and describe its appearance. (c) Repeat, using a larger number of bins and different bin limits. (d) Did your visual impression of the data change when you increased the number of bins? Explain. *Note:* You may use software such as MegaStat or Minitab if your instructor agrees. 📁 **NoodlesSpending**

6.54	6.58	6.58	6.62	6.66	6.70	6.71	6.73	6.75	6.75	6.76	6.76
6.76	6.77	6.77	6.79	6.81	6.81	6.82	6.84	6.85	6.89	6.90	6.91
6.91	6.92	6.93	6.93	6.94	6.95	6.95	6.95	6.96	6.96	6.98	6.99
7.00	7.00	7.00	7.02	7.03	7.03	7.03	7.04	7.05	7.05	7.07	7.07
7.08	7.11	7.11	7.13	7.13	7.16	7.17	7.18	7.21	7.25	7.28	7.28
7.30	7.33	7.33	7.35	7.37	7.38	7.45	7.56	7.57	7.58	7.64	7.65
7.87	7.97										

3.3 EFFECTIVE EXCEL CHARTS

Making your own charts in Excel is something you will have to learn by experience. Professionals who use Excel say that they learn new things every day. Excel offers a vast array of charts besides histograms. The chart icons in Figure 3.6 are designed to be visually self-explanatory. We will discuss those that are most useful in business and economics.

Figure 3.6

Common Excel Chart Types

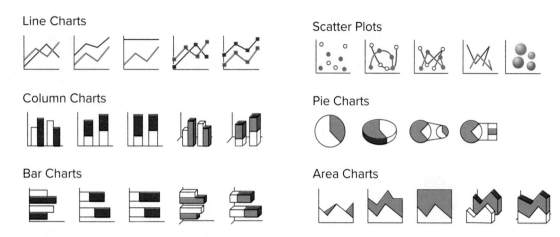

Excel's default charts tend to be very plain. But business charts need not be dull. You can customize any graph to your taste. For example, you can

- Improve the titles (main, *X*-axis, *Y*-axis).
- Change the axis scales (minimum, maximum, ticks).
- Display the data values (not always a good idea).
- Add a data table underneath the graph (if there is room).
- Change color in the plot area or chart area.
- Format decimals to make axis scales more readable.
- Edit the gridlines and borders (dotted or solid, colors).
- Alter the appearance of bars (color, pattern, gap width).
- Choose different data markers and lines (size, color).

Click on the chart to select it, and the Format and Design ribbons will be highlighted at the top of the screen, as illustrated in Figure 3.7. Alternatively, you can click on a graph and use its *Chart Elements* menus (see Figure 3.8) to add or edit specific features of your chart. Although certain features are unique to each chart type, these ribbons are similar for all chart types. If you don't like the result of your edits, just click the ⤺ Undo icon (or Ctrl-Z).

Figure 3.7

Chart Tool Ribbons

Source: Microsoft Excel.

Figure 3.8

Chart Elements Menus

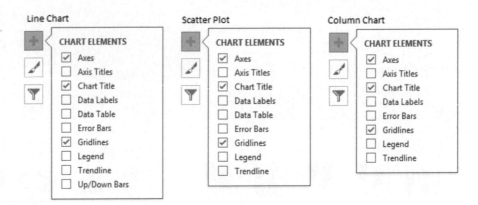

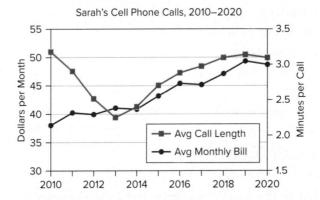

3.4 LINE CHARTS

LO 3-5

Make an effective line chart.

A **line chart** is used to display a time series, to spot trends, or to compare time periods. Line charts can be used to display several variables at once. If two variables are displayed, the right and left scales can differ, using the right scale for one variable and the left scale for the other. Excel's *two-scale line chart,* illustrated in Figure 3.9, lets you compare variables that *differ in magnitude* or are measured in *different units.* But keep in mind that someone who only glances at the chart may mistakenly conclude that both variables are of the same magnitude.

Figure 3.9

Two Scales

📁 **CellPhones**

How many variables can be displayed at once on a line graph? Too much clutter ruins any visual display. If you try to display half a dozen time series variables at once, no matter how cleverly you choose symbols and graphing techniques, the result is likely to be unsatisfactory. You will have to use your judgment.

A line graph usually has no vertical gridlines. What about horizontal gridlines? While gridlines do add background clutter, they make it easier to establish the Y value for a given year. One compromise is to use lightly colored dashed or dotted gridlines to minimize the clutter and to increase gridline spacing. If the intent is to convey only a general sense of the data magnitudes, gridlines may be omitted.

Making an Excel Line Chart

Step 1:   Highlight the data that you want to display in the line chart, click on the Insert ribbon, click on the Line icon, and choose a line chart style. *Hint:* Do *not* highlight the X-axis labels (if any). You can add X-axis labels later. The default line plot is quite basic, so you may wish to customize it.

Step 2:   To customize your graph, click on it. Its border will change to show that it has been selected, and the *Chart Elements* control will appear (just to the right of the graph).

Step 3:   If you wish to add a fitted trend, right-click on the data series on the line chart and choose Add Trendline. By default, the fitted trend will be linear. There is an option to display the trend equation and its R^2 statistic (a measure of "fit" of the line).

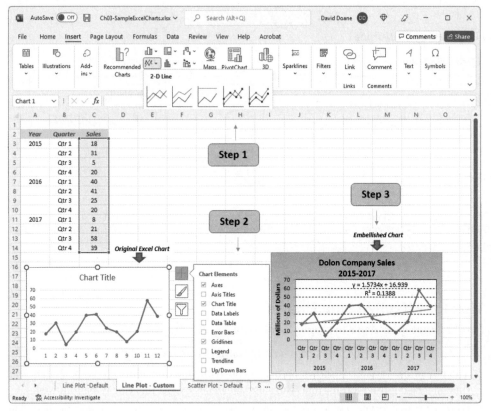

Source: Microsoft Excel.

Log Scales

On the customary **arithmetic scale**, distances on the *Y*-axis are proportional to the magnitude of the variable being displayed. But on a **logarithmic scale**, equal distances represent equal *ratios* (for this reason, a log scale is sometimes called a *ratio scale*). When data vary over a wide range, say, by more than an order of magnitude (e.g., from 6 to 60), we might prefer a *log scale* for the vertical axis to reveal more detail for small data values. A log graph reveals whether the quantity is growing at an *increasing percent* (convex function), *constant percent* (straight line), or *declining percent* (concave function). On a log scale, *equal distances* represent *equal ratios.* That is, the distance from 100 to 1,000 is the same as the distance from 1,000 to 10,000 (because both have the same 10:1 ratio). Because logarithms are undefined for negative or zero values (try it on your calculator), a log scale is only suited for positive data values.

A log scale is useful for time series data that might be expected to grow at a compound annual percentage rate (e.g., GDP, the national debt, or your future income). Log scales are common in financial charts that cover long periods of time or for data that grow rapidly (e.g., revenues for a start-up company). Some experts feel that corporate annual reports and stock prospectuses should avoid ratio scales, on the grounds that they may be misleading to uninformed individuals. But then how can we fairly portray data that vary by orders of magnitude? Should investors become better informed? The bottom line is that business students must understand log scales because they are sure to run into them.

Example 3.2

U.S. Trade   USTrade

Figure 3.10 shows the U.S. balance of trade. Although exports and imports are increasing in absolute terms, the log graph suggests that the *growth rate* in both series may be slowing because the log graph is slightly concave. On the log graph, the recently increasing trade deficit is not *relatively* as large. Regardless of how it is displayed, the trade deficit remains a concern for policymakers.

Figure 3.10

Comparison of Arithmetic and Log Scales 📄 USTrade

Source: *Economic Report of the President*, 2018, Table B24.

Tips for Effective Line Charts

Here are some general tips to keep in mind when creating line charts:

1. Line charts are used for *time series data* (never for cross-sectional data).

2. The numerical variable is shown on the *Y*-axis, while the time units go on the *X*-axis with time increasing from left to right. Business audiences expect this rule to be followed.

3. Except for log scales, use a zero origin on the *Y*-axis unless more detail is needed. The zero-origin rule is mandatory for a corporate annual report or investor stock prospectus.

4. To avoid graph clutter, numerical labels usually are *omitted* on a line chart, especially when the data cover many time periods. Use gridlines to help the reader read data values.

5. Data markers (squares, triangles, circles) are helpful. But when the series has many data values or when many variables are being displayed, they clutter the graph.

Section Exercises

Mc Graw Hill **connect**

3.11 (a) Use Excel to prepare a line chart to display the data on housing starts. Modify the default colors, fonts, etc., to make the display effective. (b) Describe the pattern, if any.

U.S. Housing Starts (thousands), 2000–2020 📄 Housing					
Year	**Starts**	**Year**	**Starts**	**Year**	**Starts**
2000	1,573	2007	1,342	2014	1,000
2001	1,601	2008	900	2015	1,107
2002	1,710	2009	554	2016	1,178
2003	1,854	2010	586	2017	1,209
2004	1,950	2011	612	2018	1,250
2005	2,073	2012	784	2019	1,267
2006	1,812	2013	928	2020	1,366

Source: https://fred.stlouisfed.org.

3.12 (a) Use Excel to prepare a line chart to display the skier/snowboarder data. Modify the default colors, fonts, etc., to make the display effective. (b) Describe the pattern, if any.

U.S. Skier/Snowboarder Visits (Millions), 2000–2021 Snowboards

Season	Visits	Season	Visits	Season	Visits
2000–01	57.3	2007–08	60.5	2014–15	53.6
2001–02	54.4	2008–09	57.4	2015–16	52.8
2002–03	57.6	2009–10	59.8	2016–17	54.8
2003–04	57.1	2010–11	60.5	2017–18	53.3
2004–05	56.9	2011–12	51.0	2018–19	59.3
2005–06	58.9	2012–13	56.9	2019–20	51.1
2006–07	55.1	2013–14	56.5	2020–21	59.0

Source: www.nsaa.org/nsaa/press.

3.13 (a) Use Excel to prepare a line chart to display the lightning death data. Modify the default colors, fonts, etc., as you judge appropriate to make the display effective. (b) Describe the pattern, if any.

U.S. Deaths by Lightning, 1940–2020 Lightning

Year	Deaths	Year	Deaths	Year	Deaths
1940	340	1970	122	2000	51
1945	268	1975	91	2005	38
1950	219	1980	74	2010	29
1955	181	1985	74	2015	27
1960	129	1990	74	2020	17
1965	149	1995	85		

Source: www.weather.gov/safety/lightning-victims.

3.14 (a) Use Excel to prepare a line chart to display the following transplant data. Modify the default colors, fonts, etc., to make the display effective. (b) Describe the pattern, if any.

U.S. Organ Transplants, 2000–2010 Transplants

Year	Heart	Liver	Kidney	Year	Heart	Liver	Kidney
2000	2,172	4,816	13,258	2006	2,192	6,650	17,094
2001	2,202	5,177	14,152	2007	2,210	6,493	16,624
2002	2,153	5,326	14,741	2008	2,163	6,318	16,517
2003	2,057	5,673	15,137	2009	2,212	6,320	16,829
2004	2,015	6,169	16,004	2010	2,333	6,291	16,898
2005	2,125	6,443	16,481				

Source: *Statistical Abstract of the United States, 2012*, p. 123.

3.5 SCATTER PLOTS

LO 3-6

Make and interpret a
scatter plot.

A **scatter plot** shows n pairs of observations (x_1, y_1), (x_2, y_2), . . ., (x_n, y_n) as dots (or some other symbol) on an X-Y graph. This type of display is so important in statistics that it deserves careful attention. A scatter plot is a starting point for bivariate data analysis. We create scatter plots to investigate the relationship between two variables. Typically, we would like to know if there is an *association* between two variables and, if so, what kind of association exists. As we did with univariate data analysis, let's look at a scatter plot to see what we can observe.

Example 3.3

Birth Rates and Life Expectancy

Source: *The World Factbook 2003.*
Central Intelligence Agency, 2003.
www.cia.gov.

Figure 3.11 shows a scatter plot with life expectancy on the X-axis and birth rates on the Y-axis. In this illustration, there seems to be an association between X and Y. That is, nations with higher birth rates tend to have lower life expectancy (and vice versa). No cause-and-effect relationship is implied because, in this example, both variables could be influenced by a third variable that is not mentioned (e.g., GDP per capita).

Figure 3.11

Scatter Plot of Birth Rates and Life Expectancy ($n = 153$ nations) 📂 **BirthLife**

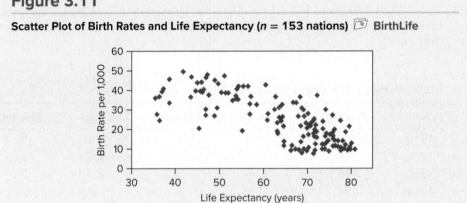

Figure 3.12 shows some scatter plot patterns similar to those that you might observe when you have a sample of (X, Y) data pairs. A scatter plot can convey patterns in data pairs that would not be apparent from a table. Compare the scatter plots in Figure 3.13 with the prototypes and use your own words to describe the patterns that you see.

Figure 3.12

**Prototype Scatter Plot
Patterns**

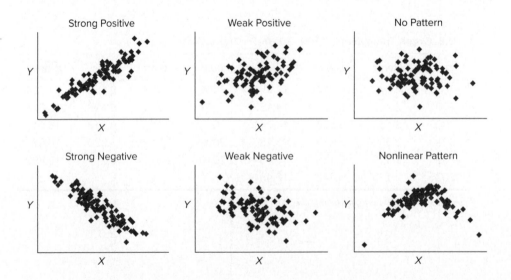

Figure 3.13

Four Scatter Plots

Very Strong Positive Association

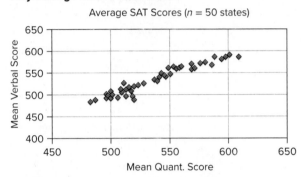

Source: *Statistical Abstract of the United States, 2001*, p. 151.

Strong Positive Association

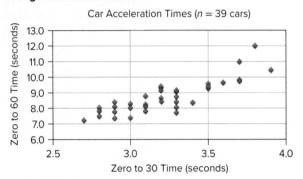

Source: Tests by a popular car magazine.

Moderate Positive Association

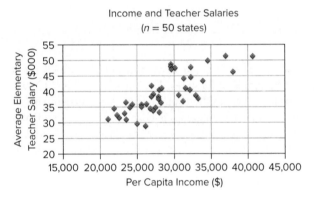

Source: *Statistical Abstract of the United States, 2001*, p. 151.

Little or No Association

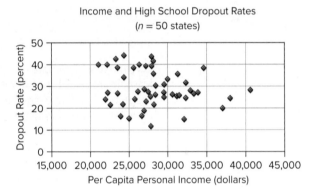

Source: *Statistical Abstract of the United States, 2001*, p. 151.

Making an Excel Scatter Plot

Making a scatter plot is easy using Excel. However, you will probably want to improve the graph after making it, as explained in the following steps. Note that Excel assumes that the first column is *X* and the second column is *Y*.

Step 1:   Highlight the (*x, y*) data pairs that you want to display in the scatter plot, click on the Insert ribbon, click on the Scatter icon, and choose a scatter plot style. The default scatter plot is quite basic, so you may wish to customize it.

Step 2:   To customize your graph, click on it. Its border will change to show that it has been selected, and the *Chart Elements* menu will appear. Select a specific feature to edit (e.g., *Chart Title, Axis Titles, Gridlines*). Alternatively, you can just right-click on any feature of your graph (e.g., chart area, plot area, *X*-axis, *Y*-axis, gridlines) and explore the options until you are satisfied with the graph's appearance. An example of a customized scatter plot can be found on the top of the next page.

Step 3:   If you wish to add a fitted trend, right-click on the data series on the scatter plot and choose Add Trendline. By default, the fitted trend will be linear. There is an option to display the trend equation and its R^2 statistic (a measure of "fit" of the line).

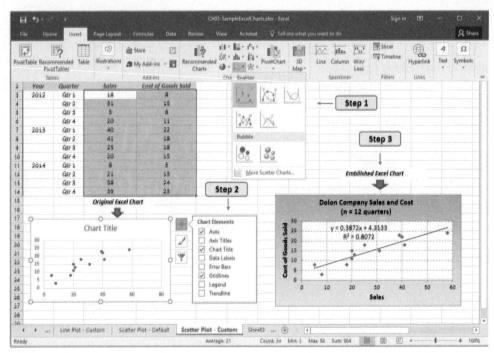

Source: Microsoft Excel.

Linear Trend Example   Figure 3.14 shows Excel's fitted linear trend for X = total gross leasable area and Y = total retail sales for a sample of 28 states. The slope of the line (0.2594) suggests that a unit change in X (each "unit" is one million square feet) is associated with an extra $0.2594 billion in retail sales, on average. The intercept is near zero, suggesting that a shopping center with no leasable area would have no sales. Later (in Chapter 12), you will learn how Excel fits a trend line, how to interpret it, and when such a line is meaningful. But because almost every student discovers this option the first time they make a scatter plot, we must mention Excel's fitted trend line here purely as a *descriptive tool* that may help you find patterns in (X, Y) data.

Figure 3.14

Excel Scatter Plot with Fitted Trend Line (n = 28 states) 📁 **RetailSales**

Source: *Statistical Abstract of the United States,* 2007, p. 660.

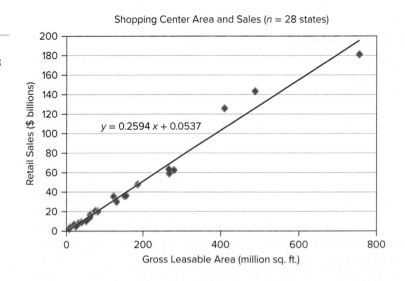

Section Exercises

📊 connect

3.15   (a) Use Excel to make a scatter plot of the data for bottled water sales for 10 weeks, placing Price on the X-axis and Units Sold on the Y-axis. Add titles and modify the default colors, fonts, etc., as you judge appropriate to make the scatter plot effective. (b) Describe the relationship (if any) between X and Y. Weak? Strong? Negative? Positive? Linear? Nonlinear? 📁 **WaterSold**

Unit Price	Units Sold
1.15	186
0.94	216
1.04	173
1.05	182
1.08	183
1.33	150
0.99	190
1.25	165
1.16	190
1.11	201

3.16 (a) Use Excel to make a scatter plot of these vehicle data, placing Weight on the *X*-axis and City MPG on the *Y*-axis. Add titles and modify the default colors, fonts, etc., as you judge appropriate to make the scatter plot effective. (b) Describe the relationship (if any) between *X* and *Y*. Weak? Strong? Negative? Positive? Linear? Nonlinear?

Weight and MPG for Selected 2020 Vehicles (*n* = 50) CityMPG

Brand and Model	City MPG	Weight (lbs.)
Acura RDX SH SWD 4dr SUV	21	4026
Acura TLX sedan	23	3505
Audi V10 2dr Quattro convertible	13	3572
. . .	. . .	. . .
Toyota Yaris L 4dr sedan	30	2385
Volvo XC40 T5 4dr SUV	22	3756
VW Tiguan SE R-Line 4dr SUV	22	3757

3.17 (a) Use Excel to make a scatter plot of the following retail store sales data, placing Coupon on the *X*-axis and Revenue on the *Y*-axis. Add titles and modify the default colors, fonts, etc., as you judge appropriate to make the scatter plot effective. (b) Describe the relationship (if any) between *X* and *Y*. Weak? Strong? Negative? Positive? Linear? Nonlinear?

Retail Store Revenue and Percent of Transactions with Coupon Redeemed (*n* = 20) Coupons

Week	Revenue ($000)	Coupon %	Week	Revenue ($000)	Coupon %
May 9	780	66.7	Jul 18	983	48.3
May 16	921	40.0	Jul 25	859	55.0
May 23	883	55.6	Aug 1	904	73.3
May 30	932	84.6	Aug 8	909	85.7
Jun 6	887	70.0	Aug 15	988	94.1
Jun 13	1,121	69.2	Aug 22	943	80.0
Jun 20	723	17.5	Aug 29	1,007	83.3
Jun 27	857	16.3	Sep 5	930	90.9
Jul 4	849	25.5	Sep 12	928	87.5
Jul 11	859	14.9	Sep 19	920	70.0

3.18 (a) Use Excel to make a scatter plot of the data, placing Floor Space on the *X*-axis and Weekly Sales on the *Y*-axis. Add titles and modify the default colors, fonts, etc., as you judge appropriate to make the scatter plot effective. (b) Describe the relationship (if any) between *X* and *Y*. Weak? Strong? Negative? Positive? Linear? Nonlinear FloorSpace

Floor Space (sq. ft.)	Weekly Sales (dollars)
6,060	16,380
5,230	14,400
4,280	13,820
5,580	18,230
5,670	14,200
5,020	12,800
5,410	15,840
4,990	16,610
4,220	13,610
4,160	10,050
4,870	15,320
5,470	13,270

3.6 COLUMN AND BAR CHARTS

LO 3-7

Make an effective column chart or bar chart.

A **column chart** is a *vertical* display of data, and a **bar chart** is a *horizontal* display of data. Figure 3.15 shows simple column and bar charts comparing trends in Instagram use.

Excel allows either vertical or horizontal display of categorical data. Each column or bar may be separated from its neighbors by a slight gap to improve legibility. You can control gap width in Excel. Most people find a column chart display easier to read, but a bar chart can be useful when the axis labels are long or when there are many categories. You can add a value label to each column (or row) to display the data value for each category. Value labels are desirable unless there are so many categories that the labels become crowded. You can use Excel's COUNTIF() function to tabulate frequencies of categorical data for a column or bar chart.

Figure 3.15

Same Data Displayed Two Ways Instagram

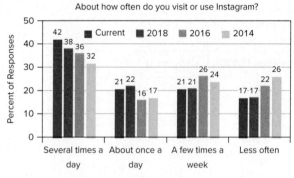
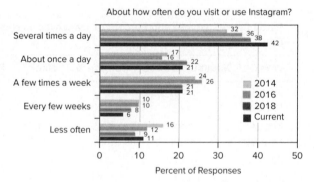

Source: Adapted from Pew Research Center.

Pareto Charts

A special type of column chart used in business is the **Pareto chart**. A Pareto chart displays categorical data, with categories displayed in descending order of frequency so that the most common categories appear first. Typically, only a few categories account for the majority of observations. This phenomenon is called the *80/20 Rule*. This rule holds true for many aspects of business. For example, in a sample of U.S. guests responding to a Vail Resorts' guest satisfaction survey, 80 percent of the respondents were visiting from just 20 percent of the states in the United States.

Pareto charts are commonly used in quality management to display the *frequency* of defects or errors of different types. The majority of quality problems can usually be traced to only a few sources or causes. Sorting the categories in descending order helps managers focus on the *vital few* causes of problems rather than the *trivial many*.

Figure 3.16 shows a Pareto chart for complaints collected from a sample of *n* = 398 customers at a concession stand. Notice that the top three categories make up 76 percent of all complaints. The owners of the concession stand can concentrate on ensuring that their food is not cold, on decreasing the time their customers spend in line, and on providing vegetarian choices.

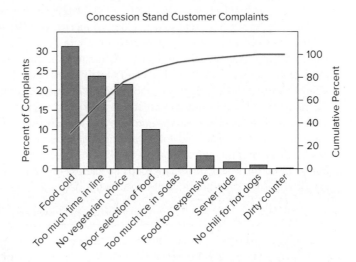

Figure 3.16

Pareto Chart
Concessions

Note: The cumulative percentage was added as a second data series on a secondary axis.

Stacked Column Chart

In a **stacked column chart** like Figure 3.17, the bar height is the sum of several subtotals. Areas may be compared by color to show patterns in the subgroups, as well as showing the total. Stacked column charts can be effective for any number of groups but work best when you have only a few. Use numerical labels if exact data values are of importance.

Tips for Effective Bar and Column Charts

The following guidelines will help you to create the most effective bar charts:

1. The numerical variable of interest usually is shown with vertical bars on the *Y*-axis.

2. If the quantity displayed is a time series, the category labels (e.g., years) are displayed on the horizontal *X*-axis with time increasing from left to right.

3. The height or length of each bar should be proportional to the quantity displayed. This is easy because most software packages default to a zero origin on a bar graph.

4. Add numerical values at the top of each bar unless they would impair legibility.

5. Bar scale position is harder to read on 3-D charts; 2-D charts are preferred.

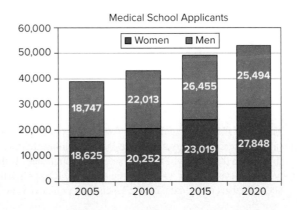

Figure 3.17

Stacked Column Chart MedSchool

Source: www.aamc.org.

Analytics in Action

Vail Resorts Epic Pass

When Vail Resorts was conducting market research for its new season ski pass, the Epic Pass, it surveyed its guests to determine which features of a season pass were most important. Features of a season pass include the number of days of skiing allowed (limited vs. unlimited), the number of resorts included on the pass, the number of blackout dates (all holidays, some holidays, or no blackouts), and the price of the pass.

A market survey was sent to a random sample of Vail Resorts guests. Vail received 1,930 responses. The respondents were sorted into groups based on the number of ski vacations they typically take each year. A summary of the responses is displayed in the clustered column chart below. The chart clearly shows that the feature considered most important to all three groups was the price of the pass. Determining the right price of the pass was critical in order to create a valuable product that a skier would purchase. An analytics tool called *conjoint analysis* was used to summarize additional data from subsequent surveys. The Epic Pass was priced at $579 after this further analysis. Charts can be effective communication tools by allowing the analyst to compare and summarize information from many different groups and across different variables. Decision makers can then see in a snapshot the areas on which to focus.

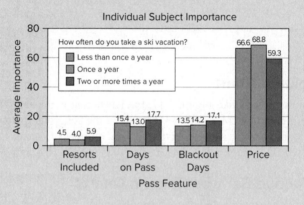

Section Exercises

3.19 (a) Use Excel to prepare a line chart to display the following gasoline price data. Modify the default colors, fonts, etc., to make the display effective. (b) Change it to a 2-D column chart. Modify the display if necessary to make the display attractive. (c) Do you prefer the line chart or bar chart? Why?

Average U.S. Retail Price of Gasoline (dollars per gallon) 🗁 **GasPrice**

Year	Price	Year	Price	Year	Price
1965	0.31	1985	1.20	2005	2.30
1970	0.36	1990	1.16	2010	3.02
1975	0.57	1995	1.15	2015	2.33
1980	1.25	2000	1.51	2020	1.77

Source: www.fueleconomy.gov and eia.doe.gov. Pre-1980 prices are for unleaded gas.

3.20 The table shows the number of TVs sold by a major retailer in California for three years. (a) Use Excel to prepare a 2-D side-by-side column chart with screen size on the horizontal axis. (b) Change the chart type to a 2-D stacked column chart. Is the stacked chart a better way to compare sales over time or among screen sizes? (c) Right-click the data series, choose Add Data Labels, and add labels to the data. Do the labels help or hinder viewer understanding?

TV Units Sold by Screen Size (inches) 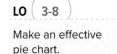 TVSales				
Year	Less than 40"	40" to 49"	50" to 59"	60" or More
2012	2,204	1,107	343	157
2014	1,802	811	508	273
2016	1,558	666	585	386

3.21 (a) Use Excel to prepare a Pareto chart of the following data. (b) Which three complaint categories account for approximately 80 percent of all complaints? (c) Which category should the telephone company focus on first?

Telephone Company Service Complaints, $n = 791$ Complaints		
Customer Complaints	Frequency	Percent
Wait too long	350	44.2%
Service person rude	187	23.6%
Difficult to reach a real person	90	11.4%
Service person not helpful	85	10.7%
Automated instructions confusing	45	5.7%
Customer service phone number hard to find	21	2.7%
Automated voice annoying	13	1.6%

3.7 PIE CHARTS

Many statisticians feel that a table or bar chart is often a better choice than a **pie chart**. But, because of their visual appeal, pie charts appear daily in company annual reports and the popular press (e.g., *USA Today, The Wall Street Journal, Scientific American*), so you must understand their uses and misuses. A pie chart can only convey a *general idea of the data* because it is hard to assess areas precisely. It should have only a few slices (typically two to five), and the slices should be labeled with data values or percents. The only correct use of a pie chart is to *portray data that sum to a total* (e.g., percent market shares). A pie chart is never used to display time series data.

Figure 3.18 shows a simple *2-D pie chart*. Excel pie charts with data labels on the slices may shrink so much that they are difficult to read. Instead of a pie chart, you can use a column or bar chart. Another alternative to display parts of a whole is Excel's hierarchy chart (Figure 3.19).

LO 3-8

Make an effective pie chart.

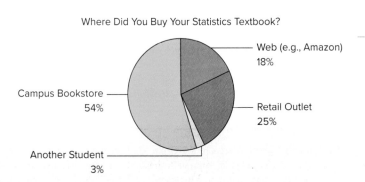

Where Did You Buy Your Statistics Textbook?

Figure 3.18

2-D Pie with Labels
Textbook

Figure 3.19

Hierarchy Chart
📄 **Textbook**

Where Did You Purchase Your Statistics Textbook?

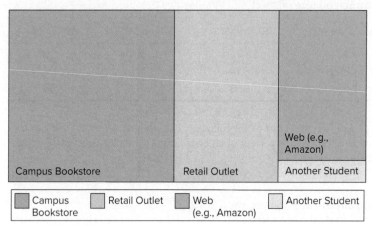

You will see 3-D or "exploded" pie charts in business publications because of their strong visual impact, but the sizes of their pie slices are harder to assess (Figure 3.20).

Figure 3.20

Exploded Pie Chart
📄 **PieCharts**

Moderately Conservative Investment Portfolio

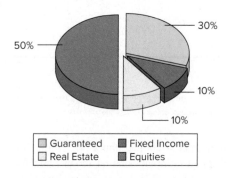

Common Pie Chart Errors

- Pie charts can only convey a general idea of the data values.
- Pie charts are ineffective when they have too many slices.
- Pie chart data must represent *parts of a whole* (e.g., percent market share).

Section Exercises

McGraw Hill **connect**

3.22 (a) Use Excel to prepare a 2-D pie chart for these web-surfing data. Modify the default colors, fonts, etc., as you judge appropriate to make the display effective. (b) Right-click the chart area, select Chart Type, and change to a bar chart. (c) Right-click the chart area, select Chart Type, and change to a hierarchy chart. Which do you prefer? Why?

Are You Concerned about Being Tracked While Web Surfing? 📄 Privacy	
Level of Concern	**Percent**
Very concerned	68
Somewhat concerned	23
Not very concerned	9
Total	100

3.23 (a) Use Excel to prepare a 2-D pie chart for the following data. Modify the default colors, fonts, etc., as you judge appropriate to make the display effective. (b) Right-click the chart area, select Chart Type, and change to a bar chart. (c) Right-click the chart area, select Chart Type, and change to a hierarchy chart. Which do you prefer? Why?

Dolona Corporation Medical Claims in 2017 Dolona	
Spent On	**Percent of Total**
Hospital services	47.5
Physicians	27.0
Pharmaceuticals	19.5
Mental health	5.0
Other	1.0
Total	100.0

3.24 (a) Use Excel to prepare a 2-D pie chart for these LCD (liquid crystal display) shipments data. Modify the default colors, fonts, etc., as you judge appropriate to make the display effective. (b) Do you feel that the chart has become too cluttered (i.e., are you displaying too many slices)? Would a bar chart be better? Explain. *Hint:* Include data labels with the percent *values*.

World Market Share of LCD Shipments LCDMarket	
Company	**Percent**
Samsung	18.0
Vizio	16.7
Sony	11.3
Sanyo	8.0
LG Electronics	7.8
Others	38.1
Total	100.0

Source: http://news.cnet.com/. Data are for first quarter of 2010. Percents may not add to 100 due to rounding.

3.8 TABLES

Tables are the simplest form of data display, yet creating effective tables is an acquired skill. By arranging numbers in rows and columns, their meaning can be enhanced so it can be understood at a glance.

A table may be preferred to a graph when the contents are complex. Table 3.7 is a summary from a bank's annual report for stockholders and potential investors. It shows time series data (columns) for six variables (rows). Figures were rounded to integers and stated in millions. (Otherwise the columns would have many digits and be harder to absorb quickly.) In its required 10K reports, financial data would be shown in full detail. Tables in documents or slide presentations can be linked dynamically to spreadsheets so that slides can be updated easily, but take care that spreadsheet data editing does not adversely affect the table layout.

LO 3-9

Make simple tables and pivot tables.

Table 3.7

XYBank Inc. Earnings Summary (millions of dollars)   XYBank

	Year				
	2020	**2019**	**2018**	**2017**	**2016**
Net interest income	2,306	2,025	1,805	1,772	1,652
Provision for credit losses	2	77	237	138	28
Noninterest income	1,046	1,069	1,174	1,052	842
Noninterest expense	1,821	1,748	1,915	1,815	1,586
Provision for income taxes	307	518	178	244	292
Net income	1,185	735	462	530	586

Tips for Effective Tables

Here are some tips for creating effective tables:

1. Keep the table simple and consistent with its purpose. The main point of the table should be clear to the reader within *10 seconds*. If not, break the table into parts or aggregate the data.

2. For presentations, round off to three or four significant digits (e.g., 142 rather than 142.213). Exceptions: when accounting requirements supersede the desire for rounding or when the numbers are used in subsequent calculations.

3. Physical table layout should guide the eye toward the comparison you wish to emphasize. Spaces or shading may be used to separate rows or columns. Use lines sparingly.

4. Within a column, use a consistent number of decimal digits. Right-justify or decimal-align the data unless all field widths are the same within the column.

Pivot Tables 📄 PivotTable

One of Excel's most popular and powerful features is the **pivot table**, which provides interactive analysis of a data matrix. The simplest kind of pivot table has rows and columns. Each of its cells shows a statistic for a row and column combination. The row and column variables must be either *categorical* or *discrete numerical,* and the variable for the table cells must be *numerical* (review Chapter 2 if you do not remember these terms). After the table is created, you can change the table by dragging variable names from the list specified in your data matrix. You can change the displayed statistic in the cells (sum, count, average, maximum, minimum, product) by right-clicking the display and selecting from the *field settings* menu. We show here the steps needed to create a pivot table for a small data matrix (25 homes, 3 variables). The first table shows the *sum* of square feet for all the homes in each category. The second table was created by copying the first table and then changing the cells to display the *average* square feet of homes in that cell.

Step 1: Select the Insert tab and specify the data range.

Source: Microsoft Excel.

Step 2:   Drag and drop desired fields for rows, columns, and the table body.

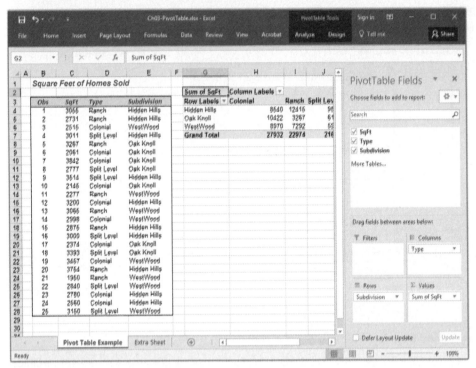

Source: Microsoft Excel.

Step 3:   Now you can format the table or right-click to choose desired field settings.

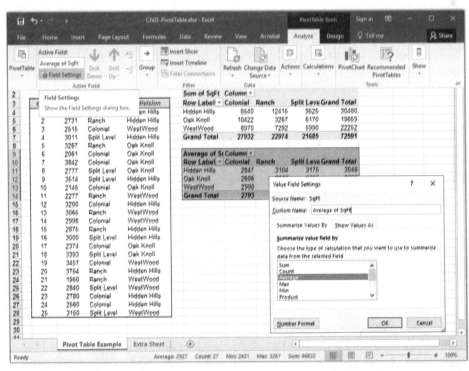

Source: Microsoft Excel.

Figure 3.21

Two Pivot Tables for U.S. Income Tax Returns (*n* = 4,801) 🗁 **Taxes**

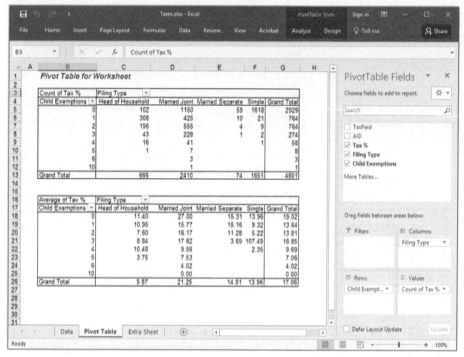

Source: Microsoft Excel.

A pivot table is especially useful when you have a large data matrix with several variables. For example, Figure 3.21 shows two pivot tables based on tax return data for *n* = 4,801 U.S. taxpayers. The first pivot table shows the number of taxpayers by filing type (single, married joint, married separate, head of household) cross-tabulated against the number of child exemptions (0, 1, 2, . . . , 10). The second pivot table shows average tax rate (percent) for each cell in the cross-tabulation. Note that some of the averages are based on small cell counts.

3.9 DECEPTIVE GRAPHS

LO 3-10

Recognize deceptive graphing techniques.

We have explained how to create *good* graphs. Now, let's turn things around. As an impartial consumer of information, you need a checklist of errors to beware of. Those who want to slant the facts may do these things deliberately, although most errors occur through ignorance. Use this list to protect yourself against ignorant or unscrupulous practitioners of the graphical arts.

Error 1: Nonzero Origin   A nonzero origin will exaggerate a trend. Measured distances would not match the stated values. The accounting profession is particularly aggressive in enforcing this rule. Although zero origins are preferred, sometimes a nonzero origin is needed to show sufficient detail. The first chart (nonzero origin) exaggerates the trend in call length.

Nonzero Origin

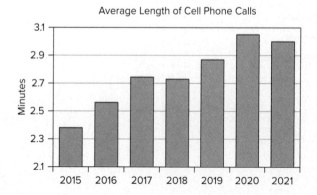

Zero Origin

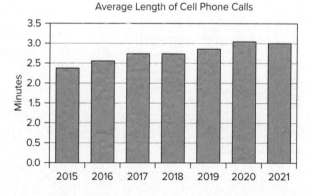

Error 2: Elastic Graph Proportions   By shortening the *X*-axis in relation to the *Y*-axis, vertical change is exaggerated. For a time series (*X*-axis representing time), this can make a sluggish sales or profit curve appear steep. Conversely, a wide *X*-axis and short *Y*-axis can downplay alarming changes. Keep the *aspect ratio* (width/height) below 2.00. Excel graphs use a default aspect ratio of about 1.68. The Golden Ratio you learned in art history suggests that 1.62 is ideal. Movies use a wide-screen format (up to 2.55), but DVDs may crop them to fit on a television screen. HDTV and multimedia computers use a 16:9 aspect ratio (about 1.78). Newer ultrawide monitors offer ratios of about 2.33. These two charts show the same data. Which *seems* to be growing faster?

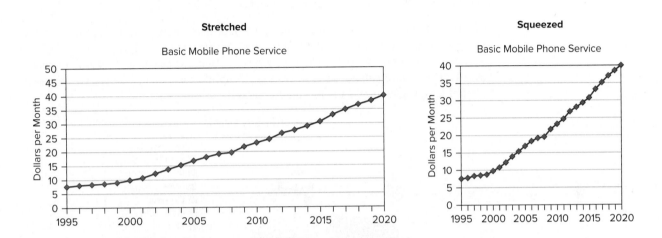

Error 3: Dramatic Titles, Distracting Art, and Perplexing Depth   In popular media, a *distracting title* often is designed more to grab the reader's attention than to convey the chart's content (*Deficit Swamps Economy*). Sometimes the title attempts to draw your conclusion for you (*Inflation Wipes Out Savings*). A title should be short but adequate for the purpose.

To add visual pizzazz, artists may superimpose *distracting pictures* or colorful figures, banners, or drawings. This can distract the reader or impart an emotional slant. Advertisements sometimes feature attractive, conservatively attired actors portraying scientists, doctors, or business leaders examining scientific-looking charts. Such displays impart credibility to self-serving commercial claims. A *3-D chart* introduces further ambiguity in reading the bar height. Excel even allows you to make a *rotated 3-D graph* that can make trends appear to dwindle into the distance or loom alarmingly toward you. Resist the temptation to use 3-D or rotated graphs. This illustration combines several errors (nonzero origin, leading title, distracting picture, vague source, rotated 3-D).

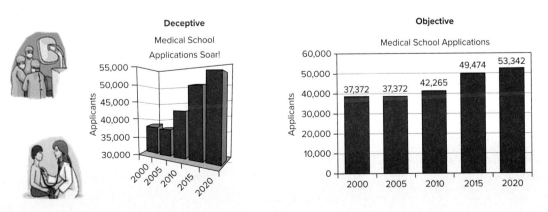

Source: www.aamc.org.

Error 4: Unclear Definitions or Scales   Missing or unclear units of measurement can render a chart useless. We must know whether the variable being plotted is sales, profits, assets, or whatever. If percent, indicate clearly *percentage of what.* Gridlines help the viewer compare magnitudes but are often omitted to avoid graph clutter. For maximum clarity in a bar graph, label each bar with its numerical value.

Error 5: Vague Sources   Large federal agencies or corporations employ thousands of people and issue hundreds of reports per year. Vague sources like "Department of Commerce" may indicate that the author lost the citation, didn't know the data source, or mixed data from several sources. Scientific publications insist on complete source citations. Rules are less rigorous for publications aimed at a general audience.

Error 6: Complex Graphs   Complicated visual displays make the reader work harder. Keep your main objective in mind. Omit "bonus" detail or put it in the appendix. Apply the *10-second rule* to graphs. If the message really is complex, can it be broken into smaller parts? This example combines several errors (silly subtitle, distracting pictures, no data labels, no definitions, vague source, too much information).

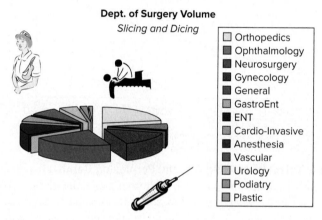

Source: Hospital reports.

Error 7: Estimated Data   In a spirit of zeal to include the "latest" figures, the last few data points in a time series are often estimated. Estimated points should be noted.

Error 8: Area Trick   One of the most pernicious visual tricks is simultaneously enlarging the width of the bars as their height increases, so the bar area misstates the true proportion (e.g., by replacing graph bars with figures like human beings, coins, or gas pumps). As figure height increases, so does width, distorting the area. This example (physician salaries) illustrates this distortion.

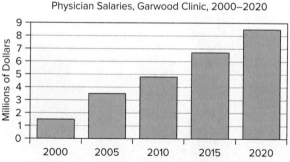

Final Advice

Can you trust any visual display (unless you created it yourself)? Be a skeptic, and be pleasantly surprised if the graph lives up to the best standards. Print media compete with TV and the web, so newspapers and magazines must use colorful charts to attract reader interest. People enjoy visual displays, so we accept some artistic liberties. Mass-readership publications like *Time, USA Today,* or even specialized business-oriented publications like *Forbes, Fortune, Bloomberg Businessweek,* and *The Wall Street Journal* should not be judged by the same standards you would apply to an academic journal. But businesses do want charts that follow the rules because a deceptive chart may have serious consequences. Decisions may be made about products or services that affect lives, market share, and jobs (including yours). So know the rules, try to follow them, and expect your peers and subordinates to do the same. Catchy graphics have a place in selling your ideas but shouldn't dominate the data.

Chapter Summary

For a set of observations on a single numerical variable, a **stem-and-leaf plot** or a **dot plot** displays the individual data values, while a **frequency distribution** classifies the data into classes called **bins** for a **histogram** of **frequencies** for each bin. The number of bins and their limits are matters left to your judgment, though **Sturges' Rule** offers advice on the number of bins. The **line chart** shows values of one or more **time series** variables plotted against time. A **log scale** is sometimes used in time series charts when data vary by orders of magnitude. The **bar chart** or **column chart** shows a **numerical** data value for each category of an **attribute.** However, a bar chart also can be used for a time series. A **scatter plot** can reveal the association (or lack of association) between two variables *X* and *Y.* The **pie chart** (showing a **numerical** data value for each category of an **attribute** if the data values are parts of a whole) is common but should be used with caution. Sometimes a **simple table** is the best visual display. Creating effective visual displays is an acquired skill. Excel offers a wide range of charts from which to choose. Consider using R (it's free) if you want to learn more about programming. Deceptive graphs are found frequently in both media and business presentations, and the consumer should be aware of common errors.

Key Terms

arithmetic scale	histogram	Pareto chart	stem-and-leaf plot
bar chart	left-skewed	pie chart	Sturges' Rule
center	line chart	pivot table	symmetric
column chart	logarithmic scale	right-skewed	trend line
dot plot	modal class	scatter plot	variability
frequency distribution	ogive	shape	
frequency polygon	outlier	stacked column chart	

Chapter Review

1. Name two attractive features and two limitations of the (a) stem-and-leaf plot and (b) dot plot.

2. (a) What is a frequency distribution? (b) What are the steps in creating one?

3. (a) What is a histogram? (b) What does it show?

4. (a) What is a bimodal histogram? (b) Explain the difference between left-skewed, symmetric, and right-skewed histograms. (c) What is an outlier?

5. (a) What is a scatter plot? (b) What do scatter plots reveal? (c) Sketch a scatter plot with a moderate positive correlation. (d) Sketch a scatter plot with a strong negative correlation.

6. For what kind of data would we use a bar chart? List three tips for creating effective bar charts.

7. For what kind of data would we use a line chart? List three tips for creating effective line charts.

8. (a) List the three most common types of charts in business, and sketch each type (no real data, just a sketch). (b) List three specialized charts that can be created in Excel, and sketch each type (no real data, just a sketch).

9. (a) For what kind of data would we use a pie chart? (b) Name two common pie chart errors. (c) Why are pie charts regarded with skepticism by some statisticians?

10. Which types of charts can be used for time series data?

11. (a) When might we need a log scale? (b) What do equal distances on a log scale represent? (c) State one drawback of a log scale graph.

12. When might we use a stacked column chart? An area chart? A Pareto chart?

13. List five deceptive graphical techniques.

14. What is a pivot table? Why is it useful?

15. Suggest pros and cons of making your graphs in R instead of Excel.

Chapter Exercises

Note: In these exercises, you may use Excel or a software package (e.g., MegaStat, Minitab, or R) if your instructor agrees.

3.25 The durations (minutes) of 26 electric power outages in the community of Sonando Heights over the past five years are shown below. (a) Make a stem-and-leaf diagram. (b) Make a histogram. (c) Describe the shape of the distribution. 📁 **Duration**

32	44	25	66	27	12	62	9	51	4	17	50	35
99	30	21	12	53	25	2	18	24	84	30	17	17

3.26 The U.S. Postal Service will ship a Priority Mail® Large Flat Rate Box (12″ × 12″ × 5½″) anywhere in the United States for a fixed price, regardless of weight. The weights (ounces) of 20 randomly chosen boxes are shown below. (a) Make a stem-and-leaf diagram. (b) Make a histogram. (c) Describe the shape of the distribution. 📁 **Weights**

72	86	28	67	64	65	45	86	31	32
39	92	90	91	84	62	80	74	63	86

3.27 A study of 40 U.S. cardiac care centers showed the following ratios of nurses to beds. (a) Prepare a dot plot. (b) Prepare a frequency distribution and histogram (you may either specify the bins yourself or use automatic bins). (c) Describe the distribution, based on these displays. 📁 **Nurses**

1.48	1.16	1.24	1.52	1.30	1.28	1.68	1.40	1.12	0.98	0.93	2.76
1.34	1.58	1.72	1.38	1.44	1.41	1.34	1.96	1.29	1.21	2.00	1.50
1.68	1.39	1.62	1.17	1.07	2.11	2.40	1.35	1.48	1.59	1.81	1.15
1.35	1.42	1.33	1.41								

3.28 The first Rose Bowl (football) was played in 1902. The next one was not played until 1916, but a Rose Bowl has been played every year since then. The margin of victory in each of the 100 Rose Bowls from 1902 through 2020 is shown below (0 indicates a tie). (a) Prepare a stem-and-leaf plot. (b) Prepare a frequency distribution and histogram (you may either specify the bins yourself or use automatic bins). (c) Describe the distribution, based on these displays. 📁 **RoseBowl**

0	0	0	1	1	1	1	1	1	1	1	1	1
2	2	3	3	3	3	3	3	3	3	4	4	4
5	5	5	5	6	6	6	7	7	7	7	7	7
7	7	7	7	7	8	8	8	8	8	8	9	9
9	9	10	10	10	10	10	11	11	11	12	13	13
13	14	14	14	14	14	14	16	16	17	17	17	18
18	20	20	20	21	21	23	24	25	25	26	27	28
28	29	29	31	32	33	33	35	36	36	39	49	49

3.29 An executive's telephone log showed the following data for the length of 60 calls initiated during the last week of July. (a) Prepare a dot plot. (b) Prepare a frequency distribution and histogram (you may either specify the bins yourself or use automatic bins). (c) Describe the distribution, based on these displays. 📁 **CallLength**

1	2	10	5	3	3	2	20	1	1
6	3	13	2	2	1	26	3	1	3
1	2	1	7	1	2	3	1	2	12
1	4	2	2	29	1	1	1	8	5
1	4	2	1	1	1	1	6	1	2
3	3	6	1	3	1	1	5	1	18

3.30 Below are batting averages of the New York Yankees players who were at bat five times or more in 2006. (a) Construct a frequency distribution. Explain how you chose the number of bins and the bin limits. (b) Make a histogram and describe its appearance. (c) Repeat, using a different number of bins and different bin limits. (d) Did your visual impression of the data change when you changed the number of bins? Explain. 📁 **Yankees**

Batting Averages for the 2006 New York Yankees

Player	Avg	Player	Avg
Derek Jeter	0.343	Craig Wilson	0.212
Johnny Damon	0.285	Bubba Crosby	0.207
Alex Rodriguez	0.290	Aaron Guiel	0.256
Robinson Cano	0.342	Kelly Stinnett	0.228
Jorge Posada	0.277	Nick Green	0.240
Melky Cabrera	0.280	Sal Fasano	0.143
Jason Giambi	0.253	Terrence Long	0.167
Bernie Williams	0.281	Kevin Thompson	0.300
Andy Phillips	0.240	Kevin Reese	0.417
Miguel Cairo	0.239	Andy Cannizaro	0.250
Bobby Abreu	0.330	Randy Johnson	0.167
Hideki Matsui	0.302	Wil Nieves	0.000
Gary Sheffield	0.298		

Source: www.thebaseballcube.com.

3.31 Download from the McGraw Hill Connect® website the full data set of measurements of cockpit noise level for an older commercial jet airliner (only the first 2 and last 2 data values are shown). (a) Use Excel to make a scatter plot, placing Airspeed on the *X*-axis and Noise Level on the *Y*-axis. Add titles and modify the default colors, fonts, etc., as you judge appropriate to make the scatter plot effective. (b) Describe the relationship (if any) between *X* and *Y*. Weak? Strong? Negative? Positive? Linear? Nonlinear? *Hint:* You may need to rescale the *X* and *Y* axes to see more detail.

Airspeed and Cockpit Noise
(*n* = 61 measurements) 🗁 **CockpitNoise**

Obs	Airspeed (knots)	Noise Level (dB)
1	250	83
2	340	89
...	...	...
60	405	93
61	250	82

Note: The decibel (dB) is a logarithmic unit that indicates the ratio of measured sound pressure to a benchmark. Some familiar examples for comparison: Vuvuzela horn at 1 m (120 dB), jack hammer at 1 m (100 dB), handheld electric mixer (65 dB).

3.32 Download the full data set from the McGraw Hill Connect® website (only the first 2 and last 2 data values are shown). (a) Use Excel to make a scatter plot, placing Dolona on the *X*-axis and ZalParm on the *Y*-axis. Add titles and modify the default colors, fonts, etc., as you judge appropriate to make the scatter plot effective. (b) Describe the relationship (if any) between *X* and *Y*. Weak? Strong? Negative? Positive? Linear? Nonlinear? (c) Make a line chart showing both time series variables on the same chart. Does the line chart support your conclusions from the scatter plot?

Week Closing Price of Two Stocks
(*n* = 52 weeks) 🗁 **StockPrices**

Week	Dolona	ZalParm
1	14.71	36.93
2	14.65	33.63
...	...	...
51	12.40	29.41
52	11.98	25.79

3.33 Download the full data set from the McGraw Hill Connect® website (only the first 2 and last 2 data values are shown). (a) Use Excel to make a scatter plot, placing GDP per Capita on the *X*-axis and Birth Rate on the *Y*-axis. Add titles and modify the default colors, fonts, etc., as you judge appropriate to make the scatter plot effective. (b) Describe the relationship (if any) between *X* and *Y*. Weak? Strong? Negative? Positive? Linear? Nonlinear?

GDP per Capita and Birth Rate
(*n* = 153 nations) 🗁 **GDPBirthRate**

Nation	GDP per Capita	Birth Rate
Afghanistan	800	41.03
Albania	3,800	18.59
...	...	...
Zambia	870	41.01
Zimbabwe	2,450	24.59

Source: *The World Factbook 2003.* Central Intelligence Agency, 2003. www.cia.gov.

3.34 (a) What kind of display is this? (b) Identify its strengths and weaknesses, using the tips and checklists shown in this chapter. (c) Can you suggest any improvements? Would a different type of display be better? 🗁 **WomenPilots**

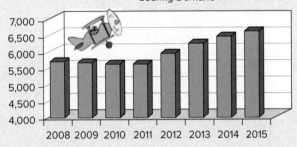

Female Air Transport Pilots
Soaring Demand

Source: www.faa.gov.

3.35 (a) What kind of display is this? (b) Identify its strengths and weaknesses, using the tips and checklists shown in this chapter. (c) Can you suggest any improvements? Would a different type of display be better? 🗁 **MedError**

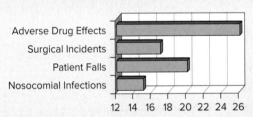

Sources of Medical Error

3.36 (a) What kind of display is this? (b) Identify its strengths and weaknesses, using the tips and checklists shown in this chapter. (c) Can you suggest any improvements? Would a different type of display be better? 🗁 **Oxnard**

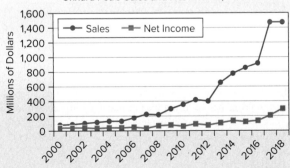

Oxnard Petro Sales and Net Income, 2000–2018

3.37 (a) What kind of display is this? (b) Identify its strengths and weaknesses, using the tips and checklists shown in this chapter. (c) Can you suggest any improvements? Would a different type of display be better?

Top Suppliers of U.S. Crude Oil Imports: 2002 (in millions of barrels)

Total imports = 3,336

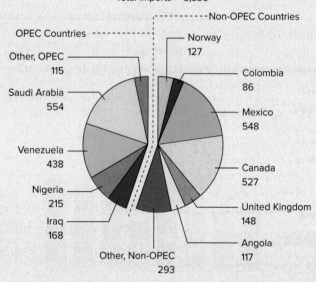

Source: *Statistical Abstract of the United States, 2003.*

3.38 (a) What kind of display is this? (b) Identify its strengths and weaknesses, using the tips and checklists shown in this chapter. (c) Can you suggest any improvements? Would a different type of display be better?

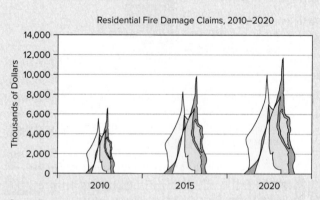

Source: Insurance company records.

3.39 (a) What kind of display is this? (b) Identify its strengths and weaknesses, using the tips and checklists shown in this chapter. (c) Can you suggest any improvements? Would a different type of display be better? 📂 **Bankruptcies**

U.S. Consumer Bankruptcies Soar

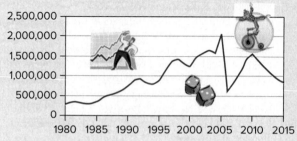

Source: *American Bankruptcy Institute*, www.abiworld.org.

3.40 (a) What kind of display is this? (b) Identify its strengths and weaknesses, using the tips and checklists shown in this chapter. (c) Can you suggest any improvements? Would a different type of display be better?

Sales of Sony Blu-Ray Disc Players at Bob's Appliance Mart

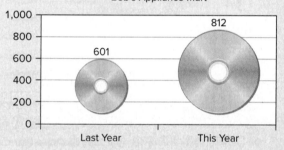

3.41 (a) Use Excel to prepare an appropriate type of chart (bar, line, pie, scatter) to display the following data. Modify the default colors, fonts, etc., as you judge appropriate to make the display effective. (b) Would more than one kind of display be acceptable? Why or why not?

Domestic Market Share, Ten Largest U.S. Airlines
📂 **AirlineMkt**

Airline	Percent
Alaska	4.5
American	18.4
Delta	16.9
Frontier	2.0
Hawaiian	1.7
JetBlue	5.4
SkyWest	2.4
Southwest	18.3
Spirit	2.7
United	14.5
Other	13.2
Total	100.0

Source: www.transtats.bts.gov. Data are based on 2015 revenue passenger miles.

3.42 (a) Use Excel to prepare an appropriate type of chart (bar, line, pie, scatter) to display the U.S. petroleum import data. Modify the default colors, fonts, etc., as you judge appropriate to make the display effective. (b) Would more than one kind of display be acceptable? Why or why not?

U.S. Petroleum Imports (millions of barrels per day) Petroleum

Year	2010	2011	2012	2013	2014	2015	2016	2017	2018
OPEC	4.91	4.56	4.27	3.72	3.24	2.89	3.45	3.37	2.89
Non-OPEC	6.89	6.88	6.33	6.14	6.00	6.55	6.61	6.78	7.05

Source: www.eia.doe.gov.

3.43 (a) Use Excel to prepare an appropriate type of chart (bar, line, pie, scatter) to display the following data. Modify the default colors, fonts, etc., as you judge appropriate to make the display effective. (b) Would more than one kind of display be acceptable? Why or why not?

U.S. Market Share for Web Browsers
WebSearch

Browser	Percent
Chrome	57.1
Safari	13.8
Firefox	10.9
Edge	10.4
Opera	3.1
Others	4.7
Total	100.0

Source: Data are for August 2016. Only desktop/laptop browser usage is included (i.e., not cell phones or similar devices).

3.44 (a) Use Excel to prepare an appropriate type of chart (bar, line, pie, scatter) to display the U.S. energy consumption data. Modify the default colors, fonts, etc., as you judge appropriate to make the display effective. (b) Would more than one kind of display be acceptable? Why or why not?

U.S. Energy Consumption by Source Energy

Source	Quad BTU	Percent
Petroleum	37.06	38.4%
Natural Gas	23.15	24.0%
Coal	20.49	21.2%
Nuclear	8.52	8.8%
Renewables	7.17	7.4%
Other	0.21	0.2%
Total	96.60	100.0%

Source: *Statistical Abstract of the United States, 2011*, p. 583.

3.45 (a) Use Excel to prepare a Pareto chart of the following data. (b) In your own words, what does this chart say?

Cell Phone Service Provider $n = 158$
CellPhone

Service Provider	Percent	Cumulative Percent
Verizon	37.3	37.3
AT&T	29.7	67.1
T-Mobile	13.3	80.4
Sprint	8.9	89.2
Other	10.8	100.0

Source: Web survey of 158 statistics students.

Related Reading

Visual Displays

Cairo, Alberto. *The Truthful Art: Data, Charts, and Maps for Communication.* Pearson, 2016.

Friendly, Michael, and Howard Wainer. *A History of Data Visualization and Graphic Communication.* Harvard University Press, 2021.

Swires-Hennessy, Ed. *Presenting Data: How to Communicate Your Message Effectively.* Wiley, 2014.

Tufte, Edward R. *The Visual Display of Quantitative Information.* 2nd ed. Graphics Press, 2004.

Wilkinson, Leland. *The Grammar of Graphics.* Springer, 2005.

CHAPTER 3   More Learning Resources    Mc Graw Hill **connect**

You can access these *LearningStats* demonstrations through Connect to help you understand visual data displays.

Topic	*LearningStats Demonstrations*
Effective visual displays	Presenting Data—I Presenting Data—II Excel Histograms R Histograms
How to make charts and tables	Excel Charts: Step-by-Step Pivot Tables: Step-by-Step Bins and Binning Stem-and-Leaf Plots Bimodal Data

Topic	LearningStats Demonstrations
Applications	⊠ Bimodal Data
	⊠ Sturges' Rule
	⊐ Stem-and-Leaf Plots
Screen Cam Tutorials	⊞ Excel Basics
	⊞ Excel Histograms
	⊞ Excel Scatter Plots

Key: ⊡ = PowerPoint ⊠ = Excel ⊐ = PDF ⊞ = Screen Cam Tutorials

Software Supplement

Histograms Using MegaStat

MegaStat® by J. B. Orris of Butler University is a simple Excel add-in that can be downloaded from www.mhhe.com/megastat and installed on your PC or Mac. MegaStat goes far beyond Excel's built-in statistical functions to offer a full range of easy-to-use statistical tools to help you analyze data, create graphs, and perform calculations. After installation, click the Excel Add-In tab. MegaStat's main menu (Figure 3.22) will appear in the upper left corner of the Excel menu bar. To get a histogram, first choose Frequency Distribution and then select Quantitative.

Figure 3.22

MegaStat Main Menu

Source: MegaStat.

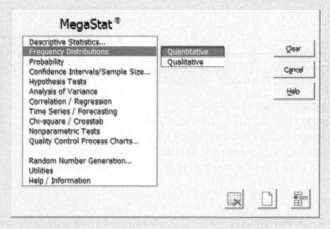

Enter your data range in the InputRange box (Figure 3.23). Or, if you click in the InputRange box and then double-click any cell in your spreadsheet data column, MegaStat will autofill the entire column data range. If you are satisfied with MegaStat's suggested values for the interval width and lower boundary for the first interval, just click OK. However, you may enter your own choices (you will be warned if your choices are inappropriate). The Option tab allows you to create unequal bin intervals. Three frequency graphs are offered: Histogram (frequency bar chart), Polygon (connected frequency bar heights), and Ogive (cumulative frequency polygon).

Figure 3.23

MegaStat Histogram Setup and Output

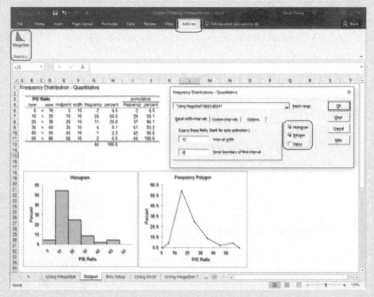

Source: MegaStat.

MegaStat provides a table with bin limits, frequencies, percent frequencies, and cumulative frequencies. Its nicely labeled histogram shows *percent* (not frequencies) on the vertical axis (the appearance of the histogram is the same either way). The histogram can be edited (e.g., title, colors, axis labels), but if you want different bins, you must revisit the MegaStat menu.

Histograms Using Minitab

Minitab™ is a comprehensive software package for statistical analysis. It has nothing to do with Excel, although you can copy data from Excel to the Minitab worksheet (and vice versa). Minitab is available at many colleges and universities or as a

download at student prices with McGraw Hill textbooks. While this textbook emphasizes Excel, it is wise to learn to use stand-alone software like Minitab to get results more easily or in a more attractive form. For example, to create a histogram, just copy your data from the Excel spreadsheet and paste it into Minitab's worksheet and choose Graphs > Histogram from the top menu bar (Figure 3.24). Let Minitab use its default options. Once the histogram has been created, you can right-click the *X*-axis to adjust the bins, axis tick marks, and so on. Here is a Minitab histogram for the price-earnings data. The graph can be copied and pasted into Excel, PowerPoint, or Word as you prepare your report.

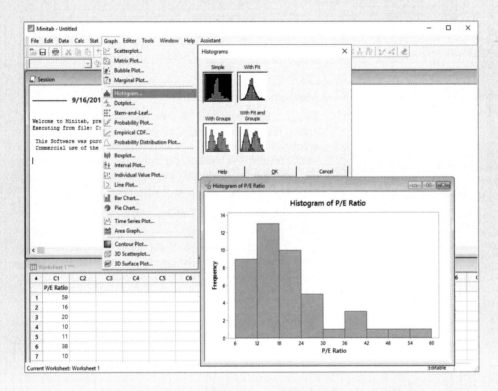

Figure 3.24

Minitab Histogram Setup and Result

Source: Minitab, Inc.

Software Supplement

Making a Histogram in R

Step 1:   Enter your data. There are several ways to enter your data into R. You can create a vector of data by entering your data values separated by commas using the R concatenate command $c(x_1, x_2, \ldots, x_n)$. However, if you have many data values, you will probably import your data from a spreadsheet. **Appendix K** explains alternative ways to import data from Excel to a data frame in R (e.g., reading the entire spreadsheet). One way is to copy our data as a single Excel data column to the clipboard and paste it into an R data frame. Our data consists of 44 price-earnings ratios for *Fortune* 500 companies. Highlight the Excel data column (including its column heading) and copy it to the clipboard using Ctrl-C (or command-C in MacOS). We will name our R data frame PERatios. The Excel column heading will become the name of the variable in R. If you have no column header, change TRUE to FALSE and R will assign a variable name:

> PERatios=read.table(file="clipboard", header=TRUE)

Step 2:   Make a histogram using the hist(x) command where x is your data vector's name. We named our data frame PERatios and our variable column was named PERatio, so we enter:

> hist(PERatios$PERatio)

Omitting all optional arguments for the hist() command, we get the default number of bins, based on Sturges' Rule (6 bins and 7 axis break points). R follows the Excel convention of using right-closed (left open) bin intervals. To set the number of bins yourself, you may include the optional argument breaks=k to create k-1 bins. The default axis label will be exactly the variable name you entered, which might look odd. To choose a better one, include the xlab argument. For example, here is how we can make 11 bins (12 axis breaks) and label the axis "Price Earnings Ratio":

> hist(PERatios$PERatio, breaks=12, xlab="Price Earnings Ratio")

Our customized histogram is shown on the right. However, its axis scale may not appear as you would expect because the actual

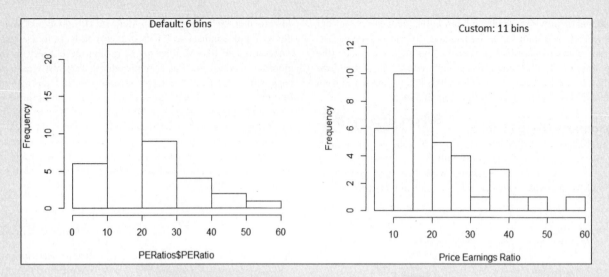

data range was 7 to 59. You can experiment with the optional R argument xlim=c(a,b) to set better axis limits a and b. But the default histogram may suffice.

Step 3: To export any graph to a file (or to the clipboard), click the Plots tab in the History pane (the lower right quadrant) and then click the Export button (use the arrow keys to scroll through whatever plots you have created). You can then paste the graph into a document.

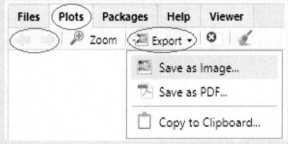

RStudio, PBC.

Making a Scatter Plot in R

Step 1: Enter your data. We have paired observations on acceleration times (in seconds) for 39 vehicles, where *x* is the time to accelerate from zero to 30 mph and *y* is the time to accelerate from zero to 60 mph. We highlight our two Excel data columns (including their column headings), copy them to the clipboard using Ctrl-C, and paste the data into an R data frame named Accel:

> Accel=read.table(file="clipboard", sep="\t", header=TRUE)

Step 2: Make a scatter plot using the plot(x,y) command where x and y are the names of your data vectors. Our variable columns were named ZeroTo30 and ZeroTo60, so we enter:

> plot(Accel$ZeroTo30, Accel$ZeroTo60)

The axis labels look a little clunky. We can create better axis labels using the xlab and ylab arguments. We can also insert the main argument into the plot() function to add a title. Our customized scatter plot is shown on the right.

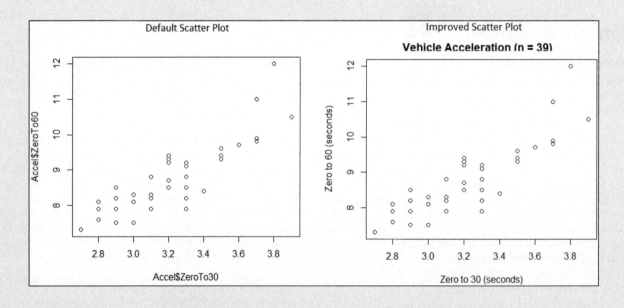

> plot(Accel$ZeroTo30, Accel$ZeroTo60, xlab="Zero to 30 (seconds)",

+ ylab="Zero to 60 (seconds)", main="Vehicle Acceleration (n = 39)")

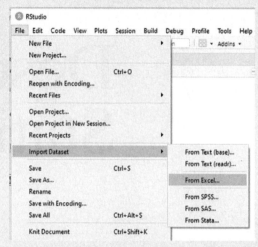

RStudio, PBC.

Importing a Spreadsheet

If you prefer to import an *entire* Excel spreadsheet, choose File>Import Dataset>From Excel in the upper left corner of the Scripts pane. When prompted, click the Browse button to locate your Excel file and follow the instructions.

Importing an Excel file works fine if your spreadsheet contains *only* the desired data in contiguous columns. However, it's unpredictable if your Excel spreadsheet has comments, pictures, or blank lines for spacing. This textbook's Excel data files (exercises, examples) are acceptable for R import if you skip the first two rows (option at the bottom of the import menu).

RStudio, PBC.

4 Descriptive Statistics

CHAPTER CONTENTS

CHAPTER LEARNING OBJECTIVES

When you finish this chapter, you should be able to

LO 4-1   Explain the concepts of center, variability, and shape.

LO 4-2   Calculate and interpret common measures of center.

LO 4-3   Calculate and interpret common measures of variability.

LO 4-4   Apply Chebyshev's theorem.

LO 4-5   Apply the Empirical Rule and recognize outliers.

LO 4-6   Transform a data set into standardized values.

LO 4-7   Calculate quartiles and other percentiles.

LO 4-8   Make and interpret box plots.

LO 4-9   Calculate and interpret a correlation coefficient and covariance.

LO 4-10   Calculate the mean and standard deviation from grouped data.

LO 4-11   Assess skewness and kurtosis in a sample.

4.1 NUMERICAL DESCRIPTION

The previous chapter explained *visual* descriptions of data (e.g., histograms, dot plots, scatter plots). This chapter explains *numerical* descriptions of data. While visual displays of data can answer many questions, a more detailed numerical description is often needed. Descriptive measures derived from a sample (*n* items) are *statistics,* while for a population (*N* items or infinite) they are *parameters.* For a sample of numerical data, we are interested in three key characteristics: center, variability, and shape. Table 4.1 summarizes the questions that we will be asking about the data.

gremlin/E+/Getty Images

Table 4.1

Characteristics of Numerical Data

Characteristic	Interpretation
Center	Where are the data values concentrated? What seems to be typical or middle data values? Is there a central tendency?
Variability	How much dispersion is there in the data? How spread out are the data values? Are there unusual values?
Shape	Are the data values distributed symmetrically? Skewed? Sharply peaked? Flat? Bimodal?

LO 4-1

Explain the concepts of center, variability, and shape.

Example 4.1

Vehicle Quality

Every year, J.D. Power and Associates issues its initial vehicle quality ratings. These ratings are of interest to consumers, dealers, and manufacturers. Table 4.2 shows defect rates for 33 vehicle brands for model year 2022. Reported defect rates are based on a sample of vehicles within each brand. Numerical statistics can be used to summarize a data set like this.

Table 4.2

Defects per 100 Vehicles 🗁 JDPower

Brand	Defects	Brand	Defects	Brand	Defects
Acura	192	GMC	162	Mazda	180
Alfa Romeo	211	Honda	183	Mercedes-Benz	189
Audi	239	Hyundai	185	MINI	168
BMW	165	Infiniti	204	Mitsubishi	226
Buick	139	Jaguar	210	Nissan	167
Cadillac	163	Jeep	199	Porsche	200
Chevrolet	147	Kia	156	Ram	186
Chrysler	265	Land Rover	193	Subaru	191
Dodge	143	Lexus	157	Toyota	172
Ford	167	Lincoln	167	Volkswagen	230
Genesis	156	Maserati	255	Volvo	256

Source: J.D. Power and Associates 2022 Initial Quality Study™. Ratings are intended for educational purposes only and should not be used as a guide to consumer decisions. Tesla and Polestar brands are not shown because they did not meet the J.D. Power study criteria.

Before calculating any statistics, we consider how the data were collected. A web search reveals that J.D. Power and Associates is a well-established independent company whose methods are widely considered to be objective. Data on defects are obtained by inspecting randomly chosen vehicles for each brand, counting the defects, and dividing the number of defects by the number of vehicles inspected. J.D. Power multiplies the result by 100 to obtain defects per 100 vehicles, rounded to the nearest integer. The underlying measurement scale is continuous (e.g., if 4 defects were found in 3 vehicles, the defect rate would be 1.333333, or 133 defects per 100 vehicles). Defect rates would vary from year to year and perhaps even within a given model year, so the timing of the study could affect the results. Because the analysis is based on sampling of vehicles within each brand, we must allow for the possibility of sampling error. With these cautions in mind, we look at the data. A good first step is to sort the data, as shown in Table 4.3.

Table 4.3

Defects per 100 Vehicles Ranked Lowest to Highest
JDPower

Brand	Defects	Brand	Defects	Brand	Defects
Buick	139	Nissan	167	Jeep	199
Dodge	143	MINI	168	Porsche	200
Chevrolet	147	Toyota	172	Infiniti	204
Genesis	156	Mazda	180	Jaguar	210
Kia	156	Honda	183	Alfa Romeo	211
Lexus	157	Hyundai	185	Mitsubishi	226
GMC	162	Ram	186	Volkswagen	230
Cadillac	163	Mercedes-Benz	189	Audi	239
BMW	165	Subaru	191	Maserati	255
Ford	167	Acura	192	Volvo	256
Lincoln	167	Land Rover	193	Chrysler	265

Except for tiny samples, sorting would be done in Excel. The sorted data reveal a wide range of data values, from 139 (Buick) to 265 (Chrysler). This suggests a high degree of *variability*. The next visual step is a histogram, shown in Figure 4.1. The modal class (largest frequency) between 150 and 170 reveals the *center*. The *shape* of the histogram is right-skewed (most data on the left, longer right tail).

Figure 4.1

Histogram of J.D. Power Data (*n* = 33) JDPower

Source: Minitab.

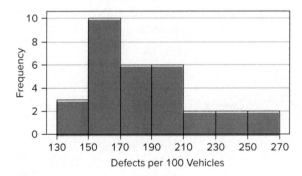

4.2 MEASURES OF CENTER

LO 4-2

Calculate and interpret common measures of center.

When we speak of the *center*, we are trying to describe the middle or typical values of a distribution. You can assess central tendency in a general way from a dot plot or histogram, but numerical statistics allow more precise statements. Table 4.4 lists six common measures of center. Each has strengths and weaknesses. We need to look at several of them to obtain a clear picture of the central tendency.

Mean

The most familiar statistical measure of center is the mean. It is the sum of the data values divided by the number of data items. For a population, we denote it μ, while for a sample we call it $\bar{x}$. We use equation 4.1 to calculate the mean of a population:

(4.1)
$$\mu = \frac{\sum_{i=1}^{N} x_i}{N} \quad \text{(population definition)}$$

Table 4.4

Six Measures of Center

Statistic	Formula	Excel*	Pro	Con
Mean	$\dfrac{1}{n}\sum\limits_{i=1}^{n} x_i$	=AVERAGE(Data)	Familiar and uses all the sample information.	Influenced by extreme values.
Median	Middle value in a sorted array	=MEDIAN(Data)	Robust when extreme data values exist.	Ignores extremes and can be affected by gaps in data values.
Mode	Most frequently occurring data value	=MODE(Data)	Useful for attribute data or discrete data with a small range.	May not be unique and is not helpful for continuous data.
Midrange	$\dfrac{x_{min} + x_{max}}{2}$	=0.5*(MIN(Data) +MAX(Data))	Easy to understand and calculate.	Influenced by extreme values and ignores most data values.
Geometric mean (G)	$\sqrt[n]{x_1 x_2 \ldots x_n}$	=GEOMEAN(Data)	Useful for growth rates and mitigates high extremes.	Less familiar and requires positive data.
Trimmed mean	Same as the mean except omit the highest and lowest k% of data values (e.g., 5%)	=TRIMMEAN(Data, Percent)	Mitigates effects of extreme values.	Excludes some data values that could be relevant.

*R offers equivalent functions mean(x), median(x), and trimmed mean mean(x,trim=p%) but not the others on this list. See **Appendix J** for a side-by-side list of R and Excel functions.

Because we rarely deal with populations, the sample notation of equation 4.2 is more commonly seen:

$$\bar{x} = \frac{\sum\limits_{i=1}^{n} x_i}{n} \quad \text{(sample definition)} \tag{4.2}$$

We calculate the mean by using Excel's function =AVERAGE(Data) where Data is an array containing the data. So for the sample of $n = 33$ car brands:

$$\bar{x} = \frac{\sum\limits_{i=1}^{n} x_i}{n} = \frac{139 + 143 + 147 + \cdots + 255 + 256 + 265}{33} = \frac{6{,}223}{33} = 188.576$$

Characteristics of the Mean

The arithmetic mean is the "average" with which most of us are familiar. The mean is affected by every sample item. It is the balancing point or fulcrum in a distribution if we view the X-axis as a lever arm and represent each data item as a physical weight, as illustrated in Figure 4.2 for the J.D. Power data.

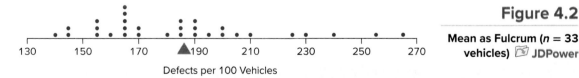

Figure 4.2

Mean as Fulcrum (*n* = 33 vehicles) 📷 JDPower

The mean is the balancing point because it has the property that distances from the mean to the data points *always* sum to zero:

$$\sum_{i=1}^{n} (x_i - \bar{x}) = 0 \tag{4.3}$$

This statement is true for *any* sample or population, regardless of its shape (skewed, symmetric, bimodal, etc.). Even when there are extreme values, the distances below the mean are *exactly* counterbalanced by the distances above the mean. The wide range and long right tail in the dot plot (Figure 4.2) suggest that the mean may not be a very representative measure of central tendency.

Median

The **median** (denoted *M*) is the 50th percentile or midpoint of the *sorted* sample data set $x_1, x_2, \ldots, x_n$. It separates the upper and lower halves of the sorted observations:

The median is the middle observation in the sorted array if *n* is odd, but the average of the middle two observations if *n* is even, as illustrated in Figure 4.3.

Figure 4.3

Illustration of the Median

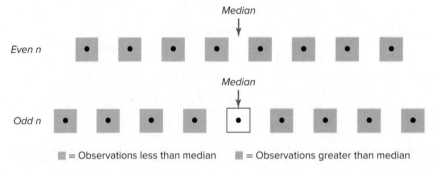

= Observations less than median   = Observations greater than median

---

Tip

If we have an even *n* (e.g., *n* = 6) then the median is halfway *between* the third and fourth observations in the sorted array:

$$Median = 16$$

| 11 | 12 | 15 | 17 | 21 | 32 |

But for an odd *n* (e.g., *n* = 7) the median is the fourth observation in this sorted array:

$$Median = 25$$

| 12 | 23 | 23 | (25) | 27 | 34 | 41 |

It is tempting to imagine that half the observations are less than the median, but this is not necessarily the case. For example:

$$Median = 78$$

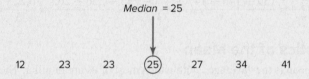

51 66 71 78 78 (78) 81 82 82 91 99

Only three data values are *below* 78, while five data values are *above* 78. This median did not provide a clean "50-50 split" in the data. This situation is not unusual. In fact, we expect clustering near the middle when there is a strong central tendency in a data set.

Excel's function for the median is =MEDIAN(Data) where Data is the data array. For the 33 vehicle quality ratings (odd *n*), the median is the 17th observation (16 below, 16 above), which is $x_{17} = 185$. For the J.D. Power data set, the median (185) does provide a clean "50-50 split" in the data.

The median is especially useful when there are extreme values. For example, government statistics use the median income because a few very high incomes will render the mean atypical. The median's insensitivity to extremes may seem advantageous or not, depending on your point of view. Consider three students' scores on five quizzes:

Tom's quiz scores: 20, 40, 70, 75, 80	Mean = 57, Median = 70
Jake's quiz scores: 60, 65, 70, 90, 95	Mean = 76, Median = 70
Mary's quiz scores: 50, 65, 70, 75, 90	Mean = 70, Median = 70

Each student has the same median quiz score (70). Tom, whose mean is pulled down by a few low scores, would rather have his grade based on the median. Jake, whose mean is pulled up by a few high scores, would prefer the mean. Mary is indifferent because her measures of center agree (she has symmetric scores).

The median lacks some of the mean's useful mathematical properties. For example, if we multiply the mean by the sample size, we get the sum of the data values. But this is not true for the median. For instance, Tom's total points on all five quizzes (285) are the product of the sample size times his mean ($5 \times 57 = 285$). But this is not true for his median ($5 \times 70 = 350$). That is one reason why instructors tend to base their semester grades on the mean. Otherwise, the lowest and highest scores would not "count."

Mode

The **mode** is the most frequently occurring data value. It may be similar to the mean and median if data values near the center of the sorted array tend to occur often. But it also may be quite different from the mean and median. A data set may have multiple modes or no mode at all.

The mode is easy to define but *not* easy to calculate (except in very small samples) because it requires tabulating the frequency of occurrence of every distinct data value. For example, the sample of defects in 33 vehicle brands has a unique mode at 167 (occurs three times), although several other data values have multiple occurrences.

139	143	147	156	156	157	162	163	165	**167**	**167**
167	168	172	180	183	185	186	189	191	192	193
199	200	204	210	211	226	230	239	255	256	265

Excel's function =MODE(Data) will return #N/A if there is no mode. If there are multiple modes, =MODE(Data) will return the first one it finds, while =MODE.MULT(Data) can return multiple modes in an array format. Sometimes the mode is far from the "middle" of the distribution and may not be "typical." For *continuous* data, the mode generally isn't useful because continuous data values rarely repeat. To assess central tendency in continuous data, we would rely on the mean or median. Yet the mode is good for describing central tendency in *categorical data* such as gender (male, female) or college major (accounting, finance, etc.). Indeed, the mode is the *only* useful measure of central tendency for categorical data. The mode is also useful to describe a *discrete* variable with a *small range* (e.g., responses to a five-point Likert scale).

Tip

The mode is most useful for discrete or categorical data with only a few distinct data values. For continuous or decimal data or data that cover a wide range, the mode is rarely useful, although the modal *class* in a histogram may help to reveal central tendency.

There may be a logical reason for the existence of modes. For example, points scored by winning college football teams on a given Saturday will tend to have modes at multiples of 7 (e.g., 7, 14, 21) because each touchdown yields 7 points (counting the extra point). Other mini-modes in football scores reflect commonly occurring combinations of scoring events. However, in business or economic data that are measured in dollars or decimal ratios, the mode is rarely useful.

A **bimodal distribution** or a **multimodal distribution** occurs when dissimilar populations are combined into one sample. For example, if the heights of 500 adult men and 500 adult women are combined into a single sample of 1,000 adults, we would get something like the second polygon in Figure 4.4.

Figure 4.4

Frequency Polygons of Heights of 1,000 Men and Women 🗁 **Heights**

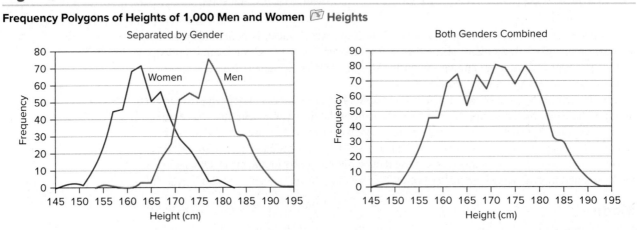

In such a case, the mean of all 1,000 adults would not represent a central tendency for either gender. When heterogeneity is known to exist, it would be better to create separate histograms or frequency polygons and carry out the analysis on each group separately. Unfortunately, we don't always know when heterogeneous populations have been combined into one sample.

Shape

The shape of a distribution may be judged by looking at the histogram or by comparing the mean and median. In **symmetric data**, the mean and median are about the same. When the data are **skewed right** (or **positively skewed**), the mean exceeds the median. When the data are **skewed left** (or **negatively skewed**), the mean is below the median. Figure 4.5 shows prototype population shapes showing varying degrees of **skewness**.

Figure 4.5

Skewness Prototype Populations

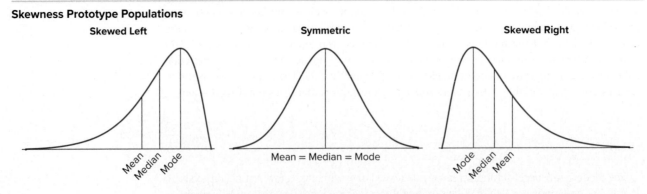

Table 4.5 summarizes the symptoms of skewness in a sample. Because few data sets are exactly symmetric, skewness is a matter of degree. Due to the nature of random sampling,

the mean and median may differ, even when a symmetric population is being sampled. Small differences between the mean and median do not indicate significant skewness and may lack practical importance.

Distribution's Shape	Histogram Appearance	Statistics
Skewed left (negative skewness)	The Long tail of histogram points left (a few low values but most data on right)	Mean < Median
Symmetric	Tails of the histogram are balanced (low/high values offset)	Mean ≈ Median
Skewed right (positive skewness)	The Long tail of the histogram points right (most data on left but a few high values)	Mean > Median

Table 4.5

Symptoms of Skewness

Address the fact that the mean is sensitive to outliers, and therefore the median is preferred as a measure of centrality.

For example, in Figure 4.6 the average spending per customer at 74 Noodles & Company restaurants appears somewhat right-skewed, so we would expect the mean to exceed the median. Actually, the difference is slight (using the spreadsheet raw data, the mean is $7.04 and the median is $7.00). The student GPA histogram in Figure 4.6 appears left-skewed, so we would expect the mean to be lower than the median. But again, the difference is slight (using the spreadsheet raw data, the mean is 3.17 and the median is 3.20). Because a histogram's appearance is affected by the way its bins are set up, its shape offers only a rough guide to skewness.

Figure 4.6

Histograms to Illustrate Skewness

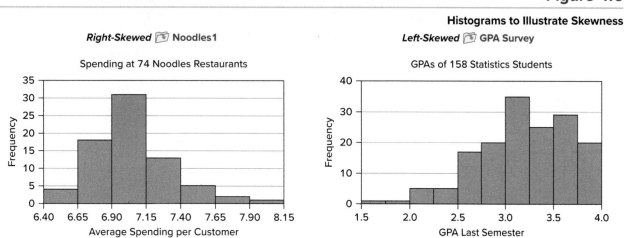

Right-Skewed 📁 Noodles1   *Left-Skewed* 📁 GPA Survey

For the sample of J.D. Power quality ratings, the mean (188.58) exceeds the median (185), which suggests right-skewness. However, this small difference between the mean and median may lack practical importance. The histogram in Figure 4.1 suggests that the skewness in IQR ratings is slight. In Section 4.8, we will introduce more precise tests for skewness.

Business data tend to be right-skewed because financial variables often are unlimited at the top but are bounded from below by zero (e.g., salaries, employees, inventory). This is also true for engineering data (e.g., time to failure, defect rates) and sports (e.g., scores in soccer). Even in a Likert scale (1, 2, 3, 4, 5), a few responses in the opposite tail can skew the mean if most replies are clustered toward the top or bottom of the scale.

Descriptive Statistics in Excel

As shown in Figure 4.7, select the Data tab and click the Data Analysis icon (on the far right side of the top menu). When the Data Analysis menu appears, select Descriptive Statistics. On the

Descriptive Statistics menu form, click anywhere inside the Input Range field, and then highlight the data block (in this case C4:C37). Specify a destination cell for the upper left corner of the output range (cell J1 in this example). Notice that we checked the Labels in the first row box because cell C4 is actually a column heading that will be used to label the output in cell K1. Check the Summary Statistics box and then click OK. The resulting statistics are shown in Figure 4.7. You probably recognize some of them (e.g., mean, median, mode), and the others will be covered later in this chapter.

Figure 4.7

Excel's Data Analysis and Descriptive Statistics
📁 **JDPower**

Note: If Data Analysis does not appear on the upper right of the Data tab, click File in the extreme upper left corner, choose Options, select Add-Ins, click Go (at the bottom), and check the box for Analysis ToolPak.

Source: Microsoft Excel.

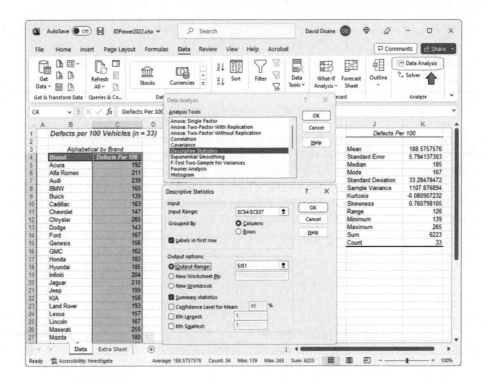

Section Exercises

Mc Graw Hill **connect**

4.1 (a) For each data set, find the mean, median, and mode. (b) Discuss anything about the data that affects the usefulness of each statistic as a measure of center.

   a. Annual campus health center visits (12 students):   0, 0, 0, 0, 0, 1, 2, 3, 3, 5, 5, 15
   b. Red Rocks ticket prices (9 concerts):   40, 40, 65, 71, 72, 75, 76, 78, 98
   c. Sodium grams in canned soup (8 varieties):   225, 255, 295, 302, 304, 337, 351, 366

4.2 For each data set, is the mode a good measure of center? Explain.

   a. Genders of 12 CEOs: M, M, F, M, F, M, M, M, F, M, M, M
   b. Ages of 10 college freshmen: 17, 17, 18, 18, 18, 18, 18, 18, 19, 20
   c. Ages of 8 MBA students: 24, 26, 27, 28, 30, 31, 33, 37

4.3 For each data set, is the mode a good measure of center? Explain.

   a. GMAT scores (8 MBA applicants):   490, 495, 542, 587, 599, 622, 630, 641
   b. Exam grades (12 students):   F, D, C, C, C, C, C, C, B, B, A, A
   c. Body mass index (7 Army recruits):   18.6, 20.2, 22.4, 23.7, 24.2, 28.8, 28.8

4.4 For each data set, which best indicates a "typical" data value (mean, median, either)?

   a. Days on campus by 11 students: 1, 1, 2, 2, 3, 3, 3, 4, 4, 5, 5
   b. P/E ratios of 6 stocks: 1.5, 6.5, 6.6, 7.3, 8.2, 9.1
   c. Textbooks in 9 backpacks: 0, 0, 0, 0, 0, 1, 2, 3, 4

4.5 For each data set, which best indicates a "typical" data value (mean, median, either)?

   a. MPG for 7 Honda Civics: 21.8, 24.1, 24.6, 26.2, 28.4, 35.2, 36.3
   b. Number of riders in 8 cars: 1, 1, 1, 1, 1, 1, 4, 6
   c. Diners at 10 restaurant tables: 1, 2, 2, 2, 3, 3, 4, 4, 4, 5

4.6 Days on the market are shown for the 36 most recent home sales in the city of Sonando Hills. (a) Calculate the mean, median, and mode. (b) Is the distribution skewed? Explain. (c) Is the mode a useful measure of center for this data set? 🖰 **Homes**

18	70	52	17	86	121	86	3	66
96	41	50	176	26	28	6	55	21
43	20	56	71	57	16	20	30	31
44	44	92	179	80	98	44	66	15

4.7 Scores are shown for the most recent state civil service exam taken by 24 applicants for positions in law enforcement. (a) Calculate the mean, median, and mode. (b) Is the distribution skewed? Explain. (c) Is the mode a useful measure of center for this data set? 🖰 **Civil**

83	93	74	98	85	82	79	78
82	68	67	82	78	83	70	99
18	96	93	62	64	93	27	58

4.8 A vehicle emissions testing facility can process up to 100 cars in an hour and the facility is open 7 a.m.–5 p.m. Monday–Thursday. Shown below are the number of cars processed each hour last week. (a) Find the mean, median, and mode for each day. (b) Do these measures of center agree? Explain. (c) For each day, note the strengths or weaknesses of each statistic of center. (d) Are the data symmetric or skewed? If skewed, in which direction? 🖰 **Emissions**

Monday	60, 60, 71, 75, 99, 73, 74, 60, 88, 60
Tuesday	79, 70, 65, 79, 65, 74, 79, 65, 65, 79
Wednesday	74, 67, 99, 72, 71, 72, 66, 74, 95, 70
Thursday	80, 98, 97, 70, 88, 85, 90, 93, 49, 10

4.9 Citibank recorded the number of customers using a downtown ATM during the noon hour on 32 consecutive workdays. (a) Find the mean, median, and mode. (b) Do these measures of center agree? Explain. (c) Make a histogram or dot plot. (d) Are the data symmetric or skewed? If skewed, in which direction? 🖰 **Citibank**

25	37	23	26	30	40	25	26
39	32	21	26	19	27	32	25
18	26	34	18	31	35	21	33
33	9	16	32	35	42	15	24

4.10 On Friday night, the owner of Chez Pierre in downtown Chicago noted the amount spent for dinner for 28 four-person tables. (a) Find the mean, median, and mode. (b) Do these measures of center agree? Explain. (c) Make a histogram or dot plot. (d) Are the data symmetric or skewed? If skewed, in which direction? 🖰 **Dinner**

95	103	109	170	114	113	107
124	105	80	104	84	176	115
69	95	134	108	61	160	128
68	95	61	150	52	87	136

4.11 An executive's telephone log showed the lengths of 60 calls initiated during the last week of July. (a) Sort the data. (b) Find the mean, median, and mode. (c) Do the measures of center agree? Explain. (d) Are the data symmetric or skewed? If skewed, in which direction? 🖰 **CallLength**

1	2	10	5	3	3	2	20	1	1
6	3	13	2	2	1	26	3	1	3
1	2	1	7	1	2	3	1	2	12
1	4	2	2	29	1	1	1	8	5
1	4	2	1	1	1	1	6	1	2
3	3	6	1	3	1	1	5	1	18

Mini Case 4.1

ATM Deposits

Table 4.6 shows a sorted random sample of 100 deposits at an ATM located in the student union on a college campus. The sample was selected at random from 1,459 deposits in one 30-day month. Deposits range from $3 to $1,341. The dot plot shown in Figure 4.8 indicates a right-skewed distribution with a few large values in the right tail and a strong clustering on the left (i.e., most ATM deposits are small). Excel's Descriptive Statistics indicate a very skewed distribution because the mean (233.89) greatly exceeds the median (135). The mode (100) is somewhat "typical," occurring five times. However, 40 and 50 each occur four times (mini-modes).

Table 4.6

100 ATM Deposits (dollars) 🗇 **ATMDeposits**

3	10	15	15	20	20	20	22	23	25	26	26
30	30	35	35	36	39	40	40	40	40	47	50
50	50	50	53	55	60	60	60	67	75	78	86
90	96	100	100	100	100	100	103	105	118	125	125
130	131	139	140	145	150	150	153	153	156	160	163
170	176	185	198	200	200	200	220	232	237	252	259
260	268	270	279	295	309	345	350	366	375	431	433
450	450	474	484	495	553	600	720	777	855	960	987
1,020	1,050	1,200	1,341								

Source: Michigan State University Federal Credit Union.

Figure 4.8

Dot Plot for ATM Deposits ($n = 100$)

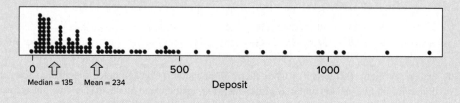

Deposit

Geometric Mean

The **geometric mean** (denoted G) is a multiplicative average, obtained by multiplying the data values and then taking the nth root of the product. This is a measure of central tendency used when all the data values are positive (greater than zero).

(4.4) $$G = \sqrt[n]{x_1 x_2 \cdots x_n} \quad \text{for the geometric mean}$$

The product of n numbers can be quite large. For the J.D. Power quality data:

$$G = \sqrt[33]{(139)(143)(147) \cdots (255)(256)(265)} = \sqrt[33]{7.67411 \times 10^{74}} = 185.88$$

The calculation is easy using Excel's function =GEOMEAN(Data). Scientific calculators have a y^x key whose inverse permits taking the nth root needed to calculate G. The geometric mean tends to mitigate the effects of high outliers.

Growth Rates

We can use a variation on the geometric mean to find the *average growth rate* for a time series (e.g., sales in a growing company):

$$GR = \sqrt[n-1]{\frac{x_n}{x_1}} - 1 \quad \text{(average growth rate of a time series)} \tag{4.5}$$

For example, from 2011 to 2015, JetBlue Airlines' revenues grew as shown in Table 4.7.

Year	Revenue
2011	4,504
2012	4,982
2013	5,441
2014	5,817
2015	6,416

Table 4.7

JetBlue Airlines Operating Revenue (millions of dollars)
🐦 **JetBlue**

Source: JetBlue's published annual Form 10K reports. Data are for December 31 of each year.

The *average growth rate* is given by taking the geometric mean of the ratios of each year's revenue to the preceding year. However, due to cancellations, only the first and last years are relevant:

$$GR = \sqrt[4]{\left(\frac{4982}{4504}\right)\left(\frac{5441}{4982}\right)\left(\frac{5817}{5441}\right)\left(\frac{6416}{5817}\right)} - 1 = \sqrt[4]{\frac{6416}{4504}} - 1 = 1.09248 - 1 = .092$$

or 9.2 percent per year. In Excel, we could use the formula =(6416/4504)^(1/4) − 1 to get the same result.

Midrange

The **midrange** is the point halfway between the lowest and highest values of *X*. It is easy to calculate but is not a robust measure of central tendency because it is sensitive to extreme data values. It is useful when you only have x_{min} and x_{max}.

$$\text{Midrange} = \frac{x_{min} + x_{max}}{2} \tag{4.6}$$

For the J.D. Power data:

$$\text{Midrange} = \frac{x_1 + x_{33}}{2} = \frac{139 + 265}{2} = 202$$

For the J.D. Power data, the midrange is higher than the mean (188.58) or median (185) and is a dubious measure of center because it is "pulled up" by a few high data values.

Trimmed Mean

The **trimmed mean** is calculated like any other mean, except that the highest and lowest *k* percent of the observations in the sorted data array are removed. The trimmed mean mitigates the effects of extreme high values on either end. For a 5 percent trimmed mean, the Excel function is =TRIMMEAN(Data, 0.10) because .05 + .05 = .10. For the J.D. Power data (*n* = 33), we would remove only one observation from each end because .05 × 33 = 1.65 = 1 (truncated to the next *lower* integer) and then take the average of the middle 31 observations.

Excel's measures of center for the J.D. Power data:

Mean:	=AVERAGE(Data)	= 188.58
Median:	=MEDIAN(Data)	= 185
Mode:	=MODE(Data)	= 167
Geo Mean:	=GEOMEAN(Data)	= 185.89
Midrange:	=(MIN(Data)+MAX(Data))/2	= 202
5% Trim Mean:	=TRIMMEAN(Data,0.10)	= 187.71

Figure 4.9

16 Percent Trimmed Mean for CPI

Source: Federal Reserve Bank of Cleveland, www.clevelandfed.org.

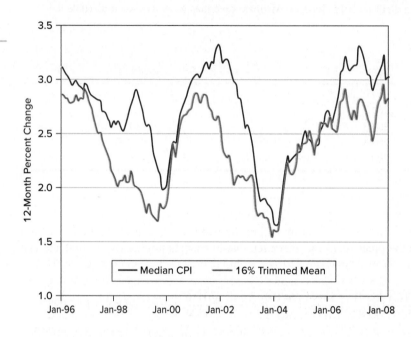

The Federal Reserve uses a 16 percent trimmed mean to mitigate the effect of extremes in its analysis of trends in the Consumer Price Index, as illustrated in Figure 4.9.

Analytics in Action

Future Job Titles?

The terms "data science" and "analytics" cover a lot of territory, but they rest on basic stuff that you are learning in this chapter: organizing and describing data succinctly to reveal their essential properties and potential relevance to specific problems or decisions. Here are some job titles you may soon encounter (for more ideas, see www.datasciencecentral.com).

Data Analyst

The data analyst "scrubs" and analyzes data looking for potential bias or concerns about their quality. Then the data are examined to reveal the structure and essential features. This is a career starting point for many newly trained individuals.

Information Security Analyst

Cyberattacks have brought applications in cybersecurity to the forefront of data analytics. The goal is to identify threats and predict their severity, using real-time data. Artificial intelligence and machine learning will play a role here. Collecting data underlies it all.

Data Engineer

Innovative companies are constantly introducing new products and services. The focus of data engineering is on linking data science to the creation of business value, preventing problems, and working with teams to troubleshoot and find solutions for problems that have been identified from data about a company's products and services.

AI Product Manager
Oversight is needed for the training and rollout of artificial intelligence systems. Particular attention is recently being paid to potential bias in training and monitoring AI performance. For example, Germany's Ethics Commission has now called for regulation of algorithmic systems (*Significance,* December 2019, p. 4) that are being proposed or adopted. Jobs will exist for individuals who become skilled in machine learning and AI.

AI Architect
Professionals are needed to design the frameworks for AI systems. This is a high-level task, but recruitment will come from the ranks of those who have gained experience starting from the bottom. Mastering basic statistics is critical.

Section Exercises

Mc Graw Hill **connect**

4.12 (a) For each data set, find the median, midrange, and geometric mean. (b) Are they reasonable measures of central tendency? Explain.

    a. Vehicle speeds on highway I-5 (9 vehicles)   42, 55, 65, 67, 68, 75, 76, 78, 94
    b. Sodium grams in canned soup (8 varieties)   225, 255, 295, 302, 304, 337, 351, 366
    c. Annual campus health center visits (12 students)   0, 0, 0, 0, 0, 1, 2, 3, 3, 5, 5, 15

4.13 (a) Write the Excel function for the 10 percent trimmed mean of a data set in cells A1:A50. (b) How many observations would be trimmed in each tail? (c) How many would be trimmed overall?

4.14 In the Excel function =TRIMMEAN(Data,.10), how many observations would be trimmed from each end of the sorted data array named Data if (a) $n = 41$, (b) $n = 66$, and (c) $n = 83$?

4.15 The city of Sonando Hills has 8 police officers. In January, the work-related medical expenses for each officer were 0, 0, 0, 0, 0, 0, 150, 650. (a) Calculate the mean, median, mode, midrange, and geometric mean. (b) Which measure of center would you use to budget the expected medical expenses for the whole year by all officers?

4.16 Spirit Airlines kept track of the number of empty seats on flight 308 (DEN–DTW) for 10 consecutive trips on each weekday except Friday. (a) Sort the data for each day. (b) Find the mean, median, mode, midrange, geometric mean, and 10 percent trimmed mean (i.e., dropping the first and last sorted observations) for each day. (c) Do the measures of center agree? Explain. (d) Note the strengths or weaknesses of each statistic of center for the data. 📁 **EmptySeats**

           Monday: 6, 1, 5, 9, 1, 1, 6, 5, 5, 1
           Tuesday: 1, 3, 3, 1, 4, 6, 9, 7, 7, 6
    Wednesday: 6, 0, 6, 0, 6, 10, 0, 0, 4, 6
      Thursday: 1, 1, 10, 1, 1, 1, 1, 1, 1, 1

4.17 Citibank recorded the number of customers using a downtown ATM during the noon hour on 32 consecutive workdays. (a) Find the mean, midrange, geometric mean, and 10 percent trimmed mean. (b) Do these measures of center agree? Explain. 📁 **Citibank**

25	37	23	26	30	40	25	26
39	32	21	26	19	27	32	25
18	26	34	18	31	35	21	33
33	9	16	32	35	42	15	24

4.18 On Friday night, the owner of Chez Pierre in downtown Chicago noted the amount spent for dinner at 28 four-person tables. (a) Find the mean, midrange, geometric mean, and 10 percent trimmed mean. (b) Do these measures of center agree? Explain. 📁 **Dinner**

95	103	109	170	114	113	107
124	105	80	104	84	176	115
69	95	134	108	61	160	128
68	95	61	150	52	87	136

4.19 An executive's telephone log showed the lengths of 60 calls initiated during the last week of July. (a) Find the mean, median, mode, midrange, geometric mean, and 10 percent trimmed mean. (b) Are the data symmetric or skewed? If skewed, in which direction? 📧 **CallLength**

1	2	10	5	3	3	2	20	1	1
6	3	13	2	2	1	26	3	1	3
1	2	1	7	1	2	3	1	2	12
1	4	2	2	29	1	1	1	8	5
1	4	2	1	1	1	1	6	1	2
3	3	6	1	3	1	1	5	1	18

4.20 The number of Internet users in Latin America grew from 77.2 million in 2009 to 139.1 million in 2016. Use the geometric mean to find the annual growth rate.

4.3 MEASURES OF VARIABILITY

LO 4-3

Calculate and interpret common measures of variability.

We can use a statistic such as the mean to describe the *center* of a distribution. But it is just as important to describe *variation* around the center. Consider possible sample distributions of study time spent by several college students taking an economics class:

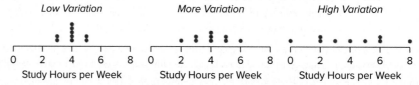

Each diagram has the same mean, but they differ in dispersion around the mean. The problem is: How do we *describe* variability in a sample? Histograms and dot plots tell us something about variation in a data set (the "spread" of data points about the center), but precise measures of dispersion are needed. Because different variables have different means and different units of measurement (dollars, pounds, yen), we want measures of variability that can be applied to many situations. Table 4.8 lists several common measures of variability. All formulas shown are for sample data sets.

Table 4.8

Five Measures of Variability for a Sample

Statistic	Formula	Excel*	Pro	Con
Range (R)	$x_{max} - x_{min}$	=MAX(Data)-MIN(Data)	Easy to calculate and easy to interpret.	Sensitive to extreme data values.
Sample variance (s^2)	$\dfrac{\sum_{i=1}^{n}(x_i - \bar{x})^2}{n-1}$	=VAR.S(Data)	Plays a key role in mathematical statistics.	Less intuitive meaning.
Sample standard deviation (s)	$\sqrt{\dfrac{\sum_{i=1}^{n}(x_i - \bar{x})^2}{n-1}}$	=STDEV.S(Data)	Most common measure. Same units as the raw data ($, £, ¥, grams, etc.).	Less intuitive meaning.
Coefficient of variation (CV)	$100 \times \dfrac{s}{\bar{x}}$	None	Expresses relative variation in *percent*, so can compare data sets with different units of measurement.	Requires nonnegative data.
Mean absolute deviation (MAD)	$\dfrac{\sum_{i=1}^{n}|x_i - \bar{x}|}{n}$	=AVEDEV(Data)	Easy to understand.	Lacks "nice" theoretical properties.

*R offers equivalent functions for range(x), var(x), and sd(x), but not the others. See **Appendix J** for a side-by-side list of R and Excel functions.

Range

The **range** is the difference between the largest and smallest observations:

$$\text{Range} = x_{max} - x_{min} \qquad (4.7)$$

For the J.D. Power data, the range is

$$\text{Range} = 265 - 139 = 126$$

Although it is easy to calculate, a drawback of the range is that it only considers the two extreme data values. It seems desirable to seek a broad-based measure of variability that is based on *all* the data values $x_1, x_2, \ldots, x_n$.

Variance and Standard Deviation

If we calculate the differences between each data value x_i and the mean, we would have both positive and negative differences. The mean is the balancing point of the distribution, so if we just sum these differences and take the average, we will always get zero, which obviously doesn't give us a useful measure of variability. One way to avoid this is to *square* the differences before we find the average. Following this logic, the **population variance** (denoted σ^2, where σ is the lowercase Greek letter "sigma") is defined as the sum of squared deviations from the mean divided by the population size:

$$\sigma^2 = \frac{\sum_{i=1}^{N}(x_i - \mu)^2}{N} \qquad (4.8)$$

If we have a sample (i.e., most of the time), we replace μ with $\bar{x}$ to get the **sample variance** (denoted s^2):

$$s^2 = \frac{\sum_{i=1}^{n}(x_i - \bar{x})^2}{n - 1} \qquad (4.9)$$

A variance is basically a mean squared deviation. But why do we divide by $n - 1$ instead of n when using sample data? This question perplexes many students. A sample contains n pieces of information, each of which can have any value, independently from the others. But once you have calculated the sample mean (as you must in order to find the variance), there are only $n - 1$ pieces of independent information left (because the sample values must add to a fixed total that gives the mean). We divide the sum of squared deviations by $n - 1$ instead of n because we have "lost" one piece of information. Otherwise, s^2 would tend to underestimate the unknown population variance σ^2.

In describing variability, we most often use the **standard deviation** (the square root of the variance). A standard deviation is a single number that helps us understand how individual values in a data set vary from the mean. Because the square root has been taken, its units of measurement are the same as X (e.g., dollars, kilograms, miles). To find the standard deviation of a population, we use

$$\sigma = \sqrt{\frac{\sum_{i=1}^{N}(x_i - \mu)^2}{N}} \qquad (4.10)$$

and for the standard deviation of a sample:

$$s = \sqrt{\frac{\sum_{i=1}^{n}(x_i - \bar{x})^2}{n - 1}} \qquad (4.11)$$

Many inexpensive calculators have built-in formulas for the standard deviation. To distinguish between the population and sample formulas, some calculators have one function key labeled σ_x and another labeled s_x. Others have one key labeled σ_n and another labeled σ_{n-1}. The only question is whether to divide the numerator by the number of data items or the number of data items minus one. Computers and calculators don't know whether your data are a sample or a population. They will use whichever formula you request. It is up to you to know which is appropriate for your data. Excel has built-in functions for the variance and standard deviation.

	Sample Formula	Population Formula
Variance	=VAR.S(Data)	=VAR.P(Data)
Std deviation	=STDEV.S(Data)	=STDEV.P(Data)

The J.D. Power data on vehicle defects is a sample, so we use =STDEV.S(Data) to obtain $s = 33.28$. Because it is not a unit-free measure, the standard deviation would only be useful (for example) to compare variation between different years.

Calculating a Standard Deviation

Table 4.9 illustrates the calculation of a standard deviation using Stephanie's scores on five quizzes (40, 55, 75, 95, 95). Her mean is 72.

Table 4.9

Worksheet for Standard Deviation Stephanie

i	X_i	$X_i - \overline{X}$	$(X_i - \overline{X})^2$	X_i^2
1	40	$40 - 72 = -32$	$(-32)^2 = 1{,}024$	$40^2 = 1{,}600$
2	55	$55 - 72 = -17$	$(-17)^2 = 289$	$55^2 = 3{,}025$
3	75	$75 - 72 = +3$	$(3)^2 = 9$	$75^2 = 5{,}625$
4	95	$95 - 72 = +23$	$(23)^2 = 529$	$95^2 = 9{,}025$
5	95	$95 - 72 = +23$	$(23)^2 = 529$	$95^2 = 9{,}025$
Sum	360	0	2,380	28,300
Mean	72			

Notice that the deviations around the mean (column three) sum to zero, an important property of the mean. Because the mean is rarely a "nice" number, such calculations typically require a spreadsheet or a calculator. Stephanie's sample standard deviation is

$$s = \sqrt{\frac{\sum_{i=1}^{n}(x_i - \overline{x})^2}{n-1}} = \sqrt{\frac{2{,}380}{5-1}} = \sqrt{595} = 24.39$$

Characteristics of the Standard Deviation

The standard deviation is nonnegative because the deviations around the mean are squared. When every observation is exactly equal to the mean, then the standard deviation is zero (i.e., there is no variation). For example, if every student received the same score on an exam, the numerators of formulas 4.8 through 4.11 would be zero because every student would be at the mean. At the other extreme, the greatest dispersion would be if the data were concentrated at x_{min} and x_{max} (e.g., if half the class scored 0 and the other half scored 100).

But the standard deviation can have any nonnegative value, depending on the unit of measurement. For example, yields on n randomly chosen investment bond funds (e.g., Westcore Plus at 7.2 percent in 2010) would have a small standard deviation compared to annual revenues of n randomly chosen *Fortune* 500 corporations (e.g., Walmart at $429 billion in 2010).

Standard deviations can be compared *only* for data sets measured in the same units. For example, the prices of hotel rooms in Tokyo (yen) cannot be compared with the prices of hotel rooms in Paris (euros). Also, standard deviations should not be compared if the means differ substantially, even when the units of measurement are the same. For instance, the weights of apples (ounces) have a smaller mean than weights of watermelons (ounces).

Standard Deviation Used to Compare Risks

"Generally speaking, the standard deviation is how much an investment's returns have varied historically from its average. The statistical measure is usually computed using monthly returns for the most recent three-year period. Standard deviation can be useful when comparing the investment risk of various mutual funds. If two funds have similar average returns but different standard deviations, the fund with the higher standard deviation is the more volatile of the two."

Source: *T. Rowe Price Investor*, September 2009, p. 8.

Coefficient of Variation

To compare dispersion in data sets with dissimilar units of measurement (e.g., kilograms and ounces) or dissimilar means (e.g., home prices in two different cities), we define the **coefficient of variation** (*CV*), which is a unit-free measure of dispersion:

$$CV = 100 \times \frac{s}{\bar{x}} \qquad\qquad (4.12)$$

The *CV* is the standard deviation expressed as a percent of the mean. For the J.D. Power data on vehicle defects, the coefficient of variation is $CV = 100 \times (33.285)/(188.576) = 17.65\%$. In some data sets, the standard deviation can exceed the mean, so the *CV* can exceed 100 percent. This can happen in skewed data sets, especially if there are outliers. The *CV* is useful for comparing variables measured in different units. For example:

Defect rates: $s = 24.94$, $\bar{x} = 134.51$ $CV = 100 \times (24.94)/(134.51) = 19\%$
ATM deposits: $s = 280.80$, $\bar{x} = 233.89$ $CV = 100 \times (280.80)/(233.89) = 120\%$
P/E ratios: $s = 14.08$, $\bar{x} = 22.72$ $CV = 100 \times (14.08)/(22.72) = 62\%$

Despite the different units of measurement, we can say that ATM deposits have much greater relative dispersion (120 percent) than either defect rates (18 percent) or P/E ratios (62 percent). The chief weakness of the *CV* is that it is undefined if the mean is zero or negative, so it is appropriate only for positive data.

Analytics in Action

Measuring Volatility

Investors make decisions based on many factors, including the volatility of a stock's price. We might measure volatility by calculating the mean and variance (or standard deviation) of historical prices of a stock over a certain time period. The coefficient of variation would provide a unit-free measure of "historic volatility" that could be used to compare volatility among different stocks. However, stock prices may violate some assumptions (e.g., normality) and may not be predictive of the future. Because an investor's objective is to predict the future (e.g., pricing of stock options), more complex methods are required.

The Black Sholes method of option contract pricing incorporates additional variables, such as contract strike price, current stock price, time to expiration, and volatility. It requires complex mathematical calculations (after all, its developers earned a Nobel Prize).

A powerful statistical tool called the VIX Index has been introduced to measure the market's expectation of future volatility. It is calculated in real-time based on S&P 500 stock prices. Calculating the VIX also entails complex algebra and statistics. While you may enjoy doing Internet research on your own to understand these terms, you had better plan on taking classes in finance if you intend to become an active investor. Your basic training in statistics provides a good starting point.

Mean Absolute Deviation

An additional measure of dispersion is the **mean absolute deviation** (*MAD*). This statistic reveals the average distance from the center. Absolute values must be used; otherwise, the deviations around the mean would sum to zero.

(4.13)
$$MAD = \frac{\sum_{i=1}^{n} |x_i - \bar{x}|}{n}$$

The *MAD* is appealing because of its simple, concrete interpretation. Using the lever analogy, the *MAD* tells us what the average distance is from an individual data point to the fulcrum. Excel's function =AVEDEV(Data) will calculate the *MAD*. For the J.D. Power defect data, Excel's formula gives *MAD* = 26.1. In other words, the *average* distance from the mean is 26.1.

Mini Case 4.2

Bear Markets

Investors know that stock prices have extended cycles of downturns ("bear markets") or upturns ("bull markets"). But how long must an investor be prepared to wait for the cycle to end? Table 4.10 shows the duration of 15 bear markets since 1929 and the decline in the S&P 500 stock index.

Table 4.10

Duration of 14 Bear Markets **BearMarkets**

Peak	Trough	Duration	S&P Loss (%)
May 1946	May 1947	12	28.8
Jun 1947	Jun 1948	12	20.6
Aug 1956	Oct 1957	15	21.6
Dec 1961	Jun 1962	7	28.0
Feb 1966	Oct 1966	8	22.2
Nov 1968	May 1970	18	36.1
Jan 1973	Oct 1974	21	48.2
Nov 1980	Aug 1982	21	27.1
Aug 1987	Dec 1987	3	33.5
Mar 2000	Sep 2001	18	36.8
Jan 2002	Oct 2002	9	33.8
Oct 2007	Nov 2008	14	51.9
Jan 2009	Mar 2009	2	27.6
Feb 2020	Mar 2020	1	33.9

Note: This table shows market corrections of 20 percent or more that lasted at least one month, consolidating various expert assessments of timing and duration (e.g., morningstar.com, tiaa.org, thebalance.com, raymondjames.com, investopedia.com, seekingalpha.com, en.wikipedia.org, and fred.stlouisfed.org).

Figure 4.10 shows that bear markets typically are short-lived (half of them last one year or less). S&P losses often are in the 25–35 percent range, but can be much larger. While bear markets are nearly symmetric, the S&P losses appear right-skewed.

Figure 4.10

Dot Plots of Bear Market Measurements

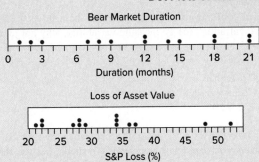

Table 4.11 shows that mean and median duration are similar (roughly symmetric) while the mean S&P loss slightly exceeds the median (due to two extremes). However, there is more relative variation in duration (58.9% CV) compared with S&P 500 losses (28.8% CV). Other measures of center and variability cannot be compared because the units of measurement differ.

Table 4.11

Statistical Summary of 14 Bear Markets

Statistic	Duration	S&P Loss (%)
Count	14	14
Min	1	20.6
Max	21	51.9
Median	12	31.2
Mean	11.50	32.15
St Dev	6.77	9.24
Coef Var	58.9%	28.8%

Section Exercises

4.21 (a) Find the mean and standard deviation for each sample. (b) What does this exercise show about the standard deviation?

Sample A: 6, 7, 8
Sample B: 61, 62, 63
Sample C: 1,000; 1,001; 1,002

4.22 For each data set: (a) Find the mean. (b) Find the standard deviation, treating the data as a sample. (c) Find the standard deviation, treating the data as a population. (d) What does this exercise show about the two formulas?

Data Set A: 6, 7, 8
Data Set B: 4, 5, 6, 7, 8, 9, 10
Data Set C: 1, 2, 3, 4, 5, 6, 7, 8, 9, 10, 11, 12, 13

4.23 In fuel economy tests in city driving conditions, a hybrid vehicle's mean was 43.2 mpg with a standard deviation of 2.2 mpg. A comparably sized gasoline vehicle's mean was 27.2 mpg with a standard deviation of 1.9 mpg. Which vehicle's mpg was more consistent in relative terms?

connect

4.24 Over the past month, Bob's bowling score mean was 182 with a standard deviation of 9.1. His bowling partner Cedric's mean was 152 with a standard deviation of 7.6. Which bowler is more consistent in relative terms?

4.25 Use Excel's AVEDEV function to find the mean absolute deviation (*MAD*) of the integers 1 through 10.

4.26 Use Excel's AVEDEV function to find the mean absolute deviation (*MAD*) of these five numbers: 12, 18, 21, 22, 27.

4.27 (a) Find the coefficient of variation for the prices of these three stocks. (b) Which stock has the greatest relative variation? (c) To measure variability, why not just compare the standard deviations?

Stock A: $\bar{X} = \$24.50, s = 5.25$
Stock B: $\bar{X} = \$147.25, s = 12.25$
Stock C: $\bar{X} = \$5.75, s = 2.08$

4.28 The asset turnover ratio (ATR) is the ratio of a company's revenues to the value of its assets (indicating its efficiency in deploying its assets). We should not use the standard deviation to compare ATR variation among industrial sectors because firms with large asset bases (e.g., utilities, financial) typically have lower mean ATR than, say, retail firms. (a) Use the sample data to calculate the coefficient of variation for each sector. (b) Which sector has the highest degree of *relative* variation? The lowest? (c) If someone (incorrectly) used the standard deviations to compare variation, would the ranking among sectors be the same?

	Retail	Utilities	Financial
Mean	2.088	0.0441	0.0743
St Dev	0.484	0.0083	0.0297

4.29 Noodles & Company tested consumer reactions to two spaghetti sauces. Each of 70 raters assessed both sauces on a scale of 1 (worst) to 10 (best) using several taste criteria. To correct for possible bias in tasting order, half the raters tasted *Sauce A* first, while the other half tasted *Sauce B* first. Actual results are shown below for "overall liking." (a) Calculate the mean and standard deviation for each sample. (b) Calculate the coefficient of variation for each sample. (c) What is your conclusion about consumer preferences for the two sauces? (Source: Noodles & Company.) 🗁 **Spaghetti**

Sauce A:

6, 7, 7, 8, 8, 6, 8, 6, 8, 7, 8, 8, 6, 8, 7, 7, 7, 8, 8, 8, 7, 7, 6, 7, 7,

8, 3, 8, 8, 7, 8, 6, 7, 8, 7, 7, 3, 6, 8, 7, 1, 8, 8, 7, 6, 7, 7, 4, 8, 8,

3, 8, 7, 7, 7, 5, 7, 7, 7, 9, 5, 7, 6, 8, 8, 8, 4, 5, 9, 8

Sauce B:

7, 7, 7, 8, 8, 7, 8, 6, 8, 7, 7, 6, 7, 7, 8, 7, 8, 7, 8, 8, 7, 8, 5, 7, 7,

9, 4, 8, 8, 7, 8, 8, 8, 8, 7, 7, 3, 7, 9, 8, 9, 7, 8, 8, 6, 7, 7, 7, 8, 8,

7, 7, 8, 6, 6, 7, 7, 9, 7, 9, 8, 8, 6, 7, 7, 9, 4, 4, 9, 8

Mini Case 4.3

Defect Rates: Numerical Statistics 🗁 JDPower

Numerical statistics are helpful when making comparisons (e.g., over time). In Table 4.12, using published J.D. Power and Associates' Initial Quality Study (IQS) data, we see that the mean and median number of defects in 2020–2022 was sharply higher than in 2018–2019. While *absolute* variation (standard deviation) has increased, *relative* variation (coefficient of variation) has remained in the 15–20 percent range. Was the sharp increase in defects due to the COVID-19 pandemic? Or did J.D. Power change its definition of "Defects per 100 Vehicles"? Further research is required.

Table 4.12

Defect Rate Statistics by Year 📁 **JDPower**

Year	N	Mean	StDev	CoefVar	Min	Median	Max
2022	33	188.58	33.28	17.65	139	185	265
2021	32	170.97	29.18	17.07	128	164	251
2020	31	171.16	24.59	14.37	136	174	228
2019	32	98.63	15.68	15.90	63	99	130
2018	31	96.87	19.82	20.46	68	95	160

The dot plots in Figure 4.11 illustrate the increase in variation. The dot plots also illustrate that there can be multiple modes (e.g., 2018). Excel's function =MODE(Data) will return only the first mode in a data column, while =MODE.MULT(Data) will display multiple modes. Even when a unique mode exists, it may not be near the perceived center.

Figure 4.11

Dot Plots of Defect Rate by Year 📁 **JDPower**

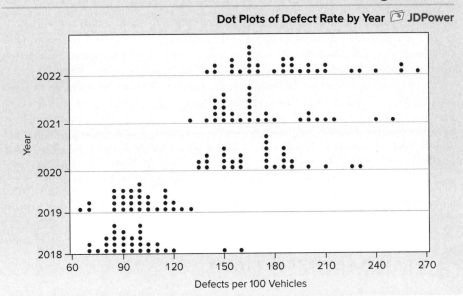

Center versus Variability

Figure 4.12 shows histograms of hole diameters drilled in a steel plate during a manufacturing process. The desired distribution is shown in red. The samples from Machine A have the

Figure 4.12

Center versus Variability

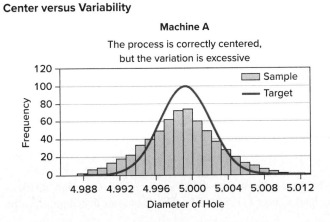
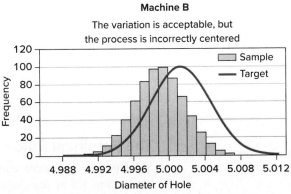

desired *mean* diameter (5 mm) but too much *variation* around the mean. It might be an older machine whose moving parts have become loose through normal wear, so there is a greater variation in the holes drilled. Samples from Machine B have acceptable *variation* in hole diameter, but the *mean* is incorrectly adjusted (less than the desired 5 mm). To monitor quality, we would take frequent samples from the output of each machine so that the process can be stopped and adjusted if the sample statistics indicate a problem.

Analytics in Action

Sharpe Ratio

Investors can use the **Sharpe Ratio** (denoted *SR*) to assess the return/risk trade-off on alternative investments. Its numerator is the expected value of the difference between the average asset return (R_a) and the average "risk-free" return (R_e) on assets such as U.S. Treasury bonds. The denominator is the standard deviation (σ_a) of returns on the asset or portfolio.

$$SR = E(R_a - R_e)/\sigma_a$$

A higher Sharpe Ratio would suggest a more favorable balance of return versus risk. A ratio of less than 1 is clearly undesirable, while a ratio above 2 is considered good. This ratio has been popularized in many ways but stems from 1966 work on the capital-asset pricing model by Nobel Prize–winning economist William Sharpe and others.

The averages and the standard deviation are calculated from observations on an asset (or portfolios of assets) over a given period of time (e.g., monthly observations) and may be done using a geometric mean or an arithmetic mean. Investors must therefore be aware that the choice of time periods and method of calculation allow a potential for manipulation. Another limitation is that a non-normal distribution of returns (e.g., right skewed) will violate an underlying assumption of the model, so *skewness* and *kurtosis* also must be considered by the individual investor or financial advisor.

See www.investopedia.com/ and https://web.stanford.edu/~wfsharpe/art/sr/sr.htm.

4.4 STANDARDIZED DATA

LO 4-4

Apply Chebyshev's theorem.

The standard deviation is an important measure of variability because of its many roles in statistics. One of its main uses is to gauge the position of items within a data array.

Chebyshev's Theorem

The French mathematician Jules Bienaymé (1796–1878) and the Russian mathematician Pafnuty Chebyshev (1821–1894) proved that, for any data set, no matter how it is distributed, the percentage of observations that lie within k standard deviations of the mean (i.e., within $\mu \pm k\sigma$) must be at least $100[1 - 1/k^2]$. Commonly called **Chebyshev's Theorem**, it says that for *any population* with mean μ and standard deviation σ:

$k = 2$ at least 75.0% will lie within $\mu \pm 2\sigma$.
$k = 3$ at least 88.9% will lie within $\mu \pm 3\sigma$.
$k = 4$ at least 93.8% will lie within $\mu \pm 4\sigma$.

For example, for an exam with $\mu = 72$ and $\sigma = 8$, at least 75 percent of the scores will be within the interval $72 \pm 2(8)$ or [56, 88] regardless of how the scores are distributed. Although applicable to any data set, these limits tend to be rather wide.

LO 4-5

Apply the Empirical Rule and recognize outliers.

The Empirical Rule

More precise statements can be made about data from a normal or Gaussian distribution, named for its discoverer, Karl Gauss (1777–1855). The Gaussian distribution is the well-known bell-shaped curve. Commonly called the **Empirical Rule**, it says that for data from a *normal distribution,* we expect the interval $\mu \pm k\sigma$ to contain a known percentage of the data:

$k = 1$	68.26% will lie within $\mu \pm 1\sigma$.
$k = 2$	95.44% will lie within $\mu \pm 2\sigma$.
$k = 3$	99.73% will lie within $\mu \pm 3\sigma$.

The Empirical Rule is illustrated in Figure 4.13. The Empirical Rule does *not* give an upper bound but merely describes what is *expected.* Rounding off a bit, we say that in samples from a normal distribution, we expect 68 percent of the data within 1 standard deviation, 95 percent within 2 standard deviations, and virtually all of the data within 3 standard deviations. Data values outside $\mu \pm 3\sigma$ are rare (less than 1%) in a normal distribution and are called **outliers**.

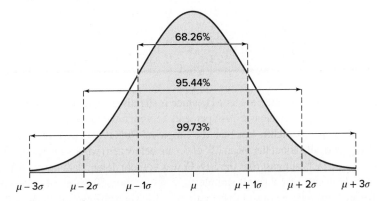

Figure 4.13

The Empirical Rule for a Normal Population

For example, suppose that 80 students take an exam. How many students will score within 2 standard deviations of the mean? Assuming that exam scores follow a normal or bell-shaped curve, we might be willing to rely on the Empirical Rule, which predicts that 95.44% × 80, or approximately 76 students, will score within 2 standard deviations from the mean. Because a normal distribution is symmetric about the mean, we expect that about 2 students will score more than 2 standard deviations above the mean and 2 below the mean. Using the Empirical Rule, we can further say that it is unlikely (though not impossible) that a student will score more than 3 standard deviations from the mean (99.73% × 80 = 79.78 ≈ 80).

Standardized Data

A general approach to identifying unusual observations is to redefine each observation in terms of its distance from the mean in standard deviations to obtain **standardized data**. We get the standardized value (called a *z*-score) by transforming each value of the observed data:

LO 4-6

Transform a data set into standardized values.

$$z_i = \frac{x_i - \mu}{\sigma} \quad \textit{for a population} \tag{4.14}$$

$$z_i = \frac{x_i - \bar{x}}{s} \quad \textit{for a sample} \tag{4.15}$$

By looking at the standardized *z*-score (z_i), we can tell at a glance how far away from the mean an observation lies. Excel's function =STANDARDIZE(XValue,Mean,StDev) makes it easy to calculate standardized values from a column of data. For the J.D. Power data, we set Mean = 188.576 and StDev = 33.284 and then use Excel to produce Table 4.13.

There are no *outliers* (beyond three standard deviations from the mean) but two data values are *unusual* (beyond two standard deviations from the mean):

$$\textit{Volvo} \quad z_i = \frac{x_i - \bar{x}}{s} = \frac{256 - 188.576}{33.285} = 2.026$$

$$\textit{Chrysler} \quad z_i = \frac{x_i - \bar{x}}{s} = \frac{265 - 188.576}{33.285} = 2.296$$

By tabulating the standardized *z*-scores, we can compare our sample with the Empirical Rule for a normal distribution. We find that 24 of 33 observations (72.73 percent) lie within the range $\bar{x} \pm 1s$

Table 4.13

Standardized z-Scores for Defect Rates by Vehicle Brand (*n* = 33) 🖻 JDPower

Brand	Defects	z-Score	Brand	Defects	z-Score	Brand	Defects	z-Score
Buick	139	−1.489	Nissan	167	−0.648	Jeep	199	0.313
Dodge	143	−1.369	MINI	168	−0.618	Porsche	200	0.343
Chevrolet	147	−1.249	Toyota	172	−0.498	Infiniti	204	0.463
Genesis	156	−0.979	Mazda	180	−0.258	Jaguar	210	0.644
KIA	156	−0.979	Honda	183	−0.168	Alfa Romeo	211	0.674
Lexus	157	−0.949	Hyundai	185	−0.107	Mitsubishi	226	1.124
GMC	162	−0.798	Ram	186	−0.077	Volkswagen	230	1.245
Cadillac	163	−0.768	Mercedes-Benz	189	0.013	Audi	239	1.515
BMW	165	−0.708	Subaru	191	0.073	Maserati	255	1.996
Ford	167	−0.648	Acura	192	0.103	Volvo	256	2.026
Lincoln	167	−0.648	LandRover	193	0.133	Chrysler	265	2.296

(*z*-score within the range −1 to +1), which is slightly more than the Empirical Rule (68.26 percent), and 31 of 33 observations (93.94 percent) lie within the range $x \pm 2s$ (*z*-score within the range −2 to +2), which is slightly less than the Empirical Rule (95.44 percent). The *z*-scores are distributed asymmetrically about the mean (27 of 33 are below zero while only six of 33 are above zero).

These facts suggest that the J.D. Power vehicle defect data do not follow a normal distribution. However, for a small sample (say, fewer than 50 observations), comparing the sample frequencies with a normal distribution is risky because we really don't have very much information about *shape*.

What to Do about Outliers?

Extreme values of a variable are vexing, but what do we do about them? For a large sample (e.g., $n = 1,000$), it would *not* be surprising to see a few data values outside the three-standard-deviation range (by the Empirical Rule, 99.73 percent of 1,000 is 997, so we would expect three outliers). But it can be tempting to discard unusual data points. Discarding an outlier would be reasonable only if we had reason to suppose the data value is erroneous. For example, a blood pressure reading of 1200/80 seems impossible (it probably was supposed to be 120/80). Perhaps the lab technician was distracted by a conversation while marking down the reading.

An outrageous observation may be invalid. But how do we guard against self-deception? More than one scientist has been convinced to disregard data that didn't fit the pattern, when in fact the weird observation was trying to say something important. For example, in the 1980s, some instruments monitoring the ozone layer over Antarctica were programmed to disregard readings more than two standard deviations from the long-term mean as likely due to measurement error (*New Scientist*, December 6, 2008, p. 32). Fortunately, the raw data were retrievable, and scientists were able to spot an increase in the number of discarded readings as the "ozone hole" grew. Strict new regulations on the release of chlorofluorocarbons have been successful in restoring the ozone in the atmosphere. At this stage of your statistical training, it suffices to *recognize* unusual data points and outliers and their potential impact and to know that there are entire books that cover the topic of outliers (see Related Reading).

Unusual Observations

Based on its standardized *z*-score, a data value is classified as

Unusual	if $|z_i| > 2$	(beyond $\mu \pm 2\sigma$)
Outlier	if $|z_i| > 3$	(beyond $\mu \pm 3\sigma$)

Estimating Sigma

For a normal distribution, essentially all the observations lie within $\mu \pm 3\sigma$, so the range is approximately 6σ (from $\mu - 3\sigma$ to $\mu + 3\sigma$). Therefore, if you know the range $x_{max} - x_{min}$, you can estimate the standard deviation as $\sigma = (x_{max} - x_{min})/6$. This rule can come in handy for approximating the standard deviation when all you know is the range. For example, the caffeine content of a cup of tea depends on the type of tea and the length of time the tea steeps, with a range of 20 to 90 mg. Knowing the range, we could estimate the standard deviation as $s = (90 - 20)/6$, or about 12 mg. This estimate assumes that the caffeine content of a cup of tea is normally distributed.

Mini Case 4.4

Presidential Ages

Table 4.14 shows the sorted ages at inauguration of the first 46 U.S. presidents. The mean is 55.52 years with a standard deviation of 7.32 years. In terms of his standardized z-score, President Joe Biden ($x = 78$, $z = +3.09$) was a clear outlier. The next oldest (Donald Trump, $x = 70$, $z = 1.98$) and the youngest (Theodore Roosevelt, ($x = 42$, $z = -1.85$) were within two standard deviations of the mean (although just barely).

Table 4.14

Ages at Inauguration of 45 U.S. Presidents (sorted) 🖾 **Presidents**

President	Age	President	Age	President	Age
T. Roosevelt	42	Carter	52	J. Q. Adams	57
Kennedy	43	Van Buren	54	Monroe	58
Grant	46	Hayes	54	Truman	60
Cleveland	47	McKinley	54	J. Adams	61
Clinton	47	Hoover	54	Jackson	61
Obama	47	G. W. Bush	54	Ford	61
Pierce	48	B. Harrison	55	Eisenhower	62
Polk	49	Cleveland	55	Taylor	64
Garfield	49	Harding	55	G. H. W. Bush	64
Fillmore	50	L. Johnson	55	Buchanan	65
Tyler	51	A. Johnson	56	W. H. Harrison	68
Arthur	51	Wilson	56	Reagan	69
Taft	51	Nixon	56	Trump	70
Coolidge	51	Washington	57	Biden	78
F. Roosevelt	51	Jefferson	57		
Lincoln	52	Madison	57		

The Minitab dot plot in Figure 4.14 shows modes at 51 and 54 (five times each). President Biden appears at the extreme right end of the scale (age 78).

Figure 4.14

Dot Plot of Presidents' Ages at Inauguration

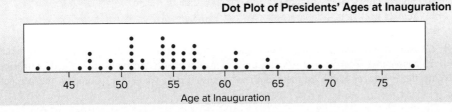

Age at Inauguration

The Minitab histogram in Figure 4.15 shows that the modal class is 50 to 55 (13 presidents are in that class). However, because the next higher class has almost as many observations, it might be more helpful to say that presidents tend to be between 50 and 59 years of age upon inauguration (25 presidents are within this range).

Figure 4.15

Histogram of Presidents' Ages at Inauguration

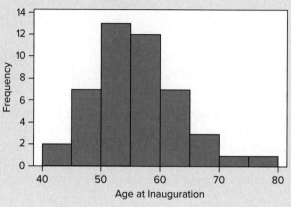

Section Exercises

4.30 (a) By Chebyshev's Theorem, at least how many students in a class of 200 would score within the range $\mu \pm 2\sigma$? (b) By the Empirical Rule, how many students in a class of 200 would score within the range $\mu \pm 2\sigma$? (c) What assumption is required in order to apply the Empirical Rule?

4.31 An exam has a mean of 70 with a standard deviation of 10. Use Chebyshev's Theorem to find a lower bound for the number of students in a class of 400 who scored between 50 and 90.

4.32 The mean monthly rent of students at Oxnard University is $875 with a standard deviation of $219. (a) John's rent is $1,325. What is his standardized z-score? (b) Is John's rent an outlier? (c) How high would the rent have to be to qualify as an outlier?

4.33 The mean collection period for accounts receivable at Ephemeral Products is 18.5 days with a standard deviation of 4.8 days. (a) What is the standardized z-score for an account that is paid in 30 days? (b) Is that account an outlier? (c) How many days (to the nearest integer) would qualify an account as an outlier?

4.34 Convert each individual data value to a standardized z-score. Is it an outlier?
   a. Ages of airline passengers: $x = 92$, $\mu = 46$, $\sigma = 13$
   b. FICO credit scores: $x = 583$, $\mu = 723$, $\sigma = 69$
   c. Condo rental vacancy days: $x = 28$, $\mu = 22$, $\sigma = 7$

4.35 Convert each individual X data value to a standardized Z value and interpret it.
   a. Student weekly grocery bills: Sam's bill is 91, $\mu = 79$, $\sigma = 5$
   b. Student GPAs: Mary's GPA is 3.18, $\mu = 2.87$, $\sigma = 0.31$
   c. Weekly study hours: Jaime studies 18 hours, $\mu = 15.0$, $\sigma = 5.0$

4.36 In a regional high school swim meet, women's times (in seconds) in the 200-yard freestyle ranged from 109.7 to 126.2. Estimate the standard deviation, using the Empirical Rule.

4.37 Find the original data value corresponding to each standardized z-score.
   a. Student GPAs: Bob's z-score is $z = +1.71$, $\mu = 2.98$, $\sigma = 0.36$
   b. Weekly work hours: Sarah's z-score is $z = +1.18$, $\mu = 21.6$, $\sigma = 7.1$
   c. Bowling scores: Dave's z-score is $z = -1.35$, $\mu = 150$, $\sigma = 40$

4.38 Citibank recorded the number of customers using a downtown ATM during the noon hour on 32 consecutive workdays. (a) Use Excel or MegaStat to sort and standardize the data. (b) Based on the Empirical Rule, are there outliers? Unusual data values? (c) Compare the percent of observations that lie within 1 and 2 standard deviations of the mean with a normal distribution. What is your conclusion?  **Citibank**

25	37	23	26	30	40	25	26
39	32	21	26	19	27	32	25
18	26	34	18	31	35	21	33
33	9	16	32	35	42	15	24

4.39 An executive's telephone log showed the lengths of 60 calls initiated during the last week of July. (a) Use Excel or MegaStat to sort and standardize the data. (b) Based on the standardized z-scores, are there outliers? Unusual data values? (c) Compare the percent of observations that lie within 1 and 2 standard deviations of the mean with a normal distribution. What is your conclusion? **CallLength**

1	2	10	5	3	3	2	20	1	1
6	3	13	2	2	1	26	3	1	3
1	2	1	7	1	2	3	1	2	12
1	4	2	2	29	1	1	1	8	5
1	4	2	1	1	1	1	6	1	2
3	3	6	1	3	1	1	5	1	18

4.5 PERCENTILES, QUARTILES, AND BOX PLOTS

Percentiles

You are familiar with percentile scores of national educational tests such as the ACT, SAT, and GMAT, which tell you where you stand in comparison with others. For example, if you are in the 83rd percentile, then 83 percent of the test-takers scored below you, and you are in the top 17 percent of all test-takers. However, only when the sample is large can we meaningfully divide the data into 100 groups (*percentiles*). Alternatively, we can divide the data into 10 groups (*deciles*), 5 groups (*quintiles*), or 4 groups (*quartiles*).

LO 4-7

Calculate quartiles and other percentiles.

Percentiles in Excel

The *p*th percentile of a sorted data array $x_1, x_2, \ldots, x_n$ is the value of x that defines the lowest p percent of the data values. Excel's formula for the *p*th percentile is =PERCENTILE .EXC(Data, Proportion), where Proportion is the *proportion* below the *p*th percentile. For example, =PERCENTILE.EXC(Data,.95) returns the 95th percentile. To find the *p*th percentile in R, use the function quantile(x, p, type=6). See **Appendix J** for details.

Percentiles generally have to be interpolated *between* two data values. For example, suppose you want the 95th percentile for a sample of $n = 73$ items. Because $.95 \times 73 = 69.35$, you would have to interpolate between the 69th and 70th observations (i.e., between x_{69} and x_{70}) to obtain the 95th percentile. The Excel formula =PERCENTILE.EXC(Data,.95) handles this interpolation automatically.

In health care, manufacturing, and banking, selected percentiles (e.g., 5, 25, 50, 75, and 95 percent) are calculated to establish *benchmarks* so that any firm can compare itself with similar firms (i.e., other firms in the same industry) in terms of profit margin, debt ratio, defect rate, or any other relevant performance measure. In finance, quartiles (25, 50, and 75 percent) are commonly used to assess financial performance of companies and stock portfolio performances. In human resources, percentiles are used in employee salary benchmarking. Occupational Employment Statistics (OES) published by the U.S. Bureau of

Labor Statistics (www.bls.gov) show the 25th, 50th, and 75th percentiles for over 800 occupations in different metropolitan areas. Individuals could use these benchmarks to assess their potential earnings in different locations, or an employer could use them to estimate the cost of hiring its employees. The number of groups depends on the task at hand and the sample size, but quartiles deserve special attention because they are meaningful even for fairly small samples.

Quartiles

The quartiles (denoted Q_1, Q_2, Q_3) are scale points that divide the sorted data into four groups of approximately equal size, that is, the 25th, 50th, and 75th percentiles, respectively.

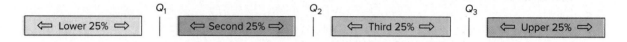

The second quartile Q_2 is the *median*. Because equal numbers of data values lie below and above the median, it is an important indicator of *center*.

The first and third quartiles Q_1 and Q_3 indicate the *center* because they define the boundaries for the middle 50 percent of the data. But Q_1 and Q_3 also indicate *variability* because the interquartile range $Q_3 - Q_1$ (denoted *IQR*) measures the degree of spread in the data (the middle 50 percent).

Conceptually, the first quartile Q_1 is the median of the data values below Q_2, and the third quartile Q_3 is the median of the data values above Q_2. Depending on n, the quartiles Q_1, Q_2, Q_3 may be members of the data set or may lie *between* two of the sorted data values. Figure 4.16 shows four possible situations.

Figure 4.16

Possible Quartile Positions

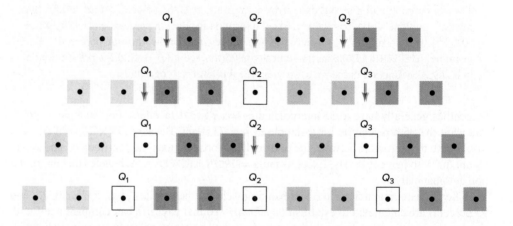

Quartiles do not always provide clean cutpoints in the sorted data, particularly in small samples or when there are repeating data values. For example:

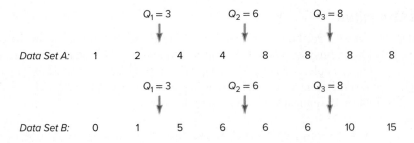

These two data sets have identical quartiles but are not really similar. Because of the small sample size and "gaps" in the data, the quartiles do not represent either data set well.

Method of Medians

For small data sets, you can find the quartiles using the **method of medians**:

- Step 1: Sort the observations.
- Step 2: Find the median Q_2.
- Step 3: Find the median of the data values that lie below Q_2.
- Step 4: Find the median of the data values that lie above Q_2.

When interpolation is necessary, we simply go halfway between the two data values as illustrated in Figure 4.17. This method is attractive because it is quick and logical (see Freund and Perles 1987 in Related Reading). However, Excel uses a different method.

Example 4.2

Method of Medians

A financial analyst has a portfolio of 12 energy equipment stocks. She has data on their recent price/earnings (P/E) ratios. To find the quartiles, she sorts the data, finds Q_2 (the median) halfway between the middle two data values, and then finds Q_1 and Q_3 (medians of the lower and upper halves, respectively).

Figure 4.17

Method of Medians

Company	Sorted P/E
Maverick Tube	7
BJ Services	22
FMC Technologies	25
Nabors Industries	29
Baker Hughes	31
Varco International	35
National-Oilwell	36
Smith International	36
Cooper Cameron	39
Schlumberger	42
Halliburton	46
Transocean	49

Q_1 is between x_3 and x_4 so
$Q_1 = (x_3 + x_4)/2 = (25 + 29)/2 = 27.0$

Q_2 is between x_6 and x_7 so
$Q_2 = (x_6 + x_7)/2 = (35 + 36)/2 = 35.5$

Q_3 is between x_9 and x_{10} so
$Q_3 = (x_9 + x_{10})/2 = (39 + 42)/2 = 40.5$

Introduce the .inc version of these formulas and clarify the difference between the .exc and .inc versions of these formulas.

Excel Quartiles

Excel does not use the method of medians, but instead uses a formula[1] to interpolate its quartiles. Excel has a function =QUARTILE.EXC(Data,k) to return the kth quartile in a data array $x_1, x_2, \ldots, x_n$. For example, =QUARTILE.EXC(Data,1) will return Q_1 and =QUARTILE.EXC (Data,3) will return Q_3. You could get the same results by using =PERCENTILE.EXC(Data,.25) and =PERCENTILE .EXC(Data,.75). Table 4.15 summarizes Excel's quartile methods.

Table 4.15

Calculating Quartiles Using Excel

Note: The functions ending in .EXC will match Minitab, R, and other statistical packages.

Quartile	Percent Below	Excel Quartile Function	Excel Percentile Function	Interpolated Position in Data Array
Q_1	25%	=QUARTILE.EXC(Data,1)	=PERCENTILE.EXC(Data,.25)	$.25n + .25$
Q_2	50%	=QUARTILE.EXC(Data,2)	=PERCENTILE.EXC(Data,.50)	$.50n + .50$
Q_3	75%	=QUARTILE.EXC(Data,3)	=PERCENTILE.EXC(Data,.75)	$.75n + .75$

Excel's Q_2 (the median) will be the same as with the method of medians. Although Excel's Q_1 and Q_3 may differ from the method of medians, the differences are usually unimportant.

Example 4.3

Excel Formula Method

Figure 4.18 illustrates Excel's quartile calculations using =QUARTILE.EXC for the same sample of P/E ratios. The resulting quartiles are similar to those using the method of medians but are not identical.

Figure 4.18

Excel's Quartile Interpolation Method

Company	Sorted P/E
Maverick Tube	7
BJ Services	22
FMC Technologies	25
Nabors Industries	29
Baker Hughes	31
Varco International	35
National-Oilwell	36
Smith International	36
Cooper Cameron	39
Schlumberger	42
Halliburton	46
Transocean	49

Q_1 is at observation $0.25n + 0.25 = (0.25)(12) + 0.25 = 3.25$, so we interpolate between x_3 and x_4 to get
$Q_1 = x_3 + (0.25)(x_4 - x_3) = 25 + (0.25)(29 - 25) = 26.00$

Q_2 is at observation $0.50n + 0.50 = (0.50)(12) + 0.50 = 6.50$, so we interpolate between x_6 and x_7 to get
$Q_2 = x_6 + (0.50)(x_7 - x_6) = 35 + (0.50)(36 - 35) = 35.50$

Q_3 is at observation $0.75n + 0.75 = (0.75)(12) + 0.75 = 9.75$, so we interpolate between x_9 and x_{10} to get
$Q_3 = x_9 + (0.75)(x_{10} - x_9) = 39 + (0.75)(42 - 39) = 41.25$

Tip

Whether you use the method of medians or Excel, your quartiles will be about the same. Small differences in calculation techniques typically do not lead to different conclusions in business applications.

---

[1]There are several acceptable ways to define the quartile position within a data array. Excel's QUARTILE.EXC and PERCENTILE.EXC functions define the Pth percentile position as $pn + p$ where $p = P/100$. See Eric Langford, "Quartiles in Elementary Statistics," *Journal of Statistics Education* 14, no. 3 (November 2006).

Box Plots

A useful tool of *exploratory data analysis* (EDA) is the **box plot** (also called a *box-and-whisker plot*) based on the **five-number summary**:

$$x_{min}, Q_1, Q_2, Q_3, x_{max}$$

The box plot is displayed visually, like this:

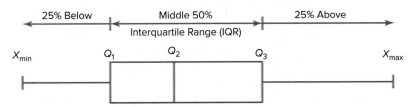

A box plot shows the *center* (position of the median Q_2). A box plot shows *variability* (width of the "box" defined by Q_1 and Q_3 and the range between x_{min} and x_{max}). A box plot shows *shape* (skewness if the whiskers are of unequal length and/or if the median is not in the center of the box). For example, right-skewness would be suggested by a longer right whisker or by a median to the left of the center of the box. Figure 4.19 shows simple box plots and histograms for samples drawn from different types of populations. A box plot gives a simple visual complement to the histogram and/or numerical statistics that we use to describe data.

Figure 4.19

Sample Box Plots from Four Populations (*n* = 1,000)

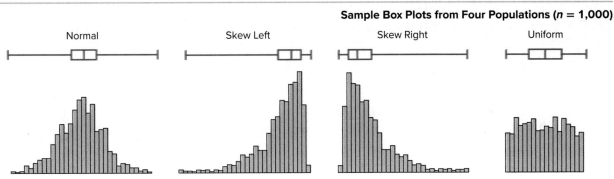

Fences and Unusual Data Values

We can use the quartiles to identify unusual data points. The idea is to detect data values that are far below Q_1 or far above Q_3. The *fences* are based on the *interquartile range (IQR)*:

$$IQR = Q_3 - Q_1 \tag{4.16}$$

To get the fences, we merely add or subtract a multiple of the *IQR* from Q_1 and Q_3.

Lower Fence	*Upper Fence*	
$Q_1 - 1.5(Q_3 - Q_1)$	$Q_3 + 1.5(Q_3 - Q_1)$	**(4.17)**

Observations outside the fences are considered as outliers. A diagram helps to visualize the fence calculations:

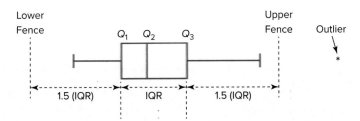

The box plot for the J.D. Power data is fairly symmetric and has no outliers.

$$Lower\ Fence \quad Q_1 - 1.5(Q_3 - Q_1) = 164 - 1.5(207 - 164) = 106.5$$

$$Upper\ Fence \quad Q_3 + 1.5(Q_3 - Q_1) = 207 + 1.5(207 - 164) = 262.5$$

To make a box plot in Excel, highlight the data and choose Insert > Statistics Chart > Box and Whisker Plot as illustrated in Figure 4.20. Excel's box plot is vertical instead of horizontal. Outliers (beyond the fences) would be displayed as dots beyond the end of the whiskers.

Figure 4.20

Excel Box Plot of J.D. Power Defects

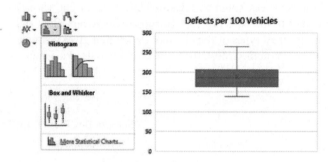

Mini Case 4.5

Defect Rates: Quartiles and Box Plots

Table 4.16 shows the five-number summary of the J.D. Power and Associates Initial Quality Study (IQS) for three years. We note that more vehicle brands were included in each successive year.

Table 4.16

Five-Number Summary for Box Plots by Year

Year	N	Min	Q_1	Q_2	Q_3	Max
2022	33	139	164	185	207	265
2021	32	128	150	164	190	251
2020	31	136	151	174	186	228
2019	32	63	90	99	113	130
2018	31	68	84	95	103	160

The Excel box plots in Figure 4.21 make it easy to compare both center and variation for these five model years. The box height (interquartile range) was greater in 2020–2022 than in 2018–2019, as were all three quartiles. Was this an effect of the COVID-19 pandemic? Labor shortages? Supply chain issues? Or is it merely due to a change in J.D. Power's definitions and sampling techniques? Statistics helps to *measure* and *describe* events but does not reveal *causes*.

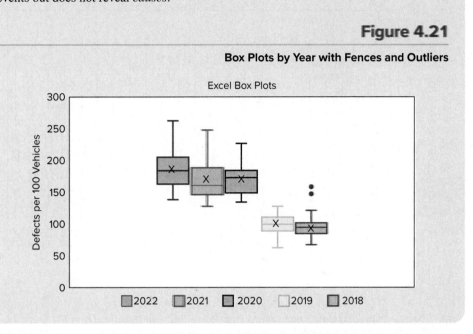

Figure 4.21

Box Plots by Year with Fences and Outliers

Midhinge

Quartiles can be used to define an additional measure of center that has the advantage of not being influenced by outliers. The midhinge is the average of the first and third quartiles:

$$\text{Midhinge} = \frac{Q_1 + Q_3}{2} \qquad (4.18)$$

The midhinge is always exactly *halfway* between Q_1 and Q_3, while the median Q_2 can be *anywhere* within the "box," which suggests a new way to describe skewness:

Median < Midhinge	⇒ Skewed right (longer right tail)
Median ≅ Midhinge	⇒ Symmetric (tails roughly equal)
Median > Midhinge	⇒ Skewed left (longer left tail)

Example 4.4

Hospital Bed Occupancy

The 24 box plots in Figure 4.22 show occupied beds in a hospital emergency room hour by hour. Each box plot is based on $n = 52$ Saturdays in one year. There are 24 hourly readings, so the total sample is $24 \times 52 = 1{,}248$ data points. Box plots are shown vertically to facilitate comparisons over time. The median (50th percentile) and quartiles (25th and 75th percentiles) change slowly and predictably over the course of the day. In several box plots, there are unusual values (very high occupancy) marked with an asterisk. Because the upper whiskers are longer than the lower whiskers, the occupancy rates are positively skewed. On several occasions (e.g., between midnight and 8 a.m.) there were Saturdays with zero bed occupancy. Similar charts can be prepared for each day of the week, aiding the hospital in planning its emergency room staffing and bed capacity.

Figure 4.22

Box Plots of Hospital Emergency Room Bed Occupancy

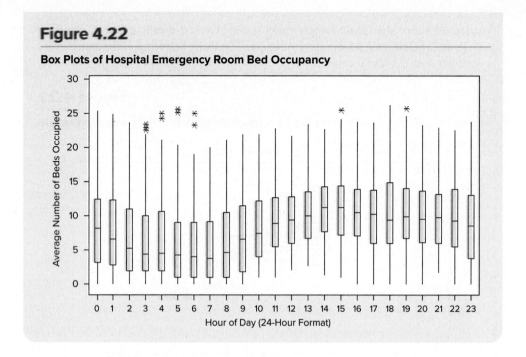

Section Exercises

4.40 Scores on an accounting exam ranged from 42 to 96, with quartiles $Q_1 = 61$, $Q_2 = 77$, and $Q_3 = 85$. (a) Sketch a simple box plot (5-number summary without fences) using a nicely scaled X-axis. (b) Describe its shape (skewed left, symmetric, skewed right).

4.41 In 2007, total compensation (in thousands of dollars) for 40 randomly chosen CEOs ranged from 790 to 192,920, with quartiles $Q_1 = 3,825$, $Q_2 = 8,890$, and $Q_3 = 17,948$. (a) Sketch a simple box plot (5-number summary without fences) using a nicely scaled X-axis. (b) Describe its shape (skewed left, symmetric, skewed right).

4.42 Waiting times (minutes) for a table at Joey's BBQ on Friday at 5:30 p.m. have quartiles $Q_1 = 21$, $Q_2 = 27$, and $Q_3 = 33$. Using the inner fences as a criterion, would a wait time of 45 minutes be considered an outlier?

4.43 Coffee temperatures (degrees Fahrenheit) at a certain restaurant have quartiles $Q_1 = 160$, $Q_2 = 165$, and $Q_3 = 170$. Using the inner fences as a criterion, would a temperature of 149 be considered an outlier?

4.44 The Comer-Correr Taco Wagon is only open from 11:00 a.m. to 2:00 p.m. on Saturdays. The owner kept track of the number of customers served on Saturdays for 60 weeks. (a) Visually estimate the quartiles Q_1, Q_2, Q_3. (b) Approximately how many customers were served on the busiest day? The slowest day? (c) Is the distribution symmetric? 🗂 **Customers**

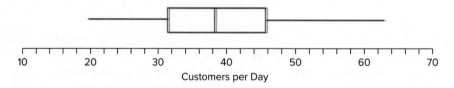

4.45 On San Martin Boulevard, embedded sensors kept track of the vehicle traffic count each hour for five weekdays, Monday through Friday, between 6 a.m. and 8 p.m. (5 weeks × 14 hours = 70 observations). (a) Visually estimate the quartiles Q_1, Q_2, Q_3. (b) Estimate x_{min} and x_{max}. (c) Is the distribution symmetric? 🗂 **Traffic**

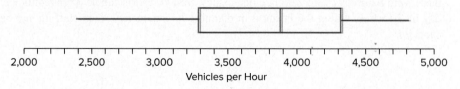

4.46 Citibank recorded the number of customers using a downtown ATM during the noon hour on 32 consecutive workdays. (a) Use Excel to find the quartiles. What do they tell you? (b) Find the midhinge. What does it tell you? (c) Make a box plot and interpret it. ⬚ **Citibank**

25	37	23	26	30	40	25	26
39	32	21	26	19	27	32	25
18	26	34	18	31	35	21	33
33	9	16	32	35	42	15	24

4.47 An executive's telephone log showed the lengths of 60 calls initiated during the last week of July. (a) Use Excel to find the quartiles. What do they tell you? (b) Find the midhinge. What does it tell you? (c) Make a box plot and interpret it. ⬚ **CallLength**

1	2	10	5	3	3	2	20	1	1
6	3	13	2	2	1	26	3	1	3
1	2	1	7	1	2	3	1	2	12
1	4	2	2	29	1	1	1	8	5
1	4	2	1	1	1	1	6	1	2
3	3	6	1	3	1	1	5	1	18

Mini Case 4.6

Airline Delays ⬚ UnitedAir

In 2005, United Airlines announced that it would award 500 frequent flier miles to every traveler on flights departing from Chicago O'Hare to seven other hub airports that arrived more than 30 minutes late (see *The Wall Street Journal,* June 14, 2005). What is the likelihood of such a delay? On a randomly chosen day (Tuesday, April 26, 2005), the Bureau of Transportation Statistics website (www.bts.gov) showed 278 United Airlines departures from O'Hare. The mean arrival delay was −7.45 minutes (i.e., flights arrived early, on average). The quartiles were $Q_1 = -19$ minutes, $Q_2 = -10$ minutes, and $Q_3 = -3$ minutes. While these statistics show that most of the flights arrive early, we must look further to estimate the probability of a frequent flier bonus.

In the box plot with fences (Figure 4.23), the "box" is entirely below zero. In the right tail, one flight was slightly above the inner fence (unusual), and eight flights were above the outer fence (outliers). An empirical estimate of the probability of a frequent flier award is 8/278, or about a 3 percent chance. A longer period of study might alter this estimate (e.g., if there were days of bad winter weather or traffic congestion).

Figure 4.23

Box Plot of Flight Arrival Delays

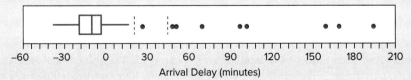

Arrival Delay (minutes)

The histogram (Figure 4.24) shows that the distribution of arrival delays is rather bell-shaped, except for the unusual values in the right tail. This is consistent with the view that "normal" flight operations are predictable, with only random variation around the mean. While it is difficult for flights to arrive earlier than planned, unusual factors could delay them by a lot.

Figure 4.24

Histogram of Flight Arrival Delays

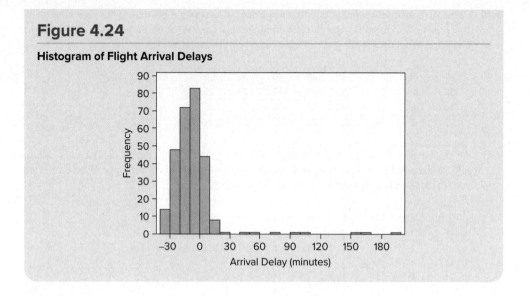

4.6 COVARIANCE AND CORRELATION

Covariance

The **covariance** of two random variables X and Y is denoted Cov(X,Y) or simply σ_{XY}. The covariance measures the degree to which the values of X and Y change together. This concept is particularly important in financial portfolio analysis. For example, if the prices of two stocks X and Y tend to move in the same direction, their covariance is positive ($\sigma_{XY} > 0$), and conversely if their prices tend to move in opposite directions ($\sigma_{XY} < 0$). If the prices of X and Y are unrelated, their covariance is zero ($\sigma_{XY} = 0$). A portfolio manager can apply this concept to reduce volatility in the overall portfolio by combining stocks in a way that reduces variation.

(4.19) *Population covariance*:

$$\sigma_{XY} = \frac{\sum_{i=1}^{N}(x_i - \mu_X)(y_i - \mu_Y)}{N}$$

(4.20) *Sample covariance*:

$$s_{XY} = \frac{\sum_{i=1}^{n}(x_i - \bar{x})(y_i - \bar{y})}{n - 1}$$

We would use the Excel function =COVARIANCE.P(XData,YData) for a population or (more commonly) the Excel function =COVARIANCE.S(XData,YData) for a sample. The units of measurement for the covariance are unpredictable because the magnitude and/or units of measurement of X and Y may differ. For this reason, analysts generally work with the correlation coefficient, which is a standardized value of the covariance that ensures a range between −1 and +1.

Correlation

Conceptually, a correlation coefficient is a covariance divided by the product of the standard deviations (denoted σ_X and σ_Y for a population or s_X and s_Y for a sample). For a population, the correlation coefficient is indicated by the lowercase Greek letter ρ (rho), while for a sample, we use the lowercase Roman letter r.

(4.21) *Population correlation coefficient*:

$$\rho = \frac{\sigma_{XY}}{\sigma_X \sigma_Y}$$

(4.22) *Sample correlation coefficient*:

$$r = \frac{s_{XY}}{s_X s_Y}$$

The **sample correlation coefficient** describes the degree of linearity between *paired* observations on two quantitative variables X and Y. The data set consists of n pairs (x_i, y_i) that are usually displayed on a *scatter plot* (review Chapter 3 if you need to refresh your memory about making scatter plots). A useful formula to calculate the sample correlation is

$$r = \frac{\sum_{i=1}^{n}(x_i - \bar{x})(y_i - \bar{y})}{\sqrt{\sum_{i=1}^{n}(x_i - \bar{x})^2} \sqrt{\sum_{i=1}^{n}(y_i - \bar{y})^2}} \tag{4.23}$$

Its range is $-1 \leq r \leq +1$. When r is near 0, there is little or no linear relationship between X and Y. An r value near $+1$ indicates a strong positive relationship, while an r value near -1 indicates a strong negative relationship. In Chapter 12, you will learn how to determine when a correlation is "significant" in a statistical sense (i.e., significantly different than zero), but for now, it is enough to recognize the correlation coefficient as a *descriptive statistic*.

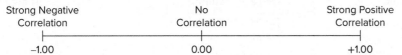

Excel's formula =CORREL(XData,YData) will return the sample correlation coefficient for two columns (or rows) of paired data. In fact, many scientific calculators will calculate r. The diagrams in Figure 4.25 will give you some idea of what various correlations look like. The correlation coefficient is a measure of the *linear relationship*. Some scatter plots may reveal a relationship, but not a *linear* one. And be wary—you will often hear the term "significant correlation" used imprecisely or incorrectly in the media to describe just about anything.

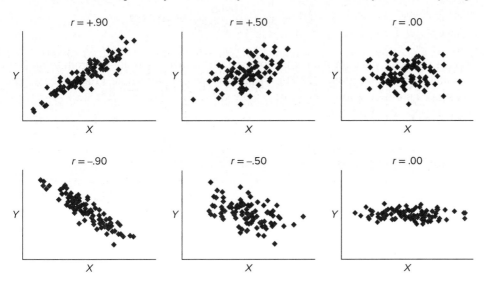

Figure 4.25

Illustration of Correlation Coefficients

Application: Stock Prices   📂 **TwoStocks** The prices of two stocks are recorded at the close of trading each Friday for 12 weeks, as displayed in Figure 4.26.

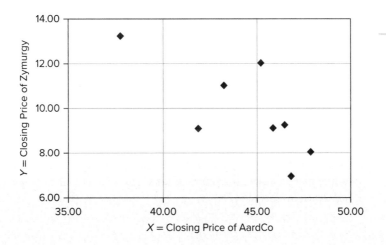

Figure 4.26

Scatter Plot of Two Stock Prices ($n = 8$)
📂 **TwoStocks**

Closing Price for Week ($n = 8$ weeks)								
Company	1	2	3	4	5	6	7	8
X (AardCo)	41.87	47.87	43.26	37.76	45.86	45.22	46.83	46.49
Y (Zymurgy)	9.11	8.07	11.02	13.24	9.14	12.04	6.96	9.27

These two stock prices seem to move in opposite directions, so we anticipate a negative covariance (and a negative correlation). Because we have only a sample of stock prices, we need the Excel function for the *sample* covariance.

Statistic	Result	Excel Function	R Function
Sample covariance:	$s_{XY} = -5.0890$	=COVARIANCE.S(XData,YData)	>cov(x,y)
Std. dev. of X:	$s_X = 3.3146$	=STDEV.S(XData)	>sd(x)
Std. dev. of Y:	$s_Y = 2.0896$	=STDEV.S(YData)	>sd(y)
Sample correlation:	$r = -0.7347$	=CORREL(XData,YData)	>cor(x,y)
Sample size:	$n = 8$ weeks	=COUNT(XData)	>length(x)

The sample correlation coefficient is

$$r = \frac{s_{XY}}{s_X s_Y} = \frac{-5.0890}{(3.3146)(2.0896)} = -0.7347$$

This is the same value for the correlation coefficient that we would get from formula 4.24 or using the Excel function =CORREL(). Using this type of information, a financial analyst can construct a portfolio whose total value is more stable, knowing that these stock prices tend to move in opposite directions.

Correlations Help Portfolio Design

"A good way to decrease the standard deviation of your portfolio is through diversification. By investing in different types of funds, you can minimize the impact that any one subasset class may have on your total holding."

Source: *T. Rowe Price Investor,* September 2009, p. 8.

Mini Case 4.7

Vail Resorts Customer Satisfaction

Figure 4.27 is a matrix showing correlations between several satisfaction variables from a sample of respondents to a Vail Resorts' satisfaction survey. The correlations are all positive, suggesting that greater satisfaction with any one of these criteria tends to be associated with greater satisfaction with the others (positive covariance). The highest correlation ($r = 0.488$) is between *SkiSafe* (attention to skier safety) and *SkiPatV* (Ski Patrol visibility). This makes intuitive sense. When a skier sees a ski patroller, you would expect an increased perception that the organization is concerned with skier safety. While many of the correlations seem small, they are all *statistically significant* (as you will learn in Chapter 12).

Figure 4.27

Correlation Matrix of Skier Satisfaction Variables (*n* = 502) 📂 VailGuestSat

	LiftOps	LiftWait	TrailVar	SnoAmt	GroomT	SkiSafe	SkiPatV
LiftOps	1.000						
LiftWait	0.180	1.000					
TrailVar	0.206	0.128	1.000				
SnoAmt	0.242	0.227	0.373	1.000			
GroomT	0.271	0.251	0.221	0.299	1.000		
SkiSafe	0.306	0.196	0.172	0.200	0.274	1.000	
SkiPatV	0.190	0.207	0.172	0.184	0.149	0.488	1.000

where

LiftOps = helpfulness/friendliness of lift operators
LiftWait = lift line wait
TrailVar = trail variety
SnoAmt = amount of snow
GroomTr = amount of groomed trails
SkiSafe = attention to skier safety
SkiPatV = Ski Patrol visibility

Section Exercises

📓 connect

4.48 For each *X–Y* data set (*n* = 12): (a) Make a scatter plot. (b) Find the sample correlation coefficient. (c) Is there a linear relationship between *X* and *Y?* If so, describe it. *Note:* Use Excel or MegaStat or Minitab if your instructor permits. 📂 **XYDataSets**

Data Set (a)

X	64.7	25.9	65.6	49.6	50.3	26.7	39.5	56.0	90.8	35.9	39.9	64.1
Y	5.8	18.1	10.6	11.9	11.4	14.6	15.7	4.4	2.2	15.4	14.7	9.9

Data Set (b)

X	55.1	59.8	72.3	86.4	31.1	41.8	40.7	36.8	42.7	28.9	24.8	16.2
Y	15.7	17.5	15.2	20.6	7.3	8.2	9.8	8.2	13.7	11.2	7.5	4.5

Data Set (c)

X	53.3	18.1	49.8	43.8	68.3	30.4	18.6	45.8	34.0	56.7	60.3	29.3
Y	10.2	6.9	14.8	13.4	16.8	9.5	16.3	16.4	1.5	11.4	10.9	19.7

4.49 Closing prices of two stocks are recorded for 50 trading days. The sample standard deviation of stock *X* is 4.638, and the sample standard deviation of stock *Y* is 9.084. The sample covariance is −36.111. (a) Calculate the sample correlation coefficient. (b) Describe the relationship between the prices of these two stocks.

4.50 For a sample of (*X, Y*) data values, the covariance is 48.724, the standard deviation of *X* is 11.724, and the standard deviation of *Y* is 8.244. (a) Calculate the sample correlation coefficient. (b) What does the sample correlation coefficient suggest about the relationship between *X* and *Y?*

4.51 (a) Make a scatter plot of the following data on *X* = home size and *Y* = selling price (thousands of dollars) for new homes (*n* = 20) in a suburb of an eastern city. (b) Calculate the sample correlation coefficient. (c) Is there a linear relationship between *X* and *Y?* If so, describe it. 📂 **HomePrice**

Square Feet	Selling Price (thousands)	Square Feet	Selling Price (thousands)	Square Feet	Selling Price (thousands)
3,570	861	3,240	809	3,160	778
3,410	740	2,660	639	3,310	760
2,690	563	3,160	778	2,930	729
3,260	698	3,460	737	3,020	720
3,130	624	3,340	806	2,320	575
3,460	737	3,240	809	3,130	785
3,340	806	2,660	639		

4.7 GROUPED DATA

LO 4-10

Calculate the mean and standard deviation from grouped data.

Weighted Mean

The **weighted mean** is a sum that assigns each data value a weight w_j that represents a fraction of the total (i.e., the k weights must sum to 1).

(4.24)
$$\bar{x} = \sum_{j=1}^{k} w_j x_j \quad \text{where} \quad \sum_{j=1}^{k} w_j = 1.00$$

For example, your instructor might give a weight of 30 percent to homework, 20 percent to the midterm exam, 40 percent to the final exam, and 10 percent to a term project (so that $.30 + .20 + .40 + .10 = 1.00$). Suppose your scores on these were 85, 68, 78, and 90. Your weighted average for the course would be

$$\bar{x} = \sum_{j=1}^{k} w_j x_j = .30 \times 85 + .20 \times 68 + .40 \times 78 + .10 \times 90 = 79.3$$

Despite a low score on the midterm exam, you are right at the borderline for an 80 (if your instructor rounds up). The weighted mean is widely used in cost accounting (weights for cost categories), finance (asset weights in investment portfolios), and other business applications.

Mini Case 4.8

What Is the DJIA? DJIA

The Dow Jones Industrial Average (commonly called the DJIA) is the oldest U.S. stock market price index, based on the prices of 30 large, widely held, and actively traded "blue chip" public companies in the United States (e.g., Coca-Cola, Microsoft, Walmart, Walt Disney). Actually, only a few of its 30 component companies are "industrial." The DJIA is not a simple average but rather a *weighted average* based on the prices of its component stocks. Originally a simple mean of stock prices, the DJIA now is calculated as the sum of the 30 stock prices divided by a "divisor" that compensates for stock splits and other changes over time. The divisor is revised as often as necessary (see *The Wall Street Journal* for the latest divisor value). Because high-priced stocks comprise a larger proportion of the sum, the DJIA is more strongly affected by changes in high-priced stocks. That is, a 10 percent price increase in a $10 stock would have less effect than a 10 percent price increase in a $50 stock, even if both companies have the same total market capitalization (the total number of shares times the price per share; often referred to as "market cap"). Broad-based market price indexes (e.g., NSDQ, AMEX, NYSE, S&P 500, Russ 2K) are widely used by fund managers, but the venerable "Dow" is still the one you see first on CNN or MSNBC.

Grouped Data

We can apply the idea of a weighted mean when we must work with observations that have been grouped. When a data set is tabulated into bins, we lose information about the location of the x values within bins but gain clarity of presentation because grouped data can be displayed more compactly than raw data. As long as the bin limits are given, we can estimate the mean and standard deviation using weights based on the bin frequencies. The accuracy of the grouped estimates will depend on the number of bins, the distribution of data within bins, and the bin frequencies.

Grouped Mean and Standard Deviation

Table 4.17 shows a frequency distribution for prices of a 90-day supply of a common cholesterol-lowering prescription drug from 47 retail pharmacies in three cities. The observations are classified into bins of equal width 5. When calculating a mean or standard deviation from grouped data, we treat all observations within a bin *as if they were located at the midpoint*. For example, in the third class (70 but less than 75), we pretend that all 11 prices were equal to $72.50 (the interval midpoint). In reality, observations may be scattered within each interval, but we assume that *on average* they are located at the class midpoint.

Table 4.17

Worksheet for Grouped Cholesterol Medication Data ($n = 47$) 📁 RxPrice

From	To	f_j	m_j	$f_j m_j$	$m_j - \bar{x}$	$(m_j - \bar{x})^2$	$f(m - \bar{x})^2$
60	65	6	62.5	375.0	−10.42553	108.69172	652.15029
65	70	11	67.5	742.5	−5.42553	29.43640	323.80036
70	75	11	72.5	797.5	−0.42553	0.18108	1.99185
75	80	13	77.5	1,007.5	4.57447	20.92576	272.03486
80	85	5	82.5	412.5	9.57447	91.67044	458.35220
85	90	0	87.5	0.0	14.57447	212.41512	0.00000
90	95	1	92.5	92.5	19.57447	383.15980	383.15980
	Sum	47	Sum	3,427.5		Sum	2,091.48936
			Mean ($\bar{x}$)	72.9255		Std Dev (s)	6.74293

Each interval j has a midpoint m_j and a frequency f_j. We calculate the estimated mean by multiplying the midpoint of each class by its class frequency, taking the sum over all k classes, and dividing by sample size n.

$$\bar{x} = \frac{\sum_{j=1}^{k} f_j m_j}{n} = \frac{3,427.5}{47} = 72.9255 \tag{4.25}$$

We then estimate the standard deviation by subtracting the estimated mean from each class midpoint, squaring the difference, multiplying by the class frequency, taking the sum over all classes to obtain the sum of squared deviations about the mean, dividing by $n - 1$, and taking the square root. *Avoid the common mistake of "rounding off" the mean before subtracting it from each midpoint.*

$$s = \sqrt{\frac{\sum_{j=1}^{k} f_j (m_j - \bar{x})^2}{n - 1}} = \sqrt{\frac{2,091.48936}{47 - 1}} = 6.74293 \tag{4.26}$$

Once we have the mean and standard deviation, we can estimate the coefficient of variation in the usual way:

$$CV = 100(s/\bar{x}) = 100(6.74293/72.9255) = 9.2\%$$

Accuracy Issues

How accurate are grouped estimates of $\bar{x}$ and s? To the extent that observations are *not* evenly spaced within the bins, accuracy would be lost. But, unless there is systematic skewness (say, clustering at the low end of each class), the effects of uneven distributions within bins will tend to average out.

Accuracy tends to improve as the number of bins increases. If the first or last class is open-ended, there will be no class midpoint and therefore no way to estimate the mean. For nonnegative data (e.g., GPA), we can assume a lower limit of zero when the first class is open-ended, although this assumption may make the first class too wide. Such an assumption may occasionally be possible for an open-ended top class (e.g., the upper limit of people's ages could be assumed to be 100), but many variables have no obvious upper limit (e.g., income). It is usually possible to estimate the median and quartiles from grouped data even with open-ended classes (the end-of-chapter *LearningStats* demonstrations give the formulas and illustrate grouped quartile calculations).

Section Exercises

4.52 Estimate the mean from this table of grouped data and frequencies.

From		To	f
0	<	20	5
20	<	40	12
40	<	60	18
60	<	80	9
		Total	44

4.53 Estimate the mean from this table of grouped data and frequencies.

From		To	f
2	<	6	7
6	<	10	12
10	<	14	3
14	<	18	2
18	<	22	1
		Total	25

4.8 SKEWNESS AND KURTOSIS

LO 4-11

Assess skewness and kurtosis in a sample.

Skewness

In a general way, *skewness* (as shown in Figure 4.28) may be judged by looking at the sample histogram, or by comparing the mean and median. However, this comparison is imprecise and does not take into account of the sample size. When more precision is needed, we look at the sample's **skewness coefficient:**

Figure 4.28

Skewness Prototype Populations

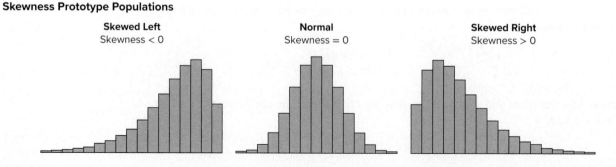

$$\text{Skewness} = \frac{n}{(n-1)(n-2)} \sum_{i=1}^{n} \left(\frac{x_i - \overline{x}}{s} \right)^3 \tag{4.27}$$

This unit-free statistic can be used to compare two samples measured in different units (say, dollars and yen) or to compare one sample with a known reference distribution such as the symmetric normal (bell-shaped) distribution. The skewness coefficient is obtained from Excel's Tools>Data Analysis>DescriptiveStatistics or by the function =SKEW(Data).

Table 4.18 shows the expected range within which the sample skewness coefficient would be expected to fall 90 percent of the time if the population being sampled were normal. A sample skewness statistic within the 90 percent range may be attributed to random variation, while coefficients outside the range would suggest that the sample came from a non-normal population. As *n* increases, the range of chance variation narrows.

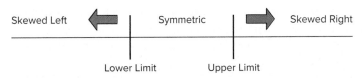

n	Lower 5%	Upper 5%	n	Lower 5%	Upper 5%
20	−0.84	+0.84	90	−0.41	+0.41
30	−0.69	+0.69	100	−0.40	+0.40
40	−0.61	+0.61	150	−0.33	+0.33
50	−0.55	+0.55	200	−0.28	+0.28
60	−0.51	+0.51	300	−0.23	+0.23
70	−0.47	+0.47	400	−0.20	+0.20
80	−0.44	+0.44	500	−0.18	+0.18

Table 4.18

90 Percent Range for Excel's Sample Skewness Coefficient

Source: Simulation of 100,000 samples using R with CRAN e1071 library.

When Excel is not available, an alternative measure of skewness is the **Pearson 2 skewness coefficient** (denoted Sk_2):

$$Sk_2 = \frac{3(\overline{x} - m)}{s} \tag{4.28}$$

where m is the sample median. This intuitively attractive statistic is easy to calculate using just three common descriptive statistics. It expresses the difference between the mean and median in terms of standard deviations. When the mean and median are nearly the same, Sk_2 will be near zero. Its sign indicates the direction of skewness. The farther Sk_2 is from zero, the more we doubt that the sample came from a symmetric normal population. A useful rule of thumb called **Schield's Rule** says that in samples of 10 or more from a normal population, we expect Sk_2 values within the range $\pm 4/\sqrt{n}$ at least 90 percent of the time. Sk_2 values outside that range would suggest that the sample did not come from a normal, symmetric population. If more precision is needed, tables (see Related Reading) are available to assess sample variation in Sk_2.

Kurtosis

Kurtosis refers to the relative length of the tails and the degree of concentration in the center. A normal bell-shaped population is called **mesokurtic** and serves as a benchmark (see Figure 4.29). A population that is flatter than a normal population (i.e., has heavier tails) is called **platykurtic**, while one that is more sharply peaked than a normal population (i.e., has thinner tails) is **leptokurtic**. Kurtosis is *not* the same thing as variability, although the two are easily confused.

A histogram is an unreliable guide to kurtosis because its scale and axis proportions may vary, so a numerical statistic is needed:

$$\text{Kurtosis} = \frac{n(n+1)}{(n-1)(n-2)(n-3)} \sum_{i=1}^{n} \left(\frac{x_i - \overline{x}}{s} \right)^4 - \frac{3(n-1)^2}{(n-2)(n-3)} \tag{4.29}$$

Figure 4.29

Kurtosis Prototype Shapes

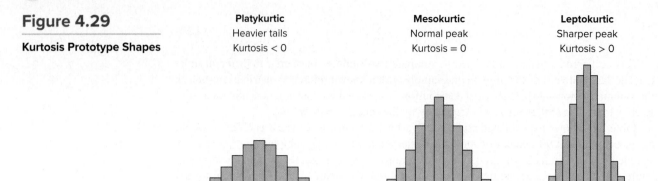

Platykurtic	Mesokurtic	Leptokurtic
Heavier tails	Normal peak	Sharper peak
Kurtosis < 0	Kurtosis = 0	Kurtosis > 0

The sample **kurtosis coefficient** is obtained from Excel's function =KURT(Data). Table 4.19 shows the expected range within which sample kurtosis coefficients would be expected to fall 90 percent of the time if the population is normal. A sample coefficient within the ranges shown may be attributed to chance variation, while a coefficient outside this range would suggest that the sample differs from a normal population. As the sample size increases, the chance range narrows. Unless you have at least 50 observations, inferences about kurtosis are risky.

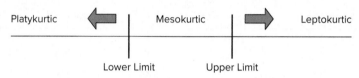

Platykurtic   Mesokurtic   Leptokurtic

Lower Limit   Upper Limit

Table 4.19

90 Percent Range for Excel's Sample Kurtosis Coefficient

Source: Simulation of 100,000 samples using R with CRAN e1071 library. Note that the lower and the upper table limits are *not* symmetric.

n	Lower 5%	Upper 5%	n	Lower 5%	Upper 5%
40	−0.89	1.35	100	−0.62	0.88
50	−0.82	1.23	150	−0.53	0.71
60	−0.76	1.13	200	−0.47	0.62
70	−0.72	1.04	300	−0.40	0.50
80	−0.68	0.98	400	−0.35	0.44
90	−0.65	0.92	500	−0.32	0.39

Mini Case 4.9

Stock Prices

An investor is tracking two stocks in the energy sector (Chevron and ExxonMobil) and two in the information technology sector (IBM and Oracle). She recorded the closing price of each stock on 55 consecutive trading days during a period of market uncertainty. The box plots in Figure 4.30 suggest negative skewness (longer left tail), except for ExxonMobil, and more price variation in the energy stocks than the information technology stocks (but note that each has a different axis scale). The Oracle box plot has an outlier, yet its variation (narrower box) *appears* less than IBM's. But is this merely an artifact of the differing axis scales? We need to look at *unit-free* statistics.

Coefficients of variation in Table 4.20 support the conclusion that the energy stocks (Chevron 8.9%, ExxonMobil 6.1%) had more *relative* variation over this time period than the information technology stocks (IBM 2.8%, Oracle 3.0%). The visual impression of a narrower "box" for Oracle is due to the scales of measurement. Skewness coefficients are within the normal range (see Table 4.18) except for ExxonMobil (+0.67) and Oracle (−0.62)). Chevron's kurtosis (−1.49) indicates a *platykurtic* shape, while Oracle's kurtosis (+0.76) implies a *leptokurtic* shape and the others are within normal range (see Table 4.19).

Figure 4.30

Box Plots for Prices of Four Stocks

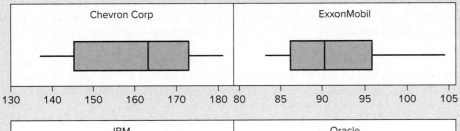

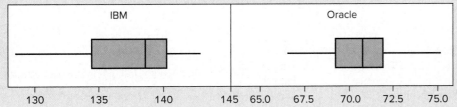

Table 4.20

Four Companies' Stock Prices

Statistic	Chevron	ExxonMobil	IBM	Oracle
Mean	159.70	90.90	136.92	70.59
StDev	14.26	5.57	3.86	2.08
Coef Var	8.9%	6.1%	2.8%	3.0%
Skewness	−0.17	0.67	−0.48	−0.62
Kurtosis	−1.49	−0.45	−0.59	0.76

Source: Stock prices are from https://finance.yahoo.com/, accessed July 21, 2022. This example is for statistical education only and not as a guide to investment decisions.

In addition to patterns in price variation, an investor would consider many other factors (e.g., prospects for growth, dividends, stability, etc.) in evaluating a portfolio, as well as sampling prices in other time periods.

Excel Hints

Hint 1: Formats When You Copy Data from Excel
Excel's dollar format (e.g., $214.07) or comma format (e.g., 12,417) will cause some statistical packages (e.g., Minitab and R) to interpret the pasted data as text (because "$" and "," are not numbers) or may cause other copy errors. In Minitab, a column heading such as C1-T indicates that the data column is text. Text cannot be analyzed numerically, so you can't get means, medians, etc. Check the format before you copy and paste.

Hint 2: Decimals When You Copy Data from Excel
Suppose you have adjusted Excel's decimal cell format to display 2.4 instead of 2.35477. When you copy this cell and paste it into Minitab, the pasted cell contains 2.4 (not 2.35477). Thus, Excel's statistical calculations (based on 2.35477) will not agree with Minitab. If you copy several columns of data (e.g., for a regression model), the differences can be serious. When you use the % format in Excel, a cell value such as 0.165 will be displayed as 16.5% but will be interpreted as text in R and will be treated as 0.165 in Minitab.

Chapter Summary

The **mean** and **median** describe a sample's **center** and also indicate **skewness**. The **mode** is useful for discrete data with a small range. The **trimmed mean** eliminates extreme values. The **geometric mean** mitigates high extremes but cannot be used when zeros or negative values are present. The **midrange** is easy to calculate but is sensitive to extremes. Variability is typically measured by the **standard deviation,** while relative dispersion is given by the **coefficient of variation** for nonnegative data. **Standardized data** reveal **outliers** or unusual data values, and the **Empirical Rule** offers a comparison with a normal distribution. In measuring dispersion, the **mean absolute deviation** or **MAD** is easy to understand but lacks nice mathematical properties. **Quartiles** are meaningful even for fairly small data sets, while **percentiles** are used only for large data sets. **Box plots** show the quartiles and data range. The **correlation coefficient** measures the degree of linearity between two variables. The **covariance** measures the degree to which two variables move together. We can estimate many common descriptive statistics from **grouped data.** Sample coefficients of **skewness** and **kurtosis** allow more precise inferences about the **shape** of the population being sampled instead of relying on histograms.

Key Terms

Center	*Variability*	*Shape*	*Other*
geometric mean	Chebyshev's Theorem	bimodal distribution	box plot
mean	coefficient of variation	kurtosis	covariance
median	Empirical Rule	kurtosis coefficient	five-number summary
midhinge	mean absolute deviation	leptokurtic	interquartile range
midrange	outliers	mesokurtic	method of medians
mode	population variance	multimodal distribution	quartiles
trimmed mean	range	negatively skewed	sample correlation coefficient
weighted mean	sample variance	Pearson 2 skewness coefficient	Sharpe's ratio
	standard deviation	platykurtic	
	standardized data	positively skewed	
	z-score	Schield's Rule	
		skewed left	
		skewed right	
		skewness	
		skewness coefficient	
		symmetric data	

Choosing the Appropriate Statistic or Visual Display

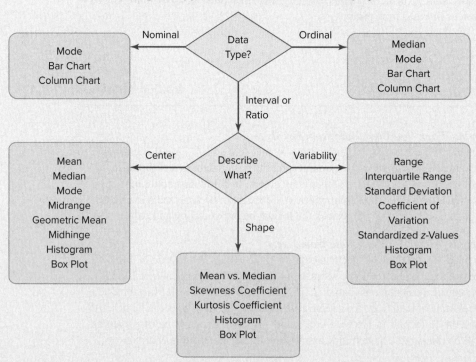

Commonly Used Formulas in Descriptive Statistics

Sample mean: $\quad \bar{x} = \dfrac{1}{n}\sum_{i=1}^{n} x_i$

Geometric mean: $\quad G = \sqrt[n]{x_1 x_2 \ldots x_n}$

Growth rate: $\quad GR = \sqrt[n-1]{\dfrac{x_n}{x_1}} - 1$

Range: $\quad \text{Range} = x_{max} - x_{min}$

Midrange: $\quad \text{Midrange} = \dfrac{x_{max} + x_{min}}{2}$

Sample standard deviation: $\quad s = \sqrt{\dfrac{\sum_{i=1}^{n}(x_i - \bar{x})^2}{n-1}}$

Coefficient of variation:

Population	Sample
$CV = 100 \times \dfrac{\sigma}{\mu}$	$CV = 100 \times \dfrac{s}{\bar{x}}$

Standardized variable:

Population	Sample
$z_i = \dfrac{x_i - \mu}{\sigma}$	$z_i = \dfrac{x_i - \bar{x}}{s}$

Midhinge: $\quad \text{Midhinge} = \dfrac{Q_1 + Q_3}{2}$

Sample correlation coefficient: $\quad r = \dfrac{\sum_{i=1}^{n}(x_i - \bar{x})(y_i - \bar{y})}{\sqrt{\sum_{i=1}^{n}(x_i - \bar{x})^2}\sqrt{\sum_{i=1}^{n}(y_i - \bar{y})^2}}$ or $r = \dfrac{s_{XY}}{s_X s_Y}$

Weighted mean: $\quad \bar{x} = \sum_{j=1}^{k} w_j x_j$ where $\sum_{j=1}^{k} w_j = 1.00$

Grouped mean: $\quad \bar{x} = \dfrac{\sum_{j=1}^{k} f_j m_j}{n}$

Chapter Review Connect

1. What are descriptive statistics? How do they differ from visual displays of data?

2. Explain each concept: (a) center, (b) variability, and (c) shape.

3. (a) Why is sorting usually the first step in data analysis? (b) Why is it useful to begin a data analysis by thinking about how the data were collected?

4. List the strengths and weaknesses of each measure of center and write its Excel function: (a) mean, (b) median, and (c) mode.

5. (a) Why must the deviations around the mean sum to zero? (b) What is the position of the median in the data array when n is even? When n is odd? (c) Why is the mode of little use in continuous data? (d) For what type of data is the mode most useful?

6. (a) What is a bimodal distribution? (b) Explain two ways to detect skewness.

7. List strengths and weaknesses of each measure of center and give its Excel function (if any): (a) midrange, (b) geometric mean, and (c) 10 percent trimmed mean.

8. (a) What is variability? (b) Name five measures of variability. List the main characteristics (strengths, weaknesses) of each measure.

9. (a) Which standard deviation formula (population, sample) is used most often? Why? (b) When is the coefficient of variation useful?

10. (a) To what kind of data does Chebyshev's Theorem apply? (b) To what kind of data does the Empirical Rule apply? (c) What is an outlier? An unusual data value?

11. (a) In a normal distribution, approximately what percent of observations are within 1, 2, and 3 standard deviations of the mean? (b) In a sample of 10,000 observations, about how many observations would you expect beyond 3 standard deviations of the mean?

12. (a) Write the mathematical formula for a standardized variable. (b) Write the Excel formula for standardizing a data value in cell F17 from an array with mean Mu and standard deviation Sigma.

13. (a) Why is it dangerous to delete an outlier? (b) When might it be acceptable to delete an outlier?

14. (a) Explain how quartiles can measure both center and variability. (b) Why don't we calculate percentiles for small samples?

15. (a) Explain the method of medians for calculating quartiles. (b) Write the Excel formula for the first quartile of an array named XData.

16. (a) What is a box plot? What does it tell us? (b) What is the role of fences in a box plot? (c) Define the midhinge and interquartile range.

17. What does a correlation coefficient measure? What is its range? Why is a correlation coefficient easier to interpret than a covariance?

18. (a) Why is some accuracy lost when we estimate the mean or standard deviation from grouped data? (b) Why do open-ended classes in a frequency distribution make it impossible to estimate the mean and standard deviation? (c) When would group data be presented instead of the entire sample of raw data?

19. (a) What is the skewness coefficient of a normal distribution? A uniform distribution? (b) Why do we need a table for sample skewness coefficients that are is based on sample size?

20. (a) What is kurtosis? (b) Sketch a platykurtic population, a leptokurtic population, and a mesokurtic population. (c) Why can't we rely on a histogram to assess kurtosis?

Chapter Exercises connect

4.54 (a) For each data set, calculate the mean, median, and mode. (b) Which, if any, of these three measures is the weakest indicator of a "typical" data value? Why?

    a. Number of e-mail accounts (12 students): 1, 1, 1, 1, 2, 2, 2, 3, 3, 3, 3, 3

    b. Number of siblings (5 students): 0, 1, 2, 2, 10

    c. Asset turnover ratio (8 retail firms): 1.85, 1.87, 2.02, 2.05, 2.11, 2.18, 2.29, 3.01

4.55 If the mean asset turnover for retail firms is 2.02 with a standard deviation of 0.22, without assuming a normal distribution, within what range will at least 75% of retail firms' asset turnover fall?

4.56 For each data set: (a) Find the mean, median, and mode. (b) Which, if any, of these three measures is the weakest indicator of a "typical" data value? Why?

    a. 100 m dash times ($n = 6$ top runners): 9.87, 9.98, 10.02, 10.15, 10.36, 10.36

    b. Number of children ($n = 13$ families): 0, 1, 1, 2, 2, 2, 2, 2, 2, 2, 2, 2, 6

    c. Number of cars in the driveway ($n = 8$ homes): 0, 0, 1, 1, 2, 2, 3, 5

4.57 During a rock concert, the noise level (in decibels) in front-row seats has a mean of 95 dB with a standard deviation of 8 dB. Without assuming a normal distribution, find the minimum percentage of noise level readings within 3 standard deviations of the mean.

4.58 Bags of jelly beans have a mean weight of 396 gm with a standard deviation of 5 gm. Use Chebyshev's Theorem to find a lower bound for the number of bags in a sample of 200 that weigh between 386 and 406 gm.

4.59 Based on experience, the Ball Corporation's aluminum can manufacturing facility in Ft. Atkinson, Wisconsin, knows that the metal thickness of incoming shipments has a mean of 0.2731 mm with a standard deviation of 0.000959 mm. (a) A certain shipment has a diameter of 0.2761. Find the standardized z-score for this shipment. (b) Is this an outlier?

4.60 SAT scores for the entering class of 2010 at Oxnard University were normally distributed with a mean of 1340 and a standard deviation of 90. Bob's SAT score was 1430.

(a) Find Bob's standardized z-score. (b) By the Empirical Rule, is Bob's SAT score unusual?

4.61 Find the data value that corresponds to each of the following z-scores.

    a. Final exam scores: Allison's z-score = 2.30, $\mu = 74$, $\sigma = 7$

    b. Weekly grocery bill: James' z-score = -1.45, $\mu = \$53$, $\sigma = \$12$

    c. Daily video game play time: Eric's z-score = -0.79, $\mu = 4.00$ hours, $\sigma = 1.15$ hours

4.62 The average time a Boulder High varsity lacrosse player plays in a game is 30 minutes with a standard deviation of 7 minutes. Nolan's playing time in last week's game against Fairview was 48 minutes. (a) Calculate the z-score for Nolan's playing time against Fairview. (b) By the Empirical Rule, was Nolan's playing time *unusual* when compared to the typical playing time?

4.63 The number of blueberries in a blueberry muffin baked by EarthHarvest Bakeries can range from 18 to 30 blueberries. (a) Use the Empirical Rule to estimate the standard deviation of the number of blueberries in a muffin. (b) What assumption did you make about the distribution of the number of blueberries?

Note: Unless otherwise instructed, you may use any desired statistical software for calculations and graphs in the following problems.

DESCRIBING DATA

4.64 Below are the monthly rents paid by 30 students who live off campus. (a) Find the mean, median, and mode. (b) Do the measures of central tendency agree? Explain. (c) Calculate the standard deviation. (d) Sort and standardize the data. (e) Are there outliers or unusual data values? (f) Using the Empirical Rule, do you think the data could be from a normal population? Rents

730	730	730	930	700	570
690	1,030	740	620	720	670
560	740	650	660	850	930
600	620	760	690	710	500
730	800	820	840	720	700

4.65 How many days in advance do travelers purchase their airline tickets? Below are data showing the advance days for a sample of 28 passengers on United Airlines Flight 815 from Chicago to Los Angeles. (a) Calculate the mean, median, mode, and midrange. (b) Calculate the quartiles and midhinge. (c) Why can't you use the geometric mean for this data set? 📁 **Days**

```
11  7 11  4 15 14 71 29  8  7 16 28 17 249
 0 20 77 18 14  3 15 52 20  0  9  9 21   3
```

4.66 The durations (minutes) of 26 electric power outages in the community of Sonando Heights over the past five years are shown below. (a) Find the mean, median, and mode. (b) Are the mean and median about the same? (c) Is the mode a good measure of center for this data set? Explain. (d) Is the distribution skewed? Explain. 📁 **Duration**

```
32 44 25 66 27 12 62  9 51  4 17 50 35
99 30 21 12 53 25  2 18 24 84 30 17 17
```

4.67 The U.S. Postal Service will ship a Priority Mail® Large Flat Rate Box (12″ × 12″ × 5½″) anywhere in the United States for a fixed price, regardless of weight. The weights (ounces) of 20 randomly chosen boxes are shown below. (a) Find the mean, median, and mode. (b) Are the mean and median about the same? If not, why not? (c) Is the mode a "typical" data value? Explain. (d) Is the distribution skewed? Explain 📁 **Weights**

```
72 86 28 67 64 65 45 86 31 32
39 92 90 91 84 62 80 74 63 86
```

4.68 A sample of size $n = 70$ showed a skewness coefficient of 0.773 and a kurtosis coefficient of 1.277. What is the distribution's shape?

4.69 The "expense ratio" is a measure of the cost of managing the portfolio. Investors prefer a low expense ratio, all else equal. Below are expense ratios for 23 randomly chosen stock funds and 21 randomly chosen bond funds. (a) Calculate the mean and median for each sample. (b) Calculate the standard deviation and coefficient of variation for each sample. (c) Which type of fund has more variability? Explain. 📁 **Funds**

23 Stock Funds
```
1.12 1.44 1.27 1.75 0.99 1.45 1.19 1.22 0.99 3.18 1.21 1.89
0.60 2.10 0.73 0.90 1.79 1.35 1.08 1.28 1.20 1.68 0.15 0.15
```

21 Bond Funds
```
1.96 0.51 1.12 0.64 0.69 0.20 1.44 0.68 0.40 0.94 0.75 1.77
0.93 1.25 0.85 0.99 0.95 0.35 0.64 0.41 0.90
```

4.70 This year, Dolon Company's website averaged 12,104 daily views with a standard deviation of 3,026. Last year, the mean number of daily page views was 6,804 with a standard deviation of 1,701. Describe the *relative* variation in page views in these two years.

4.71 At Chipotle Mexican Grill, the number of calories in an order of chips and salsa is normally distributed with a mean of 620 and a standard deviation of 12. Bob's order had only 580 calories. Was this an outlier? Explain.

4.72 A plumbing supplier's mean monthly demand for vinyl washers is 24,212 with a standard deviation of 6,053. The mean monthly demand for steam boilers is 6.8 with a standard deviation of 1.7. Which demand pattern has more relative variation? Explain.

4.73 The table below shows average daily sales of Rice Krispies Treats in the month of June in 74 Noodles & Company restaurants. (a) Make a histogram for the data. (b) Would you say the distribution is skewed? (c) Calculate the mean and standard deviation. (d) Are there any outliers? 📁 **RiceKrispies**

```
32  8 14 20 28 19 37 31 16 16
16 29 11 34 31 18 22 17 27 16
24 49 25 18 25 21 15 16 20 11
21 29 14 25 10 15  8 12 12 19
21 28 27 26 12 24 18 19 24 16
17 20 23 13 17 17 19 36 16 34
25 15 16 13 20 13 13 23 17 22
11 17 17  9
```

4.74 Analysis of investment portfolio returns over a 20-year period showed the statistics below. (a) Calculate and compare the coefficients of variation. (b) Why would we use a coefficient of variation? Why not just compare the standard deviations? (c) What do the data tell you about risk and return? 📁 **Returns**

Investment Type	Mean Return	Standard Deviation	Coefficient of Variation
Venture funds	19.2	14.0	
Common stocks	15.6	14.0	
Real estate	11.5	16.8	
Federal short-term paper	6.7	1.9	

4.75 Analysis of annualized returns over a 10-year period showed that prepaid tuition plans had a mean return of 6.3 percent with a standard deviation of 2.7 percent, while the Standard & Poor's 500 stock index had a mean return of 12.9 percent with a standard deviation of 15.8 percent. (a) Calculate and compare the coefficients of variation. (b) Why would we use a coefficient of variation? Why not just compare the standard deviations?

4.76 Caffeine content in a 5-ounce cup of brewed coffee ranges from 60 to 180 mg, depending on brew time, coffee bean type, and grind. (a) Use the midrange as a measure of center. (b) Use the Empirical Rule to estimate the standard deviation. (c) Why is the assumption of a normal, bell-shaped distribution important in making these estimates? (d) Why might caffeine content of coffee *not* be normal?

4.77 Chlorine is added to city water to kill bacteria. In a certain year, chlorine content in water from the Lake Huron Water Treatment plant ranged from 0.79 ppm (parts per million) to 0.92 ppm. (a) Use the midrange as a measure of center. (b) Use the Empirical Rule to estimate the standard deviation.

4.78 A sample of size $n = 100$ showed a skewness coefficient of −0.35 and a kurtosis coefficient of +0.75. What is the distribution's shape?

4.79 A sample of size $n = 200$ showed a skewness coefficient of −0.35 and a kurtosis coefficient of +0.75. What is the distribution's shape?

THINKING ABOUT DISTRIBUTIONS

4.80 At the Midlothian Independent Bank, a study shows that the mean ATM transaction takes 74 seconds, the median 63 seconds, and the mode 51 seconds. (a) Sketch the distribution based on these statistics. (b) What factors might cause the distribution to be like this?

4.81 At the Eureka library, the mean time a book is checked out is 13 days, the median is 10 days, and the mode is 7 days. (a) Sketch the distribution based on these statistics. (b) What factors might cause the distribution to be like this?

4.82 On Professor Hardtack's last cost accounting exam, the mean score was 71, the median was 77, and the mode was 81. (a) Sketch the distribution based on these statistics. (b) What factors might cause the distribution to be like this?

4.83 The median life span of a mouse is 118 weeks. (a) Would you expect the mean to be higher or lower than 118? (b) Would you expect the life spans of mice to be normally distributed? Explain.

4.84 The median waiting time for a liver transplant in the United States is 321 days. Would you expect the mean to be higher or lower than 321 days? Explain. (See www.livermd.org.)

4.85 A small suburban community agreed to purchase police services from the county sheriff's department. The newspaper said, "In the past, the charge for police protection from the Sheriff's Department has been based on the median cost of the salary, fringe benefits, etc. That is, the cost per deputy was set halfway between the most expensive deputy and the least expensive." (a) Is this the median? If not, what is it? (b) Which would probably cost the city more, the midrange or the median? Why?

4.86 A company's contractual "trigger" point for a union absenteeism penalty is a certain distance above the *mean* days missed by all workers. Now the company wants to switch the trigger to a certain number of days above the *median* days missed for all workers. (a) Visualize the distribution of missed days for all workers (symmetric, skewed left, skewed right). (b) Discuss the probable effect on the trigger point of switching from the mean to the median. (c) What position would the union be likely to take on the company's proposed switch?

EXCEL PROJECTS

4.87 (a) Use Excel functions to calculate the mean and standard deviation for weekend occupancy rates (percent) in nine resort hotels during the off-season. (b) What conclusion would a casual observer draw about center and variability based on your statistics? (c) Now calculate the median for each sample. (d) Make a dot plot for each sample. (e) What did you learn

Observation	Week 1	Week 2	Week 3	Week 4
1	32	33	38	37
2	41	35	39	42
3	44	45	39	45
4	47	50	40	46
5	50	52	56	47
6	53	54	57	48
7	56	58	58	50
8	59	59	61	67
9	68	64	62	68

from the medians and dot plots that was not apparent from the means and standard deviations? 🗂 **Occupancy**

4.88 (a) In Excel, enter =ROUND(NORM.INV(RAND(),70,10),0) in cells B1:B100. This will create 100 random data points from a normal distribution using parameters $\mu = 70$ and $\sigma = 10$. Think of these numbers as exam scores for 100 students. (b) Use the Excel functions =AVERAGE(B1:B100) and =STDEV.S(B1:B100) to calculate the sample mean and standard deviation for your data array. (c) Write down the sample mean and standard deviation. (d) Compare the sample statistics with the desired parameters $\mu = 70$ and $\sigma = 10$. Do Excel's random samples have approximately the desired characteristics?

GROUPED DATA

Note: In each of the following tables, the upper bin limit is excluded from that bin but is included as the lower limit of the next bin.

4.89 A random sample of individuals who filed their own income taxes were asked how much time (hours) they spent preparing last year's federal income tax forms. (a) Estimate the mean. (b) Estimate the standard deviation. (c) Do you think the observations would be distributed uniformly within each interval? Why would that matter? (d) Why do you imagine that unequal bin sizes (interval widths) were used? 🗂 **Taxes**

From	To (not incl)	f
0	2	7
2	4	42
4	8	33
8	16	21
16	32	11
32	64	6

4.90 This table shows the distribution of winning times in the Kentucky Derby (a horse race) over 87 years. (a) From the grouped data, calculate the mean. Show your calculations clearly in a worksheet. (b) What additional information would you have gained by having the raw data? (c) Do you think it likely that the distribution of times within each interval might not be uniform? Why would that matter?

Kentucky Derby Winning Times, 1930–2019 (seconds) 🗂 **Derby**

From	To	f
119	120	1
120	121	5
121	122	20
122	123	27
123	124	16
124	125	11
125	126	5
126	127	3
127	128	2
	Total	90

4.91 The self-reported number of hours worked per week by 204 top executives is given below. (a) Estimate the mean, standard deviation, and coefficient of variation using an Excel worksheet to organize your calculations. (b) Do the unequal class sizes hamper your calculations? Why do you suppose that was done?

Weekly Hours of Work by Top Executives ⬚ Work

From	To	f
40	50	12
50	60	116
60	80	74
80	100	2
	Total	204

4.92 How long does it take to fly from Denver to Atlanta on Delta Airlines? The table below shows 56 observations on flight times (in minutes) for the first week of March 2005. (a) Use the grouped data formula to estimate the mean and standard deviation. (b) Using the ungrouped data (not shown), the *ungrouped* sample mean is 161.63 minutes and the ungrouped standard deviation is 8.07 minutes. How close did your *grouped* estimates come? (c) Why might flight times *not* be uniformly distributed within the second and third class intervals? (See www.bts.gov.)

Flight Times DEN to ATL (minutes) ⬚ DeltaAir

From		To	Frequency
140	<	150	1
150	<	160	25
160	<	170	24
170	<	180	4
180	<	190	2
		Total	56

DO-IT-YOURSELF SAMPLING PROJECTS

4.93 (a) Record the points scored by the winning team in 30 college football games played last weekend (if it is not football season, do the same for basketball or another sport of your choice). If you can't find 30 scores, do the best you can. (b) Make a frequency distribution and histogram. Describe the histogram. (c) Calculate the mean, median, and mode. Which is the best measure of center? Why? (d) Calculate the standard deviation and coefficient of variation. (e) Standardize the data. Are there any outliers? (f) Make a box plot. What does it tell you?

4.94 On the web, look up the defects per 100 for all vehicles in the latest J.D. Power Initial Quality Ratings. (a) Make a histogram and describe it. (b) Calculate descriptive statistics of center, variation, and shape. State in words what they tell you. (c) Make a box plot. What does it tell you?

SCATTER PLOTS AND CORRELATION

Note: Data files for Exercises 4.95 through 4.98 may be downloaded from Connect or may be obtained from your instructor.

4.95 (a) Make an Excel scatter plot of X = 1990 assault rate per 100,000 population and Y = 2004 assault rate per 100,000 population for the 50 U.S. states. (b) Use Excel's =CORREL function to find the correlation coefficient. (c) What do the graph and correlation coefficient say about assault rates by state for these two years? (d) Use MegaStat or Excel to find the mean, median, and standard deviation of assault rates in these two years. What does this comparison tell you? ⬚ **Assault**

4.96 (a) Make an Excel scatter plot of X = curb weight of vehicle and Y = vehicle miles per gallon in city driving for 50 selected new vehicles. (b) Describe the relationship (if any). Weak? Strong? Negative? Positive? (c) Why might a relationship exist? (d) Calculate the sample correlation. What does it tell you? ⬚ **Vehicles**

4.97 (a) Make an Excel scatter plot of X = percent change in U.S. real GDP and Y = percent change in U.S. personal consumption expenditures 2000–2015. (b) Describe the relationship (if any). Weak? Strong? Negative? Positive? (c) Why might a relationship exist? (d) Calculate the sample correlation using the Excel function =CORREL. What does it tell you? ⬚ **Consumption**

4.98 (a) Make an Excel scatter plot of X = weekly closing price of Dolana stock and Y = weekly closing price of ZalParm stock over 52 consecutive weeks. (b) Describe the scatter plot. *Hint:* You may need to rescale the X and Y axes to see more detail. (c) Calculate the sample correlation using the Excel function =CORREL. What does it tell you? (d) Select both data columns (including column headings) and make a line chart displaying both series. What does the line chart say about the relationship? ⬚ **TwoStocks2**

MINI-PROJECTS

Note that in the following data sets, only the first three and last three observations are shown. You may download the complete data files from the problems in Connect or through your instructor. Choose a data set and prepare a brief, descriptive report. Refer to Appendix I for tips on report writing. You may use any computer software you wish (e.g., Excel, MegaStat, Minitab). Include relevant worksheets or graphs in your report. If some questions do not apply to your data set, explain why not.

4.99 (a) Sort the data and find X_{min} and X_{max}. (b) Make a histogram. Describe its shape. (c) Calculate the mean and median. Are the data skewed? (d) Calculate the standard deviation. (e) Standardize the data. Are there outliers? If so, list them along with their z-scores. (f) Calculate the quartiles and make a box plot. Describe its appearance.

DATA SET A Advertising Dollars as Percent of Sales in Selected Industries (*n* = 30) 🗂 Ads

Industry	Percent
Accident and health insurance	0.9
Apparel and other finished products	5.5
Beverages	7.4
⋮	⋮
Steel works and blast furnaces	1.9
Tires and inner tubes	1.8
Wine, brandy, and spirits	11.3

Source: George E. Belch and Michael A. Belch, *Advertising and Promotion*, pp. 219–220. McGraw Hill Education, 2004.

DATA SET B Maximum Rate of Climb for Selected Piston Aircraft (*n* = 54) 🗂 ClimbRate

Manufacturer/Model	Year	Climb (ft./min.)
AMD CH 2000	2000	820
Beech Baron 58	1984	1,750
Beech Baron 58P	1984	1,475
⋮	⋮	⋮
Sky Arrow 650 TC	1998	750
Socata TB20 Trinidad	1999	1,200
Tiger AG-5B	2002	850

Source: *Flying Magazine* (various issues from 1997 to 2002).

DATA SET C December Heating Degree-Days for Selected U.S. Cities (*n* = 35) 🗂 Heating

City	Degree-Days
Albuquerque	911
Baltimore	884
Bismarck	1,538
⋮	⋮
St. Louis	955
Washington, D.C.	809
Wichita	949

Note: A degree-day is the sum over all days in the month of the difference between 65 degrees Fahrenheit and the daily mean temperature of each city.
Source: U.S. Bureau of the Census, *Statistical Abstract of the United States, 2012.*

DATA SET D Deposits as Percent of Assets for Selected U.S. Banks (*n* = 41) 🗂 Banks

Bank	Percent
Ally Financial Inc	56.7
American Express Corp	36.7
Bank of America Corp	57.9
⋮	⋮
U.S. Bancorp	67.9
Wells Fargo & Co	66.8
Zions Bancorp	81.1

Source: 2018 company annual reports, www.wikipedia.org, and www.federalreserve.gov.

DATA SET E Caffeine Content of Randomly Selected Beverages (*n* = 32) 🗂 Caffeine

Company/Brand	mg/oz.
Barq's Root Beer	1.83
Coca-Cola Classic	2.83
Cool from Nestea	1.33
⋮	⋮
Snapple Sweet Tea	1.00
Sunkist Orange Soda	3.42
Vanilla Coke	2.83

Note: Caffeine content may have changed since sample was taken. For current data, see American Beverage Association (www.ameribev.org).

DATA SET F Super Bowl Scores 1967–2020 (*n* = 54 games) 🗂 SuperBowl

Year	Teams and Scores
1967	Green Bay 35, Kansas City 10
1968	Green Bay 33, Oakland 14
1969	NY Jets 16, Baltimore 7
⋮	⋮
2018	Philadelphia 41, New England 33
2019	New England 13, Los Angeles 3
2020	Kansas City 31, San Francisco 20

Source: en.wikipedia.org.

DATA SET G Property Crimes per 100,000 Residents (*n* = 68 cities) 🗂 Crime

City and State	Crime
Albuquerque, NM	8,515
Anaheim, CA	2,827
Anchorage, AK	4,370
⋮	⋮
Virginia Beach, VA	3,438
Washington, DC	6,434
Wichita, KS	5,733

Source: *Statistical Abstract of the United States, 2002.*

DATA SET H Size of Whole Foods Stores (*n* = 171) 🗂 WholeFoods

Location (Store Name)	Sq. Ft.
Albuquerque, NM (Academy)	33,000
Alexandria, VA (Annandale)	29,811
Ann Arbor, MI (Washtenaw)	51,300
⋮	⋮
Winter Park, FL	20,909
Woodland Hills, CA	28,180
Wynnewood, PA	14,000

Source: www.wholefoodsmarket.com/stores/.

Related Reading

Doane, David P., and Lori E. Seward. "Measuring Skewness: A Forgotten Statistic?" *Journal of Statistics Education* 19, no. 2 (2011).

Freund, John E., and Benjamin M. Perles. "A New Look at Quartiles of Ungrouped Data." *The American Statistician* 41, no. 3 (August 1987), pp. 200–203.

Loftus, Stephen. *Basic Statistics with R: Reaching Decisions with Data*. 1st ed. Elsevier, 2021.

Pukelsheim, Friedrich. "The Three Sigma Rule." *The American Statistician* 48, no. 2 (May 1994), pp. 88–91.

Verzani, John. *Using R for Introductory Statistics*. 2nd ed. Chapman-Hall, 2014.

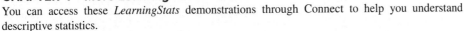

CHAPTER 4   More Learning Resources

You can access these *LearningStats* demonstrations through Connect to help you understand descriptive statistics.

Topic	LearningStats Demonstrations
Overview	Describing Data Using MegaStat Using Minitab Using R
Descriptive statistics	Basic Statistics Quartiles Box Plot Simulation Grouped Data Significant Digits
ScreenCam Tutorials	Using MegaStat Excel Descriptive Statistics Excel Scatter Plots

Key: = PowerPoint = Excel = PDF = ScreenCam Tutorials

Software Supplement

Descriptive Statistics Using Megastat

You can obtain descriptive statistics (and more) from MegaStat, as illustrated in Figure 4.31. Click the Add-Ins tab on the top menu, and then click on the MegaStat icon (left side of the top menu in this example). On the list of MegaStat procedures, click Descriptive Statistics. On the new menu, enter the data range (in this case C4:C37) in the Input range field (or highlight the data block on the worksheet). MegaStat offers you various statistics and visual displays, including a dot plot and stem-and-leaf. Compare Excel and MegaStat to see similarities and differences in their interfaces and results.

Figure 4.31

MegaStat's Descriptive Statistics JDPower

Source: MegaStat.

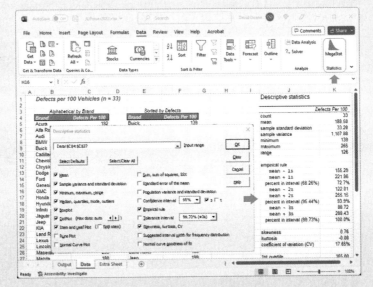

MegaStat output normally appears on a separate worksheet, but these results have been copied to the data worksheet so that you can see everything on one screen. Because we checked the Empirical Rule option, MegaStat has tabulated the sample frequencies for the J.D. Power data within each interval ($\bar{x} \pm 1s$, $\bar{x} \pm 2s$, $\bar{x} \pm 3s$) based on z-scores.

Descriptive Statistics Using Minitab

A quick way to obtain some basic descriptive statistics in Minitab is to click on Stat > Basic Statistics > Graphical Summary. If you need more detail, choose Display Descriptive Statistics. Figure 4.32 shows Minitab's menus and its graphical summary for the J.D. Power data. Note that Minitab (like MegaStat) likes to round off its decimals to improve legibility.

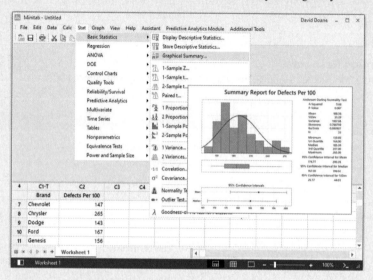

Figure 4.32

Minitab's Basic Statistics > Graphical Summary.

Source: Minitab.

Software Supplement

Descriptive Statistics Using R

R offers many functions to calculate descriptive statistics or make visual displays. For example:

```
> mean(x)        # calculate the mean
> mean(x,p)      # calculate trim mean p%
> median(x)      # calculate the median
> range(x)       # calculate the range
> sd(x)          # calculate the standard deviation
> summary(x)     # calculate several basic statistics at once
> hist(x)        # make a histogram
> boxplot(x)     # make a boxplot
> plot(x,y)      # make a scatter plot of y against x
```

For quartiles in R, use the function quantile(x, p, type=6) where p = .25, .50, or .75. If you have only a few observations, you can create a data vector by entering your data values separated by commas using the R concatenate command c() (see **Appendix K**). However, when you have many data values and/or many variables, you'll probably import your data from a spreadsheet. **Appendix K** explains alternative ways to import data from Excel to a data frame in R. An easy way is to highlight a data block in Excel and copy it to the clipboard (Ctrl-C in Windows or command-C in MacOS). For example, suppose we have a spreadsheet with data on 50 new vehicles. We will highlight numerical data in columns F through M including the column headings, which will become the variable names in R:

	A	B	C	D	E	F	G	H	I	J	K	L	M
1	Obs	Brand and Model	Style	Drive	Doors	HP	Engine	CityMPG	HwyMPG	Weight	Length	Width	Height
2	1	Acura RDX SH SWD 4dr SUV	SUV	AWD	4	272	2000	21	27	4026	186.7	74.8	65.7
3	2	Acura TLX sedan	Sedan	FWD	4	206	2400	23	33	3505	190.7	73.0	57.0
4	3	Audi V10 2dr quattro convertible	Other	AWD	2	602	5200	13	20	3572	174.3	76.4	49.0
5	...	...	...	...	...	...	...	...	...	...	...	...	...
6	47	Toyota Sienna XLE 4dr minivan	Other	AWD	4	296	3500	19	26	4590	200.6	78.1	70.7
7	48	Toyota Yaris L 4dr sedan	Sedan	FWD	4	106	1500	30	39	2385	171.2	66.7	58.5
8	49	Volvo XC40 T5 4dr SUV	SUV	AWD	4	248	2000	22	30	3756	174.2	75.2	65.3
9	50	VW Tiguan SE R-Line 4dr SUV	SUV	FWD	4	184	2000	22	29	3757	185.1	72.4	66.3

Source: Microsoft Excel.

We can create a data frame that we will call VehicleData:

In Windows:

```
> VehicleData = read.table(file="clipboard", sep="\t", header=TRUE)
```

In MacOS:

```
> VehicleData = read.table(pipe("pbpaste", sep="\t", header=TRUE)
```

Once the data are imported to VehicleData, we can view the data, calculate statistics, and so on. For example, to calculate the average city miles per gallon for our 50 vehicles (the column named CityMPG in the imported data), we use the function mean() and name the column in the data frame followed by $ (the dollar symbol) as follows:

> mean(VehicleData$CityMPG)
[1] 21.98

If our data frame columns have names (as they do here), we can get summary statistics for variables of interest using the summary() command. For example:

> summary(VehicleData[c("Weight","Length")])

	Weight		Length
Min.	:2385	Min.	:151.1
1st Qu.	:3356	1st Qu.	:181.9
Median	:3662	Median	:192.2
Mean	:3954	Mean	:190.9
3rd Qu.	:4661	3rd Qu.	:198.7
Max.	:5917	Max.	:231.9

Be careful of the syntax, e.g., the square brackets [ ]. You can also refer to columns in your data frame by number, separated by commas (or by a colon : instead of a comma if your columns are contiguous).

> summary(VehicleData[c(6:8)])

	HP		Engine		CityMPG
Min.	:106.0	Min.	:1400	Min.	:13.00
1st Qu.	:182.5	1st Qu.	:2000	1st Qu.	:18.00
Median	:249.0	Median	:2500	Median	:22.00
Mean	:262.3	Mean	:2898	Mean	:21.98
3rd Qu.	:308.0	3rd Qu.	:3500	3rd Qu.	:26.00
Max.	:602.0	Max.	:6200	Max.	:31.00

You can create a graph and export it from the Plots tab (lower right pane in R) to paste it into your written report. For example, we can create a simple box plot and histogram with optional labels for the axes and graph titles:

> boxplot(VehicleData$Weight, ylab="Pounds", main="Vehicle Weight")
> hist(VehicleData$Weight, xlab="Pounds", main="Vehicle Weight")

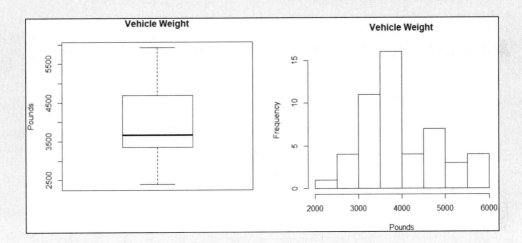

Exam Review Questions for Chapters 1–4

1. Which type of statistic (descriptive, inferential) is each of the following?

   a. Estimating the default rate on all U.S. mortgages from a random sample of 500 loans.

   b. Reporting the percentage of students in your statistics class who use Verizon.

   c. Using a sample of 50 iPhones to predict the average battery life in typical usage.

2. Which is *not* an ethical obligation of a statistician? Explain.

   a. To know and follow accepted procedures.

   b. To ensure data integrity and accurate calculations.

   c. To support the client's wishes in concluding from the data.

3. "Driving without a seat belt is not risky. I've done it for 25 years without an accident." This *best* illustrates which fallacy?

   a. Unconscious bias.

   b. Conclusion from a small sample.

   c. *Post hoc* reasoning.

4. Which data type (categorical, numerical) is each of the following?

   a. Your current credit card balance.

   b. Your college major.

   c. Your car's odometer mileage reading today.

5. Give the type of measurement (nominal, ordinal, interval, ratio) for each variable.

   a. Length of time required for a randomly chosen vehicle to cross a toll bridge.

   b. Student's ranking of five cell phone service providers.

   c. The type of charge card used by a customer (Visa, Mastercard, AmEx, Other).

6. Tell if each variable is continuous or discrete.

   a. Tonnage carried by an oil tanker at sea.

   b. Wind velocity at 7 o'clock this morning.

   c. Number of text messages you received yesterday.

7. To choose a sample of 12 students from a statistics class of 36 students, which type of sample (simple random, systematic, cluster, convenience) is each of these?

a. Picking every student who was wearing blue that day.
b. Using Excel's =RANDBETWEEN(1,36) to choose students from the class list.
c. Selecting every 3rd student starting from a randomly chosen position.

8. Which of the following is *not* a reason for sampling? Explain.

a. The destructive nature of some tests.
b. High cost of studying the entire population.
c. The expense of obtaining random numbers.

9. Which statement is *correct?* Why not the others?

a. Likert scales are interval if scale distances are meaningful.
b. Cross-sectional data are measured over time.
c. A census is always preferable to a sample.

10. Which statement is *false?* Explain.

a. Sampling error can be reduced by using appropriate data coding.
b. Selection bias means that respondents are not typical of the target population.
c. Simple random sampling requires a list of the population.

11. The management of a theme park obtained a random sample of the ages of 36 riders of its Space Adventure Simulator. (a) Make a nice histogram. (b) Did your histogram follow Sturges' Rule? If not, why not? (c) Describe the distribution of sample data. (d) Make a dot plot of the data. (e) What can be learned from each display (dot plot and histogram)?

39	46	15	38	39	47	50	61	17
40	54	36	16	18	34	42	10	16
16	13	38	14	16	56	17	18	53
24	17	12	21	8	18	13	13	10

12. Which one of the following is *true?* Why not the others?

a. Histograms are useful for visualizing correlations.
b. Pyramid charts are generally preferred to bar charts.
c. A correlation coefficient can be negative.

13. Which data would be most suitable for a pie chart? Why not the others?

a. Presidential vote in the last election by party (Democratic, Republican, Other).
b. Retail prices of six major brands of color laser printers.
c. Labor cost per vehicle for 10 major world automakers.

14. Find the mean, standard deviation, and coefficient of variation for $X = 5, 10, 20, 10, 15$.

15. Here are the ages of a random sample of 20 CEOs (chief executive officers) of *Fortune* 500 U.S. corporations. (a) Find the mean, median, and mode. (b) Discuss the advantages and disadvantages of each of these measures of center for this data set. (c) Find the quartiles and interpret them. (d) Sketch a box plot and describe it.

| 57 | 56 | 58 | 46 | 70 | 62 | 55 | 60 | 59 | 64 |
| 62 | 67 | 61 | 55 | 53 | 58 | 63 | 51 | 52 | 77 |

16. A consulting firm used a random sample of 12 CIOs (chief information officers) of large businesses to examine the relationship (if any) between salary (in thousands) and years of service in the firm. (a) Make a scatter plot and describe it. (b) Calculate a correlation coefficient and interpret it.

| Years (X) | 4 | 15 | 15 | 8 | 11 | 5 | 5 | 8 | 10 | 1 | 6 | 17 |
| Salary (Y) | 133 | 129 | 143 | 132 | 144 | 61 | 128 | 79 | 140 | 116 | 88 | 170 |

17. Which statement is *true?* Why not the others?

a. We expect the median to exceed the mean in positively skewed data.
b. The geometric mean is not possible when there are negative data values.
c. The midrange is resistant to outliers.

18. Which statement is *false?* Explain.

a. If $\mu = 52$ and $\sigma = 15$, then $X = 81$ would be an outlier.
b. If the data are from a normal population, about 68 percent of the values will be within $\mu \pm \sigma$.
c. If $\mu = 640$ and $\sigma = 128$, then the coefficient of variation is 20 percent.

19. Which is *not* a characteristic of using a log scale to display time series data? Explain.

a. A log scale helps if we are comparing changes in two-time series of dissimilar magnitude.
b. General business audiences find it easier to interpret a log scale.
c. If you display data on a log scale, equal distances represent equal ratios.

CHAPTER

5 Probability

CHAPTER LEARNING OBJECTIVES

When you finish this chapter, you should be able to

LO 5-1   Describe the sample space of a random experiment.

LO 5-2   Distinguish among the three views of probability.

LO 5-3   Apply the definitions and rules of probability.

LO 5-4   Calculate odds from given probabilities.

LO 5-5   Determine when events are independent.

LO 5-6   Apply the concepts of probability to contingency tables.

LO 5-7   Interpret a tree diagram.

LO 5-8   Use Bayes' Theorem to calculate revised probabilities.

LO 5-9   Apply counting rules to calculate possible event arrangements.

ARICAN/E+/Getty Images

You've learned that a statistic is a measurement that describes a sample data set of observations. Descriptive statistics allow us to describe a business process that we have already observed. But how will that process behave in the future? Nothing makes a businessperson more nervous than not being able to anticipate customer demand, supplier delivery dates, or their employees' output. Businesses want to be able to quantify the *uncertainty* of future events. What are the chances that revenue next month will exceed last year's average? How likely is it that our new production system will help us decrease our product defect rate? Businesses also want to understand how they can increase the chance of positive future events (increasing market share) and decrease the chance of negative future events (failing to meet forecasted sales). The field of study called *probability* allows us to understand and quantify the uncertainty about the future. We use the rules of probability to bridge the gap between what we know now and what is unknown about the future.

5.1 RANDOM EXPERIMENTS

Sample Space

A **random experiment** is an observational process whose results cannot be known in advance. For example, when a customer enters a Lexus dealership, will the customer buy a car or not? How much will the customer spend? The set of all possible *outcomes* (denoted S) is the **sample space** for the experiment. Some sample spaces can be enumerated easily, while others may be immense or impossible to enumerate. For example, when Citibank makes a consumer loan, we might define a sample space with only two outcomes:

$$S = \{\text{default, no default}\}$$

The sample space describing a Walmart customer's payment method might have four outcomes:

$$S = \{\text{cash, debit card, credit card, check}\}$$

The sample space to describe rolling a die has six outcomes:

$$S = \{ \text{} \}$$

A sample space could be so large that it is impractical to enumerate the possibilities (e.g., the 12-digit UPC bar code on a product in your local Walmart could have 1 trillion values). If the outcome of the experiment is a *continuous* measurement, the sample space cannot be listed, but it can be described by a rule. For example, the sample space for the length of a randomly chosen cell phone call would be

$$S = \{\text{all } X \text{ such that } X \geq 0\}$$

and the sample space to describe a randomly chosen student's GPA would be

$$S = \{\text{all } X \text{ such that } 0.00 \leq X \leq 4.00\}$$

Similarly, some *discrete* measurements are best described by a rule. For example, the sample space for the number of hits on a YouTube website on a given day is

$$S = \{\text{all } X \text{ such that } X = 0, 1, 2, \ldots\}$$

Event

An **event** is any subset of outcomes in the sample space. A **simple event**, or *elementary event*, is a single outcome. A discrete sample space S consists of all the simple events, denoted $E_1, E_2, \ldots, E_n$.

(5.1) $$S = \{E_1, E_2, \ldots, E_n\}$$

Consider the random experiment of tossing a balanced coin. The sample space for this experiment would be $S = \{\text{heads, tails}\}$. The chance of observing a head is the same as the chance of observing a tail. We say that these two elementary events are *equally likely*. When you buy a lottery ticket, the sample space $S = \{\text{win, lose}\}$ also has two elementary events; however, these events are not equally likely.

Simple events are the building blocks from which we can define a **compound event** consisting of two or more simple events. For example, when shopping for jazz music on Amazon Prime, there are 19 categories from which one might choose: $S = \{\text{Acid, Avant Garde, Bebop, Brazilian, Cool, Dixieland, European, Fusion, Jam, Jive, Latin, Post-bop, New Orleans, Orchestral, Smooth, Soul, Swing, Ragtime, Vocal}\}$. Within this sample space, we could define compound events "contemporary" as $A = \{\text{Acid, Avant Garde, Cool, Fusion, Jam, Post-bop}\}$ and "traditional" as $B = \{\text{Bebop, Brazilian, Dixieland, European, Jive, Latin, New Orleans, Orchestral, Smooth, Soul, Swing, Ragtime, Vocal}\}$.

Section Exercises

Mc Graw Hill **connect**

5.1 A credit card customer at Barnes & Noble can use Visa (*V*), Mastercard (*M*), or American Express (*A*). The merchandise may be books (*B*), electronic media (*E*), or other (*O*). (a) Enumerate the elementary events in the sample space describing a customer's purchase. (b) Would each elementary event be equally likely? Explain.

5.2 A survey asked tax accounting firms their business form (*S* = sole proprietorship, *P* = partnership, *C* = corporation) and type of risk insurance they carry (*L* = liability only, *T* = property loss only, *B* = both liability and property). (a) Enumerate the elementary events in the sample space. (b) Would each elementary event be equally likely? Explain.

5.3 A baseball player bats either left-handed (*L*) or right-handed (*R*). The player either gets on base (*B*) or does not get on base (*B'*). (a) Enumerate the elementary events in the sample space. (b) Would these elementary events be equally likely? Explain.

5.4 A die is thrown (1, 2, 3, 4, 5, 6), and a coin is tossed (*H, T*). (a) Enumerate the elementary events in the sample space for the die/coin combination. (b) Are the elementary events equally likely? Explain.

5.5 Each undergraduate student at Streeling University is enrolled in one of five colleges: arts & sciences (*A*), business (*B*), engineering (*E*), health sciences (*H*), or music (*M*). The student is either a freshman (*F*), sophomore (*S*), junior (*J*), or senior (*R*). (a) Enumerate the elementary events for this sample space. (b) Are the events equally likely? Explain.

5.6 Sal's pizza has pepperoni, veggie, margarita, and custom pizza options. There are small, medium, and large sizes. (a) Enumerate the elementary events for this sample space. (b) Are the events equally likely? Explain.

5.2 PROBABILITY

LO 5-2

Distinguish among the three views of probability.

The concept of probability is so familiar to most people that it can easily be misused. Therefore, we begin with some precise definitions and a few rules.

Definitions

The **probability** of an event is a number that measures the relative likelihood that the event will occur. The probability of an event A, denoted $P(A)$, must lie within the interval from 0 to 1:

$$0 \leq P(A) \leq 1 \tag{5.2}$$

$P(A) = 0$ means the event cannot occur (e.g., a naturalized citizen becoming president of the United States), while $P(A) = 1$ means the event is certain to occur (e.g., rain occurring in Hilo, Hawaii, sometime this year). In a discrete sample space, the probabilities of all simple events must sum to 1 because it is certain that one of them will occur:

$$P(S) = P(E_1) + P(E_2) + \cdots + P(E_n) = 1 \tag{5.3}$$

For example, if 32 percent of purchases are made by credit card, 15 percent by debit card, 35 percent by cash, and 18 percent by check, then

$$P(\text{credit card}) + P(\text{debit card}) + P(\text{cash}) + P(\text{check}) = .32 + .15 + .35 + .18 = 1$$

What Is "Probability"?

There are three distinct ways of assigning probability, listed in Table 5.1. Many people mix them up or use them interchangeably; however, each approach must be considered separately.

Table 5.1

Three Views of Probability

Approach	How Assigned?	Example
Empirical	Estimated from observed outcome frequency	There is a 3.2 percent chance of twins in a randomly chosen birth.
Classical	Known *a priori* by the nature of the experiment	There is a 50 percent chance of heads on a coin flip.
Subjective	Based on informed opinion or judgment	There is a 60 percent chance that Toronto will bid for the 2030 Winter Olympics.

Empirical Approach

Empirical data are data that are collected through observation. We can use the **empirical** or **relative frequency approach** to assign probabilities by dividing the frequency of observed outcomes (f) in our experimental sample space by the number of observations (n). The estimated probability is f/n. For example, we could estimate the reliability of a bar code scanner as

$$P(\text{a missed scan}) = \frac{\text{number of missed scans}}{\text{number of items scanned}}$$

or the default rate on student loans as

$$P(\text{a student defaults}) = \frac{\text{number of defaults}}{\text{number of loans}}$$

As we increase the number of times we perform the experiment (n), our estimate will become more and more accurate. We use the ratio f/n to represent the probability. Here are some examples of empirical probabilities:

- An industrial components manufacturer interviewed 280 production workers before hiring 70 of them.

H = event that a randomly chosen interviewee is hired

$$P(H) = f/n = \frac{70}{280} = .25$$

- Over 20 years, a medical malpractice insurer saw only one claim for "wrong-site" surgery (e.g., amputating the wrong limb) in 112,994 malpractice claims.

M = event that malpractice claim is for wrong-site surgery

$$P(M) = f/n = \frac{1}{112,994} = .00000885$$

- On average, 2,118 out of 100,000 Americans live to age 100 or older.

C = event that a randomly chosen American lives to 100 or older

$$P(C) = f/n = \frac{2,118}{100,000} = .02118$$

An important probability theorem is the **law of large numbers**, which says that as the number of trials increases, any empirical probability approaches its theoretical limit. Imagine flipping a coin 50 times. You know that the proportion of heads should be near .50. But in any finite sample it will probably not be .50 but something like 7 out of 13 (.5385) or 28 out of 60 (.4667). Coin flip experiments show that a large n may be needed before the proportion gets really close to .50, as illustrated in Figure 5.1.

A common variation on the law of large numbers says that as the sample size increases, there is an increased probability that any event (even an unlikely one) will occur. For example, the same five winning numbers [4, 21, 23, 34, 39] came up twice in a three-day period in the North Carolina Cash-5 game.[1] Was something wrong with the lottery? The odds on this event are indeed small (191,919 to 1), but in a large number of drawings, an event like this becomes less surprising. Similarly, the odds that one ticket will win the lottery are low; yet the probability is high that *someone* will win the lottery because so many people buy lottery tickets. Gamblers tend to misconstrue this principle as implying that a streak of bad luck is "bound to change" if they just keep betting. Unfortunately, the necessary number of tries is likely to exhaust the gambler's bankroll.

Law of Large Numbers

The law of large numbers states that as the number of trials of a random experiment increases, the average of the outcomes will approach the expected value.

You may know that **actuarial science** is a high-paying career that involves estimating empirical probabilities. Actuaries help companies calculate payout rates on life insurance, pension plans, and health care plans. Actuaries created the tables that guide IRA withdrawal rates for individuals from age 70 to 99. Here are a few challenges that actuaries face:

- Is n "large enough" to say that f/n has become a good approximation to the probability of the event of interest? Data collection costs money, and decisions must be made. The sample should be large enough but not larger than necessary for a given level of precision.

---

[1] See Leonard A. Stefanski, "The North Carolina Lottery Coincidence," *The American Statistician* 62, no. 2 (May 2008), pp. 130–134.

Figure 5.1

Law of Large Numbers Illustrated 📁 **CoinFlips**

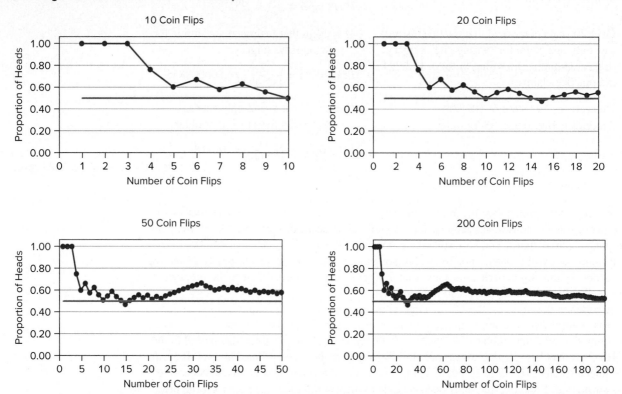

- Was the experiment repeated identically? Subtle variations may exist in the experimental conditions and data collection procedures.
- Is the underlying process stable over time? For example, default rates on 2020 student loans may not apply in 2025 due to changes in attitudes and interest rates.
- Do nonstatistical factors override data collection? Clinical trials of a drug should last long enough to ascertain its adverse side effects, yet ethical considerations forbid withholding a drug that could be beneficial. Given the severity of the health risks, Moderna's COVID-19 vaccine was approved by the FDA under an Emergency Use Authorization (EUA) to stem the spread of the dangerous SARS-CoV II virus.
- What if repeated trials are impossible? A good example occurred when Lloyd's of London was asked to insure a traveling exhibition of Monet paintings that was sent on a tour of the United States. Such an event only occurs once, so we have no *f/n* to help us.

Classical Approach

Statisticians use the term *a priori* to refer to the process of assigning probabilities before we actually observe the event or try an experiment. When flipping a coin or rolling a pair of dice, we do not actually have to perform an experiment because the nature of the process allows us to envision the entire sample space. Instead, we can use deduction to determine $P(A)$. This is the **classical approach** to probability. For example, in the sample space for the two-dice experiment shown in Figure 5.2, there are 36 possible outcomes. Rolling a seven is shown by the shaded outcomes below.

$$P(\text{rolling a seven}) = \frac{\text{number of possible outcomes with 7 dots}}{\text{number of outcomes in sample space}} = \frac{6}{36} = .1667$$

The probability is obtained *a priori* without actually doing an experiment. We can apply pure reason to cards, lottery numbers, and roulette. Also, in some physical situations, we can assume that the probability of an event such as a defect (leak, blemish) occurring in a particular unit of area, volume, or length is proportional to the ratio of that unit's size to the total area, volume, or length. Examples would be pits on rolled steel, stress fractures in concrete, or leaks in pipelines. These are *a priori* or *classical* probabilities if they are based on logic or theory, not on experience. Such calculations are rarely possible in business situations.

(1,1)	(1,2)	(1,3)	(1,4)	(1,5)	**(1,6)**
(2,1)	(2,2)	(2,3)	(2,4)	**(2,5)**	(2,6)
(3,1)	(3,2)	(3,3)	**(3,4)**	(3,5)	(3,6)
(4,1)	(4,2)	**(4,3)**	(4,4)	(4,5)	(4,6)
(5,1)	**(5,2)**	(5,3)	(5,4)	(5,5)	(5,6)
(6,1)	(6,2)	(6,3)	(6,4)	(6,5)	(6,6)

Figure 5.2

Sample Space for Rolling Two Dice 🗎 **DiceRolls**

Subjective Approach

A *subjective* probability reflects someone's informed judgment about the likelihood of an event. The **subjective approach** to probability is needed when there is no repeatable random experiment. For example:

- What is the probability that Ford's new supplier of plastic fasteners will be able to meet the September 23 shipment deadline?
- What is the probability that a new truck product program will show a return on investment of at least 10 percent?
- What is the probability that the price of Ford's stock will rise within the next 30 days?

In such cases, we rely on personal judgment or expert opinion. However, such a judgment is not random because it is typically based on experience with similar events and knowledge of the underlying causal processes. Assessing the New York Knicks' chances of an NBA title next year would be an example. Thus, subjective probabilities have something in common with empirical probabilities, although their empirical basis is informal and not quantified.

Instructions for Exercises 5.7–5.16: Which kind of probability is it (empirical, classical, subjective)?

Section Exercises

Mc Graw Hill **connect**

5.7 There is a 75 percent chance that Streeling University's fall enrollment will exceed last fall's.

5.8 Twelve percent of students will ace the first test in this class.

5.9 There is a 20 percent chance that a new stock offered in an initial public offering (IPO) will reach or exceed its target price on the first day.

5.10 There is a 25 percent chance that AT&T Wireless and Verizon will merge.

5.11 Commercial rocket launches have a 95 percent success rate.

5.12 The probability of rolling three sevens in a row with dice is .0046.

5.13 The probability that a randomly selected student in your class is celebrating a birthday today is 1/365.

5.14 More than 30 percent of the results from major search engines for the keyword phrase "ring tone" are fake pages created by spammers.

5.15 Based on the reported experience of climbers from a given year, a climber who attempts Everest has a 31 percent chance of success.

5.16 An entrepreneur who plans to open a Cuban restaurant in Nashville has a 20 percent chance of success.

LO 5-3

Apply the definitions and rules of probability.

5.3 RULES OF PROBABILITY

The field of probability has a distinct vocabulary that is important to understand. This section reviews the definitions of probability terms and illustrates how to use them.

Complement of an Event

The **complement** of an event A is denoted A' and consists of everything in the sample space S except event A, as illustrated in the **Venn diagram** in Figure 5.3.

Since A and A' together comprise the sample space, their probabilities sum to 1:

$$P(A) + P(A') = 1 \qquad (5.4)$$

The probability of the complement of A is found by subtracting the probability of A from 1:

$$P(A') = 1 - P(A) \qquad (5.5)$$

For example, if 33 percent of all new small businesses fail within the first 2 years, then the probability that a new small business will survive at least 2 years is

$$P(\text{survival}) = 1 - P(\text{failure}) = 1 - .33 = .67, \text{ or } 67\%$$

Figure 5.3

Complement of Event A

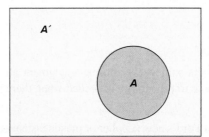

Union or Intersection?

The **union** of two events consists of all outcomes in the sample space S that are either in event A or in event B or in both. The union of A and B is sometimes denoted $A \cup B$ or "A or B," as illustrated on the left side of Figure 5.4. The symbol $\cup$ may be read "or" because it means that either or both events occur. For example, when we choose a card at random from a deck of playing cards, if Q is the event that we draw a queen and R is the event that we draw a red card, $Q \cup R$ consists of getting *either* a queen (4 possibilities in 52) *or* a red card (26 possibilities in 52) or *both* a queen and a red card (2 possibilities in 52).

Figure 5.4

Union and Intersection of Two Events

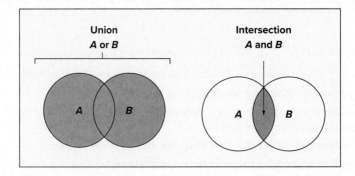

The **intersection** of two events A and B is the event consisting of all outcomes in the sample space S that are contained in both event A and event B. The intersection of A and B is denoted $A \cap B$ or "A and B," as illustrated on the right side of Figure 5.4. The probability of $A \cap B$ is called the **joint probability** and is denoted $P(A \cap B)$. The symbol $\cap$ may be read "and" because the intersection means that both events occur. For example, if Q is the event that we draw a queen and R is the event that we draw a red card, then $Q \cap R$ is the event that we get a card that is both a queen and red. That is, the intersection of sets Q and R consists of two cards ($Q^\heartsuit$ and $Q^\diamondsuit$).

General Law of Addition

The **general law of addition** says that the probability of the union of two events A and B is the sum of their probabilities less the probability of their intersection.

General Law of Addition

(5.6)
$$P(A \cup B) = P(A) + P(B) - P(A \cap B)$$

The rationale for this formula is apparent from an examination of Figure 5.4. If we just add the probabilities of A and B, we would count the intersection twice, so we must subtract the probability of $A \cap B$ to avoid overstating the probability of $A \cup B$. For the card example:

Queen: $P(Q) = 4/52$ (there are 4 queens in a deck)

Red: $P(R) = 26/52$ (there are 26 red cards in a deck)

Queen and Red: $P(Q \cap R) = 2/52$ (there are 2 red queens in a deck)

Therefore,

Queen or Red: $P(Q \cup R) = P(Q) + P(R) - P(Q \cap R)$

$$= 4/52 + 26/52 - 2/52$$

$$= 28/52 = .5385, \text{ or a } 53.85\% \text{ chance}$$

A survey of introductory statistics students showed that 29.7 percent have AT&T wireless service (event A), 73.4 percent have a Visa card (event B), and 20.3 percent have both (event $A \cap B$). The probability that a student uses AT&T *or* has a Visa card is

$$P(A \cup B) = P(A) + P(B) - P(A \cap B) = .297 + .734 - .203 = .828$$

Example 5.1

Cell Phones and Credit Cards 🗀 **WebSurvey**

Mutually Exclusive Events

Events A and B are **mutually exclusive** (or **disjoint**) if their intersection is the **empty set** (a set that contains no elements). In other words, one event precludes the other from occurring. The empty set is often called the null set and is denoted ϕ.

(5.7)
$$\text{If } A \cap B = \phi, \text{ then } P(A \cap B) = 0$$

As illustrated in Figure 5.5, the probability of $A \cap B$ is zero when the events do not overlap. For example, if A is the event that an Applebee's customer finishes her lunch in less than 30 minutes and B is the event that she takes 30 minutes or more, then $P(A \cap B) = P(\phi) = 0$. Here are examples of events that are mutually exclusive (cannot be in both categories):

- *Customer age:* A = under 21, B = over 65
- *Purebred dog breed:* A = border collie, B = golden retriever
- *Business form:* A = corporation, B = sole proprietorship

Figure 5.5

Mutually Exclusive Events

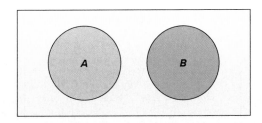

These events may not cover all of the possibilities (e.g., a business could also be a partnership or S-corporation). The only issue is whether the categories overlap. Here are examples of events that are *not* mutually exclusive (can be in both categories):

- *Student's major:* A = marketing major, B = economics major
- *Bank account:* A = Bank of America, B = JPMorgan Chase Bank
- *Credit card held:* A = Visa, B = Mastercard, C = American Express

Special Law of Addition

If *A* and *B* are mutually exclusive events, then $P(A \cap B) = 0$ and the general addition law can be simplified to the sum of the individual probabilities for *A* and *B*, the **special law of addition**.

> ### Special Law of Addition
>
> $$P(A \cup B) = P(A) + P(B) \quad \text{(addition law for mutually exclusive events)} \qquad \textbf{(5.8)}$$

For example, if we look at a person's age, then $P(\text{under } 21) = .28$ and $P(\text{over } 65) = .12$, so $P(\text{under } 21 \text{ or over } 65) = .28 + .12 = .40$ because these events do not overlap.

Collectively Exhaustive Events

Events are **collectively exhaustive** if their union is the entire sample space *S* (i.e., all the events that can possibly occur). Two mutually exclusive, collectively exhaustive events are **binary events**. For example, a car repair either is covered by the warranty (*A*) or is not covered by the warranty (*A'*). There can be more than two mutually exclusive, collectively exhaustive events. For example, Google maps gives you directions by car (*C*), public transit (*P*), walking (*W*), car hire (*T*), or bike (*B*).

Conditional Probability

The probability of event *A* *given* that event *B* has occurred is a **conditional probability**, denoted $P(A \mid B)$, which is read "the probability of *A* given *B*." The vertical line is read as "given." The conditional probability is the joint probability of *A* and *B* divided by the probability of *B*.

$$P(A \mid B) = \frac{P(A \cap B)}{P(B)} \quad \text{for} \quad P(B) > 0 \qquad \textbf{(5.9)}$$

The logic of formula 5.9 is apparent by looking at the Venn diagram in Figure 5.6. The sample space is restricted to *B*, an event that we know has occurred (the green shaded circle). The intersection, $A \cap B$, is the part of *B* that is also in *A* (the blue shaded area). The ratio of the relative size of set ($A \cap B$) to set *B* is the conditional probability $P(A \mid B)$.

Figure 5.6

Conditional Probability

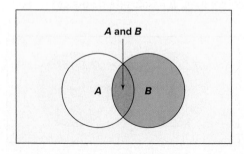

Example 5.2

High School Dropouts

Of the population age 16–21 and not in college, 13.50 percent are unemployed, 29.05 percent are high school dropouts, and 5.32 percent are unemployed high school dropouts. What is the conditional probability that a member of this population is unemployed, given that the person is a high school dropout? To answer this question, define

U = the event that the person is unemployed

D = the event that the person is a high school dropout

This "story problem" contains three facts:

$$P(U) = .1350 \qquad P(D) = .2905 \qquad P(U \cap D) = .0532$$

So by formula 5.9, the conditional probability of an unemployed youth given that the person dropped out of high school is

$$P(U \mid D) = \frac{P(U \cap D)}{P(D)} = \frac{.0532}{.2905} = .1831, \text{ or } 18.31\%$$

The *conditional* probability of being unemployed is $P(U \mid D) = .1831$ (18.31 percent), which is greater than the *unconditional probability* of being unemployed $P(U) = .1350$ (13.50 percent). In other words, knowing that someone is a high school dropout alters the probability that the person is unemployed.

Using algebra, we can rewrite formula 5.9, resulting in a **general law of multiplication** for the joint probability of two events.

General Law of Multiplication

(5.10) $$P(A \cap B) = P(A \mid B)P(B)$$

Example 5.3

Videos on Demand

Last year, 70 percent of new movies were available as video on demand (VOD) within 14 days of their release. Of those movies, 50 percent were released as TV pay-per-view (as opposed through a subscription streaming service). What is the joint probability that a new movie was available as VOD within 14 days of release for TV pay-per-view? To answer this question, define

V = the event that a new movie was available as VOD within 14 days of release
T = the event that a new movie was available as TV pay-per-view

This "story problem" contains two facts:

$$P(V) = .70 \qquad P(T \mid V) = .50$$

By formula 5.10, the joint probability that a new movie was available as VOD within 14 days of release and as TV pay-per-view is

$$P(V \cap T) = P(T \mid V)\, P(V) = .50 \times .70 = .35$$

Therefore, 35 percent of new movies last year were available as video on demand within 14 days of their release and as TV pay-per-view.

LO 5-4

Calculate odds from given probabilities.

Odds of an Event

Statisticians usually speak of probabilities rather than odds, but in sports and games of chance, we often hear **odds** quoted. We define the *odds in favor* of an event A as the ratio of the probability that event A will occur to the probability that event A will not occur. Its reciprocal is the *odds against* event A.

Odds in *favor* of A:

$$\frac{P(A)}{P(A')} = \frac{P(A)}{1 - P(A)}$$

Odds *against* A:

$$\frac{P(A')}{P(A)} = \frac{1 - P(A)}{P(A)}$$

If a probability is expressed as a percentage, you can easily convert it to odds. For example, suppose the IRS tax audit rate is 1.41 percent among taxpayers earning between $100,000 and $199,999. Let A = the event that the taxpayer is audited and set $P(A) = .0141$. The odds against an audit are

$$\frac{P(\text{no audit})}{P(\text{audit})} = \frac{1 - P(A)}{P(A)} = \frac{1 - .0141}{.0141} = 70 \text{ to } 1 \text{ } against \text{ being audited}$$

In horse racing and other sports, odds usually are quoted *against* winning. If the odds against event A are quoted as b to a, then the implied probability of event A is:

$$P(A) = \frac{a}{a + b}$$

For example, if a race horse has 4 to 1 odds *against* winning, this is equivalent to saying that the odds-makers assign the horse a 20 percent chance of winning:

$$P(\text{win}) = \frac{a}{a + b} = \frac{1}{1 + 4} = \frac{1}{5} = .20, \text{ or } 20\%$$

Section Exercises

connect

5.17 Are these characteristics of a student at your university mutually exclusive or not? Explain.

    a. A = works 20 hours or more, B = majoring in accounting

    b. A = born in the United States, B = born in Canada

    c. A = owns a Toyota, B = owns a Honda

5.18 Are these events collectively exhaustive or not? Explain.

    a. A = college grad, B = some college, C = no college

    b. A = born in the United States, B = born in Canada, C = born in Mexico

    c. A = full-time student, B = part-time student, C = not enrolled as a student

5.19 Given $P(A) = .40$, $P(B) = .50$, and $P(A \cap B) = .05$, find (a) $P(A \cup B)$, (b) $P(A \mid B)$, and (c) $P(B \mid A)$.

5.20 Given $P(A) = .70$, $P(B) = .30$, and $P(A \cap B) = .00$, find (a) $P(A \cup B)$, (b) $P(A \mid B)$, and (c) $P(B \mid A)$.

5.21 Given $P(A) = .70$, $P(B \mid A) = .40$, and $P(B) = .30$, find (a) $P(A \cap B)$, (b) $P(A \cup B)$, and (c) $P(A \mid B)$.

5.22 Given $P(A) = .40$, $P(B \mid A) = .20$, and $P(B) = .20$, find (a) $P(A \cap B)$, (b) $P(A \cup B)$, and (c) $P(A \mid B)$.

5.23 Sixty-two percent of freshmen are taking a foreign language course. Of those freshmen taking a foreign language course, 80 percent are taking Spanish. What is the probability that a randomly selected freshman is taking Spanish as a foreign language?

5.24 On any given day in the summer, 54 percent of guests at Vail Resorts are visiting from outside Colorado. Twenty-two percent of guests are visiting from outside Colorado and are visiting Vail for the first time. What is the probability that a guest visiting from outside the state of Colorado is visiting for the first time?

5.25 Currently Samsung ships 21.7 percent of the organic light-emitting diode (OLED) displays in the world. Let S be the event that a randomly selected OLED display was made by Samsung. Find (a) $P(S)$, (b) $P(S')$, (c) the odds *in favor* of event S, and (d) the odds *against* event S.

5.26 The probability of an IRS audit is 1.7 percent for U.S. taxpayers who file form 1040 and who earned $100,000 or more. (a) What are the odds that such a taxpayer will be audited? (b) What are the odds *against* such a taxpayer being audited?

5.27 Let S be the event that a randomly chosen female aged 18–24 is a smoker. Let C be the event that a randomly chosen female aged 18–24 is a Caucasian. Given $P(S) = .246$, $P(C) = .830$, and $P(S \cap C) = .232$, find each probability.

a. $P(S')$.
b. $P(S \cup C)$.
c. $P(S \mid C)$.
d. $P(S \mid C')$.

5.28 Let C be the event that a randomly chosen adult has some college education. Let M be the event that a randomly chosen adult is married. Given $P(C) = .4$, $P(M) = .5$, and $P(C \cap M) = .24$, find each probability.

a. $P(C')$.
b. $P(C \cup M)$.
c. $P(M \mid C)$.
d. $P(C \mid M)$.

Analytics in Action

Climate Change

How can local governments, financial institutions, and businesses assess risks of climate change? Heat, rainfall, and flooding impact crop yields, costs of air-conditioning, construction plans, and even business location. Climate risks can affect credit ratings (e.g., Moody's) when firms or governments seek to borrow for growth or capital improvements. Analytics firms (e.g., 427mt.com and jupiterintel.com) are springing up to offer climate intelligence services for clients such as governments, pension funds, and banks. Some individual firms are beginning to train their in-house staff to improve forecasts of impacts of climate change. These tech start-ups seek to encourage data-driven preparation for climate change. (See *The Economist* 433, no. 9170 (November 23, 2019), pp. 68–69.)

5.4 INDEPENDENT EVENTS

LO 5-5

Determine when events are independent.

When $P(A)$ differs from $P(A \mid B)$, the events are **dependent**. You can easily think of examples of dependence. For example, cell phone text messaging is more common among young people, while arteriosclerosis is more common among older people. Therefore, knowing a person's age would affect the *probability* that the individual uses text messaging or has arteriosclerosis. Dependent events may be causally related, but statistical dependence does *not* prove cause and effect. It only means that knowing that event B has occurred will affect the *probability* that event A will occur.

When knowing that event B has occurred does *not* affect the probability that event A will occur, then events A and B are **independent**. In other words, event A is independent of event B if the conditional probability $P(A \mid B)$ is the same as the unconditional probability $P(A)$; that is, if the probability of event A is the same whether event B occurs or not. For example, if text messaging among high school students is *independent* of gender, this means that knowing whether a student is a male or female does not *change* the probability that the student uses text messaging.

Independent Events

(5.11) Event A is independent of event B if and only if $P(A \mid B) = P(A)$.

If A and B are independent events, we can simplify the general law of multiplication (formula 5.10) to show that the joint probability of events A and B is the product of their individual probabilities. Recall the general law of multiplication:

$$P(A \cap B) = P(A \mid B)P(B)$$

If A and B are independent, then we can substitute $P(A)$ for $P(A \mid B)$. The result is shown in formula 5.12, the **special law of multiplication**.

Special Law of Multiplication

If events A and B are independent, then

$$P(A \cap B) = P(A)P(B) \qquad\qquad (5.12)$$

Example 5.4

Restaurant Orders

Based on past data, the probability that a customer at a certain Noodles & Company restaurant will order a dessert (event D) with the meal is .08. The probability that a customer will order a bottled beverage (event B) is .14. The joint probability that a customer will order both a dessert *and* a bottled beverage is .0112. Is ordering a dessert independent of ordering a bottled beverage?

$$P(D) = .08 \qquad P(B) = .14 \qquad P(D \cap B) = .0112$$
$$P(D) \times P(B) = .08 \times .14 = .0112 = P(D \cap B)$$

We see that $P(D \cap B) = P(D) \times P(B)$, so D and B are independent of each other. If we know that a customer has ordered a bottled beverage, does this information change the probability that the customer will also order a dessert? No, because the events are independent.

Example 5.5

Television Ads

The target audience is 2,000,000 viewers. Ad A reaches 500,000 viewers, ad B reaches 300,000 viewers, and both ads reach 100,000 viewers. That is:

$$P(A) = \frac{500{,}000}{2{,}000{,}000} = .25 \quad P(B) = \frac{300{,}000}{2{,}000{,}000} = .15 \quad P(A \cap B) = \frac{100{,}000}{2{,}000{,}000} = .05$$

Applying the definition of conditional probability from formula 5.9, the conditional probability that ad A reaches a viewer *given* that ad B reaches the viewer is

$$P(A \mid B) = \frac{P(A \cap B)}{P(B)} = \frac{.05}{.15} = .3333, \text{ or } 33.3\%$$

We see that A and B are not independent because $P(A) = .25$ is not equal to $P(A \mid B) = .3333$. That is, knowing that ad B reached the viewer raises the probability that ad A reached the viewer from $P(A) = .25$ to $P(A \mid B) = .3333$. Alternatively, because $P(A)P(B) = (.25)(.15) = .0375$ is not equal to $P(A \cap B) = .05$, we know that events A and B are not independent.

Applications of the Multiplication Law

The probability of more than two independent events occurring simultaneously is the product of their separate probabilities, as shown in formula 5.13 for n independent events $A_1, A_2, \ldots, A_n$.

$$P(A_1 \cap A_2 \cap \ldots \cap A_n) = P(A_1)P(A_2) \ldots P(A_n) \quad \text{if the events are independent} \quad (5.13)$$

The special multiplication law for independent events can be applied to system reliability. To illustrate, suppose a website has two independent file servers (i.e., no shared power or other components). Each server has 99 percent reliability (i.e., is "up" 99 percent of the time). What is the total system reliability? Let F_1 be the event that server 1 fails and F_2 be the event that server 2 fails. Then

$$P(F_1) = 1 - 0.99 = .01$$
$$P(F_2) = 1 - 0.99 = .01$$

Applying the rule of independence:

$$P(F_1 \cap F_2) = P(F_1)P(F_2) = (.01)(.01) = .0001$$

The probability that at least one server is up is 1 minus the probability that both servers are down, or $1 - .0001 = .9999$. Dual file servers dramatically improve reliability to 99.99 percent.

When individual components have a low reliability, high reliability can still be achieved with massive redundancy. For example, an experimental HP supercomputer called Teramac had over 7 million components. About 3 percent were defective (nanodevices are extremely difficult to manufacture). Yet, programs could run reliably because there was significant redundancy in the interconnect circuitry, so a valid path could almost always be found (see *Scientific American* 293, no. 5 [November 2005], p. 75).

Example 5.6

Space Shuttle

Redundancy can increase system reliability even when individual component reliability is low. For example, the NASA space shuttle has three flight computers. Suppose that they function independently but that each has an unacceptable .03 chance of failure (3 failures in 100 missions). Let F_j = event that computer j fails. Then

$$
\begin{aligned}
P(\text{all 3 fail}) &= P(F_1 \cap F_2 \cap F_3) \\
&= P(F_1)P(F_2)P(F_3) \quad \text{(presuming that failures are independent)} \\
&= (.03)(.03)(.03) \\
&= .000027, \text{ or 27 in 1,000,000 missions.}
\end{aligned}
$$

Triple redundancy can reduce the probability of computer failure to .000027 (27 failures in 1,000,000 missions). Of course, in practice, it is very difficult to have truly independent computers because they may share electrical buses or cables. On one shuttle mission, two of the three computers actually did fail, which proved the value of redundancy. Another example of space shuttle redundancy is the four independent fuel gauges that prevent the shuttle's main engines from shutting down too soon. Initial launch rules allowed the shuttle to fly as long as two of them were functional, but after the *Challenger* launch explosion, the rules were modified to require that 3 of 4 be functional, and they were modified again after the *Columbia* accident to require that all 4 be functional. (See https://history.nasa.gov/sts511.html.)

The Five Nines Rule

How high must reliability be? Prime business customers expect public carrier-class telecommunications data links to be available 99.999 percent of the time. This so-called five nines rule implies only 5 minutes of downtime per year. Such high reliability is needed not only in telecommunications but also for mission-critical systems such as airline reservation systems or banking funds transfers. Table 5.2 shows some expected system reliabilities in contemporary applications.

Suppose a certain network web server is up only 94 percent of the time (i.e., its probability of being down is .06). How many independent servers are needed to ensure that the system

Table 5.2

Typical System Reliabilities in Various Applications

Type of System	Typical Reliability (%)
Commercial fiber-optic cable systems	99.999
Cellular-radio base stations with mobile switches connected to public-switched telephone networks	99.99
Private-enterprise networking (e.g., connecting two company offices)	99.9
Airline luggage systems	99

is up at least 99.99 percent of the time? This is equivalent to requiring that the probability of all the servers being down is .0001 (i.e., $1 - .9999$). Four servers will accomplish the goal[2]:

2 servers: $P(F_1 \cap F_2) = (.06)(.06) = .0036$
3 servers: $P(F_1 \cap F_2 \cap F_3) = (.06)(.06)(.06) = .000216$
4 servers: $P(F_1 \cap F_2 \cap F_3 \cap F_4) = (.06)(.06)(.06)(.06) = .00001296$

Applications of Redundancy

The principle of redundancy is found in many places. Basketball teams have more than five players, even though only five can play at once. You set two alarm clocks in case the first doesn't wake you up. The Embraer Legacy 13-passenger jet ($21.2 million) has two identical generators on each of its two engines to allow the plane to be used as a commercial regional jet (requiring 99.5 percent dispatch reliability) as well as for private corporate travel. With four generators, plus an auxiliary power unit that can be started and run in flight, the Legacy can fly even after the failure of a generator or two (*Flying* 131, no. 9 [September 2004], p. 50).

Older airliners (e.g., Boeing 747) had four engines, not only because older engine designs were less powerful but also because they were less reliable. Particularly for transoceanic flights, four-engine planes could fly even if one engine failed (or maybe even two). Modern airliners (e.g., Boeing 777) have only two engines because newer engines are more powerful and more reliable.

It is not just a matter of individual component reliability but also of cost and consequence. Cars have only one battery because the consequence of battery failure (walking home or calling AAA) does not justify the expense of having a backup battery. But spare tires are cheap enough that most cars carry one (maybe two, if you are driving in Alaska, even though they may never be used.).

Redundancy is not required when components are highly reliable, cost per component is high, and consequences of system failure are tolerable (e.g., cell phone, alarm clock). Unfortunately, true component independence is difficult to achieve. The same catastrophe (fire, flood, etc.) that damages one component may well damage the backup system. On August 24, 2001, a twin-engine Air Transat Airbus A330 transiting the Atlantic Ocean did have a double-engine shutdown with 293 passengers aboard. Fortunately, the pilot was able to glide 85 miles to a landing in the Azores, resulting in only minor injuries.

Dependent Events in Business

Banks and credit unions know that the probability that a customer will default on a car loan is dependent on the customer's past record of unpaid credit obligations. That is why lenders consult credit bureaus (e.g., Equifax, Experian, and TransUnion) before they make a loan. Your credit score is based on factors such as the ratio of your credit card balance to your credit limit, length of your credit history, number of accounts with balances, and frequency of requests for credit. Your score can be compared with actuarial data and national averages to see what percentile you are in. The lender can then decide whether your loan is worth the risk.

Automobile insurance companies (e.g., AAA, Allstate, State Farm) know that the probability that a driver will be involved in an accident depends on the driver's age, past traffic

---

[2]In general, if p is the probability of failure, we can set $p^k = .0001$, plug in $p = .06$, take the log of both sides, and solve for k. In this case, $k = 3.27$, so we can then round up to the next higher integer.

convictions, and similar factors. This actuarial information is used in deciding whether to accept you as a new customer and in setting your insurance premium. The situation is similar for life insurance. Can you think of factors that might affect a person's life insurance premium?

In each of these loan and insurance examples, knowing B will affect our estimate of the likelihood of A. Obviously, bankers and insurance companies need to quantify these conditional probabilities precisely. An *actuary* studies conditional probabilities empirically, using accident statistics, mortality tables, and insurance claims records. Although few people undergo the extensive training to become actuaries, many businesses rely on actuarial services, so a business student needs to understand the concepts of conditional probability and statistical independence.

5.29 Given $P(J) = .26$, $P(K) = .48$. If J and K are independent, find $P(J \cup K)$.

5.30 Given $P(A) = .40$, $P(B) = .50$. If A and B are independent, find $P(A \cap B)$.

5.31 Given $P(A) = .40$, $P(B) = .50$, and $P(A \cap B) = .05$. (a) Find $P(A \mid B)$. (b) In this problem, are A and B independent?

5.32 Which pairs of events are independent?
   a. $P(A) = .60$, $P(B) = .40$, $P(A \cap B) = .24$.
   b. $P(A) = .90$, $P(B) = .20$, $P(A \cap B) = .18$.
   c. $P(A) = .50$, $P(B) = .70$, $P(A \cap B) = .25$.

5.33 Given $P(J) = .2$, $P(K) = .4$, and $P(J \cap K) = .15$. (a) Find $P(J \mid K)$. (b) In this problem, are J and K independent?

5.34 Which pairs of events are independent?
   a. $P(J) = .50$, $P(K) = .40$, $P(J \cap K) = .3$.
   b. $P(J) = .60$, $P(K) = .20$, $P(J \cap K) = .12$.
   c. $P(J) = .15$, $P(K) = .5$, $P(J \cap K) = .1$.

5.35 The probability that a student has a Visa card (event V) is .73. The probability that a student has a Mastercard (event M) is .18. The probability that a student has both cards is .03. (a) Find the probability that a student has either a Visa card or a Mastercard. (b) In this problem, are V and M independent? Explain.

5.36 Bob sets two alarm clocks (battery-powered) to be sure he arises for his Monday 8:00 a.m. accounting exam. There is a 75 percent chance that either clock will wake Bob. (a) What is the probability that Bob will oversleep? (b) If Bob had three clocks, would he have at least a 99 percent chance of waking up?

5.37 A hospital's backup power system has three independent emergency electrical generators, each with uptime averaging 95 percent (some downtime is necessary for maintenance). Any of the generators can handle the hospital's power needs. Does the overall reliability of the backup power system meet the five nines test?

5.38 Over 1,000 people try to climb Mt. Everest every year. Of those who try to climb Everest, 31 percent succeed. The probability that a climber is at least 60 years old is .04. The probability that a climber is at least 60 years old and succeeds in climbing Everest is .005. (a) Find the probability of success, given that a climber is at least 60 years old. (b) Is success in climbing Everest independent of age?

5.39 Suppose 50 percent of the customers at Pizza Palooza order a square pizza, 80 percent order a soft drink, and 40 percent order both a square pizza and a soft drink. Is ordering a soft drink independent of ordering a square pizza? Explain.

5.5 CONTINGENCY TABLES

LO 5-6

Apply the concepts of probability to contingency tables.

In Chapter 3, you saw how Excel's pivot tables can be used to display the frequency of co-occurrence of data values (e.g., how many taxpayers in a sample are filing as "single" and also have at least one child?). Because a probability usually is estimated as a relative frequency, we can use tables of frequency data to learn about relationships (e.g., dependent events or conditional probabilities) that are useful in business planning. Frequency data for the table are typically responses from a survey or summary qualitative data from an empirical study.

What Is a Contingency Table?

A contingency table is a cross-tabulation of frequencies into rows and columns. The intersection of each row and column is a cell that shows a count of data values that lie in each cell. A contingency table is similar to a frequency distribution for a single variable, except that it has two variables (rows and columns). A contingency table with r rows and c columns has rc cells and is called an $r \times c$ table. For example, Table 5.3 shows a cross-tabulation of MBA program type versus the level of college engagement of its graduates for a large business college at a public university. The empirical study was based on a sample of 360 MBA alumni. Engagement level is measured by the MBA alumni's number of contacts with the college (e.g., classroom speaking, student mentoring, or case competition judging).

Table 5.3

Contingency Table of Frequencies ($n = 360$ alumni) 📖 Alumni

| | Engagement Level | | | |
MBA Program	Low (E_1)	Medium (E_2)	High (E_3)	Row Total
Full Time (M_1)	210	20	5	235
Part Time (M_2)	64	17	7	88
Executive (M_3)	6	16	15	37
Column Total	280	53	27	360

From this contingency table, we can calculate relative frequencies to answer questions such as "What is the probability that an MBA graduate is engaged at the lowest level?" or "What percent of our MBA alumni are from our part-time program and are medium level engagers?" or "What is the chance that an executive MBA alumnus is engaged at the highest level?"

Marginal Probabilities

The marginal probability of an event is a relative frequency that is found by dividing a row or column total by the total sample size. For example, using the column totals, 280 out of 360 alumni were engaged at the lowest level. The marginal probability of a low engagement level is $P(E_1) = 280/360 = .7778$. In other words, about 77.78 percent of the MBA alumni were engaged at the lowest level. These values are highlighted in Table 5.4.

Table 5.4

Marginal Probability of Event E_1

| | Engagement Level | | | |
MBA Program	Low (E_1)	Medium (E_2)	High (E_3)	Row Total
Full Time (M_1)	210	20	5	235
Part Time (M_2)	64	17	7	88
Executive (M_3)	6	16	15	37
Column Total	280	53	27	360

Similarly, using the *row* totals, we see that 88 of the 360 engaged alumni were graduates from the part-time program, so the marginal probability of a part-time alumnus is $P(M_2) = 88/360 = .2444$, or a 24.44 percent chance.

Joint Probabilities

A joint probability represents the intersection of two events. For example, the middle cell highlighted in Table 5.5 is the joint event that a graduate is from the part-time MBA program (M_2) and is also engaged at a medium level (E_2). We can write this probability either as $P(M_2$ and $E_2)$ or as $P(M_2 \cap E_2)$. Because 17 out of 360 alumni are in this category, we calculate the joint probability as $P(M_2 \cap E_2) = 17/360 = .0472$. In other words, there is less than a 5 percent chance that an MBA alumnus is both a graduate of the part-time program and engaged at a medium level.

Table 5.5

Joint Probability of Event $M_2 \cap E_2$

MBA Program	Engagement Level			Row Total
	Low (E_1)	Medium (E_2)	High (E_3)	
Full Time (M_1)	210	20	5	235
Part Time (M_2)	64	17	7	88
Executive (M_3)	6	16	15	37
Column Total	280	53	27	360

Table 5.5

Joint Probability of Event $M_2 \cap E_2$

Conditional Probabilities

If we restrict ourselves to a single row or column (the condition), we can calculate conditional probabilities. For example, if we restrict ourselves to the 37 executive alumni in the third row, we can calculate the **conditional probability** that the MBA graduate is engaged at a high level (E_3) given that the alumnus is an executive MBA (M_3). These cells are highlighted in Table 5.6. This conditional probability may be written $P(E_3 \mid M_3)$. There were 15 highly engaged alumni out of the 37 executive MBAs, so $P(E_3 \mid M_3) = 15/37 = .4054$. There is about a 40 percent chance that an executive MBA will be highly engaged.

MBA Program	Engagement Level			Row Total
	Low (E_1)	Medium (E_2)	High (E_3)	
Full Time (M_1)	210	20	5	235
Part Time (M_2)	64	17	7	88
Executive (M_3)	6	16	15	37
Column Total	280	53	27	360

Table 5.6

Conditional Probability of Event $E_3 \mid M_3$

Independence

To check whether events in a contingency table are independent, we can look at conditional probabilities. For example, if a high level of engagement (E_3) were independent of being an alumnus from the executive MBA program (M_3), then the *conditional* probability $P(E_3 \mid M_3)$ would be the same as the *marginal* probability $P(E_3)$. But this is not the case here:

Conditional Probability *Marginal Probability*

$P(E_3 \mid M_3) = 15/37 = .4054$ $P(E_3) = 27/360 = .0750$

Thus, a high level of involvement (E_3) is not independent of being a graduate of the executive MBA program (M_3). Graduates of the executive program are much more likely to be highly engaged with the school. We have shown that these are *dependent* events.

There is another way to show that E_3 and M_3 are not independent events. If two events are independent, their joint probability would be the product of their marginal probabilities. But this is not the case here:

Joint Probability *Product of Marginal Probabilities*

$P(E_3 \text{ and } M_3) = 15/360 = .0417$ $P(E_3)P(M_3) = (27/360)(37/360) = .0077$

Therefore, using an alternative approach, we verify that high engagement (E_3) is not independent of executive MBA graduation (M_3).

Relative Frequencies

To facilitate certain kinds of probability calculations, we divide each cell frequency f_{ij} by the total sample size ($n = 360$) to get the relative frequencies shown in Table 5.7. For example, the upper left-hand cell becomes $210/360 = .5833$. Each cell is a *joint* probability. The cells sum to 1.0000 because these are all the possible event combinations.

Table 5.7

Relative Frequencies: Cell Frequencies Divided by Sample Size

MBA Program	Engagement Level			Row Total
	Low (E_1)	Medium (E_2)	High (E_3)	
Full Time (M_1)	.5833	.0556	.0139	.6528
Part Time (M_2)	.1778	.0472	.0194	.2444
Executive (M_3)	.0167	.0444	.0417	.1028
Column Total	.7778	.1472	.0750	1.0000

Summing the *joint* probabilities across a row or down a column gives the *marginal* probability for that row or column. You can also verify that the marginal row or column probabilities sum to 1.0000.

Example 5.7

Payment Method and Purchase Quantity 📂 **Payment**

A small grocery store would like to know if the number of items purchased by a customer is independent of the type of payment method the customer chooses to use. Having this information can help the store manager determine how to set up the various checkout lanes. The manager collected a random sample of 368 customer transactions. The results are shown in Table 5.8.

Table 5.8

Contingency Table for Payment Method by Number of Items Purchased

Number of Items Purchased	Payment Method			Row Total
	Cash	Check	Credit/Debit Card	
5 or Fewer	30	15	43	88
6 to 9	46	23	66	135
10 to 19	31	15	43	89
20 or More	19	10	27	56
Column Total	126	63	179	368

Looking at the frequency data presented in the table, we can calculate the marginal probability that a customer will use cash to make the payment. Let C be the event that the customer chose cash as the payment method.

$$P(C) = \frac{126}{368} = .3424$$

Is $P(C)$ the same if we condition on number of items purchased?

$$P(C \mid 5 \text{ or fewer}) = \frac{30}{88} = .3409 \quad P(C \mid 6 \text{ to } 9) = \frac{46}{135} = .3407$$

$$P(C \mid 10 \text{ to } 19) = \frac{31}{89} = .3483 \quad P(C \mid 20 \text{ or more}) = \frac{19}{56} = .3393$$

Notice that there is little difference in these probabilities. If we perform the same type of analysis for the next two payment methods, we find that *payment method* and *number of items purchased* are essentially independent. Based on this study, the manager might decide to offer a cash-only checkout lane that is *not* restricted to the number of items purchased.

Mini Case 5.1

Business Ownership: Gender and Company Size

Table 5.9 is a contingency table showing the number of employer businesses in the United States based on ownership gender and number of employees. A business is considered female-owned if 51 percent or more of the ownership is female. An employer-based business is one in which the business has a payroll.

Table 5.9

Business Ownership by Gender and Number of Employees Ownership

Ownership	Number of Employees			Total
	1 to 4	5 to 99	100+	
Female-Owned (F)	507,400	294,700	7,700	809,800
Male- or Equally Owned (M)	3,110,364	1,908,813	101,155	5,120,332
Total	3,617,764	2,203,513	108,855	5,930,132

Source: www.census.gov/econ.

Marginal probabilities can be found by calculating the ratio of the column or row totals to the grand total. We see that the $P(F) = 809,800/5,930,132 = .1366$. In other words, 13.66 percent of employer businesses are female-owned. Likewise, $P(1 \text{ to } 4 \text{ employees}) = 3,617,764/5,930,132 = .6101$, so 61.01 percent of employer businesses have only 1 to 4 employees on their payroll.

Conditional probabilities can be found from Table 5.9 by restricting ourselves to a single row or column (the *condition*). For example, for businesses with 100 or more employees:

$P(F \mid 100+) = 7,700/108,855 = .0707$ There is a 7.07 percent chance that a business is female-owned *given* that the business has 100 or more employees.

$P(M \mid 100+) = 101,155/108,855 = .9293$ There is a 92.93 percent chance that a business is male- or equally owned *given* that the business has 100 or more employees.

These conditional probabilities show that large businesses are much less likely to be female-owned than male- or equally owned $(.0707 < .9293)$. We also can conclude that the chance of a large business being female-owned is about half that when considering employer businesses of all sizes $(.0707 < .1366)$.

Table 5.10 shows the *relative frequency* obtained by dividing each table frequency by the grand total $(n = 5,930,132)$.

Table 5.10

Business Ownership by Gender and Number of Employees

Ownership	Number of Employees			Total
	1 to 4	5 to 99	100+	
Female-Owned (F)	.0856	.0497	.0013	0.1366
Male- or Equally Owned (M)	.5245	.3219	.0171	0.8634
Total	.6101	.3716	.0184	1.0000

For example, the joint probability $P(F \cap 1 \text{ to } 4 \text{ employees}) = .0856$ (i.e., about 8.56 percent of employer businesses are female-owned with fewer than 5 employees). The six joint probabilities sum to 1.0000, as they should (except for rounding).

Section Exercises

5.40   The contingency table below shows the results of a survey of video viewing habits by age. Find the following probabilities or percentages: 📓 **Videos**

   a.   Probability that a viewer is aged 18–34.
   b.   Probability that a viewer prefers watching videos on a TV screen.
   c.   Percentage of viewers who are 18–34 and prefer videos on a mobile or laptop device.
   d.   Percentage of viewers aged 18–34 who prefer videos on a mobile or laptop device.
   e.   Percentage of viewers who are 35–54 or prefer videos on a mobile or laptop device.

Viewer Age	Video Viewing Platform Preferred		Row Total
	Mobile/Laptop Device	TV Screen	
18–34	39	30	69
35–54	10	10	20
55+	3	8	11
Column Total	52	48	100

5.41   The contingency table below summarizes a survey of 1,000 bottled beverage consumers. Find the following probabilities or percentages: 📓 **Recycling**

   a.   Probability that a consumer recycles beverage bottles.
   b.   Probability that a consumer who lives in a state with a deposit law does not recycle.
   c.   Percentage of consumers who recycle and live in a state with a deposit law.
   d.   Percentage of consumers in states with a deposit law who recycle.

	Lives in a State with a Deposit Law	Lives in a State with No Deposit Law	Row Total
Recycles Beverage Bottles	154	186	340
Does Not Recycle Beverage Bottles	66	594	660
Column Total	220	780	1,000

5.42   A survey of 158 introductory statistics students showed the following contingency table. Find each event probability. 📓 **WebSurvey**

   a. $P(V)$
   b. $P(A)$
   c. $P(A \cap V)$
   d. $P(A \cup V)$
   e. $P(A \mid V)$
   f. $P(V \mid A)$

Cell Phone Provider	Visa Card (V)	No Visa Card (V′)	Row Total
AT&T (A)	32	15	47
Other (A′)	84	27	111
Column Total	116	42	158

5.43 A survey of 156 introductory statistics students showed the following contingency table. Find each event probability. 📂 **WebSurvey**

a. $P(D)$
b. $P(R)$
c. $P(D \cap R)$
d. $P(D \cup R)$
e. $P(R \mid D)$
f. $P(R \mid P)$

Newspaper Read	Living Where?			Row Total
	Dorm (D)	Parents (P)	Apt (A)	
Never (N)	13	6	6	25
Occasionally (O)	58	30	21	109
Regularly (R)	8	7	7	22
Column Total	79	43	34	156

5.44 This contingency table describes 200 business students. Find each probability and interpret it in words. 📂 **GenderMajor**

a. $P(A)$
b. $P(M)$
c. $P(A \cap M)$
d. $P(F \cap S)$
e. $P(A \mid M)$
f. $P(A \mid F)$
g. $P(F \mid S)$
h. $P(E \cup F)$

Gender	Major			Row Total
	Accounting (A)	Economics (E)	Statistics (S)	
Female (F)	44	30	24	98
Male (M)	56	30	16	102
Column Total	100	60	40	200

5.45 Based on the previous problem, is major independent of gender? Explain the basis for your conclusion.

5.46 The following contingency table shows average yield (rows) and average duration (columns) for 38 bond funds. For a randomly chosen bond fund, find the probability that: 📂 **Duration**

a. The bond fund is of long duration.
b. The bond fund has high yield.
c. The bond fund has high yield given that it is of short duration.
d. The bond fund is of short duration given that it has high yield.

Yield	Average Portfolio Duration			Row Total
	Short (D_1)	Intermediate (D_2)	Long (D_3)	
Low (Y_1)	8	2	0	10
Medium (Y_2)	1	6	6	13
High (Y_3)	2	4	9	15
Column Total	11	12	15	38

LO 5-7

Interpret a tree diagram.

5.6 TREE DIAGRAMS

What Is a Tree?

Events and probabilities can be displayed in the form of a **tree diagram** or *decision tree* to help visualize all possible outcomes. This is a common business planning activity. We begin with a contingency table. Table 5.11 shows a cross-tabulation of expense ratios (low, medium, high) by fund type (bond, stock) for a sample of 21 bond funds and 23 stock funds. For purposes of this analysis, a fund's expense ratio is defined as "low" if it is in the lowest $\frac{1}{3}$ of the sample, "medium" if it is in the middle $\frac{1}{3}$ of the sample, and "high" if it is in the upper $\frac{1}{3}$ of the sample.

Table 5.11

Frequency Tabulation of Expense Ratios by Fund Type 📁 BondFund

Expense Ratio	Fund Type		Row Total
	Bond Fund (*B*)	Stock Fund (*S*)	
Low (*L*)	11	3	14
Medium (*M*)	7	9	16
High (*H*)	3	11	14
Column Total	21	23	44

To label the tree, we need to calculate *conditional probabilities.* Table 5.12 shows conditional probabilities by fund type (i.e., dividing each cell frequency by its column total). For example, $P(L \mid B) = 11/21 = .5238$. This says there is about a 52 percent chance that a fund has a low expense ratio if it is a bond fund. In contrast, $P(L \mid S) = 3/23 = .1304$. This says there is about a 13 percent chance that a fund has a low expense ratio if it is a stock fund.

Table 5.12

Conditional Probabilities by Fund Type 📁 BondFund

Expense Ratio	Fund Type	
	Bond Fund (*B*)	Stock Fund (*S*)
Low (*L*)	$= P(L \mid B) = 11/21 = 0.5238$	$= P(L \mid S) = 3/23 = 0.1304$
Medium (*M*)	$= P(M \mid B) = 7/21 = 0.3333$	$= P(M \mid S) = 9/23 = 0.3913$
High (*H*)	$= P(H \mid B) = 3/21 = 0.1429$	$= P(H \mid S) = 11/23 = 0.4783$
Column Total	1.0000	1.0000

The tree diagram in Figure 5.7 shows all events along with their marginal, conditional, and joint probabilities. To illustrate the calculation of joint probabilities, we make a slight modification of the formula for conditional probability for events *A* and *B:*

$$P(A \mid B) = \frac{P(A \cap B)}{P(B)} \qquad \text{or} \qquad P(A \cap B) = P(B)P(A \mid B)$$

Figure 5.7

Tree Diagram for Fund Type and Expense Ratios BondFund

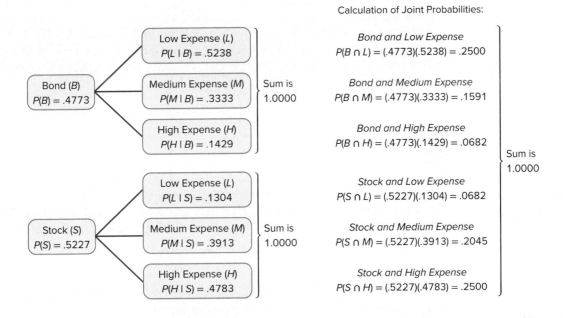

Calculation of Joint Probabilities:

Thus, the joint probability of each terminal event on the tree can be obtained by multiplying the probabilities along its branches. For example, following the top branch of the tree, the joint probability of a bond fund (B) with low expenses (L) is

$$P(B \cap L) = P(B)P(L \mid B) = (.4773)(.5238) = .2500$$

The conditional probabilities sum to 1 *within* each branch, and the joint probabilities also sum to 1 *for all six terminal events.*

Section Exercises

connect

5.47 Of grocery shoppers who have a shopping cart, 70 percent pay by credit/debit card (event C_1), 20 percent pay cash (event C_2), and 10 percent pay by check (event C_3). Of shoppers without a shopping cart, 50 percent pay by credit/debit card (event C_1), 40 percent pay cash (event C_2), and 10 percent pay by check (event C_3). On Saturday morning, 80 percent of the shoppers take a shopping cart (event S_1) and 20 percent do not (event S_2). (a) Sketch a tree based on these data. (b) Calculate the probability of all joint probabilities (e.g., $S_1 \cap C_1$). (c) Verify that the joint probabilities sum to 1.

5.48 A study showed that 60 percent of *The Wall Street Journal* subscribers watch CNBC every day. Of these, 70 percent watch it outside the home. Only 20 percent of those who don't watch CNBC every day watch it outside the home. Let D be the event "watches CNBC daily" and O be the event "watches CNBC outside the home." (a) Sketch a tree based on these data. (b) Calculate the probability of all joint probabilities (e.g., $D \cap O$). (c) Verify that the joint probabilities sum to 1.

Analytics in Action

Can Amazon Read Your Mind?

You go to Amazon.com to search for a Roomba robot vacuum for your apartment. Amazon returns over 3,000 results from your search (new, used, various vendors and models). But Amazon also recommends that you consider buying an iTouchless automatic stainless steel trashcan or Tossits car trash bags. How did it decide on these suggestions? The answer is that Amazon has a matrix (like an Excel pivot table) that keeps track of the frequency of *copurchased* items (e.g., books, music, DVDs) for web shoppers. Probabilities derived from the cells in this contingency table are used to recommend products that are likely to be of interest to you, assuming that you are "like" other buyers. While such predictions of your behavior are only probabilistic, even a modest chance of landing extra sales can make a difference in bottom-line profit. There are even more sophisticated logic engines that can track your web clicks. Is this an invasion of your privacy? Does it bother you to think that you may be predictable? Interestingly, many consumers don't seem to mind and actually find value in this kind of statistical information system.

LO 5-8

Use Bayes' Theorem to calculate revised probabilities.

5.7 BAYES' THEOREM

An important theorem published by Thomas Bayes (1702–1761) provides a method of revising probabilities to reflect new information. The **prior** (unconditional) **probability** of an event B is revised after event A has occurred to yield a **posterior** (conditional) **probability**. We begin with a formula slightly different from the standard definition of conditional probability:

$$P(B \mid A) = \frac{P(A \mid B)P(B)}{P(A)} \qquad (5.14)$$

Unfortunately, in some situations $P(A)$ is not given. The most useful and common form of **Bayes' Theorem** replaces $P(A)$ with an expanded formula:

$$P(B \mid A) = \frac{P(A \mid B)P(B)}{P(A \mid B)P(B) + P(A \mid B')P(B')} \qquad (5.15)$$

How Bayes' Theorem Works

Bayes' Theorem is best understood by example. Suppose that 10 percent of the women who purchase over-the-counter pregnancy testing kits are actually pregnant. For a particular brand of kit, if a woman is pregnant, the test will yield a positive result 96 percent of the time and a negative result 4 percent of the time (called a "false negative"). If she is not pregnant, the test will yield a positive result 5 percent of the time (called a "false positive") and a negative result 95 percent of the time. Suppose the test comes up positive. What is the probability that she is really pregnant?

We can solve this problem intuitively. If 1,000 women use this test, the results should look like Figure 5.8. Of the 1,000 women, 100 will actually be pregnant and 900 will not. The test yields 4 percent false negatives (.04 × 100 = 4) and 5 percent false positives (.05 × 900 = 45). Therefore, of the 141 women who will test positive (96 + 45), only 96 will actually be pregnant, so P(pregnant | positive test) = 96/141 = .6809.

Figure 5.8

Bayes' Theorem—Intuitive Method for 1,000 Women

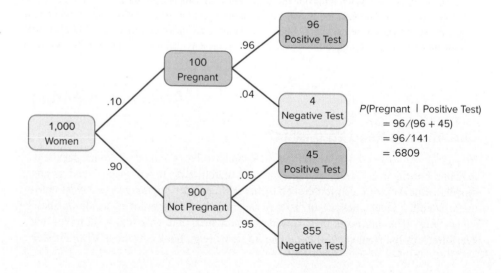

If the test is positive, there is a 68.09 percent chance that a woman is actually pregnant. A common error is to think that the probability should be 96 percent. But that is the probability of a positive test *if* the woman is pregnant, while many of the women who take the test are *not* pregnant. This intuitive calculation (assuming a large number of individuals) also can be arranged in a contingency like Table 5.13.

Table 5.13

Bayes' Theorem—Intuitive Method for 1,000 Women

	Positive Test	Negative Test	Row Total
Pregnant	96	4	100
Not Pregnant	45	855	900
Column Total	141	859	1,000

Bayes' Theorem allows us to derive this result more formally. We define these events:

A = positive test B = pregnant

A' = negative test B' = not pregnant

The given facts may be stated as follows:

$$P(A \mid B) = .96 \qquad P(A \mid B') = .05 \quad P(B) = .10$$

The complement of each event is found by subtracting from 1:

$$P(A' \mid B) = .04 \qquad P(A' \mid B') = .95 \quad P(B') = .90$$

Applying Bayes' Theorem:

$$P(B \mid A) = \frac{P(A \mid B)P(B)}{P(A \mid B)P(B) + P(A \mid B')P(B')} = \frac{(.96)(.10)}{(.96)(.10) + (.05)(.90)}$$

$$= \frac{.096}{.096 + .045} = \frac{.096}{.141} = .6809$$

There is a 68.09 percent chance that a woman is pregnant, given that the test is positive. This is the same result that we obtained using the intuitive method.

What Bayes' Theorem does is to show us how to revise our *prior* probability of pregnancy (10 percent) to get the *posterior* probability (68.09 percent) after the results of the pregnancy test are known:

Prior (before the test)   *Posterior (after positive test result)*

$P(B) = .10$ $P(B \mid A) = .6809$

The given information did not permit a direct calculation of $P(B \mid A)$ because we only knew the conditional probabilities $P(A \mid B)$ and $P(A \mid B')$. Bayes' Theorem is useful in situations like this.

A probability tree diagram (Figure 5.9) is helpful in visualizing the situation. Only branches 1 and 3 have a positive test, and only in branch 1 is the woman actually pregnant, so $P(B \mid A) = (.096)/(.096 + .045) = .6809$. The conditional probabilities for each branch sum to 1.

The result is the same using either method. While the formal method is somewhat less intuitive, the intuitive method has the drawback that a large number of individuals must be

Figure 5.9

Tree Diagram for Home Pregnancy Test
Pregnancy

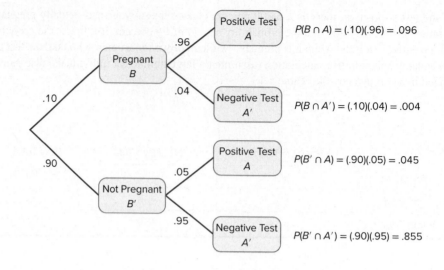

assumed if the resulting frequencies are to be integers. Use whichever method makes sense for the problem at hand.

General Form of Bayes' Theorem

A generalization of Bayes' Theorem allows event B to have as many mutually exclusive and collectively exhaustive categories as we wish $(B_1, B_2, \ldots, B_n)$ rather than just the two dichotomous categories B and B':

$$P(B_i \mid A) = \frac{P(A \mid B_i)P(B_i)}{P(A \mid B_1)P(B_1) + P(A \mid B_2)P(B_2) + \ldots + P(A \mid B_n)P(B_n)} \tag{5.16}$$

Example 5.8

Hospital Trauma Centers

Based on historical data, three hospital trauma centers have 50, 30, and 20 percent of the cases, respectively. The probability of a case resulting in a malpractice suit in each of the three hospitals is .001, .005, and .008, respectively. If a malpractice suit is filed, what is the probability that it originated in hospital 1? This problem is solved as follows. We define

Event A = event that a malpractice suit is filed by a patient
Event B_i = event that the patient was treated at trauma center i $(i = 1, 2, 3)$

The given information can be presented in a table like Table 5.14.

Table 5.14

Given Information for Hospital Trauma Centers Malpractice

Hospital	Marginal	Conditional: Suit Filed
1	$P(B_1) = 0.50$	$P(A \mid B_1) = 0.001$
2	$P(B_2) = 0.30$	$P(A \mid B_2) = 0.005$
3	$P(B_3) = 0.20$	$P(A \mid B_3) = 0.008$

Applying formula 5.16, we can find $P(B_1 \mid A)$ as follows:

$$
\begin{aligned}
P(B_1 \mid A) &= \frac{P(A \mid B_1)P(B_1)}{P(A \mid B_1)P(B_1) + P(A \mid B_2)P(B_2) + P(A \mid B_3)P(B_3)} \\
&= \frac{(.001)(.50)}{(.001)(.50) + (.005)(.30) + (.008)(.20)} \\
&= \frac{.0005}{.0005 + .0015 + .0016} = \frac{.0005}{.0036} = .1389
\end{aligned}
$$

The probability that the malpractice suit was filed in hospital 1 is .1389, or 13.89 percent. Although hospital 1 sees 50 percent of the trauma patients, it is expected to generate less than half the malpractice suits because the other two hospitals have a much higher incidence of malpractice suits.

There is nothing special about $P(B_1 \mid A)$. In fact, it is easy to calculate *all* the posterior probabilities at once by using a worksheet, as shown in Table 5.15:

$P(B_1 \mid A) = .1389$ (probability that a malpractice lawsuit originated in hospital 1)

$P(B_2 \mid A) = .4167$ (probability that a malpractice lawsuit originated in hospital 2)

$P(B_3 \mid A) = .4444$ (probability that a malpractice lawsuit originated in hospital 3)

Table 5.15

Worksheet for Bayesian Probability of Malpractice Suit 🖰 Malpractice

Hospital	Prior (Given) $P(B_i)$	Given $P(A \mid B_i)$	$P(B_i \cap A) = P(A \mid B_i)P(B_i)$	Posterior (Revised) $P(B_i \mid A) = P(B_i \cap A)/P(A)$
1	.50	.001	$(.001)(.50) = .0005$	$.0005/.0036 = .1389$
2	.30	.005	$(.005)(.30) = .0015$	$.0015/.0036 = .4167$
3	.20	.008	$(.008)(.20) = .0016$	$.0016/.0036 = \underline{.4444}$
Total	$\overline{1.00}$		$P(A) = .0036$	1.0000

We could also approach the problem intuitively by imagining 10,000 patients, as shown in Table 5.16. First, calculate each hospital's expected number of patients (50, 30, and 20 percent of 10,000). Next, find each hospital's expected number of malpractice suits by multiplying its malpractice rate by its expected number of patients:

Table 5.16

Malpractice Frequencies for 10,000 Hypothetical Patients 🖰 Malpractice

Hospital	Malpractice Suit Filed	No Malpractice Suit Filed	Total
1	5	4,995	5,000
2	15	2,985	3,000
3	16	1,984	2,000
Total	36	9,964	10,000

Hospital 1: $.001 \times 5{,}000 = 5$ (expected malpractice suits at hospital 1)

Hospital 2: $.005 \times 3{,}000 = 15$ (expected malpractice suits at hospital 2)

Hospital 3: $.008 \times 2{,}000 = 16$ (expected malpractice suits at hospital 3)

Refer to Table 5.16. Adding down, the total number of malpractice suits filed is 36. Hence, $P(B_1 \mid A) = 5/36 = .1389$, $P(B_2 \mid A) = 15/36 = .4167$, and $P(B_3 \mid A) = 16/36 = .4444$. These three probabilities add to 1. Overall, there are 36 malpractice suits, so we can also calculate $P(A) = 36/10{,}000 = .0036$. Many people find the table method easier to understand than the formulas. Do you agree?

We could visualize this situation as shown in Figure 5.10. The initial sample space consists of three mutually exclusive and collectively exhaustive events (hospitals B_1, B_2, B_3). As indicated by their relative areas, B_1 is 50 percent of the sample space, B_2 is 30 percent of the sample space, and B_3 is 20 percent of the sample space. But *given* that a malpractice case has been filed (event A), then the relevant sample space is *reduced* to that of event A.

Figure 5.10

Illustration of Hospital Trauma Center Example

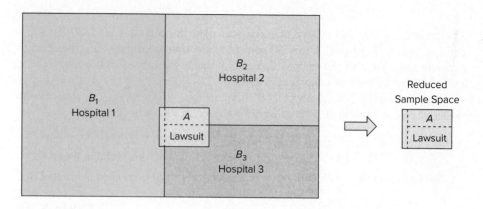

The revised (posterior) probabilities are the relative areas *within* event *A:*

$P(B_1 \mid A)$ is the proportion of *A* that lies within $B_1 = 13.89\%$

$P(B_2 \mid A)$ is the proportion of *A* that lies within $B_2 = 41.67\%$

$P(B_3 \mid A)$ is the proportion of *A* that lies within $B_3 = 44.44\%$

These percentages were calculated in Table 5.15. A worksheet is still needed to calculate $P(A)$ for the denominator.

Section Exercises

Mc Graw Hill connect

5.49 A drug test for athletes has a 5 percent false positive rate and a 10 percent false negative rate. Of the athletes tested, 4 percent have actually been using the prohibited drug. If an athlete tests positive, what is the probability that the athlete has actually been using the prohibited drug? Explain your reasoning clearly.

5.50 Half of a set of the parts are manufactured by machine *A* and half by machine *B*. Four percent of all the parts are defective. Six percent of the parts manufactured on machine *A* are defective. Find the probability that a part was manufactured on machine *A*, given that the part is defective. Explain your reasoning clearly.

5.51 An airport gamma ray luggage scanner coupled with a neural net artificial intelligence program can detect a weapon in suitcases with a false positive rate of 2 percent and a false negative rate of 2 percent. Assume a .001 probability that a suitcase contains a weapon. If a suitcase triggers the alarm, what is the probability that the suitcase contains a weapon? Explain your reasoning.

5.52 An analyst has predicted that the probability of stocks 1, 2, and 3 going down are .2, .4, and .4, respectively. The same analyst predicts the chances of stock 4 going up when each of these stocks go down are .1, .3, and .4, respectively. Given this analyst's predictions, what is the probability that stock 1 will go down if stock 4 goes up?

5.53 Thirty percent of the finance majors at Streeling University take a class in R, compared with 15 percent of the nonfinance business majors. If 20 percent of the business students are finance majors, what is the probability that a randomly selected business student taking a class in R is a finance major?

Mini Case 5.2

Bayes' Theorem and Decision Analysis

Businesses must make decisions that involve uncertainty about future events. For example, a company might want to build a new manufacturing facility in anticipation of sales growth, but the demand for its product and the amount of future sales are uncertain. If it builds the facility and demand goes down, it will have built more capacity than needed. However, if it doesn't build the new facility and demand goes up, then it will lose out on the increased revenue. Tree diagrams and Bayes' Theorem are used to model the uncertainty of future events and allow business managers to quantify how this uncertainty will affect the financial outcome of their decisions.

Baxter Inc. is a medium-sized maker of lawn furniture. Its business is doing well, and it is considering building a new manufacturing plant to handle potential increased sales next year. Will next year's demand for lawn furniture be up (U) or down (D)? Baxter's own financial staff estimates a 60 percent chance that demand for lawn furniture will be up next year. These are their *prior* probabilities.

$$P(U) = .6 \text{ and } P(D) = 1 - .6 = .4$$

But Baxter's financial staff are not experienced in macroeconomic forecasting. Baxter is considering whether to engage an expert market analyst who has wide experience in forecasting demand for consumer goods. In terms of predicting directional (+ or −) changes in demand, she has been correct 80 percent of the time. If the expert is hired, Baxter could revise its estimate, using two conditional probabilities:

$$P(\text{Expert says "Up" } | U) = .8 \text{ and } P(\text{Expert says "Down" } | D) = .8$$

Using the complement rule, Baxter also could define two more conditional probabilities:

$$P(\text{Expert says "Down" } | U) = .2 \text{ and } P(\text{Expert says "Up" } | D) = .2$$

The tree diagram in Figure 5.11 shows the marginal, conditional, and joint probabilities for these events.

What Baxter really wants to know is the chance the demand will go up *if* the expert says the demand will go up (or conversely). Bayes' Theorem could be used to update probabilities. For example:

$$P(U \mid \text{Expert says "Up"}) = \frac{P(U)P(\text{Expert says "Up" } | U)}{P(U)P(\text{Expert says "Up" } | U) + P(D)P(\text{Expert says "Up" } | D)}$$

$$= \frac{(.60)(.80)}{(.60)(.80) + (.40)(.20)} = \frac{.48}{.48 + .08} = \frac{.48}{.56} = .8571$$

From this calculation, if the expert predicts a rise in demand, the *posterior* probability of demand going up (.8571) would be greater than Baxter's *prior* probability (.6000). Therefore, the expert's added information would be valuable to Baxter.

Figure 5.11

Tree Diagram for Demand and Expert Predictions

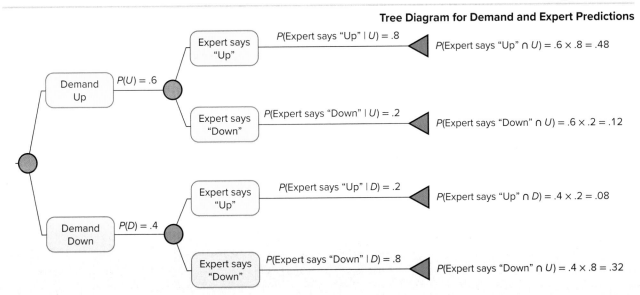

5.8 COUNTING RULES

Fundamental Rule of Counting

If event A can occur in n_1 ways and event B can occur in n_2 ways, then events A and B can occur in $n_1 \times n_2$ ways. In general, the number of ways that m events can occur is $n_1 \times n_2 \times \ldots \times n_m$.

Example 5.9

Stockkeeping Labels

How many unique stockkeeping unit (SKU) labels can a chain of hardware stores create by using two letters (ranging from *AA* to *ZZ*) followed by four numbers (digits 0 through 9)? For example:

AF1078: hex-head 6 cm bolts—box of 12

RT4855: Lime-A-Way cleaner—16 ounce

LL3119: Rust-Oleum Professional primer—gray 15 ounce

This problem may be viewed as filling six empty boxes, as shown in Figure 5.12.

Figure 5.12

Creating SKU Labels

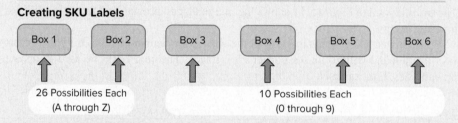

There are 26 ways (letters *A* through *Z*) to fill either the first or the second box. There are 10 ways (digits 0 through 9) to fill the third through sixth boxes. The number of unique inventory labels is therefore $26 \times 26 \times 10 \times 10 \times 10 \times 10 = 6{,}760{,}000$. Such a system should suffice for a moderately large retail store.

Example 5.10

Shirt Inventory

The number of possibilities can be large, even for a very simple counting problem. For example, the L.L.Bean men's cotton chambray shirt comes in six colors (blue, stone, rust, green, plum, indigo), five sizes (*S, M, L, XL, XXL*), and two styles (short sleeve, long sleeve). Its stock, therefore, might include $6 \times 5 \times 2 = 60$ possible shirts. The number of shirts of each type to be stocked will depend on prior demand experience. Counting the outcomes is easy with the counting formula, but even for this simple problem, a tree diagram would be impossible to fit on one page, and the enumeration of them all would be tedious (but necessary for L.L.Bean).

Factorials

The number of unique ways that n items can be arranged in a particular order is n **factorial**, the product of all integers from 1 to n.

$$n! = n(n - 1)(n - 2) \cdots 1 \tag{5.17}$$

This rule is useful for counting the possible arrangements of any n items. There are n ways to choose the first item, $n - 1$ ways to choose the second item, and so on, until we reach the last item. By definition, $0! = 1$.

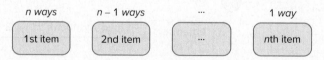

Example 5.11

Truck Routing

In very small problems, we can actually count the possibilities. For example, a home appliance service truck must make three stops (*A, B, C*). In how many ways could the three stops be arranged? There are six possible arrangements: {*ABC, ACB, BAC, BCA, CAB, CBA*}. But if all we want is the *number of possibilities* without listing them all:

$$3! = 3 \times 2 \times 1 = 6$$

Even in moderate-sized problems, listing all the possibilities is not feasible. For example, the number of possible arrangements of nine baseball players in a batting order rotation is:

$$9! = 9 \times 8 \times 7 \times 6 \times 5 \times 4 \times 3 \times 2 \times 1 = 362{,}880$$

You can calculate factorials with your calculator using the key labeled n!. Alternatively, if you have Excel, use the Excel function =FACT(n). If you have Internet access, you can enter n! in the Google search window. For example, 38! would be

Excel Function	*Google Search Window*
=FACT(38) = 5.23023E+44	38! = 5.23022612E+44

Permutations

Choose *r* items at random without replacement from a group of *n* items. In how many ways can the *r* items be arranged, treating each arrangement as a different event (i.e., treating the three-letter sequence *XYZ* as different from the three-letter sequence *ZYX*)? A **permutation** is an arrangement of the *r* sample items *in a particular order*. The number of possible permutations of *n* items taken *r* at a time is denoted $_nP_r$.

(5.18)
$$_nP_r = \frac{n!}{(n-r)!}$$

Permutations are used when we are interested in finding how many possible arrangements there are when we select *r* items from *n* items, when each possible arrangement of items is a distinct event.

Example 5.12

Appliance Service Calls

Five home appliance customers (*A, B, C, D, E*) need service calls, but the field technician can service only three of them before noon. The order in which they are serviced is important (to the customers, anyway), so each possible arrangement of three service calls is different. The dispatcher must assign the sequence. The number of possible permutations is

$$_nP_r = \frac{n!}{(n-r)!} = \frac{5!}{(5-3)!} = \frac{5 \cdot 4 \cdot 3 \cdot 2 \cdot 1}{2!} = \frac{120}{2} = 60$$

This may seem a surprisingly large number, but it can be enumerated. There are 10 distinct groups of three customers (two customers must be omitted):

ABC ABD ABE ACD ACE ADE BCD BCE BDE CDE

In turn, each group of three customers can be arranged in six possible orders. For example, the first distinct set of customers {*A, B, C*} could be arranged in six distinct ways:

ABC ACB CAB CBA BAC BCA

We could do the same for each of the other nine groups of three customers. Because there are 10 distinct groups of three customers and six possible arrangements per group, there are $10 \times 6 = 60$ permutations. Clearly, we would prefer not to enumerate sequences like this very often.

As long as n and r are not too large, you can use your calculator's permutation key labeled $_nP_r$. The equivalent Excel function is =PERMUT(n, r). For example, the number of permutations of 49 things taken 9 at a time is

Excel Function	*Calculator*
=PERMUT(49,9) = 7.45521E+14	$_{49}P_9 = 7.455208604^{14}$

Combinations

A combination is a collection of r items chosen at random without replacement from n items where the order of the selected items is *not* important (i.e., treating the three-letter sequence *XYZ* as being the same as the three-letter sequence *ZYX*). The number of possible combinations of r items chosen from n items is denoted $_nC_r$.

$$_nC_r = \frac{n!}{r!(n-r)!} \tag{5.19}$$

We use combinations when the only thing that matters is which r items are chosen, regardless of how they are arranged.

Example 5.13

Appliance Service Calls Revised

Suppose that five customers (*A, B, C, D, E*) need service calls and the maintenance worker can only service three of them this morning. The customers don't care when they are serviced as long as it's before noon, so the dispatcher does not care who is serviced first, second, or third. In other words, the dispatcher regards *ABC, ACB, BAC, BCA, CAB,* or *CBA* as being the same event because the same three customers (*A, B, C*) get serviced. The number of combinations is

$$_nC_r = \frac{n!}{r!(n-r)!} = \frac{5!}{3!(5-3)!} = \frac{5 \cdot 4 \cdot 3 \cdot 2 \cdot 1}{(3 \cdot 2 \cdot 1)(2 \cdot 1)} = \frac{120}{12} = 10$$

This is much smaller than the number of permutations in the previous example, where order was important. In fact, the possible combinations can be enumerated easily because there are only 10 distinct groups of three customers:

ABC ABD ABE ACD ACE ADE BCD BCE BDE CDE

For combinations, you can try your calculator's combination key labeled *nCr*. Alternatively, use the Excel function =COMBIN(n,r) or enter n choose r in the Google search window. For example, the number of combinations of 52 things taken 24 at a time is

Excel Function	*Calculator*	*Google Search Window*
=COMBIN(52,24) = 4.26385E+14	$_{52}C_{24} = 4.26384982^{14}$	52 choose 24 = 4.2638498E+14

Permutations or Combinations?

Permutations and combinations both calculate the number of ways we could choose r items from n items. But in permutations, *order is important*, while in combinations, *order does not matter*. The number of permutations $_nP_r$ always is at least as great as the number of combinations $_nC_r$ in a sample of r items chosen at random from n items.

Section Exercises

5.54 (a) Find 8! without a calculator. Show your work. (b) Use your calculator to find 32!. (c) Find 32! by typing "32!" in the Google search window. (d) Which method would you use most often? Why?

5.55 (a) Find $_{20}C_5$ without a calculator. Show your work. (b) Use your calculator to find $_{20}C_5$. (c) Find $_{20}C_5$ by entering "20 choose 5" in the Google search window. (d) Which method would you use most often? Why?

5.56   In the Minnesota Northstar Cash Drawing, you pick five different numbers between 1 and 31. What is the probability of picking the winning combination (order does not matter)? *Hint:* Count how many ways you could pick the first number, the second number, and so on, and then divide by the number of permutations of the five numbers.

5.57   American Express Business Travel uses a six-letter record locator number (RLN) for each client's trip (e.g., KEZLFS). (a) How many different RLNs can be created using capital letters (A–Z)? (b) What if it allows any mixture of capital letters (A–Z) and digits (0–9)? (c) What if it allows capital letters and digits but excludes the digits 0 and 1 and the letters O and I because they look too much alike?

5.58   At Oxnard University, a student ID consists of two letters (26 possibilities) followed by four digits (10 possibilities). (a) How many unique student IDs can be created? (b) Would one letter followed by three digits suffice for a university with 40,000 students? (c) Why is extra capacity in student IDs a good idea?

5.59   Until 2005, the UPC bar code had 12 digits (0–9). The first six digits represent the manufacturer, the next five represent the product, and the last is a check digit. (a) How many different manufacturers could be encoded? (b) How many different products could be encoded? (c) In 2005, the EAN bar code replaced the UPC bar code, adding a 13th digit. If this new digit is used for product identification, how many different products could now be encoded?

5.60   Bob has to study for four final exams: accounting (*A*), biology (*B*), communications (*C*), and drama (*D*). (a) If he studies one subject at a time, in how many different ways could he arrange them? (b) List the possible arrangements in the sample space.

5.61   (a) In how many ways could you arrange seven books on a shelf? (b) Would it be feasible to list the possible arrangements?

5.62   Find the following permutations $_nP_r$:

a. $n = 8$ and $r = 3$.
b. $n = 8$ and $r = 5$.
c. $n = 8$ and $r = 1$.
d. $n = 8$ and $r = 8$.

5.63   Find the following combinations $_nC_r$:

a. $n = 8$ and $r = 3$.
b. $n = 8$ and $r = 5$.
c. $n = 8$ and $r = 1$.
d. $n = 8$ and $r = 8$.

5.64   A real estate office has 10 sales agents. Each of four new customers must be assigned an agent. (a) Find the number of agent arrangements where order *is* important. (b) Find the number of agent arrangements where order is *not* important. (c) Why is the number of combinations smaller than the number of permutations?

Chapter Summary

The **sample space** for a **random experiment** contains all possible outcomes. **Simple events** in a **discrete** sample space can be enumerated, while outcomes of a **continuous** sample space can only be described by a rule. An **empirical** probability is based on relative frequencies, a **classical** probability can be deduced from the nature of the experiment, and a **subjective** probability is based on judgment. An event's **complement** is every outcome except the event. The **odds** are the ratio of an event's probability to the probability of its complement. The **union** of two events is all outcomes in either or both, while the intersection is only those events in both. **Mutually exclusive** events cannot both occur, and **collectively exhaustive** events cover all possibilities. The **conditional probability** of an event is its probability given that another event has occurred. Two events are **independent** if the conditional probability of one is the same as its **unconditional** probability. The **joint probability** of independent events is the product of their probabilities. A **contingency table** is a cross-tabulation of frequencies for two variables with categorical outcomes and can be used to calculate probabilities. A **tree** visualizes events in a sequential diagram. **Bayes' Theorem** shows how to revise a **prior probability** to obtain a **conditional** or **posterior probability** when another event's occurrence is known. The number of arrangements of sampled items drawn from a population is found with the formula for **permutations** (if order is important) or **combinations** (if order does not matter).

Key Terms

actuarial science	dependent	joint probability	redundancy
Bayes' Theorem	disjoint	law of large numbers	relative frequency approach
binary events	empirical approach	marginal probability	sample space
classical approach	empty set	mutually exclusive	simple event
collectively exhaustive	event	odds	special law of addition
combination	factorial	permutation	special law of multiplication
complement	general law of addition	posterior probability	subjective approach
compound event	general law of multiplication	prior probability	tree diagram
conditional probability	independent	probability	union
contingency table	intersection	random experiment	Venn diagram

Commonly Used Formulas in Probability

$$\text{Odds:} \quad \underset{\text{Odds for } A}{\dfrac{P(A)}{1 - P(A)}} \qquad \underset{\text{Odds against } A}{\dfrac{1 - P(A)}{P(A)}}$$

General Law of Addition: $\quad P(A \cup B) = P(A) + P(B) - P(A \cap B)$

Special Law of Addition: $\quad P(A \cup B) = P(A) + P(B)$

Conditional probability: $\quad P(A \mid B) = \dfrac{P(A \cap B)}{P(B)}$

General Law of Multiplication: $\quad P(A \cap B) = P(A \mid B)P(B)$

Special Law of Multiplication: $\quad P(A \cap B) = P(A)P(B)$

Bayes' Theorem: $\quad P(B \mid A) = \dfrac{P(A \mid B)P(B)}{P(A \mid B)P(B) + P(A \mid B')P(B')}$

Permutation: $\quad {}_nP_r = \dfrac{n!}{(n - r)!}$

Combination: $\quad {}_nC_r = \dfrac{n!}{r!(n - r)!}$

Chapter Review

1. Define (a) random experiment, (b) sample space, (c) simple event, and (d) compound event.

2. What are the three approaches to determining probability? Explain the differences among them.

3. Sketch a Venn diagram to illustrate (a) complement of an event, (b) union of two events, (c) intersection of two events, (d) mutually exclusive events, and (e) dichotomous events.

4. Define *odds*. What does it mean to say that odds are usually quoted against an event?

5. (a) State the general addition law. (b) Why do we subtract the intersection?

6. (a) Write the formula for conditional probability. (b) When are two events independent?

7. (a) What is a contingency table? (b) How do we convert a contingency table into a table of relative frequencies?

8. In a contingency table, explain the concepts of (a) marginal probability and (b) joint probability.

9. Why are tree diagrams useful? Why are they not always practical?

10. What is the main point of Bayes' Theorem?

11. Define (a) fundamental rule of counting, (b) factorial, (c) permutation, and (d) combination.

Chapter Exercises

Note: Explain answers and show your work clearly. Problems marked * are more difficult.

EMPIRICAL PROBABILITY EXPERIMENTS

5.65 (a) Make your own empirical estimate of the probability that a car is parked "nose first" (as opposed to "backed in"). Choose a local parking lot, such as a grocery store. Let *A* be the event that a car is parked nose first. Out of *n* cars examined, let *f* be the number of cars parked nose first. Then $P(A) = f/n$. (b) Do you feel your sample is large enough to have a reliable empirical probability? (c) If you had chosen a different parking lot (such as a church or a police station), would you expect the estimate of $P(A)$ to be similar? That is, would $P(A \mid church) = P(A \mid police station)$? Explain.

5.66 (a) Make your own empirical estimate of the probability that an item offered for sale on ebay.com in the category "women's leggings" will be shipped for free. For *n* items sampled from this category (items chosen using random numbers or some other random method), let *f* be the number of items with free shipping. Then $P(A) = f/n$. (b) Do you feel your sample is large enough to have a reliable empirical probability? (c) If you had chosen a different category, would you expect $P(A)$ to be similar or different? Explain.

5.67 (a) Make your own empirical estimate of the probability that a YouTube music video will be shorter than two minutes. For *n* music videos sampled from YouTube (items chosen using random numbers or some other random method), let *f* be the number of music videos that are shorter than two minutes. Then $P(A) = f/n$. (b) Do you feel your sample is large enough to have a reliable empirical probability? (c) If you had chosen sports videos, would you expect $P(A)$ to be similar or different? Explain.

5.68 M&Ms are blended in a ratio of 13 percent brown, 14 percent yellow, 13 percent red, 24 percent blue, 20 percent orange, and 16 percent green. Suppose you choose a sample of two M&Ms at random from a large bag. (a) Show the sample space. (b) What is the probability that both are brown? (c) Both blue? (d) Both green? (e) Find the probability of one brown and one green M&M. (f) Actually take 100 samples of two M&Ms (with replacement) and record the frequency of each outcome listed in (b) and (c) above. How close did your empirical results come to your predictions? (g) Which definition of probability applies in this situation?

(Source: www.mmmars.com.)

PROBLEMS

5.69 For male high school athletes, a news article reported that the probability of receiving a college scholarship is .0139 for basketball players, .0324 for swimmers/divers, and .0489 for lacrosse players. Which type of probabilities (classical, empirical, subjective) do you think these are?

5.70 A judge concludes that there is a 20 percent chance that a certain defendant will fail to appear in court if he is released after paying a full cash bail deposit. Which type of probability (classical, empirical, subjective) do you think this is?

5.71 A survey showed that 44 percent of online Internet shoppers experience some kind of technical failure at checkout (e.g., when submitting a credit card) after loading their shopping cart. Which type of probability (classical, empirical, subjective) do you think this is?

5.72 The new marketing director says we have a 75 percent chance that our online sales will increase significantly by offering free shipping. Which type of probability (classical, empirical, subjective) do you think this is?

5.73 In a typical 6 out of 49 lottery game, the probability of winning is approximately 7×10^{-8}. Which type of probability (classical, empirical, subjective) do you think this is?

5.74 The reported ACL injury rate for female high school soccer players is 20 percent. Which type of probability (classical, empirical, subjective) do you think this is?

5.75 A recent article states that there is a 2 percent chance that an asteroid 100 meters or more in diameter will strike the Earth before 2100. Which type of probability (classical, empirical, subjective) do you think this is?

5.76 If Punxsutawney Phil sees his shadow on February 2, then legend says that winter will last 6 more weeks. In 118 years, Phil has seen his shadow 104 times. (a) What is the probability that Phil will see his shadow on a randomly chosen Groundhog Day? (b) Which type of probability (classical, empirical, subjective) do you think this is?

5.77 On Los Angeles freeways during the rush hour, there is an 18 percent probability that a driver is using a handheld cell phone. Which type of probability (classical, empirical, subjective) do you think this is?

5.78 Bob owns two stocks. There is an 80 percent probability that stock *A* will rise in price, while there is a 60 percent chance that stock *B* will rise in price. There is a 40 percent chance that both stocks will rise in price. Are the stock prices independent?

5.79 To run its network, the Ramjac Corporation wants to install a system with dual independent servers. Employee Bob grumbled, "But that will double the chance of system failure." Is Bob right? Explain your reasoning with an example.

5.80 A study showed that trained police officers can detect a lie 65 percent of the time based on controlled studies of videotapes with real-life lies and truths. What are the odds in favor of a lie being detected?

5.81 The probability that a stolen vehicle is not recovered is .436. Find the odds *in favor* of a stolen vehicle being recovered.

5.82 The probability of being struck by lightning is .00016. Find the odds *against* being struck by lightning. Round your answer to the nearest whole number.

5.83 Prior to the start of the 2016 NCAA Men's Basketball playoffs, the reported odds that Butler University would *not* make it to the final game were 125 to 1. What was the implied probability that Butler would make it to the finals in 2016?

5.84 A certain model of remote-control Stanley garage door opener has nine binary (off/on) switches. The homeowner can set any code sequence. (a) How many separate codes can be programmed? (b) If you try to use your door opener on 1,000 other garages, how many times would you expect to succeed?

5.85 (a) In a certain state, license plates consist of three letters (A–Z) followed by three digits (0–9). How many different plates can be issued? (b) If the state allows any six-character mix (in any order) of 26 letters and 10 digits, how many unique plates are possible? (c) Why might some combinations of digits and letters be disallowed? *(d) Would the system described in (b) permit a unique license number for every car in the United States? For every car in the world? Explain your assumptions. *(e) If the letters O and I are not used because they look too much like the numerals 0 and 1, how many different plates can be issued?

5.86   Bob, Mary, and Jen go to dinner. Each orders a different meal. The waiter forgets who ordered which meal, so he randomly places the meals before the three diners. Let C be the event that a diner gets the correct meal and let N be the event that a diner gets an incorrect meal. Enumerate the sample space and then find the probability that

a.   No diner gets the correct meal.

b.   Exactly one diner gets the correct meal.

c.   Exactly two diners get the correct meal.

d.   All three diners get the correct meal.

5.87   An MBA program offers seven concentrations: accounting (A), finance (F), human resources (H), information systems (I), international business (B), marketing (M), and operations management (O). Students in the capstone business policy class are assigned to teams of three. In how many different ways could a team contain exactly one student from each concentration?

5.88   The probability that a customer at a grocery store pays for their items with a credit or debit card (event C) is .78. The probability that the grocery bill for a customer is greater than \$100 (event H) is .37. The probability that a customer pays with a credit or debit card given that their grocery bill was greater than \$100 is .95. (a) What is the probability that a customer pays with a credit or debit card and their grocery bill was greater than \$100? (b) What is the probability that a customer's grocery bill was over \$100 given that they paid with a credit or debit card? (c) In this problem, are C and H independent? Explain.

5.89   The probability that an international flight leaving the United States is delayed in departing (event D) is .25. The probability that an international flight leaving the United States is a transpacific flight (event P) is .57. The probability that an international flight leaving the United States is a transpacific flight and is delayed in departing is .12. (a) What is the probability that an international flight leaving the United States is delayed in departing given that the flight is a transpacific flight? (b) In this problem, are D and P independent? Explain.

5.90   A certain airplane has two independent alternators to provide electrical power. The probability that a given alternator will fail on a one-hour flight is .02. What is the probability that (a) both will fail? (b) Neither will fail? (c) One or the other will fail? Show all steps carefully.

5.91   There is a 30 percent chance that a bidding firm will get contract A and a 40 percent chance it will get contract B. There is a 5 percent chance that it will get both. Are the events independent?

5.92   In clinical trials, 60 percent of new cancer drugs advance from Phase I to Phase II, 25 percent advance from Phase II to Phase III, and 80 percent advance from Phase III to Phase IV. If 100 new drugs are entered into clinical trials, how many are expected to make it to Phase IV?

5.93   A turboprop aircraft has two attitude gyroscopes, driven from independent electrical sources. On a six-hour flight, assume the probability of failure of each attitude gyroscope is .0008. Does this achieve "five nines" reliability (i.e., a probability of at least .99999 that not all gyroscopes will fail)?

5.94   Which are likely to be independent events? For those you think are not, suggest reasons why.

a.   Gender of two babies born consecutively in a hospital.

b.   Car accident rates and the driver's gender.

c.   Phone call arrival rates at a university admissions office and time of day.

5.95   In child-custody cases, about 70 percent of the fathers win the case if they contest it. In the next three custody cases, what is the probability that all three win? What assumption(s) are you making?

*5.96   A web server hosting company advertises 99.999 percent guaranteed network uptime. (a) How many independent network servers would be needed if each has 99 percent reliability? (b) If each has 90 percent reliability?

*5.97   Fifty-six percent of American adults eat at a table-service restaurant at least once a week. Suppose that four American adults are asked if they ate at table-service restaurants last week. What is the probability that all of them say yes?

*5.98   The probability is 1 in 4,000,000 that a single auto trip in the United States will result in a fatality. Over a lifetime, an average U.S. driver takes 50,000 trips. (a) What is the probability of a fatal accident over a lifetime? Explain your reasoning carefully. *Hint:* Assume independent events. (b) Why might the assumption of independence be violated? (c) Why might a driver be tempted not to use a seat belt "just on this trip"?

*5.99   If there are two riders on a city bus, what is the probability that no two have the same birthday? What if there are 10 riders? 20 riders? 50 riders? *Hint:* Use the *LearningStats* demonstration "Birthday Problem."

*5.100   How many riders would there have to be on a bus to yield (a) a 50 percent probability that at least two will have the same birthday? (b) A 75 percent probability? *Hint:* Use the *LearningStats* demonstration "Birthday Problem."

5.101   Four students divided the task of surveying the types of vehicles in parking lots of four different shopping malls. Each student examined 100 cars in each of four large suburban malls, resulting in the 5×4 contingency table shown below. (a) Calculate each probability (i–vi) and explain in words what it means. (b) Do you see evidence that vehicle type is not independent of mall location? Explain. (Data are from an independent project by MBA students Steve Bennett, Alicia Morais, Steve Olson, and Greg Corda.) **Malls**

i. $P(C)$

ii. $P(G)$

iii. $P(V \mid S)$

iv. $P(C \mid J)$

v. $P(C \text{ and } G)$

vi. $P(T \text{ and } O)$

Number of Vehicles of Each Type in Four Shopping Malls

Vehicle Type	Somerset (S)	Oakland (O)	Great Lakes (G)	Jamestown (J)	Row Total
Car (C)	44	49	36	64	193
Minivan (M)	21	15	18	13	67
Full-size van (F)	2	3	3	2	10
SUV (V)	19	27	26	12	84
Truck (T)	14	6	17	9	46
Column Total	100	100	100	100	400

5.102 Refer to the contingency table shown below. (a) Calculate each probability (i–vi) and explain in words what it means. (b) Do you see evidence that smoking and race are *not* independent? Explain. (c) Do the smoking rates shown here correspond to your experience? (d) Why might public health officials be interested in this type of data? **Smoking2**

i. $P(S)$
ii. $P(W)$
iii. $P(S \mid W)$
iv. $P(S \mid B)$
v. $P(S \text{ and } W)$
vi. $P(N \text{ and } B)$

Smoking by Race for Males Aged 18–24

	Smoker (S)	Nonsmoker (N)	Row Total
White (W)	290	560	850
Black (B)	30	120	150
Column Total	320	680	1,000

5.103 Researchers examined forecasters' interest rate predictions for 34 quarters to see whether the predictions corresponded to what actually happened. The 2×2 contingency table below shows the frequencies of actual and predicted interest rate movements. Calculate each probability (i–vi) and explain in words what it means. **Forecasts**

i. $P(F-)$
ii. $P(A+)$
iii. $P(A- \mid F-)$
iv. $P(A+ \mid F+)$
v. $P(A+ \text{ and } F+)$
vi. $P(A- \text{ and } F-)$

Interest Rate Forecast Accuracy

	Actual Change		
Forecast Change	Decline (A−)	Rise (A+)	Row Total
Decline (F−)	7	12	19
Rise (F+)	9	6	15
Column Total	16	18	34

5.104 High levels of cockpit noise in an aircraft can damage the hearing of pilots who are exposed to this hazard for many hours. Cockpit noise in a jet aircraft is mostly due to airflow at hundreds of miles per hour. This 3×3 contingency table shows 61 observations of data collected by an airline pilot using a handheld sound meter in a certain aircraft cockpit.

Noise level is defined as "low" (under 88 decibels), "medium" (88 to 91 decibels), or "high" (92 decibels or more). There are three flight phases (climb, cruise, descent). (a) Calculate each probability (i–vi) and explain in words what it means. (b) Do you see evidence that noise level depends on flight phase? Explain. (c) Where else might ambient noise be an ergonomic issue? (*Hint:* Search the web.) ⌨ **Cockpit**

i. $P(B)$
ii. $P(L)$
iii. $P(H \mid C)$
iv. $P(H \mid D)$
v. $P(L \text{ and } B)$
vi. $P(L \text{ and } C)$

Cockpit Noise

Noise Level	Flight Phase			Row Total
	Climb (*B*)	Cruise (*C*)	Descent (*D*)	
Low (*L*)	6	2	6	14
Medium (*M*)	18	3	8	29
High (*H*)	1	3	14	18
Column Total	25	8	28	61

5.105 Suppose a recent sample of 400 employed U.S. adults found that 45 percent currently put the maximum amount possible in a 401(k). Sixty-three percent of the respondents say they are planning to retire by age 65. Sixty percent of those who plan to retire by age 65 put the maximum amount possible in a 401(k).

(a) Fill in the contingency table below using the information above.

(b) What percentage of adults who put the maximum in a 401(k) plan to retire after age 65? (c) Are planning to retire by age 65 and putting the maximum in a 401(k) independent of each other? Show your calculations.

401(k) and Retirement

	Retire by Age 65	Retire After Age 65	Row Total
Contribute Maximum			
Contribute Less Than Maximum			
Column Total			400

BAYES' THEOREM

5.106 A test for ovarian cancer has a 5 percent rate of false positives and a 0 percent rate of false negatives. On average, 1 in every 2,500 American females over age 35 actually has ovarian cancer. If a female over 35 tests positive, what is the probability that she actually has cancer? *Hint:* Make a contingency table for a hypothetical sample of 100,000 females. Explain your reasoning.

5.107 A biometric security device using fingerprints erroneously refuses to admit 1 in 1,000 authorized persons from a facility containing classified information. The device will erroneously admit 1 in 1,000,000 unauthorized persons. Assume that 95 percent of those who seek access are authorized. If the alarm goes off and a person is refused admission, what is the probability that the person was really authorized?

5.108 Dolon Web Security Consultants requires all job applicants to submit to a test for illegal drugs. If the applicant has used illegal drugs, the test has a 90 percent chance of a positive result. If the applicant has not used illegal drugs, the test has an 85 percent chance of a negative result. Actually, 4 percent of the job applicants have used illegal drugs. If an applicant has a positive test, what is the probability that they have actually used illegal drugs? *Hint:* Make a 2 × 2 contingency table of frequencies, assuming 500 job applicants.

Related Reading

Albert, James H. "College Students' Conceptions of Probability." *The American Statistician* 57, no. 1 (February 2001), pp. 37–45.

CHAPTER 5   More Learning Resources

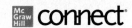

You can access these *LearningStats* demonstrations through Connect to help you understand probability.

Topic	LearningStats Demonstrations
Contingency tables	Contingency Tables Cross-Tabulations Independent Events
Probability	Probability Using Excel Probability Using R Birthday Problem System Reliability
Random processes	Law of Large Numbers Dice Rolls Pick a Card Random Names
Bayes' Theorem	Bayes' Theorem

Key: ▒ = Excel   ▣ = PowerPoint

Software Supplement

Making a Contingency Table in R   WebSurvey

We have results of a web survey of students with several *categorical* variables (i.e., data whose values are nonnumerical or qualitative). There are 158 responses, recorded in an Excel spreadsheet. The first few rows are shown:

	A	B	C	D	E	F
1		Cell Phone	Political Views	Religious Attend	Second Language Skill	Newspaper Reading
2		Verizon	Conservative	Very Often	None	Never
3		Verizon	Very Liberal	Often	Slight	Occasionally
4		AT&T	Conservative	Often	Slight	Occasionally
5		AT&T	Liberal	Rarely	Moderate	Occasionally
6		AT&T	Conservative	Never	Moderate	Regularly
7		...	...	...	...	...

Source: Microsoft Excel.

We would like to make cross-tabulations (i.e., to create one or more contingency tables). To import our data from Excel to a data frame in R, we can highlight a data block in Excel and copy it to the clipboard (Ctrl-c in Windows or command-c in MacOS). We can create a data frame in R that we might call WebSurvey:

In Windows:
```
> WebSurvey = read.table(file="clipboard", sep="\t", header=TRUE)
```

In MacOS:
```
> WebSurvey = read.table(pipe("pbpaste"), sep="\t", header=TRUE)
```

Because there are spaces in the column headings, R will name each variable by replacing the space with a period (e.g., Political.Views and Newspaper.Reading). Making a contingency table is easy. For example, to cross-tabulate Political.Views against Newspaper.Reading:

```
> MyTable=table(WebSurvey$Political.Views, WebSurvey$Newspaper.Reading)
> MyTable
```

```
                        Newspaper.Reading
Political.Views   Never Occasionally Regularly
  Conservative       7           30         8
  Liberal            8           34         3
  Very Conservative  2            6         4
  Very Liberal       9           40         7
```

Note that there are no row or column totals. You can calculate them by using table() for each variable separately, but they will just be printed (not as part of the table). It may be more convenient to copy and paste the contingency table into Word or Excel (it is a text table) to add your own row and column totals and format the table as desired. You could get the same table by referring to the desired column numbers in the data frame (2 and 5 in this example). If you do this, be careful about square brackets and parentheses:

```
> MyTable =table(WebSurvey[c(2,5)])
```

CHAPTER CONTENTS

CHAPTER LEARNING OBJECTIVES

When you finish this chapter, you should be able to

LO 6-1 Define a discrete random variable and its probability distribution.

LO 6-2 Solve problems using expected value and variance.

LO 6-3 Define and apply the uniform discrete model.

LO 6-4 Find binomial probabilities using tables, formulas, or Excel.

LO 6-5 Find Poisson probabilities using tables, formulas, or Excel.

LO 6-6 Use the Poisson approximation to the binomial (optional).

LO 6-7 Find hypergeometric probabilities using Excel.

LO 6-8 Use the binomial approximation to the hypergeometric (optional).

LO 6-9 Calculate geometric probabilities (optional).

LO 6-10 Apply rules for transformations of random variables (optional).

This chapter shows how probability can be used to analyze business activities or processes that generate random data. Many business processes create data that can be thought of as random. For example, consider cars being serviced in a quick oil change shop or calls arriving at the L.L.Bean order center. Each car or call can be thought of as an experiment with random outcomes. The variable of interest for the car might be service time. The variable of interest for the call might be amount of the order. Service time and amount of order will vary randomly for each car or call.

A **probability model** assigns a probability to each outcome in the sample space defined by a random (or stochastic) process. We use probability models to depict the essential characteristics of a stochastic process, to guide decisions or make predictions. How many service technicians do we need from noon to 1 p.m. on Friday afternoon? To answer this, we need to model the process of servicing cars during the lunch hour. Can L.L.Bean predict its total order amount from the next 50 callers? To answer this question, L.L.Bean needs to model the process of call orders to its call center. Probability models must be reasonably realistic yet simple enough to be analyzed.

Many random processes can be described by using common probability models whose properties are well known. To correctly use these probability models, it is important that you understand their development. In the following sections, we will explain how probability models are developed and describe several commonly used models.

6.1 DISCRETE PROBABILITY DISTRIBUTIONS

Random Variables

A **random variable** is a function or rule that assigns a numerical value to each outcome in the sample space of a random experiment. We use X when referring to a random variable in general, while specific values of X are shown in lowercase (e.g., x_1). The random variable often is a direct result of an observational experiment (e.g., counting the number of takeoffs in a given hour at O'Hare International Airport). A **discrete random variable** has a countable number of distinct values. Some random variables have a clear upper limit (e.g., number of absences in a class of 40 students), while others do not (e.g., number of text messages you receive in a given hour). Here are some examples of decision problems involving discrete random variables.

LO 6-1

Define a discrete random variable and its probability distribution.

Decision Problem

- Oxnard University has space in its MBA program for 65 new students. In the past, 75 percent of those who were admitted actually enrolled. The decision is made to admit 80 students. What is the probability that more than 65 admitted students will actually enroll?

Discrete Random Variable

- X = number of admitted MBA students who actually enroll ($X = 0, 1, 2, \ldots, 80$)

Decision Problem	*Discrete Random Variable*
• On the late morning (9 to 12) work shift, L.L.Bean's order processing center staff can handle up to 5 orders per minute. The mean arrival rate is 3.5 orders per minute. What is the probability that more than 5 orders will arrive in a given minute?	• X = number of phone calls that arrive in a given minute at the L.L.Bean order processing center ($X = 0, 1, 2, \ldots$)
• Rolled steel from a certain supplier averages 0.01 defect per linear meter. Toyota will reject a shipment of 500 linear meters if inspection reveals more than 10 defects. What is the probability that the order will be rejected?	• X = number of defects in 500 meters of rolled steel ($X = 0, 1, 2, \ldots$)

Probability Distributions

A **discrete probability distribution** assigns a probability to each value of a discrete random variable X. The distribution must follow the rules of probability defined in Chapter 5. If X has n distinct values $x_1, x_2, \ldots, x_n$, then

(6.1) $\qquad 0 \le P(x_i) \le 1 \qquad$ (the probability for any given value of X)

(6.2) $\qquad \displaystyle\sum_{i=1}^{n} P(x_i) = 1 \qquad$ (the sum over all values of X)

Discrete probability distributions follow the rules of functions. More than one sample space outcome can be assigned to the same number, but you cannot assign one outcome to two different numbers. Likewise, more than one random variable value can be assigned to the same probability, but one random variable value cannot have two different probabilities. The probabilities must sum to 1. Figure 6.1 illustrates the relationship between the sample space, the random variable, and the probability distribution function for a simple experiment of rolling a die.

Figure 6.1

Random Experiment: Rolling a Die

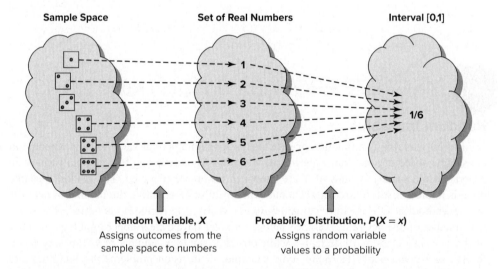

Sample Space	Set of Real Numbers	Interval [0,1]

Random Variable, X
Assigns outcomes from the sample space to numbers

Probability Distribution, P(X = x)
Assigns random variable values to a probability

Example 6.1

Coin Flips

ThreeCoins

When you flip a fair coin three times, the sample space has eight equally likely simple events: {HHH, HHT, HTH, THH, HTT, THT, TTH, TTT}. If X is the number of heads, then X is a random variable whose probability distribution is shown in Table 6.1 and Figure 6.2.

Table 6.1   **Probability Distribution for Three Coin Flips**

Possible Events	x	P(x)
TTT	0	1/8
HTT, THT, TTH	1	3/8
HHT, HTH, THH	2	3/8
HHH	3	1/8
Total		1

The values of X need not be equally likely. In this example, $X = 1$ and $X = 2$ are more likely than $X = 0$ or $X = 3$. However, the probabilities sum to 1, as in any probability distribution.

Figure 6.2   **Probability Distribution for Three Coin Flips**

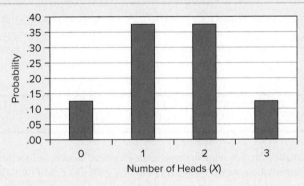

What Is a PDF or a CDF?

A known distribution can be described either by its **probability distribution function** (PDF) or by its **cumulative distribution function** (CDF). The PDF and CDF are defined either by a list of X-values and their probabilities or by mathematical equations. A discrete PDF shows the probability of each X-value, $P(X = x_i)$. The CDF shows the cumulative sum of probabilities, $P(X \le x_i)$, adding from the smallest to the largest X-value. Figure 6.3 illustrates a discrete PDF and the corresponding CDF. Notice that the CDF approaches 1, and the PDF values of X will sum to 1.

Figure 6.3

Illustration of PDF and CDF

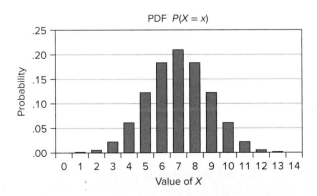

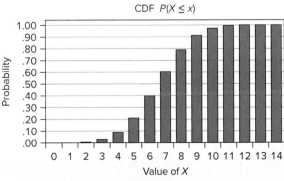

Random variables and their distributions are described by their **parameters**. The equations for the PDF, the CDF, and the characteristics of the distribution (such as the mean and standard deviation) will depend on the parameters of the process. The rest of this chapter explains several well-known discrete distributions and their applications. Many random business processes can be described by these common distributions.

Section Exercises

6.1 Which of the following could *not* be probability distributions? Explain.

Example A		Example B		Example C	
x	P(x)	x	P(x)	x	P(x)
0	.80	1	.05	50	.30
1	.20	2	.15	60	.60
		3	.25	70	.40
		4	.40		
		5	.10		

6.2 On hot, sunny summer days, Jane rents inner tubes by the river that runs through her town. Based on her past experience, she has assigned the following probability distribution to the number of tubes she will rent on a randomly selected day. (a) Find $P(X = 75)$. (b) Find $P(X \leq 75)$. (c) Find $P(X > 50)$. (d) Find $P(X < 100)$. (e) Which of the probability expressions in parts (a)–(d) is a value of the CDF?

x	25	50	75	100	Total
P(x)	.20	.40	.30	.10	1.00

6.3 On the midnight shift, the number of patients with head trauma in an emergency room has the probability distribution shown below. (a) Find $P(X \geq 3)$. (b) Find $P(X \leq 2)$. (c) Find $P(X < 4)$. (d) Find $P(X = 1)$. (e) Which of the probability expressions in parts (a)–(d) is a value of the CDF?

x	0	1	2	3	4	5	Total
P(x)	.05	.30	.25	.20	.15	.05	1.00

6.2 EXPECTED VALUE AND VARIANCE

Recall that a discrete probability distribution is defined only at specific points on the X-axis. The **expected value** $E(X)$ of a discrete random variable is the sum of all X-values weighted by their respective probabilities. It is a measure of *center*. If there are N distinct values of X (x_1, x_2, ..., x_N), the expected value is

(6.3)
$$E(X) = \mu = \sum_{i=1}^{N} x_i P(x_i)$$

The expected value is a *weighted* average because outcomes can have different probabilities. Because it is an average, we usually call $E(X)$ the *mean* and use the symbol μ.

Example 6.2

Service Calls

ServiceCalls

The distribution of Sunday emergency service calls by Ace Appliance Repair is shown in Table 6.2. The probabilities sum to 1, as must be true for any probability distribution.

The mode (most likely value of X) is 2, but the *expected* number of service calls $E(X)$ is 2.75, that is, $\mu = 2.75$. In other words, the "average" number of service calls is 2.75 on Sunday:

$$E(X) = \mu = \sum_{i=1}^{N} x_i P(x_i) = 0P(0) + 1P(1) + 2P(2) + 3P(3) + 4P(4) + 5P(5)$$

$$= 0(.05) + 1(.10) + 2(.30) + 3(.25) + 4(.20) + 5(.10) = 2.75$$

Table 6.2   **Probability Distribution of Service Calls**

x	P(x)	xP(x)
0	0.05	0.00
1	0.10	0.10
2	0.30	0.60
3	0.25	0.75
4	0.20	0.80
5	0.10	0.50
Total	1.00	2.75

In Figure 6.4, we see that this particular probability distribution is not symmetric around the mean $\mu = 2.75$. However, the mean $\mu = 2.75$ is still the balancing point, or fulcrum.

Note that $E(X)$ need not be an observable event. For example, you could have 2 service calls or 3 service calls but not 2.75 service calls. This makes sense because $E(X)$ is an *average*. It is like saying that "the average American family has 2.1 children" (even though families come only in integer sizes) or "Albert Pujols's batting average is .312" (even though the number of hits by Pujols in a particular game must be an integer).

Figure 6.4   **Probability Distribution for Service Calls**

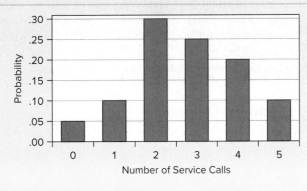

Application: Life Insurance   Expected value is the basis of life insurance, a purchase that almost everyone makes. For example, based on U.S. mortality statistics, the probability that a 30-year-old white female will die within the next year is .000642, so the probability of living another year is $1 - .000642 = .999358$. What premium should a life insurance company charge to break even on a \$500,000 one-year term insurance policy (i.e., to achieve zero expected payout)? This situation is shown in Table 6.3. Let X be the amount paid by the company to settle the policy. The expected payout is \$321, so the premium should be \$321 plus whatever return the company needs to cover its administrative overhead and profit.

Event	x	P(x)	xP(x)
Live	0	.999358	.00
Die	500,000	.000642	321.00
Total		1.000000	321.00

Table 6.3

Expected Payout for a One-Year Term Life Policy

Source: Centers for Disease Control and Prevention, *National Vital Statistics Reports* 58, no. 19 (2010).

The mortality rate shown here is for *all* 30-year-old females. An insurance quote (e.g., from the web) is likely to yield a lower premium, as long as you are a healthy, educated non-smoker in a nonrisky occupation. Insurance companies make money by knowing the actuarial probabilities and using them to set their premiums. The task is difficult because actuarial probabilities must be revised as life expectancies change over time.

Application: Raffle Tickets   Expected value can be applied to raffles and lotteries. If it costs $2 to buy a ticket in a raffle to win a new luxury automobile worth $55,000 and 29,346 raffle tickets are sold, the expected value of a lottery ticket is

$$E(X) = (\text{value if you win})P(\text{win}) + (\text{value if you lose})P(\text{lose})$$

$$= (55,000)\left(\frac{1}{29,346}\right) + (0)\left(\frac{29,345}{29,346}\right)$$

$$= (55,000)(.000034076) + (0)(.999965924) = \$1.87$$

The raffle ticket is actually worth $1.87. So why would you pay $2.00 for it? Partly because you hope to beat the odds, but also because you know that your ticket purchase helps the charity. Because the idea of a raffle is to raise money, the sponsor tries to sell enough tickets to push the expected value of the ticket below its price (otherwise, the charity would lose money on the raffle). If the raffle prize is donated (or partially donated) by a well-wisher, the break-even point may be much less than the full value of the prize.

Like a lottery, an **actuarially fair** insurance program must collect as much in overall revenue as it pays out in claims. This is accomplished by setting the premiums to reflect empirical experience with the insured group. Individuals may gain or lose, but if the pool of insured persons is large enough, the total payout is predictable. Of course, many insurance policies have exclusionary clauses for war and natural disaster (e.g., Hurricane Katrina), to deal with cases where the events are not independent. Actuarial analysis is critical for corporate pension fund planning. Group health insurance is another major application.

Variance and Standard Deviation

The **variance** Var(X) of a discrete random variable is the sum of the squared deviations about its expected value, weighted by the probability of each X-value. If there are N distinct values of X, the variance is

(6.4)
$$\text{Var}(X) = \sigma^2 = \sum_{i=1}^{N}[x_i - \mu]^2 P(x_i)$$

Just as the expected value $E(X)$ is a weighted average that measures *center,* the variance Var(X) is a weighted average that measures variability about the mean. And just as we interchangeably use μ or $E(X)$ to denote the mean of a distribution, we use either σ^2 or Var(X) to denote its variance.

The *standard deviation* is the square root of the variance and is denoted σ:

(6.5)
$$\sigma = \sqrt{\sigma^2} = \sqrt{\text{Var}(X)}$$

Example 6.3

Bed and Breakfast

🖆 **RoomRent**

The Bay Street Inn is a seven-room bed-and-breakfast in the sunny California coastal city of Santa Theresa. Demand for rooms generally is strong during February, a prime month for tourists. However, experience shows that demand is quite variable. The probability distribution of room rentals during February is shown in the first two columns of Table 6.4, where X = the number of rooms rented (X = 0, 1, 2, 3, 4, 5, 6, 7). The worksheet shows the calculation of $E(X)$ and Var(X).

Table 6.4

Worksheet for E(X) and Var(X) for February Room Rentals

x	P(x)	xP(x)	x − μ	[x − μ]²	[x − μ]²P(x)
0	.05	0.00	−4.71	22.1841	1.109205
1	.05	0.05	−3.71	13.7641	0.688205
2	.06	0.12	−2.71	7.3441	0.440646
3	.10	0.30	−1.71	2.9241	0.292410
4	.13	0.52	−0.71	0.5041	0.065533
5	.20	1.00	+0.29	0.0841	0.016820
6	.15	0.90	+1.29	1.6641	0.249615
7	.26	1.82	+2.29	5.2441	1.363466
Total	1.00	μ = 4.71			σ² = 4.225900

The formulas are:

$$E(X) = \mu = \sum_{i=1}^{N} x_i P(x_i) = 4.71$$

$$Var(X) = \sigma^2 = \sum_{i=1}^{N} [x_i - \mu]^2 P(x_i) = 4.2259$$

$$\sigma = \sqrt{4.2259} = 2.0557$$

This distribution is skewed to the left and bimodal. The mode (most likely value) is 7 rooms rented, but the average is only 4.71 room rentals in February. The standard deviation of 2.06 indicates that there is considerable variation around the mean, as seen in Figure 6.5.

Figure 6.5   Probability Distribution of Room Rentals

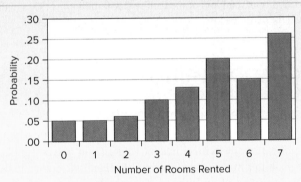

The car insurance industry estimates probabilities of vehicle accidents using big data sets of historical events. Insurance rates are based on *expected values* calculated using conditional probabilities on factors such as type of vehicle, age, income, and gender of driver, as well as geographic location. These all play a role in determining the risk of a traffic accident that causes vehicle damage, injury, or death. Because more than 90% of vehicle accidents are attributed to drivers, traditional risk assessment models place heavy weight on driver characteristics.

Analytics in Action

Car Insurance, Risk, and Driverless Vehicles

Artificial intelligence and analytics have now made driverless vehicles a reality. This transition from manned vehicles to autonomous vehicles will shift the bulk of the risk from the driver to the vehicle manufacturers and software designers. Liability in the case of damage, injury, or a fatality will now be shared between the vehicle owner and the AI software. Personal car insurance is being redefined to account for this shift and adapt to new technology.

Section Exercises

connect

6.4 On hot, sunny summer days, Jane rents inner tubes by the river that runs through her town. Based on her past experience, she has assigned the following probability distribution to the number of tubes she will rent on a randomly selected day. (a) Calculate the expected value and standard deviation of this random variable X by using the PDF shown. (b) Describe the shape of this distribution.

x	25	50	75	100	Total
P(x)	.20	.40	.30	.10	1.00

6.5 On the midnight shift, the number of patients with head trauma in an emergency room has the probability distribution shown below. (a) Calculate the mean and standard deviation. (b) Describe the shape of this distribution.

x	0	1	2	3	4	5	Total
P(x)	.05	.30	.25	.20	.15	.05	1.00

6.6 Pepsi and Mountain Dew products sponsored a contest giving away a Lamborghini sports car worth $215,000. The probability of winning from a single bottle purchase was .00000884. Find the expected value. Show your calculations clearly.

6.7 Student Life Insurance Company wants to offer an insurance plan with a maximum claim amount of $5,000 for dorm students to cover theft of certain items. Past experience suggests that the probability of a maximum claim is .01. What premium should be charged if the company wants to make a profit of $25 per policy? Assume any student who files a claim files for the maximum amount and there is no deductible. Show your calculations clearly.

6.8 A lottery ticket has a grand prize of $28 million. The probability of winning the grand prize is .000000023. Based on the expected value of the lottery ticket, would you pay $1 for a ticket? Show your calculations and reasoning clearly.

6.9 Oxnard Petro Ltd. is buying hurricane insurance for its off-coast oil drilling platform. During the next five years, the probability of total loss of only the above-water superstructure ($250 million) is .30, the probability of total loss of the facility ($950 million) is .30, and the probability of no loss is .40. Find the expected loss.

6.3 UNIFORM DISTRIBUTION

LO 6-3

Define and apply the uniform discrete model.

Characteristics of the Uniform Distribution

The **uniform distribution** is one of the simplest discrete models. It describes a random variable with a finite number of consecutive integer values from a to b. That is, the entire distribution depends only on the two parameters a and b. Each value is equally likely. Table 6.5 summarizes the characteristics of the uniform discrete distribution.

Table 6.5

Uniform Discrete Distribution

Parameters	a = lower limit b = upper limit
PDF	$P(X = x) = \dfrac{1}{b - a + 1}$
CDF	$P(X \leq x) = \dfrac{x - a + 1}{b - a + 1}$
Domain	$x = a, a + 1, a + 2, \ldots, b$
Mean	$\dfrac{a + b}{2}$
Standard deviation	$\sqrt{\dfrac{[(b - a) + 1]^2 - 1}{12}}$
Random data in Excel	=RANDBETWEEN(a, b)
Comments	Generate random integers for sampling or simulation models.

Figure 6.6

PDF and CDF for Rolling a Die

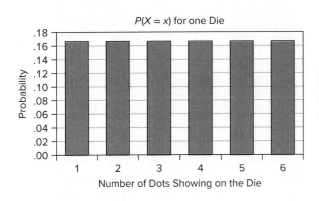

$P(X = x)$ for one Die

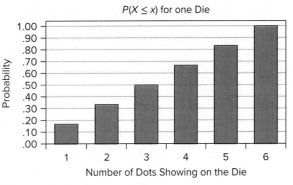

$P(X \leq x)$ for one Die

When you roll one die, the number of dots forms a uniform discrete random variable with six equally likely integer values 1, 2, 3, 4, 5, 6, shown in the PDF and CDF in Figure 6.6. You can see that the mean (3.5) must be halfway between 1 and 6, but there is no way you could anticipate the standard deviation without using a formula.

For this example, the mean and standard deviation are

Mean $\qquad \mu = \dfrac{a+b}{2} = \dfrac{1+6}{2} = 3.5$

Std. Dev. $\qquad \sigma = \sqrt{\dfrac{[(b-a)+1]^2 - 1}{12}} = \sqrt{\dfrac{[(6-1)+1]^2 - 1}{12}} = 1.708$

Consider another example of the discrete uniform distribution. When filling your car with gas, the last two digits (pennies) showing on the display of gallons dispensed will be a uniform random integer (assuming you don't "top off" but just let the pump stop automatically) ranging from $a = 00$ to $b = 99$. You could verify the predicted mean and standard deviation shown here by looking at a large sample of fill-ups on your own car:

PDF $\qquad P(X = x) = \dfrac{1}{b - a + 1} = \dfrac{1}{99 - 0 + 1} = \dfrac{1}{100} = .010$ for all x

Mean $\qquad \mu = \dfrac{a+b}{2} = \dfrac{0+99}{2} = 49.5$

Std. Dev. $\qquad \sigma = \sqrt{\dfrac{[(b-a)+1]^2 - 1}{12}} = \sqrt{\dfrac{[(99-0)+1]^2 - 1}{12}} = 28.87$

The discrete uniform distribution is often used to generate random integers that are then used to randomly sample from a population. To accomplish this, we can use the Excel function =RANDBETWEEN(a, b). For example, to generate a random integer from 5 through 10, the Excel function would be =RANDBETWEEN(5, 10). The same integer may come up more than once, so to obtain *n* distinct random integers, you would have to generate a few extras and then eliminate the duplicates if you are sampling without replacement (see Chapter 2). This method is useful in accounting and auditing (e.g., to allow the auditor to choose numbered invoices at random).

Mini Case 6.1

The "Daily 3" Lottery

Many states have a "daily 3" lottery. The daily 3 is a uniformly distributed discrete random variable whose values range from 000 through 999. There are 1,000 equally likely outcomes, so the probability of any given three-digit number is 1/1,000. The theoretical characteristics of this lottery are

$$P(X = x) = \frac{1}{b - a + 1} = \frac{1}{999 - 0 + 1} = \frac{1}{1,000} = .001$$

$$\mu = \frac{a + b}{2} = \frac{0 + 999}{2} = 499.5$$

$$\sigma = \sqrt{\frac{(b - a + 1)^2 - 1}{12}} = \sqrt{\frac{(999 - 0 + 1)^2 - 1}{12}} = 288.67$$

In a large sample of three-digit lottery numbers, you would expect the sample mean and standard deviation to be very close to 499.5 and 288.67, respectively. For example, in Michigan's daily three-digit lottery, from January 1, 2010, through December 31, 2010, there were 364 evening drawings. The mean of all the three-digit numbers drawn over that period was 497.1 with a standard deviation of 289.5. These sample results are extremely close to what would be expected. It is the nature of random samples to vary, so no sample is expected to yield statistics identical with the population parameters.

In Michigan, randomization is achieved by drawing a numbered ping-pong ball from each of three bins. Within each bin, the balls are agitated using air flow. Each bin contains 10 ping-pong balls. Each ball has a single digit (0, 1, 2, 3, 4, 5, 6, 7, 8, 9). The drawing is televised, so there is no possibility of bias or manipulation. Lotteries are studied frequently to make sure that they are truly random using statistical comparisons like these as well as tests for overall shape and patterns over time.

Section Exercises

6.10 The number of tickets purchased by an individual for Beckham College's holiday music festival is a uniformly distributed random variable ranging from 2 to 8. Find the mean and standard deviation of this random variable.

6.11 The ages of Python programmers at SynFlex Corp. range from 20 to 60. (a) If their ages are uniformly distributed, what would be the mean and standard deviation? (b) What is the probability that a randomly selected programmer's age is at least 40? At least 30? *Hint:* Treat employee ages as integers.

6.12 Use Excel to generate 100 random integers from (a) 1 through 2, inclusive; (b) 1 through 5, inclusive; and (c) 0 through 99, inclusive. (d) In each case, write the Excel formula. (e) In each case, calculate the mean and standard deviation of the sample of 100 integers you generated, and compare them with their theoretical values.

6.4 BINOMIAL DISTRIBUTION

Bernoulli Experiments

LO **6-4**

Find binomial probabilities using tables, formulas, or Excel.

A random experiment that has only two outcomes is called a **Bernoulli experiment**, named after Jakob Bernoulli (1654–1705). To create a random variable, we arbitrarily call one outcome a "success" (denoted $X = 1$) and the other a "failure" (denoted $X = 0$). The probability of success is denoted π (the Greek letter "pi," *not* to be confused with the mathematical constant 3.14159).[1] The probability of failure is $1 - \pi$, so the probabilities sum to 1, that is, $P(0) + P(1) = (1 - \pi) + \pi = 1$. The probability of success, π, remains the same for each trial.

The examples in Table 6.6 show that a success ($X = 1$) may in fact represent something undesirable but of main interest. Metallurgists look for signs of metal fatigue. Auditors look for expense voucher errors. Bank loan officers look for loan defaults. A success, then, is merely an event of interest.

The probability of success π can be any value between 0 and 1. In flipping a fair coin, π is .50. But in other applications π could be close to 1 (e.g., the probability that a customer's

---

[1] Some textbooks denote the probability of success *p*. However, in this textbook, we prefer to use Greek letters for population parameters. In Chapters 8 and 9, the symbol *p* will be used to denote a *sample* estimate of π.

Table 6.6

Bernoulli Experiment	Possible Outcomes	Probability of "Success"
Flip a coin	1 = heads 0 = tails	$\pi = .50$
Inspect a jet turbine blade	1 = crack found 0 = no crack found	$\pi = .001$
Purchase a tank of gas	1 = pay by credit card 0 = do not pay by credit card	$\pi = .78$
Do a mammogram test	1 = positive test 0 = negative test	$\pi = .0004$

Examples of Bernoulli Experiments

Visa purchase will be approved) or close to 0 (e.g., the probability that a purchased lightbulb is defective). Table 6.6 is only intended to suggest that π depends on the situation. The definitions of success and failure are arbitrary and can be switched, although we usually define success as the less likely outcome so that π is less than .50.

The only parameter needed to define a Bernoulli process is π. A Bernoulli experiment has mean π and variance $\pi(1 - \pi)$, as we see from the definitions of $E(X)$ and $\text{Var}(X)$:

$$E(X) = \sum_{i=1}^{2} x_i P(x_i) = (0)(1 - \pi) + (1)(\pi) = \pi \qquad \text{(Bernoulli mean)} \qquad \textbf{(6.6)}$$

$$\text{Var}(X) = \sum_{i=1}^{2} [x_i - E(X)]^2 P(x_i)$$
$$= (0 - \pi)^2(1 - \pi) + (1 - \pi)^2(\pi) = \pi(1 - \pi) \qquad \text{(Bernoulli variance)} \qquad \textbf{(6.7)}$$

There are many business applications that can be described by the Bernoulli model, and it is an important building block for more complex models. We will use the Bernoulli distribution to develop the next model.

Binomial Distribution

Bernoulli experiments lead to an important and more interesting model. The **binomial distribution** arises when a Bernoulli experiment is repeated n times. Each Bernoulli trial is independent so that the probability of success π remains constant on each trial. In a binomial experiment, we are interested in X = the number of successes in n trials, so the binomial random variable X is the sum of n independent Bernoulli random variables:

$$X = X_1 + X_2 + \cdots + X_n$$

We can add the n identical Bernoulli means ($\pi + \pi + \cdots + \pi$) to get the binomial mean $n\pi$. Because the n Bernoulli trials are independent, we can add[2] the n identical Bernoulli variances $\pi(1 - \pi) + \pi(1 - \pi) + \ldots + \pi(1 - \pi)$ to obtain the binomial variance $n\pi(1 - \pi)$ and hence its standard deviation $\sqrt{n\pi(1 - \pi)}$. The domain of the binomial is $x = 0, 1, 2, \ldots, n$. The binomial probability of a particular number of successes $P(X = x)$ is determined by the two parameters n and π. The characteristics of the binomial distribution are summarized in Table 6.7. The binomial probability function is

$$P(X = x) = \frac{n!}{x!(n - x)!}\pi^x(1 - \pi)^{n-x}, \quad \text{for } X = 0, 1, 2, 3, 4, \ldots, n. \qquad \textbf{(6.8)}$$

Application: Uninsured Patients 📄 **Uninsured** On average, 20 percent of the emergency room patients at Greenwood General Hospital lack health insurance. In a random sample of four patients, what is the probability that two will be uninsured? Define X = number of uninsured patients and set $\pi = .20$ (i.e., a 20 percent chance that a given patient will be uninsured)

---

[2]The last section in this chapter (optional) explains the rules for transforming and summing random variables.

Table 6.7

Binomial Distribution

Parameters	n = number of trials π = probability of success
PDF	$P(X=x) = \dfrac{n!}{x!(n-x)!}\pi^{x}(1-\pi)^{n-x}$
Excel* PDF	=BINOM.DIST(x, n, π, 0)
Excel* CDF	=BINOM.DIST(x, n, π, 1)
Domain	$x = 0, 1, 2, \ldots, n$
Mean	$n\pi$
Standard deviation	$\sqrt{n\pi(1-\pi)}$
Random data generation in Excel	=BINOM.INV(n, π, RAND()) or use Excel's Data Analysis Tools
Comments	Skewed right if $\pi < .50$, skewed left if $\pi > .50$, and symmetric if $\pi = .50$.

*See the **R Software Supplement** at the end of this chapter for a list of R's binomial probability functions.

and $1 - \pi = .80$ (i.e., an 80 percent chance that a patient will be insured). The domain is $X = 0, 1, 2, 3, 4$ patients. Applying the binomial formulas, the mean and standard deviation are

$$\text{Mean} = \mu = n\pi = (4)(.20) = 0.8 \text{ patient}$$
$$\text{Standard deviation} = \sigma = \sqrt{n\pi(1-\pi)} = \sqrt{(4)(.20)(1-.20)} = 0.8 \text{ patient}$$

Application: Quick Oil Change   Consider a shop that specializes in quick oil changes. It is important to this type of business to ensure that a car's service time is not considered "late" by the customer. We can define service times as being either *late* or *not late* and define the random variable X to be the number of cars that are late out of the total number of cars serviced. We assume that cars are independent of each other and the chance of a car being late stays the same for each car. Suppose that historically $P(\text{car is late}) = \pi = .10$.

Now, think of each car as a Bernoulli experiment and let's apply the binomial distribution. Suppose we would like to know the probability that exactly 2 of the next 12 cars serviced are late. In this case, $n = 12$, and we want to know $P(X = 2)$:

$$P(X = 2) = \frac{12!}{2!(12-2)!}(.10)^{2}(1-10)^{12-2} = .2301$$

Alternatively, we could calculate this by using the Excel function =BINOM.DIST(2, 12, .1, 0). The fourth parameter, 0, means that we want Excel to calculate $P(X = 2)$ rather than $P(X \leq 2)$.

Binomial Shape

A binomial distribution is skewed right if $\pi < .50$, skewed left if $\pi > .50$, and symmetric only if $\pi = .50$. However, skewness decreases as n increases, regardless of the value of π, as

Figure 6.7

Binomial Distributions

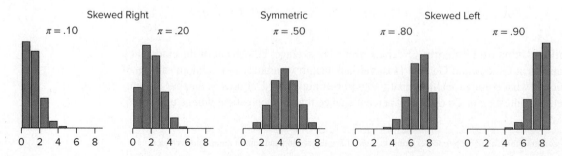

illustrated in Figure 6.7. Notice that $\pi = .20$ and $\pi = .80$ have the same shape, except reversed from left to right. This is true for any values of π and $1 - \pi$.

Recognizing Binomial Applications

Can you recognize a binomial situation? The binomial distribution has five main characteristics.

- The number of trials (n) is fixed.
- There are only two outcomes for each trial: *success* or *failure.*
- The probability of success for each trial (π) remains constant.
- The trials are independent of each other.
- The random variable (X) is the number of successes out of n trials.

Ask yourself if the five characteristics above make sense in the following examples.

In a sample of 20 friends, how many are left-handed?

In a sample of 50 cars in a parking lot, how many have hybrid engines?

In a sample of 10 emergency patients with chest pain, how many will be uninsured?

Even if you don't know π, you may have a binomial experiment. In practice, the value of π would be estimated from experience, but in this chapter it will be given.

Using the Binomial Formula

The PDF and CDF are shown in Table 6.8. We can calculate binomial probabilities by using Excel's binomial formula =BINOM.DIST(x, n, π, cumulative), where cumulative is 0 (if you want a PDF) or 1 (if you want a CDF). We also can use a calculator to work it out from the mathematical formula with $n = 4$ and $\pi = .20$.

x	PDF $P(X = x)$	CDF $P(X \leq x)$
0	0.4096	0.4096
1	0.4096	0.8192
2	0.1536	0.9728
3	0.0256	0.9984
4	0.0016	1.0000

Table 6.8

Binomial Distribution for $n = 4$, $\pi = .20$

PDF Formula   *Excel Function*

$$P(X = 0) = \frac{4!}{0!(4 - 0)!}(.20)^0(1 - .20)^{4-0} = 1 \times .20^0 \times .80^4 = .4096$$
=BINOM.DIST(0, 4, .20, 0)

$$P(X = 1) = \frac{4!}{1!(4 - 1)!}(.20)^1(1 - .20)^{4-1} = 4 \times .20^1 \times .80^3 = .4096$$
=BINOM.DIST(1, 4, .20, 0)

$$P(X = 2) = \frac{4!}{2!(4 - 2)!}(.20)^2(1 - .20)^{4-2} = 6 \times .20^2 \times .80^2 = .1536$$
=BINOM.DIST(2, 4, .20, 0)

CDF Calculation   *Excel Function*

$$P(X \leq 2) = P(X = 0) + P(X = 1) + P(X = 2) = .9728$$
=BINOM.DIST(2, 4, .2, 1)

As for any discrete probability distribution, the probabilities sum to one. That is, $P(0) + P(1) + P(2) + P(3) + P(4) = .4096 + .4096 + .1536 + .0256 + .0016 = 1.0000$. Figure 6.8 illustrates the PDF and CDF. Because $\pi < .50$, the distribution is right-skewed. The mean $\mu = n\pi = 0.8$ would be the balancing point or fulcrum of the PDF.

Figure 6.8

Binomial Distribution for $n = 4$, $\pi = .20$

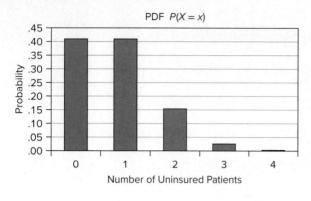
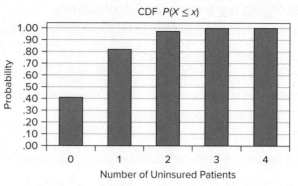

Compound Events

A compound event is expressed using an inequality. Consider the event that the sample of four patients will contain *at most* two uninsured patients. The probability of this event would be expressed as $P(X \leq 2)$. As shown previously, we can add the PDF values:

$$P(X \leq 2) = P(X = 0) + P(X = 1) + P(X = 2) = .4096 + .4096 + .1536 = .9728$$

Alternatively, we can calculate this probability using Excel's CDF function:

$$P(X \leq 2) = \text{BINOM.DIST}(2, 4, .2, 1) = .9728$$

The probability that *fewer than* two patients have insurance is the same as *at most* one patient is uninsured:

$$P(X < 2) = P(X \leq 1) = \text{BINOM.DIST}(1, 4, .2, 1) = .8192$$

Suppose we want to calculate the probability that *at least* two patients are uninsured: $P(X \geq 2)$. Because $P(X \geq 2)$ and $P(X \leq 1)$ are complementary events, we can obtain $P(X \geq 2)$ using our result from above:

$$P(X \geq 2) = 1 - P(X \leq 1) = 1 - \text{BINOM.DIST}(1, 4, .2, 1) = 1 - .8192 = .1808$$

To interpret phrases such as "more than" or "at least," it is helpful to sketch a diagram:

				$P(X \geq 2)$	
"At least two"	0	1	2	3	4

	$P(X < 2)$				
"Fewer than two"	0	1	2	3	4

	$P(X < 2)$			$P(X > 2)$	
"Fewer than 2 or more than 2"	0	1	2	3	4

Using Tables: Appendix A

The binomial formula is cumbersome, even for small *n,* so we prefer to use a computer program (Excel, Minitab, or MegaStat) or a calculator with a built-in binomial function. When you have no access to a computer (e.g., taking an exam), you can use Appendix A to look up binomial probabilities for selected values of *n* and π. An abbreviated portion of Appendix A is shown in Figure 6.9. The probabilities for $n = 4$ and $\pi = .20$ are highlighted. Probabilities in Appendix A are rounded to four decimal places, so the values may differ slightly from Excel.

Figure 6.9

Binomial Probabilities from Appendix A

n	X	.01	.02	.05	.10	.15	.20	.30	.40	.50
2	0	.9801	.9604	.9025	.8100	.7225	.6400	.4900	.3600	.2500
	1	.0198	.0392	.0950	.1800	.2550	.3200	.4200	.4800	.5000
	2	.0001	.0004	.0025	.0100	.0225	.0400	.0900	.1600	.2500
3	0	.9703	.9412	.8574	.7290	.6141	.5120	.3430	.2160	.1250
	1	.0294	.0576	.1354	.2430	.3251	.3840	.4410	.4320	.3750
	2	.0003	.0012	.0071	.0270	.0574	.0960	.1890	.2880	.3750
	3	—	—	.0001	.0010	.0034	.0080	.0270	.0640	.1250
4	0	.9606	.9224	.8145	.6561	.5220	.4096	.2401	.1296	.0625
	1	.0388	.0753	.1715	.2916	.3685	.4096	.4116	.3456	.2500
	2	.0006	.0023	.0135	.0486	.0975	.1536	.2646	.3456	.3750
	3	—	—	.0005	.0036	.0115	.0256	.0756	.1536	.2500
	4	—	—	—	.0001	.0005	.0016	.0081	.0256	.0625

The column heading over the probabilities is π.

Using Excel

Figure 6.10 shows Excel's Formulas > Insert Function menu to calculate the probability of $x = 67$ successes in $n = 1{,}024$ trials with success probability $\pi = .048$. Alternatively, you could just enter the formula =BINOM.DIST(67, 1024, 0.048, 0) in the spreadsheet cell.

MegaStat will compute an entire binomial PDF (not just a single point probability) for any n and π that you specify, and also show you a graph of the PDF. This is even easier than entering your own Excel functions.

If you need binomial random data, Excel's Data Analysis or MegaStat will generate random binomial values. A third option is to use the function =BINOM.INV(4, .20, RAND()) to create a binomial random variable value.

Figure 6.10

Excel Binomial Probability Function

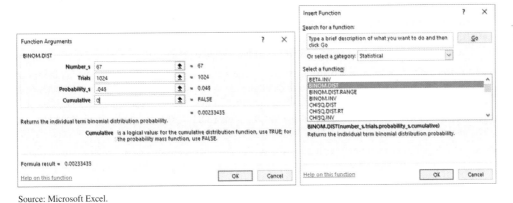

Source: Microsoft Excel.

Section Exercises

6.13 List the *X* values that are included in each italicized event.

   a. You can miss *at most 2* quizzes out of 16 quizzes (*X* = number of missed quizzes).

   b. You go to Starbucks *at least 4* days a week (*X* = number of Starbucks visits).

   c. You are penalized if you have *more than 3* absences out of 10 lectures (*X* = number of absences).

6.14 Write the probability of each italicized event in symbols (e.g., $P(X \geq 5)$).

   a. *At least* 7 correct answers on a 10-question quiz (*X* = number of correct answers).

   b. *Fewer than* 4 "phishing" e-mails out of 20 e-mails (*X* = number of phishing e-mails).

   c. *At most* 2 no-shows at a party where 15 guests were invited (*X* = number of no-shows).

6.15 Find the mean and standard deviation for each binomial random variable:
a. $n = 8, \pi = .10$
b. $n = 10, \pi = .40$
c. $n = 12, \pi = .50$

6.16 Find the mean and standard deviation for each binomial random variable:
a. $n = 30, \pi = .90$
b. $n = 80, \pi = .70$
c. $n = 20, \pi = .80$

6.17 Calculate each binomial probability:
a. $X = 5, n = 9, \pi = .90$
b. $X = 0, n = 6, \pi = .20$
c. $X = 9, n = 9, \pi = .80$

6.18 Calculate each binomial probability:
a. $X = 2, n = 8, \pi = .10$
b. $X = 1, n = 10, \pi = .40$
c. $X = 3, n = 12, \pi = .70$

6.19 Calculate each compound event probability:
a. $X \leq 3, n = 8, \pi = .20$
b. $X > 7, n = 10, \pi = .50$
c. $X < 3, n = 6, \pi = .70$

6.20 Calculate each compound event probability:
a. $X \leq 10, n = 14, \pi = .95$
b. $X > 2, n = 5, \pi = .45$
c. $X \leq 1, n = 10, \pi = .15$

6.21 Calculate each binomial probability:
a. More than 10 successes in 16 trials with an 80 percent chance of success.
b. At least 4 successes in 8 trials with a 40 percent chance of success.
c. No more than 2 successes in 6 trials with a 20 percent chance of success.

6.22 Calculate each binomial probability:
a. Fewer than 4 successes in 12 trials with a 10 percent chance of success.
b. At least 3 successes in 7 trials with a 40 percent chance of success.
c. At most 9 successes in 14 trials with a 60 percent chance of success.

6.23 In the Ardmore Hotel, 20 percent of the customers pay by American Express credit card. (a) Of the next 10 customers, what is the probability that none pay by American Express? (b) At least two? (c) Fewer than three? (d) What is the expected number who pay by American Express? (e) Find the standard deviation. (f) Construct the probability distribution (using Excel or Appendix A). (g) Make a graph of its PDF, and describe its shape.

6.24 Historically, 5 percent of a mail-order firm's repeat charge-account customers have an incorrect current address in the firm's computer database. (a) What is the probability that none of the next 12 repeat customers who call will have an incorrect address? (b) One customer? (c) Two customers? (d) Fewer than three? (e) Construct the probability distribution (using Excel or Appendix A), make a graph of its PDF, and describe its shape.

6.25 At a Noodles & Company restaurant, the probability that a customer will order a nonalcoholic beverage is .38. Use Excel to find the probability that in a sample of 5 customers, (a) none of the 5 will order a nonalcoholic beverage, (b) at least 2 will, (c) fewer than 4 will, (d) all 5 will order a nonalcoholic beverage.

6.26 J.D. Power and Associates says that 60 percent of car buyers now use the Internet for research and price comparisons. (a) Find the probability that in a sample of 8 car buyers, all 8 will use the Internet; (b) at least 5; (c) more than 4. (d) Find the mean and standard deviation of the probability distribution. (e) Sketch the PDF (using Excel or Appendix A) and describe its appearance (e.g., skewness).

6.27 There is a 70 percent chance that an airline passenger will check bags. In the next 16 passengers who check in for their flight at Denver International Airport, find the probability that (a) all will check bags; (b) fewer than 10 will check bags; (c) at least 10 will check bags.

6.28 Police records in the town of Saratoga show that 15 percent of the drivers stopped for speeding have invalid licenses. If 12 drivers are stopped for speeding, find the probability that (a) none will have an invalid license; (b) exactly one will have an invalid license; (c) at least 2 will have invalid licenses.

6.5 POISSON DISTRIBUTION

Named for the French mathematician Siméon-Denis Poisson (1781–1840), the **Poisson distribution** describes the number of occurrences within a randomly chosen unit of time (e.g., minute, hour) or space (e.g., square foot, linear mile). For the Poisson distribution to apply, the events must occur randomly and independently over a continuum of time or space, as illustrated below. We will call the continuum "time" because the most common Poisson application is modeling **arrivals** *per unit of time*. Each dot (•) is an occurrence of the event of interest.

Random Unit of Time	Random Unit of Time		Random Unit of Time
$X = 3$	$X = 1$		$X = 5$

Flow of Time ⟶

Let X = the number of events per unit of time. Then the value of X is a random variable. The figure above shows that we could get $X = 3$ or $X = 1$ or $X = 5$ events, for randomly chosen units of time.

We often call the Poisson distribution the *model of arrivals* (customers, defects, accidents). Arrivals can reasonably be regarded as Poisson events if each event is **independent** (i.e., each arrival has no effect on the probability of other arrivals). Some situations lack this characteristic. For example, computer users know that a power interruption often presages another within seconds or minutes. But, as a practical matter, the Poisson assumptions often are met sufficiently to make it a useful model of reality. For example:

- X = number of customers arriving at a bank ATM in a given minute.

- X = number of file server virus infections at a data center during a 24-hour period.

- X = number of asthma patient arrivals in a given hour at a walk-in clinic.

- X = number of Airbus 330 aircraft engine shutdowns per 100,000 flight hours.

- X = number of blemishes per sheet of white bond paper.

The Poisson model has only one parameter, denoted λ (the lowercase Greek letter "lambda"), representing the *mean number of events per unit of time or space*. The unit of time usually is chosen to be short enough that the mean arrival rate is not large (typically $\lambda < 20$). For this reason, the Poisson distribution is sometimes called the *model of* **rare events**. If the mean is large, we can reformulate the time units to yield a smaller mean. For example, $\lambda = 90$ events per hour is the same as $\lambda = 1.5$ events per minute. However, the Poisson model works for any λ as long as its assumptions are met.

Characteristics of the Poisson Distribution

All characteristics of the Poisson model are determined by its mean λ, as shown in Table 6.9. The constant e (the base of the natural logarithm system) is approximately 2.71828 (to see a more precise value of e, use your calculator's e^x function with $x = 1$). The mean of the Poisson distribution is λ, and its standard deviation is the square root of the mean. The simplicity of the Poisson formulas makes it an attractive model (e.g., easier than the binomial). Unlike the binomial, X has no obvious limit, that is, the number of events that can occur in a given unit of time is not bounded. However, Poisson probabilities taper off toward zero as X increases, so the effective range is usually small.

Table 6.9

Poisson Distribution

Parameter	λ = mean arrivals per unit of time or space
PDF	$P(X = x) = \dfrac{\lambda^x e^{-\lambda}}{x!}$
Excel* PDF	=POISSON.DIST($x, \lambda, 0$)
Excel* CDF	=POISSON.DIST($x, \lambda, 1$)
Domain	$x = 0, 1, 2, \ldots$ (no obvious upper limit)
Mean	λ
Standard deviation	$\sqrt{\lambda}$
Comments	Always right-skewed, but less so for larger λ.

*See the **R Software Supplement** at the end of this chapter for a list of R's Poisson probability functions.

Example 6.4

Coffee Shop Customers

📂 CoffeeShop

On Thursday morning between 9 a.m. and 10 a.m., customers arrive at a mean rate of 1.7 customers per minute at the campus coffee shop and enter the queue (if any) for a coffee. Using the Poisson formulas with $\lambda = 1.7$, the equations for the PDF, mean, and standard deviation are:

$$\text{PDF: } P(X = x) = \frac{\lambda^x e^{-\lambda}}{x!} = \frac{(1.7)^x e^{-1.7}}{x!}$$

$$\text{Mean: } \lambda = 1.7$$

$$\text{Standard deviation: } \sigma = \sqrt{\lambda} = \sqrt{1.7} = 1.304$$

Suppose we would like to know the probability that exactly 3 customers will enter the queue for a coffee in a minute. Using $\lambda = 1.7$ and $x = 3$:

$$P(X = 3) = \frac{1.7^3 e^{-1.7}}{3!} = .1496$$

Alternatively, we could have calculated this using the Excel function =POISSON.DIST (3, 1.7, 0). The third parameter, 0, means that we want Excel to calculate $P(X = 3)$ rather than $P(X \leq 3)$.

Table 6.10 shows values for the Poisson PDF for different values of λ. Going down each column, the probabilities must sum to 1.0000 (except for rounding because these probabilities are only accurate to four decimals). The Poisson probability function is

(6.9)
$$P(X = x) = \frac{\lambda^x e^{-\lambda}}{x!}, \text{ for } X = 1, 2, 3, 4, \ldots$$

Table 6.10

Poisson Probabilities for Various Values of λ

x	$\lambda = 0.1$	$\lambda = 0.5$	$\lambda = 0.8$	$\lambda = 1.6$	$\lambda = 2.0$
0	.9048	.6065	.4493	.2019	.1353
1	.0905	.3033	.3595	.3230	.2707
2	.0045	.0758	.1438	.2584	.2707
3	.0002	.0126	.0383	.1378	.1804
4	—	.0016	.0077	.0551	.0902
5	—	.0002	.0012	.0176	.0361
6	—	—	.0002	.0047	.0120
7	—	—	—	.0011	.0034
8	—	—	—	.0002	.0009
9	—	—	—	—	.0002
Sum	1.0000	1.0000	1.0000	1.0000	1.0000

Note: Probabilities less than .0001 have been omitted. Columns may not sum to 1 due to rounding.

Poisson distributions are always right-skewed (long right tail) but become less skewed and more bell-shaped as λ increases, as illustrated in Figure 6.11.

Figure 6.11

Poisson Becomes Less Skewed for Larger λ

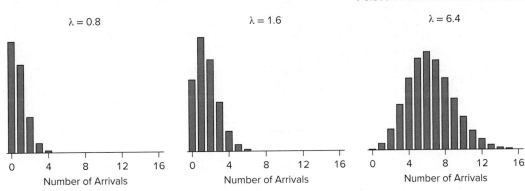

Recognizing Poisson Applications

Can you recognize a Poisson situation? The Poisson distribution has four main characteristics.

- An event of interest occurs randomly over time or space.
- The average arrival rate (λ) remains constant.
- The arrivals are independent of each other.
- The random variable (X) is the number of events within an observed time interval.

Ask yourself if the four characteristics above make sense in the following examples.

In the last week, how many credit card applications did you receive by mail?

In the last week, how many checks did you write?

In the last week, how many e-mail viruses did your firewall deflect?

It may be a Poisson process, even if you don't know the mean (λ). In business applications, the value of λ would have to be estimated from experience, but in this chapter λ will be given.

Using the Poisson Formula

Table 6.11 shows the PDF and CDF for each value of X. The probabilities for individual X-values can be calculated by inserting λ = 1.7 into the Poisson PDF or by using Excel's Poisson function =POISSON.DIST(x, λ, cumulative) where cumulative is 0 (if you want a PDF) or 1 (if you want a CDF).

Table 6.11

Probability Distribution for λ = 1.7

x	P(X = x)	P(X ≤ x)
0	.1827	.1827
1	.3106	.4932
2	.2640	.7572
3	.1496	.9068
4	.0636	.9704
5	.0216	.9920
6	.0061	.9981
7	.0015	.9996
8	.0003	.9999
9	.0001	1.0000

PDF Formula	*Excel Function*
$P(X=0) = \dfrac{1.7^0 e^{-1.7}}{0!} = .1827$	=POISSON.DIST(0, 1.7, 0)
$P(X=1) = \dfrac{1.7^1 e^{-1.7}}{1!} = .3106$	=POISSON.DIST(1, 1.7, 0)
$P(X=2) = \dfrac{1.7^2 e^{-1.7}}{2!} = .2640$	=POISSON.DIST(2, 1.7, 0)
$\vdots$	$\vdots$
$P(X=8) = \dfrac{1.7^8 e^{-1.7}}{8!} = .0003$	=POISSON.DIST(8, 1.7, 0)
$P(X=9) = \dfrac{1.7^9 e^{-1.7}}{9!} = .0001$	=POISSON.DIST(9, 1.7, 0)

Poisson probabilities must sum to 1 (except due to rounding) as with any discrete probability distribution. Beyond $X = 9$, the probabilities are below .0001. Graphs of the PDF and CDF are shown in Figure 6.12. The most likely event is one arrival (probability .3106, or a 31.1 percent chance), although two arrivals are almost as likely (probability .2640, or a 26.4 percent chance). This PDF would help the coffee shop schedule for the Thursday morning work shift.

Figure 6.12

Poisson PDF and CDF for $\lambda = 1.7$

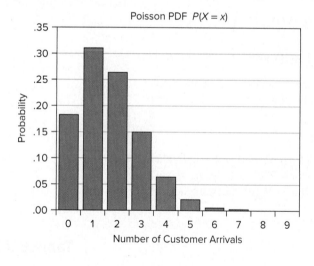

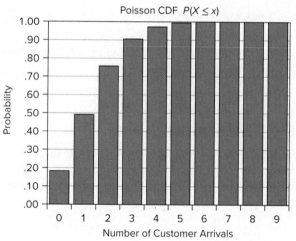

Compound Events

Cumulative probabilities can be evaluated by summing individual X probabilities. For example, the probability that *at most* two customers will arrive in a given minute is the sum of probabilities for several events:

$$P(X \leq 2) = P(X=0) + P(X=1) + P(X=2) = .1827 + .3106 + .2640 = .7572$$

Alternatively, we could calculate this probability using Excel's CDF function:

$$P(X \leq 2) = \text{POISSON.DIST}(2, 1.7, 1) = .7572$$

We could then calculate the probability of *at least* three customers (the complementary event):

$$P(X \geq 3) = 1 - P(X \leq 2) = 1 - \text{POISSON.DIST}(2, 1.7, 1) = 1 - .7572 = .2428$$

Using Tables (Appendix B)

Appendix B facilitates Poisson calculations, as illustrated in Figure 6.13 with highlighted probabilities for the terms in the sum for $P(X \geq 3)$. Appendix B doesn't go beyond $\lambda = 20$, partly because the table would become huge, but mainly because we have Excel.

X	λ					
	1.6	**1.7**	**1.8**	**1.9**	**2.0**	**2.1**
0	.2019	.1827	.1653	.1496	.1353	.1225
1	.3230	.3106	.2975	.2842	.2707	.2572
2	.2584	.2640	.2678	.2700	.2707	.2700
3	.1378	.1496	.1607	.1710	.1804	.1890
4	.0551	.0636	.0723	.0812	.0902	.0992
5	.0176	.0216	.0260	.0309	.0361	.0417
6	.0047	.0061	.0078	.0098	.0120	.0146
7	.0011	.0015	.0020	.0027	.0034	.0044
8	.0002	.0003	.0005	.0006	.0009	.0011
9	—	.0001	.0001	.0001	.0002	.0003
10	—	—	—	—	—	.0001
11	—	—	—	—	—	—

Figure 6.13

Poisson Probabilities for $P(X \geq 3)$ from Appendix B

Using Excel

Tables are helpful for taking statistics exams (when you may not have access to Excel). However, tables contain only selected λ values, and in real-world problems, we cannot expect λ always to be a nice round number. Excel's menus are illustrated in Figure 6.14. In this example, Excel calculates =POISSON.DIST(11, 17, 0) as .035544812, which is more accurate than Appendix B.

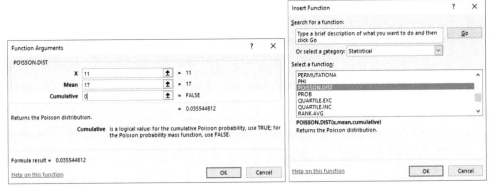

Source: Microsoft Excel.

Figure 6.14

Excel Poisson Probability Function

Section Exercises

McGraw Hill connect

6.29 Find the mean and standard deviation for each Poisson:

  a. $\lambda = 1.0$
  b. $\lambda = 2.0$
  c. $\lambda = 4.0$

6.30 Find the mean and standard deviation for each Poisson:

  a. $\lambda = 9.0$
  b. $\lambda = 12.0$
  c. $\lambda = 7.0$

6.31 Calculate each Poisson probability:

  a. $P(X = 6), \lambda = 4.0$
  b. $P(X = 10), \lambda = 12.0$
  c. $P(X = 4), \lambda = 7.0$

6.32   Calculate each Poisson probability:

    a. $P(X = 2)$, $\lambda = 0.1$
    b. $P(X = 1)$, $\lambda = 2.2$
    c. $P(X = 3)$, $\lambda = 1.6$

6.33   Calculate each compound event probability:

    a. $P(X \le 3)$, $\lambda = 4.3$
    b. $P(X > 7)$, $\lambda = 5.2$
    c. $P(X < 3)$, $\lambda = 2.7$

6.34   Calculate each compound event probability:

    a. $P(X \le 10)$, $\lambda = 11.0$
    b. $P(X > 3)$, $\lambda = 5.2$
    c. $P(X < 2)$, $\lambda = 3.7$

6.35   Calculate each Poisson probability:

    a. More than 10 arrivals with $\lambda = 8.0$.
    b. No more than 5 arrivals with $\lambda = 4.0$.
    c. At least 2 arrivals with $\lambda = 5.0$

6.36   Calculate each Poisson probability:

    a. Fewer than 4 arrivals with $\lambda = 5.8$.
    b. At least 3 arrivals with $\lambda = 4.8$.
    c. At most 9 arrivals with $\lambda = 7.0$.

6.37   According to J.D. Power and Associates' 2006 Initial Quality Study, consumers reported on average 1.7 problems per vehicle with new 2006 Volkswagens. In a randomly selected new Volkswagen, find the probability of (a) at least one problem; (b) no problems; (c) more than three problems. (d) Construct the probability distribution using Excel or Appendix B, make a graph of its PDF, and describe its shape. (Data are from J.D. Power and Associates 2006 Initial Quality Study[SM].)

6.38   At an outpatient mental health clinic, appointment cancellations occur at a mean rate of 1.5 per day on a typical Wednesday. Let X be the number of cancellations on a particular Wednesday. (a) Justify the use of the Poisson model. (b) What is the probability that no cancellations will occur on a particular Wednesday? (c) One? (d) More than two? (e) Five or more?

6.39   The average number of items (such as a drink or dessert) ordered by a Noodles & Company customer in addition to the meal is 1.4. These items are called *add-ons*. Define X to be the number of add-ons ordered by a randomly selected customer. (a) Justify the use of the Poisson model. (b) What is the probability that a randomly selected customer orders at least 2 add-ons? (c) No add-ons? (d) Construct the probability distribution using Excel or Appendix B, make a graph of its PDF, and describe its shape.

6.40   (a) Why might the number of yawns per minute by students in a warm classroom not be a Poisson event? (b) Give two additional examples of events per unit of time that might violate the assumptions of the Poisson model, and explain why.

LO 6-6

Use the Poisson approximation to the binomial (optional).

Poisson Approximation to Binomial (Optional)

The binomial and Poisson are close cousins. The Poisson distribution may be used to approximate a binomial by setting $\lambda = n\pi$. This approximation is helpful when the binomial calculation is difficult (e.g., when n is large) and when Excel is not available. For example, suppose 1,000 women are screened for a rare type of cancer that has a nationwide incidence of 6 cases per 10,000 (i.e., $\pi = .0006$). What is the probability of finding two or fewer cases? The number of cancer cases would follow a binomial distribution with $n = 1,000$ and $\pi = .0006$. However, the binomial formula would involve awkward factorials. To use a Poisson approximation, we set the Poisson mean (λ) equal to the binomial mean ($n\pi$):

$$\lambda = n\pi = (1000)(.0006) = 0.6$$

To calculate the probability of x successes, we can then use Appendix B or the Poisson PDF $P(X = x) = \lambda^x e^{-\lambda}/x!$ which is simpler than the binomial PDF $P(X = x) = \dfrac{n!}{x!(n-x)!}\pi^x (1 - \pi)^{n-x}$. The Poisson approximation of the desired probability is $P(X \le 2) = P(0) + P(1) + P(2) = .5488 + .3293 + .0988 = .9769$.

Poisson Approximation

$P(X = 0) = 0.6^0 e^{-0.6}/0! = .5488$

$P(X = 1) = 0.6^1 e^{-0.6}/1! = .3293$

$P(X = 2) = 0.6^2 e^{-0.6}/2! = .0988$

Actual Binomial Probability

$P(X = 0) = \dfrac{1000!}{0!(1000 - 0)!} .0006^0 (1 - .0006)^{1000-0} = .5487$

$P(X = 1) = \dfrac{1000!}{1!(1000 - 1)!} .0006^1 (1 - .0006)^{1000-1} = .3294$

$P(X = 2) = \dfrac{1000!}{2!(1000 - 2)!} .0006^2 (1 - .0006)^{1000-2} = .0988$

The Poisson calculations are easy, and (at least in this example) the Poisson approximation is accurate. The Poisson approximation does a good job in this example, but when is it "good enough" in other situations? The general rule for a good approximation is that n should be "large" and π should be "small." A common rule of thumb says the approximation is adequate if $n \geq 20$ and $\pi \leq .05$.

6.41 An experienced order taker at the L.L.Bean call center has a .003 chance of error on each keystroke (i.e., $\pi = .003$). In 500 keystrokes, find the approximate probability of (a) at least two errors and (b) fewer than four errors. (c) Is the Poisson approximation justified?

6.42 The probability of a manufacturing defect in an aluminum beverage can is .00002. If 100,000 cans are produced, find the approximate probability of (a) at least one defective can and (b) two or more defective cans. (c) Is the Poisson approximation justified?

6.43 Three percent of the letters placed in a certain postal drop box have incorrect postage. Suppose 200 letters are mailed. (a) For this binomial, what is the expected number with incorrect postage? (b) For this binomial, what is the standard deviation? (c) What is the approximate probability that at least 10 letters will have incorrect postage? (d) Fewer than five? (e) Is the Poisson approximation justified?

6.44 In a string of 100 Christmas lights, there is a .01 chance that a given bulb will fail within the first year of use (if one bulb fails, it does not affect the others). Find the approximate probability that two or more bulbs will fail within the first year.

6.45 The probability that a passenger's bag will be mishandled on a U.S. airline is .0046. If 500 passenger bags are checked, (a) what is the expected number of mishandled bags? (b) What is the approximate probability of no mishandled bags? More than two? (c) Is a Poisson approximation justified?

6.6 HYPERGEOMETRIC DISTRIBUTION

LO 6-7

Find hypergeometric probabilities using Excel.

The **hypergeometric distribution** is similar to the binomial except that sampling is *without replacement* from a finite population of N items. Therefore, the trials are not independent and the probability of success is *not* constant from trial to trial. The hypergeometric distribution has three parameters: N (the number of items in the population), n (the number of items in the sample), and s (the number of successes in the population). The distribution of X (the number of successes in the sample) is hypergeometric, with the characteristics shown in Table 6.12. The hypergeometric distribution may be skewed right or left and is symmetric only if $s/N = .50$ (i.e., if the proportion of successes in the population is 50 percent).

Table 6.12

Hypergeometric Distribution

Parameters	N = number of items in the population n = sample size s = number of successes in population
PDF	$P(X = x) = \dfrac{{}_sC_x \, {}_{N-s}C_{n-x}}{{}_N C_n}$
Excel* PDF	=HYPGEOM.DIST(x, n, s, N, 0)
Excel* CDF	=HYPGEOM.DIST(x, n, s, N, 1)
Domain	$\max(0, n - N + s) \leq X \leq \min(s, n)$
Mean	$n\pi$, where $\pi = s/N$
Standard deviation	$\sqrt{n\pi(1 - \pi)} \sqrt{\dfrac{N - n}{N - 1}}$
Comments	Similar to binomial, but sampling is without replacement from a finite population. It can be approximated by a binomial with $\pi = s/N$ if $n/N < 0.05$ and is symmetric if $s/N = 0.50$.

*See the **R Software Supplement** at the end of this chapter for a list of R's hypergeometric probability functions.

The hypergeometric PDF, shown in formula 6.10, uses the formula for combinations:

(6.10)
$$P(X = x) = \frac{{}_sC_x \, {}_{N-s}C_{n-x}}{{}_NC_n}$$

where

$${}_sC_x = \text{the number of ways to choose } x$$
$$\text{successes from } s \text{ successes in the population}$$

$${}_{N-s}C_{n-x} = \text{the number of ways to choose } n - x$$
$$\text{failures from } N - s \text{ failures in the population}$$

$${}_NC_n = \text{the number of ways to choose } n \text{ items}$$
$$\text{from } N \text{ items in the population}$$

and $N - s$ is the number of failures in the population, x is the number of successes in the sample, and $n - x$ is the number of failures in the sample.

Example 6.5

Damaged iPads

In a shipment of 10 iPads, 2 are damaged and 8 are good. The receiving department at Best Buy tests a sample of 3 iPads at random to see if they are defective. The number of damaged iPads in the sample is a random variable X. The problem description is

$N = 10$ (number of iPads in the shipment)

$n = 3$ (sample size drawn from the shipment)

$s = 2$ (number of damaged iPads in the shipment, i.e., population "successes")

$N - s = 8$ (number of nondamaged iPads in the shipment)

$x = ?$ (number of damaged iPads in the sample, i.e., sample "successes")

$n - x = ?$ (number of nondamaged iPads in the sample)

It is tempting to think of this as a binomial problem with $n = 3$ and $\pi = s/N = 2/10 = .20$. But π is not constant. On the first draw, the probability of a damaged iPad is indeed $\pi_1 = 2/10$. But on the second draw, the probability of a damaged iPad could be $\pi_2 = 1/9$ (if the first draw contained a damaged iPad) or $\pi_2 = 2/9$ (if the first draw did not contain a damaged iPad). On the third draw, the probability of a damaged iPad could be $\pi_3 = 0/8$, $\pi_3 = 1/8$, or $\pi_3 = 2/8$ depending on what happened in the first two draws.

Recognizing Hypergeometric Applications

Look for a finite population (N) containing a known number of successes (s) and *sampling without replacement* (n items in the sample) where the probability of success is not constant for each sample item drawn. For example:

- Forty automobiles are to be inspected for California emissions compliance. Thirty-two are compliant, but 8 are not. A sample of 7 cars is chosen at random. What is the probability that all are compliant? At least 5?

- A law enforcement agency must process 500 background checks for firearms purchasers. Fifty applicants are convicted felons. Through a computer error, 10 applicants are approved without a background check. What is the probability that none is a felon? At least two are?

- A medical laboratory receives 40 blood specimens to check for HIV. Eight actually contain HIV. A worker is accidentally exposed to 5 specimens. What is the probability that none contained HIV?

Using the Hypergeometric Formula

For the iPad example, the only possible values of x are 0, 1, and 2 because there are only 2 damaged iPads in the population. The probabilities are

PDF Formula   *Excel Function*

$$P(X = 0) = \frac{{}_2C_0\,{}_8C_3}{{}_{10}C_3} = \frac{\left(\frac{2!}{0!2!}\right)\left(\frac{8!}{3!5!}\right)}{\left(\frac{10!}{3!7!}\right)} = \frac{56}{120} = \frac{7}{15} = .4667 \qquad \text{=HYPGEOM.DIST(0, 3, 2, 10, 0)}$$

$$P(X = 1) = \frac{{}_2C_1\,{}_8C_2}{{}_{10}C_3} = \frac{\left(\frac{2!}{1!1!}\right)\left(\frac{8!}{2!6!}\right)}{\left(\frac{10!}{3!7!}\right)} = \frac{56}{120} = \frac{7}{15} = .4667 \qquad \text{=HYPGEOM.DIST(1, 3, 2, 10, 0)}$$

$$P(X = 2) = \frac{{}_2C_2\,{}_8C_1}{{}_{10}C_3} = \frac{\left(\frac{2!}{2!0!}\right)\left(\frac{8!}{1!7!}\right)}{\left(\frac{10!}{3!7!}\right)} = \frac{8}{120} = \frac{1}{15} = .0667 \qquad \text{=HYPGEOM.DIST(2, 3, 2, 10, 0)}$$

The values of $P(X)$ sum to 1, as they should: $P(0) + P(1) + P(2) = 7/15 + 7/15 + 1/15 = 1$. We also can find the probability of compound events. For example, the probability of at least one damaged iPad is $P(X \geq 1) = P(1) + P(2) = 7/15 + 1/15 = 8/15 = .533$, or 53.3 percent. The PDF and CDF are illustrated in Figure 6.15.

Figure 6.15

Hypergeometric PDF and CDF Illustrated with $N = 10$, $n = 3$, and $S = 2$

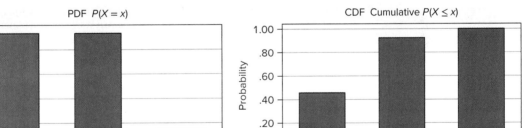

Using Excel

The hypergeometric formula is tedious and tables are impractical because there are three parameters, so we prefer Excel's hypergeometric function =HYPGEOM.DIST(x, n, s, N, 0). For example, using $x = 5$, $n = 10$, $s = 82$, $N = 194$, the formula =HYPGEOM.DIST(5, 10, 82, 194, 0) gives 0.222690589, as illustrated in Figure 6.16. You also can get hypergeometric probabilities from MegaStat or Minitab (menus not shown).

Figure 6.16

Excel Hypergeometric Probability Function

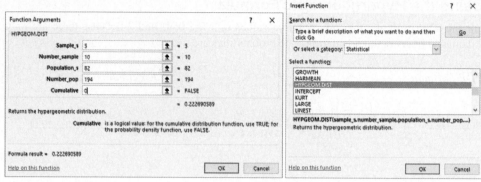

Source: Microsoft Excel.

Section Exercises

![McGraw Hill connect]

6.46 (a) State the values that X can assume in each hypergeometric scenario. (b) Use the hypergeometric PDF formula to find the probability requested. (c) Check your answer by using Excel.

    i. $N = 10$, $n = 3$, $s = 4$, $P(X = 3)$
    ii. $N = 20$, $n = 5$, $s = 3$, $P(X = 2)$
    iii. $N = 36$, $n = 4$, $s = 9$, $P(X = 1)$
    iv. $N = 50$, $n = 7$, $s = 10$, $P(X = 3)$

6.47 ABC Warehouse has eight refrigerators in stock. Two are side-by-side models, and six are top-freezer models. (a) Using Excel, calculate the entire hypergeometric probability distribution for the number of top-freezer models in a sample of four refrigerators chosen at random. (b) Make an Excel graph of the PDF for this probability distribution and describe its appearance.

6.48 A statistics textbook chapter contains 60 exercises, 6 of which are essay questions. A student is assigned 10 problems. Define X to be the number of essay questions the student receives. (a) Use Excel to calculate the entire hypergeometric probability distribution for X. (b) What is the probability that none of the questions are essay? (c) That at least one is essay? (d) That two or more are essay? (e) Make an Excel graph of the PDF of the hypergeometric distribution and describe its appearance.

6.49 Fifty employee travel expense reimbursement vouchers were filed last quarter in the finance department at Ramjac Corporation. Of these, 20 contained errors. A corporate auditor inspects five vouchers at random. Let X be the number of incorrect vouchers in the sample. (a) Use Excel to calculate the entire hypergeometric probability distribution. (b) Find $P(X = 0)$. (c) Find $P(X = 1)$. (d) Find $P(X \geq 3)$. (e) Make an Excel graph of the PDF of the hypergeometric distribution and describe its appearance.

6.50 A medical laboratory receives 40 blood specimens to check for HIV. Eight actually contain HIV. A worker is accidentally exposed to five specimens. (a) Use Excel to calculate the entire hypergeometric probability distribution. (b) What is the probability that none contained HIV? (c) Fewer than three? (d) At least two? (e) Make an Excel graph of the PDF of the hypergeometric distribution and describe its appearance.

LO 6-8

Use the binomial approximation to the hypergeometric (optional).

Binomial Approximation to the Hypergeometric (Optional)

There is a strong similarity between the binomial and hypergeometric models. Both involve samples of size n, and both treat X as the number of successes in the sample. If you replaced each item after it was selected, then you would have a binomial distribution instead of a hypergeometric distribution. If the size of the sample (n) is small in relation to the population (N), then the probability of success is nearly constant on each draw, so the two models are almost the same if we set $\pi = s/N$. A common *rule of thumb* is that the binomial is a safe approximation to the hypergeometric whenever $n/N < 0.05$. In other words, if we sample less than 5 percent of the population, π will remain essentially constant, even if we sample without replacement. For example, suppose we want $P(X = 6)$ for a hypergeometric with $N = 400$, $n = 10$, and $s = 200$. Because $n/N = 10/400 = .025$, the binomial approximation would be acceptable. Set $\pi = s/N = 200/400 = .50$ and use Appendix A to obtain $P(X = 6) = .2051$. But in the iPad example, the binomial approximation would be unacceptable because we sampled more than 5 percent of the population (i.e., $n/N = 3/10 = .30$ for the iPad problem).

> ### Rule of Thumb
>
> If $n/N < .05$, it is safe to use the binomial approximation to the hypergeometric, using sample size n and success probability $\pi = s/N$.

6.51 (a) Check whether the binomial approximation is acceptable in each of the following hypergeometric situations. (b) Find the binomial approximation (using Appendix A) for each probability requested. (c) Check the accuracy of your approximation by using Excel to find the actual hypergeometric probability.

    a. $N = 100$, $n = 3$, $s = 40$, $P(X = 3)$
    b. $N = 200$, $n = 10$, $s = 60$, $P(X = 2)$
    c. $N = 160$, $n = 12$, $s = 16$, $P(X = 1)$
    d. $N = 500$, $n = 7$, $s = 350$, $P(X = 5)$

6.52 Two hundred employee travel expense reimbursement vouchers were filed last year in the finance department at Ramjac Corporation. Of these, 20 contained errors. A corporate auditor audits a sample of five vouchers. Let X be the number of incorrect vouchers in the sample. (a) Justify the use of the binomial approximation. (b) Find the probability that the sample contains no erroneous vouchers. (c) Find the probability that the sample contains at least two erroneous vouchers.

6.53 A law enforcement agency processes 500 background checks for firearms purchasers. Fifty applicants are convicted felons. Through a clerk's error, 10 applicants are approved without checking for felony convictions. (a) Justify the use of the binomial approximation. (b) What is the probability that none of the 10 is a felon? (c) That at least 2 of the 10 are convicted felons? (d) That fewer than 4 of the 10 are convicted felons?

6.54 Four hundred automobiles are to be inspected for California emissions compliance. Of these, 320 actually are compliant, but 80 are not. A random sample of 6 cars is chosen. (a) Justify the use of the binomial approximation. (b) What is the probability that all are compliant? (c) At least 4?

6.7 GEOMETRIC DISTRIBUTION (OPTIONAL)

The **geometric distribution** is related to the binomial. It describes the number of Bernoulli trials until the first success is observed. But the number of trials is not fixed. We define X as the number of trials until the first success and π as the constant *probability* of a success on each trial. The geometric distribution depends only on the parameter π (i.e., it is a one-parameter model). The domain of X is $\{1, 2, \ldots\}$ because we must have at least one trial to obtain our first success, but there is no limit on how many trials may be necessary. The characteristics of the geometric distribution are shown in Table 6.13. It is always skewed to the right. It can be shown that the geometric probabilities sum to 1 and that the mean and standard deviation are nearly the same when π is small. Probabilities diminish as X increases, but not rapidly.

LO 6-9

Calculate geometric probabilities (optional).

Table 6.13

Geometric Distribution

Parameter	π = probability of success
PDF	$P(X = x) = \pi (1 - \pi)^{x-1}$
CDF	$P(X \le x) = 1 - (1 - \pi)^{x}$
Domain	$x = 1, 2, \ldots$ where x = number of trials before the first success
Mean	$1/\pi$
Standard deviation	$\sqrt{\dfrac{1 - \pi}{\pi^{2}}}$
Random data in Excel*	=1+INT(LN(1−RAND())/LN(1 − π))
Comments	Highly skewed.

*Excel offers no geometric PDF or CDF (perhaps because the formulas are simple). But see the **R Software Supplement** at the end of this chapter for a list of R's geometric probability functions.

Example 6.6

Telefund Calling

At Faber University, 15 percent of the alumni (the historical percentage) make a donation or pledge during the annual telefund. What is the probability that the first donation will come within the first five calls? To calculate this geometric probability, we would set $\pi = .15$, apply the PDF formula $P(X = x) = \pi(1 - \pi)x^{-1}$, and then sum the probabilities:

$$P(1) = (.15)(1 - .15)^{1-1} = (.15)(.85)^0 = .1500$$
$$P(2) = (.15)(1 - .15)^{2-1} = (.15)(.85)^1 = .1275$$
$$P(3) = (.15)(1 - .15)^{3-1} = (.15)(.85)^2 = .1084$$
$$P(4) = (.15)(1 - .15)^{4-1} = (.15)(.85)^3 = .0921$$
$$P(5) = (.15)(1 - .15)^{5-1} = (.15)(.85)^4 = .0783$$

Then $P(X \le 5) = P(1) + P(2) + P(3) + P(4) + P(5) = .5563$. Alternately, we can use the CDF to get $P(X \le 5) = 1 - (1 - .15)^5 = 1 - .4437 = .5563$. The CDF is a much easier method when sums are required. The PDF and CDF are illustrated in Figure 6.17.

The expected number of phone calls until the first donation is

$$\mu = 1/\pi = 1/(.15) = 6.67 \text{ calls}$$

On the average, we expect to call between 6 and 7 alumni until the first donation. However, the standard deviation is rather large:

$$\sigma = [(1 - \pi)/\pi^2]^{1/2} = [(1 - .15)/(.15)^2]^{1/2} = 6.15 \text{ calls}$$

The large standard deviation is a signal that it would be unwise to regard the mean as a good prediction of how many trials will be needed until the first donation.

Figure 6.17

Geometric PDF and CDF for $\pi = .15$

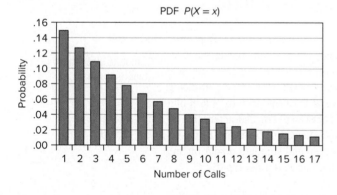

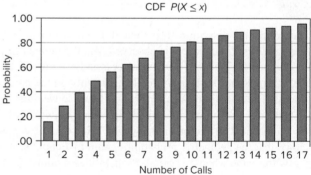

Section Exercises

6.55   Find each geometric probability.

    a. $P(X = 5)$ when $\pi = .50$
    b. $P(X = 3)$ when $\pi = .25$
    c. $P(X = 4)$ when $\pi = .60$

6.56   In the Ardmore Hotel, 20 percent of the guests (the historical percentage) pay by American Express credit card. (a) What is the expected number of guests until the next one pays by American Express credit card? (b) What is the probability that the first guest to use an American Express credit card is within the first 10 to check out?

6.57   In a certain Kentucky Fried Chicken franchise, half of the customers request "crispy" instead of "original," on average. (a) What is the expected number of customers before the next customer requests "crispy"? (b) What is the probability of serving more than 10 customers before the first request for "crispy"?

6.8 TRANSFORMATIONS OF RANDOM VARIABLES (OPTIONAL)

Linear Transformation

A **linear transformation** of a random variable X is performed by adding a constant, multiplying by a constant, or both. Below are two useful rules about the mean and variance of a transformed random variable $aX + b$, where a and b are any constants ($a \geq 0$). Adding a constant shifts the mean but does not affect the standard deviation. Multiplying by a constant affects both the mean and the standard deviation.

> ***Rule 1:*** $\mu_{aX+b} = a\mu_X + b$ (mean of a transformed variable)
>
> ***Rule 2:*** $\sigma_{aX+b} = a\sigma_X$ (standard deviation of a transformed variable)

Application: Exam Scores   Professor Hardtack gave a tough exam whose raw scores had $\mu = 40$ and $\sigma = 10$, so he decided to raise the mean by 20 points. One way to increase the mean to 60 is to shift the curve by adding 20 points to every student's score. Rule 1 says that adding a constant to all X-values will *shift the mean* but will leave the standard deviation unchanged, as illustrated in the left-side graph in Figure 6.18 using $a = 1$ and $b = 20$.

Figure 6.18

Effect of Adding a Constant to X or Multiplying X by a Constant

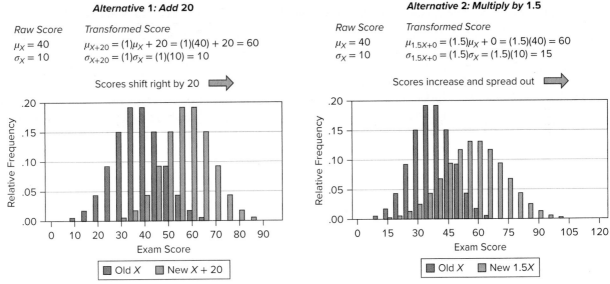

Alternative 1: Add 20

Raw Score	Transformed Score
$\mu_X = 40$	$\mu_{X+20} = (1)\mu_X + 20 = (1)(40) + 20 = 60$
$\sigma_X = 10$	$\sigma_{X+20} = (1)\sigma_X = (1)(10) = 10$

Scores shift right by 20 ⟹

Alternative 2: Multiply by 1.5

Raw Score	Transformed Score
$\mu_X = 40$	$\mu_{1.5X+0} = (1.5)\mu_X + 0 = (1.5)(40) = 60$
$\sigma_X = 10$	$\sigma_{1.5X+0} = (1.5)\sigma_X = (1.5)(10) = 15$

Scores increase and spread out ⟹

Alternatively, Professor Hardtack could multiply every exam score by 1.5, which also would accomplish the goal of raising the mean from 40 to 60. However, Rule 2 says that the standard deviation would rise from 10 to 15, thereby also *increasing the dispersion*. In other words, this policy would "spread out" the students' exam scores. Some scores might even exceed 100, as illustrated in the right-side graph in Figure 6.18.

Application: Total Cost   A linear transformation useful in business is the calculation of total cost as a function of quantity produced: $C = vQ + F$, where C is total cost, v is variable cost per unit, Q is the number of units produced, and F is fixed cost. Sonoro Ltd. is a small firm that manufactures kortholts. Its variable cost per unit is $v = \$35$, its annual fixed cost is $F = \$24{,}000$, and its monthly order quantity Q is a random variable with mean $\mu_Q = 500$ units and standard deviation $\sigma_Q = 40$. Total cost C is a random variable $C = vQ + F$, so we can apply Rules 1 and 2:

$$\text{Mean of total cost: } \mu_{vQ+F} = v\mu_Q + F = (35)(500) + 24{,}000 = \$41{,}500$$
$$\text{Std. Dev. of total cost: } \sigma_{vQ+F} = v\sigma_Q = (35)(40) = \$1{,}400$$

Sums of Random Variables

Below are two useful rules that apply to **sums of random variables**. For example, if a firm has k different products, each with a random demand, then total revenue R is the sum of the revenue for each of the k products: $R = R_1 + R_2 + R_3 + \cdots + R_k$. Rule 3 says that the means can be added. Rule 4 says that the variances can be added *if* the variables are independent. These rules apply to the sum of any number of variables and are useful for analyzing situations where we must combine random variables.

> **Rule 3:** $\mu_{X+Y} = \mu_X + \mu_Y$ (mean of sum of two random variables X and Y)

> **Rule 4:** $\sigma_{X+Y} = \sqrt{\sigma_X^2 + \sigma_Y^2}$ (standard deviation of sum if X and Y are independent)

Note that for the *difference* of two random variables X and Y, $\mu_{X-Y} = \mu_X - \mu_Y$ and $\sigma_{X-Y} = \sqrt{\sigma_X^2 + \sigma_Y^2}$. The variance in both X and Y still contribute to the variance in $X - Y$.

Application: Gasoline Expenses   The daily gasoline expense of Apex Movers, Inc., is a random variable with mean $\mu = \$125$ and standard deviation $\sigma = \$35$ ($\sigma^2 = 1{,}225$). If we define Y to be the gasoline expense *per year,* and there are 250 working days per year at Apex Movers, then

$$\mu_Y = \mu + \cdots + \mu = \underbrace{\$125 + \cdots + \$125}_{250 \text{ times}} = \$31{,}250$$

$$\sigma_Y = \sqrt{\sigma^2 + \cdots + \sigma^2} = \underbrace{\sqrt{1225 + \cdots + 1225}}_{250 \text{ times}} = \$553.399$$

This assumes that daily gasoline expenses are independent of each other.

Application: Project Scheduling   The initial phase of a construction project entails three activities that must be undertaken sequentially (i.e., the second activity cannot begin until the first is complete, and so on) and the time (in days) to complete each activity is a random variable with a known mean and standard deviation:

Excavation	*Foundations*	*Structural Steel*
$\mu_1 = 25$ days	$\mu_2 = 14$ days	$\mu = 58$ days
$\sigma_1 = 3$ days	$\sigma_2 = 2$ days	$\sigma = 7$ days

By Rule 3, the mean time to complete the whole project is the sum of the task means (even if the task times are not independent):

$$\mu = \mu_1 + \mu_2 + \mu_3 = 25 + 14 + 58 = 97 \text{ days}$$

By Rule 4, if the times to complete each activity are *independent,* the overall variance for the project is the sum of the task variances, so the standard deviation for the entire project is:

$$\sigma = \sqrt{\sigma_1^2 + \sigma_2^2 + \sigma_3^2} = \sqrt{3^2 + 2^2 + 7^2} = \sqrt{62} = 7.874 \text{ days}$$

From this information, we can construct $\mu \pm 1\sigma$ or $\mu \pm 2\sigma$ intervals for the entire project:

$$97 \pm (1)(7.874), \text{ or between } 89.1 \text{ and } 104.9 \text{ days}$$
$$97 \pm (2)(7.874), \text{ or between } 81.3 \text{ and } 112.8 \text{ days}$$

So the *Empirical Rule* would imply that there is about a 95 percent chance that the project will take between 81.3 and 112.8 days. This calculation could help the construction firm estimate upper and lower bounds for the project completion time. Of course, if the distribution is not normal, the Empirical Rule may not apply.

Covariance

If X and Y are *not* independent (i.e., if X and Y are *correlated*), then we cannot use Rule 4 to find the standard deviation of their sum. Recall from Chapter 4 that the **covariance** of two random variables, denoted $\text{Cov}(X, Y)$ or σ_{XY}, describes how the variables vary in relation to each other. A positive covariance indicates that the two variables tend to move in the same direction while a negative covariance indicates that the two variables move in opposite directions. We use both the covariance and the variances of X and Y to calculate the standard deviation of the sum of X and Y.

$$\textit{Rule 5:} \quad \sigma_{X+Y} = \sqrt{\sigma_X^2 + \sigma_Y^2 + 2\sigma_{XY}}$$

Finance is one area where the covariance measure has important applications because a portfolio's income is the sum of the incomes of the assets in the portfolio. If two asset returns have a positive covariance, the variance in total return will be *greater* than the sum of the two variances (i.e., risk is increased) while if the two asset returns have a negative covariance, the variance in total return will be *less* (i.e., risk is decreased). If the asset returns are uncorrelated, then their variances can just be summed (Rule 4). The point is that investors can reduce portfolio risk (i.e., smaller variance) by choosing assets appropriately.

Application: Centralized versus Decentralized Warehousing The design of an optimal distribution network is critical for the operations of a firm. High demand variation means there will be a high need for inventory safety stock. Lower demand variation decreases the amount of safety stock needed at a distribution center (DC). A firm can centralize its distribution centers so that one DC services all markets, or it can decentralize its network so each market has its own DC. The demand patterns will dictate whether a centralized or decentralized network is best. Suppose a firm has two markets to service. The parameters of the two distribution centers on demand for each market are given below as well as their covariance.

Market X Demand
$\mu_X = 500$ units/day
$\sigma_X = 45$ units

Market Y Demand
$\mu_Y = 250$ units/day
$\sigma_Y = 35$ units

$\sigma_{XY} = -882$ units

Applying Rule 3, the average *aggregate* demand for both markets combined would be

$$\mu_D = \mu_X + \mu_Y = 500 + 250 = 750 \text{ units}$$

Given that the two markets have a negative covariance, by Rule 5 the standard deviation in aggregate demand would be

$$\sigma_D \sqrt{\sigma_X^2 + \sigma_Y^2 + 2\sigma_{XY}} = \sqrt{45^2 + 35^2 + 2(-882)} = \sqrt{1486} = 38.55 \text{ units}$$

Note that the variation in aggregate demand is less than the variation for market X demand and close to the variation for market Y demand. This means a centralized DC could be the best solution because the firm would need less safety inventory on hand. (Of course, the cost savings would need to be compared to possibly increased transportation costs.)

Section Exercises

connect

6.58 The height of a Los Angeles Lakers basketball player averages 6 feet 7.6 inches (i.e., 79.6 inches) with a standard deviation of 3.24 inches. To convert from inches to centimeters, we multiply by 2.54. (a) In centimeters, what is the mean? (b) In centimeters, what is the standard deviation? (c) Which rules did you use?

6.59 July sales for Melodic Kortholt, Ltd., average $\mu_1 = \$9,500$ with $\sigma_1^2 = \$1,250$. August sales average $\mu_2 = \$7,400$ with $\sigma_2^2 = \$1,425$. September sales average $\mu_3 = \$8,600$ with $\sigma_3^2 = \$1,610$. (a) Find the mean and standard deviation of total sales for the third quarter. (b) What assumptions are you making?

6.60 The mean January temperature in Fort Collins, CO, is 37.1° F with a standard deviation of 10.3° F. Express these Fahrenheit parameters in degrees Celsius using the transformation C = (5/9)F − 17.78.

6.61 There are five accounting exams. Bob's typical score on each exam is a random variable with a mean of 80 and a standard deviation of 5. His final grade is based on the sum of his exam scores. (a) Find the mean and standard deviation of Bob's point total assuming his performances on exams are independent of each other. (b) By the Empirical Rule (see Chapter 4), would you expect that Bob would earn at least 450 points (the required total for an "A")?

Chapter Summary

A **random variable** assigns a numerical value to each outcome in the sample space of a random process. A **discrete random variable** has a countable number of distinct values. Probabilities in a **discrete probability distribution** must be between zero and one and must sum to one. The **expected value** is the mean of the distribution, measuring center, and its **variance** is a measure of variability. A known distribution is described by its **parameters,** which imply its **probability distribution function** (PDF) and its **cumulative distribution function** (CDF).

As summarized in Table 6.14, the **uniform distribution** has two parameters (a, b) that define its domain. The **Bernoulli distribution** has one parameter (π, the probability of success) and two outcomes (0 or 1). The **binomial distribution** has

two parameters (n, π). It describes the sum of n independent Bernoulli random experiments with constant probability of success. It may be skewed left (π > .50) or right (π > .50) or be symmetric (π = .50) but becomes less skewed as n increases. The **Poisson distribution** has one parameter (λ, the mean arrival rate). It describes arrivals of independent events per unit of time or space. It is always right-skewed, becoming less so as λ increases. The **hypergeometric distribution** has three parameters (N, n, s). It is like a binomial, except that sampling of n items is without replacement from a finite population of N items containing s successes. The **geometric distribution** is a one-parameter model (π, the probability of success) that describes the number of trials until the first success. Figure 6.19 shows the relationships among these five discrete models.

Table 6.14

Comparison of Models

Model	Parameters	Mean	Variance	Characteristics
Bernoulli	π	π	π(1 − π)	Used to generate the binomial.
Binomial	n, π	nπ	nπ(1 − π)	Skewed right if π < .50, left if π > .50.
Geometric	π	1/π	(1 − π)/π²	Always skewed right and leptokurtic.
Hypergeometric	N, n, s	nπ where π = s/N	nπ(1 − π)[(N − n)/(N − 1)]	Like binomial except sampling without replacement from a finite population.
Poisson	λ	λ	λ	Always skewed right and leptokurtic.
Uniform	a, b	(a + b)/2	[(b − a + 1)² − 1]/12	Always symmetric and platykurtic.

Figure 6.19

Relationships among Discrete Models

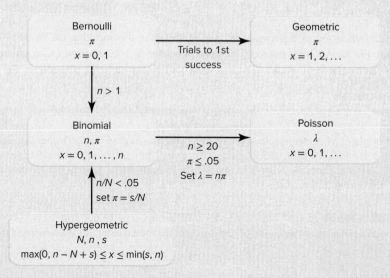

Rules for **linear transformations** of random variables say that adding a constant to a random variable shifts the distribution but does not change its variance, while multiplying a random variable by a constant changes both its mean and its variance. Rules for summing random variables permit adding of their means, but their variances can be summed only if the random variables are independent.

Key Terms

actuarially fair	discrete probability	linear transformation (of a	random variable
arrivals (Poisson)	distribution	random variable)	rare events (Poisson)
Bernoulli experiment	discrete random variable	parameters	sums of random variables
binomial distribution	expected value	Poisson distribution	uniform distribution
covariance	geometric distribution	probability distribution	variance
cumulative distribution	hypergeometric distribution	function	
function	independent	probability model	

Commonly Used Formulas in Discrete Distributions

Total probability: $\displaystyle\sum_{i=1}^{N} P(x_i) = 1$

Expected value: $E(X) = \mu = \displaystyle\sum_{i=1}^{N} x_i P(x_i)$ if there are N distinct values $x_1, x_2, \ldots, x_N$

Variance: $\mathrm{Var}(X) = \sigma^2 = \displaystyle\sum_{i=1}^{N} [x_i - \mu]^2 P(x_i)$

Uniform PDF: $P(X = x) = \dfrac{1}{b - a + 1}$ $x = a, a + 1, \ldots, b$

Binomial PDF: $P(X = x) = \dfrac{n!}{x!(n - x)!} \pi^x (1 - \pi)^{n-x}$ $x = 0, 1, 2, \ldots, n$

Poisson PDF: $P(X = x) = \dfrac{\lambda^x e^{-\lambda}}{x!}$ $x = 0, 1, 2, \ldots$

Hypergeometric PDF: $P(X = x) = \dfrac{{}_sC_x \, {}_{N-s}C_{n-x}}{{}_NC_n}$ $\max(0, n - N + s) \le x \le \min(s, n)$

Geometric PDF: $P(X = x) = \pi(1 - \pi)^{x-1}$ $x = 1, 2, \ldots$

Chapter Review

Note: Questions marked * are harder or rely on optional material from this chapter.

1. Define (a) random process, (b) random variable, (c) discrete random variable, and (d) probability distribution.

2. Without using formulas, explain the meaning of (a) expected value of a random variable, (b) actuarial fairness, and (c) variance of a random variable.

3. What is the difference between a PDF and a CDF? Sketch a picture of each.

4. (a) What are the two parameters of a uniform distribution? (b) Why is the uniform distribution the first one considered in this chapter?

5. (a) Describe a Bernoulli experiment and give two examples. (b) What is the connection between a Bernoulli experiment and a binomial distribution?

6. (a) What are the parameters of a binomial distribution? (b) What is the mean of a binomial distribution? The standard deviation? (c) When is a binomial skewed right? Skewed left? Symmetric? (d) Suggest a data-generating situation that might be binomial.

7. (a) What are the parameters of a Poisson distribution? (b) What is the mean of a Poisson distribution? The standard deviation? (c) Is a Poisson ever symmetric? (d) Suggest a data-generating situation that might be Poisson.

8. In the binomial and Poisson models, why is the assumption of independent events important?

*9. (a) When are we justified in using the Poisson approximation to the binomial? (b) Why would we want to do this approximation?

10. (a) Explain a situation when we would need the hypergeometric distribution. (b) What are the three parameters of the hypergeometric distribution? (c) How does it differ from a binomial distribution?

*11. When are we justified in using (a) the Poisson approximation to the binomial? (b) The binomial approximation to the hypergeometric?

12. Name a situation when we would need the (a) hypergeometric distribution; *(b) geometric distribution; (c) uniform distribution.

*13. What do Rules 1 and 2 say about transforming a random variable?

*14. What do Rules 3 and 4 say about sums of several random variables?

*15. In Rule 5, what does the covariance measure? What happens to the variance of the sum of two random variables when the covariance is (a) positive? (b) negative? (c) zero?

Chapter Exercises

Mc Graw Hill **connect**

Note: Show your work clearly. Problems marked * are harder or rely on optional material from this chapter.

6.62 The probability that a 30-year-old white male will live another year is .99863. What premium would an insurance company charge to break even on a one-year, $1 million term life insurance policy?

6.63 Your grandmother is mailing you a gift card worth $250 for your birthday. There is a 2 percent chance it will be lost in the mail. She could insure the gift card for $4. Should she? Explain, using the concept of expected value.

6.64 Jane is planning to offer a Groupon for inner tube rentals that she will distribute on hot, sunny summer days by the river that runs through her town. Based on her past experience with Groupon, she has assigned the following probability distribution to the number of tubes she will rent on a randomly selected day. If Jane would like her expected revenue to be at least $300 per day, what should the Groupon price be? Round the price to the nearest dollar.

x	30	60	120	180
P(x)	.10	.40	.40	.10

6.65 Consider the Bernoulli model. What would be a typical probability of success (π) for (a) free throw shooting by a good college basketball player? (b) Hits by a good baseball batter? (c) Passes completed by a good college football quarterback? (d) Incorrect answers on a five-part multiple choice exam if you are guessing? (e) Can you suggest reasons why independent events might not be assumed in some of these situations? Explain.

6.66 There is a 14 percent chance that a Noodles & Company customer will order bread with the meal. Use Excel to find the probability that in a sample of 10 customers, (a) more than five will order bread; (b) no more than two will; 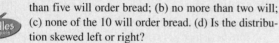 (c) none of the 10 will order bread. (d) Is the distribution skewed left or right?

6.67 In a certain year, on average 10 percent of the vehicles tested for emissions failed the test. Suppose that five vehicles are tested. (a) What is the probability that all pass? (b) All but one pass? (c) Sketch the probability distribution and discuss its shape.

6.68 The probability that an American CEO can transact business in a foreign language is .20. Ten American CEOs are chosen at random. (a) What is the probability that none can transact business in a foreign language? (b) That at least two can? (c) That all 10 can?

6.69 In a certain Kentucky Fried Chicken franchise, half of the customers typically request "crispy" instead of "original." (a) What is the probability that none of the next four customers will request "crispy"? (b) At least two? (c) At most two?

(d) Construct the probability distribution (Excel or Appendix A), make a graph of its PDF, and describe its shape.

6.70 On average, 40 percent of U.S. beer drinkers order light beer. (a) What is the probability that none of the next eight customers who order beer will order light beer? (b) That one customer will? (c) Two customers? (d) Fewer than three? (e) Construct the probability distribution (Excel or Appendix A), make a graph of its PDF, and describe its shape.

6.71 Write the Excel binomial formula for each probability.

a. Three successes in 20 trials with a 30 percent chance of success.

b. Seven successes in 50 trials with a 10 percent chance of success.

c. Six or fewer successes in 80 trials with a 5 percent chance of success.

d. At least 30 successes in 120 trials with a 20 percent chance of success.

6.72 Tired of careless spelling and grammar, a company decides to administer a test to all job applicants. The test consists of 20 sentences. Applicants must state whether each sentence contains any grammar or spelling errors. Half the sentences contain errors. The company requires a score of 14 or more. (a) If an applicant guesses randomly, what is the probability of passing? (b) What minimum score would be required to reduce the probability of "passing by guessing" to 5 percent or less?

6.73 The default rate on government-guaranteed student loans at a certain private 4-year institution is 7 percent. The college extends 10 such loans. (a) What is the probability that none of them will default? (b) That at least three will default? (c) What is the expected number of defaults?

6.74 Experience indicates that 8 percent of the pairs of men's trousers dropped off for dry cleaning will have an object in the pocket that should be removed before cleaning. Suppose that 14 pairs of pants are dropped off and the cleaner forgets to check the pockets. What is the probability that no pair has an object in the pocket?

6.75 A study by the Parents' Television Council showed that 80 percent of movie commercials aired on network television between 8 and 9 p.m. (the prime family viewing hour) were for R-rated films. (a) Find the probability that in 16 commercials during this time slot at least 10 will be for R-rated films. (b) Find the probability of fewer than 8 R-rated films.

6.76 Write the Excel formula for each Poisson probability, using a mean arrival rate of 10 arrivals per hour.

a. Seven arrivals.

b. Three arrivals.

c. Fewer than five arrivals.

d. At least 11 arrivals.

6.77 A small feeder airline knows that the probability is .10 that a reservation holder will not show up for its daily 7:15 a.m. flight into a hub airport. The flight carries 10 passengers. (a) If the flight is fully booked, what is the probability that all those with reservations will show up? (b) If the airline overbooks by selling 11 seats, what is the probability that no one will have to be bumped? (c) That more than one passenger will be bumped? *(d) The airline wants to overbook the flight by enough seats to ensure a 95 percent chance that the flight will be full, even if some passengers may be bumped. How many seats would it sell?

6.78 Although cable modems are tested before they are placed in the installer's truck, the installer knows that 20 percent of them won't work properly. The driver must install eight modems today in a large apartment building. (a) Ten modems are placed in the truck. What is the probability that the driver will have enough working modems *(b) How many should the driver load to ensure a 95 percent probability of having enough working modems?

6.79 (a) Why might the number of calls received per minute at a fire station not be a Poisson event? (b) Name two other events per unit of time that might violate the assumptions of the Poisson model.

6.80 Software filters rely heavily on "blacklists" (lists of known "phishing" URLs) to detect fraudulent e-mails. But such filters typically catch only 20 percent of phishing URLs. Jason receives 16 phishing e-mails. (a) What is the expected number that would be caught by such a filter? (b) What is the chance that such a filter would detect none of them?

6.81 Lunch customers arrive at a Noodles & Company restaurant at an average rate of 2.8 per minute. Define X to be the number of customers to arrive during a randomly selected minute during the lunch hour and assume X has a Poisson distribution. (a) Calculate the probability that exactly five customers will arrive in a minute during the lunch hour. (b) Calculate the probability that no more than five customers will arrive in a minute. (c) What is the average customer arrival rate for a 5-minute interval? (d) What property of the Poisson distribution did you use to find this arrival rate?

6.82 In a Major League Baseball game, the average is 1.0 broken bat per game. Find the probability of (a) no broken bats in a game; (b) at least 2 broken bats.

6.83 In the last 50 years, the average number of deaths due to alligators in Florida is 0.3 death per year. Assuming no change in this average, in a given year find the probability of (a) no alligator deaths; (b) at least 2 alligator deaths.

6.84 In a recent year, potentially dangerous commercial aircraft incidents (e.g., near collisions) averaged 1.2 per 100,000 flying hours. Let X be the number of incidents in a 100,000-hour period. (a) Justify the use of the Poisson model. (b) What is the probability of at least one incident? (c) More than three incidents? (d) Construct the probability distribution (Excel or Appendix B) and make a graph of its PDF.

6.85 At an outpatient mental health clinic, appointment cancellations occur at a mean rate of 1.5 per day on a typical Wednesday. Let X be the number of cancellations on a particular Wednesday. (a) Justify the use of the Poisson model. (b) What is the probability that no cancellations will occur on a particular Wednesday? (c) That one will? (d) More than two? (e) Five or more?

6.86 Car security alarms go off at a mean rate of 3.8 per hour in a large Costco parking lot. Find the probability that in an hour there will be (a) no alarms; (b) fewer than four alarms; and (c) more than five alarms.

6.87 In a certain automobile manufacturing paint shop, paint defects on the hood occur at a mean rate of 0.8 defect per square meter. A hood on a certain car has an area of 3 square meters. (a) Justify the use of the Poisson model. (b) If a customer inspects a hood at random, what is the probability that there will be no defects? (c) One defect? (d) Fewer than two defects?

6.88 Past insurance company audits have found that 2 percent of dependents claimed on an employee's health insurance actually are ineligible for health benefits. An auditor examines a random sample of 7 claimed dependents. (a) What is the probability that all are eligible? (b) That at least one is ineligible?

*6.89 A "rogue wave" (one far larger than others surrounding a ship) can be a threat to oceangoing vessels (e.g., naval vessels, container ships, oil tankers). The European Centre for Medium-Range Weather Forecasts issues a warning when such waves are likely. The average for this rare event is estimated to be .0377 rogue wave per hour in the South Atlantic. Find the probability that a ship will encounter at least one rogue wave in a 5-day South Atlantic voyage (120 hours).

6.90 In Northern Yellowstone Lake, earthquakes occur at a mean rate of 1.2 quakes per year. Let X be the number of quakes in a given year. (a) Justify the use of the Poisson model. (b) What is the probability of fewer than three quakes? (c) More than five quakes? (d) Construct the probability distribution (Excel or Appendix B) and make a graph of its PDF.

6.91 On New York's Verrazano Narrows bridge, traffic accidents occur at a mean rate of 2.0 crashes per day. Let X be the number of crashes in a given day. (a) Justify the use of the Poisson model. (b) What is the probability of at least one crash? (c) Fewer than five crashes? (d) Construct the probability distribution (Excel or Appendix B), make a graph of its PDF, and describe its shape.

APPROXIMATIONS

*6.92 Leaks occur in a pipeline at a mean rate of 1 leak per 1,000 meters. In a 2,500-meter section of pipe, what is the probability of (a) no leaks? (b) Three or more leaks? (c) What is the expected number of leaks?

*6.93 Among live deliveries, the probability of a twin birth is .02. (a) In 200 live deliveries, how many would be expected to have twin births? (b) What is the probability of no twin births? (c) One twin birth? (d) Calculate these probabilities both with and without an approximation. (e) Is the approximation justified? Discuss fully.

*6.94 The probability is .03 that a passenger on United Airlines Flight 9841 is a Platinum flyer (50,000 miles per year). If 200 passengers take this flight, use Excel to find the binomial probability of (a) no Platinum flyers, (b) one Platinum flyer, and (c) two Platinum flyers. (d) Calculate the same probabilities using a Poisson approximation. (e) Is the Poisson approximation justified? Explain.

*6.95 The probability of being "bumped" (voluntarily or involuntarily) on a U.S. airline was .00128. The average number of passengers traveling through Denver International Airport each hour is 5,708. (a) What is the expected number of bumped passengers per hour? (b) What is the approximate Poisson probability of fewer than 10 bumped passengers? More than 5? (c) Would you expect the approximation likely to be accurate (cite a rule of thumb)?

*6.96 On average, 2 percent of all persons who are given a breathalyzer test by the state police pass the test (blood alcohol under .08 percent). Suppose that 500 breathalyzer tests are given. (a) What is the expected number who pass the test? (b) What is the approximate Poisson probability that 5 or fewer will pass the test?

*6.97 The probability of an incorrect call by an NFL referee is .025 (e.g., calling a pass complete, but the decision reversed on instant replays). In a certain game, there are 150 plays. (a) What is the probability of at least 4 incorrect calls by the referees? (b) Justify any assumptions that you made.

*6.98 In CABG surgery, there is a .00014 probability of a retained foreign body (e.g., a sponge or a surgical instrument) left inside the patient. (a) In 100,000 CABG surgeries, what is the expected number of retained foreign bodies? (b) What is the Poisson approximation to the binomial probability of five or fewer retained foreign bodies in 100,000 CABG surgeries? (c) Look up CABG on the Internet if you are unfamiliar with the acronym.

GEOMETRIC

*6.99 The probability of a job offer in a given interview is .25. (a) What is the expected number of interviews until the first job offer? (b) What is the probability the first job offer occurs within the first six interviews?

*6.100 The probability that a bakery customer will order a birthday cake is .04. (a) What is the expected number of customers until the first birthday cake is ordered? (b) What is the probability the first cake order occurs within the first 20 customers?

*6.101 In a certain city, 8 percent of the cars have a burned-out headlight. (a) What is the expected number that must be inspected before the first one with a burned-out headlight is found? (b) What is the probability of finding the first one within the first five cars? *Hint:* Use the CDF.

*6.102 For patients aged 81 to 90, the probability is .07 that a coronary bypass patient will die soon after the surgery. (a) What is the expected number of operations until the first fatality? (b) What is the probability of conducting 20 or more operations before the first fatality? *Hint:* Use the CDF.

*6.103 Historically, 5 percent of a mail-order firm's regular charge-account customers have an incorrect current address in the firm's computer database. (a) What is the expected number of customer orders until the first one with an incorrect current address places an order? (b) What is the probability of mailing 30 bills or more until the first one is returned with a wrong address? *Hint:* Use the CDF.

*6.104 At a certain clinic, 2 percent of all Pap smears show signs of abnormality. What is the expected number of Pap smears that must be inspected before the first abnormal one is found?

TRANSFORMATIONS AND COVARIANCE

*6.105 The weight of a Los Angeles Lakers basketball player averages 233.1 pounds with a standard deviation of 34.95 pounds. To express these measurements in terms a European would understand, we could convert from pounds to kilograms by multiplying by .4536. (a) In kilograms, what is the mean? (b) In kilograms, what is the standard deviation?

*6.106 The Rejuvo Corp. manufactures a granite countertop cleaner and polish. Quarterly sales Q is a random variable with a mean of 25,000 bottles and a standard deviation of 2,000 bottles. Variable cost is $8 per unit, and fixed cost is $150,000. (a) Find the mean and standard deviation of Rejuvo's total cost. (b) If all bottles are sold, what would the selling price have to be to break even, on average? To make a profit of $20,000?

*6.107 A manufacturer fills one-gallon cans (3,785 ml) on an assembly line in two independent steps. First, a high-volume spigot injects most of the paint rapidly. Next, a more precise but slower spigot tops off the can. The fill amount in each step is a normally distributed random variable. For step one, $\mu_1 = 3,420$ ml and $\sigma_1 = 10$ ml, while for step two $\mu_2 = 390$ ml and $\sigma_2 = 2$ ml. Find the mean and standard deviation of the total fill $X_1 + X_2$.

*6.108 A manufacturing project has five independent phases whose completion must be sequential. The time to complete each phase is a random variable. The mean and standard deviation of the time for each phase are shown below. (a) Find the expected completion time. (b) Make a 2-sigma interval around the mean completion time ($\mu \pm 2\sigma$).

Phase	Mean (hours)	Std. Dev. (hours)
Set up dies and other tools	20	4
Milling and machining	10	2
Finishing and painting	14	3
Packing and crating	6	2
Shipping	48	6

*6.109 In September, demand for industrial furnace boilers at a large plumbing supply warehouse has a mean of 7 boilers with a standard deviation of 2 boilers. The warehouse pays a unit cost of $2,225 per boiler plus a fee of $500 per month to act as dealer for these boilers. Boilers are sold for $2,850 each. (a) Find the mean and standard deviation of September profit (revenue minus cost). (b) Which rules did you use?

*6.110 A certain outpatient medical procedure has five steps that must be performed in sequence. (a) Assuming that the time (in minutes) required for each step is an independent

Step	Mean (minutes)	Std. Dev. (minutes)
Patient check-in	15	4
Pre-op preparation	30	6
Medical procedure	25	5
Recovery	45	10
Checkout and discharge	20	5

random variable, find the mean and standard deviation for the total time. (b) Why might the assumption of independence be doubtful?

*6.111 Malaprop Ltd. sells two products. Daily sales of product *A* have a mean of $70 with a standard deviation of $10, while sales of product *B* have a mean of $200 with a standard deviation of $30. Sales of the products tend to rise and fall at the same time, having a positive covariance of 400. (a) Find the mean daily sales for both products together. (b) Find the standard deviation of total sales of both products. (c) Is the variance of the total sales greater than or less than the sum of the variances for the two products?

Related Reading

Forbes Catherine, Merran Evans, Nicholas Hastings, and Brian Peacock. *Statistical Distributions.* 4th ed. John Wiley & Sons, 2011.

CHAPTER 6 More Learning Resources

You can access these *LearningStats* demonstrations through Connect to help you understand visual data displays.

Topic	LearningStats Demonstrations
Discrete distributions	⊠ Probability Calculator ⊠ Random Discrete Data ⊠ Binomial/Poisson Approximation
Table	⊠ Table A–Binomial Probabilities ⊠ Table B–Poisson Probabilities
Applications	⊠ Covariance in Asset Portfolios: A Simulation ⊠ Expected Value: Life Expectancy

Key: ⊠ = Excel

Software Supplement

Discrete Distributions in R

Discrete Probability Distribution	Excel Function	R Function
Binomial distribution (*n* trials, prob π)		
PDF: Returns probability $P(X = x)$	BINOM.DIST(x, n, π, 0)	dbinom(x, n, π)
CDF: Returns probability $P(X \le x)$	BINOM.DIST(x, n, π, 1)	pbinom(x, n, π)
Inverse CDF: Returns *x* for $P(X \le x) = \alpha$	BINOM.INV(n, π, α)	qbinom(α, n, π)
. . .Random data (makes *k* samples in R)	BINOM.INV(n, π, RAND())	rbinom (k, n, π)
Poisson distribution (with mean λ)		
PDF: Returns probability $P(X = x)$	POISSON.DIST(x, λ, 0)	dpois(x, λ)
CDF: Returns probability $P(X \le x)$	POISSON.DIST(x, λ, 1)	ppois(x, λ)
Inverse CDF: Returns *x* for $P(X \le x) = \alpha$	----------------	qpois(α, λ)
. . .Random data (makes *k* samples in R)	----------------	rpois(x, λ)
Hypergeometric distribution (*N* pop size)		
PDF: Returns probability $P(X = x)$	HYPGEOM.DIST(x, n, s, N, 0)	dhyper(x, s, $N−s$, n)
CDF: Returns probability $P(X \le x)$	HYPGEOM.DIST(x, n, s, N, 1)	phyper(x, s, $N−s$, n)
Inverse CDF: Returns *x* for $P(X \le x) = \alpha$	----------------	qhyper(α, s, $N−s$, n)
. . .Random data (makes *k* samples in R)	----------------	rhyper(k, s, $N−s$, n)
Geometric distribution (success prob π)	*x* = num *trials* before success	*x* = num *failures* before success
PDF: Returns probability $P(X = x)$	$\pi^*(1 − \pi)^\wedge(x − 1)$	dgeom(x, λ)
CDF: Returns probability $P(X \le x)$	$1 − (1 − \pi)^\wedge x$	pgeom(x, λ)
Inverse CDF: Returns *x* for $P(X \le x) = \alpha$	----------------	qgeom(α, λ)
. . .Random data (makes *k* samples in R)	----------------	rgeom(k, λ)

CHAPTER CONTENTS

CHAPTER LEARNING OBJECTIVES

When you finish this chapter, you should be able to

LO 7-1   Define a continuous random variable.

LO 7-2   Calculate uniform probabilities.

LO 7-3   Know the form and parameters of the normal distribution.

LO 7-4   Find the normal probability for a given z or x using tables or Excel.

LO 7-5   Solve for z or x for a given normal probability using tables or Excel.

LO 7-6   Use the normal approximation to a binomial or a Poisson.

LO 7-7   Find the exponential probability for a given x.

LO 7-8   Solve for x for a given exponential probability.

LO 7-9   Use the triangular distribution for "what-if" analysis (optional).

CONTINUOUS PROBABILITY DISTRIBUTIONS

Continuous Random Variables

In Chapter 6, you learned about probability models and discrete random variables. We will now expand our discussion of probability models to include models that describe **continuous random variables**. Recall that a discrete random variable usually arises from *counting* something such as the number of customer arrivals in the next minute. In contrast, a continuous random variable usually arises from *measuring* something such as the waiting time until the next customer arrives. Unlike a discrete variable, a continuous random variable can have noninteger (decimal) values.

Probability for a discrete variable is defined at a point such as $P(X = 3)$ or as a sum over a series of points such as $P(X \leq 2) = P(0) + P(1) + P(2)$. But when X is a continuous variable (e.g., waiting time), it does not make sense to speak of probability "at" a particular X-value (e.g., $X = 54$ seconds) because the values of X are not a set of discrete points. Rather, probabilities are defined as *areas under a curve* called the *probability density function* (PDF). Probabilities for a continuous random variable are defined on intervals such as $P(53.5 \leq X \leq 54.5)$ or $P(X < 54)$ or $P(X \geq 53)$. Figure 7.1 illustrates the differences between discrete and continuous random variables. This chapter explains how to recognize data-generating situations that produce continuous random variables, how to calculate event probabilities, and how to interpret the results.

Sorbis/Shutterstock

LO 7-1

Define a continuous random variable.

Discrete Variable: Defined at Each Point

0 1 2 3 4 5

Continuous Variable: Defined over an Interval

53   53.5   54   54.5   55

Figure 7.1

Discrete and Continuous Events

PDFs and CDFs

A continuous probability distribution can be described either by its **probability density function** (PDF) or by its **cumulative distribution function** (CDF). For a continuous random variable, the PDF is an equation that shows the height of the curve $f(x)$ at each possible value of X. Any continuous PDF must be nonnegative, and the area under the entire PDF must be 1. The mean, variance, and shape of the distribution depend on the PDF and its *parameters*. The CDF is denoted $F(x)$ and shows $P(X \leq x)$, the cumulative *area* to the left of a given value of X. The CDF is useful for probabilities, while the PDF reveals the *shape* of the distribution. There are Excel functions for many common PDFs or CDFs.

For example, Figure 7.2 shows a hypothetical PDF and CDF for a distribution of freeway speeds. The random variable *miles per hour* is a continuous variable that can be expressed with any level of precision we choose. The curves are smooth, with the PDF showing the probability density at points along the X-axis. The CDF shows the *cumulative* probability of speeds, gradually approaching 1 as X approaches 90. In this illustration, the distribution is symmetric and bell-shaped (normal or Gaussian) with a mean of 75 and a standard deviation of 5.

Figure 7.2

Freeway Speed Examples

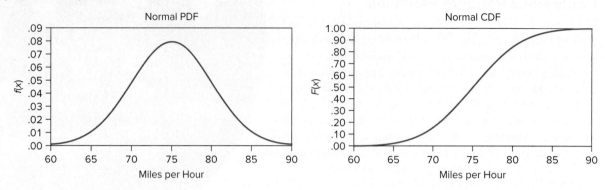

Probabilities as Areas

With discrete random variables, we take sums of probabilities over groups of points. But continuous probability functions are smooth curves, so the area *at* any point would be zero. Instead of taking sums of probabilities, we speak of *areas under curves*. In calculus terms, we would say that $P(a < X < b)$ is the **integral** of the probability density function $f(x)$ over the interval from a to b. Because $P(X = a) = 0$ the expression $P(a < X < b)$ is equal to $P(a \leq X \leq b)$. Figure 7.3 shows the area under a continuous PDF. The entire area under any PDF must be 1.

Figure 7.3

Probability as an Area

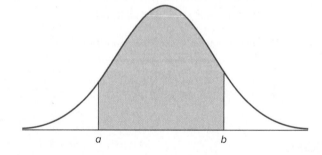

Expected Value and Variance

The mean and variance of a continuous random variable are analogous to $E(X)$ and $\text{Var}(X)$ for a discrete random variable, except that the integral sign $\int$ replaces the summation sign Σ. Integrals are taken over all X-values. The mean is still the balancing point or fulcrum for the entire distribution, and the variance is still a measure of dispersion about the mean. The mean is still the average of all X-values weighted by their probabilities, and the variance is still the weighted average of all squared deviations around the mean. The standard deviation is still the square root of the variance.

		Continuous Random Variable	*Discrete Random Variable*
(7.1)	Mean	$E(X) = \mu = \displaystyle\int_{-\infty}^{+\infty} x f(x)\, dx$	$E(X) = \mu = \displaystyle\sum_{\text{all } x} x P(x)$
(7.2)	Variance	$\text{Var}(X) = \sigma^2 = \displaystyle\int_{-\infty}^{+\infty} (x - \mu)^2 f(x)\, dx$	$\text{Var}(X) = \sigma^2 = \displaystyle\sum_{\text{all } x} [x - \mu]^2 P(x)$

Calculus notation is used here for the benefit of those who have studied it. But statistics can be learned without calculus, if you are willing to accept that others have worked out the details by using calculus. If you decide to become an actuary, you *will* use calculus (so don't sell your calculus book). However, in this chapter, the means and variances are presented *without* proof for the distributions that you are most likely to see applied to business situations.

Section Exercises

connect

7.1 Flight 202 is departing Los Angeles. Is each random variable discrete (D) or continuous (C)?

    a. Number of airline passengers traveling with children under age 3.
    b. Proportion of passengers traveling without checked luggage.
    c. Weight of a randomly chosen passenger on Flight 202.

7.2 It is Saturday morning at Starbucks. Is each random variable discrete (D) or continuous (C)?

    a. Temperature of the coffee served to a randomly chosen customer.
    b. Number of customers who order only coffee with no food.
    c. Waiting time before a randomly chosen customer is handed the order.

7.3 Which of the following could *not* be probability density functions for a continuous random variable? Explain. *Hint:* Find the area under the function $f(x)$.

    a. $f(x) = .25$ for $0 \leq x \leq 1$
    b. $f(x) = .25$ for $0 \leq x \leq 4$
    c. $f(x) = x$ for $0 \leq x \leq 2$

7.4 For a continuous PDF, why can't we sum the probabilities of all x-values to get the total area under the curve?

7.2 UNIFORM CONTINUOUS DISTRIBUTION

Characteristics of the Uniform Distribution

The **uniform continuous distribution** is perhaps the simplest model one can imagine. If X is a random variable that is uniformly distributed between a and b, its PDF has constant height, as shown in Figure 7.4. The uniform continuous distribution is sometimes denoted $U(a, b)$ for short. Its mean and standard deviation are shown in Table 7.1.

LO 7-2

Calculate uniform probabilities.

Figure 7.4

Uniform Distribution

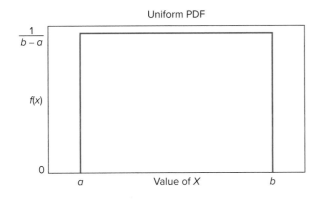
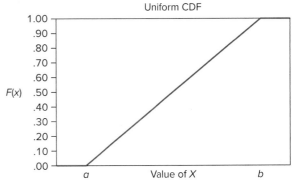

Because the PDF is rectangular, you can easily verify that the area under the curve is 1 by multiplying its base $(b - a)$ by its height $1/(b - a)$. Its CDF increases linearly to 1, as shown in Figure 7.4. Because events can easily be shown as rectangular areas, we rarely need to refer to the CDF, whose formula is just $P(X \leq x) = (x - a)/(b - a)$.

The continuous uniform distribution is similar to the discrete uniform distribution if the x-values cover a wide range. For example, three-digit lottery numbers ranging from 000 to 999 would closely resemble a continuous uniform with $a = 0$ and $b = 999$.

Table 7.1

Uniform Continuous Distribution

Parameters	a = lower limit
	b = upper limit
PDF	$f(x) = \dfrac{1}{b-a}$
CDF	$P(X \leq x) = \dfrac{x-a}{b-a}$
Domain	$a \leq x \leq b$
Mean	$\dfrac{a+b}{2}$
Standard deviation	$\sqrt{\dfrac{(b-a)^2}{12}}$
Shape	Symmetric with no mode.
Random data in Excel	=a+(b−a)*RAND() (returns 1 value)
Random data in R	runif(k, a, b) (returns k values)

Example 7.1

Anesthesia Effectiveness

An oral surgeon injects a painkiller prior to extracting a tooth. Given the varying characteristics of patients, the dentist views the time for anesthesia effectiveness as a uniform random variable that takes between 15 and 30 minutes. In short notation, we could say that X is $U(15, 30)$. Setting $a = 15$ and $b = 30$, we obtain the mean and standard deviation:

$$\mu = \frac{a+b}{2} = \frac{15+30}{2} = 22.5 \text{ minutes}$$

$$\sigma = \sqrt{\frac{(b-a)^2}{12}} = \sqrt{\frac{(30-15)^2}{12}} = 4.33 \text{ minutes}$$

An event probability is simply an interval width expressed as a proportion of the total. Thus, the probability of taking between c and d minutes is

(7.3) $P(c < X < d) = (d-c)/(b-a)$ (area between c and d in a uniform model)

For example, the probability that the anesthetic takes between 20 and 25 minutes is

$$P(20 < X < 25) = (25-20)/(30-15) = 5/15 = 0.3333, \text{ or } 33.33\%.$$

This situation is illustrated in Figure 7.5.

Figure 7.5

Uniform Probability $P(20 < X < 25)$

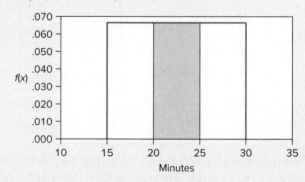

A special case of the continuous uniform distribution, denoted $U(0, 1)$, with limits $a = 0$ and $b = 1$, is shown in Figure 7.6. Using the formulas for the mean and standard deviation, you can easily show that this distribution has $\mu = 0.5000$ and $\sigma = 0.2887$. This special case is important because Excel's function =RAND() uses this distribution. If you create random numbers by using =RAND(), you know what their mean and standard deviation should be. This important distribution is discussed in more detail in later chapters on simulation and goodness-of-fit tests.

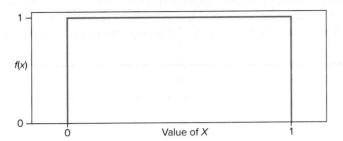

Figure 7.6

Unit $U(0, 1)$ Distribution

Uses of the Uniform Model

The uniform model $U(a, b)$ is used only when you have no reason to imagine that any X-values are more likely than others. The uniform distribution can be useful in business for what-if analysis, in situations where you know the "worst" and "best" range but don't want to make any assumptions about the distribution in between. That may sound like a conservative approach. But bear in mind that if the data-generating situation has any central tendency at all, the assumption of a uniform distribution would lead to a higher standard deviation than might be appropriate.

Section Exercises

connect

7.5   Find the mean and standard deviation for each uniform continuous model.

   a. $U(0, 10)$
   b. $U(100, 200)$
   c. $U(1, 99)$

7.6   Find each uniform continuous probability and sketch a graph showing it as a shaded area.

   a. $P(X < 10)$ for $U(0, 50)$
   b. $P(X > 500)$ for $U(0, 1,000)$
   c. $P(25 < X < 45)$ for $U(15, 65)$

7.7   For a continuous uniform distribution, why is $P(25 < X < 45)$ the same as $P(25 \leq X \leq 45)$?

7.8   Assume the weight of a randomly chosen American passenger car is a uniformly distributed random variable ranging from 2,500 pounds to 4,500 pounds. (a) What is the mean weight of a randomly chosen vehicle? (b) The standard deviation? (c) What is the probability that a vehicle will weigh less than 3,000 pounds? (d) More than 4,000 pounds? (e) Between 3,000 and 4,000 pounds?

7.9   The actual arrival time of the scheduled 10:20 a.m. bus at the Alpine and Broadway stop is a uniformly distributed random variable ranging from 10:18 to 10:23. (a) What is the average arrival time of the 10:20 a.m. bus? (b) The standard deviation? (c) What is the probability that the bus is early? (d) What is the probability that the bus arrives between 10:19 and 10:21 a.m.?

7.10 Jill's resting heart rate (bpm) is a uniformly distributed random variable ranging between 74 and 77 bpm. (a) What is her average resting heart rate? (b) The standard deviation? (c) What is the probability that her bpm on a randomly chosen morning is below 75.5? (d) What is the probability that her bpm on a randomly chosen morning is greater than 76.3?

7.3 NORMAL DISTRIBUTION

LO 7-3

Know the form and parameters of the normal distribution.

Characteristics of the Normal Distribution

The **normal** or **Gaussian distribution**, named for German mathematician Karl Gauss (1777–1855), has already been mentioned several times. Its importance gives it a major role in our discussion of continuous models. A normal probability distribution is defined by two parameters,

μ and σ. It is often denoted $N(\mu, \sigma)$. The domain of a normal random variable is $-\infty < x < +\infty$. However, as a practical matter, the interval $[\mu - 3\sigma, \mu + 3\sigma]$ includes almost all the area (as you know from the Empirical Rule in Chapter 4). Besides μ and σ, the normal probability density function $f(x)$ depends on the constants e (approximately 2.71828) and π (approximately 3.14159). The expected value of a normal random variable is μ, and its variance is σ^2. The normal distribution is always symmetric. Table 7.2 summarizes its main characteristics.

Table 7.2

Normal Distribution

Parameters	μ = population mean
	σ = population standard deviation
PDF	$f(x) = \dfrac{1}{\sigma\sqrt{2\pi}} e^{-\frac{1}{2}\left(\frac{x-\mu}{\sigma}\right)^2}$
Domain	$-\infty < x < +\infty$
Mean	μ
Std. Dev.	σ
Shape	Symmetric, mesokurtic, and bell-shaped.
PDF in Excel*	=NORM.DIST(x, μ, σ, 0)
CDF in Excel*	=NORM.DIST(x, μ, σ, 1)
Random data in Excel	=NORM.INV(RAND(), μ, σ)

*See the **R Software Supplement** at the end of this chapter for a list of R's normal probability functions.

The normal probability density function $f(x)$ reaches a maximum at μ and has points of inflection at $\mu \pm \sigma$, as shown in the left chart in Figure 7.7. Despite its appearance, $f(x)$ does not reach the X-axis beyond $\mu \pm 3\sigma$ but is merely asymptotic to it. Its single peak and symmetry cause some observers to call it "mound-shaped" or "bell-shaped." Its CDF has a "lazy-S" shape, as shown in the right chart in Figure 7.7. It approaches, but never reaches, 1.

Figure 7.7

Normal PDF and CDF

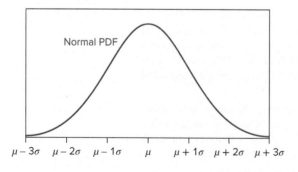
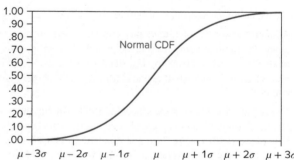

A normal distribution with mean μ and standard deviation σ is sometimes denoted $N(\mu, \sigma)$ for short. All normal distributions have the same general shape, differing only in the axis scales. For example, the left chart in Figure 7.8 shows the distribution of diameters of golf balls from a manufacturing process that produces normally distributed diameters with a mean diameter of $\mu = 42.70$ mm and a standard deviation $\sigma = 0.01$ mm, or $N(42.70, 0.01)$ in short notation. The right chart in Figure 7.8 shows the distribution of scores on the CPA theory exam, assumed to be normal with a mean of $\mu = 70$ and a standard deviation $\sigma = 10$, or $N(70, 10)$ in short notation. Although the shape of each PDF is the same, notice that the horizontal and vertical axis scales differ.

It is a common misconception that $f(x)$ must be smaller than 1, but in the left chart in Figure 7.8 you can see that this is not the case. Because the area under the entire curve must be 1, when X has a small range (e.g., the golf ball diameter range is about 0.06 mm), the height of $f(x)$ is large (about 40 for the golf ball diameters). Conversely, as shown in the right chart in Figure 7.8, when X has a large range (e.g., the CPA exam range is about 60 points), the height of $f(x)$ is small (about 0.40 for the exam scores).

Figure 7.8

All Normal Distributions Look Alike Except for Scaling

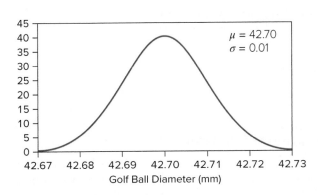
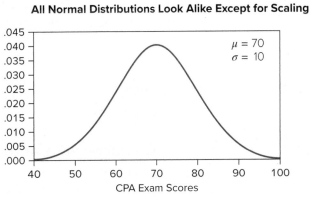

What Is Normal?

Many physical measurements in engineering and the sciences resemble normal distributions. Normal random variables also can be found in economic and financial data, behavioral measurement scales, marketing research, and operations analysis. The normal distribution is especially important as a sampling distribution for estimation and hypothesis testing. To be regarded as a candidate for normality, a random variable should

- Be measured on a continuous scale.
- Possess a clear center.
- Have only one peak (unimodal).
- Exhibit tapering tails.
- Be symmetric about the mean (equal tails).

When the range is large, we often treat a discrete variable as continuous. For example, exam scores are discrete (range from 0 to 100) but are often treated as continuous data. Here are some random variables that *might* be expected to be approximately normally distributed:

- $X =$ quantity of beverage in a 2-liter bottle of Diet Pepsi.
- $X =$ cockpit noise level in a Boeing 777 at the captain's left ear during cruise.
- $X =$ diameter in millimeters of a manufactured steel ball bearing.

Each of these variables would tend toward a certain mean but would exhibit random variation. For example, even with excellent quality control, not every bottle of a soft drink will have exactly the same fill (even if the variation is only a few milliliters). The mean and standard deviation depend on the nature of the data-generating process. Precision manufacturing can achieve very small σ in relation to μ (e.g., steel ball bearing diameter), while other data-generating situations produce relatively large σ in relation to μ (e.g., your driving fuel mileage). Thus, each normally distributed random variable may have a different coefficient of variation, even though they may share a common shape.

There are statistical tests to see whether a sample came from a normal population. In Chapter 4, for example, you saw that a histogram can be used to assess normality in a general way. More precise tests will be discussed in Chapter 15. For now, our task is to learn more about the normal distribution and its applications.

Section Exercises

7.11 If all normal distributions have the same shape, how do they differ?

7.12 (a) At what x value does $f(x)$ reach a maximum for a normal distribution $N(75, 5)$? (b) Does $f(x)$ touch the X-axis at $\mu \pm 3\sigma$?

7.13 State the Empirical Rule for a normal distribution (see Chapter 4).

7.14 Discuss why you would or would not expect each of the following variables to be normally distributed. *Hint:* Would you expect a single central mode and tapering tails? Would the distribution be roughly symmetric? Would one tail be longer than the other?

    a. Shoe sizes of adult males.
    b. Years of higher education of 30-year-old employed women.
    c. Days from mailing home utility bills to receipt of payment.
    d. Time to process insurance claims for residential fire damage.

7.4 STANDARD NORMAL DISTRIBUTION

Characteristics of the Standard Normal

LO 7-4

Find the normal probability for a given z or x using tables or Excel.

Because there is a different normal distribution for every pair of values of μ and σ, we often transform the variable by subtracting the mean and dividing by the standard deviation to produce a *standardized variable,* just as in Chapter 4, except that now we are talking about a population distribution instead of sample data. This important transformation is shown in formula 7.4.

(7.4) $$z = \frac{x - \mu}{\sigma} \qquad \text{(transformation of each } x\text{-value to a } z\text{-value)}$$

If X is normally distributed $N(\mu, \sigma)$, the standardized variable Z has a **standard normal distribution**. Its mean is 0 and its standard deviation is 1, denoted $N(0, 1)$. The maximum height of $f(z)$ is at 0 (the mean) and its points of inflection are at ± 1 (the standard deviation). The shape of the distribution is unaffected by the z transformation. Table 7.3 summarizes the main characteristics of the standard normal distribution.

Table 7.3

Standard Normal Distribution

Parameters	μ = population mean σ = population standard deviation
PDF	$f(z) = \dfrac{1}{\sqrt{2\pi}} e^{-z^2/2}$ where $z = \dfrac{x - \mu}{\sigma}$
Domain	$-\infty < z < +\infty$
Mean	0
Standard deviation	1
Shape	Symmetric, mesokurtic, and bell-shaped.
PDF in Excel*	=NORM.S.DIST(z,0)
CDF in Excel*	=NORM.S.DIST(z,1)
Random data in Excel*	=NORM.S.INV(RAND())
Comment	There is no simple formula for normal areas so we need tables or Excel.

*See the **R Software Supplement** at the end of this chapter for a list of R's normal probability functions.

Notation

Use an uppercase variable name like Z or X when speaking in general and a lowercase variable name like z or x to denote a particular value of Z or X.

Because every transformed normal distribution will look the same, we can use a common scale, usually labeled from -3 to $+3$, as shown in Figure 7.9. Because $f(z)$ is a probability density function, the entire area under the curve is 1, as you can approximately verify by treating it as a triangle (area = ½ base × height). As a rule, we are not interested in the height of the function $f(z)$ but rather in areas under the curve (although Excel will provide either). The probability of an event $P(z_1 < Z < z_2)$ is a definite integral of $f(z)$. Although there is no simple integral for $f(z)$, a normal area can be approximated to any desired degree of accuracy using various methods [e.g., covering the area from 0 to $f(z)$ with many narrow rectangles and summing their areas]. You do not need to worry about this because tables or Excel functions are available.

Figure 7.9

Standard Normal PDF and CDF

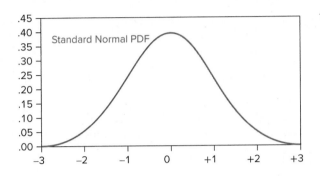
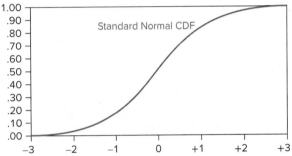

Normal Areas from Appendix C-1

Tables of normal probabilities have been prepared so you can look up any desired normal area. Such tables have many forms. Table 7.4 illustrates Appendix C-1, which shows areas from 0 to z using increments of 0.01 from $z = 0$ to $z = 3.69$ (beyond this range, areas are very small). For example, to calculate $P(0 < Z < 1.96)$, you select the row for $z = 1.9$ and the column for 0.06 (because $1.96 = 1.90 + 0.06$). This row and column are shaded in Table 7.4. At the intersection of the shaded row and column, we see $P(0 < Z < 1.96) = .4750$. This area is illustrated in Figure 7.10. Because half the total area under the curve lies to the right of the mean, we can find a right-tail area by subtraction. For example, $P(Z > 1.96) = .5000 - P(0 < Z < 1.96) = .5000 - .4750 = .0250$.

Suppose we want a middle area such as $P(-1.96 < Z < +1.96)$. Because the normal distribution is symmetric, we also know that $P(-1.96 < Z < 0) = .4750$. Adding these areas, we get

$$P(-1.96 < Z < +1.96) = P(-1.96 < Z < 0) + P(0 < Z < 1.96)$$
$$= .4750 + .4750 = .9500$$

Table 7.4

Normal Area from 0 to z (from Appendix C-1)

z	0.00	0.01	0.02	0.03	0.04	0.05	0.06	0.07	0.08	0.09
0.0	.0000	.0040	.0080	.0120	.0160	.0199	.0239	.0279	.0319	.0359
0.1	.0398	.0438	.0478	.0517	.0557	.0596	.0636	.0675	.0714	.0753
0.2	.0793	.0832	.0871	.0910	.0948	.0987	.1026	.1064	.1103	.1141
⋮	⋮	⋮	⋮	⋮	⋮	⋮	⋮	⋮	⋮	⋮
1.6	.4452	.4463	.4474	.4484	.4495	.4505	.4515	.4525	.4535	.4545
1.7	.4554	.4564	.4573	.4582	.4591	.4599	.4608	.4616	.4625	.4633
1.8	.4641	.4649	.4656	.4664	.4671	.4678	.4686	.4693	.4699	.4706
1.9	.4713	.4719	.4726	.4732	.4738	.4744	.4750	.4756	.4761	.4767
2.0	.4772	.4778	.4783	.4788	.4793	.4798	.4803	.4808	.4812	.4817
2.1	.4821	.4826	.4830	.4834	.4838	.4842	.4846	.4850	.4854	.4857
2.2	.4861	.4864	.4868	.4871	.4875	.4878	.4881	.4884	.4887	.4890
2.3	.4893	.4896	.4898	.4901	.4904	.4906	.4909	.4911	.4913	.4916
⋮	⋮	⋮	⋮	⋮	⋮	⋮	⋮	⋮	⋮	⋮
3.6	.49984	.49985	.49985	.49986	.49986	.49987	.49987	.49988	.49988	.49989
3.7	.49989	.49990	.49990	.49990	.49991	.49991	.49992	.49992	.49992	.49992

Inequalities—Strict (<) or Inclusive (≤)?

Because a point has no area in a continuous distribution, the probability $P(-1.96 \leq Z \leq +1.96)$ is the same as $P(-1.96 < Z < +1.96)$, so, for simplicity, we sometimes omit the equality.

Figure 7.10

Finding Areas Using Appendix C-1

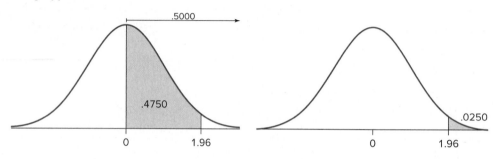

The interval $-1.96 < Z < 1.96$ encloses 95 percent of the area under the normal curve. Figure 7.11 illustrates this calculation.

Figure 7.11

Finding Areas Using Appendix C-1

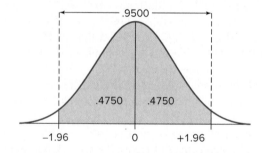

From Appendix C-1 we can see the basis for the Empirical Rule, illustrated in Figure 7.12. These are the "k-sigma" intervals mentioned in Chapter 4 and used by statisticians for quick reference to the normal distribution. Thus, it is *approximately* correct to say that a "2-sigma interval" contains 95 percent of the area (actually $z = 1.96$ would yield a 95 percent area):

$$P(-1.00 < Z < +1.00) = 2 \times P(0 < Z < 1.00) = 2 \times .3413 = .6826, \text{ or } 68.26\%$$

$$P(-2.00 < Z < +2.00) = 2 \times P(0 < Z < 2.00) = 2 \times .4772 = .9544, \text{ or } 95.44\%$$

$$P(-3.00 < Z < +3.00) = 2 \times P(0 < Z < 3.00) = 2 \times .49865 = .9973, \text{ or } 99.73\%$$

Normal Areas from Appendix C-2

Table 7.5 illustrates another kind of table. Appendix C-2 shows cumulative normal areas from the left to z. You can think of Appendix C-2 as the CDF for the normal distribution. This corresponds to the way Excel calculates normal areas. Using this approach, we see that $P(Z < -1.96) = .0250$ and $P(Z < +1.96) = .9750$. By subtraction, we get

$$P(-1.96 < Z < +1.96) = P(Z < +1.96) - P(Z < -1.96) = .9750 - .0250 = .9500$$

The result is identical to that obtained previously. The interval $-1.96 < Z < 1.96$ encloses 95 percent of the area under the normal curve. This calculation is illustrated in Figure 7.13.

Figure 7.12

Normal Areas within $\mu \pm k\sigma$

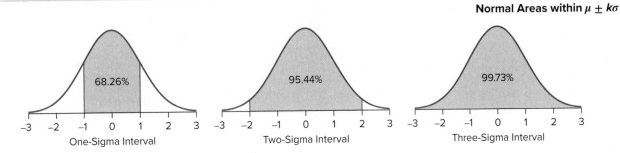

Table 7.5

Cumulative Normal Area from Left to *z* (from Appendix C-2)

z	0.00	0.01	0.02	0.03	0.04	0.05	0.06	0.07	0.08	0.09
−3.7	.00011	.00010	.00010	.00010	.00009	.00009	.00008	.00008	.00008	.00008
−3.6	.00016	.00015	.00015	.00014	.00014	.00013	.00013	.00012	.00012	.00011
⋮	⋮	⋮	⋮	⋮	⋮	⋮	⋮	⋮	⋮	⋮
−2.3	.0107	.0104	.0102	.0099	.0096	.0094	.0091	.0089	.0087	.0084
−2.2	.0139	.0136	.0132	.0129	.0125	.0122	.0119	.0116	.0113	.0110
−2.1	.0179	.0174	.0170	.0166	.0162	.0158	.0154	.0150	.0146	.0143
−2.0	.0228	.0222	.0217	.0212	.0207	.0202	.0197	.0192	.0188	.0183
−1.9	.0287	.0281	.0274	.0268	.0262	.0256	.0250	.0244	.0239	.0233
−1.8	.0359	.0351	.0344	.0336	.0329	.0322	.0314	.0307	.0301	.0294
−1.7	.0446	.0436	.0427	.0418	.0409	.0401	.0392	.0384	.0375	.0367
−1.6	.0548	.0537	.0526	.0516	.0505	.0495	.0485	.0475	.0465	.0455
⋮	⋮	⋮	⋮	⋮	⋮	⋮	⋮	⋮	⋮	⋮
0.0	.5000	.5040	.5080	.5120	.5160	.5199	.5239	.5279	.5319	.5359
0.1	.5398	.5438	.5478	.5517	.5557	.5596	.5636	.5675	.5714	.5753
0.2	.5793	.5832	.5871	.5910	.5948	.5987	.6026	.6064	.6103	.6141
⋮	⋮	⋮	⋮	⋮	⋮	⋮	⋮	⋮	⋮	⋮
1.6	.9452	.9463	.9474	.9484	.9495	.9505	.9515	.9525	.9535	.9545
1.7	.9554	.9564	.9573	.9582	.9591	.9599	.9608	.9616	.9625	.9633
1.8	.9641	.9649	.9656	.9664	.9671	.9678	.9686	.9693	.9699	.9706
1.9	.9713	.9719	.9726	.9732	.9738	.9744	.9750	.9756	.9761	.9767
2.0	.9772	.9778	.9783	.9788	.9793	.9798	.9803	.9808	.9812	.9817
2.1	.9821	.9826	.9830	.9834	.9838	.9842	.9846	.9850	.9854	.9857
2.2	.9861	.9864	.9868	.9871	.9875	.9878	.9881	.9884	.9887	.9890
2.3	.9893	.9896	.9898	.9901	.9904	.9906	.9909	.9911	.9913	.9916
⋮	⋮	⋮	⋮	⋮	⋮	⋮	⋮	⋮	⋮	⋮
3.6	.99984	.99985	.99985	.99986	.99986	.99987	.99987	.99988	.99988	.99989
3.7	.99989	.99990	.99990	.99990	.99991	.99991	.99992	.99992	.99992	.99992

Because Appendix C-1 and Appendix C-2 yield identical results, you should use whichever table is easier for the area you are trying to find. Appendix C-1 is often easier for "middle areas." It also has the advantage of being more compact (it fits on one page), which is one reason why it has traditionally been used for statistics exams and in other textbooks (e.g., marketing). But Appendix C-2 is easier for left-tail areas and some complex areas. Further, Appendix C-2 corresponds to the way Excel calculates normal areas. When subtraction is required for a right-tail or middle area, either table is equally convenient.

Figure 7.13

Finding Areas by Using Appendix C-2

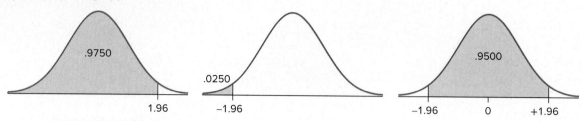

Section Exercises

Note: Use Appendix C-1 or C-2 for these exercises.

7.15 Find the standard normal area for each of the following, showing your reasoning clearly and indicating which table you used.

    a. $P(0 < Z < 0.50)$
    b. $P(-0.50 < Z < 0)$
    c. $P(Z > 0)$
    d. $P(Z = 0)$

7.16 Find the standard normal area for each of the following, showing your reasoning clearly and indicating which table you used.

    a. $P(1.22 < Z < 2.15)$
    b. $P(2.00 < Z < 3.00)$
    c. $P(-2.00 < Z < 2.00)$
    d. $P(Z > 0.50)$

7.17 Find the standard normal area for each of the following, showing your reasoning clearly and indicating which table you used.

    a. $P(-1.22 < Z < 2.15)$
    b. $P(-3.00 < Z < 2.00)$
    c. $P(Z < 2.00)$
    d. $P(Z = 0)$

7.18 Find the standard normal area for each of the following. Sketch the normal curve and shade in the area represented below.

    a. $P(Z < -1.96)$
    b. $P(Z > 1.96)$
    c. $P(Z < 1.65)$
    d. $P(Z > -1.65)$

7.19 Find the standard normal area for each of the following. Sketch the normal curve and shade in the area represented below.

    a. $P(Z < -1.28)$
    b. $P(Z > 1.28)$
    c. $P(-1.96 < Z < 1.96)$
    d. $P(-1.65 < Z < 1.65)$

7.20 Bob's exam score was 2.17 standard deviations above the mean. The exam was taken by 200 students. Assuming a normal distribution, how many scores were higher than Bob's?

7.21 Joan's finishing time for the Bolder Boulder 10K race was 1.75 standard deviations faster than the women's average for her age group. There were 405 women who ran in her age group. Assuming a normal distribution, how many women ran faster than Joan?

LO 7-5

Solve for *z* or *x* for a given normal probability using tables or Excel.

Finding *z* for a Given Area

We also can use the tables to find the *z*-value that corresponds to a given area. For example, what *z*-value defines the top 1 percent of a normal distribution? Because half the area lies above the mean, an upper area of 1 percent implies that 49 percent of the area must lie between 0 and *z*. Searching Appendix C-1 for an area of .4900, we see that $z = 2.33$ yields

Table 7.6

**Normal Area from 0 to z
(from Appendix C-1)**

z	0.00	0.01	0.02	0.03	0.04	0.05	0.06	0.07	0.08	0.09
0.0	.0000	.0040	.0080	.0120	.0160	.0199	.0239	.0279	.0319	.0359
0.1	.0398	.0438	.0478	.0517	.0557	.0596	.0636	.0675	.0714	.0753
0.2	.0793	.0832	.0871	.0910	.0948	.0987	.1026	.1064	.1103	.1141
⋮	⋮	⋮	⋮	⋮	⋮	⋮	⋮	⋮	⋮	⋮
1.6	.4452	.4463	.4474	.4484	.4495	.4505	.4515	.4525	.4535	.4545
1.7	.4554	.4564	.4573	.4582	.4591	.4599	.4608	.4616	.4625	.4633
1.8	.4641	.4649	.4656	.4664	.4671	.4678	.4686	.4693	.4699	.4706
1.9	.4713	.4719	.4726	.4732	.4738	.4744	.4750	.4756	.4761	.4767
2.0	.4772	.4778	.4783	.4788	.4793	.4798	.4803	.4808	.4812	.4817
2.1	.4821	.4826	.4830	.4834	.4838	.4842	.4846	.4850	.4854	.4857
2.2	.4861	.4864	.4868	.4871	.4875	.4878	.4881	.4884	.4887	.4890
2.3	.4893	.4896	.4898	.4901	.4904	.4906	.4909	.4911	.4913	.4916
⋮	⋮	⋮	⋮	⋮	⋮	⋮	⋮	⋮	⋮	⋮
3.6	.49984	.49985	.49985	.49986	.49986	.49987	.49987	.49988	.49988	.49989
3.7	.49989	.49990	.49990	.49990	.49991	.49991	.49992	.49992	.49992	.49992

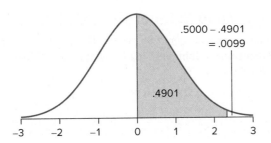

.5000 − .4901
= .0099

.4901

Figure 7.14

**Finding Areas by Using
Appendix C-1**

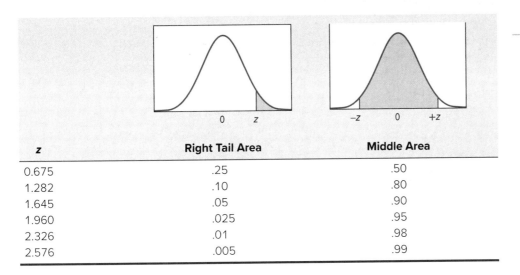

Table 7.7

Important Normal Areas

z	Right Tail Area	Middle Area
0.675	.25	.50
1.282	.10	.80
1.645	.05	.90
1.960	.025	.95
2.326	.01	.98
2.576	.005	.99

an area of .4901. Without interpolation, that is as close as we can get to 49 percent. This is illustrated in Table 7.6 and Figure 7.14.

We can find other important areas in the same way. Because we are often interested in the top 25 percent, 10 percent, 5 percent, 1 percent, etc., or the middle 50 percent, 90 percent, 95 percent, 99 percent, and so forth, it is convenient to record these important z-values for quick reference. Table 7.7 summarizes some important normal areas. For greater accuracy, these z-values are shown to three decimals (they were obtained from Excel).

Section Exercises

Note: If you are using Appendix C-1 or C-2, you may only be able to approximate the actual area unless you interpolate.

7.22 Find the associated z-score for each of the following standard normal areas.

    a. Highest 10 percent   b. Lowest 50 percent   c. Highest 7 percent

7.23 Find the associated z-score for each of the following standard normal areas.

    a. Lowest 6 percent   b. Highest 40 percent   c. Lowest 7 percent

7.24 Find the associated z-score or scores that represent the following standard normal areas.

    a. Middle 50 percent   b. Lowest 5 percent   c. Middle 90 percent

7.25 Find the associated z-score or scores that represent the following standard normal areas.

    a. Middle 60 percent   b. Highest 2 percent   c. Middle 95 percent

7.26 High school students across the nation compete in a financial capability challenge each year by taking a National Financial Capability Challenge Exam. Students who score in the top 20 percent are recognized publicly for their achievement by the Department of the Treasury. Assuming a normal distribution, how many standard deviations above the mean does a student have to score to be publicly recognized?

7.27 The fastest 10 percent of runners who complete the Nosy Neighbor 5K race win a gift certificate to a local running store. Assuming a normal distribution, how many standard deviations below the mean must a runner's time be in order to win the gift certificate?

Finding Areas by Using Standardized Variables

John took an economics exam and scored 86 points. The class mean was 75 with a standard deviation of 7. What percentile is John in? That is, what is $P(X < 86)$? We need first to calculate John's standardized Z-score:

$$z_{John} = \frac{x_{John} - \mu}{\sigma} = \frac{86 - 75}{7} = \frac{11}{7} = 1.57$$

This says that John's score is 1.57 standard deviations above the mean. From Appendix C-2 we get $P(X < 86) = P(Z < 1.57) = .9418$, so John is approximately in the 94th percentile. That means that his score was better than 94 percent of the class, as illustrated in Figure 7.15. The table gives a slightly different value from Excel due to rounding.

Figure 7.15

Two Equivalent Areas

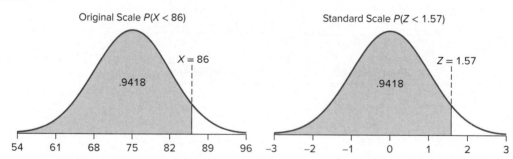

On this exam, what is the probability that a randomly chosen test-taker would have a score of at least 65? We begin by standardizing:

$$z = \frac{x - \mu}{\sigma} = \frac{65 - 75}{7} = \frac{-10}{7} = -1.43$$

Using Appendix C-1 we can calculate $P(X \geq 65) = P(Z \geq -1.43)$ as

$$P(Z \geq -1.43) = P(-1.43 < Z < 0) + .5000$$

$$= .4236 + .5000 = .9236, \text{or } 92.4\%$$

Using Appendix C-2 we can calculate $P(X \geq 65) = P(Z \geq -1.43)$ as

$$P(Z \geq -1.43) = 1 - P(Z < -1.43) = 1 - .0764 = .9236, \text{ or } 92.4\%$$

Using either method, there is a 92.4 percent chance that a student scores 65 or above on this exam. These calculations are illustrated in Figure 7.16.

Figure 7.16

Two Ways to Find an Area

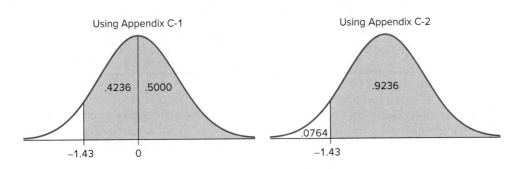

Finding Normal Areas with Excel

Excel offers several functions for the normal and standard normal distributions, as shown in Figure 7.17.

Figure 7.17

Inserting Excel Normal Functions

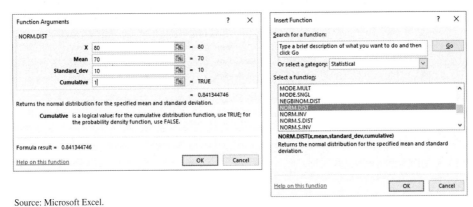

Source: Microsoft Excel.

Table 7.8 illustrates Excel functions that return left-tail normal areas for a given value of x or z. Excel is more accurate than a table; however, you must be careful of syntax. It is a good idea to sketch a normal curve and shade the desired area to help you *visualize* the answer you expect so that you will recognize if you are getting an unreasonable answer from Excel. In Table 7.8, note that the cumulative argument is set to 1 (or TRUE) because we want the CDF (left-tail area) rather than the PDF (height of the normal curve).

Excel's NORM.DIST and NORM.INV functions let us evaluate areas and inverse areas *without* standardizing. For example, let X be the diameter of a manufactured steel ball bearing whose mean diameter is $\mu = 2.040$ cm and whose standard deviation is $\sigma = .001$ cm. What is the probability that a given steel bearing will have a diameter between 2.039 and 2.042 cm? We use Excel's function =NORM.DIST(x,μ,σ,cumulative), where cumulative is TRUE.

Because Excel gives left-tail areas, we first calculate $P(X < 2.039)$ and $P(X < 2.042)$ as in Figure 7.18. We then obtain the area between by subtraction, as illustrated in Figure 7.19.

Table 7.8

Excel Normal CDF Functions

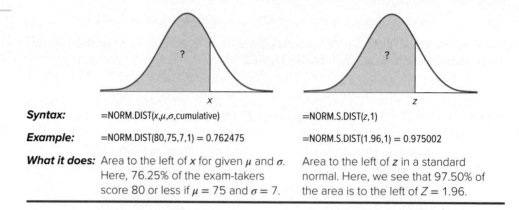

Syntax: =NORM.DIST(x,μ,σ,cumulative)   =NORM.S.DIST(z,1)

Example: =NORM.DIST(80,75,7,1) = 0.762475   =NORM.S.DIST(1.96,1) = 0.975002

What it does: Area to the left of x for given μ and σ. Here, 76.25% of the exam-takers score 80 or less if $\mu = 75$ and $\sigma = 7$.   Area to the left of z in a standard normal. Here, we see that 97.50% of the area is to the left of $Z = 1.96$.

The desired area is approximately 81.9 percent. Of course, we could do exactly the same thing by using Appendix C-2:

$$P(2.039 < X < 2.042) = P(X < 2.042) - P(X < 2.039)$$
$$= \text{NORM.DIST}(2.042,2.04,.001,1) - \text{NORM.DIST}(2.039,2.04,.001,1)$$
$$= .9772 - .1587 = .8185, \text{ or } 81.9\%$$

Figure 7.18

Left-Tail Areas Using Excel Normal CDF

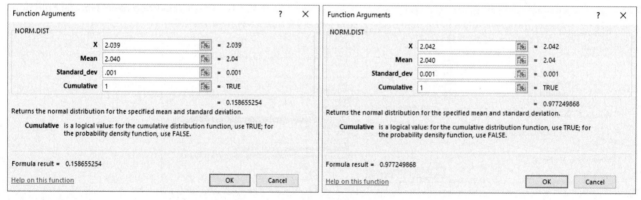

Source: Microsoft Excel.

Figure 7.19

Cumulative Areas from Excel's NORM.DIST

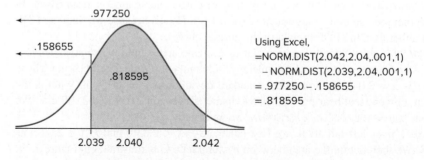

Note: Use Excel for the following exercises.

7.28 Suppose $X \sim N(56, 4)$. Write the Excel function to calculate the following probabilities:

    a. $P(X < 52.3)$ b. $P(X \geq 58.0)$ c. $P(50.0 \leq X < 63.7)$

7.29 Suppose the volume of liquid soap (X) in a randomly chosen bottle of dish soap is normally distributed with $\mu = 600$ ml and $\sigma = 5$ ml. Write the Excel function to calculate the following probabilities: a. $P(X > 592)$ b. $P(X \leq 603)$ c. $P(592 < X \leq 603)$

7.30 Daily output of Marathon's Garyville, Louisiana, refinery is normally distributed with a mean of 232,000 barrels of crude oil per day with a standard deviation of 7,000 barrels. (a) What is the probability of producing at least 232,000 barrels? (b) Between 232,000 and 239,000 barrels? (c) Less than 239,000 barrels? (d) Less than 245,000 barrels? (e) More than 225,000 barrels?

7.31 Assume that the number of calories in a McDonald's Egg McMuffin is a normally distributed random variable with a mean of 290 calories and a standard deviation of 14 calories. (a) What is the probability that a particular serving contains fewer than 300 calories? (b) More than 250 calories? (c) Between 275 and 310 calories? Show all work clearly. (Data are from McDonalds.com.)

7.32 The weight of a miniature Tootsie Roll is normally distributed with a mean of 3.30 grams and standard deviation of 0.13 gram. (a) Within what weight range will the middle 95 percent of all miniature Tootsie Rolls fall? (b) What is the probability that a randomly chosen miniature Tootsie Roll will weigh more than 3.50 grams? (Data are from a project by MBA student Henry Scussel.)

7.33 The pediatrics unit at Carver Hospital has 24 beds. The number of patients needing a bed at any point in time is $N(19.2, 2.5)$. What is the probability that the number of patients needing a bed will exceed the pediatrics unit's bed capacity?

7.34 The cabin of a business jet has a height 5 feet 9 inches high. If a business traveler's height is $N(5'10'', 2.7'')$, what percentage of the business travelers will have to stoop?

7.35 On January 1, 2011, a new standard for baseball bat "liveliness" called BBCOR (Ball-Bat Coefficient of Restitution) was adopted for teams playing under NCAA rules. A higher BBCOR allows the ball to travel farther when hit, so bat manufacturers want a high BBCOR. The maximum allowable BBCOR is 0.500. BigBash Inc. produces bats whose BBCOR is $N(0.480, 0.008)$. What percentage of its bats will exceed the BBCOR standard?

7.36 Last year's freshman class at Big State University totaled 5,324 students. Of those, 1,254 received a merit scholarship to help offset tuition costs their freshman year (although the amount varied per student). The amount a student received was $N(\$3456, \$478)$. If the cost of full tuition was \$4,200 last year, what percentage of students who received a merit scholarship did *not* receive enough to cover full tuition?

Inverse Normal

How can we find the various normal percentiles (5th, 10th, 25th, 75th, 90th, 95th, etc.) known as the **inverse normal**? That is, how can we find X for a given area? We simply turn the standardizing transformation around:

$$x = \mu + z\sigma \quad \left(\text{solving for } x \text{ in } z = \frac{x - \mu}{\sigma} \right) \quad\quad (7.5)$$

Recall that John's economics class exam had a mean equal to 75 and a standard deviation equal to 7. Using Table 7.7 (or looking up the areas in Excel), we obtain the results shown in Table 7.9. Note that to find a lower tail area (such as the lowest 5 percent), we must use negative Z-values.

Percentile	z	$x = \mu + z\sigma$	x (to nearest integer)
95th (highest 5%)	1.645	$x = 75 + (1.645)(7)$	86.52, or 87 (rounded)
90th (highest 10%)	1.282	$x = 75 + (1.282)(7)$	83.97, or 84 (rounded)
75th (highest 25%)	0.675	$x = 75 + (0.675)(7)$	79.73, or 80 (rounded)
25th (lowest 25%)	−0.675	$x = 75 - (0.675)(7)$	70.28, or 70 (rounded)
10th (lowest 10%)	−1.282	$x = 75 - (1.282)(7)$	66.03, or 66 (rounded)
5th (lowest 5%)	−1.645	$x = 75 - (1.645)(7)$	63.49, or 63 (rounded)

Table 7.9

Percentiles for Desired Normal Area

Using the two Excel functions =NORM.INV() and =NORM.S.INV() shown in Table 7.10, we can solve for the value of x or z that corresponds to a given normal area. Also, for a given area and x, we might want to solve for μ or σ.

For example, suppose that John's economics professor has decided that any student who scores below the 10th percentile must retake the exam. The exam scores are normal with

Table 7.10

Excel Inverse Normal
Functions

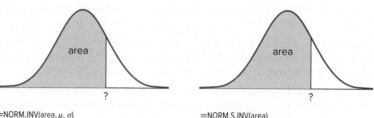

Syntax:	=NORM.INV(area, μ, σ)	=NORM.S.INV(area)
Example:	=NORM.INV(0.99,75,7) = 91.2844	=NORM.S.INV(0.75) = 0.674490
What it does:	Value of x for given left-tail area. If $\mu = 75$ and $\sigma = 7$, the 99th percentile for exam-takers is a score of 91.28 or 91 to the nearest integer.	Value of z corresponding to a given left-tail area. Here, the 75th percentile (third quartile) of a standard normal is $z = 0.675$.

$\mu = 75$ and $\sigma = 7$. What is the score that would require a student to retake the exam? We need to find the value of x that satisfies $P(X < x) = .10$. The approximate z-score for the 10th percentile is $z = -1.28$. The steps to solve the problem are

- Use Appendix C or Excel to find $z = -1.28$ to satisfy $P(Z < -1.28) = .10$.

- Substitute the given information into $z = \dfrac{x - \mu}{\sigma}$ to get $-1.28 = \dfrac{x - 75}{7}$.

- Solve for x to get $x = 75 - (1.28)(7) = 66.04$ (or 66 after rounding)

Students who score below 66 points on the economics exam will be required to retake the exam.

Analytics in Action

Describing Business Uncertainty Using a Simulation Model

Most business analyses involve uncertainty. Managers know they must account for uncertainty in their analysis, but how? To incorporate uncertainty in complex business problems, analysts often perform a computer simulation because it is a highly flexible, practical, and intuitive tool.

Simulation combines mathematical and logical concepts to imitate real-world processes or systems, allowing users to better understand how different factors affect the outcome. Instead of requiring mastery of the mathematics necessary for developing simulation models from scratch, most simulation software packages nowadays simplify this task by allowing users to describe uncertainty by choosing a **probability distribution** that closely represents the scenario. For example:

- To illustrate an uncertain cost, managers may choose a **uniform distribution** when it is reasonable to estimate the lowest and highest possible costs, and any outcome (cost) in between is equally likely.

- To describe an uncertain activity duration in a project, if managers can estimate the shortest, most likely, and longest duration, the **triangular** distribution may be a good candidate.

- To characterize uncertain return on investment (%), managers may depict it with a **normal distribution** if the outcome follows a bell-shaped curve (most likely around the average, symmetric and tapering tails, etc.).

Understanding key characteristics of probability distributions and choosing the appropriate distribution for their business process enables analysts to fully leverage the power of simulation modeling. Simulations have incredible range of applications across industry. More discussion on simulation can be found in Chapter 18 of this textbook.

Note: Use Excel for the following exercises.

7.37 Suppose $X \sim N(56, 4)$. Write the Excel function to find the X values associated with the following percentages: a. Highest 20 percent   b. Middle 60 percent   c. Lowest 30 percent

7.38 Suppose the volume of liquid soap (X) in a randomly chosen bottle of dish soap is normally distributed with $\mu = 600$ ml and $\sigma = 5$ ml. Write the Excel function to find the volume associated with the following percentages: a. Lowest 60 percent   b. Middle 40 percent   c. Highest 80 percent

7.39 The time required to verify and fill a common prescription at a neighborhood pharmacy is normally distributed with a mean of 10 minutes and a standard deviation of 3 minutes. Find the time for each event. Show your work.

    a. Highest 10 percent   b. Middle 50 percent
    c. Highest 80 percent   d. Lowest 10 percent

7.40 The time required to cook a pizza at a neighborhood pizza joint is normally distributed with a mean of 12 minutes and a standard deviation of 2 minutes. Find the time for each event. Show your work.

    a. Highest 5 percent   b. Lowest 50 percent
    c. Middle 95 percent   d. Lowest 80 percent

7.41 The weight of a McDonald's cheeseburger is normally distributed with a mean of 114 ounces and a standard deviation of 7 ounces. Find the weight that corresponds to each event. Show your work.

    a. Highest 5 percent   b. Lowest 50 percent
    c. Middle 95 percent   d. Lowest 80 percent

7.42 The weight of a small Starbucks coffee is a normally distributed random variable with a mean of 360 grams and a standard deviation of 9 grams. Find the weight that corresponds to each event. Show your work.

    a. Highest 10 percent   b. Middle 50 percent
    c. Highest 80 percent   d. Lowest 10 percent

7.43 The weights of newborn babies in Foxboro Hospital are normally distributed with a mean of 6.9 pounds and a standard deviation of 1.2 pounds. (a) How unusual is a baby weighing 8.0 pounds or more? (b) What would be the 90th percentile for birth weight? (c) Within what range would the middle 95 percent of birth weights lie?

7.44 The credit scores of 35-year-olds applying for a mortgage at Ulysses Mortgage Associates are normally distributed with a mean of 600 and a standard deviation of 100. (a) Find the credit score that defines the upper 5 percent. (b) Seventy-five percent of the customers will have a credit score higher than what value? (c) Within what range would the middle 80 percent of credit scores lie?

7.45 The number of patients needing a bed at any point in time in the pediatrics unit at Carver Hospital is $N(19.2, 2.5)$. Find the middle 50 percent of the number of beds needed (round to the next higher integer because a "bed" is indivisible).

7.46 A ski resort pays its part-time seasonal employees on an hourly basis. At a certain mountain, the hourly rates have a normal distribution with $\sigma = \$3.00$. If 20 percent of all part-time seasonal employees make more than $13.16 an hour, what is the average hourly pay rate at this mountain?

7.47 The average cost of an IRS Form 1040 tax filing at Thetis Tax Service is $157.00. Assuming a normal distribution, if 70 percent of the filings cost less than $171.00, what is the standard deviation?

Example 7.2

Service Times in a Quick Oil Change Shop: Four Worked Problems

After studying the process of changing oil, the shop's manager has found that the distribution of service times, X, is normal with a mean $\mu = 28$ minutes and a standard deviation $\sigma = 5$ minutes, that is, $X \sim N(28, 5)$. This information can now be used to answer questions about normal probabilities.

To answer these types of questions, it is helpful to follow a few basic steps. (1) Draw a picture and label the picture with the information you know. (2) Shade in the area that will answer your question. (3) Standardize the random variable. (4) Find the area by using one of the tables or Excel.

Worked Problem #1   *What proportion of cars will be finished in less than half an hour?*

- **Steps 1 and 2:**   Draw a picture and shade the area to the left of 30 minutes.

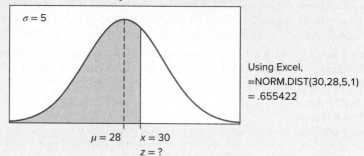

$\sigma = 5$

$\mu = 28$ $x = 30$
$z = ?$

Using Excel,
=NORM.DIST(30,28,5,1)
= .655422

- **Step 3:** $z = \dfrac{30 - 28}{5} = 0.40$

- **Step 4:**   Using Appendix C-2 or Excel, we find that $P(Z \leq 0.40) = .6554$.

Approximately 66 percent of the cars will be finished in less than half an hour.

Worked Problem #2   *What is the chance that a randomly selected car will take longer than 40 minutes to complete?*

- **Steps 1 and 2:**   Draw a picture and shade the area to the right of 40 minutes.

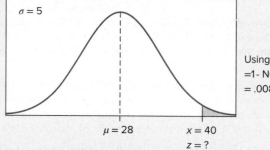

$\sigma = 5$

$\mu = 28$ $x = 40$
$z = ?$

Using Excel,
=1- NORM.DIST(40,28,5,1)
= .008198

- **Step 3:** $z = \dfrac{40 - 28}{5} = 2.4$

- **Step 4:**   Using Appendix C-2 or Excel we find that $P(Z > 2.4) = 1 - P(Z \leq 2.4) = 1 - .9918 = .0082$.

There is less than a 1 percent chance that a car will take longer than 40 minutes to complete.

Worked Problem #3   *What service time corresponds to the 90th percentile?*

- **Steps 1 and 2:**   Draw a picture and shade the desired area.

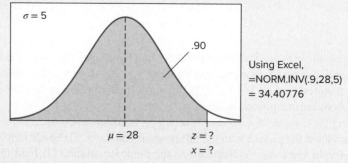

$\sigma = 5$

.90

$\mu = 28$ $z = ?$
$x = ?$

Using Excel,
=NORM.INV(.9,28,5)
= 34.40776

In this case, steps 3 and 4 need to be reversed.

- **Step 3:**   Find $z = 1.28$ by using the tables or Excel.

- **Step 4:** $1.28 = \dfrac{x - 28}{5}$, so $x = 28 + 5(1.28) = 34.4$ minutes.

Ninety percent of the cars will be finished in 34.4 minutes or less.

Worked Problem #4 *The manager wants to be able to service 80 percent of the vehicles within 30 minutes. What must the mean service time be to accomplish this goal?*

- **Steps 1 and 2:** Draw a curve and shade the desired area.

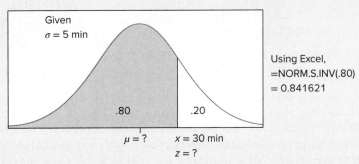

Given
$\sigma = 5$ min

.80   .20

$\mu = ?$ $x = 30$ min
         $z = ?$

Using Excel,
=NORM.S.INV(.80)
= 0.841621

- **Step 3:** Use tables or Excel to find $z = 0.84$ (approximately) for an upper tail area of .20 (lower tail area of .80).

- **Step 4:** Substitute into $z = \dfrac{x - \mu}{\sigma}$ to get $0.84 = \dfrac{30 - \mu}{5}$ and solve for $\mu = 30 - 0.84(5) = 25.8$.

The mean service time would have to be 25.8 minutes to ensure that 80 percent are serviced within 30 minutes.

7.48 Use Excel to find each probability.

   a. $P(X < 110)$ for $N(100, 15)$ b. $P(X < 2.00)$ for $N(0, 1)$
   c. $P(X < 5000)$ for $N(6000, 1000)$ d. $P(X < 450)$ for $N(600, 100)$

7.49 Use Excel to find each probability.

   a. $P(80 < X < 110)$ for $N(100, 15)$ b. $P(1.50 < X < 2.00)$ for $N(0, 1)$
   c. $P(4500 < X < 7000)$ for $N(6000, 1000)$ d. $P(225 < X < 450)$ for $N(600, 100)$

7.50 The weight of a small Starbucks coffee is a normal random variable with a mean of 360 g and a standard deviation of 9 g. Use Excel to find the weight corresponding to each percentile of weight.

   a. 10th percentile   b. 32nd percentile   c. 75th percentile
   d. 90th percentile   e. 99.9th percentile   f. 99.99th percentile

7.51 A study found that the mean waiting time to see a physician at an outpatient clinic was 40 minutes with a standard deviation of 28 minutes. Use Excel to find each probability. (a) What is the probability of more than an hour's wait? (b) Less than 20 minutes? (c) At least 10 minutes?

7.52 High-strength concrete is supposed to have a compressive strength greater than 6,000 pounds per square inch (psi). A certain type of concrete has a mean compressive strength of 7,000 psi, but due to variability in the mixing process, it has a standard deviation of 420 psi. Assume a normal distribution. Using Excel, what is the probability that a given pour of concrete from this mixture will fail to meet the high-strength criterion? In your judgment, does this mixture provide an adequate margin of safety?

7.5 NORMAL APPROXIMATIONS

Normal Approximation to the Binomial

We have seen that (unless we are using Excel) binomial probabilities may be difficult to calculate when *n* is large, particularly when many terms must be summed. Instead, we can use a normal approximation. The logic of this approximation is that as *n* becomes large, the discrete binomial

LO 7-6

Use the normal approximation to a binomial or a Poisson.

Figure 7.20

Binomial Approaches Normal as *n* Increases

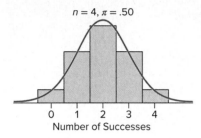

$n = 4, \pi = .50$

Number of Successes

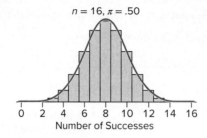

$n = 16, \pi = .50$

Number of Successes

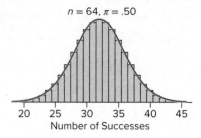

$n = 64, \pi = .50$

Number of Successes

bars become more like a smooth, continuous, normal curve. Figure 7.20 illustrates this idea for 4, 16, and 64 flips of a fair coin with *X* defined as the number of heads in *n* tries. As sample size increases, it becomes easier to visualize a smooth, bell-shaped curve overlaid on the bars.

As a rule of thumb, when $n\pi \geq 10$ and $n(1 - \pi) \geq 10$, it is safe to use the normal approximation to the binomial, setting the normal μ and σ equal to the binomial mean and standard deviation:

(7.6)
$$\mu = n\pi$$

(7.7)
$$\sigma = \sqrt{n\pi(1 - \pi)}$$

Example 7.3

Coin Flips

What is the probability of more than 17 heads in 32 flips of a fair coin? In binomial terms, this would be $P(X \geq 18) = P(18) + P(19) + \ldots + P(32)$, which would be a tedious sum even if we had a table. Could the normal approximation be used? With $n = 32$ and $\pi = .50$, we clearly meet the requirement that $n\pi \geq 10$ and $n(1 - \pi) \geq 10$. However, when translating a discrete scale into a continuous scale, we must be careful about individual points. The event "more than 17" actually falls halfway *between* 17 and 18 on a discrete scale, as shown in Figure 7.21.

Figure 7.21

Normal Approximation to $P(X \geq 18)$

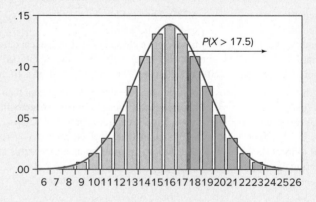

You don't need to draw the entire distribution. All you need is a little diagram (ignoring the low and high ends of the scale because they are not relevant) to show the event "more than 17" visually:

$$\ldots 14 \quad 15 \quad 16 \quad 17 \quad \textit{18} \quad \textit{19} \quad \textit{20} \quad \textit{21} \quad \textit{22} \quad \textit{23} \ldots$$

If you make a diagram like this, you can *see* the correct cutoff point. Because the cutoff point for "more than 17" is halfway between 17 and 18, the normal approximation is

$P(X > 17.5)$. The 0.5 is an adjustment called the **continuity correction**. The normal parameters are

$$\mu = n\pi = (32)(0.5) = 16$$
$$\sigma = \sqrt{n\pi(1 - \pi)} = \sqrt{(32)(0.5)(1 - 0.5)} = 2.82843$$

We then perform the usual standardizing transformation with the continuity-corrected X-value:

$$z = \frac{x - \mu}{\sigma} = \frac{17.5 - 16}{2.82843} = .53$$

From Appendix C-1 we find $P(Z > .53) = .5000 - P(0 < Z < .53) = .5000 - .2019 = .2981$. Alternately, we could use Appendix C-2 to get $P(Z > .53)$, which, by the symmetry of the normal distribution, is the same as $P(Z < -.53) = .2981$. The calculations are illustrated in Figure 7.22.

Figure 7.22

Normal Area for $P(Z > .53)$

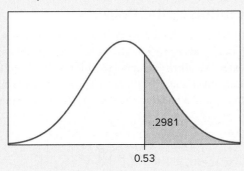

How accurate is this normal approximation to the binomial $P(X \geq 18)$ in our coin flip example? We can check it by using Excel. Because Excel's function is cumulative to the left, we find $P(X \leq 17)$ with the Excel function =BINOM.DIST(17, 32, 0.5, 1) and then subtract from 1:

$$P(X \geq 18) = 1 - P(X \leq 17) = 1 - .7017 = .2983$$

In this case, the normal approximation (.2981) is very close to the binomial probability (.2983), partly because this binomial is roughly symmetric (π is near .50). The rule of thumb results in an error of less than .01 in the approximation. (For a demonstration of this result, see *LearningStats* and other Learning Resources at the end of this chapter.)

When a binomial distribution is badly skewed (π near 0 or 1), the normal approximation is less accurate, *ceteris paribus*. But when n is large, the normal approximation improves, regardless of π. In a *right-skewed* binomial (when $\pi < .50$), the rule $n\pi \geq 10$ ensures that the mean $\mu = n\pi$ is far enough above 0 to prevent severe truncation. In a *left-skewed* binomial distribution (when $\pi > .50$), the rule that $n(1 - \pi) \geq 10$ guards against severe truncation at the upper end of the scale by making sure that the mean is well below n. That is why both rules are needed.

To be sure you understand the continuity correction, consider the events in the table below. We sketch a diagram to find the correct cutoff point to approximate a discrete model with a continuous one.

Event	Relevant Values of X	Normal Cutoff
At least 17	. . . 14 15 16 *17 18 19 20* . . .	Use $x = 16.5$
More than 15	. . . 14 15 *16 17 18 19 20* . . .	Use $x = 15.5$
Fewer than 19	. . . *14 15 16 17 18* 19 20 . . .	Use $x = 18.5$

Section Exercises

connect

Note: Use Excel for the following exercises.

7.53 The default rate on government-guaranteed student loans at a certain public four-year institution is 7 percent. (a) If 1,000 student loans are made, what is the approximate normal probability of fewer than 50 defaults? (b) More than 100? Show your work carefully.

7.54 In a certain store, there is a .03 probability that the scanned price in the bar code scanner will not match the advertised price. The cashier scans 800 items. (a) What is the expected number of mismatches? The standard deviation? (b) What is the approximate normal probability of at least 20 mismatches? (c) Of more than 30 mismatches? Show your calculations clearly.

7.55 The probability is .90 that a vending machine in the Oxnard University Student Center will dispense the desired item when correct change is inserted. If 200 customers try the machine, what is the approximate normal probability that (a) at least 175 will receive the desired item? (b) That fewer than 190 will receive the desired item? Explain.

7.56 When confronted with an in-flight medical emergency, pilots and crew can consult staff physicians at a global response center located in Arizona. If the global response center is called, there is a 4.8 percent chance that the flight will be diverted for an immediate landing. Suppose the response center is called 8,465 times in a given year. (a) What is the expected number of diversions? (b) What is the approximate normal probability of at least 400 diversions? (c) Fewer than 450 diversions? Show your work carefully.

Normal Approximation to the Poisson

The normal approximation for the Poisson works best when λ is fairly large. If you can't find λ in Appendix B (which only goes up to $\lambda = 20$), you are reasonably safe in using the normal approximation. Some textbooks allow the approximation when $\lambda \geq 10$, which is comparable to the rule that the binomial mean must be at least 10. To use the normal approximation to the Poisson, we set the normal μ and σ equal to the Poisson mean and standard deviation:

(7.8)
$$\mu = \lambda$$

(7.9)
$$\sigma = \sqrt{\lambda}$$

Example 7.4

Utility Bills

On Wednesday between 10 a.m. and noon, customer billing inquiries arrive at a mean rate of 42 inquiries per hour at Consumers Energy. What is the probability of receiving more than 50 calls? Call arrivals presumably follow a Poisson model, but the mean $\lambda = 42$ is too large to use Appendix B. The formula would entail an infinite sum $P(51) + P(52) + \ldots$ whose terms gradually become negligible (recall that the Poisson has no upper limit), but the calculation would be tedious at best. However, the normal approximation is simple. We set

$$\mu = \lambda = 42$$
$$\sigma = \sqrt{\lambda} = \sqrt{42} = 6.48074$$

The continuity-corrected cutoff point for $X \geq 51$ is $X = 50.5$ (halfway between 50 and 51):

$$\ldots 46 \ 47 \ 48 \ 49 \ 50 \ \overrightarrow{\mathit{51} \ \mathit{52} \ \mathit{53} \ldots}$$

The standardized Z-value for the event "more than 50" is $P(X > 50.5) = P(Z > 1.31)$ because

$$z = \frac{x - \mu}{\sigma} = \frac{50.5 - 42}{6.48074} \cong 1.31$$

Using Appendix C-2 we look up $P(Z < -1.31) = .0951$, which is the same as $P(Z > 1.31)$ because the normal distribution is symmetric. We can check the actual Poisson probability by using Excel's cumulative function =POISSON.DIST(50, 42, 1) and subtracting from 1:

$$P(X \geq 51) = 1 - P(X \leq 50) = 1 - .9025 = .0975$$

In this case, the normal approximation (.0951) comes fairly close to the Poisson result (.0975). Of course, if you have access to Excel, you don't need the approximation at all.

Note: Use Excel for the following exercises.

7.57 Small power surges (5–10 volts) occur in a residence an average of 50 times in a 24-hour period. (a) What is the approximate normal probability of at least 60 small power surges in a randomly chosen 24-hour period? (b) Of fewer than 35 small power surges? (c) Use Excel to calculate the actual Poisson probabilities. How close were your approximations?

7.58 The average defect rate on a 2010 Volkswagen vehicle was reported to be 1.35 defects per vehicle. Suppose that we inspect 100 Volkswagen vehicles at random. (a) What is the approximate normal probability of finding at least 150 defects? (b) Of finding fewer than 100 defects? (c) Use Excel to calculate the actual Poisson probabilities. How close were your approximations?

7.59 On average, 28 patients per hour arrive in the Foxboro 24-Hour Walk-in Clinic on Friday between 6 p.m. and midnight. (a) What is the approximate normal probability of more than 35 arrivals? (b) Of fewer than 25 arrivals? (c) Is the normal approximation justified? Show all calculations. (d) Use Excel to calculate the actual Poisson probabilities. How close were your approximations?

7.60 For a large Internet service provider (ISP), web virus attacks occur at a mean rate of 150 per day. (a) What is the approximate normal probability of at least 175 attacks in a given day? (b) Of fewer than 125 attacks? (c) Is the normal approximation justified? Show all calculations. (d) Use Excel to calculate the actual Poisson probabilities. How close were your approximations?

7.6 EXPONENTIAL DISTRIBUTION

In Chapter 6 we introduced the idea of a *random process*. For example, consider the process of customers arriving at various points of time at a restaurant, illustrated in Figure 7.23. There are two different variables that could be used to describe this process. We could count the number of customers who arrive in a randomly selected minute, or we could measure the time between two customer arrivals. As you learned in Chapter 6, the *count* of customer arrivals is a discrete random variable that typically has a Poisson distribution. If the count of customer arrivals has a Poisson distribution, the distribution of the time between two customer arrivals will have an **exponential distribution**, whose characteristics are detailed in Table 7.11. In the exponential model, the focus is on the *waiting time* until the next event, a continuous variable.

LO 7-7

Find the exponential probability for a given x.

Customer Arrivals

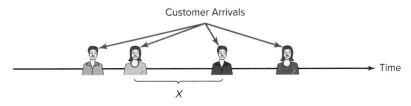

X = time between two customer arrivals

Figure 7.23

Customer Arrival Process at a Restaurant

Parameter	λ = mean arrival rate per unit of time or space (same as Poisson mean)
PDF	$f(x) = \lambda e^{-\lambda x}$
CDF	$P(X \le x) = 1 - e^{-\lambda x}$
Domain	$x \ge 0$
Mean	$1/\lambda$
Standard deviation	$1/\lambda$
Shape	Always right-skewed.
PDF in Excel*	=EXPON.DIST($x, \lambda, 0$)
CDF in Excel*	=EXPON.DIST($x, \lambda, 1$)
Random data in Excel*	=−LN(RAND())/λ (returns 1 value)
Comments	We often know $1/\lambda$ (the mean time between events) instead of λ (the arrival rate).

Table 7.11

Exponential Distribution

*See the **R Software Supplement** at the end of this chapter for a list of R's exponential probability functions.

Figure 7.24

Family of Exponential Distribution PDFs for Selected λ Values

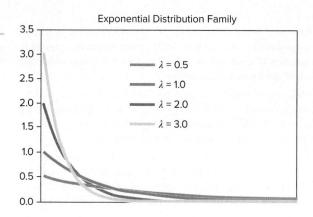

FIGURE 7.25

Exponential Areas Using CDF

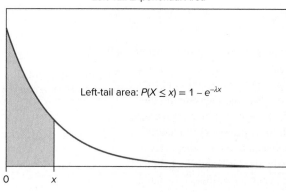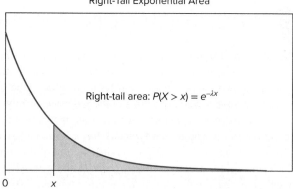

The exponential distribution has one parameter, the mean arrival rate λ. Each value of λ gives a different PDF, as illustrated in Figure 7.24. However, all curves in the exponential PDF family have the same shape. The vertical axis intercept is always λ because inserting $x = 0$ in the PDF in Table 7.11 gives $f(0) = \lambda$.

The exponential probability density function approaches zero as x increases and is right-skewed. We usually are not interested in the height of the function $f(x)$ but rather in areas under the curve, as illustrated in Figure 7.25. Fortunately, the CDF is simple. No tables are needed, just a calculator that has the e^x function key. The probability of waiting more than x units of time until the next arrival is $e^{-\lambda x}$, while the probability of waiting x units of time or less is $1 - e^{-\lambda x}$.

(7.10)	$P(X \leq x) = 1 - e^{-\lambda x}$	(probability of waiting x or less)
(7.11)	$P(X > x) = e^{-\lambda x}$	(probability of waiting more than x)
(7.12)	$P(x_1 \leq X < x_2) = e^{-\lambda x_1} - e^{-\lambda x_2}$	(probability of waiting between x_1 and x_2)

Recall that $P(X \leq x)$ is the same as $P(X < x)$ because the point x has no area. For this reason, we could use either $<$ or $\leq$ in formula 7.10 and formula 7.12.

Example 7.5

Customer Waiting Time

Between 2 p.m. and 4 p.m. on Wednesday, patient insurance inquiries arrive at Blue Choice insurance at a mean rate of 2.2 calls per minute. What is the probability of waiting more than 30 seconds for the next call? We set $\lambda = 2.2$ events per minute and $x = 0.50$ minute. Note that we must convert 30 seconds to 0.50 minute because λ is expressed in minutes, and the units of measurement must be the same. We have

$$P(X > 0.50) = e^{-\lambda x} = e^{-(2.2)(0.50)} = .3329, \text{ or } 33.29\%$$

There is about a 33 percent chance of waiting more than 30 seconds before the next call arrives. Because $x = 0.50$ is a *point* that has no area in a continuous model, $P(X \geq 0.50)$ and $P(X > 0.50)$ refer to the same event (unlike, say, a binomial model, in which a point *does* have a probability). The probability that 30 seconds or less (0.50 minute) will be needed before the next call arrives is

$$P(X \leq 0.50) = 1 - e^{-(2.2)(0.50)} = 1 - .3329 = .6671$$

These calculations are illustrated in Figure 7.26.

Figure 7.26

Exponential Tail Areas for $\lambda = 2.2$

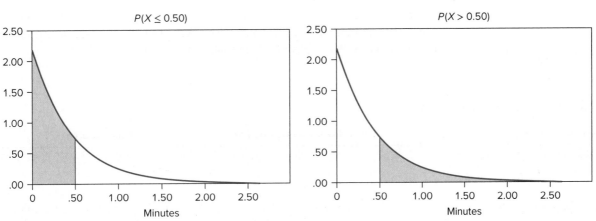

Section Exercises

7.61 In Santa Theresa, false alarms are received at the downtown fire station at a mean rate of 0.3 per day. (a) What is the probability that more than 7 days will pass before the next false alarm arrives? (b) Less than 2 days? (c) Explain fully.

7.62 Between 11 p.m. and midnight on Thursday night, Mystery Pizza gets an average of 4.2 telephone orders per hour. Find the probability that (a) at least 30 minutes will elapse before the next telephone order; (b) less than 15 minutes will elapse; and (c) between 15 and 30 minutes will elapse.

7.63 A passenger metal detector at Chicago's Midway Airport gives an alarm 2.1 times a minute. What is the probability that (a) less than 60 seconds will pass before the next alarm? (b) More than 30 seconds? (c) At least 45 seconds?

7.64 The Johnson family uses a propane gas grill for cooking outdoors. During the summer they need to replace their tank on average every 30 days. At a randomly chosen moment, what is the probability that they can grill out (a) at least 40 days before they need to replace their tank; (b) no more than 20 days?

7.65 At a certain Noodles & Company restaurant, customers arrive during the lunch hour at a rate of 2.8 per minute. What is the probability that (a) at least 30 seconds will pass before the next customer walks in; (b) no more than 15 seconds; (c) more than 1 minute?

Inverse Exponential

We can use the exponential area formula in reverse. If the mean arrival rate is 2.2 calls per minute, we want the 90th percentile for waiting time (the top 10 percent of waiting time) as illustrated in Figure 7.27. We want to find the x-value that defines the upper 10 percent.

LO 7-8

Solve for x for a given exponential probability.

Because $P(X \leq x) = .90$ implies $P(X > x) = .10$, we set the right-tail area to .10, take the natural logarithm of both sides, and solve for x:

$$P(X \leq x) = 1 - e^{-\lambda x} = .90$$
$$\text{so} \quad e^{-\lambda x} = .10$$
$$-\lambda x = \ln(.10)$$
$$-(2.2)x = -2.302585$$
$$x = 2.302585/2.2$$
$$x = 1.0466 \text{ minutes}$$

Figure 7.27

Finding x for the Upper 10 Percent

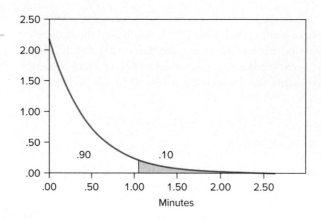

So 90 percent of the calls will arrive within 1.0466 minutes (or 62.8 seconds). We can find any percentile in the same way. For example, Table 7.12 illustrates similar calculations to find the quartiles (25 percent, 50 percent, 75 percent) of waiting time.

Table 7.12

Quartiles for Exponential with $\lambda = 2.2$

First Quartile Q_1	Second Quartile Q_2 (median)	Third Quartile Q_3
$P(X \leq x) = 1 - e^{-\lambda x} = .25$	$P(X \leq x) = 1 - e^{-\lambda x} = .50$	$P(X \leq x) = 1 - e^{-\lambda x} = .75$
so $e^{-\lambda x} = .75$	so $e^{-\lambda x} = .50$	so $e^{-\lambda x} = .25$
$-\lambda x = \ln(.75)$	$-\lambda x = \ln(.50)$	$-\lambda x = \ln(.25)$
$-(2.2)x = -0.2876821$	$-(2.2)x = -0.6931472$	$-(2.2)x = -1.386294$
$x = 0.2876821/2.2$	$x = 0.6931472/2.2$	$x = 1.386294/2.2$
$x = 0.1308$ minute, or 7.9 seconds	$x = 0.3151$ minute, or 18.9 seconds	$x = 0.6301$ minute, or 37.8 seconds

The calculations in Table 7.12 show that the mean waiting time is $1/\lambda = 1/2.2 = 0.4545$ minute, or 27 seconds. It is instructive to note that the median waiting time (18.9 seconds) is less than the mean. Because the exponential distribution is highly right-skewed, we would expect the mean waiting time to be above the median, which it is.

Mean Time between Events

Exponential waiting times are often described in terms of the **mean time between events** (MTBE) rather than in terms of Poisson arrivals per unit of time. In other words, we might be given $1/\lambda$ instead of λ.

$$\text{MTBE} = 1/\lambda = \textit{mean time between events} \quad \text{(units of time per event)}$$

$$1/\text{MTBE} = \lambda = \textit{mean events per unit of time} \quad \text{(events per unit of time)}$$

If we know MTBE, we can calculate its reciprocal to get λ. For example, if the mean time between patient arrivals in an emergency room is 20 minutes, then $\lambda = 1/20 = 0.05$ arrival per minute (or $\lambda = 3.0$ arrivals per hour). We could work the problem using either hours or minutes, as long as we are careful to make sure that x and λ are expressed in the same units when we calculate $e^{-\lambda x}$. For example, $P(X > 12 \text{ minutes}) = e^{-(0.05)(12)} = e^{-0.60}$ is the same as $P(X > 0.20 \text{ hour}) = e^{-(3)(0.20)} = e^{-0.60}$.

Example 7.6

Flat-Panel Displays

The NexGenCo color flat-panel display in an aircraft cockpit has a mean time between failures (MTBF) of 22,500 flight hours. What is the probability of a failure within the next 10,000 flight hours? Because 22,500 hours per failure implies $\lambda = 1/22,500$ failure per hour, we calculate:

$$P(X < 10,000) = 1 - e^{-\lambda x} = 1 - e^{-(1/22,500)(10,000)} = 1 - e^{-0.4444} = 1 - .6412 = .3588$$

There is a 35.88 percent chance of failure within the next 10,000 hours of flight. This assumes that failures follow the Poisson model.

Example 7.7

Warranty Period

A manufacturer of GPS navigation receivers for boats knows that their mean life under typical maritime conditions is 7 years. What warranty should be offered in order that no more than 30 percent of the GPS units will fail before the warranty expires? The situation is illustrated in Figure 7.28.

Figure 7.28

Finding *x* for the Lower 30 Percent

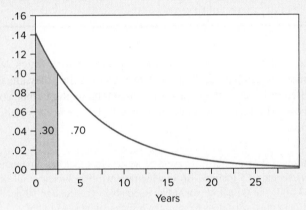

Let *x* be the length of the warranty. To solve this problem, we note that if 30 percent fail before the warranty expires, 70 percent will fail afterward. That is, $P(X > x) = 1 - P(X \le x) = 1 - 0.30 = .70$.

We set $P(X > x) = e^{-\lambda x} = .70$ and solve for *x* by taking the natural log of both sides of the equation:

$$e^{-\lambda x} = .70$$
$$-\lambda x = \ln(.70)$$
$$-\lambda x = -0.356675$$
$$x = 0.356675/\lambda$$

But in this case, we are not given λ but rather its reciprocal, MTBF = $1/\lambda$. Seven years *mean time between failures* is the same as saying $\lambda = 1/7$ *failure per year*. So we plug in $\lambda = 1/7 = 0.1428571$ to finish solving for *x:*

$$x = 0.356675/0.142857 = 2.497 \text{ years}$$

Thus, the firm would offer a 30-month warranty.

It may seem paradoxical that such a short warranty would be offered for something that lasts 7 years. However, the right tail is very long. A few long-lived GPS units will pull up the mean. This is typical of electronic equipment, which helps explain why your laptop computer may have only a 1-year warranty when we know that laptops often last for many years. Similarly, automobiles typically outlast their warranty period (although competitive pressures have recently led to warranties of 5 years or more, even though it may result in a loss on a few warranties). In general, warranty periods are a policy tool used by business to balance costs of expected claims against the competitive need to offer contract protection to consumers.

Using Excel

The Excel function =EXPON.DIST(x, Lambda, 1) will return the left-tail area $P(X \le x)$. The "1" indicates a cumulative area. If you enter 0 instead of 1, you will get the height of the PDF instead of the left-tail area for the CDF. Every situation with Poisson arrivals over time is associated with an exponential waiting time. Both models depend solely on the parameter λ = mean arrival rate per unit of time. These two closely related distributions are summarized in Table 7.13.

Table 7.13   **Relation between Exponential and Poisson Models**

Model	Random Variable	Parameter	Mean	Domain	Variable Type
Poisson	X = number of arrivals per unit of time	$\lambda = \dfrac{\text{mean arrivals}}{\text{unit of time}}$	λ	$x = 0, 1, 2, \ldots$	Discrete
Exponential	X = waiting time until next arrival	$\lambda = \dfrac{\text{mean arrivals}}{\text{unit of time}}$	$1/\lambda$	$x \ge 0$	Continuous

The exponential model may remind you of the geometric model, which describes the number of items that must be sampled until the first success. In spirit, they are similar. However, the models are different because the geometric model tells us the number of *discrete* events until the next success, while the exponential model tells the *continuous* waiting time until the next arrival of an event.

Section Exercises

7.66 The time it takes a ski patroller to respond to an accident call has an exponential distribution with an average equal to 5 minutes. (a) In what time will 90 percent of all ski accident calls be responded to? (b) If the ski patrol would like to be able to respond to 90 percent of the accident calls within 10 minutes, what does the average response time need to be?

7.67 Between 11 p.m. and midnight on Thursday night, Mystery Pizza gets an average of 4.2 telephone orders per hour. (a) Find the median waiting time until the next telephone order. (b) Find the upper quartile of waiting time before the next telephone order. (c) What is the upper 10 percent of waiting time until the next telephone order? Show all calculations clearly.

7.68 The shoplifting sensor at a certain Best Buy exit gives an alarm 0.5 time a minute. (a) Find the median waiting time until the next alarm. (b) Find the first quartile of waiting times before the next alarm. (c) Find the 30th percentile of waiting time until the next alarm. Show all calculations clearly.

7.69 Between 2 a.m. and 4 a.m. at an all-night pizza parlor, the mean time between the arrival of telephone pizza orders is 20 minutes. (a) Find the median wait for pizza order arrivals. (b) Explain why the median is not equal to the mean. (c) Find the upper quartile.

7.70 The mean life of a certain computer hard disk in continual use is 8 years. (a) How long a warranty should be offered if the vendor wants to ensure that no more than 10 percent of the hard disks will fail within the warranty period? (b) No more than 20 percent?

7.7 TRIANGULAR DISTRIBUTION (OPTIONAL)

Characteristics of the Triangular Distribution

LO 7-9

Use the triangular distribution for "what-if" analysis (optional).

Table 7.14 shows the characteristics of the **triangular distribution**. Visually, it is a simple distribution, as you can see in Figure 7.29. It can be symmetric or skewed. Its X values must lie within the interval $[a, c]$. But unlike the uniform, it has a mode or "peak." The peak is reminiscent of a normal, which also has a single maximum. But unlike the normal, the triangular does not go on forever because its X values are confined by a and c. The triangular distribution is sometimes denoted $T(a, b, c)$ or $T(\text{min}, \text{mode}, \text{max})$.

		Table 7.14
		Triangular Distribution

Parameters

a = lower limit
b = mode
c = upper limit

PDF

$$f(x) = \frac{2(x - a)}{(b - a)(c - a)} \quad \text{for} \quad a \leq x \leq b$$

$$f(x) = \frac{2(c - x)}{(c - a)(c - b)} \quad \text{for} \quad b \leq x \leq c$$

CDF

$$P(X \leq x) = \frac{(x - a)^2}{(b - a)(c - a)} \quad \text{for} \quad a \leq x \leq b$$

$$P(X \leq x) = 1 - \frac{(c - x)^2}{(c - a)(c - b)} \quad \text{for} \quad b \leq x \leq c$$

Domain

$a \leq x \leq c$

Mean

$$\frac{a + b + c}{3}$$

Standard deviation

$$\sqrt{\frac{a^2 + b^2 + c^2 - ab - ac - bc}{18}}$$

Shape

Positively skewed if $b < (a + c)/2$.
Negatively skewed if $b > (a + c)/2$.

Comments

Practical model, useful in business what-if analysis.
A symmetric triangular is the sum of two identically
distributed uniform variates.

Figure 7.29

Triangular PDFs

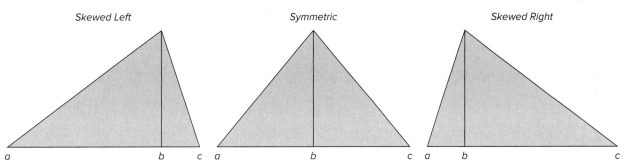

An oral surgeon injects a painkiller prior to extracting a tooth. Given the varying characteristics of patients, the dentist views the time for anesthesia effectiveness as a triangular
random variable that takes between 15 minutes and 30 minutes, with 20 minutes as the
most likely time. Setting $a = 15$, $b = 20$, and $c = 30$, we obtain

$$\mu = \frac{a + b + c}{3} = \frac{15 + 20 + 30}{3} = 21.7 \text{ minutes}$$

$$\sigma = \sqrt{\frac{a^2 + b^2 + c^2 - ab - ac - bc}{18}}$$

$$= \sqrt{\frac{15^2 + 20^2 + 30^3 - (15)(20) - (15)(30) - (20)(30)}{18}}$$

$$= 3.12 \text{ minutes}$$

Example 7.8

*Anesthesia Effectiveness
Using Triangular
Distribution*

Using the cumulative distribution function or CDF, we can calculate the probability of taking less than x minutes:

$$(7.13) \qquad P(X \le x) = \frac{(x - a)^2}{(b - a)(c - a)} \quad \text{for } a \le x \le b$$

$$(7.14) \qquad P(X \le x) = 1 - \frac{(c - x)^2}{(c - a)(c - b)} \quad \text{for } b \le x \le c$$

For example, the probability that the anesthetic takes less than 25 minutes is

$$P(X \le 25) = 1 - \frac{(30 - 25)^2}{(30 - 15)(30 - 20)} = .8333$$

Basically, we are finding the small triangle's area (½ base × height) and then subtracting from 1. This situation is illustrated in Figure 7.30. In contrast, assuming a uniform distribution with parameters $a = 15$ and $b = 30$ would yield $P(X \le 25) = .6667$. Why is it different? Because the triangular, with mode 20, has more probability on the low end, making it more likely that a patient will be fully anesthetized within 25 minutes. Assuming a uniform distribution may seem conservative, but it could lead to patients sitting around longer waiting to be sure the anesthetic has taken effect. Only experience could tell us which model is more realistic.

Figure 7.30

Triangular $P(X \le 25)$

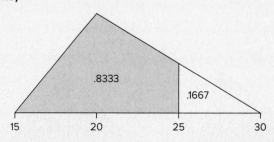

Special Case: Symmetric Triangular

An interesting special case is a **symmetric triangular distribution** centered at 0, whose lower limit is identical to its upper limit except for sign (e.g., from $-c$ to $+c$) with mode 0 (halfway between $-c$ and $+c$). If you set $c = 2.45$, the distribution $T(-2.45, 0, +2.45)$ closely resembles a standard normal distribution $N(0, 1)$.

Figure 7.31 compares these two distributions. Unlike the normal $N(0, 1)$, the triangular distribution $T(-2.45, 0, +2.45)$ always has values within the range $-2.45 \le X \le +2.45$. Yet over much of the range, the distributions are alike, and random samples from $T(-2.45, 0, +2.45)$ are surprisingly similar to samples from a normal $N(0, 1)$ distribution. It is easy to generate symmetric triangular random data in Excel by summing two $U(0, 1)$ random variables using the function =245*(RAND()+RAND()−1).

Uses of the Triangular

The triangular distribution is a way of thinking about variation that corresponds rather well to what-if analysis in business. It is not surprising that business analysts are attracted to the triangular model. Its finite range and simple form are more understandable than a normal distribution.

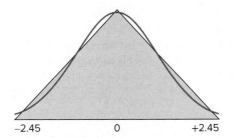

Figure 7.31

Symmetric Triangular Is Approximately Normal

It is more versatile than a normal because it can be skewed in either direction. Yet it has some of the nice properties of a normal, such as a distinct mode. The triangular model is especially handy for what-if analysis when the business case depends on predicting a stochastic variable (e.g., the price of a raw material, an interest rate, a sales volume). If the analyst can anticipate the range (*a* to *c*) and most likely value (*b*), it will be possible to calculate probabilities of various outcomes. Many times, such distributions will be skewed, so a normal wouldn't be much help. For example, a triangular distribution of completion times was used by planners for each construction phase in building the Port Miami Tunnel, a 4,200-foot undersea tunnel in Florida. The triangular distribution is often used in simulation modeling software such as Arena. In Chapter 18, we will explore what-if analysis using the triangular $T(a, b, c)$ model in simulations.

7.71 Suppose that the distribution of order sizes (in dollars) at L.L.Bean has a distribution that is $T(0, 25, 75)$. (a) Find the mean. (b) Find the standard deviation. (c) Find the probability that an order will be less than \$25. (d) Sketch the distribution and shade the area for the event in part (c).

7.72 Suppose that the distribution of oil prices (\$/bbl) is forecast to be $T(50, 65, 105)$. (a) Find the mean. (b) Find the standard deviation. (c) Find the probability that the price will be greater than \$75. (d) Sketch the distribution and shade the area for the event in part (c).

Section Exercises

connect

Chapter Summary

The **probability density function** (PDF) of a **continuous random variable** is a smooth curve, and probabilities are *areas* under the curve. The area under the entire PDF is 1. The **cumulative distribution function** (CDF) shows the area under the PDF to the left of *X*, approaching 1 as *X* increases. The mean $E(X)$ and variance $Var(X)$ are integrals, rather than sums, as for a discrete random variable. The **uniform continuous distribution,** denoted $U(a, b)$, has two parameters *a* and *b* that enclose the range. It is a simple what-if model with applications in simulation. The

normal distribution, denoted $N(\mu, \sigma)$, is symmetric and bell-shaped. It has two parameters, the mean μ and standard deviation σ. It serves as a benchmark. Because there is a different normal distribution for every possible μ and σ, we apply the transformation $z = (x - \mu)/\sigma$ to get a new random variable that follows a **standard normal distribution,** denoted $N(0, 1)$, with mean 0 and standard deviation 1. There is no simple formula for normal areas, but tables or Excel functions are available to find an area under the curve for given *z*-values or to find *z*-values that give a

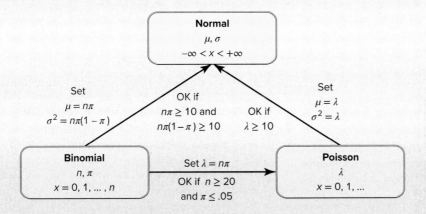

Figure 7.32

Relationships among Three Models

specified area (the "inverse normal"). As shown in Figure 7.32, a **normal approximation** for a binomial or Poisson probability is acceptable when the mean is at least 10. The **exponential distribution** describes *waiting time* until the next Poisson arrival. Its one parameter is λ (the mean arrival rate), and its right tail area is $e^{-\lambda x}$ (the probability of waiting at least x time units for the next arrival). It is strongly right-skewed and is used to predict warranty claims or to schedule facilities. The **triangular distribution** $T(a, b, c)$ has three parameters (a and c enclose the range, and b is the mode). It may be symmetric or skewed in either direction. It is easy to visualize and is a useful model for what-if simulation. Table 7.15 compares these five models.

Table 7.15 Comparison of Models

Model	Parameters	Mean	Variance	Characteristics
Uniform	a, b	$(a+b)/2$	$(b-a)^2/12$	Always symmetric.
Normal	μ, σ	μ	σ^2	Symmetric. Useful as reference benchmark.
Standard normal	μ, σ	0	1	Special case of the normal with $z = (x - \mu)/\sigma$.
Exponential	λ	$1/\lambda$	$1/\lambda^2$	Always skewed right. Right-tail area is $e^{-\lambda x}$ for waiting times.
Triangular	a, b, c	$(a+b+c)/3$	$(a^2+b^2+c^2-ab-ac-bc)/18$	Useful for what-if business modeling.

Key Terms

continuity correction	Gaussian distribution	probability density function	triangular distribution
continuous random variable	integral	standard normal	uniform continuous
cumulative distribution	inverse normal	distribution	distribution
function	mean time between events	symmetric triangular	
exponential distribution	normal distribution	distribution	

Commonly Used Formulas in Continuous Distributions

$$\text{Uniform CDF:} \quad P(X \le x) = \frac{x-a}{b-a} \quad \text{for } a \le x \le b$$

$$\text{Standard Normal Transformation:} \quad z = \frac{x-\mu}{\sigma} \quad \text{for } -\infty < x < +\infty$$

$$\text{Normal-Binomial Approximation:} \quad \mu = n\pi \quad \sigma = \sqrt{n\pi(1-\pi)}$$
$$\text{for } n\pi \ge 10 \text{ and } n(1-\pi) \ge 10$$

$$\text{Normal-Poisson Approximation:} \quad \mu = \lambda \quad \sigma = \sqrt{\lambda} \quad \text{for } \lambda \ge 0$$

$$\text{Exponential CDF:} \quad P(X \le x) = 1 - e^{-\lambda x} \quad \text{for } x \ge 0$$

Chapter Review

1. (a) Why does a point have zero probability in a continuous distribution? (b) Why are probabilities areas under curves in a continuous distribution?

2. Define (a) parameter, (b) PDF, and (c) CDF.

3. For the uniform distribution: (a) tell how many parameters it has; (b) indicate what the parameters represent; (c) describe its shape; and (d) explain when it would be used.

4. For the normal distribution: (a) tell how many parameters it has; (b) indicate what the parameters represent; (c) describe its shape; and (d) explain why all normal distributions are alike despite having different μ and σ.

5. (a) What features of a stochastic process might lead you to anticipate a normal distribution? (b) Give two examples of random variables that might be considered normal.

6. (a) What is the transformation to standardize a normal random variable? (b) Why do we standardize a variable to find normal areas? (c) How does a standard normal distribution differ from any other normal distribution, and how is it similar?

7. (a) Explain the difference between Appendix C-1 and Appendix C-2. (b) List advantages of each type of table. (c) Which table do you expect to use, and why? (d) Why not always use Excel?

8. Write an example of each of the four normal functions in Excel and tell what each function does.

9. List the standard normal *z*-values for several common areas (tail and/or middle). *You will use them often.*

10. For the exponential distribution: (a) tell how many parameters it has; (b) indicate what the parameters represent; (c) describe its shape; and (d) explain when it would be used.

11. When does the normal give an acceptable approximation (a) to a binomial and (b) to a Poisson? (c) Why might you never need these approximations? (d) When might you need them?

12. For the triangular distribution: (a) tell how many parameters it has; (b) indicate what the parameters represent; (c) describe its shape in a general way (e.g., skewness); and (d) explain when it would be used.

Chapter Exercises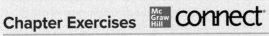

Note: Show your work clearly. Problems with * are harder or based on optional material.

7.73 Which of the following is a continuous random variable?

a. Number of Honda Civics sold in a given day at a car dealership.

b. Amount of gasoline used for a 200-mile trip in a Honda Civic.

c. Distance driven on a particular Thursday by the owner of a Honda Civic.

7.74 Which of the following could be probability density functions for a continuous random variable? Explain.

a. $f(x) = .50$ for $0 \le x \le 2$
b. $f(x) = 2 - x$ for $0 \le x \le 2$
c. $f(x) = .5x$ for $0 \le x \le 2$

7.75 Applicants for a night caretaker position are uniformly distributed in age between 25 and 65. (a) What is the mean age of an applicant? (b) The standard deviation? (c) What is the probability that an applicant will be over 45? (d) Over 55? (e) Between 30 and 60?

7.76 Passengers using New York's MetroCard system must swipe the card at a rate between 10 and 40 inches per second, or else the card must be reswiped through the card reader. Research shows that actual swipe rates by subway riders are uniformly distributed between 5 and 50 inches per second. (a) What is the mean swipe rate? (b) What is the standard deviation of the swipe rate? (c) What are the quartiles? (d) What percentage of subway riders must reswipe the card because they were outside the acceptable range?

7.77 Discuss why you would or would not expect each of the following variables to be normally distributed. *Hint:* Would you expect a single central mode and tapering tails? Would the distribution be roughly symmetric? Would one tail be longer than the other?

a. Time for households to complete the U.S. Census short form.

b. Size of automobile collision damage claims.

c. Diameters of randomly chosen circulated quarters.

d. Weight of contents of 16-ounce boxes of elbow macaroni.

7.78 Why might the following *not* be normally distributed? (a) The time it takes you to drive to the airport. (b) The annual income for a randomly chosen Major League Baseball player. (c) The annual hurricane losses suffered by homeowners in Florida.

7.79 Scores on a certain accounting exam were normally distributed with a mean of 75 and a standard deviation of 7. Find the percentile for each individual using Excel's =NORM.S.DIST function. (a) Bob's score was 82; (b) Phyllis's score was 93; (c) Tom's score was 63.

7.80 Chlorine concentration in a municipal water supply is a uniformly distributed random variable that ranges between 0.74 ppm and 0.98 ppm. (a) What is the mean chlorine concentration? (b) The standard deviation? (c) What is the probability that the chlorine concentration will exceed 0.80 ppm on a given day? (d) Will be under 0.85 ppm? (e) Will be between 0.80 ppm and 0.90 ppm?

7.81 The weekly demand for Lay's Baked potato chips at a certain Subway sandwich shop is a random variable with mean 450 and standard deviation 80. Find the value(s) of *X* for each event. Show your work.

a. Highest 50 percent
b. Lowest 25 percent
c. Middle 80 percent
d. 5th percentile

7.82 The weekly demand for Papa Chubby's pizzas on a Friday night is a random variable with mean 235 and standard deviation 10. Find the value(s) of *X* for each event. Show your work.

a. Lowest 50 percent
b. Highest 25 percent
c. 90th percentile
d. Middle 80 percent

7.83 The amounts spent by customers at a Noodles & Company restaurant during lunch are normally distributed with a mean equal to $7.00 and a standard deviation equal to $0.35. (a) What amount is the first quartile? (b) The second quartile? (c) The 90th percentile?

7.84 The length of a Colorado brook trout is normally distributed. (a) What is the probability that a brook trout's length exceeds the mean? (b) Exceeds the mean by at least 1 standard deviation? (c) Exceeds the mean by at least 2 standard deviations? (d) Is within 2 standard deviations?

7.85 The caffeine content of a cup of home-brewed coffee is a normally distributed random variable with a mean of 115 mg and a standard deviation of 20 mg. (a) What is the probability that a randomly chosen cup of home-brewed coffee will have more than 130 mg of caffeine? (b) Less than 100 mg? (c) A very strong cup of tea has a caffeine content of 91 mg. What is the probability that a cup of coffee will have less caffeine than a very strong cup of tea?

7.86 The fracture strength of a certain type of manufactured glass is normally distributed with a mean of 579 MPa with a standard deviation of 14 MPa. (a) What is the probability that a randomly chosen sample of glass will break at less than 579 MPa? (b) More than 590 MPa? (c) Less than 600 MPa?

7.87 Tire pressure monitoring systems (TPMS) warn the driver when the tire pressure of the vehicle is 25% below the target pressure. Suppose the target tire pressure of a certain car is 30 psi (pounds per square inch). (a) At what psi will the TPMS trigger a warning for this car? (b) Suppose tire pressure is a normally distributed random variable with a standard deviation equal to 2 psi. If the car's average tire pressure is on target, what is the probability that the TPMS will trigger a warning? (c) The manufacturer's recommended correct inflation range is 28 psi to 32 psi. Assume the tires' average psi is on target. If a tire on the car is inspected at random, what is the probability that the tire's inflation is within the recommended range?

7.88 In a certain microwave oven on the high power setting, the time it takes a randomly chosen kernel of popcorn to pop is normally distributed with a mean of 140 seconds and a standard deviation of 25 seconds. What percentage of the kernels will fail to pop if the popcorn is cooked for (a) 2 minutes? (b) Three minutes? (c) If you wanted 95 percent of the kernels to pop, what time would you allow? (d) If you wanted 99 percent to pop?

7.89 Procyon Manufacturing produces tennis balls. Their manufacturing process has a mean ball weight of 2.035 ounces with a standard deviation of 0.03 ounce. Regulation tennis balls are required to have a weight between 1.975 ounces and 2.095 ounces. What proportion of Procyon's production will fail to meet these specifications?

7.90 Shower temperature at the Oxnard Health Club showers is regulated automatically. The heater kicks in when the temperature falls to 99°F and shuts off when the temperature reaches 107°F. Water temperature then falls slowly until the heater kicks in again. At a given moment, the water temperature is a uniformly distributed random variable $U(99,107)$. (a) Find the mean temperature. (b) Find the standard deviation of the temperature. (c) Find the 75th percentile for water temperature.

7.91 Tests show that, on average, the Li-ion Hitachi stick driver can drive 207 drywall screws on a single charge. Bob needs to drive 230 drywall screws. If the standard deviation is 14 screws, find the probability that Bob can finish his job without recharging. *Hint:* Assume a normal distribution and treat the data as continuous.

7.92 The time it takes to give a man a shampoo and haircut is normally distributed with mean 22 minutes and standard deviation 3 minutes. Customers are scheduled every 30 minutes. (a) What is the probability that a male customer will take longer than the allotted time? *(b) If three male customers are scheduled sequentially on the half-hour, what is the probability that all three will be finished within their allotted half-hour times?

7.93 The length of a time-out during a televised professional football game is normally distributed with a mean of 84 seconds and a standard deviation of 10 seconds. If the network runs consecutive commercials totaling 90 seconds, what is the probability that play will resume before the commercials are over?

7.94 If the weight (in grams) of cereal in a box of Lucky Charms is $N(470, 5)$, what is the probability that the box will contain less than the advertised weight of 453 g?

7.95 Demand for residential electricity at 6:00 p.m. on the first Monday in October in Santa Theresa County is normally distributed with a mean of 4,905 MW (megawatts) and a standard deviation of 355 MW. Due to scheduled maintenance and unexpected system failures in a generating station, the utility can supply a maximum of 5,200 MW at that time. What is the probability that the utility will have to purchase electricity from other utilities or allow brownouts?

7.96 Jim's systolic blood pressure is a random variable with a mean of 145 mmHg and a standard deviation of 20 mmHg. For Jim's age group, 140 is the threshold for high blood pressure. (a) If Jim's systolic blood pressure is taken at a randomly chosen moment, what is the probability that it will be 135 or less? (b) 175 or more? (c) Between 125 and 165? (d) Discuss the implications of variability for physicians who are trying to identify patients with high blood pressure.

7.97 A statistics exam was given. Calculate the percentile for each of the following four students.

a. John's z-score was -1.62.
b. Mary's z-score was 0.50.
c. Zak's z-score was 1.79.
d. Frieda's z-score was 2.48.

7.98 Are the following statements true or false? Explain your reasoning.

a. "If we see a standardized z-value beyond ± 3, the variable cannot be normally distributed."
b. "If X and Y are two normally distributed random variables measured in different units (e.g., X is in pounds and Y is in kilograms), then it is not meaningful to compare the standardized z-values."
c. "Two machines fill 2-liter soft drink bottles by using a similar process. Machine A has $\mu = 1,990$ ml and $\sigma = 5$ ml while machine B has $\mu = 1,995$ ml and $\sigma = 3$ ml. The variables cannot both be normally distributed because they have different standard deviations."

*7.99 John can take either of two routes (A or B) to LAX airport. At midday on a typical Wednesday, the travel time on either route is normally distributed with parameters $\mu_A = 54$ minutes, $\sigma_A = 6$ minutes, $\mu_B = 60$ minutes, and $\sigma_B = 3$ minutes. (a) Which route should he choose if he must be at the airport in 54 minutes to pick up his spouse? (b) Sixty minutes? (c) Sixty-six minutes? Explain carefully.

7.100 The amount of fill in a half-liter (500 ml) soft drink bottle is normally distributed. The process has a standard deviation of 5 ml. The mean is adjustable. (a) Where should the mean be set to ensure a 95 percent probability that a half-liter bottle will not be underfilled? (b) A 99 percent probability? (c) A 99.9 percent probability? Explain.

7.101 The length of a certain kind of Colorado brook trout is normally distributed with a mean of 12.5 inches and a standard deviation of 1.2 inches. What minimum size limit should the Department of Natural Resources set if it wishes to allow people to keep 80 percent of the trout they catch?

7.102* Times for a surgical procedure are normally distributed. There are two methods. Method A has a mean of 28 minutes and a standard deviation of 4 minutes, while method B has a mean of 32 minutes and a standard deviation of 2 minutes. (a) Which procedure is preferred if the procedure must be completed within 28 minutes? (b) Thirty-eight minutes? (c) Thirty-six minutes? Explain your reasoning fully.

7.103* The length of a brook trout is normally distributed. Two brook trout are caught. (a) What is the probability that both exceed the mean? (b) Neither exceeds the mean? (c) One is above the mean and one is below? (d) Both are equal to the mean?

APPROXIMATIONS

7.104 Among live deliveries, the probability of a twin birth is .02. (a) In 2,000 live deliveries, what is the probability of at least 50 twin births? (b) Fewer than 35?

7.105 Nationwide, the probability that a rental car is from Hertz is 25 percent. In a sample of 100 rental cars, what is the probability that fewer than 20 are from Hertz?

7.106 The probability of being in a car accident when driving more than 10 miles over the speed limit in a residential neighborhood is .06. Of the next 1,000 cars that pass through a particular neighborhood, what are the first and third quartiles for the number of car accidents in this neighborhood?

7.107 A multiple-choice exam has 100 questions. Each question has four choices. (a) What minimum score should be required to reduce the chance of passing by random guessing to 5 percent? (b) To 1 percent? (c) Find the quartiles for a guesser.

7.108 The probability that a certain kind of flower seed will germinate is .80. (a) If 200 seeds are planted, what is the probability that fewer than 150 will germinate? (b) That at least 150 will germinate?

7.109 On a cold morning, the probability is .02 that a given car will not start in the small town of Eureka. Assume that 1,500 cars are started each cold morning. (a) What is the probability that at least 25 cars will not start? (b) More than 40?

7.110 At a certain fire station, false alarms are received at a mean rate of 0.2 per day. In a year, what is the probability that fewer than 60 false alarms are received?

EXPONENTIAL DISTRIBUTION

7.111 The HP dvd1040i 20X Multiformat DVD Writer has an MTBF of 70,000 hours. (a) Assuming continuous operation, what is the probability that the DVD writer will last more than 100,000 hours? (b) Less than 50,000 hours? (c) At least 50,000 hours but not more than 80,000 hours? (Source: www.hp.com.)

7.112 Automobile warranty claims for engine mount failure in a Troppo Malo 2000 SE are rare at a certain dealership, occurring at a mean rate of 0.1 claim per month. (a) What is the probability that the dealership will wait at least 6 months until the next claim? (b) At least a year? (c) At least 2 years? (d) At least 6 months but not more than 1 year?

7.113 Suppose the average time to service a Noodles & Company customer at a certain restaurant is 3 minutes and the service time follows an exponential distribution. (a) What is the probability that a customer will be serviced in less than 3 minutes? (b) Why is your answer more than 50 percent? Shouldn't exactly half the area be below the mean?

7.114 Systron Donner Inertial manufactures inertial subsystems for automotive, commercial/industrial, and aerospace and defense applications. The sensors use a one-piece, micromachined inertial sensing element to measure angular rotational velocity or linear acceleration. The MTBF for a single axis sensor is 400,000 hours. (a) Find the probability that a sensor lasts at least 30 years, assuming continuous operation. (b) Would you be surprised if a sensor has failed within the first 3 years? Explain.

TRIANGULAR DISTRIBUTION

*7.115 The price (dollars per 1,000 board feet) of Douglas fir from western Washington and Oregon varies according to a triangular distribution $T(300, 350, 490)$. (a) Find the mean. (b) Find the standard deviation. (c) What is the probability that the price will exceed 400?

*7.116 The distribution of scores on a statistics exam is $T(50, 60, 95)$. (a) Find the mean. (b) Find the standard deviation. (c) Find the probability that a score will be less than 75. (d) Sketch the distribution and shade the area for the event in part (c).

*7.117 The distribution of beach condominium prices in Santa Theresa ($ thousands) is $T(500, 700, 2100)$. (a) Find the mean. (b) Find the standard deviation. (c) Find the probability that a condo price will be greater than $750K. (d) Sketch the distribution and shade the area for the event in part (c).

PROJECTS AND DISCUSSION

7.118 (a) Write an Excel formula to generate a random normal deviate from $N(0, 1)$ and copy the formula into 10 cells. (b) Find the mean and standard deviation of your sample of 10 random data values. Are you satisfied that the random data have the desired mean and standard deviation? (c) Press F9 to generate 10 more data values and repeat question (b).

7.119 (a) Write an Excel formula to generate a random normal deviate from $N(4000, 200)$ and copy the formula into 100 cells. (b) Find the mean and standard deviation of your sample of 100 random data values. Are you satisfied that the random data have the desired mean and standard deviation? (c) Make a histogram of your sample. Does it appear normal?

7.120 On a police sergeant's examination, the historical mean score was 80 with a standard deviation of 20. Four officers who were alleged to be cronies of the police chief scored 195, 171, 191, and 189, respectively, on the test. This led to allegations of irregularity in the exam. (a) Convert these four officers' scores to standardized z-values. (b) Do you think there was sufficient reason to question these four exam scores? What assumptions are you making?

Related Reading

Forbes, Catherine, Merran Evans, Nicholas Hastings, and Brian Peacock. *Statistical Distributions.* 4th ed., Wiley, 2011.

CHAPTER 7 More Learning Resources

You can access these *LearningStats* demonstrations through Connect to help you understand probability.

Topic	LearningStats Demonstrations
Calculations	⬛ = Continuous Distributions Using Excel ⬛ = Continuous Distributions Using R ⬛ Normal Areas ⬛ Probability Calculator
Normal approximations	⬛ Evaluating Rules of Thumb 🔺 Why the Rule of 10?
Random data	⬛ Random Continuous Data ⬛ Visualizing Random Normal Data
Tables	⬛ Table C—Normal Probabilities

Key: ⬛ = PowerPoint ⬛ = Excel 🔺 = PDF

Software Supplement

Continuous Distributions in R

Continuous Probability Distributions	Excel Function	R Function
Normal distribution		
PDF: Returns height of $f(x)$	NORM.DIST(x, μ, σ, 0)	dnorm(x, μ, σ)
CDF: Returns probability $P(X \le x)$	NORM.DIST(x, μ, σ, 1)	pnorm(x, μ, σ)
Inverse CDF: Returns x for $P(X \le x) = \alpha$	NORM.INV(α, μ, σ)	qnorm(α, μ, σ)
Random data (makes k samples in R)	NORM.INV(RAND(), μ, σ)	rnorm(k, μ, σ)
Standard normal distribution		
PDF: Returns height of $f(z)$	NORM.S.DIST(z, 0)	dnorm(z)
CDF: Returns probability $P(Z \le z)$	NORM.S.DIST(z, 1)	pnorm(z)
Inverse CDF: Returns z for $P(Z \le z) = \alpha$	NORM.S.INV(α)	qnorm(α)
Random data (makes k samples in R)	NORM.S.INV(RAND())	rnorm(k)
Uniform distribution $a < x < b$		
PDF: Returns height of $f(x)$	$1/(b-a)$	dunif(x, a, b)
CDF: Returns probability $P(X \le x)$	$(x-a)/(b-a)$	punif(x, a, b)
Inverse CDF: Returns x for $P(X \le x) = \alpha$	$a + a*(b-a)$	qunif(α, a, b)
Random data (makes k samples in R)	$a + (b-a)*$RAND()	runif(k, a, b)
Exponential distribution		
PDF: Returns height of $f(x)$	EXPON.DIST(x, λ, 0)	dexp(x, λ)
CDF: Returns probability $P(X \le x)$	EXPON.DIST(x, λ, 1)	pexp(x, λ)
Inverse CDF: Returns x for $P(X \le x) = \alpha$	----------------	qexp(α, λ)
Random data (makes k samples in R)	----------------	rexp(k, λ)

Exam Review Questions for Chapters 5–7

1. Which type of probability (empirical, classical, subjective) is each of the following?

   a. On a given Friday, the probability that Flight 277 to Chicago is on time is 23.7%.

   b. Your chance of going to Disney World next year is 10%.

   c. The chance of rolling a 3 on two dice is 1/8.

2. For the following contingency table, find (a) $P(H \cap T)$; (b) $P(S|G)$; (c) $P(S)$

	R	S	T	Row Total
G	10	50	30	90
H	20	50	40	110
Col Total	30	100	70	200

3. If $P(A) = .30$, $P(B) = .70$, and $P(A \cap B) = .25$, are A and B independent events? Explain.

4. Which statement is *false?* Explain.

   a. If $P(A) = .05$, then the odds against event A's occurrence are 19 to 1.
   b. If A and B are mutually exclusive events, then $P(A \cup B) = 0$.
   c. The number of permutations of 5 things taken 2 at a time is 20.

5. Which statement is *true?* Why not the others?

   a. The Poisson distribution has two parameters.
   b. The binomial distribution assumes dependent random trials.
   c. The uniform distribution has two parameters.

6. If the payoff of a risky investment has three possible outcomes ($1,000, $2,000, $5,000) with probabilities .60, .30, and .10, respectively, find the expected value.

   a. $1,500
   b. $2,300
   c. $1,700

7. Assuming independent arrivals with a mean of 2.5 arrivals per minute, find the probability that in a given minute there will be (a) exactly 2 arrivals; (b) at least 3 arrivals; (c) fewer than 4 arrivals. (d) Which probability distribution did you use, and why?

8. If a random experiment whose success probability is .20 is repeated 8 times, find the probability of (a) exactly 3 successes; (b) more than 3 successes; (c) at most 2 successes. (d) Which probability distribution did you use, and why?

9. In a random experiment with 50 independent trials with constant probability of success .30, find the mean and standard deviation of the number of successes.

10. Which probability distribution (uniform, binomial, Poisson) is most nearly appropriate to describe each situation (assuming you knew the relevant parameters)?

    a. The number of dimes older than 10 years in a random sample of 8 dimes.
    b. The number of hospital patients admitted during a given minute on Tuesday morning.
    c. The last digit of a randomly chosen student's Social Security number.

11. Which statement is *false?* Explain.

    a. In the hypergeometric distribution, sampling is done without replacement.
    b. The mean of the uniform distribution is always $(a + b)/2$.
    c. We use the geometric distribution to find probabilities of arrivals per unit of time.

12. Which statement is *false?* Explain.

    a. To find probabilities in a continuous distribution, we add up the probabilities at each point.
    b. A uniform continuous model $U(5, 21)$ has mean 13 and standard deviation 4.619.
    c. A uniform PDF is constant for all values within the interval $a \leq X \leq b$.

13. Which statement is *true* for a normal distribution? Why not the others?

    a. The shape of the PDF is always symmetric regardless of μ and σ.
    b. The shape of the CDF resembles a bell-shaped curve.
    c. When no tables are available, areas may be found by a simple formula.

14. If freeway speeds are normally distributed with a mean of $\mu = 70$ mph and $\sigma = 7$ mph, find the probability that the speed of a randomly chosen vehicle (a) exceeds 78 mph; (b) is between 65 and 75 mph; (c) is less than 70 mph.

15. In the previous problem, calculate (a) the 95th percentile of vehicle speeds (i.e., 95 percent below); (b) the lowest 10 percent of speeds; (c) the highest 25 percent of speeds (3rd quartile).

16. Which of the following Excel formulas would be a correct way to calculate $P(X < 450)$ given that X is $N(500, 60)$?

    a. =NORM.DIST(450, 500, 60, 1)
    b. =NORM.S.DIST(450, 60)
    c. =1 − NORM.DIST(450, 500, 60, 0)

17. If arrivals follow a Poisson distribution with mean 1.2 arrivals per minute, find the probability that the waiting time until the next arrival will be (a) less than 1.5 minutes; (b) more than 30 seconds; (c) between 1 and 2 minutes.

18. In the previous problem, find (a) the 95th percentile of waiting times (i.e., 95 percent below); (b) the first quartile of waiting times; (c) the mean time between arrivals.

19. Which statement is *correct* concerning the normal approximation? Why not the others?

    a. The normal Poisson approximation is acceptable when $\lambda \geq 10$.
    b. The normal binomial approximation is better when n is small and π is large.
    c. Normal approximations are needed because Excel lacks discrete probability functions.

20. Which statement is *incorrect?* Explain.

    a. The triangular always has a single mode.
    b. The mean of the triangular is $(a + b + c)/3$.
    c. The triangular cannot be skewed left or right.

8 Sampling Distributions and Estimation

CHAPTER LEARNING OBJECTIVES

When you finish this chapter, you should be able to

LO 8-1   Define a sampling distribution.

LO 8-2   Explain the desirable properties of estimators.

LO 8-3   State and apply the Central Limit Theorem for a mean.

LO 8-4   Explain how sample size affects the standard error.

LO 8-5   Construct a confidence interval for a population mean using z.

LO 8-6   Know when and how to use Student's t instead of z to estimate a mean.

LO 8-7   Construct a confidence interval for a population proportion.

LO 8-8   Know how to modify confidence intervals when the population is finite.

LO 8-9   Calculate sample size to estimate a mean.

LO 8-10 Calculate sample size to estimate a proportion.

LO 8-11 Construct a confidence interval for a variance (optional).

n Chapter 1, we described two kinds of statistics: descriptive and inferential. Chapters 2–7 presented the statistical and probability tools for describing data sets and business models. Beginning with Chapter 8, we will present statistical tools for making inferences about populations. Inferential statistics involve two basic tasks: *estimating parameters* and *evaluating hypotheses* about populations using sample data. Inferential statistical methods are the foundation used by today's business analytics professionals. We begin with sampling concepts and estimation.

SAMPLING DISTRIBUTIONS

The value of a *sample statistic,* calculated from a *random sample,* will vary depending on which population items happen to be included in the sample. Some samples may represent the population well, while other samples could differ greatly from the population (particularly if the sample size is small). To illustrate **sampling variation**, let's draw some random samples from a large population of GMAT scores for MBA applicants. Figure 8.1 shows a dot plot of the entire population.

LO 8-1

Define a sampling distribution.

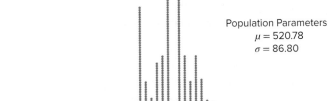

Population Parameters
$\mu = 520.78$
$\sigma = 86.80$

Population of 2,637 GMAT Scores

Figure 8.1

Dot Plot of GMAT Population 📂 **GMAT**

Source: Data for 2,637 MBA applicants at a medium-sized public university located in the Midwest.

Figure 8.2 shows several random samples of $n = 5$ from this population. The *individual* items that happen to be included in the samples vary. While sampling variation is inevitable, there is a tendency for the sample *means* to be close to the population mean ($\mu = 520.78$), shown as a dashed line in Figure 8.2. We see that the sample *means* (red markers) have much less variation than the *individual* sample items (blue dots). This is because the mean is an *average.* To see this more clearly, we can remove the *individual* sample items (blue dots) and just look at the sample *means* relative to the population mean (lower diagram). In larger samples, the sample means would tend to be even closer to μ.

Figure 8.2

Dot Plot of Eight Sample Means

📂 GMAT

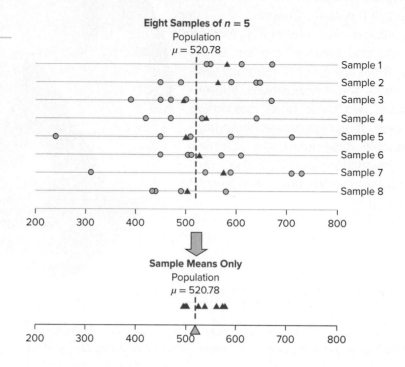

Sampling Error

The **sampling error** is the difference between the statistic and the corresponding population **parameter**. For example, for the population mean:

(8.1)
$$\text{Sampling Error} = \bar{x} - \mu$$

As demonstrated in Figure 8.2, sampling error exists because different samples will yield different values for $\bar{x}$ depending on which population items happen to be included in the sample. Because random samples and sample statistics vary, a sample statistic is a *random variable,* and random variables have probability distributions. A probability distribution that describes the values of a sample statistic calculated from all possible samples of size n is called a **sampling distribution**. Common statistics, such as the mean or proportion, have well-understood sampling distributions, derived from statistical theory and confirmed by observation. This chapter explains how to use the *sampling distribution* of a statistic to estimate unknown population parameters, taking into account four factors:

- Sample variation (uncontrollable).
- Population variation (uncontrollable).
- Sample size (controllable).
- Desired confidence in the estimate (controllable).

LO 8-2

Explain the desirable properties of estimators.

8.2 ESTIMATION

An **estimator** is a statistic calculated from a sample to estimate the value of a population parameter. An **estimate** is the value of the estimator in a particular sample. Table 8.1 and Figure 8.3 show some common estimators.

Usually, the parameter we are estimating is unknown, so we cannot calculate the sampling error. However, we can choose estimators with desirable properties when we make inferences about unknown parameters. For example, the sample mean $\bar{X}$ is a random variable that

Table 8.1

Examples of Estimators

Estimator	Formula	Parameter
Sample mean	$\bar{x} = \dfrac{1}{n}\sum_{i=1}^{n} x_i$ where x_i is the ith data value and n is the sample size	μ
Sample proportion	$p = x/n$ where x is the number of successes in the sample and n is the sample size	π
Sample standard deviation	$s = \sqrt{\dfrac{\sum_{i=1}^{n}(x_i - \bar{x})^2}{n-1}}$ where x_i is the ith data value and n is the sample size	σ

Table 8.1

Examples of Estimators

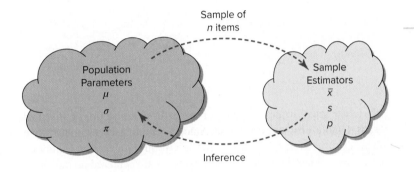

Figure 8.3

Sample Estimators of Population Parameters

correctly estimates μ on average because the sample means that overestimate μ will tend to be offset by those that underestimate μ. Four desirable properties of estimators, that are the basis for **statistical estimation**, are described next.

Properties of Estimators

The **bias** is the difference between the expected value (i.e., the average value) of the estimator and the true parameter. For the mean:

$$\text{Bias} = E(\overline{X}) - \mu \qquad (8.2)$$

An estimator is *unbiased* if its expected value is the parameter being estimated. We say $\overline{X}$ is an unbiased estimate of μ because $E(\overline{X}) = \mu$. There can be sampling error in a particular sample, but an **unbiased estimator** neither overstates nor understates the true parameter *on average*. Bias can be studied mathematically or by simulation experiments.

The sample mean ($\bar{x}$) and sample proportion (p) are unbiased estimators of μ and π, respectively. But we can find examples of biased estimators. For example, if you use Excel's population standard deviation formula =STDEV.P(Data) instead of its sample standard deviation formula =STDEV.S(Data) to estimate σ, you will get a biased estimate that underestimates the true value of σ *on average*. Sampling error is an inevitable risk in statistical sampling. You cannot *know* how much sampling error you have without knowing the population parameter (and if you knew it, you wouldn't be taking a sample). Our responsibility is to understand that a large sample will yield a more reliable estimate and to take the sample scientifically.

Sampling error is *random,* whereas bias is *systematic.* Consider an analogy with target shooting, illustrated in Figure 8.4. An expert whose rifle sights are correctly aligned will produce a target pattern like the one on the left. The same expert shooting a rifle with misaligned sights might produce the pattern on the right. Both targets show sampling variation, but the unbiased estimator is correctly *aimed.* An unbiased estimator avoids *systematic* error.

Figure 8.4

Illustration of Bias

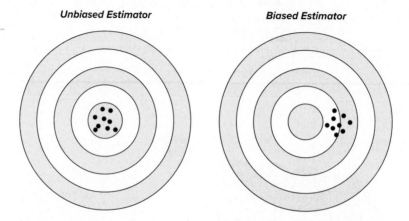

Unbiased Estimator   Biased Estimator

Efficiency refers to the variance of the estimator's sampling distribution. Smaller variance means a more efficient estimator. Among all unbiased estimators, we prefer the **minimum variance estimator**, referred to as MVUE (minimum variance unbiased estimator). Figure 8.5 shows two unbiased estimators. Both patterns are centered on the bull's-eye, but the estimator on the left has less variation. A more efficient estimator is closer *on average* to the true value of the parameter. You cannot assess efficiency from one sample, but it can be studied either mathematically or by simulation. Statisticians have proved that, for a normal distribution, $\bar{x}$ and s^2 are minimum variance estimators of μ and σ^2, respectively (i.e., no other estimators can have smaller variance). Similarly, the sample proportion p is an MVUE of the population proportion π.

Figure 8.5

Illustration of Efficiency

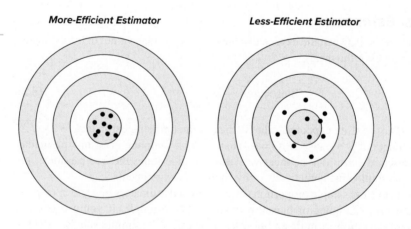

More-Efficient Estimator   Less-Efficient Estimator

A **consistent estimator** converges toward the parameter being estimated as the sample size increases. That is, the sampling distribution variation decreases around the true parameter value, as illustrated in Figure 8.6. It seems logical that in larger samples $\bar{x}$ ought to be closer to μ, p ought to be closer to π, and s ought to be closer to σ. In fact, it can be shown that the variances of these three estimators diminish as n increases, so all are consistent estimators. Figure 8.6 illustrates the importance of a large sample because in a large sample your estimated mean is likely to be closer to μ.

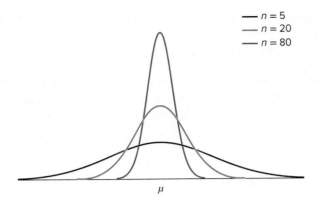

Figure 8.6

Illustration of Consistency

8.3 CENTRAL LIMIT THEOREM FOR A MEAN

LO 8-3

State and apply the Central Limit Theorem for a mean.

Consider the sample mean $\overline{X}$ used to estimate the population mean μ. Our objective is to use the sampling distribution of $\overline{X}$ to say something about the population that we are studying. To describe the sampling distribution, we need to know the mean, variance, and shape of the distribution. As we've already learned, the sample mean is an unbiased estimator for μ.

LO 8-4

Explain how sample size affects the standard error.

$$E(\overline{X}) = \mu \qquad \text{(expected value of the mean)} \tag{8.3}$$

We've also learned that $\overline{X}$ is a *random variable* whose value will change whenever we take a different sample. And as long as our samples are *random samples,* the only type of error we will have in our estimating process is *sampling error.* The sampling error of the sample mean is described by its standard deviation. This value has a special name, the **standard error of the mean.** Notice that the standard error of the mean decreases as the sample size increases:

$$\sigma_{\overline{x}} = \frac{\sigma}{\sqrt{n}} \qquad \text{(standard error of the mean)} \tag{8.4}$$

For example, if the average spending by a salon customer for a facial treatment is $\mu = \$80.00$ with standard deviation $\sigma = \$10.00$, then the standard error of the mean for a random sample of 20 facial treatments is much smaller than $10:

$$\sigma_{\overline{x}} = \frac{\$10.00}{\sqrt{20}} = \$2.236$$

If the population is normal, the sample mean has a normal distribution centered at μ, with a standard error $\sigma/\sqrt{n}$. As the sample size n increases, the distribution of the sample means converges to the population mean μ as the standard error of the mean, $\sigma/\sqrt{n}$, gets smaller (Figure 8.7).

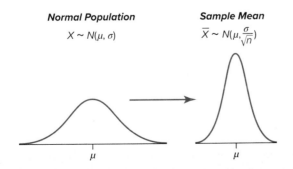

Normal Population

$X \sim N(\mu, \sigma)$

Sample Mean

$\overline{X} \sim N(\mu, \frac{\sigma}{\sqrt{n}})$

Figure 8.7

Distribution of *X* and *X̄* When *X* is Normal

However, the population may not have a normal distribution, or we may simply not know *what* the population distribution looks like. What can we do in these circumstances? We can rely on one of the most fundamental laws of statistics, the *Central Limit Theorem*. The **Central Limit Theorem** (CLT) is a powerful result that allows us to approximate the shape of the sampling distribution of $\overline{X}$ when we don't know what the population distribution looks like. Even if your population is *not* normal, by the CLT, if the sample size is large enough, the sample mean will have approximately a normal distribution. Figure 8.8 illustrates this.

Figure 8.8

Illustration of the Central Limit Theorem

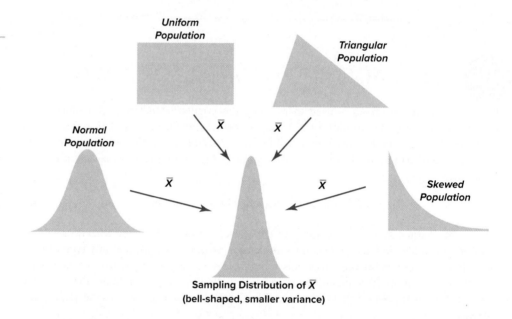

Central Limit Theorem for a Mean

If a random sample of size n is drawn from a population with mean μ and standard deviation σ, the distribution of the sample mean $\overline{X}$ approaches a normal distribution with mean μ and standard deviation $\sigma_{\overline{x}} = \sigma/\sqrt{n}$ as the sample size increases.

The Central Limit Theorem and Sample Size

A rule of thumb often used is that $n \geq 30$ is required to assume a normal sampling distribution for the sample mean, but a much smaller n can suffice for symmetric populations. The left side of Figure 8.9 shows histograms of the means of samples taken from a uniform population. There is sampling variation, but as sample size increases, the means and standard deviations of the sample means quickly approach a clear, bell-shaped histogram with decreasing variance.

A symmetric, uniform population does not pose much of a challenge for the Central Limit Theorem. But what if the population is severely skewed? For example, consider a strongly skewed population (e.g., waiting times at airport security screening). The Central Limit Theorem predicts that *for any population,* the distribution of sample means drawn from this population will approach normality.

The right side of Figure 8.9 shows histograms of means drawn from a skewed population. Despite the skewness, the sample means start to exhibit a bell-shaped curve with decreasing variation. Even for $n = 8$ the histograms begin to look normal. Samples larger than this are common, so we can rely on the CLT in most situations.

Figure 8.9

Illustrations of the Central Limit Theorem

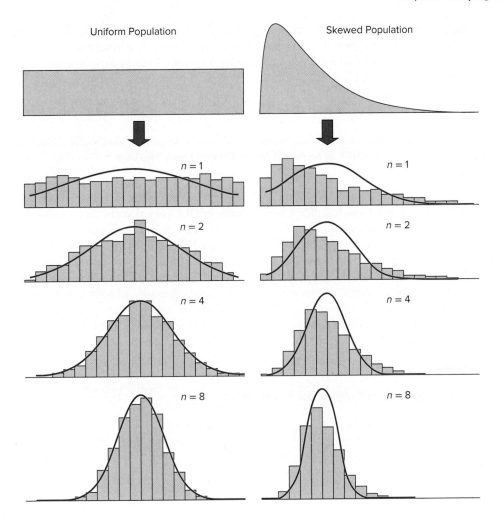

Uniform Population   Skewed Population

Range of Sample Means

The Central Limit Theorem permits us to define an interval within which the sample means are expected to fall. As long as the sample size n is large enough, we can use the normal distribution regardless of the population shape (or any n if the population is normal to begin with).

Expected Range of Sample Means

$$\mu \pm z \frac{\sigma}{\sqrt{n}} \qquad \text{where } \frac{\sigma}{\sqrt{n}} \text{ is the standard error of the mean} \qquad (8.5)$$

We use the familiar z-values for the standard normal distribution. If we know μ and σ, the CLT allows us to predict the range of sample means for samples of size n. Because the standard error decreases as n increases, you can make the interval as small as you want by increasing n.

90% Interval
$$\mu \pm 1.645 \frac{\sigma}{\sqrt{n}}$$

95% Interval
$$\mu \pm 1.960 \frac{\sigma}{\sqrt{n}}$$

99% Interval
$$\mu \pm 2.576 \frac{\sigma}{\sqrt{n}}$$

For example, within what interval would we expect GMAT sample means to fall for samples of $n = 5$ applicants (see Figure 8.1). The population is approximately normal with parameters $\mu = 520.78$ and $\sigma = 86.80$, so the predicted range for 95 percent of the sample means is

$$520.78 \pm 1.960\frac{86.80}{\sqrt{5}} \quad \text{or} \quad 520.78 \pm 76.08 \quad \text{or} \quad [444.70, 596.86]$$

Our eight sample means for $n = 5$ (see Figure 8.2) drawn from this population fall within this interval (roughly 444 to 597), as predicted by the Central Limit Theorem.

Example 8.1

Bottle Filling: Variation in $\overline{X}$

The amount of liquid in a half-liter (500 ml) bottle of Diet Coke is normally distributed with mean $\mu = 505$ ml and standard deviation $\sigma = 1.2$ ml. Because the population is normal, the sample mean $\overline{X}$ will be a normally distributed random variable for any sample size. If we sample a single bottle (i.e., $n = 1$) and measure its fill, the sample "mean" is just X, which should lie within the ranges shown in Table 8.2. It appears that the company has set the mean far enough above 500 ml that essentially all bottles contain at least the advertised half-liter quantity. If we increase the sample size to $n = 4$ bottles, we expect the sample means to lie within a narrower range, as shown in Table 8.2 because when we average four items, we *reduce the variability* in the sample mean.

Table 8.2

Expected 95 Percent Range of the Sample Mean

$n = 1$	$n = 4$
$\mu \pm 1.960\dfrac{\sigma}{\sqrt{n}}$	$\mu \pm 1.960\dfrac{\sigma}{\sqrt{n}}$
$505 \pm 1.960\dfrac{1.2}{\sqrt{1}}$	$505 \pm 1.960\dfrac{1.2}{\sqrt{4}}$
505 ± 2.352	505 ± 1.176
$[502.6, 507.4]$	$[503.8, 506.2]$

If this experiment were repeated a large number of times, we would expect that the sample means would lie within the limits shown above. For example, if we took 1,000 samples and computed the mean for each sample, we would expect that approximately 950 of the sample means would lie within the 95 percent limits. But we don't really take 1,000 samples (except in a computer simulation). We actually take only *one* sample. The importance of the CLT is that it *predicts* what will likely happen with that *one* sample.

Section Exercises

McGraw Hill **connect**

8.1 Find the interval $\left[\mu - z\dfrac{\sigma}{\sqrt{n}}, \mu + z\dfrac{\sigma}{\sqrt{n}}\right]$ within which 90 percent of the sample means would be expected to fall, assuming that each sample is from a normal population.

a. $\mu = 100, \sigma = 12, n = 36$
b. $\mu = 2,000, \sigma = 150, n = 9$
c. $\mu = 500, \sigma = 10, n = 25$

8.2 Find the interval $\left[\mu - z\dfrac{\sigma}{\sqrt{n}}, \mu + z\dfrac{\sigma}{\sqrt{n}}\right]$ within which 95 percent of the sample means would be expected to fall, assuming that each sample is from a normal population.

a. $\mu = 200, \sigma = 12, n = 36$
b. $\mu = 1,000, \sigma = 15, n = 9$
c. $\mu = 50, \sigma = 1, n = 25$

8.3 The diameter of bushings turned out by a manufacturing process is a normally distributed random variable with a mean of 4.035 mm and a standard deviation of 0.005 mm. A sample of 25 bushings is taken once an hour. (a) Within what interval should 95 percent of the bushing diameters fall? (b) Within what interval should 95 percent of the sample *means* fall? (c) What conclusion would you reach if you saw a sample mean of 4.020? A sample mean of 4.055?

8.4 Concerns about climate change and CO_2 reduction have initiated the commercial production of blends of biodiesel (e.g., from renewable sources) and petrodiesel (from fossil fuel). Random samples of 35 blended fuels are tested in a lab to ascertain the bio/total carbon ratio. (a) If the true mean is .9480 with a standard deviation of 0.0060, within what interval will 95 percent of the sample means fall? (b) What is the sampling distribution of $\overline{X}$? In other words, state the shape, center, and variability of the distribution of $\overline{X}$. (c) What theorem did you use to answer part (b)?

8.5 (a) Find the standard error of the mean for each sampling situation (assuming a normal population). (b) What happens to the standard error each time you quadruple the sample size?

    a. $\sigma = 32, n = 4$
    b. $\sigma = 32, n = 16$
    c. $\sigma = 32, n = 64$

8.6 (a) Find the standard error of the mean for each sampling situation (assuming a normal population). (b) What happens to the standard error each time you quadruple the sample size?

    a. $\sigma = 24, n = 9$
    b. $\sigma = 24, n = 36$
    c. $\sigma = 24, n = 144$

8.7 The fat content of a pouch of Keebler Right Bites© Fudge Shoppe© Mini-Fudge Stripes is normally distributed with a mean of 3.50 grams. Assume a known standard deviation of 0.25 gram. (a) What is the standard error of $\overline{X}$, the mean weight from a random sample of 10 pouches of cookies? (b) Within what interval would you expect the sample mean to fall, with 95 percent probability? (Source: www.keebler.com.)

8.8 The combined city/highway fuel economy of a 2023 Lexus UX 250 Hybrid is a normally distributed random variable with a mean of $\mu = 42.0$ mpg and a standard deviation of $\sigma = 0.33$ mpg. (a) What is the standard error of $\overline{X}$, the mean from a random sample of 16 fill-ups by one driver? (b) Within what interval would you expect the sample mean to fall, with 90 percent probability? (Source: www.fueleconomy.gov.)

8.4 CONFIDENCE INTERVAL FOR A MEAN (μ) WITH KNOWN σ

What Is a Confidence Interval?

A sample mean $\overline{x}$ calculated from a random sample $x_1, x_2, \ldots, x_n$ is a **point estimate** of the unknown population mean μ. Because samples vary, we need to indicate our uncertainty about the true value of μ. Based on our knowledge of the sampling distribution of $\overline{X}$, we can create an **interval estimate** for μ. We construct a **confidence interval** for the unknown mean μ by adding and subtracting a **margin of error** from $\overline{x}$, the mean of our random sample. The **confidence level** for this interval is expressed as a percentage such as 90, 95, or 99 percent.

LO 8-5

Construct a confidence interval for a population mean using z.

Confidence Interval for a Mean μ with Known σ

$$\overline{x} \pm z_{\alpha/2} \frac{\sigma}{\sqrt{n}} \qquad \text{where } \frac{\sigma}{\sqrt{n}} \text{ is the standard error of the mean} \qquad (8.6)$$

Because the standard error of the mean $\sigma_{\bar{x}} = \sigma/\sqrt{n}$ decreases as n increases, we can make the confidence interval as narrow as we wish by taking a large enough sample. If samples are drawn from a normal population (or if the sample is large enough that $\overline{X}$ is approximately normal by the Central Limit Theorem) and σ is known, then the margin of error is calculated using the standard normal distribution. The value $z_{\alpha/2}$ is determined by the desired level of confidence, which we call $1 - \alpha$. Because the sampling distribution is symmetric, $\alpha/2$ is the area in each tail of the normal distribution, as shown in Figure 8.10.

Figure 8.10

Confidence Level Using z

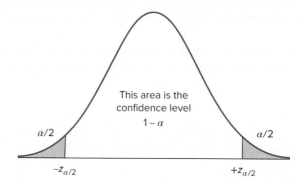

The middle area shows the confidence level. The remaining area is divided into two symmetrical tails, each having area equal to $\alpha/2$. The value of $z_{\alpha/2}$ will depend on the level of confidence desired. For example, if the chosen confidence level is 90 percent ($1 - \alpha = .90$ and $\alpha = .10$), we would use $z_{\alpha/2} = z_{.10/2} = z_{.05} = 1.645$ for an upper tail area of .05. Similarly, for a 95 percent confidence level ($1 - \alpha = .95$ and $\alpha = .05$), we would use $z_{\alpha/2} = z_{.05/2} = z_{.025} = 1.960$ for an upper tail area of .025, as illustrated in Figure 8.11. Notice that the 95 percent confidence interval is *wider* than the 90 percent confidence interval.

Figure 8.11

90 Percent and 95 Percent Confidence Levels Compared

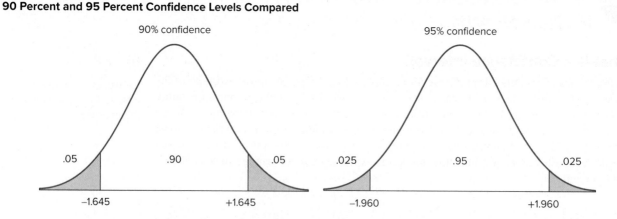

Choosing a Confidence Level

You might be tempted to assume that a higher confidence level gives a "better" estimate. However, confidence is not free—there is a trade-off that must be made. A higher confidence level leads to a wider confidence interval, as illustrated in Example 8.2. The 95 percent confidence interval is *wider* than the 90 percent confidence interval. In order to gain confidence, we must accept a wider range of possible values for μ. Greater confidence implies loss of precision (i.e., a greater margin of error). Table 8.3 shows several common confidence levels and their associated z-values. A 95 percent confidence level is often used because it is a reasonable compromise between confidence and precision.

Confidence Level	$1 - \alpha$	α	$\alpha/2$	$z_{\alpha/2}$
90%	.90	.10	.05	$z_{.05} = 1.645$
95%	.95	.05	.025	$z_{.025} = 1.960$
99%	.99	.01	.005	$z_{.005} = 2.576$

Table 8.3

Common Confidence Levels and z-Values

Example 8.2

Bottle Filling: Confidence Intervals for μ

The volume of liquid in a half-liter bottle of Diet Coke is a normally distributed random variable. The standard deviation of volume is known to be $\sigma = 1.20$ ml. A sample of 10 bottles gives a sample mean volume $\bar{x} = 503.4$ ml. Because the population is normal, the sample mean is a normally distributed random variable for any sample size, so we can use the z distribution to construct a confidence interval for μ using formula 8.6 and z values from Table 8.3.

90% confidence interval: $\quad \bar{x} \pm z_{\alpha/2} \dfrac{\sigma}{\sqrt{n}} \quad$ or $\quad 503.4 \pm 1.645 \dfrac{1.20}{\sqrt{10}} \quad$ or $\quad [502.78, 504.02]$

95% confidence interval: $\quad \bar{x} \pm z_{\alpha/2} \dfrac{\sigma}{\sqrt{n}} \quad$ or $\quad 503.4 \pm 1.960 \dfrac{1.20}{\sqrt{10}} \quad$ or $\quad [502.66, 504.14]$

99% confidence interval: $\quad \bar{x} \pm z_{\alpha/2} \dfrac{\sigma}{\sqrt{n}} \quad$ or $\quad 503.4 \pm 2.576 \dfrac{1.20}{\sqrt{10}} \quad$ or $\quad [502.42, 504.38]$

Lower bounds for all three of the confidence intervals are well above 500, indicating that the mean of the bottle-filling process is safely above the required minimum half liter (500 ml).

How to Interpret a Confidence Interval

We can think of the confidence level $1 - \alpha$ as a *probability on the procedure* used to calculate the confidence interval.

$$P\left(\overline{X} - z_{\alpha/2} \frac{\sigma}{\sqrt{n}} < \mu < \overline{X} + z_{\alpha/2} \frac{\sigma}{\sqrt{n}} \right) = 1 - \alpha \qquad (8.7)$$

This is a statement about the *random variable* $\overline{X}$. It is not a statement about a *specific* sample mean $\bar{x}$. If you took 100 random samples from the same population and used exactly this procedure to construct 100 confidence intervals using a 95 percent confidence level, approximately 95 of the intervals would contain the true mean μ, while approximately 5 intervals would not. Figure 8.12 illustrates this idea using 20 samples (note that one interval does not contain μ).

Once you've taken a random sample and have calculated $\bar{x}$ and the corresponding confidence interval, the confidence level $1 - \alpha$ no longer is thought of as a probability. The confidence interval you've calculated either does or does not contain μ. Because you only do it once, you won't know if your specific interval contains the true mean μ or not. You can only say that $1 - \alpha$ is now your level of *confidence* that the interval contains μ.

Figure 8.12

Twenty 95 Percent Confidence Intervals for μ

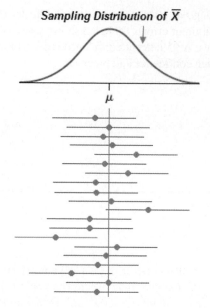

Sampling Distribution of $\overline{X}$

When Can We Assume Normality?

If σ is known and the population is normal, then we can safely use formula 8.6 to construct the confidence interval for μ. If σ is known but we do not know whether the population is normal, a common rule of thumb is that $n \geq 30$ is sufficient to assume a normal sampling distribution for $\overline{X}$ (by the CLT) as long as the population is reasonably symmetric and has no outliers. However, a larger n may be needed to assume normality if you are sampling from a strongly skewed population or one with outliers. When σ is *unknown,* a different approach is used, as described in the next section.

Is σ Ever Known?

Yes, but not very often. In quality control applications with ongoing manufacturing processes, it may be reasonable to assume that σ stays the same over time. The type of confidence interval just seen is therefore important because it is used to construct *control charts* to track the mean of a process (such as bottle filling) over time. However, the case of unknown σ is more typical and will be examined in the next section.

Section Exercises

8.9 Construct a confidence interval for μ assuming that each sample is from a normal population.

    a. $\overline{x} = 14$, $\sigma = 4$, $n = 5$, 90 percent confidence
    b. $\overline{x} = 37$, $\sigma = 5$, $n = 15$, 99 percent confidence
    c. $\overline{x} = 121$, $\sigma = 15$, $n = 25$, 95 percent confidence

8.10 Construct a confidence interval for μ assuming that each sample is from a normal population.

    a. $\overline{x} = 24$, $\sigma = 3$, $n = 10$, 90 percent confidence
    b. $\overline{x} = 125$, $\sigma = 8$, $n = 25$, 99 percent confidence
    c. $\overline{x} = 12.5$, $\sigma = 1.2$, $n = 50$, 95 percent confidence

8.11 Use the sample information $\overline{x} = 2.4$, $\sigma = 0.15$, $n = 9$ to calculate the following confidence intervals for μ assuming the sample is from a normal population: (a) 90 percent confidence; (b) 95 percent confidence; (c) 99 percent confidence. (d) Describe how the intervals change as you increase the confidence level.

8.12 Use the sample information $\overline{x} = 37$, $\sigma = 5$, $n = 15$ to calculate the following confidence intervals for μ assuming the sample is from a normal population: (a) 90 percent confidence; (b) 95 percent confidence; (c) 99 percent confidence. (d) Describe how the intervals change as you increase the confidence level.

8.13 A random sample of 25 items is drawn from a population whose standard deviation is known to be $\sigma = 40$. The sample mean is $\bar{x} = 270$.

    a. Construct an interval estimate for μ with 95 percent confidence.
    b. Repeat part a assuming that $n = 50$.
    c. Repeat part a assuming that $n = 100$.
    d. Describe how the confidence interval changes as n increases.

8.14 A random sample of 100 items is drawn from a population whose standard deviation is known to be $\sigma = 50$. The sample mean is $\bar{x} = 850$.

    a. Construct an interval estimate for μ with 95 percent confidence.
    b. Repeat part a assuming that $\sigma = 100$.
    c. Repeat part a assuming that $\sigma = 200$.
    d. Describe how the confidence interval changes as σ increases.

8.15 The combined city/highway fuel economy of a 2023 Lexus UX 250 Hybrid is normally distributed with a known standard deviation of 0.33 mpg. If a random sample of 10 trips yields a mean of 41.5 mpg, find the 95 percent confidence interval for the true mean mpg. (Source: www.fueleconomy.gov.)

8.16 Guest ages at a ski mountain typically have a right-skewed distribution. Assume the standard deviation (σ) of *age* is 14.5 years. (a) Even though the population distribution of age is right-skewed, what will be the shape of the distribution of $\bar{X}$, the average age, in a random sample of 40 guests? (b) From a random sample of 40 guests, the sample mean is 36.4 years. Calculate a 99 percent confidence interval for μ, the true mean age of ski mountain guests.

8.17 The Ball Corporation's beverage can manufacturing plant in Fort Atkinson, Wisconsin, uses a metal supplier that provides metal with a known thickness standard deviation $\sigma = .000959$ mm. If a random sample of 58 sheets of metal resulted in $\bar{x} = 0.2731$ mm, calculate the 99 percent confidence interval for the true mean metal thickness.

8.5 CONFIDENCE INTERVAL FOR A MEAN (μ) WITH UNKNOWN σ

Student's *t* Distribution

In situations where the population is normal but its standard deviation σ is unknown, the **Student's *t* distribution** should be used instead of the normal *z* distribution. This is particularly important when the sample size is small. When σ is unknown, the formula for a confidence interval resembles the formula for known σ except that *t* replaces *z* and *s* replaces σ.

LO 8-6

Know when and how to use Student's *t* instead of *z* to estimate a mean.

Confidence Interval for a Mean μ with Unknown σ

$$\bar{x} \pm t_{\alpha/2} \frac{s}{\sqrt{n}} \qquad \text{where} \; \frac{s}{\sqrt{n}} \; \text{is the } \textit{estimated} \text{ standard error of the mean} \qquad \textbf{(8.8)}$$

The interpretation of the confidence interval is the same as when σ is known, as illustrated in Figure 8.13. However, the confidence intervals will be wider (other things being the same) because $t_{\alpha/2}$ is always greater than $z_{\alpha/2}$. Intuitively, our confidence interval will be wider because we face added uncertainty when we use the sample standard deviation *s* to estimate the unknown population standard deviation σ.

The Student's *t* distributions were proposed by a Dublin brewer named W. S. Gosset (1876–1937), who published his research under the name "Student" because his employer did not approve of publishing research based on company data. The *t* distributions are symmetric and shaped very much like the standard normal distribution, except they are somewhat less peaked and have thicker tails. Note that the *t* distributions are a class of distributions, each of which is dependent on the size of the sample we are using. Figure 8.14 shows how the tails of

Figure 8.13

Confidence Level Using Student's *t*

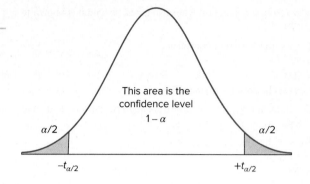

This area is the confidence level
$1 - \alpha$

$\alpha/2$ $\alpha/2$

$-t_{\alpha/2}$ $+t_{\alpha/2}$

Figure 8.14

Comparison of Normal and Student's *t*

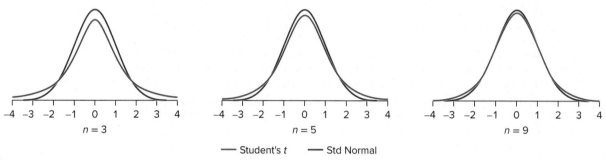

$n = 3$ $n = 5$ $n = 9$

— Student's *t*   — Std Normal

the distributions change as the sample size increases. A closer look reveals that the *t* distribution's tails lie *above* the normal (i.e., the *t* distribution always has longer tails).

Degrees of Freedom

Knowing the sample size allows us to calculate a parameter called **degrees of freedom** (sometimes abbreviated *d.f.*). This parameter is used to determine the value of the *t* statistic used in the confidence interval formula. The degrees of freedom tell us how many observations we used to calculate *s*, the sample standard deviation, less the number of intermediate estimates we used in our calculation. Recall that the formula for *s* uses all *n* individual values from the sample and also $\bar{x}$, the sample mean. Therefore, the degrees of freedom are equal to the sample size minus 1.

(8.9) $d.f. = n - 1$ (degrees of freedom for a confidence interval for μ)

For large degrees of freedom, the *t* distribution approaches the shape of the normal distribution, as illustrated in Figure 8.14. However, in small samples, the difference is important. For example, in Figure 8.14 the lower axis scale range extends out to ±4, while a range of ±3 would cover most of the area for a standard normal distribution. We have to go out further into the tails of the *t* distribution to enclose a given area, so for a given confidence level, *t is always larger than z,* so the confidence interval is always *wider* than if *z* were used.

Comparison of *z* and *t*

Table 8.4 shows that for very small samples the *t*-values differ substantially from the normal *z* values. But for a given confidence level, as degrees of freedom increase, the *t*-values approach the familiar normal *z*-values (shown at the bottom of each column corresponding to an infinitely large sample). For example, for $n = 31$, we would have degrees of freedom $d.f. = 31 - 1 = 30$, so for a 90 percent confidence interval, we would use $t = 1.697$, which is only slightly larger than $z = 1.645$. It might seem tempting to use the *z*-values to avoid having to look up the correct degrees of freedom, but this would not be conservative (because the resulting confidence interval would be too narrow).

Table 8.4

Student's *t*-values for Selected Degrees of Freedom

	Confidence Level				
d.f.	80%	90%	95%	98%	99%
1	3.078	6.314	12.706	31.821	63.656
2	1.886	2.920	4.303	6.965	9.925
3	1.638	2.353	3.182	4.541	5.841
4	1.533	2.132	2.776	3.747	4.604
5	1.476	2.015	2.571	3.365	4.032
10	1.372	1.812	2.228	2.764	3.169
20	1.325	1.725	2.086	2.528	2.845
30	1.310	1.697	2.042	2.457	2.750
40	1.303	1.684	2.021	2.423	2.704
60	1.296	1.671	2.000	2.390	2.660
100	1.290	1.660	1.984	2.364	2.626
∞	1.282	1.645	1.960	2.326	2.576

Note: The bottom row shows the *z*-values for each confidence level.

Using Excel

The Excel function =T.INV.2T(probability, degrees of freedom) gives the value of $t_{\alpha/2}$, where probability is α. For example, for a 95 percent confidence interval, $\alpha = 1 - .95 = .05$. With 60 degrees of freedom, the function =T.INV.2T(0.05, 60) yields $t_{.025} = 2.000298$.

Example 8.3

GMAT Scores 📋

GMATScores

Consider a random sample of GMAT scores submitted by 20 applicants to an MBA program. A dot plot of this sample is shown in Figure 8.15.

530	450	600	570	360
550	640	490	460	550
480	440	530	470	560
500	430	640	420	530

Figure 8.15

Dot Plot and 90 Percent Confidence Interval (*n* = 20 Scores) 📋 **GMATScores**

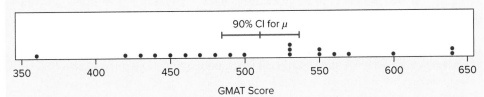

GMAT Score

We will construct a 90 percent confidence interval for the mean GMAT score of all MBA applicants. The sample mean is $\bar{x} = 510$, and the sample standard deviation is $s = 73.77$. Because the population standard deviation σ is unknown, we will use the Student's *t* for our confidence interval with 19 degrees of freedom:

$$d.f. = n - 1 = 20 - 1 = 19 \quad \text{(degrees of freedom for } n = 20\text{)}$$

For a 90 percent confidence interval, we use the two-tailed Excel function =T.INV.2T(0.10,19) = 1.729 or consult Appendix D and find $t_{\alpha/2} = t_{.05} = 1.729$:

	Confidence Level $(1 - \alpha)$				
	0.80	0.90	0.95	0.98	0.99
			Upper Tail Area $(\alpha/2)$		
d.f.	0.10	0.05	0.025	0.01	0.005
1	3.078	6.314	12.706	31.821	63.656
2	1.886	2.920	4.303	6.965	9.925
⋮	⋮	⋮	⋮	⋮	⋮
18	1.330	1.734	2.101	2.552	2.878
19	1.328	1.729	2.093	2.539	2.861
20	1.325	1.725	2.086	2.528	2.845

The 90 percent confidence interval is

$$\bar{x} \pm t_{\alpha/2} \frac{s}{\sqrt{n}} \quad \text{or} \quad 510 \pm (1.729)\frac{73.77}{\sqrt{20}} \quad \text{or} \quad 510 \pm 28.52$$

We are 90 percent confident that the true mean GMAT score is within the interval [481.48, 538.52], as shown in Figure 8.15. If we wanted a narrower interval with the same level of confidence, we would need a larger sample size to reduce the right-hand side of $\bar{x} \pm t_{\alpha/2} \frac{s}{\sqrt{n}}$.

Example 8.4

Hospital Stays

📁 **Maternity**

During a certain period of time, Balzac Hospital had 8,261 maternity cases. Hospital management needs to know the mean length of stay (LOS) so it can plan the maternity unit bed capacity and schedule the nursing staff. Each case is assigned a code called a DRG (which stands for "Diagnostic Related Group"). The most common DRG was 373 (simple delivery without complicating diagnoses), accounting for 4,409 cases during the study period. Using DRG 373, for a random sample of hospital records for $n = 25$ births, the mean length of stay was $\bar{x} = 39.144$ hours with a standard deviation of $s = 16.204$ hours. What is the 95 percent confidence interval for the true mean?

To justify using the Student's t distribution, we will assume that the population is normal (we will examine this assumption later). Because the population standard deviation is unknown, we use the Student's t for our confidence interval with 24 degrees of freedom:

$$d.f. = n - 1 = 25 - 1 = 24 \quad \text{(degrees of freedom for } n = 25)$$

For a 95 percent confidence interval, use the two-tailed Excel function =T.INV.2T(0.05,24) = 2.0639 or Appendix D and find $t_{\alpha/2} = t_{.025} = 2.064$. The 95 percent confidence interval is

$$\bar{x} \pm t_{\alpha/2} \frac{s}{\sqrt{n}} \quad \text{or} \quad 39.144 \pm (2.064)\frac{16.204}{\sqrt{25}} \quad \text{or} \quad 39.144 \pm 6.689$$

With 95 percent confidence, the true mean LOS is within the interval [32.455, 45.833], so our estimate is that a simple maternity stay averages between 32.5 hours and 45.8 hours. A dot plot of this sample and confidence interval are shown in Figure 8.16.

Figure 8.16

Dot Plot and 95 Percent Confidence Interval ($n = 25$ Births) 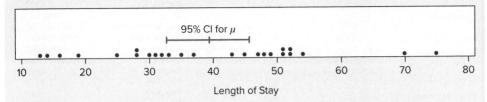 **Maternity**

Length of Stay

Our confidence interval width reflects the sample size, the confidence level, and the standard deviation. If we wanted a narrower interval (i.e., more precision), we could either increase the sample size or lower the confidence level (e.g., to 90 percent or even 80 percent). But we cannot do anything about the standard deviation because it is an aspect of the sample. In fact, some samples could have larger standard deviations than this one.

Outliers and Messy Data

Outliers and messy data are common. Managers often encounter large databases containing unruly data and must decide how to treat atypical observations. What if we had a patient in our LOS example who stayed in the hospital for 254 hours? If a handful of maternity patients stay 10 days (240 hours) instead of 2 days (48 hours), real resources will be required to treat them. Would this be common or an outlier? Even if we decide that 10 days is unusual or an outlier, we cannot simply ignore it. We should investigate what might have happened to make this patient's stay longer than usual. It's possible that the diagnostic code DRG 373 (simple delivery without complicating diagnoses) was assigned incorrectly. We should ask if there are multiple outliers that might invalidate the assumption of normality. If the data are highly skewed, we might consider taking a larger sample so that the Central Limit Theorem is more applicable. Health care managers in hospitals, clinics, insurers, and state and federal agencies all spend significant time working with messy data just like these. In the United States, health care spending is more than one-sixth of the GDP, suggesting that one job out of every six (perhaps yours) is tied directly or indirectly to health care, so examples like this are not unusual. You need to be ready to deal with messy data.

Must the Population Be Normal?

The *t* distribution assumes a normal population. While this assumption is often in doubt, simulation studies have shown that confidence intervals using Student's *t* are reliable as long as the population is not badly skewed and if the sample size is not too small (see Chapter 8 Connect® supplements for simulation demonstrations).

Using Appendix D

Beyond *d.f.* = 50, Appendix D shows *d.f.* in steps of 5 or 10. If Appendix D does not show the exact degrees of freedom that you want, use the *t*-value for the *next lower d.f.* For example, if *d.f.* = 54, you would use *d.f.* = 50. Using the next lower degrees of freedom is a conservative procedure because it overestimates the margin of error. Because *t*-values change very slowly as *d.f.* rises beyond *d.f.* = 50, rounding down will make little difference.

Figure 8.17

Excel's Confidence Interval for μ

Source: Microsoft Excel.

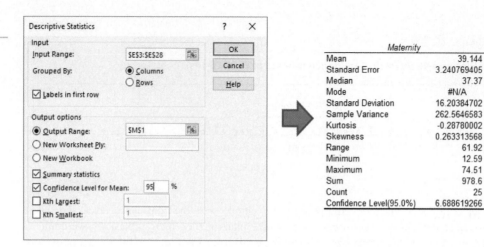

Maternity	
Mean	39.144
Standard Error	3.240769405
Median	37.37
Mode	#N/A
Standard Deviation	16.20384702
Sample Variance	262.5646583
Kurtosis	-0.28780002
Skewness	0.263313568
Range	61.92
Minimum	12.59
Maximum	74.51
Sum	978.6
Count	25
Confidence Level(95.0%)	6.688619266

Can I Ever Use z Instead of t?

In large samples, z and t give similar results. But a conservative statistician always uses the t distribution for confidence intervals when σ is unknown because using z would underestimate the margin of error. Because t tables are easy to use (or we can get t-values from Excel), there isn't much justification for using z when σ is unknown.

Excel's Descriptive Statistics

The output from Excel's Data Analysis > Descriptive Statistics does not give the confidence interval limits, but it does give the standard error and margin of error for a specified confidence level $t_{\alpha/2} s/\sqrt{n}$ (the oddly labeled last line in the table). Figure 8.17 shows Excel's results for Example 8.4 (maternity LOS).

Section Exercises

📘 **connect**

8.18 Find a confidence interval for μ assuming that each sample is from a normal population.

   a. $\bar{x} = 24$, $s = 3$, $n = 7$, 90 percent confidence
   b. $\bar{x} = 42$, $s = 6$, $n = 18$, 99 percent confidence
   c. $\bar{x} = 119$, $s = 14$, $n = 28$, 95 percent confidence

8.19 For each value of $d.f.$ (degrees of freedom), look up the value of Student's t in Appendix D for the stated level of confidence. Then use Excel to find the value of Student's t to four decimal places. Which method (Appendix D or Excel) do you prefer, and why?

   a. $d.f. = 9$, 95 percent confidence
   b. $d.f. = 15$, 98 percent confidence
   c. $d.f. = 47$, 90 percent confidence

8.20 For each value of $d.f.$, look up the value of Student's t in Appendix D for the stated level of confidence. How close is the t-value to the corresponding z-value (at the bottom of the column for $d.f. = \infty$)?

   a. $d.f. = 40$, 95 percent confidence
   b. $d.f. = 80$, 95 percent confidence
   c. $d.f. = 100$, 95 percent confidence

8.21 A random sample of 10 items is drawn from a population whose standard deviation is unknown. The sample mean is $\bar{x} = 270$ and the sample standard deviation is $s = 20$. Use Appendix D to find the values of Student's t.

   a. Construct an interval estimate for μ with 95 percent confidence.
   b. Repeat part a assuming that $n = 20$.

c. Repeat part a assuming that $n = 40$.

d. Describe how the confidence interval changes as n increases.

8.22   A random sample of 25 items is drawn from a population whose standard deviation is unknown. The sample mean is $\bar{x} = 850$, and the sample standard deviation is $s = 15$. Use Appendix D to find the values of Student's t.

a. Construct an interval estimate of μ with 95 percent confidence.

b. Repeat part a assuming that $s = 30$.

c. Repeat part a assuming that $s = 60$.

d. Describe how the confidence interval changes as s increases.

8.23   A sample of 21 minivan electrical warranty repairs for "loose, not attached" wires (one of several electrical failure categories the dealership mechanic can select) showed a mean repair cost of $45.66 with a standard deviation of $27.79. (a) Construct a 95 percent confidence interval for the true mean repair cost. (b) How could the confidence interval be made narrower?

8.24   A random sample of 16 pharmacy customers showed the waiting times below (in minutes). Find a 90 percent confidence interval for μ, assuming that the sample is from a normal population.
📁 **Pharmacy**

21	22	22	17	21	17	23	20
20	24	9	22	16	21	22	21

8.25   A random sample of monthly rent paid by 12 college seniors living off campus gave the results below (in dollars). Find a 99 percent confidence interval for μ, assuming that the sample is from a normal population. 📁 **Rent1**

1900	1810	1770	1860	1850	1790
1810	1800	1890	1720	1910	1640

8.26   A random sample of 10 shipments of stick-on labels showed the following order sizes. (a) Construct a 95 percent confidence interval for the true mean order size. (b) How could the confidence interval be made narrower? (Data are from a project by MBA student Henry Olthof Jr.) 📁 **OrderSize**

12,000	18,000	30,000	60,000	14,000	10,500	52,000	14,000	15,700	19,000

8.27   Prof. Green gave three exams last semester. Scores were normally distributed on each exam. Below are scores for 10 randomly chosen students on each exam. (a) Find the 95 percent confidence interval for the mean score on each exam. (b) Do the confidence intervals overlap? What inference might you draw by comparing the three confidence intervals? 📁 **Exams2**

> *Exam 1:* 81, 79, 88, 90, 82, 86, 80, 92, 86, 86
> *Exam 2:* 87, 76, 81, 83, 100, 95, 93, 82, 99, 90
> *Exam 3:* 77, 79, 74, 75, 82, 69, 74, 80, 74, 76

8.6   CONFIDENCE INTERVAL FOR A PROPORTION (π)

Proportions Are Important in Business

Proportions occur frequently in surveys and sampling, most often expressed as percents. Happily, estimating a proportion is simpler than estimating a mean because we are just counting things. Using surveys, businesses study many aspects of their customers' satisfaction to determine where they are doing well and where they need to improve. For example, firms often track the proportion of customers who say they are likely to recommend the company to a friend or colleague. The higher this percentage is, the larger the group of "promoters" a company has among its customers. However, the analysts also must account for sampling error to determine how reliable their estimate is. This is where the science of statistics can help.

LO   8-7

Construct a confidence interval for a population proportion.

Central Limit Theorem for Proportions

The Central Limit Theorem (CLT) applies to a sample proportion because a proportion is just a mean of data whose only values are 0 or 1. The distribution of a sample proportion

$p = x/n$ tends toward normality as n increases. The distribution is centered at the population proportion π. Its standard error σ_p will decrease as n increases, as in the case of the standard error for $\overline{X}$. In other words, the sample proportion $p = x/n$ is a *consistent* estimator of π (see Figure 8.6).

Central Limit Theorem for a Proportion

As sample size increases, the distribution of the sample proportion $p = x/n$ approaches a normal distribution with mean π and standard error

$$\sigma_p = \sqrt{\frac{\pi(1 - \pi)}{n}}.$$

The Central Limit Theorem permits us to define an interval within which the sample proportions are expected to fall. As long as the sample size n is large enough, we can use the normal distribution. For example, suppose that 20 percent of Holiday Inn Express reservations are made by Internet (i.e., $\pi = .20$). In a particular random sample of reservations, it is unlikely that we will see exactly 20 percent Internet reservations. If we take repeated random samples, many sample proportions (p) will be close to $\pi = .20$, while some will not due to sampling variation. But, on average, we expect that p should tend toward the true value π and should have approximately a normal distribution.

$$p = \frac{x}{n} = \frac{\text{number of Internet reservations}}{\text{number of items in the sample}}$$

If we could actually take repeated samples, we could empirically study the *sampling distribution* of $p = x/n$. This can be done in a computer simulation. Figure 8.18 shows histograms of 1,000 sample proportions of various sample sizes taken from a population with $\pi = .20$.

Figure 8.18

1,000 Sample Proportions from Population with $\pi = .20$ 🗂 Hotel

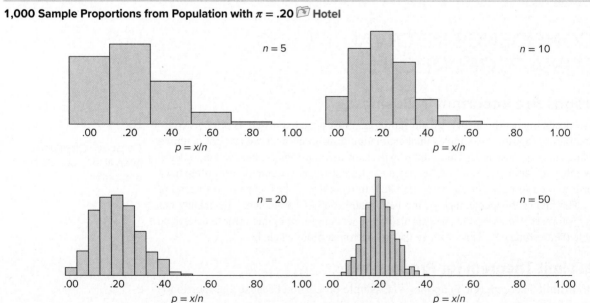

You can see that a sample proportion near .20 is most likely, but there is considerable variation. As n increases, the range of the statistic $p = x/n$ narrows, and its distribution more closely resembles a continuous random variable. For larger n, the distribution also appears more nearly symmetric and bell-shaped, as the CLT predicts. As n increases, the discreteness of $p = x/n$ becomes less noticeable, and the distribution looks more like a smooth normal distribution.

In real life, we only get *one* sample. Yet even for a single sample, we can apply the CLT to *predict* the behavior of p. In Chapter 6, you learned that the binomial model describes the number of successes in a sample of n items from a population with constant probability of success π. Using this knowledge and the CLT, we can construct a confidence interval for π from a single sample proportion.

Standard Error of the Proportion

In mathematical terms, the range of the sample proportion narrows because n appears in the denominator of the **standard error of the proportion** (σ_p):

$$\sigma_p = \sqrt{\frac{\pi(1 - \pi)}{n}} \qquad \text{(standard error of the sample proportion)} \qquad \textbf{(8.10)}$$

The standard error of the proportion depends on π, as well as on n, being largest when the population proportion is near $\pi = .50$ and becoming smaller when π is near 0 or 1. For example:

If $\pi = .50$, then   for $n = 50$ $\sigma_p = \sqrt{\dfrac{\pi(1 - \pi)}{n}} = \sqrt{\dfrac{.50(1 - .50)}{50}} = \sqrt{\dfrac{.25}{50}} = .0707$

for $n = 200$ $\sigma_p = \sqrt{\dfrac{\pi(1 - \pi)}{n}} = \sqrt{\dfrac{.50(1 - .50)}{200}} = \sqrt{\dfrac{.25}{200}} = .0354$

If $\pi = .10$, then   for $n = 50$ $\sigma_p = \sqrt{\dfrac{\pi(1 - \pi)}{n}} = \sqrt{\dfrac{.10(1 - .10)}{50}} = \sqrt{\dfrac{.09}{50}} = .0424$

for $n = 200$ $\sigma_p = \sqrt{\dfrac{\pi(1 - \pi)}{n}} = \sqrt{\dfrac{.10(1 - .10)}{200}} = \sqrt{\dfrac{.09}{200}} = .0212$

The formula is symmetric (i.e., $\pi = .10$ would give the same standard error as $\pi = .90$). Figure 8.19 shows that as n increases, the standard error σ_p decreases, but at a slower rate. Thus, if the sample size is quadrupled, the standard error is only halved.

When Is It Safe to Assume Normality of p?

The statistic $p = x/n$ may be assumed normally distributed when the sample is "large." How large must n be? Table 8.5 illustrates a conservative rule of thumb that normality may be assumed whenever $n\pi \geq 10$ and $n(1 - \pi) \geq 10$. To assume normality, the sample should have at least 10 "successes" and at least 10 "failures" (i.e., we want $x \geq 10$ and $n - x \geq 10$).[1]

Rule of Thumb

The sample proportion $p = x/n$ may be assumed normal when the sample has at least 10 "successes" and at least 10 "failures," i.e., when $x \geq 10$ and $n - x \geq 10$.

---

[1] A good alternative rule is to require $n > 9(1 - \pi)/\pi$ and $n > 9\pi/(1 - \pi)$.

Figure 8.19

Effect of n on σ_p

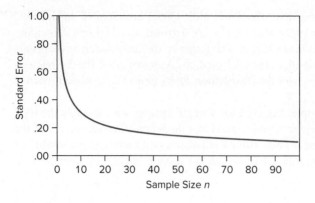

Table 8.5 shows that a very large sample may be needed to assume normality of the sample proportion when π differs greatly from .50.

Table 8.5

Minimum Sample Size to Assume Normality of $p = x/n$

π	n
.50	20
.40 or .60	25
.30 or .70	33
.20 or .80	50
.10 or .90	100
.05 or .95	200
.02 or .98	500
.01 or .99	1,000
.005 or .995	2,000
.002 or .998	5,000
.001 or .999	10,000

If your sample is too small to assume a normal distribution for p, you would need to construct a confidence interval using the binomial distribution. To do this in Excel is possible, but difficult. However, other software packages (R, Minitab, JMP) can handle this situation (see the end of chapter **Software Supplement**).

Confidence Interval for π

By the Central Limit Theorem, we can state the probability that a sample proportion will fall within a given interval. For example, there is a 95 percent chance that p will fall within the interval $\pi \pm z_{.025} \sqrt{\dfrac{\pi(1-\pi)}{n}}$, where $z_{.025} = 1.96$, and similarly for other values of z. This is the basis for a confidence interval estimate of π. Replacing π with $p = x/n$ (because π is unknown) and assuming a large sample (to justify the assumption of normality), we obtain formula 8.11 for the confidence interval for π. Table 8.6 shows $z_{\alpha/2}$ values for common confidence levels.

> **Confidence Interval for π**
>
> **(8.11)**
> $$p \pm z_{\alpha/2} \sqrt{\frac{p(1-p)}{n}}$$

Table 8.6

Common Confidence Levels and z-Values

Confidence Level	$1 - \alpha$	α	$\alpha/2$	$z_{\alpha/2}$
90%	.90	.10	.05	$z_{.05} = 1.645$
95%	.95	.05	.025	$z_{.025} = 1.960$
99%	.99	.01	.005	$z_{.005} = 2.576$

Example 8.5

Auditing

A sample of 75 retail in-store purchases showed that 24 were paid in cash. The sample proportion is

$$p = x/n = 24/75 = .32 \qquad \text{(proportion of in-store cash transactions)}$$

We can assume that p is normally distributed[2] because np and $n(1 - p)$ exceed 10. That is, $np = (75)(.32) = 24$ and $n(1 - p) = (75)(.68) = 51$. The 95 percent confidence interval is

$$p \pm z_{\alpha/2} \sqrt{\frac{p(1 - p)}{n}} \quad \text{or} \quad .32 \pm 1.960 \sqrt{\frac{.32(1 - .32)}{75}} \quad \text{or} \quad .32 \pm .106 \quad \text{or} \quad [.214, .426]$$

We cannot know with certainty whether the true proportion lies within the interval [.214, .426]. Either it does or it does not. And it is true that different samples could yield different intervals. But we can say that, on average, 95 percent of the intervals constructed in this way would contain the true population proportion π. Therefore, we are 95 percent confident that the true proportion π is between .214 and .426.

Reducing the Margin of Error

The width of the confidence interval for π depends on

- Sample size.
- Confidence level.
- Sample proportion p.

We cannot do anything about p because it is calculated from the sample. If we want a narrower interval (i.e., improved precision), we could either increase the sample size or reduce the confidence level (e.g., from 95 percent to 90 percent). But once our confidence level is chosen, our only option is to increase n. Of course, larger samples are more costly (or maybe impossible). The next example illustrates how the width of the interval becomes narrower when using a larger sample and a lower confidence level.

Example 8.6

Online Ordering

A random sample of 200 restaurants listed on OpentTable.com for Denver, Colorado, revealed that 30 of the restaurants did not provide a link to order online. What is the 90 percent confidence interval for the proportion of all restaurants listed that did not provide a link to order online? The sample proportion is

$$p = x/n = 30/200 = .15 \qquad \text{(proportion of pages without an online order link)}$$

The normality test is easily met because $np = (200)(.15) = 30$ and $n(1 - p) = (200)(.85) = 170$. The 90 percent confidence interval requires $z = 1.645$:

$$p \pm z_{\alpha/2} \sqrt{\frac{p(1 - p)}{n}} \quad \text{or} \quad .15 \pm 1.645 \sqrt{\frac{.15(1 - .15)}{200}} \quad \text{or} \quad .15 \pm .042 \quad \text{or} \quad [.108, .192].$$

We are 90 percent confident that between 10.8 percent and 19.2 percent of the pages did not have an online order link. Excel's Data Analysis does not offer a confidence interval for a proportion, but the calculations are easy:

=0.15−NORM.S.INV(.95)*SQRT(0.15*(1−0.15)/200) = .108   (lower 90% confidence limit)

=0.15+NORM.S.INV(.95)*SQRT(0.15*(1−0.15)/200) = .192   (upper 90% confidence limit)

---

[2]When constructing a confidence interval, we use p instead of π in our rule of thumb to test whether n is large enough to assure normality because π is unknown. The test is therefore equivalent to asking if $x \geq 10$ and $n - x \geq 10$.

Polls and Margin of Error

In polls and survey research, the margin of error is typically based on a 95 percent confidence level and the initial assumption that $\pi = .50$. This is a conservative assumption because σ_p is at its maximum when $\pi = .50$. Table 8.7 shows the margin of error for various sample sizes. The law of diminishing returns is apparent. Greater accuracy is possible, but each reduction in the margin of error requires a disproportionately larger sample size. Results in the presidential election of 2016 seemed to defy media projections, but this was partly because many media reports did not discuss the margin of error in the polls and the effect of sample size.

Table 8.7

Margin of Error for 95 Percent Confidence Interval Assuming $\pi = .50$

$n = 100$	$n = 200$	$n = 400$	$n = 800$	$n = 1,200$	$n = 1,600$
±9.8%	±6.9%	±4.9%	±3.5%	±2.8%	±2.5%

Example 8.7

Election Polls

A hotly contested race for a Colorado senate seat took place in November 2010 between Michael Bennet and Ken Buck. A poll by Public Policy Polling on November 1, 2010, showed Buck leading Bennet 49 percent to 48 percent. The sample size was 1,059 likely adult voters, and the reported accuracy was ±3 percent. On the same day, a FOX News poll, conducted by Pulse Opinion Research, showed Buck leading Bennet 50 percent to 46 percent, based on 1,000 likely voters, with a reported accuracy of ±3 percent. These are typical sample sizes for opinion polls on major issues such as state and national elections, foreign policy, or a Supreme Court decision. Tracking polls do vary, but if several different independent polls show the same candidate ahead and if the margin is stable over time, they usually get it right. In this case, the margins of error for both polls made the outcome too close to call before all votes were tallied. Bennet ended up winning with 48 percent of the vote, while Buck garnered 47 percent of the vote.

The margin of error is sometimes referred to as the *sample accuracy*. Popular media sometimes use statistical terminology loosely, but the idea is the same. Statewide political polls, such as a gubernatorial race, typically have 800 respondents (margin of error ±3.5 percent), while a mayoral or local political poll might have 400 respondents (margin of error ±4.9 percent). Private market research or customer mail surveys may rely on even smaller samples, while Internet surveys can yield very large samples. In spite of the large samples possible from an Internet survey, it is important to consider *nonresponse bias* (see Chapter 2).

Rule of Three

A useful quick rule is the *Rule of Three*. If in n independent trials no events occur, the upper 95 percent confidence bound is approximately $3/n$. For example, if no medical complications arise in 17 prenatal fetal surgeries, the upper bound on such complications is roughly $3/17 = .18$, or about 18 percent. This rule is sometimes used when limited data are available. This rule is especially useful because the formula for the standard error σ_p breaks down when $p = 0$. The rule of three is a conservative approach with an interesting history.[3]

Section Exercises

Mc Graw Hill **connect**

8.28 Calculate the standard error of the sample proportion.
  a. $n = 30, \pi = .50$
  b. $n = 50, \pi = .20$
  c. $n = 100, \pi = .10$
  d. $n = 500, \pi = .005$

---

[3]For further details, see B. D. Jovanovic and P. S. Levy, "A Look at the Rule of Three," *The American Statistician* 51, no. 2 (May 1997), pp. 137–139.

8.29 Calculate the standard error of the sample proportion.
   a. $n = 40, \pi = .30$
   b. $n = 200, \pi = .10$
   c. $n = 30, \pi = .40$
   d. $n = 400, \pi = .03$

8.30 Should p be assumed normal?
   a. $n = 200, \pi = .02$
   b. $n = 100, \pi = .05$
   c. $n = 50, \pi = .50$

8.31 Should p be assumed normal?
   a. $n = 25, \pi = .50$
   b. $n = 60, \pi = .20$
   c. $n = 100, \pi = .08$

8.32 Find the margin of error for a poll, assuming that $\pi = .50$.
   a. $n = 50$
   b. $n = 200$
   c. $n = 500$
   d. $n = 2,000$

8.33 A car dealer is taking a customer satisfaction survey. Find the margin of error (i.e., assuming 95% confidence and $\pi = .50$) for (a) 250 respondents, (b) 125 respondents, and (c) 65 respondents.

8.34 In a sample of 500 new websites registered on the Internet, 24 were anonymous (i.e., they shielded their name and contact information). (a) Construct a 95 percent confidence interval for the proportion of all new websites that were anonymous. (b) May normality of p be assumed? Explain.

8.35 From a list of stock mutual funds, 52 funds were selected at random. Of the funds chosen, it was found that 19 required a minimum initial investment under $1,000. (a) Construct a 90 percent confidence interval for the true proportion requiring an initial investment under $1,000. (b) May normality of p be assumed? Explain.

8.36 Of 43 bank customers depositing a check, 18 received some cash back. (a) Construct a 90 percent confidence interval for the proportion of all depositors who ask for cash back. (b) Check the normality assumption of p.

8.37 A survey showed that 4.8 percent of the 250 Americans surveyed had suffered some kind of identity theft in the past 12 months. (a) Construct a 99 percent confidence interval for the true proportion of Americans who had suffered identity theft in the past 12 months. (b) May normality of p be assumed? Explain.

Mini Case 8.1

Airline Water Quality

Is the water on your airline flight safe to drink? It isn't feasible to analyze the water on every flight, so sampling is necessary. In a recent study, the Environmental Protection Agency (EPA) found bacterial contamination in water samples from the lavatories and galley water taps on 20 of 158 randomly selected U.S. flights (12.7 percent of the flights). Alarmed by the data, the EPA ordered sanitation improvements and then tested water samples again later that year. In the second sample, bacterial contamination was found in 29 of 169 randomly sampled flights (17.2 percent of the flights).

First sample: $p = 20/158 = .12658$, or 12.7% contaminated

Second sample: $p = 29/169 = .17160$, or 17.2% contaminated

So, was the problem getting worse instead of better? From these samples, we can construct confidence intervals for the true proportion of flights with contaminated water. We begin with the 95 percent confidence interval for π based on the first water sample:

$$p \pm z_{\alpha/2} \sqrt{\frac{p(1-p)}{n}} = .12658 \pm 1.96 \sqrt{\frac{.12658(1 - .12658)}{158}}$$

$$= .12658 \pm .05185, \text{ or } 7.5\% \text{ to } 17.8\%$$

Next we determine the 95 percent confidence interval for π based on the second water sample:

$$p \pm z_{\alpha/2} \sqrt{\frac{p(1-p)}{n}} = .17160 \pm 1.96 \sqrt{\frac{.17160(1 - .17160)}{169}}$$

$$= .17160 \pm .05684, \text{ or } 11.5\% \text{ to } 22.8\%$$

The margin of error exceeds 5 percent in each sample. Because the confidence intervals overlap, we cannot rule out the possibility that there was no change in water contamination; that is, the difference could be due to sampling variation. Since then, the EPA has continued to take steps to encourage airlines to improve water quality.

See www.epa.gov/dwreginfo/aircraft-drinking-water-rule.

LO 8-8

Know how to modify confidence intervals when the population is finite.

8.7 ESTIMATING FROM FINITE POPULATIONS

In Chapter 2, we discussed infinite and finite populations and the implication of a finite population when sampling without replacement. If the sample size n is less than 5 percent of the population and we are sampling without replacement, then we consider the size of the population to be effectively infinite. However, on occasion we will take samples without replacement where n is greater than 5 percent of the population. When this happens, our margin of error on the interval estimate is actually less than when the sample size is "small" relative to the population size. As we sample more of the population, we get more precise estimates. We need to account for the fact that we are sampling a larger percentage of the population.

Finite Population Correction Factor

(8.12) $\sqrt{\dfrac{N - n}{N - 1}}$ is the finite population correction factor (FPCF)

where N = the number of items in the population
          n = the number of items in the sample

The **finite population correction factor** (FPCF) reduces the margin of error and provides a more precise interval estimate. The use of the FPCF is shown in formulas 8.13–8.15. The FPCF can be omitted when the population is infinite (e.g., when we are sampling from an ongoing production process) or *effectively* infinite (when the population is at least 20 times as large as the sample). When $n/N < .05$, the FPCF is almost equal to 1 and will have a negligible effect on the confidence interval. See Chapters 2 and 6 for further discussion of finite populations.

Confidence Intervals for Finite Populations

$$\bar{x} \pm z_{\alpha/2} \frac{\sigma}{\sqrt{n}} \sqrt{\frac{N-n}{N-1}} \qquad \text{estimating } \mu \text{ with known } \sigma \qquad \textbf{(8.13)}$$

$$\bar{x} \pm t_{\alpha/2} \frac{s}{\sqrt{n}} \sqrt{\frac{N-n}{N-1}} \qquad \text{estimating } \mu \text{ with unknown } \sigma \qquad \textbf{(8.14)}$$

$$p \pm z_{\alpha/2} \sqrt{\frac{p(1-p)}{n}} \sqrt{\frac{N-n}{N-1}} \qquad \text{estimating } \pi \qquad \textbf{(8.15)}$$

Illustration   Streeling Pharmaceutical has a staff of 1,000 employees. The human resources department sent a survey to a random sample of 75 employees to estimate the average number of hours per week its employees use the company's on-site exercise facility. The sample results showed $\bar{x} = 3.50$ hours and $s = 0.83$ hour. Because more than 5 percent of the population was sampled ($n/N = 75/1000 = 7.5\%$), a finite population correction is suggested. The FPCF would be

$$\sqrt{\frac{N-n}{N-1}} = \sqrt{\frac{1,000-75}{1,000-1}} = \sqrt{.9259} = .9623$$

The 95 percent confidence interval estimate, with $t_{.025} = 1.993$ ($d.f. = 74$), would be

$$\bar{x} \pm t_{\alpha/2} \sqrt{\frac{N-n}{N-1}} \frac{s}{\sqrt{n}} \quad \text{or} \quad 3.50 \pm 1.993(.9623)\left(\frac{0.83}{\sqrt{75}}\right) \quad \text{or} \quad 3.50 \pm 0.184$$

If the FPCF had been omitted, the confidence interval would have been:

$$\bar{x} \pm t \frac{s}{\sqrt{n}} \quad \text{or} \quad 3.50 \pm 1.993 \frac{0.83}{\sqrt{75}} \quad \text{or} \quad 3.50 \pm 0.191$$

In this example, the FPCF narrowed the confidence interval only slightly, by reducing the margin of error from ± 0.191 to ± 0.184. Because it reduces the margin of error, using the FPCF is desirable, even when its benefit is modest.

Section Exercises

Mc Graw Hill **connect**

8.38 Calculate the FPCF for each sample and population size. Can the population be considered effectively infinite in each case?

    a. $N = 450$, $n = 10$
    b. $N = 300$, $n = 25$
    c. $N = 1,800$, $n = 280$

8.39 Use $\bar{x} = 50$, $\sigma = 15$, $n = 90$, $N = 1,000$ to calculate confidence intervals for μ assuming the sample is from a normal population: (a) 90 percent confidence; (b) 95 percent confidence; (c) 99 percent confidence.

8.40 Use $\bar{x} = 3.7$, $s = 0.2$, $n = 1,200$, $N = 5,800$ to calculate confidence intervals for μ assuming the sample is from a normal population: (a) 90 percent confidence; (b) 95 percent confidence; (c) 99 percent confidence.

8.41 A random survey of 500 students was conducted from a population of 2,300 students to estimate the proportion who had part-time jobs. The sample showed that 245 had part-time jobs. Calculate the 90 percent confidence interval for the true proportion of students who had part-time jobs.

8.8 SAMPLE SIZE DETERMINATION FOR A MEAN

LO 8-9

Calculate sample size to estimate a mean.

Sample Size to Estimate μ

Suppose we wish to estimate a population mean with a maximum allowable margin of error of $\pm E$. What sample size is required? We start with the general form of the confidence interval:

General Form	*What We Want*
$\bar{x} \pm z \dfrac{\sigma}{\sqrt{n}}$	$\bar{x} \pm E$

In this confidence interval, we use z instead of t because we are going to solve for n, and degrees of freedom cannot be determined unless we know n. Equating the maximum error E to half of the confidence interval width and solving for n,

$$E = z\frac{\sigma}{\sqrt{n}} \quad \rightarrow \quad E^2 = z^2\frac{\sigma^2}{n} \quad \rightarrow \quad n = z^2\frac{\sigma^2}{E^2}$$

Thus, the formula for the sample size can be written:

(8.16) $\qquad\qquad n = \left(\dfrac{z\sigma}{E}\right)^2 \qquad$ (sample size to estimate μ)

Always round n to the next higher integer to be conservative.

A Myth

Many people believe that when the population is large, you need a larger sample to obtain a given level of precision in the estimate. This is incorrect. For a given level of precision, it is only the sample size that matters, even if the population is a million or a billion. This is apparent from the confidence interval formula, which includes n but not N.

How to Estimate σ?

We can plug our desired precision E and the appropriate z for the desired confidence level into formula 8.11. However, σ poses a problem because it is usually unknown. Table 8.8 shows several ways to approximate the value of σ. You can always try more than one method and see how much difference it makes. But until you take the sample, you will not know for sure if you have achieved your goal (i.e., the desired precision E).

Method 1: Take a Preliminary Sample

Take a small preliminary sample and use the sample estimate s in place of σ. This method is common, though its logic is somewhat circular (i.e., we must take a sample to plan a sample).

Method 2: Assume Uniform Population

Estimate upper and lower limits a and b and set $\sigma = [(b - a)^2/12]^{1/2}$. For example, we might guess the weight of a light-duty truck to range from 1,500 pounds to 3,500 pounds, implying a standard deviation of $\sigma = [(3,500 - 1,500)^2/12]^{1/2} = 577$ pounds. Because a uniform distribution has no central tendency, the actual σ is probably smaller than our guess, so we get a larger n than necessary (a conservative result).

Method 3: Assume Normal Population

Estimate upper and lower bounds a and b, and set $\sigma = (b - a)/6$. This assumes normality with most of the data within $\mu + 3\sigma$ and $\mu - 3\sigma$, so the estimated range is 6σ. For example, we might guess the weight of a light truck to range from 1,500 pounds to 3,500 pounds, implying $\sigma = (3,500 - 1,500)/6 = 333$ pounds. Recent research suggests that this method may not be conservative enough (see Related Reading).

Method 4: Poisson Arrivals

In the special case when λ is a Poisson arrival rate, then $\sigma = \sqrt{\lambda}$. For example, if you think the arrival rate is about 20 customers per hour, then you would estimate $\sigma = \sqrt{20} = 4.47$.

Table 8.8

Four Ways to Estimate σ

Example 8.8

Onion Weight

A produce manager wants to estimate the mean weight of Spanish onions being delivered by a supplier, with 95 percent confidence and an error of ± 1 ounce. A preliminary sample of 12 onions shows a sample standard deviation of 3.60 ounces. For a 95 percent confidence interval, we will set $z = 1.96$. We use $s = 3.60$ in place of σ and set the desired error $E = 1$ to obtain the required sample size:

$$n = [(1.96)(3.60)/(1)]^2 = 49.79, \text{ or } 50 \text{ onions}$$

We would round to the next higher integer and take a sample of 50 Spanish onions. This should ensure an estimate of the true mean weight with an error not exceeding ± 1 ounce.

A seemingly modest change in E can have a major effect on the sample size because E is squared. Suppose we reduce the maximum error to $E = 0.5$ ounce to obtain a more precise estimate. The required sample size would then be

$$n = [(1.96)(3.60)/(0.5)]^2 = 199.1, \text{ or } 200 \text{ onions}$$

Practical Advice

When estimating a mean, the maximum error E is expressed in the same units as X and σ. For example, E would be expressed in dollars when estimating the mean order size for mail-order customers (e.g., $E = \$2$) or in minutes to estimate the mean wait time for patients at a clinic (e.g., $E = 10$ minutes). To estimate last year's starting salaries for MBA graduates from a university, the maximum error could be large (e.g., $E = \$2,000$) because a $2,000 error in estimating μ might still be a reasonably accurate estimate.

Using z in the sample size formula for a mean is necessary but not conservative. Because t always exceeds z for a given confidence level, your actual interval may be wider than $\pm E$ as intended. As long as the required sample size is large (say 30 or more), the difference will be acceptable.

The sample size formulas for a mean are not conservative, that is, they tend to underestimate the required sample size (see the Kupper and Hafner article in Related Reading). Therefore, the sample size formulas for a mean should be regarded only as a minimum guideline. Whenever possible, samples should exceed this minimum.

> **Tip**
>
> If your required sample size (n) exceeds 5 percent of the population size (N), you can adjust the required sample size downward to $n' = \dfrac{nN}{n + (N - 1)}$. This adjustment allows a smaller sample yet will also guarantee that the sample size never exceeds the population size.

Section Exercises

8.42 For each level of precision, find the required sample size to estimate the mean starting salary for a new CPA with 95 percent confidence, assuming a population standard deviation of $7,500 (same as last year).

  a. $E = \$2,000$
  b. $E = \$1,000$
  c. $E = \$500$

8.43 Last year, a study showed that the average ATM cash withdrawal took 65 seconds with a standard deviation of 10 seconds. The study is to be repeated this year. How large a sample would be needed to estimate this year's mean with 95 percent confidence and an error of ±4 seconds?

8.44 The EPA city/hwy mpg range for a Saturn Vue FWD automatic 5-speed transmission is 20 to 28 mpg. (a) Estimate σ using Method 3 from Table 8.8. (b) If you owned this vehicle, how large a sample (e.g., how many tanks of gas) would be required to estimate your mean mpg with an error of ±1 mpg and 90 percent confidence? (Source: www.fueleconomy.gov.).

8.45 Popcorn kernels are believed to take between 100 and 200 seconds to pop in a certain microwave. (a) Estimate σ using Method 3 from Table 8.8. (b) What sample size (number of kernels) would be needed to estimate the true mean seconds to pop with an error of ±5 seconds and 95 percent confidence?

8.46 Analysis showed that the mean arrival rate for vehicles at a certain Shell station on Friday afternoon last year was 4.5 vehicles per minute. How large a sample would be needed to estimate this year's mean arrival rate with 98 percent confidence and an error of ±0.5?

8.47 Noodles & Company wants to estimate the mean spending per customer at a certain restaurant with 95 percent confidence and an error of ±$0.25. What is the required sample size, assuming a standard deviation of $2.50 (based on similar restaurants elsewhere)?

8.48 In an intra-squad swim competition, men's freestyle 100 swim times at a certain university ranged from 43.89 seconds to 51.96 seconds. (a) Estimate the standard deviation using Method 3 (the Empirical Rule for a normal distribution) from Table 8.8. (b) What sample size is needed to estimate the mean for all swimmers with 95 percent confidence and an error of ±0.50 second?

8.49 The combined city/highway fuel economy of a 2016 Toyota 4Runner 2WD 6-cylinder 4-L automatic 5-speed using regular gas is a normally distributed random variable with a range 17 mpg to 22 mpg. (a) Estimate the standard deviation using Method 3 (the Empirical Rule for a normal distribution) from Table 8.8. (b) What sample size is needed to estimate the mean with 90 percent confidence and an error of ±0.25 mpg? (Source: www.fueleconomy.gov.)

8.9 SAMPLE SIZE DETERMINATION FOR A PROPORTION

LO 8-10

Calculate sample size to estimate a proportion.

Suppose we wish to estimate a population proportion with a precision (maximum error) of ± E. What sample size is required? We start with the general form of the confidence interval:

General Form	*What We Want*
$p \pm z \sqrt{\dfrac{\pi(1 - \pi)}{n}}$	$p \pm E$

We equate the maximum error E to half of the confidence interval width and solve for n:

$$E = z\sqrt{\frac{\pi(1 - \pi)}{n}} \quad \rightarrow \quad E^2 = z^2\frac{\pi(1 - \pi)}{n} \quad \rightarrow \quad n = z^2\frac{\pi(1 - \pi)}{E^2}$$

Thus, the formula for the sample size for a proportion can be written:

$$n = \left(\frac{z}{E}\right)^2 \pi(1 - \pi) \qquad \text{(sample size to estimate } \pi) \qquad \textbf{(8.17)}$$

Always round n to the next higher integer.

Because a proportion is a number between 0 and 1, the maximum error E is also between 0 and 1. For example, if we want an error of ± 7 percent, we would specify $E = 0.07$.

Because π is unknown (that's why we are taking the sample), we need to make an assumption about π to plan our sample size. If we have a prior estimate of π (e.g., from last year or a comparable application), we can plug p into the formula. Or we could take a small preliminary sample to obtain an initial value of p. Some experts recommend using $\pi = .50$ because the resulting sample size will guarantee the desired precision for any π. However, this conservative assumption may lead to a larger sample than necessary because π is not always equal to .5. Sampling costs money, so if a prior estimate of π is available, it might be advisable to use it, especially if you think that π differs greatly from .50. For example, in estimating the proportion of home equity loans that result in default, we would expect π to be much smaller than .50, while in estimating the proportion of motorists who use seat belts, we would hope that π would be much larger than .50. Table 8.9 details three ways to estimate π.

Table 8.9

Three Ways to Estimate π

Method 1: Assume That $\pi = .50$
This method is conservative and ensures the desired precision. It is therefore a sound choice and is often used. However, the sample may end up being larger than necessary.

Method 2: Take a Preliminary Sample
Take a small preliminary sample and insert p into the sample size formula in place of π. This method is appropriate if π is believed to differ greatly from .50, as is often the case, though its logic is somewhat circular (i.e., we must take a sample to plan our sample).

Method 3: Use a Prior Sample or Historical Data
A reasonable approach, but how often are such data available? And might π have changed enough to make it a questionable assumption?

Example 8.9

ATM Withdrawals

A university credit union wants to know the proportion of cash withdrawals that exceed $50 at its ATM located in the student union building. With an error of ± 2 percent and a confidence level of 95 percent, how large a sample is needed to estimate the proportion of withdrawals exceeding $50? The z-value for 95 percent confidence is $z = 1.960$. Using $E = 0.02$ and assuming conservatively that $\pi = .50$, the required sample size is

$$n = \left(\frac{z}{E}\right)^2 \pi(1 - \pi) = \left(\frac{1.960}{0.02}\right)^2 (.50)(1 - .50) = 2,401$$

We would need to examine $n = 2,401$ withdrawals to estimate π within ± 2 percent and with 95 percent confidence. In this case, last year's proportion of ATM withdrawals over $50 was 27 percent. If we had used this estimate in our calculation, the required sample size would be

$$n = \left(\frac{z}{E}\right)^2 p(1 - p) = \left(\frac{1.960}{0.02}\right)^2 (.27)(1 - .27) = 1,893 \text{ (rounded to next higher integer)}$$

We would need to examine $n = 1,893$ withdrawals to estimate π within ± 0.02. The required sample size is smaller than when we make the conservative assumption $\pi = .50$.

Alternatives

Suppose that our research budget will not permit a large sample. In the previous example, we could reduce the confidence interval level from 95 to 90 percent and increase the maximum error to ± 4 percent. Assuming $\pi = .50$, the required sample size is

$$n = \left(\frac{z}{E}\right)^2 \pi(1 - \pi) = \left(\frac{1.645}{0.04}\right)^2 (.50)(1 - .50) = 423 \quad \text{(rounded to next higher integer)}$$

These seemingly modest changes make a huge difference in the sample size.

Practical Advice

Choosing a sample size is a common problem. Clients who take samples are constrained by time and money. Naturally, they prefer the highest possible confidence level and the lowest possible error. But when a statistical consultant shows them the required sample size, they may find it infeasible. A better way to look at it is that the formula for sample size provides a structure for a dialogue between statistician and client. A good consultant can propose several possible confidence levels and errors and let the client choose the combination that best balances the need for accuracy against the available time and budget. The statistician can offer advice about these trade-offs so the client's objectives are met. Other issues include nonresponse rates, dropout rates from ongoing studies, and possibly incorrect assumptions used in the calculation.

A common error is to insert $E = 2$ in the formula when you want an error of ± 2 percent. Because we are dealing with a *proportion,* a 2% error is $E = 0.02$. In other words, when estimating a proportion, E is always between 0 and 1.

Section Exercises

8.50 What sample size would be required to estimate the true proportion of American female business executives who prefer the title "Ms.," with an error of ± 0.025 and 98 percent confidence?

8.51 What sample size would be needed to estimate the true proportion of American households that still have a telephone landline, with 90 percent confidence and an error of ± 0.02?

8.52 What sample size would be needed to estimate the true proportion of students at your college (if you are a student) who are wearing backpacks, with 95 percent confidence and an error of ± 0.04?

8.53 What sample size would be needed to estimate the true proportion of American adults who know their cholesterol level, using 95 percent confidence and an error of ± 0.02?

8.54 How large a sample size would be needed to estimate the percentage of wireless routers in San Francisco that use data encryption, with an error of ± 2 percent and 95 percent confidence?

8.55 Inspection of a random sample of 19 aircraft showed that 15 needed repairs to fix a wiring problem that might compromise safety. How large a sample would be needed to estimate the true proportion of jets with the wiring problem, with 90 percent confidence and an error of ± 6 percent?

Analytics in Action

Margin of Error in the Era of Big Data

We want estimates of parameters from sample data to be both unbiased and efficient (see Figures 8.4 and 8.5). We expect big data to lead to more efficient estimates because sampling error decreases as the sample size increases. In the era of big data, the margin of error around an estimate of a population parameter can be very small, which means the estimates can be very precise. But if you recall from Chapter 2, sampling error is not the only type of error in sampling.

Quality of data is as important as *quantity* of data. Selection bias, response error, and interviewer error are not a function of sample size. When large amounts of non-random data are collected, bias can be systemic, and conclusions from big data can be misleading. Precise estimates with small margins of error cannot correct for systemic bias.

For example, data and statistics collected by police departments are dependent on where police are deployed. Crimes committed in neighborhoods without a consistent police presence might not be tracked as closely or be included in crime statistics data sets. Witness accounts may reflect ethnic or racial bias. Such factors may lead to poor quality data and biased conclusions about crime despite precise estimates.

8.10 CONFIDENCE INTERVAL FOR A POPULATION VARIANCE, σ^2 (OPTIONAL)

Chi-Square Distribution

In our previous discussions of means and differences in means, we have indicated that many times we do not know the population variance σ^2. A variance estimate can be useful information for many business applications. If the population is normal, we can construct a confidence interval for the population variance σ^2 using the **chi-square distribution** (the Greek letter χ is pronounced "kye") with degrees of freedom equal to $d.f. = n - 1$. Lower-tail and upper-tail percentiles for the chi-square distribution (denoted χ_L^2 and χ_U^2, respectively) can be found in Appendix E. Alternatively, we can use the Excel functions =CHISQ.INV($\alpha/2$, df) and =CHISQ.INV.RT($\alpha/2$, df) to find χ_L^2 and χ_U^2, respectively. Using the sample variance s^2, the confidence interval is

LO 8-11

Construct a confidence interval for a variance (optional).

$$\frac{(n-1)s^2}{\chi_U^2} \le \sigma^2 \le \frac{(n-1)s^2}{\chi_L^2} \quad \text{(confidence interval for } \sigma^2 \text{ from sample variance } s^2\text{)} \qquad \textbf{(8.18)}$$

On a particular Friday night, the charges for 40 pizza delivery orders from Mama Frida's Pizza showed a mean of $\bar{x} = 24.76$ with a sample variance $s^2 = 12.77$.

Example 8.10

Pizza Delivery 🖿 **Pizza**

29.51	21.09	29.98	29.95	21.07	29.52	21.07	24.95	21.07	24.95
24.98	29.95	24.95	21.07	25.30	25.30	29.95	29.99	29.95	24.95
24.95	21.07	24.98	21.09	21.07	24.98	21.07	25.30	29.95	25.30
25.30	24.98	16.86	25.30	25.30	16.86	24.95	24.98	25.30	21.07

The sample data were nearly symmetric (median $24.98) with no outliers. Normality of the prices will be assumed. From Appendix E, using 39 degrees of freedom ($d.f. = n - 1 = 40 - 1 = 39$), we obtain bounds for the 95 percent middle area, as illustrated in Figures 8.20 and 8.21.

$\chi_L^2 = 23.65$ (lower 2.5 percent)   or χ_L^2 =CHISQ.INV(.025,39)

$\chi_U^2 = 58.12$ (upper 2.5 percent)   or χ_U^2 =CHISQ.INV.RT(.025,39)

The 95 percent confidence interval for the population variance σ^2 is

Lower bound: $\dfrac{(n-1)s^2}{\chi_U^2} = \dfrac{(40-1)(12.77)}{58.12} = 8.569$

Upper bound: $\dfrac{(n-1)s^2}{\chi_L^2} = \dfrac{(40-1)(12.77)}{23.65} = 21.058$

With 95 percent confidence, we believe that $8.569 \le \sigma^2 \le 21.058$. If you want a confidence interval for the standard deviation, just take the square root of the interval bounds. In this example, that would give $2.93 \le \sigma \le 4.59$.

Figure 8.20

Chi-Square Values for 95 Percent Confidence with _d.f._ = 39

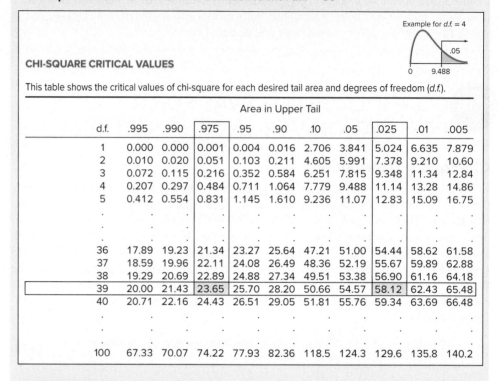

CHI-SQUARE CRITICAL VALUES

Example for _d.f._ = 4

This table shows the critical values of chi-square for each desired tail area and degrees of freedom (_d.f._).

d.f.					Area in Upper Tail					
	.995	.990	.975	.95	.90	.10	.05	.025	.01	.005
1	0.000	0.000	0.001	0.004	0.016	2.706	3.841	5.024	6.635	7.879
2	0.010	0.020	0.051	0.103	0.211	4.605	5.991	7.378	9.210	10.60
3	0.072	0.115	0.216	0.352	0.584	6.251	7.815	9.348	11.34	12.84
4	0.207	0.297	0.484	0.711	1.064	7.779	9.488	11.14	13.28	14.86
5	0.412	0.554	0.831	1.145	1.610	9.236	11.07	12.83	15.09	16.75
⋮	⋮	⋮	⋮	⋮	⋮	⋮	⋮	⋮	⋮	⋮
36	17.89	19.23	21.34	23.27	25.64	47.21	51.00	54.44	58.62	61.58
37	18.59	19.96	22.11	24.08	26.49	48.36	52.19	55.67	59.89	62.88
38	19.29	20.69	22.89	24.88	27.34	49.51	53.38	56.90	61.16	64.18
39	20.00	21.43	23.65	25.70	28.20	50.66	54.57	58.12	62.43	65.48
40	20.71	22.16	24.43	26.51	29.05	51.81	55.76	59.34	63.69	66.48
⋮	⋮	⋮	⋮	⋮	⋮	⋮	⋮	⋮	⋮	⋮
100	67.33	70.07	74.22	77.93	82.36	118.5	124.3	129.6	135.8	140.2

Figure 8.21

Chi-Square Tail Values for _d.f._ = 39

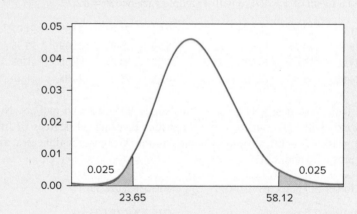

Caution: Assumption of Normality

The methods just described for confidence interval estimation of the variance and standard deviation are highly dependent on the population having a normal distribution. There is no Central Limit Theorem that can be used for the statistic s^2. If the population does not have a normal distribution, the confidence interval for the variance should not be considered accurate. What can we do about it?

An excellent alternative is the **bootstrap method**. It avoids having to assume normality when constructing a confidence interval. It can be used for estimating many different parameters. Although it requires special software, the bootstrap method is easy to explain. It rests on the principle that the sample reflects everything we know about the population. From our sample of *n* observations, we simply take repeated samples of *n* items (with replacement) and calculate the statistic of interest for each sample. The average of these statistics is the *bootstrap estimator*. The distribution of these repeated estimates is the *bootstrap distribution* (which generally is not normal). Percentiles of the bootstrap distribution provide a *bootstrap confidence interval*. For example, a 90 percent confidence interval would be formed by the 5th and 95th percentiles.

Bootstrap accuracy increases with the number of resamples (typically hundreds or thousands). Many statistical packages (e.g., Minitab) offer bootstrap estimators for most common parameters. Resampling is in the mainstream of statistics because it produces robust estimates (see *LearningStats* for details and a spreadsheet simulation).

Section Exercises

Mc Graw Hill connect

8.56 Find the 95 percent confidence interval for the population variance from these samples.
  a. $n = 15$ commuters, $s = 10$ miles driven
  b. $n = 18$ students, $s = 12$ study hours

8.57 The weights of 20 oranges (in ounces) are shown below. Construct a 95 percent confidence interval for the population standard deviation. *Note:* Scale was only accurate to the nearest ¼ ounce. (Data are from a project by statistics student Julie Gillman.) 🗁 **Oranges**

5.50	6.25	6.25	6.50	6.50	7.00	7.00	7.00	7.50	7.50
7.75	8.00	8.00	8.50	8.50	9.00	9.00	9.25	10.00	10.50

8.58 A pediatrician's records showed the mean height of a random sample of 25 girls at age 12 months to be 29.530 inches with a standard deviation of 1.0953 inches. Construct a 95 percent confidence interval for the population variance. (Data are from a project by statistics students Lori Bossardet, Shannon Wegner, and Stephanie Rader.)

8.59 Find the 90 percent confidence interval for the standard deviation of gasoline mileage mpg for these 16 San Francisco commuters driving hybrid gas–electric vehicles. 🗁 **Hybrid**

38.8	48.9	28.5	40.0	38.8	29.2	29.1	38.5
34.4	46.1	51.8	30.7	36.9	25.6	42.7	38.3

Chapter Summary

An **estimator** is a sample statistic ($\bar{x}$, s, p) that is used to estimate an unknown population **parameter** (μ, σ, π). A desirable estimator is **unbiased** (correctly centered), **efficient** (minimum variance), and **consistent** (variance goes to zero as *n* increases). **Sampling error** (the difference between an estimator and its parameter) is inevitable, but a larger sample size yields estimates that are closer to the unknown parameter. The **Central Limit Theorem** (CLT) states that the sample mean $\bar{x}$ is centered at μ and follows a normal **sampling distribution** if *n* is large, regardless of the population shape. A **confidence interval** for μ consists of lower and upper bounds that have a specified **confidence level** enclosing μ. Any confidence level may be used, but 90, 95, and 99 percent are common. If the population variance is unknown, we replace *z* in the confidence interval formula for μ with **Student's *t*** using $n - 1$ degrees of freedom. The CLT also applies to the sample proportion (*p*) as an estimator of π, using a rule of thumb to decide if normality may be assumed. The **margin of error** is the half-width of the confidence interval. The **finite population correction factor** adjusts the margin of error to reflect better estimation precision when sample size is close to the population size. Formulas exist for the required **sample size** for a given level of precision in a confidence interval for μ or π, although they entail assumptions and are only approximate. Confidence intervals and sample sizes may be adjusted for finite populations, but often the adjustments are not material. Confidence intervals may be created for a variance using the **chi-square distribution.**

Key Terms

bias	degrees of freedom	margin of error	standard error of the mean
bootstrap method	efficiency (efficient estimator)	minimum variance estimator	standard error of the proportion
Central Limit Theorem	estimate	parameter	statistical estimation
chi-square distribution	estimator	point estimate	Student's *t* distribution
confidence interval	finite population correction	sampling distribution	unbiased estimator
confidence level	factor	sampling error	
consistent estimator	interval estimate	sampling variation	

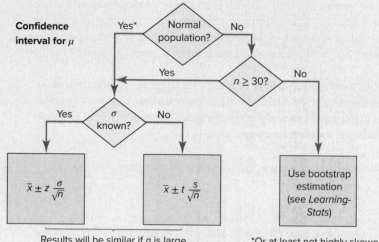

Results will be similar if n is large   *Or at least not highly skewed

Commonly Used Formulas in Sampling Distributions and Estimation

Sample proportion: $p = \dfrac{x}{n}$

Standard error of the sample mean: $\sigma_{\bar{x}} = \dfrac{\sigma}{\sqrt{n}}$

Confidence interval for μ, known σ: $\bar{x} \pm z_{\alpha/2} \dfrac{\sigma}{\sqrt{n}}$

Confidence interval for μ, unknown σ: $\bar{x} \pm t_{\alpha/2} \dfrac{s}{\sqrt{n}}$ with $d.f. = n - 1$

Standard error of the sample proportion: $\sigma_p = \sqrt{\dfrac{\pi(1 - \pi)}{n}}$

Confidence interval for π: $p \pm z_{\alpha/2} \sqrt{\dfrac{p(1 - p)}{n}}$

Finite population correction factor (FPCF): $\sqrt{\dfrac{N - n}{N - 1}}$

Confidence interval for μ, known σ, finite population: $\bar{x} \pm z_{\alpha/2} \dfrac{\sigma}{\sqrt{n}} \sqrt{\dfrac{N - n}{N - 1}}$

Confidence interval for μ, unknown σ, finite population: $\bar{x} \pm t_{\alpha/2} \dfrac{s}{\sqrt{n}} \sqrt{\dfrac{N - n}{N - 1}}$

Confidence interval for π, finite population: $p \pm z_{\alpha/2} \sqrt{\dfrac{p(1 - p)}{n}} \sqrt{\dfrac{N - n}{N - 1}}$

Sample size to estimate μ: $n = \left(\dfrac{z\sigma}{E}\right)^2$

Sample size to estimate π: $n = \left(\dfrac{z}{E}\right)^2 \pi(1 - \pi)$

Chapter Review

1. Define (a) parameter, (b) estimator, (c) sampling error, and (d) sampling distribution.

2. Explain the difference between sampling error and bias. Can they be controlled?

3. Name three estimators. Which ones are unbiased?

4. Explain what it means to say an estimator is (a) unbiased, (b) efficient, and (c) consistent.

5. State the main points of the Central Limit Theorem for a mean.

6. Why is population shape of concern when estimating a mean? What does sample size have to do with it?

7. (a) Define the standard error of the mean. (b) What happens to the standard error as sample size increases? (c) How does the law of diminishing returns apply to the standard error?

8. Define (a) point estimate, (b) margin of error, (c) confidence interval, and (d) confidence level.

9. List some common confidence levels. What happens to the margin of error as you increase the confidence level, all other things being equal?

10. List differences and similarities between Student's *t* and the standard normal distribution.

11. Give an example to show that (a) for a given confidence level, the Student's *t* confidence interval for the mean is wider than if we use a *z*-value, and (b) it makes little difference in a large sample whether we use Student's *t* or *z*.

12. Why do outliers and skewed populations pose a problem for estimating a sample mean?

13. (a) State the Central Limit Theorem for a proportion. (b) When is it safe to assume normality for a sample proportion?

14. (a) Define the standard error of the proportion. (b) What happens to the standard error as sample size increases? (c) Why does a larger sample improve a confidence interval?

15. When would you use the FPCF and what does it do to the margin of error?

16. (a) Why does σ pose a problem for sample size calculation for a mean? (b) How can σ be approximated when it is unknown?

17. When calculating a sample size for a proportion, why is it conservative to assume that $\pi = .50$?

18. Why would we be interested in a confidence interval for a variance? Give an example.

Chapter Exercises

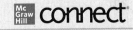

Note: Explain answers and show your work clearly. Problems marked * rely on optional material from this chapter.

8.60 A random sample of 30 lunch orders at Noodles & Company showed a mean bill of $10.36 with a standard deviation of $5.31. Find the 95 percent confidence interval for the mean bill of all lunch orders. 🗂 **NoodlesOrderSize**

8.61 A random sample of 21 nickels measured with a very accurate micrometer showed a mean diameter of 0.834343 inch with a standard deviation of 0.001886 inch. (a) Why would nickel diameters vary? (b) Construct a 99 percent confidence interval for the true mean diameter of a nickel. (c) Discuss any assumptions that are needed. (d) What sample size would ensure an error of ±0.0005 inch with 99 percent confidence? (Source: Data are from a project by MBA student Bob Tindall.)

8.62 A random sample of 10 miniature Tootsie Rolls was taken from a bag. Each piece was weighed on a very accurate scale. The results in grams were

3.087 3.131 3.241 3.241 3.270 3.353
3.400 3.411 3.437 3.477

(a) Construct a 90 percent confidence interval for the true mean weight. (b) What sample size would be necessary to estimate the true weight with an error of ±0.03 gram with 90 percent confidence? 🗂 **Tootsie**

8.63 Statistics students were asked to go home and fill a 1-cup measure with raisin bran, tap the cup lightly on the counter three times to settle the contents, if necessary add more raisin bran to bring the contents exactly up to the 1-cup line, spread the contents on a large plate, and count the raisins. For the 13 students who chose Kellogg's brand, the reported results were

23 33 44 36 29 42 31 33
61 36 34 23 24

Construct a 90 percent confidence interval for the mean number of raisins per cup. Show your work clearly. 🗂 **Raisins**

8.64 A sample of 20 pages was taken without replacement from a Yellow Pages (yp.com) business directory that has 1,591 pages. (a) Calculate the FPCF for this sample. (b) Should the population be considered effectively infinite?

8.65 Twenty-five blood samples were selected by taking every seventh blood sample from racks holding 187 blood samples from the morning draw at a medical center. (a) Calculate the FPCF for this sample. (b) Should the population be considered effectively infinite?

8.66 A sample of 20 pages was taken from a Yellow Pages (yp.com) business directory. On each page, the mean area devoted to display ads was measured (a display ad is a large block of multicolored illustrations, maps, and text). The data (in square millimeters) are shown below:

0   260 356 403 536 0   268
369 428 536 268 396 469 536
162 338 403 536 536 130

(a) Construct a 95 percent confidence interval for the true mean. (b) What sample size would be needed to obtain an error of ±20 square millimeters with 95 percent confidence? 🗂 **DisplayAds**

8.67 Sixteen owners of a 2020 Toyota Tacoma kept track of their average fuel economy for a month. The results are shown below. (a) Construct a 95 percent confidence interval for the mean. (b) What factor(s) limit the conclusions that can be drawn about the true mean? 📊 **MPG** (Source: www.fueleconomy.gov.)

20.8	20.0	19.4	19.7	21.1	22.6	18.3	20.1
20.5	19.5	17.4	22.4	18.9	20.2	19.6	19.0

8.68 Twenty-five blood samples were selected by taking every seventh blood sample from racks holding 187 blood samples from the morning draw at a medical center. The white blood count (WBC) was measured using a Coulter Counter Model S. The mean WBC was 8.636 with a standard deviation of 3.9265. (a) Construct a 90 percent confidence interval for the true mean using the FPCF. (b) What sample size would be needed for an error of ± 1.5 with 90 percent confidence? (Source: Data are from a project by MBA student Wendy Blomquist.)

8.69 Twenty-one warranty repairs were selected from a population of 126 by selecting every sixth item. The population consisted of "loose, not attached" minivan electrical wires (one of several electrical failure categories the dealership mechanic can select). The mean repair cost was $145.664 with a standard deviation of $27.793. (a) Construct a 95 percent confidence interval for the true mean repair cost. (b) What sample size would be needed to obtain an error of $\pm\$5$ with 95 percent confidence?* (c) Construct a 95 percent confidence interval for the true standard deviation.

8.70 Dave the jogger runs the same route every day (about 2.2 miles). On 18 consecutive days, he recorded the number of steps on his Fitbit. The results were

3,450	3,363	3,228	3,360	3,304	3,407	3,324
3,365	3,290	3,289	3,346	3,252	3,237	3,210
3,140	3,220	3,103	3,129			

(a) Construct a 95 percent confidence interval for the true mean number of steps Dave takes on his run. (b) What sample size would be needed to obtain an error of ± 20 steps with 95 percent confidence? (c) Using Excel, plot a line chart of the data. What do the data suggest about the pattern over time? 📊 **DaveSteps**

8.71 A pediatrician's records showed the mean height of a random sample of 25 girls at age 12 months to be 29.530 inches with a standard deviation of 1.0953 inches. (a) Construct a 95 percent confidence interval for the true mean height. (b) What sample size would be needed for 95 percent confidence and an error of $\pm.20$ inch?

*8.72 A random sample of 10 exam scores showed a standard deviation of 7.2 points. Find the 95 percent confidence interval for the population standard deviation. Use Appendix E to obtain the values of χ_L^2 and χ_U^2.

*8.73 A random sample of 30 lunch orders at Noodles & Company showed a standard deviation of $5.31. Find the 90 percent confidence interval for the population standard deviation. Use Excel to obtain χ_L^2 =CHISQ.INV(α/2,d.f.) and χ_U^2 =CHISQ.INV.RT((α/2),d.f.)

8.74 During the Rose Bowl, the length (in seconds) of 12 randomly chosen commercial breaks during timeouts (following touchdown, turnover, field goal, or punt) were

65	75	85	95	80	100	90	80
85	85	60	65				

Assuming a normal population, construct a 90 percent confidence interval for the mean length of a commercial break during the Rose Bowl. 📊 **TimeOuts**

8.75 A sample of 40 songs from a student's Spotify playlist showed a mean length of 3.542 minutes with a standard deviation of 0.311 minute. (a) Construct a 95 percent confidence interval for the mean. (b) Why might the normality assumption be an issue here? (c) What sample size would be needed to estimate μ with 95 percent confidence and an error of ± 6 seconds?

8.76 The Environmental Protection Agency (EPA) requires that cities monitor over 80 contaminants in their drinking water. Samples from the Lake Huron Water Treatment Plant gave the results shown here. All observations were below the allowable maximum, as shown by the reported range of contaminant levels. For each substance, estimate the *standard deviation σ* by using one of the methods shown in Table 8.8 in Section 8.8.

Substance	Range Detected	Allowable Maximum	Origin of Substance
Chromium	0.47 to 0.69	100	Discharge from steel and pulp mills, natural erosion
Barium	0.004 to 0.019	2	Discharge from drilling wastes, metal refineries, natural erosion
Fluoride	1.07 to 1.17	4.0	Natural erosion, water additive, discharge from fertilizer and aluminum factories

8.77 In a sample of 100 Planter's Mixed Nuts, 19 were found to be almonds. (a) Construct a 90 percent confidence interval for the true proportion of almonds. (b) May normality be assumed? Explain. (c) What sample size would be needed for 90 percent confidence and an error of ± 0.03? (d) Why would a quality control manager at Planter's need to understand sampling?

8.78 A study showed that 14 of 180 publicly traded business services companies failed a test for compliance with Sarbanes–Oxley requirements for financial records and fraud protection. Assuming that these are a random sample of all publicly traded companies, construct a 95 percent confidence interval for the overall noncompliance proportion.

8.79 How "decaffeinated" is decaffeinated coffee? If a researcher wants to estimate the mean caffeine content of a cup of Starbucks' decaffeinated espresso with 98 percent confidence and an error of ± 0.1 mg, what is the required number of cups that must be tested? Assume a standard deviation of 0.5 mg, based on a small preliminary sample of 12 cups.

8.80 Noodles & Company wants to estimate the percent of customers who order dessert, with 95 percent confidence and an error of ±10%. What is the required sample size?

8.81 Junior Achievement and Deloitte commissioned a "teen ethics poll" of 787 students aged 13–18, finding that 29 percent felt inadequately prepared to make ethical judgments. (a) Assuming that this was a random sample, find the 95 percent confidence interval for the true proportion of U.S. teens who feel inadequately prepared to make ethical judgments. (b) Is the sample size large enough to assume normality?
(Source: http://ja.org/.)

8.82 A random sample of 30 cans of Del Vino crushed tomatoes revealed a mean weight of 798.3 grams (excluding the juice). The can-filling process for Del Vino crushed tomatoes has a known standard deviation of 3.1 grams. Construct the 95 percent confidence interval for the mean weight of cans, assuming that the weight of cans is a normally distributed random variable.

8.83 A poll of 125 college students who watch *Friends* showed that 83 of them usually watch on a mobile device (e.g., laptop). (a) Assuming that this was a random sample, construct a 90 percent confidence interval for the proportion of all college students who usually watch this show on a mobile device. (b) Would a finite population correction be required? Explain.

8.84 A survey of 4,581 U.S. households that owned a mobile phone found that 58 percent are satisfied with the coverage of their cellular phone provider. Assuming that this was a random sample, construct a 90 percent confidence interval for the true proportion of satisfied U.S. mobile phone owners.

8.85 A "teen ethics poll" was commissioned by Junior Achievement and Deloitte. The survey by Harris Interactive surveyed 787 students aged 13–18. (a) Assuming that this was a random sample of all students in this age group, find the margin of error of the poll. (b) Would the margin of error be greater or smaller for the subgroup consisting only of male students? Explain.
(Source: http://ja.org/.)

8.86 Biting an unpopped kernel of popcorn hurts! As an experiment, a self-confessed connoisseur of cheap popcorn carefully counted 773 kernels and put them in a popper. After popping, the unpopped kernels were counted. There were 86. (a) Construct a 90 percent confidence interval for the proportion of all kernels that would not pop. (b) Check the normality assumption for p.

8.87 A sample of 213 newspaper tire ads from several Sunday papers showed that 98 contained a low-price guarantee (offer to "meet or beat any price"). (a) Assuming that this was a random sample, construct a 95 percent confidence interval for the proportion of all Sunday newspaper tire ads that contain a low-price guarantee. (b) Is the criterion for normality of p met?

8.88 A physician's billing office conducted a random check of patient records and found that 36 of 50 patients had changed insurance plans within the past year. Construct a 90 percent confidence interval for the true proportion.

8.89 Of 250 college students taking a statistics class, 4 reported an allergy to peanuts. (a) Is the criterion for normality of p met?

(b) Assuming that this was a random sample, use Minitab to construct a 95 percent confidence interval for the proportion of all college statistics students with a peanut allergy.

8.90 (a) A poll of 2,277 likely voters was conducted on the president's performance. Approximately what margin of error would the approval rating estimate have? (b) The poll showed that 44 percent approved the president's performance. Construct a 90 percent confidence interval for the true proportion. (c) Would you say that the percentage of all voters who approve of the president's performance could be 50 percent?

8.91 To determine the proportion of taxpayers who use an online tax preparation service, a survey of 600 taxpayers was conducted. Calculate the margin of error used to estimate this proportion. What assumptions are required to find the margin of error?

8.92 A sample of 40 songs from a student's Spotify playlist showed a mean length of 3.542 minutes with a standard deviation of 0.311 minute. Construct a 95 percent confidence interval for the population standard deviation.

8.93 Ouranos Resorts would like to send a survey to its guests asking about their satisfaction with the new website design. It would like to have a margin of error of ±5 percent on responses with 95 percent confidence. (a) Using the conservative approach, what sample size is needed to ensure this level of confidence and margin of error? (b) If Ouranos Resorts wanted the margin of error to be only ±2.5 percent, what would happen to the required sample size?

MINI-PROJECTS

8.94 This is an exercise using Excel. (a) Use =RANDBETWEEN(0,99) to create 20 samples of size $n = 4$ by choosing two-digit random numbers between 00 and 99. (b) For each sample, calculate the mean. (c) Make a histogram of the 80 *individual x-values* using bins 10 units wide (i.e., 0, 10, 20, . . . , 100). Describe the shape of the histogram. (d) Make a histogram of your 20 *sample means* using bins 10 units wide. (e) Discuss the histogram shape. Does the Central Limit Theorem seem to be working? (f) Find the mean of your 20 sample means. Was it what you would expect? Explain. (g) Find the standard deviation of your 20 sample means. Was it what you would expect?

8.95 For 10 tanks of gas for your car, calculate the miles per gallon. (a) Construct a 95 percent confidence interval for the true mean mpg for your car. (b) How many tanks of gas would you need to obtain an error of ±0.2 mpg with 95 percent confidence? Use the value of s from your sample.

8.96 (a) Look at 50 vehicles in a grocery store parking lot near you. Count the number that are SUVs (state your definition of SUV). Use any sampling method you like (e.g., the first 50 you see). (b) Construct a 95 percent confidence interval for the true population proportion of SUVs. (c) What sample size would be needed to ensure an error of ±0.025 with 98 percent confidence? (d) Would the proportion be the same if this experiment were repeated in a university parking lot?

8.97 (a) From the New York Stock Exchange website, take a random sample of 30 companies that are on the IPO (Initial Public Offering) Showcase list for last year. (b) Calculate a 90 percent confidence interval for the average closing stock price on the day of the IPO.

Related Reading

Boos, Dennis D., and Jacqueline M. Hughes-Oliver. "How Large Does *n* Have to Be for *z* and *t* Intervals?" *The American Statistician* 54, no. 2 (May 2000), pp. 121–128.

Kupper, Lawrence L., and Kerry B. Hafner. "How Appropriate Are Popular Sample Size Formulas?" *The American Statistician* 43, no. 2 (May 1989), pp. 101–105.

Lenth, Russell V. "Some Practical Guidelines for Effective Sample Size Determination." *The American Statistician* 55, no. 3 (August 2001), pp. 187–193.

Parker, Robert A. "Sample Size: More Than Calculations." *The American Statistician* 57, no. 3 (August 2003), pp. 166–170.

van Belle, Gerald. *Statistical Rules of Thumb.* 2nd ed. Wiley, 2008.

CHAPTER 8   More Learning Resources

You can access these *LearningStats* demonstrations through Connect to help you understand sampling distributions and confidence intervals.

Topic	*LearningStats Demonstrations*
Central Limit Theorem	CLT Demonstration: Simulation CLT Demonstration: Finite Population
Sampling distributions	Sampling Distributions Critical Values (z, t, χ^2) Sample Proportion Demonstration
Confidence intervals	Confidence Interval: Means Confidence Interval: Proportions Confidence Interval: Variances Confidence Interval: Simulation Confidence Interval: Bootstrap Useful Equations Estimation Using Excel Estimation Using R Finite Populations Bootstrap Explained
Sample size	Sample Size Calculator
Tables	Appendix D—Student's *t* Appendix E—Chi-Square

Key: = PowerPoint   = Excel   = PDF

Software Supplement

Confidence Interval for μ: MegaStat  Maternity

If you already have the sample mean and standard deviation, MegaStat does all the calculations for you, as illustrated in Figure 8.22 for a sample ($n = 25$) of hospital lengths of stay for maternity patients (see Example 8.4). MegaStat gives you a choice of *z* or *t*. Notice the Preview button. If you click OK, you also will see the *t*-value and other details.

Figure 8.22

MegaStat's Confidence Interval for μ

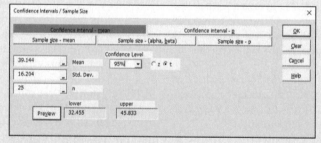

Source: MegaStat.

Confidence Interval for μ: Minitab 📁 Maternity

Use Minitab's Stat > Basic Statistics > Graphical Summary to get confidence intervals, as well as a histogram and box plot. Minitab uses Student's *t* for the confidence interval for the mean. Minitab also gives confidence intervals for the median and standard deviation. Figure 8.23 shows Minitab's Graphical Summary for the maternity length of stay data ($n = 25$).

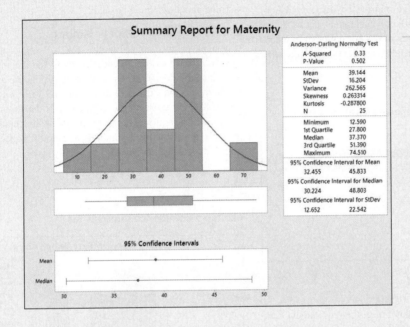

Figure 8.23

Minitab's Confidence Intervals and Graphical Summary

Source: Minitab.

Bootstrap Confidence Interval for μ: Minitab

Minitab offers a modern approach to constructing confidence intervals for a mean using resampling (the bootstrap method). We need not assume that the underlying population is normally distributed. We need not estimate the standard error. Our confidence limits are simply percentiles (e.g., .025 and .975 for 95 percent confidence) of the distribution of many resampled means. Figure 8.24 shows a 95 percent bootstrap confidence interval and histogram for the mean hospital stay based on 1,000 resamples of a random sample of 25 hospital maternity stays (see Example 8.4). Compare the bootstrap confidence interval [32.87, 45.40] with the Student's *t* confidence interval [32.455, 45.833] from Example 8.4. Each time you resample, you will get a different result. The larger the resample, the more stable the result (Minitab's limit is 1,000).

Figure 8.24   **Minitab Bootstrap Confidence Interval 📁 Maternity**

Source: Minitab.

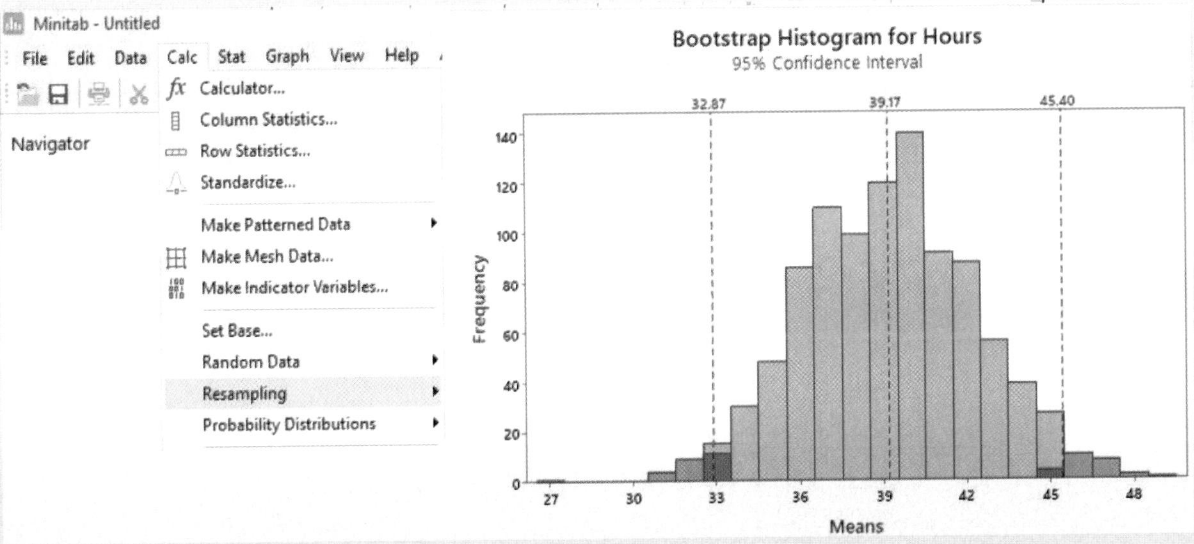

Confidence Interval for π: MegaStat

Excel's Data Analysis does not offer a confidence interval for a proportion, presumably because the calculations are easy. However, MegaStat does offer a confidence interval for a proportion, as shown in Figure 8.25. You only need to enter p and n. A convenient feature is that, if you enter p larger than 1, MegaStat assumes that it is the x-value in $p = x/n$ so you don't even have to calculate p. Click the Preview button to see the confidence interval, or click OK for additional details. MegaStat always assumes normality, even when it is not justified, so you should check this assumption for yourself.

Figure 8.25

MegaStat's 90 Percent Confidence Interval for π

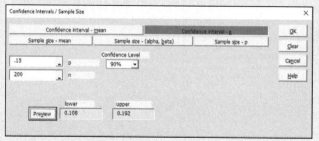

Source: MegaStat.

Confidence Interval for π: Minitab

If the sample is small (i.e., if we cannot meet the requirement that $n\pi \geq 10$ and $n(1 - \pi) \geq 10$), the distribution of p may not be well approximated by the normal. In that case, confidence limits around p can be constructed by using the binomial distribution. Minitab uses this method, which works for any n (unless you choose the Options to assume normality). Although the underlying calculations are a bit complex, Minitab does all the work and the resulting interval is correct for any n and p. For example, in a sample of 14 purchasers of the 2019 DVD movie *Star Wars: The Rise of Skywalker*, 11 watched only the film and never even looked at the "extras." The sample proportion is $p = 11/14$. We have $np = 11$, but $n(1 - p) = 3$ is less than 10, so we should not assume normality in constructing a confidence interval for the proportion of purchasers who never viewed the "extras." Figure 8.26 shows Minitab's confidence interval using the binomial distribution with Stat > Basic Statistics > One Proportion. Minitab's exact binomial confidence interval [.492, .953] is quite different from the normal confidence interval [.571, 1.000]. In this example, Minitab gives a warning about assuming normality.

Confidence Interval for σ: Minitab

If you have raw data, Minitab's Stat > Basic Statistics > Graphical Summary gives nice confidence intervals for the mean, median, and standard deviation, as well as a histogram and box plot. Minitab calculates the confidence interval for σ, as illustrated in Figure 8.27. In estimating σ, Minitab uses the chi-squared distribution, which in turn assumes normality of the population. In this example, the normality assumption is doubtful, based on the histogram's appearance and the AD normality test.

Figure 8.27

Minitab's Confidence Intervals for σ

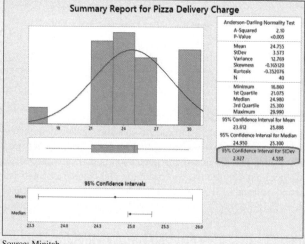

Source: Minitab.

Figure 8.26

Minitab's 95 Percent Confidence Interval for π

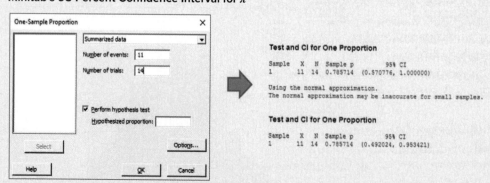

Source: Minitab.

Confidence Intervals in R Maternity

Unlike Excel, Minitab, MegaStat, and other menu-driven software packages, constructing a confidence interval requires several steps in R. We will demonstrate using a random sample of 25 randomly chosen hospital maternity stays (see Example 8.4) measured in hours (actual data).

Confidence interval for μ

What is the 95 percent confidence interval for the population mean? We can highlight the Excel data column (including its column heading), copy the data to the clipboard, and paste it into a data frame in R:

```
> MaternityStay=read.table(file="clipboard",
header=TRUE)
```

The Excel header was Hours, so our variable is named MaternityStay$Hours. For brevity, we will assign a shorter variable name x. It's a good idea that we check the mean, standard deviation, and sample size to make sure the data look OK:

```
> x=MaternityStay$Hours    #assign short variable name for brevity
> mean(x)
[1] 39.144
> sd(x)
[1] 16.20385
> length(x)
[1] 25
```

We name our confidence limits LowerLim and UpperLim (you can call them anything you want). The population variance is unknown, so our 95 percent confidence interval requires the lower .025 and upper .025 quantiles of the Student's *t* distribution using the R function qt(). Note that R provides the negative sign for the lower .025 quantile.

```
> LowerLim=mean(x)+qt(.025,length(x)−1)*sd(x)/sqrt(length(x))
> UpperLim=mean(x)+qt(.975,length(x)−1)*sd(x)/sqrt(length(x))
> LowerLim
[1] 32.45538
> UpperLim
[1] 45.83262
```

If we are careful with parentheses, we can do this all in one step:

```
> mean(x)+qt(c(.025,.975),length(x)−1)*sd(x)/sqrt(length(x))
[1] 32.45538 45.83262
```

These confidence limits agree with Example 8.4 (and with the Minitab output shown above). A sneaky way to get quick confidence limits is to use the t.test() function, but the output has a lot of extra stuff because it is intended for hypothesis tests (Chapter 9):

```
> t.test(MaternityStay$Hours,conf.level=.95)
One Sample t-test
data: MaternityStay$Hours
t = 12.079, df = 24, p-value = 1.092e-11
alternative hypothesis: true mean is not equal to 0
95 percent confidence interval:
32.45538 45.83262
sample estimates:
mean of x
39.144
```

Confidence interval for σ²

If we want a confidence interval for the variance, we must pretty much do it ourselves, as follows (note reversal of the order of quantiles—see textbook formula):

```
> (length(x)−1)*var(x)/qchisq(c(.975,.025),length(x)−1)
[1] 160.0838 508.1425
```

You may be able to identify and download other CRAN tools that will make these tasks easier (unless your university has already done that). For example, the boot() package allows you to create confidence intervals without the assumptions required by the normal or *t* distributions. However, learning to use these tools is not a goal of this brief introduction to R.

CHAPTER CONTENTS

CHAPTER LEARNING OBJECTIVES

When you finish this chapter, you should be able to

LO 9-1 Know the steps in testing hypotheses and define H_0 and H_1.

LO 9-2 Define Type I error, Type II error, and power.

LO 9-3 Formulate null and alternative hypotheses for μ or π.

LO 9-4 Explain decision rules, critical values, and rejection regions.

LO 9-5 Perform a hypothesis test for a mean with known σ using z.

LO 9-6 Use tables or Excel to find the p-value in tests of μ.

LO 9-7 Perform a hypothesis test for a mean with unknown σ using t.

LO 9-8 Perform a hypothesis test for a proportion and find the p-value.

LO 9-9 Check whether normality may be assumed in testing a proportion.

LO 9-10 Interpret a power curve or OC curve (optional).

LO 9-11 Perform a hypothesis test for a variance (optional).

D
ata are used in business every day to support marketing claims, help managers make decisions, and measure business improvement. Whether the business is small or large, profit or non-profit, the use of data allows businesses to find the best answers to their questions.

- Should Ball Corporation's aluminum can division change suppliers of metal?

- Did the proportion of defective products decrease after a new manufacturing process was introduced?

- Has the average service time at a Noodles & Company restaurant decreased since last year?

- Has a ski resort decreased its average response time to accidents?

- Did the proportion of satisfied car repair customers increase after providing more training for the employees?

michaeljung/Shutterstock

Savvy businesspeople use data and many of the statistical tools that you've already learned to answer these types of questions. We will build on these tools in this chapter and learn about one of the most widely used statistical tools—**hypothesis testing**. Hypothesis testing is used in science and business to test assumptions and theories and guide managers when facing decisions. We will first explain the logic behind hypothesis testing and then show how *statistical hypothesis testing* helps businesses make decisions.

9.1 LOGIC OF HYPOTHESIS TESTING

LO 9-1

Know the steps in testing hypotheses and define H_0 and H_1.

The business analyst asks questions, makes assumptions, and proposes testable theories about the values of key parameters of the business operating environment. Each assumption is tested against observed data. If an assumption has not been disproved, in spite of rigorous efforts to do so, the business may operate under the belief that the statement is true. The analyst states the assumption, called a **null hypothesis**, in a format that can be tested using well-known statistical procedures. The null hypothesis is compared with sample data to determine if the data are consistent or inconsistent with the hypothesis. When the data are found to be inconsistent (i.e., in conflict) with the null hypothesis, that hypothesis is discarded in favor of an alternative hypothesis.

The process of hypothesis testing can be an iterative process, as illustrated in Figure 9.1. Not every hypothesis has to be retested continuously. Some questions are relevant only at a point in time, such as asking whether consumers under age 25 prefer a new Coke flavor to an existing flavor. If a carefully designed marketing study provides a clear-cut answer, no more testing is required. On the other hand, clinical testing of new drugs may go on for years, at different hospitals and with different types of patients. The efficacy and side effects of new drugs may involve subtle effects that only show up in very large samples and over longer periods of time, sometimes with profound financial implications for the company. Automobile safety testing is ongoing because of changes in car technology and driver habits. There is plenty of work for data analysts, who continually look for changes in customer satisfaction, shifts in buying patterns, and trends in warranty claims.

317

Figure 9.1

Hypothesis Testing as an Ongoing Process

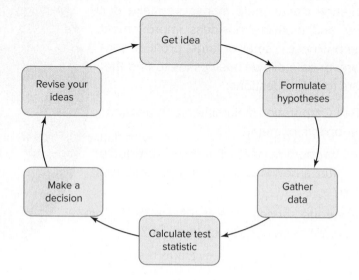

Who Uses Hypothesis Testing?

The innovative vigor of our economy is largely based on technology: new materials, new manufacturing methods, new distribution systems, new information strategies. All business managers need at least a basic understanding of hypothesis testing because managers often interact with specialists, read technical reports, and then make recommendations on key financial or strategic decisions based on statistical evidence. A confidence interval sometimes gives enough information to make a decision. Knowing the 95 percent range of likely values for a key decision parameter (e.g., the proportion of repeat customers under age 30) may be all you need. This chapter extends the idea of confidence intervals by showing how to test a sample against a benchmark and how to assess the risk of incorrect decisions.

Steps in Hypothesis Testing

Step 1: State the null and alternative hypotheses.

Step 2: Specify what level of inconsistency with the data will lead to rejection of the null hypothesis. This is called a *decision rule.*

Step 3: Collect data and calculate necessary statistics to test the hypothesis.

Step 4: Make a decision. Should the null hypothesis be rejected or not?

Step 5: Take action based on the decision.

Step 1: State the Hypotheses   Formulate a pair of mutually exclusive statements. One statement or the other must be true, but they cannot both be true.

H_0: Null Hypothesis

H_1: Alternative Hypothesis

The two statements are *hypotheses* because the truth is unknown. Efforts will be made to reject the **null hypothesis**. H_0 must be stated in a precise way so that it can be tested against empirical evidence from a sample. If H_0 happens to be an established theory, we might not expect to reject it, but we might be monitoring a situation to see if a key parameter has changed. If we reject H_0, we tentatively conclude that the **alternative hypothesis** H_1 is the case. H_0 represents the *status quo* (e.g., the current state of affairs), while H_1 is sometimes called the *action alternative* because action may be required if we reject H_0 in favor of H_1. For example:

Criminal Trial   In a criminal trial, the hypotheses are

H_0: The defendant is not guilty

H_1: The defendant is guilty

Our legal system assumes a defendant is not guilty *unless the evidence gathered by the prosecutor is sufficient to reject this assumption.*

Drug Testing   When an Olympic athlete is tested for performance-enhancing drugs ("doping"), the presumption is that the athlete is in compliance with the rules. The hypotheses are

H_0: No banned substance was used

H_1: Banned substance was used

Samples of urine or blood are taken as evidence and *used only to disprove the null hypothesis because we assume the athlete is free of banned substances.*

Biometric Security   To identify authorized and unauthorized persons for computer access, ATM withdrawals, and entry into secure facilities, there is increasing interest in using the person's physical characteristics (e.g., fingerprints, facial structure, or iris patterns) instead of paper and plastic IDs, which can be forged. The hypotheses are

H_0: User is legitimate

H_1: User is not legitimate

The system assumes the user is legitimate *unless the physical characteristic being presented is inconsistent with the biometric profile of that individual.*

Step 2: Specify the Decision Rule   Before collecting data to compare against the null hypothesis, the researcher must specify *how* the evidence will be used to reach a decision. In our legal system, the evidence presented by the prosecutor must convince a jury "beyond a reasonable doubt" that the defendant is guilty. In steroid testing, the lab that analyzes the urine or blood sample must conduct tests to decide whether the sample exceeds the agreed-upon benchmark or threshold. With biometric screening, the designer of the security system determines how many discrepancies on a fingerprint would indicate an unauthorized user.

Steps 3 and 4: Data Collection and Decision Making   Much of the critical work in hypothesis testing takes place during steps 1 and 2. Once the hypotheses and decision rule have been clearly articulated, the process of data collection, while occasionally time-consuming, is straightforward. We compare the data with the hypothesis, using the decision rule, and decide to reject or not reject the null hypothesis.

Step 5: Take Action Based on Decision   This last step—taking action—requires experience and expertise on the part of the decision maker. Suppose the evidence presented at a trial convinces a jury that the defendant is not innocent. What punishment should the judge impose? Or suppose the blood sample of an athlete shows steroid use. What fine should the athletic commission impose? Should the athlete be banned from competing? If the fingerprint presented for authentication has been rejected, should an alarm go off? Should a security breach be recorded in the system? Appropriate action for the decision should relate back to the purpose of conducting the hypothesis test in the first place.

Can a Null Hypothesis Be Proved?

No, we cannot prove a null hypothesis—we can only *fail to reject* it. A null hypothesis that survives repeated tests without rejection is "true" only in the limited sense that it has been thoroughly scrutinized and tested. Today's "true" hypothesis could be disproved tomorrow if new data are found. If we fail to reject H_0, the same hypothesis may be retested. That is how scientific inquiry works. Einstein's theories, for example, are over 100 years old but are still being subjected to rigorous tests. Few scientists really think that Einstein's theories are "wrong." Yet it's in the nature of science to keep trying to refute accepted theories, especially when a new test is possible or when new data become available. Similarly, the safety of commonly used prescription drugs is continually being studied. Sometimes, "safe" drugs are revealed to have serious side effects only after large-scale, long-term use by consumers.

Analytics in Action

Walmart, Big Data, and Retail Analytics

Walmart processes over a million customer transactions each hour, which translates into two to three petabytes of data each hour. (A petabyte is a million gigabytes!) What to do with all these data? Walmart is practicing *data democratization*. This term means making large amounts of data available to everyone in the organization so that employees can quickly react to changes in their customers' behaviors.

Walmart operates a Data Café that can be accessed by everyone in the company. When store managers notice changes in sales for particular products, they can go to the Data Café and look at data across all their stores to figure out why the changes are happening. For example, a sharp decrease in sales of one product turned out to be caused by incorrect pricing for that product. The problem can be fixed within a day rather than having to wait a week or two—which is what typically happened in the past. Differences in sales between stores can highlight mistakes such as Halloween products not making it to the shelf. Identifying the mistake and fixing it in time for customers to purchase before the big night prevents a loss in profit. Big data mean that real-time decisions can be made that positively impact business performance.

Source: www.forbes.com/sites/bernardmarr/2017/01/23/really-big-data-at-walmart-real-time-insights-from-their-40-petabyte-data-cloud/#6c7d848a6c10.

9.2 TYPE I AND TYPE II ERRORS

LO 9-2

Define Type I error, Type II error, and power.

Our ability to collect evidence can be limited by our tools and by time and financial resources. On occasion we will be making a decision about the null hypothesis that could be wrong. Consequently, our decision rule will be based on the levels of risk of making a wrong decision. We can allow more risk or less risk by changing the threshold of the decision rule.

It is possible to make an incorrect decision regarding the null hypothesis. As illustrated in the table below, either the null hypothesis is true or it is false. We have two possible choices concerning the null hypothesis. We either reject H_0 or fail to reject H_0.

	H_0 is true	H_0 is false
Reject H_0	Type I error	Correct decision
Fail to reject H_0	Correct decision	Type II error

The true situation determines whether our decision was correct. If the decision about the null hypothesis matches the true situation, the decision was correct. Rejecting the null hypothesis when it is true is a **Type I error** (also called a **false positive**). Failure to reject the null hypothesis when it is false is a **Type II error** (also called a **false negative**). In either case, an incorrect decision was made.

Because we rarely have perfect information about the true situation, we can't always know whether we have committed a Type I or Type II error. But we can calculate the *probability* of making an incorrect decision. We can minimize the chance of error by collecting as much sample evidence as our resources allow and by choosing proper testing procedures.

Consequences of Type I and Type II Errors

The consequences of these two errors are quite different, and the costs are borne by different parties. Depending on the situation, decision makers may fear one error more than the other. It would be nice if both types of error could be avoided. Unfortunately, when making a decision based on a fixed body of sample evidence, reducing the risk of one type of error often increases the risk of the other.

Criminal Trial   Type I error is convicting an innocent defendant, so the costs are borne by the defendant. Type II error is failing to convict a guilty defendant, so the costs are borne by society if the guilty person returns to the streets. Concern for the rights of the accused and stricter rules of evidence may suggest rules that would reduce the risk of Type I error. But concern over the social costs of crime and victims' rights could lead to rules that would reduce Type II error, presumably at the expense of Type I error. Both risks can be reduced only by devoting more effort to gathering evidence and strengthening the legal process (expediting trials, improving jury quality, increasing investigative work).

H_0: Defendant is not guilty
H_1: Defendant is guilty

Drug Testing   Type I error (a false positive) is unfairly disqualifying an athlete who is "clean." Type II error (a false negative) is letting the drug user get away with it and have an unfair competitive advantage. The costs of Type I error are hard feelings, unnecessary embarrassment, and possible loss of lifetime earnings for professional athletes. The costs of Type II error are tarnishing the Olympic image and rewarding those who break the rules. Over time, improved tests have reduced the risk of both types of error. However, for a given technology, the threshold can be set lower or higher, balancing Type I and Type II errors.

H_0: No illegal doping
H_1: Illegal doping

Biometric Security   Type I error means denying a legitimate user access to a facility or funds. Type II error is letting an unauthorized user have access to facilities or a financial account. Technology has progressed to the point where Type II errors have become very rare, though Type I errors remain a problem. The error rates depend, among other things, on how much is spent on the equipment and software.

H_0: User is legitimate
H_1: User is not legitimate

Probability of Type I and Type II Errors

The *probability* of a Type I error (rejecting a true null hypothesis) is denoted α (the lowercase Greek letter "alpha"). Statisticians refer to α as the **level of significance**. The probability of a Type II error (not rejecting a false hypothesis) is denoted β (the lowercase Greek letter "beta"), as shown in Table 9.1.

Key Term	What Is It?	Symbol	Definition	Also Called
Type I error	Reject a true hypothesis	α	$P(\text{reject } H_0 \mid H_0 \text{ is true})$	False positive
Type II error	Fail to reject a false hypothesis	β	$P(\text{fail to reject } H_0 \mid H_0 \text{ is false})$	False negative
Power	Correctly reject a false hypothesis	$1 - \beta$	$P(\text{reject } H_0 \mid H_0 \text{ is false})$	Sensitivity

Table 9.1

Key Terms in Hypothesis Testing

The **power** of a test is the probability that a false hypothesis *will* be rejected (as it should be). Power equals $1 - \beta$ and is the complement of Type II error. Reducing β will increase power (usually accomplished by increasing the sample size). More powerful tests are more likely to detect false hypotheses. For example, if a new weight-loss drug actually is effective, we would want to reject the null hypothesis that the drug has no effect. Larger samples lead to increased power, which is why clinical trials often involve thousands of people.

Relationship between α and β

We desire tests that avoid false negatives (small β), yet we also want to avoid false positives (small α). Given two equivalent tests, we will choose the more powerful test (smaller β). But for a given type of test and fixed sample size, there is a trade-off between α and β. The larger critical value needed to reduce α makes it harder to reject H_0, thereby increasing β. The proper balance between α and β can be elusive. Consider these examples:

- If your household carbon monoxide detector's sensitivity threshold is increased to reduce the risk of overlooking danger (reduced β), there will be more false alarms (increased α).

- A doctor who is conservative about admitting patients with symptoms of heart attack to the ICU (reduced β) will admit more patients with no heart attack (increased α).

- More sensitive airport weapons detectors (reduced β) will inconvenience more safe passengers (increased α).

The level of significance, α, is not calculated using a formula. Instead, we choose a value based on our willingness to risk making a Type I error. Choosing a smaller value of α reflects a greater concern with the consequences of a Type I error. Typical values are .01, .05, or .10. To reduce β, without changing our chosen α, we would need to increase the sample size (gathering more evidence), which is not always feasible or cost-effective.

Mini Case 9.1

Biometric Security

If your ATM could recognize your physical characteristics (e.g., fingerprint, face, palm, iris), you wouldn't need an ATM card or a PIN. A reliable biometric ID system also could reduce the risk of ID theft, eliminate computer passwords, and speed airport security screening. The hypotheses are

H_0: User is legitimate

H_1: User is not legitimate

Fujitsu Laboratories has tested a palm ID system on 700 people, ranging from children to seniors. It achieved a false rejection rate of 1 percent and a false acceptance rate of 0.5 percent. Bank of Tokyo-Mitsubishi introduced palm-scanning at its ATMs in 2004. DigitalPersona of Redwood City, California, has developed a fingerprint scanner (called *U.Are.U*) that is able to recognize fingerprints in 200 milliseconds with a 1 percent false rejection rate and a 0.002 percent false acceptance rate. Some high-end devices achieve false acceptance rates as low as 25 per million. These low rates of Type II error are encouraging because they mean that others cannot easily impersonate you. Fingerprint scanning is popular (e.g., at the entrance gates of Disney World or Universal Studios) because it is cheaper and easier to implement, though some experts believe that iris scanning or several combined tests have better long-run potential to reduce both error rates. Any such system requires a stored database of biometric data.

Product Safety

Firms are increasingly wary of Type II error (failing to recall a product as soon as sample evidence begins to indicate potential problems):

H_0: Product is performing safely

H_1: Product is not performing safely

They even may order a precautionary product recall before the statistical evidence has become convincing (e.g., Verizon's 2004 recall of 50,000 cell phone batteries after one exploded and another caused a car fire) or even *before* anything bad happens (e.g., Intel's

2004 recall of its 915 G/P and 925X chip sets from original equipment manufacturers [OEMs], before the chips actually reached any consumers). Failure to act swiftly can generate liability and adverse publicity, as with the spate of Ford Explorer rollover accidents and eventual recall of certain 15-inch Firestone radial tires. Ford and Firestone believed they had found an engineering work-around to make the tire design safe until accumulating accident data, lawsuits, and NHTSA pressure forced recognition that there was a problem. In 2004, certain COX_2 inhibitor drugs that had previously been thought effective and safe, based on extensive clinical trials, were found to be associated with increased risk of heart attack. The makers' stock price plunged (e.g., Merck). The courts often must use statistical evidence to adjudicate product liability claims.

Section Exercises

connect

9.1 If you repeated a hypothesis test 1,000 times (in other words, 1,000 different samples from the same population), how many times would you expect to commit a Type I error, assuming the null hypothesis were true, if (a) $\alpha = .05$; (b) $\alpha = .01$; or (c) $\alpha = .001$?

9.2 Define Type I and Type II errors for each scenario, and identify the cost(s) of each type of error.

   a. A 25-year-old ER patient in Minneapolis complains of chest pain. Heart attacks in 25-year-olds are rare, and beds are scarce in the hospital. The null hypothesis is that there is no heart attack (perhaps muscle pain due to shoveling snow).

   b. Approaching O'Hare for landing, a British Air flight from London has been in a holding pattern for 45 minutes due to bad weather. Landing is expected within 15 minutes. The flight crew could declare an emergency and land immediately, but an FAA investigation would be launched, and other flights might be endangered. The null hypothesis is that there is enough fuel to stay aloft for 15 more minutes.

   c. You are trying to finish a lengthy statistics report and print it for your evening class. Your color printer is very low on ink, and you just have time to get to Staples for a new cartridge. But it is snowing and you need every minute to finish the report. The null hypothesis is that you have enough ink.

9.3 A firm decides to test its employees for illegal drugs. (a) State the null and alternative hypotheses. (b) Define Type I and II errors. (c) What are the consequences of each type of error, and to whom?

9.4 A hotel installs smoke detectors with adjustable sensitivity in all public guest rooms. (a) State the null and alternative hypotheses. (b) Define Type I and II errors. (c) What are the consequences of each type of error, and to whom?

9.5 What is the consequence of a false negative in an inspection of your car's brakes? *Hint:* The null hypothesis is the status quo (things are OK).

9.6 What is the consequence of a false positive in a weekly inspection of a nuclear plant's cooling system? *Hint:* The null hypothesis is the status quo (things are OK).

9.3 DECISION RULES AND CRITICAL VALUES

LO 9-3

Formulate null and alternative hypotheses for μ or π.

A **statistical hypothesis** is a statement about the value of a population parameter that we are interested in. For example, the parameter could be a mean, a proportion, or a variance. A **hypothesis test** is a decision between two competing, mutually exclusive, and collectively exhaustive hypotheses about the value of the parameter.

The hypothesized value of the parameter is the center of interest. For example, if the true value of μ is 5, then the sample mean should not differ greatly from 5. We rely on our knowledge of the *sampling distribution* and the *standard error of the estimate* to decide if the sample estimate is far enough away from 5 to contradict the assumption that $\mu = 5$. We can calculate the likelihood of an observed sample outcome. If the sample outcome is very unlikely, we would reject the claimed mean $\mu = 5$.

The null hypothesis states a benchmark value that we denote with the subscript "0," as in μ_0 or π_0. The hypothesized value μ_0 or π_0 does not come from a sample but is based on past performance, an industry standard, a target, or a product specification.

> ## Where Do We Get μ_0 (or π_0)?
>
> For a mean (or proportion), the value of μ_0 (or π_0) that we are testing is a *benchmark* based on past experience, an industry standard, a target, or a product specification. The value of μ_0 (or π_0) does *not* come from a sample.

One-Tailed and Two-Tailed Tests

For a mean, the null hypothesis H_0 states the value(s) of μ_0 that we will try to reject. There are three possible alternative hypotheses:

Left-Tailed Test	*Two-Tailed Test*	*Right-Tailed Test*
$H_0: \mu \geq \mu_0$	$H_0: \mu = \mu_0$	$H_0: \mu \leq \mu_0$
$H_1: \mu < \mu_0$	$H_1: \mu \neq \mu_0$	$H_1: \mu > \mu_0$

The application will dictate which of the three alternatives is appropriate. The *direction of the test* is indicated by which way the inequality symbol points in H_1:

$<$ indicates a **left-tailed test**

$\neq$ indicates a **two-tailed test**

$>$ indicates a **right-tailed test**

Example 9.1

Right-Tailed Test for μ

EPA guidelines for the maximum safe level of radon (a naturally occurring radioactive gas) in a home is 4.0 pCi/L (picocuries per liter of air). When a home is tested, a number of measurements are taken and averaged. A homeowner would be concerned if the average radon level is too high. In this case, the homeowner would choose a *right-tailed test*. The inequality in H_1 indicates the tail of the test.

$H_0: \mu \leq 4.0$ pCiL $\Leftarrow$ Assume H_0 is true unless evidence says otherwise

$H_1: \mu > 4.0$ pCiL $\Leftarrow$ The $>$ inequality in H_1 points right (right-tailed test)

The mean radon level is assumed safe (i.e., H_0 is assumed to be true) unless the sample mean indicates an average radon level too far *above* 4.0 to be due to chance variation.

Example 9.2

Left-Tailed Test for μ

The Diehard Platinum automobile battery has a claimed CCA (cold cranking amperes) of 880 amps. A consumer testing agency chooses a sample of several batteries and finds the average CCA for the batteries tested. The consumer testing agency would mainly be concerned if the battery delivers less than the claimed CCA, so it would choose a *left-tailed test*. The inequality in H_1 indicates the tail of the test.

$H_0: \mu \geq 880$ CCA $\Leftarrow$ Assume H_0 is true unless evidence says otherwise

$H_1: \mu < 880$ CCA $\Leftarrow$ The $<$ inequality in H_1 points left (left-tailed test)

The CCA mean is assumed at or above 880 amps (i.e., H_0 is assumed to be true) unless the sample mean indicates that CCA is too far *below* 880 to be due to chance variation.

The width of a sheet of standard-size copier paper should be $\mu = 216$ mm (i.e., 8.5 inches). There is variation in paper width due to the nature of the production process, so the width of a sheet of paper is a random variable. Samples are taken from the production process, and the mean width is calculated. If the paper is too narrow, the pages might not be well centered in the sheet feeder. If the pages are too wide, sheets could jam in the feeder or paper trays. Either violation poses a quality problem, so the manufacturer might choose a *two-tailed test*.

$H_0: \mu = 216$ mm ⇐ Assume H_0 is true unless evidence says otherwise
$H_1: \mu \neq 216$ mm ⇐ The $\neq$ in H_1 points to both tails (two-tailed test)

If the null hypothesis is rejected, action is required to adjust the manufacturing process. The action depends on the severity of the departure from H_0.

Example 9.3

Two-Tailed Test for μ

Decision Rule

When performing a statistical hypothesis test, we compare a sample statistic to the hypothesized value of the population parameter stated in the null hypothesis. Extreme outcomes occurring in the left tail would cause us to reject the null hypothesis in a left-tailed test; extreme outcomes occurring in the right tail would cause us to reject the null hypothesis in a right-tailed test. Extreme values in *either* the left or right tail would cause us to reject the null hypothesis in a two-tailed test.

We rely on our knowledge of the *sampling distribution* and the *standard error of the estimate* to decide if the sample statistic is far enough away from μ_0 to contradict the assumption that $\mu = \mu_0$. The area under the sampling distribution curve that defines an extreme outcome is called the **rejection region**. You may visualize the level of significance (α) as an area in the tail(s) of a distribution (e.g., normal) far enough from the center that it represents an unlikely outcome if our null hypothesis is true. We will calculate a **test statistic** that measures the difference between the sample statistic and the hypothesized parameter. A test statistic that falls in the shaded region will cause rejection of H_0, as illustrated in Figure 9.2. The area of the nonrejection region (white area) is $1 - \alpha$.

LO 9-4

Explain decision rules, critical values, and rejection regions.

Figure 9.2

Tests for $H_0: \mu = \mu_0$

Left-Tailed Test   Two-Tailed Test   Right-Tailed Test

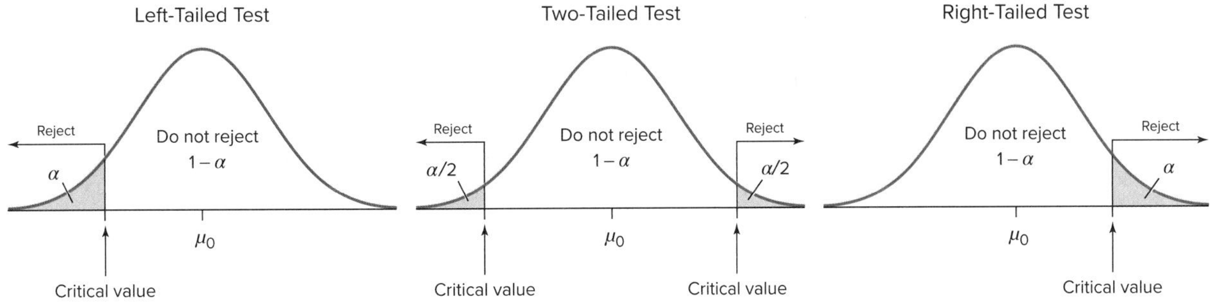

Critical Value

The **critical value** is the boundary between the two regions (reject H_0, do not reject H_0). The **decision rule** states what the critical value of the test statistic would have to be in order to reject H_0 at the chosen level of significance (α). The decision maker specifies the value of α for the test (common choices for α are .10, .05, or .01). The level of significance is usually expressed as a percent (e.g., 10 percent, 5 percent, or 1 percent). A small value of α is desirable to ensure a low probability of Type I error. For example, if we specify a decision rule based on $\alpha = .05$, we would expect to commit a Type I error about 5 times in 100 samples. In a two-tailed test, the risk is split with $\alpha/2$ in each tail because there are two ways to reject H_0. For example, in a two-tailed test using $\alpha = .10$, we would put $\alpha/2 = .05$ in each tail. We can look up the critical value from a table or from Excel (e.g., z or t).

By choosing a small α (say $\alpha = .01$), the decision maker can make it harder to reject the null hypothesis. By choosing a larger α (say $\alpha = .05$), it is easier to reject the null hypothesis. This raises the possibility of manipulating the decision. For this reason, the choice of α should precede the calculation of the test statistic, thereby minimizing the temptation to select α so as to favor one conclusion over the other.

In a left-tailed or right-tailed test, we can only test *one* value of the hypothesized parameter at a time, so we test the null hypothesis ($H_0: \mu \leq \mu_0$ or $H_0: \mu \geq \mu_0$) *only* at the point of equality $\mu = \mu_0$. If we reject $\mu = \mu_0$ in favor of the alternative, we implicitly reject the *entire class* of H_0 possibilities. For example, suppose the sample mean in the radon test is far enough above 4.0 to cause us to reject the null hypothesis $H_0: \mu = 4.0$ in favor of the alternative hypothesis $H_1: \mu > 4.0$. The same sample would also permit rejection of any value of μ *less than* $\mu = 4.0$, so we actually can reject $H_0: \mu \leq 4.0$.

In quality control, any deviation from specifications indicates that something may be wrong, so a two-tailed test is common. If the decision maker has no *a priori* reason to expect rejection in one direction, it is reasonable to use a two-tailed test. Rejection in a two-tailed test guarantees rejection in a one-tailed test at the same α.

When the consequences of rejecting H_0 are asymmetric or where one tail is of special importance to the researcher, we might perform a one-tailed test. For example, suppose that a machine is supposed to bore holes with a 3.5-mm diameter in a piece of sheet metal. Although any deviation from 3.5 mm is a violation of the specification, the consequences of rejecting H_0 may be different. Suppose an attachment pin is to be inserted into the hole. If the hole is too small, the pin cannot be inserted, but the metal piece might be reworked to enlarge the hole so the pin does fit. On the other hand, if the hole is too large, the pin will fit too loosely and may fall out. The piece may have to be discarded because an oversized hole cannot be made smaller. This is illustrated in Figure 9.3.

Figure 9.3

Asymmetric Effects of Nonconformance

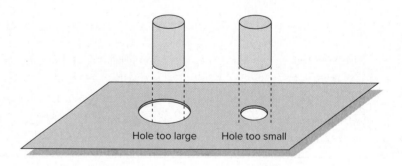

Hole too large   Hole too small

Section Exercises

9.7 A manufacturer claims that its compact fluorescent bulbs contain an average of 2.5 mg of mercury. Write the hypotheses for a two-tailed test, using the manufacturer's claim about the mean as the null hypothesis.

9.8 Noodles & Company is interested in testing whether its new menu design helps reduce the average order time for its customers. Suppose that the average order time prior to the introduction of its new menu was 1.2 minutes. Write the hypotheses for a left-tailed test, using its previous average order time as the claim about the mean for the null hypothesis.

9.9 The Scottsdale fire department aims to respond to fire calls in 4 minutes or less, on average. State the hypotheses you would use if you had reason to believe that the fire department's claim is not being met. *Hint:* Remember that sample data are used as evidence *against* the null, not to prove the null is true.

9.10   The average age of a part-time seasonal employee at a Vail Resorts ski mountain has historically been 37 years. State the hypotheses one would use to test whether this average has decreased since the last season.

9.11   Sketch a diagram of the decision rule for each pair of hypotheses.

   a. $H_0: \mu \geq 80$ versus $H_1: \mu < 80$
   b. $H_0: \mu = 80$ versus $H_1: \mu \neq 80$
   c. $H_0: \mu \leq 80$ versus $H_1: \mu > 80$

9.12   The Ball Corporation's aluminum can manufacturing facility in Fort Atkinson, Wisconsin, wants to use a sample to perform a two-tailed test to see whether the mean incoming metal thickness is at the target of 0.2731 mm. A deviation in either direction would pose a quality concern. State the hypotheses it should test.

9.4   TESTING A MEAN: KNOWN POPULATION VARIANCE

We will first explain how to test a population mean, μ. The sample statistic used to estimate μ is the random variable $\overline{X}$. The sampling distribution for $\overline{X}$ depends on whether the population variance σ^2 is known. We begin with the case of known σ^2. We learned in Chapter 8 that the sampling distribution of $\overline{X}$ will be a normal distribution provided that we have a normal population (or, by the Central Limit Theorem, if the sample size is large). In Section 9.5 we will turn to the more common case when σ^2 is unknown.

LO   9-5

Perform a hypothesis test for a mean with known σ using *z*.

Test Statistic

A *test statistic* measures the difference between a given sample mean $\overline{x}$ and a benchmark μ_0 in terms of the standard error of the mean. The test statistic is the "standardized score" of the sample statistic. When testing μ with a known σ, the test statistic is a *z* score. Once we have collected our sample, we calculate a value of the test statistic using the sample mean and then compare it against the critical value of *z*. We will refer to the calculated value of the test statistic as z_{calc}.

Test Statistic for a Mean: Known σ

Sample mean Hypothesized mean

$$z_{calc} = \frac{\overline{x} - \mu_0}{\sigma_{\overline{x}}} = \frac{\overline{x} - \mu_0}{\frac{\sigma}{\sqrt{n}}} \qquad (9.1)$$

Standard error of the sample mean

If the true mean of the population is μ_0, then the value of a particular sample mean $\overline{x}$ calculated from our sample should be near μ_0, and therefore the test statistic should be near zero.

Critical Value

The test statistic is compared with a *critical value* from a table. The critical value is the boundary between two regions (reject H_0, do not reject H_0) in the decision rule. For a two-tailed test (but *not* for a one-tailed test), the hypothesis test is equivalent to asking whether the confidence interval for μ includes zero. In a two-tailed test, half the risk of Type I error (i.e., $\alpha/2$) goes in each tail, as shown in Table 9.2, so the z-values are the same as for a confidence interval. You can verify these z-values from Excel.

Table 9.2

Some Common z-Values

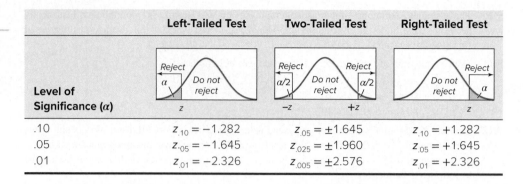

Level of Significance (α)	Left-Tailed Test	Two-Tailed Test	Right-Tailed Test
.10	$z_{.10} = -1.282$	$z_{.05} = \pm 1.645$	$z_{.10} = +1.282$
.05	$z_{.05} = -1.645$	$z_{.025} = \pm 1.960$	$z_{.05} = +1.645$
.01	$z_{.01} = -2.326$	$z_{.005} = \pm 2.576$	$z_{.01} = +2.326$

Example 9.4

Paper Manufacturing

The Hammermill Company produces paper for laser printers. Standard paper width is 216 mm, or 8.5 inches. Suppose that the actual width is a random variable that is normally distributed with a target average of 216 mm and a known standard deviation of 0.023 mm based on the manufacturing technology currently in use. Variation arises during manufacturing because of slight differences in the paper stock, vibration in the rollers and cutting tools, and wear and tear on the equipment. The cutters can be adjusted if the paper width drifts from the target mean. Suppose that a quality control inspector chooses 50 sheets at random and measures them with a precise instrument, obtaining a sample mean width of 216.0070 mm. Using a 5 percent level of significance ($\alpha = .05$), does this sample show that the paper average width is greater than the target? 🖾 **Paper**

Step 1: State the Hypotheses   The question indicates a right-tailed test, so the hypotheses would be

H_0: $\mu \leq 216$ mm (product mean is not greater than target width)

H_1: $\mu > 216$ mm (product mean is greater than the target width)

From the null hypothesis, we see that $\mu_0 = 216$ mm, which is the target width.

Step 2: Specify the Decision Rule   We use the *level of significance* to find the *critical value* of the z statistic that determines the threshold for rejecting the null hypothesis to be $\alpha = .05$. The critical value of z that accomplishes this is $z_{.05} = 1.645$. As illustrated in Figure 9.4, the decision rule is

Reject H_0 if $z_{\text{calc}} > 1.645$

Otherwise, do not reject H_0

Figure 9.4

Right-Tailed z Test for $\alpha = .05$

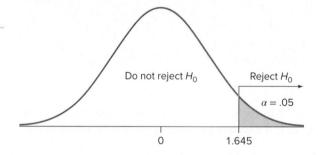

Step 3: Collect Sample Data and Calculate the Test Statistic   If H_0 is true, then the test statistic should be near 0 because $\bar{x}$ should be near μ_0. The value of the test statistic is

$$z_{\text{calc}} = \frac{\bar{x} - \mu_0}{\dfrac{\sigma}{\sqrt{n}}} = \frac{216.0070 - 216.0000}{\dfrac{0.0230}{\sqrt{50}}} = \frac{0.0070}{0.00325269} = 2.152$$

Step 4: Make the Decision   The test statistic falls in the right rejection region, so we reject the null hypothesis H_0: $\mu \leq 216$ and conclude the alternative hypothesis H_1: $\mu > 216$ at the 5 percent level of significance. Although the difference is slight, it is statistically significant.

Step 5: Take Action   Now that we have concluded that the process is producing paper with an average width *greater* than the target, it is time to adjust the manufacturing process to bring the average width back to target. Our course of action could be to readjust the machine settings, or it could be time to resharpen the cutting tools. At this point it is the responsibility of the process engineers to determine the best course of action.

p-Value Method

LO 9-6

Use tables or Excel to find the *p*-value in tests of μ.

The critical value method described above requires that you specify your rejection criterion in terms of the test statistic before you take a sample. The **p-value method** is a more flexible approach that is often preferred by statisticians over the critical value method. It requires that you express the strength of your evidence (i.e., your sample) against the null hypothesis in terms of a probability. The *p*-value is a direct measure of the likelihood of the observed sample under H_0. The *p*-value answers the following question: If the null hypothesis is true, what is the probability that we would observe our particular sample mean (or one even farther away from μ_0)? The *p*-value gives us more information than a test using one particular value of α because the observer can choose any α that is appropriate for the problem.

We compare the *p*-value with the level of significance. If the *p*-value is smaller than α, the sample contradicts the null hypothesis, and so we reject H_0. For a right-tailed test, the decision rule using the *p*-value approach is stated as

Reject H_0 if $P(Z > z_{calc}) < \alpha$; otherwise, fail to reject H_0.

Whether we use the critical value approach or the *p*-value approach, our decision about the null hypothesis will be the same.

What Is a *p*-Value?

A sample statistic is a random variable that may differ from the hypothesized value merely by chance, so we do not expect the sample to agree *exactly* with H_0. The *p*-value is the probability of obtaining a test statistic as extreme as the one observed, assuming that the null hypothesis is true. A large *p*-value (near 1.00) tends to support H_0, while a small *p*-value (near 0.00) tends to contradict H_0. If the *p*-value is less than the chosen level of significance (α), then we conclude that the null hypothesis is false.

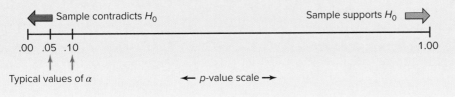

In order to calculate the *p*-value, we first need to calculate the test statistic z_{calc}. In this example, $z_{calc} = 2.152$, so the *p*-value for a right-tailed test is $P(Z > 2.152)$. The direction of the inequality in the *p*-value is the same as in the alternative hypothesis: H_1: $\mu > 216$ mm.

To find the *p*-value, we can use Excel's function =NORM.S.DIST(2.152,1) to obtain the left-tail area for the cumulative Z distribution (see Figure 9.5). Because $P(Z < 2.152) = .9843$, the right-tail area is $P(Z > 2.152) = 1 - .9843 = .0157$. This is the *p*-value for the right-tailed test, as illustrated in Figure 9.5. The *p*-value of .0157 indicates that in a right-tailed test, a test statistic of $z_{calc} = 2.152$ (or a more extreme test statistic) would happen by chance about 1.57 percent of the time if the null hypothesis were true.

Figure 9.5

p-Value for a Right-Tailed Test with $z_{calc} = 2.152$, Using Excel

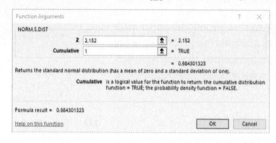
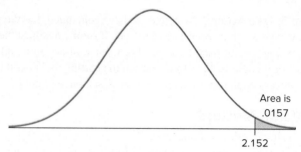

Area is
.0157

2.152

Source: Microsoft Excel.

We also could obtain the *p*-value from Appendix C-2, which shows cumulative standard normal areas less than *z*, as illustrated in Table 9.3. The cumulative area is not exactly the same as Excel because Appendix C-2 requires that we round the test statistic to two decimals ($z = 2.15$).

Table 9.3

Finding the *p*-Value for *z* = 2.15 in Appendix C-2

z	.00	.01	.02	.03	.04	.05	.06	.07	.08	.09
−3.7	.00011	.00010	.00010	.00010	.00009	.00009	.00008	.00008	.00008	.00008
−3.6	.00016	.00015	.00015	.00014	.00014	.00013	.00013	.00012	.00012	.00011
−3.5	.00023	.00022	.00022	.00021	.00020	.00019	.00019	.00018	.00017	.00017
.	.	.	.	.	.	.	.	.	.	.
.	.	.	.	.	.	.	.	.	.	.
2.0	.97725	.97778	.97831	.97882	.97932	.97982	.98030	.98077	.98124	.98169
2.1	.98214	.98257	.98300	.98341	.98382	.98422	.98461	.98500	.98537	.98574
2.2	.98610	.98645	.98679	.98713	.98745	.98778	.98809	.98840	.98870	.98899

Two-Tailed Test

In this case, the manufacturer decided that a two-tailed test would be more appropriate because the objective is to detect a deviation from the desired mean in *either* direction.

Step 1: State the Hypotheses   For a two-tailed test, the hypotheses are

H_0: $\mu = 216$ mm (product mean is on target)

H_1: $\mu \neq 216$ mm (product mean is not on target)

Step 2: Specify the Decision Rule   We will use the same $\alpha = .05$ as in the right-tailed test. But for a two-tailed test, we split the risk of Type I error by putting $\alpha/2 = .05/2 = .025$ in each tail. For $\alpha = .05$ in a two-tailed test, the critical value is $z_{.025} = \pm 1.96$, so the decision rule is

Reject H_0 if $z_{calc} > +1.96$ or if $z_{calc} < -1.96$

Otherwise, do not reject H_0

The decision rule is illustrated in Figure 9.6.

Step 3: Calculate the Test Statistic   The test statistic is *unaffected by the hypotheses or the level of significance.* The value of the test statistic is the same as for the one-tailed test:

$$z_{calc} = \frac{\bar{x} - \mu_0}{\frac{\sigma}{\sqrt{n}}} = \frac{216.0070 - 216.0000}{\frac{0.0230}{\sqrt{50}}} = \frac{0.0070}{0.0032527} = 2.152$$

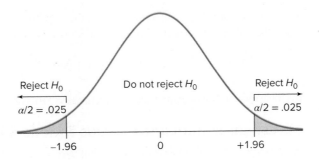

Figure 9.6

Two-Tailed *z* Test
for $\alpha = .05$

Step 4: Make the Decision   Because the test statistic falls in the right tail of the rejection region, we reject the null hypothesis $H_0: \mu = 216$ and conclude $H_1: \mu \neq 216$ at the 5 percent level of significance. Another way to say this is that the sample mean *differs significantly* from the desired target width at $\alpha = .05$ in a two-tailed test.

Step 5: Take Action   An adjustment is needed, such as changing the cutting tool settings. Now it is up to the process engineers to choose the best course of action.

Using the *p*-Value Approach

In a two-tailed test, the decision rule using the *p*-value is the same as in a one-tailed test.

Reject H_0 if *p*-value $< \alpha$. Otherwise, do not reject H_0.

The difference between a one-tailed and a two-tailed test is how we obtain the *p*-value. Because we allow rejection in either the left or the right tail in a two-tailed test, the level of significance, α, is divided equally between the two tails to establish the rejection region. In order to fairly evaluate the *p*-value against α, we must now double the tail area. The *p*-value in this two-tailed test is $2 \times P(z_{calc} > 2.152) = 2 \times .0157 = .0314$ (see Figure 9.7). This says that in a two-tailed test, a result as extreme as 2.152 would arise about 3.14 percent of the time by chance alone *if the null hypothesis were true.* The *p*-value method provides more information than a simple "yes-no" decision.

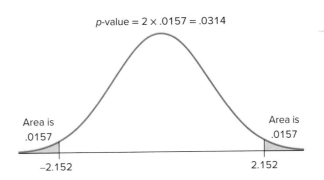

Figure 9.7

Two-Tailed *p*-Value for
$z = 2.152$

Although the sample mean 216.007 might seem very close to 216, it is more than two standard deviations from the desired mean. This example shows that even a small difference can be significant. It all depends on σ and n, that is, on the standard error of the mean in the denominator of the test statistic. In this case, there is a high degree of precision in the manufacturing process ($\sigma = 0.023$ is very small), so the standard error (and hence the allowable variation) is extremely small. Such a tiny difference in means would not be noticeable to consumers, but stringent quality control standards are applied to ensure that no shipment goes out with any noticeable nonconformance.

Analogy to Confidence Intervals

A two-tailed hypothesis test at the 5 percent level of significance ($\alpha = .05$) is equivalent to asking whether the 95 percent confidence interval for the mean includes the hypothesized mean. If the confidence interval includes the hypothesis mean $H_0: \mu = 216$, then we cannot reject the null hypothesis. In this case, the 95 percent confidence interval would be

$$\bar{x} \pm z_{\alpha/2} \frac{\sigma}{\sqrt{n}} \quad \text{or} \quad 216.0070 \pm 1.960 \frac{0.0230}{\sqrt{50}} \quad \text{or} \quad [216.001, 216.013]$$

Because this confidence interval does not include 216, we reject the null hypothesis $H_0: \mu = 216$.

Statistical Significance versus Practical Importance

Suppose that a redesigned engine could improve a truck's mpg rating by 0.5 mpg but requires spending $1.2 billion on new plant and equipment. Is it worth doing? Such a decision requires understanding not only statistical **significance** but also the **importance** of the potential improvement: the magnitude of the effect and its implications for product durability, customer satisfaction, budgets, cash flow, and staffing.

Because the standard error of most sample estimators approaches zero as sample size increases (if they are consistent estimators), even a small difference between the sample statistic and the hypothesized parameter may be significant if the sample size is large enough. Researchers who deal with large samples must expect "significant" effects, even when an effect is too slight to have practical importance. Is an improvement of 0.2 mpg in fuel economy *important* to Toyota buyers? Is a 0.2 percent loss of market share *important* to Hertz? Is a 15-minute increase in laptop battery life *important* to Dell customers?

Such questions require a cost/benefit calculation. Because resources are always scarce, a dollar spent on a quality improvement has an opportunity cost (the forgone alternative). If we spend money to make a certain product improvement, then some other project may have to be shelved. We can't do everything, so we must ask whether a proposed product improvement is the best use of our scarce resources. These are questions that must be answered by experts in medicine, product safety, or engineering rather than by statisticians.

Section Exercises

McGraw Hill **connect**

9.13 Find the z_{calc} test statistic for each hypothesis test.

    a. $\bar{x} = 242, \mu_0 = 230, \sigma = 18, n = 20$
    b. $\bar{x} = 3.44, \mu_0 = 3.50, \sigma = 0.24, n = 40$
    c. $\bar{x} = 21.02, \mu_0 = 20.00, \sigma = 2.52, n = 30$

9.14 Use Excel to find the critical value of z for each hypothesis test.

    a. 10 percent level of significance, two-tailed test
    b. 1 percent level of significance, right-tailed test
    c. 5 percent level of significance, left-tailed test

9.15 Use Excel to find the critical value of z for each hypothesis test.

    a. $\alpha = .05$, two-tailed test
    b. $\alpha = .10$, right-tailed test
    c. $\alpha = .01$, left-tailed test

9.16 Find the z_{calc} test statistic for each hypothesis test.

    a. $\bar{x} = 423, \mu_0 = 420, \sigma = 6, n = 9$
    b. $\bar{x} = 8{,}330, \mu_0 = 8{,}344, \sigma = 48, n = 36$
    c. $\bar{x} = 3.102, \mu_0 = 3.110, \sigma = .250, n = 25$

9.17 GreenBeam Ltd. claims that its compact fluorescent bulbs average no more than 3.50 mg of mercury. A sample of 25 bulbs shows a mean of 3.59 mg of mercury. (a) Write the hypotheses for a right-tailed test, using GreenBeam's claim as the null hypothesis about the mean. (b) Assuming a known standard deviation of 0.18 mg, calculate the z test statistic to test the manufacturer's claim. (c) At the 1 percent level of significance ($\alpha = .01$), does the sample exceed the manufacturer's claim? (d) Find the p-value.

9.18 The mean potassium content of a popular sports drink is listed as 140 mg in a 32-oz bottle. Analysis of 20 bottles indicates a sample mean of 139.4 mg. (a) Write the hypotheses for a two-tailed test of the claimed potassium content. (b) Assuming a known standard deviation of 2.00 mg, calculate the z test statistic to test the manufacturer's claim. (c) At the 10 percent level of significance ($\alpha = .10$), does the sample contradict the manufacturer's claim? (d) Find the p-value.

9.19   Calculate the test statistic and *p*-value for each sample.

    a. H_0: $\mu = 60$ versus H_1: $\mu \neq 60$, $\alpha = .025$, $\bar{x} = 63$, $\sigma = 8$, $n = 16$
    b. H_0: $\mu \geq 60$ versus H_1: $\mu < 60$, $\alpha = .05$, $\bar{x} = 58$, $\sigma = 5$, $n = 25$
    c. H_0: $\mu \leq 60$ versus H_1: $\mu > 60$, $\alpha = .05$, $\bar{x} = 65$, $\sigma = 8$, $n = 36$

9.20   Determine the *p*-value for each test statistic.

    a.   Right-tailed test, $z = +1.34$
    b.   Left-tailed test, $z = -2.07$
    c.   Two-tailed test, $z = -1.69$

9.21   Procyon Mfg. produces tennis balls. Weights are supposed to be normally distributed with a mean of 2.035 ounces and a standard deviation of 0.002 ounce. A sample of 25 tennis balls shows a mean weight of 2.036 ounces. At $\alpha = .025$ in a right-tailed test, is the mean weight heavier than it is supposed to be?

9.22   The mean arrival rate of flights at O'Hare Airport in marginal weather is 195 flights per hour with a historical standard deviation of 13 flights. To increase arrivals, a new air traffic control procedure is implemented. In the next 30 days of marginal weather, the mean arrival rate is 200 flights per hour. (a) Set up a right-tailed decision rule at $\alpha = .025$ to decide whether there has been a significant increase in the mean number of arrivals per hour. (b) Carry out the test and make the decision. Is it close? Would the decision be different if you used $\alpha = .01$? (c) What assumptions are you making, if any? 📷 **Flights**

210	215	200	189	200	213	202	181	197	199
193	209	215	192	179	196	225	199	196	210
199	188	174	176	202	195	195	208	222	221

9.23   An airline serves bottles of Galena Spring Water that are supposed to contain an average of 10 ounces. The filling process follows a normal distribution with process standard deviation 0.07 ounce. Twelve randomly chosen bottles had the weights shown below (in ounces). (a) Set up a two-tailed decision rule to detect quality control violations using the 5 percent level of significance. (b) Carry out the test. (c) What assumptions are you making, if any? 📷 **BottleFill**

    10.02 9.95 10.11 10.10 10.08 10.04 10.06 10.03 9.98 10.01 9.92 9.89

9.24   The Scottsdale fire department aims to respond to fire calls in 4 minutes or less, on average. Response times are normally distributed with a standard deviation of 1 minute. Would a sample of 18 fire calls with a mean response time of 4 minutes 30 seconds provide sufficient evidence to show that the goal is not being met at $\alpha = .01$? What is the *p*-value?

9.25   The lifespan of xenon metal halide arc-discharge bulbs for aircraft landing lights is normally distributed with a mean of 3,000 hours and a standard deviation of 500 hours. If a new ballast system shows a mean life of 3,515 hours in a test on a sample of 10 prototype new bulbs, would you conclude that the new lamp's mean life exceeds the current mean life at $\alpha = .01$? What is the *p*-value? (For more information, see www.xevision.com.)

9.26   Discuss the issues of *statistical significance* and *practical importance* in each scenario.

    a.   A process for producing I-beams of oriented strand board used as main support beams in new houses has a mean breaking strength of 2,000 lbs./ft. A sample of boards from a new process has a mean breaking strength of 2,150 lbs./ft. The improvement is statistically significant, but the per-unit cost is higher.
    b.   Under continuous use, the mean battery life in a certain cell phone is 45 hours. In tests of a new type of battery, the sample mean battery life is 46 hours. The improvement is statistically significant, but the new battery costs more to produce.
    c.   For a wide-screen HDTV LCD unit, the mean half-life (i.e., to lose 50 percent of its brightness) is 32,000 hours. A new process is developed. In tests of the new display, the sample mean half-life is 35,000 hours. The improvement is statistically significant, though the new process is more costly.

9.27   The target activation force of the buttons on a keyless entry clicker is 1.967 newtons. Variation exists in activation force due to the nature of the manufacturing process. A sample of 9 clickers showed a mean activation force of 1.88 newtons. The standard deviation is known to be 0.145 newton. Too much force makes the keys hard to click, while too little force means the keys might be clicked accidentally. Therefore, the manufacturer's quality control engineers use a two-tailed hypothesis test for samples taken from each production batch to detect excessive deviations in either direction. At $\alpha = .05$, does the sample indicate a significant deviation from the target?

LO 9-7

Perform a hypothesis test for a mean with unknown σ using t.

9.5 TESTING A MEAN: UNKNOWN POPULATION VARIANCE

If the population variance σ^2 must be estimated from the sample, the hypothesis testing procedure is modified. There is a loss of information when s replaces σ in the formulas, and it is no longer appropriate to use the normal distribution. However, the basic hypothesis testing steps are the same.

Using Student's t

When the population standard deviation σ is unknown (as it usually is) and the population may be assumed normal (or generally symmetric with no outliers), the test statistic follows the Student's t distribution with $n - 1$ degrees of freedom. Because σ is rarely known, we generally expect to use Student's t instead of z, as you saw for confidence intervals in the previous chapter.

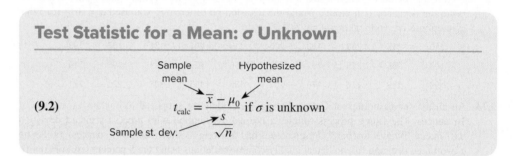

Test Statistic for a Mean: σ Unknown

(9.2)

$$t_{\text{calc}} = \frac{\overline{x} - \mu_0}{\dfrac{s}{\sqrt{n}}} \text{ if } \sigma \text{ is unknown}$$

Sample mean — Hypothesized mean — Sample st. dev.

Example 9.5

Hot Chocolate

In addition to its core business of bagels and coffee, Bruegger's Bagels also sells hot chocolate for the non-coffee crowd. Customer research shows that the ideal temperature for hot chocolate is 142°F ("hot" but not "too hot"). A random sample of 24 cups of hot chocolate is taken at various times, and the temperature of each cup is measured using an ordinary kitchen thermometer that is accurate to the nearest whole degree. **HotChoc**

140	140	141	145	143	144	142	140
145	143	140	140	141	141	137	142
143	141	142	142	143	141	138	139

The sample mean is 141.375 with a sample standard deviation of 1.99592. At $\alpha = .10$, does this sample evidence show that the true mean differs from 142?

Step 1: State the Hypotheses   We use a two-tailed test. The null hypothesis is in conformance with the desired standard.

$H_0: \mu = 142$ (mean temperature is correct)

$H_1: \mu \neq 142$ (mean temperature is incorrect)

Step 2: Specify the Decision Rule   For $\alpha = .10$, using the Excel function =T.INV.2T(α, $d.f.$) =T.INV.2T(0.10,23) = 1.714 gives the two-tailed critical value for $d.f. = n - 1 = 24 - 1 = 23$ degrees of freedom. The same value can be obtained from Appendix D, shown here in abbreviated form:

		Upper Tail Area			
d.f.	**.10**	**.05**	**.025**	**.01**	**.005**
1	3.078	6.314	12.706	31.821	63.657
2	1.886	2.920	4.303	6.965	9.925
3	1.638	2.353	3.182	4.541	5.841
⋮	⋮	⋮	⋮	⋮	⋮
21	1.323	1.721	2.080	2.518	2.831
22	1.321	1.717	2.074	2.508	2.819
23	1.319	1.714	2.069	2.500	2.807
24	1.318	1.711	2.064	2.492	2.797
25	1.316	1.708	2.060	2.485	2.787

We will reject H_0 if $t_{calc} > 1.714$ or if $t_{calc} < -1.714$, as illustrated in Figure 9.8.

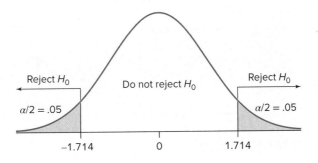

Figure 9.8

Two-Tailed Test for a Mean Using *t* for *d.f.* = 23

Step 3: Calculate the Test Statistic   Inserting the sample information, the test statistic is

$$t_{calc} = \frac{\bar{x} - \mu_0}{\dfrac{s}{\sqrt{n}}} = \frac{141.375 - 142}{\dfrac{1.99592}{\sqrt{24}}} = \frac{-0.6250}{0.40742} = -1.534$$

Step 4: Make the Decision   Because the test statistic lies within the range of chance variation, we cannot reject the null hypothesis H_0: $\mu = 142$.

Effect of α

How sensitive is our conclusion to the choice of level of significance? Table 9.4 shows several critical values of Student's *t*. At $\alpha = .20$ we could reject H_0, but not at the other α values shown. This table is not to suggest that experimenting with various α values is desirable but merely to illustrate that our decision is affected by our choice of α.

	$\alpha = .20$	$\alpha = .10$	$\alpha = .05$	$\alpha = .01$
Critical value	$t_{.10} = \pm 1.319$	$t_{.05} = \pm 1.714$	$t_{.025} = \pm 2.069$	$t_{.005} = \pm 2.807$
Decision	Reject H_0	Don't reject H_0	Don't reject H_0	Don't reject H_0

Table 9.4

Effect of α on the Decision (Two-Tailed *t* Test with *d.f.* = 23)

Using the *p*-Value

A more general approach favored by researchers is to find the *p*-value. After the *p*-value is calculated, different analysts can compare it to the level of significance (α) that is appropriate for the task. We want to determine the tail area less than $t = -1.534$ or greater than $t = +1.534$. However, from Appendix D, we can only get a range for the *p*-value. From Appendix D, we see that the two-tail *p*-value must lie between .20 and .10 (it's a two-tailed test, so we double the right-tail area). It is easier and more precise to use Excel's function =T.DIST.2T(t test statistic, degrees of freedom)=T.DIST.2T(1.534,23) to get the two-tailed *p*-value of .13867. The area of each tail is half that, or .06934, as shown in Figure 9.9. A sample mean as extreme in either tail would occur by chance about 139 times in 1,000 two-tailed tests if H_0 were true. Because the *p*-value $> \alpha$, we cannot reject H_0.

Figure 9.9

Two-Tailed *p*-Value for *t* = 1.534

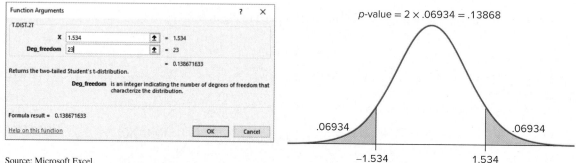

Source: Microsoft Excel.

Interpretation

It is doubtful whether a consumer could tell the difference in hot chocolate temperature within a few degrees of 142°F, so a tiny difference in means might lack *practical importance* even if it were *statistically significant*. Importance must be judged by management, not by the statistician.

In the hot chocolate example, there are no outliers and something of a bell shape, as shown in the dot plot below. The *t* test is reasonably robust to mild non-normality. However, outliers or extreme skewness can affect the test, just as when we construct confidence intervals.

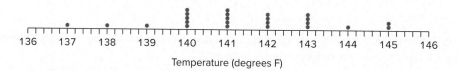

Temperature (degrees F)

Who Uses *p*-Values?

"Executives at Noodles & Company may not perform a statistical analysis themselves, but they do understand the *p*-value associated with the results of an analysis. The *p*-value allows us to objectively consider the data and statistical results when we make an important strategic decision."

Dave Boennighausen, chief executive officer at Noodles & Company

Confidence Interval versus Hypothesis Test

The two-tailed test at the 10 percent level of significance is equivalent to a two-tailed 90 percent confidence interval. If the confidence interval does not contain μ_0, we reject H_0.

For the hot chocolate, the sample mean is 141.375 with a sample standard deviation of 1.99592. Using Appendix D we find $t_{.05} = 1.714$, so the 90 percent confidence interval for μ is

$$\bar{x} \pm t_{\alpha/2} \frac{s}{\sqrt{n}} \quad \text{or} \quad 141.375 \pm (1.714)\frac{1.99592}{\sqrt{24}} \quad \text{or} \quad 141.375 \pm 0.6983$$

Because $\mu = 142$ lies within the 90 percent confidence interval [140.677, 142.073], we cannot reject the hypothesis H_0: $\mu = 142$ at $\alpha = .10$ in a two-tailed test. Many decisions can be handled either as hypothesis tests or using confidence intervals. The confidence interval has the appeal of providing a graphic feeling for the location of the hypothesized mean within the confidence interval, as shown in Figure 9.10.

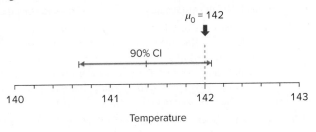

Figure 9.10

90 Percent Confidence Interval for μ

Large Samples

From Appendix D you can verify that when n is large, there is little difference between critical values of t and z (the last line in Appendix D, for $d.f. = \infty$). For this reason, it is unlikely that harm will result if you use z instead of t, as long as the sample size is not small. The test statistic is

$$z_{calc} = \frac{\bar{x} - \mu_0}{\dfrac{s}{\sqrt{n}}} \quad \text{(large sample, unknown } \sigma) \tag{9.3}$$

However, using z instead of t is not conservative because it will increase Type I error somewhat. Therefore, statisticians recommend that we always apply t when σ is unknown. We then can use Excel or Appendix D to get the critical value.

Section Exercises

Mc Graw Hill connect

9.28 Find the t_{calc} test statistic for each hypothesis test.

   a. $\bar{x} = 14.7$, $\mu_0 = 13.0$, $s = 1.8$, $n = 12$
   b. $\bar{x} = 241$, $\mu_0 = 250$, $s = 12$, $n = 8$
   c. $\bar{x} = 2,102$, $\mu_0 = 2,000$, $s = 242$, $n = 17$

9.29 Find the critical value of Student's t for each hypothesis test.

   a. 10 percent level of significance, two-tailed test, $n = 21$
   b. 1 percent level of significance, right-tailed test, $n = 9$
   c. 5 percent level of significance, left-tailed test, $n = 28$

9.30 Find the critical value of Student's t for each hypothesis test.

   a. Two-tailed test, $n = 18$, $\alpha = .05$
   b. Right-tailed test, $n = 15$, $\alpha = .10$
   c. Left-tailed test, $n = 31$, $\alpha = .01$

9.31 Find the t_{calc} test statistic for each hypothesis test.

   a. $\bar{x} = 347$, $\mu_0 = 349$, $s = 1.8$, $n = 9$
   b. $\bar{x} = 45$, $\mu_0 = 50$, $s = 12$, $n = 16$
   c. $\bar{x} = 4.103$, $\mu_0 = 4.004$, $s = 0.245$, $n = 25$

9.32 Estimate the p-value *as a range* using Appendix D (*not* Excel):

   a. $t = 1.457$, $d.f. = 14$, right-tailed test
   b. $t = 2.601$, $d.f. = 8$, two-tailed test
   c. $t = -1.847$, $d.f. = 22$, left-tailed test

9.33 Find the p-value using Excel (*not* Appendix D):

   a. $t = 1.457$, $d.f. = 14$, right-tailed test
   b. $t = 2.601$, $d.f. = 8$, two-tailed test
   c. $t = -1.847$, $d.f. = 22$, left-tailed test

9.34 Use Excel to find the *p*-value for each test statistics.

    a. Right-tailed test, $t = \pm 1.677$, $n = 13$
    b. Left-tailed test, $t = -2.107$, $n = 5$
    c. Two-tailed test, $t = -1.865$, $n = 34$

9.35 Calculate the test statistic and *p*-value for each sample. State the conclusion for the specified α.

    a. $H_0: \mu = 200$ versus $H_1: \mu \neq 200$, $\alpha = .025$, $\bar{x} = 203$, $s = 8$, $n = 16$
    b. $H_0: \mu \geq 200$ versus $H_1: \mu < 200$, $\alpha = .05$, $\bar{x} = 198$, $s = 5$, $n = 25$
    c. $H_0: \mu \leq 200$ versus $H_1: \mu > 200$, $\alpha = .05$, $\bar{x} = 205$, $s = 8$, $n = 36$

9.36 The manufacturer of an airport baggage scanning machine claims it can handle an average of 530 bags per hour. At $\alpha = .05$ in a left-tailed test, would a sample of 16 randomly chosen hours with a mean of 510 and a standard deviation of 50 indicate that the manufacturer's claim is overstated?

9.37 The manufacturer of Glo-More flat white interior latex paint claims one-coat coverage of 400 square feet per gallon on interior walls. A painter keeps careful track of 6 gallons and finds coverage (in square feet) of 360, 410, 380, 360, 390, 400. (a) At $\alpha = .10$ does this evidence contradict the claim? State your hypotheses and decision rule. (b) Is this conclusion sensitive to the choice of α? (c) Use Excel to find the *p*-value. Interpret it. (d) Discuss the distinction between importance and significance in this example. 📁 **Paint**

9.38 The average weight of a package of rolled oats is supposed to be at least 18 ounces. A sample of 18 packages shows a mean of 17.78 ounces with a standard deviation of 0.41 ounce. (a) At the 5 percent level of significance, is the true mean smaller than the specification? Clearly state your hypotheses and decision rule. (b) Is this conclusion sensitive to the choice of α? (c) Use Excel to find the *p*-value. Interpret it.

9.39 A consumer survey found that the mean purchase price of a smartphone device (such as an iPhone or Galaxy) in 2019 was $818. In 2020, a random sample of 20 business managers who owned a smartphone device showed a mean purchase price of $841 with a sample standard deviation of $23. (a) At $\alpha = .05$, has the mean purchase price increased? State the hypotheses and decision rule clearly. (b) Use Excel to find the *p*-value and interpret it.

9.40 The average age of a part-time seasonal employee at a Vail Resorts ski mountain has historically been 37 years. A random sample of 50 part-time seasonal employees in 2020 had a sample mean age of 38.5 years with a sample standard deviation equal to 16 years. At the 10 percent level of significance, does this sample show that the average age was different in 2020?

9.41 The number of entrees purchased in a single order at a Noodles & Company restaurant has had a historical average of 1.60 entrees per order. On a particular Saturday afternoon, a random sample of 40 Noodles orders had a mean number of entrees equal to 1.80 with a standard deviation equal to 1.11. At the 5 percent level of significance, does this sample show that the average number of entrees per order was greater than expected?

9.42 A small dealership leased 21 Subaru Outbacks on 2-year leases. When the cars were returned at the end of the lease, the mileage was recorded (see below). Is the dealer's mean significantly greater than the national average of 30,000 miles for 2-year leased vehicles, using the 10 percent level of significance? 📁 **Mileage**

40,060	24,960	14,310	17,370	44,740	44,550	20,250
33,380	24,270	41,740	58,630	35,830	25,750	28,910
25,090	43,380	23,940	43,510	53,680	31,810	36,780

9.43 At Oxnard University, a sample of 18 senior accounting majors showed a mean cumulative GPA of 3.35 with a standard deviation of 0.25. (a) At $\alpha = .05$ in a two-tailed test, does this differ significantly from 3.25 (the mean GPA for all business school seniors at the university)? (b) Use the sample to construct a 95 percent confidence interval for the mean. Does the confidence interval include 3.25? (c) Explain how the hypothesis test and confidence interval are equivalent.

Mini Case 9.2

Beauty Products and Small Business

Lisa has been working at a beauty counter in a department store for 5 years. In her spare time she's also been creating lotions and fragrances using all-natural products. After receiving positive feedback from her friends and family about her beauty products, Lisa decides to open her own store. Lisa knows that convincing a bank to help fund her new business will require more than a few positive testimonials from family. Based on her experience working at the department store, Lisa believes women in her area spend more than the national average on fragrance products. This fact could help make her business successful.

Lisa would like to be able to support her belief with data to include in a business plan proposal that she would then use to obtain a small business loan. Lisa took a business statistics course while in college and decides to use the hypothesis testing tool she learned. After conducting research, she learns that the national average spending by women on fragrance products is \$59 every 3 months. The hypothesis test is based on this survey result:

$H_0: \mu \le \$59$

$H_1: \mu > \$59$

In other words, she will assume the average spending in her town is the same as the national average *unless she has strong evidence that says otherwise.* Lisa takes a random sample of 25 women and finds that the sample mean $\bar{x}$ is \$68 and the sample standard deviation s is \$15. Lisa uses a t statistic because she doesn't know the population standard deviation. Her calculated t statistic is

$$t_{calc} = \frac{68 - 59}{\frac{15}{\sqrt{25}}} = 3.00 \quad \text{with 24 degrees of freedom}$$

Using the Excel formula =T.DIST.RT(3,24), Lisa finds that the right-tail p-value is .003103. This p-value is quite small and she can safely reject her null hypothesis. Lisa now has strong evidence to conclude that over a 3-month period, women in her area spend more than \$59 on average.

Lisa also would like to include an estimate for the average amount women in her area *do* spend. Calculating a confidence interval would be her next step. Lisa chooses a 95 percent confidence level and finds the t value to use in her calculations by using the Excel formula =T.INV(0.05,24). The result is $t_{.025} = 2.0639$. Her 95 percent confidence interval for μ is

$$\bar{x} \pm t_{\alpha/2}\frac{s}{\sqrt{n}} \quad \text{or} \quad \$68 \pm 2.0639\frac{15}{\sqrt{25}} \quad \text{or} \quad \$68 \pm \$6.19 \quad \text{or} \quad [61.81, 74.19]$$

Lisa's business plan proposal can confidently claim that women in her town spend more than the national average on fragrance products and that she estimates the average spending is between \$62 and \$74 every 3 months. Hopefully the bank will see not only that Lisa creates excellent beauty products, but also that she is a smart businessperson!

9.6   TESTING A PROPORTION

Proportions are used frequently in business situations, and collecting proportion data is straightforward. It is easier for customers to say whether they like or dislike this year's new automobile color than it is for customers to quantify their degree of satisfaction with the new color. Also, many business performance indicators such as market share, employee retention rates, and employee accident rates are expressed as proportions.

LO 9-8

Perform a hypothesis test for a proportion and find the p-value.

Example 9.6

Testing a Proportion

Annually, U.S. airlines mishandle 6 out of every 1,000 bags checked for air travel. One airline, FlyMoon, found that 54 percent of its mishandled bag incidents happened while transferring baggage to a connecting flight. FlyMoon recently installed a radio-frequency identification (RFID) system with the goal of decreasing the proportion of mistakes due to transfer errors and ultimately reducing the overall proportion of lost bags. After operating its new system for several months, FlyMoon would like to know if the wireless system has been effective. FlyMoon could use a hypothesis test to answer this question. The benchmark for the null hypothesis is its proportion of mistakes due to transfer errors using its old system ($\pi_0 = .54$). The possible set of statistical hypotheses would be:

Left-Tailed Test	*Two-Tailed Test*	*Right-Tailed Test*
$H_0: \pi \geq .54$	$H_0: \pi = .54$	$H_0: \pi \leq .54$
$H_1: \pi < .54$	$H_1: \pi \neq .54$	$H_1: \pi > .54$

Which set of hypotheses would be most logical for FlyMoon to use? Because FlyMoon believes the RFID system will reduce the proportion of transfer errors, it might use a left-tailed test. It would assume there has been no improvement, unless its evidence shows otherwise. If it can reject H_0 in favor of H_1 in a left-tailed test, FlyMoon would be able to say that its data provide evidence that the proportion of transfer errors has decreased since the RFID tagging system was implemented.

The steps we follow for testing a hypothesis about a population proportion, π, are the same as the ones we follow for testing a mean. The difference is that we now calculate a sample proportion, p, to calculate the test statistic. We know from Chapter 8 that for a sufficiently large sample, the sample proportion can be assumed to follow a normal distribution. Our rule is to assume normality if $n\pi_0 \geq 10$ and $n(1 - \pi_0) \geq 10$. If we can assume a normal sampling distribution, then the test statistic would be the z-score. Recall that the sample proportion is

(9.4)
$$p = \frac{x}{n} = \frac{\text{number of successes}}{\text{sample size}}$$

The test statistic, calculated from sample data, is the difference between the sample proportion p and the hypothesized proportion π_0 divided by the *standard error of the proportion* (sometimes denoted σ_p):

Test Statistic for a Proportion

(9.5)
$$z_{calc} = \frac{p - \pi_0}{\sigma_p} = \frac{p - \pi_0}{\sqrt{\dfrac{\pi_0(1 - \pi_0)}{n}}}$$

The value of π_0 we are testing is a **benchmark**, such as past performance, an industry standard, or a product specification. The value of π_0 does *not* come from a sample.

Example 9.7

Return Policy

Retailers such as H&M, Office Depot, Uniqlo, and Dick's Sporting Goods are employing new technology to crack down on return fraud—customers who abuse their return and exchange policies. For example, some customers buy an outfit, wear it once or twice, and then return it. The Retail Equation, a California-based company, licenses software that tracks a shopper's record of bringing back items and alerts the retailer when a customer's behavior is deemed risky. The goal is to reduce the number of fraudulent returns, which should reduce the proportion of returns overall. In brick-and-mortar department stores, the historical return rate for merchandise is 13.0 percent. At one department store, after implementing the new software, there were 22 returns in a sample of 250 purchases. At $\alpha = .05$, does this sample indicate that the true return rate is below the historical rate?

Step 1: State the Hypotheses   The hypotheses are

H_0: $\pi \geq .13$ (return rate is the same or greater than the historical rate)
H_1: $\pi < .13$ (return rate has fallen below the historical rate)

Step 2: Specify the Decision Rule   For $\alpha = .05$ in a left-tailed test, the critical value is $z_{.05} = -1.645$, so the decision rule is

Reject H_0 if $z_{\text{calc}} < -1.645$
Otherwise, do not reject H_0

This decision rule is illustrated in Figure 9.11.

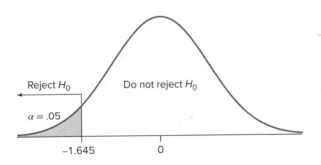

Figure 9.11

**Left-Tailed *z*
Test Using $\alpha = .05$**

Before using z we should check the normality assumption. To assume normality, we require that $n\pi_0 \geq 10$ and $n(1 - \pi_0) \geq 10$. Note that we use the hypothesized proportion π_0 (not p) to check normality because we are assuming H_0 to be the truth about the population. Inserting $\pi_0 = .13$ and $n = 250$, we see that these conditions are easily met: $(250)(.13) = 32.5$ and $(250)(1 - .13) = 217.5$.

Step 3: Calculate the Test Statistic   Because $p = x/n = 22/250 = .088$, the sample seems to favor H_1. But we will assume that H_0 is true and see if the test statistic contradicts this assumption. We test the hypothesis at $\pi = .13$. If we can reject $\pi = .13$ in favor of $\pi < .13$, then we implicitly reject the class of hypotheses $\pi \geq .13$. The test statistic is the difference between the sample proportion $p = x/n$ and the hypothesized parameter π_0 divided by the standard error of p:

$$z_{\text{calc}} = \frac{p - \pi_0}{\sqrt{\dfrac{\pi_0(1 - \pi_0)}{n}}} = \frac{.088 - .13}{\sqrt{\dfrac{.13(1 - .13)}{250}}} = \frac{-.042}{.02127} = -1.975$$

Step 4: Make the Decision   Because the test statistic falls in the left-tail rejection region, we reject H_0. We conclude that the return rate is less than .13 after implementing the new software.

Step 5: Take Action   The rate of returns seems to be reduced, so the store might want to try using the software at its other locations.

Calculating the p-Value

For our test statistic $z_{calc} = -1.975$, the p-value (.02413) can be obtained from Excel's cumulative standard normal =NORM.S.DIST(−1.975). Alternatively, if we round the test statistic to two decimals, we can use the cumulative normal table in Appendix C-2. Depending on how we round the test statistic, we might obtain two possible p-values, as shown in Table 9.5. Using the p-value, we reject H_0 at $\alpha = .05$, but the decision would be very close if we had used $\alpha = .025$. Figure 9.12 illustrates the p-value.

Table 9.5

Finding the p-Value for z = 1.975 in Appendix C-2

z	.00	.01	.02	.03	.04	.05	.06	.07	.08	.09
−3.7	.00011	.00010	.00010	.00010	.00009	.00009	.00008	.00008	.00008	.00008
−3.6	.00016	.00015	.00015	.00014	.00014	.00013	.00013	.00012	.00012	.00011
−3.5	.00023	.00022	.00022	.00021	.00020	.00019	.00019	.00018	.00017	.00017
.	.	.	.	.	.	.	.	.	.	.
.	.	.	.	.	.	.	.	.	.	.
−2.0	.02275	.02222	.02169	.02118	.02068	.02018	.01970	.01923	.01876	.01831
−1.9	.02872	.02807	.02743	.02680	.02619	.02559	.02500	.02442	.02385	.02330
−1.8	.03593	.03515	.03438	.03362	.03288	.03216	.03144	.03074	.03005	.02938

Figure 9.12

p-Value for a Left-Tailed Test with $z_{calc} = -1.975$

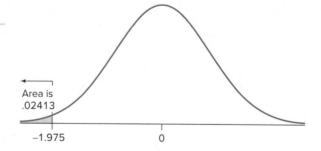

Area is
.02413

−1.975   0

The *smaller* the p-value, the more we are inclined to *reject* H_0. Does this seem backward? You might think a large p-value would be "more significant" than a small one. But the p-value is the likelihood of the sample if H_0 is true, so a small p-value makes us doubt H_0. The p-value is a direct measure of the level of significance at which we could reject H_0, so *a smaller p-value is more convincing*. For the left-tailed test, the p-value tells us that there is a .02413 probability of getting a sample proportion of .088 or less if the true proportion is .13; that is, such a sample would arise by chance only about 24 times in 1,000 tests if the null hypothesis is true. In our left-tailed test, we would reject H_0 because the p-value (.02413) is smaller than α (.05). In fact, we could reject H_0 at *any* α greater than .02413.

Two-Tailed Test

What if we used a two-tailed test? This might be appropriate if the objective is to detect a change in the return rate in *either* direction. In fact, two-tailed tests are used more often because rejection in a two-tailed test always implies rejection in a one-tailed test, other things being equal. The same sample can be used for either a one-tailed or two-tailed test. The type of hypothesis test is up to the statistician.

Step 1: State the Hypotheses   The hypotheses are

H_0: $\pi = .13$ (return rate is the same as the historical rate)

H_1: $\pi \neq .13$ (return rate is different from the historical rate)

Step 2: Specify the Decision Rule   For a two-tailed test, we split the risk of Type I error by putting $\alpha/2 = .05/2 = .025$ in each tail (as we would for a confidence interval). For $\alpha = .05$ in a two-tailed test, the critical value is $z_{.025} = 1.96$, so the decision rule is

Reject H_0 if $z_{\text{calc}} > +1.96$ or if $z_{\text{calc}} < -1.96$

Otherwise, do not reject H_0.

The decision rule is illustrated in Figure 9.13.

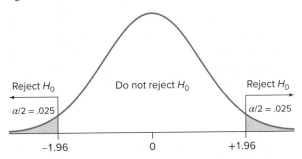

Figure 9.13

Two-Tailed *z* Test for $\alpha = .05$

Step 3: Calculate the Test Statistic   The test statistic is *unaffected by the hypotheses or the level of significance.* The value of the test statistic is the same as for the one-tailed test:

$$z_{\text{calc}} = \frac{p - \pi_0}{\sqrt{\dfrac{\pi_0(1 - \pi_0)}{n}}} = \frac{.088 - .13}{\sqrt{\dfrac{.13(1 - .13)}{250}}} = \frac{-.042}{.02127} = -1.975$$

Step 4: Make the Decision   Because the test statistic falls in the left tail of the rejection region, we reject the null hypothesis H_0: $\pi = .13$ and conclude H_1: $\pi \neq .13$ at the 5 percent level of significance. Another way to say this is that the sample proportion *differs significantly* from the historical return rate at $\alpha = .05$ in a two-tailed test. Note that this decision is rather a close one because the test statistic just barely falls into the rejection region.

The rejection was stronger in a one-tailed test, that is, the test statistic was farther from the critical value. *Holding α constant, rejection in a two-tailed test always implies rejection in a one-tailed test.* This reinforces the logic of choosing a two-tailed test unless there is a specific reason to prefer a one-tailed test.

Calculating a *p*-Value for a Two-Tailed Test

In a two-tailed test, we divide the risk into equal tails, one on the left and one on the right, to allow for the possibility that we will reject H_0 whenever the sample statistic is very small or very large. With the *p*-value approach in a two-tailed test, we find the tail area associated with our sample test statistic, multiply this by two, and then compare that probability to α. Our z statistic was calculated to be -1.975. The *p*-value would then be

$$2 \times P(Z < -1.975) = 2 \times .02413 = .04826$$

We would reject the null hypothesis because the *p*-value .04826 is less than α (.05).

Effect of α

Would the decision be the same if we had used a different level of significance? While the test statistic $z_{\text{calc}} = -1.975$ is the same regardless of our choice of α, our choice of α *does* affect the decision. Referring to Table 9.6 we see that we can reject the null hypothesis at $\alpha = .10$ ($z_{\text{crit}} = \pm1.645$) or $\alpha = .05$ ($z_{\text{crit}} = \pm1.960$), but we cannot reject at $\alpha = .01$ ($z_{\text{crit}} = \pm2.576$).

α	Test Statistic	Two-Tailed Critical Values	Decision
.10	$z_{\text{calc}} = -1.975$	$z_{.05} = \pm1.645$	Reject H_0
.05	$z_{\text{calc}} = -1.975$	$z_{.025} = \pm1.960$	Reject H_0
.01	$z_{\text{calc}} = -1.975$	$z_{.005} = \pm2.576$	Don't reject H_0

Table 9.6

Effect of Varying α

Which level of significance is the "right" one? They all are. It depends on how much Type I error we are willing to allow. Before concluding that $\alpha = .01$ is "better" than the others because it allows less Type I error, you should remember that smaller Type I error leads to increased Type II error. In this case, Type I error would imply that there has been a change in return rates when in reality nothing has changed, while Type II error implies that the software had no effect on the return rate, when in reality the software did decrease the return rate.

Example 9.8

Length of Hospital Stay

A hospital is comparing its performance against an industry benchmark that no more than 50 percent of normal births should result in a hospital stay exceeding 2 days (48 hours). Thirty-one births in a sample of 50 normal births had a length of stay (LOS) greater than 48 hours. At $\alpha = .025$, does this sample prove that the hospital exceeds the benchmark? This question requires a right-tailed test.

Step 1: State the Hypotheses   The hypotheses are

$H_0: \pi \le .50$ (the hospital is compliant with the benchmark)

$H_1: \pi > .50$ (the hospital is exceeding the benchmark)

Step 2: Specify the Decision Rule   For $\alpha = .025$ in a right-tailed test, the critical value is $z_{.025} = 1.96$, so the decision rule is

Reject H_0 if $z > 1.960$

Otherwise, do not reject H_0

This decision rule is illustrated in Figure 9.14.

Figure 9.14

Right-Tailed z Test Using $\alpha = .025$

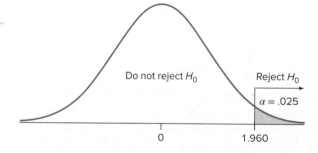

To assume normality of the sample proportion p, we require that $n\pi_0 \ge 10$ and $n(1 - \pi_0) \ge 10$. Inserting $\pi_0 = .50$ and $n = 50$, we see that the normality conditions are easily met: $(50)(.50) = 25$ and $(50)(1 - .50) = 25$.

Step 3: Calculate the Test Statistic   Because $p = x/n = 31/50 = .62$, the sample seems to favor H_1. But we will assume that H_0 is true and see if the test statistic contradicts this assumption. We test the hypothesis at $\pi = .50$. If we can reject $\pi = .50$ in favor of $\pi > .50$, then we can reject the class of hypotheses $\pi \le .50$. The test statistic is the difference between the sample proportion $p = x/n$ and the hypothesized parameter π_0 divided by the standard error of p:

$$z_{calc} = \frac{p - \pi_0}{\sqrt{\dfrac{\pi_0(1 - \pi_0)}{n}}} = \frac{.62 - .50}{\sqrt{\dfrac{.50(1 - .50)}{50}}} = \frac{.12}{.07071068} = 1.697$$

Step 4: Make the Decision   The test statistic does not fall in the right-tail rejection region, so we cannot reject the hypothesis that $\pi \leq .50$ at the 2.5 percent level of significance. In other words, the test statistic is within the realm of chance at $\alpha = .025$.

Step 5: Take Action   The hospital appears compliant with the benchmark, so no action is required at this time.

Calculating the *p*-Value

In this case, the *p*-value can be obtained from Excel's cumulative standard normal function =1-NORM.S.DIST(1.697, 1) = .04485 or from Appendix C-2 (using $z = 1.70$, we get $p = 1 - .9554 = .0446$). Excel's accuracy is greater because $z = 1.697$ is not rounded to $z = 1.70$. Because we want a right-tail area, we must subtract the cumulative distribution function from 1. The *p*-value is greater than .025, so we fail to reject the null hypothesis in a right-tailed test. We could (barely) reject at $\alpha = .05$. This demonstrates that the level of significance can affect our decision. The advantage of the *p*-value is that it tells you exactly the point of indifference between rejecting or not rejecting H_0. The *p*-value is illustrated in Figure 9.15.

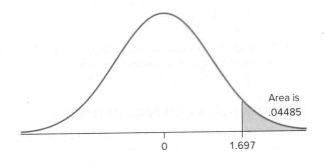

Figure 9.15

p-Value for a
Right-Tailed Test
with $z_{calc} = 1.697$

Area is
.04485

0   1.697

9.44 Interpret each *p*-value in your own words:

   a. *p*-value = .387, H_0: $\pi \geq .20$, H_1: $\pi < .20$, $\alpha = .10$
   b. *p*-value = .043, H_0: $\pi \leq .90$, H_1: $\pi > .90$, $\alpha = .05$
   c. *p*-value = .0012, H_0: $\pi = .50$, H_1: $\pi \neq .50$, $\alpha = .01$

9.45 Calculate the test statistic and *p*-value for each sample.

   a. H_0: $\pi = .20$ versus H_1: $\pi \neq .20$, $\alpha = .025$, $p = .28$, $n = 100$
   b. H_0: $\pi \leq .50$ versus H_1: $\pi > .50$, $\alpha = .025$, $p = .60$, $n = 90$
   c. H_0: $\pi \leq .75$ versus H_1: $\pi > .75$, $\alpha = .10$, $p = .82$, $n = 50$

9.46 Calculate the test statistic and *p*-value for each sample.

   a. H_0: $\pi \leq .60$ versus H_1: $\pi > .60$, $\alpha = .05$, $x = 56$, $n = 80$
   b. H_0: $\pi = .30$ versus H_1: $\pi \neq .30$, $\alpha = .05$, $x = 18$, $n = 40$
   c. H_0: $\pi \geq .10$ versus H_1: $\pi < .10$, $\alpha = .01$, $x = 3$, $n = 100$

9.47 May normality of the sample proportion p be assumed? Show your work.

   a. H_0: $\pi = .30$ versus H_1: $\pi \neq .30$, $n = 20$
   b. H_0: $\pi = .05$ versus H_1: $\pi \neq .05$, $n = 50$
   c. H_0: $\pi = .10$ versus H_1: $\pi \neq .10$, $n = 400$

9.48 In a recent survey, 10 percent of the participants rated Pepsi as being "concerned with my health." PepsiCo's response included a new "Smart Spot" symbol on its products that meet certain nutrition criteria to help consumers who seek more healthful eating options. At $\alpha = .05$, would a follow-up survey showing that 18 of 100 persons now rate Pepsi as being "concerned with my health" provide sufficient evidence that the percentage has increased?

9.49 In a hospital's shipment of 3,500 insulin syringes, 14 were unusable due to defects. (a) At $\alpha = .05$, is this sufficient evidence to reject future shipments from this supplier if the hospital's quality standard requires 99.7 percent of the syringes to be acceptable? State the hypotheses and decision rule. (b) May normality of the sample proportion p be assumed? (c) Explain the effects of Type I error and Type II error. (d) Find the p-value.

9.50 To combat antibiotic resistance, the Quality Improvement Consortium recommends a throat swab to confirm strep throat before a physician prescribes antibiotics to children under age 5. Nationally, 40 percent of children under 5 who received antibiotics did not have a throat swab. The Colorado Department of Health took a random sample of 60 children under the age of 5 who received antibiotics for throat infections and found that 18 did not have a throat swab. (a) At $\alpha = .05$, is this a statistically significant decrease from the national rate of 40 percent? (b) Is it safe to assume normality of the sample proportion p? Explain.

9.51 To encourage telephone efficiency, a catalog call center issues a guideline that at least half of all telephone orders should be completed within 2 minutes. Subsequently, a random sample of 64 telephone call orders showed that 24 calls lasted 2 minutes or less. Does this sample show that fewer than half of all orders are completed within 2 minutes? (a) State the appropriate hypotheses assuming π is the proportion of all calls that are completed within 2 minutes. (b) Find the p-value. (c) State the conclusion using $\alpha = .05$. (d) Is the difference important (as opposed to significant)?

9.52 The recent default rate on all student loans is 5.2 percent. In a recent random sample of 300 loans at private universities, there were 9 defaults. (a) Does this sample show sufficient evidence that the private university loan default rate is below the rate for all universities, using a left-tailed test at $\alpha = .01$? (b) Calculate the p-value. (c) Verify that the assumption of normality of the sample proportion p is justified.

9.53 A poll of 702 frequent and occasional fliers found that 442 respondents favored a ban on cell phones in flight, even if technology permits their use. At $\alpha = .05$, can we conclude that more than half the sampled population supports a ban?

Small Samples and Non-normality

LO (9-9)

Check whether normality may be assumed in testing a proportion.

In random tests by the FAA, 12.5 percent of all passenger flights failed the agency's test for bacterial count in water served to passengers. Airlines now are trying to improve their compliance with water quality standards. Random inspection of 16 recent flights showed that only 1 flight failed the water quality test. Has overall compliance improved?

H_0: $\pi \geq .125$ (failure rate has not improved)

H_1: $\pi < .125$ (failure rate has declined)

Our rule is to assume normality if $n\pi_0 \geq 10$ and $n(1 - \pi_0) \geq 10$. This sample is clearly too small to assume normality because $n\pi_0 = (16)(.125) = 2$. Instead, we can find the exact binomial left-tail probability of observing this sample under the assumption that $\pi = .125$. The binomial test is easy and is always correct because no assumption of normality is required. We can use Excel to calculate the cumulative binomial probability of the observed sample result under the null hypothesis as $P(X \leq 1 \mid n = 16, \pi = .125) =$ BINOM.DIST(1,16,.125,1) $= .38793$. The p-value of .388 does not permit rejection of H_0 at any of the usual levels of significance (e.g., 5 percent). We conclude that the failure rate has not improved.

Mini Case 9.3

Every Minute Counts

As more company business is transacted by telephone or Internet, there is a considerable premium to reduce customer time spent with human operators. Verizon recently installed a new speech recognition system for its repair calls. In the old system, the user had to press keys on the numeric keypad to answer questions, which led many callers to opt to talk to an operator instead. Under the old system, 94 percent of the customers had to talk to an operator to get their needs met. Suppose that, using the new system, a sample of 150 calls showed that 120 required an operator. The hypotheses are

H_0: $\pi \geq .94$ (the new system is no better than the old system)

H_1: $\pi < .94$ (the new system has reduced the proportion of operator calls)

These hypotheses call for a left-tailed test. Using $\alpha = .01$, the left-tail critical value is $z_{.01} = -2.326$, as illustrated in Figure 9.16.

Figure 9.16

Decision Rule for Left-Tailed Test

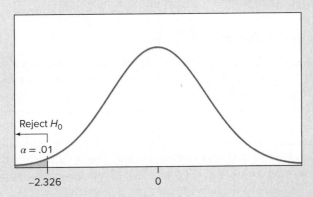

For normality of the sample proportion p, we want $n\pi_0 \geq 10$ and $n(1 - \pi_0) \geq 10$. The condition for normality is not quite met because $(150)(.94) = 141$ but $(150)(.06) = 9$. We will proceed, bearing in mind this possible concern. The sample proportion is $p = 120/150 = .80$, so the test statistic is

$$z_{calc} = \frac{p - \pi_0}{\sqrt{\frac{\pi_0(1 - \pi_0)}{n}}} = \frac{.80 - .94}{\sqrt{\frac{.94(1 - .94)}{150}}} = \frac{-.14}{.01939} = -7.22$$

The test statistic is far below the critical value, so we conclude that the percentage of customers who require an operator has declined. We can confirm that this conclusion would have been the same using the binomial probability formula in Excel: $P(X \leq 120 \mid n = 150, \pi = .94)$ =BINOM.DIST(120,150,.94,1) = 5.6×10^{-9}. Because this p-value is approximately zero, it supports our conclusion that the percentage of customers who require an operator has declined.

Besides being significant, such savings are important. For example, Boston Financial Data Services, a company that provides record-keeping services for mutual funds, shaved a minute off the mean time to process a customer request. Its call centers process 1.7 million calls a year, so the savings are very large.

Section Exercises

 connect

9.54 A Realtor claims that no more than half of the homes he sells are sold for less than the asking price. When reviewing a random sample of 12 sales over the past year, he found that actually 10 were sold below the asking price. (a) Would we be justified in assuming that the sample proportion p is normally distributed? Explain. (b) Calculate a p-value for the observed sample outcome, using the normal distribution. At the .05 level of significance in a right-tailed test, is the proportion of homes sold for less than the asking price greater than 50%? (c) Use Excel to calculate the binomial probability $P(X \geq 10 \mid n = 12, \pi = .50) = 1 - P(X \leq 9 \mid n = 12, \pi = .50)$.

9.55 BriteScreen, a manufacturer of 19-inch LCD computer screens, requires that on average 99.9 percent of all LCDs conform to its quality standard. In a day's production of 2,000 units, 4 are defective. (a) Assuming this is a random sample, is the standard being met, using the

10 percent level of significance? *Hint:* Use Excel to find the binomial probability $P(X \geq 4 \mid n = 2{,}000, \pi = .001) = 1 - P(X \leq 3 \mid n = 2{,}000, \pi = .001)$. (b) Show that normality of the sample proportion p should not be assumed.

9.56 Perfect pitch is the ability to identify musical notes correctly without hearing another note as a reference. The probability that a randomly chosen person has perfect pitch is .0005. (a) If 20 students at Julliard School of Music are tested and 2 are found to have perfect pitch, would you conclude at the .01 level of significance that Julliard students are more likely than the general population to have perfect pitch? *Hint:* Use Excel to find the right-tailed binomial probability $P(X \geq 2 \mid n = 20, \pi = .0005)$. (b) Show that normality of the sample proportion p should not be assumed.

Analytics in Action

Will Big Data Kill Statistics?

Large, complex, and unstructured data sets can arrive in real time from multiple sources in such volume and speed that it can be difficult to process the data into meaningful information. This is where sampling comes in. And while sample sizes can be quite large when there are terabytes (even petabytes) of data, significance tests still have a role to play.

Analysts must use sampling to pick a manageable subset of the incoming data torrent or to study things that are difficult to measure in real time. Virologists in San Diego test samples of wastewater for COVID-19 counts to estimate the current population infection rates (see *Union Tribune*, July 19, 2022, p. 1). Plant pathologists in Italy sample twigs from olive trees to develop models of pathogens that threaten agriculture [see *Scientific American* 327, no. 2 (August 2022), p. 22].

Companies anticipate customer demand by analyzing key attributes of past and current products or services to develop predictive models of customer behavior. They use statistics to combine data from social media, test markets, focus groups, and early store rollouts to plan, produce, and launch new products (see www.oracle.com/big-data/what-is-big-data/). Statistics allow analysts to winnow through vast numbers of predictor variables to select those that show significant and important effects. Data scientists "curate" data from many sources using data analysis, spatial analysis, and visualization.

Statistics is alive and well!

9.7 POWER CURVES AND OC CURVES (OPTIONAL)

LO 9-10

Interpret a power curve or OC curve (optional).

Recall that *power* is the probability of correctly rejecting a false null hypothesis. While we cannot always attain the power we desire in a statistical test, we can at least calculate what the power would be in various possible situations. We will show step by step how to calculate power for tests of a mean or proportion and how to draw *power curves* that show how power depends on the true value of the parameter we are estimating.

Power Curve for a Mean: An Example

Power depends on how far the true value of the parameter is from the null hypothesis value. The test will have low power if the population mean (μ) is almost the same as the hypothesized mean (μ_0). The only way to increase power would be to increase the sample size. To illustrate the calculation of β risk and power, consider a utility that is installing underground PVC pipe as a cable conduit. The specifications call for a mean strength of 12,000 psi (pounds per square inch). A sample of 25 pieces of pipe is tested under laboratory conditions to ascertain the compressive pressure that causes the pipe to collapse. The standard deviation is known from past experience to be $\sigma = 500$ psi. If the pipe proves stronger than the specification, there is no problem, so the utility requires a left-tailed test:

$H_0: \mu \geq 12,000$

$H_1: \mu < 12,000$

If the true mean strength is 11,900 psi, what is the probability that the utility will fail to reject the null hypothesis and mistakenly conclude that $\mu = 12,000$? At $\alpha = .05$, what is the power of the test? Recall that β is the risk of Type II error, the probability of incorrectly accepting a false hypothesis. Type II error is bad, so we want β to be small.

$$\beta = P(\text{accept } H_0 \mid H_0 \text{ is false}) \tag{9.6}$$

In this example, $\beta = P(\text{conclude } \mu = 12,000 \mid \mu = 11,900)$.

Conversely, power is the probability that we correctly reject a false hypothesis. More power is better, so we want power to be as close to 1 as possible:

$$\text{Power} = P(\text{reject } H_0 \mid H_0 \text{ is false}) = 1 - \beta \tag{9.7}$$

The values of β and power will vary, depending on the difference between the true mean μ and the hypothesized mean μ_0, the standard deviation σ, the sample size n, and the level of significance α.

$$\text{Power} = f(\mu - \mu_0, \sigma, n, \alpha) \quad \text{(determinants of power for a mean)} \tag{9.8}$$

Table 9.7 summarizes their effects. While we cannot change μ and σ, the sample size and level of significance often are under our control. We can get more power by increasing α, but would we really want to increase Type I error in order to reduce Type II error? Probably not, so the way we usually increase power is by choosing a larger sample size.

Parameter	If . . .	then . . .
True mean (μ)	$\lvert \mu - \mu_0 \rvert \uparrow$	Power $\uparrow$
True standard deviation (σ)	$\sigma \uparrow$	Power $\downarrow$
Sample size (n)	$n \uparrow$	Power $\uparrow$
Level of significance (α)	$\alpha \uparrow$	Power $\uparrow$

Table 9.7

Determinants of Power in Testing One Mean

Calculating Power

To calculate β and power, we follow a simple sequence of steps for any given values of μ, σ, n, and α. We assume a normal population (or a large sample) so that the sample mean $\overline{X}$ may be assumed to be normally distributed.

Step 1 Find the left-tail *critical value* for the sample mean. At $\alpha = .05$ in a left-tailed test, we know that $z_{.05} = -1.645$. Using the formula for a z-score,

$$z_{\text{critical}} = \frac{\overline{x}_{\text{critical}} - \mu_0}{\dfrac{\sigma}{\sqrt{n}}}$$

we can solve algebraically for $\overline{x}_{\text{critical}}$:

$$\overline{x}_{\text{critical}} = \mu_0 + z_{\text{critical}} \frac{\sigma}{\sqrt{n}} = 12,000 - 1.645\left(\frac{500}{\sqrt{25}}\right) = 11,835.5$$

In terms of the data units of measurement (pounds per square inch), the decision rule is

Reject $H_0: \mu \geq 12,000$ if $\overline{X} < 11,835.5$ psi

Otherwise, do not reject H_0

Now suppose that the true mean is $\mu = 11,900$. Then the sampling distribution of $\overline{X}$ would be centered at 11,900 instead of 12,000, as we hypothesized. The probability of β error is the area to the right of the critical value $\overline{x}_{critical} = 11,835.5$ (the nonrejection region) representing $P(\overline{X} > \overline{x}_{critical} \mid \mu = 11,900)$. Figure 9.17 illustrates this situation.

Figure 9.17

Finding β When $\mu = 11,900$

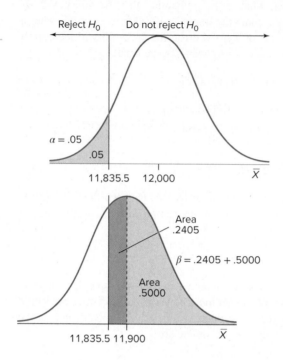

Step 2 Express the difference between the critical value $\overline{x}_{critical}$ and the true mean μ as a z-value:

$$z = \frac{\overline{x}_{critical} - \mu}{\dfrac{\sigma}{\sqrt{n}}} = \frac{11,835.5 - 11,900}{\dfrac{500}{\sqrt{25}}} = -0.645$$

Step 3 Find the β risk and power as areas under the normal curve, using Appendix C-2 or Excel:

Calculation of β

$\beta = P(\overline{X} > \overline{x}_{critical} \mid \mu = 11,900)$

$\quad = P(Z > -0.645)$

$\quad = 0.2405 + 0.5000$

$\quad = 0.7405$, or 74.1%

Calculation of Power

$\text{Power} = P(\overline{X} < \overline{x}_{critical} \mid \mu = 11,900)$

$\quad = 1 - \beta$

$\quad = 1 - 0.7405$

$\quad = 0.2595$, or 26.0%

This calculation shows that if the true mean is $\mu = 11,900$, then there is a 74.05 percent chance that we will commit β error by failing to reject $\mu = 12,000$. Because 11,900 is not very far from 12,000 in terms of the standard error, our test has relatively low power. Although our test may not be sensitive enough to reject the null hypothesis reliably if μ is only *slightly* less than 12,000, if μ is *far* below 12,000, our test would be more likely to lead to rejection of H_0. Although we cannot know the true mean, we *can* repeat our power calculation for as many values of μ and n as we wish. These calculations may appear tedious, but they are straightforward in a spreadsheet. Table 9.8 shows β and power for samples of $n = 25$, 50, and 100 for selected μ values from 12,000 down to 11,600.

Table 9.8

β **and Power for**
$\mu_0 = 12{,}000$

True μ	n = 25			n = 50			n = 100		
	z	β	Power	z	β	Power	z	β	Power
12,000	−1.645	.9500	.0500	−1.645	.9500	.0500	−1.645	.9500	.0500
11,900	−0.645	.7405	.2595	−0.231	.5912	.4088	0.355	.3612	.6388
11,800	0.355	.3612	.6388	1.184	.1183	.8817	2.355	.0093	.9907
11,700	1.355	.0877	.9123	2.598	.0047	.9953	4.355	.0000	1.0000
11,600	2.355	.0093	.9907	4.012	.0000	1.0000	6.355	.0000	1.0000

Notice that β drops toward 0 and power approaches 1 when the true value μ is far from the hypothesized mean $\mu_0 = 12{,}000$. When $\mu = 12{,}000$, there can be no β error because β error can only occur if H_0 is false. Power is then equal to $\alpha = .05$, the lowest power possible.

Effect of Sample Size

Table 9.8 also shows that, other things being equal, if sample size were to increase, β risk would decline and power would increase because the critical value $\overline{x}_{critical}$ would be closer to the hypothesized mean μ. For example, if the sample size were increased to $n = 50$, then

$$\overline{x}_{critical} = \mu_0 + z_{critical}\frac{\sigma}{\sqrt{n}} = 12{,}000 - 1.645\left(\frac{500}{\sqrt{50}}\right) = 11{,}883.68$$

$$z = \frac{\overline{x}_{critical} - \mu}{\frac{\sigma}{\sqrt{n}}} = \frac{11{,}883.68 - 11{,}900}{\frac{500}{\sqrt{25}}} = -0.231$$

$$\text{Power} = P(\overline{X} < \overline{x}_{critical} \mid \mu = 11{,}900) = P(Z < -.231) = .4088, \text{ or } 40.9\%$$

Relationship of the Power and OC Curves

Power is much easier to understand when it is made into a graph. A **power curve** is a graph whose Y-axis shows the power of the test $(1 - \beta)$ and whose X-axis shows the various possible true values of the parameter while holding the sample size constant. Figure 9.18 shows the power curve for this example, using three different sample sizes. Power increases as the departure of μ from 12,000 becomes greater. Larger sample size creates a higher power curve. Because the power curve approaches $\alpha = .05$ as the true mean approaches the hypothesized mean of 12,000, we can see that α also affects the power curve. If we increase α, the power curve will shift up. Although it is not illustrated here, power also rises if the standard deviation is smaller because a small σ gives the test more precision.

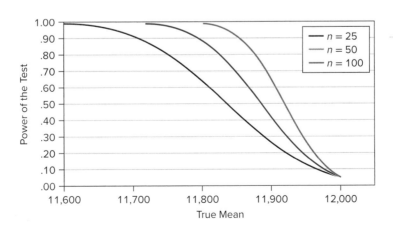

The graph of β risk against this same X-axis is called the operating characteristic or **OC curve**. Figure 9.19 shows the OC curve for this example. It is simply the converse of the power curve, so it is redundant if you already have the power curve.

Figure 9.19

OC Curves for
$H_0: \mu \geq 12,000$;
$H_1: \mu < 12,000$

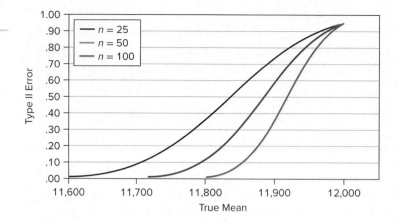

Power Curve for Tests of a Proportion

For tests of a proportion, power depends on the true proportion π, the hypothesized proportion π_0, the sample size n, and the level of significance α. Table 9.9 summarizes their effects on power. As with a mean, enlarging the sample size is the most common method of increasing power unless we are willing to raise the level of significance (i.e., reduce α).

Table 9.9

Determinants of Power in
Testing a Proportion

Parameter	If . . .	then . . .
True proportion π	$\lvert \pi - \pi_0 \rvert \uparrow$	Power $\uparrow$
Sample size n	$n \uparrow$	Power $\uparrow$
Level of significance α	$\alpha \uparrow$	Power $\uparrow$

Example 9.9

*Length of Hospital Stay:
Power Curve*

A sample is taken of 50 births in a major hospital. We are interested in knowing whether at least half of all mothers have a length of stay (LOS) less than two days. We will do a right-tailed test using $\alpha = .10$. The hypotheses are

$H_0: \pi \leq .50$
$H_1: \pi > .50$

To find the power curve, we follow the same procedure as for a mean—actually, it is easier than a mean because we don't have to worry about σ. For example, what is the power of the test if the true proportion is $\pi = .60$ and the sample size is $n = 50$?

Step 1: Find the Critical Value   Find the right-tail *critical value* for the sample proportion. At $\alpha = .10$ in a right-tailed test, we would use $z_{.10} = 1.282$ (actually, $z_{.10} = 1.28155$ if we use Excel), so

$$p_{critical} = \pi_0 + 1.28155 \sqrt{\frac{\pi_0(1 - \pi_0)}{n}} = .50 + 1.28155 \sqrt{\frac{(.50)(1 - .50)}{50}} = .590619$$

Step 2: Convert to z Value   Express the difference between the critical value $p_{critical}$ and the true proportion π as a z-value:

$$z = \frac{p_{critical} - \pi}{\sqrt{\frac{\pi(1 - \pi)}{n}}} = \frac{.590619 - .600000}{\sqrt{\frac{(.60)(1 - .60)}{50}}} = -0.1354$$

Step 3: Calculate Power   Find the β risk and power as areas under the normal curve:

Calculation of β

$\beta = P(p < p_{critical} \mid \pi = .60)$

$\quad = P(Z < -0.1354)$

$\quad = 0.4461$, or 44.61%

Calculation of Power

$Power = P(p > p_{critical} \mid \pi = .60)$

$\quad = 1 - \beta$

$\quad = 1 - 0.4461$

$\quad = 0.5539$, or 55.39%

We can repeat these calculations for any values of π and n. Table 9.10 illustrates power for selected values of π ranging from .50 to .70, at which point power is near its maximum, and for sample sizes of $n = 50$, 100, and 200. As expected, power increases sharply as sample size increases and as π differs more from $\pi_0 = .50$.

Figure 9.20 presents these results visually. Power curves for the larger sample sizes are higher, and the power is lowest when π is near the hypothesized value of $\pi_0 = .50$. The lowest point on the curve has power equal to $\alpha = .10$. Thus, if we increase α, the power curve would shift up. In other words, we can decrease β (and thereby raise power) by increasing the chance of Type I error, a trade-off we might not wish to make.

	$n = 50$			$n = 100$			$n = 200$		
π	z	β	Power	z	β	Power	z	β	Power
.50	1.282	.9000	.1000	1.282	.9000	.1000	1.282	.9000	.1000
.55	0.577	.7181	.2819	0.283	.6114	.3886	−0.133	.4470	.5530
.60	−0.135	.4461	.5539	−0.733	.2317	.7683	−1.579	.0572	.9428
.65	−0.880	.1893	.8107	−1.801	.0358	.9642	−3.104	.0010	.9990
.70	−1.688	.0457	.9543	−2.966	.0015	.9985	−4.774	.0000	1.0000

Table 9.10

β and Power for
$\pi_0 = .50$

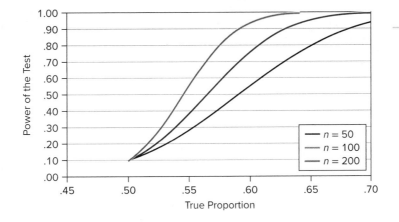

Figure 9.20

**Power Curve Families
for H_0: $\pi \le .50$;
H_1: $\pi > .50$**

Two-Tailed Power Curves and OC Curves

Both previous examples used one-tailed tests. But if we choose a two-tailed hypothesis test, we will see both sides of the power curve and/or OC curve. Figure 9.21 shows the two-tailed power and OC curves for the previous two examples (H_0: $\mu = 12{,}000$ and H_0: $\pi = .50$). Each power curve resembles an inverted normal curve, reaching its minimum value when $\mu = \mu_0$ (for a mean) or at $\pi = \pi_0$ (for a proportion). The minimum power is equal to the value of α that we select. In our examples, we chose $\alpha = .05$ for testing μ and $\alpha = .10$ for testing π. If we change α, we will raise or lower the entire power curve. You can download power curve spreadsheet demonstrations for μ and π (see McGraw Hill Connect® resources at the end of this chapter) if you want to try your own experiments with power curves (with automatic calculations).

Figure 9.21

Two-Tailed Power and OC Curves

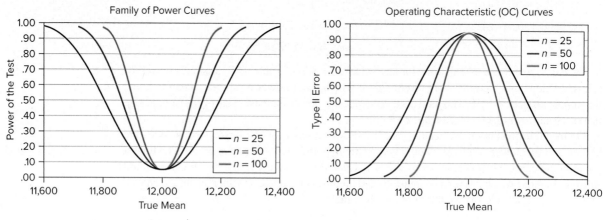
$H_0: \mu = 12{,}000, H_1: \mu \neq 12{,}000, \sigma = 500, \alpha = .05$

$H_0: \pi = .50, H_1: \pi \neq .50, \alpha = .10$

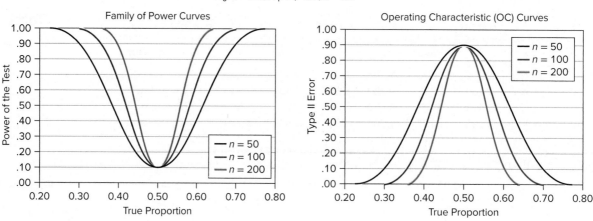

Section Exercises

connect

Hint: Check your answers using *LearningStats* (from the McGraw Hill Connect® downloads at the end of this chapter).

9.57 A quality expert inspects 400 items to test whether the population proportion of defectives exceeds .03, using a right-tailed test at $\alpha = .10$. (a) What is the power of this test if the true proportion of defectives is $\pi = .04$? (b) If the true proportion is $\pi = .05$? (c) If the true proportion of defectives is $\pi = .06$?

9.58 Repeat the previous exercise using $\alpha = .05$. For each true value of π, is the power higher or lower?

9.59 For a certain wine, the mean pH (a measure of acidity) is supposed to be 3.50 with a known standard deviation of $\sigma = .10$. The quality inspector examines 25 bottles at random to test whether the pH is too low, using a left-tailed test at $\alpha = .01$. (a) What is the power of this test if the true mean is $\mu = 3.48$? (b) If the true mean is $\mu = 3.46$? (c) If the true mean is $\mu = 3.44$?

9.60 Repeat the previous exercise, using $\alpha = .05$. For each true value of μ, is the power higher or lower?

9.8 TESTS FOR ONE VARIANCE (OPTIONAL)

LO 9-11

Perform a hypothesis test for a variance (optional).

Not all business hypothesis tests involve proportions or means. In quality control, for example, it is important to compare the variance of a process with a historical benchmark, σ_0^2, to see whether variance reduction has been achieved or to compare a process standard deviation with an engineering specification.

Example 9.10

Attachment Times

Historical statistics show that the standard deviation of attachment times for an instrument panel in an automotive assembly line is $\sigma = 7$ seconds. Observations on 20 randomly chosen attachment times are shown in Table 9.11. At $\alpha = .05$, does the variance in attachment times differ from the historical variance ($\sigma^2 = 7^2 = 49$)?

Table 9.11

Panel Attachment Times (seconds) ▧ Attachment				
120	143	136	126	122
140	133	133	131	131
129	128	131	123	119
135	137	134	115	122

FIGURE 9.22

Decision Rule for Chi-Square Test

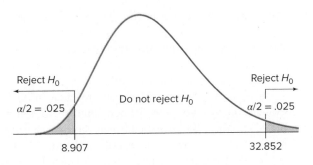

The sample mean is $\bar{x} = 129.400$ with a standard deviation of $s = 7.44382$. We ignore the sample mean because it is irrelevant to this test. For a two-tailed test, the hypotheses are

H_0: $\sigma^2 = 49$
H_1: $\sigma^2 \neq 49$

For a test of one variance, assuming a normal population, the test statistic follows the **chi-square distribution** with degrees of freedom equal to $d.f. = n - 1 = 20 - 1 = 19$. Denoting the hypothesized variance as σ_0^2, the test statistic is

$$\chi_{calc}^2 = \frac{(n-1)s^2}{\sigma_0^2} \qquad \text{(test for one variance)} \qquad \text{(9.9)}$$

For a two-tailed test, the decision rule based on the upper and lower critical values of chi-square is

Reject H_0 if $\chi_{calc}^2 < \chi_{lower}^2$ or if $\chi_{calc}^2 > \chi_{upper}^2$
Otherwise, do not reject H_0

We can use the Excel function =CHISQ.INV to get the critical values:

$$\chi_{lower}^2 = \text{CHISQ.INV}(\alpha/2, d.f.) = \text{CHISQ.INV}(0.025, 19) = 8.907$$

$$\chi_{upper}^2 = \text{CHISQ.INV}(1 - \alpha/2, d.f.) = \text{CHISQ.INV}(0.975, 19) = 32.852$$

The decision rule is illustrated in Figure 9.22. The value of the test statistic is

$$\chi_{calc}^2 = \frac{(n-1)s^2}{\sigma^2} = \frac{(20-1)(7.44382)^2}{7^2} = 21.49$$

Because the test statistic is within the middle range, we conclude that the population variance does not differ significantly from 49; that is, the assembly process variance is unchanged.

When to Use Tests for One Variance

In general, we would be interested in a test of variances when it is not the *center* of the distribution, but rather the *variability* of the process, that matters. More variation implies a more erratic data-generating process. For example, variance tests are important in manufacturing processes because increased variation around the mean can be a sign of wear and tear on equipment or other problems that would require attention.

The chi-square test for a variance is not robust to non-normality of the population. If normality cannot be assumed (e.g., if the data set has outliers or severe skewness), you might need to use a bootstrap method (see *LearningStats* Unit 08) to test the hypothesis using specialized software. In such a situation, it is best to consult a statistician.

Mini Case 9.4

Ball Corporation Metal Thickness

Ball Corporation is the largest supplier of beverage cans in the world. Metal beverage containers are lightweight, fully recyclable, quickly chilled, and easy to store. For these reasons, the metal beverage container is the package of choice in homes, vending machines, and coolers around the world. Ball's many manufacturing sites throughout the world turn out more than 100 million cans each day.

At its Fort Atkinson, Wisconsin, manufacturing facility, Ball's quality group must evaluate the metal of potential new suppliers. Because of the large quantity of cans produced each day, Ball has established very precise specifications on the metal characteristics. One characteristic that is critical to production is the variation in metal thickness. Inconsistent thickness across a sheet of metal can create serious problems for Ball's manufacturing process. Ball's current supplier has a known thickness standard deviation $\sigma = 0.000959$ mm. To qualify a potential new metal supplier, Ball conducted a two-tailed hypothesis test to determine if the potential supplier's thickness variance was consistent with that of the current supplier. The null and alternative hypotheses are

$$H_0: \sigma^2 = (0.000959)^2$$
$$H_1: \sigma^2 \neq (0.000959)^2$$

Ball received a sample of 168 sheets of metal from the potential supplier. Therefore, the degrees of freedom for the chi-square test statistic are 167. Using $\alpha = .10$, the decision rule states:

If $\chi^2_{calc} < 138.1$ or $\chi^2_{calc} > 198.2$, reject H_0

The sample standard deviation was calculated to be $s = 0.00106$ mm. The calculated test statistic was then found to be

$$\chi^2_{calc} = \frac{(n-1)s^2}{\sigma^2} = \frac{(168-1)(0.00106)^2}{(0.000959)^2} = 204.03$$

Because $\chi^2_{calc} > 198.2$, the decision was to reject H_0. Ball concluded that the variation in metal thickness for the potential supplier was not equivalent to that of their primary supplier. The potential supplier was not a candidate for supplying metal to Ball.

Section Exercises

9.61 A sample of size $n = 15$ has variance $s^2 = 35$. At $\alpha = .01$ in a left-tailed test, does this sample contradict the hypothesis that $\sigma^2 = 50$?

9.62 A sample of size $n = 10$ has variance $s^2 = 16$. At $\alpha = .10$ in a two-tailed test, does this sample contradict the hypothesis that $\sigma^2 = 24$?

9.63 A sample of size $n = 19$ has variance $s^2 = 1.96$. At $\alpha = .05$ in a right-tailed test, does this sample contradict the hypothesis that $\sigma^2 = 1.21$?

9.64   pH is a measure of acidity that winemakers must watch. A "healthy wine" should have a pH in the range 3.1 to 3.7. The target standard deviation is $\sigma = 0.10$ (i.e., $\sigma^2 = 0.01$). The pH measurements for a sample of 16 bottles of wine are shown below. At $\alpha = .05$ in a two-tailed test, is the sample variance either too high or too low? Show all steps, including the hypotheses and critical values from Appendix E. *Hint:* Ignore the mean. (See www.winemakermag.com.) 🗀 **WinePH**

3.49	3.54	3.58	3.57	3.54	3.34	3.48	3.60
3.48	3.27	3.46	3.32	3.51	3.43	3.56	3.39

9.65   In U.S. hospitals, the average length of stay (LOS) for a diagnosis of pneumonia is 137 hours with a standard deviation of 25 hours. The LOS (in hours) for a sample of 12 pneumonia patients at Santa Theresa Memorial Hospital is shown below. In a two-tailed test at $\alpha = .05$, is this sample variance consistent with the national norms? Show all steps, including the hypotheses and critical values from Appendix E. *Hint:* Ignore the mean. 🗀 **Pneumonia**

132	143	143	120	124	116
130	165	100	83	115	141

Chapter Summary

The **null hypothesis (H_0)** represents the status quo or a benchmark. We try to reject H_0 in favor of the **alternative hypothesis (H_1)** on the basis of the sample evidence. The alternative hypothesis points to the tail of the test (> for a left-tailed test, < for a right-tailed test, ≠ for a two-tailed test). Rejecting a true H_0 is **Type I error** while failing to reject a false H_0 is **Type II error**. The **power** of the test is the probability of correctly rejecting a false H_0. The probability of Type I error is denoted α (often called the **level of significance**) and can be set by the researcher. The probability of Type II error is denoted β and is dependent on the true parameter value, sample size, and α. In general, lowering α increases β, and vice versa. The **test statistic** compares the sample statistic with the hypothesized parameter. For a mean, the **decision rule** tells us whether to reject H_0 by comparing the test statistic with the **critical value** of z (known σ) or t (unknown σ) from a table

or from Excel. Tests of a proportion are based on the normal distribution (if the sample is large enough, according to a rule of thumb), although in small samples the binomial is required. In any hypothesis test, the **p-value** shows the probability that the test statistic (or one more extreme) would be observed by chance, assuming that H_0 is true. If the p-value is smaller than α, we reject H_0 (i.e., a small p-value indicates a **significant** departure from H_0). A two-tailed test is analogous to a confidence interval seen in the last chapter. Power is greater when the true parameter is farther from the null hypothesis value. A **power curve** is a graph that plots the power of the test $(1 - \beta)$ against possible values of the true parameter, while the **OC curve** is a plot of the probability of a false negative (β). Tests of a variance use the **chi-square distribution** and suffer if the data are badly skewed.

Key Terms

alternative hypothesis	hypothesis	OC curve	statistical hypothesis
benchmark	hypothesis test	power	test statistic
chi-square distribution	hypothesis testing	power curve	two-tailed test
critical value	importance	p-value method	Type I error
decision rule	left-tailed test	rejection region	Type II error
false negative	level of significance	right-tailed test	
false positive	null hypothesis	significance	

Commonly Used Formulas in One-Sample Hypothesis Tests

Type I error: $\alpha = P(\text{reject } H_0 \mid H_0 \text{ is true})$

Type II error: $\beta = P(\text{fail to reject } H_0 \mid H_0 \text{ is false})$

Power: $1 - \beta = P(\text{reject } H_0 \mid H_0 \text{ is false})$

Test statistic for sample mean, σ known: $z_{\text{calc}} = \dfrac{\bar{x} - \mu_0}{\dfrac{\sigma}{\sqrt{n}}}$

Test statistic for sample mean, σ unknown:

$$t_{\text{calc}} = \frac{\bar{x} - \mu_0}{\frac{s}{\sqrt{n}}} \quad \text{with } d.f. = n - 1$$

Test statistic for sample proportion:

$$z_{\text{calc}} = \frac{p - \pi_0}{\sqrt{\frac{\pi_0(1 - \pi_0)}{n}}}$$

Test statistic for sample variance:

$$\chi^2_{\text{calc}} = \frac{(n - 1)s^2}{\sigma_0^2} \quad \text{with } d.f. = n - 1$$

Chapter Review

Note: Questions labeled * are based on optional material from this chapter.

1. (a) List the steps in testing a hypothesis. (b) Why can't a hypothesis ever be proven?

2. (a) Explain the difference between the null hypothesis and the alternative hypothesis. (b) How is the null hypothesis chosen (why is it "null")?

3. (a) Why do we say "fail to reject H_0" instead of "accept H_0"? (b) What does it mean to "provisionally accept a hypothesis"?

4. (a) Define Type I error and Type II error. (b) Give an example to illustrate.

5. (a) Explain the difference between a left-tailed test, two-tailed test, and right-tailed test. (b) When would we choose a two-tailed test? (c) How can we tell the direction of the test by looking at a pair of hypotheses?

6. (a) What is a test statistic? (b) Explain the meaning of the rejection region in a decision rule. (c) Why do we need to know the sampling distribution of a statistic before we can do a hypothesis test?

7. (a) Define level of significance. (b) Define power.

8. (a) Why do we prefer low values for α and β? (b) For a given sample size, why is there a trade-off between α and β? (c) How could we decrease both α and β?

9. (a) Why is a "statistically significant difference" not necessarily a "practically important difference"? Give an illustration. (b) Why do statisticians play only a limited role in deciding whether a significant difference requires action?

10. (a) In a hypothesis test for a proportion, when can normality be assumed? *(b) If the sample is too small to assume normality, what can we do?

11. (a) In a hypothesis test of one mean, when do we use t instead of z? (b) When is the difference between z and t small?

12. (a) Explain what a p-value means. Give an example and interpret it. (b) Why is the p-value method an attractive alternative to specifying α in advance?

13. Why is a confidence interval similar to a two-tailed test?

*14. (a) What does a power curve show? (b) What factors affect power for a test of a mean? (c) What factors affect power for a proportion? (d) What is the most commonly used method of increasing power?

*15. (a) In testing a hypothesis about a variance, what distribution do we use? (b) When would a test of a variance be needed? (c) If the population is not normal, what can we do?

Chapter Exercises

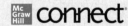

Note: Explain answers and show your work clearly. Problems marked * rely on optional material from this chapter.

HYPOTHESIS FORMULATION AND TYPE I AND II ERRORS

9.66 Suppose you always reject the null hypothesis, regardless of any sample evidence. (a) What is the probability of Type II error? (b) Why is this a bad policy?

9.67 Suppose the judge decides to acquit all defendants, regardless of the evidence. (a) What is the probability of Type I error? (b) Why is this a bad policy?

9.68 High blood pressure, if untreated, can lead to increased risk of stroke and heart attack. A common definition of hypertension is diastolic blood pressure of 90 or more. (a) State the null and alternative hypotheses for a physician who checks your blood pressure. (b) Define Type I and II errors. What are the consequences of each?

9.69 A nuclear power plant replaces its facility access system ID cards with a biometric security system that scans the iris pattern of the employee and compares it with a data bank. Users are classified as authorized or unauthorized. (a) State the null and alternative hypotheses. (b) Define Type I and II errors. What are the consequences of each?

9.70 A test-preparation company advertises that its training program raises SAT scores by an average of at least 30 points. A random sample of test-takers who had completed the training showed a mean increase smaller than 30 points. (a) Write the hypotheses for a left-tailed test of the mean. (b) Explain the consequences of a Type I error in this context.

9.71 Telemarketers use a predictive dialing system to decide whether a person actually answers a call (as opposed to an answering machine). If so, the call is routed to a telemarketer. If no telemarketer is free, the software must automatically hang up the phone within two seconds to comply with

FCC regulations against tying up the line. The SmartWay company says that its new system is smart enough to hang up on no more than 2 percent of the answered calls. Write the hypotheses for a right-tailed test, using SmartWay's claim about the proportion as the null hypothesis.

9.72 If the true mean is 50 and we reject the hypothesis that $\mu = 50$, what is the probability of Type II error? *Hint:* This is a trick question.

9.73 If we fail to reject the null hypothesis that $\pi = .50$ even though the true proportion is .60, what is the probability of Type I error? *Hint:* This is a trick question.

9.74 Pap smears are a test for abnormal cancerous and pre-cancerous cells taken from the cervix. (a) State a pair of hypotheses and then explain the meaning of a false negative and a false positive. (b) Why is the null hypothesis "null"? (c) Who bears the cost of each type of error?

9.75 In a commercially available fingerprint scanner (e.g., for your home or office PC), false acceptances are 1 in 25 million with false rejection rates of 3 percent. (a) Define Type I and II errors. (b) Why do you suppose the false rejection rate is so high compared with the false acceptance rate?

9.76 When told that over a 10-year period a mammogram test has a false positive rate of 50 percent, Bob said, "That means that about half the women tested actually have no cancer." Correct Bob's mistaken interpretation.

TESTS OF MEANS AND PROPORTIONS

9.77 Malcheon Health Clinic claims that the average waiting time for a patient is 20 minutes or less. A random sample of 15 patients shows a mean wait time of 24.77 minutes with a standard deviation of 7.26 minutes. (a) Write the hypotheses for a right-tailed test, using the clinic's claim as the null hypothesis. (b) Calculate the t test statistic to test the claim. (c) At the 5 percent level of significance ($\alpha = .05$), does the sample contradict the clinic's claim? (d) Use Excel to find the p-value and compare it to the level of significance. Did you come to the same conclusion as you did in part (c)?

9.78 The sodium content of a popular sports drink is listed as 220 mg in a 32-oz bottle. Analysis of 10 bottles indicates a sample mean of 228.2 mg with a sample standard deviation of 18.2 mg. (a) Write the hypotheses for a two-tailed test of the claimed sodium content. (b) Calculate the t test statistic to test the manufacturer's claim. (c) At the 5 percent level of significance ($\alpha = .05$), does the sample contradict the manufacturer's claim? (d) Use Excel to find the p-value and compare it to the level of significance. Did you come to the same conclusion as you did in part (c)?

9.79 A can of peeled whole tomatoes is supposed to contain an average of 19 ounces of tomatoes (excluding the juice). The actual weight is a normally distributed random variable whose standard deviation is known to be 0.25 ounce. (a) In quality control, would a one-tailed or two-tailed test be used? Why? (b) Explain the consequences of departure from the mean in either direction. (c) Which sampling distribution would you use if samples of four cans are weighed? Why? (d) Set up a two-tailed decision rule for $\alpha = .01$.

9.80 At Ajax Spring Water, a half-liter bottle of soft drink is supposed to contain a mean of 520 ml. The filling process follows a normal distribution with a known process standard

deviation of 4 ml. (a) Which sampling distribution would you use if random samples of 10 bottles are to be weighed? Why? (b) Set up hypotheses and a two-tailed decision rule for the correct mean using the 5 percent level of significance. (c) If a sample of 16 bottles shows a mean fill of 515 ml, does this contradict the hypothesis that the true mean is 520 ml?

9.81 On eight Friday quizzes, Bob received scores of 80, 85, 95, 92, 89, 84, 90, 92. He tells Prof. Hardtack that he is really a 90+ performer, but this sample just happened to fall below his true performance level. (a) State an appropriate pair of hypotheses. (b) State the formula for the test statistic and show your decision rule using the 1 percent level of significance. (c) Carry out the test. Show your work. (d) What assumptions are required? (e) Use Excel to find the p-value and interpret it. 🖻 **BobQuiz**

9.82 Pre-pandemic, the average cost of a wedding reception for 100 people was \$8,100 in the state of Colorado. A sample of 35 Colorado weddings during the summer of 2023 showed a mean cost of \$8,350 with a standard deviation equal to \$692. (a) At the .10 level of significance, has the true mean increased post-pandemic? State your hypotheses and decision rule. (b) Calculate the p-value and interpret it.

9.83 A certain ski boot rivet has a target weight of 2.268 grams. A random sample of 15 rivets from a manufacturer showed a mean weight of 2.256 grams with a standard deviation of .026 gram. (a) Using $\alpha = .05$, is the mean weight of all rivets from this manufacturer lower than the target weight? State your hypotheses and decision rule. (b) Calculate the p-value and interpret it.

9.84 A retail store had 60 credit card transactions over a two-hour period. Of those 60 transactions, 38 of them had to be re-swiped due to an error on the first try. (a) At the .10 level of significance, do more than half of the credit card transactions need to be swiped again? State your hypotheses and decision rule. (b) Calculate the p-value and interpret it.

9.85 A sample of 100 one-dollar bills from the Subway cash register revealed that 16 had something written on them besides the normal printing (e.g., "Bob ♥ Mary"). (a) At $\alpha = .05$, is this sample evidence consistent with the hypothesis that 10 percent or fewer of all dollar bills have anything written on them besides the normal printing? Include a sketch of your decision rule and show all calculations. (b) Is your decision sensitive to the choice of α? (c) Find the p-value.

9.86 A sample of 100 mortgages approved during the current year showed that 31 were issued to a single-earner family or individual. The historical percentage is 25 percent. (a) At the .05 level of significance in a right-tailed test, has the percentage of single-earner or individual mortgages risen? Include a sketch of your decision rule and show all work. (b) Is this a close decision? (c) State any assumptions that are required.

9.87 A quality control standard requires that no more than 5 percent of bags of Halloween candy be underweight. A random sample of 200 bags showed that 16 were underweight. (a) At $\alpha = .025$, is the standard being violated? Use a right-tailed test and show your work. (b) Find the p-value.

9.88 Ages for the 2022 Boston Red Sox pitchers are shown below. (a) Assuming this is a random sample of Major League Baseball pitchers, at the 5 percent level of significance, does this sample show that the true mean age of all American League pitchers is over 30 years? State your hypotheses and decision rule and show all work. (b) If there is a difference, is it important? (c) Find the *p*-value and interpret it. 📄 **RedSox**

Ages of Boston Red Sox Pitchers, 2022

Barnes	32	Paxton	32
Bello	23	Pivetta	29
Brasier	35	Rodriguez	31
Crawford	26	Sale	33
German	25	Schreiber	28
Hernandez	25	Seabold	26
Houck	26	Taylor	29
Kelly	27	Walter	26
Matta	27	Whitlock	26
Murphy	32	Winckowski	24
Ort	30		

Source: www.espn.com.

9.89 The EPA is concerned about the quality of drinking water served on airline flights. A sample of 112 flights found unacceptable bacterial contamination on 14 flights. (a) At $\alpha = .05$, does this sample show that more than 10 percent of all flights have contaminated water? (b) Find the *p*-value.

9.90 A web-based company has a goal of processing 95 percent of its orders on the same day they are received. If 485 out of the next 500 orders are processed on the same day, would this prove that they are exceeding their goal, using $\alpha = .025$?

9.91 In a major football conference, a sample showed that only 267 out of 584 freshmen players graduated within 6 years. (a) At $\alpha = .05$, does this sample contradict the claim that at least half graduate within 6 years? State your hypotheses and decision rule. (b) Calculate the *p*-value and interpret it. (c) Do you think the difference is important, as opposed to significant?

9.92 An auditor reviewed 25 oral surgery insurance claims from a particular surgical office, determining that the mean out-of-pocket patient billing above the reimbursed amount was $275.66 with a standard deviation of $78.11. (a) At the 5 percent level of significance, does this sample prove a violation of the guideline that the average patient should pay no more than $250 out-of-pocket? State your hypotheses and decision rule. (b) Is this a close decision?

9.93 The average service time at a Noodles & Company restaurant was 3.5 minutes in the previous year. Noodles implemented some time-saving measures and would like to know if they have been effective. They sample 20 service times and find the sample average is 3.2 minutes with a sample standard deviation of .4 minute. Using $\alpha = .05$, were the measures effective?

9.94 A ski tuning shop has set a goal not to exceed an average of 5 working days from the time the skis are brought in to the time they are finished. A random sample of 12 pairs of skis brought in for tuning showed the following service times (in days): 9, 2, 5, 1, 5, 4, 7, 5, 11, 3, 7, 2. At $\alpha = .05$, is the goal being met? 📄 **SkiTunes**

9.95 A recent study by the Government Accountability Office found that consumers got correct answers about Medicare only 67 percent of the time when they called 1-800-

MEDICARE. (a) At $\alpha = .05$, would a subsequent audit of 50 randomly chosen calls with 40 correct answers suffice to show that the percentage had risen? What is the *p*-value? (b) Is the normality criterion for the sample proportion met?

9.96 Beer shelf life is a problem for brewers and distributors because when beer is stored at room temperature, its flavor deteriorates. When the average furfuryl ether content reaches 6 μg per liter, a typical consumer begins to taste an unpleasant chemical flavor. (a) At $\alpha = .05$, would the following sample of 12 randomly chosen bottles stored for a month convince you that the mean furfuryl ether content exceeds the taste threshold? (b) What is the *p*-value? 📄 **BeerTaste**

6.53, 5.68, 8.10, 7.50, 6.32, 8.75, 5.98, 7.50, 5.01, 5.95, 6.40, 7.02

9.97 (a) A statistical study reported that a drug was effective with a *p*-value of .042. Explain in words what this tells you. (b) How would that compare to a drug that had a *p*-value of .087?

9.98 Bob said, "Why is a small *p*-value significant when a large one isn't? That seems backward." Try to explain it to Bob, giving an example to make your point.

9.99 Sarpedon Corp. claims that its car batteries average at least 880 CCA (cold-cranking amps). Tests on a sample of 9 batteries yield a mean of 871 CCA with a standard deviation of 15.6 CCA. (a) State the hypotheses to test Sarpedon's claim against this sample evidence. (b) What is the critical value at the 5 percent level of significance? (c) Should we reject Sarpedon's claim?

9.100 Thetis Mfg. says that its outboard watercraft engine's noise level at 75 percent throttle averages 100 decibels or less from the operator's seating position. A random sample of 8 noise level measurements showed a mean of 106 decibels with a standard deviation of 7.2 decibels. At the 5 percent level of significance, should we reject Thetis's claim?

PROPORTIONS: SMALL SAMPLES

9.101 An automaker states that its cars equipped with electronic fuel injection and computerized engine controls will start on the first try (hot or cold) 99 percent of the time. A survey of 100 new car owners revealed that 3 had not started on the first try during a recent cold snap. (a) At $\alpha = .025$, does this demonstrate that the automaker's claim is incorrect? (b) Calculate the *p*-value and interpret it. *Hint:* Use Excel to calculate the cumulative binomial probability $P(X \geq 3 \mid n = 100, \pi = .01) = 1 - P(X \leq 2 \mid n = 100, \pi = .01)$.

9.102 A quality standard says that no more than 2 percent of the eggs sold in a store may be cracked (not broken, just cracked). In 3 cartons (12 eggs each carton), 2 eggs are cracked. (a) At the .10 level of significance, does this prove that the standard is exceeded? (b) Calculate a *p*-value for the observed sample result. *Hint:* Use Excel to calculate the cumulative binomial probability $P(X \geq 2 \mid n = 36, \pi = .02) = 1 - P(X \leq 1 \mid n = 36, \pi = .02)$.

9.103 An experimental medication is administered to 16 people who suffer from migraines. After an hour, 10 say they feel better. Is the medication effective (i.e., is the percent who feel better greater than 50 percent)? Use $\alpha = .10$, explain fully, and show all steps.

9.104 The historical on-time percentage for Amtrak's Sunset Limited is 10 percent. In July, the train was on time 0 times in 31 runs. At the .10 level of significance, has the on-time percentage fallen? Explain clearly. *Hint:* Use Excel to calculate the cumulative binomial probability $P(X \leq 0 \mid n = 31, \pi = .10)$.

9.105 After 7 months, none of 238 angioplasty patients who received a drug-coated stent to keep their arteries open had experienced restenosis (re-blocking of the arteries). (a) Use Minitab to construct a 95 percent binomial confidence interval for the proportion of all angioplasty patients who experience restenosis. (b) Why is it necessary to use a binomial in this case? (c) If the goal is to reduce the occurrence of restenosis to 5 percent or less, does this sample show that the goal is being achieved?

POWER

Hint: In the power problems, use *LearningStats* (from the McGraw Hill Connect® downloads) to check your answers.

9.106 A certain brand of flat white interior latex paint claims one-coat coverage of 400 square feet per gallon. The standard deviation is known to be 20. A sample of 16 gallons is tested. (a) At $\alpha = .05$ in a left-tailed test, find the β risk and power assuming that the true mean is really 380 square feet per gallon. (b) Construct a left-tailed power curve, using increments of 5 square feet (400, 395, 390, 385, 380).

9.107 A process is normally distributed with standard deviation 12. Samples of size 4 are taken. Suppose that you wish to test the hypothesis that $\mu = 500$ at $\alpha = .05$ in a left-tailed test. (a) What is the β risk if the true mean is 495? If the true mean is 490? If the true mean is 485? If the true mean is 480? (b) Calculate the power for each of the preceding values of μ and sketch a power curve. (c) Repeat the previous exercises using $n = 16$.

TESTS OF VARIANCES

Hint: Use MegaStat or a similar software tool to check your work.

9.108 Is this sample of 25 exam scores inconsistent with the hypothesis that the true variance is 64 (i.e., $\sigma = 8$)? Use the 5 percent level of significance in a two-tailed test. Show all steps, including the hypotheses and critical values from Appendix E. ⬚ **Exams**

80	79	69	71	74
73	77	75	65	52
81	84	84	79	70
78	62	77	68	77
88	70	75	85	84

9.109 Hammermill Premium Inkjet 24 lb. paper has a specified brightness of 106. (a) At $\alpha = .005$, does this sample of 24 randomly chosen test sheets from a day's production run show that the mean brightness exceeds the specification? (b) Does the sample show that $\sigma^2 < 0.0025$? State the hypotheses and critical value for the left-tailed test from Appendix E. ⬚ **Brightness**

106.98	107.02	106.99	106.98	107.06	107.05	107.03	107.04
107.01	107.00	107.02	107.04	107.00	106.98	106.91	106.93
107.01	106.98	106.97	106.99	106.94	106.98	107.03	106.98

Related Reading

Vickers, Andrew J. What Is a p-Value Anyway? 34 Stories to Help You Actually Understand Statistics. Pearson, 2010.

Wasserstein, Ronald L., Allen L. Schirm, and Nicole A. Lazar. (2019), "Moving to a World Beyond '$p < 0.05$.'" *The American Statistician* 73, no. 51 (2019), pp. 1–19.

CHAPTER 9 More Learning Resources

You can access these *LearningStats* demonstrations through Connect to help you understand one-sample hypothesis tests.

Topic	*LearningStats Demonstrations*
Common hypothesis tests	⬚ Overview of Hypothesis Tests ⬚ One Sample Tests Using R ⬚ Do-It-Yourself Simulation ⬚ Sampling Distribution Examples
Type I error and power	⬚ Type I Error ⬚ *p*-Value Illustration ⬚ Power Curves: Examples ⬚ Power Curves: Do-It-Yourself ⬚ Power Curve Families ⬚ Finite Populations ⬚ Useful Formulas
Tables	⬚ Appendix C—Normal ⬚ Appendix D—Student's *t* ⬚ Appendix E—Chi-Square

Key: ⬚ = Excel ⬚ = PowerPoint ⬚ = PDF

Software Supplement

Using MegaStat to Test a Mean

Excel does not offer a one-sample t-test, but you can get tests for one mean, including a confidence interval, using MegaStat. Figure 9.23 shows its setup screen and output for the test of one mean for the hot chocolate data from Example 9.5. You enter the data range and everything else is automatic. You have a choice of z or t, but to use z, you must know σ.

Using MegaStat to Test a Variance

Excel has no test for a variance, but you can use Minitab or Mega-Stat. Figure 9.24 shows MegaStat's setup screen and output for the variance test in Example 9.10, including a confidence interval for σ^2.

Figure 9.23   MegaStat Test for One Mean

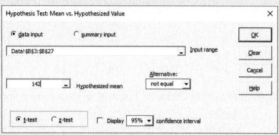
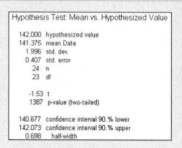

Source: MegaStat.

Figure 9.24   MegaStat Test for One Variance

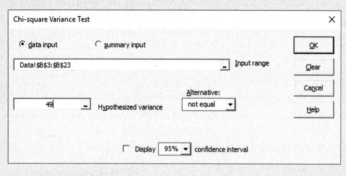

Source: MegaStat.

Software Supplement

One-Sample Test for a Mean (μ) 🖻 HotChoc

Customer research shows that the ideal temperature for hot chocolate is 142°F ("hot" but not "too hot"). A random sample of 24 cups of hot chocolate shows these temperatures:

> 140 140 141 145 143 144 142 140 145 143 140 140
> 141 141 137 142 143 141 142 142 143 141 138 139

At $\alpha = .10$, does the true mean differ from 142? The population variance is unknown, so we use the Student's t distribution with the R function t.test(). The problem asks for a two-tailed test and 90 percent confidence (where confidence $= 1 - \alpha$). Input the data any way you wish (see **Appendix K**). Our sample is small, so we will use c() to enter the data.

```
> Temp=c(140,140,141,145,143,144,142,
    140,145,143,140,140,141,141,137,142,143, 141,142,142,143,141,138,139)
> t.test(Temp,alternative="two.sided", mu=142,conf.level=.90)

One Sample t-test
data: Temp
t = −1.5341, df = 23, p-value = 0.1387
alternative hypothesis:
true mean is not equal to 142
90 percent confidence interval:
140.6767 142.0733
sample estimates:
mean of x 141.375
```

We are unable to reject the null hypothesis that $\mu = 142$, which agrees with the textbook (see Example 9.5). The *p*-value (.1387) also agrees with the textbook. Choose alternative="less" if you want a left-tailed test or alternative="greater" for a right-tailed test.

One-Sample Test for a Proportion (π)

In brick-and-mortar department stores, the historical return rate for merchandise is 13.0 percent. At one department store, there were 22 returns in the most recent 250 purchases. At $\alpha = .05$, does this sample indicate that their true return rate now is below the historical return rate? The problem asks for a left-tailed test and 95 percent confidence (where confidence = $1 - \alpha$). Because R's default confidence level is 95 percent, we can omit the confidence=.95 argument. We also omit the continuity correction argument correct=FALSE because that is the default.

```
> prop.test(x=22, n=250, alternative="less", p=0.13, correct=FALSE)
1-sample proportions test without continuity
correction
data: 22 out of 250, null probability 0.13
X-squared = 3.8992, df = 1, p-value = 0.02415
alternative hypothesis: true p is less than 0.13
95 percent confidence interval:
0.0000000 0.1220539
sample estimates:
p 0.088
```

The z_{calc} test statistic (1.9746) is the square root of the chi-square test statistic (3.8992), which agrees with the textbook (see Example 9.7). The *p*-value (.0242) also agrees with the textbook. We reject H_0: $\pi \geq .13$ in favor of H_1: $\pi < .13$.

10 Two-Sample Hypothesis Tests

CHAPTER CONTENTS

CHAPTER LEARNING OBJECTIVES

When you finish this chapter, you should be able to

LO 10-1 Recognize and perform a test for two means.

LO 10-2 Explain the assumptions underlying the two-sample test of means.

LO 10-3 Construct a confidence interval for $\mu_1 - \mu_2$.

LO 10-4 Recognize paired data and be able to perform a paired t test.

LO 10-5 Perform a test to compare two proportions using z.

LO 10-6 Check whether normality may be assumed for two proportions.

LO 10-7 Construct a confidence interval for $\pi_1 - \pi_2$.

LO 10-8 Carry out a test of two variances using the F distribution.

10.1 TWO-SAMPLE TESTS

The logic and applications of hypothesis testing that you learned in Chapter 9 will continue here, but now we consider two-sample tests. The two-sample test is used to make inferences about the two populations from which the samples were drawn. The use of these techniques is widespread in science and engineering as well as social sciences. For example, drug companies use clinical trials to determine the effectiveness of new drugs, agricultural science uses these methods to compare crop yields, and marketing firms use them to contrast purchase patterns in different demographic groups.

Gorodenkoff/Shutterstock

What Is a Two-Sample Test?

Two-sample tests compare two sample estimates *with each other,* whereas one-sample tests compare a sample estimate with a nonsample benchmark or target (a claim or prior belief about a population parameter). Two-sample tests are especially useful because they possess a built-in point of comparison. You can think of many situations where two groups are to be compared:

- Before *versus* after
- Old *versus* new
- Experimental *versus* control

Sometimes we don't really care about the actual value of the population parameter, but only whether the parameter is the same for both populations. Usually, the null hypothesis is that both samples were drawn from populations with the same parameter value, but we also can test for a given degree of difference.

The logic of two-sample tests is based on the fact that two samples drawn from the *same population* may yield *different estimates* of a parameter due to chance. For example, exhaust emission tests could yield different results for two vehicles of the same type. Only if the two sample statistics differ by more than the amount attributable to chance can we conclude that the samples came from populations with different parameter values, as illustrated in Figure 10.1. Two-sample tests occur in all types of business applications. Several examples are shown here:

Environment Samples of methane gas emissions (in standard cubic feet per day) from two different manufacturers of low-bleed pneumatic controllers were analyzed. Manufacturer A's sample mean was 510.5 with a standard deviation of 147.2 in 18 tests, compared with Manufacturer B's mean of 628.9 with a standard deviation of 237.9 in 17 tests. Is the difference statistically significant?

Safety In Dallas, some fire trucks were painted yellow (instead of red) to heighten their visibility. During a test period, the fleet of red fire trucks made 153,348 runs and had 20 accidents, while the fleet of yellow fire trucks made 135,035 runs and had 4 accidents. Is the difference in accident rates significant?

Figure 10.1

Same Population or Different?

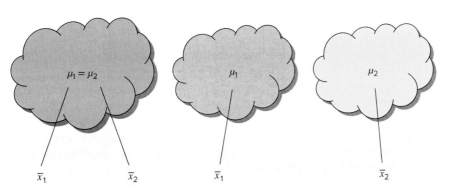

Samples came from the same population. Any differences are due to sampling variation.

Samples came from populations with different parameter values.

Education   In a certain college class, 20 randomly chosen students were given a tutorial, while 20 others used a self-study computer simulation. On the same 20-point quiz, the tutorial students' mean score was 16.7 with a standard deviation of 2.5, compared with a mean of 14.5 and a standard deviation of 3.2 for the simulation students. Did the tutorial students do better, or is it just due to chance? Is there any significant difference in the degree of variation in the two groups?

Automotive   A new bumper is installed on selected vehicles in a corporate fleet. During a 1-year test period, 12 vehicles with the new bumper were involved in accidents, incurring mean damage of $1,101 with a standard deviation of $696. During the same year, 9 vehicles with the old bumpers were involved in accidents, incurring mean damage of $1,766 with a standard deviation of $838. Did the new bumper significantly reduce damage? Did it significantly reduce variation?

Marketing   At a University of Colorado women's home basketball game, a random sample of 25 concession purchases showed a mean of $7.12 with a standard deviation of $2.14. For the next week's home game, the admission ticket had a discount coupon for popcorn printed on the back. A random sample of 25 purchases from that week showed a mean of $8.29 with a standard deviation of $3.02. Was there a statistically significant increase in the average concession stand purchases with the coupon?

Mini Case 10.1

Early Intervention Saves Lives

Statistics is helping U.S. hospitals prove the value of innovative organizational changes to deal with medical crisis situations. At the Pittsburgh Medical Center, "SWAT teams" were shown to reduce patient mortality by cutting red tape for critically ill patients. They formed a Rapid Response Team (RRT) consisting of a critical care nurse, intensive care therapist, and respiratory therapist, empowered to make decisions without waiting until the patient's doctor could be paged. Statistics were collected on cardiac arrests for 2 months before and after the RRT concept was implemented. The sample data revealed more than a 50 percent reduction in total cardiac deaths and a 46 percent decline in average ICU days after cardiac arrest from 2.59 days to only 1.50 days after RRT. These improvements were both *statistically significant* and of *practical importance* because of the medical benefits and the large cost savings in hospital care. Statistics played a similar role at the University of California San Francisco Medical Center in demonstrating the value of a new method of expediting treatment of heart attack emergency patients. (See "How Statistics Can Save Failing Hearts," *The New York Times*, March 7, 2007, p. C1.)

Test Procedure

The testing procedure is like that of one-sample tests. We state our hypotheses, set up a decision rule, insert the sample statistics, and make a decision. Because the true parameters are unknown, we rely on statistical theory to help us reach a defensible conclusion about our hypotheses. Our decision could be wrong—we could commit a **Type I** or **Type II error**—but at least we can specify our risk of making an error. Larger samples are always desirable because they permit us to reduce the chance of making either a Type I error or Type II error (i.e., increase the power of the test). However, larger samples take time and cost money, so we often must work with the available data.

Analytics in Action

A/B Testing: Old Technique, New Application

A/B testing is used to compare customers' reactions to differences in website design. The basic concept of A/B testing uses two-sample statistical hypothesis testing based on techniques developed a hundred years ago by a statistician named Ronald Fisher. While the fundamentals of the statistics haven't changed, what has changed is the technology as well as *how* the data are collected and the *quantity* of data collected. The Internet has made it possible to collect millions of customer responses on different website layouts in real time.

When designing an online retail web page, developers consider text differences (e.g., font and color), pay button differences (e.g., placement and size), user device choice (e.g., mobile or desktop), and many others. The number of combinations to consider is extremely large. To deal with the size of this problem, statisticians use multivariate A/B tests comparing customer responses from two, three, or even four website designs at a time. Results are analyzed using significance levels and margins of error—statistical terms that you've already learned about. A/B testing helps retailers design websites that increase the chance that a customer will buy their products. Examples of changes companies have made after testing include:

- Offering health service packages with a quoted package price rather than individual service pricing.

- Removing discount codes when customers only want to make a purchase.

- Simplifying text on a landing page rather than including detailed information.

- Posting photos that show customers in action versus photos with static poses.

Source: https://hbr.org/2017/06/a-refresher-on-ab-testing.

10.2 COMPARING TWO MEANS: INDEPENDENT SAMPLES

Comparing two population means is a common business problem. Is there a difference between the average customer purchase at Starbucks on Saturday and Sunday mornings? Is there a difference between the average satisfaction scores from a taste test for two versions of a new menu item at Noodles & Company? Is there a difference between the average age of full-time and part-time seasonal employees at a Vail Resorts ski mountain?

LO 10-1

Recognize and perform a test for two means.

Format of Hypotheses

The process of comparing two means starts by stating null and alternative hypotheses, just as we did in Chapter 9. To test for a difference in means of magnitude D_0, the possible pairs of null and alternative hypotheses are

Left-Tailed Test	*Two-Tailed Test*	*Right-Tailed Test*
$H_0: \mu_1 - \mu_2 \geq D_0$	$H_0: \mu_1 - \mu_2 = D_0$	$H_0: \mu_1 - \mu_2 \leq D_0$
$H_1: \mu_1 - \mu_2 < D_0$	$H_1: \mu_1 - \mu_2 \neq D_0$	$H_1: \mu_1 - \mu_2 > D_0$

For example, we might ask if the difference between the average number of years worked at a ski resort for full-time and part-time seasonal employees is greater than 2 years. In this situation, we would formulate the null hypothesis as: $H_0: \mu_1 - \mu_2 \leq 2$, where $D_0 = 2$ years. If a company is simply interested in knowing whether a difference exists between two populations, they would want to test the null hypothesis $H_0: \mu_1 - \mu_2 = 0$, where $D_0 = 0$.

Test Statistic

The test compares the observed difference in means $\overline{X}_1 - \overline{X}_2$ with the hypothesized difference D_0. The test assumes that both $\overline{X}_1$ and $\overline{X}_2$ are calculated from independent random samples taken from normal populations or if the populations are not normal but with $n_1 >= 30$ and $n_2 >= 30$ (Central Limit Theorem). The **test statistic** is the difference between the sample statistic and D_0 divided by the standard error of the sample statistic. There are three cases to consider.

For the case where we know the values of the population variances, σ_1^2 and σ_2^2, the test statistic is a *z*-score. We would use the standard normal distribution to find **p-values** or critical values of z_α. Critical values of *z*, not z_α. The critical values of *z* can be z_α or $z_{(\alpha/2)}$ depending if the problem is single- or double-tailed.

Case 1: Known Variances

(10.1)
$$z_{\text{calc}} = \frac{(\overline{x}_1 - \overline{x}_2) - D_0}{\sqrt{\dfrac{\sigma_1^2}{n_1} + \dfrac{\sigma_2^2}{n_2}}}$$

For the more common case where we *don't* know the values of the population variances but we have reason to believe they are equal, we would use the Student's *t* distribution. We would need to rely on sample estimates s_1^2 and s_2^2 for the population variances σ_1^2 and σ_2^2. By assuming that the population variances are equal, we are allowed to *pool* the sample variances by taking a weighted average of s_1^2 and s_2^2 to calculate an estimate of the common population variance. Weights are assigned to s_1^2 and s_2^2 based on their respective degrees of freedom $(n_1 - 1)$ and $(n_2 - 1)$. Because we are pooling the sample variances, the common variance estimate is called the **pooled variance** and is denoted s_p^2. Case 2 is often called the *pooled* t *test*. Degrees of freedom for the pooled *t* test will be the sum of the degrees of freedom for each individual sample.

Case 2: Unknown Variances Assumed Equal

(10.2)
$$t_{\text{calc}} = \frac{(\overline{x}_1 - \overline{x}_2) - D_0}{\sqrt{\dfrac{s_p^2}{n_1} + \dfrac{s_p^2}{n_2}}} \quad \text{where the pooled variance is}$$

$$s_p^2 = \frac{(n_1 - 1)s_1^2 + (n_2 - 1)s_2^2}{n_1 + n_2 - 2} \quad \text{and} \quad d.f. = n_1 + n_2 - 2$$

If the unknown variances σ_1^2 and σ_2^2 are assumed *unequal*, we do not pool the variances. Instead, we replace σ_1^2 and σ_2^2 with the sample variances s_1^2 and s_2^2. This is a more conservative assumption than Case 2. Under these conditions, the distribution of the random variable $\overline{X}_1 - \overline{X}_2$ is no longer certain, a difficulty known as the **Behrens-Fisher problem**. However, the comparison of means can reliably be performed using a Student's *t* test with **Welch's adjusted degrees of freedom**.

Case 3: Unknown Variances Assumed Unequal

$$t_{calc} = \frac{(\bar{x}_1 - \bar{x}_2) - D_0}{\sqrt{\dfrac{s_1^2}{n_1} + \dfrac{s_2^2}{n_2}}} \quad \text{with } d.f. = \frac{\left(\dfrac{s_1^2}{n_1} + \dfrac{s_2^2}{n_2}\right)^2}{\dfrac{\left(\dfrac{s_1^2}{n_1}\right)^2}{n_1 - 1} + \dfrac{\left(\dfrac{s_2^2}{n_2}\right)^2}{n_2 - 1}} \quad \textbf{(10.3)}$$

Finding Welch's degrees of freedom requires a tedious calculation, but this is easily handled by Excel. The formulas for Case 2 and Case 3 will usually yield the same decision about the hypotheses unless the sample sizes and variances differ greatly.

For the common situation of testing for a zero difference ($D_0 = 0$) in two population means, the possible pairs of null and alternative hypotheses are

Left-Tailed Test	*Two-Tailed Test*	*Right-Tailed Test*
$H_0: \mu_1 - \mu_2 \geq 0$	$H_0: \mu_1 - \mu_2 = 0$	$H_0: \mu_1 - \mu_2 \leq 0$
$H_1: \mu_1 - \mu_2 < 0$	$H_1: \mu_1 - \mu_2 \neq 0$	$H_1: \mu_1 - \mu_2 > 0$

Table 10.1 summarizes the formulas for the test statistic in each of the three cases described above when we assume that $\mu_1 - \mu_2 = 0$. We have simplified the formulas based on the assumption that we will usually be testing for equal population means. We have left off the expression D_0 because we are assuming that $D_0 = 0$. All of these test statistics presume independent random samples from normal populations, although in practice they are robust to non-normality as long as the samples are not too small i.e., $n_1 >= 30$ and $n_2 >= 30$ and the populations are not too skewed.

Case 1	Case 2	Case 3
Known Variances	Unknown Variances, Assumed Equal	Unknown Variances, Assumed Unequal
$z_{calc} = \dfrac{\bar{x}_1 - \bar{x}_2}{\sqrt{\dfrac{\sigma_1^2}{n_1} + \dfrac{\sigma_2^2}{n_2}}}$	$t_{calc} = \dfrac{\bar{x}_1 - \bar{x}_2}{\sqrt{\dfrac{s_p^2}{n_1} + \dfrac{s_p^2}{n_2}}}$ where $s_p^2 = \dfrac{(n_1 - 1)s_1^2 + (n_2 - 1)s_2^2}{n_1 + n_2 - 2}$	$t_{calc} = \dfrac{\bar{x}_1 - \bar{x}_2}{\sqrt{\dfrac{s_1^2}{n_1} + \dfrac{s_2^2}{n_2}}}$
For critical value, use standard normal distribution	For critical value, use Student's t with $d.f. = n_1 + n_2 - 2$	For critical value, use Student's t with Welch's adjusted degrees of freedom

Table 10.1

Test Statistic for Zero Difference of Means ($D_0 = 0$)

Most of the time you will be using a computer. As long as you have raw data (i.e., the original samples of n_1 and n_2 observations), Excel's Data Analysis menu handles all three cases, as shown in Figure 10.2. Both MegaStat and Minitab also perform these tests and will do so for summarized data as well (i.e., when you have $\bar{x}_1$, $\bar{x}_2$, s_1, s_2 instead of the n_1 and n_2 data columns).

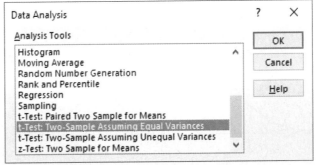

Figure 10.2

Excel's Data Analysis Menu

Source: Microsoft Excel.

Example 10.1

Drug Prices in Two States

The price of prescription drugs is an ongoing national issue in the United States. Zocor is a common prescription cholesterol-reducing drug prescribed for people who are at risk for heart disease. Table 10.2 shows Zocor prices from randomly selected pharmacies in two states. At $\alpha = .05$, is there a difference in the mean for all pharmacies in Colorado and Texas?

Table 10.2

Zocor Prices (30-Day Supply) in Two States 🖅 RxPrice

Colorado Pharmacies		Texas Pharmacies	
City	**Price ($)**	**City**	**Price ($)**
Alamosa	125.05	Austin	145.32
Avon	137.56	Austin	131.19
Broomfield	142.50	Austin	151.65
Buena Vista	145.95	Austin	141.55
Colorado Springs	117.49	Austin	125.99
Colorado Springs	142.75	Dallas	126.29
Denver	121.99	Dallas	139.19
Denver	117.49	Dallas	156.00
Eaton	141.64	Dallas	137.56
Fort Collins	128.69	Houston	154.10
Gunnison	130.29	Houston	126.41
Pueblo	142.39	Houston	114.00
Pueblo	121.99	Houston	144.99
Pueblo	141.30		
Sterling	153.43		
Walsenburg	133.39		

$$\bar{x}_1 = \$133.994$$
$$s_1 = \$11.015$$
$$n_1 = 16 \text{ pharmacies}$$

$$\bar{x}_2 = \$138.018$$
$$s_2 = \$12.663$$
$$n_2 = 13 \text{ pharmacies}$$

Source: Public Research Interest Group (www.pirg.org). Surveyed pharmacies were chosen from the telephone directory in 2004.

Step 1: State the Hypotheses To check for a significant difference ($D_0 = 0$) without regard for its direction, we choose a two-tailed test. The hypotheses to be tested are

$$H_0: \mu_1 - \mu_2 = 0$$
$$H_1: \mu_1 - \mu_2 \neq 0$$

Step 2: Specify the Decision Rule We will assume equal variances. For the pooled-variance t test, degrees of freedom are $d.f. = n_1 + n_2 - 2 = 16 + 13 - 2 = 27$. From Appendix D we get the two-tail critical value $t_{\alpha/2}$ or $t_{.025} = \pm 2.052$. The decision rule is illustrated in Figure 10.3.

Figure 10.3

Two-Tailed Decision Rule for Student's *t* with $\alpha = .05$ and $d.f. = 27$

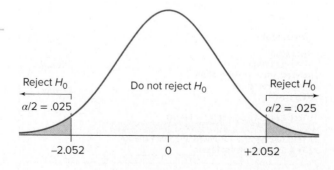

Reject H_0 Do not reject H_0 Reject H_0

$\alpha/2 = .025$ $\alpha/2 = .025$

−2.052   0   +2.052

Step 3: Calculate the Test Statistic   The sample statistics are

$$\bar{x}_1 = 133.994 \qquad \bar{x}_2 = 138.018$$

$$s_1 = 11.015 \qquad s_2 = 12.663$$

$$n_1 = 16 \qquad n_2 = 13$$

Because we are assuming equal variances, we use the formulas for Case 2. The pooled variance s_p^2 is

$$s_p^2 = \frac{(n_1 - 1)s_1^2 + (n_2 - 1)s_2^2}{n_1 + n_2 - 2} = \frac{(16 - 1)(11.015)^2 + (13 - 1)(12.663)^2}{16 + 13 - 2} = 138.6737$$

Using s_p^2, the test statistic is

$$t_{\text{calc}} = \frac{\bar{x}_1 - \bar{x}_2}{\sqrt{\dfrac{s_p^2}{n_1} + \dfrac{s_p^2}{n_2}}} = \frac{133.994 - 138.018}{\sqrt{\dfrac{138.6737}{16} + \dfrac{138.6737}{13}}} = \frac{-4.024}{4.39708} = -0.915$$

The pooled standard deviation is $s_p = \sqrt{138.6737} = 11.776$. Notice that s_p always lies between s_1 and s_2 (if not, you made an arithmetic error). This is because s_p^2 is a weighted average of s_1^2 and s_2^2.

Step 4: Make the Decision   The test statistic $t_{\text{calc}} = -0.915$ does not fall in the rejection region, so we cannot reject the hypothesis of equal means. Excel's menu and output are shown in Figure 10.4.

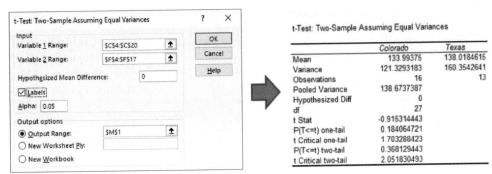

Figure 10.4

Excel's Data Analysis with Unknown but Equal Variances

Source: Microsoft Excel.

The Hypothesized Mean Difference is "0" because we are testing for $D_0 = 0$. Check Labels when the data column label is selected. The Excel default value for α is .05. Note that the Excel output provides both the one-tailed and two-tailed critical values and allows the user to determine when the one-tail critical value should be negative for a left-tailed test.

Step 5: Take Action   Because the difference in sample means is within the realm of chance assuming that H_0 is true, no further investigation is required.

The *p*-value can be verified using the two-tailed Excel function =T.DIST.2T(.915,27) = .3681. This large *p*-value says that a difference of sample means of this magnitude would happen by chance about 37 percent of the time if $\mu_1 = \mu_2$. The observed difference in sample means seems to be well within the realm of chance.

The sample variances in this example are similar, so the assumption of equal variances is reasonable. But if we did use the formulas for Case 3 (assuming *unequal* variances), the test statistic would be

$$t_{\text{calc}} = \frac{\bar{x}_1 - \bar{x}_2}{\sqrt{\dfrac{s_1^2}{n_1} + \dfrac{s_2^2}{n_2}}} = \frac{133.994 - 138.018}{\sqrt{\dfrac{(11.015)^2}{16} + \dfrac{(12.663)^2}{13}}} = \frac{-4.024}{4.4629} = -0.902$$

Welch's adjusted degrees of freedom will be smaller than in Case 2 (24 instead of 27):

$$d.f. = \frac{\left[\frac{s_1^2}{n_1} + \frac{s_2^2}{n_2}\right]^2}{\frac{\left(\frac{s_1^2}{n_1}\right)^2}{n_1 - 1} + \frac{\left(\frac{s_2^2}{n_2}\right)^2}{n_2 - 1}} = \frac{\left[\frac{(11.015)^2}{16} + \frac{(12.663)^2}{13}\right]^2}{\frac{\left((11.015)^2\right)^2}{16 - 1} + \frac{\left((12.663)^2\right)^2}{13 - 1}} = 24$$

The degrees of freedom are truncated to the next lower integer to be conservative.

For the unequal-variance t test with $d.f. = 24$, Appendix D or Excel gives the two-tail critical value $t_{.025} = \pm 2.064$. The decision rule is illustrated in Figure 10.5.

Figure 10.5

Two-Tail Decision Rule for Student's t with $\alpha = .05$ and $d.f. = 24$

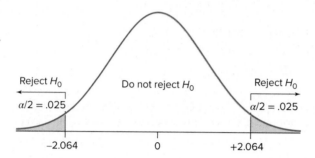

Reject H_0
$\alpha/2 = .025$

Do not reject H_0

Reject H_0
$\alpha/2 = .025$

−2.064   0   +2.064

The calculations are best done by computer. Excel's menu and output are shown in Figure 10.6. Excel shows one-tailed and two-tailed tests.

For the Zocor data, either assumption about the variances would lead to the same conclusion:

Case 2 (equal variances) $t_{.025} = \pm 2.052$ ($d.f. = 27$) $t_{calc} = -0.915$ ($p = .3681$)
Do not reject at $\alpha = .05$

Case 3 (unequal variances) $t_{.025} = \pm 2.064$ ($d.f. = 24$) $t_{calc} = -0.902$ ($p = .3761$)
Do not reject at $\alpha = .05$

Figure 10.6

Excel's Data Analysis with Unknown and Unequal Variances

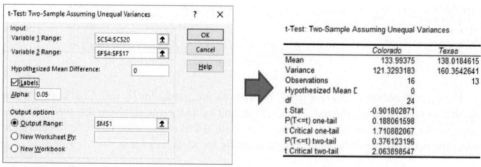

	Colorado	Texas
Mean	133.99375	138.0184615
Variance	121.3293183	160.3542641
Observations	16	13
Hypothesized Mean D	0	
df	24	
t Stat	-0.901802871	
P(T<=t) one-tail	0.188061598	
t Critical one-tail	1.710882067	
P(T<=t) two-tail	0.376123196	
t Critical two-tail	2.063898547	

t-Test: Two-Sample Assuming Unequal Variances

Source: Microsoft Excel.

LO 10-2

Explain the assumptions underlying the two-sample test of means.

Which Assumption Is Best?

If the *sample sizes are equal,* the Case 2 and Case 3 test statistics will always be identical, but the degrees of freedom (and hence the critical values) may differ. If you have no information about the population variances, then the best choice is Case 3. The fewer assumptions you make about your populations, the less likely you are to make a mistake in your conclusions. Case 1 (known population variances) is not explored further here because it is uncommon in business.

Must Sample Sizes Be Equal?

Unequal sample sizes are common, and the formulas still apply. However, there are advantages to equal sample sizes. We avoid unbalanced sample sizes when possible to increase the *power* of the test. But many times, we have to take the samples as they come.

Large Samples

For unknown variances, if both samples are large ($n_1 \geq 30$ and $n_2 \geq 30$) and you have reason to think the population isn't badly skewed (look at the histograms or dot plots of the samples), it is common to use formula 10.4 with Appendix C. Although it usually gives results very close to the "proper" t tests, this approach is not conservative (i.e., it may increase Type I risk).

$$z_{calc} = \frac{\overline{x}_1 - \overline{x}_2}{\sqrt{\dfrac{s_1^2}{n_1} + \dfrac{s_2^2}{n_2}}} \qquad \text{(large samples, symmetric populations)} \qquad \textbf{(10.4)}$$

Caution: Three Issues

Bear in mind three questions when you are comparing two sample means:

- Are the populations skewed? Are there outliers?
- Are the sample sizes large ($n \geq 30$)?
- Is the difference *important* as well as significant?

The first two questions refer to the assumption of normal populations, upon which the tests are based. Fortunately, the t test is robust to non-normality as long as the samples are not too small and the populations are not too skewed. Skewness or outliers can usually be seen in a histogram or dot plot of each sample. The t tests (Case 2 and Case 3) are OK in the face of moderate skewness, especially if the samples are large (e.g., sample sizes of at least 30). Outliers are more serious and might require consultation with a statistician. In such cases, you might ask yourself whether a test of means is appropriate. With small samples or skewed data, the mean may not reflect central tendency, and your test may lack power. In such situations, it may be better merely to describe the samples and skip the formal t tests.

A small difference in means or proportions could be *statistically* significant if the sample size is large because the standard error gets smaller as the sample size gets larger. So, we must separately ask if the difference is *important*. The answer depends on the data magnitude and the consequences to the decision maker. For example, individuals who took an experimental weight-loss drug called Lorcaserin lost 5.8 percent of their weight after a year, compared to 2.5 percent for those who took a placebo. The difference was statistically significant. However, in rejecting the obesity drug, the FDA said that the 3.3 percent difference was below the 5 percent efficacy criterion set by the FDA. Although *significant,* the difference was not *important* enough to approve the new medication, given the possibility of additional health risks. (See *The New York Times,* September 17, 2010, p. B1.)

Mini Case 10.2

Methane Gas Emissions of Pneumatic Controllers

Certain types of equipment can be controlled pneumatically through the use of pressurized gas such as natural gas. Such equipment is often used in the oil and gas production industry because electricity is not always available in remote locations. As part of normal operations, pneumatic controllers release methane gas to the atmosphere. Older-technology controllers, called high-bleed controllers, typically release more methane gas than newer-technology controllers, called low-bleed controllers. Measuring the amount of gas emissions reduction is a critical factor when evaluating the benefits of switching out the high-bleed for the low-bleed controllers.

The American Carbon Registry, www.americancarbonregistry.org, provides approved measurement methodologies for energy companies that wish to show greenhouse gas emission reduction through retrofitting high-bleed pneumatic controllers with low-bleed controllers. One step is determining whether different manufacturers of high-bleed controllers have different baseline gas emissions. Independent samples of gas emissions from two different manufacturers were analyzed using a two-tailed two-sample hypothesis test. Summary sample data from the two manufacturers are shown in Table 10.3.

Table 10.3   Pneumatic Controller Methane Gas Emission (scfd)

Cemco	Invalco
$\bar{x}_1 = 510.5$	$\bar{x}_2 = 628.9$
$s_1 = 147.2$	$s_2 = 237.9$
$n_1 = 18$	$n_2 = 17$

Note: scfd: standard cubic foot per day.
Source: www.americancarbonregistry.org/carbonaccounting/.

We will do a two-tailed test at $\alpha = .10$. The hypotheses are

$$H_0: \mu_1 - \mu_2 = 0$$
$$H_1: \mu_1 - \mu_2 \neq 0$$

Because the variances are unknown and the sample standard deviations appear very different, we will assume the population variances are unequal, using formula 10.3 for the t statistic. (Later on, in Section 10.7, we will learn how to actually test the equality of variances.)

$$t_{calc} = \frac{\bar{x}_1 - \bar{x}_2}{\sqrt{\frac{s_1^2}{n_1} + \frac{s_2^2}{n_2}}} = \frac{510.5 - 628.9}{\sqrt{\frac{147.2^2}{18} + \frac{237.9^2}{17}}} = -1.759$$

Welch's adjusted degrees of freedom are

$$d.f. = \frac{\left[\frac{s_1^2}{n_1} + \frac{s_2^2}{n_2}\right]^2}{\frac{\left(\frac{s_1^2}{n_1}\right)^2}{n_1 - 1} + \frac{\left(\frac{s_2^2}{n_2}\right)^2}{n_2 - 1}} = \frac{\left[\frac{147.2^2}{18} + \frac{237.9^2}{17}\right]^2}{\frac{\left(\frac{147.2^2}{18}\right)^2}{18 - 1} + \frac{\left(\frac{237.9^2}{17}\right)^2}{17 - 1}} = 26$$

The critical value is =T.INV.2T(.10,26) = $t_{.05}$ = 1.706. Because |−1.759| > 1.706, we reject H_0 and conclude that the average emissions from the two manufacturers' controllers are not equal. Alternatively, we could compare the p-value to α. The p-value for this test is found using Excel: =T.DIST.2T(1.759,26) = .0903. Because .0903 < .10, our decision is to reject the assumption that the means are equal and conclude that the means are not equal. The implication is that baseline emissions must be calculated separately for each manufacturer in order to accurately measure emissions reductions when retrofitting high-bleed controllers with low-bleed controllers.

Section Exercises

connect

Hint: Calculate the p-values using Excel, and show each Excel formula you used. Excel has no t test for summarized data (i.e., given $\bar{x}_1, s_1, n_1, \bar{x}_2, s_2, n_2$), but you can download a spreadsheet calculator for summarized data from McGraw Hill's Connect®, or you can use MegaStat.

10.1   Do a two-sample test for equality of means assuming equal variances. Calculate the p-value.

a. Comparison of GPA for randomly chosen college juniors and seniors: $\bar{x}_1 = 3.05$, $s_1 = .20$, $n_1 = 15$, $\bar{x}_2 = 3.25$, $s_2 = .30$, $n_2 = 15$, $\alpha = .025$, left-tailed test.

b. Comparison of average commute miles for randomly chosen students at two community colleges: $\bar{x}_1 = 15$, $s_1 = 5$, $n_1 = 22$, $\bar{x}_2 = 18$, $s_2 = 7$, $n_2 = 19$, $\alpha = .05$, two-tailed test.

c. Comparison of credits at time of graduation for randomly chosen accounting and economics students: $\bar{x}_1 = 139$, $s_1 = 2.8$, $n_1 = 12$, $\bar{x}_2 = 137$, $s_2 = 2.7$, $n_2 = 17$, $\alpha = .05$, right-tailed test.

10.2   Repeat the previous exercise, assuming unequal variances. Calculate the *p*-value using Excel, and show the Excel formula you used.

10.3   Is there a difference in the average number of years' seniority between returning part-time seasonal employees and returning full-time seasonal employees at a ski resort? From a random sample of 191 returning part-time employees, the average seniority, $\bar{x}_1$, was 4.9 years with a standard deviation, s_1, equal to 5.4 years. From a random sample of 833 returning full-time employees, the average seniority, $\bar{x}_2$, was 7.9 years with a standard deviation, s_2, equal to 8.3 years. Assume the population variances are not equal. (a) Test the hypothesis of equal means using $\alpha = .01$. (b) Calculate the *p*-value using Excel.

10.4   The average mpg usage for a 2022 Toyota Prius for a sample of 10 tanks of gas was 53.4 with a standard deviation of 1.8. For a 2022 Ford Fusion, the average mpg usage for a sample of 10 tanks of gas was 42.0 with a standard deviation of 2.3. (a) Assuming equal variances, at $\alpha = .01$, is the difference between the Toyota Prius's true mean and the Ford Fusion's true mean more than 10 mpg? (b) Calculate the *p*-value using Excel.

10.5   When the background music tempo was slow, the mean amount of bar purchases for a sample of 17 restaurant patrons was $30.47 with a standard deviation of $15.10. When the background music tempo was fast, the mean amount of bar purchases for a sample of 14 patrons in the same restaurant was $21.62 with a standard deviation of $9.50. (a) Assuming equal variances, at $\alpha = .01$, is the true mean higher when the music is slow? (b) Calculate the *p*-value using Excel.

10.6   Are women's feet getting bigger? Retailers in the past 20 years have had to increase their stock of larger sizes. Walmart Stores, Inc., and Payless ShoeSource, Inc., have been aggressive in stocking larger sizes, and Nordstrom's reports that its larger sizes typically sell out first. Assuming equal variances, at $\alpha = .025$, do these random shoe size samples of 12 randomly chosen women in each age group show that women's shoe sizes have increased? 🖭 **ShoeSize1**

Born in 1980:	8	7.5	8.5	8.5	8	7.5	9.5	7.5	8	8	8.5	9
Born in 1960:	8.5	7.5	8	8	7.5	7.5	7.5	8	7	8	7	8

10.7   Researchers analyzed 12 samples of two kinds of Stella's decaffeinated coffee. The caffeine in a cup of decaffeinated espresso had a mean of 9.4 mg with a standard deviation of 3.2 mg, while brewed decaffeinated coffee had a mean of 12.7 mg with a standard deviation of 0.35 mg. Assuming unequal population variances, is there a significant difference in caffeine content between these two beverages at $\alpha = .01$?

10.8   On a random basis, Bob buys a small take-out coffee from one of two restaurants. As a statistics project in the month of May, he measured the temperature of each cup immediately after purchase, using an analog cooking thermometer. Assuming equal variances, is the mean temperature higher at Panera than at Bruegger's at $\alpha = .01$? *Note:* Ideal coffee temperature is a matter of individual preference (see www.coffeedetective.com). 🖭 **Coffee**

Panera	171	161	169	179	171	166	169	178	171	165	172	172
Bruegger's	168	165	172	151	162	158	157	160	158	160	158	164

10.9   For a marketing class term project, Bob is investigating whether college seniors eat less frequently in fast-food chains than college freshmen. He asked 11 freshmen and 11 seniors to keep track of how many times they ate in a fast-food restaurant during the month of October. Assuming equal variances, can you conclude that the mean is significantly smaller for college seniors at the 5 percent level of significance? 🖭 **FastFood**

Seniors	10	5	15	13	5	7	18	8	19	9	8
Freshmen	16	9	17	14	15	11	18	12	7	16	20

10.3 CONFIDENCE INTERVAL FOR THE DIFFERENCE OF TWO MEANS, $\mu_1 - \mu_2$

LO 10-3

Construct a confidence interval for $\mu_1 - \mu_2$.

There may be occasions when we want to estimate the difference between two unknown population means. The point estimate for $\mu_1 - \mu_2$ is $\bar{X}_1 - \bar{X}_2$, where $\bar{X}_1$ and $\bar{X}_2$ are calculated from independent random samples. We can use a confidence interval estimate to find a range within which the true difference might fall. If the confidence interval for the **difference of two means** includes zero, we could conclude that there is no significant difference in means.

When the population variances are unknown (the usual situation), the procedure for constructing a confidence interval for $\mu_1 - \mu_2$ depends on our assumption about the unknown variances. If both populations are normal and the population variances can be assumed equal, the difference of means follows a Student's t distribution with $(n_1 - 1) + (n_2 - 1)$ degrees of freedom. Assuming *equal variances,* the pooled variance is a weighted average of the sample variances with weights $n_1 - 1$ and $n_2 - 1$ (the respective degrees of freedom for each sample):

(10.5)
$$(\bar{x}_1 - \bar{x}_2) \pm t_{\alpha/2} \sqrt{\frac{(n_1 - 1)s_1^2 + (n_2 - 1)s_2^2}{n_1 + n_2 - 2}} \sqrt{\frac{1}{n_1} + \frac{1}{n_2}}$$

with $d.f. = (n_1 - 1) + (n_2 - 1)$

If the population variances are unknown and are assumed to be unequal, we should not pool the variances. For *unequal variances,* a practical alternative is to use the t distribution, adding the variances and using Welch's formula for the degrees of freedom:

(10.6)
$$(\bar{x}_1 - \bar{x}_2) \pm t_{\alpha/2} \sqrt{\frac{s_1^2}{n_1} + \frac{s_2^2}{n_2}} \quad \text{with} \quad d.f. = \frac{(s_1^2/n_1 + s_2^2/n_2)^2}{\dfrac{(s_1^2/n_1)^2}{n_1 - 1} + \dfrac{(s_2^2/n_2)^2}{n_2 - 1}}$$

Example 10.2

Marketing Teams

Do teams that collaborate virtually feel they get along better than teams that collaborate face-to-face? A study was conducted with senior marketing majors at a large business school. Students were randomly assigned to a team that collaborated online or a team that collaborated face-to-face. Both teams were given five cases to analyze. At the end of the study, each team member was asked to rate how well they felt the team got along by responding to the statement "I felt our members got along well together." The response scale was a 1–5 Likert scale with "1" = strongly disagree and "5" = strongly agree.

Table 10.4 shows the means and standard deviations for the two groups. The population variances are unknown but will be assumed equal (note the similar standard deviations). For a confidence level of 90 percent, we use Student's t with $d.f. = 44 + 42 - 2 = 84$. From Appendix D we obtain $t_{.05} = 1.664$ (using 80 degrees of freedom, the next lower value). The confidence interval is

$$(\bar{x}_1 - \bar{x}_2) \pm t_{\alpha/2} \text{ or } t_{.05} \sqrt{\frac{(n_1 - 1)s_1^2 + (n_2 - 1)s_2^2}{n_1 + n_2 - 2}} \sqrt{\frac{1}{n_1} + \frac{1}{n_2}}$$

$$= (3.58 - 2.93) \pm (1.664) \sqrt{\frac{(44 - 1)(0.76)^2 + (42 - 1)(0.82)^2}{44 + 42 - 2}} \sqrt{\frac{1}{44} + \frac{1}{42}}$$

$$= 0.65 \pm 0.284 \quad \text{or} \quad [0.366, 0.934]$$

Table 10.4 **Means and Standard Deviations for the Two Marketing Teams**

Statistic	Virtual Team	Face-to-Face Team
Sample Mean	$\bar{x}_1 = 3.58$	$\bar{x}_2 = 2.93$
Sample Std. Dev.	$s_1 = 0.76$	$s_2 = 0.82$
Sample Size	$n_1 = 44$	$n_2 = 42$

Because this confidence interval does not include zero, we can say with 90 percent confidence that there is a difference between the means (i.e., the virtual team's mean differs from the face-to-face team's mean). If we had not assumed equal variances, the results would be the same in this case because the samples are large and of similar size, and the variances do not differ greatly. But when you have small, unequal sample sizes or unequal variances, the methods can yield different conclusions.

Because the calculations for the comparison of two sample means are time-consuming, it is helpful to use software. See the Software Supplement at the end of this chapter for an illustration of MegaStat's menu for comparing two sample means.

Should Sample Sizes Be Equal?

Many people instinctively try to choose equal sample sizes for tests of means. It is preferable to avoid unbalanced sample sizes to increase the power of the test, but it is not necessary. Unequal sample sizes are common, and the formulas still apply.

10.10 A special bumper was installed on selected vehicles in a large fleet. The dollar cost of body repairs was recorded for all vehicles that were involved in accidents over a 1-year period. Those with the special bumper are the test group, and the other vehicles are the control group, shown below. Each "repair incident" is defined as an invoice (which might include more than one separate type of damage).

Statistic	Test Group	Control Group
Mean Damage	$\bar{x}_1 = \$1,101$	$\bar{x}_2 = \$1,766$
Sample Std. Dev.	$s_1 = \$696$	$s_2 = \$838$
Repair Incidents	$n_1 = 12$	$n_2 = 9$

(a) Construct a 90 percent confidence interval for the true difference of the means assuming equal variances. Show all work clearly. (b) Repeat, using the assumption of unequal variances with Welch's formula for *d.f.* (c) Did the assumption about variances change the conclusion? (d) Construct separate confidence intervals for each mean. Do they overlap?

10.11 In trials of an Internet-based method of learning statistics, post-tests were given to two groups: traditional instruction (22 students) and Internet-based (17 students). On the post-test, the first group (traditional instruction) had a mean score of 8.64 with a standard deviation of 1.88, while the second group (experimental instruction) had a mean score of 8.82 with a standard deviation of 1.70. (a) Construct a 90 percent confidence interval for the true difference of the means assuming equal variances. (b) Repeat, using the assumption of unequal variances with Welch's formula for *d.f.* (c) Did the assumption about variances change the conclusion? (d) Construct separate confidence intervals for each mean. Do they overlap?

10.12 Construct a 95 percent confidence interval for the difference of mean monthly rent paid by undergraduates and graduate students. What do you conclude? 📂 **Rent2**

Undergraduate Student Rents ($n = 10$)

1,820	1,780	1,870	1,670	1,800
1,790	1,810	1,180	2,000	1,730

Graduate Student Rents ($n = 12$)

2,130	1,920	1,930	1,880	1,780	1,910
1,790	1,840	1,930	1,910	1,860	1,850

10.4 MEAN DIFFERENCE IN PAIRED SAMPLES

Paired Data

When sample data consist of n matched pairs, a different approach is required. If the *same* individuals are observed twice but under different circumstances, we have a **paired comparison**. For example:

LO 10-4

Recognize paired data and be able to perform a paired *t* test.

- Fifteen retirees with diagnosed hypertension are assigned a program of diet, exercise, and meditation. A baseline measurement of blood pressure is taken *before* the program begins and again *after* 2 months. Was the program effective in reducing blood pressure?

- Ten cutting tools use lubricant A for 10 minutes. The blade temperatures are taken. When the machine has cooled, it is run with lubricant B for 10 minutes, and the blade temperatures are again measured. Which lubricant makes the blades run cooler?

- Weekly sales of Snapple at 12 Walmart stores are compared *before* and *after* installing a new eye-catching display. Did the new display increase sales?

Paired data typically come from a *before-after* experiment, but not always. For example, paired data might also arise in marketing studies that use one focus group to rate two different products. When we use *one* focus group of n individuals to compare two products, we have *paired* ratings. If we treat the data as two independent samples, ignoring the *dependence* between the data pairs, the test is less powerful.

Paired *t* Test

In the **paired *t* test** we define a new variable $d = X_1 - X_2$ as the *difference* between X_1 and X_2. The *two* samples are reduced to *one* sample of n differences $d_1, d_2, \ldots, d_n$. We usually present the n observed differences in column form:

Obs	Sample 1	Sample 2	Differences
1	x_{11}	x_{12}	$d_1 = x_{11} - x_{12}$
2	x_{21}	x_{22}	$d_2 = x_{21} - x_{22}$
3	x_{31}	x_{32}	$d_3 = x_{31} - x_{32}$
. . .	. . .	. . .	. . .
n	x_{n1}	x_{n2}	$d_n = x_{n1} - x_{n2}$

The same sample data also could be presented in row form:

Obs	1	2	3	. . .	n
Sample 1	x_{11}	x_{21}	x_{31}	. . .	x_{n1}
Sample 2	x_{12}	x_{22}	x_{32}	. . .	x_{n2}
Difference	$d_1 = x_{11} - x_{12}$	$d_2 = x_{21} - x_{22}$	$d_3 = x_{31} - x_{32}$	. . .	$d_n = x_{n1} - x_{n2}$

We calculate the mean $\overline{d}$ and standard deviation s_d of the sample of n differences $d_1, d_2, \ldots, d_n$ with the usual formulas for a mean and standard deviation. We call the mean $\overline{d}$ instead of $\overline{x}$ merely to remind ourselves that we are dealing with *differences*.

$$(10.7) \qquad \overline{d} = \frac{\sum_{i=1}^{n} d_i}{n} \quad \text{(mean of } n \text{ differences)}$$

$$(10.8) \qquad s_d = \sqrt{\sum_{i=1}^{n} \frac{(d_i - \overline{d})^2}{n - 1}} \quad \text{(std. dev. of } n \text{ differences)}$$

Because the population variance of d is unknown, we will do a paired t test using Student's t with $n - 1$ degrees of freedom to compare the sample mean difference $\overline{d}$ with a hypothesized difference μ_d (usually $\mu_d = 0$). The test statistic is really a one-sample t test, just like those in Chapter 9.

$$(10.9) \qquad t_{calc} = \frac{\overline{d} - \mu_d}{\dfrac{s_d}{\sqrt{n}}} \quad \text{(test statistic for } \textbf{paired samples)}$$

Example 10.3

Repair Estimates

Repair

An insurance company's procedure in settling a claim under $10,000 for fire or water damage to a homeowner is to require two estimates for cleanup and repair of structural damage before allowing the insured to proceed with the work. The insurance company compares estimates from two contractors who most frequently handle this type of work in this geographical area. Table 10.5 shows the 10 most recent claims for which damage estimates were provided by both contractors. At the .05 level of significance, is there a difference between the two contractors?

Step 1: State the Hypotheses We will choose a two-tailed test using these hypotheses:

$H_0: \mu_d = 0$

$H_1: \mu_d \neq 0$

Table 10.5

Damage Repair Estimates ($) for 10 Claims

🖾 Repair

Claim	X_1 Contractor A	X_2 Contractor B	$d = X_1 - X_2$ Difference
1. Jones, C.	5,500	6,000	−500
2. Smith, R.	1,000	900	100
3. Xia, Y.	2,500	2,500	0
4. Gallo, J.	7,800	8,300	−500
5. Carson, R.	6,400	6,200	200
6. Petty, M.	8,800	9,400	−600
7. Tracy, L.	600	500	100
8. Barnes, J.	3,300	3,500	−200
9. Rodriguez, J.	4,500	5,200	−700
10. Van Dyke, P.	6,500	6,800	−300
			$\overline{d} = -240.00$
			$s_d = 327.28$
			$n = 10$

Step 2: Specify the Decision Rule Our test statistic will follow a Student's t distribution with $d.f. = n - 1 = 10 - 1 = 9$, so from Appendix D with $\alpha = .05$ the two-tail critical value is $t_{.025} = \pm 2.262$, as illustrated in Figure 10.7. The decision rule is

Reject H_0 if $t_{calc} < -2.262$ or if $t_{calc} > +2.262$

Otherwise do not reject H_0

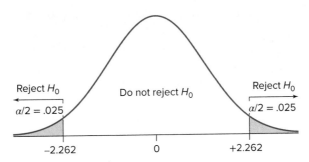

Figure 10.7

Decision Rule for Two-Tailed Paired t Test at $\alpha = .05$

Step 3: Calculate the Test Statistic The mean and standard deviation are calculated in the usual way, as shown in Table 10.5, so the test statistic is

$$t_{calc} = \frac{\overline{d} - \mu_d}{\dfrac{s_d}{\sqrt{n}}} = \frac{-240 - 0}{\left(\dfrac{327.28}{\sqrt{10}}\right)} = \frac{-240}{103.495} = -2.319$$

Step 4: Make the Decision Because $t_{calc} = -2.319$ falls in the left-tail critical region (below -2.262), we reject the null hypothesis and conclude that there is a significant difference between the two contractors. However, it is a *very* close decision.

Step 5: Take Action Because the difference is significant in a two-tailed test, it also would be significant in a left-tailed test. The insurance company might talk to Contractor B to see why its estimates are higher.

Excel's Paired Difference Test

The calculations for our repair estimates example are easy in Excel, as illustrated in Figure 10.8. Excel gives you the option of choosing either a one-tailed or two-tailed test and also shows the p-value. For a two-tailed test, the p-value is $p = .0456$, which would barely lead to rejection of the hypothesis of zero difference of means at $\alpha = .05$. The borderline *p*-value reinforces our conclusion that the decision is sensitive to our choice of α. MegaStat and Minitab also provide a paired *t* test.

Figure 10.8

Results of Excel's Paired *t* Test at $\alpha = .05$

📂 **Repair**

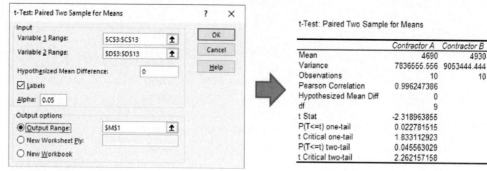

Source: Microsoft Excel.

Analogy to Confidence Interval

A two-tailed test for a zero difference is equivalent to asking whether the confidence interval for the true mean difference μ_d includes zero.

(10.10)
$$\overline{d} \pm t_{\alpha/2} \frac{s_d}{\sqrt{n}} \qquad \text{(confidence interval for mean difference)}$$

It depends on the confidence level:

90% confidence ($t_{\alpha/2} = 1.833$): $[-429.72, -50.28]$
95% confidence ($t_{\alpha/2} = 2.262$): $[-474.12, -5.88]$
99% confidence ($t_{\alpha/2} = 3.250$): $[-576.34, +96.34]$

As Figure 10.9 shows, the 99 percent confidence interval includes zero, but the 90 percent and 95 percent confidence intervals do not.

Figure 10.9

Confidence Intervals for Mean Difference

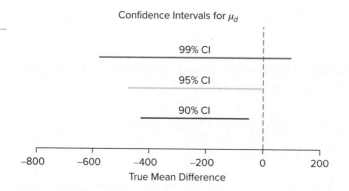

Why Not Treat Paired Data as Independent Samples?

When observations are matched pairs, the paired *t* test is more powerful because it considers variation between the paired values in each column as well as variation within each column of data. The difference in individual paired values is ignored if we treat samples independently. To show this, let's revisit Example 10.3 and treat each column of repair data as an **independent sample**. The summary statistics are

$$\overline{x}_1 = 4{,}690.00 \qquad \overline{x}_2 = 4{,}930.00$$
$$s_1 = 2{,}799.38 \qquad s_2 = 3{,}008.89$$
$$n_1 = 10 \qquad n_2 = 10$$

Assuming equal variances, we get the results shown in Figure 10.10. The *p*-values (one-tail or two-tail) are not even close to being significant at the usual α levels. By ignoring the dependence between the samples, we unnecessarily *sacrifice the power of the test*. Therefore, if the two data columns are paired, we should not treat them independently.

Best Test : Paired Samples

t-Test: Paired Two Sample for Means

	Contractor A	Contractor B
Mean	4690	4930
Variance	7836555.556	9053444.444
Observations	10	10
Pearson Correlation	0.996247386	
Hypothesized Mean Diff	0	
df	9	
t Stat	-2.318963855	
P(T<=t) one-tail	0.022781515	
t Critical one-tail	1.833112923	
P(T<=t) two-tail	0.045563029	
t Critical two-tail	2.262157158	

Less Power : Independent Samples

t-Test: Two-Sample Assuming Equal Variances

	Contractor A	Contractor B
Mean	4690	4930
Variance	7836555.556	9053444.444
Observations	10	10
Pooled Variance	8445000	
Hypothesized Mean Differe	0	
df	18	
t Stat	-0.184670029	
P(T<=t) one-tail	0.427776285	
t Critical one-tail	1.734063592	
P(T<=t) two-tail	0.855552569	
t Critical two-tail	2.100922037	

Figure 10.10

Excel's Paired Sample and Independent Sample *t* Tests
📁 **Repair**

Mini Case 10.3

Price-Earnings Ratios

During the first six months of 2022, the U.S. stock market was rattled by fears of inflation, recession, and the Russian invasion of Ukraine. Table 10.6 shows P/E ratios for a random sample of 25 S&P 500 companies at the beginning and end of this six-month period (January 3 to June 30).

Table 10.06

Price-Earnings Ratios for 25 Companies 📁 **PERatios**

Obs	Company	Ticker	Before	After	Difference
1	Align Technology	ALGN	67.82	30.30	37.52
2	Alphabet (Class C)	GOOG	25.79	20.34	5.45
3	American Electric Power	AEP	17.64	18.85	−1.21
⋮	⋮	⋮	⋮	⋮	⋮
23	Realty Income Corporation	O	80.26	68.72	11.54
24	RoperTechnologies	ROP	45.27	14.67	30.60
25	TE Connectivity	TEL	21.72	14.77	6.95

Source: Individual company P/E ratios are from www.macrotrends.net, accessed August 16, 2022.

Figure 10.11 shows a downward shift in P/E ratios over this period of time, as well as fewer extreme values. This shift reflects the overall market decline in S&P 500 stock prices, but also includes possible changes in company earnings.

Figure 10.11

Price Earnings Ratios for 25 Companies

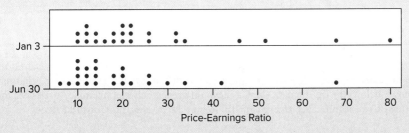

The S&P 500 stock price index fell by an average of 13 percent over this period. Some analysts had predicted an even sharper decline in price-earnings ratios. Did that occur? On January 3 (the first trading day of the year), the average P/E ratio for the S&P 500 was reported as 23.11 (www.multpl.com). We can use Excel's paired t-test for the hypotheses $H_0: \mu_d \leq 3.00$ and $H_1: \mu_d > 3.00$ (i.e., 13 percent of 23.11) to decide whether P/E ratios did decline more than stock prices.

t-Test: Paired Two Sample for Means

Statistic	Jan 3	Jun 30
Mean	26.362	19.688
Variance	314.63	179.72
Observations	25	25
Hypothesized Mean Difference	3.000	
Sample Mean Difference	6.674	
df	24	
t Stat	1.995	
P(T<=t) one-tail	0.029	
t Critical one-tail	1.711	

The test statistic is $t_{calc} = 1.995$ ($p = 0.029$), so at $\alpha = 0.05$ our sample suggests that P/E ratios did in fact decline more than stock prices for this period of time. This change not only would be *significant* but also would be *important* for investors who base their portfolio decisions on P/E ratios.

Section Exercises

10.13 (a) At $\alpha = .05$, does the following sample show that daughters are taller than their mothers? (b) Is the decision close? (c) Why might daughters tend to be taller than their mothers? Why might they not? **Height**

Family	Daughter's Height (cm)	Mother's Height (cm)
1	167	172
2	166	162
3	176	157
4	171	159
5	165	157
6	181	177
7	173	174

10.14 An experimental surgical procedure is being studied as an alternative to the old method. Both methods are considered safe. Five surgeons perform the operation on two patients matched by age, sex, and other relevant factors, with the results shown. The time to complete the surgery (in minutes) is recorded. (a) At the 5 percent significance level, is the new way faster? State your hypotheses and show all steps clearly. (b) Is the decision close? **Surgery**

	Surgeon 1	Surgeon 2	Surgeon 3	Surgeon 4	Surgeon 5
Old way	36	55	28	40	62
New way	29	42	30	32	56

10.15 Pregabalin and gabapentin are the generic names for two common anticonvulsant drugs used to treat epilepsy and other seizure disorders. GoodRx prices for a sample of 10 pharmacies are shown below. (a) Do the data show a mean difference between pharmacy prices for these two drugs? Use a significance level of .05. (b) Report the p-value. **GoodRxPrices**

Pharmacy	Pregabalin	Gabapentin
1	0.88	2.54
2	4.25	4.73
3	3.52	4.44
4	3.78	6.89
5	6.20	3.08
6	0.67	2.35
7	2.56	7.56
8	4.48	3.45
9	4.26	1.68
10	6.57	9.45

10.16 The U.S. government's "Cash for Clunkers" program encouraged individuals to trade in their old gas-guzzlers for new, more efficient vehicles. At $\alpha = .05$, do the data below support the hypothesis that the gain in mpg was more than 5 mpg? *Hint:* The null hypothesis is $H_0: \mu_d \leq 5$ mpg. 📷 **Guzzlers**

Buyer	New Car	Old Car	Buyer	New Car	Old Car
Buyer 1	23.0	20.1	Buyer 8	26.3	16.1
Buyer 2	20.3	15.8	Buyer 9	27.2	19.2
Buyer 3	24.6	19.1	Buyer 10	28.0	19.8
Buyer 4	25.3	17.4	Buyer 11	18.9	17.7
Buyer 5	20.6	13.9	Buyer 12	23.8	21.8
Buyer 6	26.8	15.6	Buyer 13	27.8	18.5
Buyer 7	22.9	17.1	Buyer 14	27.1	19.4

10.17 The coach told the high school swim team that times in the 200-yard individual medley in the Division I-AA Swim Championships typically are more than 1/2 second faster than their seed times going into the meet. At $\alpha = .05$, do these times (in seconds) for 26 male competitors (actual data with altered names) support the coach's assertion? *Hint:* The null hypothesis is $H_0: \mu_d \leq 0.5$ second. 📷 **Swimmers**

Swimmer	Seed	Prelims	Swimmer	Seed	Prelims
Jason	122.06	117.49	Wolfgang	122.33	122.35
Walter	116.23	116.62	Jin	120.68	117.56
Fred	122.88	121.93	Li	121.81	121.7
Tom	123.33	123.81	Brian	121.08	120.58
Andy	122.33	122.52	Pete	122.38	123.16
David	120.41	118.75	Otto	120.87	120.64
Jon	122.44	123.57	Joon	122.46	124.82
Bruce	120.42	123.27	Jeffrey	115.16	110.88
Victor	118.19	117.15	Roger	119.68	114.16
Karl	123.27	121.28	Bill	115.94	113.3
Kurt	120.23	119.21	John	122.98	119.93
Frank	117.66	113.1	Preston	123.28	120.7
Cedric	122.61	122.40	Steve	116.50	113.15

10.18 Below is a random sample of shoe sizes for 12 mothers and their daughters. (a) At $\alpha = .01$, does this sample show that women's shoe sizes have increased? State your hypotheses and show all steps clearly. (b) Is the decision close? (c) Are you convinced? (d) Why might shoe sizes change over time? 📷 **ShoeSize2**

	1	2	3	4	5	6	7	8	9	10	11	12
Daughter	8	8	7.5	8	9	9	8.5	9	9	8	7	8
Mother	8	7	7.5	8	8.5	8.5	7.5	7.5	6	8	7	7

10.19 A newly installed automatic gate system was being tested to see if the number of failures in 1,000 entry attempts was the same as the number of failures in 1,000 exit attempts. A random sample of eight delivery trucks was selected for data collection. Do these sample results show that there is a significant difference between entry and exit gate failures? Use $\alpha = .01$.　📁 **Gates**

	Truck 1	Truck 2	Truck 3	Truck 4	Truck 5	Truck 6	Truck 7	Truck 8
Entry failures	43	45	53	56	61	51	48	44
Exit failures	48	51	60	58	58	45	55	50

10.5 COMPARING TWO PROPORTIONS

LO 10-5

Perform a test to compare two proportions using z.

The test for two proportions is perhaps the most commonly used two-sample test because percents are ubiquitous. Is the president's approval rating greater than, less than, or the same as last month? Is the proportion of satisfied Dell customers greater than HP's? Is the annual nursing turnover percentage at Mayo Clinic higher than, lower than, or the same as that at Johns Hopkins? To answer such questions, we would compare two sample proportions.

Testing for Zero Difference: $\pi_1 - \pi_2 = 0$

Let the true proportions in the two populations be denoted π_1 and π_2. When testing the difference between two proportions, we typically assume the population proportions are equal and set up our hypotheses using the null hypothesis $H_0: \pi_1 - \pi_2 = 0$. This is similar to our approach when testing the difference between two means. The research question will determine the format of our alternative hypothesis. The three possible pairs of hypotheses are

Left-Tailed Test	*Two-Tailed Test*	*Right-Tailed Test*
$H_0: \pi_1 - \pi_2 \geq 0$	$H_0: \pi_1 - \pi_2 = 0$	$H_0: \pi_1 - \pi_2 \leq 0$
$H_1: \pi_1 - \pi_2 < 0$	$H_1: \pi_1 - \pi_2 \neq 0$	$H_1: \pi_1 - \pi_2 > 0$

Sample Proportions

The sample proportion p_1 is a point estimate of π_1, and the sample proportion p_2 is a point estimate of π_2. A "success" is any event of interest (not necessarily something desirable).

(10.11)
$$p_1 = \frac{x_1}{n_1} = \frac{\text{number of "successes" in sample 1}}{\text{number of items in sample 1}}$$

(10.12)
$$p_2 = \frac{x_2}{n_2} = \frac{\text{number of "successes" in sample 2}}{\text{number of items in sample 2}}$$

Pooled Proportion

If H_0 is true, there is no difference between π_1 and π_2, so the samples can logically be *pooled* into one "big" sample to estimate the *combined* population proportion p_c:

(10.13) $\quad p_c = \frac{x_1 + x_2}{n_1 + n_2} = \frac{\text{number of successes in combined samples}}{\text{combined sample sizes}} = \textbf{pooled proportion}$

Test Statistic

If the samples are large, the difference of proportions $p_1 - p_2$ may be assumed normally distributed. The *test statistic* is the difference of the sample proportions $p_1 - p_2$ minus the hypothesized difference $\pi_1 - \pi_2$ divided by the standard error of the difference $p_1 - p_2$. The test statistic for testing the difference between two proportions is

(10.14)
$$z_{\text{calc}} = \frac{(p_1 - p_2) - (\pi_1 - \pi_2)}{\sqrt{\dfrac{p_1(1 - p_1)}{n_1} + \dfrac{p_2(1 - p_2)}{n_2}}}$$

If we are testing the hypothesis that $\pi_1 - \pi_2 = 0$, the standard error is calculated by using the pooled proportion p_c. We can then substitute p_c for both p_1 and p_2 in formula 10.14. The result is shown in formula 10.15.

Test Statistic for Equality of Proportions

(10.15)
$$z_{calc} = \frac{p_1 - p_2}{\sqrt{p_c(1 - p_c)\left(\dfrac{1}{n_1} + \dfrac{1}{n_2}\right)}}$$

Example 10.4

Active Promoters, Vail Resorts

In order to measure the level of satisfaction with Vail Resorts' websites, the Vail Resorts marketing team periodically surveys a random sample of guests and asks them to rate their likelihood of recommending the website to a friend or colleague. An *active promoter* is a guest who responds that they are highly likely to recommend the website. From a random sample of 2,386 Vail ski mountain guests in last year's ski season, there were 2,014 active promoters, and from a random sample of 2,309 Vail ski mountain guests in this year's ski season, there were 2,048 active promoters. Results from the survey are shown in Table 10.7. At the .01 level of significance, did the proportion of active promoters increase?

Table 10.7 **Website Satisfaction Survey**

Statistic	This Year	Last Year
Number of active promoters	$x_1 = 2{,}048$	$x_2 = 2{,}014$
Number of guests surveyed	$n_1 = 2{,}309$	$n_2 = 2{,}386$
Active promoter proportion	$p_1 = \dfrac{2{,}048}{2{,}309} = .8870$	$p_2 = \dfrac{2{,}014}{2{,}386} = .8441$

Step 1: State the Hypotheses Because Vail Resorts had redesigned its ski mountain websites for this year's season, it was interested in seeing if the proportion of active promoters had increased. Therefore, we will do a right-tailed test for equality of proportions.

$H_0: \pi_1 - \pi_2 \leq 0$
$H_1: \pi_1 - \pi_2 > 0$

Step 2: Specify the Decision Rule Using $\alpha = .01$, the right-tail critical value is $z_{.01} = 2.326$, which yields the decision rule

Reject H_0 if $z_{calc} > 2.326$
Otherwise, do not reject H_0

The decision rule is illustrated in Figure 10.12. Because Excel uses cumulative left-tail areas, the right-tail critical value $z_{.01} = 2.326$ is obtained using =NORM.S.INV(.99).

Step 3: Calculate the Test Statistic The sample proportions indicate that this year's season had a higher proportion of active promoters than last year's season. We assume that $\pi_1 - \pi_2 = 0$ and see if a contradiction stems from this assumption. Assuming that the proportions are equal, we can pool the two samples to obtain a **pooled estimate** of the common proportion by dividing the combined number of active promoters by the combined sample size.

$$p_c = \frac{x_1 + x_2}{n_1 + n_2} = \frac{2{,}048 + 2{,}014}{2{,}309 + 2{,}386} = \frac{4{,}062}{4{,}695} = .8652, \text{ or } 86.52\%$$

Figure 10.12

Right-Tailed Test for Two Proportions

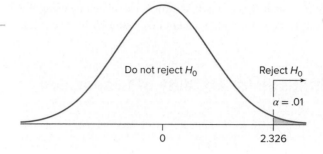

Assuming normality (i.e., large samples), the test statistic is

$$z_{calc} = \frac{p_1 - p_2}{\sqrt{p_c(1 - p_c)\left(\dfrac{1}{n_1} + \dfrac{1}{n_2}\right)}} = \frac{.8870 - .8441}{\sqrt{.8652(1 - .8652)\left(\dfrac{1}{2309} + \dfrac{1}{2386}\right)}} = 4.303$$

Step 4: Make the Decision If H_0 were true, the test statistic should be near zero. Because the test statistic (z_{calc} = 4.303) exceeds the critical value ($z_{.01}$ = 2.326), we reject the null hypothesis and conclude that $\pi_1 - \pi_2 > 0$. If we were to use the *p*-value approach, we would find the *p*-value by using the function =1–NORM.S.DIST(4.313,1) in Excel. This function returns a value so small (.00000807) it is, for all practical purposes, equal to zero. Because the *p*-value is less than .01, we would reject the null hypothesis.

Whether we use the critical value approach or the *p*-value approach, we would reject the null hypothesis of equal proportions. In other words, the proportion of active promoters this year is significantly greater than in the previous year. The new website design appeared to be attractive to Vail Resorts' guests.

Step 5: Take Action Keep the new website design because it appears to increase the number of promoters.

LO 10-6

Check whether normality may be assumed for two proportions.

Checking Normality

We have assumed a normal distribution for the statistic $p_1 - p_2$. This assumption can be checked. For a test of two proportions, the criterion for normality is $n\pi \geq 10$ and $n(1 - \pi) \geq 10$ for *each* sample, using each sample proportion in place of π:

$$n_1 p_1 = (2,309)(2,048/2,309) = 2,048 \quad n_1(1 - p_1) = (2,309)(1 - 2,048/2,309) = 261$$
$$n_2 p_2 = (2,386)(2,014/2,386) = 2,014 \quad n_2(1 - p_2) = (2,386)(1 - 2,014/2,386) = 372$$

The normality requirement is comfortably fulfilled in this case. Ideally, these numbers should exceed 10 by a comfortable margin, as they do in this example. Note that when using sample data, the sample size rule of thumb is equivalent to requiring that each sample contains at least 10 "successes" and at least 10 "failures." In this example, this requirement is easily met (there are 2,048 and 2,014 promoters and 261 and 372 nonpromoters).

If sample sizes do not justify the normality assumption, each sample should be treated as a binomial experiment. Unless you have good computational software, this may not be worthwhile. If the samples are small, the test is likely to have low power anyway.

Must Sample Sizes Be Equal?

While most analysts will instinctively choose the same sample sizes, equal sample sizes are not necessary. The formulas still apply even if sample sizes are not equal.

Mini Case 10.4

How Does Noodles & Company Provide Value to Customers?

Value perception is an important concept for all companies but is especially relevant for consumer-oriented industries such as retail and restaurants. Most retailers and restaurant concepts periodically make price increases to reflect changes in inflationary items such as cost of goods and labor costs. Noodles & Company took the opposite approach when it evaluated its value perception through its consumers.

Through rigorous statistical analysis, Noodles recognized that a significant percentage of current customers would increase their frequency of visits if the menu items were priced slightly lower. The company evaluated the trade-offs that a price decrease would represent and determined that they should be able to increase revenue by reducing price. Despite not advertising this price decrease, the company did in fact see an increase in frequency of visits after the change. To measure the impact, the company statistically evaluated both the increase in frequency and the customer evaluations of Noodles & Company's value perception. Within a few months, the statistical analysis showed not only that had customer frequency increased by 2 to 3 percent but also that improved value perception led to an increase in average party size of 2 percent. They concluded that the price decrease of roughly 2 percent led to a total revenue increase of 4 to 5 percent.

Section Exercises

10.20 Calculate the test statistic and *p*-value for a test of equal population proportions. What is your conclusion?

    a. Right-tailed test, $\alpha = .10$, $x_1 = 228$, $n_1 = 240$, $x_2 = 703$, $n_2 = 760$
    b. Left-tailed test, $\alpha = .05$, $x_1 = 36$, $n_1 = 80$, $x_2 = 66$, $n_2 = 120$
    c. Two-tailed test, $\alpha = .05$, $x_1 = 52$, $n_1 = 80$, $x_2 = 56$, $n_2 = 70$

10.21 Calculate the test statistic and *p*-value for a test of equal population proportions. What is your conclusion?

    a. Left-tailed test, $\alpha = .10$, $x_1 = 28$, $n_1 = 336$, $x_2 = 14$, $n_2 = 112$
    b. Right-tailed test, $\alpha = .05$, $x_1 = 276$, $n_1 = 300$, $x_2 = 440$, $n_2 = 500$
    c. Two-tailed test, $\alpha = .10$, $x_1 = 35$, $n_1 = 50$, $x_2 = 42$, $n_2 = 75$

10.22 Find the sample proportions and test statistic for equal proportions. Find the *p*-value.

    a. Dissatisfied workers in two companies: $x_1 = 40$, $n_1 = 100$, $x_2 = 30$, $n_2 = 100$, $\alpha = .05$, two-tailed test.
    b. Rooms rented at least a week in advance at two hotels: $x_1 = 24$, $n_1 = 200$, $x_2 = 12$, $n_2 = 50$, $\alpha = .01$, left-tailed test.
    c. Home equity loan default rates in two banks: $x_1 = 36$, $n_1 = 480$, $x_2 = 26$, $n_2 = 520$, $\alpha = .05$, right-tailed test.

10.23 Find the test statistic and do the two-sample test for equality of proportions.

    a. Repeat buyers at two car dealerships: $p_1 = .30$, $n_1 = 50$, $p_2 = .54$, $n_2 = 50$, $\alpha = .01$, left-tailed test.
    b. Honor roll students in two sororities: $p_1 = .45$, $n_1 = 80$, $p_2 = .25$, $n_2 = 48$, $\alpha = .10$, two-tailed test.
    c. First-time Hawaii visitors at two hotels: $p_1 = .20$, $n_1 = 80$, $p_2 = .32$, $n_2 = 75$, $\alpha = .05$, left-tailed test.

10.24 During the period 1990–1998 there were 46 Atlantic hurricanes, of which 19 struck the United States. During the period 1999–2006 there were 70 hurricanes, of which 45 struck the United States. (a) State the hypotheses to test whether the percentage of hurricanes that strike the United States is increasing. (b) Calculate the test statistic. (c) State the critical value at $\alpha = .01$. (d) What is your conclusion? (e) Can normality of $p_1 - p_2$ be assumed?

10.25 In 2022, a sample of 200 in-store shoppers showed that 66 paid by debit card. In 2023, a sample of the same size showed that 86 paid by debit card. (a) Formulate appropriate hypotheses to test whether the percentage of debit card shoppers increased. (b) Carry out the test at $\alpha = .01$. (c) Find the *p*-value. (d) Test whether normality of $p_1 - p_2$ may be assumed.

10.26 A survey of 100 mayonnaise purchasers showed that 65 were loyal to one brand. For 100 bath soap purchasers, only 53 were loyal to one brand. Perform a two-tailed test comparing the proportion of brand-loyal customers at $\alpha = .05$.

10.27 A 20-minute consumer survey mailed to 500 adults aged 25–34 included a $5 Starbucks gift certificate. The same survey was mailed to 500 adults aged 25–34 without the gift certificate. There were 65 responses from the first group and 45 from the second group. Perform a two-tailed test comparing the response rates (proportions) at $\alpha = .05$.

10.28 Is the water on your airline flight safe to drink? It is not feasible to analyze the water on every flight, so sampling is necessary. The Environmental Protection Agency (EPA) found bacterial contamination in water samples from the lavatories and galley water taps on 20 of 158 randomly selected U.S. flights. Alarmed by the data, the EPA ordered sanitation improvements and then tested water samples again. In the second sample, bacterial contamination was found in 29 of 169 randomly sampled flights. (a) Use a left-tailed test at $\alpha = .05$ to check whether the percent of all flights with contaminated water was lower in the first sample. (b) Find the p-value. (c) Discuss the question of significance versus importance in this specific application. (d) Test whether normality of $p_1 - p_2$ may be assumed.

10.29 When tested for compliance with Sarbanes-Oxley requirements for financial records and fraud protection, 14 of 180 publicly traded business services companies failed, compared with 7 of 67 computer hardware, software, and telecommunications companies. (a) Is this a statistically significant difference at $\alpha = .05$? (b) Can normality of $p_1 - p_2$ be assumed?

Testing for Nonzero Difference (Optional)

Testing for equality of π_1 and π_2 is a special case of testing for a specified difference D_0 between the two proportions:

Left-Tailed Test	Two-Tailed Test	Right-Tailed Test
$H_0: \pi_1 - \pi_2 \geq D_0$	$H_0: \pi_1 - \pi_2 = D_0$	$H_0: \pi_1 - \pi_2 \leq D_0$
$H_1: \pi_1 - \pi_2 < D_0$	$H_1: \pi_1 - \pi_2 \neq D_0$	$H_1: \pi_1 - \pi_2 > D_0$

We have shown how to test for $D_0 = 0$, that is, $\pi_1 = \pi_2$. If the hypothesized difference D_0 is nonzero, we do not pool the sample proportions but instead use the test statistic shown in formula 10.16.

$$(10.16) \qquad z_{calc} = \frac{p_1 - p_2 - D_0}{\sqrt{\dfrac{p_1(1 - p_1)}{n_1} + \dfrac{p_2(1 - p_2)}{n_2}}} \qquad \text{(test statistic for nonzero differences } D_0\text{)}$$

Example 10.5

Magazine Ads

A sample of 111 magazine advertisements in *Good Housekeeping* showed 70 that listed a website. In *Fortune,* a sample of 145 advertisements showed 131 that listed a website. At $\alpha = .025$, does the *Fortune* proportion differ from the *Good Housekeeping* proportion by more than 20 percent? Table 10.8 shows the data.

Table 10.8   **Magazine Ads with Websites**

Statistic	Fortune	Good Housekeeping
Number with websites	$x_1 = 131$ with website	$x_2 = 70$ with website
Number of ads examined	$n_1 = 145$ ads	$n_2 = 111$ ads
Proportion	$p_1 = \dfrac{131}{145} = .90345$	$p_2 = \dfrac{70}{111} = .63063$

We will do a right-tailed test for $D_0 = .20$. The hypotheses are

$H_0: \pi_1 - \pi_2 \leq .20$

$H_1: \pi_1 - \pi_2 > .20$

The test statistic is

$$z_{calc} = \frac{p_1 - p_2 - D_0}{\sqrt{\dfrac{p_1(1 - p_1)}{n_1} + \dfrac{p_2(1 - p_2)}{n_2}}}$$

$$= \frac{.90345 - .63063 - .20}{\sqrt{\dfrac{.90345(1 - .90345)}{145} + \dfrac{.63063(1 - .63063)}{111}}} = 1.401$$

At $\alpha = .025$, the right-tail critical value is $z_{.025} = 1.960$, so the difference of proportions is insufficient to reject the hypothesis that the difference is .20 or less. The decision rule is illustrated in Figure 10.13.

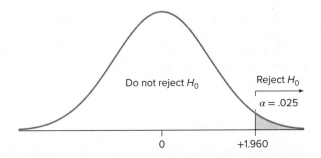

Figure 10.13

Right-Tailed Test for Magazine Ads at $\alpha = .025$

Source: Project by MBA students Frank George, Karen Orso, and Lincy Zachariah.

Calculating the *p*-Value

Using the *p*-value approach, we would insert the test statistic $z_{calc} = 1.401$ into Excel's cumulative normal =1-NORM.S.DIST(1.401,1) to obtain a right-tail area of .0806, as shown in Figure 10.14. Because the *p*-value $>.025$, we would not reject H_0. The conclusion is that the difference in proportions is not greater than .20.

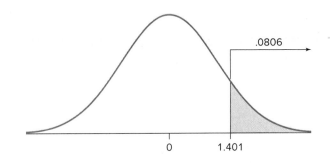

Figure 10.14

***p*-Value for Magazine Proportions Differing by $D_0 = .20$**

Note: Use software of your choice for calculations.

Section Exercises

Mc Graw Hill **connect**

10.30 In 2022, a sample of 200 in-store shoppers showed that 66 paid by debit card. In 2023, a sample of the same size showed that 86 paid by debit card. (a) Formulate appropriate hypotheses to test whether the percentage of debit card shoppers increased by more than 5 percent, using $\alpha = .10$. (b) Find the *p*-value.

10.31 From a telephone log, an executive finds that 36 of 128 incoming telephone calls last week lasted at least 5 minutes. She vows to make an effort to reduce the length of time spent on calls. The phone log for the next week shows that 14 of 96 incoming calls lasted at least 5 minutes. (a) At $\alpha = .05$, has the proportion of 5-minute phone calls declined by more than 10 percent? (b) Find the *p*-value.

10.32 A 30-minute consumer survey mailed to 500 adults aged 25–34 included a $10 gift certificate to Barnes & Noble. The same survey was mailed to 500 adults aged 25–34 without the gift certificate. There were 185 responses from the first group and 45 from the second group. (a) At $\alpha = .025$, did the gift certificate increase the response rate by more than 20 percent? (b) Find the *p*-value.

Mini Case 10.5

Automated Parking Lot Entry/Exit Gate System

Large universities have many different parking lots. Delivery trucks travel between various buildings all day long to deliver food, mail, and other items. Automated entry/exit gates make travel time much faster for the trucks and cars entering and exiting the different parking lots because the drivers do not have to stop to activate the gate manually. The gate is electronically activated as the truck or car approaches the parking lot.

One large university with two campuses recently negotiated with a company to install a new automated system. One requirement of the contract stated that the proportion of failed gate activations on one campus would be no different from the proportion of failed gate activations on the second campus. (A failed activation was one in which the driver had to manually activate the gate.) The university facilities operations manager designed and conducted a test to establish whether the gate company had violated this requirement of the contract. The university could renegotiate the contract if there was significant evidence showing that the two proportions were different.

The test was set up as a two-tailed test, and the hypotheses tested were

$$H_0: \pi_1 - \pi_2 = 0$$
$$H_1: \pi_1 - \pi_2 \neq 0$$

Both the university and the gate company agreed on a 5 percent level of significance. Random samples from each campus were collected. The data are shown in Table 10.9.

Table 10.9   **Proportion of Failed Gate Activations**

Statistic	Campus 1	Campus 2
Number of failed activations	$x_1 = 52$	$x_2 = 63$
Sample size (number of entry/exit attempts)	$n_1 = 1{,}000$	$n_2 = 1{,}000$
Proportion	$p_1 = \dfrac{52}{1{,}000} = .052$	$p_2 = \dfrac{63}{1{,}000} = .063$

The pooled proportion is

$$p_c = \frac{x_1 + x_2}{n_1 + n_2} = \frac{52 + 63}{1{,}000 + 1{,}000} = \frac{115}{2{,}000} = .0575$$

The test statistic is

$$z_{calc} = \frac{p_1 - p_2}{\sqrt{p_c(1 - p_c)\left(\dfrac{1}{n_1} + \dfrac{1}{n_2}\right)}} = \frac{.052 - .063}{\sqrt{.0575(1 - .0575)\left(\dfrac{1}{1{,}000} + \dfrac{1}{1{,}000}\right)}} = -1.057$$

Using the 5 percent level of significance, the critical value is $z_{.025} = 1.96$, so it is clear that there is no significant difference between these two proportions. This conclusion is reinforced by Excel's cumulative normal function =NORM.S.DIST(–1.057,1), which gives the area to the left of -1.057 as .1453. Because this is a two-tailed test, the p-value is .2906.

Was it reasonable to assume normality of the test statistic? Yes, because we have at least 10 successes and 10 failures in each sample:

$$n_1 p_1 = 1{,}000(52/1{,}000) = 52 \qquad n_1(1 - p_1) = 1{,}000(1 - 52/1{,}000) = 948$$
$$n_2 p_2 = 1{,}000(63/1{,}000) = 63 \qquad n_2(1 - p_2) = 1{,}000(1 - 63/1{,}000) = 937$$

Based on this sample, the university had no evidence to refute the gate company's claim that the failed activation proportions were the same for each campus.

Source: This case was based on a real contract negotiation between a large western university and a private company.

10.6 CONFIDENCE INTERVAL FOR THE DIFFERENCE OF TWO PROPORTIONS, $\pi_1 - \pi_2$

A confidence interval for the **difference of two population proportions**, $\pi_1 - \pi_2$, is given by

$$(p_1 - p_2) \pm z_{\alpha/2} \sqrt{\frac{p_1(1 - p_1)}{n_1} + \frac{p_2(1 - p_2)}{n_2}} \qquad (10.17)$$

LO 10-7

Construct a confidence interval for $\pi_1 - \pi_2$.

This formula assumes that both samples are large enough to assume normality. The rule of thumb for assuming normality is that $np \geq 10$ and $n(1 - p) \geq 10$ for each sample.

Example 10.6

Hospital Cost

Hospital emergency department visits are a major contributor to health care costs. Moreover, such visits often are followed by another visit soon after discharge. Researchers wanted to know if extra counseling of emergency patients prior to discharge would reduce the likelihood of a return emergency visit within 30 days. Following treatment, randomly selected emergency patients were divided into two groups in a double-blind experiment. The first group received normal counseling prior to discharge, while the second group received extra counseling, education, and follow-up phone calls. The results are shown in Table 10.10.

Table 10.10   Return Emergency Visits within 30 Days

Statistic	Usual Counseling	Extra Counseling
Number of return visits	$x_1 = 90$	$x_2 = 61$
Number of patients in group	$n_1 = 370$	$n_2 = 368$
Rate of return visits	$p_1 = \frac{90}{370}$	$p_2 = \frac{61}{368}$
	$= .24324$	$= .16576$

The 95 percent confidence interval for the difference between the proportions is

$$(p_1 - p_2) \pm z_{\alpha/2} \sqrt{\frac{p_1(1 - p_1)}{n_1} + \frac{p_2(1 - p_2)}{n_2}}$$

$$= (.24324 - .16576) \pm 1.960 \sqrt{\frac{.24324(1 - .24324)}{370} + \frac{.16576(1 - .16576)}{368}}$$

$$= .07748 \pm .05792 \quad \text{or} \quad [.01956, .13540].$$

Because the confidence interval [.01956, .13540] does not include zero, there is a significant difference in the rate of return visits. This suggests that extra counseling would be helpful in terms of patient safety and also might reduce cost as long as the per-patient cost of extra counseling is less than the cost of a return emergency visit.

Section Exercises

connect

10.33 The American Bankers Association reported that in a sample of 120 consumer purchases in France, 60 were made with cash, compared with 26 in a sample of 50 consumer purchases in the United States. Construct a 90 percent confidence interval for the difference in proportions.

10.34 A study showed that 36 of 72 cell phone users with a headset missed their exit, compared with 12 of 72 talking to a passenger. Construct a 95 percent confidence interval for the difference in proportions.

10.35 A survey of 100 cigarette smokers showed that 71 were loyal to one brand, compared to 122 of 200 toothpaste users. Construct a 90 percent confidence interval for the difference in proportions.

LO 10-8

Carry out a test of two variances using the F distribution.

10.7 COMPARING TWO VARIANCES

Comparing the *variances* may be as important as comparing the *means* of two populations. In manufacturing, smaller variation around the mean would indicate a more reliable product. In finance, smaller variation around the mean would indicate less volatility in asset returns. In services, smaller variation around the mean would indicate more consistency in customer treatment. For example, is the *variance* in Ford Mustang assembly times the same this month as last month? Is the *variability* in customer waiting times the same at two Tim Horton's franchises? Is the *variation* the same for customer concession purchases at a movie theater on Friday and Saturday nights?

Format of Hypotheses

We may test the null hypothesis against a left-tailed, two-tailed, or right-tailed alternative:

Left-Tailed Test	*Two-Tailed Test*	*Right-Tailed Test*
$H_0: \sigma_1^2 \geq \sigma_2^2$	$H_0: \sigma_1^2 = \sigma_2^2$	$H_0: \sigma_1^2 \leq \sigma_2^2$
$H_1: \sigma_1^2 < \sigma_2^2$	$H_1: \sigma_1^2 \neq \sigma_2^2$	$H_1: \sigma_1^2 > \sigma_2^2$

An equivalent way to state these hypotheses is to look at the *ratio* of the two variances. A ratio near 1 would indicate equal variances.

Left-Tailed Test	*Two-Tailed Test*	*Right-Tailed Test*
$H_0: \dfrac{\sigma_1^2}{\sigma_2^2} \geq 1$	$H_0: \dfrac{\sigma_1^2}{\sigma_2^2} = 1$	$H_0: \dfrac{\sigma_1^2}{\sigma_2^2} \leq 1$
$H_1: \dfrac{\sigma_1^2}{\sigma_2^2} < 1$	$H_1: \dfrac{\sigma_1^2}{\sigma_2^2} \neq 1$	$H_1: \dfrac{\sigma_1^2}{\sigma_2^2} > 1$

The F Test

In a left-tailed or right-tailed test, we actually test only at the equality, with the understanding that rejection of H_0 would imply rejecting values more extreme. The test statistic is the ratio of the sample variances. Assuming the populations are normal, the test statistic follows the **F distribution**, named for Ronald A. Fisher (1890–1962), one of the most famous statisticians of all time.

(10.18)
$$F_{\text{calc}} = \frac{s_1^2}{s_2^2} \qquad \begin{array}{l} df_1 = n_1 - 1 \\ \\ df_2 = n_2 - 1 \end{array}$$

If the null hypothesis of equal variances is true, this ratio should be near 1:

$$F_{\text{calc}} \cong 1 \qquad (\text{if } H_0 \text{ is true})$$

If the test statistic F is much less than 1 or much greater than 1, we would reject the hypothesis of equal population variances. The numerator s_1^2 has degrees of freedom $df_1 = n_1 - 1$, while the denominator s_2^2 has degrees of freedom $df_2 = n_2 - 1$. The F distribution is skewed. Its mean is always greater than 1, and its mode (the "peak" of the distribution) is always less than 1, but both the mean and mode tend to be near 1 for large samples. F cannot be negative because s_1^2 and s_2^2 cannot be negative.

Critical Values for the F Test

Critical values for the **F test** are denoted F_L (left tail) and F_R (right tail). The rejection regions and Excel formulas for the critical values are shown in Figure 10.15. Notice that the rejection regions for the two-tailed test are asymmetric.

Figure 10.15

Critical Values for the *F* Test for Equal Variances

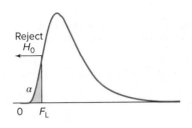

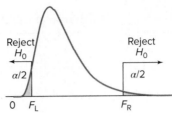

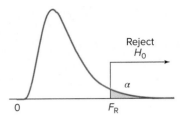

	Left-Tailed Test	**Two-Tailed Test**	**Right-Tailed Test**
Critical Value	$F_L =$ F.INV(α, df_1, df_2)	$F_L =$ F.INV$(\alpha/2, df_1, df_2)$ $F_R =$ F.INV.RT$(\alpha/2, df_1, df_2)$	$F_R =$ F.INV.RT(α, df_1, df_2)
Decision Rule	Reject H_0 if $F_{calc} < F_L$	Reject H_0 if $F_{calc} < F_L$ or if $F_{calc} > F_R$	Reject H_0 if $F_{calc} > F_R$

A right-tail critical value F_R may be found from Appendix F using df_1 and df_2 degrees of freedom. It is written

$$F_R = F_{df_1, df_2} \qquad \text{(right-tail critical } F) \qquad\qquad \textbf{(10.19)}$$

To obtain a left-tail critical value F_L from Appendix F, we reverse the numerator and denominator degrees of freedom, find the critical value from the table, and take its reciprocal:

$$F_L = \frac{1}{F_{df_2, df_1}} \qquad \text{(left-tail critical } F \text{ with reversed } df_1 \text{ and } df_2) \qquad \textbf{(10.20)}$$

Example 10.7
Collision Damage

An experimental bumper was designed to reduce damage in low-speed collisions. This bumper was installed on an experimental group of vans in a large fleet, but not on a control group. At the end of a trial period, accident data showed 12 repair incidents (a "repair incident" is a repair invoice) for the experimental vehicles and 9 repair incidents for the control group vehicles. Table 10.11 shows the dollar cost of the repair incidents.

A dot plot of the two samples, shown in Figure 10.16, suggests that the new bumper may have reduced the *mean* damage. However, the firm was also interested in whether the *variance* in damage had changed. We use the *F* test to test the hypothesis of equal variances. Do the sample variances support the idea of equal variances in the population? We will perform a two-tailed test.

Step 1: State the Hypotheses   For a two-tailed test for equality of variances, the hypotheses are

$$H_0: \sigma_1^2 = \sigma_2^2 \qquad \text{or} \qquad H_0: \sigma_1^2/\sigma_2^2 = 1$$
$$H_1: \sigma_1^2 \neq \sigma_2^2 \qquad\qquad\quad H_1: \sigma_1^2/\sigma_2^2 \neq 1$$

Step 2: Specify the Decision Rule   Degrees of freedom for the *F* test are

Numerator: $df_1 = n_1 - 1 = 12 - 1 = 11$

Denominator: $df_2 = n_2 - 1 = 9 - 1 = 8$

For a two-tailed test, we split the α risk and put $\alpha/2$ in each tail. Using Excel, the left-tail critical value for $\alpha/2 = .025$ is $F_L =$ F.INV$(.025, 11, 8) = 0.273$, and the right-tail critical value is $F_R =$ F.INV$(.975, 11, 8) = 4.243$. Alternatively, we could use Appendix F. To avoid interpolating,

Table 10.11

Repair Cost ($) for Accident Damage **Damage**

	Experimental Vehicles	Control Vehicles
	1,973	1,185
	403	885
	509	2,955
	2,103	815
	1,153	2,852
	292	1,217
	1,916	1,762
	1,602	2,592
	1,559	1,632
	547	
	801	
	359	
	$\bar{x}_1 = \$1,101.42$	$\bar{x}_2 = \$1,766.11$
	$s_1 = \$696.20$	$s_2 = \$837.62$
	$n_1 = 12$ incidents	$n_2 = 9$ incidents

Figure 10.16

Dot Plots for Collision Repair Costs **Damage**

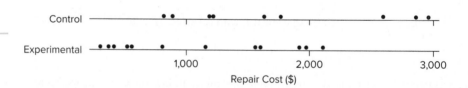

we use the next lower degrees of freedom when the required entry is not found in Appendix F. This conservative practice will not increase the probability of Type I error. For example, because $F_{11,8}$ is not in the table, we use $F_{10,8}$, as shown in Figure 10.17.

Figure 10.17

Critical Values of F for $\alpha/2 = .025$

CRITICAL VALUES OF $F_{.025}$

This table shows the 2.5 percent right-tail critical values of F for the stated degrees of freedom.

Denominator Degrees of Freedom (df_2)	Numerator Degrees of Freedom (df_1)										
	1	2	3	4	5	6	7	8	9	10	12
1	647.8	799.5	864.2	899.6	921.8	937.1	948.2	956.6	963.3	968.6	976.7
2	38.51	39.00	39.17	39.25	39.30	39.33	39.36	39.37	39.39	39.40	39.41
3	17.44	16.04	15.44	15.10	14.88	14.73	14.62	14.54	14.47	14.42	14.34
4	12.22	10.65	9.98	9.60	9.36	9.20	9.07	8.98	8.90	8.84	8.75
5	10.01	8.43	7.76	7.39	7.15	6.98	6.85	6.76	6.68	6.62	6.52
6	8.81	7.26	6.60	6.23	5.99	5.82	5.70	5.60	5.52	5.46	5.37
7	8.07	6.54	5.89	5.52	5.29	5.12	4.99	4.90	4.82	4.76	4.67
8	7.57	6.06	5.42	5.05	4.82	4.65	4.53	4.43	4.36	4.30	4.20
9	7.21	5.71	5.08	4.72	4.48	4.32	4.20	4.10	4.03	3.96	3.87
10	6.94	5.46	4.83	4.47	4.24	4.07	3.95	3.85	3.78	3.72	3.62
11	6.72	5.26	4.63	4.28	4.04	3.88	3.76	3.66	3.59	3.53	3.43
12	6.55	5.10	4.47	4.12	3.89	3.73	3.61	3.51	3.44	3.37	3.28
13	6.41	4.97	4.35	4.00	3.77	3.60	3.48	3.39	3.31	3.25	3.15
14	6.30	4.86	4.24	3.89	3.66	3.50	3.38	3.29	3.21	3.15	3.05
15	6.20	4.77	4.15	3.80	3.58	3.41	3.29	3.20	3.12	3.06	2.96

$$F_R = F_{df_1, df_2} = F_{11,8} \approx F_{10,8} = 4.30 \qquad \text{(right-tail critical value)}$$

To find the left-tail critical value, we reverse the numerator and denominator degrees of freedom, find the critical value from Appendix F, and take its reciprocal:

$$F_L = \frac{1}{F_{df_2, df_1}} = \frac{1}{F_{8,11}} = \frac{1}{3.66} = 0.273 \qquad \text{(left-tail critical value)}$$

As shown in Figure 10.18, the two-tailed decision rule is

Reject H_0 if $F_{calc} < 0.273$ or if $F_{calc} > 4.30$

Otherwise, do not reject H_0

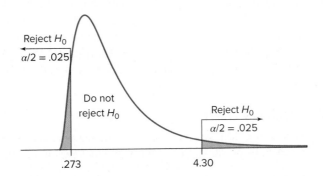

Figure 10.18

Two-Tailed *F* Test
at $\alpha = .05$

Step 3: Calculate the Test Statistic   The test statistic is

$$F_{calc} = \frac{s_1^2}{s_2^2} = \frac{(696.20)^2}{(837.62)^2} = 0.691$$

Step 4: Make the Decision   Because $F_{calc} = 0.691$, we cannot reject the hypothesis of equal variances in a two-tailed test at $\alpha = .05$. In other words, the ratio of the sample variances does not differ significantly from 1. The *p*-value will depend on the value of F_{calc}:

If $F_{calc} > 1$ Two-tailed *p*-value is =2*F.DIST.RT(F_{calc}, df_1, df_2)

If $F_{calc} < 1$ Two-tailed *p*-value is =2*F.DIST(F_{calc}, df_1, df_2, 1)

For the bumper data, $F_{calc} = 0.691$, so Excel's two-tailed *p*-value is =2*F.DIST(0.691,11,8,1) = .5575.

Folded *F* Test

We can make the two-tailed test for equal variances into a right-tailed test, so it is easier to look up the critical values in Appendix F. This method requires that we put the *larger observed variance* in the numerator and then look up the critical value for $\alpha/2$ instead of the chosen α. The test statistic for the folded *F* test is

$$F_{calc} = \frac{s_{max}^2}{s_{min}^2} \qquad \text{Reject } H_0 \text{ if } F_{calc} > F_{\alpha/2} \qquad (10.21)$$

The larger variance goes in the numerator and the smaller variance in the denominator. *"Larger" refers to the variance (not to the sample size).* But the hypotheses are the same as for a two-tailed test:

H_0: $\sigma_1^2/\sigma_2^2 = 1$

H_1: $\sigma_1^2/\sigma_2^2 \neq 1$

For the bumper data, the second sample variance ($s_2^2 = 837.62$) is larger than the first sample variance ($s_1^2 = 696.20$), so the folded *F* test statistic is

$$F_{calc} = \frac{s_{max}^2}{s_{min}^2} = \frac{s_2^2}{s_1^2} = \frac{(837.62)^2}{(696.20)^2} = 1.448$$

We must be careful that the degrees of freedom match the variances in the modified F statistic. In this case, the second sample variance is larger (it goes in the numerator), so we must reverse the degrees of freedom:

Numerator: $n_2 - 1 = 9 - 1 = 8$
Denominator: $n_1 - 1 = 12 - 1 = 11$

Now we look up the critical value for $F_{8,11}$ in Appendix F using $\alpha/2 = .05/2 = .025$:

$$F_{.025} = 3.66$$

Alternatively, we can use the Excel function =F.INV.RT(0.025,8,11) = 3.66. Because the test statistic $F_{calc} = 1.448$ does not exceed the critical value $F_{.025} = 3.66$, we cannot reject the hypothesis of equal variances. This is the same conclusion that we reached in the two-tailed test. Because $F_{calc} < 1$, Excel's two-tailed p-value is =2*F.DIST.RT(1.448,8,11) = .5569, which is the same as in the previous result except for rounding. Anytime you want a two-tailed F test, you may use the folded F test if you think it is easier.

One-Tailed *F* Test

Suppose that the firm was interested in knowing whether the new bumper had *reduced* the variance in collision damage cost. We would then perform a left-tailed test.

Step 1: State the Hypotheses   The hypotheses for a left-tailed test are

$H_0: \sigma_1^2/\sigma_2^2 \geq 1$
$H_1: \sigma_1^2/\sigma_2^2 < 1$

Step 2: Specify the Decision Rule   Degrees of freedom for the F test are the same as for a two-tailed test (the hypothesis doesn't affect the degrees of freedom):

Numerator: $df_1 = n_1 - 1 = 12 - 1 = 11$
Denominator: $df_2 = n_2 - 1 = 9 - 1 = 8$

However, now the entire $\alpha = .05$ goes in the left tail. Using Excel, the left-tail critical value is F_L =F.INV(.05,11,8) = 0.339. Alternatively, we can find the left-tail critical value from Appendix F by reversing the degrees of freedom and calculating the reciprocal of the table value, as illustrated in Figures 10.19 and 10.20. Notice that the asymmetry of the F distribution causes the left-tail area to be compressed in the horizontal direction.

$$F_L = \frac{1}{F_{df_2,df_1}} = \frac{1}{F_{8,11}} = \frac{1}{2.95} = 0.339 \qquad \text{(left-tail critical value)}$$

The decision rule is

Reject H_0 if $F_{calc} < 0.339$
Otherwise, do not reject H_0

Step 3: Calculate the Test Statistic   The test statistic is the same as for a two-tailed test (the hypothesis doesn't affect the test statistic):

$$F_{calc} = \frac{s_1^2}{s_2^2} = \frac{(696.20)^2}{(837.62)^2} = 0.691$$

Step 4: Make the Decision   Because the test statistic $F = 0.691$ is not in the critical region, we cannot reject the hypothesis of equal variances in a one-tailed test. The bumpers did not significantly decrease the variance in collision repair cost.

CRITICAL VALUES OF $F_{.05}$

Figure 10.19

Right-Tail F_R for $\alpha = .05$

This table shows the 5 percent right-tail critical values of F for the stated degrees of freedom.

Denominator Degrees of Freedom (df_2)	Numerator Degrees of Freedom (df_1)										
	1	2	3	4	5	6	7	8	9	10	12
1	161.4	199.5	215.7	224.6	230.2	234.0	236.8	238.9	240.5	241.9	243.9
2	18.51	19.00	19.16	19.25	19.30	19.33	19.35	19.37	19.38	19.40	19.41
3	10.13	9.55	9.28	9.12	9.01	8.94	8.89	8.85	8.81	8.79	8.74
4	7.71	6.94	6.59	6.39	6.26	6.16	6.09	6.04	6.00	5.96	5.91
5	6.61	5.79	5.41	5.19	5.05	4.95	4.88	4.82	4.77	4.74	4.68
6	5.99	5.14	4.76	4.53	4.39	4.28	4.21	4.15	4.10	4.06	4.00
7	5.59	4.74	4.35	4.12	3.97	3.87	3.79	3.73	3.68	3.64	3.57
8	5.32	4.46	4.07	3.84	3.69	3.58	3.50	3.44	3.39	3.35	3.28
9	5.12	4.26	3.86	3.63	3.48	3.37	3.29	3.23	3.18	3.14	3.07
10	4.96	4.10	3.71	3.48	3.33	3.22	3.14	3.07	3.02	2.98	2.91
11	4.84	3.98	3.59	3.36	3.20	3.09	3.01	2.95	2.90	2.85	2.79
12	4.75	3.89	3.49	3.26	3.11	3.00	2.91	2.85	2.80	2.75	2.69
13	4.67	3.81	3.41	3.18	3.03	2.92	2.83	2.77	2.71	2.67	2.60
14	4.60	3.74	3.34	3.11	2.96	2.85	2.76	2.70	2.65	2.60	2.53
15	4.54	3.68	3.29	3.06	2.90	2.79	2.71	2.64	2.59	2.54	2.48

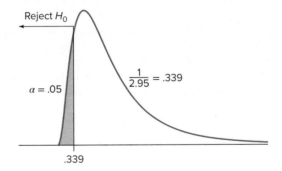

Figure 10.20

Left-Tail F_L for $\alpha = .05$

$\frac{1}{2.95} = .339$

Reject H_0

$\alpha = .05$

.339

Excel's *F* Test

Figure 10.21 shows Excel's left-tailed test. For the bumper data, the *p*-value of .279 indicates that a sample variance ratio as extreme as $F = 0.691$ would occur by chance about 28 percent of the time if the population variances were in fact equal. Because the *p*-value is not less than $\alpha = .05$, we conclude that there is no significant difference in variances.

Figure 10.21

Excel's *F* Test of Variances

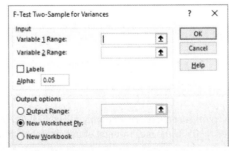

F-Test Two-Sample for Variances

	Experimental	Control
Mean	1101.416667	1766.111111
Variance	484700.8106	701611.1111
Observations	12	9
df	11	8
F	0.690839702	
P(F<=f) one-tail	0.278623089	
F Critical one-tail	0.33921414	

Source: Microsoft Excel.

Excel's Data Analysis > F Test Two Sample for Variances always calculates the test statistic $F_{calc} = s_1^2/s_2^2$ and then performs a *one-tailed* test in the direction of the observed inequality. For the bumper data, s_1^2 is less than s_2^2, so Excel performs a left-tailed test. If you want a two-tailed test, you can double Excel's p-value.

Significance versus Importance

The test of means showed a mean difference of $665 per repair incident. That is large enough that it might be important. The incremental cost per vehicle of the new bumper would have to be compared with the discounted present value of the expected annual savings per vehicle over its useful life. In a large fleet of vehicles, the payback period could be calculated. Most firms require that a change pay for itself in a fairly short period of time. *Importance* is a question to be answered ultimately by financial experts, not statisticians.

Section Exercises

Mc Graw Hill **connect**

Hint: Use software of your choice to perform calculations.

10.36 Which samples show unequal variances? Use $\alpha = .10$ in all tests. Show the critical values and degrees of freedom clearly and illustrate the decision rule.

a. $s_1 = 10.2$, $n_1 = 22$, $s_2 = 6.4$, $n_2 = 16$, two-tailed test
b. $s_1 = 0.89$, $n_1 = 25$, $s_2 = 0.67$, $n_2 = 18$, right-tailed test
c. $s_1 = 124$, $n_1 = 12$, $s_2 = 260$, $n_2 = 10$, left-tailed test

10.37 Which samples show unequal variances? Use $\alpha = .05$ in all tests. Show the critical values and degrees of freedom clearly and illustrate the decision rule.

a. $s_1 = 5.1$, $n_1 = 11$, $s_2 = 3.2$, $n_2 = 8$, two-tailed test
b. $s_1 = 221$, $n_1 = 8$, $s_2 = 445$, $n_2 = 8$, left-tailed test
c. $s_1 = 67$, $n_1 = 10$, $s_2 = 15$, $n_2 = 13$, right-tailed test

10.38 Researchers at the Mayo Clinic studied the effect of sound levels on patient healing and found a significant association (louder hospital ambient sound level is associated with slower postsurgical healing). Based on the Mayo Clinic's experience, Ardmore Hospital installed a new vinyl flooring that is supposed to reduce the mean sound level (decibels) in the hospital corridors. The sound level is measured at five randomly selected times in the main corridor. (a) At $\alpha = .05$, has the mean been reduced? Show the hypotheses, decision rule, and test statistic. (b) At $\alpha = .05$, has the variance changed? Show the hypotheses, decision rule, and test statistic. 🖫 **Decibels**

New Flooring	Old Flooring
42	48
41	51
40	44
37	48
44	52

10.39 A manufacturing process drills holes in sheet metal that are supposed to be 0.5000 cm in diameter. Before and after a new drill press is installed, the hole diameter is carefully measured (in cm) for 12 randomly chosen parts. At $\alpha = .05$, do these independent random samples prove that the new process has smaller variance? Show the hypotheses, decision rule, and test statistic. *Hint:* Use Excel =F.INV(α,n_1-1,n_2-1) to get F_L. 🖫 **Diameter**

New drill:	.5005	.5010	.5024	.4988	.4997	.4995
	.4976	.5042	.5014	.4995	.4988	.4992
Old drill:	.5052	.5053	.4947	.4907	.5031	.4923
	.5040	.5035	.5061	.4956	.5035	.4962

10.40 Examine the data below showing the weights (in pounds) of randomly selected checked bags for an airline's flights on the same day. (a) At $\alpha = .05$, is the mean weight of an international bag greater?

Show the hypotheses, decision rule, and test statistic. (b) At $\alpha = .05$, is the variance greater for bags on an international flight? Show the hypotheses, decision rule, and test statistic. 📄 **Luggage**

International (10 bags)			Domestic (15 bags)		
39	47		29	37	43
54	48		36	33	42
46	28		33	29	32
39	54		34	43	35
69	62		38	39	39

Chapter Summary

A **two-sample test** compares samples with each other rather than comparing with a benchmark, as in a one-sample test. For independent samples, the comparison of means generally utilizes the Student's t distribution because the population variances are almost always unknown. If the unknown variances are **assumed equal,** we use a **pooled variance** estimate and **add the degrees of freedom.** If the unknown variances are **assumed unequal, we** do not pool the variances and we reduce the degrees of freedom by using **Welch's formula.** The test statistic is the difference of means divided by their standard error. For tests of means or proportions, **equal sample sizes** are desirable but not necessary. The t **test for paired samples** uses the differences of n paired

observations, thereby being a **one-sample t test.** For two proportions, the samples may be **pooled** if the population proportions are assumed equal, and the test statistic is the difference of proportions divided by the standard error, the square root of the sum of the sample variances. For proportions, **normality** of $p_1 - p_2$ may be assumed if both samples are large, that is, if each contains at least 10 successes and 10 failures. The **F test** for equality of **two variances** is named after Sir Ronald Fisher. Its test statistic is the **ratio** of the sample variances. We want to see if the ratio differs significantly from 1. The F table shows critical values based on both **numerator** and **denominator** degrees of freedom.

Key Terms

Behrens-Fisher problem	F test	pooled estimate	two-sample tests
difference of two means	independent sample	pooled proportion	Type I error
difference of two population proportions	paired comparison	pooled variance	Type II error
	paired samples	p-values	Welch's adjusted degrees of freedom
F distribution	paired t test	test statistic	

Commonly Used Formulas in Two-Sample Hypothesis Tests

Test Statistic (Difference of Means, Assuming Equal Variances):

$$t_{calc} = \frac{\overline{x}_1 - \overline{x}_2}{\sqrt{\frac{s_p^2}{n_1} + \frac{s_p^2}{n_1}}}, \quad \text{with } d.f. = n_1 + n_2 - 2 \text{ and } s_p^2 = \frac{(n_1 - 1)s_1^2 + (n_2 - 1)s_2^2}{n_1 + n_2 - 2}$$

Test Statistic (Difference of Means, Assuming Unequal Variances):

$$t_{calc} = \frac{\overline{x}_1 - \overline{x}_2}{\sqrt{\frac{s_1^2}{n_1} + \frac{s_2^2}{n_2}}}, \quad \text{with } d.f. = \frac{(s_1^2/n_1 + s_2^2/n_2)^2}{\frac{(s_1^2/n_1)^2}{n_1 - 1} + \frac{(s_2^2/n_2)^2}{n_2 - 1}}$$

Confidence Interval for $\mu_1 - \mu_2$:

$$(\overline{x}_1 - \overline{x}_2) \pm t_{\alpha/2}\sqrt{\frac{s_1^2}{n_1} + \frac{s_2^2}{n_2}}, \quad \text{with } d.f. = \frac{\left(\frac{s_2^1}{n_1} + \frac{s_2^2}{n_2}\right)^2}{\frac{\left(\frac{s_1^2}{n_1}\right)^2}{n_1 - 1} + \frac{\left(\frac{s_2^2}{n_2}\right)^2}{n_2 - 1}}$$

Test Statistic (Paired Differences): $t_{calc} = \dfrac{\bar{d} - \mu_d}{\dfrac{s_d}{\sqrt{n}}}$, with $d.f. = n - 1$

Test Statistic (Equality of Proportions): $z_{calc} = \dfrac{p_1 - p_2}{\sqrt{p_c(1 - p_c)\left(\dfrac{1}{n_1} + \dfrac{1}{n_2}\right)}}$, with $p_c = \dfrac{x_1 + x_2}{n_1 + n_2}$

Confidence Interval for $\pi_1 - \pi_2$: $(p_1 - p_2) \pm z_{\alpha/2}\sqrt{\dfrac{p_1(1 - p_1)}{n_1} + \dfrac{p_2(1 - p_2)}{n_2}}$

Test Statistic (Two Variances): $F_{calc} = \dfrac{s_1^2}{s_2^2}$, with $df_1 = n_1 - 1, df_2 = n_2 - 1$

Test Statistic (Folded F Test): $F_{calc} = \dfrac{s_{max}^2}{s_{min}^2}$ (use $df_1 = n_2 - 1$ and $df_2 = n_1 - 1$ if $s_2^2 > s_1^2$)

Chapter Review

1. (a) Explain why two samples from the same population could differ. (b) Why do we say that two-sample tests have a built-in point of reference?

2. (a) In a two-sample test of proportions, what is a pooled proportion? (b) Why is the test for normality important for a two-sample test of proportions? (c) What is the criterion for assuming normality of the test statistic?

3. (a) Is it necessary that sample sizes be equal for a two-sample test of proportions? Is it desirable? (b) Explain the analogy between overlapping confidence intervals and testing for equality of two proportions.

4. List the three cases for a test comparing two means. Explain how they differ.

5. Consider *Case 1* (known variances) in the test comparing two means. (a) Why is *Case 1* unusual and not used very often? (b) What distribution is used for the test statistic? (c) Write the formula for the test statistic.

6. Consider *Case 2* (unknown but equal variances) in the test comparing two means. (a) What distribution is used for the test statistic? (b) State the degrees of freedom used in this test. (c) Write the formula for the pooled variance and interpret it. (d) Write the formula for the test statistic.

7. Consider *Case 3* (unknown and unequal variances) in the test comparing two means. (a) What complication arises in degrees of freedom for *Case 3*? (b) What distribution is used for the test statistic? (c) Write the formula for the test statistic.

8. (a) Is it ever acceptable to use a normal distribution in a test of means with unknown variances? (b) If we assume normality, what is gained? What is lost?

9. Why is it a good idea to use a computer program like Excel to do tests of means?

10. (a) Explain why the paired t test for dependent samples is really a one-sample test. (b) State the degrees of freedom for the paired t test. (c) Why not treat two paired samples as if they were independent?

11. Explain how a difference in means could be statistically *significant* but not *important*.

12. (a) Why do we use an F test? (b) When two population variances are equal, what value would you expect of the F test statistic?

13. (a) In an F test for two variances, explain how to obtain left- and right-tail critical values. (b) What are the assumptions underlying the F test?

Chapter Exercises

Note: For tests on two proportions, two means, or two variances, it is a good idea to check your work by using Minitab, MegaStat, or the *LearningStats* two-sample calculators in Unit 10.

10.41 The top food snacks consumed by adults aged 18–54 are gum, chocolate candy, fresh fruit, potato chips, breath mints/candy, ice cream, nuts, cookies, bars, yogurt, and crackers. Out of a random sample of 25 men, 15 ranked fresh fruit in their top five snack choices. Out of a random sample of 32 women, 22 ranked fresh fruit in their top five snack choices. Is there a difference in the proportion of men and women who rank fresh fruit in their top five list of snacks? (a) State the hypotheses and a decision rule for $\alpha = .10$. (b) Calculate the sample proportions. (c) Find the

test statistic and its p-value. What is your conclusion? (d) Is normality of $p_1 - p_2$ assured?

10.42 In an early home game, an NBA team made 66 of its 94 free throw attempts. In one of its last home games, the team made 68 of 89 attempts. (a) At $\alpha = .10$, did the team significantly improve its free throw percentage (left-tailed test)? (b) Use Excel to calculate the p-value and interpret it.

10.43 Are college students more likely than young children to eat cereal? Researchers surveyed both age groups to find the answer. The results are shown in the table below. (a) State the hypotheses used to answer the question. (b) Using $\alpha = .05$, state the decision rule and sketch it. (c) Find the sample proportions and z statistic. (d) Make a decision. (e) Find the *p*-value and interpret it. (f) Is the normality assumption fulfilled? Explain.

Statistic	College Students (ages 18 to 25)	Young Children (ages 6 to 11)
Number who eat cereal	$x_1 = 833$	$x_2 = 692$
Number surveyed	$n_1 = 850$	$n_2 = 740$

10.44 A recent study found that 202 women held board seats out of a total of 1,195 seats in the *Fortune* 100 companies. A previous study found that 779 women held board seats out of a total of 5,727 seats in the *Fortune* 500 companies. Treating these as random samples (board seat assignments change often), can we conclude that *Fortune* 100 companies have a greater proportion of women board members than the *Fortune* 500? (a) State the hypotheses. (b) Calculate the sample proportions. (c) Find the test statistic and its *p*-value. What is your conclusion at $\alpha = .05$? (d) If statistically significant, can you suggest factors that might explain the increase?

10.45 A study of the *Fortune* 100 board of director members showed that there were 36 minority women holding board seats out of 202 total female board members. There were 142 minority men holding board seats out of 993 total male board members. (a) Treating the findings from this study as samples, calculate the sample proportions. (b) Find the test statistic and its *p*-value. (c) At the 5 percent level of significance, is there a difference in the percentage of minority women board directors and minority men board directors?

10.46 To test his hypothesis that students who finish an exam first get better grades, a professor kept track of the order in which papers were handed in. Of the first 25 papers, 10 received a B or better, compared with 8 of the last 24 papers handed in. Is the first group better, at $\alpha = .10$? (a) State your hypotheses and obtain a test statistic and *p*-value. Interpret the results. (b) Are the samples large enough to assure normality of $p_1 - p_2$? (c) Make an argument that early finishers should do better. Then make the opposite argument. Which is more convincing?

10.47 How many full-page advertisements are found in a magazine? In an October issue of *Muscle and Fitness,* there were 252 ads, of which 97 were full-page. For the same month, the magazine *Glamour* had 342 ads, of which 167 were full-page. (a) Is the difference significant at $\alpha = .01$? (b) Find the *p*-value. (c) Is normality assured? (d) Based on what you know of these magazines, why might the proportions of full-page ads differ?

10.48 eShopNet, an online clothing retailer, is testing a new e-mail campaign by sending one version of the e-mail with the word "free" in the subject line (version A) to a group of 1,500 customers and another version of the e-mail with the word "discount" in the subject line (version B) to a different group of 1,500 customers. This type of test is called A/B testing or split testing. After tracking the responses to the two different versions of the e-mail advertisement, eShopNet finds that 90 responded to version A (with the word "free") and 129 responded to version B (with the word "discount"). Using $\alpha = .01$, was the response rate to version B significantly higher than the response rate to version A?

10.49 After John F. Kennedy Jr. was killed in an airplane crash at night, a survey was taken, asking whether a noninstrument-rated pilot should be allowed to fly at night. Of 409 New York State residents, 61 said yes. Of 70 aviation experts who were asked the same question, 40 said yes. (a) At $\alpha = .01$, did a larger proportion of experts say yes compared with the general public, or is the difference within the realm of chance? (b) Find the *p*-value and interpret it. (c) Is normality of $p_1 - p_2$ assured?
Source: www.siena.edu/sri.

10.50 A ski company in Vail owns two ski shops, one on the east side and one on the west side. Sales data showed that at the eastern location there were 56 pairs of large gloves sold out of 304 total pairs sold. At the western location there were 145 pairs of large gloves sold out of 562 total pairs sold. (a) Calculate the sample proportion of large gloves for each location. (b) At $\alpha = .05$, is there a significant difference in the proportion of large gloves sold? (c) Can you suggest any reasons why a difference might exist? (*Note:* Problem is based on actual sales data.)

10.51 At a University of Colorado women's home basketball game, a random sample of 25 concession purchases showed a mean of $7.12 with a standard deviation of $2.14. For the next week's home game, the admission ticket had a discount coupon for popcorn printed on the back. A random sample of 25 purchases from that week showed a mean of $8.29 with a standard deviation of $3.02. Was there an increase in the average concession stand purchases with the coupon? Assume unequal variances and use $\alpha = .05$.

10.52 A ski resort tracks the proportion of seasonal employees who are rehired each season. Rehiring a seasonal employee is beneficial in many ways, including lowering the costs incurred during the hiring process such as training costs. A random sample of 833 full-time and 386 part-time seasonal employees showed that 434 full-time employees were rehired compared with 189 part-time employees. (a) Is there a significant difference in the proportion of rehires between the full-time and part-time seasonal employees? Use $\alpha = .10$ for the level of significance. (b) Use Excel to calculate the *p*-value.

10.53 Does a "follow-up reminder" increase the online review rate for a local business? The business sent out requests for an online review to 760 customers (without a reminder) and got 703 reviews. As an experiment, it sent out 240 review requests (with a reminder) and got 228 reviews. (a) At $\alpha = .05$, was the online review rate higher in the experimental group? (b) Can normality be assumed?

10.54 A study revealed that the 30-day readmission rate was 31.5 percent for 400 patients who received after-hospital care instructions (e.g., how to take their medications) compared to a readmission rate of 38.5 percent for 400 patients who did not receive such information. (a) Set up the hypotheses to see whether the readmission rate was lower for those who received the information. (b) Find the *p*-value for the test. (c) What is your conclusion at $\alpha = .05$? At $\alpha = .01$?

10.55 In a marketing class, 44 student members of virtual (Internet) project teams (group 1) and 42 members of face-to-face project teams (group 2) were asked to respond on a 1–5 scale to the question: "As compared to other teams, the members helped each other." For group 1 the mean was 2.73 with a standard deviation of 0.97, while for group 2 the mean was 1.90 with a standard deviation of 0.91. At $\alpha = .01$, is the virtual team mean significantly higher?

10.56 In San Francisco, a sample of 3,200 wireless routers showed that 1,312 used encryption (to prevent hackers from intercepting information). In Seattle, a sample of 1,800 wireless routers showed that 684 used encryption. (a) Set up hypotheses to test whether the population proportion of encryption is higher in San Francisco than Seattle. (b) Test the hypotheses at $\alpha = .05$.

10.57 Researchers at a medical center implanted defibrillators in 742 patients after a heart attack and compared them with 490 similar patients without the implant. Over the next 2 years, 104 of those with defibrillators had died, compared with 98 of those without defibrillators. (a) State the hypotheses for a one-tailed test to see if the defibrillators reduced the death rate. (b) Obtain a test statistic and p-value. (c) Is normality assured? (d) Why might such devices not be widely implanted in heart attack patients?

10.58 Noodles & Company introduced spaghetti and meatballs to its menu. Before putting it on the menu, it performed taste tests to determine the best-tasting spaghetti sauce. In a paired comparison, 70 tasters were asked to rate their satisfaction with two different sauces on a scale of 1–10, with 10 being the highest. The mean difference in ratings was $\bar{d} = -0.385714$ with a standard deviation of $s_d = 1.37570$. (a) State the hypotheses for a two-tailed test. (b) Calculate the test statistic. (c) Calculate the critical value for $\alpha = .05$. (d) Calculate the p-value. (e) State your conclusion.

10.59 Has the cost to outsource a standard employee background check changed? A random sample of 10 companies in spring 2019 showed a sample average of $105 with a sample standard deviation equal to $32. A random sample of 10 different companies in spring 2020 resulted in a sample average of $75 with a sample standard deviation equal to $45. (a) Conduct a hypothesis test to test the difference in sample means with a level of significance equal to .05. Assume the population variances are not equal. (b) Discuss why a paired sample design might have made more sense in this case.

10.60 From her firm's computer telephone log, an executive found that the mean length of 64 telephone calls during July was 4.48 minutes with a standard deviation of 5.87 minutes. She vowed to make an effort to reduce the length of calls. The August phone log showed 48 telephone calls whose mean was 2.396 minutes with a standard deviation of 2.018 minutes. (a) State the hypotheses for a right-tailed test. (b) Obtain a test statistic and p-value assuming unequal variances. Interpret these results using $\alpha = .01$. (c) Why might the sample data *not* resemble a normal, bell-shaped curve? If not, how might this affect your conclusions?

10.61 An experimental bumper was designed to reduce damage in low-speed collisions. This bumper was installed on an experimental group of vans in a large fleet but not on a control group. At the end of a trial period, accident data showed 12 repair incidents for the experimental group and

9 repair incidents for the control group. Vehicle downtime (in days per repair incident) is shown below. At $\alpha = .05$, did the new bumper reduce downtime? (a) Make stacked dot plots of the data (a sketch is OK). (b) State the hypotheses. (c) State the decision rule and sketch it. (d) Find the test statistic. (e) Make a decision. (f) Find the p-value and interpret it. (g) Do you think the difference is large enough to be important? Explain. ⌨ **DownTime**

New bumper (12 repair incidents):
9, 2, 5, 12, 5, 4, 7, 5, 11, 3, 7, 1

Control group (9 repair incidents):
7, 5, 7, 4, 18, 4, 8, 14, 13

10.62 Based on the sample data below, is the average Medicare spending in the northern region significantly less than the average spending in the southern region at the 1 percent level? (a) State the hypotheses and decision rule. (b) Find the test statistic assuming unequal variances. (c) State your conclusion. Is this a strong conclusion? (d) Can you suggest reasons why a difference might exist?

Medicare Spending per Patient (adjusted for age, sex, and race)		
Statistic	**Northern Region**	**Southern Region**
Sample mean	$3,123	$8,456
Sample standard deviation	$1,546	$3,678
Sample size	14 patients	16 patients

10.63 In a 15-day survey of air pollution in two European capitals, the mean particulate count (micrograms per cubic meter) in Athens was 39.5 with a standard deviation of 3.75, while in London the mean was 31.5 with a standard deviation of 2.25. (a) Assuming equal population variances, does this evidence convince you that the mean particulate count is higher in Athens, at $\alpha = .05$? (b) Are the variances equal or not, at $\alpha = .05$?

10.64 One group of accounting students took a distance learning class, while another group took the same course in a traditional classroom. At $\alpha = .10$, is there a significant difference in the mean scores listed below? (a) State the hypotheses. (b) State the decision rule and sketch it. (c) Find the test statistic. (d) Make a decision. (e) Use Excel to find the p-value and interpret it.

Exam Scores for Accounting Students		
Statistic	**Distance**	**Classroom**
Mean scores	$\bar{x}_1 = 9.1$	$\bar{x}_2 = 10.3$
Sample std. dev.	$s_1 = 2.4$	$s_2 = 2.5$
Number of students	$n_1 = 20$	$n_2 = 20$

10.65 Do male and female school superintendents earn the same pay? Salaries for 20 males and 17 females in a certain metropolitan area are shown below. At $\alpha = .01$, were the mean superintendent salaries greater for men than for women? (a) State the hypotheses. (b) State the decision rule and sketch it. (c) Find the test statistic. (d) Make a decision. (e) Estimate the p-value and interpret it. ⌨ **Paycheck**

School Superintendent Pay

Men (n = 20)		Women (n = 17)	
114,000	121,421	94,675	96,000
115,024	112,187	123,484	112,455
115,598	110,160	99,703	120,118
108,400	128,322	86,000	124,163
109,900	128,041	108,000	76,340
120,352	125,462	94,940	89,600
118,000	113,611	83,933	91,993
108,209	123,814	102,181	
110,000	111,280	86,840	
151,008	112,280	85,000	

10.66 The average take-out order size for Ashoka Curry House restaurant is shown. Assuming equal variances, at $\alpha = .05$, is there a significant difference in the order sizes? (a) State the hypotheses. (b) State the decision rule and sketch it. (c) Find the test statistic. (d) Make a decision. (e) Use Excel to find the p-value and interpret it.

Customer Order Size ($)

Statistic	Friday Night	Saturday Night
Mean order size	$\bar{x}_1 = 22.32$	$\bar{x}_2 = 25.56$
Standard deviation	$s_1 = 4.35$	$s_2 = 6.16$
Number of orders	$n_1 = 13$	$n_2 = 18$

10.67 Cash withdrawals from a college credit union for a random sample of 30 Fridays and 30 Mondays are shown. At $\alpha = .01$, is there a difference in the mean withdrawal on Monday and Friday? (a) Make stacked dot plots of the data (a sketch is OK). (b) State the hypotheses. (c) State the decision rule and sketch it. (d) Find the test statistic. (e) Make a decision. (f) Find the p-value and interpret it. ATM

Randomly Chosen Cash Withdrawals ($)

Friday			Monday		
250	10	10	40	30	10
20	10	30	100	70	370
110	20	10	20	20	10
40	20	40	30	50	30
70	10	10	200	20	40
20	20	400	20	30	20
10	20	10	10	20	100
50	20	10	30	40	20
100	20	20	50	10	20
20	60	70	60	10	20

10.68 In Mini Case 10.2, we found that the mean methane gas emissions for the two pneumatic controller manufacturers were not equal. When choosing formula 10.3 to calculate the t statistic, we assumed that their variances were not equal. Was this a valid assumption? Use Cemco's

sample $s_1 = 147.2$ scfd ($n_1 = 18$) and Invalco's sample $s_2 = 237.9$ scfd ($n_2 = 17$) to find out. (a) State the hypotheses for a two-tailed test. (b) Calculate the critical value for $\alpha = .10$. (c) Calculate the test statistic. (d) What is your conclusion?

10.69 A ski company in Vail owns two ski shops, one on the west side and one on the east side of Vail. Is there a difference in daily average goggle sales between the two stores? Assume equal variances. (a) State the hypotheses for a two-tailed test. (b) State the decision rule for a level of significance equal to 5 percent and sketch it. (c) Find the test statistic and state your conclusion.

Sales Data for Ski Goggles

Statistic	East Side Shop	West Side Shop
Mean sales	$328	$435
Sample std. dev.	$104	$147
Sample size	28 days	29 days

10.70 A ski company in Vail owns two ski shops, one on the west side and one on the east side of Vail. Ski hat sales data (in dollars) for a random sample of 5 Saturdays during the ski season showed the following results. Is there a significant difference in sales dollars of hats between the west side and east side stores at the 10 percent level of significance? (a) State the hypotheses. (b) State the decision rule and sketch it. (c) Find the test statistic and state your conclusion. Hats

Saturday Sales Data ($) for Ski Hats

Saturday	East Side Shop	West Side Shop
1	548	523
2	493	721
3	609	695
4	567	510
5	432	532

10.71 Emergency room arrivals in a large hospital showed the statistics below for 2 months. At $\alpha = .05$, has the variance changed? Show all steps clearly, including an illustration of the decision rule.

Statistic	October	November
Mean arrivals	177.032	171.733
Standard deviation	13.482	15.427
Days	31	30

10.72 Concerned about graffiti, mayors of nine suburban communities instituted a citizen Community Watch program. (a) State the hypotheses to see whether the number of graffiti incidents declined. (b) Find the test statistic. (c) State the critical value for $\alpha = .05$. (d) Find the p-value. (e) State your conclusion. 📁 **Graffiti**

Community	Monthly Incidents After	Monthly Incidents Before
Burr Oak	8	12
Elgin Corners	3	6
Elm Grove	7	8
Greenburg	0	1
Huntley	4	2
North Lyman	0	4
Pin Oak	4	3
South Lyman	4	4
Victorville	0	3

10.73 A certain company will purchase the house of any employee who is transferred out of state and will handle all details of reselling the house. The purchase price is based on two assessments, one assessor being chosen by the employee and one by the company. Based on the sample of eight assessments shown, do the two assessors agree? Use the .01 level of significance, state hypotheses clearly, and show all steps. 📁 **HomeValue**

10.74 Nine homes are chosen at random from real estate listings in two suburban neighborhoods, and the square footage of each home is noted in the following table. At the .10 level of significance, is there a difference between the average sizes of homes in the two neighborhoods? State your hypotheses and show all steps clearly. 📁 **HomeSize**

10.75 Two labs produce 1280 × 1024 LCD displays. At random, records are examined for 12 independently chosen hours of production in each lab, and the number of bad pixels per thousand displays is recorded. (a) Assuming equal variances, at the .01 level of significance, is there a difference in the defect rate between the two labs? State your hypotheses and show all steps clearly. (b) At the .01 level of significance, can you reject the hypothesis of equal variances? State your hypotheses and show all steps clearly. 📁 **LCDDefects**

Defects in Randomly Inspected LCD Displays

Lab A 422, 319, 326, 410, 393, 368, 497, 381, 515, 472, 423, 355

Lab B 497, 421, 408, 375, 410, 489, 389, 418, 447, 429, 404, 477

10.76 A cognitive retraining clinic assists outpatient victims of head injury, anoxia, or other conditions that result in cognitive impairment. Each incoming patient is evaluated to establish an appropriate treatment program and estimated length of stay. To see if the evaluation teams are consistent, 12 randomly chosen patients are separately evaluated by two expert teams (A and B) as shown. At the .10 level of significance, is there a difference between the evaluator teams' estimated length of stay? State your hypotheses and show all steps clearly. 📁 **LengthStay**

Assessments of Eight Homes ($ thousands) 📁 **HomeValue**

Assessed by	Home 1	Home 2	Home 3	Home 4	Home 5	Home 6	Home 7	Home 8
Company	328	350	455	278	290	285	535	745
Employee	318	345	470	285	310	280	525	765

Size of Homes in Two Subdivisions 📁 **HomeSize**

Subdivision	Square Footage								
Greenwood	2,320	2,450	2,270	2,200	2,850	2,150	2,400	2,800	2,430
Pinewood	2,850	2,560	2,300	2,100	2,750	2,450	2,550	2,750	3,150

Estimated Length of Stay in Weeks 📁 **LengthStay**

Team	Patient											
	1	2	3	4	5	6	7	8	9	10	11	12
A	24	24	52	30	40	30	18	30	18	40	24	12
B	24	20	52	36	36	24	36	16	52	24	24	16

10.77 Rates of return (annualized) in two investment portfolios are compared over the past 12 quarters. They are considered similar in safety, but portfolio B is advertised as being "less volatile." (a) At $\alpha = .025$, does the sample show that portfolio A has significantly greater variance in rates of return than portfolio B? (b) At $\alpha = .025$, is there a significant difference in the means? 📄 **Portfolio**

Portfolio A	Portfolio B	Portfolio A	Portfolio B
5.23	8.96	7.89	7.68
10.91	8.60	9.82	7.62
12.49	7.61	9.62	8.71
4.17	6.60	4.93	8.97
5.54	7.77	11.66	7.71
8.68	7.06	11.49	9.91

10.78 Is there a difference between the variance in ages for full-time seasonal employees and part-time seasonal employees at a ski resort? A sample of 62 full-time employees had an $s_1^2 = 265.69$. A sample of 78 part-time employees had an $s_2^2 = 190.44$. (a) Test for equal variances with $\alpha = .05$. (b) If you were to then test for equal mean ages between the two groups, would you use the pooled t statistic for the test statistic? Why or why not?

10.79 A survey of 100 mayonnaise purchasers showed that 65 were loyal to one brand. For 100 bath soap purchasers, only 53 were loyal to one brand. Form a 95 percent confidence interval for the difference of proportions. Does it include zero?

10.80 A 20-minute consumer survey mailed to 500 adults aged 25–34 included a $5 Starbucks gift card. The same survey was mailed to 500 adults aged 25–34 without the gift card. There were 65 responses from the first group and 45 from the second group. Form a 95 percent confidence interval for the difference of proportions. Does it include zero?

10.81 One group of accounting students used simulation programs, while another group received a tutorial. Scores on an exam were compared. (a) Construct a 90 percent confidence interval for the true difference in mean scores, explaining any assumptions that are necessary. (b) Do you think the learning methods have significantly different results? Explain.

Statistic	Simulation	Tutorial
Mean score	$\bar{x}_1 = 9.1$	$\bar{x}_2 = 10.3$
Sample std. dev.	$s_1 = 2.4$	$s_2 = 2.5$
Number of students	$n_1 = 20$	$n_2 = 20$

10.82 Advertisers fear that users of digital video recorders (DVRs) will fast-forward past commercials when they watch a recorded program. A leading British pay television company told its advertisers that this effect might be offset because DVR users watch more TV. A sample of 15 DVR users showed a daily mean screen time of 2 hours and 26 minutes with a standard deviation of 14 minutes, compared with a daily mean of 2 hours and 7 minutes with a standard deviation of 12 minutes for a sample of 15 non-DVR users. (a) Construct a 95 percent confidence interval for the difference in mean TV watching. Would this sample support the company's claim (i.e., zero within the confidence interval for the mean difference)? (b) Discuss any assumptions that are needed.

10.83 In preliminary tests of a vaccine that may help smokers quit by reducing the "rush" from tobacco, 64 subjects who wanted to quit smoking were given either a placebo or the vaccine. Of the 32 in the placebo group, only 3 quit smoking for 30 days (the U.S. Food and Drug Administration's criterion for smoking cessation) compared with 11 for the vaccine group. (a) Assuming equal sample sizes, find the 95 percent confidence interval for the difference in proportions. What does it suggest? (b) Why is the sample size a problem here?

10.84 Do positive emotions reduce susceptibility to colds? Healthy volunteers were divided into two groups based on their emotional profiles, and each group was exposed to rhinovirus (the common cold). Of those who reported mostly positive emotions, 14 of 50 developed cold symptoms, compared with 23 of 56 who reported mostly negative emotions. (a) Find the 95 percent confidence interval for the difference in proportions. What does it suggest? (b) Is the criterion for normality met?

10.85 Male and female students in a finance class were asked how much their last tank of gas cost. Can you conclude that on average males spent more on gas than females? (a) State the hypotheses for this test. (b) Show the calculation of the test statistic, assuming unequal population variances. (c) State the decision rule, using $\alpha = .01$. (d) Draw the conclusion. (e) Is it reasonable to assume unequal variances? Explain.

Males	Females
$\bar{x}_1 = \$43.20$	$\bar{x}_2 = \$36.60$
$s_1 = \$8.30$	$s_2 = \$3.10$
$n_1 = 13$	$n_2 = 9$

10.86 Students in nutrition classes at two high schools were asked to keep track of the number of times during the past month that they ordered from a fast-food chain restaurant. (a) The research hypothesis is that Sonando High School students choose fast-food restaurants more often. State the hypotheses for this test. (b) Show the calculation of the test statistic, using $\alpha = .01$ and assuming equal population variances. (c) State the decision rule. (d) Draw the conclusion. (e) Is it reasonable to assume equal variances? Explain.

Sonando High School	Gedacht High School
$\bar{x}_1 = 14.51$	$\bar{x}_2 = 11.88$
$s_1 = 2.69$	$s_2 = 2.66$
$n_1 = 11$	$n_2 = 16$

10.87 A retailer compared the frequency of customer merchandise returns at two locations. Last month, store A had 57 returns on 760 purchases, while store B had 62 returns on 1,240 purchases. At $\alpha = .01$, was the return rate significantly higher at store A?

10.88 Streeling University surveyed a random sample of employees to estimate the frequency of sexually inappropriate comments they had heard at work during the last month.

Of 80 respondents, 32 said they had heard such comments. The survey was repeated after all employees had attended a required training seminar on appropriate workplace behavior. After the seminar, 36 of 120 respondents said they had heard such comments. At $\alpha = .10$, was the proportion reporting inappropriate comments lower after the seminar?

10.89 The Fischer Theatre compared attendance at its Saturday and Sunday matinee performances of a major Broadway musical. At $\alpha = .05$, is the Sunday matinee attendance significantly greater than that for the Saturday matinee?  **Matinee**

Date	Sunday	Saturday
Oct 1–2	4,897	4,833
Oct 8–9	4,846	4,710
Oct 15–16	4,848	4,759
Oct 22–23	4,822	4,862
Oct 29–30	4,924	4,898

10.90 Random samples of tires being replaced by a car dealer showed the tire life (miles) below, based on whether the owner had checked once a month for recommended tire inflation. Use $\alpha = .05$ in the following questions. (a) Are the population variances equal? (b) Do the population means differ? (c) What is the advantage of answering (a) before (b)? **TireLife**

Checked	Not Checked	Checked	Not Checked
46,540	28,260	42,780	32,200
36,970	47,450	40,330	33,880
40,430	44,300	37,670	22,650
40,120	37,870	46,210	37,310

DO-IT-YOURSELF

10.91 Count the number of two-door vehicles among 50 vehicles from a college or university student parking lot. Use any sampling method you like (e.g., the first 50 you see). Do the same for a grocery store that is not very close to the college or university. At $\alpha = .10$, is there a significant difference in the proportion of two-door vehicles in these two locations? (a) State the hypotheses. (b) State the decision rule and sketch it. (c) Find the sample proportions and z test statistic. (d) Make a decision. (e) Find the p-value and interpret it. (f) Is the normality assumption fulfilled? Explain.

Related Reading

Best, D. J., and J. C. W. Rayner. "Welch's Approximate Solution for the Behrens-Fisher Problem." *Technometrics* 29 (1987), pp. 205–10.

Posten, H. O. "Robustness of the Two-Sample *t*-Test under Violations of the Homogeneity of Variance Assumption, Part II." *Communications in Statistics—Theory and Methods* 21 (1995), pp. 2169–84.

Scheffè, H. "Practical Solutions of the Behrens-Fisher Problem." *Journal of the American Statistical Association* 65 (1970), pp. 1501–8.

Shoemaker, Lewis F. "Fixing the *F* Test for Equal Variances." *The American Statistician* 57, no. 2 (May 2003), pp. 105–14.

CHAPTER 10   More Learning Resources

You can access these *LearningStats* demonstrations through Connect to help you understand two-sample hypothesis tests.

Mc Graw Hill **connect**

Topic	LearningStats Demonstrations
Common hypothesis tests	Two-Sample Tests Using Excel Two-Sample Tests Using R Calculator for Two Means Calculator for Two Proportions Useful Formulas
	Appendix F–*F* Distribution

Key: = Excel   = PowerPoint   = PDF

Software Supplement

Using MegaStat to Compare Two Means

MegaStat's menu for comparing two sample means is shown in Figure 10.22. One can choose the *t* statistic assuming equal or unequal variances. As an option, MegaStat also will calculate the confidence interval for the difference between the means and will test for equal variances. The output shown refers to Example 10.2.

Using Minitab to Test Equality of Variances

The *F* test assumes that the populations being sampled are normal. Unfortunately, that test is rather sensitive to non-normality of the sampled populations. To test for equal variances, Minitab reports both the *F* test and a more robust alternative known as *Levene's test* along with their *p*-values. As long as you know how to interpret a *p*-value, you really don't need to know the details of Levene's test. An attractive feature of Minitab's *F* test is its graphical display of a confidence interval for each population standard deviation, shown in Figure 10.23.

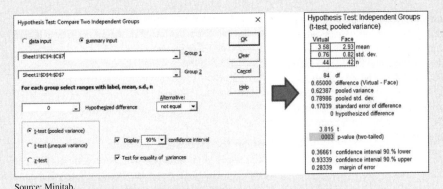

Figure 10.22

MegaStat's Test for Equal Means

Source: Minitab.

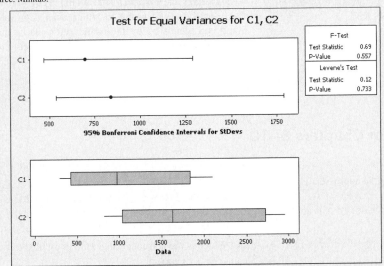

Figure 10.23

Minitab's Test for Equal Variances

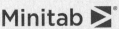

Software Supplement

Two-Sample Test for Means ($\mu_1 - \mu_2$) 📁 RxPrice

Prices were recorded for a 30-day supply of a common prescription drug in 16 Colorado cities and 13 Texas cities (see textbook Example 10.1). At $\alpha = .05$, can we conclude that the mean price is different in the two states? The population variances are unknown, so we use the Student's t distribution and the R function t.test(). We copy the two Excel data columns (with headings) to the clipboard and paste the data into an R data frame RxPrice:

```
> RxPrice=read.table(file="clipboard",sep="\t",header=TRUE)
```

We request a two-tailed test with 95 percent confidence $(1 - \alpha)$ and unequal variances (Welch's test). Because the file lengths differ (Texas has three fewer observations), R imported the last three Texas data values as NA. However, we can use the na.omit() function to omit NA from the Texas data (that would also work on Colorado even though it is not needed).

```
> t.test(RxPrice$Colorado,na.omit(RxPrice$Texas),var.equal=FALSE,
    alternative="two.sided",conf.level=.95)
    Welch Two Sample t-test
data: RxPrice$Colorado and na.omit(RxPrice$Texas)
t = -0.9018, df = 24.025, p-value = 0.3761
alternative hypothesis: true difference in means is not equal to 0
95 percent confidence interval:
    -13.235294  5.185871
sample estimates:
mean of x mean of y
    133.9938 138.0185
```

The p-value (.3761) indicates that we would not reject the null hypothesis $\mu_1 = \mu_2$ (also because the confidence interval for the difference in means $\mu_1 - \mu_2$ includes zero). This agrees with textbook Example 10.1.

Two-Sample Test for Proportions ($\pi_1 - \pi_2$)

Electronic gate entry activations by delivery vehicles in university parking lots failed 52 times in 1,000 attempts at Campus 1 compared with 63 failures in 1,000 attempts at Campus 2. At $\alpha = .05$, do the failure proportions differ? We use a two-tailed test and 95 percent confidence (confidence = $1 - \alpha$). We include correct=FALSE because we do not want a continuity correction.

```
> prop.test(x=c(52,63),n=c(1000,1000),alternative="two.sided",conf.
    level=.95,correct=FALSE)
2-sample test for equality of proportions without continuity correction
data: c(52, 63) out of c(1000, 1000)
X-squared = 1.1164, df = 1, p-value = 0.2907
alternative hypothesis: two.sided
95 percent confidence interval:
    -0.031399356 0.009399356
sample estimates:
prop 1 prop 2
0.052  0.063
```

The z_{calc} test statistic (1.057) is the square root of the chi-square test statistic (1.1164). The p-value (.2907) indicates that the proportions do not differ significantly. These results agree with textbook Example 10.4 because sample sizes are the same. However, when sample sizes are unequal, R results would differ somewhat because R uses a chi-square test instead of a pooled sample z test as illustrated in the textbook.

Two-Sample Test for Variances (σ_1^2/σ_2^2) 📁 Damage

An experimental bumper was installed on a group of vans in a large fleet but not on a control group. At the end of a trial period, there were 12 repair incidents for the experimental vehicles and 9 repair incidents for the control group vehicles. The cost of damage repair (dollars) was recorded. We create two data vectors and perform a two-tailed F test to see whether the variance differs between the two groups. These results agree with textbook Example 10.7. The p-value indicates that we cannot reject the hypothesis of equal population variances.

```
> Experimental=c(1973,403,509,2103,1153,292,
   1916,1602,1559,547,801,359)
> Control=c(1185,885,2955,815,2852,1217,1762,2592,1632)
> var.test(Experimental,Control,Ratio=1,Tails="two-sided")
   F test to compare two variances
data: Experimental and Control
```

```
F = 0.69084, num df = 11, denom df = 8, p-value = 0.5572
alternative hypothesis: true ratio of variances is not equal to 1
95 percent confidence interval:
   0.1628029 2.5311116
sample estimates:
ratio of variances
   0.6908397
```

Option for Two-Sample Tests

Both t.test(x,y) and var.test(x,y) will work with stacked data, but the syntax is then t.test(x~y) and var.test(x~y), where x is a column vector of *data values* and y is a *factor* (a column vector of the same length as x identifying the group to which each data value belongs). The group identifiers in the factor vector can be numbers (e.g., "1" or "2") or characters ("A" or "B") or labels ("Group 1" or "Group 2").

Exam Review Questions for Chapters 8–10

1. Which statement is *not* correct? Explain.
   a. The sample data $x_1, x_2, \ldots, x_n$ will be approximately normal if the sample size n is large.
   b. For a skewed population, the distribution of $\overline{X}$ is approximately normal if n is large.
   c. The expected value of $\overline{X}$ is equal to the true mean μ even if the population is skewed.

2. Match each statement to the correct property of an estimator (unbiased, consistent, efficient):
   a. The estimator "collapses" on the true parameter as n increases.
   b. The estimator has a relatively small variance.
   c. The expected value of the estimator is the true parameter.

3. Concerning confidence intervals, which statement is *most nearly* correct? Why not the others?
   a. We should use z instead of t when n is large.
   b. We use the Student's t distribution when σ is unknown.
   c. Using the Student's t distribution instead of z narrows the confidence interval.

4. A sample of 9 customers in the "quick" lane in a supermarket showed a mean purchase of $14.75 with a standard deviation of $2.10. (a) Find the 95 percent confidence interval for the true mean. (b) Why should you use t instead of z in this case?

5. A sample of 200 customers at a supermarket showed that 28 used a debit card to pay for their purchases. (a) Find the 95 percent confidence interval for the population proportion. (b) Why is it OK to assume normality in this case? (c) What sample size would be needed to estimate the population proportion with 90 percent confidence and an error of $\pm.03$?

6. Which statement is *incorrect?* Explain.
   a. If $p = .50$ and $n = 100$, the estimated standard error of the sample proportion is .05.
   b. In a sample size calculation for estimating π, it is conservative to assume $\pi = .50$.
   c. If $n = 250$ and $p = .07$, it is not safe to assume normality in a confidence interval for π.

7. Given H_0: $\mu \geq 18$ and H_1: $\mu < 18$, we would commit Type I error if we
   a. conclude that $\mu \geq 18$ when the truth is that $\mu < 18$.
   b. conclude that $\mu < 18$ when the truth is that $\mu \geq 18$.
   c. fail to reject $\mu \geq 18$ when the truth is that $\mu < 18$.

8. Which is the correct z value for a two-tailed test at $\alpha = .05$?
   a. $z = \pm 1.645$
   b. $z = \pm 1.960$
   c. $z = \pm 2.326$

9. The process that produces Sonora Bars (a type of candy) is intended to produce bars with a mean weight of 56 grams (g). The process standard deviation is known to be 0.77 g. A random sample of 49 candy bars yields a mean weight of 55.82 g. (a) State the hypotheses to test whether the mean is smaller than it is supposed to be. (b) What is the test statistic? (c) At $\alpha = .05$, what is the critical value for this test? (d) What is your conclusion?

10. A sample of 16 ATM transactions shows a mean transaction time of 67 seconds with a standard deviation of 12 seconds. (a) State the hypotheses to test whether the mean transaction time exceeds 60 seconds. (b) Find the test statistic. (c) At $\alpha = .025$, what is the critical value for this test? (d) What is your conclusion?

11. Which statement is *correct?* Why not the others?
   a. The level of significance α is the probability of committing Type I error.
   b. As the sample size increases, critical values of $t_{.05}$ increase, gradually approaching $z_{.05}$.
   c. When σ is unknown, it is conservative to use $z_{.05}$ instead of $t_{.05}$ in a hypothesis test for μ.

12. Last month, 85 percent of the visitors to the Sonora Candy Factory made a purchase in the on-site candy shop after taking the factory tour. This month, a random sample of 500 such visitors showed that 435 purchased candy after the tour. The manager said, "Good, the percentage of candy-buyers has risen significantly." (a) At $\alpha = .05$, do you agree? (b) Why is it reasonable to assume normality in this test?

13. Weights of 12 randomly chosen Sonora Bars (a type of candy) from assembly line 1 had a mean weight of 56.25 grams (g) with a standard deviation of 0.65 g, while the weights of 12 randomly chosen Sonora Bars from assembly line 2 had a mean weight of 56.75 g with a standard deviation of 0.55 g. (a) Find the test statistic to test whether the mean population weights are the same for both assembly lines (i.e., that the difference is due to random variation). (b) State the critical value for $\alpha = .05$ and degrees of freedom that you are using. (c) State your conclusion.

14. In a random sample of 200 Colorado residents, 150 had skied at least once last winter. A similar sample of 200 Utah residents revealed that 140 had skied at least once last winter. At $\alpha = .025$, is the percentage significantly greater in Colorado? Explain fully and show calculations.

15. Five students in a large lecture class compared their scores on two exams. "Looks like the class mean was higher on the second exam," Bob said. (a) What kind of test would you use? (b) At $\alpha = .10$, what is the critical value? (c) Do you agree with Bob? Explain.

	Bill	Mary	Sam	Sarah	Megan
Exam 1	75	85	90	65	86
Exam 2	86	81	90	71	89

16. Which statement is *not* correct concerning a *p*-value? Explain.

    a. *Ceteris paribus,* a larger *p*-value makes it more likely that H_0 will be rejected.
    b. The *p*-value shows the risk of Type I error if we reject H_0 when H_0 is true.
    c. In making a decision, we compare the *p*-value with the desired level of significance.

17. Given $n_1 = 8$, $s_1 = 14$, $n_2 = 12$, $s_2 = 7$. (a) Find the test statistic for a test for equal population variances. (b) At $\alpha = .05$ in a two-tailed test, state the critical value and degrees of freedom.

11 Analysis of Variance

CHAPTER CONTENTS

CHAPTER LEARNING OBJECTIVES

When you finish this chapter, you should be able to

LO 11-1 Use basic ANOVA terminology correctly.

LO 11-2 Explain the assumptions of ANOVA and why they are important.

LO 11-3 Recognize from data format when one-factor ANOVA is appropriate.

LO 11-4 Interpret sums of squares and calculations in an ANOVA table.

LO 11-5 Use Excel or other software for ANOVA calculations.

LO 11-6 Use a table or Excel to find critical values for the F distribution.

LO 11-7 Understand and perform Tukey's test for paired means.

LO 11-8 Use Hartley's test for equal variances in c treatment groups.

LO 11-9 Recognize from data format when two-factor ANOVA is needed.

LO 11-10 Interpret results in a two-factor ANOVA without replication.

LO 11-11 Interpret main effects and interaction effects in two-factor ANOVA.

LO 11-12 Recognize the need for experimental design and GLM (optional).

11.1 OVERVIEW OF ANOVA

You have already learned to compare the means of two samples. In this chapter, you will learn to compare more than two means *simultaneously* and how to trace sources of variation to potential explanatory factors by using **analysis of variance** (commonly referred to as **ANOVA**). Proper *experimental design* can make efficient use of limited data to draw the strongest possible inferences. Although analysis of variance has a relatively short history, it is one of the richest and most thoroughly explored fields of statistics. Originally developed by the English statistician Ronald A. Fisher (1890–1962) in connection with agricultural research (factors affecting crop growth), it was quickly applied in biology and medicine. Because of its versatility, it is now used in engineering, psychology, marketing, and many other areas. In this chapter, we will only illustrate a few kinds of problems where ANOVA may be utilized (see Related Reading if you need to go further).

Juice Images/Glow Images

LO 11-1

Use basic ANOVA terminology correctly.

The Goal: Explaining Variation

Analysis of variance seeks to identify *sources of variation* in a numerical *dependent* variable Y (the **response variable**). Variation in the response variable about its mean either is **explained** by one or more categorical *independent* variables (the **factors**) or is **unexplained** (random error):

$$\begin{array}{ccc} \text{Variation } Y & = \text{ Explained Variation} & + \text{ Unexplained Variation} \\ \text{(around its mean)} & \text{(due to factors)} & \text{(random error)} \end{array}$$

ANOVA is a *comparison of means*. Each possible value of a factor or combination of factors is a **treatment**. Sample observations within each treatment are viewed as coming from populations with possibly different means. We test whether each factor has a significant effect on Y, and sometimes we test for interaction between factors. The test uses the F distribution, which was introduced in Chapter 10. ANOVA can handle any number of factors, but the researcher often is interested only in a few. Also, data collection costs may impose practical limits on the number of factors or treatments we can choose. This chapter concentrates on ANOVA models with one or two factors, although more complex models are briefly mentioned at the end of the chapter. We will begin with some illustrations of **one-factor ANOVA**.

Illustration: Manufacturing Defect Rates

Figure 11.1 shows a dot plot of daily defect rates for automotive computer chips manufactured at four plant locations. Samples of 10 days' production were taken at each plant. Are the observed differences in the plants' sample mean defect rates merely due to random variation? Or are the observed differences between the plants' defect rates too great to be attributed to chance? This is the kind of question that one-factor ANOVA is designed to answer.

Figure 11.1

Chip Defect Rates at Four Plants

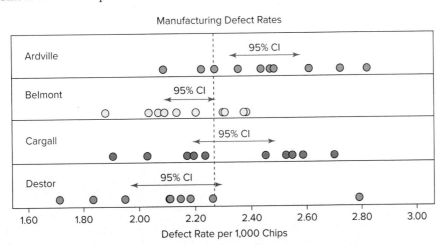

A simple way to state the one-factor ANOVA hypothesis is

$H_0: \mu_1 = \mu_2 = \mu_3 = \mu_4$ (mean defect rates are the same at all four plants)

H_1: Not all the means are equal (at least one mean differs from the others)

If we cannot reject H_0, then we conclude that the observations within each treatment or group actually have a common mean μ (represented by a dashed line in Figure 11.1). This one-factor ANOVA model may be visualized as in Figure 11.2.

Figure 11.2

ANOVA Model for Chip Defect Rates

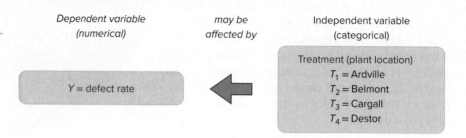

Illustration: Hospital Length of Stay

To allocate resources and fixed costs correctly, hospital management needs to test whether a patient's length of a stay (LOS) depends on the diagnostic-related group (DRG) code. Consider the case of a bone fracture. LOS is a *numerical* response variable (measured in hours). The hospital organizes the data by using five diagnostic codes for type of fracture (facial, radius or ulna, hip or femur, other lower extremity, all other), as illustrated in Figure 11.3. Type of fracture is a *categorical* variable.

Figure 11.3

ANOVA Model for Hospital Length of Stay

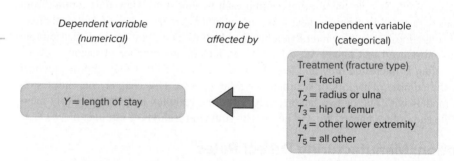

Illustration: Automobile Painting

Paint quality is a major concern of car makers. A key characteristic of paint is its viscosity, a continuous *numerical* variable. Viscosity is to be tested for dependence on application temperature (low, medium, high), as illustrated in Figure 11.4. Although temperature is a numerical variable, it has been coded into *categories* that represent the test conditions of the experiment because the car maker did not want to assume that viscosity was linearly related to temperature.

Figure 11.4

ANOVA Model for Paint Viscosity

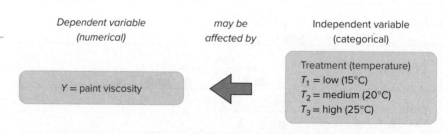

ANOVA Assumptions

Analysis of variance assumes that the

LO 11-2

Explain the assumptions of ANOVA and why they are important.

- Observations on Y are independent.
- Populations being sampled are normal.
- Populations being sampled have equal variances.

Fortunately, ANOVA is somewhat robust to departures from the normality and equal variance assumptions. Later in this chapter, you will see tests for equal variances and advice on handling non-normality.

ANOVA Calculations

ANOVA calculations usually are too tedious to do by calculator, so after we choose an ANOVA model and collect the data, we rely on software (e.g., Excel, MegaStat, Minitab, R) to do the calculations. In some applications (accounting, finance, human resources, marketing), large samples can easily be taken from existing records, while in others (engineering, manufacturing, computer systems), experimental data collection is so expensive that small samples are used. Large samples increase the power of the test, but power also depends on the degree of variation in Y. Specialized software is needed to calculate power for ANOVA experiments.

Analytics in Action

Experiments or Big Data?

Experimental design, developed for agriculture and manufacturing before the advent of big data, is still an important tool for business. When trying to find the root cause of a product defect or customer problem, sifting through lots of data for clues can be time-consuming and sometimes lead to incorrect conclusions. A thoughtfully designed experiment that collects the *right* data, even if it's a small amount, can often provide insight faster than taking a scattershot approach to a big data set.

Boeing's 787 Dreamliner fleet was grounded after experiencing several electrical fires during its first year in service. A series of designed experiments found the root cause of the fires to be a short circuit in the lithium ion battery installed on the aircraft. The solution to correct it required a redesign of the battery enclosure by the Japanese company GS Yuasa, which manufactured the batteries. The Dreamliner was back operating after three months. Experimental design has proven to benefit all types of industries such as automotive, electronics, health, and IT (see https://hbr.org/).

11.2 ONE-FACTOR ANOVA (COMPLETELY RANDOMIZED MODEL)

LO 11-3

Recognize from data format when one-factor ANOVA is appropriate.

Data Format

If we are only interested in comparing the means of c groups (*treatments* or *factor levels*), we have a one-factor ANOVA. If subjects (or individuals) are assigned randomly to treatments, then we call this the **completely randomized model**. This is by far the most common ANOVA model that covers many business problems. The one-factor ANOVA is usually viewed as a comparison between several columns of data, although the data also could be presented in rows. Table 11.1 illustrates the data format for a one-factor ANOVA with treatments $T_1, T_2, \ldots, T_c$ with group means $\overline{y}_1, \overline{y}_2, \ldots, \overline{y}_c$

Table 11.1

Format of One-Factor ANOVA Data for c Treatments $T_1, T_2, \ldots, T_c$.

Data in Columns				Data in Rows						
T_1	T_2		T_c							
y_{11}	y_{12}	$\cdots$	y_1c	T_1	y_{11}	y_{21}	y_{31}	$\cdots$	etc.	n_1 obs. $\quad \bar{y}_1$
y_{21}	y_{22}	$\cdots$	y_2c	T_2	y_{12}	y_{22}	y_{32}	$\cdots$	etc.	n_2 obs. $\quad \bar{y}_2$
y_{31}	y_{32}	$\cdots$	y_3c	$\cdots$					$\cdots$	$\cdots \qquad \cdots$
$\cdots$	$\cdots$	$\cdots$	$\cdots$	T_c	y_1c	y_2c	y_3c	$\cdots$	etc.	n_c obs. $\quad \bar{y}_c$
etc.	etc.	$\cdots$	etc.							
n_1 obs.	n_2 obs.		n_c obs.							
$\bar{y}_1$	$\bar{y}_2$	$\cdots$	$\bar{y}_c$							

Within each treatment j we have n_j observations on Y. Sample sizes within each treatment do *not* need to be equal, although there are advantages to having balanced sample sizes. Equal sample size (1) ensures that each treatment contributes equally to the analysis, (2) reduces problems arising from violations of the assumptions (e.g., nonindependent Y values, unequal variances or nonidentical distributions within treatments, or non-normality of Y), and (3) increases the power of the test (i.e., the ability of the test to detect differences in treatment means). The total number of observations is the sum of the sample sizes for each treatment:

(11.1)
$$n = n_1 + n_2 + \cdots + n_c$$

Hypotheses to Be Tested

The question of interest is whether the mean of Y varies from treatment to treatment. The hypotheses to be tested are

$H_0: \mu_1 = \mu_2 = \cdots = \mu_c$ (all the treatment means are equal)

$H_1:$ Not all the means are equal (at least one pair of treatment means differs)

Because one-factor ANOVA is a generalization of the test for equality of two means, why not just compare all possible pairs of means by using repeated two-sample t tests (as in Chapter 10)? Consider our experiment comparing the four manufacturing plant average defect rates. To compare pairs of plant averages, we would have to perform six different t tests. If each t test has a Type I error probability equal to .05, then the probability that at least one of those tests results in a Type I error is $1 - (.95)^6 = .2649$. ANOVA tests all the means *simultaneously* and therefore does not inflate our Type I error.

One-Factor ANOVA as a Linear Model

An equivalent way to express the one-factor model is to say that observations in treatment j came from a population with a common mean (μ) plus a treatment effect (T_j) plus random error (ε_{ij}):

(11.2) $\qquad y_{ij} = \mu + T_j + \varepsilon_{ij} \qquad j = 1, 2, \ldots, c \quad$ and $\quad i = 1, 2, \ldots, n_j$

The random error is assumed to be normally distributed with zero mean and the same variance for all treatments. If we are interested only in what happens to the response for the particular *levels* of the factor that were selected (a **fixed-effects model**), then the hypotheses to be tested are

$H_0: T_1 = T_2 = \cdots = T_c = 0$ (all treatment effects are zero)

$H_1:$ Not all A_j are zero (some treatment effects are nonzero)

If the null hypothesis is true ($T_j = 0$ for all j), then knowing that an observation x came from treatment j does not help explain the variation in Y, and the ANOVA model collapses to

(11.3)
$$y_{ij} = \mu + \varepsilon_{ij}$$

If the null hypothesis is false, then at least some of the T_j must be nonzero. In that case, the T_j that are negative (below μ) must be offset by the T_j that are positive (above μ) when weighted by sample size.

Group Means

The *mean of each group* is calculated in the usual way by summing the observations in the treatment and dividing by the sample size:

$$\bar{y}_j = \frac{1}{n_j} \sum_{j=1}^{n_j} y_{ij} \tag{11.4}$$

The *overall sample mean* or *grand mean* $\bar{y}$ can be calculated either by summing *all* the observations and dividing by n or by taking a weighted average of the c sample means:

$$\bar{y} = \frac{1}{n} \sum_{j=1}^{c} \sum_{i=1}^{n_j} y_{ij} = \frac{1}{n} \sum_{j=1}^{c} n_j \bar{y}_j \tag{11.5}$$

Partitioned Sum of Squares

To understand the logic of ANOVA, consider that for a given observation y_{ij}, the following relationship must hold (on the right-hand side we just add and subtract $\bar{y}_j$):

$$(y_{ij} - \bar{y}) = (\bar{y}_j - \bar{y}) + (y_{ij} - \bar{y}_j) \tag{11.6}$$

LO 11-4

Interpret sums of squares and calculations in an ANOVA table.

This says that any deviation of an observation from the grand mean $\bar{y}$ may be expressed in two parts: the deviation of the column mean ($\bar{y}_j$) from the grand mean ($\bar{y}$), or *between* treatments, and the deviation of the observation (y_{ij}) from its own column mean ($\bar{y}_j$), or *within* treatments. We can show that this relationship also holds for *sums* of squared deviations, yielding the **partitioned sum of squares:**

$$\sum_{j=1}^{c} \sum_{i=1}^{n_j} (y_{ij} - \bar{y})^2 = \sum_{j=1}^{c} n_j (\bar{y}_j - \bar{y})^2 + \sum_{j=1}^{c} \sum_{i=1}^{n_j} (y_{ij} - \bar{y}_j)^2 \tag{11.7}$$

This important relationship may be expressed simply as

$$SST = SSB + SSE \qquad \text{(partitioned sum of squares)} \tag{11.8}$$

Partitioned Sum of Squares

Sum of Squares Total (*SST*)	=	Sum of Squares between Treatments (*SSB*)	+	Sum of Squares within Treatments (*SSE*)
		↑ Explained by Treatments		↑ Unexplained Random Error

If the treatment means do not differ greatly from the grand mean, *SSB* will be relatively small and *SSE* will be relatively large (and conversely). The sums *SSB* and *SSE* may be used to test the hypothesis that the treatment means differ from the grand mean. However, we first divide each sum of squares by its *degrees of freedom* (to adjust for group sizes). The *F test statistic* is the ratio of the resulting **mean squares**. These calculations can be arranged in a worksheet like Table 11.2.

Source of Variation	Sum of Squares	Degrees of Freedom	Mean Square	F Statistic
Treatment (between groups)	$SSB = \sum_{j=1}^{c} n_j (\bar{y}_j - \bar{y})^2$	$c - 1$	$MSB = \dfrac{SSB}{c - 1}$	$F = \dfrac{MSB}{MSE}$
Error (within groups)	$SSE = \sum_{j=1}^{c} \sum_{i=1}^{n_j} (y_{ij} - \bar{y}_j)^2$	$n - c$	$MSE = \dfrac{SSE}{n - c}$	
Total	$SST = \sum_{j=1}^{c} \sum_{i=1}^{n_j} (y_{ij} - \bar{y})^2$	$n - 1$		

Table 11.2

One-Factor ANOVA Table

Using Excel

The ANOVA calculations are mathematically simple but involve tedious sums. These calculations are almost always done on a computer.[1] For example, Excel's one-factor ANOVA menu using Data Analysis is shown in Figure 11.5. MegaStat uses a similar menu.

Figure 11.5

Excel's ANOVA Menu

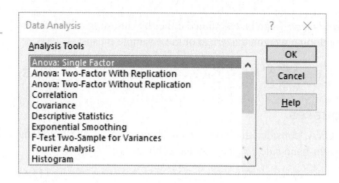

Test Statistic

At the beginning of this chapter, we described the variation in Y as consisting of explained variation and unexplained variation. To test whether the independent variable explains a significant proportion of the variation in Y, we need to compare the explained (due to treatments) and unexplained (due to error) variation. Recall that the F distribution describes the *ratio of two variances*. Therefore, it makes sense that the ANOVA test statistic is an *F test statistic*. The F statistic is the ratio of the variance due to treatments to the variance due to error. *MSB* is the mean square between treatments, and *MSE* is the mean square *within* treatments. Formula 11.9 shows the F statistic and its degrees of freedom.

(11.9)

Between groups (explained)

Within groups (unexplained)

$$F \equiv \frac{MSB}{MSE} = \frac{\left(\dfrac{SSB}{c-1}\right)}{\left(\dfrac{SSE}{n-c}\right)}$$

$df_1 = c - 1$ (numerator)

$df_2 = n - c$ (denominator)

The test statistic $F = MSB/MSE$ cannot be negative (it's based on sums of squares—see Table 11.2), so the F test for equal treatment means is always a right-tailed test. If there is little difference among treatments, we would expect *MSB* to be near zero because the treatment means $\bar{y}_j$ would be near the overall mean $\bar{y}$. Thus, when F is near zero, we would not expect to reject the hypothesis of equal group means. The larger the F statistic, the more we are inclined to reject the hypothesis of equal means. But how large must F be to convince us that the means differ? Just as with a z test or a t test, we need a *decision rule*.

Decision Rule

The F distribution is a right-skewed distribution that starts at zero (F cannot be negative because variances are sums of squares) and has no upper limit (because the variances could be of any magnitude). For ANOVA, the F test is a right-tailed test. For a given level of significance α, we can use Appendix F to obtain the right-tail critical value of F. Alternatively, we can use Excel's function =F.INV.RT(α, df_1, df_2). The decision rule is illustrated in Figure 11.6. This critical value is denoted F_{df_1, df_2} or $F_{c-1, n-c}$ or simply F_{critical}. If our sample F statistic exceeds F_{critical}, we reject the hypothesis of equal group means.

[1]Detailed step-by-step examples of all ANOVA calculations can be downloaded from the McGraw Hill Connect® case studies in *LearningStats* Unit 11.

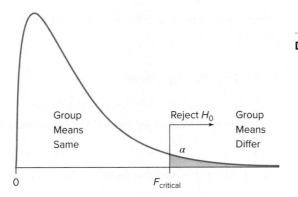

Figure 11.6

Decision Rule for an *F* Test

Example 11.1

Carton Packing

A cosmetics manufacturer's regional distribution center has four workstations that are responsible for packing cartons for shipment to small retailers. Each workstation is staffed by two workers. The task involves assembling each order, placing it in a shipping carton, inserting packing material, taping the carton, and placing a computer-generated shipping label on each carton. Generally, each station can pack 200 cartons a day, and often more. However, there is variability due to differences in orders, labels, and workers. Table 11.3 shows the number of cartons packed per day for the past 5 days. Is the variation among stations within the range attributable to chance, or do these samples indicate actual differences in the station means?

Table 11.3

Number of Cartons Packed 🗁 **Cartons**

	Station 1	Station 2	Station 3	Station 4
	236	238	220	241
	250	239	236	233
	252	262	232	212
	233	247	243	231
	239	246	213	213
Sum	1,210	1,232	1,144	1,130
Mean	242.0	246.4	228.8	226.0
St. Dev.	8.515	9.607	12.153	12.884
n	5	5	5	5

As a preliminary step, we plot the data (Figure 11.7) to check for any time pattern and just to visualize the data. We see some potential differences in means but no obvious time pattern

Figure 11.7

Plot of the Data

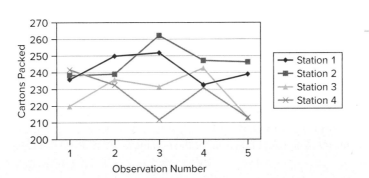

(otherwise we would have to consider observation order as a second factor). We proceed with the hypothesis test.

Step 1: State the Hypotheses   The hypotheses to be tested are

$H_0: \mu_1 = \mu_2 = \mu_3 = \mu_4$ (the means are the same)

H_1: Not all the means are equal (at least one mean is different)

LO 11-6

Use a table or Excel to find critical values for the F distribution.

Step 2: State the Decision Rule   There are $c = 4$ groups and $n = 20$ observations, so degrees of freedom for the F test are

Numerator: $df_1 = c - 1 = 4 - 1 = 3$ (between treatments, factor)

Denominator: $df_2 = n - c = 20 - 4 = 16$ (within treatments, error)

We will use $\alpha = .05$ for the test. The 5 percent right-tail critical value from Appendix F is $F_{3,16} = 3.24$. Instead of Appendix F we could use Excel's function =F.INV.RT(0.05,3,16), which yields $F_{.05} = 3.238872$. This decision rule is illustrated in Figure 11.8.

Figure 11.8

F Test Using $\alpha = .05$ with $F_{3,16}$

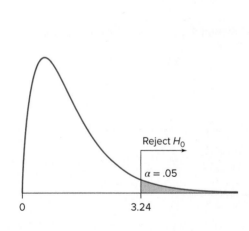

Appendix F: Critical Values for $F_{.05}$

This table shows the critical value that defines the 5% right-tail area for the stated degrees of freedom.

			Numerator Degrees of Freedom (v_1)								
	1	2	3	4	5	6	7	8	9	10	12
1	161.4	199.5	215.7	224.6	230.2	234.0	236.8	238.9	240.5	241.9	243.9
2	18.51	19.00	19.16	19.25	19.30	19.33	19.35	19.37	19.38	19.40	19.41
3	10.13	9.55	9.28	9.12	9.01	8.94	8.89	8.85	8.81	8.79	8.74
4	7.71	6.94	6.59	6.39	6.26	6.16	6.09	6.04	6.00	5.96	5.91
5	6.61	5.79	5.41	5.19	5.05	4.95	4.88	4.82	4.77	4.74	4.68
6	5.99	5.14	4.76	4.53	4.39	4.28	4.21	4.15	4.10	4.06	4.00
7	5.59	4.74	4.35	4.12	3.97	3.87	3.79	3.73	3.68	3.64	3.57
8	5.32	4.46	4.07	3.84	3.69	3.58	3.50	3.44	3.39	3.35	3.28
9	5.12	4.26	3.86	3.63	3.48	3.37	3.29	3.23	3.18	3.14	3.07
10	4.96	4.10	3.71	3.48	3.33	3.22	3.14	3.07	3.02	2.98	2.91
11	4.84	3.98	3.59	3.36	3.20	3.09	3.01	2.95	2.90	2.85	2.79
12	4.75	3.89	3.49	3.26	3.11	3.00	2.91	2.85	2.80	2.75	2.69
13	4.67	3.81	3.41	3.18	3.03	2.92	2.83	2.77	2.71	2.67	2.60
14	4.60	3.74	3.34	3.11	2.96	2.85	2.76	2.70	2.65	2.60	2.53
15	4.54	3.68	3.29	3.06	2.90	2.79	2.71	2.64	2.59	2.54	2.48
16	4.49	3.63	3.24	3.01	2.85	2.74	2.66	2.59	2.54	2.49	2.42
17	4.45	3.59	3.20	2.96	2.81	2.70	2.61	2.55	2.49	2.45	2.38
18	4.41	3.55	3.16	2.93	2.77	2.66	2.58	2.51	2.46	2.41	2.34
19	4.38	3.52	3.13	2.90	2.74	2.63	2.54	2.48	2.42	2.38	2.31
20	4.35	3.49	3.10	2.87	2.71	2.60	2.51	2.45	2.39	2.35	2.28

Denominator Degrees of Freedom (v_2)

Reject H_0

$\alpha = .05$

0   3.24

Step 3: Perform the Calculations   Using Excel for the calculations, we obtain the results shown in Figure 11.9. You can specify the desired level of significance. Excel's default is $\alpha = .05$. Note that Excel labels SSB "between groups" and SSE "within groups." This is an intuitive and attractive way to describe the variation.

Step 4: Make the Decision   Because the test statistic $F = 4.12$ exceeds the critical value $F_{.05} = 3.24$, we can reject the hypothesis of equal means. Because Excel gives the p-value, you don't actually need Excel's critical value. The p-value ($p = .024124$) is less than the level of significance ($\alpha = .05$), which confirms that we should reject the hypothesis of equal treatment means. The Excel function for the p-value is =F.DIST.RT(4.121769,3,16).

Figure 11.9

Excel's One-Factor ANOVA Results
📂 Cartons

Anova: Single Factor

SUMMARY

Groups	Count	Sum	Average	Variance
Station 1	5	1210	242.0	72.5
Station 2	5	1232	246.4	92.3
Station 3	5	1144	228.8	147.7
Station 4	5	1130	226.0	166.0

ANOVA

Source of Variation	SS	df	MS	F	p-Value	F crit
Between Groups	1479.2	3	493.0667	4.121769	0.024124	3.238872
Within Groups	1914.0	16	119.6250			
Total	3393.2	19				

Step 5: Take Action As there is a significant difference in means, the distributor will undertake further analysis to see which stations are most efficient and identify possible reasons. The dot plot of observations by group, shown in Figure 11.10, includes group means (shown as short horizontal tick marks) and the overall mean (shown as a dashed line). The dot plot suggests that stations 3 and 4 have means below the overall mean, while stations 1 and 2 are above the overall mean.

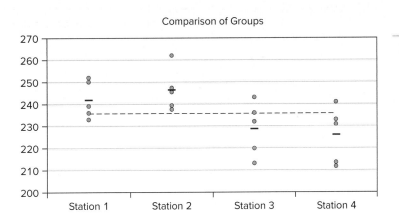

Comparison of Groups

Figure 11.10

Dot Plot of Four Samples
📂 Cartons

11.1 Using the following Excel results: (a) What was the overall sample size? (b) How many groups were there? (c) Write the hypotheses. (d) Find the critical value of F for $\alpha = .05$. (e) Calculate the test statistic. (f) Do the population means differ at $\alpha = .05$?

ANOVA

Source of Variation	SS	df	MS	F
Between Groups	119.9816	4	29.9954	
Within Groups	583.2201	35	16.6634	
Total	703.2016	39		

11.2 Using the following Excel results: (a) What was the overall sample size? (b) How many groups were there? (c) Write the hypotheses. (d) Find the critical value of F for $\alpha = .10$. (e) Calculate the test statistic. (f) Do the population means differ at $\alpha = .10$?

ANOVA

Source of Variation	SS	df	MS	F
Between Groups	120.8706	3	40.29018	
Within Groups	325.0202	20	16.25101	
Total	445.8907	23		

11.3 In a one-factor ANOVA with sample sizes $n_1 = 5$, $n_2 = 7$, $n_3 = 6$, $n_4 = 7$, $n_5 = 5$, the test statistic was $F_{calc} = 2.447$. (a) State the hypotheses. (b) State the degrees of freedom for the test. (c) What is the critical value of F for $\alpha = .10$? (d) What is your conclusion? (e) Write the Excel function for the p-value.

11.4 In a one-factor ANOVA with sample sizes $n_1 = 8$, $n_2 = 5$, $n_3 = 6$, $n_4 = 6$, the test statistic was $F_{calc} = 3.251$. (a) State the hypotheses. (b) State the degrees of freedom for the test. (c) What is the critical value of F for $\alpha = .05$? (d) What is your conclusion? (e) Write the Excel function for the p-value.

Instructions for Exercises 11.5 through 11.8: Use Excel's Data Analysis (or other software) to perform the one-factor ANOVA using $\alpha = .05$. For each exercise: (a) State the hypotheses. (b) What are the degrees of freedom for the test? (c) What is the critical value of F at the 5 percent level of significance? (d) State your conclusion about the population means. (e) Interpret the p-value. *Optional:* (f*) Make a plot of the data for each group (e.g., using MegaStat) or confidence intervals for the group means (e.g., using Minitab). What do the plots show?

11.5 Scrap rates per thousand (parts whose defects cannot be reworked) are compared for five randomly selected days at three plants. Do the data show a significant difference in mean scrap rates? 📁 **ScrapRate**

Scrap Rate (per thousand units)		
Plant A	**Plant B**	**Plant C**
11.4	11.1	10.2
12.5	14.1	9.5
10.1	16.8	9.0
13.8	13.2	13.3
13.7	14.6	5.9

11.6 One particular morning, the length of time spent in the examination rooms is recorded for each patient seen by each physician at an orthopedic clinic. Do the data show a significant difference in mean times? 📁 **Physicians**

Time in Examination Rooms (minutes)			
Physician 1	**Physician 2**	**Physician 3**	**Physician 4**
34	33	17	28
25	35	30	33
27	31	30	31
31	31	26	27
26	42	32	32
34	33	28	33
21		26	40
		29	

11.7   Semester GPAs are compared for seven randomly chosen students in each class level at Oxnard University. Do the data show a significant difference in mean GPAs? 🖺 **GPA1**

GPA for Randomly Selected Students in Four Business Majors			
Accounting	**Finance**	**Human Resources**	**Marketing**
2.48	3.16	2.93	3.54
2.19	3.01	2.89	3.71
2.62	3.07	3.48	2.94
3.15	2.88	3.33	3.46
3.56	3.33	3.53	3.50
2.53	2.87	2.95	3.25
3.31	2.85	3.58	3.20

11.8   Sales of *People* magazine are compared over a 5-week period at four Barnes and Noble stores in the Chicago area. Do the data show a significant difference in mean weekly sales? 🖺 **Magazines**

Weekly Sales			
Store 1	**Store 2**	**Store 3**	**Store 4**
102	97	89	100
106	77	91	116
105	82	75	87
115	80	106	102
112	101	94	100

11.3 MULTIPLE COMPARISONS

Tukey's Test

Besides performing an *F* test to compare the *c* means *simultaneously,* we also could ask whether *pairs* of means differ. You might expect to do a *t* test for two independent means (Chapter 10) or check whether there is overlap in the confidence intervals for each group's mean. But the null hypothesis in ANOVA is that *all* the means are the same, so to maintain the desired overall probability of Type I error, we need to create a *simultaneous confidence interval* for the difference of means based on the *pooled* variances for all *c* groups at once and then see which pairs exclude zero. For *c* groups, there are $c(c - 1)/2$ distinct pairs of means to be compared.

Several **multiple comparison** tests are available. Their logic is similar. We will discuss only one, called **Tukey's studentized range test** (sometimes called the *HSD* or "honestly significant difference" test). It has good power and is widely used. We will refer to it as *Tukey's test,* named for statistician John Wilder Tukey (1915–2000). This test is available in most statistical packages (but not in Excel's Data Analysis). It is a two-tailed test for equality of paired means from *c* groups compared simultaneously and is a natural follow-up when the results of the one-factor ANOVA test show a significant difference in at least one mean. The hypotheses to compare group *j* with group *k* are

$H_0: \mu_j = \mu_k$

$H_1: \mu_j \neq \mu_k$

Tukey's test statistic is

$$T_{calc} = \frac{|\bar{y}_j - \bar{y}_k|}{\sqrt{MSE\left[\dfrac{1}{n_j} + \dfrac{1}{n_k}\right]}} \tag{11.10a}$$

LO 11-7

Understand and perform Tukey's test for paired means.

We would reject H_0 if $T_{calc} > T_{c,n-c}$ where $T_{c,n-c}$ is a critical value for the desired level of significance. Table 11.4 shows 5 percent critical values of $T_{c,n-c}$. If the desired degrees of freedom cannot be found, we could interpolate or rely on a computer package to provide the exact critical value. We take *MSE* directly from the ANOVA calculations (see Table 11.2). The *MSE* is the *pooled variance* for all *c* samples combined (rather than pooling just two sample variances as in Chapter 10). Therefore, Tukey's test statistic also could be written as

$$\textbf{(11.10b)} \quad T_{calc} = \frac{|\bar{x}_j - \bar{x}_k|}{\sqrt{\dfrac{s_p^2}{n_j} + \dfrac{s_p^2}{n_k}}} \quad \text{where } s_p^2 = \frac{(n_1 - 1)s_1^2 + (n_2 - 1)s_2^2 + \cdots + (n_c - 1)s_c^2}{(n_1 - 1) + (n_2 - 1) + \cdots + (n_c - 1)}$$

This alternative formula reveals similarities to the two-sample *t* test (e.g., pooled variance) yet also reveals dissimilarities (e.g., pooling of *c* variances instead of just two variances). *MSE* is available from the ANOVA, so we will use the first formula.

Table 11.4

Five Percent Critical Values of Tukey Test Statistic*

*Table shows studentized range divided by $\sqrt{2}$ to obtain $T_{c,n-c}$. Entries are calculated using the R function qtukey(0.95, nmeans, df)/sqrt(2) where nmeans = c and df = n − c. The last line in the table uses df = 10000.

n − *c*	Number of Groups (c)								
	2	**3**	**4**	**5**	**6**	**7**	**8**	**9**	**10**
5	2.57	3.25	3.69	4.01	4.27	4.48	4.65	4.81	4.95
6	2.45	3.07	3.46	3.75	3.98	4.17	4.33	4.47	4.59
7	2.36	2.95	3.31	3.58	3.79	3.96	4.11	4.24	4.35
8	2.31	2.86	3.20	3.45	3.65	3.82	3.96	4.08	4.18
9	2.26	2.79	3.12	3.36	3.55	3.71	3.84	3.96	4.06
10	2.23	2.74	3.06	3.29	3.47	3.62	3.75	3.86	3.96
15	2.13	2.60	2.88	3.09	3.25	3.38	3.49	3.59	3.68
20	2.09	2.53	2.80	2.99	3.14	3.27	3.37	3.46	3.54
30	2.04	2.47	2.72	2.90	3.04	3.16	3.25	3.34	3.41
40	2.02	2.43	2.68	2.86	2.99	3.10	3.20	3.28	3.35
60	2.00	2.40	2.64	2.81	2.94	3.05	3.14	3.22	3.29
120	1.98	2.37	2.61	2.77	2.90	3.00	3.09	3.16	3.22
∞	1.96	2.34	2.57	2.73	2.85	2.95	3.03	3.10	3.16

We will illustrate Tukey's test for the carton-packing data. We assume that a one-factor ANOVA has already been performed and the results showed that at least one mean was significantly different. We will use the *MSE* from the ANOVA. For the carton-packing data, there are 4 groups and 20 observations, so $c = 4$ and $n - c = 20 - 4 = 16$. From Table 11.4 we must interpolate between $T_{4,15} = 2.88$ and $T_{4,20} = 2.80$ to get $T_{4,16} = 2.86$. The decision rule for any pair of means is therefore

$$\text{Reject } H_0 \text{ if } T_{calc} = \frac{|\bar{y}_i - \bar{y}_k|}{\sqrt{MSE\left[\dfrac{1}{n_j} + \dfrac{1}{n_k}\right]}} > 2.86$$

There may be a different decision rule for every pair of stations unless the sample sizes n_j and n_k are identical (in our example, the group sizes are the same). For example, to compare groups 2 and 4, the test statistic is

$$T_{calc} = \frac{|\bar{y}_2 - \bar{y}_4|}{\sqrt{MSE\left[\dfrac{1}{n_2} + \dfrac{1}{n_4}\right]}} = \frac{|246.4 - 226.0|}{\sqrt{119.625\left[\dfrac{1}{5} + \dfrac{1}{5}\right]}} = 2.95$$

Because $T_{calc} = 2.95$ exceeds 2.86, we reject the hypothesis of equal means for stations 2 and 4. We conclude that there is a significant difference between the mean output of stations 2 and 4. MegaStat shows the critical values $T_{.05} = 2.86$ and $T_{.01} = 3.67$ for the experimentwise error rate in the Tukey test, so you do not need to refer to Table 11.4.

A similar test must be performed for every possible pair of means. All six possible comparisons of means are shown in Figure 11.11. Only stations 2 and 4 differ at $\alpha = .05$. An

Tukey simultaneous comparison *t* values (d.f. = 16)

		Station 4 226.0	Station 3 228.8	Station 1 242.0	Station 2 246.4
Station 4	226.0				
Station 3	228.8	0.40			
Station 1	242.0	2.31	1.91		
Station 2	246.4	2.95	2.54	0.64	

critical values for experimentwise error rate:

0.05	2.86
0.01	3.67

Figure 11.11

MegaStat's Tukey Tests and Independent Sample *t* Tests 🖾 **Cartons**

attractive feature of MegaStat's Tukey test is that it highlights significant results using color-coding for $\alpha = .05$ and $\alpha = .01$, as seen in Figure 11.11.

We use Tukey's *T* test instead of comparing pairs of means using repeated independent *t* tests (Chapter 10). Performing *c* independent *t* tests on the same data would increase our risk of finding a difference of means even if none exists in the population. Tukey's *T* test prevents this buildup of the Type I error probability (see Section 11.2).

11.9 Consider a one-factor ANOVA with $n_1 = 9$, $n_2 = 10$, $n_3 = 7$, $n_4 = 8$. (a) How many possible comparisons of means are there? (b) State the degrees of freedom for Tukey's *T*. (c) Find the critical value of Tukey's *T* for $\alpha = .05$.

11.10 Consider a one-factor ANOVA with $n_1 = 6$, $n_2 = 5$, $n_3 = 4$, $n_4 = 6$, $n_5 = 4$. (a) How many possible comparisons of means are there? (b) State the degrees of freedom for Tukey's *T*. (c) Find the critical value of Tukey's *T* for $\alpha = .05$.

Instructions for Exercises 11.11 through 11.14: (a) How many possible comparisons of means are there? (b) State the degrees of freedom for Tukey's *T*. (c) Find the critical value of Tukey's *T* for $\alpha = .05$ from Table 11.4. (d) State your conclusions about pairs of means. *Hint:* Use MegaStat, Minitab, or other software to perform the Tukey test.

11.11 Refer to Exercise 11.5. Which pairs of mean scrap rates differ significantly (3 plants)? 🖾 **ScrapRate**

11.12 Refer to Exercise 11.6. Which pairs of mean examination times differ significantly (4 physicians)? 🖾 **Physicians**

11.13 Refer to Exercise 11.7. Which pairs of mean GPAs differ significantly (4 majors)? 🖾 **GPA1**

11.14 Refer to Exercise 11.8. Which pairs of mean weekly sales differ significantly (4 stores)? 🖾 **Magazines**

Section Exercises

Mc
Graw
Hill **connect**

11.4 TESTS FOR HOMOGENEITY OF VARIANCES

ANOVA Assumptions

Analysis of variance assumes that observations on the response variable are from normally distributed populations that have the same variance. We have noted that few populations meet these requirements perfectly, and unless the sample is quite large, a test for normality may be impractical. However, we can test the assumption of **homogeneous** (equal) **variances**. Although the one-factor ANOVA test is only slightly affected by inequality of variance when group sizes are equal or nearly so, it is still a good idea to test this assumption. In general, surprisingly large differences in variances must exist to conclude that the population variances are unequal.

LO 11-8

Use Hartley's test for equal variances in *c* treatment groups.

Hartley's Test

If we had only two groups, we could use the *F* test you learned in Chapter 10 to compare the variances. But for *c* groups, a more general test is required. One such test is **Hartley's test**, named for statistician H. O. Hartley (1912–1980). The hypotheses are

$H_0: \sigma_1^2 = \sigma_2^2 = \cdots = \sigma_c^2$ (equal variances)

$H_1:$ The σ_j^2 are not all equal (unequal variances)

Hartley's test statistic is the ratio of the largest sample variance to the smallest sample variance:

$$(11.11) \qquad H_{calc} = \frac{s_{max}^2}{s_{min}^2} = \frac{max(s_1^2, s_2^2, \ldots, s_c^2)}{min(s_1^2, s_2^2, \ldots, s_c^2)}$$

The calculation is simple. Just look at Excel's list of variances. Choose the largest and smallest and calculate their ratio. If the variances in each population group are the same, we would expect $H_{calc} \approx 1$. Hartley's test is always a right-tailed test because the larger variance is always in the numerator. The decision rule is

Reject H_0 if $H_{calc} > H_{critical}$

Critical values of $H_{critical}$ may be found in Table 11.5 using degrees of freedom given by

Numerator: $df_1 = c$

Denominator: $df_2 = \dfrac{n}{c} - 1$

where n is the total number of observations. This test assumes equal group sizes, so df_2 would be an integer. For group sizes that are not drastically unequal, this procedure will still be approximately correct, using the next lower integer if df_2 is not an integer.

Table 11.5

Critical 5 Percent Values of Hartley's $H = s_{max}^2/s_{min}^2$

Source: E. S. Pearson and H. O. Hartley, *Biometrika Tables for Statisticians,* 3rd ed. (Oxford University Press, 1970), p. 202.

Denominator df_2	Numerator df_1								
	2	**3**	**4**	**5**	**6**	**7**	**8**	**9**	**10**
2	39.0	87.5	142	202	266	333	403	475	550
3	15.4	27.8	39.2	50.7	62.0	72.9	83.5	93.9	104
4	9.60	15.5	20.6	25.2	29.5	33.6	37.5	41.1	44.6
5	7.15	10.8	13.7	16.3	18.7	20.8	22.9	24.7	26.5
6	5.82	8.38	10.4	12.1	13.7	15.0	16.3	17.5	18.6
7	4.99	6.94	8.44	9.7	10.8	11.8	12.7	13.5	14.3
8	4.43	6.00	7.18	8.12	9.03	9.78	10.5	11.1	11.7
9	4.03	5.34	6.31	7.11	7.80	8.41	8.95	9.45	9.91
10	3.72	4.85	5.67	6.34	6.92	7.42	7.87	8.28	8.66
12	3.28	4.16	4.79	5.30	5.72	6.09	6.42	6.72	7.00
15	2.86	3.54	4.01	4.37	4.68	4.95	5.19	5.40	5.59
20	2.46	2.95	3.29	3.54	3.76	3.94	4.10	4.24	4.37
30	2.07	2.40	2.61	2.78	2.91	3.02	3.12	3.21	3.29
60	1.67	1.85	1.96	2.04	2.11	2.17	2.22	2.26	2.30
∞	1.00	1.00	1.00	1.00	1.00	1.00	1.00	1.00	1.00

Example 11.2

Carton Packing: Hartley's Test 🖼 **Cartons**

Using the carton-packing data in Table 11.3, there are 4 groups and 20 total observations, so we have

Numerator: $df_1 = c = 4$

Denominator: $df_2 = n/c - 1 = 20/4 - 1 = 5 - 1 = 4$

From Table 11.5 we choose the critical value $H_{critical} = 20.6$ using $df_1 = 4$ and $df_2 = 4$. The sample statistics (from Excel) for our workstations are

Work Station	n	Mean	Variance
Station 1	5	242.0	$s_{min}^2 \rightarrow 72.5$
Station 2	5	246.4	92.3
Station 3	5	228.8	147.7
Station 4	5	226.0	$s_{max}^2 \rightarrow 166.0$

The test statistic is

$$H_{calc} = \frac{s_{max}^2}{s_{min}^2} = \frac{166.0}{72.5} = 2.29$$

In this case, we cannot reject the hypothesis of equal variances. Indeed, Table 11.5 makes it clear that unless the sample size is very large, the variance ratio would have to be quite large to reject the hypothesis of equal population variances.

Levene's Test

Hartley's test relies on the assumption of normality in the populations from which the sample observations are drawn. A more robust alternative is **Levene's test**, which does not assume a normal distribution. This test requires a computer package other than Excel. It is not necessary to discuss the computational procedure except to say that Levene's test is based on the distances of the observations from their sample *medians* rather than their sample *means*. As long as you know how to interpret a *p*-value, Levene's test is easy to use. Figure 11.12 shows Minitab's test of homogeneity of variance for the carton-packing data using Levene's test, with the added attraction of confidence intervals for each population standard deviation. Because the confidence intervals overlap and the *p*-value (.823) is large, we cannot reject the hypothesis of equal population variances. This confirms that the one-factor ANOVA procedure was appropriate for the carton-packing data.

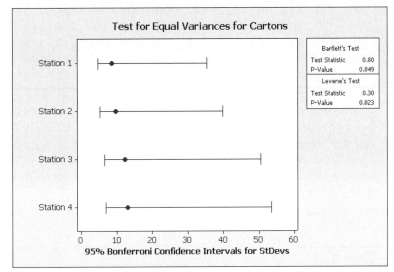

Source: Minitab.

Figure 11.12

Minitab's Equal-Variance Test Cartons

Section Exercises

connect

11.15 In a one-factor ANOVA with $n_1 = 6$, $n_2 = 4$, and $n_3 = 5$, the sample variances were $s_1^2 = 121$, $s_2^2 = 929$, and $s_3^2 = 456$. For Hartley's test: (a) State the hypotheses. (b) Calculate the degrees of freedom. (c) Find the critical value at the 5 percent level of significance. (d) Calculate Hartley's test statistic. (e) What is your conclusion?

11.16 In a one-factor ANOVA with $n_1 = 7$, $n_2 = 6$, $n_3 = 5$, $n_4 = 5$, and $n_5 = 7$, the sample standard deviations were $s_1 = 12$, $s_2 = 24$, $s_3 = 16$, $s_4 = 46$, and $s_5 = 27$. For Hartley's test: (a) State the hypotheses. (b) Calculate the degrees of freedom. (c) Find the critical value at the 5 percent level of significance. (d) Calculate Hartley's test statistic. (e) What is your conclusion?

Instructions for Exercises 11.17 through 11.20: Use the table of critical values from Table 11.5 and the sample variances from your previous ANOVA (see Exercises 11.5–11.8). (a) State the degrees of freedom for Hartley's test. (b) Find Hartley's critical value at the 5 percent level of significance. (c) Calculate Hartley's test statistic. (d) What is your conclusion about group variances? *Optional challenge:* (e*) If you have access to Minitab or another software package, perform Levene's test for equal group variances at the 5 percent level of significance. Does Levene's test agree with Hartley's test?

11.17 Refer to Exercise 11.5. Are the population variances the same for scrap rates (3 plants)? 📸 **ScrapRate**

11.18 Refer to Exercise 11.6. Are the population variances the same for examination times (4 physicians)? 📸 **Physicians**

11.19 Refer to Exercise 11.7. Are the population variances the same for the GPAs (4 majors)? 📸 **GPA1**

11.20 Refer to Exercise 11.8. Are the population variances the same for weekly sales (4 stores)? 📸 **Magazines**

Mini Case 11.1

Hospital Emergency Arrivals

To plan its staffing schedule, a large urban hospital examined the number of arrivals per day over a 13-week period, as shown in Table 11.6. Data are shown in rows rather than in columns to make a more compact table.

Table 11.6

Number of Emergency Arrivals by Day of the Week 📸 **Emergency**

Mon	188	175	208	176	179	184	191	194	174	191	198	213	217
Tue	174	167	165	164	169	164	150	175	178	164	202	175	191
Wed	177	169	180	173	182	181	168	165	174	175	174	177	182
Thu	170	164	190	169	164	170	153	150	156	173	177	183	208
Fri	177	167	172	185	185	170	170	193	212	171	175	177	209
Sat	162	184	173	175	144	170	163	157	181	185	199	203	198
Sun	182	176	183	228	148	178	175	174	188	179	220	207	193

We perform a one-factor ANOVA to test the model *Arrivals = f(Weekday)*. The single factor (*Weekday*) has seven treatments. The Excel results, shown in Figure 11.13, indicate that *Weekday* does have a significant effect on *Arrivals* because the test statistic $F = 3.270$ exceeds the 5 percent critical value $F_{6,84} = 2.209$. The p-value (.006) indicates that a test statistic this large would arise by chance only about 6 times in 1,000 samples if the hypothesis of equal daily means were true.

Figure 11.13

One-Factor ANOVA for Emergency Arrivals and Sample Plot

SUMMARY

Groups	Count	Sum	Average	Variance
Mon	13	2488	191.385	206.423
Tue	13	2238	172.154	173.141
Wed	13	2277	175.154	29.808
Thu	13	2227	171.308	250.564
Fri	13	2363	181.769	216.692
Sat	13	2294	176.462	308.769
Sun	13	2431	187.000	445.667

ANOVA

Source of Variation	SS	df	MS	F	P-value	F crit
Between Groups	4570.989	6	761.8315	3.26953	0.00613	2.20855
Within Groups	19572.769	84	233.0092			
Total	24143.758	90				

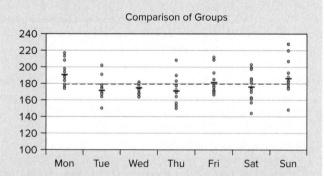

The Tukey multiple comparison test (Figure 11.14) shows that the only pairs of *significantly* different means at $\alpha = .05$ are (*Mon, Tue*) and (*Mon, Thu*). In testing for equal variances, we get conflicting conclusions, depending on which test we use. Hartley's test gives $H_{calc} = 445.667/29.808 = 14.95$, which exceeds the critical value $H_{7,12} = 6.09$ (note that *Wed* has a *very* small variance). But Levene's test for homogeneity of variances

(Figure 11.15) has a *p*-value of .221, which at $\alpha = .05$ does not allow us to reject the equal-variance assumption that underlies the ANOVA test. When it is available, we prefer Levene's test because it does not depend on the assumption of normality.

Figure 11.14

MegaStat's Tukey Test for $\mu_j = \mu_k$

Tukey simultaneous comparison *t*-values (d.f. = 84)

		Thu 171.3	Tue 172.2	Wed 175.2	Sat 176.5	Fri 181.8	Sun 187.0	Mon 191.4
Thu	171.3							
Tue	172.2	0.14						
Wed	175.2	0.64	0.50					
Sat	176.5	0.86	0.72	0.22				
Fri	181.8	1.75	1.61	1.10	0.89			
Sun	187.0	2.62	2.48	1.98	1.76	0.87		
Mon	191.4	3.35	3.21	2.71	2.49	1.61	0.73	

critical values for experimentwise error rate:

0.05	3.03
0.01	3.59

Figure 11.15

Minitab Test for Equal Variances 🖅 **Emergency**

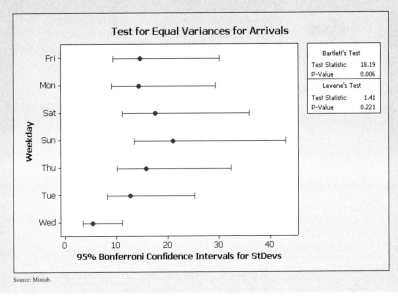

Source: Mintab.

11.5 TWO-FACTOR ANOVA WITHOUT REPLICATION (RANDOMIZED BLOCK MODEL)

Data Format

Suppose that two factors *A* and *B* may affect *Y*. One way to visualize this is to imagine a data matrix with *r* rows and *c* columns. Each row is a level of factor *A*, while each column is a level of factor *B*. Initially, we will consider the case where all levels of both factors occur, and each cell contains only one observation. In this **two-factor ANOVA without replication** (or *non-repeated measures design*), each factor combination is observed exactly once. The mean of *Y* can be computed either across the rows or down the columns, as shown in Table 11.7. The grand mean $\bar{y}$ is the sum of all data values divided by the sample size *rc*.

LO 11-9

Recognize from data format when two-factor ANOVA is needed.

Table 11.7

Format of Two-Factor
ANOVA Data Set without
Replication

Levels of Factor A	Levels of Factor B				Row Mean
	B_1	B_2	...	B_c	
A_1	y_{11}	y_{12}	...	y_{1c}	$\overline{y}_{1.}$
A_2	y_{21}	y_{22}	...	y_{2c}	$\overline{y}_{2.}$
...	...	...	...	...	...
A_r	y_{r1}	yr_2	...	y_{rc}	$\overline{y}_{r.}$
Col Mean	$\overline{y}_{.1}$	$\overline{y}_{.2}$	...	$\overline{y}_{.c}$	$\overline{y}$

Figure 11.16 illustrates a two-factor ANOVA model in which a numerical response variable (paint viscosity) may vary both by temperature (*Factor A*) and by paint supplier (*Factor B*). Three different temperature settings (A_1, A_2, A_3) were tested on shipments from three different suppliers (B_1, B_2, B_3), yielding a table with $3 \times 3 = 9$ cells. Each factor combination is a *treatment*. With only one observation per treatment, no interaction between the two factors is included.[2]

Figure 11.16

Two Factor Model of Paint Viscosity

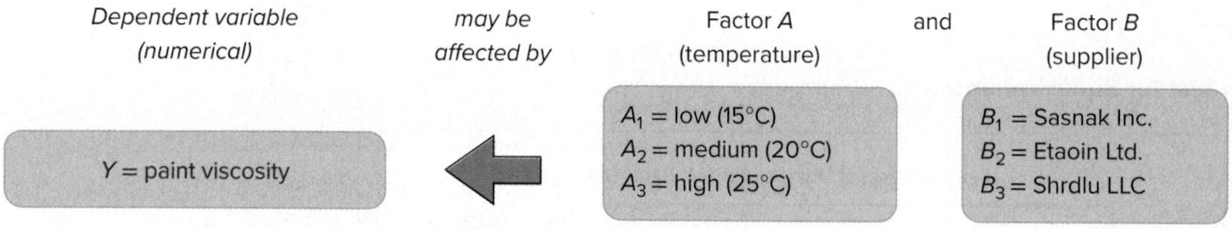

Dependent variable (numerical)	may be affected by	Factor A (temperature)	and	Factor B (supplier)
Y = paint viscosity		A_1 = low (15°C) A_2 = medium (20°C) A_3 = high (25°C)		B_1 = Sasnak Inc. B_2 = Etaoin Ltd. B_3 = Shrdlu LLC

Two-Factor ANOVA Model

Expressed in linear form, the two-factor ANOVA model is

(11.12)
$$y_{jk} = \mu + A_j + B_k + \varepsilon_{jk}$$

where

y_{jk} = observed data value in row j and column k

μ = common mean for all treatments

A_j = effect of row factor A ($j = 1, 2, \ldots, r$)

B_k = effect of column factor B ($k = 1, 2, \ldots, c$)

ε_{jk} = random error

The random error is assumed to be normally distributed with zero mean and the same variance for all treatments.

Hypotheses to Be Tested

If we are interested only in what happens to the response for the particular levels of the factors that were selected (a *fixed-effects model*), then the hypotheses to be tested are

Factor A

$H_0: A_1 = A_2 = \cdots = A_r = 0$ (row means are the same)

$H_1:$ Not all the A_j are equal to zero (row means differ)

---

[2]There are not enough degrees of freedom to estimate an interaction unless the experiment is replicated.

Factor B

$H_0: B_1 = B_2 = \cdots = B_c = 0$ (column means are the same)

H_1: Not all the B_k are equal to zero (column means differ)

If we are unable to reject either null hypothesis, all variation in Y is just a random disturbance around the mean μ:

$$y_{jk} = \mu + \varepsilon_{jk} \qquad \textbf{(11.13)}$$

Randomized Block Model

A special terminology is used when only one factor is of research interest and the other factor is merely used to control for potential confounding influences. In this case, the two-factor ANOVA model with one observation per cell is sometimes called the **randomized block model**. In the randomized block model, it is customary to call the column effects *treatments* (as in one-factor ANOVA to signify that they are the effect of interest), while the row effects are called *blocks*.[3] For example, a North Dakota agribusiness might want to study the effect of four kinds of fertilizer (F_1, F_2, F_3, F_4) in promoting wheat growth (Y) on three soil types (S_1, S_2, S_3). To control for the effects of soil type, we could define three blocks (rows), each containing one soil type, as shown in Table 11.8. Subjects within each block (soil type) would be randomly assigned to the treatments (fertilizer).

Table 11.8

Format of Randomized Block Experiment: Two Factors

Block (Soil Type)	Treatment (Fertilizer)			
	F_1	F_2	F_3	F_4
S_1				
S_2				
S_3				

A randomized block model looks like a two-factor ANOVA and is computed exactly like a two-factor ANOVA. However, its interpretation by the researcher may resemble a one-factor ANOVA because only the column effects (treatments) are of interest. The blocks exist only to reduce variance. The effect of the blocks will show up in the hypothesis test but is of no interest to the researcher as a separate factor. In short, the difference between a randomized block model and a standard two-way ANOVA model lies in the mind of the researcher. Because calculations for a randomized block design are identical to the two-factor ANOVA with one observation per cell, we will not call the row factor a "block" and the column factor a "treatment." Instead, we just call them *factor A* and *factor B*. Interpretation of the factors is not a mathematical issue. If only the column effect is of interest, you may call the column effect the "treatment."

Calculation of Unreplicated Two-Factor ANOVA

Calculations for the unreplicated two-factor ANOVA may be arranged as in Table 11.9. Degrees of freedom sum to $n - 1$. For a data set with r rows and c columns, notice that $n = rc$. The total sum of squares shown in Table 11.9 has three components:

$$SST = SSA + SSB + SSE \qquad \textbf{(11.14)}$$

where

SST = total sum of squared deviations about the mean

SSA = between rows sum of squares (effect of factor A)

SSB = between columns sum of squares (effect of factor B)

SSE = error sum of squares (residual variation)

LO 11-10

Interpret results in a two-factor ANOVA without replication.

---

[3]In principle, either rows or columns could be the blocking factor, but it is customary to put the blocking factor in rows.

SSE is a measure of unexplained variation. If *SSE* is relatively high, we suspect that the factor effects do not differ significantly from zero. Conversely, if *SSE* is relatively small, it is a sign that at least one factor is a relevant predictor of *Y*, and we would expect either *SSA* or *SSB* (or both) to be relatively large. Before doing the *F* test, each sum of squares must be divided by its degrees of freedom to obtain the *mean square*. Calculations are almost always done by a computer (see *LearningStats* Unit 11).

Table 11.9

Format of Two-Factor ANOVA with One Observation per Cell

Source of Variation	Sum of Squares	Degrees of Freedom	Mean Square	F Ratio
Factor *A* (row effect)	$SSA = c\sum_{j=1}^{r}(\bar{y}_{j.} - \bar{y})^2$	$r - 1$	$MSA = \dfrac{SSA}{r-1}$	$F_A = \dfrac{MSA}{MSE}$
Factor *B* (column effect)	$SSB = r\sum_{k=1}^{c}(\bar{y}_{.k} - \bar{y})^2$	$c - 1$	$MSB = \dfrac{SSB}{c-1}$	$F_B = \dfrac{MSB}{MSE}$
Error	$SSE = \sum_{j=1}^{r}\sum_{k=1}^{c}(y_{jk} - \bar{y}_{j.} - \bar{y}_{.k} + \bar{y})^2$	$(r-1)(c-1)$	$MSE = \dfrac{SSE}{(c-1)(r-1)}$	
Total	$SST = \sum_{j=1}^{r}\sum_{k=1}^{c}(y_{jk} - \bar{y})^2$	$rc - 1$		

Example 11.3

Vehicle Acceleration

A driver steps down on the vehicle accelerator pedal in order to speed up the vehicle. What is the peak acceleration to a final speed of 80 mph? Tests were carried out on one vehicle at four different initial speeds (10, 25, 40, and 55 mph) and three different levels of pedal rotation (5, 8, and 10 degrees). The acceleration results are shown in Table 11.10. At $\alpha = .05$, does this sample show that the two experimental factors (pedal rotation, initial speed) are significant predictors of acceleration? This is an *unreplicated* experiment.

Table 11.10

Maximum Acceleration Under Test Conditions 🖼 Acceleration

| Pedal Rotation | Initial Speed | | | |
	10 mph	25 mph	40 mph	55 mph
5 degrees	0.35	0.19	0.14	0.10
8 degrees	0.37	0.28	0.19	0.19
10 degrees	0.42	0.30	0.29	0.23

Note: Maximum acceleration is measured as a fraction of acceleration due to gravity (32 ft./sec.^{2}).

Step 1: State the Hypotheses It is helpful to assign short, descriptive variable names to each factor. The general form of the model is

$$Acceleration = f(PedalRotation, InitialSpeed)$$

Stated as a linear model:

$$y_{jk} = \mu + A_j + B_k + \varepsilon_{jk}^{\cdot}$$

The hypotheses are

Factor A (PedalRotation)

H_0: $A_1 = A_2 = A_3 = 0$ (pedal rotation has no effect)
H_1: Not all the A_j are equal to zero

Factor B (InitialSpeed)

H_0: $B_1 = B_2 = B_3 = B_4 = 0$ (initial speed has no effect)
H_1: Not all the B_k are equal to zero

Step 2: State the Decision Rule   Each F test may require a different right-tail critical value because the numerator degrees of freedom depend on the number of factor levels, while denominator degrees of freedom (error *SSE*) are the same for all three tests:

Factor A: $df_1 = r - 1 = 3 - 1 = 2$ ($r = 3$ pedal rotations)
Factor B: $df_1 = c - 1 = 4 - 1 = 3$ ($c = 4$ initial speeds)
Error: $df_2 = (r - 1)(c - 1) = (3 - 1)(4 - 1) = 6$

From Appendix F, the 5 percent critical values in a right-tailed test (all ANOVA tests are right-tailed tests) are

$F_{2,6} = 5.14$ for factor A or using Excel: =F.INV.RT(0.05,2,6) = 5.1432
$F_{3,6} = 4.76$ for factor B or using Excel: =F.INV.RT(0.05,3,6) = 5.7571

We will reject the null hypothesis (no factor effect) if the F test statistic exceeds the critical value.

Step 3: Perform the Calculations   Calculations are done by using Excel's Data Analysis. The menu and results are shown in Figure 11.17. There is a table of means and variances, followed by the ANOVA table.

Figure 11.17

Excel's ANOVA: Two-Factor without Replication Acceleration

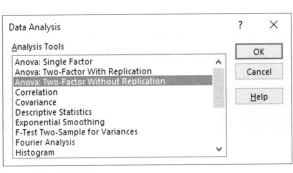

Anova: Two-Factor Without Replication

SUMMARY	Count	Sum	Average	Variance
5 degrees	4	0.78	0.1950	0.0120
8 degrees	4	1.03	0.2575	0.0074
10 degrees	4	1.24	0.3100	0.0063
10 mph	3	1.14	0.3800	0.0013
25 mph	3	0.77	0.2567	0.0034
40 mph	3	0.62	0.2067	0.0058
55 mph	3	0.52	0.1733	0.0044

ANOVA						
Source of Variation	SS	df	MS	F	P-value	F crit
Rows	0.026517	2	0.013258	22.83732	0.001565	5.143253
Columns	0.073892	3	0.024631	42.42584	0.000196	4.757063
Error	0.003483	6	0.000581			
Total	0.103892	11				

Source: Microsoft Excel.

Step 4: Make the Decision   Because $F_A = 22.84$ (rows) exceeds $F_{2,6} = 5.14$, we see that factor A (pedal rotation) has a significant effect on acceleration. The p-value for pedal rotation is very small ($p = .001565$), which says that the F statistic is not due to chance at $\alpha = .05$. Similarly, $F_B = 42.43$ exceeds $F_{3,6} = 4.76$, so we see that factor B (initial speed) also has a significant effect on acceleration. Its tiny p-value (.000196) is unlikely to be a chance result. In short, we conclude that

- Acceleration is significantly affected by pedal rotation ($p = .001565$).
- Acceleration is significantly affected by initial speed ($p = .000196$).

Step 5: Take Action   The *p*-values suggest that initial speed is a more significant predictor than pedal rotation, although both are highly significant. These results conform to your own experience. Maximum acceleration ("pushing you back in your seat") from a low speed or standing stop is greater than when you are driving down the freeway, and, of course, the harder you press the accelerator pedal, the faster you will accelerate. In fact, you might think of the pedal rotation as a blocking factor because its relationship to acceleration is tautological and of little research interest. Nonetheless, omitting pedal rotation and using a one-factor model would not be a correct model specification. Further, the engineers who did this experiment were actually interested in both effects.

Visual Display of Data

Figure 11.18 shows MegaStat's dot plot and ANOVA table. The dot plot shows the column factor (presumed to be the factor of research interest) on the horizontal axis, while the row factor (presumed to be a blocking factor) is only used to define the line graphs. MegaStat rounds its ANOVA results more than Excel and highlights significant *p*-values. MegaStat does not provide critical *F* values, which are basically redundant because you have the *p*-values.

Figure 11.18

MegaStat's Two-Factor ANOVA (Randomized Block Model) 📋 **Acceleration**

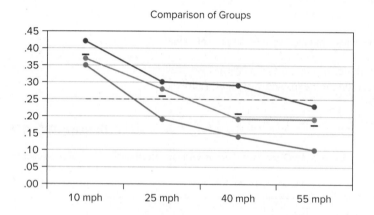

Comparison of Groups

Randomized blocks ANOVA

Mean	n	Std. Dev	
0.38000	3	0.03606	10 mph
0.25667	3	0.05859	25 mph
0.20667	3	0.07638	40 mph
0.17333	3	0.06658	55 mph
0.19500	4	0.10970	5 degrees
0.25750	4	0.08617	8 degrees
0.31000	4	0.07958	10 degrees
0.25417	12	0.09718	Total

ANOVA table

Source	SS	df	MS	F	p-value
Treatments	0.0739	3	0.02463	42.43	.0002
Blocks	0.0265	2	0.01326	22.84	.0016
Error	0.0035	6	0.00058		
Total	0.1039	11			

Limitations of Two-Factor ANOVA without Replication

When replication is impossible or extremely expensive, two-factor ANOVA without replication must suffice. For example, crash-testing of automobiles to estimate collision damage is very costly. However, whenever possible, there is a strong incentive to replicate the experiment to add power to the tests. Would different results have been obtained if the car had been tested not once but several times at each speed? Or if several different cars had been tested? For testing acceleration, there would seem to be no major cost impediment to replication except the time and effort required to take the measurements. Of course, it could be argued that if the measurements of acceleration were careful and precise the first time, replication would be a waste of time. And yet some random variation is found in any experiment. These are matters to ponder. But two-factor ANOVA *with replication* does offer advantages, as you will see.

Section Exercises

Instructions: Use Excel's Data Analysis (or MegaStat, or Minitab, or JMP) to perform each two-factor ANOVA without replication at $\alpha = .05$. For each data set: (a) State the hypotheses. If you are viewing this data set as a randomized block, which is the blocking factor, and why? (b) State your conclusions about the treatment means. (c) Interpret the *p*-values carefully. *Optional challenge:* (d*) Make a plot of the data for each group (e.g., using MegaStat) or individual value plots (e.g., using Minitab). What do the plots show?

11.21 Concerned about Friday absenteeism, management examined absenteeism rates for the last three Fridays in four assembly plants. Does this sample provide sufficient evidence to conclude that there is a significant difference in treatment means? 📂 **Absences**

	Plant 1	Plant 2	Plant 3	Plant 4
March 4	19	18	27	22
March 11	22	20	32	27
March 18	20	16	28	26

11.22 Engineers are testing company fleet vehicle fuel economy (miles per gallon) performance by using different types of fuel. One vehicle of each size is tested. Does this sample provide sufficient evidence to conclude that there is a significant difference in treatment means? 📂 **MPG2**

	87 Octane	89 Octane	91 Octane	Ethanol 5%	Ethanol 10%
Compact	27.2	30.0	30.3	26.8	25.8
Mid-Size	23.0	25.6	28.6	26.6	23.3
Full-Size	21.4	22.5	22.2	18.9	20.8
SUV	18.7	24.1	22.1	18.7	17.4

11.23 Driving time (minutes) from the airport to each of four downtown hotels is recorded for five Uber drivers. Does this sample provide sufficient evidence to conclude that there is a significant difference in treatment means? 📂 **DriveTime**

	Driver A	Driver B	Driver C	Driver D	Driver E
Hotel 1	30.9	22.3	34.9	31.2	20.9
Hotel 2	25.5	24.6	28.7	26.5	20.3
Hotel 3	29.0	26.0	29.6	25.0	23.7
Hotel 4	19.9	28.0	27.8	24.1	23.9

11.24 A beer distributor is comparing quarterly sales of Coors Light (number of six-packs sold) at three convenience stores. Does this sample provide sufficient evidence to conclude that there is a significant difference in treatment means? 📂 **BeerSales**

	Store 1	Store 2	Store 3
Qtr 1	1,521	1,298	1,708
Qtr 2	1,396	1,492	1,382
Qtr 3	1,178	1,052	1,132
Qtr 4	1,730	1,659	1,851

Mini Case 11.2

Automobile Interior Noise Level

Most consumers prefer quieter cars. Table 11.11 shows interior noise level for five vehicles selected from tests performed by a popular magazine. Noise level (in decibels) was measured at idle, at 60 miles per hour, and under hard acceleration from 0 to 60 mph.

For reference, 60 dB is a normal conversation, 75 dB is a typical vacuum cleaner, 85 dB is city traffic, 90 dB is a typical hair dryer, and 110 dB is a chain saw. Two questions may be asked: (1) Does noise level vary significantly among the vehicles? (2) Does noise level vary significantly with speed? If you wish to think of this as a randomized block experiment, the column variable (*vehicle type*) is the research question, while the row variable (*speed*) is the blocking factor.

Table 11.11

Interior Noise Levels in Five Randomly Selected Vehicles 📄 NoiseLevel

	Full-Size Sedan	Sport Wagon	Full-Size SUV	Mid-Size Sedan	Small SUV
Idle	41	45	44	45	46
60 mph	65	67	66	66	76
0–60 mph	76	72	76	77	64

The general form of the model is *NoiseLevel* = *f*(*CarSpeed*, *CarType*). Degrees of freedom for *CarSpeed* (rows) will be $r - 1 = 3 - 1 = 2$, while degrees of freedom for *CarType* (columns) will be $c - 1 = 5 - 1 = 4$. Denominator degrees of freedom will be the same for both factors because *SSE* has degrees of freedom $(r - 1)(c - 1) = (3 - 1)(5 - 1) = 8$. Excel's ANOVA results (using Excel's default level of significance $\alpha = .05$) and MegaStat's dot plot are shown in Figure 11.19.

Figure 11.19

Two-Factor ANOVA without Replication for Car Noise

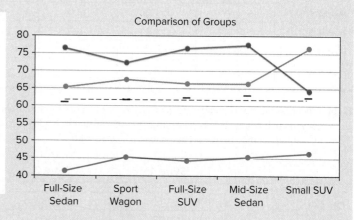

Because $F = 46.01$ exceeds $F_{2,8} = 4.46$, we see that *CarSpeed* (row factor) does have a highly significant effect on noise level. Its very small *p*-value ($p = .00004$) is unlikely to be a chance result. But *CarType* (column factor) has no significant effect on noise level because $F = 0.07$ does not exceed $F_{4,8} = 3.84$. The *p*-value for *CarType* ($p = .99006$) says that its *F* statistic could easily have arisen by chance. In short, we conclude that

- Interior noise *is* significantly affected by car speed ($p = .00004$).

- Interior noise *is not* significantly affected by car type ($p = .9901$).

We do not bother with Tukey multiple comparisons of means because we know that car type has no significant effect on noise level (the research hypothesis) and the effect of initial speed is of less research interest (a blocking factor).

11.6 TWO-FACTOR ANOVA WITH REPLICATION (FULL FACTORIAL MODEL)

What Does Replication Accomplish?

LO 11-11

Interpret main effects and interaction effects in two-factor ANOVA.

In a two-factor model, suppose that each factor combination is observed m times. With multiple observations within each cell, we can do more detailed statistical tests. With an equal number of observations in each cell (*balanced data*), we have a two-factor ANOVA model *with* replication. Replication allows us to test not only the factors' **main effects** but also an **interaction effect**. This model is often called the **full factorial** model. In linear model format it may be written

$$y_{ijk} = \mu + A_j + B_k + AB_{jk} + \varepsilon_{ijk} \qquad (11.15)$$

where

y_{ijk} = observation i for row j and column k ($i = 1, 2, \ldots, m$)

μ = common mean for all treatments

A_j = effect attributed to factor A in row j ($j = 1, 2, \ldots, r$)

B_k = effect attributed to factor B in column k ($k = 1, 2, \ldots, c$)

AB_{jk} = effect attributed to interaction between factors A and B

ε_{ijk} = random error (normally distributed, zero mean, same variance for all treatments)

Interaction effects can be important. For example, an agribusiness researcher might postulate that corn yield is related to seed type (A), soil type (B), interaction between seed type and soil type (AB), or all three. An interaction effect can be significant even if the main effects are not. In the absence of any factor effects, all variation about the mean μ is purely random.

Format of Hypotheses

For a fixed-effects ANOVA model, the hypotheses that could be tested in the two-factor ANOVA model with replicated observations are

Factor *A:* Row Effect

$H_0: A_1 = A_2 = \cdots = A_r = 0$ (row means are the same)

$H_1:$ Not all the A_j are equal to zero (row means differ)

Factor *B:* Column Effect

$H_0: B_1 = B_2 = \cdots = B_c = 0$ (column means are the same)

$H_1:$ Not all the B_k are equal to zero (column means differ)

Interaction Effect

$H_0:$ All the AB_{jk} are equal to zero (there is no interaction effect)

$H_1:$ Not all AB_{jk} are equal to zero (there is an interaction effect)

If none of the proposed factors has anything to do with Y, then the model collapses to

$$y_{ijk} = \mu + \varepsilon_{ijk} \qquad (11.16)$$

Format of Data

Table 11.12 shows the format of a data set with two factors and a balanced (equal) number of observations per treatment (each row/column intersection is a treatment). To avoid needless subscripts, the m observations in each treatment are represented simply as *yyy*. Except for the replication within cells, the format is the same as the unreplicated two-factor ANOVA.

Table 11.12

Data Format of Replicated Two-Factor ANOVA

Levels of Factor A	Levels of Factor B				Row Mean
	B_1	B_2	...	B_c	
A_1	yyy yyy ... yyy	yyy yyy ... yyy	... ...	yyy yyy ... yyy	$\bar{y}_{1.}$
A_2	yyy yyy ... yyy	yyy yyy ... yyy	... ...	yyy yyy ... yyy	$\bar{y}_{2.}$
...	...	...	...	...	...
A_r	yyy yyy ... yyy	yyy yyy ... yyy	... ...	yyy yyy ... yyy	$\bar{y}_{r.}$
Col Mean	$\bar{y}_{.1}$	$\bar{y}_{.2}$	...	$\bar{y}_{.c}$	$\bar{\bar{y}}$

Sources of Variation

There are now three F tests that could be performed: one for each main effect (factors A and B) and a third F test for interaction. The total sum of squares is partitioned into four components:

(11.17)
$$SST = SSA + SSB + SSI + SSE$$

where

SST = total sum of squared deviations about the mean

SSA = between rows sum of squares (effect of factor A)

SSB = between columns sum of squares (effect of factor B)

SSI = interaction sum of squares (effect of AB)

SSE = error sum of squares (residual variation)

For an experiment with r rows, c columns, and m replications per treatment, the sums of squares and ANOVA calculations may be presented in a table, shown in Table 11.13.

Table 11.13

Two-Factor ANOVA with Replication

Source of Variation	Sum of Squares	Degrees of Freedom	Mean Square	F Ratio
Factor A (row effect)	$SSA = cm\sum_{j=1}^{r}(\bar{y}_{j.} - \bar{\bar{y}})^2$	$r - 1$	$MSA = \dfrac{SSA}{r-1}$	$F_A = \dfrac{MSA}{MSE}$
Factor B (column effect)	$SSB = rm\sum_{k=1}^{c}(\bar{y}_{.k} - \bar{\bar{y}})$	$c - 1$	$MSB = \dfrac{SSB}{c-1}$	$F_B = \dfrac{MSB}{MSE}$
Interaction ($A \times B$)	$SSI = m\sum_{j=1}^{r}\sum_{k=1}^{c}(\bar{y}_{jk} - \bar{y}_{j.} - \bar{y}_{.k} + \bar{\bar{y}})^2$	$(r-1)(c-1)$	$MSI = \dfrac{SSI}{(r-1)(c-1)}$	$F_I = \dfrac{MSI}{MSE}$
Error	$SSE = \sum_{i=1}^{m}\sum_{j=1}^{r}\sum_{k=1}^{c}(\bar{y}_{ijk} - \bar{y}_{jk})^2$	$rc(m-1)$	$MSE = \dfrac{SSE}{rc(m-1)}$	
Total	$SST = \sum_{i=1}^{m}\sum_{j=1}^{r}\sum_{k=1}^{c}(y_{ijk} - \bar{\bar{y}})^2$	$rcm - 1$		

If *SSE* is relatively high, we expect that we would fail to reject H_0 for the various hypotheses. Conversely, if *SSE* is relatively small, it is likely that at least one of the factors (row effect, column effect, or interaction) is a relevant predictor of *Y*. Before doing the *F* test, each sum of squares must be divided by its degrees of freedom to obtain its *mean square*. Degrees of freedom sum to $n - 1$ (note that $n = rcm$).

Example 11.4

Delivery Time

A health maintenance organization orders weekly medical supplies for its four clinics from five different suppliers. Delivery times (in days) for 4 recent weeks are shown in Table 11.14. Do the means differ between clinics or suppliers? Is there interaction?

Table 11.14

Delivery Times (in days) 📁 **Deliveries**

	Supplier 1	Supplier 2	Supplier 3	Supplier 4	Supplier 5
Clinic A	8	14	10	8	17
	7	11	15	7	12
	10	14	10	13	9
	12	11	7	10	10
Clinic B	14	9	12	6	15
	14	7	10	10	12
	13	7	10	12	12
	13	8	11	8	10
Clinic C	11	8	12	10	14
	10	9	10	11	13
	12	11	13	7	10
	14	12	10	10	12
Clinic D	7	8	7	8	14
	10	13	5	5	13
	10	9	6	11	8
	13	12	5	4	11

Using short variable names, the two-factor ANOVA model has the general form

$$DeliveryTime = f(Clinic, Supplier, Clinic \times Supplier)$$

The effects are assumed additive. The linear model is

$$y_{ijk} = \mu + A_j + B_k + AB_{jk} + \varepsilon_{ijk}$$

Step 1: State the Hypotheses   The hypotheses are

Factor *A*: Row Effect (*Clinic*)

$H_0: A_1 = A_2 = \cdots = A_r = 0$ (clinic means are the same)
$H_1:$ Not all the A_j are equal to zero (clinic means differ)

Factor *B*: Column Effect (*Supplier*)

$H_0: B_1 = B_2 = \cdots = B_c = 0$ (supplier means are the same)
$H_1:$ Not all the B_k are equal to zero (supplier means differ)

Interaction Effect (*Clinic* × *Supplier*)

$H_0:$ All the AB_{jk} are equal to zero (there is no interaction effect)
$H_1:$ Not all AB_{jk} are equal to zero (there is an interaction effect)

Step 2: State the Decision Rule   Each F test may require a different right-tail critical value because the numerator degrees of freedom depend on the number of factor levels, while denominator degrees of freedom (error SSE) are the same for all three tests:

> $Factor\ A$: $df_1 = r - 1 = 4 - 1 = 3$ ($r = 4$ clinics)
> $Factor\ B$: $df_1 = c - 1 = 5 - 1 = 4$ ($c = 5$ suppliers)
> $Interaction\ (AB)$: $df_1 = (r - 1)(c - 1) = (4 - 1)(5 - 1) = 12$
> $Error$: $df_2 = rc(m - 1) = 4 \times 5 \times (4 - 1) = 60$

Excel provides the right-tail F critical values for $\alpha = .05$, which we can verify using Appendix F:

> $F_{3,60} = 2.76$ for Factor A
> $F_{4,60} = 2.53$ for Factor B
> $F_{12,60} = 1.92$ for Factor AB

We reject the null hypothesis if an F test statistic exceeds its critical value.

Step 3: Perform the Calculations   Excel provides tables of row and column sums and means (not shown here because they are lengthy). The ANOVA table in Figure 11.20 summarizes the partitioning of variation into its component sums of squares, degrees of freedom, mean squares, F test statistics, p-values, and critical F-values for $\alpha = .05$.

Figure 11.20

Excel's Two-Factor ANOVA with Replication 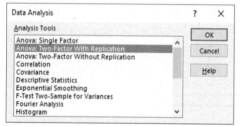 **Deliveries**

ANOVA						
Source of Variation	SS	df	MS	F	P-value	F crit
Sample	51.350	3	17.1167	3.5784	0.018939	2.758078
Columns	104.425	4	26.1063	5.4578	0.000815	2.525215
Interaction	134.775	12	11.2313	2.3480	0.015167	1.917396
Within	287.000	60	4.78333			
Total	577.550	79				

Source: Microsoft Excel.

Step 4: Make the Decision   For the row variable ($Clinic$), the test statistic $F = 3.578$ and its p-value ($p = .0189$) lead us to conclude that the mean delivery times among clinics are not the same at $\alpha = .05$. For the column variable ($Supplier$), the test statistic $F = 5.458$ and its p-value ($p = .0008$) lead us to conclude that the mean delivery times from suppliers are not the same at $\alpha = .05$. For the interaction effect, the test statistic $F = 2.348$ and its p-value ($p = .0152$) indicate significance at $\alpha = .05$. The p-values permit a more flexible interpretation because α need not be specified in advance. In summary:

Variable	p-Value	Interpretation
Clinic	.0189	Clinic means differ (significant at $\alpha = .05$)
Supplier	.0008	Supplier means differ (significant at $\alpha = .01$)
Clinic × Supplier	.0152	Interaction exists (significant at $\alpha = .05$)

Interaction Effect

The statistical test for interaction is just like any other F test. But you might still wonder, What *is* an interaction, anyway? You may be familiar with the idea of drug interaction. If you consume a few ounces of vodka, it has an effect on you. If you take an allergy pill, it has an effect on you. But if you combine the two, the effect may be different (and possibly dangerous) compared with using either drug by itself. That is why many medications carry a warning like "Avoid alcohol while using this medication."

To visualize an interaction, we plot the treatment means for one factor against the levels of the other factor. Within each factor level, we connect the means. In the absence of an interaction, the lines will be roughly parallel or will tend to move in the same direction at the same time. If there is a strong interaction, the lines will have differing slopes and will tend to cross one another.

Figure 11.21 shows that the interaction plot lines for delivery times are *not* parallel (crossing lines). A significant *interaction effect* says that suppliers have different mean delivery times for different clinics. For example, for clinic *B*, average delivery times by supplier 2 are quicker than by supplier 1. However, supplier 3 gives quicker average delivery times for clinic *D* than for clinic *C*. Therefore, there is not one preferred supplier for all clinics. We can make other comparisons of this sort, confirming visually that there is a significant interaction effect in this ANOVA. This visual conclusion is consistent with the interaction *p*-value ($p = .0152$) for the *F* test of $A \times B$. Figure 11.22 illustrates some prototype interaction patterns, using a simplified hypothetical two-factor ANOVA model in which factor *A* has three levels and factor *B* has two levels.

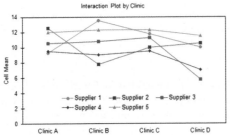
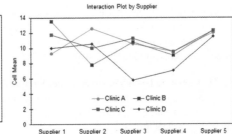

Figure 11.21

**Interaction Plots
for Delivery Times
📂 Deliveries**

Figure 11.22

Prototype Interaction Patterns

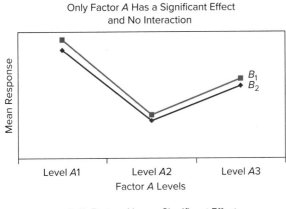

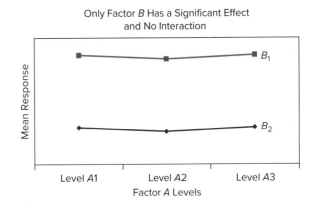

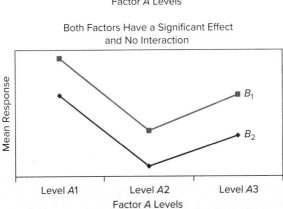

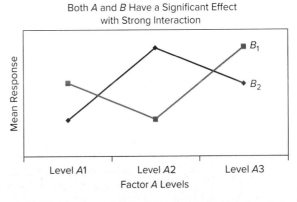

Significance versus Importance

A table of treatment means (Figure 11.23) allows us to explore these differences further. For clinics *A* and *C,* the supplier mean delivery times differ by 2–3 days, which might not matter unless inventories were low. But for clinics *B* and *D,* the mean delivery times differ by 4–5 days. This suggests that differences in average supplier times (and the interaction effect) may be *important* as well as *significant.* You can construct your own table like this if it is not supplied by software. A Tukey test (not shown) can be performed by software to see which means differ significantly.

Figure 11.23

Mean Delivery Times
📁 **Deliveries**

	Supplier 1	Supplier 2	Supplier 3	Supplier 4	Supplier 5	
Clinic A	9.25	12.50	10.50	9.50	12.00	10.75
Clinic B	13.50	7.75	10.75	9.00	12.25	10.65
Clinic C	11.75	10.00	11.25	9.50	12.25	10.95
Clinic D	10.00	10.50	5.75	7.00	11.50	8.95
	11.13	10.19	9.56	8.75	12.00	10.33

Section Exercises

📕 **connect**

Instructions: For each data set: (a) State the hypotheses. (b) Use Excel's Data Analysis (or other software) to perform the two-factor ANOVA with replication, using $\alpha = .05$. (c) State your conclusions about the main effects and interaction effects. (d) Interpret the *p*-values carefully. *Ambitious students:* (e*) Create interaction plots and interpret them.

11.25 A small independent stock broker has created four sector portfolios for her clients. Each portfolio always has five stocks that may change from year to year. The volatility (coefficient of variation) of each stock is recorded for each year. Are the main effects significant? Is there an interaction? 📁 **Volatility**

Year	Stock Portfolio Type			
	Health	**Energy**	**Retail**	**Leisure**
2007	14.5	23.0	19.4	17.6
	18.4	19.9	20.7	18.1
	13.7	24.5	18.5	16.1
	15.9	24.2	15.5	23.2
	16.2	19.4	17.7	17.6
2009	21.6	22.1	21.4	25.5
	25.6	31.6	26.5	24.1
	21.4	22.4	21.5	25.9
	26.6	31.3	22.8	25.5
	19.0	32.5	27.4	26.3
2011	12.6	12.8	22.0	12.9
	13.5	14.4	17.1	11.1
	13.5	13.1	24.8	4.9
	13.0	8.1	13.4	13.3
	13.6	14.7	22.2	12.7

11.26 Oxnard Petro, Ltd., has three interdisciplinary project development teams that function on an ongoing basis. Team members rotate from time to time. Every 4 months (three times a year) each department head rates the performance of each project team (using a 0–100 scale, where 100 is the best rating). Are the main effects significant? Is there an interaction? 📁 **Ratings**

Year	Marketing	Engineering	Finance
2007	90	69	96
	84	72	86
	80	78	86
2009	72	73	89
	83	77	87
	82	81	93
2011	92	84	91
	87	75	85
	87	80	78

11.27 A market research firm is testing consumer reaction to a new shampoo on four age groups in four regions. There are five consumers in each test panel. Each consumer completes a 10-question product satisfaction instrument with a 5-point scale (5 is the highest rating), and the average score is recorded. Are the main effects significant? Is there an interaction? 📁 **Satisfaction**

	Northeast	Southeast	Midwest	West
Youth (under 18)	3.9	3.9	3.6	3.9
	4.0	4.2	3.9	4.4
	3.7	4.4	3.9	4.0
	4.1	4.1	3.7	4.1
	4.3	4.0	3.3	3.9
College (18–25)	4.0	3.8	3.6	3.8
	4.0	3.8	3.6	3.8
	3.7	3.7	3.8	3.6
	3.8	3.6	3.9	3.6
	3.8	3.7	4.0	4.1
Adult (26–64)	3.2	3.5	3.5	3.8
	3.8	3.3	3.8	3.6
	3.7	3.4	3.8	3.4
	3.4	3.5	4.0	3.7
	3.4	3.4	3.7	3.1
Senior (65+)	3.4	3.6	3.3	3.4
	2.9	3.4	3.3	3.2
	3.6	3.6	3.1	3.5
	3.7	3.6	3.1	3.3
	3.5	3.4	3.1	3.4

11.28 Oxnard Petro, Ltd., has three suppliers of catalysts. Orders are placed with each supplier every 15 working days, or about once every 3 weeks. The delivery time (days) is recorded for each order over 1 year. Are the main effects significant? Is there an interaction? 📁 **Deliveries2**

	Supplier 1	Supplier 2	Supplier 3
Qtr 1	12	10	16
	15	13	13
	11	11	14
	11	9	14
Qtr 2	13	10	14
	11	10	11
	13	13	12
	12	11	12
Qtr 3	12	11	13
	8	9	8
	8	8	13
	13	6	6
Qtr 4	8	8	11
	10	10	11
	13	10	10
	11	10	11

Mini Case 11.3

Turbine Engine Thrust

Engineers testing turbofan aircraft engines wanted to know if oil pressure and turbine temperature are related to engine thrust (pounds). They chose four levels for each factor and observed each combination five times, using the two-factor replicated ANOVA model $Thrust = f(OilPres, TurbTemp, OilPres \times TurbTemp)$. The test data are shown in Table 11.15.

Table 11.15

Turbofan Engine Thrust Test Results   Turbines

	Turbine Temperature			
Oil Pressure	**T1**	**T2**	**T3**	**T4**
P1	1,945.0	1,942.3	1,934.2	1,916.7
	1,933.0	1,931.7	1,930.0	1,943.0
	1,942.4	1,946.0	1,944.0	1,948.8
	1,948.0	1,959.0	1,941.0	1,928.0
	1,930.0	1,939.9	1,942.0	1,946.0
P2	1,939.4	1,922.0	1,950.6	1,929.6
	1,952.8	1,936.8	1,947.9	1,930.0
	1,940.0	1,928.0	1,950.0	1,934.0
	1,948.0	1,930.7	1,922.0	1,923.0
	1,925.0	1,939.0	1,918.0	1,914.0
P3	1,932.0	1,939.0	1,952.0	1,960.4
	1,955.0	1,932.0	1,963.0	1,946.0
	1,949.7	1,933.1	1,923.0	1,931.0
	1,933.0	1,952.0	1,965.0	1,949.0
	1,936.5	1,943.0	1,944.0	1,906.0
P4	1,960.2	1,937.0	1,940.0	1,924.0
	1,909.3	1,941.0	1,984.0	1,906.0
	1,950.0	1,928.2	1,971.0	1,925.8
	1,920.0	1,938.9	1,930.0	1,923.0
	1,964.9	1,919.0	1,944.0	1,916.7

Source: Research project by three engineering students enrolled in an MBA program. Data are disguised.

The ANOVA results in Figure 11.24 indicate that only turbine temperature is significantly related to thrust at $\alpha = .05$. The table of means suggests that because mean thrust varies only over a tiny range, the effect may not be very important. The lack of interaction is revealed by the nearly parallel **interaction plots**. Levene's test for equal variances (not shown) shows a p-value of $p = .42$, indicating that variances may be assumed equal, as is desirable for an ANOVA test.

Figure 11.24

MegaStat Two-Factor ANOVA Results

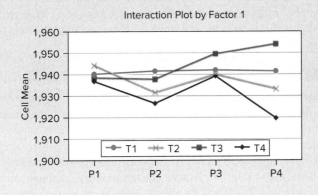

Means:		Temperature				
		T1	T2	T3	T4	
Pressure	P1	1,939.68	1,943.78	1,938.24	1,936.50	1,939.55
	P2	1,941.04	1,931.30	1,937.70	1,926.12	1,934.04
	P3	1,941.24	1,939.82	1,949.40	1,938.48	1,942.24
	P4	1,940.88	1,932.82	1,953.80	1,919.10	1,936.65
		1,940.71	1,936.93	1,944.79	1,930.05	1,938.12

ANOVA table: Two-Factor with Replication

Source	SS	df	MS	F	p-value
Pressure	755.708	3	251.9028	1.32	.2756
Temperature	2,353.426	3	784.4755	4.11	.0099
Interaction	1,989.095	9	221.0106	1.16	.3367
Error	12,212.052	64	190.8133		
Total	17,310.282	79			

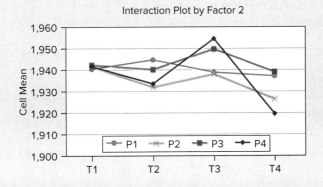

11.7 HIGHER-ORDER ANOVA MODELS (OPTIONAL)

Higher-Order ANOVA Models

Why limit ourselves to two factors? Although a three-factor data set cannot be shown in a two-dimensional table, the idea of a three-factor ANOVA is not difficult to grasp. Consider the paint viscosity problem introduced at the beginning of this chapter. Figure 11.25 adds a third factor (solvent ratio) to the paint viscosity model.

LO 11-12

Recognize the need for experimental design and GLM (optional).

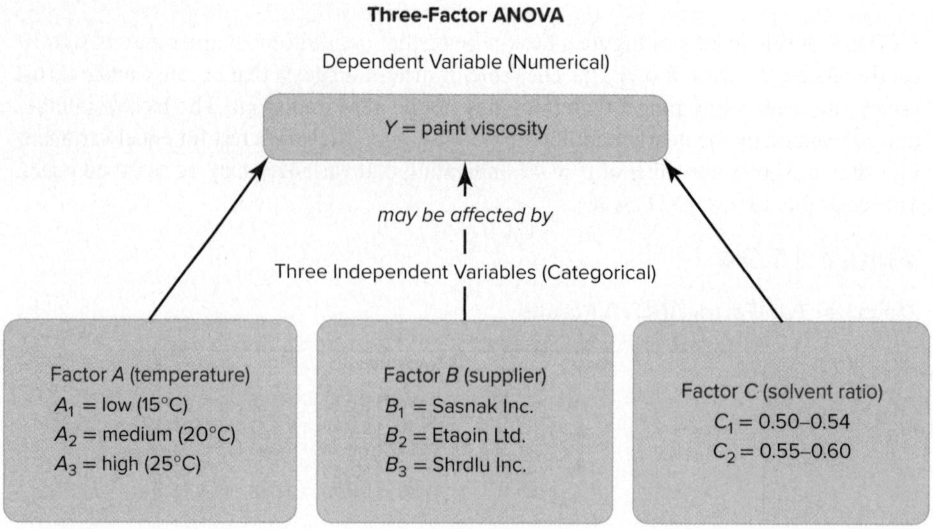

Figure 11.25

Three-Factor ANOVA Model for Paint Viscosity

Three-Factor ANOVA

Dependent Variable (Numerical)

Y = paint viscosity

may be affected by

Three Independent Variables (Categorical)

Factor A (temperature)
A_1 = low (15°C)
A_2 = medium (20°C)
A_3 = high (25°C)

Factor B (supplier)
B_1 = Sasnak Inc.
B_2 = Etaoin Ltd.
B_3 = Shrdlu Inc.

Factor C (solvent ratio)
C_1 = 0.50–0.54
C_2 = 0.55–0.60

A three-factor ANOVA allows more two-factor interactions ($A \times B$, $A \times C$, $B \times C$) and even a three-factor interaction ($A \times B \times C$). However, because the computations are already done by computer, the analysis would be no harder than a two-factor ANOVA. The "catch" is that higher-order ANOVA models are beyond Excel's capabilities, so you will need fancier software. Fortunately, any general-purpose statistical package (e.g., Minitab, SPSS, SAS, R) can handle ANOVA with *any* number of factors with *any* number of levels (subject to computer software limitations).

What Is GLM?

The **general linear model** (GLM) is a versatile tool for estimating large and complex ANOVA models. Besides allowing more than two factors, GLM permits unbalanced data (unequal sample size within treatments) and any desired subset of interactions among factors (including three-way interactions or higher) as long as you have enough observations (i.e., enough degrees of freedom) to compute the effects. GLM also can provide predictions and identify unusual observations. GLM does not require equal variances, although care must be taken to avoid sparse or empty cells in the data matrix. Data are expected to be in stacked format (one column for Y and one column for each factor A, B, C, etc.). The output of GLM is easily understood by anyone who is familiar with ANOVA, as you can see in Mini Case 11.4.

Mini Case 11.4

Hospital Maternity Stay MaternityLOS

The data set consists of 4,409 maternity hospital visits whose DRG (diagnostic-related group) code is 373 (simple delivery without complicating diagnoses). The dependent variable of interest is the length of stay (LOS) in the hospital. The model contains one discrete numerical factor and two categorical factors: the number of surgical stops (*NumStops*), the CCS diagnostic code (*CCSDiag*), and the CCS procedure code (*CCSProc*). CCS codes are a medical classification scheme developed by the American Hospital Research Council to help hospitals and researchers organize medical information. The proposed model has three factors and one interaction:

$$LOS = f(NumStops, CCSDiag, CCSProc, CCDiag \times CCSProc)$$

Before starting the GLM analysis, a frequency tabulation was prepared for each factor. The tabulation (not shown) revealed that some factor levels were observed too rarely to be useful. Cross-tabulations (not shown) also revealed that some treatments would be empty or very sparse. Based on this preliminary data screening, the factors were recoded to avoid GLM estimation problems. *NumStops* was recoded as a binary variable (2 if there were one or two stops, 3 if there were three or more stops). *CCSDiag* codes with a frequency less than 100 were recoded as 999. Patients whose *CCSProc* code occurred less than 10 times (19 patients) were deleted from the sample, leaving a sample of 4,390 patients.

Minitab's menu and GLM results are shown in Figure 11.26. The first thing shown is the number of levels for each factor and the discrete values of each factor. Frequencies of the factor values are not shown but can be obtained from Minitab's Tables command.

Figure 11.26

Minitab GLM Menu and Results

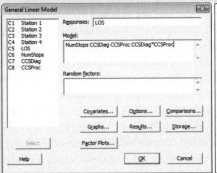

Source: Minitab.

The *p*-values from the ANOVA table suggest that *NumStops* is significant at $\alpha = .10$ ($p = .072$) while the other two main effects, *CCSDiag* and *CCSProc* ($p = .000$ and $p = .001$), are highly significant (such small *p*-values would arise less than once in 1,000 samples if there were no relationship). The interaction *CCSDiag* × *CCSProc* is also highly significant ($p = .000$). Because the sample size is large, even slight effects could be *significant*, so further analysis may be needed to see if the effects are also *important*.

What Is Experimental Design?

Experimental design refers to the number of factors under investigation, the number of levels assigned to each factor, the way factor levels are defined, and the way observations are obtained. *Fully crossed* or *full factorial* designs include all possible combinations of factor levels. **Fractional factorial designs**, for reasons of economy, limit data collection to a subset of possible factor combinations. If all levels of one factor are fully contained in another, the design is **nested** or **hierarchical**. **Balanced designs** are characterized by an equal number of observations for each factor combination. In a *fixed-effects model,* the levels of each factor are predetermined, which implies that our inferences are valid only for the specified factor levels. For example, if a firm has only three paint suppliers (S_1, S_2, S_3), these would be our factor levels. In a *random effects model,* the factor levels would be chosen randomly from a population of potential factor levels. For example, if a firm has 20 paint suppliers (S_1 through S_{20}), but we only want to study three of them, we might choose three at random (say, S_7, S_{11}, and S_{18}) from the 6,840 possible ways to choose 3 items from 20. Fixed effects are by far the most common models used in business analysis, where randomization and controlled experiments are not practical.

Experimental design is a specialized topic that goes far beyond this textbook. However, you may need to interact professionally with engineers or quality improvement teams that are working on product design, reliability, and product performance. It is therefore helpful to have a general idea of what experimental design is all about and to learn some of the basic terminology. If you become more deeply involved, you can ask your employer to send you to a 3-day training class in experimental design to boost your skills.

2^k Models

When there are k factors, each with two levels, we have a 2^k *factorial design.* Reducing a factor to two levels is a useful simplification that reduces the data requirements in a replicated experiment because the data matrix will have fewer cells. Even a continuous factor (e.g., *Pressure*) can be "binarized" into roughly equal groups (*Low, High*) by cutting the data array at the median. The 2^k design is especially useful when the number of factors is very large. In automotive engineering, for example, it is not uncommon to study more than a dozen factors that are predictive of exhaust emissions. Even when each factor is limited to only two levels, full factorial 2^k experiments with replication can require substantial data-collection effort.

Fractional Factorial Designs

Unlike a full factorial design, a *fractional factorial* design, for reasons of economy, limits data collection to a subset of the possible factor combinations. Fractional factorial designs are important in real-life situations where many factors exist. For example, suppose that automobile combustion engineers are investigating 10 factors, each with two levels, to determine their effect on emissions. This would yield $2^{10} = 1,024$ possible factor combinations. It would be impractical and uneconomical to gather data for all 1,024 factor combinations.

By excluding some factor combinations, a fractional factorial model necessarily sacrifices some of the interaction effects. But if the most important objective is to study the *main effects* (which is frequently the case or is at least an acceptable compromise), it is possible to get by with a much smaller number of observations. It often is possible to estimate some, though not all, interaction effects in a fractional factorial experiment. Templates are published to guide experimenters in choosing the correct design and sample size for the desired number of factors (see Related Reading) to make efficient use of available data.

Nested or Hierarchical Design

If all levels of one factor are fully contained within another, the design is *nested* or *hierarchical.* Using most computer packages, nested designs can be represented using simple notation like

$$Defects = f(Experience, Method(Machine))$$

In this model, *Machine* is nested within *Method,* so the effect of *Machine* cannot appear as a main effect. Presumably the nature of the manufacturing process dictates that *Machine* depends on *Method.* Although the model is easy to state, this example is not intended to suggest that estimates of nested models are easy to interpret.

Random Effects Models

In a *fixed-effects model,* the levels of each factor are predetermined, which implies that our inferences are valid only for the specified factor levels. In a *random effects model,* the factor levels are chosen randomly from a population of potential factor levels. Computation and interpretation of random effects are more complicated, and not all tests may be feasible. Novices are advised that estimation of random effects models should be preceded by further study (see Related Reading).

Chapter Summary

ANOVA tests whether a numerical dependent variable (**response variable**) is associated with one or more categorical independent variables (**factors**) with several **levels.** Each level or combination of levels is a treatment. A **one-factor** ANOVA compares means in c columns of data. It is a generalization of a two-tailed t test for two independent sample means. Fisher's **F statistic** is a ratio of two variances (treatment vs. error). It is compared with a right-tailed critical value from an F table or from Excel for appropriate numerator and denominator degrees of freedom. Alternatively, we can compare the p-value for the F test statistic with the desired level of significance (any *p*-value less than α is significant). An **unreplicated two-factor ANOVA** can be viewed as a **randomized block model** if only one factor is of research interest. A **replicated two-factor ANOVA** (or full factorial model) has more than one observation per treatment, permitting inclusion of an interaction test in addition to tests for the **main effects. Interaction effects** can be seen as crossing lines on plots of factor means. The **Tukey test** compares individual treatment means. We test for homogeneous variances (an assumption of ANOVA) using **Hartley's test** or **Levene's test**. The **general linear model** (GLM) can be used when there are more than two factors. **Experimental design** helps make efficient use of limited data. Other general advice:

- ANOVA may be helpful even if those who collected the data did not utilize a formal experimental design (often the case in real-world business situations).

- ANOVA calculations are tedious because of the sums required, so computers are generally used.

- One-factor ANOVA is the most common and suffices for many business situations.

- ANOVA is an overall test. To tell which specific pairs of treatment means differ, use the Tukey test.

- Although real-life data may not perfectly meet the normality and equal-variance assumptions, ANOVA is reasonably robust (and alternative tests do exist).

Key Terms

analysis of variance (ANOVA)	full factorial	main effects	response variable
balanced designs	general linear model	mean squares	treatment
completely randomized model	Hartley's test	multiple comparison	Tukey's studentized range test
experimental design	hierarchical design	nested design	two-factor ANOVA without
explained variance	homogeneous variances	one-factor ANOVA	replication
factors	interaction effect	partitioned sum of squares	unexplained variance
fixed-effects model	interaction plots	randomized block model	
fractional factorial designs	Levene's test	replication	

Chapter Review

Note: Questions labeled * are based on optional material from this chapter.

1. Explain each term: (a) explained variation; (b) unexplained variation; (c) factor; (d) treatment.

2. (a) Explain the difference between one-factor and two-factor ANOVA. (b) Write the linear model form of one-factor ANOVA. (c) State the hypotheses for a one-factor ANOVA in two different ways. (d) Why is one-factor ANOVA used a lot?

3. (a) State three assumptions of ANOVA. (b) What do we mean when we say that ANOVA is fairly robust to violations of these assumptions?

4. (a) Sketch the format of a one-factor ANOVA data set (completely randomized model). (b) Must group sizes be the same for one-factor ANOVA? Is it better if they are? (c) Explain the concepts of variation *between treatments* and variation *within treatments*. (d) What is the *F* statistic? (e) State the degrees of freedom for the *F* test in one-factor ANOVA.

5. (a) Sketch the format of a two-factor ANOVA data set without replication. (b) State the hypotheses for a two-factor ANOVA without replication. (c) What is the difference between a randomized block model and a two-factor ANOVA without replication? (d) What do the two *F* statistics represent in a two-factor ANOVA without replication? (e) What are their degrees of freedom?

6. (a) Sketch the format of a two-factor ANOVA data set with replication. (b) What is gained by replication? (c) State the hypotheses for a two-factor ANOVA with replication. (d) What do the three *F* statistics represent in a two-factor ANOVA with replication? (e) What are their degrees of freedom?

7. (a) What is the purpose of the Tukey test? (b) Why can't we just compare all possible pairs of group means using the two-sample *t* test?

8. (a) What does a test for homogeneity of variances tell us? (b) Why should we test for homogeneity of variances? (c) Explain what Hartley's test measures. (d) Why might we use Levene's test instead of Hartley's test?

*9. What is the general linear model and why is it useful?

*10. (a) What is a 2^k design, and what are its advantages? (b) What is a fractional factorial design, and what are its advantages? (c) What is a nested or hierarchical design? (d) How is a random effects model different than a fixed-effects model?

Chapter Exercises

Instructions: You may use Excel, MegaStat, Minitab, JMP, or another computer package of your choice. Attach appropriate copies of the output or capture the screens, tables, and relevant graphs and include them in a written report. Try to state your conclusions succinctly in language that would be clear to a decision maker who is a nonstatistician. Exercises marked * are based on optional material. Answer the following questions, or those your instructor assigns.

a. Choose an appropriate ANOVA model. State the hypotheses to be tested.

b. Display the data visually (e.g., dot plots or line plots by factor). What do the displays show?

c. Do the ANOVA calculations using the computer.

d. State the decision rule for $\alpha = .05$ and make the decision. Interpret the *p*-value.

e. In your judgment, are the observed differences in treatment means (if any) large enough to be of practical importance?

f. Given the nature of the data, would more data collection be practical?

g. Perform Tukey multiple comparison tests and discuss the results.

h. Perform a test for homogeneity of variances. Explain fully.

11.29 Below are grade point averages for 25 randomly chosen university business students during a recent semester. *Research question:* Are the mean grade point averages the same for students in these four class levels? 🗀 **GPA2**

Grade Point Averages of 25 Business Students

Freshman (5 students)	Sophomore (7 students)	Junior (7 students)	Senior (6 students)
1.91	3.89	3.01	3.32
2.14	2.02	2.89	2.45
3.47	2.96	3.45	3.81
2.19	3.32	3.67	3.02
2.71	2.29	3.33	3.01
	2.82	2.98	3.17
	3.11	3.26	

11.30 The XYZ Corporation is interested in possible differences in days worked by salaried employees in three departments in the financial area. A survey of 23 randomly chosen employees reveals the data shown below. Because of the casual sampling methodology in this survey, the sample sizes are unequal. *Research question:* Are the mean annual attendance rates the same for employees in these three departments? 🗀 **DaysWorked**

Days Worked Last Year by 23 Employees

Department	Days Worked
Budgets (5 workers)	278 260 265 245 258
Payables (10 workers)	205 270 220 240 255 217 266 239 240 228
Pricing (8 workers)	240 258 233 256 233 242 244 249

11.31 Mean output of solar cells of three types are measured six times under random light intensity over a period of 5 minutes, yielding the results shown. *Research question:* Is the mean solar cell output the same for all cell types? 🗀 **SolarWatts**

Solar Cell Output (watts)

Cell Type		Output (watts)				
A	123	121	123	124	125	127
B	125	122	122	121	122	126
C	126	128	125	129	131	128

11.32 In a bumper test, three types of autos were deliberately crashed into a barrier at 5 mph, and the resulting damage (in dollars) was estimated. Five test vehicles of each type were crashed,

with the results shown below. *Research question:* Are the mean crash damages the same for these three vehicles? 🗀 **Crash1**

Crash Damage ($)

Goliath	Varmint	Weasel
1,600	1,290	1,090
760	1,400	2,100
880	1,390	1,830
1,950	1,850	1,250
1,220	950	1,920

11.33 The waiting time (in minutes) for emergency room patients with non-life-threatening injuries was measured at four hospitals for all patients who arrived between 6:00 and 6:30 p.m. on a certain Wednesday. The results are shown below. *Research question:* Are the mean waiting times the same for emergency patients in these four hospitals? 🗀 **ERWait**

Emergency Room Waiting Time (minutes)

Hospital A (5 patients)	Hospital B (4 patients)	Hospital C (7 patients)	Hospital D (6 patients)
10	8	5	0
19	25	11	20
5	17	24	9
26	36	16	5
11		18	10
		29	12
		15	

11.34 The results shown below are mean productivity measurements (average number of assemblies completed per hour) for a random sample of workers at each of three plants. *Research question:* Are the mean hourly productivity levels the same for workers in these three plants? 🗀 **Productivity**

Hourly Productivity of Assemblers in Plants

Plant	Finished Units Produced per Hour
A (9 workers)	3.6 5.1 2.8 4.6 4.7 4.1 3.4 2.9 4.5
B (6 workers)	2.7 3.1 5.0 1.9 2.2 3.2
C (10 workers)	6.8 2.5 5.4 6.7 4.6 3.9 5.4 4.9 7.1 8.4

11.35 Below are results of braking tests of the Ford Explorer on glare ice, packed snow, and split traction (one set of wheels on ice, the other on dry pavement), using three braking methods. *Research question:* Is the mean stopping distance affected by braking method and/or by surface type? 🗀 **Braking**

Stopping Distance from 40 mph to 0 mph

Method	Ice	Split Traction	Packed Snow
Pumping	441	223	149
Locked	455	148	146
ABS	460	183	167

11.36 An MBA director examined GMAT scores for the first 10 MBA applicants (assumed to be a random sample of early applicants) for 4 academic quarters. *Research question:* Do the mean GMAT scores for early applicants differ by quarter? ☞ **GMAT**

GMAT Scores of First 10 Applicants

Fall	490	580	440	580	430	420	640	470	530	640
Winter	310	590	730	710	540	450	670	390	500	470
Spring	500	450	510	570	610	490	450	590	640	650
Summer	450	590	710	240	510	670	610	550	540	540

11.37 An ANOVA study was conducted to compare dental offices in five small towns. The response variable was the number of days each dental office was open last year. *Research question:* Is there a difference in the means among these five towns? ☞ **DaysOpen**

Dental Clinic Days Open during the Last Year in Five Towns

Chalmers	Greenburg	Villa Nueve	Ulysses	Hazeltown
230	194	206	198	214
215	193	200	186	196
221	208	208	206	194
205	198	206	189	190
232		232	181	203
210		208		

11.38 The Environmental Protection Agency (EPA) advocates a maximum arsenic level in water of 10 micrograms per liter. Below are results of EPA tests on randomly chosen wells in a suburban Michigan county. *Research question:* Is the mean arsenic level affected by well depth and/or age of well? ☞ **Arsenic**

Arsenic Level in Wells (micrograms per liter)

Well Depth	Age of Well (years)		
	Under 10	10 to 19	20 and Over
Shallow	5.4	6.1	6.8
	4.3	4.1	5.4
	6.1	5.8	5.7
Medium	3.4	5.1	4.5
	3.7	3.7	5.5
	4.3	4.4	4.6
Deep	2.4	3.8	3.9
	2.9	2.7	2.9
	2.7	3.4	4.0

11.39 Is a state's income related to its high school dropout rate? *Research question:* Do the high school dropout rates differ among the five income quintiles? ☞ **Dropout**

State High School Dropout Rates by Income Groups

Lowest Income Quintile		2nd Income Quintile		3rd Income Quintile		4th Income Quintile		Highest Income Quintile	
State	Dropout %	State	Dropout %	State	Dropout %	State	Dropout %	State	Dropout %
Mississippi	40.0	Kentucky	34.3	N. Carolina	39.5	Oregon	26.0	Minnesota	15.3
W. Virginia	24.2	S. Carolina	44.5	Wyoming	23.3	Ohio	30.5	Illinois	24.6
New Mexico	39.8	N. Dakota	15.5	Missouri	27.6	Pennsylvania	25.1	California	31.7
Arkansas	27.3	Arizona	39.2	Kansas	25.5	Michigan	27.2	Colorado	28.0
Montana	21.5	Maine	24.4	Nebraska	12.1	Rhode Island	31.3	N. Hampshire	27.0
Louisiana	43.0	S. Dakota	28.1	Texas	39.4	Alaska	33.2	Maryland	27.4
Alabama	39.0	Tennessee	40.1	Georgia	44.2	Nevada	26.3	New York	39.0
Oklahoma	26.9	Iowa	16.8	Florida	42.2	Virginia	25.7	New Jersey	20.4
Utah	16.3	Vermont	19.5	Hawaii	36.0	Delaware	35.9	Massachusetts	25.0
Idaho	22.0	Indiana	28.8	Wisconsin	21.9	Washington	25.9	Connecticut	28.2

Source: *Statistical Abstract of the United States, 2002.*

11.40 In a bumper test, three test vehicles of each of three types of autos were crashed into a barrier at 5 mph, and the resulting damage was estimated. Crashes were from three angles: head-on, slanted, and rear-end. The results are shown to the right. *Research questions:* Is the mean repair cost affected by crash type and/or vehicle type? Are the observed effects (if any) large enough to be of practical importance (as opposed to statistical significance)? ☞ **Crash2**

5 mph Collision Damage ($)

Crash Type	Goliath	Varmint	Weasel
Head-on	700	1,700	2,280
	1,400	1,650	1,670
	850	1,630	1,740
Slanted	1,430	1,850	2,000
	1,740	1,700	1,510
	1,240	1,650	2,480
Rear-end	700	860	1,650
	1,250	1,550	1,650
	970	1,250	1,240

11.41 As a volunteer for a consumer research group, LaShonda was assigned to analyze the freshness of three brands of tortilla chips. She examined four randomly chosen bags of chips for four brands of chips from three different stores. She recorded the number of days from the current date until the "fresh until" expiration date printed on the package. *Research question:* Do mean days until the expiration date differ by brand or store? *Note:* Some data values are negative. 🖭 **Freshness**

Days until Expiration Date on Package

	Store 1	Store 2	Store 3
Brand A	−1	25	17
	−1	24	18
	20	10	21
	22	27	6
Brand B	−7	15	29
	30	−8	40
	24	6	24
	23	31	50
Brand C	16	11	41
	7	16	17
	16	30	27
	19	21	18
Brand D	21	42	31
	11	32	30
	10	38	39
	19	28	45

11.42 Three samples of each of three types of PVC pipe of equal wall thickness are tested to failure under three temperature conditions, yielding the results shown below. *Research questions:* Is mean burst strength affected by temperature and/or by pipe type? Is there a "best" brand of PVC pipe? Explain. 🖭 **PVCPipe**

Burst Strength of PVC Pipes (psi)

Temperature	PVC1	PVC2	PVC3
Hot (70°C)	250	301	235
	273	285	260
	281	275	279
Warm (40°C)	321	342	302
	322	322	315
	299	339	301
Cool (10°C)	358	375	328
	363	355	336
	341	354	342

11.43 Below are data on truck production (number of vehicles completed) during the second shift at five truck plants for each day in a randomly chosen week. *Research question:* Are the mean production rates the same by plant and by day? 🖭 **Trucks**

Trucks Produced during Second Shift

	Mon	Tue	Wed	Thu	Fri
Plant A	130	157	208	227	216
Plant B	204	230	252	250	196
Plant C	147	208	234	213	179
Plant D	141	200	288	260	188

11.44 To check pain-relieving medications for potential side effects on blood pressure, it is decided to give equal doses of each of four medications to test subjects. To control for the potential effect of weight, subjects are classified by weight groups. Subjects are approximately the same age and are in general good health. Two subjects in each category are chosen at random from a large group of male prison volunteers. Subjects' blood pressures 15 minutes after the dose are shown below. *Research question:* Is mean blood pressure affected by body weight and/or by medication type? 🖭 **Systolic**

Systolic Blood Pressure of Subjects (mmHg)

Ratio of Subject's Weight to Normal Weight	Medication M1	Medication M2	Medication M3	Medication M4
Under 1.1	131	146	140	130
	135	136	132	125
1.1 to 1.3	136	138	134	131
	145	145	147	133
1.3 to 1.5	145	149	146	139
	152	157	151	141

11.45 To assess the effects of instructor and student gender on student course scores, an experiment was conducted in 11 sections of managerial accounting classes ranging in size from 25 to 66 students. The factors were instructor gender (*M, F*) and student gender (*M, F*). There were 11 instructors (7 male, 4 female). Steps were taken to eliminate subjectivity in grading, such as common exams and sharing exam grading responsibility among all instructors so no one instructor could influence exam grades unduly. (a) What type of ANOVA is this? (b) What conclusions can you draw? (c) Discuss sample size and raise any questions you think may be important.

Analysis of Variance for Students' Course Scores

Source of Variation	Sum of Squares	Degrees of Freedom	Mean Square	F Ratio	p-Value
Instructor gender (*I*)	97.84	1	97.84	0.61	0.43
Student gender (*S*)	218.23	1	218.23	1.37	0.24
Interaction (*I* × *S*)	743.84	1	743.84	4.66	0.03
Error	63,358.90	397	159.59		
Total	64,418.81	400			

See Marlys Gascho Lipe, "Further Evidence on the Performance of Female versus Male Accounting Students," *Issues in Accounting Education* 4, no. 1 (Spring 1989), pp. 144–50.

11.46 In a market research study, members of a consumer test panel are asked to rate the visual appeal (on a 1 to 10 scale) of the texture of dashboard plastic trim in a mockup of a new fuel cell car. The manufacturer is testing four finish textures. Panelists are assigned randomly to evaluate each texture. The test results are shown below. Each cell shows the average rating by panelists who evaluated each texture. *Research question:* Is mean rating affected by age group and/or by surface type? 🖭 **Texture**

Mean Ratings of Dashboard Surface Texture

Age Group	Shiny	Satin	Pebbled	Pattern
Youth (under 21)	6.7	6.6	5.5	4.3
Adult (21 to 39)	5.5	5.3	6.2	5.9
Middle Age (40 to 61)	4.5	5.1	6.7	5.5
Senior (62 and over)	3.9	4.5	6.1	4.1

11.47 This table shows partial results for a one-factor ANOVA. (a) Calculate the F test statistic. (b) Calculate the p-value using Excel's function =F.DIST.RT(F, DF1, DF2). (c) Find the critical value $F_{.05}$ from Appendix F or using Excel's function =F.INV.RT(.05, DF1, DF2). (d) Interpret the results.

ANOVA

Source of Variation	SS	df	MS	F	p-value	$F_{.05}$
Between groups	3207.5	3	1069.17			
Within groups	441730	36	12270.28			
Total	444937.5	39				

11.48 (a) What kind of ANOVA is this (one-factor, two-factor, or two-factor with replication)? (b) Calculate each F test statistic. (c) Calculate the p-value for each F test using Excel's function =F.DIST.RT(F, DF1, DF2). (d) Interpret the results.

ANOVA

Source of Variation	SS	df	MS	F	p-value
Factor A	36,598.56	3	12,199.52		
Factor B	22,710.29	2	11,355.15		
Interaction	177,015.38	6	29,502.56		
Error	107,561.25	36	2,987.81		
Total	343,885.48	47			

11.49 Here is an Excel ANOVA table for an experiment to assess the effects of ambient noise level and plant location on worker productivity. (a) What kind of ANOVA is this (one-factor, two-factor, two-factor replicated)? (b) How many plant locations were there? How many noise levels were examined? How many observations were there for each cell? (c) At $\alpha = 0.05$, what are your conclusions?

ANOVA

Source of Variation	SS	df	MS	F	p-value	F_{05}
Plant location	3.0075	3	1.0025	2.561	0.1200	3.863
Noise level	8.4075	3	2.8025	7.16	0.0093	3.863
Error	3.5225	9	0.3914			
Total	14.9375					

11.50 Several friends go bowling several times per month. They keep track of their scores over several months. An ANOVA was performed. (a) What kind of ANOVA is this (one-factor, two-factor, etc.)? (b) How many friends were there? How many months were observed? How many observations per bowler per month? Explain how you know. (c) At $\alpha = .01$, what are your conclusions about bowling scores? Explain, referring to either the F tests or p-values.

ANOVA

Source of Variation	SS	df	MS	F	p-value	F crit
Month	1702.389	2	851.194	11.9793	0.0002	3.4028
Bowler	4674.000	3	1558.000	21.9265	0.0000	3.0088
Interaction	937.167	6	156.194	2.1982	0.0786	2.5082
Within	1705.333	24	71.056			
Total	9018.889	35				

11.51 Air pollution (micrograms of particulate per milliliter of air) was measured along four freeways at each of five different times of day, with the results shown below. (a) What kind of ANOVA is this (one-factor, two-factor, etc.)? (b) What is your conclusion about air pollution? Explain, referring to either the F tests or p-values. (c) Do you think the variances should be assumed equal? Explain your reasoning. Why does it matter?

SUMMARY	Count	Sum	Average	Variance
Chrysler	5	1584	316.8	14333.7
Davidson	5	1047	209.4	3908.8
Reuther	5	714	142.8	2926.7
Lodge	5	1514	302.8	11947.2
12:00A-6:00A	4	505	126.25	872.9
6:00A-10:00A	4	1065	266.25	11060.3
10:00A-3:00P	4	959	239.75	5080.3
3:00P-7:00P	4	1451	362.75	14333.6
7:00P-12:00A	4	879	219.75	7710.9

ANOVA

Source of Variation	SS	df	MS	F	p-value	F crit
Freeway	100957.4	3	33652.45	24.903	0.000	3.490
Time of Day	116249.2	4	29062.3	21.506	0.000	3.259
Error	16216.4	12	1351.367			
Total	233423	19				

11.52 A company has several suppliers of office supplies. It receives several shipments each quarter from each supplier. The time (days) between order and delivery was recorded for several randomly chosen shipments from each supplier in each quarter, and an ANOVA was performed. (a) What kind of ANOVA is this (one-factor, two-factor, etc.)? (b) How many suppliers were there? How many quarters? How many observations per supplier per quarter? Explain how you know. (c) At $\alpha = .01$, what are your conclusions about shipment time? Explain, referring to either the F tests or p-values.

ANOVA

Source of Variation	SS	df	MS	F	p-value	F crit
Quarter	148.04	3	49.34667	6.0326	0.0009	2.7188
Supplier	410.14	4	102.535	12.5348	0.0000	2.4859
Interaction	247.06	12	20.5883	2.5169	0.0073	1.8753
Within	654.40	80	8.180			
Total	1459.64	99				

11.53 Several friends go bowling several times per month. They keep track of their scores over several months. An ANOVA was performed. (a) What kind of ANOVA is this (one-factor, two-factor, etc.)? (b) How could you tell how many friends there were in the sample just from the ANOVA table? Explain. (c) What are your conclusions about bowling scores? Explain, referring to either the F test or p-value. *Ambitious students:* (d*) State the critical value for Hartley's test for unequal variances. (e*) Calculate Hartley's test statistic. (f*) Should group variances be assumed equal?

SUMMARY

Bowler	Count	Sum	Average	Variance
Mary	15	1856	123.733	77.067
Bill	14	1599	114.214	200.797
Sally	12	1763	146.917	160.083
Robert	15	2211	147.400	83.686
Tom	11	1267	115.182	90.164

ANOVA

Source of Variation	SS	df	MS	F	p-value	F crit
Between Groups	14465.63	4	3616.408	29.8025	0.0000	2.5201
Within Groups	7523.444	62	121.3459			
Total	21989.07	66				

11.54 Are large companies more profitable *per dollar of assets?* The largest 500 companies in the world were ranked according to their number of employees, with groups defined as follows: Small = Under 25,000 employees, Medium = 25,000 to 49,999 employees, Large = 50,000 to 99,999 employees, Huge = 100,000 employees or more. An ANOVA was performed using the company's profit-to-assets ratio (percent) as the dependent variable. (a) What kind of ANOVA is this (one-factor, two-factor, etc.)? (b) What is your conclusion about the research question? Explain, referring to either the F test or p-value. (c) What can you learn from the plots that compare the groups? (d) Do you think the variances can be assumed equal? Explain your reasoning. (e) Perform Hartley's test to test for unequal variances. (f) Which groups of companies have significantly different means? Explain.

Mean	n	Std. Dev	
1.054	119	2.8475	Small
3.058	113	4.8265	Medium
3.855	147	5.8610	Large
3.843	120	4.3125	Huge
3.004	499	4.7925	Total

ANOVA table

Source	SS	df	MS	F	p-value
Treatment	643.9030	3	214.63433	9.84	2.58E-06
Error	10,794.2720	495	21.80661		
Total	11,438.1750	498			

Post hoc analysis
Tukey simultaneous comparison t-values (d.f. = 495)

		Small 1.054	Medium 3.058	Huge 3.843	Large 3.855
Small	1.054				
Medium	3.058	3.27			
Huge	3.843	4.62	1.28		
Large	3.855	4.86	1.37	0.02	

critical values for experimentwise error rate:

0.05	2.60
0.01	3.18

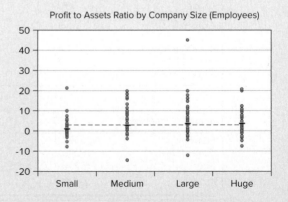

Profit to Assets Ratio by Company Size (Employees)

Related Reading

Montgomery, Douglas C. *Design and Analysis of Experiments.* 10th ed. John Wiley & Sons, 2019.

CHAPTER 11   More Learning Resources

You can access these *LearningStats* demonstrations through Connect to help you understand analysis of variance.

Topic	LearningStats Demonstrations
Overview of ANOVA	What Is ANOVA? Examples: ANOVA Tests ANOVA Simulation ANOVA Using R
Tables	Appendix F—Critical Values of *F*

Key: = Excel   = PowerPoint   = PDF

Software Supplement

MegaStat ANOVA

Figure 11.27 shows MegaStat's one-factor ANOVA table for the carton-packing data. The results are the same as with Excel, although MegaStat rounds things off, highlights significant *p*-values, and gives standard deviations (instead of variances) for each treatment. MegaStat also provides a dot plot of observations by group and Tukey tests for pairs of means.

Figure 11.27

MegaStat's One-Factor ANOVA Results   Cartons

One-Factor ANOVA

Mean	n	Std. Dev	
242.0	5	8.51	Station 1
246.4	5	9.61	Station 2
228.8	5	12.15	Station 3
226.0	5	12.88	Station 4
235.8	20	13.36	Total

ANOVA table

Source	SS	df	MS	F	p-value
Treatment	1,479.20	3	493.067	4.12	.0241
Error	1,914.00	16	119.625		
Total	3,393.20	19			

MegaStat's two-factor ANOVA results, shown in Figure 11.28, are similar to Excel's except that the table of treatment means is more compact, the results are rounded, and significant *p*-values are highlighted (bright yellow for $\alpha = .01$, light green for $\alpha = .05$). As an option, Tukey tests also are reported for each pair of treatment means.

Figure 11.28

MegaStat's Two-Factor ANOVA   Deliveries

ANOVA table

Source	SS	df	MS	F	p-value
Factor 1	51.35	3	17.117	3.43	.0224
Factor 2	104.43	4	26.106	5.24	.0011
Interaction	102.78	12	8.565	1.72	.0852
Error	299.00	60	4.983		
Total	557.56	79			

Minitab ANOVA

Minitab and most other statistical packages want the data in *stacked* format (see Figure 11.29). Each variable has its own column (e.g., column one lists all the *Y* values, while column two is a list of category labels like "Station 1"). Minitab will convert unstacked data (Excel format) to stacked data for one-factor ANOVA, but not for other ANOVA models.

Figure 11.29

Minitab ANOVA Menu

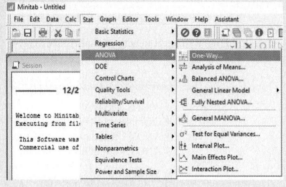

Source: Minitab.

Minitab's one-factor ANOVA output, shown in Figure 11.30, is the same as Excel's except that Minitab rounds off the results and displays a confidence interval for each group mean, an attractive feature. Minitab also gives a plot of the confidence intervals. In this carton-packing example, the confidence intervals overlap, except possibly stations 2 and 4.

Figure 11.30

Minitab's One-Factor ANOVA

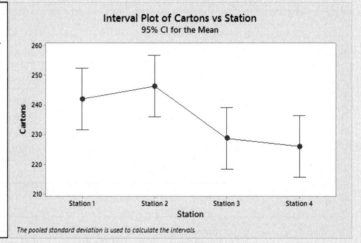

Analysis of Variance

Source	DF	Adj SS	Adj MS	F-Value	P-Value
Station	3	1479	493.1	4.12	0.024
Error	16	1914	119.6		
Total	19	3393			

Means

Station	N	Mean	StDev	95% CI
Station 1	5	242.00	8.51	(231.63, 252.37)
Station 2	5	246.40	9.61	(236.03, 256.77)
Station 3	5	228.80	12.15	(218.43, 239.17)
Station 4	5	226.00	12.88	(215.63, 236.37)

Pooled StDev = 10.9373

The pooled standard deviation is used to calculate the intervals.

Source: Minitab.

Software Supplement

One-Factor ANOVA 🗐 Cartons

A cosmetics manufacturer's regional distribution center has four workstations that pack cartons for shipment to small retailers. There is variability in the number of cartons packed due to differences in orders, labels, and workers. Does this sample of daily output for the most recent five working days show only random variation or is there a difference in the station means?

A rectangular data array is easy to understand, but R requires *stacked* data, so the input data must be in two columns (abbreviated above). Input the data any way you wish (see **Appendix K**). We will copy the two data columns from Excel (including headings "Station" and "Cartons" that will become our R variable names) to the clipboard and create an R data frame we call CartonData. We designate Cartons as the response and Station as a factor. Because Station looks like a number, we use as.factor() to tell R that Station is a categorical variable (this would be unnecessary if we

▲	A	B	C	D	E	F	G	H
1		*Number of Cartons Packed*						
2								
3		*Station 1*	*Station 2*	*Station 3*	*Station 4*		*Station*	*Cartons*
4		236	238	220	241		1	236
5		250	239	236	233		2	238
6		252	262	232	212		3	220
7		233	247	243	231		4	241
8		239	246	213	213		1	250
9							2	239
10							3	236
11							...	...

Source: Microsoft Excel.

had coded Station as text). Use the aov() function with the ~ symbol to define the model:

```
> CartonData=read.table(file="clipboard", sep="\t",header=TRUE)
> result1=aov(Cartons~as.factor(Station),data=CartonData)
> summary(result1)
```

	Df	Sum Sq	Mean Sq	F value	Pr(>F)
as.factor(Station)	3	1479	493.1	4.122	0.0241*
Residuals	16	1914	119.6		

Signif. codes: 0 '***' 0.001 '**' 0.01 '*' 0.05 '.' 0.1 ' ' 1

The results agree with textbook Example 11.1. Note that we store the ANOVA output as a temporary object result1 and then display it using the summary() command, which gives a nicer presentation. In this case, we can reject the null hypothesis of equal means at $\alpha = .05$. To see which station means differ, we can use the function TukeyHSD() with its default 95 percent confidence level. Only Stations 4 and 2 differ significantly at $\alpha = .05$, which agrees with textbook Figure 11.11.

```
> TukeyHSD(result)
Tukey multiple comparisons of means
95% family-wise confidence level
Fit: aov(formula = Cartons ~ as.factor(Station), data = CartonData)
$`as.factor(Station)`
```

	diff	lwr	upr	p adj
2-1	4.4	-15.39073	24.1907311	0.9188131
3-1	-13.2	-32.99073	6.5907311	0.2635407
4-1	-16.0	-35.79073	3.7907311	0.1364778
3-2	-17.6	-37.39073	2.1907311	0.0904731
4-2	-20.4	-40.19073	-0.6092689	0.0421928
4-3	-2.8	-22.59073	16.9907311	0.9768149

Two-Factor ANOVA Unreplicated

We want to see if a vehicle's initial speed and pedal position affect acceleration, based on the test results shown below. To perform the ANOVA in R, we must express the data in column format, as in columns G and H.

Source: Microsoft Excel.

We copy the columns to the clipboard (including headings) and paste the data into an R data frame AccelData. The file headings become the variable names, although R will replace the space with a period. The response Acceleration is a function of two factors, separated by a + symbol in the model specification. Both factors are text (not numbers), so we can omit the as.factor() function (although using it would be OK).

```
> AccelData=read.table(file="clipboard",sep="\ t",header=TRUE)
> result2=aov(Acceleration~Pedal.Position+Initial.Speed,data=AccelData)
> summary(result2)
```

	Df	Sum Sq	Mean Sq	F value	Pr(>F)
Pedal.Position	2	0.02652	0.013258	22.84	0.001565**
Initial.Speed	3	0.07389	0.024631	42.43	0.000196***
Residuals	6	0.00348	0.000581		

Signif. codes: 0 '***'0.001 '**'0.01 '*'0.05 '.'0.1 ' '1

Both factors are highly significant, as indicated by the tiny *p*-values. These results agree with the textbook (Example 11.3).

Two-Factor ANOVA Replicated (Full Factorial)

A health maintenance organization orders weekly medical supplies for its four clinics from five different suppliers. Based on this sample of four recent weeks, do mean delivery times (in days) differ by supplier or clinic? Is there interaction between these two factors?

Source: Microsoft Excel.

R requires *stacked* data, so we copy the three data columns (abbreviated above) from Excel (including headings) to the clipboard and create an R data frame we will call Delivery. The summary() function displays a table like Excel. Statisticians are skeptical of the * notation for significance, but it is widely used.

```
> Delivery=read.table(file="clipboard", sep="\t", header=TRUE)
> result3=aov(Days~Supplier+Clinic+Supplier*Clinic,data=Delivery)
> summary(result3)
```

	Df	Sum Sq	Mean Sq	F value	Pr(>F)
Supplier	4	104.42	26.106	5.458	0.000815***
Clinic	3	51.35	17.117	3.578	0.018939*
Supplier:Clinic	12	134.78	11.231	2.348	0.015167*
Residuals	60	287.00	4.783		

Signif. codes: 0 '***'0.001 '**'0.01 '*'0.05 '.'0.1 ' '1

Design element credits: MegaStat logo: MegaStat; Minitab logo: MINITAB® and all other trademarks and logos for the Company's products and services are the exclusive property of Minitab, LLC. All other marks referenced remain the property of their respective owners. See minitab.com for more information. Noodles & Company logo: Kristoffer Tripplaar/Alamy Stock Photo; R logo: The R Foundation; Microsoft logo: Microsoft.

CHAPTER CONTENTS

CHAPTER LEARNING OBJECTIVES

When you finish this chapter, you should be able to

LO 12-1 Calculate and test a correlation coefficient for significance.

LO 12-2 Interpret a regression equation and use it to make predictions.

LO 12-3 Explain the form and assumptions of a simple regression model.

LO 12-4 Explain the least squares method, apply formulas for coefficients, and interpret R^2.

LO 12-5 Construct confidence intervals and test hypotheses for the slope and intercept.

LO 12-6 Interpret the ANOVA table and use it to calculate F, R^2, and standard error.

LO 12-7 Distinguish between confidence and prediction intervals for Y.

LO 12-8 Calculate residuals and perform tests of regression assumptions.

LO 12-9 Identify unusual residuals and tell when they are outliers.

LO 12-10 Define leverage and identify high-leverage observations.

LO 12-11 Improve data conditioning and use transformations if needed (optional).

LO 12-12 Identify when logistic regression is appropriate and calculate predictions for a binary response variable.

VISUAL DISPLAYS AND CORRELATION ANALYSIS

Up to this point, our study of the discipline of statistical analysis has primarily focused on learning how to describe and make inferences about single variables. It is now time to learn how to describe and summarize relationships *between* variables. Businesses of all types can be quite complex. Understanding how different variables in our business processes are related to each other helps us predict and, hopefully, improve our business performance.

Examples of quantitative variables that might be related to each other include spending on advertising and sales revenue, produce delivery time and percentage of spoiled produce, premium and regular gas prices, and preventive maintenance spending and manufacturing productivity rates. It may be that with some of these pairs there is one variable that we would like to be able to *predict* such as sales revenue, percentage of spoiled produce, and productivity rates. But first we must learn how to *visualize, describe,* and *quantify* the relationships between variables such as these.

Jecapix/E+/Getty Images

Visual Displays

Analysis of **bivariate data** (i.e., two variables) typically begins with a **scatter plot** that displays each observed data pair (x_i, y_i) as a dot on an *X-Y* grid. This diagram provides a visual indication of the strength of the relationship or association between the two random variables. This simple display requires no assumptions or computation. A scatter plot is typically the precursor to more complex analytical techniques.

Figure 12.1 shows a scatter plot comparing the annual revenue for the Star Golf Shoe Company at 29 retail locations compared to the number of competitive brands carried at those locations.

LO 12-1

Calculate and test a correlation coefficient for significance.

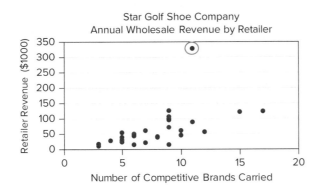

Figure 12.1

Golf Shoes (*n* = 29 retailers) GolfShoes

We look at scatter plots to get an initial idea of the relationship between two random variables. Is there an evident pattern to the data? Is the pattern linear or nonlinear? Are there data points that are not part of the overall pattern? We would characterize the relationship between revenue and number of competitive brands as linear (although not perfectly linear) and as positive (as number of competitive brands at a retailer increases, the revenue for Star Golf Shoes increases). We see one pair of values set apart from the rest, above and slightly to the right. This retailer generated $328,000 for Star with only 11 competitive brands, a much greater revenue than retailers with more competitive brands.

In contrast to the positive relationship seen in Figure 12.1, Figure 12.2 shows a negative relationship. This scatter plot compares the median home price to the unemployment rate for a sample of zip codes. We would characterize this relationship as fairly linear and negative (as the unemployment rate for a zip code increases, the median home price decreases). It seems we have one unusual value, circled, with median price at slightly over $900,000. Yet this unusual pair of values is consistent with the negative trend.

Figure 12.2

Median Home Prices
(n = 206 zip codes) 📁
MedianHomePrices

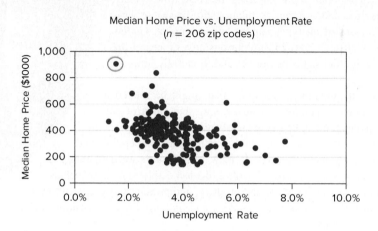

Median Home Price vs. Unemployment Rate
(n = 206 zip codes)

Correlation Coefficient

A visual display is a good first step in analysis, but we would also like to quantify the strength of the association between two variables. Therefore, accompanying the scatter plot is the **sample correlation coefficient** (also called the Pearson correlation coefficient). This statistic measures the degree of linearity in the relationship between two random variables X and Y and is denoted r. Its value will fall in the interval $[-1, 1]$.

Strong Negative Correlation	No Correlation	Strong Positive Correlation
├	┼	┤
−1.00	0.00	+1.00

When r is near 0, there is little or no linear relationship between X and Y. An r-value near $+1$ indicates a strong positive relationship, while an r-value near -1 indicates a strong negative relationship.

$$(12.1) \qquad r = \frac{\sum_{i=1}^{n}(x_i - \bar{x})(y_i - \bar{y})}{\sqrt{\sum_{i=1}^{n}(x_i - \bar{x})^2}\sqrt{\sum_{i=1}^{n}(y_i - \bar{y})^2}} \qquad \text{(sample correlation coefficient)}$$

The product will be negative, that is, a negative correlation, when x_i tends to be *above* its mean while the associated y_i is below its mean. Conversely, the correlation coefficient will be positive when x_i and the associated y_i tend to be above their means at the same time or below their means at the same time. To simplify the notation here and elsewhere in this chapter, we define three terms called **sums of squares**:

$$(12.2) \qquad SS_{xx} = \sum_{i=1}^{n}(x_i - \bar{x})^2 \qquad SS_{yy} = \sum_{i=1}^{n}(y_i - \bar{y})^2 \qquad SS_{xy} = \sum_{i=1}^{n}(x_i - \bar{x})(y_i - \bar{y})$$

Using this notation, the formula for the sample correlation coefficient can be written

$$(12.3) \qquad r = \frac{SS_{xy}}{\sqrt{SS_{xx}}\sqrt{SS_{yy}}} \qquad \text{(sample correlation coefficient)}$$

The correlation coefficient for the variables shown in Figure 12.1 is $r = .5604$, which includes the outlier point circled. Without that data point, $r = .7392$. This higher value is not surprising because we observe a fairly strong positive relationship between retail revenue and competitive brands carried. The correlation coefficient for the variables shown in Figure 12.2 is $r = -.4467$. We observed a fairly negative linear relationship between unemployment rate and median home price.

Figure 12.3 shows prototype scatter plots. We see that a correlation of .500 implies a great deal of random variation, and even a correlation of .900 is far from "perfect" linearity. The last scatter plot shows $r = .00$ despite an obvious curvilinear relationship between X and Y. This illustrates the fact that a correlation coefficient only measures the degree of *linear* relationship between X and Y.

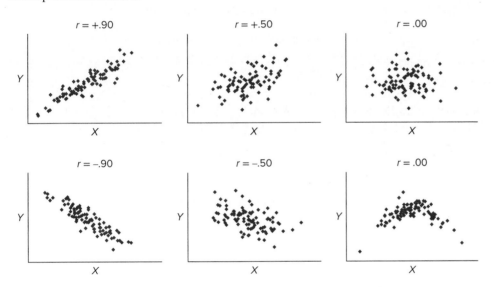

Figure 12.3

Scatter Plots Showing
Various Correlation
Coefficient Values
($n = 100$)

Correlation analysis has many business applications. For example:

- Financial planners study correlations between asset classes over time in order to help their clients diversify their portfolios.
- Marketing analysts study correlations between customer online purchases in order to develop new web advertising strategies.
- Human resources experts study correlations between measures of employee performance in order to devise new job-training programs.

Tests for Significant Correlation Using Student's *t*

The sample correlation coefficient r is an estimate of the **population correlation coefficient** ρ (the Greek letter rho). There is no flat rule for a "high" correlation because sample size must be taken into consideration. To test the hypothesis $H_0: \rho = 0$, the test statistic is

$$t_{calc} = r \sqrt{\frac{n-2}{1-r^2}} \quad \text{(test for zero correlation)} \tag{12.4}$$

We compare this *t* test statistic with a critical value of *t* for a one-tailed or two-tailed test from Appendix D using *d.f.* = *n* − 2 degrees of freedom and any desired *α*. Recall that we lose a degree of freedom for each parameter that we estimate when we calculate a statistic. Because both $\bar{x}$ and $\bar{y}$ are used to calculate *r*, we lose 2 degrees of freedom and so *d.f.* = *n* − 2. After calculating the *t* statistic, we can find its *p*-value by using Excel's function =T.DIST.2T(t, deg_freedom).

Example 12.1

MBA Applicants

📁 **MBA**

In its admission decision process, a university's MBA program examines an applicant's score on the Graduate Management Aptitude Test (GMAT), which has both verbal and quantitative components. Figure 12.4 shows the scatter plot with the sample correlation coefficient for 30 MBA applicants randomly chosen from 1,961 MBA applicant records at a public university in the Midwest. Is the correlation (*r* = .4356) between verbal and quantitative GMAT scores statistically significant? It is not clear from the scatter plot shown in Figure 12.4 that there is a statistically significant linear relationship.

Figure 12.4 **Scatter Plot for 30 MBA Applicants** 📁 **MBA**

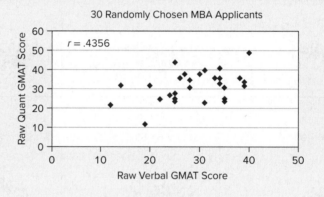

Step 1: State the Hypotheses The hypotheses are

$H_0: \rho = 0$

$H_0: \rho \neq 0$

Step 2: Specify the Decision Rule We will use a two-tailed test for significance at *α* = .05. For a two-tailed test using *d.f.* = *n* − 2 = 30 − 2 = 28 degrees of freedom, Appendix D gives $t_{.025} = 2.048$. The decision rule is

Reject H_0 if $t_{calc} > 2.048$ or if $t_{calc} < -2.048$.

Step 3: Calculate the Test Statistic To calculate the test statistic, we first need to calculate the value for *r*. Using Excel's function =CORREL(array1, array2), we find *r* = .4356 for the variables *Quant GMAT* and *Verbal GMAT*. We must then calculate t_{calc}.

$$t_{calc} = r\sqrt{\frac{n-2}{1-r^2}} = .4356 \times \sqrt{\frac{30-2}{1-(.4356)^2}} = 2.561$$

Step 4: Make a Decision The test statistic value ($t_{calc} = 2.561$) exceeds the critical value $t_{.025} = 2.048$, so we reject the hypothesis of zero correlation at $\alpha = .05$. We also can find the p-value using the Excel function =T.DIST.2T(t, deg_freedom). The two-tailed p-value for GMAT score is =T.DIST.2T(2.561,28) = .0161. We would reject $\rho = 0$ because the p-value < .05.

Step 5: Take Action The admissions officers recognize that these scores tend to vary together for applicants.

Critical Value for Correlation Coefficient

An equivalent approach is to calculate a critical value for the correlation coefficient. First, look up the critical value of t from Appendix D with $d.f. = n - 2$ degrees of freedom for either a one-tailed or two-tailed test, with the α you choose. Then the critical value of the correlation coefficient, $r_{critical}$, is

$$r_{critical} = \frac{t}{\sqrt{t^2 + n - 2}} \quad \text{(critical value for a correlation coefficient)} \quad \textbf{(12.5)}$$

An advantage of this method is that you get a benchmark for the correlation coefficient. Its disadvantage is that there is no p-value and it is inflexible if you change your mind about α. MegaStat uses this method, giving two-tail critical values for $\alpha = .05$ and $\alpha = .01$.

Table 12.1 shows that, as sample size increases, the critical value of r becomes smaller. Thus, in very large samples, even very small correlations could be "significant." In a larger sample, smaller values of the sample correlation coefficient can be considered "significant." While a larger sample does give a better estimate of the true value of ρ, a larger sample does *not* mean that the correlation is stronger, nor does its increased *significance* imply increased *importance*.

Sample Size	$r_{.05}$	$r_{.01}$
$n = 25$	.395	.504
$n = 50$	.279	.361
$n = 100$	.197	.256
$n = 200$	.139	.182

Table 12.1

Values for $r_{critical}$ for Different Sample Sizes

Significance versus Importance

In large samples, small correlations may be significant, even if the scatter plot shows little evidence of linearity. Thus, a *significant* correlation may lack practical *importance*.

12.1 For each sample, do a test for zero correlation. (a) Use Excel to find the critical value of t_α. (b) State the hypotheses about ρ. (c) Perform the t test and report your decision.

a. $r = +.45$, $n = 20$, $\alpha = .05$, two-tailed test
b. $r = -.35$, $n = 30$, $\alpha = .10$, two-tailed test
c. $r = +.60$, $n = 7$, $\alpha = .05$, right-tailed test
d. $r = -.30$, $n = 61$, $\alpha = .01$, left-tailed test

Section Exercises

Mc Graw Hill **connect**

Instructions for exercises 12.2 and 12.3: (a) Make a scatter plot. What does it suggest about the population correlation between *X* and *Y*? (b) Make an Excel worksheet to calculate SS_{xx}, SS_{yy}, and SS_{xy}. Use these sums to calculate the sample correlation coefficient. Check your work by using Excel's function =CORREL(array1, array2). (c) Use Excel to find $t_{.025}$ for a two-tailed test for zero correlation at $\alpha = .05$. (d) Calculate the *t* test statistic. Can you reject $\rho = 0$? (e) Use Excel's function =T.DIST.2T(t, deg_freedom) to calculate the two-tail *p*-value.

12.2 **College Student Weekly Earnings in Dollars ($n = 5$)** 📂 **WeekPay**

Hours Worked (X)	Weekly Pay (Y)
10	93
15	171
20	204
20	156
35	261

12.3 **Wait Time for Vehicle Registration in Seconds ($n = 5$)** 📂 **DMVWait**

Clerk Windows Open (X)	Wait Time (Y)
4	385
5	335
6	383
7	344
8	288

Instructions for exercises 12.4–12.6: (a) Make a scatter plot of the data. What does it suggest about the correlation between *X* and *Y*? (b) Use Excel to calculate the correlation coefficient. (c) Use Excel to find $t_{.025}$ for a two-tailed test at $\alpha = .05$. (d) Calculate the *t* test statistic. (e) Can you reject $\rho = 0$?

12.4 **Moviegoer Snack Spending ($n = 10$)** 📂 **Movies**

Age (X)	Spent (Y)	Age (X)	Spent (Y)
30	6.85	33	10.75
50	10.50	36	7.60
34	5.50	26	10.10
12	10.35	18	12.35
37	10.20	46	8.35

12.5 **Annual Percent Return on Mutual Funds ($n = 17$)** 📂 **Portfolio**

Last Year (X)	This Year (Y)
11.9	15.4
19.5	26.7
11.2	18.2
14.1	16.7
14.2	13.2
5.2	16.4
20.7	21.1
11.3	12.0
−1.1	12.1
3.9	7.4
12.9	11.5
12.4	23.0
12.5	12.7
2.7	15.1
8.8	18.7
7.2	9.9
5.9	18.9

12.6 **Order Size and Shipping Cost ($n = 12$)** 📂 **ShipCost**

Orders (X)	Ship Cost (Y)
1,068	4,489
1,026	5,611
767	3,290
885	4,113
1,156	4,883
1,146	5,425
892	4,414
938	5,506
769	3,346
677	3,673
1,174	6,542
1,009	5,088

Mini Case 12.1

Do Loyalty Cards Promote Sales Growth?

A business can achieve sales growth by increasing the number of new customers. Another way is by increasing business from existing customers. Loyal customers visit more often, thus contributing to sales growth. Loyalty cards are used by many companies to foster positive relationships with their customers. Customers carry a card that records the number of purchases or visits they make. They are rewarded with a free item or discount after so many visits. But do these loyalty cards provide incentive to repeat customers to visit more often? Surprisingly, Noodles & Company found out that this wasn't happening in some markets. After several years of running a loyalty card program without truly measuring its impact on the business, Noodles performed a correlation analysis on the variables "Sales Growth Percentage" and "Loyalty Card Sales Percentage." The results showed that in some markets there was no significant correlation, meaning the loyalty cards weren't associated with increased sales revenue. However, in other markets there was actually *a statistically significant negative correlation.* In other words, loyalty cards were associated with a decrease in sales growth. Why? Ultimately, the free visits that customers had earned were replacing visits that they would have otherwise paid full price for. Moreover, the resources the company was devoting to the program were taking away from more proven sales-building techniques, such as holding nonprofit fund-raisers or tastings for local businesses. Based on this analysis, Noodles & Company made the decision to discontinue its loyalty card program and focused on other approaches to building loyal customers. See Chapter Exercise 12.70.

12.2 SIMPLE REGRESSION

What Is Simple Regression?

Correlation coefficients and scatter plots provide clues about relationships among variables and may suffice for some purposes. But often, the analyst would like to mathematically model the relationship for prediction purposes. For example, a business might hypothesize that

LO 12-2

Interpret a regression equation and use it to make predictions.

- Advertising expenditures predict quarterly sales revenue.
- Number of dependents predicts employee prescription drug expenses.
- Apartment size predicts monthly rent.
- Engine horsepower predicts miles per gallon.

The hypothesized relationship may be linear, quadratic, or some other form. For now we will focus on the simple linear model in slope-intercept form: $Y = \text{slope} \times X + y - \text{intercept}$. In statistics this straight-line model is often referred to as a **simple regression equation**. The slope and intercept of the simple regression equation are used to describe the relationship between the two variables.

We define the Y variable as the **response variable** (the *dependent variable*) and the X variable as the **predictor variable** (the *independent variable*). If the relationship can be estimated, a business can explore policy questions such as

- How much extra sales will be generated, on average, by a \$1 million increase in advertising expenditures? What would expected sales be with no advertising?
- How much do prescription drug costs per employee rise, on average, with each extra dependent? What would be the expected cost if the employee had no dependents?
- How much extra rent, on average, is paid per extra square foot?
- How much fuel efficiency, on average, is lost when the engine horsepower is increased?

Response or Predictor?

The *response* variable is the *dependent* variable. This is the Y variable. The *predictor* variable is the *independent* variable. This is the X variable. Only the dependent variable (not the independent variable) is treated as a random variable.

Interpreting an Estimated Regression Equation

The intercept and slope of an estimated regression can provide useful information. The slope tells us how much and in which direction the response variable will change for each one-unit increase in the explanatory variable. However, it is important to interpret the intercept with caution because it is meaningful only if the explanatory variable would reasonably have a value equal to zero. For example:

$Sales = 268 + 7.37\ Ads$	Each extra \$1 million of advertising will generate \$7.37 million of sales on average. The firm would average \$268 million of sales with zero advertising. However, the intercept may not be meaningful because $Ads = 0$ may be outside the range of observed data.
$DrugCost = 410 + 550\ Dependents$	Each extra dependent raises the mean annual prescription drug cost by \$550. An employee with zero dependents averages \$410 in prescription drugs.
$Rent = 150 + 1.05\ SqFt$	Each extra square foot adds \$1.05 to monthly apartment rent. The intercept is not meaningful because no apartment can have $SqFt = 0$.
$MPG = 49.22 - 0.079\ Horsepower$	Each unit increase in engine horsepower decreases the fuel efficiency by 0.079 mile per gallon. The intercept is not meaningful because a zero horsepower engine does not exist.

The proposed relationship between the explanatory variable and response variable is not an assumption of causation. One cannot conclude that the explanatory variable *causes* a change in the response variable. Consider a regression equation with unemployment rate per capita as the explanatory variable and crime rate per capita as the response variable.

Crime Rate $= 0.125 + 0.031$ *Unemployment Rate*

The slope value, 0.031, means that for each one-unit increase in the unemployment rate, we expect to see an increase of .031 in the crime rate. Does this mean being out of work causes crime to increase? No, there are many lurking variables that could further explain the change in crime rates (e.g., poverty rate, education level, or police presence).

Cause and Effect?

When we propose a regression model, we might have a causal mechanism in mind, but cause and effect are not proven by a simple regression. We cannot assume that the explanatory variable is "causing" the variation we see in the response variable.

Prediction Using Regression

One of the main uses of regression is to make predictions. Once we have a fitted regression equation that shows the estimated relationship between X (the independent variable) and Y (the dependent variable), we can plug in any value of X to obtain the prediction for Y. For example:

Sales = 268 + 7.37 *Ads*

If the firm spends $10 million on advertising, its predicted sales would be $341.7 million; that is, *Sales* = 268 + 7.37(10) = 341.7.

DrugCost = 410 + 550 *Dependents*

If an employee has four dependents, the predicted annual drug cost would be $2,610; that is, *DrugCost* = 410 + 550(4) = 2,610.

Rent = 150 + 1.05 *SqFt*

The predicted rent on an 800-square-foot apartment is $990; that is, *Rent* = 150 + 1.05(800) = 990.

MPG = 49.22 − 0.079 *Horsepower*

If an engine has 200 horsepower, the predicted fuel efficiency is 33.42 mpg; that is, *MPG* = 49.22 − 0.079(200) = 33.42.

Extrapolation Outside the Range of *X*

Predictions from our fitted regression model are stronger within the range of our sample *x* values. The relationship seen in the scatter plot may not be true for values far outside our observed *x* range. Extrapolation outside the observed range of *x* is always tempting but should be approached with caution.

Section Exercises

Mc Graw Hill **connect**

12.7 (a) Interpret the slope of the fitted regression *HomePrice* = 125,000 + 150 *SquareFeet*. (b) What is the prediction for *HomePrice* if *SquareFeet* = 2,000? (c) Would the intercept be meaningful if this regression applies to home sales in a certain subdivision?

12.8 (a) Interpret the slope of the fitted regression *Sales* = 842 − 37.5 *Price*. (b) If *Price* = 20, what is the prediction for Sales? (c) Would the intercept be meaningful if this regression represents coffee mug sales at REI?

12.9 (a) Interpret the slope of the fitted regression *CarTheft* = 1,667 − 35.3 *MedianAge*, where *CarTheft* is the number of car thefts per 100,000 people by state and *MedianAge* is the median age of the population. (b) What is the prediction for *CarTheft* if *MedianAge* is 40? (c) Would the intercept be meaningful if this regression applies to car thefts per 100,000 people by state?

12.10 (a) Interpret the slope of the fitted regression *Computer power dissipation* = 15.73 + 0.032 *Microprocessor speed,* where *Computer power dissipation* is measured in watts and *Microprocessor speed* is measured in MHz. (b) What is the prediction for *Computer power dissipation* if *Microprocessor speed* is 3,000 MHz? (c) Is this intercept meaningful?

12.11 (a) Interpret the slope of the fitted regression *Number of International Franchises* = −47.5 + 1.75 *PowerDistanceIndex*. The *PowerDistanceIndex* is a measure on a scale of 0–100 of the wealth gap between the richest and poorest in a country. (b) What is the prediction for number of international franchises in a country that has a *PowerDistanceIndex* of 85? (c) Is this intercept meaningful?

12.3 REGRESSION MODELS

Model and Parameters

The regression model's *unknown population parameters* are denoted by Greek letters β_0 (the **intercept**) and β_1 (the **slope**). The *population model* for a linear relationship is

$$y = \beta_0 + \beta_1 x + \epsilon \qquad \text{(population regression model)} \qquad (12.6)$$

This relationship is assumed to be true for all (x_i, y_i) pairs in the population. Inclusion of a random error ϵ is necessary because other unspecified variables also may affect *Y* and because there may be measurement error in *Y*.

Each value of *X* may be paired with many different values of *Y* (e.g., there are many 800-square-foot apartments near a university, but each may rent for a different amount). The

LO 12-3

Explain the form and assumptions of a simple regression model.

regression model without the error term represents the expected value of Y for a given x value. This is called the *simple regression equation:*

(12.7) $E(Y|x) = \beta_0 + \beta_1 x$ (simple regression equation)

Even though the error term ϵ is not observable, we assume that the error is a normally distributed random variable with mean 0 and standard deviation σ. We also assume that σ is the same for each x and that the errors are independent of each other. These are called the **regression assumptions**. Figure 12.5 illustrates the regression assumptions.

- Assumption 1: The errors are normally distributed.
- Assumption 2: The errors have constant variance, σ^2.
- Assumption 3: The errors are independent of each other.

Figure 12.5

Regression Parameters and Error Assumptions

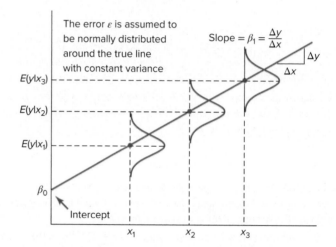

From the sample, we estimate the regression equation and use it to *predict* the expected value of Y for a given value of X:

(12.8) $\hat{y} = b_0 + b_1 x$ (estimated regression equation)

Roman letters denote the *coefficients* b_0 (the estimated intercept) and b_1 (the estimated slope). For a given value x_i the estimated value of the dependent variable is $\hat{y}_i$. (You can read this as "y-hat.") The difference between the observed value y_i and its estimated value $\hat{y}_i$ is called a **residual** and is denoted e_i. The residual is the vertical distance between each y_i and the estimated regression line on a scatter plot of (x_i, y_i) values.

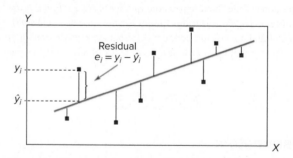

What Is a Residual?

A residual is calculated as the observed value of y minus the estimated value of y:

(12.9) $e_i = y_i - \hat{y}_i$ (residual)

The n residuals $e_1, e_2, \ldots, e_n$ are used to estimate σ, the standard deviation of the errors.

Fitting a Regression on a Scatter Plot

From a scatter plot, we could visually estimate the slope and intercept. Although this method is inexact, experiments suggest that people are pretty good at "eyeball" line fitting. We instinctively try to adjust the line to ensure that the line passes through the "center" of the scatter of data points, to match the data as closely as possible. In other words, we try to minimize the vertical distances between the *fitted* line and the observed y values.

A more precise method is to let Excel calculate the estimates. We enter observations on the independent variable $x_1, x_2, \ldots, x_n$ and the dependent variable $y_1, y_2, \ldots, y_n$ into separate columns and let Excel fit the regression equation, as illustrated in Figures 12.6 and 12.7.[1] Excel will choose the regression coefficients so as to produce a good fit.

- Step 1: Highlight the data columns.
- Step 2: Click on Insert and choose Scatter to create a graph.
- Step 3: Click on the scatter plot points to select the data.
- Step 4: Right-click and choose Add Trendline.
- Step 5: Choose Options and check Display Equation on chart.

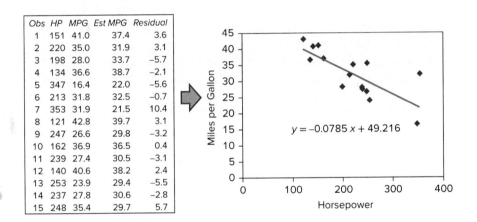

Obs	HP	MPG	Est MPG	Residual
1	151	41.0	37.4	3.6
2	220	35.0	31.9	3.1
3	198	28.0	33.7	−5.7
4	134	36.6	38.7	−2.1
5	347	16.4	22.0	−5.6
6	213	31.8	32.5	−0.7
7	353	31.9	21.5	10.4
8	121	42.8	39.7	3.1
9	247	26.6	29.8	−3.2
10	162	36.9	36.5	0.4
11	239	27.4	30.5	−3.1
12	140	40.6	38.2	2.4
13	253	23.9	29.4	−5.5
14	237	27.8	30.6	−2.8
15	248	35.4	29.7	5.7

$y = -0.0785\,x + 49.216$

Figure 12.6

Excel's Trendline

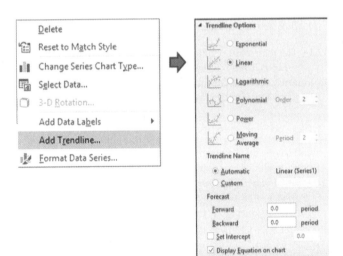

Figure 12.7

Excel's Trendline Menus

Source: Microsoft Excel.

---

[1] Excel calls its regression equation a "trendline," although we usually use that term for a time series trend.

Illustration: Miles per Gallon and Horsepower

Figure 12.6 shows a sample of miles per gallon and horsepower for 15 engines. The Excel graph and its fitted regression equation also are shown.

Slope Interpretation The fitted regression is $\hat{y} = 49.216 - 0.0785x$. The slope ($b_1 = -0.0785$) says that for each additional unit of engine horsepower, the miles per gallon decreases by 0.0785 mile. This estimated slope is a *statistic* because a different sample might yield a different estimate of the slope.

Intercept Interpretation The intercept ($b_0 = 49.216$) suggests that when the engine has no horsepower ($x = 0$), the fuel efficiency would be quite high. However, the intercept has little meaning in this case, not only because zero horsepower makes no logical sense but also because extrapolating to $x = 0$ is beyond the range of the observed data.

Regression Caveats

- The "fit" of the regression does *not* depend on the sign of its slope. The sign of the fitted slope merely tells whether X has a positive or negative association with Y.
- View the intercept with skepticism unless $x = 0$ is logically possible and is within the observed range of X.
- Regression does not demonstrate cause and effect between X and Y. A good fit only shows that X and Y vary together. Both could be affected by another variable or by the way the data are defined.

Section Exercises

12.12 The regression equation *NetIncome* $= 2,277 + 0.0307$ *Revenue* was estimated from a sample of 100 leading world companies (variables are in millions of dollars). (a) Interpret the slope. (b) Is the intercept meaningful? Explain. (c) Make a prediction of *NetIncome* when *Revenue* = 20,000. 🖻 **Global100**

12.13 The regression equation *HomePrice* $= 51.3 + 2.61$ *Income* was estimated from a sample of 34 cities in the eastern United States. Both variables are in thousands of dollars. *HomePrice* is the median selling price of homes in the city and *Income* is median family income for the city. (a) Interpret the slope. (b) Is the intercept meaningful? Explain. (c) Make a prediction of *HomePrice* when *Income* = 50 and also when *Income* = 100. 🖻 **HomePrice1**

12.14 The regression equation *Credits* $= 15.4 - 0.07$ *Work* was estimated from a sample of 21 statistics students. *Credits* is the number of college credits taken and *Work* is the number of hours worked per week at an outside job. (a) Interpret the slope. (b) Is the intercept meaningful? Explain. (c) Make a prediction of *Credits* when *Work* = 0 and when *Work* = 40. What do these predictions tell you? 🖻 **Credits**

12.15 Below are fitted regressions for Y = asking price of a used vehicle and X = the age of the vehicle. The observed range of X was 1 to 8 years. The sample consisted of all vehicles listed for sale in a particular week. (a) Interpret the slope of each fitted regression. (b) Interpret the intercept of each fitted regression. Does the intercept have meaning? (c) Predict the price of a 5-year-old Chevy Blazer. (d) Predict the price of a 5-year-old Chevy Silverado. 🖻 **CarPrices**

Chevy Blazer: *Price* $= 16,189 - 1,050$ *Age* ($n = 21$ vehicles, observed X range was 1 to 8 years).
Chevy Silverado: *Price* $= 22,591 - 1,339$ *Age* ($n = 24$ vehicles, observed X range was 1 to 10 years).

12.16 Refer back to the regression equation in exercise 12.12: *NetIncome* $= 2,277 + 0.0307$ *Revenue*. Recall that the variables are both in millions of dollars. (a) Calculate the residual for the *x, y* pair ($41,078, $8,301). Did the regression equation underestimate or overestimate the net income? (b) Calculate the residual for the *x, y* pair ($61,768, $893). Did the regression equation underestimate or overestimate the net income?

12.17 Refer back to the regression equation in exercise 12.14: *Credits* = 15.4 − 0.07 *Work*. (a) Calculate the residual for the *x*, *y* pair (14, 18). Did the regression equation underestimate or overestimate the credits? (b) Calculate the residual for the *x*, *y* pair (30, 6). Did the regression equation underestimate or overestimate the credits?

12.4 ORDINARY LEAST SQUARES FORMULAS

Slope and Intercept

The **ordinary least squares** method (or **OLS** method for short) is used to estimate a regression so as to ensure the best fit. "Best" fit in this case means that we have selected the slope and intercept so that our residuals are as small as possible. Recall that a residual $e_i = y_i - \hat{y}_i$ is the difference between the observed *y* and the estimated *y*. Residuals can be either positive or negative. It is a characteristic of the OLS estimation method that the residuals around the regression line always sum to zero. That is, the positive residuals exactly cancel the negative ones:

LO 12-4

Explain the least squares method, apply formulas for coefficients, and interpret R^2.

$$\sum_{i=1}^{n} e_i = \sum_{i=1}^{n} (y_i - \hat{y}_i) = 0 \quad \text{(OLS residuals always sum to zero)} \quad \textbf{(12.10)}$$

Therefore, to work with an equation that has a nonzero sum, we square the residuals, just as we squared the deviations from the mean when we developed the equation for variance back in Chapter 4. The fitted coefficients b_0 and b_1 are chosen so that the fitted linear model $\hat{y}_i = b_0 + b_1 x$ has the smallest possible sum of squared residuals (*SSE*):

$$SSE = \sum_{i=1}^{n} e_i^2 = \sum_{i=1}^{n} (y_i - \hat{y}_i)^2 = \sum_{i=1}^{n} (y_i - b_0 - b_1 x_i)^2 \quad \text{(sum to be minimized)} \quad \textbf{(12.11)}$$

This is an optimization problem that can be solved for b_0 and b_1 by using Excel's Solver Add-In. However, we also can use calculus to solve for b_0 and b_1.

$$b_1 = \frac{\sum_{i=1}^{n} (x_i - \bar{x})(y_i - \bar{y})}{\sum_{i=1}^{n} (x_i - \bar{x})^2} \quad \text{(OLS estimator for slope)} \quad \textbf{(12.12)}$$

$$b_0 = \bar{y} - b_1 \bar{x} \quad \text{(OLS estimator for intercept)} \quad \textbf{(12.13)}$$

If we use the notation for sums of squares (see formula 12.2), then the OLS formula for the slope can be written

$$b_1 = \frac{SS_{xy}}{SS_{xx}} \quad \text{(OLS estimator for slope)} \quad \textbf{(12.14)}$$

These formulas require only a few spreadsheet operations to find the means, deviations around the means, and their products and sums. They are built into Excel and many calculators. Their Excel formulas are

$$b_0 = \text{INTERCEPT(YData, XData)}$$
$$b_1 = \text{SLOPE(YData, XData)}$$

The OLS formulas give unbiased and consistent estimates[2] of β_0 and β_1. The OLS regression line always passes through the point $(\bar{x}, \bar{y})$ for any data, as illustrated in Figure 12.8.

Illustration: Exam Scores and Study Time

What is the relationship between the number of hours a student studies and his or her exam score? We can estimate the regression line for these two variables using a sample of 10 students. The worksheet in Table 12.2 shows the calculations of the sums needed for the slope and intercept. Figure 12.9 shows a fitted regression line. The vertical line segments in the scatter

---

[2]Recall from Chapter 8 that an unbiased estimator's expected value is the true parameter and that a consistent estimator approaches ever closer to the true parameter as the sample size increases.

Figure 12.8

OLS Regression Line Always Passes through $(\bar{x}, \bar{y})$

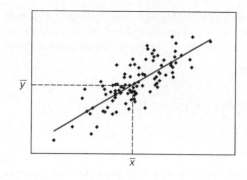
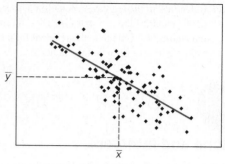

Table 12.2

Worksheet for Slope and Intercept Calculations

ExamScores

Student	Hours x_i	Score y_i	$x_i - \bar{x}$	$y_i - \bar{y}$	$(x_i - \bar{x})(y_i - \bar{y})$	$(x_i - \bar{x})^2$
Tom	1	53	−9.5	−17.1	162.45	90.25
Mary	5	74	−5.5	3.9	−21.45	30.25
Sarah	7	59	−3.5	−11.1	38.85	12.25
Oscar	8	43	−2.5	−27.1	67.75	6.25
Cullyn	10	56	−0.5	−14.1	7.05	0.25
Jaime	11	84	0.5	13.9	6.95	0.25
Theresa	14	96	3.5	25.9	90.65	12.25
Knut	15	69	4.5	−1.1	−4.95	20.25
Jin-Mae	15	84	4.5	13.9	62.55	20.25
Courtney	19	83	8.5	12.9	109.65	72.25
Sum	105	701	0	0	$SS_{xy} = 519.50$	$SS_{xx} = 264.50$
Mean	$\bar{x} = 10.5$	$\bar{y} = 70.1$				

Figure 12.9

Scatter Plot with Fitted Line and Residuals Shown as Vertical Line Segments

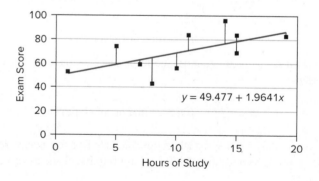

plot show the differences between the actual and fitted exam scores (i.e., residuals). The OLS residuals always sum to zero. We have:

$$b_1 = \frac{SS_{xy}}{SS_{xx}} = \frac{519.50}{264.50} = 1.9641 \qquad \text{(fitted slope)}$$

$$b_0 = \bar{y} - b_1\bar{x} = 70.1 - (1.9641)(10.5) = 49.477 \qquad \text{(fitted intercept)}$$

The fitted regression *Score* $= 49.477 + 1.9641$ *Study* says that, on average, each additional hour of study yields a little less than 2 additional exam points (the slope). A student who did not study (*Study* $= 0$) would expect a score of about 49 (the intercept). In this example, the intercept is meaningful because zero study time not only is possible (though hopefully uncommon) but also was almost within the range of observed data. The scatter plot shows an imperfect fit because not all of the variation in exam scores can be explained by study time. The remaining *unexplained* variation in exam scores reflects other factors (e.g., previous night's sleep, class attendance, test anxiety). We can use the fitted regression equation $\hat{y} = 1.9641x + 49.477$ to find each student's *expected* exam score. Each prediction is a *conditional mean,* given the student's study hours. For example:

Student and Study Time	Expected Exam Score
Oscar, $x = 8$ hours	$\hat{y} = 49.48 + 1.964(8) = 65.19$ (65 to nearest integer)
Theresa, $x = 14$ hours	$\hat{y} = 49.48 + 1.964(14) = 76.98$ (77 to nearest integer)
Courtney, $x = 19$ hours	$\hat{y} = 49.48 + 1.964(19) = 86.79$ (87 to nearest integer)

Oscar's actual exam score was only 43, so he did worse than his predicted score of 65. Theresa scored 96, far above her predicted score of 77. Courtney, who studied the longest (19 hours), scored 83, fairly close to her predicted score of 87. These examples show that study time is not a perfect predictor of exam scores.

Sources of Variation in *Y*

In a regression, we seek to explain the variation in the dependent variable around its mean. We express the *total variation* as a sum of squares (denoted *SST*):

$$SST = \sum_{i=1}^{n}(y_i - \overline{y})^2 \quad \text{(total sum of squares)} \tag{12.15}$$

We can split the total variation into two parts:

$$\begin{matrix} SST & = & SSR & + & SSE \\ (\textit{total}\text{ variation} & & (\text{variation explained} & & (\text{unexplained} \\ \text{around the mean}) & & \text{by the }\textit{regression}) & & \text{or }\textit{error}\text{ variation}) \end{matrix}$$

The *explained variation* in Y (denoted *SSR*) is the sum of the squared differences between the conditional mean $\hat{y}_i$ (conditioned on a given value x_i) and the unconditional mean $\overline{y}$ (same for all x_i):

$$SSR = \sum_{i=1}^{n}(\hat{y}_i - \overline{y})^2 \quad (\textit{regression}\text{ sum of squares, explained}) \tag{12.16}$$

The *unexplained variation* in y (denoted *SSE*) is the sum of *squared* residuals, sometimes referred to as the **error sum of squares.**

$$SSE = \sum_{i=1}^{n}(y_i - \hat{y}_i)^2 \quad (\textit{error}\text{ sum of squares, unexplained}) \tag{12.17}$$

If the fit is good, *SSE* will be relatively small compared to *SST.* If each observed data value y_i is exactly the same as its estimate $\hat{y}_i$ (i.e., a perfect fit), then *SSE* will be zero. There is no upper limit on *SSE.* Table 12.3 shows the calculation of *SSE* for the exam scores.

Table 12.3

Calculations of Sums of Squares   ExamScores

Student	Hours x_i	Score y_i	Estimated Score $\hat{y}_i = 1.9641x_i + 49.477$	Residual $e_i = y_i - \hat{y}_i$	$(y_i - \hat{y}_i)^2$	$(\hat{y}_i - \overline{y})^2$	$(y_i - \overline{y})^2$
Tom	1	53	51.441	1.559	2.43	348.15	292.41
Mary	5	74	59.298	14.702	216.15	116.68	15.21
Sarah	7	59	63.226	−4.226	17.86	47.25	123.21
Oscar	8	43	65.190	−22.190	492.40	24.11	734.41
Cullyn	10	56	69.118	−13.118	172.08	0.96	198.81
Jaime	11	84	71.082	12.918	166.87	0.96	193.21
Theresa	14	96	76.974	19.026	361.99	47.25	670.81
Knut	15	69	78.939	−9.939	98.78	78.13	1.21
Jin-Mae	15	84	78.939	5.061	25.61	78.13	193.21
Courtney	19	83	86.795	−3.795	14.40	278.72	166.41
					SSE = 1,568.57	*SSR* = 1,020.34	*SST* = 2,588.90

Assessing Fit: Coefficient of Determination

Because the magnitude of *SSE* is dependent on sample size and on the units of measurement (e.g., dollars, kilograms, ounces), we want a *unit-free* benchmark to assess the fit of the regression equation. We can obtain a measure of *relative fit* by comparing *SST* to *SSR*. Recall that total variation in *y* can be expressed as

$$SST = SSR + SSE$$

By dividing both sides by *SST*, we now have the sum of two proportions on the right-hand side.

$$\frac{SST}{SST} = \frac{SSR}{SST} + \frac{SSE}{SST} \quad \text{or} \quad 1 = \frac{SSR}{SST} + \frac{SSE}{SST}$$

The first proportion, *SSR/SST*, has a special name: **coefficient of determination** or R^2. You can calculate this statistic in two ways.

(12.18)
$$R^2 = \frac{SSR}{SST} \quad \text{or} \quad R^2 = 1 - \frac{SSE}{SST}$$

The range of the coefficient of determination is $0 \leq R^2 \leq 1$. The highest possible R^2 is 1 because, if the regression gives a perfect fit, then $SSE = 0$:

$$R^2 = 1 - \frac{SSE}{SST} = 1 - \frac{0}{SST} = 1 - 0 = 1 \quad \text{if } SSE = 0 \text{ (perfect fit)}$$

The lowest possible R^2 is 0 because, if knowing the value of *X* does not help predict the value of *Y*, then $SSE = SST$:

$$R^2 = 1 - \frac{SSE}{SST} = 1 - \frac{SST}{SST} = 1 - 1 = 0 \quad \text{if } SSE = SST \text{ (worst fit)}$$

For the exam scores, the coefficient of determination is

$$R^2 = 1 - \frac{SSE}{SST} = 1 - \frac{1,568.57}{2,588.90} = 1 - .6059 = .3941$$

Because a coefficient of determination always lies in the range $0 \leq R^2 \leq 1$, it is often expressed as a *percent of variation explained.* Because the exam score regression yields $R^2 = .3941$, we could say that *X* (hours of study) "explains" 39.41 percent of the variation in *y* (exam scores). On the other hand, 60.59 percent of the variation in exam scores is *not* explained by study time. The *unexplained variation* reflects factors not included in our model (e.g., reading skills, hours of sleep, hours of work at a job, physical health, etc.) or just plain random variation. Although the word "explained" does not necessarily imply causation, in this case we have a *priori* reason to believe that causation exists, that is, that increased study time improves exam scores.

R^2 and r

In a bivariate regression, R^2 is the square of the correlation coefficient *r*. Thus, if $r = .50$, then $R^2 = .25$. For this reason, MegaStat (and some textbooks) denotes the coefficient of determination as r^2 instead of R^2. In this textbook, the uppercase notation R^2 is used to indicate the difference in their definitions. It is tempting to think that a low R^2 indicates that the model is not useful. Yet in some applications (e.g., predicting crude oil future prices), even a slight improvement in predictive power can translate into millions of dollars.

Section Exercises

Instructions for exercises 12.18 and 12.19: (a) Make an Excel worksheet to calculate SS_{xx}, SS_{yy}, and SS_{xy} (the same worksheet you used in exercises 12.2 and 12.3). (b) Use the formulas to calculate the slope and intercept. (c) Use your estimated slope and intercept to make a worksheet to calculate *SSE*, *SSR*, and *SST*. (d) Use these sums to calculate the R^2. (e) To check your answers, make an Excel scatter plot of *X* and *Y*, select the data points, right-click, select Add Trendline, select the Options tab, and choose Display equation on chart and Display R-squared value on chart.

12.18 College Student Weekly Earnings in Dollars ($n = 5$)
📁 WeekPay

Hours Worked (X)	Weekly Pay (Y)
10	93
15	171
20	204
20	156
35	261

12.19 Wait Time for Vehicle Registration in Seconds ($n = 5$)
📁 DMVWait

Clerk Windows Open (X)	Wait Time (Y)
4	385
5	335
6	383
7	344
8	288

Instructions for exercises 12.20–12.22: (a) Use Excel to make a scatter plot of the data. (b) Select the data points, right-click, select Add Trendline, select the Options tab, and choose Display equation on chart and Display R-squared value on chart. (c) Interpret the fitted slope. (d) Is the intercept meaningful? Explain. (e) Interpret the R^2.

12.20 Moviegoer Snack Spending ($n = 10$)
📁 Movies

Age (X)	Spent (Y)
30	6.85
50	10.50
34	5.50
12	10.35
37	10.20
33	10.75
36	7.60
26	10.10
18	12.35
46	8.35

12.21 Annual Percent Return on Mutual Funds ($n = 17$)
📁 Portfolio

Last Year (X)	This Year (Y)
11.9	15.4
19.5	26.7
11.2	18.2
14.1	16.7
14.2	13.2
5.2	16.4
20.7	21.1
11.3	12.0
−1.1	12.1
3.9	7.4
12.9	11.5
12.4	23.0
12.5	12.7
2.7	15.1
8.8	18.7
7.2	9.9
5.9	18.9

12.22 Order Size and Shipping Cost ($n = 12$)
📁 ShipCost

Orders (X)	Ship Cost (Y)
1,068	4,489
1,026	5,611
767	3,290
885	4,113
1,156	4,883
1,146	5,425
892	4,414
938	5,506
769	3,346
677	3,673
1,174	6,542
1,009	5,088

12.5 TESTS FOR SIGNIFICANCE

Standard Error of Regression

A measure of overall fit is the **standard error** of the estimate, denoted s_e:

$$s_e = \sqrt{\frac{\sum_{i=1}^{n}(y_i - \hat{y}_i)^2}{n-2}} = \sqrt{\frac{SSE}{n-2}} \quad \text{(standard error)} \quad \textbf{(12.19)}$$

LO 12-5

Construct confidence intervals and test hypotheses for the slope and intercept.

Recall that SSE is the sum of the squared residuals, $\sum e_i^2$. If the model's predictions are perfect, then the residuals will be zero, or $SSE = 0$, and the standard error s_e will be zero. In general, a smaller value of s_e indicates a better fit. For the exam scores, we can use SSE from Table 12.3 to find s_e:

$$s_e = \sqrt{\frac{SSE}{n-2}} = \sqrt{\frac{1,568.57}{10-2}} = \sqrt{\frac{1,568.57}{8}} = 14.002$$

The standard error s_e is an estimate of σ (the standard deviation of the unobservable errors). Because it measures overall fit, the standard error s_e serves somewhat the same function as the coefficient of determination. However, unlike R^2, the magnitude of s_e depends on the units of measurement of the dependent variable (e.g., dollars, kilograms, ounces) and on the data magnitude. For this reason, R^2 is often the preferred measure of overall fit because its scale is always 0 to 1. The main use of the standard error s_e is to construct confidence intervals.

Confidence Intervals for Slope and Intercept

Once we have the standard error s_e we construct confidence intervals for the coefficients from the formulas shown below.

(12.20) $$s_{b_1} = \frac{s_e}{\sqrt{\sum_{i=1}^{n} (x_i - \bar{x})^2}} \qquad \text{(standard error of slope)}$$

(12.21) $$s_{b_0} = s_e \sqrt{\frac{1}{n} + \frac{\bar{x}^2}{\sum_{i=1}^{n} (x_i - \bar{x})^2}} \qquad \text{(standard error of intercept)}$$

For the exam score data, plugging in the sums from Table 12.2, we get

$$s_{b_1} = \frac{s_e}{\sqrt{\sum_{i=1}^{n} (x_i - \bar{x})^2}} = \frac{14.002}{\sqrt{264.50}} = 0.86095$$

$$s_{b_0} = s_e \sqrt{\frac{1}{n} + \frac{\bar{x}^2}{\sum_{i=1}^{n} (x_i - \bar{x})^2}} = 14.002 \sqrt{\frac{1}{10} + \frac{(10.5)^2}{264.50}} = 10.006$$

These standard errors are used to construct confidence intervals for the true slope and intercept, using Student's t with $d.f. = n - 2$ degrees of freedom and any desired confidence level.

(12.22) $$b_1 - t_{\alpha/2} s_{b_1} \leq \beta_1 \leq b_1 + t_{\alpha/2} s_{b_1} \qquad \text{(CI for true slope)}$$

(12.23) $$b_0 - t_{\alpha/2} s_{b_0} \leq \beta_0 \leq b_0 + t_{\alpha/2} s_{b_0} \qquad \text{(CI for true intercept)}$$

For the exam scores, degrees of freedom are $n - 2 = 10 - 2 = 8$, so from Appendix D we get $t_{.025} = 2.306$ for 95 percent confidence. The 95 percent confidence intervals for the coefficients are

Slope

$$b_1 - t_{.025} s_{b_1} \leq \beta_1 \leq b_1 + t_{.025} s_{b_1}$$
$$1.9641 - (2.306)(0.86095) \leq \beta_1 \leq 1.9641 + (2.306)(0.86095)$$
$$-0.0213 \leq \beta_1 \leq 3.9495$$

Intercept

$$b_0 - t_{.025} s_{b_0} \leq \beta_0 \leq b_0 + t_{.025} s_{b_0}$$
$$49.477 - (2.306)(10.066) \leq \beta_0 \leq 49.477 + (2.306)(10.066)$$
$$26.26 \leq \beta_0 \leq 72.69$$

These confidence intervals are fairly wide. The width of any confidence interval can be reduced by obtaining a larger sample, partly because the t-value would decrease (toward the normal z-value) but mainly because the standard errors decrease as n increases. For the exam scores, the confidence interval for the slope includes zero, suggesting that the true slope could be zero. Some software packages (e.g., Excel and MegaStat) provide confidence intervals automatically, while others do not (e.g., Minitab).

Hypothesis Tests

Is the true slope different from zero? This is an important question because if $\beta_1 = 0$, then X is not associated with Y and the regression model collapses to a constant β_0 plus a random error term:

Proposed Model	*If $\beta_1 = 0$*	*Then*
$y = \beta_0 + \beta_1 x + \varepsilon$	$y = \beta_0 + (0)x + \varepsilon$	$y = \beta_0 + \varepsilon$

For testing either coefficient, we use a t test with $d.f. = n - 2$ degrees of freedom. Usually we are interested in testing whether the parameter is equal to zero, as shown here, but you may substitute another value in place of 0 if you wish. The hypotheses and their test statistics are

Coefficient	*Hypotheses*	*Test Statistic*	
Slope	$H_0: \beta_1 = 0$ $H_1: \beta_1 \neq 0$	$t_{calc} = \dfrac{\text{estimated slope } - \text{ hypothesized slope}}{\text{standard error of the slope}} = \dfrac{b_1 - 0}{s_{b_1}}$	**(12.24)**
Intercept	$H_0: \beta_0 = 0$ $H_1: \beta_0 \neq 0$	$t_{calc} = \dfrac{\text{estimated intercept } - \text{ hypothesized intercept}}{\text{standard error of the intercept}} = \dfrac{b_0 - 0}{s_{b_0}}$	**(12.25)**

The critical value of t is obtained from Appendix D or from Excel's function =T.INV.2T(α, d.f.). Often, the researcher uses a two-tailed test as the starting point because rejection in a two-tailed test always implies rejection in a one-tailed test (but not vice versa). In practice, we rarely test the intercept because its value typically has no meaning within the observed range of the explanatory variable. However, it's important to include the intercept in the model when making predictions.

Slope versus Correlation

The test for zero slope is the same as the test for zero correlation. That is, the t test for zero slope (formula 12.24) will always yield *exactly* the same t_{calc} as the t test for zero correlation (formula 12.4).

Test for Zero Slope: Exam Scores 📄 ExamScores

For the exam scores, we would anticipate a positive slope (i.e., more study hours should improve exam scores), so we will use a right-tailed test.

Step 1: State the Hypotheses

$$H_0: \beta_1 \leq 0$$
$$H_1: \beta_1 > 0$$

Step 2: Specify the Decision Rule   For a right-tailed test with $\alpha = .05$ and $d.f. = 10 - 2 = 8$, $t_{.05} = 1.860$. Our decision rule states:

$$\text{Reject } H_0 \text{ if } t_{calc} > 1.860$$

Step 3: Calculate the Test Statistic   To calculate the test statistic, we use the slope estimate ($b_1 = 1.9641$) and the standard error ($s_{b_1} = 0.86095$) we calculated previously:

$$t_{calc} = \frac{b_1 - 0}{s_{b_1}} = \frac{1.9641 - 0}{0.86095} = 2.281$$

Step 4: Make a Decision   Because $t_{calc} > t_{.05}$ (2.281 > 1.860), we can reject the hypothesis of a zero slope in a right-tailed test. (We would be unable to do so in a two-tailed test because the critical value of our t statistic would be 2.306.) Once we calculate the test statistic for the slope or intercept, we can find the p-value by using Excel's function =T.DIST.RT(2.281, 8) = .025995. Because .025995 < .05, we would reject H_0. We conclude the slope is positive.

Using Excel: Exam Scores 📁 ExamScores

These calculations are normally done by computer (we have demonstrated the calculations only to illustrate the formulas). The Excel menu to accomplish these tasks can be found in Excel's Data Analysis Tool Pak and is shown in Figure 12.10. The resulting output, shown in Figure 12.11, can be used to verify our calculations. Excel always does two-tailed tests, so you must halve the p-value if you need a one-tailed test. You may specify the confidence level, but Excel's default is 95 percent confidence.

Figure 12.10

Excel's Regression Menus

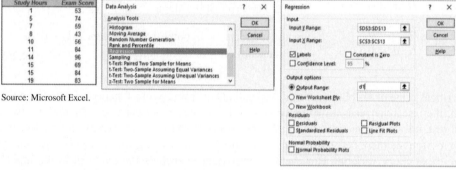

Source: Microsoft Excel.

Source: Microsoft Excel.

Figure 12.11

Excel's Regression Output

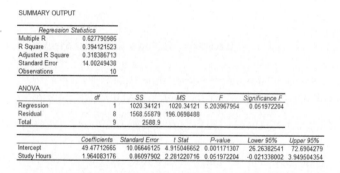

SUMMARY OUTPUT

Regression Statistics	
Multiple R	0.627790986
R Square	0.394121523
Adjusted R Square	0.318386713
Standard Error	14.00249438
Observations	10

ANOVA

	df	SS	MS	F	Significance F
Regression	1	1020.34121	1020.34121	5.203967954	0.051972204
Residual	8	1568.55879	196.0698488		
Total	9	2588.9			

	Coefficients	Standard Error	t Stat	P-value	Lower 95%	Upper 95%
Intercept	49.47712665	10.06646125	4.915046652	0.001171307	26.26382541	72.6904279
Study Hours	1.964083176	0.86097902	2.281220716	0.051972204	-0.021338002	3.949504354

Constant Is Zero?

Avoid checking the Constant is Zero box in Excel's menu. This would force the intercept through the origin, changing the model drastically. Leave this option to the experts.

Application: Retail Sales 📁 RetailSales

Is there a positive association between gross leasable area (X) and retail sales (Y) in shopping malls? We will assume a linear relationship between X and Y:

$$Sales = \beta_0 + \beta_1 Area + \varepsilon$$

We anticipate a positive slope (more leasable area permits more retail sales) and an intercept near zero (zero leasable space would imply no retail sales). Because retail sales do not depend solely on leasable area, the random error term will reflect all other factors that influence retail sales as well as possible measurement error. The regression line is estimated using a sample of $n = 24$ shopping malls randomly chosen from 24 different U.S. states.

Based on the scatter plot and Excel's fitted linear regression, displayed in Figure 12.12, the linear model seems justified. The very high R^2 says that *Area* "explains" about 98 percent of the variation in *Sales*. Although it is reasonable to assume causation between *Area* and *Sales* in this model, the high R^2 alone does not prove cause and effect.

For a more detailed look, we examine the regression output for these data, shown in Figure 12.13. On average, each extra million square feet of leasable space yields an extra $259.4 billion in retail sales ($b_1 = .2590$). The slope is nonzero in the two-tail test ($t = 30.972$), as indicated by its tiny p-value (1.22×10^{-19}). The yellow highlight indicates that the slope differs significantly from zero at $\alpha = .01$, and the narrow confidence interval for the slope [0.2417 to 0.2764] does not enclose zero. We conclude that this sample result (nonzero slope) did not arise by chance—rarely will you see such small p-values (except perhaps in time series data). But the intercept ($b_0 = 0.3852$) does not differ significantly from zero (p-value = .8479, $t = 0.194$), and the confidence interval for the intercept [−3.7320, 4.5023] includes zero. These conclusions are in line with our prior expectations.

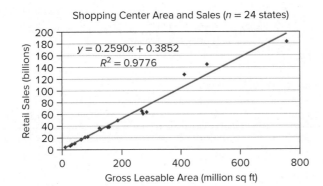

Shopping Center Area and Sales ($n = 24$ states)

$y = 0.2590x + 0.3852$
$R^2 = 0.9776$

Retail Sales (billions)
Gross Leasable Area (million sq ft)

Figure 12.12

Leasable Area and Retail Sales RetailSales

Figure 12.13

Regression Results for Retail Sales RetailSales

| Regression Output | | | | | Confidence Interval | |
Variables	Coefficients	Std. Error	t Stat	p-Value	Lower 95%	Upper 95%
Intercept	0.3852	1.9853	0.194	.8479	−3.7320	4.5023
Area	0.2590	0.0084	30.972	1.22E-19	0.2417	0.2764

Tip

The test for zero slope always yields a t statistic that is identical to the test for zero correlation coefficient. Therefore, it is not necessary to do both tests. Because regression output always includes a t test for the slope, that is the test we usually use.

Mini Case 12.2

Does Per Person Spending Predict Weekly Sales? NoodlesRevenue

Can Noodles & Company predict its average weekly sales at a restaurant from the average amount a person spends when visiting its restaurant? A random sample of data from 74 restaurants was used to answer this question. The scatter plot in Figure 12.14 shows the relationship between *average spending per person* and *average weekly sales*.

The scatter plot shows almost no relationship between the two variables. This observation is also supported by the regression results shown in Figure 12.15.

Figure 12.14

Noodles Weekly Sales

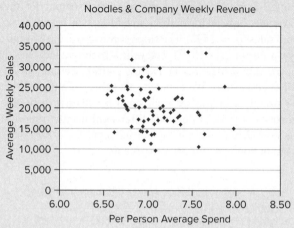

Noodles & Company Weekly Revenue

Figure 12.15

Regression Output for Weekly Revenue

Regression Output					Confidence Interval	
Variables	Coefficients	Std. Error	t Stat	p-Value	Lower 95%	Upper 95%
Intercept	32,710.56	14,987.28	2.183	.0323	2,833.97	62,587.15
Per Person Spend	−1,751.45	2,125.68	−0.824	.4127	−5,988.92	2,486.02

The regression results show that b_1, the estimate for the slope β_1, is −$1,751.45. It would appear that, on average, for each additional dollar an individual spends in a restaurant, weekly sales would decrease by $1,751.45—seemingly a large number. But notice that the confidence interval for the slope [−$5,988.92, $2,486.02] contains zero and the standard error ($2,125.68) is larger than the estimated coefficient. When we perform a two-tailed test for zero slope with $H_0: \beta_1 = 0$ and $H_1: \beta_1 \neq 0$, the p-value for this test is .4127. Because this p-value is much greater than any value of α we might choose, we *fail to reject* the null hypothesis of zero slope. Both the hypothesis for zero slope and the confidence interval show that the slope is not significantly different from zero. Our conclusion is that average weekly sales *should not be* predicted by per person average spending. Based on the information we have here, our best prediction of average weekly sales at a Noodles & Company restaurant is simply the mean ($\bar{y} = \$20,373$).

Section Exercises

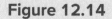

Instructions for exercises 12.23 and 12.24: (a) Perform a regression using MegaStat or Excel. (b) State the null and alternative hypotheses for a two-tailed test for a zero slope. (c) Report the p-value and the 95 percent confidence interval for the slope shown in the regression results. (d) Is the slope significantly different from zero? Explain your conclusion.

12.23 College Student Weekly Earnings in Dollars ($n = 5$) 📄 WeekPay

Hours Worked (X)	Weekly Pay (Y)
10	93
15	171
20	204
20	156
35	261

12.24 Wait Time for Vehicle Registration in Seconds ($n = 5$) 📄 DMVWait

Clerk Windows Open (X)	Wait Time (Y)
4	385
5	335
6	383
7	344
8	28

12.25 A regression was performed using data on 32 NFL teams. The variables were Y = current value of team (millions of dollars) and X = total debt held by the team owners (millions of dollars). (a) Write the fitted regression equation. (b) Construct a 95 percent confidence interval for the slope. (c) Perform a right-tailed t test for zero slope at $\alpha = .05$. State the hypotheses clearly. (d) Use Excel to find the p-value for the t statistic for the slope. 📧 **NFL**

Variables	Coefficients	Std. Error
Intercept	557.4511	25.3385
Debt	3.0047	0.8820

12.26 A regression was performed using data on 16 randomly selected charities. The variables were Y = expenses (millions of dollars) and X = revenue (millions of dollars). (a) Write the fitted regression equation. (b) Construct a 95 percent confidence interval for the slope. (c) Perform a right-tailed t test for zero slope at $\alpha = .05$. State the hypotheses clearly. (d) Use Excel to find the p-value for the t statistic for the slope. 📧 **Charities**

Variables	Coefficients	Std. Error
Intercept	7.6425	10.0403
Revenue	0.9467	0.0936

12.6 ANALYSIS OF VARIANCE: OVERALL FIT

Decomposition of Variance

A regression seeks to explain variation in the dependent variable around its mean. A simple way to see this is to express the deviation of y_i from its mean $\bar{y}$ as the sum of the deviation of y_i from the regression estimate $\hat{y}_i$ plus the deviation of the regression estimate $\hat{y}_i$ from the mean $\bar{y}$:

LO 12-6

Interpret the ANOVA table and use it to calculate F, R^2, and standard error.

$$y_i - \bar{y} = (y_i - \hat{y}_i) + (\hat{y}_i - \bar{y}) \quad \text{(adding and subtracting } \hat{y}_i) \tag{12.26}$$

It can be shown (see Kutner et al. in Related Reading) that this same decomposition also holds for the *sums of squares:*

$$\sum_{i=1}^{n}(y_i - \bar{y})^2 = \sum_{i=1}^{n}(y_i - \hat{y}_i)^2 + \sum_{i=1}^{n}(\hat{y}_i - \bar{y})^2 \quad \text{(sums of squares)} \tag{12.27}$$

As we have already seen, this *decomposition of variance* may be written as

SST	$=$	SSE	$+$	SSR
(*total* variation around the mean)		(unexplained or *error* variation)		(variation explained by the *regression*)

F Statistic for Overall Fit

To test a regression for overall significance, we use an F test to compare the explained (SSR) and unexplained (SSE) sums of squares. We divide each sum by its respective degrees of freedom to obtain *mean squares* (*MSR* and *MSE*). The *F statistic* is the ratio of these two mean squares. Calculations of the F statistic are arranged in a table called the *analysis of variance* or ANOVA table (see Table 12.4).

Table 12.4

ANOVA Table for a Simple Regression

Source of Variation	Sum of Squares	df	Mean Square	F	Excel p-Value
Regression (explained)	$SSR = \sum_{i=1}^{n}(\hat{y}_i - \bar{y})^2$	1	$MSR = \dfrac{SSR}{1}$	$F_{calc} = \dfrac{MSR}{MSE}$	=F.DIST.RT(F_{calc}, 1, n−2)
Residual (unexplained)	$SSE = \sum_{i=1}^{n}(y_i - \hat{y}_i)^2$	n − 2	$MSE = \dfrac{SSE}{n-2}$		
Total	$SST = \sum_{i=1}^{n}(y_i - \bar{y})^2$	n − 1			

The ANOVA table also contains the sums required to calculate $R^2 = SSR/SST$. An ANOVA table is provided automatically by any regression software (e.g., Excel, MegaStat). The formula for the F test statistic is

$$\text{(12.28)} \qquad F_{calc} = \frac{MSR}{MSE} = \frac{SSR/1}{SSE/(n-2)} = (n-2)\frac{SSR}{SSE} \qquad (F \text{ statistic for simple regression})$$

The F statistic reflects both the sample size and the ratio of SSR to SSE. For a given sample size, a larger F statistic indicates a better fit (larger SSR relative to SSE), while F close to zero indicates a poor fit (small SSR relative to SSE). The F statistic must be compared with a critical value $F_{1,\,n-2}$ from Appendix F for whatever level of significance is desired, and we can find the p-value by using Excel's function =F.DIST(F_{calc}, 1, n−2). Software packages provide the p-value automatically.

Example 12.2

Exam Scores: F *Statistic*

📄 **ExamScores**

Figure 12.16 shows an ANOVA table for the exam scores. The F statistic is

$$F_{calc} = \frac{MSR}{MSE} = \frac{1020.3412}{196.0698} = 5.20$$

From Appendix F, the critical value of $F_{1,8}$ at the 5 percent level of significance would be 5.32, so the exam score regression is not quite significant at $\alpha = .05$. The p-value of .052 says a sample such as ours would be expected about 52 times in 1,000 samples if X and Y were unrelated. In other words, if we reject the hypothesis of no relationship between X and Y, we face a Type I error risk of 5.2 percent. This p-value might be called *marginally significant.*

Figure 12.16   ANOVA Table for Exam Data

ANOVA table

Source	SS	df	MS	F	p-Value
Regression	1,020.3412	1	1,020.3412	5.20	.0520
Residual	1,568.5588	8	196.0698		
Total	2,588.9000	9			

From the ANOVA table, we can calculate the standard error from the mean square for the residuals:

$$s_e = \sqrt{MSE} = \sqrt{196.0698} = 14.002 \quad \text{(standard error for exam scores)}$$

F Test p-Value and t Test p-Value

In a simple regression, the F test always yields the same p-value as a two-tailed t test for zero slope, which in turn always gives the same p-value as a two-tailed test for zero correlation. The relationship between the test statistics is $F_{calc} = t^2_{calc}$.

Section Exercises

connect

12.27 Below is a regression using X = home price (000), y = annual taxes (000), n = 12 homes. (a) Write the fitted regression equation. (b) Write the formula for each t statistic and verify the t statistics shown below. (c) State the degrees of freedom for the t tests and find the two-tail critical value for t using Excel. (d) Use Excel's function =T.DIST.2T(t, d.f.) to verify the p-value shown for each t statistic (slope, intercept). (e) Verify that $F = t^2$ for the slope. (f) In your own words, describe the fit of this regression.

R²	0.452
Std. Error	0.454
n	12

ANOVA table

Source	SS	df	MS	F	p-Value
Regression	1.6941	1	1.6941	8.23	.0167
Residual	2.0578	10	0.2058		
Total	3.7519	11			

Regression output

Variables	Coefficients	Std. Error	t Stat	p-Value	Lower 95%	Upper 95%
Intercept	1.8064	0.6116	2.954	.0144	0.4438	3.1691
Slope	0.0039	0.0014	2.869	.0167	0.0009	0.0070

12.28 Below is a regression using X = average price, Y = units sold, n = 20 stores. (a) Write the fitted regression equation. (b) Write the formula for each t statistic and verify the t statistics shown below. (c) State the degrees of freedom for the t tests and find the two-tail critical value for t using Excel. (d) Use Excel's function =T.DIST.2T(t, d.f.) to verify the p-value shown for each t statistic (slope, intercept). (e) Verify that $F = t^2$ for the slope. (f) In your own words, describe the fit of this regression.

R²	0.200
Std. Error	26.128
n	20

ANOVA table

Source	SS	df	MS	F	p-Value
Regression	3,080.89	1	3,080.89	4.51	.0478
Residual	12,288.31	18	682.68		
Total	15,369.20	19			

Regression output

Variables	Coefficients	Std. Error	t Stat	p-Value	Lower 95%	Upper 95%
Intercept	614.9300	51.2343	12.002	.0000	507.2908	722.5692
Slope	−109.1120	51.3623	−2.124	.0478	−217.0202	−1.2038

Instructions for exercises 12.29–12.31: (a) Use Excel's Data Analysis > Regression (or MegaStat or Minitab) to obtain regression estimates. (b) Interpret the 95 percent confidence interval for the slope. Does it contain zero? (c) Interpret the *t* test for the slope and its *p*-value. (d) Interpret the *F* statistic. (e) Verify that the *p*-value for *F* is the same as for the slope's *t* statistic, and show that $r^2 = F$. (f) Describe the fit of the regression.

12.29   Moviegoer Snack Spending ($n = 10$) Movies

Age (X)	Spent (Y)	Age (X)	Spent (Y)
30	6.85	33	10.75
50	10.50	36	7.60
34	5.50	26	10.10
12	10.35	18	12.35
37	10.20	46	8.35

12.30   Annual Percent Return on Mutual Funds ($n = 17$) Portfolio

Last Year (X)	This Year (Y)
11.9	15.4
19.5	26.7
11.2	18.2
14.1	16.7
14.2	13.2
5.2	16.4
20.7	21.1
11.3	12.0
−1.1	12.1
3.9	7.4
12.9	11.5
12.4	23.0
12.5	12.7
2.7	15.1
8.8	18.7
7.2	9.9
5.9	18.9

12.31   Order Size and Shipping Cost ($n = 12$) ShipCost

Orders (X)	Ship Cost (Y)
1,068	4,489
1,026	5,611
767	3,290
885	4,113
1,156	4,883
1,146	5,425
892	4,414
938	5,506
769	3,346
677	3,673
1,174	6,542
1,009	5,088

Mini Case 12.3

Airplane Cockpit Noise Cockpit

Career airline pilots face the risk of progressive hearing loss due to the noisy cockpits of most jet aircraft. Much of the noise comes not from engines but from air roar, which increases at high speeds. To assess this workplace hazard, a pilot measured cockpit noise at randomly selected points during the flight by using a handheld meter. Noise level (in decibels) was measured in seven different aircraft at the first officer's left ear position using a handheld meter. For reference, 60 dB is a normal conversation, 75 is a typical vacuum cleaner, 85 is city traffic, 90 is a typical hair dryer, and 110 is a chain saw. Figure 12.17 shows 61 observations on cockpit noise (decibels) and airspeed (knots indicated air speed, KIAS) for a Boeing 727, an older type of aircraft lacking design improvements in newer planes.

The scatter plot in Figure 12.17 suggests that a linear model provides a reasonable description of the data. The fitted regression shows that each additional knot of airspeed

Figure 12.17   Scatter Plot of Cockpit Noise Data Cockpit

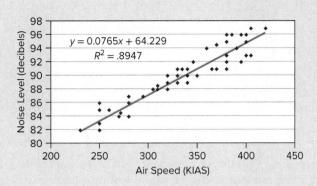

increases the noise level by 0.0765 dB. Thus, a 100-knot increase in airspeed would add about 7.65 dB of noise. The intercept of 64.229 suggests that if the plane were not flying ($KIAS = 0$), the noise level would be only slightly greater than a normal conversation.

The regression results in Figure 12.18 show that the fit is very good ($R^2 = .895$) and that the regression is highly significant ($F = 501.16$, p-value $< .001$). Both the slope and intercept have p-values below .001, indicating that the true parameters are nonzero. Thus, the regression is significant, as well as having practical value.

Figure 12.18   Regression Results of Cockpit Noise

Regression Analysis

r^2	0.895	n	61
r	0.946	k	1
Std. Error	1.292	Dep. Var.	Noise

ANOVA table

Source	SS	df	MS	F	p-Value
Regression	836.9817	1	836.9817	501.16	1.60E-30
Residual	98.5347	59	1.6701		
Total	935.5164	60			

Regression output · confidence interval

Variables	Coefficients	Std. Error	t Stat	p-Value	Lower 95%	Upper 95%
Intercept	64.2294	1.1489	55.907	8.29E-53	61.9306	66.5283
Speed	0.0765	0.0034	22.387	1.60E-30	0.0697	0.0834

12.7   CONFIDENCE AND PREDICTION INTERVALS FOR Y

The regression line is an estimate of the *conditional mean* of Y, that is, the expected value of Y for a given value of X, denoted $E(Y \mid x_i)$. Because the *point estimate* is most likely not equal to the true value, we need an *interval estimate* to show a range of likely values. To do this, we insert the x_i value into the fitted regression equation, calculate the estimated $\hat{y}_i$ and use the formulas shown below. The first formula gives a **confidence interval** for the conditional mean of Y, while the second is a **prediction interval** for individual values of Y. The formulas

LO 12-7

Distinguish between confidence and prediction intervals for Y.

are similar, except that prediction intervals are wider because *individual Y* values vary more than the *mean* of *Y*.

(12.29)
$$\hat{y}_i \pm t_{\alpha/2} s_e \sqrt{\frac{1}{n} + \frac{(x_i - \bar{x})^2}{\sum_{i=1}^{n}(x_i - \bar{x})^2}} \qquad \text{(confidence interval for mean of } Y\text{)}$$

(12.30)
$$\hat{y}_i \pm t_{\alpha/2} s_e \sqrt{1 + \frac{1}{n} + \frac{(x_i - \bar{x})^2}{\sum_{i=1}^{n}(x_i - \bar{x})^2}} \qquad \text{(prediction interval for individual } Y\text{)}$$

Interval width depends on the value of x_i. Intervals will be narrower when x_i is near its mean. (Note that if $x_i = \bar{x}$, the last term under the square root disappears completely.) Conversely, predictions based on x_i values far from $\bar{x}$ will be less precise.

To understand the difference between these two formulas, let's first calculate the 95 percent confidence interval for the *expected* exam score (i.e., the mean score) of students who study 4 hours. Using the regression model from Section 12.4, the predicted exam score would be $\hat{y} = 1.9641(4) + 49.477 = 57.333$. For 95 percent confidence with $d.f. = n - 2 = 10 - 2 = 8$, we use $t_{.025} = 2.306$. Using formula 12.29 and the sums from Table 12.3, the 95 percent confidence interval is

$$57.333 \pm (2.306)(14.002)\sqrt{\frac{1}{10} + \frac{(4 - 10.5)^2}{264.5}} \quad \text{or} \quad 57.33 \pm 16.46$$

The *average* exam score for students who study 4 hours is estimated to be between 40.87 and 73.79.

How does this result differ from the 95 percent *prediction* interval for the exam score of an individual student who studies 4 hours? Using the same $\hat{y} = 57.333$ we calculated above and formula 12.30, the 95 percent prediction interval is

$$57.333 \pm (2.306)(14.002)\sqrt{1 + \frac{1}{10} + \frac{(4 - 10.5)^2}{264.5}} \quad \text{or} \quad 57.33 \pm 36.24$$

As expected, the margin of error for the *predicted* individual score is much greater than the margin of error for the estimated *average* score when study hours is 4 hours: 36.24 versus 16.46. Given that the fit for the exam score data was not very high ($R^2 = .3941$), it is not surprising that the prediction interval is wide. Prediction intervals are more precise when R^2 is high.

Two Illustrations: Exam Scores and Retail Sales   ExamScores   RetailSales

Because there will be a different interval for every X value, it is helpful to display confidence and prediction intervals over the entire range of X (an option on many software packages). Figure 12.19 shows confidence and prediction intervals for exam scores and retail sales. The contrast between the two graphs is striking. Confidence and prediction intervals for exam scores are wide but become narrower for X values near the mean. The prediction bands for exam scores for large X values (e.g., $x = 20$ hours of study) even extend above 100 points (presumably the upper limit for an exam score). In contrast, the intervals for retail sales appear narrow and only slightly wider for X values below or above the mean. While the prediction bands for retail sales seem narrow, they still represent billions of dollars (e.g., for $x = 500$, the retail sales prediction interval has a width of about \$33 billion). This shows that a high R^2 does not guarantee precise predictions.

Figure 12.19 Confidence and Prediction Intervals Illustrated

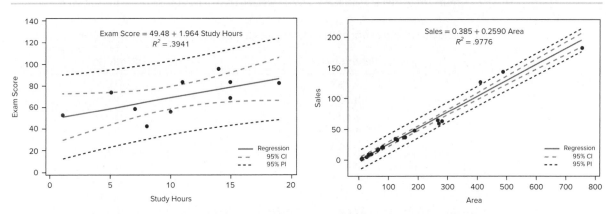

Analytics in Action

Predictive Maintenance and Machine Learning

The ideal time to perform a maintenance task (e.g., replace a part, adjust equipment settings, or add coolant) is just *before* equipment failure. Regression models are used to predict when a failure is about to happen, which allows preventive maintenance rather than reactionary maintenance. But choosing the best regression model can be difficult when there is a large quantity of operating data.

Machine learning can help. Equipment sensors are used to automatically track many different operating characteristics of mechanical and electrical components. Different regression models are analyzed, and each model's error is recorded. The model(s) with the lowest prediction error are used to predict how many operating cycles remain before a component will fail. Process engineers and maintenance crews can monitor the output in real time to know when they should shut down a piece of equipment and perform a preventive maintenance task. This type of predictive maintenance is used in the airline industry, NASA, and a multitude of manufacturing applications. Companies save millions of dollars by avoiding costly equipment breakdowns and reducing maintenance activities.

Section Exercises

Mc Graw Hill connect

12.32 Refer to the Weekly Earnings data set below. (a) Calculate confidence and prediction intervals for *Y* using the following set of *x* values: 12, 17, 21, 25, and 30. (b) Report the 95 percent confidence interval and prediction interval for $x = 17$. (c) Calculate the 95 percent confidence interval for μ_y using the appropriate method from Chapter 8. (d) Compare the result from part (c) to the confidence interval you reported in part (b). How are they different?

College Student Weekly Earnings ($n = 5$) 📄 WeekPay

Hours Worked (*X*)	Weekly Pay (*Y*)
10	93
15	171
20	204
20	156
35	261

12.33 Refer to the Revenue and Profit data set below. Data are in billions of dollars. (a) Calculate confidence and prediction intervals for *Y* using the following set of *x* values: 1.8, 15, and 30. (b) Report the 95 percent confidence interval and prediction interval for $x = 15$. (c) Calculate the 95 percent confidence interval for μ_y using the appropriate method from Chapter 8. (d) Compare the result from part (c) to the confidence interval you reported in part (b). How are they different?

Revenue and Profit of Entertainment Companies ($n = 9$) 🖫 Entertainment	
Revenue (X)	**Profit (Y)**
1.792	−0.020
8.931	1.146
2.446	−0.978
1.883	−0.162
2.490	0.185
43.877	2.639
1.311	0.155
26.585	1.417
27.061	1.267

12.8 RESIDUAL TESTS

Three Important Assumptions

LO 12-8

Calculate residuals and perform tests of regression assumptions.

Recall that the dependent variable is a random variable that has an error component, ε. In Section 12.3 we discussed three assumptions that the OLS method makes about the random error term ε. The regression assumptions are restated here:

- Assumption 1: The errors are *normally* distributed.
- Assumption 2: The errors have *constant* variance.
- Assumption 3: The errors are *independent*.

Because we cannot observe the error ε, we must rely on the residuals $e_1, e_2, \ldots, e_n$ from the estimated regression for clues about possible violations of these assumptions. While formal tests exist for identifying assumption violations, many analysts rely on simple visual tools to help them determine when an assumption has not been met and how serious the violation is.

In this chapter we will discuss the consequences of violating each assumption and explain the visual tools used for examining the residuals. In Chapter 13 we will discuss in more detail how to remedy an assumption violation and demonstrate a more formal method for examining assumption 3.

Violation of Assumption 1: Non-normal Errors

The main consequence of non-normality of errors is that confidence intervals for the parameters may be untrustworthy because the normality assumption is used to justify using Student's t to construct confidence intervals. If the sample size is large (say, $n > 30$), the confidence intervals should still be informative. An exception would be if extreme outliers exist, posing a serious estimation problem. Non-normality of errors is usually considered a mild violation because the regression parameter estimates b_0 and b_1 and their variances remain unbiased and consistent. The hypotheses are

H_0: Errors are normally distributed

H_1: Errors are not normally distributed

A simple way to check for non-normality is to make a histogram of the residuals. You can use either plain residuals or **standardized residuals**. A *standardized residual* is obtained by dividing each residual by its standard error. Histogram shapes will be the same, but standardized residuals offer the advantage of a predictable scale (between −3 and +3 unless there are outliers). A simple "eyeball test" can usually reveal outliers or serious asymmetry. Figure 12.20 shows a standardized residual histogram for Mini Case 12.3 (cockpit noise). There are no outliers and the histogram is roughly symmetric, albeit possibly platykurtic (i.e., flatter than normal).

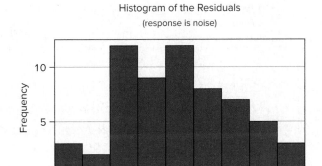

Figure 12.20

Cockpit Noise Residuals (Histogram) Cockpit

Another visual test for normality is the **normal probability plot** (also known as a *quantile-quantile plot* or simply **q-q plot**). It plots observed cumulative data values against those that would be expected under the assumption of normality. It is produced as an option by many software regression tools. Ideally, the residual normal probability plot should be linear. For example, in Figure 12.21 we see only slight deviations from linearity at the lower and upper ends of the residual probability plot. These residuals seem reasonably consistent with the null hypothesis of normality. There are more precise tests for normality, but the histogram and normal probability plot suffice for most purposes.

Normal Probability Plot of the Residuals
(response is noise)

Figure 12.21

Cockpit Noise Residuals (Normal Probability Plot)

Normality Assumption Violation

Non-normality is not considered a major violation because the parameter estimates remain unbiased and consistent. Don't worry too much about it *unless* you spot major outliers.

Violation of Assumption 2: Nonconstant Variance

The regression should fit equally well for all values of X. If the error magnitude is constant for all X, the errors are **homoscedastic** (the ideal condition). If the error magnitudes increase or decrease as X changes, they are **heteroscedastic**. Although the OLS regression parameter estimates b_0 and b_1 are still unbiased and consistent, their estimated variances are biased and are neither efficient nor asymptotically efficient. In the most common form of heteroscedasticity, the variances of the estimators are likely to be understated, resulting in overstated t statistics and artificially narrow confidence intervals. Your regression results may thus seem more significant than is warranted.

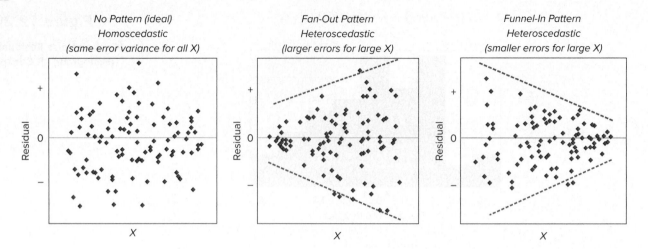

For a simple regression, you can see heteroscedasticity on the *XY* scatter plot, but a more general visual test is to plot the residuals against *X* or against $\hat{Y}$. Ideally, there is no pattern in the residuals as we move along the horizontal axis, as shown in the illustration above.

Although many patterns of nonconstant variance might exist, the "fan-out" pattern (increasing residual variance) is most common. Less frequently, we might see a "funnel-in" pattern, which shows decreasing residual variance. The residuals *always* have a mean of zero, whether the residuals exhibit homoscedasticity or heteroscedasticity.

Residual plots provide a fairly sensitive "eyeball test" for heteroscedasticity. The residual plot is therefore considered an important tool in the statistician's diagnostic kit. The hypotheses are

H_0: Errors have constant variance (homoscedastic)

H_1: Errors have nonconstant variance (heteroscedastic)

Figure 12.22 shows a residual plot for Mini Case 12.3 (cockpit noise). In this residual plot, the magnitude of the residuals does not increase or decrease as we look from left to right. A pattern like this is consistent with the null hypothesis of homoscedasticity (constant variance). When using visual tools, we are looking for *obvious* departures from the assumptions of normality and constant variance. Consider the overall shape of the plot and don't focus on one or two residuals that might be considered outliers.

Figure 12.22

Cockpit Noise Residual Plot (Homoscedastic)

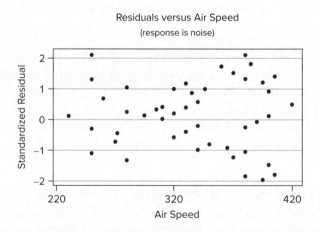

Figure 12.23 shows a heteroscedastic residual plot from a regression of home prices versus square feet. The magnitude of the residuals increases as the home size increases. That is, the model predicts better for small homes than for large homes. A pattern like this is consistent with heteroscedasticity (nonconstant variance). In Chapter 13 you will learn how we might remedy the violation of constant error variance.

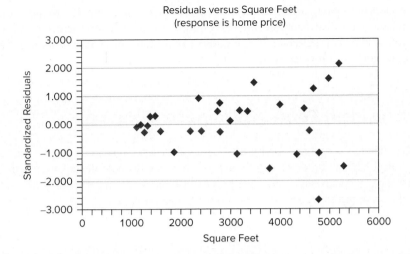

Figure 12.23

**Home Price Residual Plot
(Heteroscedastic)**

Homoscedastic Errors Assumption Violation

Although it can widen the confidence intervals for the coefficients, heteroscedasticity does not bias the estimates. At this stage of your training, it is sufficient just to recognize its existence.

Violation of Assumption 3: Autocorrelated Errors

Autocorrelation is a pattern of nonindependent errors, mainly found in time-series data. When variable observations are collected in a specific order such as in a time series, it is possible that the errors $e_1, e_2, \ldots, e_n$ are related to each other. In a regression, each residual e_t should be independent of its predecessors $e_{t-1}, e_{t-2}, \ldots, e_1$. Violations of this assumption can show up in different ways. In the simple model of *first-order autocorrelation,* we would find that *et* is correlated with the prior residual et_{-1}. The OLS estimators b_0 and b_1 are still unbiased and consistent, but their estimated variances are biased in a way that typically leads to confidence intervals that are too narrow and *t* statistics that are too large. Thus, the model's fit may be overstated.

Positive autocorrelation is indicated by runs of residuals with the *same* sign, while *negative autocorrelation* is indicated by runs of residuals with *alternating* signs. Such patterns can sometimes be seen in a plot of the residuals against the order of data entry. In the *runs test,* we count the number of sign reversals (i.e., how often does the residual plot cross the zero centerline?). If the pattern is random, the number of sign changes should be approximately $n/2$. Fewer than $n/2$ centerline crossings would suggest positive autocorrelation, while more than $n/2$ centerline crossings would suggest negative autocorrelation. For example, if $n = 50$, we would expect about 25 centerline crossings. In the first illustration, there are only 11 crossings

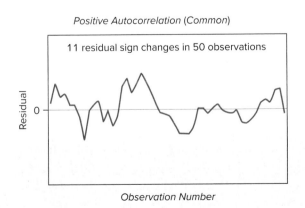

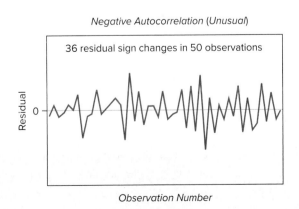

(positive autocorrelation), while in the second illustration there are 36 crossings (negative autocorrelation). Positive autocorrelation is common in economic time-series regressions due to the cyclical nature of the economy. It is harder to envision logical reasons for negative auto-correlation, and in fact it is rarely observed.

Independent Errors Assumption Violation

Autocorrelation is a concern with time series data but rarely in cross-sectional data, where data have no time order. Although it can widen the confidence intervals for the coefficients, autocorrelation does not bias the estimates. At this stage of your training, it is sufficient just to recognize when you have autocorrelation.

Mini Case 12.4

Exports and Imports ⬚ Exports

We often see headlines about the persistent imbalance in U.S. foreign trade. Yet when U.S. imports increase, other nations acquire dollar balances that should eventually lead to increased purchases of U.S. goods and services, thereby increasing U.S. exports (i.e., trade imbalances are supposed to be self-correcting). What do the data tell us? Figure 12.24 shows a regression of the percent change in U.S. exports (Y) against the percent change in U.S. imports (X) for the period 1960–2018. Percent changes are used instead of dollars in order to reduce heteroscedasticity (changing data magnitude) and autocorrelation (time-series data).

Figure 12.24

Excel Scatter Plot and Regression

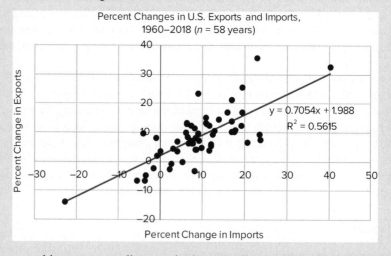

Exports and imports generally move in the same direction (mostly growing), but the slope (0.7054) indicates that exports have grown less than imports. Why? An econo-mist would examine other factors such as tariffs, domestic labor costs, trade agreements, and sources of exchange rate rigidity. Simple regression would not suffice. What about possible violations of OLS assumptions? In Figure 12.25 the residual histogram is not perfectly bell-shaped, yet the probability plot of residuals is roughly linear with no extreme outliers (normality). There is no pattern in the plot of residuals against fitted

Y (homoscedastic), but the residual runs plot (in order of time) has only 15 centerline crossings (sign changes). We would expect about $n/2 = 58/2 = 29$ centerline crossings (suggesting positive autocorrelation).

Figure 12.25

Residual Plots from Minitab

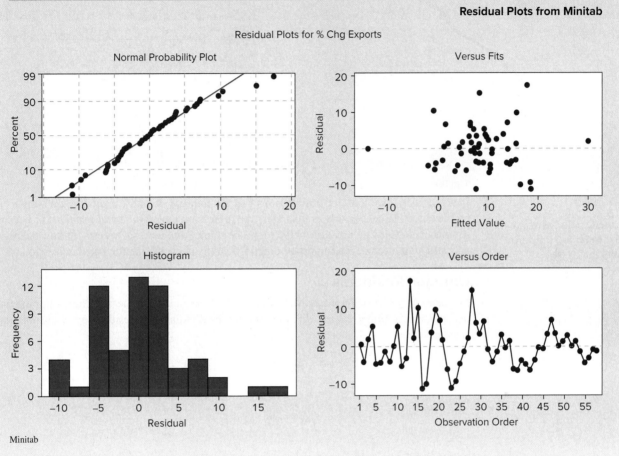

Minitab

12.34 Review the two residual plots below. Do either of these show evidence that the regression error assumptions of normality and constant variation have been violated? Explain.

Section Exercises

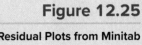

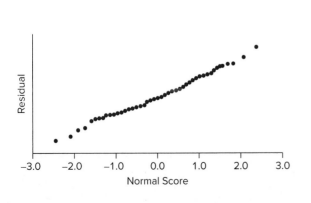

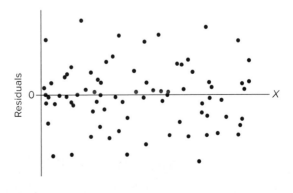

12.35 Review the two residual plots below. Do either of these show evidence that the regression error assumptions of normality and constant variation have been violated? Explain.

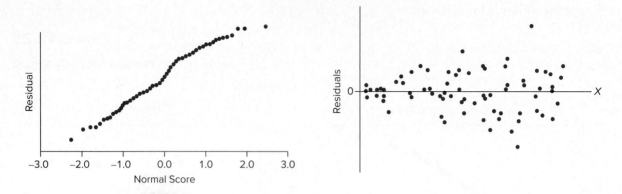

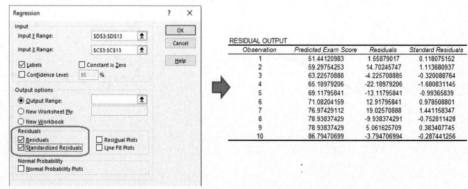

12.9 UNUSUAL OBSERVATIONS

LO 12-9

Identify unusual residuals and tell when they are outliers.

In a regression, we look for observations that are unusual. An observation could be unusual because its *Y*-value is poorly predicted by the regression model (*unusual residual*) or because its unusual *X*-value greatly affects the regression line (*high leverage*). Tests for unusual residuals and high leverage are important diagnostic tools in evaluating the fitted regression.

Unusual Residuals

Because every regression may have different *Y* units (e.g., stock price in dollars, shipping time in days), it is helpful to *standardize* the residuals by dividing each residual, e_i, by its individual standard error, s_{e_i}.

(12.31) $$e_i^* = \frac{e_i}{s_{e_i}}$$ (standardized residual for observation *i*)

where

$$s_{e_i} = s_e \sqrt{1 - h_i} \quad \text{and} \quad h_i = \frac{1}{n} + \frac{(x_i - \overline{x})^2}{\sum(x_i - \overline{x})^2}$$

Notice that this calculation requires a unique adjustment for each residual based on the observation's distance from the mean. We will refer to this value e_i^* as a *standardized residual.* An equivalent name for this value is **studentized residual**, which is used by many software packages.

Using the Empirical Rule as a rule of thumb, any standardized residual whose absolute value is 2 or more is unusual, and any residual whose absolute value is 3 or more would be considered an outlier.

Excel's Data Analysis > Regression provides residuals as an option, as shown in Figure 12.26. Excel calculates its "standardized residuals" by dividing each residual by the standard

Figure 12.26

Excel's Exam Score Residuals

RESIDUAL OUTPUT

Observation	Predicted Exam Score	Residuals	Standard Residuals
1	51.44120983	1.55879017	0.118075152
2	59.29754253	14.70245747	1.113680937
3	63.22570888	-4.225708885	-0.320088764
4	65.18979206	-22.18979206	-1.680831145
5	69.11795841	-13.11795841	-0.99365839
6	71.08204159	12.91795841	0.978508801
7	76.97429112	19.02570888	1.441158347
8	78.93837429	-9.938374291	-0.752811428
9	78.93837429	5.061625709	0.383407745
10	86.79470699	-3.794706994	-0.287441256

Source: Microsoft Excel.

deviation of the column of residuals. This procedure is not quite correct but generally suffices to identify unusual residuals. Using the Empirical Rule, there are no unusual residuals in Figure 12.26.

High Leverage

A high **leverage** statistic indicates that the observation is far from the mean of X. Such observations have great influence on the regression estimates because they are at the "end of the lever." Figure 12.27 illustrates this concept. One individual worked 65 hours, while the others worked between 12 and 42 hours. This individual will have a big effect on the slope estimate because he is so far above the mean of X.

The leverage for observation i is denoted h_i and is calculated as

$$h_i = \frac{1}{n} + \frac{(x_i - \overline{x})^2}{\sum\limits_{i=1}^{n}(x_i - \overline{x})^2}$$

(12.32)

LO 12-10

Define leverage and identify high-leverage observations.

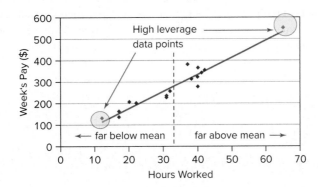

Figure 12.27

Illustration of High Leverage 🖥 **Leverage**

High Leverage

As a rule of thumb for a simple regression, a leverage statistic that exceeds $4/n$ is unusual (if $x_i = \overline{x}$, the leverage statistic h_i is $1/n$, so the rule of thumb is just four times this value).

What to Do about Unusual Observations?

Unless they are erroneous, keep all the data points in the analysis and note them in the report. If these data pairs were omitted, the regression results might change noticeably. Valid observations are not to be discarded except in extreme circumstances. Merely note that they are unusual, comment on whether they are influential, and leave them be.

We see from Figure 12.28 that two data points (Tom and Courtney) are likely to have high leverage because Tom studied for only 1 hour (far below the mean) while Courtney studied for 19 hours (far above the mean). Using the sums from Table 12.3, we can calculate their leverages:

$$h_{\text{Tom}} = \frac{1}{10} + \frac{(1 - 10.5)^2}{264.50} = .441 \qquad \text{(Tom's leverage)}$$

$$h_{\text{Courtney}} = \frac{1}{10} + \frac{(19 - 10.5)^2}{264.50} = .373 \qquad \text{(Courtney's leverage)}$$

Example 12.3

Exam Scores: Leverage and Influence 🖥

ExamScores

Figure 12.28 Scatter Plot for Exam Data 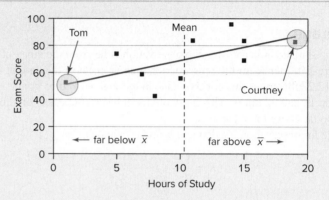 ExamScores

Tom's leverage exceeds $4/n = 4/10 = .400$, so it appears that his observation is *influential*. Yet, despite his high leverage, the regression fits Tom's actual exam score well, so his residual is not unusual. This illustrates that *high leverage* and *unusual residuals* are two different concepts.

Mini Case 12.5

Body Fat 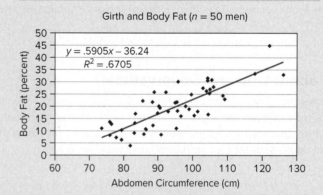 BodyFat

Is waistline a good predictor of body fat? A random sample of 50 men's body fat (percent) and girths (centimeters) was collected. Figure 12.29 suggests that a linear regression is appropriate, and the MegaStat output in Figure 12.30 shows that the regression is highly significant ($F = 97.68$, $t = 9.883$, p-value $= .0000$).

Figure 12.29 Body Fat Regression

Girth and Body Fat ($n = 50$ men)

$y = .5905x - 36.24$
$R^2 = .6705$

The table of residuals, shown in Figure 12.31, highlights three unusual observations. Observations 5, 45, and 50 have high leverage values (exceeding $4/n = 4/50 = .08$) because their abdomen measurements are far from the mean. Observation 37 is included because it has a fairly high studentized residual equal to 1.96 (actual body fat of 30.20 percent is much greater than the predicted 20.33 percent). It does not quite meet our criteria of an unusual residual because its value is still less than 2.0. "Well-behaved" observations are not shown here because they are not unusual according to any of the diagnostic criteria (leverage or studentized residual).

Figure 12.30   Body Fat Regression

Regression Analysis

r^2	0.671		n	50
r	0.819		k	1
Std. Error	5.086		Dep. Var.	Fat% 1

ANOVA table

Source	SS	df	MS	F	p-Value
Regression	2,527.1190	1	2,527.1190	97.68	3.71E-13
Residual	1,241.8162	48	25.8712		
Total	3,768.9352	49			

Regression output   *Confidence Interval*

Variables	coefficients	Std. Error	t Stat	p-value	Lower 95%	Upper 95%
Intercept	−36.2397	5.6690	−6.393	6.28E-08	−47.6379	−24.8415
Abdomen	0.5905	0.0597	9.883	3.71E-13	0.4704	0.7107

Figure 12.31   Unusual Body Fat Residuals

Observation	Fat%	Predicted	Residual	Leverage	Studentized Residual
5	33.60	33.44	0.16	0.099	0.033
37	30.20	20.33	9.87	0.020	1.960
45	33.10	38.28	−5.18	0.162	−1.114
50	45.10	35.86	9.24	0.128	1.945

Section Exercises

12.36 Calculate the standardized residual e_i^* and determine whether it is unusual or an outlier. (a) $e_i = 25$, $s_e = 8$; (b) $e_i = -15$, $s_e = 7$; (c) $e_i = 112$, $s_e = 77$

12.37 Given $s_e = 14$, what is the raw residual if the standardized residual is (a) $e_i^* = 2.00$; (b) $e_i^* = -1.50$; (c) $e_i^* = 0.50$.

12.38 An estimated regression for a random sample of observations on an assembly line is *Defects* = 3.2 + 0.045 *Speed*, where *Defects* is the number of defects per million parts and *Speed* is the number of units produced per hour. The estimated standard error is $s_e = 1.07$. Suppose that 100 units per hour are produced and the actual (observed) defect rate is *Defects* = 4.4. (a) Calculate the predicted *Defects*. (b) Calculate the residual. (c) Standardize the residual using s_e. (d) Is this observation an outlier?

12.39 An estimated regression for a random sample of vehicles is *MPG* = 49.22 − 0.081 *Horsepower*, where *MPG* is miles per gallon and *Horsepower* is the engine's horsepower. The standard error is $s_e = 2.03$. Suppose an engine has 200 horsepower and its actual (observed) fuel efficiency is *MPG* = 38.15. (a) Calculate the predicted *MPG*. (b) Calculate the residual. (c) Standardize the residual using s_e. (d) Is this engine an outlier?

12.40 Is leverage high? (a) $h_i = .13$, $n = 16$; (b) $h_i = .13$, $n = 50$; (c) $h_i = .36$, $n = 100$.

12.41 State the h_i cutoff point for "high leverage" when (a) $n = 25$; (b) $n = 40$; (c) $n = 200$.

12.42 A sample of season performance measures for 29 NBA teams was collected for a season. A regression analysis was performed on two of the variables with $Y = $ *total number of free throws made* and $X = $ *total number of free throws attempted*. Calculate the leverage statistic for the following three teams and state whether the leverage would be considered high. Given: $SS_{xx} = 999{,}603$ and $\bar{x} = 2004$.

a. The Golden State Warriors attempted 2,382 free throws.

b. The New Jersey Nets attempted 2,125 free throws.

c. The New York Knicks attempted 1,620 free throws.

12.43 A sample of 74 Noodles & Company restaurants was used to perform a regression analysis with $Y = \%$ *Annual Revenue Growth* and $X = \%$ *Revenue Due to Loyalty Card Use.* Calculate the leverage statistic for the following three restaurants and state whether the leverage would be considered high. Given: $SS_{xx} = 22.285$ and $\bar{x} = 2.027$ percent.

a. Restaurant 21 earned .072 percent of revenue from loyalty card use.

b. Restaurant 29 earned 1.413 percent of revenue from loyalty card use.

c. Restaurant 64 earned 3.376 percent of revenue from loyalty card use.

12.10 OTHER REGRESSION PROBLEMS (OPTIONAL)

Outliers

We have mentioned outliers under the discussion of non-normal residuals. However, outliers are the source of many other woes, including loss of fit. What causes outliers? An outlier may be an error in recording the data. If so, the observation should be deleted. But how can you tell? Impossible or bizarre data values are *prima facie* reasons to discard a data value. For example, in a sample of body fat data, one adult man's weight was reported as 205 pounds and his height as 29.5 inches (probably a typographical error that should have been 69.5 inches). It is reasonable to discard the observation on grounds that it represents a population different from the other men. An outlier may be an observation that has been influenced by an unspecified "lurking" variable that should have been controlled but wasn't. If so, we should try to identify the lurking variable and formulate a *multiple* regression model that includes the lurking variable(s) as predictors.

Model Misspecification

If a relevant predictor has been omitted, then the model is *misspecified.* Instead of simple regression, you should use *multiple regression.* Such a situation is so common that it is almost a warning against relying on bivariate regression because we usually can think of more than one explanatory variable. As you will see in the next chapter, multiple regression is computationally easy because the computer does all the work. In fact, most computer packages just call it "regression" regardless of the number of predictors.

Ill-Conditioned Data

Variables in the regression should be of the same general order of magnitude, and most people take steps intuitively to make sure this is the case (**well-conditioned data**). Unusually large or small data (called **ill-conditioned**) can cause loss of regression accuracy or can create awkward formatting. This may occur in business applications with accounting data expressed in millions of dollars with many digits of accuracy. For example, Figure 12.32 shows two scatter plots of net income and revenue of 30 large U.S. companies. Their appearance is the same, but the first graph has disastrously crowded axis labels. The graphs have the same slope and R^2,

Figure 12.32

Ill-Conditioned versus Well-Conditioned Data Global30

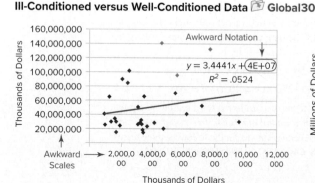
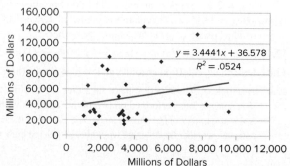

but the first regression has an unintelligible intercept (4E+07), yet the only difference is the position of the decimal point in the data (thousands versus millions).

Awkwardly small numbers also may require adjustment. For example, the number of automobile thefts per capita in the United States in 2018 was 0.00229. However, this statistic is easier to work with if we report it as "229 thefts per 100,000 population." It is also bad to mix large numbers with tiny numbers. For example, in 2018 the average U.S. family income was $89,930 while the number of active physicians per capita was 0.00295. To avoid mixing magnitudes, we might rescale income as $89.93 (thousands) and physicians as 295 (active physicians per 100,000 population). Decimal rescaling does not alter relationships among the variables.

> ## Scaling Data
>
> Adjust the magnitude of your data *before* running the regression.

Spurious Correlation 🗂 Prisoners

In a **spurious correlation**, two variables appear related because of the way they are defined. For example, consider the hypothesis that a state's spending on education is a linear function of its prison population. Such a hypothesis seems absurd, and we would expect the regression to be insignificant. But if the variables are defined as *totals* without adjusting for population, we will observe significant correlation. This phenomenon is called the *size effect* or the *problem of totals*. The first graph in Figure 12.33 shows that, contrary to expectation, the regression on totals gives a very strong fit to the data. Yet the graph on the right side of Figure 12.33 shows that if we divide both variables by population and adjust the decimals, the fit is nil and the slope is indistinguishable from zero. This spurious correlation arises merely because both variables reflect the size of a state's population. For example, Texas, New York, and California lie far to the upper right on the first scatter plot because they are populous states, while less populous states like South Dakota and Delaware are clustered near the origin.

Figure 12.33 Spurious Correlation: Totals versus Per Capita

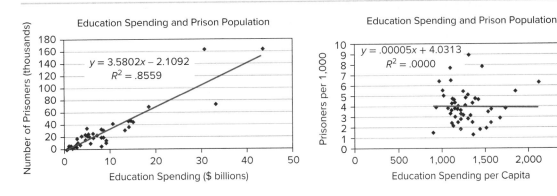

Model Form and Variable Transforms 🗂 MPG1

Sometimes a relationship cannot be modeled using a linear regression. For example, Figure 12.34 plots fuel efficiency (city MPG) on engine size (horsepower) for a sample of 93 vehicles. On the left is a nonlinear model form fitted by Excel. This is one of several nonlinear forms offered by Excel (there are also logarithmic and exponential functions). On the right is a linear regression after taking *logarithms* of each variable. These logarithms are in base 10, but any base will do (scientists prefer natural logs in base *e*). This is an example of a **variable transform**. An advantage of the **log transformation** is that it reduces heteroscedasticity and improves the normality of the residuals, especially when dealing with totals (the size problem mentioned earlier). However, log transforms will not work if any data values are zero or negative.

Figure 12.34   Nonlinear Regression versus Log Transform

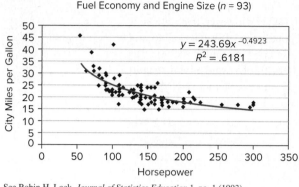

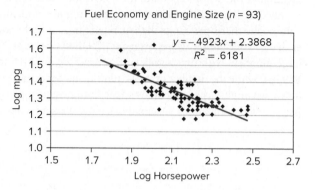

See Robin H. Lock, *Journal of Statistics Education* 1, no. 1 (1993).

Excel makes it easy to fit all sorts of regression models. But fit is only one criterion for evaluating a regression model. Because nonlinear or transformed models might be hard to justify or explain to others, the principle of *Occam's Razor* (choosing the simplest explanation that fits the facts) favors linear regression, unless there are other compelling factors.

Mini Case 12.6

Bank Assets and Market Capitalization 📊 MktCap

Market capitalization (or "market cap") is the total dollar market value of a company's outstanding shares of stock, calculated by multiplying the number of shares by the current share market price. Investors use market cap to assess a company's size (versus alternative measures such as total revenue or total assets). For example, Figure 12.35 shows the relationship between total assets and market capitalization for 75 U.S. banks.

Figure 12.35   Bank Assets and Market Cap

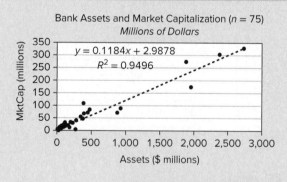

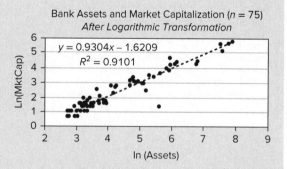

Because banks vary enormously in size (e.g., JPMorganChase is nearly 20 times as large as the smallest bank in our sample), it is difficult to see what is happening near the origin in the first scatter plot. Only a few banks show up as distinct points on the scatter plot. Rescaling will not help, as the majority of U.S. banks have assets below \$100m (https://en.wikipedia.org/). In the second scatter plot, after a log transformation, more detail is visible, and yet both graphs have comparable fit ($R^2 = .9496$ versus $R^2 = .9101$). We can obtain predictions from the fitted log model by exponentiating (e.g., with Excel's =EXP() function).

12.11 LOGISTIC REGRESSION (OPTIONAL)

Binary Response Variable

Sometimes we need to predict something that has only two possible values (a binary *dependent* variable). For example, will a Chase bank customer use online banking ($Y = 1$) or not ($Y = 0$)? Will a Verizon customer switch cell phone providers when the current contract expires ($Y = 1$) or remain with Verizon ($Y = 0$)? Will an Amazon customer make another purchase within the next six months ($Y = 1$) or not ($Y = 0$)? Such research questions would seem to be candidates for regression modeling because we could define possible predictors such as a customer's age, gender, length of time as an existing customer, or past transaction history.

LO 12-12

Identify when logistic regression is appropriate and calculate predictions for a binary response variable.

Why Not Use Least Squares?

Unfortunately, if you perform an ordinary least-squares regression with a binary response variable, there will be complications. While the *actual* value of Y can only be 1 (if the event occurs) or 0 (if the event does not occur), the *predicted* value of Y should be a number between 0 and 1, denoting the *probability* of the event of interest. Using Excel's linear regression, the predicted Y values could be greater than one or less than zero, which would be illogical. Another issue is that your regression errors will violate the assumption of homoscedasticity (constant variance) because as the predicted Y values vary from .50 (in either direction), the variance of the errors will decrease and approach zero. Finally, significance tests assume normally distributed errors, which cannot be the case when Y has only two values ($Y = 0$ or $Y = 1$). Therefore, tests for significance would be in doubt if you used linear regression with a binary response variable.

The solution is to choose **logistic regression** using the *nonlinear* regression model shown below. This equation predicts the *probability* that $Y = 1$ for any specified value of the independent variable. This model form ensures predictions within the range $0 < \hat{y} < 1$.

$$\hat{y} = \frac{e^{b_0+b_1x}}{1 + e^{b_0+b_1x}} \qquad (12.33)$$

The logistic regression model has an *S*-shaped form, as illustrated in Figure 12.36. The logistic function approaches 1 as the value of the independent variable increases.

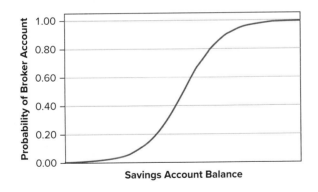

Figure 12.36

Form of the Logistic Model

Estimating a Logistic Regression Model

The underlying model is the Bernoulli (binary) distribution. The event of interest either occurs (probability π) or does not occur (probability $1 - \pi$). Instead of the least squares method, we estimate the parameters using the method of **maximum likelihood**. This method chooses values of the regression parameters that will maximize the probability of obtaining the observed sample data. While easy to state in words, the computational procedure requires specialized software. Any major statistical package will safely perform logistic regression (sometimes called *logit* for short) and will provide *p*-values for the estimated coefficients and predictions for Y. An iterative process is required because there is no simple formula for the parameter estimates. What is important at this stage of training is for you to recognize the need for a specialized tool when Y is a binary (0, 1) variable. In the next chapter, this topic will be developed more fully and will allow for more than one predictor.

Example 12.4

Brokerage Accounts

We might expect that the size of a bank customer's savings balance would predict whether ($Y = 1$) or not ($Y = 0$) the customer also would have a brokerage account with the bank to facilitate making investments. A sample of 20 bank customers is shown in Table 12.5.

Table 12.5

Savings Balance and Existence of Broker Account ($n = 20$)
📂 **Broker**

Savings	Broker?	Savings	Broker?	Savings	Broker?	Savings	Broker?
1,100	0	9,900	0	12,500	1	30,900	0
1,600	0	10,000	0	20,500	0	32,100	1
3,400	0	11,300	0	20,500	0	36,000	1
8,600	0	11,400	0	20,700	1	56,000	1
9,400	0	12,300	0	22,600	0	72,800	1

Linear regression is inappropriate, as can be seen in Figure 12.37, because Y has only two values ($Y = 0$ or 1). Instead, we use software (such as Minitab) to fit a nonlinear logistic regression.

Figure 12.37

Scatter Plot and Fitted Logistic Regression Model ($n = 20$)

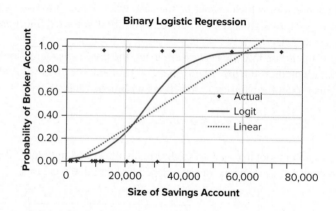

The estimated logistic equation is

$$\hat{y} = P(Broker) = \frac{exp(-4.000768 + .00014792 \, Savings)}{1 + exp(-4.000768 + .00014792 \, Savings)}$$

We can use the fitted model to estimate the probability that a given customer will have a broker account. For example, the predicted probability is .2607 that a customer whose savings balance is \$20,000 will have a broker account, compared with a probability of .8717 for a customer whose savings balance is \$40,000:

If *Savings* = 20,000: $P(Broker) = \dfrac{exp(-4.000768 + .00014792 \times 20,000)}{1 + exp(-4.000768 + .00014792 \times 20,000)} = .2607$

If *Savings* = 40,000: $P(Broker) = \dfrac{exp(-4.000768 + .00014792 \times 40,000)}{1 + exp(-4.000768 + .00014792 \times 40,000)} = .8717$

Section Exercises

📕 **connect**

12.44 A binary logistic regression was fitted using a sample of employees of a large corporation to investigate if the employee's monthly salary (*Salary*) predicts whether (1) or not (0) the employee has a 401(k) retirement plan (*401K*). Use the fitted equation to predict the probability of a 401(k) plan for an employee whose monthly salary is (a) \$3,000; (b) \$5,000; (c) \$8,000.

$$P(401K) = \frac{exp(-6.13 + .000881 \, Salary)}{1 + exp(-6.13 + .000881 \, Salary)}$$

12.45 Does the number of address changes (*Moves*) in the last decade predict whether (1) or not (0) a college graduate lacks health insurance (*Uninsured*)? Use the fitted binary logistic regression to predict the probability that a college graduate will lack insurance if the number of address changes is (a) 0; (b) 3; (c) 6.

$$P(Uninsured) = \frac{\exp(-5.88 + 1.055\ Moves)}{1 + \exp(-5.88 + 1.055\ Moves)}$$

Chapter Summary

The **sample correlation coefficient** r measures linear association between X and Y, with values near 0 indicating a lack of linearity while values near -1 (negative correlation) or $+1$ (positive correlation) suggest linearity. The **t test** is used to test hypotheses about the **population correlation** ρ. In **simple regression** there is an assumed linear relationship between the independent variable X (the **predictor**) and the dependent variable Y (the **response**). The slope (β_1) and intercept (β_0) are unknown **parameters** that are estimated from a sample. **Residuals** are the differences between **observed** and **fitted** Y-values. The **ordinary least squares** (OLS) method yields **regression coefficients** for the slope (b_1) and intercept (b_0) that minimize the sum of squared residuals. The **coefficient of determination** (R^2) measures the overall fit of the regression, with R^2 near 1 signifying a good fit and R^2 near 0 indicating a poor fit. The **F statistic** in the ANOVA table is used to test for significant overall regression, while the **t statistics** (and their p-values) are used to test hypotheses about the slope and intercept. The **standard error** of the regression is used to create **confidence intervals** or **prediction intervals** for Y. Regression assumes that the errors are normally distributed, independent random variables with constant variance σ^2. **Residual tests** identify possible **violations** of assumptions (**non-normality, autocorrelation, heteroscedasticity**). Data values with high **leverage** (unusual X-values) have strong influence on the regression. Unusual **standardized residuals** indicate cases where the regression gives a poor fit. **Ill-conditioned** data may lead to **spurious** correlation or other problems. **Data transforms** may help, but they also change the **model specification**. **Logistic regression** is used when creating a model used to predict a **binary response** variable.

Key Terms

autocorrelation	log transformation	R^2	standard error
bivariate data	logistic regression	regression assumptions	standardized residuals
coefficient of determination	maximum likelihood	residual	studentized residuals
confidence interval	normal probability plot	response variable	sums of squares
error sum of squares	ordinary least squares (OLS)	sample correlation	t statistic
heteroscedastic	population correlation	coefficient, r	variable transform
homoscedastic	coefficient, ρ	scatter plot	well-conditioned data
ill-conditioned data	prediction interval	simple regression equation	
intercept	predictor variable	slope	
leverage	q-q plot	spurious correlation	

Commonly Used Formulas in Simple Regression

Sample correlation coefficient:
$$r = \frac{\sum_{i=1}^{n}(x_i - \bar{x})(y_i - \bar{y})}{\sqrt{\sum_{i=1}^{n}(x_i - \bar{x})^2}\sqrt{\sum_{i=1}^{n}(y_i - \bar{y})^2}}$$

Test statistic for zero correlation:
$$t_{calc} = r\sqrt{\frac{n-2}{1-r^2}}\quad \text{with}\quad d.f. = n - 2$$

True regression line:
$$y = \beta_0 + \beta_1 x + \epsilon$$

Fitted regression line:
$$\hat{y} = b_0 + b_1 x$$

Slope of fitted regression:
$$b_1 = \frac{\sum_{i=1}^{n}(x_i - \bar{x})(y_i - \bar{y})}{\sum_{i=1}^{n}(x_i - \bar{x})^2}\quad \text{or}\quad b_1 = \frac{SS_{xy}}{SS_{xx}}$$

Intercept of fitted regression:
$$b_0 = \bar{y} - b_1\bar{x}$$

Sum of squared residuals:
$$SSE = \sum_{i=1}^{n}(y_i - \hat{y}_i)^2 = \sum_{i=1}^{n}(y_i - b_0 - b_1 x_i)^2$$

Coefficient of determination:
$$R^2 = 1 - \frac{\sum_{i=1}^{n}(y_i - \hat{y}_i)^2}{\sum_{i=1}^{n}(y_i - \bar{y})^2} = 1 - \frac{SSE}{SST} \quad \text{or} \quad R^2 = \frac{SSR}{SST}$$

Standard error of the estimate:
$$s_e = \sqrt{\frac{\sum_{i=1}^{n}(y_i - \hat{y}_i)^2}{n - 2}} = \sqrt{\frac{SSE}{n - 2}}$$

Standard error of the slope:
$$s_{b1} = \frac{s_e}{\sqrt{\sum_{i=1}^{n}(x_i - \bar{x})^2}} \quad \text{with } d.f. = n - 2$$

t test for zero slope:
$$t_{calc} = \frac{b_1 - 0}{s_{b_1}} \quad \text{with } d.f. = n - 2$$

Confidence interval for true slope:
$$b_1 - t_{\alpha/2}s_{b_1} \leq \beta_1 \leq b_1 + t_{\alpha/2}s_{b_1} \quad \text{with } d.f. = n - 2$$

Confidence interval for conditional mean of Y:
$$\hat{y}_i \pm t_{\alpha/2}s_e\sqrt{\frac{1}{n} + \frac{(x_i - \bar{x})^2}{\sum_{i=1}^{n}(x_i - \bar{x})^2}}$$

Prediction interval for Y:
$$\hat{y}_i \pm t_{\alpha/2}s_e\sqrt{1 + \frac{1}{n} + \frac{(x_i - \bar{x})^2}{\sum_{i=1}^{n}(x_i - \bar{x})^2}}$$

Chapter Review

1. (a) How does correlation analysis differ from regression analysis? (b) What does a correlation coefficient reveal? (c) What sums are needed to calculate a correlation coefficient? (d) What are the two ways of testing a correlation coefficient for significance?

2. (a) What is a simple regression model? (b) State three caveats about regression. (c) What does the random error component in a regression model represent? (d) What is the difference between a regression residual and the true random error?

3. (a) Explain how you fit a regression to an Excel scatter plot. (b) What are the limitations of Excel's scatter plot fitted regression?

4. (a) Explain the logic of the ordinary least squares (OLS) method. (b) How are the least squares formulas for the slope and intercept derived? (c) What sums are needed to calculate the least squares estimates?

5. (a) Why can't we use the sum of the residuals to assess fit? (b) What sums are needed to calculate R^2? (c) Name an advantage of using the R^2 statistic instead of the standard error s_e to measure fit. (d) Why do we need the standard error s_e?

6. (a) Explain why a confidence interval for the slope or intercept would be equivalent to a two-tailed hypothesis test. (b) Why is it especially important to test for a zero slope?

7. (a) What does the F statistic show? (b) What is its range? (c) What is the relationship between the F test and the t tests for the slope and correlation coefficient?

8. (a) For a given X, explain the distinction between a confidence interval for the conditional mean of Y and a prediction interval for an individual Y-value. (b) Why is the individual prediction interval wider? (c) Why are these intervals narrowest when X is near its mean?

9. (a) What is a residual? (b) What is a standardized residual, and why is it useful? (c) Name two alternative ways to identify unusual residuals.

10. (a) When does a data point have high leverage (refer to the scatter plot)? (b) Name one test for unusual leverage.

11. (a) Name three assumptions about the random error term in the regression model. (b) Why are the residuals important in testing these assumptions?

12. (a) What are the consequences of non-normal errors? (b) Explain two tests for non-normality.

13. (a) What is heteroscedasticity? Identify its two common forms. (b) What are its consequences? (c) How do we test for it?

14. (a) What is autocorrelation? Identify two main forms of it. (b) What are its consequences? (c) How do we test for it?

15. (a) Why might there be outliers in the residuals? (b) What actions could be taken?

16. (a) What are ill-conditioned data? How can they be avoided? (b) What is spurious correlation? How can it be avoided?

17. (a) What is a log transform? (b) What are its advantages and disadvantages?

18. (a) When is logistic regression needed? (b) Why not use Excel for logistic regression?

Chapter Exercises

Note: Problem marked * relies on optional material from this chapter.

Instructions: Choose one or more of the data sets A–J below, or as assigned by your instructor. The first column is the X, or independent, variable and the second column is the Y, or dependent, variable. Use Excel or a statistical package (e.g., MegaStat or Minitab) to obtain the simple regression and required graphs. Write your answers to exercises 12.46 through 12.61 (or those assigned by your instructor) in a concise report, labeling your answers to each question. Insert tables and graphs in your report as appropriate. You may work with a partner if your instructor allows it.

12.46 Are the variables cross-sectional data or time-series data?

12.47 How do you imagine the data were collected?

12.48 Is the sample size sufficient to yield a good estimate? If not, do you think more data could easily be obtained, given the nature of the problem?

12.49 State your *a priori* hypothesis about the sign of the slope. Is it reasonable to suppose a cause-and-effect relationship?

12.50 Make a scatter plot of Y against X. Discuss what it tells you.

12.51 Use Excel's Add Trendline feature to fit a linear regression to the scatter plot. Is a linear model credible?

12.52 Interpret the slope. Does the intercept have meaning, given the range of the data?

12.53 Use Excel, MegaStat, or Minitab to fit the regression model, including residuals and standardized residuals.

12.54 (a) Does the 95 percent confidence interval for the slope include zero? If so, what does this tell you? If not, what does it mean? (b) Do a two-tailed *t* test for zero slope at $\alpha = .05$. State the hypotheses, degrees of freedom, and critical value for your test. (c) Interpret the *p*-value for the slope. (d) Did the sample support your hypothesis about the sign of the slope?

12.55 (a) Based on the R^2 and ANOVA table for your model, how would you assess the fit? (b) Interpret the *p*-value for the *F* statistic. (c) Would you say that your model's fit is good enough to be of practical value?

12.56 Study the table of residuals. Identify as *outliers* any standardized residuals that exceed 3 and as *unusual* any that exceed 2. Can you suggest any reasons for these unusual residuals?

12.57 (a) Make a histogram (or normal probability plot) of the residuals and discuss its appearance. (b) Do you see evidence that your regression may violate the assumption of normal errors?

12.58 Inspect the residual plot to check for heteroscedasticity and report your conclusions.

12.59 Is an autocorrelation test appropriate for your data? If so, perform an eyeball inspection of residual plot against observation order or a runs test.

12.60 Use MegaStat or Minitab to generate 95 percent confidence and prediction intervals for various X-values.

12.61 Use MegaStat or Minitab to identify observations with high leverage.

DATA SET A   **Median Income and Median Home Prices by State (n = 51)** 🖅 **HomePrice3**

State	Median Income	Median Price
AK	$40,933	$241,750
AL	57,848	128,969
AR	46,896	120,560
⋮	⋮	⋮
WI	42,777	153,935
WV	50,351	129,369
WY	52,201	183,202

Source: http://www.fhfa.gov and http://www.census.gov.
Note: n = 51 refers to 50 states plus DC.

DATA SET B   **Midterm and Final Exam Scores for Business Statistics Students, Fall Semester 2011 (n = 58 students)** 🖅 **ExamScores2**

Midterm Exam Score	Final Exam Score
80	78
87	85
72	81
⋮	⋮
80	82
68	70
70	69

DATA SET C   **Estimated and Actual Length of Stay in Months (n = 16 patients)** 🖅 **Hospital**

Patient	ELOS	ALOS
1	10.5	10
2	4.5	2
3	7.5	4
⋮	⋮	⋮
14	6	10
15	7.5	7
16	3	5.5

Source: Records of a hospital outpatient cognitive retraining clinic.
Note: ELOS used a 42-item assessment instrument combined with expert team judgment. Patients had suffered head trauma, stroke, or other medical conditions affecting cognitive function.

DATA SET D Single-Engine Aircraft Performance (n = 52 airplanes) ⌨ Airplanes

Mfgr/Model	TotalHP	Cruise
AMD CH 2000	116	100
Beech Baron 58	600	200
Beech Baron 58	650	241
⋮	⋮	⋮
Sky Arrow 650 TC	81	98
Socata TB20 Trinidad	250	163
Tiger AG-5B	180	143

Source: New and used airplane reports in *Flying* (various issues).

Note: Cruise is in knots (nautical miles per hour). Data are for educational purposes only and should not be used as a guide to aircraft performance. *TotalHP* is total horse power.

DATA SET E Microprocessor Speed and Power Dissipation (n = 14 chips) ⌨ Microprocessors

Chip	Speed (MHz)	Power (watts)
1989 Intel 80486	20	3
1993 Pentium	100	10
1997 Pentium II	233	35
1998 Intel Celeron	300	20
1999 Pentium III	600	42
1999 AMD Athlon	600	50
2000 Pentium 4	1300	51
2004 Celeron D	2100	73
2004 Pentium 4	3800	115
2005 Pentium D	3200	130
2007 AMD Phenom	2300	95
2008 Intel Core 2	3200	136
2009 Intel Core i7	2900	95
2009 AMD Phenom II	3200	125

See wikipedia.org and *New Scientist* 208, no. 2780 (October 2, 2010), p. 41.

DATA SET F Restaurant Weekly Revenue and Weekly Website Hits (n = 10 restaurants) ⌨ WebSiteHits

Restaurant	Website Hits	Weekly Revenue
John's Café	1,213	$12,113
Buccan	1,490	11,409
New City Diner	1,365	14,579
Black Pearl	1,455	11,605
Saratoga	1,269	12,308
Burnt Toast	1,632	12,320
University Seat	1,323	13,225
Jimmy's	1,865	13,652
Maroon and Orange	1,590	13,893
Burger Palace	1,878	13,896

DATA SET G Mileage and Vehicle Weight (n = 73 vehicles) ⌨ MPG2

Vehicle	Weight	City MPG
Acura TL	3,968	20
Audi A5	3,583	22
BMW 4 Series 428i	3,470	22
⋮	⋮	⋮
Volkswagen Passat SE	3,230	24
Volvo S60 T5	3,528	21
Volvo XC90	4,667	16

Sources: Acura.com; AudiUSA.com; VW.com; Volvo.com. Vehicles are a random sample of 2014 vehicles sold in the U.S. All are gas or flex-fuel (no hybrids or electrics). Data are intended for statistical education and should not be viewed as a guide to vehicle performance.

DATA SET H Pasta Sauce Total Calories and Fat Calories (n = 20 products) ⌨ Pasta

Product	Fat Cal/gm	Cal/gm
Barilla Roasted Garlic & Onion	0.20	0.64
Barilla Tomato & Basil	0.12	0.56
Classico Tomato & Basil	0.08	0.40
⋮	⋮	⋮
Ragu Roasted Garlic	0.19	0.70
Ragu Traditional	0.20	0.56
Sutter Home Tomato & Garlic	0.16	0.64

Source: Independent project by statistics students Donna Bennett, Nicole Cook, Latrice Haywood, and Robert Malcolm.
Note: Data are intended for educational purposes only and should not be viewed as a nutrition guide.

DATA SET I Temperature and Energy Usage for a Residence (n = 24 months) ⌨ Electric

Month	Avg Temp (F°)	Usage (kWh)
1	62	436
2	71	464
3	76	446
⋮	⋮	⋮
22	25	840
23	38	867
24	48	606

Sources: Electric bills for a residence and NOAA weather data.

DATA SET J U.S. Annual Percent Inflation in Prices of Medical Care and Apparel (n = 44 years) ⌨ Inflation

Year	Medical Care%	Apparel%
1975	9.8	2.4
1976	10.0	4.6
1977	8.9	4.3
⋮	⋮	⋮
2016	4.1	−0.1
2017	1.8	−1.6
2018	2.0	−0.1

Source: *Economic Report of the President,* 2018.

12.62 Researchers found a correlation coefficient of $r = .50$ on personality measures for identical twins. A reporter interpreted this to mean that "the environment orchestrated one-half of their personality differences." Do you agree with this interpretation? Discuss.

12.63 A study of the role of spreadsheets in planning in 55 small firms defined $Y =$ "satisfaction with sales growth" and $X =$ "executive commitment to planning." Analysis yielded an overall correlation of $r = .3043$. Do a two-tailed test for zero correlation at $\alpha = .025$.

12.64 In a study of stock prices from 1970 to 1994, the correlation between Nasdaq closing prices on successive days (i.e., with a 1-day lag) was $r = .13$ with a t statistic of 5.47. Interpret this result.

(Source: David Nawrocki, "The Problems with Monte Carlo Simulation," *Journal of Financial Planning* 14, no. 11 [November 2001], p. 96.)

12.65 Regression analysis of free throws by 29 NBA teams during the 2002–2003 season revealed the fitted regression $Y = 55.2 + .73X$ ($R^2 = .874$, $s_e = 53.2$), where $Y =$ total free throws made and $X =$ total free throws attempted. The observed range of X was from 1,620 (New York Knicks) to 2,382 (Golden State Warriors). (a) Find the expected number of free throws made for a team that shoots 2,000 free throws.

(b) Do you think that the intercept is meaningful? *Hint:* Make a scatter plot and let Excel fit the line. (c) Make a 95 percent prediction interval for Y when $X = 2,000$. ⎙ **FreeThrows**

12.66 Exhibit "A" below shows a regression using $X =$ weekly pay, $Y =$ income tax withheld, and $n = 35$ McDonald's employees. (a) Write the fitted regression equation. (b) State the degrees of freedom for a two-tailed test for zero slope, and use Appendix D to find the critical value at $\alpha = .05$. (c) What is your conclusion about the slope? (d) Interpret the 95 percent confidence limits for the slope. (e) Verify that $F = t^2$ for the slope. (f) In your own words, describe the fit of this regression.

12.67 Exhibit "B" below shows a regression using $X =$ monthly maintenance spending (dollars), $Y =$ monthly machine downtime (hours), and $n = 15$ copy machines. (a) Write the fitted regression equation. (b) State the degrees of freedom for a two-tailed test for zero slope, and use Appendix D to find the critical value at $\alpha = .05$. (c) What is your conclusion about the slope? (d) Interpret the 95 percent confidence limits for the slope. (e) Verify that $F = t^2$ for the slope. (f) In your own words, describe the fit of this regression.

EXHIBIT "A"
⎙ **Tax Withheld**

R^2	0.202
Std. Error	6.816
n	35

ANOVA table

Source	SS	df	MS	F	p-value
Regression	387.6959	1	387.6959	8.35	.0068
Residual	1,533.0614	33	46.4564		
Total	1,920.7573	34			

Regression output — Confidence Interval

Variables	Coefficients	Std. Error	t Stat	p-value	Lower 95%	Upper 95%
Intercept	30.7963	6.4078	4.806	.0000	17.7595	43.8331
Slope	0.0343	0.0119	2.889	.0068	0.0101	0.0584

EXHIBIT "B"
⎙ **Downtime**

R^2	0.370
Std. Error	286.793
n	15

ANOVA table

Source	SS	df	MS	F	p-value
Regression	628,298.2	1	628,298.2	7.64	.0161
Residual	1,069,251.8	13	82,250.1		
Total	1,697,550.0	14			

Regression output — Confidence Interval

Variables	Coefficients	Std. Error	t Stat	p-value	Lower 95%	Upper 95%
Intercept	1,743.57	288.82	6.037	.0000	1,119.61	2,367.53
Slope	-1.2163	0.4401	-2.764	.0161	-2.1671	-0.2656

12.68 Exhibit "C" below shows a regression using X = total assets ($ billions), Y = total revenue ($ billions), and n = 64 large banks. (a) Write the fitted regression equation. (b) State the degrees of freedom for a two-tailed test for zero slope, and use Appendix D to find the critical value at α = .05. (c) What is your conclusion about the slope? (d) Interpret the 95 percent confidence limits for the slope. (e) Verify that $F = t^2$ for the slope. (f) In your own words, describe the fit of this regression.

12.69 Do stock prices of competing companies move together? Below are daily closing prices for 42 consecutive trading days of two computer services firms (IBM = International Business Machines Corporation, HPQ = Hewlett-Packard Company. (a) Calculate the sample correlation coefficient (e.g., using Excel or MegaStat). (b) At α = .01 can you conclude that the true correlation coefficient is not equal to zero? (c) Make a scatter plot of the data. What does it say? **StockPrices**

(Source: finance.yahoo.com.)

Date	IBM	HPQ
10/3/22	121.51	25.89
10/4/22	125.50	26.64
10/5/22	125.74	26.78
⋮	⋮	⋮
11/28/22	146.18	29.21
11/29/22	146.49	28.88
11/30/22	148.90	30.04

12.70 Following are percentages for *annual sales growth* and *net sales attributed to loyalty card usage* at 74 Noodles & Company restaurants. (a) Make a scatter plot. (b) Find the correlation coefficient and interpret it. (c) Test the correlation coefficient for significance, clearly stating the degrees of freedom. (d) Does it appear that loyalty card usage is associated with increased sales growth? **LoyaltyCard**

Store	Growth%	Loyalty%
1	−8.3	2.1
2	−4.0	2.5
3	−3.9	1.7
⋮	⋮	⋮
72	20.8	1.1
73	25.5	0.6
74	28.8	1.8

Source: Noodles & Company.

12.71 Below are fertility rates (average children born per woman) in 15 EU nations for 2 years. (a) Make a scatter plot. (b) Find the correlation coefficient and interpret it. (c) Test the correlation coefficient for significance, clearly stating the degrees of freedom. **Fertility**

Nation	1990	2000
Austria	1.5	1.3
Belgium	1.6	1.5
Denmark	1.6	1.7
⋮	⋮	⋮
Spain	1.4	1.1
Sweden	2.0	1.4
U.K.	1.8	1.7

12.72 Consider the Excel regression (next page) of perceived sound quality as a function of price for 27 stereo speakers. (a) Is the coefficient of *Price* significantly different from zero at α = .05? (b) What does the R^2 tell you? (c) Given these results, would you conclude that a higher price implies higher sound quality? **Speakers**

EXHIBIT "C"
Bank Revenue

R^2	0.519
Std. Error	6.977
n	64

ANOVA table

Source	SS	df	MS	F	p-value
Regression	3,260.0981	1	3,260.0981	66.97	1.90E-11
Residual	3,018.3339	62	48.6828		
Total	6,278.4320	63			

Regression output | | | | | Confidence Interval | |

Variables	Coefficients	Std. Error	t Stat	p-value	Lower 95%	Upper 95%
Intercept	6.5763	1.9254	3.416	.0011	2.7275	10.4252
X1	0.0452	0.0055	8.183	1.90E-11	0.0342	0.0563

Regression Statistics	
R Square	0.01104
Standard Error	4.02545
Observations	27

Variables	Coefficients	Std Error	t Stat
Intercept	88.4902	1.67814	52.731
Price	−0.00239	0.00453	−0.528

12.73 Choose *one* of these three data sets (see Exhibit "D" below). (a) Make a scatter plot. (b) Let Excel estimate the regression line, with fitted equation and R^2. (c) Describe the fit of the regression. (d) Write the fitted regression equation and interpret the slope. (e) Do you think that the estimated intercept is meaningful? Explain.

12.74 Simple regression was employed to establish the effects of childhood exposure to lead (see Exhibit "E" below). The effective sample size was about 122 subjects. The independent variable was the level of dentin lead (parts per million). Below are regressions using various dependent variables. (a) Calculate the t statistic for each slope. (b) From the p-values, which slopes differ from zero at $\alpha = .01$?

(c) Do you feel that cause and effect can be assumed? *Hint:* Do a web search for information about effects of childhood lead exposure.

12.75 Below are recent financial ratios for a random sample of 20 integrated health care systems. *Operating Margin* is total revenue minus total expenses divided by total revenue plus net operating profits. *Equity Financing* is fund balance divided by total assets. (a) Make a scatter plot of $Y =$ operating margin and $X =$ equity financing (both variables are in percent). (b) Use Excel to fit the regression, with fitted equation and R^2. (c) In your own words, describe the fit. 📄 **HealthCare**

Operating Margin	Equity Financing
3.89	35.58
8.23	59.68
2.56	40.48
⋮	⋮
4.75	54.21
0.00	59.73
10.79	46.21

EXHIBIT "D"
📄 **Three Data Sets**

Commercial Real Estate Y = assessed value, $000 X = floor space, sq. ft. (n = 15) 📄 **Assessed**		Employee Salaries Y = employee salary, $000 X = employee age (n = 23) 📄 **Salaries**		New Home Sales Y = selling price, $000 X = home size, sq. ft. (n = 20) 📄 **HomePrice2**	
Size	**Assessed**	**Age**	**Salary**	**SqFt**	**Price**
1,790	1,796	23	28.6	3,570	861
4,720	1,544	31	53.3	3,410	740
5,940	2,094	44	73.8	2,690	563
⋮	⋮	⋮	⋮	⋮	⋮
4,880	1,678	54	75.8	3,020	720
1,620	710	44	79.8	2,320	575
1,820	678	36	70.2	3,130	785

EXHIBIT "E"
📄 **Lead Exposure**

Dependent Variable	R^2	Estimated Slope	Std. Error	p-Value
Highest grade achieved	.061	−0.027	0.009	.008
Reading grade equivalent	.121	−0.070	0.018	.000
Class standing	.039	−0.006	0.003	.048
Absence from school	.071	4.8	1.7	.006
Grammatical reasoning	.051	0.159	0.062	.012
Vocabulary	.108	−0.124	0.032	.000
Hand-eye coordination	.043	0.041	0.018	.020
Reaction time	.025	11.8	6.66	.080
Minor antisocial behavior	.025	−0.639	0.36	.082

Source: H. L. Needleman et al., *The New England Journal of Medicine* 322, no. 2 (January 1990), p. 86.

12.76 Consider the following data on 20 chemical reactions, with Y = chromatographic retention time (seconds) and X = molecular weight (gm/mole). (a) Make a scatter plot. (b) Use Excel to fit the regression, with fitted equation and R^2. (c) In your own words, describe the fit. 🖉 **Chemicals**

Chemical Name	Retention Time	Molecular Weight
alpha-pinene	34.50	136.24
cyclopentene	95.27	68.12
p-diethylbenzene	284.00	134.22
⋮	⋮	⋮
pentane	78.00	72.15
isooctane	136.90	114.23
hexane	106.00	86.18

Source: Data provided by John Seeley of Oakland University.

12.77 SAT scores are often used as a proxy for how well a state educates the students who reside in that state. However, not all students in a state actually take the SAT and SAT participation rates vary across states. For example, some states require the SAT as an assessment test in high school. Those states will have higher participation rates than states that do not require the test in high school. If a student attends a high school that does not require the SAT, but the student plans to apply to a college that requires the SAT for admission, then that student will choose to take the exam. Regression results from a regression using Y = state average SAT and X = state SAT participation are shown below. (a) What do these regression results tell you about the relationship between state average SAT score and state SAT participation? (b) Do you consider this a strong relationship? Explain. (c) Can you suggest reasons why this relationship might exist?

Regression Statistics	
R Square	0.8419
Standard Error	59.94
Observations	51

Variables	Coefficients	Std Error	t Stat
Intercept	1752.44	12.916	135.683
Participation	−388.08	24.024	−16.154

Source: www.collegeboard.org.

12.78 Below are revenue and profit (both in $ billions) for nine large entertainment companies. (a) Make a scatter plot of profit as a function of revenue. (b) Use Excel to fit the regression, with fitted equation and R^2. (c) In your own words, describe the fit. 🖉 **Entertainment**

Company	Revenue	Profit
AMC Entertainment	1.792	−0.020
Clear Channel	8.931	1.146
Liberty Media	2.446	−0.978
⋮	⋮	⋮
Univision	1.311	0.155
Viacom	26.585	1.417
Walt Disney	27.061	1.267

Source: *Fortune* 149, no. 7 (April 5, 2005), p. F-50.

12.79 Below are fitted regressions based on used vehicle ads. The assumed regression model is $AskingPrice = f(VehicleAge)$. (a) Interpret the slopes. (b) Are the intercepts meaningful? Explain. (c) Assess the fit of each model. (d) Is a bivariate model adequate to explain vehicle prices? If not, what other predictors might be considered?

Vehicle	n	Intercept	Slope	R^2
Ford Explorer	31	22,252	−2,452	.643
Ford F-150	43	26,164	−2,239	.713
Ford Mustang	33	21,308	−1,691	.328
Ford Taurus	32	13,160	−906	.679

Source: *Detroit's AutoFocus* 4, no. 38 (September 17–23, 2004).

12.80 Exhibit "F" (below) shows results of a regression of Y = average stock returns (in percent) as a function of X = average price/earnings ratios for the period 1949–1997 (49 years). Separate regressions were done for various holding periods (sample sizes are therefore variable). (a) Summarize what the regression results tell you. (b) Would you anticipate autocorrelation in this type of data? Explain.

12.81* Does a new car owner's monthly car payment (*CarPmt*) predict whether (1) or not (0) the owner subscribes to satellite radio (*SatRad*)? Use the fitted logistic model below to predict the probability of subscribing to satellite radio for a car owner whose monthly car payment is (a) $350; (b) $450; (c) $550.

$$P(SatRad) = \frac{exp(-11.96 + .02421 CarPmt)}{1 + exp(-11.96 + .02421 CarPmt)}$$

EXHIBIT "F"
🖉 **Stock Returns**

Holding Period	Intercept	Slope	t	R^2	p
1 year	28.10	−0.92	1.86	.0688	.0686
2 years	26.11	−0.86	2.57	.1252	.0136
5 years	20.67	−0.57	2.99	.1720	.0046
8 years	24.73	−0.94	6.93	.5459	.0000
10 years	24.51	−0.95	8.43	.6516	.0000

Source: Ruben Trevino and Fiona Robertson, "P/E Ratios and Stock Market Returns," *Journal of Financial Planning* 15, no. 2 (February 2002), p. 78.

DO-IT-YOURSELF MINI-PROJECT

12.82 Adult height is somewhat predictable from average height of both parents. For females, a commonly used equation is *YourHeight = ParentHeight* − 2.5, while for males the equation is *YourHeight = ParentHeight* + 2.5. (a) Test these equations on yourself and 9 friends. (b) How well did the equations predict height?

Related Reading

Kutner, Michael H.; Chris Nachtsheim; and John Neter. *Applied Linear Regression Models.* McGraw Hill, 2004.

Osborne, Jason W. *Best Practices in Logistic Regression.* Sage, 2014.

Vigen, Tyler. *Spurious Correlations.* Hachette Books, 2015.

CHAPTER 12   More Learning Resources

You can access these *LearningStats* demonstrations through Connect to help you understand probability.

Topic	LearningStats Demonstrations
Correlation	▣ Overview of Correlation ▣ Scatter Plot Simulation ▣ Zero Correlation Demonstration
Regression	▣ Overview of Simple Regression ▣ Simple Regression Using Excel ▣ Simple Regression Using R
Ordinary least squares estimators	▣ Least Squares Method Demonstration ▣ Derivation of OLS Estimators ▣ Regression Calculations ▣ Effect of *X* Range and Model Form
Confidence and prediction intervals	▣ Confidence versus Prediction Intervals ▣ Regression Calculations ▣ Superimposing Many Fitted Regressions
Violations of assumptions	▣ Non-Normal Errors ▣ Heteroscedastic Errors ▣ Autocorrelated Errors

Key: ▣ = PowerPoint   ▣ = Excel   ▣ = PDF

Software Supplement

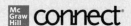

Using MegaStat: Exam Scores ExamScores

Figure 12.38 shows MegaStat's regression menu. The output format is similar to Excel's, except that MegaStat highlights coefficients that differ significantly from zero at $\alpha = .05$ in a two-tailed test.

Figure 12.38

MegaStat's Regression Menus

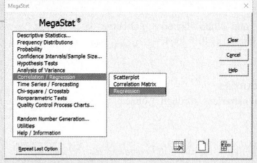

Source: MegaStat.

Using Minitab: Exam Scores ExamScores

Figure 12.39 shows Minitab's regression menus. Minitab gives you the same general output as Excel but with strongly rounded results.

You may have noticed that both Excel and Minitab calculated something called "adjusted R-Square." For a bivariate regression, this statistic is of little interest, but in the next chapter it becomes important.

Figure 12.39

Minitab's Regression Menus

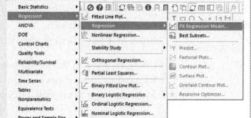

Source: MegaStat.

Using MegaStat: Calculating Confidence and Prediction Intervals

Both MegaStat and Minitab will let you type in the x_i values and will give both confidence and prediction intervals *only* for that x_i value, but you must make your own graphs. Shown below in Figure 12.40 is MegaStat's menu.

Figure 12.40

MegaStat Regression Predictions

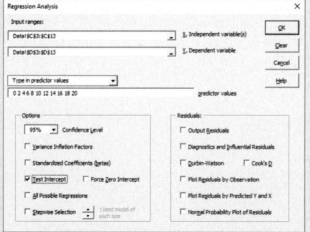

Source: MegaStat.

Using MegaStat: Calculating Residuals

Although there are subtle differences in the way Excel, Mega-Stat, and Minitab calculate and display standardized residuals, MegaStat gives the same general output as Excel. Its regression menu is shown in Figure 12.41.

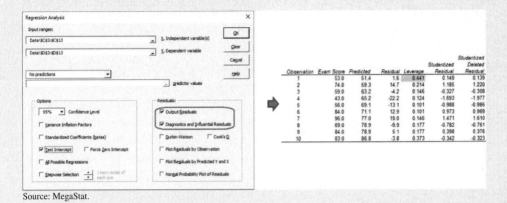

Figure 12.41

MegaStat's Exam Score Residuals

Source: MegaStat.

MegaStat calculates both studentized residuals and *studentized deleted residuals*. The calculation for studentized deleted residuals is equivalent to rerunning the regression *n* times, with each observation omitted in turn, and recalculating the studentized residuals. Further calculation details are reserved for an advanced statistics class, but interpretation is simple. A studentized deleted residual whose absolute value is 2 or more is unusual and one whose absolute value is 3 or more is typically considered an outlier. To make the output more readable, MegaStat rounds off the values and highlights unusual and outlier standardized residuals.

Using Minitab: Calculating Exam Score Residuals

Minitab gives the same general output as Excel and MegaStat but with rounded results and more detailed residual information. Its menus are shown in Figure 12.42. Minitab reports standardized residuals, which usually are close in value to Excel's "standardized" residuals.

Minitab's results confirm that there are no unusual residuals in the exam score regression. An attractive feature of Minitab is that the actual and fitted *Y*-values are displayed (Excel shows only the fitted *Y*-values). Minitab also gives the standard error for the mean of *Y* (the column labeled SE Fit), which you can multiply by $t_{a/2}$ to get the confidence interval width.

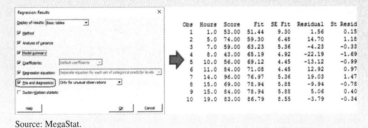

Figure 12.42

Minitab's Exam Score Residuals

Source: MegaStat.

Simple Regression in R

Here are some R functions to estimate a linear regression $MyY = b_0 + b_1 MyX$ and display the results (we are using generic names of all objects):

Fit model and view key results:

```
> My.fit=lm(MyY~MyX,MyData)          # Fit regression and save results
> summary(My.fit)                    # display essential regression results
> coef(My.fit)                       # show only estimated coefficients
> anova(My.fit)                      # anova table for the regression
> plot(MyData$MyX,MyData$MyY)        # scatter plot of data
```

View residuals and perform diagnostic tests:

```
> residuals(My.fit)                        # list of residuals
> rstandard(My.fit)                        # list of standardized residuals
> hist(residuals(My.fit))                  # histogram of residuals
> hist(rstandard(Exam.fit))                # histogram of standardized residuals
> qqnorm(rstandard(My.fit))                # normal probability plot of residuals
> plot(fitted.values(My.fit),residuals(My.fit))    # heteroscedasticity scan
> hatvalues(My.fit)                        # leverage statistic for each observation
```

Example Using Exam Scores 🗁 ExamScores

We have 10 paired observations of study hours and exam scores. Copy two data columns to data frame ExamData using the clipboard (or any other method). List the data. Estimate the regression using the linear model function lm(). View the results using summary() and anova(). Make scatter plots (Exam.Scores versus Study.Hours, residuals versus fits). Perform residual diagnostics (histograms of residuals, normal probability plot of residuals). View leverage statistics.

```
> ExamData=read.table(file="clipboard", sep="\t", header=TRUE)
> ExamData
        Study.Hours     Exam.Score
1           1               53
2           5               74
3           7               59
4           8               43
5           10              56
6           11              84
7           14              96
8           15              69
9           15              84
10          19              83
> Exam.fit=lm(Exam.Score~Study.Hours,ExamData)
> summary(Exam.fit)
Residuals:
```

Min	1Q	Median	3Q	Max
−22.190	−8.510	−1.118	10.954	19.026

Coefficients:

	Estimate Std.	Error	t value	Pr(>|t|)	
(Intercept)	49.477	10.066	4.915	0.00117	**
Study.Hours	1.964	0.861	2.281	0.05197	.

```
Signif. codes: 0 '***'  0.001 '**'  0.01 '*'  0.05  '.'  0.1  ' '  1
Residual standard error: 14.00 on 8 degrees of freedom
Multiple R-squared: 0.3941,   Adjusted R-squared:  0.3184
F-statistic: 5.204 on 1 and 8 DF,   p-value: 0.05197

> coef(Exam.fit)
(Intercept)    Study.Hours
49.477127      1.964083

> anova(Exam.fit)
Analysis of Variance Table
Response: Exam.Score
```

	Df	Sum Sq	Mean Sq	F value	Pr(>F)	
Study.Hours	1	1020.3	1020.34	5.204	0.05197	.
Residuals	8	1568.6	196.07			

```
Signif. codes:  0 '***' 0.001 '**' 0.01 '*' 0.05 '.' 0.1 ' ' 1

> plot(ExamData$Study.Hours,ExamData$Exam.Score)   # scatter plot of data
> plot(residuals(Exam.fit),fitted.values(Exam.fit))      # heteroskedastic?
> hist(residuals(Exam.fit)                               # residual histogram for normality check
> qqnorm(rstandard(Exam.fit))                            # normal probability plot of residuals
> hatvalues(Exam.fit)                                    #leverage statistics
```

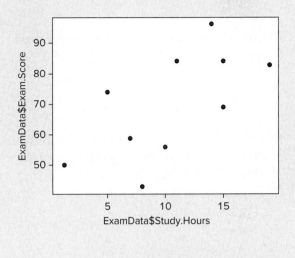

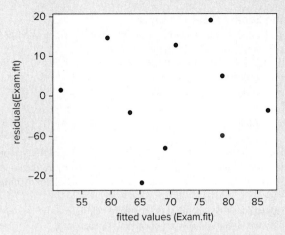

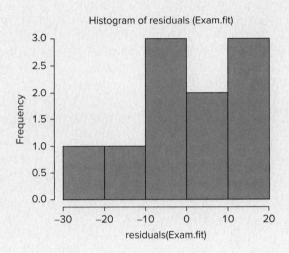

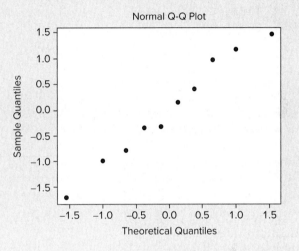

1	2	3	4	5	6	7
0.4412098	0.2143667	0.1463138	0.1236295	0.1009452	0.1009452	0.1463138

8	9	10
0.1765595	0.1765595	0.3731569

```
> # Create object RES to display residual statistics in nice columns
> RES=data.frame     (residuals(Exam.fit),    rstandard(Exam.fit),    hatvalues(Exam.fit))
> RES
      residuals.Exam.fit.    rstandard.Exam.fit.    hatvalues.Exam.fit.1
1         1.558790               0.1489217               0.4412098
2        14.702457               1.1846078               0.2143667
3        -4.225709              -0.3266217               0.1463138
4       -22.189792              -1.6927934               0.1236295
5       -13.117958              -0.9880246               0.1009452
6        12.917958               0.9729609               0.1009452
7        19.025709               1.4705721               0.1463138
8        -9.938374              -0.7821568               0.1765595
9         5.061626               0.3983534               0.1765595
10       -3.794707              -0.3422894               0.3731569
```

Example Using Broker Data 📂 Broker

Does the size of a bank customer's savings account balance predict whether (1) or not (0) the customer has a brokerage account for investments? We have a sample of 20 bank clients:

Client	Savings	Broker
1	1100	0
2	1600	0
. . .	. . .	. . .
19	56000	1
20	72800	1

Copy the two data columns to a data frame SavingData using the clipboard (or any other method). Estimate the logistic regression using the general linear model function glm(). Save the result in an object named MyLogit. Display the results using summary(). Plot the fitted probabilities against Savings to view the S-shaped curve (see textbook Example 12.4).

```
> SavingData=read.table(file="clipboard", sep="\t", header=TRUE)
> MyLogit.fit=glm(Broker~Savings,family="binomial",data=SavingData)
> summary(MyLogit.fit)
Deviance Residuals:
    Min       1Q    Median        3Q       Max
-1.42697  -0.44122  -0.36617   0.07737   2.12688
Coefficients:
              Estimate    Std.Error   z value   Pr(>|z|)
(Intercept)  -4.001e+00   1.663e+00   -2.406    0.0161 *
Savings       1.479e-04   7.156e-05    2.067    0.0387 *
---
Signif. codes: 0 '***'  0.001 '**'  0.01 '*'  0.05 '.'  0.1 ' ' 1
> plot(SavingData$Savings,fitted.values(MyLogit.fit))
```

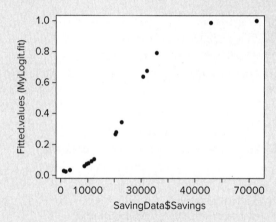

CHAPTER
13 Multiple Regression

CHAPTER CONTENTS

13.1   Multiple Regression

13.2   Assessing Overall Fit

13.3   Predictor Significance

13.4   Confidence Intervals for Y

13.5   Categorical Variables

13.6   Tests for Nonlinearity and Interaction

13.7   Multicollinearity

13.8   Regression Diagnostics

13.9   Other Regression Topics

13.10 Logistic Regression (Optional)

CHAPTER LEARNING OBJECTIVES

When you finish this chapter, you should be able to

LO 13-1   Use a fitted multiple regression equation to make predictions.

LO 13-2   Use the ANOVA table to perform an F test for overall significance.

LO 13-3   Construct confidence intervals for coefficients and test predictors for significance.

LO 13-4   Calculate the standard error and construct approximate confidence and prediction intervals for Y.

LO 13-5   Incorporate categorical variables into a multiple regression model.

LO 13-6   Perform basic tests for nonlinearity and interaction.

LO 13-7   Detect multicollinearity and assess its effects.

LO 13-8   Analyze residuals to check for violations of residual assumptions.

LO 13-9   Identify unusual residuals and tell when they are outliers.

LO 13-10 Identify high leverage observations and their possible causes.

LO 13-11 Explain the purpose of data conditioning and stepwise regression.

LO 13-12 Identify when logistic regression is appropriate and calculate predictions for a binary response variable.

fstop123/Getty Images

13.1 MULTIPLE REGRESSION

Suppose you own a home and need to sell. How do you predict the selling price? It would be naïve to use a *simple regression* model based on one independent variable (e.g., square footage) to predict your home's value. Home values are affected by economic conditions as well as the physical characteristics of your house. Economic conditions might include mortgage interest rates and job opportunities in your city. Lower interest rates tend to be associated with higher home prices because buyers can afford to borrow more to purchase a house. Better job opportunities can increase the demand for homes in your area, which can increase home prices. Physical characteristics such as the size of your home, the number of bathrooms and bedrooms, and the age of the home also are used to determine the price of your home. Because *multiple* variables affect your home's value, we need a model that uses multiple independent variables for predicting.

Multiple regression extends simple regression to include several independent variables (called *predictors* or *explanatory variables*). Multiple regression is required when a single-predictor model does not adequately explain the variation in the dependent variable Y (the response variable). Expanding the regression model to include additional predictors ($X_1, X_2, X_3, \ldots$) can help explain variation and improve predictions. The interpretation of multiple regression is similar to simple regression because simple regression is a special case of multiple regression. In fact, statisticians make no distinction between simple and multiple regression—they just call it *regression*.

Some of the proposed predictors may be useful, while others may not. The regression analysis will tell us whether each variable is useful. One of our objectives in regression modeling is to know whether we have a *parsimonious* model. A parsimonious regression model is a *lean* model, that is, one that has only useful predictors. If an estimated coefficient has a positive (+) sign, then higher X values are associated with higher Y values, and conversely if an estimated coefficient has a negative sign.

Calculations are done by computer; therefore, there is no extra computational burden. However, there are additional interpretations to consider when analyzing a multiple regression. Using multiple predictors is more than a matter of "improving the fit." Rather, it is a question of specifying a correct model. A low R^2 in a simple regression model does not necessarily mean that X and Y are unrelated but may simply indicate that the model is incorrectly specified. Omission of relevant predictors (*model misspecification*) can cause biased estimates and misleading results.

Limitations of Simple Regression

- Multiple relationships usually exist.
- Biased estimates if relevant predictors are omitted.
- Lack of fit does not show that X is unrelated to Y if the true model is multivariate.

Regression Terminology

The **response variable** (Y) is assumed to be related to the k **predictors** ($X_1, X_2, \ldots, X_k$) by a linear equation called the *population regression model*:

(13.1) $$y = \beta_0 + \beta_1 x_1 + \beta_2 x_2 + \cdots + \beta_k x_k + \varepsilon$$

A *random error* ε represents everything that is not part of the model. The unknown regression coefficients $\beta_0, \beta_1, \beta_2, \ldots, \beta_k$ are *parameters* and are denoted by Greek letters. Each coefficient

β_j shows the change in the expected value of Y for a unit change in X_j while holding everything else constant (*ceteris paribus*). The errors are assumed to be unobservable, independent random disturbances that are normally distributed with zero mean and constant variance, that is, $\varepsilon \sim N(0, \sigma^2)$. Under these assumptions, the ordinary least squares (OLS) estimation method yields unbiased, consistent, efficient estimates of the unknown parameters.

The *sample estimates* of the regression coefficients are denoted by Roman letters b_0, b_1, b_2, . . . , b_k. The *predicted* value of the response variable is denoted $\hat{y}$ and is calculated by inserting the values of the predictors into the *estimated regression equation:*

$$\hat{y} = b_0 + b_1 x_1 + b_2 x_2 + \cdots + b_k x_k \quad \text{(predicted value of } Y) \tag{13.2}$$

In this chapter, we will not show formulas for the estimated coefficients b_0, b_1, b_2, . . . , b_k because they entail matrix algebra. Regression equations are estimated by computer software (Excel, MegaStat, Minitab, R) utilizing the appropriate formulas.

In a simple regression (one predictor), the fitted regression is a *line,* while in multiple regression (e.g., two predictors), the fitted regression is a *surface* or *plane,* as illustrated in Figure 13.1. If there are more than two predictors, no diagram can be drawn because the fitted regression is represented by a hyperplane.

Data Format

To obtain a fitted regression, we need n observed values of the response variable Y and its proposed predictors X_1, X_2, . . . , X_k. A multivariate data set is a single column of Y-values and k columns of X-values. The form of this $n \times k$ matrix of observations is shown in Figure 13.2.

In Excel's Data Analysis > Regression, you are required to have the X data in contiguous columns. However, other software packages permit nonadjacent columns of X data. Flexibility in choosing data columns is useful if you decide to omit one or more X data columns and rerun the regression (e.g., to seek parsimony).

Figure 13.1

Fitted Regression: Bivariate versus Multivariate

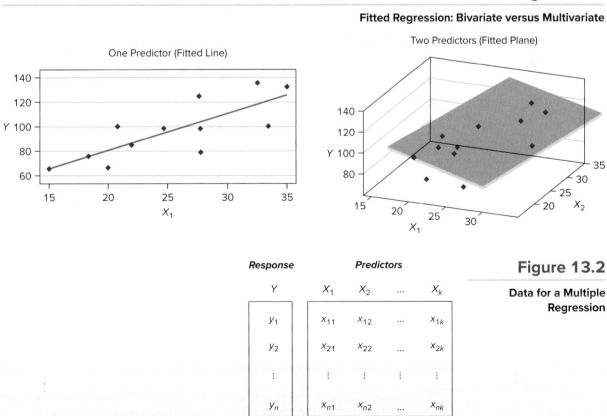

Figure 13.2

Data for a Multiple Regression

Illustration: Home Prices

Table 13.1 shows sales of 30 new homes in an upscale development. Although the selling price of a home (the *response variable*) may depend on many factors, we will examine three potential *explanatory variables.*

Definition of Variable	Short Name
Y = selling price of a home (thousands of dollars)	*Price*
X_1 = home size (square feet)	*SqFt*
X_2 = lot size (thousand square feet)	*LotSize*
X_3 = number of bathrooms	*Baths*

Table 13.1 Characteristics of 30 New Homes NewHomes

Home	Price	SqFt	LotSize	Baths	Home	Price	SqFt	LotSize	Baths
1	505.5	2,192	16.4	2.5	16	675.1	3,076	19.8	3.0
2	784.1	3,429	24.7	3.5	17	710.4	3,259	20.8	3.5
3	649.0	2,842	17.7	3.5	18	674.7	3,162	19.4	4.0
4	689.8	2,987	20.3	3.5	19	663.6	2,885	23.2	3.0
5	709.8	3,029	22.2	3.0	20	606.6	2,550	20.2	3.0
6	590.2	2,616	20.8	2.5	21	758.9	3,380	19.6	4.5
7	643.3	2,978	17.3	3.0	22	723.7	3,131	22.5	3.5
8	789.7	3,595	22.4	3.5	23	621.8	2,754	19.2	2.5
9	683.0	2,838	27.4	3.0	24	622.4	2,710	21.6	3.0
10	544.3	2,591	19.2	2.0	25	631.3	2,616	20.8	2.5
11	822.8	3,633	26.9	4.0	26	574.0	2,608	17.3	3.5
12	637.7	2,822	23.1	3.0	27	863.8	3,572	29.0	4.0
13	618.7	2,994	20.4	3.0	28	652.7	2,924	21.8	2.5
14	619.3	2,696	22.7	3.5	29	844.2	3,614	25.5	3.5
15	490.5	2,134	13.4	2.5	30	629.9	2,600	24.1	3.5

Using short variable names instead of Y and X, we may write the regression model in an intuitive form:

$$Price = \beta_0 + \beta_1 SqFt + \beta_2 LotSize + \beta_3 Baths + \varepsilon$$

Logic of Variable Selection

Before doing the estimation, it is desirable to state our hypotheses about the sign of the coefficients in the model. In so doing, we force ourselves to think about our motives for including each predictor instead of just throwing predictors into the model willy-nilly. In the home price example, each predictor is expected to contribute positively to the selling price.

Predictor	Anticipated Sign	Reasoning
SqFt	>0	Larger homes cost more to build and give greater utility to the buyer.
LotSize	>0	Larger lots are desirable for privacy, gardening, and play.
Baths	>0	Additional baths give more utility to the purchaser with a family.

Explicit *a priori* reasoning about cause and effect permits us to compare the regression estimates with our expectation and to recognize any surprising results that may occur.

Estimated Regression

A regression equation can be estimated using Excel or any other statistical package. Using the sample of $n = 30$ home sales, we obtain the fitted regression and its statistics of fit (R^2 is the coefficient of determination and *SE* is the standard error):

$$Price = -28.85 + 0.171 \; SqFt + 6.78 \; LotSize + 15.53 \; Baths \quad (R^2 = .956, \; SE = 20.31)$$

The intercept is not meaningful because there can be no home with $SqFt = 0$, $LotSize = 0$, and $Baths = 0$. Each additional square foot seems to add about 0.171 (i.e., \$171 because *Price* is measured in thousands of dollars) to the average selling price, *ceteris paribus*. The coefficient of *LotSize* implies that, on average, each additional thousand square feet of lot size adds 6.78 (i.e., \$6,780) to the selling price. The coefficient of *Baths* says that, on average, each additional bathroom adds 15.53 (i.e., \$15,530) to the selling price. Although the fit ($R^2 = .956$) is good, its standard error (20.31 or \$20,310) suggests that prediction intervals will be wide.

Predictions from a Fitted Regression

We can use the fitted regression model to make predictions for various assumed predictor values. For example, what would be the expected selling price of a 2,800-square-foot home with 2½ baths on a lot with 18,500 square feet? In the fitted regression equation, we simply plug in $SqFt = 2800$, $LotSize = 18.5$, and $Baths = 2.5$ to get the predicted selling price:

$$SqFt = 2800 \quad LotSize = 18.5 \quad Baths = 2.5$$

$$Price = -28.85 + 0.171(2800) + 6.78(18.5) + 15.53(2.5) = 614.23, \text{ or } \$614,230$$

Although we could plug in any desired values of the predictors (*SqFt, LotSize, Baths*), it is risky to use predictor values outside the predictor value ranges in the data set used to estimate the fitted regression. For example, it would be risky to choose $SqFt = 4000$ because no home this large was seen in the original data set. Although the prediction might turn out to be reasonable, we would be extrapolating beyond the range of observed data.

Common Misconceptions about Fit

A common mistake is to assume that the equation that best fits our observed data is preferred. Sometimes a model with a low R^2 may give useful predictions, while a model with a high R^2 may conceal problems. Fit is only one criterion for assessing a regression. For example, a bivariate model using only *SqFt* as a predictor does a pretty good job of predicting *Price* and has an attractive simplicity:

$$Price = 15.47 + 0.222 \; SqFt \quad (R^2 = .914, \; s = 27.28)$$

Should we perhaps prefer the simpler model? The principle of **Occam's Razor** says that a complex model that is only slightly better may not be preferred if a simpler model will do the job. However, in this case, the three-predictor model is not very complex and is based on solid *a priori* logic.

Principle of Occam's Razor

When two explanations are otherwise equivalent, we prefer the simpler, more parsimonious one.

Also, a high R^2 only indicates a good fit for the observed data set ($i = 1, 2, \ldots, n$). If we wanted to use the fitted regression equation to predict Y from a different set of X's, the fit might not be the same. For this reason, if the sample is large enough, a statistician likes to use half the data to *estimate* the model and the other half to *test* the model's predictions.

Regression Modeling

The choices of predictors and model form (e.g., linear or nonlinear) are tasks of *regression modeling*. To begin with, we restrict our attention to predictors that meet the test of *a priori* logic to avoid endless "data shopping." Naturally, we want predictors that are significant in "explaining" the variation in *Y* (i.e., predictors that improve the "fit"). But we also prefer predictors that add new information rather than mirroring one another.

For example, we would expect that *LotSize* and *SqFt* are related (a bigger house may require a bigger lot) and likewise *SqFt* and *Baths* (a bigger house is likely to require more baths). If so, there may be overlap in their contributions to explaining *Price*. Closely related predictors can introduce instability in the regression estimates. If we include too many predictors, we violate the principle of Occam's Razor, which favors simple models, *ceteris paribus*. In this chapter, you will see how these criteria can be used to develop and assess regression models.

Four Criteria for Regression Assessment

- **Logic**   Is there an *a priori* reason to expect a causal relationship between the predictors and the response variable?

- **Fit**   Does the *overall* regression show a significant relationship between the predictors and the response variable?

- **Parsimony**   Does *each predictor* contribute significantly to the explanation? Are some predictors not worth the trouble?

- **Stability**   Are the predictors related to one another so strongly that regression estimates become erratic?

Section Exercises

13.1 Observations are taken on net revenue from sales of a certain LCD TV at 50 retail outlets. The regression model was Y = net revenue (thousands of dollars), X_1 = shipping cost (dollars per unit), X_2 = expenditures on print advertising (thousands of dollars), X_3 = expenditure on electronic media ads (thousands), X_4 = rebate rate (percent of retail price). (a) Write the fitted regression equation. (b) Interpret each coefficient. (c) Would the intercept be likely to have meaning in this regression? (d) Use the fitted equation to make a prediction for *NetRevenue* when *ShipCost* = 10, *PrintAds* = 50, *WebAds* = 40, and *Rebate%* = 15. ✉ **LCDTV**

Predictor	Coefficient
Intercept	4.306
ShipCost	−0.082
PrintAds	2.265
WebAds	2.498
Rebate%	16.697

13.2 Observations are taken on sales of a certain mountain bike in 30 sporting goods stores. The regression model was Y = total sales (thousands of dollars), X_1 = display floor space (square meters), X_2 = competitors' advertising expenditures (thousands of dollars), X_3 = advertised price (dollars per unit). (a) Write the fitted regression equation. (b) Interpret each coefficient. (c) Would the intercept seem to have meaning in this regression? (d) Make a prediction for *Sales* when *FloorSpace* = 80, *CompetingAds* = 100, and *Price* = 1,200. ✉ **Bikes**

Predictor	Coefficient
Intercept	1225.44
FloorSpace	11.52
CompetingAds	−6.935
Price	−0.1496

13.3 A ski resort asked a random sample of guests to rate their satisfaction on various attributes of their visit on a scale of 1–5 with 1 = very unsatisfied and 5 = very satisfied. The estimated regression model was Y = overall satisfaction score, X_1 = lift line wait, X_2 = amount of ski trail grooming, X_3 = safety patrol visibility, and X_4 = friendliness of guest services. (a) Write the fitted regression equation. (b) Interpret each coefficient. (c) Would the intercept seem to have meaning in this regression? (d) Make a prediction for *Overall Satisfaction* when a guest's satisfaction in all four areas is rated a 5. 🖅 **ResortGuestSat**

Predictor	Coefficient
Intercept	2.8931
LiftWait	0.1542
AmountGroomed	0.2495
SkiPatrolVisibility	0.0539
FriendlinessHosts	−0.1196

13.4 A regression model to predict Y, the state-by-state burglary crime rate per 100,000 people, used the following four state predictors: X_1 = median age in years, X_2 = number of bankruptcies per 1,000 people, X_3 = federal expenditures per capita, and X_4 = high school graduation percentage. (a) Write the fitted regression equation. (b) Interpret each coefficient. (c) Would the intercept seem to have meaning in this regression? (d) Make a prediction for *Burglary* when X_1 = 35 years, X_2 = 7.0 bankruptcies per 1,000, X_3 = $6,000, and X_4 = 80 percent. 🖅 **Burglary**

Predictor	Coefficient
Intercept	4,198.5808
AgeMed	−27.3540
Bankrupt	17.4893
FedSpend	−0.0124
HSGrad%	−29.0314

13.2 ASSESSING OVERALL FIT

As in simple regression, there is one residual for every observation in a multiple regression:

$$e_i = y_i - \hat{y}_i \qquad \text{for } i = 1, 2, \ldots, n$$

Figure 13.3 illustrates the residual for one data value in a two-predictor regression. Each expected value of Y is a point on the fitted regression plane for a given pair of X-values (x_1, x_2). The residual is the vertical distance from the actual y_i value for those particular X-values (x_1, x_2) to $\hat{y}_i$. Just as in simple regression, we use the sum of squared residuals (SSE) as a measure of "fit" of the model.

LO 13-2

Use the ANOVA table to perform an *F* test for overall significance.

In a multiple regression, each residual is a vertical distance from the actual *Y*-value to the expected (predicted) *Y* on the regression plane

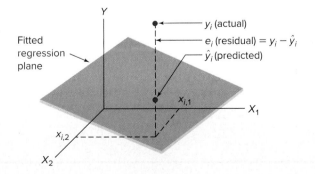

Figure 13.3

Residual in Two-Predictor Model

F Test for Significance

Before determining which, if any, of the individual predictors are significant, we perform a *global test* for overall fit using the **F test**. For a regression with k predictors, the hypotheses to be tested are

H_0: All the true coefficients are zero ($\beta_1 = \beta_2 = \cdots = \beta_k = 0$)

H_1: At least one of the coefficients is nonzero

Recall from Chapters 10 and 11 that the F statistic is the ratio of two variances. The basis for the regression F test is the **ANOVA table**, which decomposes variation of the response variable around its mean into two parts:

(13.3)

$$\underset{\substack{\text{Total} \\ \text{variation}}}{SST} = \underset{\substack{\text{Explained by} \\ \text{regression}}}{SSR_i} + \underset{\substack{\text{Unexplained} \\ \text{error}}}{SSE}$$

$$\sum_{i=1}^{n}(y_i - \overline{y})^2 = \sum_{i=1}^{n}(\hat{y}_i - \overline{y})^2 + \sum_{i=1}^{n}(y_i - \hat{y}_i)^2$$

The OLS method of estimation will minimize the sum of the squared residuals represented by the *SSE* term, in formula 13.3 above, where *SSE* is the *unexplained* variation in *Y*. Each predicted value $\hat{y}_i$ is based on a fitted regression equation with k predictors. The ANOVA calculations for a k-predictor model can be summarized in a table like Table 13.2.

When F_{calc} is close to 1, the values of *MSR* and *MSE* are close in magnitude. This suggests that *none* of the predictors provides a good predictive model for *Y* (i.e., all β_j are equal to 0). When the value of *MSR* is much greater than *MSE*, this suggests that at least one of the predictors in the regression model is significant (i.e., at least one β_j is not equal to 0).

Table 13.2

ANOVA Table Format

Source of Variation	Sum of Squares	df	Mean Square	F	Excel p-Value
Regression (explained)	$SSR = \sum_{i=1}^{n}(\hat{y}_i - \overline{y})^2$	k	$MSR = \dfrac{SSR}{k}$	$F_{calc} = \dfrac{MSR}{MSE}$	=F.DIST.RT(F_{calc}, k, n−k−1)
Residual (unexplained)	$SSE_i = \sum_{i=1}^{n}(y_i - \hat{y}_i)^2$	$n - k - 1$	$MSE = \dfrac{SSE}{n-k-1}$		
Total	$SST = \sum_{i=1}^{n}(y_i - \overline{y})^2$	$n - 1$			

After simplifying the ratios in Table 13.2, the formula for the F test statistic is

(13.4)

$$F_{calc} = \frac{MSR}{MSE} = \frac{\sum_{i=1}^{n}(\hat{y}_i - \overline{y})^2}{\sum_{i=1}^{n}(y_i - \hat{y}_i)^2}\left(\frac{n - k - 1}{k}\right)$$

Table 13.3 shows the ANOVA table for the home price regression with $n = 30$ observations and $k = 3$ predictors.

Table 13.3

ANOVA Results for Three-Predictor Home Price Regression

Source	Sum of Squares	df	Mean Square	F	p-Value
Regression	232,450	3	77,483	187.92	.0000
Error	10,720	26	412.32		
Total	243,170	29			

The hypotheses to be tested are

H_0: All the coefficients are zero ($\beta_1 = \beta_2 = \beta_3 = 0$)

H_1: At least one coefficient is nonzero

Calculation of the sums *SSR, SSE,* and *SST* would be tedious without the computer. The *F* test statistic is $F_{calc} = MSR/MSE = 77{,}483/412.32 = 187.92$. Degrees of freedom are $k = 3$ for the numerator and $n - k - 1 = 30 - 3 - 1 = 26$ for the denominator. For $\alpha = .05$, Appendix F gives a critical value of $F_{3,26} = 2.98$, so the regression clearly is significant overall. We also can use Excel's function =F.DIST.RT(187.92,3,26) to verify the *p*-value (.000).

Coefficient of Determination (R^2)

The most common measure of overall fit is the **coefficient of determination** or R^2, which is based on the ANOVA table's sums of squares. It can be calculated in two ways by using the error sum of squares (*SSE*), regression sum of squares (*SSR*), and total sum of squares (*SST*). The formulas are illustrated using the three-predictor regression of home prices.

$$R^2 = 1 - \frac{SSE}{SST} = 1 - \frac{\sum_{i=1}^{n}(y_i - \hat{y}_i)^2}{\sum_{i=1}^{n}(y_i - \bar{y})^2} = 1 - \frac{10{,}720}{243{,}170} = 1 - .044 = .956 \qquad \textbf{(13.5)}$$

or equivalently

$$R^2 = \frac{SSR}{SST} = \frac{\sum_{i=1}^{n}(\hat{y}_i - \bar{y})^2}{\sum_{i=1}^{n}(y_i - \bar{y})^2} = \frac{232{,}450}{243{,}170} = .956 \qquad \textbf{(13.6)}$$

For the home price data, the R^2 statistic indicates that 95.6 percent of the variation in selling price is "explained" by our three predictors. While this indicates a very good fit, there is still some unexplained variation. Adding more predictors can *never* decrease the R^2. However, when R^2 already is high, there is not a lot of room for improvement.

Adjusted R^2

In multiple regression, it is possible to raise the coefficient of determination R^2 by including additional predictors. This may tempt you to imagine that we should always include many predictors to get a "better fit." To discourage this tactic (called *overfitting* the model), an adjustment can be made to the R^2 statistic to penalize the inclusion of useless predictors. The **adjusted coefficient of determination** using *n* observations and *k* predictors is

$$R^2_{adj} = 1 - \frac{\left(\dfrac{SSE}{n - k - 1}\right)}{\left(\dfrac{SST}{n - 1}\right)} \qquad \text{(adjusted } R^2) \qquad \textbf{(13.7)}$$

R^2_{adj} is always less than R^2. As you add predictors, R^2 will not decrease. But R^2_{adj} may increase, remain the same, or decrease, depending on whether the added predictors increase R^2 sufficiently to offset the penalty. If R^2_{adj} is substantially smaller than R^2, it suggests that the model contains useless predictors. For the home price data with three predictors, both statistics are similar ($R^2 = .956$ and $R^2_{adj} = .951$), which suggests that the model does not contain useless predictors.

$$R^2_{adj} = 1 - \frac{\left(\dfrac{10{,}720}{26}\right)}{\left(\dfrac{243{,}170}{29}\right)} = .951$$

There is no fixed rule of thumb for comparing R^2 and R^2_{adj}. A smaller gap between R^2 and R^2_{adj} indicates a more parsimonious model. A large gap would suggest that if some weak predictors were deleted, a leaner model would be obtained without losing very much predictive power.

How Many Predictors?

One way to prevent overfitting the model is to limit the number of predictors based on the sample size. A conservative rule (**Evans' Rule**) suggests that n/k should be at least 10 (i.e., at least 10 observations per predictor). A more relaxed rule (**Doane's Rule**) suggests that n/k be at least 5 (i.e., at least 5 observations per predictor).[1] For the home price regression with $n = 30$ and $k = 3$ example, $n/k = 30/3 = 10$, so either guideline is met.

> Evans' Rule (*conservative*): $n/k \geq 10$ (at least 10 observations per predictor)
> Doane's Rule (*relaxed*): $n/k \geq 5$ (at least 5 observations per predictor)

These rules are merely suggestions. Technically, a regression is possible as long as the sample size exceeds the number of predictors. But when n/k is small, the R^2 no longer gives a reliable indication of fit. Sometimes, researchers must work with small samples that cannot be enlarged. For example, a start-up business selling a new type of insulin-monitoring device might have only 12 observations on quarterly sales. Should it attempt a regression model to predict sales using two predictors (advertising, product price)? Although $n = 12$ and $k = 2$ would violate the conservative guideline ($n/k = 12/2 = 6$), the firm might feel that an imperfect analysis is better than none at all. On the other hand, in modern data mining applications, enormous sample sizes are possible. When predictors can be included at will, the analyst may face the opposite problem: distilling a parsimonious model from a huge set of potential predictors.

Section Exercises

connect

13.5 Refer to the ANOVA table below. (a) State the degrees of freedom for the F test for overall significance. (b) Use Appendix F to look up the critical value of F for $\alpha = .05$. (c) Calculate the F statistic. Is the regression significant overall? (d) Calculate R^2 and R^2_{adj}, showing your formulas clearly. **LCDTV**

Source	df	SS	MS
Regression	4	259,412	64,853
Error	45	224,539	4,990
Total	49	483,951	

13.6 Refer to the ANOVA table below. (a) State the degrees of freedom for the F test for overall significance. (b) Use Appendix F to look up the critical value of F for $\alpha = .05$. (c) Calculate the F statistic. Is the regression significant overall? (d) Calculate R^2 and R^2_{adj}, showing your formulas clearly. **Bikes**

Source	df	SS	MS
Regression	3	1,196,410	398,803
Error	26	379,332	14,590
Total	29	1,575,742	

13.7 Refer to the ANOVA table below. (a) State the degrees of freedom for the F test for overall significance. (b) Use Appendix F to look up the critical value of F for $\alpha = .05$. (c) Calculate the F statistic. Is the regression significant overall? (d) Calculate R^2 and R^2_{adj}, showing your formulas clearly. **ResortGuestSat**

Source	df	SS	MS
Regression	4	33.0730	8.2682
Residual	497	317.9868	0.6398
Total	501	351.0598	

---

[1] If information is assumed proportional to the logarithm of sample size, a sample twice as large *(Evans' Rule)* would offer only 43 percent more information *(Doane's Rule)* because $\log(10)/\log(5) = 1.43$.

13.8   Refer to the ANOVA table below. (a) State the degrees of freedom for the F test for overall significance. (b) Use Appendix F to look up the critical value of F for $\alpha = .05$. (c) Calculate the F statistic. Is the regression significant overall? (d) Calculate R^2 and R^2_{adj}, showing your formulas clearly. **Burglary**

Source	df	SS	MS
Regression	4	1,182,733	295,683
Residual	45	1,584,952	35,221
Total	49	2,767,685	

13.3   PREDICTOR SIGNIFICANCE

Hypothesis Tests

Each estimated coefficient shows the change in the conditional mean of Y associated with a one-unit change in an explanatory variable, holding the other explanatory variables constant. If a predictor coefficient β_j is equal to zero, it means that the explanatory variable X_j does not help explain variation in the response variable Y. We are usually interested in testing each fitted coefficient to see whether it is significantly different from zero. If there is an *a priori* reason to anticipate a particular direction of association, we could choose a right-tailed or left-tailed test. For example, we would expect *SqFt* to have a positive effect on *Price,* so a right-tailed test might be used. However, the default choice is a two-tailed test because, if the null hypothesis can be rejected in a two-tailed test, it also can be rejected in a one-tailed test at the same level of significance.

LO 13-3

Construct confidence intervals for coefficients and test predictors for significance.

Hypothesis Tests for Coefficient of Predictor X_j

Left-Tailed Test	*Two-Tailed Test*	*Right-Tailed Test*
$H_0: \beta_j = 0$	$H_0: \beta_j = 0$	$H_0: \beta_j = 0$
$H_1: \beta_j < 0$	$H_1: \beta_j \neq 0$	$H_1: \beta_j > 0$

p-Values and Software

Software packages, including Excel, report only two-tailed *p*-values because if you can reject H_0 in a two-tailed test, you also can reject H_0 in a one-tailed test at the same α.

If we cannot reject the hypothesis that a coefficient is zero, then the corresponding predictor does not significantly contribute to the prediction of Y. For example, consider a three-predictor model:

$$y = \beta_0 + \beta_1 x_1 + \beta_2 x_2 + \beta_3 x_3 + \varepsilon$$

Does X_2 help us to predict Y? To find out, we might choose a two-tailed test:

$H_0: \beta_2 = 0$ (X_2 is *not* related to Y)
$H_1: \beta_2 \neq 0$ (X_2 *is* related to Y)

If we are unable to reject H_0, the term involving x_2 will drop out:

$$y = \beta_0 + \beta_1 x_1 + \boxed{0 x_2} + \beta_3 x_3 + \varepsilon \qquad (x_2 \text{ term drops out if } \beta_2 = 0)$$

and the regression will collapse to a *two-variable* model:

$$y = \beta_0 + \beta_1 x_1 + \beta_3 x_3 + \varepsilon$$

Test Statistic

Rarely would a fitted coefficient be *exactly* zero, so we use a t test to test whether the difference from zero[2] is *significant*. For predictor X_j, the test statistic for k predictors is Student's t with $n - k - 1$ degrees of freedom. To test for a zero coefficient, we take the ratio of the fitted coefficient b_j to its standard error s_j:

(13.8)
$$t_{\text{calc}} = \frac{b_j - 0}{s_j} \quad \text{(test statistic for coefficient of predictor } X_j \text{)}$$

where

(13.9)
$$s_j = \sqrt{\frac{MSE}{SS_{X_j}(1 - R_j^2)}}$$

The value of s_j is usually not calculated by the analyst but rather taken from the regression output because the calculation is tedious. The value of *MSE* comes from the ANOVA table on the regression output (see Table 13.2). SS_{X_j} is the sum of the squared deviations of X_j about its mean and R_j^2 is the coefficient of determination when predictor j is regressed against *all* the other predictors (excluding Y).

We can use Appendix D to find a critical value of t for a chosen level of significance α, or we could find the p-value for the t statistic using Excel's function =T.DIST.2T(t, deg_freedom). All computer packages report the t statistic and the p-value for each predictor, so we actually do not need tables. To test for a zero coefficient, we could alternatively construct a confidence interval for the true coefficient β_j and see whether the interval includes zero. Formula 13.10 shows the confidence interval formula for a coefficient:

(13.10)
$$b_j - t_{\alpha/2} s_j \le \beta_j \le b_j + t_{\alpha/2} s_j \quad \text{(confidence interval for coefficient } \beta_j \text{)}$$

When using Excel to perform the regression analysis, you can enter any confidence level you wish. All calculations are provided by Excel, so you only have to know how to interpret the results.

Confidence Intervals versus Hypothesis Tests

Checking to see whether the confidence interval includes zero is equivalent to a two-tailed test of H_0: $\beta_j = 0$.

Example 13.1

Home Prices

📁 **NewHomes**

Figure 13.4 shows the fitted regression for the three-predictor model of home prices, including a table of estimated coefficients, standard errors, t statistics, and p-values. Excel computes two-tailed p-values, as do most statistical packages. Notice that 0 is within the 95 percent confidence interval for *Baths*, while the confidence intervals for *SqFt* and *LotSize* do not include 0. This suggests that the hypothesis of a zero coefficient can be rejected for *SqFt* and *LotSize* but not for *Baths*.

Figure 13.4   **Regression for Home Prices (three predictors)**

Regression output					Confidence Interval	
Variables	Coefficients	Std. Error	t Stat	p-Value	Lower 95%	Upper 95%
Intercept	−28.8477	29.7115	−0.971	0.3405	−89.9206	32.2251
SqFt	0.1709	0.0154	11.064	0.0000	0.1392	0.2027
LotSize	6.7777	1.4213	4.769	0.0001	3.8562	9.6992
Baths	15.5347	9.2083	1.687	0.1036	−3.3932	34.4626

---

[2]While $\beta_j = 0$ is the default null hypothesis in Excel and other statistical packages, you could specify a nonzero value. For example, to test whether an extra square foot adds at least $200 to a home's selling price, you would use 200 instead of 0 in the formula for the test statistic.

There are four estimated coefficients (counting the intercept). For reasons stated previously, the intercept is of no interest. For the three predictors, each t test uses $n - k - 1$ degrees of freedom. Because we have $n = 30$ observations and $k = 3$ predictors, we have $n - k - 1 = 30 - 3 - 1 = 26$ degrees of freedom. From Appendix D we can obtain two-tailed critical values of t for α equal to .10, .05, or .01 ($t_{.05} = 1.706$, $t_{.025} = 2.056$, and $t_{.005} = 2.779$). However, because p-values are provided, we do not really need these critical values.

$$SqF: t_{calc} = 0.1709/0.01545 = 11.06 \ (p\text{-value} = .0000)$$

$$LotSize: t_{calc} = 6.778/1.421 = 4.77 \ (p\text{-value} = .0001)$$

$$Baths: t_{calc} = 15.535/9.208 = 1.69 \ (p\text{-value} = .1036)$$

The coefficients of *SqFt* and *LotSize* differ significantly from zero at any common α because their p-values are practically zero. The coefficient of *Baths* is not quite significant at $\alpha = .10$. Based on the t-values, we conclude that *SqFt* is a very significant predictor of *Price,* followed closely by *LotSize,* while *Baths* is of marginal significance.

It is important to consider the coefficients of the predictor variables as well. Our three predictors have the following coefficients: 0.1709 for *SqFt,* 6.7777 for *LotSize,* and 15.5347 for *Baths.* Are these coefficient values logical? Recall that the response variable has been scaled by $1000, so each coefficient should be multiplied by 1000 in order to understand its impact on *Price.* The coefficient of *SqFt* says that the average price per square foot for a home in this neighborhood is $170.9 ($0.1709 \times 1000$). The average price per thousand square feet of lot size is $6777.7, and the average price for an additional bathroom is $15,534.7. Each of these coefficient values is reasonable and helps validate the regression model. One might choose to keep the predictor *Baths* in the model because its p-value is very close to .10 and its coefficient is logical.

Regression output contains many statistics, but some of them are especially important in getting the "big picture." Figure 13.5 shows a typical regression printout ($Y =$ car theft rate in the 50 states) with certain key features circled and comments that a statistician might make.

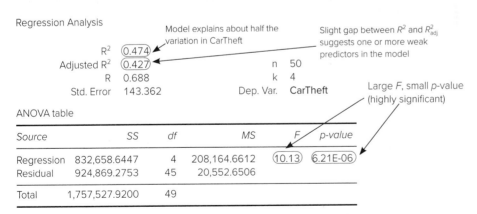

Figure 13.5

Typical Regression Output

Regression Analysis

Model explains about half the variation in CarTheft

Slight gap between R^2 and R^2_{adj} suggests one or more weak predictors in the model

R^2	(0.474)
Adjusted R^2	(0.427)
R	0.688
Std. Error	143.362

n	50
k	4
Dep. Var.	CarTheft

Large F, small p-value (highly significant)

ANOVA table

Source	SS	df	MS	F	p-value
Regression	832,658.6447	4	208,164.6612	(10.13)	(6.21E-06)
Residual	924,869.2753	45	20,552.6506		
Total	1,757,527.9200	49			

Regression output   *confidence interval*

variables	coefficients	std. error	t (df = 45)	p-value	95% lower	95% upper
Intercept	−1,097.5258	254.7116	−4.309	.0001	−1,610.5413	−584.5103
Income	0.0230	0.0063	3.630	.0007	0.0103	0.0358
Unem	40.6282	18.3963	2.208	.0323	3.5762	77.6801
Pupil/Tea	37.6388	9.7047	3.878	.0003	18.0926	57.1851
Divorce	26.3106	20.1711	(1.304)	(1987)	(−14.3161	66.9374)

This predictor is not significant even at $\alpha = .10$

Zero is included in 95% CI, so coefficient for Divorce does not differ from 0

13.9 Observations are taken on net revenue from sales of a certain LCD TV at 50 retail outlets. The regression model was Y = net revenue (thousands of dollars), X_1 = shipping cost (dollars per unit), X_2 = expenditures on print advertising (thousands of dollars), X_3 = expenditure on electronic media ads (thousands), X_4 = rebate rate (percent of retail price). (a) Calculate the t statistic for each coefficient to test for $\beta = 0$. (b) Look up the critical value of Student's t in Appendix D for a two-tailed test at $\alpha = .01$. Which coefficients differ significantly from zero? (c) Use Excel to find the p-value for each coefficient. 📊 **LCDTV**

Predictor	Coefficient	SE
Intercept	4.310	70.82
ShipCost	−0.0820	4.678
PrintAds	2.265	1.050
WebAds	2.498	0.8457
Rebate	16.697	3.570

13.10 Observations are taken on sales of a certain mountain bike in 30 sporting goods stores. The regression model was Y = total sales (thousands of dollars), X_1 = display floor space (square meters), X_2 = competitors' advertising expenditures (thousands of dollars), X_3 = advertised price (dollars per unit). (a) Calculate the t statistic for each coefficient to test for $\beta = 0$. (b) Look up the critical value of Student's t in Appendix D for a two-tailed test at $\alpha = .01$. Which coefficients differ significantly from zero? (c) Use Excel to find the p-value for each coefficient. 📊 **Bikes**

Predictor	Coefficient	SE
Intercept	1225.4	397.3
FloorSpace	11.522	1.330
CompetingAds	−6.935	3.905
Price	−0.14955	0.08927

13.11 A random sample of 502 ski resort guests were asked to rate their satisfaction on various attributes of their visit on a scale of 1–5 with 1 = very unsatisfied and 5 = very satisfied. The regression model was Y = overall satisfaction score, X_1 = lift line wait, X_2 = amount of ski trail grooming, X_3 = ski patrol visibility, and X_4 = friendliness of guest services. (a) Calculate the t statistic for each coefficient to test for $\beta_j = 0$. (b) Look up the critical value of Student's t in Appendix D for a two-tailed test at $\alpha = .01$. Which coefficients differ significantly from zero? (c) Use Excel to find a p-value for each coefficient. 📊 **ResortGuestSat**

Predictor	Coefficient	SE
Intercept	2.8931	0.3680
LiftWait	0.1542	0.0440
AmountGroomed	0.2495	0.0529
SkiPatrolVisibility	0.0539	0.0443
FriendlinessHosts	−0.1196	0.0623

13.12 A regression model to predict Y, the state burglary rate per 100,000 people, used the following four state predictors: X_1 = median age in years, X_2 = number of bankruptcies per 1,000 population, X_3 = federal expenditures per capita (a *leading* predictor), and X_4 = high school graduation percentage. (a) Calculate the t statistic for each coefficient to test for $\beta_j = 0$. (b) Look up the critical value of Student's t in Appendix D for a two-tailed test at $\alpha = .01$. Which coefficients differ significantly from zero? (c) Use Excel to find a p-value for each coefficient. 📊 **Burglary**

Predictor	Coefficient	SE
Intercept	4,198.5808	799.3395
AgeMed	−27.3540	12.5687
Bankrupt	17.4893	12.4033
FedSpend	−0.0124	0.0176
HSGrad%	−29.0314	7.1268

13.4 CONFIDENCE INTERVALS FOR *Y*

Standard Error

Another important measure of fit is the **standard error** (s_e) of the regression, derived from the sum of squared residuals (*SSE*) for *n* observations and *k* predictors:

$$s_e = \sqrt{\frac{\sum_{i=1}^{n}(y_i - \hat{y}_i)^2}{n - k - 1}} = \sqrt{\frac{SSE}{n - k - 1}} \qquad \text{(standard error of the regression)} \quad \textbf{(13.11)}$$

LO 13-4

Calculate the standard error and construct approximate confidence and prediction intervals for *Y*.

The standard error is measured in the same units as the response variable *Y* (dollars, square feet, etc.). A smaller s_e indicates a better fit. If all predictions were perfect (i.e., if $y_i = \hat{y}_i$ for all observations), then s_e would be zero. However, perfect predictions are unlikely.

> From the ANOVA table for the three-predictor home price model, we obtain $SSE = 10{,}720$, so
>
> $$s_e = \sqrt{\frac{SSE}{n - k - 1}} = \sqrt{\frac{10{,}720}{30 - 3 - 1}} = 20.31$$
>
> $s_e = 20.31$ (i.e., \$20,310 because *Y* is measured in thousands of dollars) suggests that the model has room for improvement despite its good fit ($R^2 = .956$). Forecasters find the standard error more useful than R^2 because s_e tells more about the *practical utility* of the forecasts, especially when it is used to make confidence or prediction intervals.

Example 13.2

Home Prices II

Approximate Confidence and Prediction Intervals for *Y*

In a multiple regression model, matrix algebra is required to construct *exact* confidence or prediction intervals for a given set of *X* values. However, we can use the standard error to create useful *approximate*[3] confidence or prediction intervals for $\mu_{Y|X}$ or *Y*. The intervals are most accurate when $X_1, X_2, \ldots, X_k$ are near their respective means (a reasonable case). Although these approximate intervals somewhat understate the interval widths, they are helpful when you only need a general idea of the accuracy of your model's predictions. Formula 13.12 gives an approximate confidence interval for the conditional mean of *Y*, while formula 13.13 gives an approximate prediction interval for individual values of *Y*. The formulas are similar, except that prediction intervals are wider because *individual Y* values vary more than the *mean* of *Y*.

$$\hat{y}_i \pm t_{\alpha/2}\frac{s_e}{\sqrt{n}} \qquad \text{(approximate confidence interval for conditional mean of } Y) \qquad \textbf{(13.12)}$$

$$\hat{y}_i \pm t_{\alpha/2}s_e \qquad \text{(approximate prediction interval for individual } Y\text{-value)} \qquad \textbf{(13.13)}$$

> For home prices using the three-predictor model ($s_e = 20.31$), the 95 percent confidence interval would require $n - k - 1 = 30 - 3 - 1 = 26$ degrees of freedom. From Appendix D we obtain $t_{.025} = 2.056$, so the *approximate* intervals are
>
> $$\hat{y}_i \pm (2.056)\frac{20.31}{\sqrt{30}} \text{ or } \hat{y}_i \pm 7.62 \qquad (95\% \text{ confidence interval for conditional mean})$$
>
> $$\hat{y}_i \pm (2.056)(20.31) \text{ or } \hat{y}_i \pm 41.76 \quad (95\% \text{ prediction interval for individual home price})$$

Example 13.3

Home Prices III

---

[3]If you need exact intervals, you should use Minitab or a similar computer package. You must specify the value of *each predictor* $X_1, X_2, \ldots, X_k$ for which the confidence interval or prediction is desired.

Exact 95 percent confidence and prediction intervals for a home with $SqFt = 2{,}950$, $Lot\text{-}Size = 21$, and $Baths = 3$ (these values are very near the predictor means for our sample) are $\hat{y}_i \pm 8.55$ and $\hat{y}_i \pm 42.61$, respectively. Thus, our *approximate* intervals are not conservative (i.e., slightly too narrow). Nonetheless, the approximate intervals provide a ballpark idea of the accuracy of the model's predictions. Despite its good fit ($R^2 = .956$), we see that the three-predictor model's predictions are far from perfect. For example, the 95 percent prediction interval for an individual home price is $\hat{y}_i \pm \$41{,}760$.

The *t*-values for a 95 percent confidence level are typically near 2 (as long as n is not too small). This suggests a quick interval, without using a t table:

(13.14) $\quad \hat{y}_i \pm 2\dfrac{s_e}{\sqrt{n}} \qquad$ (quick 95% confidence interval for conditional mean of Y)

(13.15) $\quad \hat{y}_i \pm 2s_e \qquad$ (quick 95% prediction interval for individual Y-value)

These quick formulas are suitable only for rough calculations when you lack access to regression software or t tables.

Section Exercises

Mc Graw Hill **connect**

13.13 A regression of accountants' starting salaries in a large firm was estimated using 40 new hires and five predictors (college GPA, gender, score on CPA exam, years' prior experience, size of graduating class). The standard error was \$3,620. Find the approximate width of a 95 percent prediction interval for an employee's salary, assuming that the predictor values for the individual are near the means of the sample predictors.

13.14 An agribusiness performed a regression of wheat yield (bushels per acre) using observations on 25 test plots with four predictors (rainfall, fertilizer, soil acidity, hours of sun). The standard error was 1.17 bushels. Find the approximate width of a 95 percent prediction interval for wheat yield, assuming that the predictor values for a test plot are near the means of the sample predictors.

Mini Case 13.1

Birth Rates and Life Expectancy 🖭 BirthRates1

Table 13.4 shows the birth rate (Y = births per 1,000 population), life expectancy (X_1 = life expectancy at birth), and literacy (X_2 = percent of population that can read and write) for a random sample of 49 world nations.

Table 13.4 Birth Rates, Life Expectancy, and Literacy in Selected World Nations

Nation	BirthRate	LifeExp	Literate
Albania	18.59	72.1	93
Algeria	22.34	70.2	62
Australia	12.71	80.0	100
⋮	⋮	⋮	⋮
Yemen	43.30	60.6	38
Zambia	41.01	37.4	79
Zimbabwe	24.59	36.5	85

From Figure 13.6, the fitted regression equation is $BirthRate = 65.9 - 0.362\ LifeExp - 0.233\ Literate$, which says, *ceteris paribus*, that one year's increase in $LifeExp$ is associated with 0.362 fewer babies per 1,000 persons, while one extra percent of $Literate$ is associated with 0.233 fewer babies per 1,000 persons. The coefficient of determination is fairly high ($R^2 = .743$), and the overall regression is significant ($F_{calc} = 66.42$, p-value $= .000$).

Figure 13.6   Regression for Birth Rates (three predictors)

Regression Analysis: Birth Rates

R^2	0.743	n	49
Adjusted R^2	0.732	k	2
Std. Error	5.190	Dep. Var.	**BirthRate**

ANOVA table

Source	SS	df	MS	F	p-Value
Regression	3,578.2364	2	1,789.1182	66.42	0.0000
Residual	1,239.1479	46	26.9380		
Total	4,817.3843	48			

Regression output   Confidence Interval

Variables	Coefficients	Std. Error	t Stat	p-Value	Lower 95%	Upper 95%
Intercept	65.8790	3.8513	17.106	0.0000	58.1268	73.6312
LifeExp	−0.3618	0.0666	−5.431	0.0000	−0.4960	−0.2277
Literate	−0.2330	0.0415	−5.610	0.0000	−0.3166	−0.1494

Because both predictors are significant ($t_{calc} = -5.431$ and $t_{calc} = -5.610$, *p*-values near .000), the evidence favors the hypothesis that birth rates tend to fall as nations achieve higher life expectancy and greater literacy. Although cause and effect are unproven, the conclusions are consistent with what we know about nutrition, health, and education.

Source: Central Intelligence Agency, *The World Factbook, 2003.*

13.5   CATEGORICAL VARIABLES

LO 13-5

Incorporate categorical variables into a multiple regression model.

We cannot directly include a **categorical variable** (qualitative data) as a predictor in a regression because regression requires *numerical* data (quantitative data). But through simple data coding, we can convert categorical data into useful predictors. We will begin our discussion with categorical variables that have only two levels, for which coding is quite simple. We achieve this by defining a binary variable. Recall that a binary variable has two values, denoting the presence or absence of a condition (usually coded 0 and 1). By coding each category as a binary variable, we have created a **binary predictor**. Statisticians like to use intuitive names for the binary predictor variable. For example:

For *n* Graduates from an MBA Program

Employed = 1   (if the individual is currently employed)

Employed = 0   (otherwise)

For *n* Quarters of Sales Data

Recession = 1   (if the sales data are for a recession year)

Recession = 0   (otherwise)

For *n* Business Schools

AACSB = 1   (if the school is accredited by the AACSB)

AACSB = 0   (otherwise)

For *n* States

West = 1   (if the state is west of the Mississippi)

West = 0   (otherwise)

Binary predictors are easy to create and are extremely important because they allow us to capture the effects of nonquantitative (categorical) variables such as gender (female, male) or stock fund type (load, no-load). Such variables are also called **dummy**, **dichotomous**, or **indicator variables**.

Naming Binary Variables

Name the binary variable for the characteristic that is present when the variable is 1 (e.g., *Male*) so that others can immediately see what the "1" stands for.

Testing a Binary for Significance

The binary predictor variable's coefficient is tested for equality to zero by using a t test, just as we test the coefficient on a quantitative predictor variable. If the binary coefficient is found to be significantly different from zero, then we conclude that the binary predictor is a significant predictor for Y. Its coefficient contributes to the predicted value of Y when the binary variable value is 1 but has no effect on Y when the binary is 0.

Effects of a Binary Predictor

A binary predictor is sometimes called a **shift variable** because it shifts the regression plane up or down. Suppose that we have a two-predictor fitted regression $y = b_0 + b_1x_1 + b_2x_2$ where x_1 is a binary predictor. Because the only values that x_1 can take on are 0 or 1, its contribution to the regression is either b_1 or nothing, as seen in this example:

If $x_1 = 0$, then $y = b_0 + b_1(0) + b_2x_2$, so $y = b_0 + b_2x_2$.

If $x_1 = 1$, then $y = b_0 + b_1(1) + b_2x_2$, so $y = (b_0 + b_1) + b_2x_2$.

The coefficient b_2 is the same regardless of the value of x_1, but the intercept either is b_0 (when $x_1 = 0$) or $b_0 + b_1$ (when $x_1 = 1$).

For example, suppose we have a fitted regression of fuel economy based on a sample of 43 cars:

$$MPG = 39.5 - 0.00463 \; Weight + 1.51 \; Manual$$

where

$$Weight = \text{vehicle curb weight as tested (pounds)}$$
$$Manual = 1 \text{ if manual transmission, 0 if automatic}$$

If *Manual* = 0, then the predicted *MPG* is:

$$MPG = 39.5 - 0.00463 \; Weight + 1.51(0)$$
$$= 39.5 - 0.00463 \; Weight$$

If *Manual* = 1, then the predicted *MPG* is:

$$MPG = 39.5 - 0.00463 \; Weight + 1.51(1)$$
$$= 41.01 - 0.00463 \; Weight$$

The binary variable shifts the intercept, leaving the slope unchanged, as illustrated in Figure 13.7. In this case, although a manual transmission raises *MPG* slightly (by 1.51 miles per gallon, on average), the change in the intercept is rather small (i.e., manual transmission did not have a very large effect). Many experts feel that the choice of automatic versus manual transmission makes very little difference in fuel economy today.

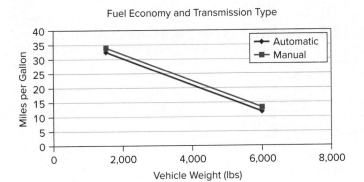

Fuel Economy and Transmission Type

Figure 13.7

Binary Shift Variable Illustrated

EXAMPLE 13.4

Subdivision Home Prices

📂 OakKnoll

We know that location is an important determinant of home price. But how can we include "location" in a regression? The answer is to code it as a binary predictor. Table 13.5 shows 20 home sales in two different subdivisions, Oak Knoll and Hidden Hills. We create a binary predictor, arbitrarily designating *OakKnoll* = 1 if the home is in the Oak Knoll subdivision and *OakKnoll* = 0 otherwise. We then do an ordinary regression, shown in Figure 13.8.

The model has a rather good fit ($R^2 = .922$) and is significant overall ($F_{calc} = 100.94$, p-value = .0000). Both predictors have a significant effect on *Price* at $\alpha = .05$, although *SqFt* ($t_{calc} = 14.008$, p-value = .0000) is a much stronger predictor than *OakKnoll* ($t_{calc} = 2.340$, p-value = .0317). The fitted coefficient of *OakKnoll* tells us that, on average, a home in the Oak Knoll subdivision sells for 33.538 more than a home in Hidden Hills (i.e., \$33,538 because *Price* is in thousands of dollars). Rounded off a bit, the fitted regression equation is *Price* = 10.6 + 0.199 *SqFt* + 33.5 *OakKnoll*. The intercept ($t_{calc} = 0.237$, p-value = .8154) does not differ significantly from zero, as also can be seen from the 95 percent confidence interval for the intercept (which includes zero).

Table 13.5   Home Prices with Binary Predictor 📂 OakKnoll

Obs	Price ($000)	SqFt	OakKnoll	Subdivision
1	$615.6	3,055	0	Hidden Hills
2	557.4	2,731	0	Hidden Hills
3	472.6	2,515	0	Hidden Hills
4	595.3	3,011	0	Hidden Hills
5	696.9	3,267	1	Oak Knoll
6	409.2	2,061	1	Oak Knoll
7	814.2	3,842	1	Oak Knoll
8	592.4	2,777	1	Oak Knoll
9	695.5	3,514	0	Hidden Hills
10	495.3	2,145	1	Oak Knoll
11	488.4	2,277	1	Oak Knoll
12	605.4	3,200	0	Hidden Hills
13	635.7	3,065	0	Hidden Hills
14	654.8	2,998	0	Hidden Hills
15	565.6	2,875	0	Hidden Hills
16	642.2	3,000	0	Hidden Hills
17	568.9	2,374	1	Oak Knoll
18	686.5	3,393	1	Oak Knoll
19	724.5	3,457	0	Hidden Hills
20	749.7	3,754	0	Hidden Hills

Figure 13.8 Oak Knoll Regression for 20 Home Sales

Regression Statistics	
Multiple R	0.9604
R Square	0.9223
R Sqr (adj)	0.9132
Std Error	29.6697
Obs	20

ANOVA

	df	SS	MS	F	Signif F
Regression	2	177706.8	88853.4	100.94	0.0000
Residual	17	14965.0	880.3		
Total	19	192671.7			

	Coefficients	Std Error	t Stat	p-Value	Lower 95%	Upper 95%
Intercept	10.6185	44.7725	0.237	0.8154	−83.8431	105.0802
SqFt	0.1987	0.0142	14.008	0.0000	0.1688	0.2286
OakKnoll	33.5383	14.3328	2.340	0.0317	3.2987	63.7779

More Than One Binary

A variable like gender (male, female) requires only one binary predictor (e.g., *Male*) because *Male* = 0 would indicate the individual is a female. But what if we need several binary predictors to code the data? This occurs when the number of categories to be coded exceeds two. For example, we might have home sales in five subdivisions or quarterly Walmart profits or student GPA by class level:

> Home sales by subdivision: *OakKnoll, HiddenHills, RockDale, Lochmoor, KingsRidge*
>
> Walmart profit by quarter: *Qtr1, Qtr2, Qtr3, Qtr4*
>
> GPA by class level: *Freshman, Sophomore, Junior, Senior, Master's, Doctoral*

Each category is a binary variable denoting the presence (1) or absence (0) of the characteristic of interest. For example:

> *Freshman* = 1 if the student is a freshman, 0 otherwise
>
> *Sophomore* = 1 if the student is a sophomore, 0 otherwise
>
> *Junior* = 1 if the student is a junior, 0 otherwise
>
> *Senior* = 1 if the student is a senior, 0 otherwise
>
> *Master's* = 1 if the student is a master's candidate, 0 otherwise
>
> *Doctoral* = 1 if the student is a PhD candidate, 0 otherwise

But if there are *c* categories (assuming they are mutually exclusive and collectively exhaustive), we need only *c* − 1 binaries to code each observation. This is equivalent to omitting any *one* of the categories. This is possible because *c* − 1 binary values uniquely determine the omitted binary. For example, Table 13.6 shows that we could omit the last binary column without losing any information. Because only one column can be 1 and the other columns must be 0, the following relation holds:

$$Freshman + Sophomore + Junior + Senior + Master's + Doctoral = 1$$

that is,

$$Doctoral = 1 - Freshman - Sophomore - Junior - Senior - Master's$$

Name	Freshman	Sophomore	Junior	Senior	Master's	Doctoral (Omitted)
Jaime	0	0	1	0	0	0
Fritz	0	1	0	0	0	0
Mary	0	0	0	0	0	1
Jean	0	0	0	1	0	0
Otto	0	0	0	0	1	0
Gail	1	0	0	0	0	0
etc.	. . .	. . .	. . .	. . .	. . .	. . .

Table 13.6

Why We Need Only
c − 1 Binaries to Code
c Categories

That Mary is a doctoral student can be inferred from the fact that 0 appears in all the other columns. Because Mary is *not* in any of the other five categories, she must be in the sixth category:

$$Doctoral = 1 - 0 - 0 - 0 - 0 - 0 = 1$$

There is nothing special about the last column; we could have omitted any other column instead. Similarly, we might omit the *KingsRidge* data column from home sales data because a home that is not in one of the first four subdivisions must be *KingsRidge*. We could omit the *Qtr4* column from the Walmart time series because if an observation is not from the first, second, or third quarter, it must be from *Qtr4*:

Home sales: *OakKnoll, HiddenHills, RockDale, Lochmoor, ~~KingsRidge~~*

Walmart profit: *Qtr1, Qtr2, Qtr3, ~~Qtr4~~*

Again, there is nothing special about omitting the last category. We can omit any single binary instead. The omitted binary becomes the base reference point for the regression; that is, it is part of the intercept. No information is lost.

What If I Forget to Exclude One Binary?

If you include all *c* binaries for *c* categories, you will have a redundant independent variable that is *collinear* with the other binary categories. When the value of one independent variable can be determined from the values of the other independent variables, this creates a serious problem for the regression estimation because one column in the *X* data matrix will then be a perfect linear combination of the other column(s). The least squares estimation would then fail because the data matrix would be singular (i.e., would have no inverse). Minitab automatically checks for such a situation and omits one of the offending predictors, while Excel merely gives an error. It is safer to decide for yourself which binary to exclude.

Mini Case 13.2

Age or Gender Bias? ▨ Oxnard

We can't use simple *t* tests to compare employee groups based on gender or age or job classification because they fail to take into account relevant factors such as education and experience. A simplistic salary equity study that fails to account for such control variables would be subject to criticism. Instead, we can use binary variables to study the effects of age, experience, gender, and education on salaries within a corporation. Gender and education can be coded as binary variables, and age can be forced into a binary variable that defines older employees explicitly rather than assuming that age has a linear effect on salary.

Table 13.7 shows salaries for 25 employees in the advertising department at Oxnard Petro, Ltd. As an initial step in a salary equity study, the human resources consultant performed a linear regression using the proposed model $Salary = \beta_0 + \beta_1\ Male + \beta_2\ Exper + \beta_3\ Ovr50 + \beta_4\ MBA$. *Exper* is the employee's experience in years; *Salary* is in thousands of dollars. Binaries are used for gender ($Male = 0, 1$), age ($Ovr50 = 0, 1$), and MBA degree ($MBA = 0, 1$). Can we reject the hypothesis that the coefficients of *Male* and *Ovr50* are zero? If so, it would suggest salary inequity based on gender and/or age.

Table 13.7 Salaries of Advertising Staff of Oxnard Petro, Ltd.

Obs	Employee	Salary	Male	Exper	Ovr50	MBA
1	Mary	28.6	0	0	0	1
2	Frieda	53.3	0	4	0	1
3	Alicia	73.8	0	12	0	0
4	Tom	26.0	1	0	0	0
5	Nicole	77.5	0	19	0	0
6	Xihong	95.1	1	17	0	0
7	Ellen	34.3	0	1	0	1
8	Bob	63.5	1	9	0	0
9	Vivian	96.4	0	19	0	0
10	Cecil	122.9	1	31	0	0
11	Barry	63.8	1	12	0	0
12	Jaime	111.1	1	29	1	0
13	Wanda	82.5	0	12	0	1
14	Sam	80.4	1	19	1	0
15	Saundra	69.3	0	10	0	0
16	Pete	52.8	1	8	0	0
17	Steve	54.0	1	2	0	1
18	Juan	58.7	1	11	0	0
19	Dick	72.3	1	14	0	0
20	Lee	88.6	1	21	0	0
21	Judd	60.2	1	10	0	0
22	Sunil	61.0	1	7	0	0
23	Marcia	75.8	0	18	0	0
24	Vivian	79.8	0	19	0	0
25	Igor	70.2	1	12	0	0

The coefficients in Figure 13.9 suggest that a male ($Male = 1$) makes $3,013 more on average than a female. However, the coefficient of *Male* does not differ significantly from zero even at $\alpha = .10$ ($t_{calc} = 0.862$, *p*-value = .3990). The evidence for age discrimination is a little stronger. Although an older employee ($Ovr50 = 1$) makes $8,598 less than others, on average, the *p*-value for *Ovr50* ($t_{calc} = -1.360$, *p*-value = .1891) is not convincing at $\alpha = .10$. The coefficient of MBA indicates that, *ceteris paribus,* MBA degree holders earn $9,587 more than others, and the coefficient differs from zero at $\alpha = .10$ ($t_{calc} = 1.916$, *p*-value = .0698). Salaries at Oxnard Petro are dominated by *Exper* ($t_{calc} = 12.08$, *p*-value = .0000). Each additional year of experience adds $3,019, on average, to an employee's salary. The regression is significant overall ($F_{calc} = 52.62$, $p = .0000$) and has a good fit ($R^2 = .913$). Although the sample fails Evans' 10:1 ratio test for *n/k*, it passes Doane's 5:1 ratio test. A more complete salary equity study might consider additional predictors.

Figure 13.9   **Regression Results for Oxnard Salary Equity Study**

Estimated regression equation:
$Salary = 28.878 + 3.013 \, Male + 3.019 \, Exper - 8.598 \, Ovr50 + 9.587 \, MBA$
$R^2 = .9480$ Std Err = 3.8818 $F = 52.62 \, (p = .0000)$

	Coefficients	Std Error	t Stat	p-Value
Intercept	28.878	4.925	5.864	0.0000
Male	3.013	3.496	0.862	0.3990
Exper	3.019	0.250	12.081	0.0000
Ovr50	−8.598	6.324	−1.360	0.1891
MBA	9.587	5.003	1.916	0.0698

Regional Binaries

One very common use of binaries is to code regions. Figure 13.10 shows how the 50 states of the United States could be divided into four regions by using these binaries:

Midwest = 1 if state is in the Midwest, 0 otherwise

Neast = 1 if state is in the Northeast, 0 otherwise

Seast = 1 if state is in the Southeast, 0 otherwise

West = 1 if state is in the West, 0 otherwise

For example, we can use regression to analyze the U.S. voting patterns in the 2000 U.S. presidential election. Binary predictors could permit us to analyze the effects of region (a qualitative variable) on voting patterns.

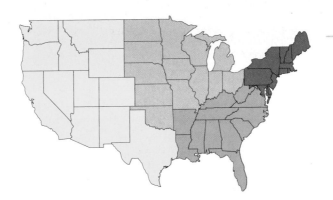

Figure 13.10

Four Regional Binaries

Mini Case 13.3

Regional Voting Patterns   🗁   Election2008

Table 13.8 shows an abbreviated data set for the 50 U.S. states. There are four regional binaries, but we only need to include three in our model. Arbitrarily, we omit the *Seast* column, which becomes the baseline for the regression to examine a hypothesis about the effects of population age, college graduation rates, home ownership, unemployment rates, and region on voting patterns in the 2008 U.S. presidential election. The dependent variable (*Obama%*) is the percentage vote for Barack Obama, and the proffered hypothesis to be investigated is

$$Obama\% = \beta_0 + \beta_1 Age65\% + \beta_2 ColGrad\% + \beta_3 HomeOwn\% + \beta_4 UnEmp\%$$
$$+ \beta_5 Midwest + \beta_6 Neast + \beta_7 West$$

Table 13.8 Characteristics of U.S. States in 2008 Election

State	Obama%	Age65%	ColGrad%	HomeOwn%	Unemp%	Midwest	Neast	West	Seast (Omitted)
AL	38.8	13.8	22.0	73.0	5.0	0	0	0	1
AK	37.9	7.3	27.3	66.4	6.7	0	0	1	0
AZ	44.9	13.3	25.1	69.1	5.5	0	0	1	0
AR	38.9	14.3	18.8	68.9	5.1	0	0	0	1
CA	60.9	11.2	29.6	57.5	7.2	0	0	1	0
CO	53.7	10.3	35.6	69.0	4.9	0	0	1	0
CT	60.6	13.7	35.6	70.7	5.7	0	1	0	0
.	.	.	.	.	.	.	.	.	.
etc.	etc.	etc.	etc.	etc.	etc.	etc.	etc.	etc.	etc.

The fitted regression shown in Figure 13.11 has four quantitative predictors and three binaries. The regression is significant overall ($F_{calc} = 12.27$, p-value = 2.04E-08). The p-values suggest that, *ceteris paribus*, the percent of voters choosing Obama was higher in states with older citizens, a higher percentage of college graduates, and higher unemployment. Home ownership was not a highly significant predictor with $t_{calc} = -1.593$ (p-value = .1187). The Obama vote was, *ceteris paribus*, significantly higher in the Northeast as compared to the Southeast ($t_{calc} = 2.037$, p-value = .048). The difference between the Western and Southeastern states was marginally significant ($t_{calc} = 1.673$, p-value = .1018), but there was *not* a significant difference in the Obama vote between the Midwestern states and the Southeastern states ($t_{calc} = 1.299$, p-value = .2009). Using regional binaries allows us to analyze the effects of these qualitative factors.

Figure 13.11

Regression Output for Voting Patterns

Regression Analysis

R^2	0.672			
Adjusted R^2	0.617		n	50
R	0.819		k	7
Std. Error	5.883		Dep. Var.	**Obama%**

ANOVA table

Source	SS	df	MS	F	p-Value
Regression	2,971.6423	7	424.5203	12.27	2.04E-08
Residual	1,453.6465	42	34.6106		
Total	4,425.2888	49			

Regression output   *confidence interval*

Variables	Coefficients	Std. Error	t Stat	p-Value	Lower 95%	Upper 95%
Intercept	0.4988	23.5964	0.021	.9832	−47.1207	48.1184
Age65%	2.2891	0.6195	3.695	.0006	1.0389	3.5393
ColGrad%	0.9787	0.2608	3.752	.0005	0.4523	1.5050
HomeOwn%	−0.3377	0.2120	−1.593	.1187	−0.7655	0.0902
Unemp%	2.6293	0.7262	3.621	.0008	1.1637	4.0949
Midwest	3.2490	2.5005	1.299	.2009	−1.7971	8.2951
Neast	6.7259	3.3019	2.037	.0480	0.0624	13.3893
West	4.6094	2.7555	1.673	.1018	−0.9514	10.1702

13.15 A regression model to predict the price of a condominium for a weekend getaway in a resort community included the following predictor variables: number of nights needed, number of bedrooms, whether the condominium complex had a swimming pool or not, and whether or not a parking garage was available. (a) Identify the quantitative predictor variable(s). (b) How many binary variables would be included in the model? (c) Write the proposed model form for predicting condominium price.

13.16 A regression model to predict the price of diamonds included the following predictor variables: the weight of the stone (in carats where 1 carat = 0.2 gram), the color rating (D, E, F, G, H, or I), and the clarity rating (IF, VVS1, VVS2, VS1, or VS2). (a) Identify the quantitative predictor variable(s). (b) How many indicator variables would be included in the model in order to prevent the least squares estimation from failing? (c) Write the proposed model form for predicting a diamond price.

13.17 Refrigerator prices are affected by characteristics such as whether or not the refrigerator is on sale, whether or not it is listed as a Sub-Zero brand, the number of doors (one door or two doors), and the placement of the freezer compartment (top, side, or bottom). The table below shows the regression output from a regression model using the natural log of price as the dependent variable. The model was developed by the Bureau of Labor Statistics. (a) Write the regression model, being careful to exclude the base indicator variable. (b) Find the *p*-value for each coefficient, using 319 degrees of freedom. Using an $\alpha = .01$, which predictor variable(s) are *not* significant predictors? (c) By how much does the natural log of refrigerator price decrease from a *two-door, side freezer model* to a *two-door, top freezer model?* (d) Which model demands a higher price: the side freezer or the one door with freezer model?

Variable	Coefficient	Standard Error	t Statistic
Intercept	5.4841	0.13081	41.92
Sale price	−0.0733	0.0234	−3.13
Sub-Zero brand	1.1196	0.1462	7.66
Total capacity (in cubic ft)	0.06956	0.005351	13.00
Two-door, freezer on bottom	0.04657	0.08085	0.58
Two-door, side freezer	Base		
Two-door, freezer on top	−0.3433	0.03596	−9.55
One door with freezer	−0.7096	0.1310	−5.42
One door, no freezer	−0.8820	0.1491	−5.92

Source: www.bls.gov/cpi/quality-adjustment/refrigerators.htm.

13.18 A model was developed to predict the length of a sentence (the response variable) for a male convicted of assault using the following predictor variables: age (in years), number of prior felony convictions, whether the criminal was married or not (1 = married), and whether the criminal was employed or not (1 = employed). The table below shows the regression output for a sample of *n* = 50 cases. (a) Write the regression model. (b) Using 45 degrees of freedom, find the *p*-value for each coefficient. Using an $\alpha = .01$, which predictor variable(s) are *not* significant predictors of length of sentence? (c) Interpret the coefficient of *Married?*. (d) How much shorter is the sentence if the criminal is employed? (e) Predict the length of sentence for an unmarried, unemployed, 25-year-old male with one prior conviction. Show your calculations. 🗂 **Sentencing**

Variable	Coefficient	Standard Error	t Statistic
Intercept	3.2563	4.3376	0.751
Age	0.5219	0.1046	4.989
Convictions	7.7412	1.0358	7.474
Married?	−6.0852	2.5809	−2.358
Employed?	−14.3402	2.5356	−5.656

13.6 TESTS FOR NONLINEARITY AND INTERACTION

Tests for Nonlinearity

Sometimes the effect of a predictor is nonlinear. A simple example would be estimating the volume of lumber to be obtained from a tree before cutting. This is a practical problem facing a timber farm. The manager can inventory the trees and measure their heights and diameters and then use this information to estimate the lumber volume. In addition to improving the accuracy of asset valuation on the balance sheet, the manager can decide the best time to cut the trees, based on their expected growth rates.

The volume of lumber that can be milled from a tree depends on the height of the tree and its radius, that is, *Volume = f(Height, Radius)*. But what is the appropriate model form? Figure 13.12 shows a scatter plot of tree *radius* versus tree *volume*. You'll see that the quadratic model actually fits the data better than the linear model.

Figure 13.12

Scatter Plot Tree Radius versus Volume

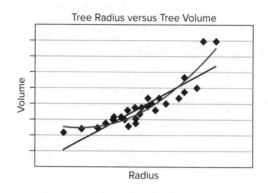

Figure 13.13 shows the regression output for two regressions of *Volume* on *Height* and *Radius:*

Model 1: $Volume = -58.0 + 0.3393\ Height + 9.4163\ Radius\ (R^2 = .948, s = 3.88)$

Model 2: $Volume = -27.5 + 0.3488\ Height + 0.6738\ Radius^2\ (R^2 = .973, s = 2.78)$

Figure 13.13

Regression Results for Tree Data

Model 1: The estimated regression equation is:
$Volume = -57.99 + 0.3393\ Height + 9.4163\ Radius$
$R^2 = .9480 \qquad SE = 3.8818$

Model 2: The estimated regression equation is
$Volume = -27.51 + .3488\ Height + .6738\ Radius^2$
$R^2 = .9729 \qquad SE = 2.7995$

Model 1: Regression Output

Variables	Coefficients	Std Err	t-Stat	p-Value
Intercept	−57.9877	8.6382	−6.713	0.0000
Height	0.3393	0.1302	2.607	0.0145
Radius	9.4163	0.5285	17.816	0.0000

Model 2: Regression Output

Variables	Coefficients	Std Err	t-Stat	p-Value
Intercept	−27.5116	6.5577	−4.195	0.0000
Height	0.3488	0.0932	3.744	0.0008
Radius^2	0.6738	0.0267	25.222	0.0000

If we regard a log as a cylinder, we would prefer the second regression because the usable volume of a cylinder is proportional to the square of its radius.[4] The *t* statistics for both *Height* and *Radius* are improved in Model 2, and the higher R^2 and reduced standard errors indicate a better fit. Although we introduced a squared term for the radius predictor variable, the model still is said to be linear because none of the parameters (i.e., β_0, β_1, or β_2) shows up as exponents nor do we divide any parameter by another.

To test for suspected nonlinearity of any *predictor,* we can include its square in the regression. For example, instead of

$$y = \beta_0 + \beta_1 x_1 + \beta_2 x_2 + \varepsilon \tag{13.16}$$

we would use

$$y = \beta_0 + \beta_1 x_1 + \beta_2 x_1^2 + \beta_3 x_2 + \beta_4 x_2^2 + \varepsilon \tag{13.17}$$

If the coefficients of the squared predictors (β_2 and β_4) do not differ significantly from zero, then the model collapses to the form in formula 13.16. On the other hand, rejection of the hypothesis H_0: $\beta_2 = 0$ would suggest a quadratic relationship between *Y* and X_1 and rejection of the hypothesis H_0: $\beta_4 = 0$ would suggest a quadratic relationship between *Y* and X_2. Some researchers include squared predictors as a matter of course in large studies. Squared predictors add model complexity and impose a cost of reduced degrees of freedom for significance tests (we lose 1 degree of freedom for each squared predictor), but the potential reward is a more appropriate model specification.

For example, rising total U.S. petroleum consumption from 1980 through 2004 can be fairly well described by a linear time trend model ($R^2 = .8691$). Yet adding a squared predictor (making it a quadratic model) gives an even better fit ($R^2 = .9183$) and the x_t^2 term is significant ($t_{calc} = 3.640$). This suggests a nonlinear trend, using $x_t = 1, 2, \ldots, 25$ to represent the year.

$$\text{Linear: } \hat{y}_t = 0.2012 \, x_t + 15.029 \; (R^2 = 0.8691)$$

$$\text{Quadratic: } \hat{y}_t = 0.0074 \, x_t^2 + 0.0078 \, x_t + 15.899 \; (R^2 = 0.9183)$$

Multiplicative Models

We can use linear regression to estimate *exponents* in a **multiplicative model** (as opposed to an additive model). The Cobb-Douglas model of production with *k* inputs $x_1, x_2, \ldots, x_k$ is an example:

$$y = \beta_0 x_1^{\beta_1} x_2^{\beta_2} \ldots x_k^{\beta_k} \tag{13.18}$$

Although the exponents in this model cannot be directly estimated using Excel, by taking logs of both sides, we obtain a transformed model that is now in a linear form. The coefficients of the log expression can be estimated using least squares:

$$\log(y) = \log(\beta_0) + \beta_1 \log(x_1) + \beta_2 \log(x_2) + \cdots + \beta_k \log(x_k) + \varepsilon \tag{13.19}$$

Example: Bank Revenue ⌫ **BankRevenue** Banks may be viewed as producers of revenue using two inputs: assets and employees.

$$Revenue = \beta_0 Assets^{\beta_1} \, Employees^{\beta_2}$$

For a sample of $n = 64$ large banks, the estimated production function (logs in base 10) is

$$\log(Revenue) = -0.2781 + 0.4973 \log(Assets) + 0.2209 \log(Employees)$$

$$(t \text{ stat } = 9.27) \qquad (t \text{ stat } = 6.84)$$

---

[4] $V = \pi h r^2$ would describe the relationship between the tree's radius (*r*), height (*h*), and volume (*V*). A logarithmic model $ln(Volume) = \beta_0 + \beta_1 \, ln(Height) + \beta_2 \, ln(Radius)$ might be more appropriate, although the resulting R^2 and s_e would not be comparable to the models shown above because the response variable would be in different units.

The t statistics indicate that both coefficients are significant and the fit is fairly good ($R^2 = .6909$). Setting $b_0 = 10^{-0.2871} = 0.5163$, the corresponding fitted multiplicative model is

$$Revenue = 0.5163 Assets^{0.4973} Employees^{0.2209}$$

Tests for Interaction

We can test for **interaction** between two predictors by including their product in the regression. For example, we might hypothesize that Y depends on X_1 and X_2, and X_1X_2. To test for interaction, we estimate the model:

(13.20) $$y = \beta_0 + \beta_1 x_1 + \beta_2 x_2 + \beta_3 x_1 x_2 + \varepsilon$$

If the t test for β_3 allows us to reject the hypothesis H_0: $\beta_3 = 0$, then we conclude that there is a significant interaction effect that transcends the roles of X_1 and X_2 separately (similar to the two-factor ANOVA tests for interaction in Chapter 11). Interaction effects require careful interpretation and cost 1 degree of freedom per interaction. However, if the interaction term improves the model specification, it is well worth the cost. For example, a bank's lost revenue (*Loss*) due to loan defaults depends on the loan size (*Size*) and the degree of risk (*Risk*). Small loans may be risky but may not contribute much to the total losses. Large loans may be less risky but potentially represent a large loss. An interaction term (*Size* × *Risk*) would be large if either predictor is large, thereby capturing the effect of both predictors. Thus, the interaction term might be a significant predictor in the model:

$$Loss = b_0 + b_1\ Size + b_2\ Risk + b_3\ Size \times Risk$$

Analytics In Action

People Analytics at Work

Workplace design is often an afterthought. But the design of working spaces can either promote or hinder productivity. How can a business know if its office environment is set up to improve performance? By collecting and analyzing the *data exhaust* from employees' Skype, e-mails, and face-to-face interactions. Data exhaust is secondary data related to employee interactions—either in person or on the Internet. Length of e-mails, frequency of e-mail exchanges between users, and time of day that Skype conversations take place are all examples of secondary data.

A group of researchers from MIT's Media Lab mined hundreds of megabytes of employee communication data to see what was related to highly productive teams. What they found was that team performance could be accurately predicted by this secondary interaction data and that personal information or communication content wasn't necessary to predict productivity. Based on this research, the scientists founded Humanyze, Inc., a software company that provides a dashboard called Elements. The software platform analyzes communication data and uses it to identify work patterns that get things done. Industries such as banking, real estate, and NASA have all benefited from analyzing the massive amounts of data generated by each employee.

Source: www.humanyze.com/.

Section Exercises

Mc Graw Hill connect

13.19 The data set below shows a sample of salaries for 39 engineers employed by the Solnar Company along with each engineer's years of experience. (a) Construct a scatter plot using *Salary* as the response variable and *Years* as the explanatory variable. Describe the shape of the scatter plot. Does it appear that a nonlinear model would be appropriate? Explain. (b) Run the regression with *Salary* as the response variable and *Years* and *YearsSq* as the explanatory variables. Report the R^2, F_{calc} statistic, and *p*-value. Is the nonlinear model significant? (c) Report the *p*-values for both *Years* and *YearsSq*. Are these variables significant predictors? Use $\alpha = .10$.

Engineer Salaries and Years of Experience at Solnar Company (n = 39) 📁 **Salaries**

Salaries ($)	Years	YearsSq
50,000	1	1
54,000	1	1
52,000	1	1
⋮	⋮	⋮
134,000	32	1,023
118,000	34	1,156
134,000	35	1,225

13.20 The same data set from exercise 13.19 also has gender information for each engineer. The binary variable *Male* = 1 indicates the engineer is male and *Male* = 0 indicates the engineer is female. Run the regression with *Salary* as the response variable and *Years, YearsSq, Male,* and *Years* × *Male* as the explanatory variables. Report the *p*-values for the binary variable *Male* and the interaction term *Years* × *Male*. Are these two variables significant? Use $\alpha = .10$.

Engineer Salaries, Years of Experience, and Gender at Solnar Company (n = 39) 📁 **Salaries**

Salaries ($)	Years	YearsSq	Male	Years × Male
48,000	1	1	0	0
50,000	1	1	0	0
52,000	1	1	1	1
⋮	⋮	⋮	⋮	⋮
131,000	20	400	1	20
134,000	32	1,023	1	32
134,000	35	1,225	1	35

13.7 MULTICOLLINEARITY

What Is Multicollinearity?

When the independent variables $X_1, X_2, \ldots, X_m$ are related to each other instead of being independent, we have a condition known as **multicollinearity**. If only two predictors are correlated, we have **collinearity**. Almost any data set will have some degree of correlation among the predictors. The depth of our concern would depend on the *degree* of multicollinearity.

LO 13-7

Detect multicollinearity and assess its effects.

Variance Inflation

Multicollinearity does not bias the least squares estimates or the predictions for *Y*, but it does induce *variance inflation*. When predictors are strongly intercorrelated, the variances of their estimated coefficients tend to become inflated, widening the confidence intervals for the true coefficients $\beta_1, \beta_2, \ldots, \beta_k$ and making the *t* statistics less reliable. It can thus be difficult to identify the separate contribution of each predictor to "explaining" the response variable, due to the entanglement of their roles. Consequences of variance inflation can range from trivial to severe. In the most extreme case, when one *X* data column is an exact linear function of one or more other *X* data columns, the least squares estimation will fail.[5] That could happen, for example, if you inadvertently included the same predictor twice, or if you forgot to omit one of the *c* binaries used to code *c* attribute categories. Some software packages will check for perfect multicollinearity and will remove one of the offending predictors, but don't count on it.

---

[5] If the *X* data matrix has no inverse, we cannot solve for the OLS estimates.

Variance inflation generally does not cause major problems, and some researchers suggest that it is best ignored except in extreme cases. However, it is a good idea to investigate the degree of multicollinearity in the regression model. There are several ways to do this.

Correlation Matrix

To check whether two predictors are correlated (collinearity), we can inspect the **correlation matrix** for the predictors using Excel's Data Analysis > Correlation. For example, Table 13.9 shows correlations among selected characteristics of 50 vehicles that might be used to predict city MPG (e.g., horsepower, engine displacement, curb weight, dimensions, drive type). The response variable (*CityMPG*) is not included because collinearity among the predictors is the condition we are investigating. Cells above the diagonal are redundant and hence are not shown. Correlations that differ from zero at $\alpha = .05$ or $\alpha = .01$ in a two-tailed test are highlighted. In this example, a majority of the predictors are significantly correlated, which is not an unusual situation in regression modeling.

Table 13.9

Correlation Matrix for Vehicle MPG Predictors 🖼 VehicleMPG

	Doors	HP	Engine	Weight	Length	Width	AWD	RWD
Doors	1.000							
HP	−.339	1.000						
Engine	−.204	.806	1.000					
Weight	.095	.654	.729	1.000				
Length	.318	.503	.567	.738	1.0000			
Width	.090	.714	.681	.821	.763	1.000		
AWD	.014	.215	.113	.253	−.069	.214	1.000	
RWD	−.055	.418	.394	.409	.503	.418	−.421	1.000

± .279 critical value of $\alpha_{.05}$ (two-tail)
± .361 critical value of $\alpha_{.01}$ (two-tail)

Significant predictor correlations do not *per se* indicate a serious problem. A common rule called **Klein's Rule** suggests that we should worry about the stability of the regression estimates or accuracy of confidence intervals only when a pairwise predictor correlation exceeds the multiple correlation coefficient (i.e., the square root of R^2). Until we actually perform the regression, we cannot say for sure whether Klein's Rule is violated. However, high correlations (e.g., *Engine* versus *HP*) suggest that some predictors may indeed be redundant.

Variance Inflation Factor (VIF)

Although the matrix scatter plots and correlation matrix are easy to understand, they only show correlations between *pairs* of predictors (e.g., X_1 and X_2). A general test for multicollinearity should reveal more complex relationships *among* predictors. For example, X_2 might be a linear function of X_1, X_3, and X_4 even though its pairwise correlation with each is not very large.

The **variance inflation factor (VIF)** for each predictor provides a more comprehensive test. For a given predictor X_j, the VIF is defined as

(13.21)
$$VIF_j = \frac{1}{1 - R_j^2}$$

where R_j^2 is the coefficient of determination when predictor j is regressed against *all* the other predictors (excluding Y).

Response Variable	Explanatory Variables	R^2
X_1	$X_2, X_3, \ldots, X_k$	R_1^2
X_2	$X_1, X_3, \ldots, X_k$	R_2^2
$\vdots$	$\vdots$	$\vdots$
X_k	$X_1, X_2, \ldots, X_{k-1}$	R_k^2

If predictor j is unrelated to the other predictors, its R_j^2 will be 0 and its VIF will be 1 (an ideal situation that will rarely be seen with actual data). Some possible situations are

R_j^2	VIF_j	Interpretation
0.00	$\dfrac{1}{1 - R_j^2} = \dfrac{1}{1 - 0.00} = 1.0$	No variance inflation
0.50	$\dfrac{1}{1 - R_j^2} = \dfrac{1}{1 - 0.50} = 2.0$	Mild variance inflation
0.90	$\dfrac{1}{1 - R_j^2} = \dfrac{1}{1 - 0.90} = 10.0$	Strong variance inflation
0.99	$\dfrac{1}{1 - R_j^2} = \dfrac{1}{1 - 0.99} = 100.0$	Severe variance inflation

What to Do about Multicollinearity?

There is no limit on the magnitude of a VIF. Some researchers suggest that when a predictor variable j has a VIF that exceeds 10, there is cause for concern, or even that one should remove predictor j from the model. But that rule of thumb is perhaps too conservative. A VIF of 10 says that the other predictors "explain" 90 percent of the variation in predictor j. While a VIF of 10 shows that predictor j is strongly related to the other predictors, it is not necessarily indicative of instability in the least squares estimates. Removing a relevant predictor is a step that should not be taken lightly, for it could result in misspecification of the model. A better way to think of it is that a large VIF is a warning to consider whether predictor j really belongs in the model.

Are Coefficients Stable?

Evidence of instability would be when X_1 and X_2 have a high pairwise correlation with Y, yet one or both predictors have insignificant t statistics in the fitted multiple regression. Another symptom would be if X_1 and X_2 are positively correlated with Y, yet one of them has a negative slope in the multiple regression. As a general test, you can try dropping a collinear predictor from the regression and watch what happens to the fitted coefficients in the reestimated model. If they do not change very much, multicollinearity was probably not a concern. If dropping one collinear predictor causes sharp changes in one or more of the remaining coefficients in the model, then your multicollinearity may be causing instability. Keep in mind that a predictor must be significantly different from zero in order to say that it "changed" in the reestimation.

Mini Case 13.4

Regional Voting Patterns Election2008

Mini Case 13.3 investigated a hypothesis about predictors for the percentage vote for Barack Obama in each of the 50 states. The proffered model was $Obama\% = \beta_0 + \beta_1 Age65\% + \beta_2 ColGrad\% + \beta_3 HomeOwn\% + \beta_4 UnEmp\% + \beta_5 Midwest + \beta_6 Neast + \beta_7 West$. Table 13.10 shows the correlation matrix for the predictors. A correlation is significant at $\alpha = .05$ if it exceeds $r_{crit} = .2187$ (using formula 12.5 with $t_{crit} = 2.0106$ for a two-tailed test). By this rule, 11 of the predictor correlations are significant at $\alpha = .05$ (shaded cells in table). However, none is close to the multiple correlation coefficient ($R = .819$), so by Klein's Rule we should not worry. It also should be mentioned that correlations between the binary variables may exist by design. For example, it is understood that if a state is in the Midwest, it *cannot* be in any of the other three regions.

Table 13.10   Correlation Matrix for 2008 Election Predictors Election2008

	Age65%	ColGrad%	HomeOwn%	Unemp%	Midwest	Neast	West
ColGrad%	−0.194						
HomeOwn%	0.153	−0.357					
Unemp%	−0.122	−0.091	−0.147				
Midwest	0.161	−0.109	0.137	−0.118			
Neast	0.227	0.548	−0.067	0.013	−0.315		
West	−0.476	0.038	−0.318	−0.078	−0.370	−0.331	
Seast	0.114	−0.460	0.259	0.191	−0.333	−0.298	−0.350

Despite the significant correlations between certain predictors, Figure 13.14 shows that for the election data, no VIF exceeds 10 and the overall mean VIF is small. Thus, the confidence intervals should be reliable. *Note:* Excel does not compute VIFs (these are from MegaStat).

Figure 13.14   VIFs for Election Study

Regression output

Variables	Coefficients	Std. Error	t Stat	p-Value	VIF
Intercept	0.4988	23.5964	0.021	.9832	
Age65%	2.2891	0.6195	3.695	.0006	1.563
ColGrad%	0.9787	0.2608	3.752	.0005	2.183
HomeOwn%	−0.3377	0.2120	−1.593	.1187	1.390
Unemp%	2.6293	0.7262	3.621	.0008	1.174
Midwest	3.2490	2.5005	1.299	.2009	1.738
Neast	6.7259	3.3019	2.037	.0480	2.703
West	4.6094	2.7555	1.673	.1018	2.211

Section Exercises

 connect

13.21 Using the "Resort Guest Satisfaction Survey" data, construct a correlation matrix of the 11 independent variables. The response variable is *ovalue*. (a) Identify the four pairs of independent variables that have the highest pairwise correlation values. Do they show significant correlation? (b) Run the regression with all 11 predictor variables and calculate the VIF for each predictor. Use software such as MegaStat or Minitab. (c) Did you see any cause for concern based on the VIF values? Why or why not? **ResortGuestSat**

13.22 Using the "Metals" data, construct a correlation matrix of the six independent variables. The response variable is *Price/lb*. (a) Identify any pairs of independent variables that have a significant pairwise correlation. (b) Run the regression with all six predictor variables and calculate the VIF for each predictor. Use software such as MegaStat or Minitab. (c) Did you see any cause for concern based on the VIF values? Why or why not? **Metals**

13.8 REGRESSION DIAGNOSTICS

LO 13-8

Analyze residuals to check for violations of residual assumptions.

Recall that the least squares method makes several assumptions about the random error ε. Although ε is unobservable, clues may be found in the residuals e_i. We routinely test three important assumptions:

- *Assumption* 1: The errors are normally distributed.
- *Assumption* 2: The errors have constant variance (i.e., they are homoscedastic).
- *Assumption* 3: The errors are independent (i.e., they are nonautocorrelated).

Regression residuals often violate one or more of these assumptions. The consequences may be mild, moderate, or severe. **Residual tests** for violations of regression assumptions are routinely provided by regression software. Visual tools for testing these assumptions were presented in Sections 12.8 and 12.9. We will review each assumption and suggest remedies if violations exist.

Non-normal Errors

Except when there are major outliers, non-normal residuals are usually considered a mild violation. The regression coefficients and their variances remain unbiased and consistent. The main ill consequence is that confidence intervals for the parameters may be unreliable because the normality assumption is used to construct them. However, if the sample size is large (say, $n > 30$), the confidence intervals generally are OK unless serious outliers exist. The hypotheses are

H_0: Errors are normally distributed

H_1: Errors are not normally distributed

A simple "eyeball test" of the *histogram of residuals* can usually reveal outliers or serious asymmetry. You can use either plain residuals or standardized (i.e., studentized) residuals. Standardized residuals offer the advantage of a predictable scale (between -3 and $+3$ unless there are outliers). Another visual test for normality is the **normal probability plot** (also called a *quantile-quantile plot* or simply **q-q plot**), which is produced as an option by many software packages. If the normality hypothesis is true, the probability plot should be approximately linear.

What can we do about non-normality? First, consider trimming outliers–but only if they clearly are mistakes. Second, can you increase the sample size? If so, it will help assure asymptotic normality of the estimates. Third, you could try a logarithmic transformation of the variables. However, this is a new model specification that may require advice from a professional statistician. Fourth, you could do nothing—just be aware of the problem.

Nonconstant Variance (Heteroscedasticity)

The regression should fit equally well for all values of X or Y. This desirable property is called **homoscedasticity**. If the error variance is constant, the errors are *homoscedastic*. If the error variance is nonconstant, we have **heteroscedasticity** and say that the errors are *heteroscedastic*. This violation is potentially serious. Although the least squares regression parameter estimates are still unbiased and consistent, their estimated variances are biased and are neither efficient nor asymptotically efficient. In the most common form of heteroscedasticity, the variances of the estimators are likely to be understated, resulting in overstated t statistics and artificially narrow confidence intervals. In a multiple regression, a visual test for constant variance can be performed by examining scatter plots of the residuals against each predictor or against the fitted Y-values. Ideally, there will be no pattern and the vertical spread (residual variance) will be similar regardless of the X-values. The hypotheses are

H_0: Errors have constant variance (homoscedastic)

H_1: Errors have nonconstant variance (heteroscedastic)

In a multiple regression, to avoid looking at all k residual plots (one for each predictor), we usually just examine the plot of residuals against the predicted Y-values. Although many patterns of nonconstant variance might exist, the "fan-out" pattern of increasing variance is most common (see Figure 13.15). The zero line appears more or less in the center of the residual plot because the residuals always sum to zero.

Figure 13.15

Homoscedastic and Heteroscedastic Residual Plots

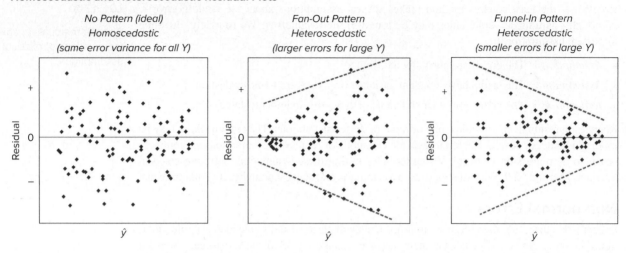

Heteroscedasticity may arise in economic time series data if X and Y increase in magnitude over time, causing the errors to increase. In financial data (e.g., GDP), heteroscedasticity can sometimes be reduced by expressing the data in constant dollars (dividing by a price index). In cross-sectional data (e.g., total crimes), heteroscedasticity may be mitigated by expressing the data in relative terms (e.g., per capita crime). A more general approach to reducing heteroscedasticity is to transform the variables (e.g., by taking logs). However, this is a new model specification, which requires a reverse transformation when making predictions of Y.

Example 13.5

NFL Free Agent Annual Contracts

📂 **NFLFreeAgent**

A multiple regression model was developed for a sample of 50 NFL free agent defensive players. The response variable was their annual contract value. The four predictor variables included their age, number of Pro Bowls they played in, which conference they were in (AFC or NFC), and the ratio of games started to games played. The model explained approximately 20% of the variation in annual contract. However, the plot of residuals against predicted annual contract shows a clear pattern, shown in Figure 13.16.

The presence of nonconstant error variance can cause the confidence intervals for the regression coefficients to be narrower than they should be because the standard errors of the estimates are possibly underestimated. Techniques for reducing the effect of heteroscedasticity include transforming either one of the X or Y variables and/or expressing the variables in relative terms. Residual plots for individual predictors might reveal more about the source of the observed pattern. It is also important to consider other explanatory variables that might better explain variation in Y than our chosen X variables. You will have a chance to examine these data in more detail in the end-of-chapter exercises.

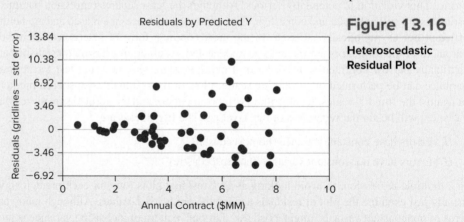

Figure 13.16

Heteroscedastic Residual Plot

Source: Data are from an MBA Class Project from students Jalal Aslam, John Hornbuckle, Jake Jesielowski, Antonio King, Samuel Rolfes, and Colin Tinley.

Mini Case 13.5

Tests for Normality and Homoscedasticity 📂 HeartDeaths

Figure 13.17 shows a histogram and normal probability plot of the residuals for a regression model of heart deaths in all 50 U.S. states for the year 2000. The dependent variable is *Heart* = heart deaths per 100,000 population and the three predictors are *Age65%* = percent of population age 65 and over, *Income* = per capita income in thousands of dollars, and *Black%* = percent of population that is African American.

Figure 13.17

Histogram and Normal Plot of Residuals for Heart Death Regression

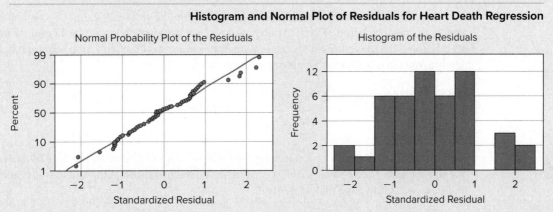

The histogram is arguably bell-shaped. Because the residuals have been standardized, we can see that there are no outliers (more than 3 standard errors from zero). The probability plot reveals slight deviations from linearity at the lower and upper ends, but overall the plot is consistent with the hypothesis of normality. For a test for heteroscedasticity, we can look at the plot of residuals against the fitted *Y*-values and the *X*-values seen in Figure 13.18. These show no pronounced consistent "fan-out" or "funnel-in" pattern, thereby favoring the hypothesis of homoscedasticity (constant variance).

Figure 13.18

Residual Plots for Heart Death Regression

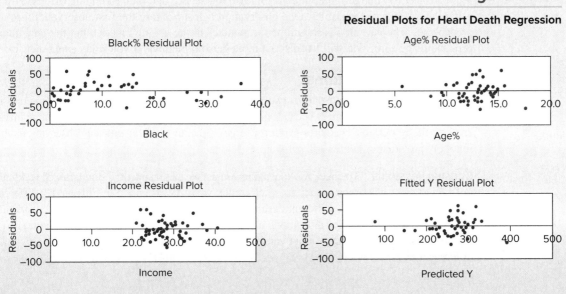

Autocorrelation

If you are working with time series data, you need to be aware of the possibility of **autocorrelation**, a pattern of nonindependent errors that violates the regression assumption that each error is independent of its predecessor. Cross-sectional data may exhibit autocorrelation, but usually it is an artifact of the order of data entry and so may be ignored. When the errors in a regression are autocorrelated, the least squares estimators of the coefficients are still unbiased and consistent. However, their estimated variances are biased in a way that typically leads to confidence intervals that are too narrow and t statistics that are too large. Thus, the model's fit may be overstated. The hypotheses are

H_0: Errors are nonautocorrelated

H_1: Errors are autocorrelated

Because the true errors are unobservable, we rely on the residuals $e_1, e_2, \ldots, e_n$ for evidence of autocorrelation. The most common test for autocorrelation is the Durbin-Watson test. Using e_t to denote the tth residual (assuming you are working with time series data), the Durbin-Watson test statistic for autocorrelation is

(13.22)
$$DW = \frac{\sum_{t=2}^{n} (e_t - e_{t-1})^2}{\sum_{t=1}^{n} e_t^2} \qquad \text{(Durbin-Waston test statistic)}$$

If you study econometrics or forecasting, you will use a special table to test the DW statistic for significance. (See downloadable *LearningStats* demonstrations at the end of this chapter.) For now, we simply note that, in general,

$DW < 2$ suggests positive autocorrelation (common).

$DW \approx 2$ suggests no autocorrelation (ideal).

$DW > 2$ suggests negative autocorrelation (rare).

What can we do about autocorrelation? First-order time series autocorrelation can be reduced by transforming the variables. A very simple transformation is the *method of first differences* in which all the variables are redefined as *changes:*

$$\Delta x_t = x_t - x_{t-1} \quad \text{(change in X from period } t-1 \text{ to period } t)$$
$$\Delta y_t = y_t - y_{t-1} \quad \text{(change in Y from period } t-1 \text{ to period } t)$$

Then we regress ΔY against all the ΔX *predictors.* This transformation can easily be done in a spreadsheet by subtracting each cell from its predecessor and then rerunning the regression using the transformed variables. One observation is lost because the first observation has no predecessor. The new slopes should be the same as in the original model, but the new intercept should be zero. You will learn about more general methods if you study econometrics.

Unusual Observations

LO 13-9

Identify unusual residuals and tell when they are outliers.

An observation may be unusual for two reasons: (1) because the fitted model's prediction is poor (*unusual residuals*) or (2) because one or more observations may be having a large influence on the regression estimates (*high leverage*). Several tests for unusual observations are routinely provided by regression software.

Unusual Residuals   To check for unusual residuals, we can inspect the standardized residuals to find instances where the model does not predict well. As explained in Chapter 12, different software packages may use different definitions for "standardized" or "studentized" residuals, but usually they give similar indications of "unusual" residuals. For example, in Figure 13.19 the last two columns of the printout can be interpreted the same even though there are slight differences in the values.

Unusual Residuals

We apply the Empirical Rule. Standardized residuals more than $2s_e$ from zero are *unusual*, while residuals more than $3s_e$ from zero are *outliers*.

Figure 13.19

Regression Output for Heart Death

Regression Analysis: Heart Deaths per 100,000

R^2	0.779		
Adjusted R^2	0.764	n	50
R	0.883	k	3
Std. Error	27.422	Dep. Var.	**Heart**

ANOVA table

Source	SS	df	MS	F	p-Value
Regression	121,809.5684	3	40,603.1895	54.00	0.0000
Residual	34,590.3558	46	751.9643		
Total	156,399.9242	49			

Regression output   *confidence interval*

Variables	Coefficients	Std. Error	t Stat	p-Value	Lower 95%	Upper 95%	VIF
Intercept	−37.0813	39.2007	−0.946	0.3491	−115.9882	41.8256	
Age65%	24.2509	2.0753	11.686	0.0000	20.0736	28.4282	1.018
Income	−1.0800	0.9151	−1.180	0.2440	−2.9220	0.7620	1.012
Black%	2.2682	0.4116	5.511	0.0000	1.4398	3.0967	1.013

Unusual Observations

Observation	Heart	Predicted	Residual	Leverage	Studentized Residual	Studentized Deleted Residual
AK	90.90	76.62	14.28	0.304	0.624	0.620
CT	278.10	274.33	3.77	0.208	0.154	0.153
FL	340.40	392.45	−52.05	0.177	−2.092	−2.176
HI	203.30	259.06	−55.76	0.037	−2.072	−2.152
MS	337.20	316.02	21.18	0.223	0.876	0.874
OK	335.40	274.87	60.53	0.047	2.261	2.372
UT	130.80	145.05	−14.25	0.172	−0.571	−0.567
WV	377.50	317.55	59.95	0.108	2.315	2.436

High Leverage   To check for high leverage, we look at the *leverage statistic* for each observation. It shows how far the predictors are from their means. As you saw in Chapter 12 (Section 12.8), such observations potentially have great influence on the regression estimates because they are at the "end of the lever."

LO **13-10**

Identify high leverage observations and their possible causes.

High Leverage

For n observations and k predictors, an observation is considered to be a high leverage observation if the leverage statistic exceeds

$$\frac{2(k+1)}{n}$$

What to Do about Unusual Observations?

Unless these observations are erroneous, keep them in the analysis and note them in the report. If they were omitted, the regression results might change noticeably. But valid observations are not to be discarded except in extreme circumstances. Merely note that they are unusual, comment on whether they are influential, and leave them be.

Mini Case 13.6

Unusual Observations ⌦ HeartDeaths

Figure 13.19 shows regression results for a regression model of heart deaths in the 50 U.S. states for the year 2000 (variables are drawn from the *LearningStats* state database). The response variable is *Heart* = heart deaths per 100,000 population, with predictors *Age65%* = percent of population age 65 and over, *Income* = per capita income in thousands of dollars, and *Black%* = percent of population classified as African American. The fitted regression is

$$Heart = -37.1 + 24.3\,Age65\% - 1.08\,Income + 2.27\,Black\%$$

Age65% has the anticipated positive sign and is highly significant, and similarly for *Black%*. Income has a negative sign but is not significant even at $\alpha = .10$. The regression overall is significant ($F_{calc} = 54.00$, *p*-value = .0000). The R^2 shows that the predictors explain 77.9 percent of the variation in *Heart* among the states. The adjusted R^2 is 76.4 percent, indicating that no unhelpful predictors are present. Because this is cross-sectional data, the DW statistic was not requested.

Figure 13.19 lists eight unusual observations (the other 42 states are not unusual). Five states (AK, CT, FL, MS, UT) are highlighted because they have unusual *leverage*—a leverage statistic that exceeds $2(k+1)/n = (2)(3+1)/50 = 0.16$. One or more predictors for these states must differ greatly from the mean of that predictor, but only by inspecting the *X* data columns (*Age65%, Income,* or *Black%*) could we identify their unusual *X*-values. Four states are highlighted because they have unusual *residuals*—a gap of at least two standard deviations (studentized residuals) between actual and predicted *Heart* values. FL and HI are more than 2 standard deviations lower than predicted, while OK and WV are more than 2 standard deviations higher than predicted. One state, FL, is unusual with respect to both residual and leverage.

Mc Graw Hill **connect**

Section Exercises

13.23 Which would be "high leverage" observations?

 a. Leverage $h_i = .15$ in a regression with 5 predictors and 72 observations.

 b. Leverage $h_i = .18$ in a regression with 4 predictors and 100 observations.

 c. Leverage $h_i = .08$ in a regression with 7 predictors and 240 observations.

13.24 Which violations of regression assumptions, if any, do you see in these residual diagnostics? Explain.

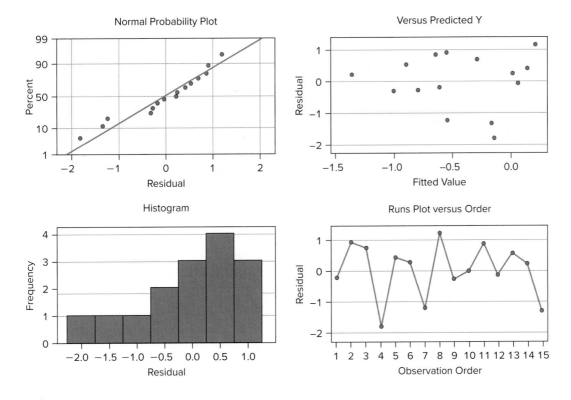

LO (13-11)

Explain the purpose of data conditioning and stepwise regression.

13.9 OTHER REGRESSION TOPICS

Outliers

An outlier may be due to an error in recording the data. If so, the observation should be deleted. But how can you tell? Impossible or truly bizarre data values are apparent reasons to discard an observation. For example, a Realtor's database of recent home sales in a million-dollar neighborhood contained this observation:

Price	BR	Bath	Basement	Built	SqFeet	Garage
95,000	4	3	Y	2001	4,335	Y

The 95,000 price is probably a typographical error. Even if the price were correct, it would be reasonable to discard the observation on grounds that it represents a different population than the other homes (e.g., a "gift" sale by a wealthy parent to a newlywed couple).

Missing Predictors

An outlier also may be an observation that has been influenced by an unspecified "lurking" variable that should have been controlled but wasn't. In this case, we should try to identify the lurking variable and formulate a multiple regression model that includes both predictors. For example, a reasonable model such as Y = home price, X_1 = square feet, and X_2 = lot size might give poor predictions unless we add a neighborhood binary predictor (you can probably think of areas where a large house on a large lot might still command a poor price). If there are unspecified "lurking" variables, our fitted regression model will not give accurate predictions.

Ill-Conditioned Data

All variables in the regression should be of the same general order of magnitude (not too small, not too large). If your coefficients come out in exponential notation (e.g., 7.3154 E+06), you probably should adjust the decimal point in one or more variables to a convenient magnitude, as long as you treat all the values in the same data column consistently.

Significance in Large Samples

Statistical significance may not imply *practical importance*. In a large sample, we can obtain very large *t* statistics with low *p*-values for our predictors when, in fact, their effect on *Y* is very slight. There is an old saying in statistics that you can make anything significant if you get a large enough sample. In medical research, where thousands of patients are enrolled in clinical trials, this is a familiar problem. It can become difficult in large data-mining models to figure out which *significant* variables are really *important*.

Model Specification Errors

If you estimate a linear model when actually a nonlinear model is required or when you omit a relevant predictor, then you have a *misspecified model.* How can you detect misspecification? You can

- Plot the residuals against estimated *Y* (should be no discernable pattern).
- Plot the residuals against actual *Y* (should be no discernable pattern).
- Plot the fitted *Y* against the actual *Y* (should be a 45-degree line).

What are the cures for misspecification? Start by looking for a missing relevant predictor, seek a model with a better theoretical basis, or try transforming your variables (e.g., using a logarithm transform, which can create a linear model from a nonlinear model). Model specification is a topic that will be covered in considerable depth if you study econometrics. For now, just remember that *residual patterns* are clues that the model may be incorrectly specified.

Missing Data

If many values in a data column are missing, we might want to discard that variable. If a *Y* data value is missing, we must discard the entire observation. If any *X* data values are missing, the conservative action is to discard the entire observation. However, because discarding an entire observation would mean losing other good information, statisticians have developed procedures for imputing missing values, such as using the mean of the *X* data column or by a regression procedure to "fit" the missing *X*-value from the complete observations. Imputing missing values requires specialized software and expert statistical advice.

Stepwise and Best Subsets Regression

It may have occurred to you that there ought to be a way to automate the task of fitting the "best" regression using *k* predictors. **Stepwise regression** uses the power of the computer to fit the best model using 1, 2, 3, . . . , *k* predictors. In *backward elimination,* we start with all predictors in the model, removing at each stage the predictor with the highest *p*-value greater than α, stopping when there are no more predictors with *p*-values greater than α to be removed. This method has the appeal of being easy to understand and implement without special software, but it tends to increase effective α and may not yield the highest R^2 for the number of predictors. In *forward selection,* we start with the single best predictor, adding at each stage the next best predictor in terms of increasing R^2, until there are no more predictors with *p*-values less than α to be added to the model. A drawback of this method is that adding a predictor may render one or more already-added predictors insignificant, and special software is needed for all but the simplest models. In *best subsets* regression, we try all possible combinations of predictors and then choose the best model for the number of predictors we feel are justified (e.g., by Evans' Rule). This is the most comprehensive method, though it may present a lot of output.

Example: Takeoff Thrust   Aerospace engineers had a large data set of 469 observations on *Thrust* (takeoff thrust of a jet turbine) along with seven potential predictors (*TurbTemp, AirFlow, TurbSpeed, OilTemp, OilPres, RunTime,* and *ThermCyc*). In the absence of a theoretical model, a stepwise regression was run using MegaStat, with the results shown in Figure 13.20. Only *p*-values are shown for each predictor, along with R^2, R^2_{adj}, and standard error. In this example, most *p*-values are tiny due to the large *n*. Note that each additional predictor adds less to the R^2 than the previous predictor.

Figure 13.20

Stepwise Regression of Turbine Data

Regression Analysis—Stepwise Selection Displaying the Best Model of Each Size

469 observations
Thrust is the dependent variable

| | | | | *p-Values for the coefficients* | | | | | | |
Nvar	TurbTemp	Airflow	TurbSpeed	OilTemp	OilPres	RunTime	ThermCyc	s	Adj R^2	R^2
1		.0000						12.370	.252	.254
2		.0000			.0004			12.219	.270	.273
3	.0003	.0000					.0005	12.113	.283	.287
4		.0000		.0081	.0000		.0039	12.041	.291	.297
5	.0010	.0000		.0009	.0003		.0006	11.914	.306	.314
6	.0010	.0000	.1440	.0037	.0005		.0010	11.899	.308	.317
7	.0008	.0000	.1624	.0031	.0007	.2049	.0006	11.891	.309	.319

While stepwise regression can help identify the "best fit" model for each number of predictors, it is only appropriate when there is no theoretical model to guide us in choosing our predictors. Such theory-free procedures are unappealing to statisticians because they ignore the analyst's knowledge about the data and may result in illogical or impractical models. Yet quasi-automated methods like these (and there are many others, e.g., decision trees, clustering, neural networks) are necessary in data mining applications when analysts face gigantic data sets with many predictors and no theoretical guidelines. Such methods can permit the analyst to spot relationships that would otherwise be overlooked and may suggest models that might merit further investigation. Business graduates are likely to encounter these methods and so must understand their strengths and weaknesses.

 13.10 LOGISTIC REGRESSION (OPTIONAL)

Logistic Regression

Sometimes we need to predict something that has only two possible values (a binary *dependent* variable). In Chapter 12 you saw **logistic regression** (sometimes called *logit*) with a single predictor. Now we can extend this model to include multiple predictors $x_1, x_2, \ldots, x_k$, which is more realistic in applied data analytics. For example, in trying to predict mortgage loan default ($Y = 0$ or 1), we'd want to examine *several* possible predictors such as a borrower's age, income, years of education, employment status, marital status, size of the mortgage, and so on. The logistic regression equation resembles a simple logistic regression (see Section 12.11) except that the exponent includes k predictors. This fitted model predicts the *probability* that $Y = 1$ for specified values of the independent variables $x_1, x_2, \ldots, x_k$. This *nonlinear* model form ensures predictions within the range $0 < \hat{y} < 1$.

LO 13-12

Identify when logistic regression is appropriate and calculate predictions for a binary response variable.

$$y = \frac{e^{b_0+b_1x_1+b_2x_2+\cdots+b_kx_k}}{1 + e^{b_0+b_1x_1+b_2x_2+\cdots+b_kx_k}} \tag{13.23}$$

The logistic function has an S-shaped form that approaches 1 as the value of the exponent increases (see Figure 13.21).

Estimating a Logistic Regression Model

We would not estimate a logit model using ordinary least squares, both because the function is nonlinear and because the assumptions of linear regression are not fulfilled (errors are heteroscedastic and non-normal). Instead, we use the method of **maximum likelihood** to choose

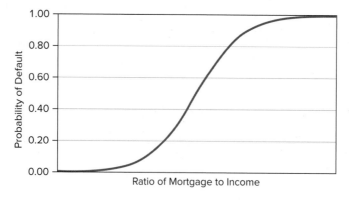

Figure 13.21

Form of the Logistic Model

the values of the regression parameters that will maximize the probability of obtaining the observed sample data. An iterative process is required because there is no simple formula for the parameter estimates $b_0, b_1, \ldots, b_k$.

Because we cannot use Excel, specialized software is required. Any major statistical package will safely perform logistic regression (sometimes called *logit* for short) and will provide *p*-values for the estimated coefficients and predictions for *Y*. You will no longer see the familiar statistics such as R^2, but equivalent measures of fit and significance (with *p*-values) will be provided. Generally, a fairly large sample is required, especially when the event of interest is rare (i.e., when there are not very many 1's in the sample). As a minimum, we want at least 10 observations per predictor, and many experts recommend even more. To achieve a parsimonious model, a common method is to start with the full model (all predictors included) and to use backward stepwise regression to eliminate the weakest predictors until further elimination would weaken the overall model (according to whatever criterion has been selected).

Example: Mortgage Loan Defaults Table 13.11 shows data on a sample of mortgage loans. The binary variable is whether the loan was in default ($Y = 1$) or not ($Y = 0$) within five years after loan approval. In this sample of 200 loans, there were 51 defaults. There are seven potential predictors: borrower's age, income, education, employment status, marital status, mortgage size, and mortgage/income ratio. We omit *Income* and *Mortgage* on grounds that *Ratio* should capture their effects (and to avoid multicollinearity).

Table 13.11

Borrower Characteristics and Mortgage Default Status ($n = 200$ Loans)
📁 Default

Obs	Default	Age	Income	YrsEduc	Employed	Married	Mortgage	Ratio
1	0	44	73.190	16	1	1	204.200	2.79
2	1	59	94.520	16	1	1	283.600	3.00
3	0	50	49.970	16	1	1	63.500	1.27
⋮	⋮	⋮	⋮	⋮	⋮	⋮	⋮	⋮
198	1	57	95.700	10	1	0	169.400	1.77
199	0	38	104.870	14	1	0	311.500	2.97
200	0	46	97.870	20	1	1	283.800	2.90

Note: Only the first three and last three observations are shown. In case of more than one borrower, data are for the first co-signer. Borrower characteristics are recorded at the time the loan was approved.

The following table (from Minitab) shows *z*-statistics for the **Wald test** of individual predictor significance. Predictors *Age* ($z = 1.17$) and *YrsEduc* ($z = -0.70$) are not significant even at $\alpha = .10$ (*p*-values 0.243 and 0.482, respectively). The other three predictors are significant and have the expected sign. The strongest predictor is *Ratio* ($z = 4.27$, *p*-value $= 0.000$), and the coefficient's positive sign implies that the odds of default increase with the ratio of mortgage to income. The predictor *Employed* ($z = -2.98$, p-value $= 0.003$) is significant at $\alpha = .01$, indicating that the probability of default is lower for borrowers who are employed full time (as we would anticipate). The predictor *Married* ($z = -2.00$, p-value $= 0.046$) is barely significant at $\alpha = .05$.

Logistic Regression Table							
					Odds	95%	CI
Predictor	Coef	SE Coef	Z	P	Ratio	Lower	Upper
Constant	-2.46612	1.63241	-1.51	0.131			
Age	0.0252382	0.0216385	1.17	0.243	1.03	0.98	1.07
YrsEduc	-0.0554601	0.0788360	-0.70	0.482	0.95	0.81	1.10
Employed	-1.68500	0.565769	-2.98	0.003	0.19	0.06	0.56
Married	-0.865059	0.432924	-2.00	0.046	0.42	0.18	0.98
Ratio	1.31807	0.308886	4.27	0.000	3.74	2.04	6.84

Log-Likelihood = −95.134

Test that all slopes are zero: G = 36.836, DF = 5, P-Value = 0.000

Where did these p-values come from? The square of the Wald z test statistic follows approximately a chi-square distribution with *d.f.* $= 1$, so each p-value can be verified using the Excel function =CHISQ.DIST.RT(z^2, 1). For overall significance, we look at the **log-likelihood ratio test** statistic $G = 36.836$ with *d.f.* $= k$, where k is the number of predictors. In this example, the right-tail p-value can be found using the Excel function =CHISQ.DIST.RT(36.836,1) = 1.285E-09, which is effectively zero, as shown here.

A simple way to assess the proposed logistic regression model is to count the correct predictions for individual cases and classify the predictions either as *concordant* (correct) or *discordant* (incorrect). Ideally, the frequencies of false positives and false negatives would be relatively small compared with the frequencies of correct predictions. For the mortgage sample data, the frequency of correct predictions was 77.3 percent. Minitab's summary measures require special tables. For more information, consult the end-of-chapter references and download the *LearningStats* demonstrations from McGraw Hill's Connect®.

Measures of Association				
(Between the Response Variable and predicated Probabilities)				
Pairs	Number	Percent	Summary Measures	
Concordant	5877	77.3	Somers' D	0.55
Discordant	1697	22.3	Goodman-Kruskal Gamma	0.55
Ties	25	0.3	Kendall's Tau-a	0.21
Total	7599	100.0		

Chapter Summary

Multivariate regression extends simple regression to include multiple **predictors** of the **response variable.** Criteria to judge a fitted regression model include **logic, fit, parsimony,** and **stability.** Using too many predictors violates the principle of **Occam's Razor,** which favors a simpler model if it is adequate. If the R^2 differs greatly from R^2_{adj}, the model may contain unhelpful predictors. The ANOVA table and **F test** measure overall significance, while the **t test** is used to test hypotheses about individual predictors. A **confidence interval** for each unknown **parameter** is equivalent to a two-tailed hypothesis test for $\beta = 0$. The **standard error** of the regression is used to create **confidence intervals** or **prediction intervals** for *Y*. A **binary predictor** (also called a **dummy variable** or an **indicator**) has value 1 if the condition of interest is present, 0 otherwise. For *c* categories, we only include $c - 1$ binaries or the regression will fail. Including a squared predictor provides a test for **nonlinearity** of the predictors. Including the product of two predictors is a test for **interaction. Collinearity** (correlation between *two* predictors) is detected in the **correlation matrix,** while **multicollinearity** (when a predictor depends on *several* other predictors) is identified from the **variance inflation factor** (VIF) for each predictor. Regression assumes that the errors are normally distributed, independent random variables with constant variance. **Residual tests** identify possible **non-normality, autocorrelation,** or **heteroscedasticity. Logistic regression** is needed when the response variable is binary (0 or 1).

Key Terms

adjusted coefficient of determination (R^2_{adj})	Doane's Rule	logistic regression	residual tests
ANOVA table	dummy variable	log-likelihood ratio test	response variable
autocorrelation	Evans' Rule	maximum likelihood	shift variable
binary predictor	F test	multicollinearity	stability
categorical variable	fit	multiple regression	standard error (s_e)
coefficient of	heteroscedasticity	multiplicative model	stepwise regression
determination (R^2)	homoscedasticity	normal probability plot	variance inflation factor (VIF)
collinearity	indicator variable	Occam's Razor	Wald test
correlation matrix	interaction	parsimony	
dichotomous variable	Klein's Rule	predictors	
	logic	q-q plot	

Commonly Used Formulas

Population regression model for k predictors: $\quad y = \beta_0 + \beta_1 x_1 + \beta_2 x_2 + \cdots + \beta_k x_k + \varepsilon$

Fitted regression equation for k predictors: $\quad \hat{y} = b_0 + b_1 x_1 + b_2 x_2 + \cdots + b_x x_k$

Residual for ith observation: $\quad e_i = y_i - \hat{y}_i \quad$ (for $i = 1, 2, \ldots, n$)

ANOVA sums: $\quad SST = SSR + SSE$

SST (total sum of squares): $\quad \displaystyle\sum_{i=1}^{n}(y_i - \bar{y})^2$

SSR (regression sum of squares): $\quad \displaystyle\sum_{i=1}^{n}(\hat{y}_i - \bar{y})^2$

SSE (error sum of squares): $\quad \displaystyle\sum_{i=1}^{n}(y_i - \hat{y}_i)^2$

MSR (regression mean square): $\quad MSR = SSR/k$

MSE (error mean square): $\quad MSE = SSE/(n - k - 1)$

F test statistic for overall significance: $\quad F_{calc} = MSR/MSE$

Coefficient of determination: $\quad R^2 = 1 - \dfrac{SSE}{SST} \text{ or } R^2 = \dfrac{SSR}{SST}$

Adjusted R^2: $\quad R^2_{adj} = 1 - \dfrac{\left(\dfrac{SSE}{n - k - 1}\right)}{\left(\dfrac{SST}{n - 1}\right)}$

Test statistic for coefficient of predictor X_j: $\quad t_{calc} = \dfrac{b_j - 0}{s_j} \text{ where } s_j \text{ is the standard error of } b_j$

Confidence interval for coefficient β_j: $\quad b_j - t_{\alpha/2}\, s_j \leq \beta_j \leq b_j + t_{\alpha/2}\, s_j \text{ with } d.f. = n - k - 1$

Estimated standard error of the regression: $\quad s_e = \sqrt{\dfrac{\displaystyle\sum_{i=1}^{n}(y_i - \hat{y}_i)^2}{n - k - 1}} = \sqrt{\dfrac{SSE}{n - k - 1}} = \sqrt{MSE}$

Approximate confidence interval for $E(Y/X)$: $\quad \hat{y}_i \pm t_{\alpha/2} \dfrac{s_e}{\sqrt{n}} \text{ with } d.f. = n - k - 1$

Approximate prediction interval for Y: $\quad \hat{y}_i \pm t_{\alpha/2}\, s_e \text{ with } d.f. = n - k - 1$

Variance inflation factor for predictor j: $\quad VIF_j = \dfrac{1}{1 - R^2_j}$

Evans' Rule (conservative): $\quad n/k \geq 10 \text{ (10 observations per predictor)}$

Doane's Rule (relaxed): $\quad n/k \geq 5 \text{ (5 observations per predictor)}$

High leverage: $\quad h_j > \dfrac{2(k + 1)}{n}$

Chapter Review

1. (a) List two limitations of simple regression. (b) Why is estimating a multiple regression model just as easy as simple regression?

2. (a) What does ε represent in the regression model? (b) What assumptions do we make about ε? What is the distinction between Greek letters (β) and Roman letters (b) in representing a regression equation?

3. (a) Describe the format of a multiple regression data set. (b) Why is it a good idea to write down our *a priori* reasoning about a proposed regression?

4. (a) Why does a higher R^2 not always indicate a good model? (b) State the principle of Occam's Razor. (c) List four criteria for assessing a regression model.

5. (a) What is the role of the F test in multiple regression? (b) How is the F statistic calculated from the ANOVA table? (c) Why are tables rarely needed for the F test?

6. (a) Why is testing H_0: $\beta = 0$ a very common test for a predictor? (b) How many degrees of freedom do we use in a t test for an individual predictor's significance?

7. (a) Explain why a confidence interval for a predictor coefficient is equivalent to a two-tailed test of significance. (b) Why are t tables rarely needed in performing significance tests?

8. (a) What does a coefficient of determination (R^2) measure? (b) When R^2 and R^2_{adj} differ considerably, what does it indicate?

9. State some guidelines to prevent inclusion of too many predictors in a regression.

10. (a) State the formula for the standard error of the regression. (b) Why is it sometimes preferred to R^2 as a measure of "fit"? (c) What is the formula for a quick prediction interval for individual Y-values? (d) When you need an exact prediction, what must you do?

11. (a) What is a binary predictor? (b) Why is a binary predictor sometimes called a "shift variable"? (c) How do we test a binary predictor for significance?

12. If we have c categories for an attribute, why do we only use $c - 1$ binaries to represent them in a fitted regression?

13. (a) Explain why it might be useful to include a quadratic term in a regression. (b) Explain why it might be useful to include an interaction term between two predictors in a regression. (c) Name a drawback to including quadratic or interaction terms in a regression.

14. (a) What is multicollinearity? (b) What are its potential consequences? (c) Why is it a matter of degree? (d) Why might it be ignored?

15. (a) How does multicollinearity differ from collinearity? (b) Explain how we can use the correlation matrix to test for collinearity. (c) State a quick rule to test for significant collinearity in a correlation matrix. (d) What is Klein's Rule?

16. (a) State the formula for a variance inflation factor (VIF) for a predictor. (b) Why does the VIF provide a more general test for multicollinearity than a correlation matrix or a matrix plot? (c) State a rule of thumb for detecting strong variance inflation.

17. If multicollinearity is severe, what might its symptoms be?

18. (a) How can we detect an unusual residual? An outlier? (b) How can we identify an influential observation?

19. (a) Name two ways to detect non-normality of the residuals. (b) What are the potential consequences of this violation? (c) What remedies might be appropriate?

20. (a) Name two ways to detect heteroscedastic residuals. (b) What are the potential consequences of this violation? (c) What remedies might be appropriate?

21. (a) Name two ways to detect autocorrelated residuals. (b) What are the potential consequences of this violation? (c) What remedies might be appropriate?

22. (a) What is a lurking variable? How might it be inferred? (b) What are ill-conditioned data?

23. (a) When is logistic regression appropriate? (b) Why do we not use Excel for logistic regression?

Chapter Exercises

Note: Exercises marked * are based on optional material.
Instructions for Data Sets: Choose one of the data sets *A–L* below or as assigned by your instructor. Only the first three and last three observations are shown for each data set. In each data set, the dependent variable (*response*) is the first variable. Choose the independent variables (*predictors*) as you judge appropriate. Use a spreadsheet or a statistical package (e.g., MegaStat or Minitab) to perform the necessary regression calculations and to obtain the required graphs. Write a concise report answering questions 13.25 through 13.41 (or a subset of these questions assigned by your instructor). Label sections of your report to correspond to the questions. Insert tables and graphs in your report as appropriate. You may work with a partner if your instructor allows it.

13.25 Are these cross-sectional data or time series data? What is the unit of observation (e.g., firm, individual, year)?

13.26 Are the X and Y data well-conditioned? If not, make any transformations that may be necessary and explain.

13.27 State your *a priori* hypotheses about the sign (+ or −) of each predictor and your reasoning about cause and effect. Would the intercept have meaning in this problem? Explain.

13.28 Does your sample size fulfill Evans' Rule ($n/k \geq 10$) or at least Doane's Rule ($n/k \geq 5$)?

13.29 Perform the regression and write the estimated regression equation (round off to 3 or 4 significant digits for clarity). Do the coefficient signs agree with your *a priori* expectations?

13.30 Does the 95 percent confidence interval for each predictor coefficient include zero? What conclusion can you draw? *Note:* Skip this question if you are using Minitab because predictor confidence intervals are not shown.

13.31 Do a two-tailed *t* test for zero slope for each predictor coefficient at $\alpha = .05$. State the degrees of freedom and look up the critical value in Appendix D (or from Excel).

13.32 (a) Which *p*-values indicate predictor significance at $\alpha = .05$? (b) Do the *p*-values support the conclusions you reached from the *t* tests? (c) Do you prefer the *t* test or the *p*-value approach? Why?

13.33 Based on the R^2 and ANOVA table for your model, how would you describe the fit?

13.34 Use the standard error to construct an *approximate* prediction interval for *Y*. Based on the width of this prediction interval, would you say the predictions are good enough to have practical value?

13.35 (a) Generate a correlation matrix for your predictors. Round the results to three decimal places. (b) Based on the correlation matrix, is collinearity a problem?

13.36 (a) If you did not already do so, rerun the regression requesting variance inflation factors (VIFs) for your predictors. (b) Do the VIFs suggest that multicollinearity is a problem? Explain.

13.37 (a) If you did not already do so, request a table of standardized residuals. (b) Are any residuals *outliers* (three standard errors) or *unusual* (two standard errors)?

13.38 If you did not already do so, request leverage statistics. Are any observations influential? Explain.

13.39 If you did not already do so, request a histogram of standardized residuals and/or a normal probability plot. Do the residuals suggest non-normal errors? Explain.

13.40 If you did not already do so, request a plot of residuals versus the fitted *Y*. Is heteroscedasticity a concern?

13.41 If you are using time series data, perform one or more tests for autocorrelation (visual inspection of residuals plotted against observation order, runs test, Durbin-Watson test). Is autocorrelation a concern?

DATA SET A Mileage and Other Characteristics of Randomly Selected Vehicles ($n = 73, k = 4$) 📷 Mileage

Obs	Vehicle	CityMPG	Length	Width	Weight	ManTran
1	Acura TL	20	109.3	74.0	3968	0
2	Audi A5	22	108.3	73.0	3583	1
3	BMW 4 Series 428i	22	182.6	71.9	3470	0
⋮	⋮	⋮	⋮	⋮	⋮	⋮
71	Volkswagen Passat SE	24	191.6	72.2	3230	0
72	Volvo S60 T5	21	182.2	73.4	3528	0
73	Volvo XC90	16	189.3	76.2	4667	0

CityMPG = EPA miles per gallon in city driving; *Length* = vehicle length (inches); *Width* = vehicle width (inches); *Weight* = weight (pounds); *ManTran* = 1 if manual shift transmission, 0 otherwise.

Sources: Acura.com; AudiUSA.com; VW.com; Volvo.com. Vehicles are a random sample of 2014 vehicles sold in the U.S. All are gas or flex-fuel (no hybrids or electrics). Data are intended for statistical education and should not be viewed as a guide to vehicle performance.

DATA SET B Noodles & Company Sales, Seating, and Demographic Data ($n = 74, k = 5$) 📷 Noodles2

Obs	Sales/SqFt	Seats-Inside	Seats-Patio	MedIncome	MedAge	BachDeg%
1	702	66	18	45.2	34.4	31
2	210	69	16	51.9	41.2	20
3	365	67	10	51.4	40.3	24
⋮	⋮	⋮	⋮	⋮	⋮	⋮
72	340	63	28	60.9	43.5	21
73	401	72	15	73.8	41.6	29
74	327	76	24	64.2	31.4	15

Sales/SqFt = sales per square foot of floor space, *Seats-Inside* = number of interior seats, *Seats-Patio* = number of outside seats. The three demographic variables refer to a three-mile radius of the restaurant: *MedIncome* = median family income, *MedAge* = median age, and *BachDeg%* = percentage of population with at least a bachelor's degree.

Source: Noodles & Company.

DATA SET C   Assessed Value of Small Medical Office Buildings (*n* = 32, *k* = 5)
📂 **Assessed**

Obs	Assessed	Floor	Offices	Entrances	Age	Freeway
1	1796	4790	4	2	8	0
2	1544	4720	3	2	12	0
3	2094	5940	4	2	2	0
⋮	⋮	⋮	⋮	⋮	⋮	⋮
30	1264	3580	3	2	27	0
31	1162	3610	2	1	8	1
32	1447	3960	3	2	17	0

Assessed = assessed value (thousands of dollars); *Floor* = square feet of floor space; *Offices* = number of offices in the building; *Entrances* = number of customer entrances (excluding service doors); *Age* = age of the building (years); *Freeway* = 1 if within one mile of freeway, 0 otherwise.

DATA SET D   Changes in Consumer Price Index, Capacity Utilization, Changes in Money Supply Components, and Unemployment (*n* = 54, *k* = 4)
📂 **Money**

Year	ChgCPI	CapUtil	ChgM1	ChgM2	Unem
1966	2.9	91.1	2.5	4.6	3.8
1967	3.1	87.2	6.6	9.3	3.8
1968	4.2	87.1	7.7	8.0	3.6
⋮	⋮	⋮	⋮	⋮	⋮
2017	2.1	76.1	8.1	4.9	4.4
2018	2.4	78.0	3.7	3.9	3.9
2019	1.8	77.4	6.2	6.7	3.5

ChCPI = percent change in the Consumer Price Index (CPI) over previous year, *CapUtil* = percent utilization of manufacturing capacity in current year, *ChgM1* = percent change in currency and demand deposits (M1) over previous year, *ChgM2* = percent change in small time deposits and other near-money (M2) over previous year, *Unem* = civilian unemployment rate in percent.

Sources: www.bls.gov and https://fred.stlouisfed.org/.

DATA SET E   College Graduation Rate and Selected Characteristics of U.S. States (*n* = 50, *k* = 8) 📂 **ColGrads**

State	ColGrad%	Dropout	EdSpend	Metro%	Age	LPRFem	Neast	Seast	West
AL	19.8	19.1	1221	89.2	37.4	55.8	0	1	0
AK	28.7	8.3	2187	74.7	33.9	65.6	0	0	1
AZ	27.9	14.3	1137	96.7	34.5	57.4	0	0	1
⋮	⋮	⋮	⋮	⋮	⋮	⋮	⋮	⋮	⋮
WV	15.1	17.6	1538	75.0	40.7	49.1	0	1	0
WI	25.1	9.5	1792	85.9	37.9	66.6	0	0	0
WY	22.0	9.1	1896	71.5	39.1	65.3	0	0	1

ColGrad% = percent of state population with a college degree; *Dropout* = percent of high school students who do not graduate; *EdSpend* = per capita spending on K–12 education; *Metro%* = percent of population living in metropolitan and micropolitan statistical areas; *Age* = median age of state's population; *LPRFem* = percent of adult females who are in the labor force; *Neast* = 1 if state is in the Northeast, 0 otherwise; *Seast* = 1 if state is in the Southeast, 0 otherwise; *West* = 1 if state is in the West, 0 otherwise. *Midwest* is the omitted fourth binary.

Source: *Statistical Abstract of the United States, 2007.*

DATA SET F   Characteristics of Selected Piston Aircraft (*n* = 55, *k* = 4) 📷 CruiseSpeed

Obs	Mfgr/Model	Cruise	Year	TotalHP	NumBlades	Turbo
1	Cessna Turbo Stationair TU206	148	1981	310	3	1
2	Cessna 310 R	194	1975	570	3	0
3	Piper 125 Tri Pacer	107	1951	125	2	0
⋮	⋮	⋮	⋮	⋮	⋮	⋮
53	OMF Aircraft Symphony	128	2002	160	2	0
54	Liberty XL-2	132	2003	125	2	0
55	Piper 6X	148	2004	300	3	0

Cruise = best cruise speed (knots indicated air speed) at 65–75 percent power; *Year* = year of manufacture; *TotalHP* = total horsepower (both engines if twin); *NumBlades* = number of propeller blades; *Turbo* = 1 if turbocharged, 0 otherwise.

Source: *Flying Magazine* (various issues). Data are for educational purposes only and not as a guide to performance.

DATA SET G   Price of Immunotherapy Drugs (mAbs) (*n* = 30, *k* = 4) 📷 mAbsPrice

Obs	Name	Price	CEOComp	Rebates	Biosimilar?	#Trials	Company
1	Erenumab	$8,243	$16.90	12.202	1	21	Amgen
2	Fremanezumab	$2,564	$32.50	4.926	1	16	Teva
3	Duplimab	$9,930	$26.51	19.085	0	2264	Regneron/Sanofi
⋮	⋮	⋮	⋮	⋮	⋮	⋮	⋮
28	Tocilizumab	$5,416	$12.30	7.588	1	195	Roche Genentech
29	Golimumab	$95,380	$29.80	19.996	0	54	J&J
30	Daratumumab	$5,241	$29.80	19.996	0	28	J&J

mAbs are monoclonal antibodies, which is a class of immunotherapy drugs. *Price* = ($) price per gram, *CEOComp* = CEO Compensation ($mil), *Rebates* = rebates and chargebacks manufacturers make to distributors, *Biosimilar?* = 1 if a generic drug exists and 0 otherwise, *#Trials* = number of clinical trials.

Source: Data collected for a class project by Avril Vermunt and Bryan Prust.

DATA SET H   Foreclosure Rates (*n* = 50, *k* = 7) 📷 Foreclosures

State	Foreclosure	MassLayoff	SubprimeShare	PriceIncomeRatio	Ownership	5YrApp	UnempChange	%HousMoved
Alabama	2.70	5.96	28%	4.04	76.6	32.51	0.00%	0.448
Alaska	4.90	2.78	18%	5.54	66.0	53.91	−4.62%	0.426
Arizona	15.20	1.56	26%	6.79	71.1	96.55	−7.32%	0.383
⋮	⋮	⋮	⋮	⋮	⋮	⋮	⋮	⋮
West Virginia	0.50	1.05	21%	4.34	81.3	35.82	−2.13%	0.502
Wisconsin	4.90	12.50	21%	4.83	71.1	36.37	4.26%	0.428
Wyoming	1.50	0.96	23%	5.06	72.8	62.56	−9.09%	0.498

Each observation shows the state foreclosure rate for the year. *MassLayoff* = mass layoff events per 100,000 people, *SubprimeShare* = share of new mortgages that were subprime, *PriceIncomeRatio* = average home prices to median household income ratio, *Ownership* = home ownership rates (%), *5YrApp* = average 5-year home price appreciation, *UnempChange* = 2007 unemployment rate percent change, *%HousMoved* = % of housing that was moved into in the past five years.

Source: MBA project by Steve Rohlwing and Rediate Eshetu.

DATA SET I   Body Fat and Personal Measurements for Males (*n* = 50, *k* = 8) 🖫 **BodyFat2**

Obs	Fat%	Age	Weight	Height	Neck	Chest	Abdomen	Hip	Thigh
1	12.6	23	154.25	67.75	36.2	93.1	85.2	94.5	59.0
2	6.9	22	173.25	72.25	38.5	93.6	83.0	98.7	58.7
3	24.6	22	154.00	66.25	34.0	95.8	87.9	99.2	59.6
⋮	⋮	⋮	⋮	⋮	⋮	⋮	⋮	⋮	⋮
48	6.4	39	148.50	71.25	34.6	89.8	79.5	92.7	52.7
49	13.4	45	135.75	68.50	32.8	92.3	83.4	90.4	52.0
50	5.0	47	127.50	66.75	34.0	83.4	70.4	87.2	50.6

Fat% = percent body fat, *Age* = age (yrs.), *Weight* = weight (lbs.), *Height* = height (in.), *Neck* = neck circumference (cm), *Chest* = chest circumference (cm), *Abdomen* = abdomen circumference (cm), *Hip* = hip circumference (cm), *Thigh* = thigh circumference (cm).

Source: Data are a subsample of 252 males analyzed in Roger W. Johnson, "Fitting Percentage of Body Fat to Simple Body Measurements," *Journal of Statistics Education* 4, no. 1 (1996).

DATA SET J   Used Vehicle Prices (*n* = 637, *k* = 4) 🖫 **Vehicles**

Obs	Model	Price	Age	Car	Truck	SUV
1	Astro GulfStream Conversion	12,988	3	0	0	0
2	Astro LS 4.3L V6	5,950	9	0	0	0
3	Astro LS V6	19,995	4	0	0	0
⋮	⋮	⋮	⋮	⋮	⋮	⋮
635	DC 300M Autostick	10,995	6	1	0	0
636	DC 300M Special Edition	22,995	1	1	0	0
637	GM 3500 4×4 w/8ft bed and plow	17,995	5	0	1	0

Price = asking price ($); *Age* = vehicle age (yrs); *Car* = 1 if passenger car, 0 otherwise; *Truck* = 1 if truck, 0 otherwise; *SUV* = 1 if sport utility vehicle, 0 otherwise (*Van* is the omitted fourth binary).

**DATA SET K   Summer National Senior Games—500-Yard Freestyle
(*n* = 198, *k* = 3)** 🖫 **Swimming**

Swim Time	Seed	Gender	Age
371.20	446.29	1	51
372.28	390.56	1	52
380.10	391.20	1	50
⋮	⋮	⋮	⋮
789.29	795.30	0	85
883.70	1021.20	0	87
1027.22	1074.70	0	86

Source: www.fastlanetek.com. Data are for the 2009 NSGA competition. Participants are age 50 and over. Participant names have been omitted.

DATA SET L   NFL Free Agent Contracts (*n* = 50, *k* = 4) 🖫 **NFLFreeAgent2**

Player Name	Annual Contract ($millions)	Age	Pro-Bowls Played	Purchasing Conference (NFC = 1)	GamesStarted/ GamesPlayed
Apple	3.75	26	0	0	0.863
Averett	4	27	0	0	0.477
Bradberry	7.25	28	1	1	0.989
⋮	⋮	⋮	⋮	⋮	⋮
Witherspoon	4	27	0	0	0.643
Woods	5	27	0	1	0.844
Young	1.365	28	0	1	0.480

Source: Data are from an MBA Class Project from students Jalal Aslam, John Hornbuckle, Jake Jesielowski, Antonio King, Samuel Rolfes, and Colin Tinley.

GENERAL EXERCISES

13.42 In a model of Ford's quarterly revenue $TotalRevenue = \beta_0 + \beta_1\ CarSales + \beta_2\ TruckSales + \beta_3\ SUVSales + \varepsilon$, the three predictors are measured in number of units sold (not dollars). (a) Interpret each slope. (b) Would the intercept be meaningful? Explain.

13.43 In a study of paint peel problems, a regression was suggested to predict defects per million (the response variable). The intended predictors were supplier (four suppliers, coded as binaries), spray pressure, drying time, and drying temperature. There were 11 observations. Explain why regression is impractical in this case, and suggest a remedy.

13.44 A hospital emergency department (ED) studied factors that it thought had an effect on the number of patients who left without being seen (*LWBS*). It analyzed $n = 164$ 24-hour observation periods using six binary predictors representing days of the week, daily occupancy rate, and number of patients presenting to the ED. The fitted regression equation was $LWBS = 1.87 + 1.19Mon - 0.187Tue - 0.785Wed - 0.580Thu - 0.451Fri - 0.267Sat + 0.078Occ + 0.025\#PatPres$ ($SE = 6.18$, $R^2 = .294$, $R^2_{adj} = .292$). (a) Why did the analyst use only six binaries for days when there are 7 days in a week? (b) Which day sees an increase in patients who leave without being seen? (c) Does the number of patients who leave without being seen increase or decrease as occupancy and number of total patients increase? (d) Assess the regression model's fit.

13.45 Using test data on 20 types of laundry detergent, an analyst fitted a regression to predict *CostPerLoad* (average cost per load in cents per load) using binary predictors *TopLoad* (1 if washer is a top-loading model, 0 otherwise) and *Powder* (if detergent was in powder form, 0 otherwise). (a) Write the proposed regression model. (b) Referring to the *F* statistic and its *p*-value (see Exhibit "A" below), what do you conclude about the overall fit of this model? (c) Referring to the predictor *p*-values, what do you conclude about each individual predictor's significance? **Laundry**

13.46 A researcher used stepwise regression (see Exhibit "B" below) to create regression models to predict *BirthRate* (births per 1,000) using five predictors: *LifeExp* (life expectancy in years), *InfMort* (infant mortality rate), *Density* (population density per square kilometer), *GDPCap* (gross domestic product per capita), and *Literate* (literacy percent). (a) Which model (Nvar 1, 2, 3, 4, or 5) best balances fit and parsimony? Explain. (b) Does the addition of *LifeExp* and *Density* improve the model with respect to the R^2_{adj}? (c) Which two variables appear to be the most significant? **BirthRates2**

13.47 A sports enthusiast created an equation to predict *Victories* (the team's number of victories in the National Basketball Association regular season play) using predictors *FGP* (team field goal percentage), *FTP* (team free throw percentage), *Points* (team average points per game), *Fouls* (team average number of fouls per game), *TrnOvr* (team average number of turnovers per game), and *Rbnds* (team average number of rebounds per game). The fitted regression was $Victories = -281 + 523\ FGP + 3.12\ FTP + 0.781\ Points - 2.90\ Fouls + 1.60\ TrnOvr + 0.649\ Rbnds$ ($R^2 = .802$, $F = 10.80$, $SE = 6.87$). The strongest predictors were *FGP* ($t = 4.35$) and *Fouls* ($t = -2.146$). The other predictors were only marginally significant, and *FTP* and *Rbnds* were not significant. The matrix of correlations is shown in Exhibit "C" below. At the time of this analysis, there were 23 NBA teams. (a) Do the signs of the regression coefficients make sense? (b) Is the intercept meaningful? Explain. (c) Is the sample size a problem (using Evans' Rule or Doane's Rule)? (d) Why might collinearity account for the lack of significance of some predictors?

EXHIBIT "A"
Laundry

R^2	0.117		
Adjusted R^2	0.006	n	19
R	0.341	k	2
Std. Error	5.915	Dep. Var.	**Cost per Load**

ANOVA table

Source	*SS*	*df*	*MS*	*F*	*p-Value*
Regression	73.8699	2	36.9350	1.06	.3710
Residual	559.8143	16	34.9884		
Total	633.6842	18			

Regression output

Variables	*Coefficients*	*Std. Error*	*t Stat*	*p-Value*	*Confidence Interval* Lower 95%	Upper 95%
Intercept	26.0000	4.1826	6.216	1.23E-05	17.1333	34.8667
Top-Load	−6.3000	4.5818	−1.375	.1881	−16.0130	3.4130
Powder	−0.2714	2.9150	−0.093	.9270	−6.4509	5.9081

EXHIBIT "B"
BirthRates2

Regression Analysis—Stepwise Selection (best model of each size)

153   observations
BirthRate is the dependent variable

p-Values for the coefficients

Nvar	LifeExp	InfMort	Density	GDPCap	Literate	s	Adj R^2	R^2
1		.0000				6.318	.722	.724
2		.0000			.0000	5.334	.802	.805
3		.0000		.0242	.0000	5.261	.807	.811
4	.5764	.0000		.0311	.0000	5.273	.806	.812
5	.5937	.0000	.6289	.0440	.0000	5.287	.805	.812

EXHIBIT "C"
NBA Wins

	FGP	FTP	Points	Fouls	TrnOvr	Rbnds
FGP	1.000					
FTP	−0.039	1.000				
Points	0.475	0.242	1.000			
Fouls	−0.014	0.211	0.054	1.000		
TrnOvr	0.276	0.028	0.033	0.340	1.000	
Rbnds	0.436	0.137	0.767	−0.032	0.202	1.000

13.48 An expert witness in a case of alleged racial discrimination in a state university school of nursing introduced a regression of the determinants of *Salary* of each professor for each year during an eight-year period ($n = 423$) with the following results, with dependent variable *Salary* and predictors *Year* (year in which the salary was observed), *YearHire* (year when the individual was hired), *Race* (1 if individual is Black, 0 otherwise), and *Rank* (1 if individual is an assistant professor, 0 otherwise). (a) Which of these predictor variables are significant predictors of *Salary?* Explain. (b) What does the coefficient on *Race* tell you about the relationship between *Salary* and *Race?* (c) What does the coefficient on *Rank* tell you about the relationship between *Salary* and *Rank?*

Variable	Coefficient	t
Intercept	−3,816,521	−29.4
Year	1,948	29.8
YearHire	−826	−5.5
Race	−2,093	−4.3
Rank	−6,438	−22.3

13.49 Analysis of a Detroit Marathon ($n = 1,015$ men, $n = 150$ women) produced the regression results shown in Exhibit "D" below, with dependent variable *Time* (the marathon time in minutes) and predictors *Age* (runner's age), *Weight* (runner's weight in pounds), *Height* (runner's height in inches), and *Exp* (1 if runner had prior marathon experience, 0 otherwise). (a) Interpret the coefficient of *Exp*. (b) Does the intercept have any meaning? (c) Why do you suppose squared predictors were included? (d) Plug in your own *Age, Height, Weight,* and *Exp* to predict your own running time. Do you believe it?

13.50 Using test data on 43 vehicles, an analyst fitted a regression (see Exhibit "E") to predict *CityMPG* (miles per gallon in city driving) using as predictors *Length* (length of car in inches), *Width* (width of car in inches), and *Weight* (weight of car in pounds). (a) Referring to the *F* statistic and its *p*-value, what do you conclude about the overall fit of this model? (b) Do you see evidence that some predictors were unhelpful? Explain. (c) Do you suspect that multicolllinearity is a problem? Explain. **CityMPG**

13.51 A researcher used stepwise regression to create regression models to predict *CarTheft* (thefts per 1,000) using four predictors: *Income* (per capita income), *Unem* (unemployment percent), *Pupil/Tea* (pupil-to-teacher ratio), and

EXHIBIT "D"
🗁 Marathon
Times

Variable	Men (n = 1,015)			Women (n = 150)	
	Coefficient	t		Coefficient	t
Intercept	−366			−2,820	
Age	−4.827	−6.1		−3.593	−2.5
Age^2	0.0767	7.1		0.0524	2.6
Weight	−1.598	−1.9		3.000	0.7
$Weight^2$	0.00896	3.4		−0.00404	−2.0
Height	24.65	1.5		96.13	1.6
$Height^2$	−0.2074	−1.7		−0.8040	−1.8
Exp	−41.74	−17.0		−28.65	−4.3
	$R^2 = 0.423$			$R^2 = 0.334$	

Source: Detroit Striders.

EXHIBIT "E"
🗁 CityMPG

R^2	0.682			
Adjusted R^2	0.658	n	43	
R	0.826	k	3	
Std. Error	2.558	Dep. Var.	**CityMPG**	

ANOVA table

Source	SS	df	MS	F	p-Value
Regression	547.3722	3	182.4574	27.90	8.35E-10
Residual	255.0929	39	6.5408		
Total	802.4651	42			

Regression output | | | | | Confidence Interval | | |

Variables	Coefficients	Std. Error	t Stat	p-Value	Lower 95%	Upper 95%	VIF
Intercept	39.4492	8.1678	4.830	.0000	22.9283	55.9701	
Length (in)	−0.0016	0.0454	−0.035	.9725	−0.0934	0.0902	2.669
Width (in)	−0.0463	0.1373	−0.337	.7379	−0.3239	0.2314	2.552
Weight (lbs)	−0.0043	0.0008	−5.166	.0000	−0.0060	−0.0026	2.836

Divorce (divorces per 1,000 population) for the 50 U.S. states (see Exhibit "F" below). (a) Which model (Nvar 1, 2, 3, or 4) best balances fit and parsimony? Explain. (b) Does the addition of *Divorce* improve the model with respect to the R^2_{adj}? (c) Which two variables appear to be the most significant? 🗁 **CarTheft**

*13.52 The following table shows a portion of a data set for 200 individuals who visited an Apple retail store on a particular weekend. Use whatever software is available (e.g., Minitab, Stata, SPSS, SAS) to estimate the binary logistic regression to see whether the three independent variables (income, prior, male) are significant predictors of the binary event (whether or not the individual made a purchase). Include all the predictors and use $\alpha = .05$ in your significance tests. 🗁 **Purchase**

Obs	Buy	Income	Prior	Male
1	1	55.5	0	0
2	1	40.0	0	1
3	1	60.5	1	0
⋮	⋮	⋮	⋮	⋮
198	0	49.8	1	1
199	0	23.3	0	1
200	0	73.0	1	1

Definitions: *Buy* = 1 if individual made a purchase during this visit to the Apple retail store, 0 otherwise; *Income* = individual's annual income last year (thousands); *Prior* = 1 if at least one Apple product already owned, 0 otherwise; *Male* = 1 if borrower was male, 0 otherwise.

EXHIBIT "F"
🗁 **Car Theft Regression**

Regression Analysis—Stepwise Selection (best model of each size)

50 observations
CarTheft is the dependent variable

	p-Values for the coefficients						
Nvar	Income	Unem	Pupil/Tea	Divorce	Std. Err	Adj R^2	R^2
1			.0004		167.482	.218	.234
2	.0018		.0000		152.362	.353	.379
3	.0013	.0157	.0001		144.451	.418	.454
4	.0007	.0323	.0003	.1987	143.362	.427	.474

Related Reading

Hosmer, David W., Stanley Lemeshow, and Rodney X. Sturdivant. *Applied Logistic Regression.* 3rd ed. Wiley, 2014.

Kutner, Michael H., Chris Nachtsheim, and John Neter. *Applied Linear Regression Models.* McGraw Hill, 2004.

Osborne, Jason W. *Best Practices in Logistic Regression.* Sage, 2014.

Tetlock, Philip E., and Dan Gardner. *Superforecasting: The Art and Science of Prediction.* Crown, 2015.

Wooldridge, Jeff. *Introductory Econometrics.* 7th ed. Southwestern, Cengage, 2020.

CHAPTER 13   More Learning Resources

You can access these *LearningStats* demonstrations through Connect to help you understand multiple regression.

Topic	LearningStats Demonstrations
Multiple regression overview	Multiple Regression Overview Violations of Assumptions Multiple Regression Using R
Simulation	Effects of Collinearity Effects of Multicollinearity
Logistic regression	Logistic Regression Binary Logistic Regression
Other supplements	Partial *F* Test Durbin-Watson Test

Key: ☒ = Excel ☐ = PowerPoint 🗏 = Adobe PDF

Software Supplement

Minitab Regression Options

Minitab offers attractive regression options, shown in Figure 13.22. For example, you can choose either regular or standardized residuals. For simplicity, select the Four in One option, or else choose only the residual plots you want. You also may request a separate residual plot against each predictor.

Figure 13.22

Minitab Residual Plot Options 🗏 CockpitNoise

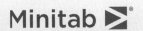

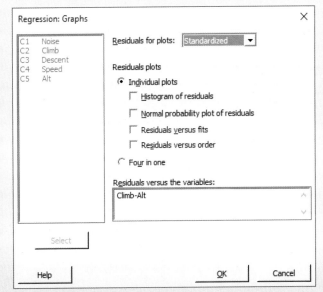

Source: Minitab.

Another useful Minitab feature is the Regression Model menu, which allows you to create squared predictors and/or interactions among predictors, as illustrated in Figure 13.23.

Figure 13.23

Minitab Regression Model Options 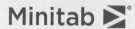 CockpitNoise

Source: Minitab.

Multiple Linear Regression in R

Regression Example Using Cockpit Noise 🖭 CockpitNoise

Multiple regression in R is like simple regression (review the **Software Supplement** in Chapter 12) except that we use a + symbol between each predictor to specify the model. We have 61 observations of cockpit noise (Noise) with predictors flight phase (Climb, Descent), airspeed (Speed), and altitude (Alt). Read data columns into a data frame NoiseData. List the data. Estimate the regression using the linear model function lm(). Save the results in Noise.fit. View the results using summary() and anova(). Calculate leverage statistics using hatvalues(). Check residuals visually for normality (histogram, normality plot) and heteroscedasticity (residual plot against each X_i or only against Y_{fit}). To test for multicollinearity, we prefer VIFs, but that requires an add-on CRAN package. We show instead a correlation matrix using cor() that may suffice to spot worrying predictor relationships.

```
> NoiseData        #display the data (truncated in this printout)
```

	Noise	Climb	Descent	Speed	Alt
1	83.0	1	0	250	10.0
2	89.0	1	0	340	15.0
3	88.0	1	0	320	18.0

```
> Noise.fit=lm(Noise~Climb+Descent+Speed+Alt,NoiseData)
> summary(Noise.fit)
Residuals:
```

Min	1Q	Median	3Q	Max
-2.15787	-0.94498	0.06266	0.89711	2.62745

Coefficients:

	Estimate	Std. Error	t value	Pr(>|t|)	
(Intercept)	64.089257	1.358709	47.169	<2e-16	***
Climb	-0.510960	0.534866	-0.955	0.344	
Descent	-1.228482	0.565320	-2.173	0.034	*
Speed	0.079330	0.003586	22.123	<2e-16	***
Alt	-0.003823	0.026686	-0.143	0.887	

```
Signif.  codes:   0 '***'   0.001 '**'   0.01 '*'   0.05 '.' 0.1 ' ' 1
```

Residual standard error: 1.247 on 56 degrees of freedom
Multiple R-squared: 0.9069,   Adjusted R-squared: 0.9003
F-statistic: 136.4 on 4 and 56 DF,   p-value: < 2.2e-16

```
> anova(Noise.fit)    #show ANOVA table for regression
Analysis of Variance Table
Response: Noise
```

	Df	Sum Sq	Mean Sq	F value	Pr(>F)
Climb	1	67.29	67.29	43.2688	1.72e-08 ***
Descent	1	4.89	4.89	3.1438	0.08166 .
Speed	1	776.22	776.22	499.1446	< 2.2e-16 ***
Alt	1	0.03	0.03	0.0205	0.88660

Residuals 56 87.09 1.56

```
---
Signif. codes:   0 '***' 0.001 '**' 0.01 '*' 0.05 '.' 0.1 ' ' 1
> cor(NoiseData)   #correlation matrix for Y and all X
```

	Noise	Climb	Descent	Speed	Alt
Noise	1.0000000	-0.2681893	0.2382587	0.94587182	0.10949230
Climb	-0.2681893	1.0000000	-0.6944444	-0.31870624	-0.17505540
Descent	0.2382587	-0.6944444	1.0000000	0.35286341	-0.28168304
Speed	0.9458718	-0.3187062	0.3528634	1.00000000	0.06334232
Alt	0.1094923	-0.1750554	-0.2816830	0.06334232	1.00000000

```
> plot(NoiseData$Noise,fitted.values(Noise.fit))    # Y(obs) vs Y(fit)
> plot(residuals(Noise.fit),fitted.values(Noise.fit))  # hetero test
> hist(residuals(Exam.fit))                          # normality test
> qqnorm(rstandard(Exam.fit))                        # normal prob plot
```

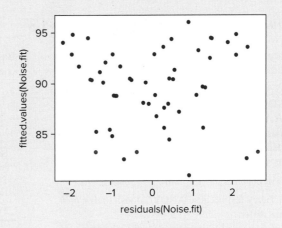

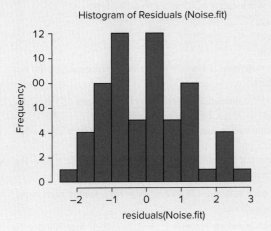

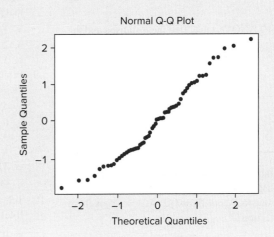

```
> # Create object ResCols to display residual stuff in adjacent columns
> ResCols=data.frame(residuals(Noise.fit),rstandard(Noise.fit),hatvalues(Noise.fit))
> ResCols  #only three rows are displayed for brevity
    residuals.Noise.fit   rstandard.Noise.fit   hatvalues.Noise.fit
1   -0.37255018          -0.31360640           0.09251380
2   -1.49312924          -1.22868330           0.05036831
3   -0.89506207          -0.73266449           0.04029816
```

Logit Example Using Loan Defaults 📁 Default

The binary variable is whether the loan was in default ($Y = 1$) or not ($Y = 0$) within five years after loan approval. In a sample of 200 loans, there were 51 defaults. There are seven potential predictors: borrower's age, income, education, employment status, marital status, mortgage size, and mortgage/income ratio. Copy data columns to a data frame DefaultData. List the data (only 3 lines shown here). Estimate the logistic regression using the general linear model function glm(). Omit Income and Mortgage on grounds that Ratio should capture both effects (and to avoid multicollinearity). Save the result in an object named DLogit. Display the results using summary(). Results are identical with the textbook example in Section 13.10.

```
> DefaultData
    Default  Age  Income  YrsEduc  Employed  Married  Mortgage  Ratio
1   0        44   73.19   16       1         1        204.2     2.79
2   1        59   94.52   16       1         1        283.6     3.00
3   0        50   49.97   16       1         1        63.5      1.27
...

> DLogit.fit=glm(Default~Age+YrsEduc+Employed+Married+Ratio,
    family="binomial",data=DefaultData)
> summary(DLogit.fit)
Deviance Residuals:
    Min       1Q      Median      3Q       Max
  -1.9345  -0.7368    -0.4810   0.4733   2.5003
Coefficients:
              Estimate   Std. Error   z value   Pr(>|z|)
(Intercept)   -2.46612   1.63241      -1.511    0.1309
Age            0.02524   0.02164       1.166    0.2435
YrsEduc       -0.05546   0.07884      -0.703    0.4818
Employed      -1.68500   0.56577      -2.978    0.0029 **
Married       -0.86506   0.43292      -1.998    0.0457 *
Ratio          1.31807   0.30889       4.267    1.98e-05 ***
---
Signif. codes: 0 `***' 0.001 `**' 0.01 `*' 0.05 `.' 0.1 ` ' 1
```

Exam Review Questions for Chapters 11–13

1. Which statement is *correct* concerning one-factor ANOVA? Why not the others?

   a. The ANOVA is a test to see whether the variances of c groups are the same.
   b. In ANOVA, the k groups are compared two at a time, not simultaneously.
   c. ANOVA depends on the assumption of normality of the populations sampled.

2. Which statement is *incorrect?* Explain.

   a. We need a Tukey test because ANOVA doesn't tell *which* group means differ.
   b. Hartley's test is needed to determine whether the means of the groups differ.
   c. ANOVA assumes equal variances in the k groups being compared.

3. Given the following ANOVA table, find the F statistic and the critical value of $F_{.05}$.

Source	SS	df	MS	F
Treatment	744.00	4		
Error	751.50	15		
Total	1,495.50	19		

4. Given the following ANOVA: (a) How many ATM locations were there? (b) What was the sample size? (c) At $\alpha = .05$, is there a significant effect due to *Day of Week?* (d) At $\alpha = .05$, is there a significant interaction?

Source	SS	df	MS	F
ATM	41926.67	2	20963.33	9.133
Day of Week	4909.52	6	818.25	0.356
Interaction	29913.33	12	2492.78	1.086
Error	433820.00	189	2295.34	
Total	510569.52	209		

5. Given a sample correlation coefficient $r = .373$ with $n = 30$, can you reject the hypothesis $\rho = 0$ for the population at $\alpha = .01$? Explain, stating the critical value you are using in the test.

6. Which statement is *incorrect?* Explain.

   a. Correlation uses a t test with $n - 2$ degrees of freedom.
   b. Correlation analysis assumes that X is independent and Y is dependent.
   c. Correlation analysis is a test for the degree of linearity between X and Y.

7. Based on the information in this partial ANOVA table, the coefficient of determination R^2 is approximately

   a. 0.499
   b. 0.501
   c. 0.382

Source	SS	df	MS
Regression	158.33	1	158.33
Residual	159.08	25	6.36
Total	317.41	26	

8. In a test of the regression model $Y = \beta_0 + \beta_1 X$ with 27 observations, what is the critical value of t to test the hypothesis that $\beta_1 = 0$ using $\alpha = .05$ in a two-tailed test?

   a. 1.960
   b. 2.060
   c. 1.708

9. Which statement is *correct* for a simple regression? Why not the others?

   a. A 95% confidence interval (CI) for the mean of Y is wider than the 95% CI for the predicted Y.
   b. A confidence interval for the predicted Y is widest when $X = \bar{x}$.
   c. The t test for zero slope always gives the same t_{calc} as the correlation test for $\rho = 0$.

10. Tell if each statement is *true* or *false* for a simple regression. If false, explain.

    a. If the standard error is $s_e = 3,207$, then a residual $e_i = 4,327$ would be an outlier.
    b. In a regression with $n = 50$, a leverage statistic $h_i = .10$ indicates unusual leverage.
    c. A decimal change is often used to improve data conditioning.

11. For a multiple regression, which statement is *true?* Why not the others?

    a. Evans' Rule suggests at least 10 observations for each predictor.
    b. The t_{calc} in a test for significance of a binary predictor can have only two values.
    c. Occam's Razor says we must prefer simple regression because it is simple.

12. For a multiple regression, which statement is *false?* Explain.

    a. If $R^2 = .752$ and $R^2_{adj} = .578$, the model probably has at least one weak predictor.
    b. R^2_{adj} can exceed R^2 if the model contains some very strong predictors.
    c. Deleting a predictor could increase the R^2_{adj} but will not increase R^2.

13. Which predictor coefficients differ significantly from zero at $\alpha = .05$?

    a. X3 and X5
    b. X5 only
    c. all but X1 and X3

	Coefficients	Lower 95%	Upper 95%
X1	−0.2430	−0.5743	0.0882
X2	0.1876	−0.3778	0.7529
X3	−0.3397	−0.4681	−0.2114
X4	0.0019	−0.0144	0.0182
X5	1.6025	0.1326	3.0724

14. Which predictors differ significantly from zero at $\alpha = .05$?

    a. X3 only
    b. X4 only
    c. both X3 and X4

	Coefficients	Std. Error	p-Value
X1	−0.2280	0.1782	0.2099
X2	0.2189	0.3008	0.4721
X3	−0.3437	0.0597	0.0000
X4	1.5884	0.7427	0.0402

15. In this regression with $n = 40$, which predictor differs significantly from zero at $\alpha = .01$?

    a. X2
    b. X3
    c. X5

	Coefficients	Std. Error
X1	−0.03472	0.02328
X2	0.02679	0.03974
X3	−0.04852	0.00902
X4	0.00027	0.00114
X5	0.22893	0.10333

CHAPTER CONTENTS

CHAPTER LEARNING OBJECTIVES

When you finish this chapter, you should be able to

LO 14-1 Define time series data and its components.

LO 14-2 Interpret a linear, exponential, or quadratic trend model.

LO 14-3 Fit any common trend model and use it to make forecasts.

LO 14-4 Know the definitions of common fit measures.

LO 14-5 Interpret a moving average and use Excel to create it.

LO 14-6 Use exponential smoothing to forecast trendless data.

LO 14-7 Interpret seasonal factors and use them to make forecasts.

LO 14-8 Use regression with seasonal binaries to make forecasts.

LO 14-9 Interpret index numbers.

14.1 TIME SERIES COMPONENTS

Time Series Data

Businesses must track their performance. By looking at their sales, costs, or profits over time, businesses can tell where they've been, whether they are performing poorly or satisfactorily, and how much improvement is needed, in both the short term and the long term. A **time series variable** (denoted Y) consists of data observed over n periods of time. Consider a clothing retailer that specializes in blue jeans. Examples of time series data this company might be interested in tracking would be the number of jeans sold and the company's market share. Or, from the manufacturing perspective, the company might track cost of raw materials over time.

Businesses also use time series data to monitor whether a particular process is stable or unstable. And they use time series data to help anticipate the future, a process we call *forecasting*. In addition to business time series data, we see economic time series data in *The Wall Street Journal* or *Bloomberg Businessweek* and also in *USA Today* or *Time* or even when we browse the web. Although business and economic time series data are most common, we can see time series data for population, health, crime, sports, and social problems. Usually, time series data are presented in a graph, like Figures 14.1 and 14.2.

It is customary to plot time series data either as a line graph or as a bar graph, with time on the horizontal X-axis and the variable of interest on the vertical Y-axis to reveal how the variable changes over time. In a line graph, the X-Y data points are connected with line segments to make it easier to see fluctuations. While anyone can understand time series graphs in a general way, this chapter explains how to interpret time series data *statistically* and to make defensible forecasts. Our analysis begins with sample observations $y_1, y_2, \ldots, y_n$ covering n time periods. The following notation is used:

- y_t is the value of the time series in period t.
- t is an index denoting the time period ($t = 1, 2, \ldots, n$).
- n is the number of time periods.
- $y_1, y_2, \ldots, y_n$ is the data set for analysis.

To distinguish time series data from cross-sectional data, we use y_t for an individual observation, with a subscript t instead of i.

JUNG YEON-JE/AFP/Getty Images

Figure 14.1

U.S. Employment (monthly, not seasonally adjusted) Labor

Source: www.bls.gov. Latest data shown are for January 1, 2020.

Figure 14.2

Exchange Rates (daily) 📁 **Exchange**

Source: www.federalreserve.gov. Latest data shown are for January 24, 2020.

Time series data may be measured *at a point in time* or *over an interval of time.* For example, in accounting, balance sheet data are measured at the end of the fiscal year, while income statement data are measured over an entire fiscal year. The gross domestic product (GDP) is a flow of goods and services measured *over an interval of time,* while the prime rate of interest is measured *at a point in time.* Your GPA is measured *at a point in time,* while your weekly pay is measured *over an interval of time.* The distinction is sometimes vague in reported data, but a little thought will usually clarify matters. For example, Canada's 2018 unemployment rate (5.6 percent) would be measured at a point in time (e.g., at year's end), while Canada's 2018 hydroelectric production (382 terawatt-hours) would be measured over the entire year (see www.statcan.gc.ca).

Periodicity

The **periodicity** is the time interval over which data are collected (decade, year, quarter, month, week, day, hour). For example, the U.S. population is measured each *decade,* your personal income tax is calculated *annually,* GDP is reported *quarterly,* the unemployment rate is estimated *monthly,* and *The Wall Street Journal* reports the closing price of Apple stock *daily* (although stock prices are also monitored continuously on the web). Firms typically report profits by quarter but pension liabilities only at year's end. Any periodicity is possible, but the principles of time series modeling can be understood with three common data types:

- Annual data (1 observation per year)
- Quarterly data (4 observations per year)
- Monthly data (12 observations per year)

Time Series Components

Time series *decomposition* seeks to separate a time series Y into four components: trend (T), cycle (C), seasonal (S), and irregular (I). Figure 14.3 illustrates these four components in a hypothetical monthly time series. The four components may be thought of as layering atop one another to produce the actual time series. In this example, the irregular component (I) is large enough to obscure the cycle (C) and seasonal (S) components but not the trend (T). However, we can usually extract the original components from the time series by using statistical methods. These components are assumed to follow either an **additive model** or a **multiplicative model**, as shown in Table 14.1.

Figure 14.3

Four Components of a Time Series

Trend

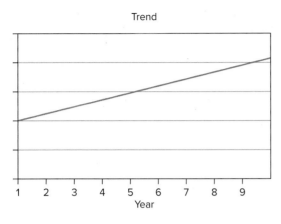

Trend + Cycle

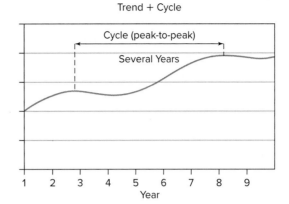
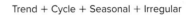

Trend + Cycle + Seasonal

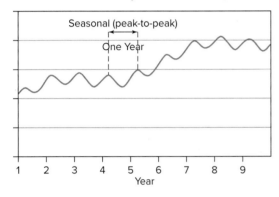

Trend + Cycle + Seasonal + Irregular

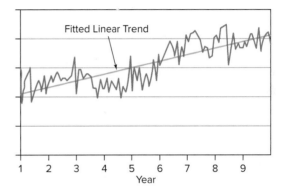

Table 14.1

Components of a Time Series

Model	Components	Used for
Additive	$Y = T + C + S + I$	Data of similar magnitude (short-run or trend-free data) with constant *absolute* growth or decline.
Multiplicative	$Y = T \times C \times S \times I$	Data of increasing or decreasing magnitude (long-run or trended data) with constant *percent* growth or decline.

The additive form is attractive for its simplicity, but the multiplicative model is often more useful for forecasting financial data, particularly when the data vary over a range of magnitudes. Especially in the short run, it may not matter greatly which form is assumed. In fact, the model forms are fundamentally equivalent because the multiplicative model becomes additive if logarithms are taken (as long as the data are nonnegative):

$$\log(Y) = \log(T \times C \times S \times I) = \log(T) + \log(C) + \log(S) + \log(I)$$

Trend

Trend (T) is a general movement over all years ($t = 1, 2, \ldots, n$). Change over a few years is not a trend. Some trends are steady and predictable. For example, the data may be steadily growing (e.g., total U.S. population), neither growing nor declining (e.g., your current car's mpg), or steadily declining (infant mortality rates in a developing nation). A mathematical trend can be fitted to any data, but its predictive value depends on the situation. For example, to predict health expenditures or Amazon's net sales (Figure 14.4), a mathematical trend might be useful, but a mathematical model might not be very helpful for predicting frequency of hurricanes or Fargo, ND, snowfall (Figure 14.5).

Most of us think of three general patterns: growth, stability, or decline. But there are sub-tler trends within each category. A time series can increase at a steady *linear* rate (e.g., the number of books you have read in your lifetime), at an *increasing* rate (e.g., Medicare costs for an aging population), or at a *decreasing* rate (e.g., live attendance at NFL football games). It can grow for a while and then level off (e.g., sales of 86-inch TVs) or grow toward an asymptote (e.g., percent of adults owning an iPhone).

Figure 14.4

Steady Trend 📁 Steady

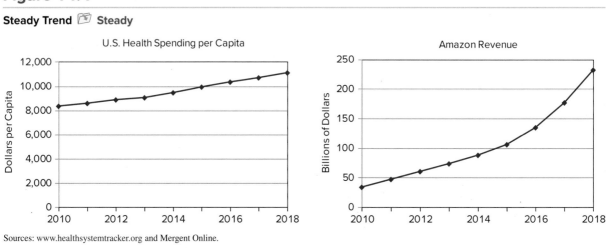

Sources: www.healthsystemtracker.org and Mergent Online.

Figure 14.5

Erratic Pattern 📁 Erratic

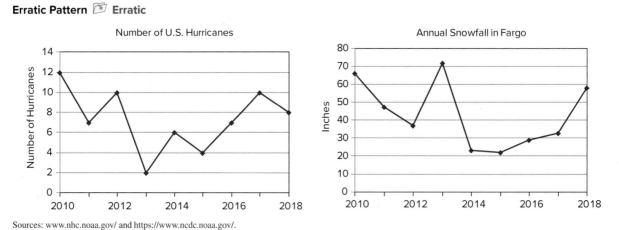

Sources: www.nhc.noaa.gov/ and https://www.ncdc.noaa.gov/.

Cycle

Cycle (*C*) is a repetitive up-and-down movement around the trend that covers *several years*. For example, industry analysts have studied cycles for sales of new automobiles, new home construction, inventories, and business investment. These cycles are based primarily on prod-uct life and replacement cycles. In any market economy, there are broad business cycles that affect employment and production. After we have extracted the trend and seasonal components of a time series, a cycle may be detected as autocorrelation in the residuals (see Chapter 12, Section 12.8). Although cycles are conceptually important, there is no general theory of cycles, and even those cycles that have been identified in specific industries have erratic timing and complex causes that defy generalization. Over a small number of time periods (a typical forecasting situation), cycles are undetectable or may resemble a trend. For this reason, cycles are not discussed further in this chapter.

Seasonal

Seasonal (*S*) is a repetitive cyclical pattern *within a year.*[1] For example, many retail businesses experience strong sales during the fourth quarter because of Christmas. Automobile sales rise when new models are released. Peak demand for airline flights to Europe occurs during summer vacation travel. Although often imagined as sine waves, seasonal patterns may not be smooth. Peaks and valleys can occur in any month or quarter, and each industry may face its own unique seasonal pattern. For example, June weddings tend to create a "spike" in bridal sales, but there is no "sine wave" pattern in bridal sales. By definition, annual data have no seasonality.

Irregular

Irregular (*I*) is a random disturbance that follows no apparent pattern. It also is called the *error* component or *random noise* reflecting all factors other than trend, cycle, and seasonality. For example, daily prices of many common stocks fluctuate greatly. When the irregular component is large, it may be difficult to isolate other individual model components. In such cases, we use special techniques (e.g., **moving average** or **exponential smoothing**) to make short-run forecasts. Faced with erratic data, experts may use their own knowledge to make *judgment forecasts.* For example, vehicle sales forecasts may combine judgment forecasts from dealers, financial staff, and economists. However, a major systemic shock such as the COVID-19 crisis may dominate other time series components and render forecasting efforts moot.

 14.2 TREND FORECASTING

There are many forecasting methods designed for specific situations. Much of this chapter deals with *trend models* because they are so common in business. You also will learn to use *decomposition* to make adjustments for *seasonality* and how to use *smoothing models.* The important topics of *ARIMA models* and *causal models* are reserved for a more specialized class in forecasting. Figure 14.6 summarizes the main categories of forecasting models.

 LO 14-2

Interpret a linear, exponential, or quadratic trend model.

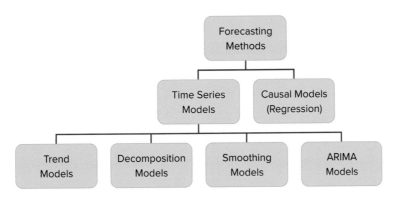

Figure 14.6

Overview of Forecasting

---

[1]Repetitive patterns within a week, day, or other time period also may be considered seasonal. For example, mail volume in the U.S. Postal Service is higher on Monday. Emergency arrivals at hospitals are lower during the first shift (between midnight and 6:00 a.m.). In this chapter, we will discuss only *monthly* and *quarterly* seasonal patterns because these are most typical of business data.

Three Trend Models

There are many possible trend models, but three of them are especially useful in business:

(14.1) $\qquad$ $y_t = b_0 + b_1 t$ $\qquad\qquad$ for $t = 1, 2, \ldots, n$ (linear trend)

(14.2) $\qquad$ $y_t = b_0 e^{b_1 t}$ $\qquad\qquad$ for $t = 1, 2, \ldots, n$ (exponential trend)

(14.3) $\qquad$ $y_t = b_0 + b_1 t + b_2 t^2$ $\quad$ for $t = 1, 2, \ldots, n$ (quadratic trend)

The linear and exponential models are widely used because they have only two parameters and are familiar to most business audiences. The quadratic model may be useful when the data have a turning point. All three can be fitted by Excel.

Linear Trend Model

The **linear trend** model has the form $y_t = b_0 + b_1 t$. It is useful for a time series that grows or declines by the same amount (b_1) in each period, as shown in Figure 14.7. It is the simplest model and may suffice for short-run forecasting. It is generally preferred in business as a baseline forecasting model unless there are compelling reasons to consider a more complex model.

Figure 14.7

Linear Trend Models

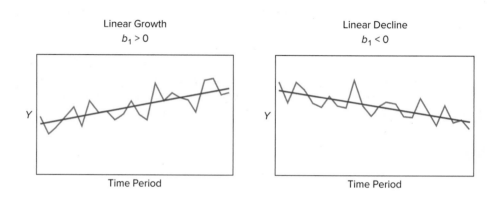

Illustration: Linear Trend

Following the population "echo boom," some U.S. states (especially in the Northeast and Midwest) began to see declines in the number of high school graduates. Concerned about its potential loss of traditional student populations aged 18–25, a Midwestern university wanted to extrapolate the recent trend in fall semester enrollments. The slope of Excel's fitted trend, shown in Figure 14.8, indicates that, on average, the university has lost 235 students per year. This decline was probably exacerbated by the COVID epidemic.

Figure 14.8

Excel's Linear Trend
📁 **Enrollment**

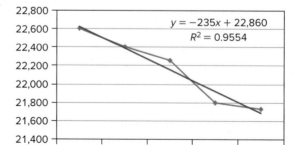

Linear Trend Calculations

The linear trend is fitted in the usual way by using the ordinary least squares formulas. Because you are already familiar with regression, we will only point out the use of the index $t = 1, 2, 3, 4, 5$ as the independent variable (instead of using the years 2018, 2019, 2020, 2021, 2022). We use this time index to simplify the calculations and keep the data magnitudes under control (Excel uses this method too).

LO 14-3

Fit any common trend model and use it to make forecasts.

$$Slope: \quad b_1 = \frac{\sum_{t=1}^{n}(t - \bar{t})(y_t - \bar{y})}{\sum_{t=1}^{n}(t - \bar{t})^2} = \frac{-2,350}{10} = -235$$

$$Intercept: \quad b_0 = \bar{y} - b_1\bar{t} = 22,155 - (-235)(3) = 22,860$$

The *slope* of the fitted trend $y_t = 22,860 - 235t$ says that, unless the university takes steps to recruit new, nontraditional student populations, it can expect to lose 235 students each year ($dy_t/dt = -235$). The *intercept* is the "starting point" for the time series in period $t = 0$; that is, $y_0 = 22,860 - 235(0) = 22,860$.

Time Labels

In fitting a trend to annual data, the years (2018, 2019, 2020, 2021, 2022) are merely used as labels for the *X*-axis. The yearly labels should *not* be used in fitting the trend or calculating the forecast. To fit a trend to annual data, use the time index ($t = 1, 2, \ldots$, etc.). To make a forecast, insert a value for the time index ($t = 1, 2, \ldots$, etc.) into Excel's fitted trend.

Forecasting a Linear Trend

We can make a forecast for any future year by using the fitted model $y_t = 22,860 - 235t$. In the enrollment example, the fitted trend equation is based on only 5 years' data, so we should be wary of extrapolating very far ahead:

For 2023 ($t = 6$): $y_6 = 22,860 - 235(6) = 21,450$
For 2024 ($t = 7$): $y_7 = 22,860 - 235(7) = 21,215$
For 2025 ($t = 8$): $y_8 = 22,860 - 235(8) = 20,980$

Linear Trend: Calculating R^2

In this illustration, the linear model gives a good fit ($R^2 = 0.9554$) to the *past* data. However, a good fit to the past data does not guarantee good *future* forecasts. A deeper analysis of underlying causes of enrollment declines is needed. Are the causal forces likely to remain the same in subsequent years? Could the current demographic decline continue indefinitely, or will enrollments approach an asymptote or even start to grow again? Will the COVID threat diminish or recur? These are questions that forecasters must ask. The forecast is simply a projection of current trend assuming that nothing changes.

$$Coefficient\ of\ determination: \quad R^2 = 1 - \frac{\sum_{t=1}^{n}(y_t - \hat{y}_t)^2}{\sum_{t=1}^{n}(y_t - \bar{y})^2} = 1 - \frac{25,750}{578,000} = 0.9554$$

Exponential Trend Model

The **exponential trend** model has the form $y_t = b_0 e^{b_1 t}$. It is useful for a time series that grows or declines at the same *rate* (b_1) in each period, as shown in Figure 14.9. When the growth rate is positive ($b_1 > 0$), then Y grows by an *increasing* amount each period (unlike the linear model, which assumes a *constant* increment each period). If the growth rate is negative ($b_1 < 0$), then Y declines by a *decreasing* amount each period (unlike the linear model, which assumes a *constant* decrement each period).

Figure 14.9

Exponential Trend Models

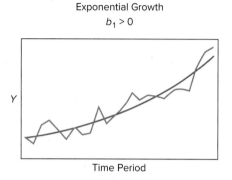
Exponential Growth
$b_1 > 0$

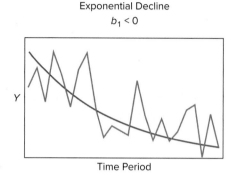
Exponential Decline
$b_1 < 0$

When to Use the Exponential Model

The exponential model is often preferred for financial data or data that cover a longer period of time. When you invest money in a commercial bank savings account, interest accrues at a given percent. Your savings grow faster than a linear rate because you earn interest on the accumulated interest. Banks use the exponential formula to calculate interest on CDs. Financial analysts often find the exponential model attractive because costs, revenue, and salaries are best projected under assumed *percent* growth rates.

Another nice feature of the exponential model is that you can compare two growth rates in two time series variables with dissimilar data units (i.e., a percent growth rate is *unit-free*). For example, between 2000 and 2020 the number of Medicare enrollees grew from 40.0 million persons to 62.6 million persons (2.3 percent annual growth rate), while Medicare payments grew from $217 billion to $830 billion (6.9 percent annual growth rate). Comparing these percents, we see that Medicare insurance payments have been growing three times as fast as the Medicare head count (see www.cms.gov). These facts underlie the ongoing debate about Medicare spending in the United States.

There may not be much difference between a linear and exponential model when the growth rate is small and the data set covers only a few time periods. For example, suppose your starting salary is $50,000. Table 14.2 compares salary increases of $2,500 each year ($y_t = 50{,}000 + 2{,}500t$) with a continuously compounded 4.879 percent salary growth ($y_t = 50{,}000e^{0.04879t}$). Over the first few years, there is little difference. But after 20 years, the difference is obvious, as shown in Figure 14.10. Despite its attractive simplicity,[2] the linear model's assumptions may be inappropriate for some financial variables.

Table 14.2

Two Models of Salary Growth

t	$y_t = 50{,}000 + 2{,}500t$ Linear	$y_t = 50{,}000e^{0.04879t}$ Exponential
0	50,000	50,000
5	62,500	63,814
10	75,000	81,445
15	87,500	103,946
20	100,000	132,665

[2]In a sense, the linear model ($y_t = b_0 + b_1 t$) and the exponential model ($y_t = b_0 e^{b_1 t}$) are equally simple because they are two-parameter models, and a log-transformed exponential model $\ln(y_t) = \ln(b_0) + b_1 t$ is actually linear.

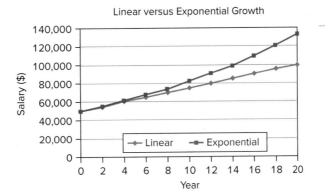

Figure 14.10

Linear and Exponential Growth Compared

Illustration: Exponential Trend

Spending on Internet security in the United States has shown explosive growth. For example, Figure 14.11 shows revenue growth for one web security company. Clearly, a linear trend (constant *dollar* growth) would be inadequate. It is more reasonable to assume a constant *percent* rate of growth and fit an exponential model. Excel's fitted exponential trend is $y_t = 3.8197e^{0.3894t}$. The value of b_1 in the exponential model $y_t = b_0 e^{b_1 t}$ is the continuously compounded growth rate, so we can say that Dolon's revenue is growing at an astonishing rate of 38.94 percent per year. A negative value of b_1 in the equation $y_t = b_0 e^{b_1 t}$ would indicate *decline* instead b_0 is the "starting point" in period $t = 0$. For example, $y_0 = 3.8197e^{0.3894(0)} = 3.8197$.

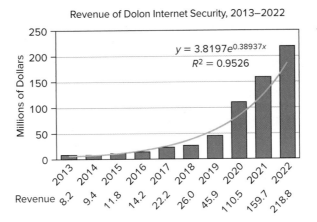

Figure 14.11

Excel's Exponential Trend
📁 **DolonCorp**

Exponential Trend Calculations

Calculations of the exponential trend are done by using a transformed variable $z_t = \ln(y_t)$, instead of y_t, to produce a linear equation so that we can use the least squares formulas.

$$\text{Slope:} \quad b_1 = \frac{\sum_{t=1}^{n}(t - \bar{t})(z_t - \bar{z})}{\sum_{t=1}^{n}(t - \bar{t})^2} = \frac{32.12329}{82.5} = 0.3893732$$

$$\text{Intercept:} \quad b_0 = \bar{z} - b_1 \bar{t} = 3.481731 - (.3893732)(5.5) = 1.340178$$

When the least squares calculations are completed, we must transform the intercept back to the original units by exponentiation to get the correct intercept $b_0 = e^{1.340178} = 3.8197$. The fitted trend equation is

$$y_t = b_0 e^{b_1 t} = 3.8197e^{0.3894t}$$

Forecasting an Exponential Trend

We can make a forecast of debit card usage for any future year by using the fitted model:[3]

For 2023 ($t = 11$): $y_{11} = 3.8197e^{0.38937(11)} = 276.8$

For 2024 ($t = 12$): $y_{12} = 3.8197e^{0.38937(12)} = 408.5$

For 2025 ($t = 13$): $y_{13} = 3.8197e^{0.38937(13)} = 603.0$

Can Dolon's revenue actually continue to grow at a rate of 38.937 percent? Typically, when a new product is introduced, its growth rate at first is very strong but eventually slows down as the market becomes saturated and/or as competitors arise. Wild card factors such as the COVID crisis add further uncertainty. However, demand for Internet security protection is expected to grow, so Dolon's high growth rate might continue if the firm is able to manage its expansion. Trend forecasts are useful only if the past trend can reasonably be assumed to provide a reliable guide to the near future.

Exponential Trend: Calculating R^2

We calculate R^2 the same way as for the linear trend, except that we replace the dependent variable y_t with $z_t = \ln(y_t)$ and the fitted value with $\hat{z}_t = 1.340178 + .389373t$. This is necessary because Excel's trend-fitting calculations are done in logarithms:

$$\text{Coefficient of determination:} \quad R^2 = 1 - \frac{\sum_{t=1}^{n}(z_t - \hat{z}_t)^2}{\sum_{t=1}^{n}(z_t - \bar{z})^2} = 1 - \frac{0.62228}{13.13021} = 0.9526$$

In this example, the exponential trend gives a very good fit ($R^2 = 0.9526$) to the past data. Although a high R^2 does not guarantee good forecasts, demand for Internet security protection is expected to grow, so Dolon's high growth rate could continue if the firm is able to manage its expansion.

Knowing only y_1 and y_t (the starting and ending values) you can estimate the compound growth rate b_1 using this formula:

(14.4)
$$b_1 = [\ln(y_t) - \ln(y_1)]/(t - 1)$$

We can apply this formula to Dolon's revenue using Excel's natural log function:

$$b_1 = (\text{LN}(218.8) - \text{LN}(8.2))/(10 - 1) = 0.3649$$

This quick estimate can be useful (e.g., for comparing investments) when you only know where you started and where you are now.

Quadratic Trend Model

The **quadratic trend** model has the form $y_t = b_0 + b_1t + b_2t^2$. The t^2 term allows a nonlinear shape. It is useful for a time series that has a turning point or that is not captured by the exponential model. If $b_2 = 0$, the quadratic model $y_t = b_0 + b_1t + b_2t^2$ becomes a linear model because the term b_2t^2 drops out of the equation (i.e., the linear model is a special case of the quadratic model). Fitting a quadratic model is a way of checking for nonlinearity. If the coefficient b_2 does not differ significantly from zero (and if the quadratic R^2 is about the same as a linear model), then the linear model would suffice. Depending on the values of b_1 and b_2, the quadratic model can assume any of four shapes, as shown in Figure 14.12.

---

[3]Excel uses the exponential formula $y_t = b_0 \exp(b_1t)$, where b_1 is the *continuously compounded* growth rate. Minitab uses $y_t = y_0 (1 + r)^t$, which you may recognize as the formula for compound interest. Although the formulas appear different, they give identical forecasts. To convert Minitab's fitted equation to Excel's, set $b_0 = y_0$ and $b_1 = \ln(1 + r)$. To convert Excel's fitted equation to Minitab's, set $y_0 = b_0$ and $r = \exp(b_1) - 1$.

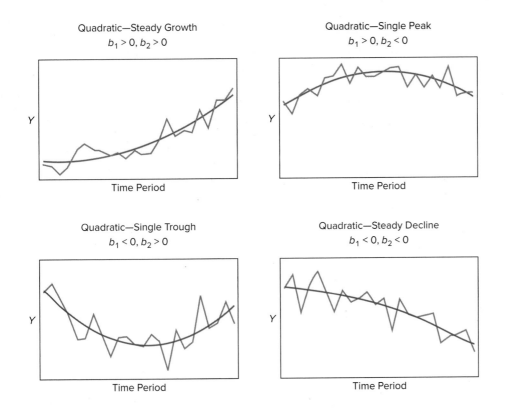

Figure 14.12

Four Quadratic Trend Models

Illustration: Quadratic Trend

The number of hospital beds in a certain rural county (Table 14.3) declined during, showed signs of leveling out, and then declined again. What trend would we choose if the objective is to make a realistic one-year forecast?

Year	Beds	Year	Beds
2013	1,081	2018	984
2014	1,062	2019	987
2015	1,035	2020	976
2016	1,013	2021	965
2017	994	2022	956

Table 14.3

Number of Hospital Beds, 2013–2022
📁 **HospitalBeds**

Figure 14.13 shows one-year projections using the linear and quadratic models. Many observers would think that the quadratic model offers a more believable prediction because the quadratic model is able to capture the slight curvature in the data pattern. But this gain in credibility must be weighed against the added complexity of the quadratic model. It appears that the forecasts would turn upward if projected more than one year ahead. We should be especially skeptical of any polynomial model that is projected more than one or two periods into the future.

Figure 14.13

Two Trend Models for Hospital Beds, 2013–2023 📁 **HospitalBeds**

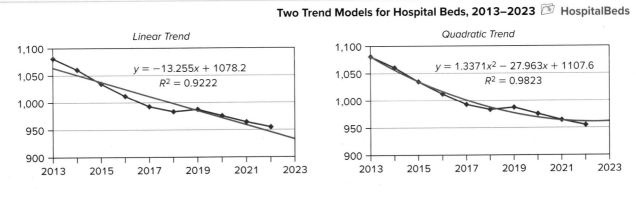

Using Excel for Trend Fitting

Plot the data, right-click on the data, and choose a trend. Figure 14.14 shows Excel's menu of six trend options. The menu includes a sketch of each trend type. Click the Options tab if you want to display the R^2 and fitted equation on the graph, or if you want to plot forecasts (trend extrapolations) on the graph. The quadratic model is a **polynomial model** of order 2. Despite Excel's many choices, some patterns cannot be captured by any of the common trend models. For women air transport pilots, the fitted quadratic (polynomial) regression predicts continued growth, but at a slowing rate. By default, Excel reports four- or five-decimal accuracy. However, you can click on Excel's fitted trend equation, choose Format Data Labels, choose Number, and set the number of decimal places you want to see.

Figure 14.14

Excel's Trend-Fitting Menus   WomenPilots

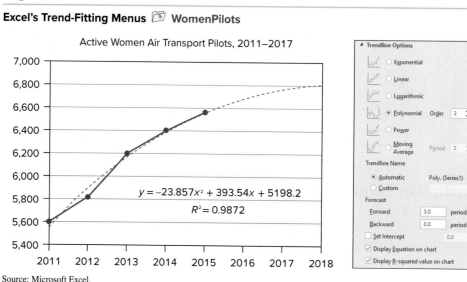

Source: Microsoft Excel.

Principle of Occam's Razor

Given two *sufficient* explanations, we prefer the simpler one.

William of Occam (1285–1347)

Trend-Fitting Criteria

It is so easy to fit a trend in Excel that it is tempting to "shop around" for the best fit. Forecasters prefer the simplest trend model that adequately matches the trend. Simple models are easier to interpret and explain to others. However, that is *not* to say that a simpler model is always preferred. Occam's Razor is merely a "tie-breaker" when we have two *equally* good models. Criteria for selecting a trend model for forecasting include

Criterion	*Ask Yourself*
• Occam's Razor	Would a simpler model suffice?
• Overall fit	How does the trend fit the past data?
• Believability	Does the extrapolated trend "look right"?
• Fit to recent data	Does the fitted trend match the last few data points?

Example 14.1

Comparing Trends

You can usually increase the R^2 by choosing a more complex model. But if you are making a *forecast,* this is not the only relevant issue because R^2 measures the fit to the *past* data. Figure 14.15 shows four fitted trends using the same data, with three-period forecasts. For this data set, the linear model may be inadequate because its fit to recent periods is marginal (we prefer the simplest model *only if* it "does the job"). The cubic trend yields the highest R^2, but the fitted equation is nonintuitive and would be hard to explain or defend. Also, its forecasts appear to be increasing too rapidly. In this example, the exponential model has the lowest R^2, yet it matches the recent data fairly well and its forecasts appear credible when projected a few periods ahead.

Figure 14.15

Four Fitted Trends Using the Same Data

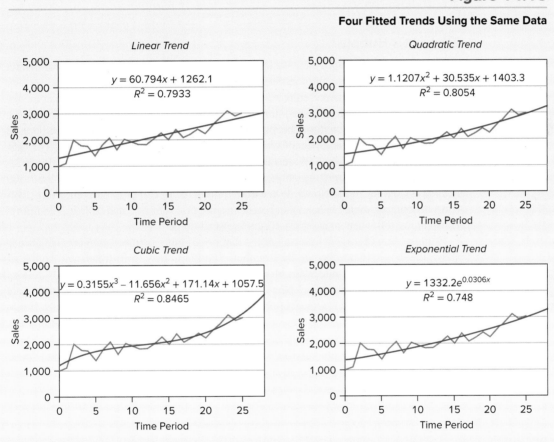

Any trend model's forecasts become less reliable as they are extrapolated farther into the future. The quadratic trend, the simplest of Excel's polynomial models, is sometimes acceptable for short-term forecasting. However, forecasters avoid higher-order polynomial models (cubic and higher) not only because they are complex but also because they can give bizarre forecasts when extrapolated more than one period ahead. Table 14.4 compares the features of common trend models.

Table 14.4

Comparison of Three Trend Models

Model	Pro	Con
Linear	1. Simple, familiar to everyone. 2. May suffice for short-run data.	1. Assumes constant slope. 2. Cannot capture nonlinear change.
Exponential	1. Familiar to financial analysts. 2. Shows compound growth rate.	1. Some managers are unfamiliar with e^x. 2. Data values must be positive.
Quadratic	1. Useful for data with a turning point. 2. Useful test for nonlinearity.	1. Complex and lacks intuitive interpretation. 2. Can give untrustworthy forecasts if extrapolated too far.

Analytics in Action

Trend? Or Bubble?

A price "bubble" is a rapid increase in the price of an asset well above its typical market value, which creates instability in the market. Rising prices may inspire investors to seek credit to finance their increased purchases of the asset. Rising prices may also convince banks and capital markets to accommodate these requests by issuing credit. However, a bubble can collapse if creditors become alarmed and stop allowing the asset to be used as collateral. A bubble is especially fragile when financing is highly leveraged. The U.S. housing bubble peaked in 2006, reached new lows by 2012, then peaked again by 2022. Are similar bubbles occurring now? A recent *Forbes* article was titled "The S&P 500 Bubble Is Coming: What Now?" When does a surge in price of cryptocurrency (e.g., Bitcoin) become a bubble? What about recent surging prices of rare metals (e.g., rhodium)? How can we tell when an asset price increase is a bubble?

Some market observers have suggested that prices more than *two standard deviations* above the longer trend may signal a bubble. You know how to fit trends and construct 95 percent confidence intervals, so you can do your own bubble hunting. Of course, everyone else can do the same. Nobel prize–winning economist Robert Shiller of Yale University has published extensively on using analytics to develop models of asset valuation. His models are far from simple. One of Shiller's insights is that in efficient markets everyone has access to the same information and analytical tools. Be forewarned—there is no simple formula for investment success. And it's psychology—not just economics.

See Robert J. Shiller, *Narrative Economics: How Stories Go Viral and Drive Major Economic Events;* (Princeton, NJ: Princeton University Press, 2019) and wikipedia.org/wiki/2000s_United_States_housing_bubble.

Section Exercises

Mc Graw Hill **connect**

14.1 In 2009, US Airways Flight 1549 made a successful emergency landing in the Hudson River, after striking birds shortly after takeoff. Are bird strikes an increasing threat to planes? (a) Make an Excel graph of the data on bird strikes. (b) Discuss the underlying causes that might explain the trend. (c) Fit three trends (linear, quadratic, and exponential) to the time series. (d) Use *each* of the three fitted trend equations to make numerical forecasts for the next three years. How much difference does the choice of model make? Which forecasts do you trust the most, and why? 🗇 **BirdStrikes**

Number of Reported Bird Strikes to Civil Aircraft in U.S., 2008–2018 🗇 BirdStrikes

Year	Strikes	Year	Strikes	Year	Strikes
2008	7,213	2012	10,918	2016	13,454
2009	8,950	2013	11,417	2017	14,664
2010	9,905	2014	13,694	2018	16,020
2011	10,119	2015	13,808		

Source: https://www.faa.gov/sites/faa.gov/files/2022-07/Wildlife-Strike-Report-1990-2021.pdf.

14.2 (a) Make an Excel graph of the data on usage of renewable energy in the United States. (b) Discuss the underlying causes that might explain the trend or pattern. (c) Fit three trends (linear, quadratic, exponential) to the time series. (d) Use *each* of the three fitted trend equations to make numerical forecasts for the next three years. How similar are the three models' forecasts? Renew

U.S. Usage of Renewable Energy (quad BTU), 2011–2018

Year	Usage	Year	Usage
2011	9.20	2015	9.72
2012	8.85	2016	10.37
2013	9.45	2017	11.18
2014	9.74	2018	11.52

Source: https://usafacts.org/data/.

14.3 (a) Make an Excel line graph of the data on employee work stoppages. (b) Discuss the underlying causes that might explain the trend or pattern. (c) Fit three trends (linear, exponential, quadratic). (d) Which trend model is best, and why? If none is satisfactory, explain. (e) Would you trust a trend forecast for 2020? Explain. Strikers

U.S. Workers Involved in Work Stoppages, 2000–2019 (thousands)

Year	Strikers	Year	Strikers	Year	Strikers	Year	Strikers
2000	397	2005	102	2010	45	2015	49
2001	102	2006	77	2011	113	2016	102
2002	47	2007	193	2012	150	2017	25
2003	131	2008	83	2013	55	2018	485
2004	232	2009	13	2014	34	2019	466

Source: http://data.bls.gov.

14.4 You want to invest $1,000. Which growth curve would yield the largest principal 5 years from now? 10 years? 20 years? Explain. *Hint:* Show all the forecasts.
   a. $y_t = 1000e^{0.039t}$
   b. $y_t = 1000 + 45t$
   c. $y_t = 1000 + 11t + 3t^2$

14.5 For each of the following fitted trends, make a prediction for period $t = 15$:
   a. $y_t = 926e^{-0.026t}$
   b. $y_t = 2{,}217 - 8t$
   c. $y_t = 447 - 29t + 7t^2$

Mini Case 14.1

U.S. Trade Deficit

The imbalance between imports and exports has been a vexing policy problem for U.S. policymakers for decades. The last time the United States had a trade surplus was in 1975, partly due to reduced dependency on foreign oil through conservation measures enacted after the oil crisis (shortages and gas lines) in the early 1970s. However, the trade deficit has become more acute over time due partly to continued oil imports and, more recently, to availability of cheaper goods from China and other emerging economies.

Prior to the recent recession, imports had been growing faster than exports. Yet over the past decade, the fitted trend equations show that exports have grown at a slightly higher compound annual rate of 6.37 percent, compared with 5.11 percent for imports (see Figure 14.16). Possible reasons would include reduced industrial production due to the recession, improved vehicle fuel economy, rising import prices, and the impact of exchange rates.

Figure 14.16

U.S. Trade, 2000–2020 **TradeDeficit**

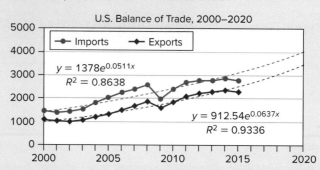

Perhaps the United States can achieve trade balance–but only far in the future, given the tiny difference in growth rates. Further, the assumption of *ceteris paribus* may not hold. Much depends on U.S. trade treaties, global challenges (e.g., climate change), how the United States and other nations handle their internal finances, and international conflicts. The disruptive COVID-19 crisis of 2020 illustrates the difficulty of making economic policy based on trend projections.

Forecasts are less a way of predicting the future than of showing where we are heading *if* nothing changes. A paradox of forecasting is that, as soon as decision makers see the implications of a distasteful forecast, they may try to take steps to ensure that the forecast is wrong!

14.3 ASSESSING FIT

Five Measures of Fit

In time series analysis, you are likely to encounter several different measures of "fit" that show how well the estimated trend model matches the observed time series. "Fit" refers to historical data, and you should bear in mind that a good fit is no guarantee of good forecasts–the usual goal. Five common measures of fit are shown in Table 14.5.

Table 14.5

Five Measures of Fit

Statistic	Description	Pro	Con
(14.5) $\quad R^2 = 1 - \dfrac{\sum_{t=1}^{n}(y_t - \hat{y}_t)^2}{\sum_{t=1}^{n}(y_t - \bar{y}_t)^2}$	**Coefficient of determination (R^2)**	1. Unit-free measure. 2. Very common.	1. Often interpreted incorrectly (e.g., "percent of correct predictions").
(14.6) $\quad MAPE = \dfrac{100}{n}\sum_{t=1}^{n}\left|\dfrac{y_t - \hat{y}_t}{y_t}\right|$	**Mean absolute percent error (*MAPE*)**	1. Unit-free measure (%). 2. Intuitive meaning.	1. Lacks nice math properties.
(14.7) $\quad MAD = \dfrac{1}{n}\sum_{t=1}^{n}\left|y_t - \hat{y}_t\right|$	**Mean absolute deviation (*MAD*)**	1. Intuitive meaning. 2. Same units as y_t.	1. Not unit-free. 2. Lacks nice math properties.
(14.8) $\quad MSD = \dfrac{1}{n}\sum_{t=1}^{n}(y_t - \hat{y}_t)^2$	**Mean squared deviation (*MSD*)**	1. Nice math properties. 2. Penalizes big errors more.	1. Nonintuitive meaning. 2. Rarely reported.
(14.9) $\quad SE = \sqrt{\dfrac{\sum_{t=1}^{n}(y_t - \hat{y}_t)^2}{n - 2}}$	**Standard error (*SE*)**	1. Same units as y_t. 2. For confidence intervals.	1. Nonintuitive meaning.

EXAMPLE 14.2

Fire Losses

Figure 14.17 shows an Excel graph with fitted linear trend and three-year forecasts for fire loss claims paid to homeowners (in millions of dollars) by an insurance company. Table 14.6 shows the calculations for these statistics of fit. Because the residuals $y_t - \hat{y}_t$ sum to zero, we see why it's necessary to sum either their absolute values or their squares to obtain a measure of fit. *MAPE, MAD, MSD,* and *SE* would be zero if the trend provided a perfect fit to the time series.

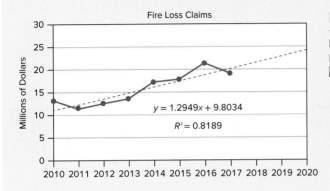

Figure 14.17

**Fire Loss Claims
Paid—Linear Model**
📄 **FireLosses**

Using the sums in Table 14.6, we can apply the formulas for each fit statistic:

$$MAPE = \frac{100}{n} \sum_{t=1}^{n} \left| \frac{y_t - \hat{y}_t}{y_t} \right| = \frac{100}{8}(0.6590) = 8.24\%$$

$$MAD = \frac{1}{n} \sum_{t=1}^{n} |y_t - \hat{y}_t| = \frac{1}{8}(9.9142) = 1.239$$

$$MSD = \frac{1}{n} \sum_{t=1}^{n} (y_t - \hat{y}_t)^2 = \frac{1}{8}(15.5683) = 1.946$$

$$SE = \sqrt{\frac{\sum_{t=1}^{n}(y_t - \hat{y}_t)^2}{n-2}} = \sqrt{\frac{15.5683}{8-2}} = 1.611$$

$$R^2 = 1 - \frac{\sum_{t=1}^{n}(y_t - \hat{y}_t)^2}{\sum_{t=1}^{n}(y_t - \bar{y})^2} = 1 - \frac{15.5683}{85.9878} = 0.8189$$

The R^2 statistic says that a linear trend alone can "explain" about 82 percent of the variation in claims paid. *MAPE* says that our fitted trend has a mean absolute error of 8.24 percent. *MAD* says that the average error is 1.239 million dollars (ignoring the sign). *MSD* lacks a simple interpretation. These fit statistics are most useful in comparing different trend models for the same data. All the statistics (especially the *MSD*) are affected by the unusual residual in 2016, when fire losses greatly exceeded the trend. The standard error is useful if we want to make a prediction interval for a forecast, using formula 14.9. It is the same formula you saw in Chapter 12. This formula widens the confidence interval when the time index t is far from its historic mean.

$$\hat{y}_t \pm t_{n-2} SE \sqrt{1 + \frac{1}{n} + \frac{(t - \bar{t})^2}{\sum_{t=1}^{n}(t - \bar{t})^2}} \quad \text{(prediction interval for future } y_t) \quad \textbf{(14.10)}$$

Table 14.6

SUMS for *MAD, MAPE, MSD,* and Standard Error 📄 FireLosses

Period	Year	y_t	$\hat{y}_t = 9.8034 + 1.2949t$	$y_t - \hat{y}_t$	$\|y_t - \hat{y}_t\|$	$\|y_t - \hat{y}_t\|/y_t$	$(y_t - \hat{y}_t)^2$
1	2010	12.940	11.0983	1.8417	1.8417	0.1423	3.3919
2	2011	11.510	12.3932	−0.8832	0.8832	0.0767	0.7800
3	2012	12.428	13.6881	−1.2601	1.2601	0.1014	1.5879
4	2013	13.457	14.9830	−1.5260	1.5260	0.1134	2.3287
5	2014	17.118	16.2779	0.8401	0.8401	0.0491	0.7058
6	2015	17.586	17.5728	0.0132	0.0132	0.0008	0.0002
7	2016	21.129	18.8677	2.2613	2.2613	0.1070	5.1135
8	2017	18.874	20.1626	−1.2886	1.2886	0.0683	1.6605
			Sum	0.000	9.9142	0.6590	15.5683
			Mean	0.000	1.2393	0.0824	1.9460

14.4 Moving Averages

LO 14-5

Interpret a moving average and use Excel to create it.

Trendless or Erratic Data

What if the time series $y_1, y_2, \ldots, y_n$ is erratic or has no consistent trend? In such cases, there may be little point in fitting a trend, and if the mean is changing over time, we cannot just "take the average" over the entire data set. Instead, a conservative approach is to calculate a **moving average**. There are two main types of moving averages: trailing or centered. We will illustrate each.

Trailing Moving Average (*TMA*)

The simplest kind of moving average is the **trailing moving average** (*TMA*) over the last m periods.

$$\textbf{(14.11)} \qquad \hat{y}_t = \frac{y_t + y_{t-1} + \cdots + y_{t-m+1}}{m} \qquad \text{(trailing moving average over } m \text{ periods)}$$

The *TMA* smooths the past fluctuations in the time series, helping us see the pattern more clearly. The choice of m depends on the situation. A larger m yields a "smoother" *TMA* but requires more data. The value of $\hat{y}_t$ also may be used as a forecast for period $t + 1$. Beyond the range of the observed data $y_1, y_2, \ldots, y_n$ there is no way to update the moving average, so it is best regarded as a *one-period-ahead forecast*.

EXAMPLE 14.3

Fuel Economy

Many drivers keep track of their fuel economy. For a given vehicle, there is likely to be little trend over time, but there is always random fluctuation. Also, current driving conditions (e.g., snow, hot weather, road trips) could temporarily affect mileage over several consecutive time periods. In this situation, a moving average might be considered. Table 14.7 shows Andrew's fuel economy data set. Column five shows a three-period *TMA*. For example, for period 6 (yellow-shaded cells), the *TMA* is

$$\hat{y}_6 = \frac{24.392 + 21.458 + 24.128}{3} = 23.326$$

It is easiest to appreciate the moving average's "smoothing" of the data when it is displayed on a graph, as in Figure 14.18. It is clear that Andrew's mean is around 23 mpg, though the moving average fluctuates over a range of approximately ±2 mpg.

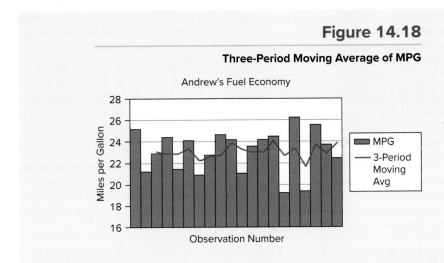

Figure 14.18

Three-Period Moving Average of MPG

Table 14.7   Andrew's Miles per Gallon ($n = 20$)   AndrewsMPG

Obs	Date	Miles Driven	Gallons	MPG	TMA	CMA	
1	5-Jan	285	11.324	25.168			
2	7-Jan	185	8.731	21.189		23.074	
3	11-Jan	250	10.934	22.864	23.074	22.815	
4	15-Jan	296	12.135	24.392	22.815	22.905	
5	19-Jan	232	10.812	21.458	22.905	23.326	
6	25-Jan	301	12.475	24.128	23.326	22.158	Example: *TMA*
7	30-Jan	285	13.645	20.887	22.158	22.581	
8	3-Feb	263	11.572	22.727	22.581	22.747	
9	7-Feb	250	10.152	24.626	22.747	23.856	
10	14-Feb	307	12.678	24.215	23.856	23.283	
11	22-Feb	242	11.520	21.007	23.283	22.942	
12	29-Feb	288	12.201	23.605	22.942	22.937	
13	5-Mar	285	11.778	24.198	22.937	24.103	
14	8-Mar	313	12.773	24.505	24.103	22.638	Example: *CMA*
15	13-Mar	283	14.732	19.210	22.638	23.330	
16	18-Mar	318	12.103	26.274	23.330	21.620	
17	22-Mar	195	10.064	19.376	21.620	23.746	
18	28-Mar	320	12.506	25.588	23.746	22.904	
19	2-Apr	270	11.369	23.749	22.904	23.910	
20	12-Apr	259	11.566	22.393	23.910		

Source: Data were collected by statistics student Andrew Fincher for his 11-year-old car.

Centered Moving Average (*CMA*)

Another moving average is the **centered moving average** (*CMA*). Formula 14.12 shows a *CMA* for $m = 3$ periods. The formula looks both forward *and* backward in time, to express the current "forecast" as the mean of the current observation *and* observations on either side of the current data.

$$\hat{y}_t = \frac{y_{t-1} + y_t + y_{t+1}}{3} \quad \text{(centered moving average over } m \text{ periods)} \quad \textbf{(14.12)}$$

This is not really a forecast at all, but merely a way of smoothing the data. In Table 14.7, column seven shows the *CMA* for Andrew's MPG data. For example, for period 14 (blue-shaded cells), the *CMA* is

$$\hat{y}_{14} = \frac{24.198 + 24.505 + 19.210}{3} = 22.638$$

When *m* is odd (*m* = 3, 5, etc.), the *CMA* is easy to calculate. When *m* is even, the formula is more complex because the mean of an even number of data points would lie *between* two data points and would not be correctly centered. Instead, we take a double moving average (yipe!) to get the resulting *CMA* centered properly. For example, for *m* = 4, we would average y_{t-2} through y_{t+1}, then average y_{t-1} through y_{t+2}, and finally average the two averages! You need not worry about this formula for now. It will be illustrated shortly in the context of seasonal data.

Using Excel for a *TMA*

Excel offers a *TMA* in its Add Trendline option when you click on a time series line graph or bar chart. Its menus are displayed in Figure 14.19. The *TMA* is a conservative choice whenever you doubt that one of Excel's five other trend models (linear, logarithmic, polynomial, power, exponential) would be appropriate. However, Excel does *not* give you the option of making any forecasts with its moving average model.

Figure 14.19

Excel's Moving Average Menus

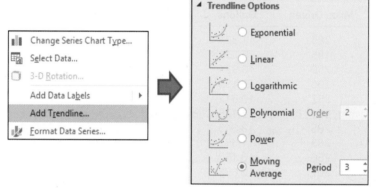

Source: Microsoft Excel.

14.6 (a) Make an Excel line graph of the exchange rate data (only first 3 and last 3 days are shown). Describe the pattern. (b) Click on the data and choose Add Trendline > Moving Average. Describe the effect of increasing *m* (e.g., *m* = 2, 4, 6, etc.). Include a copy of each graph with your answer. (c) Discuss how this moving average might help a currency speculator. 📄 **DollarEuro**

Daily Dollar/Euro Exchange Rate (*n* = 61 days)						
Day	1	2	3	. . .	59	60	61
Date	11/1/19	11/4/19	11/5/19	. . .	1/29/20	1/30/20	1/31/20
Rate	1.1169	1.1144	1.1070	. . .	1.1004	1.1032	1.1082

Source: www.federalreserve.gov.

14.5 EXPONENTIAL SMOOTHING

Forecast Updating

The **exponential smoothing** model is a special kind of moving average. It is used for ongoing one-period-ahead forecasting for data that have up-and-down movements but no consistent trend. For example, a retail outlet may place orders for thousands of different stock-keeping units (SKUs) each week so as to maintain its inventory of each item at the desired level (to avoid emergency calls to warehouses or suppliers). For such forecasts, many firms choose exponential smoothing, a simple forecasting model with only two inputs and one constant. The updating formula for the forecasts is

$$F_{t+1} = \alpha y_t + (1 - \alpha) F_t \quad \text{(Smoothing update)} \qquad \text{(14.13)}$$

where

F_{t+1} = the forecast for the next period

α = the "smoothing constant" $(0 \leq \alpha \leq 1)$

y_t = the actual data value in period t

F_t = the previous forecast for period t

Smoothing Constant (α)

The next forecast F_{t+1} is a weighted average of y_t (the current data) and F_t (the previous forecast). The value of α, called the **smoothing constant**, is the weight given to the latest data. A small value of α would give low weight to the most recent observation and heavy weight $1 - \alpha$ to the previous forecast (a "heavily smoothed" series). The larger the value of α, the more quickly the forecasts adapt to recent data. For example,

If $\alpha = .05$, then $F_{t+1} = .05y_t + .95F_t$ (heavy smoothing, slow adaptation)

If $\alpha = .20$, then $F_{t+1} = .20y_t + .80F_t$ (moderate smoothing, moderate adaptation)

If $\alpha = .50$, then $F_{t+1} = .50y_t + .50F_t$ (little smoothing, quick adaptation)

Choosing the Value of α

If $\alpha = 1$, there is no smoothing at all, and the forecast for next period is the same as the latest data point, which basically defeats the purpose of exponential smoothing. Minitab uses $\alpha = .20$ (i.e., moderate smoothing) as its default, which is a fairly common choice of α. The fit of the forecasts to the data will change as you try different values of α. Most computer packages can, as an option, solve for the "best" α using a criterion such as minimum *SSE*.

Over time, earlier data values have less effect on the exponential smoothing forecasts than more recent y-values. To see this, we can replace F_t in formula 14.12 with the prior forecast F_{t-1}, and repeat this type of substitution indefinitely to obtain this result:

$$F_{t+1} = \alpha y_t + \alpha(1 - \alpha)y_{t-1} + \alpha(1 - \alpha)^2 y_{t-2} + \alpha(1 - \alpha)^3 y_{t-3} + \cdots \qquad \text{(14.14)}$$

We see that the next forecast F_{t+1} depends on *all* the prior data (y_{t-1}, y_{t-2}, etc.). However, as long as $\alpha < 1$, as we go farther into the past, each prior data value has less and less impact on the current forecast.

Initializing the Process

From formula 14.12, we see that F_{t+1} depends on F_t, which in turn depends on F_{t-1}, and so on, all the way back to F_1. But where do we get F_1 (the initial forecast)? There are many ways to initialize the forecasting process. For example, Excel simply sets the initial forecast equal to the first actual data value:

Method A

Set $F_1 = y_1$ (use the first data value)

This method has the advantage of simplicity, but if y_1 happens to be unusual, it could take a few iterations for the forecasts to stabilize. Another approach is to set the initial forecast equal to the average of the first several observed data values. This method tends to iron out the effects of unusual y-values. For example, Minitab uses the first six data values:

Method B

Set $F_1 = \dfrac{y_1 + y_2 + y_3 + y_4 + y_5 + y_6}{6}$ (average of first 6 data values)

EXAMPLE 14.4

Weekly Sales Data

Table 14.8 shows weekly sales of deck sealer (a paint product sold in gallon containers) at a large do-it-yourself warehouse-style retailer. For exponential smoothing forecasts, the company uses $\alpha = .10$. Its choice of α is based on experience. Because α is fairly small, it will provide strong smoothing. The last two columns compare the two methods of initializing the forecasts. Unusually high sales in week 5 have a strong effect on method B's starting point. At first, the difference in forecasts is striking, but over time the methods converge.

Using Method A:

$$F_2 = \alpha y_1 + (1 - \alpha)F_1 = (.10)(106) + (.90)(106) = 106$$
$$F_3 = \alpha y_2 + (1 - \alpha)F_2 = (.10)(110) + (.90)(106) = 106.4$$
$$F_4 = \alpha y_3 + (1 - \alpha)F_3 = (.10)(108) + (.90)(106.4) = 106.56$$
$$\vdots$$
$$F_{19} = \alpha y_{18} + (1 - \alpha)F_{18} = (.10)(120) + (.90)(130.908) = 129.82$$

Table 14.8 **Deck Sealer Sales: Exponential Smoothing ($n = 18$ weeks)**
📁 DeckSealer

Week	Actual Sales	Method A: $F_1 = y_1$	Method B: F_1 = Average (1st 6)
1	106	106.000	127.833
2	110	106.000	125.650
3	108	106.400	124.085
4	97	106.560	122.477
5	210	105.604	119.929
6	136	116.044	128.936
7	128	118.039	129.642
8	134	119.035	129.478
9	107	120.532	129.930
10	123	119.179	127.637
11	139	119.561	127.174
12	140	121.505	128.356
13	144	123.354	129.521
14	94	125.419	130.969
15	108	122.277	127.272
16	168	120.849	125.344
17	179	125.564	129.610
18	120	130.908	134.549

Smoothed forecasts using $\alpha = .10$.

Using Method B:

$$F_2 = \alpha y_1 + (1 - \alpha)F_1 = (.10)(106) + (.90)(127.833) = 125.650$$
$$F_3 = \alpha y_2 + (1 - \alpha)F_2 = (.10)(110) + (.90)(125.650) = 124.085$$
$$F_4 = \alpha y_3 + (1 - \alpha)F_3 = (.10)(108) + (.90)(124.085) = 122.477$$
$$\vdots$$
$$F_{19} = \alpha y_{18} + (1 - \alpha)F_{18} = (.10)(120) + (.90)(134.549) = 133.094$$

Despite their different starting points, the forecasts for period 19 do not differ greatly. Rounding to the next higher integer, for week 19, the firm would order 130 gallons (using method *A*) or 134 gallons (using method *B*). Figure 14.20 shows the similarity in *patterns* of the forecasts, although the *level* of forecasts is always higher in method *B* because of its higher initial value. This demonstrates that the choice of starting values *does* affect the forecasts.

Figure 14.20

Initializing Methods Compared

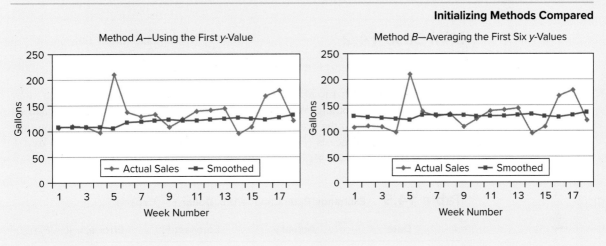

Using Excel

Excel has an exponential smoothing option. It is found in the Data Analysis menu. Instead of the smoothing constant α, Excel asks for a *damping factor,* which is equal to $1 - \alpha$. Excel uses method *A* (setting $F_1 = y_1$) to initialize its forecasts. Figure 14.21 shows Excel's exponential smoothing dialogue box and its output chart of actual and forecast values. Excel's chart is difficult to read, so you may wish to make your own "improved" line chart, like the one shown in Figure 14.21. Excel makes no *future* forecasts past period $t = 18$, but you can do it yourself (see the period $t = 19$ forecast calculations in Example 14.4). Exponential smoothing forecasts can't be updated beyond one period ahead because there are no more actual y_t values to plug into the updating formula.

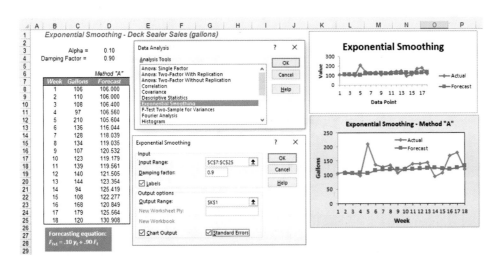

Figure 14.21

Excel's Exponential Smoothing

Source: Microsoft Excel.

Smoothing with Trend and Seasonality

Single exponential smoothing is intended for *trendless* data. If your data have a trend, you can try *Holt's method* with *two* smoothing constants (one for *trend,* one for *level*). If you have both trend and seasonality, you can try *Winters's method* with *three* smoothing constants (one for *trend,* one for *level,* one for *seasonality*). These advanced methods are similar to single smoothing in that they use simple formulas to update the forecasts, and you may use them without special caution. These topics are usually reserved for a class in forecasting, so they will not be explained here.

Mini Case 14.2

Exchange Rates

We have data for March 1 to March 30 and want to forecast 1 day ahead to March 31 by using exponential smoothing. We choose a smoothing constant value of $\alpha = .20$ and set the initial forecast F_1 to the average of the first six actual values. Table 14.9 shows the actual data (y_t) and forecasts (F_t) for each date. The March 31 forecast is $F_{23} = \alpha y_{22} + (1 - \alpha)F_{22} = (.20)(1.2164) + (.80)(1.21395) = 1.2144$.

Table 14.9 **Exchange Rate: Canada/U.S. Dollar** 📷 Canada

t	Date	Actual y_t	Forecast F_t	Error $e_t = y_t - F_t$
1	01-Mar	1.2425	1.23450	0.0080
2	02-Mar	1.2395	1.23610	0.0034
3	03-Mar	1.2463	1.23678	0.0095
.	.	.	.	.
.	.	.	.	.
21	29-Mar	1.2135	1.21406	−0.0006
22	30-Mar	1.2164	1.21395	0.0024
23	31-Mar		**1.21444**	

Source: Data from www.federalreserve.gov.

The column of errors (e_t) shown in Table 14.9 can be used to calculate measures of fit (e.g., *MAPE, MAD*). Figure 14.22 shows that the forecasts adapt, but always with a lag. Exponential smoothing is really a kind of moving average, so its "forecasts" are mainly of short-term value.

Figure 14.22

Excel's Exponential Smoothing ($\alpha = .20$)

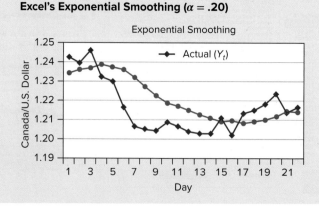

14.7 (a) Make an Excel line graph of the following bond yield data (only the first and last three data values are shown). Describe the pattern. Is there a consistent trend? (b) Perform exponential smoothing with $\alpha = .20$. Initialize the forecast using the first value in the time series (Method *A*). Record the statistics of fit. (c) Do the smoothing again with $\alpha = .10$ and then with $\alpha = .30$, recording the statistics of fit. (d) Compare the statistics of fit for the three values of α. (e) Make a one-period forecast (i.e., $t = 53$) using each of the three α values. How did α affect your forecasts? 🖎 **BondYield**

U.S. Treasury 10-Year Bond Yields at Week's End ($n = 51$ weeks)

Date	4-Jan	11-Jan	18-Jan	. . .	13-Dec	20-Dec	27-Dec
Rate	2.67	2.71	2.79	. . .	1.82	1.92	1.88

14.6 SEASONALITY

When and How to Deseasonalize

When the data periodicity is monthly or quarterly, we should calculate a seasonal index and use it to **deseasonalize** the data (annual data have no seasonality). This process is called **decomposition** of a time series. For a multiplicative model (the usual assumption), a seasonal index is a *ratio*. For example, if the seasonal index for July is 1.25, it means that July is 125 percent of the monthly average. If the seasonal index for January is 0.84, it means that January is 84 percent of the monthly average. If the seasonal index for October is 1.00, it means that October is an average month. The seasonal indexes must sum to 12 for monthly data or 4 for quarterly data. The following steps are used to deseasonalize data for time series observations:

- Step 1. Calculate a centered moving average (*CMA*) for each month (quarter).
- Step 2. Divide each observed y_t value by the *CMA* to obtain seasonal ratios.
- Step 3. Average the seasonal ratios by month (quarter) to get raw seasonal indexes.
- Step 4. Adjust the raw seasonal indexes so they sum to 12 (monthly) or 4 (quarterly).
- Step 5. Divide each y_t by its seasonal index to get deseasonalized data.

In step 1, we lose 12 observations (monthly data) or 4 observations (quarterly data) because of the centering process. We will illustrate this technique for quarterly data.

LO 14-7

Interpret seasonal factors and use them to make forecasts.

Illustration of Calculations

Table 14.10 shows six years' data on quarterly revenue from sales of carpeting, tile, wood, and vinyl flooring by a floor-covering retailer. The data have an upward trend (see Figure 14.23), perhaps due to a boom in consumer spending on home improvement and new homes. There also appears to be seasonality, with lower sales in the third quarter (summer) and higher sales in the first quarter (winter). Excel has no seasonal decomposition feature, but you can perform your own calculations as shown in Table 14.11.

Quarter	2012	2013	2014	2015	2016	2017
1	259	306	379	369	515	626
2	236	300	262	373	373	535
3	164	189	242	255	339	397
4	222	275	296	374	519	488

Table 14.10

Sales of Floor Covering Materials ($ thousands)
🖎 **FloorSales**

Table 14.11

Deseasonalized Sales
($n = 24$ quarters)
📁 **FloorSales**

Obs	Year	Quarter	Sales	CMA	Sales/CMA	Seasonal Index	Deseasonalized
1	2012	1	259			1.252	206.9
2		2	236			1.021	231.1
3		3	164	**226.125**	0.725	0.740	221.7
4		4	222	240.000	0.925	0.987	224.9
5	2013	1	306	251.125	1.219	1.252	244.4
6		2	300	260.875	1.150	1.021	293.8
7		3	189	276.625	0.683	0.740	255.5
8		4	275	281.000	0.979	0.987	278.6
9	2014	1	379	282.875	1.340	1.252	302.7
10		2	262	292.125	0.897	1.021	256.6
11		3	242	293.500	0.825	0.740	327.2
12		4	296	306.125	0.967	0.987	299.8
13	2015	1	369	321.625	1.147	1.252	294.7
14		2	373	333.000	1.120	1.021	365.3
15		3	255	361.000	0.706	0.740	344.7
16		4	374	379.250	0.986	0.987	378.8
17	2016	1	515	389.750	1.321	1.252	411.3
18		2	373	418.375	0.892	1.021	365.3
19		3	339	450.375	0.753	0.740	458.3
20		4	519	484.500	1.071	0.987	525.7
21	2017	1	626	512.000	1.223	1.252	500.0
22		2	535	515.375	1.038	1.021	524.0
23		3	397			0.740	536.7
24		4	488			0.987	494.3

Figure 14.23

Deseasonalized
Trend 📁 **FloorSales**

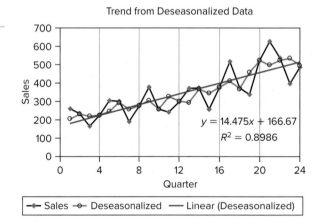

Trend from Deseasonalized Data

$y = 14.475x + 166.67$
$R^2 = 0.8986$

— Sales — Deseasonalized — Linear (Deseasonalized)

Table 14.12

Calculation of Seasonal
Indexes 📁 **FloorSales**

Quarter	2012	2013	2014	2015	2016	2017	Mean	Adjusted
1		1.219	1.340	1.147	1.321	1.223	1.250	1.252
2		1.150	0.897	1.120	0.892	1.038	1.019	1.021
3	0.725	0.683	0.825	0.706	0.753		0.738	0.740
4	0.925	0.979	0.967	0.986	1.071		0.986	0.987
							3.993	4.000

Due to rounding, details may not yield the result shown.

Because the number of subperiods (quarters) is even ($m = 4$), each value of the *CMA* is the average of two averages. For example, the first *CMA* value, 226.125, is the average of $(259 + 236 + 164 + 222)/4$ and $(236 + 164 + 222 + 306)/4$. Table 14.12 shows how the indexes are averaged. The *CMA* loses two quarters at the beginning and two quarters at the end, so each seasonal index is an average of only five quarters (instead of six). Each mean is then adjusted to force the sum to be 4.000, and these are the seasonal indexes. If we had monthly data, the indexes would be adjusted so that their sum would be 12.000. Calculations are ordinarily performed with software (e.g., MegaStat, Minitab, or R).

After the data have been deseasonalized, the trend (Figure 14.23) is fitted based on deseasonalized data. The sharper peaks and valleys in the original time series (Y) have been smoothed by removing the seasonality (S). Any remaining variation about the trend (T) is irregular (I) or "random noise." To make a forecast k periods ahead, multiply the *deseasonalized* trend estimate $\hat{y}_{t+k}$ by the seasonal index for period $t+k$.

Seasonal Forecasts Using Binary Predictors

Another way to address seasonality is to estimate a regression model using **seasonal binaries** as predictors. For quarterly data, for example, the data set would look as shown in Table 14.13. When we have four binaries (i.e., four quarters), we must exclude one binary to prevent perfect multicollinearity (see Chapter 13, Section 13.5). Arbitrarily, we exclude the fourth quarter binary *Qtr4* (it will be a portion of the intercept when *Qtr1* = 0 and *Qtr2* = 0 and *Qtr3* = 0).

LO 14-8

Use regression with seasonal binaries to make forecasts.

Table 14.13

Sales Data with Seasonal Binaries FloorSales

Year	Quarter	Sales	Time	Qtr1	Qtr2	Qtr3
2012	1	259	1	1	0	0
	2	236	2	0	1	0
	3	164	3	0	0	1
	4	222	4	0	0	0
2013	1	306	5	1	0	0
	2	300	6	0	1	0
	3	189	7	0	0	1
	4	275	8	0	0	0
2014	1	379	9	1	0	0
	2	262	10	0	1	0
	3	242	11	0	0	1
	4	296	12	0	0	0
2015	1	369	13	1	0	0
	2	373	14	0	1	0
	3	255	15	0	0	1
	4	374	16	0	0	0
2016	1	515	17	1	0	0
	2	373	18	0	1	0
	3	339	19	0	0	1
	4	519	20	0	0	0
2017	1	626	21	1	0	0
	2	535	22	0	1	0
	3	397	23	0	0	1
	4	488	24	0	0	0

We assume a linear trend, and specify the regression model *Sales* = *f*(*Time, Qtr1, Qtr2, Qtr3*). The estimated regression is shown in Figure 14.24. This is an additive model of the form $Y = T + S + I$ (recall that we omit the cycle C in practice). The fitted equation (rounded) is

$$Sales = 161 + 14.4\ Time + 89.8\ Qtr1 + 12.9\ Qtr2 - 83.6\ Qtr3$$

Figure 14.24

Fitted Regression for Seasonal Binaries

	Coefficient	Standard Error	t Stat	p-Value
Intercept	161.208	24.334	6.625	0.0000
Time	14.366	1.244	11.549	0.0000
Qtr1	89.765	24.324	3.690	0.0016
Qrt2	12.899	24.164	0.534	0.5997
Qrt3	−83.634	24.068	−3.475	0.0025

$R^2 = 0.9001$ $SE = 41.63$ $F = 42.78$

Time is a significant predictor ($p = .0000$), indicating significant linear trend. Two of the binaries are significant: *Qtr1* ($p = .0016$) and *Qtr3* ($p = .0025$). The second-quarter binary *Qtr2* ($p = .5997$) is not significant. The model gives a good overall fit ($R^2 = 0.9001$). The main virtue of the seasonal regression model is its versatility. We can plug in future values of *Time* and the seasonal binaries to create forecasts as far ahead as we wish. For example:

Period 25: Sales $= 161 + 14.4(25) + 89.8(1) + 12.9(0) − 83.6(0) = 610.8$

Period 26: Sales $= 161 + 14.4(26) + 89.8(0) + 12.9(1) − 83.6(0) = 548.3$

Period 27: Sales $= 161 + 14.4(27) + 89.8(0) + 12.9(0) − 83.6(1) = 466.2$

Period 28: Sales $= 161 + 14.4(28) + 89.8(0) + 12.9(0) − 83.6(0) = 564.2$

Section Exercises

connect

14.8 (a) Plot the PepsiCo data. Is there a trend? (b) Do you see evidence of seasonality? (c) Use software of your choice (e.g., MegaStat, Minitab, or R) to deseasonalize the data and calculate quarterly seasonal indexes. (d) If there is seasonality, suggest possible reasons. (e*) Perform a regression using seasonal binaries. Interpret the results. 📊 **PepsiCo**

PepsiCo Revenues ($ billions), 2014–2019						
Quarter	**2014**	**2015**	**2016**	**2017**	**2018**	**2019**
Qtr1	12.62	12.22	11.86	12.05	12.56	12.88
Qtr2	16.89	15.92	15.40	15.71	16.09	16.45
Qtr3	17.22	16.33	16.03	16.24	16.48	17.19
Qtr4	19.95	18.58	19.52	19.53	19.52	20.64

Source: Form 10-K reports for PepsiCo, Inc. and online earnings announcements. Data are for December 31 of each year.

14.9 (a) Plot the Corvette data. Is there a trend? (b) Do you see evidence of seasonality? (c) Use software of your choice (e.g., MegaStat, Minitab, or R) to deseasonalize the data and calculate monthly seasonal indexes. (d) If there is seasonality, suggest possible reasons. (e*) Perform a regression using seasonal binaries. Interpret the results. 📊 **Corvette**

U.S. Corvette Sales, 2004–2007 (number of cars sold)				
Month	**2004**	**2005**	**2006**	**2007**
Jan	2,986	2,382	2,579	2,234
Feb	2,382	2,365	3,058	2,784
Mar	3,033	3,215	3,655	3,158
Apr	3,169	3,177	3,516	3,227
May	3,420	3,078	3,317	3,300
Jun	3,398	2,417	2,938	3,055
Jul	3,492	1,872	2,794	2,377
Aug	2,067	2,202	2,990	2,877
Sep	3,705	2,372	3,056	2,837
Oct	2,607	2,981	2,761	2,484
Nov	2,120	3,157	2,773	2,438
Dec	2,897	3,271	3,081	2,914
Total	35,276	32,489	36,518	33,685

Source: *Ward's Automotive Yearbook, 2005–2008.*

Mini Case 14.3

Using Seasonal Binaries 🗁 Beer

Figure 14.25 shows monthly U.S. shipments of bottled beer for six years. A strong seasonal pattern is evident, presumably because people drink more beer in the warmer months. How can we describe the pattern statistically?

We create a regression data set with linear trend (Time = 1, 2, ..., 72) and 11 seasonal binaries (Feb–Dec). The January binary is omitted to prevent perfect multicollinearity. The regression results, shown in Figure 14.26, indicate a good fit ($R^2 = 0.857$), significant upward trend ($p = 0.000$ for Time), and several seasonal binaries that differ significantly from zero (p-values near zero). Binary predictor coefficients indicate that shipments are above the January average during the spring and summer (Mar–Aug), below the January average in the winter (Nov–Feb), and near the January average in the fall (Sep–Oct). The fitted regression equation can be used to forecast any future months' shipments.

Figure 14.25

U.S. Bottled Beer Shipments, 2001–2006

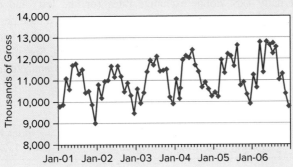

Source: www.census.gov.

Figure 14.26

Minitab's Fitted Regression for Seasonal Binaries

Regression Equation

Beer = 10164 + 16.90 Time − 484 Feb + 768 Mar + 579 Apr
     + 1311 May + 1182 Jun + 975 Jul + 892 Aug − 99 Sep
     + 107 Oct − 644 Nov − 1089 Dec

Coefficients

Term	Coef	SE Coef	T-Value	P-Value	VIF
Constant	10164	161	62.97	0.000	
Time	16.90	2.08	8.11	0.000	1.03
Feb	−484	209	−2.31	0.024	1.83
Mar	768	209	3.67	0.001	1.83
Apr	579	209	2.77	0.008	1.83
May	1311	209	6.26	0.000	1.84
Jun	1182	209	5.64	0.000	1.84
Jul	975	210	4.65	0.000	1.84
Aug	892	210	4.26	0.000	1.84
Sep	−99	210	−0.47	0.638	1.84
Oct	107	210	0.51	0.613	1.85
Nov	−644	210	−3.06	0.003	1.85
Dec	−1089	210	−5.18	0.000	1.86

Model Summary

S	R-sq	R-sq(adj)	R-sq(pred)
362.302	85.67%	82.76%	78.81%

14.7 INDEX NUMBERS

A simple way to measure changes over time (and especially to compare two or more variables) is to convert time series data into **index numbers**. The idea is to create an index that starts at 100 in a *base period,* so we can see *relative changes* in the data regardless of the original data units. Indexes are most often used for financial data (e.g., prices, wages, costs) but can be used with any numerical data (e.g., number of units sold, warranty claims, computer spam).

Relative Indexes

To convert a time series $y_1, y_2, \ldots, y_n$ into a *relative index* (sometimes called a *simple index*), we divide each data value y_t by the data value y_1 in a base period and multiply by 100. The relative index I_t for period t is

$$(14.15) \qquad\qquad I_t = 100 \times \frac{y_t}{y_1}$$

The index in the base period is always $I_1 = 100$, so the index $I_1, I_2, \ldots, I_n$ makes it easy to see *relative changes* in the data, regardless of the original data units. For example, Table 14.14 shows 60 days of daily U.S. dollar exchange rates (on the left) and the corresponding index numbers (on the right) using November 1, 2019 = 100 as a base period. Because each index starts at the same point (100), we can easily see fluctuations and trends in Figure 14.27. We could fit a moving average, if we wanted to smooth the data. Speculators who engage in currency arbitrage would use even more-sophisticated tools to analyze movements in currency indexes.

Table 14.14

U.S. Foreign Exchange Rates  **Currency**

	Foreign Currency per Dollar			Index Numbers (Nov. 1, 2019 = 100)		
Date	U.K.	Mexico	Canada	U.K.	Mexico	Canada
01-Nov-19	1.3145	19.0990	1.2950	100.0	100.0	100.0
04-Nov-19	1.3145	19.1660	1.2906	100.0	100.4	99.7
05-Nov-19	1.3176	19.2190	1.2870	100.2	100.6	99.4
⋮	⋮	⋮	⋮	⋮	⋮	⋮
28-Jan-20	1.3174	18.7800	1.2996	100.2	98.3	100.4
29-Jan-20	1.3201	18.6762	1.3012	100.4	97.8	100.5
30-Jan-20	1.3216	18.7990	1.3106	100.5	98.4	101.2

Source: Federal Reserve Bank of St. Louis (https://fred.stlouisfed.org).

Figure 14.27

U.S. Foreign Exchange Rates  **Currency**

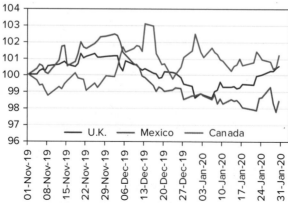

Indexes of Foreign Currency per Dollar (Nov. 11, 2019 = 100)

Weighted Indexes

A different calculation is required for a *weighted index* such as the Consumer Price Index for all urban consumers (CPI-U). The CPI-U is a measure of the relative prices paid by urban consumers for a market basket of goods and services, based on prices of hundreds of goods and services in eight major groups. The goal is to make the CPI-U representative of the prices paid for all goods and services purchased by all urban consumers. This requires assigning weights to each consumer good or service to reflect its importance relative to all the other goods and services in the market basket (e.g., housing gets a higher weight because it is a larger proportion of total spending). The basic formula for a simple weighted price index is

$$I_t = 100 \times \frac{\sum_{i=1}^{m} p_{it} q_i}{\sum_{i=1}^{m} p_{i1} q_i} = 100 \times \frac{p_{1t} q_1 + p_{2t} q_2 + \cdots + p_{mt} q_m}{p_{11} q_1 + p_{21} q_2 + \cdots + p_{m1} q_m} \qquad (14.16)$$

where

I_t = weighted index for period t $(t = 1, 2, \ldots, n)$

p_{it} = price of good i in period t $(i = 1, 2, \ldots, m; t = 1, 2, \ldots, n)$

q_i = weight assigned to good i $(i = 1, 2, \ldots, m)$

The numerator is the cost of buying a given market basket of goods and services at today's prices (period t) relative to the cost of the same market basket in the base period (period 1). The weight q_i represents the relative quantity of the item in the consumer's budget. For example, suppose there is a price increase of 5 percent for food and beverages and a 10 percent increase for medical care costs, with no price changes for the other expenditure categories. This would result in an increase of 1.4 percent in the CPI, as shown in Table 14.15.

From Table 14.15, the price index rose from 100.0 to 101.4, or a 1.4 percent increase:

$$I_2 = 100 \times \frac{\sum_{i=1}^{m} p_{i2} q_i}{\sum_{i=1}^{m} p_{i1} q_i} = 100 \times \frac{101.4}{100.0} = 101.4$$

Formula 14.16 is called a *Laspeyres index*. It treats the base year quantity weights as constant. Weights are based on the *Survey of Consumer Expenditures*. In your economics classes, you may learn more sophisticated methods that take into account the fact that expenditure weights do change over time. One such method is the *Paasche index*, which uses a formula similar to the Laspeyres index, except that quantity weights are adjusted for each period.

Table 14.15

Illustrative Calculation of Price Index

Expenditure Category	Base Year ($t = 1$)				Current Year ($t = 2$)		
	Weight (q_i)	Relative Price (p_{i1})	Relative Spending ($p_{i1}q_i$)		Weight (q_i)	Relative Price (p_{i2})	Relative Spending ($p_{i2}q_i$)
Food and beverages	15.7 ×	1.00 =	15.7		15.7 ×	1.05 =	16.5
Housing	40.9 ×	1.00 =	40.9		40.9 ×	1.00 =	40.9
Apparel	4.4 ×	1.00 =	4.4		4.4 ×	1.00 =	4.4
Transportation	17.1 ×	1.00 =	17.1		17.1 ×	1.00 =	17.1
Medical care	5.8 ×	1.00 =	5.8		5.8 ×	1.10 =	6.4
Recreation	6.0 ×	1.00 =	6.0		6.0 ×	1.00 =	6.0
Education/communication	5.8 ×	1.00 =	5.8		5.8 ×	1.00 =	5.8
Other goods and services	4.3 ×	1.00 =	4.3		4.3 ×	1.00 =	4.3
Sum	100.0	$\sum_{i=1}^{m} p_{i1} q_i =$	100.0		100.0	$\sum_{i=1}^{m} p_{i2} q_i =$	101.4

Importance of Index Numbers

The CPI affects nearly all Americans because it is used to adjust things like retirement benefits, food stamps, school lunch benefits, alimony, and tax brackets. The CPI-U could be compared with an index of salary growth for workers, or to measure current-dollar salaries in "real dollars." The Bureau of Labor Statistics (www.bls.gov) publishes CPI historical statistics for 31 categories. The most widely used CPI-U uses 1982–84 as a reference. That is, the Bureau of Labor Statistics sets the CPI-U (the average price level) for the years 1982, 1983, and 1984 equal to 100, and then measures changes in relation to that figure. As of December 2019 for example, the CPI-U was 257.0, meaning that, on average, prices had more than doubled over the previous 36 years (about a 2.7 percent annual increase, applying the geometric mean formula 4.5 with $n = 36$). The CPI is based on the buying habits of the "average" consumer, so it may not be a perfect reflection of anyone's individual price experience.

Other familiar price indexes, such as the Dow Jones Industrial Average (DJIA), have their own unique methodologies. Originally a simple arithmetic mean of stock prices, the DJIA now is the sum of the 30 stock prices divided by a "divisor" to compensate for stock splits and other changes over time. The divisor is revised periodically. Because high-priced stocks comprise a larger proportion of the sum, the DJIA is more strongly affected by changes in high-priced stocks. A little web research can tell you a lot about how stock price indexes are calculated, their strengths and weaknesses, and some alternative indexes that finance experts have invented.

 ## 14.8 FORECASTING: FINAL THOUGHTS

Role of Forecasting

In many ways, forecasting resembles planning. *Forecasting* is an analytical way to describe a "what-if" future that might confront the organization. *Planning* is the organization's attempt to determine actions it will take under each foreseeable contingency. Forecasts help decision makers become aware of trends or patterns that require a response. Actions taken by the decision makers may actually head off the contingency envisioned in the forecast. Thus, forecasts tend to be self-defeating because they trigger homeostatic organizational responses. So-called black swan events (https://en.wikipedia.org/wiki/Black_swan_theory) such as the COVID-19 crisis pose a vast challenge in making projections of many time series variables.

Behavioral Aspects of Forecasting

Forecasts can facilitate organizational communication. The forecast (or even just a nicely prepared time series chart) lets everyone examine the same facts concurrently and perhaps argue with the data or the assumptions that underlie the forecast or its relevance to the organization. A quantitative forecast helps *make assumptions explicit*. Those who prepare the forecast must explain and defend their assumptions, while others must challenge them. In the process, everyone gains understanding of the data, the underlying realities, and the imperfections in the data. Forecasts *focus the dialogue* and can make it more productive.

Of course, this assumes a certain maturity among the individuals around the table. Strong leaders (or possibly meeting facilitators) can play a role in guiding the discourse to produce a positive result. The danger is that people may try to find scapegoats (yes, they do tend to blame the forecaster), deny facts, or avoid responsibility for tough decisions. But one premise of this book is that statistics, when done well, can strengthen any dialogue and lead to better decisions.

Forecasts Are Always Wrong

We discussed several measures to use to determine if a forecast model fits the time series. Successful forecasters understand that a forecast is never precise. There is always some error, but we can *use* the error measures to track forecast accuracy. Many companies use several different forecasting models and rely on the model that has had the least error over some time period. We have described simple models in this chapter. You may take a class specifically focusing on forecasting in which you will learn about other time series models including AR (autoregressive) and ARIMA (autoregressive integrated moving average) models. Such models take advantage of the dependency that might exist between values in the time series. To ensure good forecast outcomes

- Maintain up-to-date databases of *relevant* data.
- Allow sufficient lead time to analyze the data.
- State several alternative forecasts or scenarios.
- Track forecast errors over time.
- State your assumptions and qualifications and consider your time horizon.
- Don't underestimate the power of a good graph.

Mini Case 14.4

How Does Noodles & Company Ensure Its Ingredients are as Fresh as Possible?

Using only fresh ingredients is key for great food and success for restaurants like Noodles & Company. To be sure that the restaurants are serving only the freshest ingredients, while also reducing food waste, Noodles & Company turned to statistical forecasting for ordering ingredients and daily food preparation. The challenge was to create a forecast that is sophisticated enough to be accurate yet simple enough for new restaurant employees to understand.

Noodles & Company uses a food management software system to forecast the demand for its menu items based on the moving average of the previous four weeks' sales. This simple forecasting technique has been very accurate. The automated process also uses the forecast of each item to estimate how many ingredients to order as well as how much to prepare each day. For example, the system might forecast that during next Wednesday's lunch, the location in Longmont, Colorado, will sell 55 Pesto Cavatappis. After forecasting the demand for each menu item, the system then specifies exactly how much of each ingredient to prepare for that lunch period.

For the restaurant teams, the old manual process of estimating and guessing how much of each ingredient to prepare is now replaced with an automated prep sheet. Noodles & Company has reduced food waste because restaurants are less likely to overorder ingredients and overprepare menu items. The restaurant teams are more efficient and customers are served meals made with the freshest ingredients possible.

Chapter Summary

A **time series** is assumed to have four components. For most business data, **trend** is the general pattern of change over all years observed, while **cycle** is a repetitive pattern of change around the trend over several years and **seasonality** is a repetitive pattern within a year. The **irregular** component is a random disturbance that follows no pattern. The **additive model** is adequate in the short run because the four components' magnitude does not change much, but for observations over longer periods of time, the **multiplicative model** is preferred. Common trend models include **linear** (constant slope and no turning point), **quadratic** (one turning point), and **exponential** (constant percent growth or decline). Higher polynomial models are untrustworthy and liable to give strange forecasts, though any trend model is less reliable the farther out it is projected. In forecasting, forecasters use fit measures besides R^2, such as mean absolute percent error (**MAPE**), mean absolute deviation (**MAD**), and mean squared deviation (**MSD**). For trendless or erratic data, we use a **moving average** over m periods or **exponential smoothing.** Forecasts adapt rapidly to changing data when the **smoothing constant** α is large (near 1) and conversely for a small α (near 0). For monthly or quarterly data, a **seasonal adjustment** is required before extracting the trend. Alternatively, regression with **seasonal binaries** can be used to capture seasonality and make forecasts. **Index numbers** are used to show changes relative to a base period.

Key Terms

additive model
centered moving average (*CMA*)
coefficient of determination
cycle
decomposition
deseasonalize
exponential smoothing

exponential trend
index numbers
irregular
linear trend
mean absolute deviation (*MAD*)
mean absolute percent error
    (*MAPE*)

mean squared deviation (*MSD*)
moving average
multiplicative model
periodicity
polynomial model
quadratic trend
seasonal

seasonal binaries
smoothing constant
standard error (*SE*)
time series variable
trailing moving average (*TMA*)
trend

Commonly Used Formulas

Additive time series model: $Y = T + C + S + I$

Multiplicative time series model: $Y = T \times C \times S \times I$

Linear trend model: $y_t = b_0 + b_1 t$

Exponential trend model: $y_t = b_0 e^{b_1 t}$

Quadratic trend model: $y_t = b_0 + b_1 t + b_2 t^2$

Coefficient of determination: $R^2 = 1 - \dfrac{\sum\limits_{t=1}^{n}(y_t - \hat{y}_t)^2}{\sum\limits_{t=1}^{n}(y_t - \bar{y})^2}$

Mean absolute percent error: $MAPE = \dfrac{100}{n}\sum\limits_{t=1}^{n}\left|\dfrac{y_t - \hat{y}_t}{y_t}\right|$

Mean absolute deviation: $MAD = \dfrac{1}{n}\sum\limits_{t=1}^{n}|y_t - \hat{y}_t|$

Mean squared deviation: $MSD = \dfrac{1}{n}\sum\limits_{t=1}^{n}(y_t - \hat{y}_t)^2$

Standard error: $SE = \sqrt{\dfrac{\sum\limits_{t=1}^{n}(y_t - \hat{y}_t)^2}{n - 2}}$

Forecast updating equation for exponential smoothing: $F_{t+1} = \alpha y_t + (1 - \alpha)F_t$

Chapter Review

1. Explain the difference between (a) stocks and flows; (b) cross-sectional and time series data; (c) additive and multiplicative models.

2. (a) What is periodicity? (b) Give original examples of data with different periodicity.

3. (a) What are the distinguishing features of each component of a time series (trend, cycle, seasonal, irregular)? (b) Why is cycle usually ignored in time series modeling?

4. Name four criteria for assessing a trend forecast.

5. Name two advantages and two disadvantages of each of the common trend models (linear, exponential, quadratic).

6. When would the exponential trend model be preferred to a linear trend model?

7. Explain how to obtain the compound percent growth rate from a fitted exponential model.

8. (a) When might a quadratic model be useful? (b) What precautions must be taken when forecasting with a quadratic model? (c) Why are higher-order polynomial models dangerous?

9. Name five measures of fit for a trend, and state their advantages and disadvantages.

10. (a) When do we use a moving average? (b) Name two types of moving averages. (c) When is a centered moving average harder to calculate?

11. (a) When is exponential smoothing most useful? (b) Interpret the smoothing constant α. What is its range? (c) What does a small α say about the degree of smoothing? A large α?

12. (a) Explain two ways to initialize the forecasts in an exponential smoothing process. (b) Name an advantage and a disadvantage of each method.

13. (a) Why is seasonality irrelevant for annual data? (b) List the steps in deseasonalizing a monthly time series. (c) What is the sum of a monthly seasonal index? A quarterly index?

14. (a) How can forecasting improve communication within an organization? (b) List five tips for ensuring effective forecasting outcomes.

15. (a) Explain how seasonal binaries can be used to model seasonal data. (b) What is the advantage of using seasonal binaries?

16. Explain the equivalency between the two forms of an exponential trend model.

17. What is the purpose of index numbers?

Chapter Exercises

McGraw Hill **connect**

Instructions: For each exercise, make an attractive, well-labeled time series line chart. Adjust the *Y*-axis scale if necessary to show more detail (because Excel usually starts the scale at zero). If a fitted trend is called for, display the equation and R^2 statistic (or *MAPE, MAD,* and *MSD* in Minitab). Include printed copies of all relevant graphs with your answers to each exercise. Exercises marked with * are based on harder material.

14.10 (a) Make a line chart for JetBlue's revenue. (b) Describe the trend (if any) and discuss possible causes. (c) Fit both a linear and an exponential trend to the data. (d) Which model is preferred? Why? 🗇 **JetBlue**

JetBlue Airlines Revenue, 2012–2019 (billions)

Year	Revenue
2012	4.98
2013	5.44
2014	5.82
2015	6.42
2016	6.63
2017	7.01
2018	7.66
2019	8.09

Source: JetBlue's published annual Form 10-K reports. Data are for December 31 of each year.

14.11 (a) Plot both Swiss watch time series on the same graph. (b) Describe the trend (if any) and discuss possible causes. (c) Fit an exponential trend to each time series. (d) Interpret each fitted trend carefully. What conclusion do you draw? 🗇 **Swiss**

Swiss Watch Exports (thousands of units), 2014–2019

Year	Mechanical	Electronic
2014	8,131	20,455
2015	7,812	20,325
2016	6,963	18,433
2017	7,238	17,068
2018	7,525	16,215
2019	7,236	13,398

Source: Fédération de L'Industrie Horlogère Suisse, Swiss Watch Exports, www.fhs.swiss/eng/statistics.html.

14.12 (a) Make a line graph of the U.S. civilian labor force data. (b) Describe the trend (if any) and discuss possible causes. (c) Fit three trend models: linear, exponential, and quadratic. Which model would offer the most believable forecasts? Explain. (d) Choose one of the fitted trend models and make forecasts for years 2020–2022. Justify your choice. 🗇 **LaborForce**

U.S. Civilian Labor Force (thousands)

Year	Labor Force
2010	153,156
2011	153,373
2012	154,904
2013	154,408
2014	155,521
2015	157,245
2016	158,968
2017	159,880
2018	162,510
2019	164,007

Source: www.bls.gov. Data are in thousands for December 31 of each year, seasonally adjusted.

14.13 (a) Plot the voter participation rate. (b) Describe the trend (if any) and discuss possible causes. (c) Fit three trend models: linear, exponential, and quadratic. (d) Which trend model is preferred? Why? (e) Make a forecast for 2024, using a trend model of your choice. (f*) *Optional challenge:* Check your forecast. How accurate was it? Discuss. *Note:* Time is in four-year increments, so use $t = 17$ for the 2024 forecast. 🗇 **Voters**

U.S. Presidential Election Voter Participation, 1960–2020

Year	Voting Age Population	Voted for President	% Voting Pres
1960	109,672	68,836	62.8
1964	114,090	70,098	61.4
1968	120,285	73,027	60.7
1972	140,777	77,625	55.1
1976	152,308	81,603	53.6
1980	163,945	86,497	52.8
1984	173,995	92,655	53.3
1988	181,956	91,587	50.3
1992	189,493	104,600	55.2
1996	196,789	96,390	49.0
2000	209,787	105,594	50.3
2004	219,553	122,349	55.7
2008	229,945	131,407	57.1
2012	235,248	129,235	53.8
2016	249,422	136,669	54.8
2020	257,605	159,690	62.0

Source: wikipedia.org.

14.14 For each of the following fitted trends, make a prediction for period $t = 17$:

a. $y_t = 2286e^{0.076t}$

b. $y_t = 1149 + 12.78t$

c. $y_t = 501 + 18.2t - 7.1t^2$

14.15 (a) Make a line graph of total consumer credit outstanding. (b) Describe the trend (if any) and discuss possible causes. (c) Fit linear, exponential, and quadratic trends. (d) Plot both revolving and nonrevolving credit on the same graph. Do the trends differ? Explain. 🖭 **Consumer**

14.16 (a) Plot the data on U.S. general aviation shipments. (b) Describe the pattern and discuss possible causes. *Hint:* What economic factors affect major capital investments? (c) Would a fitted trend be helpful? Explain. (d) Fit a moving average (e.g., period 2) to the data. Is it useful? (e) Would trend forecasts for 2020 and beyond be appropriate? Explain. 🖭 **Airplanes**

14.17 For each of the following fitted trends, make a prediction for period $t = 12$:

a. $y_t = 372e^{-0.041t}$

b. $y_t = 719 + 10t$

c. $y_t = 1299 - 51t + 7t^2$

14.18 (a) Plot *either* receipts and outlays *or* federal debt and GDP (plot both time series on the same graph). (b) Describe the two trends and discuss possible causes. (c) Fit an exponential trend to each. (d) Compare the growth rates. Explain the implications. (e) Plot the ratio of debt to GDP. What does it show? 🖭 **FedBudget**

U.S. Consumer Credit Outstanding, 2000–2019 ($ billions) 🖭 Consumer

Year	Total	Revolving	Nonrevolving
2000	1,722	683	1,039
2001	1,868	715	1,153
2002	1,972	751	1,221
.	.	.	.
.	.	.	.
2017	3,828	1,022	2,806
2018	4,010	1,054	2,956
2019	4,176	1,086	3,090

Source: www.federalreserve.gov. Data are for December 31 of each year. Units are billions of dollars, seasonally adjusted. Total is short and intermediate term credit to individuals, the sum of revolving credit (mostly credit card and home equity loans) and nonrevolving credit (for a specific purchase such as a car, mobile home, education, boats, trailers, or vacations).

U.S. Manufactured General Aviation Shipments, 2002–2019 🖭 Airplanes

Year	Planes	Year	Planes	Year	Planes	Year	Planes
2002	2,207	2007	3,279	2012	1,516	2017	1,595
2003	2,137	2008	3,079	2013	1,615	2018	1,746
2004	2,355	2009	1,585	2014	1,631	2019	1,771
2005	2,857	2010	1,334	2015	1,592		
2006	3,147	2011	1,323	2016	1,531		

Source: U.S. Manufactured General Aviation Shipments, *Statistical Databook*, General Aviation Manufacturers Association.

U.S. Federal Finances, 2001–2019 ($ billions current) 🖭 FedBudget

Year	Receipts	Outlays	Fed Debt	GDP	Debt/GDP
2001	1,991	1,863	5,770	10,565	0.546
2002	1,853	2,011	6,198	10,877	0.570
2003	1,782	2,160	6,760	11,332	0.597
.	.	.	.	.	.
.	.	.	.	.	.
2017	3,316	3,982	20,206	19,272	1.048
2018	3,330	4,109	21,462	20,236	1.061
2019	3,462	4,447	22,668	21,220	1.068

Source: *Economic Report of the President, 2019.*

14.19 (a) Plot both men's and women's winning times (in minutes) on the same graph. (b) Fit a linear trend model to each series. Ask Excel for forecasts 20 years ahead. From the fitted trends, will the times eventually converge? Explain. (c) Make a copy of your graph, click each fitted trend, and change it to a moving average trend type. (d) Is a moving average a reasonable approach to modeling these data sets? *Note:* The data file ⌧ **Boston** has the data converted to decimal minutes. Only the first three and last three lines are shown here.

14.20 (a) Plot the data on leisure and hospitality employment. (b) Describe the trend (if any) and discuss possible causes. (c) Fit the linear, exponential, and quadratic trends. Would any of these trend models give credible forecasts for 2020 and beyond? Explain. ⌧ **Leisure**

Leisure and Hospitality Employment, 2006–2019 (thousands)

Year	Employees	Year	Employees
2006	13,292	2013	14,454
2007	13,550	2014	14,892
2008	13,256	2015	15,407
2009	12,944	2016	15,845
2010	13,158	2017	16,195
2011	13,538	2018	16,555
2012	13,978	2019	16,942

Source: http://data.bls.gov.

14.21 (a) Plot the data on law enforcement officers killed. (b) Describe the trend (if any) and discuss possible causes. (c) Would a fitted trend be helpful? Explain. ⌧ **LawOfficers**

U.S. Law Enforcement Officers Killed, 2006–2017

Year	Killed	Year	Killed
2006	114	2012	97
2007	141	2013	76
2008	109	2014	96
2009	96	2015	86
2010	128	2016	118
2011	125	2017	94

Source: https://ucr.fbi.gov.

14.22 (a) Plot the data on lightning deaths. (b) Describe the trend (if any) and discuss possible causes. (c) Fit an exponential trend to the data. Interpret the fitted equation. (d) Make a forecast for 2020, using a trend model of your choice (or a judgment forecast). Explain the basis for your forecast. *Note:* Time is in five-year increments, so use $t = 17$ for your 2020 forecast. ⌧ **Lightning**

U.S. Lightning Deaths, 1940–2015

Year	Deaths	Year	Deaths
1940	340	1980	74
1945	268	1985	74
1950	219	1990	74
1955	181	1995	85
1960	129	2000	51
1965	149	2005	38
1970	122	2010	29
1975	91	2015	28

Sources: *Statistical Abstract of the United States, 2011*, p. 234, and www.nws.noaa.gov.

Boston Marathon Champions, 1980–2019 ⌧ Boston

Men			Women		
Year	Name of Winner	Time	Year	Name of Winner	Time
1980	Bill Rodgers	2:12:11	1980	Jacqueline Gareau	2:34:28
1981	Toshihiko Seko	2:09:26	1981	Allison Roe	2:26:46
1982	Alberto Salazar	2:08:52	1982	Charlotte Teske	2:29:33
⋮	⋮	⋮	⋮	⋮	⋮
2017	Geoffrey Kirui	2:09:37	2017	Edna Kiplagat	2:21:52
2018	Yuki Kawauchi	2:15:58	2018	Desi Linden	2:39:54
2019	Lawrence Cherono	2:07:57	2019	Worknesh Degefa	2:23:31

Source: www.wikipedia.org.

14.23 (a) Plot the data on skier/snowboard visits. (b) Would a fitted trend be helpful? Explain. 📁 **SnowBoards**

14.24 (a) Plot both men's and women's 100-meter dash winning times on the same graph. (b) Fit a linear trend model to each series (men, women). (c) Use Excel's option to forecast each trend graphically to 2040 (i.e., up to period $t = 27$ periods because observations are in four-year increments). From these projections, does it appear that the times will eventually converge? *Optional challenge:* (d*) Set the fitted trends equal, solve for x (the time period when the trends will cross), and convert x to a year. Is the result plausible? Explain. *Note:* Only the first three and last three years are displayed here. 📁 **Olympic**

14.25 (a) Plot U.S. petroleum imports on a graph. (b) Describe the trend (if any) and discuss possible causes. (c) Fit linear, exponential, and quadratic trends. (d) Do any of these trends seem appropriate to make forecasts? Explain. (e) Make a projection for 2020 using any method (including judgment or a moving average). Do you believe it? Explain. *Note:* Time increments are five years, so use $t = 13$ for the 2020 forecast. 📁 **Petroleum**

U.S. Annual Petroleum Imports, 1960–2015 (billions of barrels)

Year	Imports	Year	Imports
1960	372	1990	2,151
1965	452	1995	2,639
1970	483	2000	3,320
1975	1,498	2005	3,696
1980	1,926	2010	3,363
1985	1,168	2015	2,687

Source: www.eia.doe.gov.

14.26 (a) Make a line chart for an m-period moving average to the exchange rate data shown below (only the first 3 and last 3 days are shown). with $m = 2, 3, 4,$ and 5 periods. For each method, state the last MA value. (b) Which value of m do you prefer? Why? (c) Is a moving average appropriate for this kind of data? 📁 **Sterling**

U.S. Skier/Snowboarder Visits, 2000–2021 (millions) 📁 SnowBoards

Season	Visits	Season	Visits	Season	Visits
2000–01	57.3	2007–08	60.5	2014–15	53.6
2001–02	54.4	2008–09	57.4	2015–16	52.8
2002–03	57.6	2009–10	59.8	2016–17	54.8
2003–04	57.1	2010–11	60.5	2017–18	53.3
2004–05	56.9	2011–12	51.0	2018–19	59.3
2005–06	58.9	2012–13	56.9	2019–20	51.1
2006–07	55.1	2013–14	56.5	2020–21	59.0

Source: www.nsaa.org/nsaa/press/industryStats.asp.

Summer Olympics 100-Meter Dash Winning Times, 1928–2016 📁 Olympic

Year	Men's 100-Meter Winner	Seconds	Women's 100-Meter Winner	Seconds
1928	Percy Williams, Canada	10.80	Elizabeth Robinson, United States	12.20
1932	Eddie Tolan, United States	10.30	Stella Walsh, Poland	11.90
1936	Jesse Owens, United States	10.30	Helen Stephens, United States	11.50
...	...	....	...	...
2008	Usain Bolt, Jamaica	9.69	Shelly-Ann Fraser, Jamaica	10.78
2012	Usain Bolt, Jamaica	9.63	Shelly-Ann Fraser, Jamaica	10.75
2016	Usain Bolt, Jamaica	9.81	Elaine Thompson, Jamaica	10.71

Source: www.wikipedia.org.

Daily Spot Exchange Rate, U.S. Dollars per Pound Sterling ($n = 60$ days) 📁 Sterling

Day	1	2	3	...	58	59	60
Date	11/1/19	11/4/19	11/5/19	...	1/28/20	1/29/20	1/30/20
Rate	1.2950	1.2906	1.2870	...	1.2996	1.3012	1.3106

Source: Federal Reserve Board of Governors.

14.27 Refer to exercise 14.26. (a) Plot the dollar/pound exchange rate data. Copy and paste the chart so that you have four copies (one for each α). (b) Perform simple exponential smoothing (using Excel's Data Analysis or other software such as Minitab) using $\alpha = .05, .10, .20,$ and $.50$. (c) Which value of α do you prefer? Why? (d) Is exponential smoothing appropriate for this kind of data? 🗂 **Sterling**

14.28 (a) Plot the data on natural gas bills. (b) Can you see seasonal patterns? Explain. (c) Use MegaStat, Minitab, or R to calculate estimated seasonal indexes. (d) Which months are the highest? The lowest? Can you explain this pattern? (e) Is there a trend in the deseasonalized data? *Optional challenge:* (f*) Perform a regression using 11 seasonal binaries. Interpret the results. 🗂 **GasBills**

14.29 (a) Plot the data on air travel delays. (b) Can you see seasonal patterns? Explain. (c) Use MegaStat, Minitab, or R to calculate estimated seasonal indexes. (d) Which months have the most delays? The fewest? Can you suggest reasons? 🗂 **Delays**

14.30 (a) Plot the data on airplane shipments. (b) Can you see seasonal patterns? Explain. (c) Use MegaStat, Minitab, or R to calculate estimated seasonal indexes. (d) In which quarters are shipments highest? Lowest? Can you suggest reasons? 🗂 **AirplanesQtr**

14.31 Ten years ago, on the last trading day of year 1, Felicia invested $1,000 in each of three stock funds. On the last trading day of year 10, their values were:

Fund A: $2,509   Fund B: $2,096   Fund C: $3,034.

Use formula 14.4 to estimate the implied rate of return for each fund.

14.32 (a) Use MegaStat, Minitab, or R to deseasonalize the quarterly data on Coca-Cola's revenues and calculate seasonal indexes. (b) Interpret the seasonal indexes. If there is seasonality, suggest possible reasons. (c*) Perform a regression using seasonal binaries. Interpret the results. 🗂 **CocaCola**

Natural Gas Bills for a California Residence, 2017–2020 🗂 **GasBills**

Month	2017	2018	2019	2020
Jan	78.98	118.86	101.44	155.37
Feb	84.44	111.31	122.20	148.77
Mar	65.54	75.62	99.49	115.12
Apr	62.60	77.47	55.85	85.89
May	29.24	29.23	44.94	46.84
Jun	18.10	17.10	19.57	24.93
Jul	91.57	16.59	15.98	20.84
Aug	6.48	27.64	14.97	26.94
Sep	19.35	28.86	18.03	34.17
Oct	29.02	48.21	56.98	88.58
Nov	94.09	67.15	115.27	100.63
Dec	101.65	125.18	130.95	174.63

Source: Homeowner's records.

U.S. Airspace Total System Delays, 2013–2017 🗂 **Delays**

Month	2013	2014	2015	2016	2017
Jan	16,240	15,385	18,571	18,035	29,548
Feb	17,031	19,755	18,553	20,989	25,607
Mar	21,697	20,227	22,326	28,237	38,291
Apr	37,117	25,912	24,416	22,683	41,977
May	35,740	35,218	31,125	28,455	49,208
Jun	46,693	43,059	41,560	39,238	52,981
Jul	46,715	37,967	38,308	43,881	49,913
Aug	31,101	34,499	32,711	41,335	47,951
Sep	21,844	28,302	25,455	27,085	32,091
Oct	21,066	31,940	21,893	26,619	31,248
Nov	16,316	20,647	21,376	23,498	20,732
Dec	21,809	28,206	29,087	25,411	25,381

Source: www.faa.gov.

14.33 (a) Use MegaStat, Minitab, or R to deseasonalize the monthly data on student pilots and calculate seasonal indexes. (b) Interpret the seasonal indexes. If there is seasonality, suggest possible reasons. (c*) Perform a regression using seasonal binaries. Interpret the results.
📁 **StudentPilots**

14.34 The following seasonal regression was fitted with quarterly seasonal binaries beginning in the first quarter ($Qtr4$ is omitted to avoid multicollinearity). Make a prediction for y_t in period (a) $t = 21$; (b) $t = 8$; (c) $t = 15$.

$$y_t = 213 + 11t - 9\,Qtr1 + 12\,Qtr2 - 15\,Qtr3.$$

14.35 The following seasonal regression was fitted with quarterly seasonal binaries beginning in the first quarter ($Qtr4$ is omitted to avoid multicollinearity). Make a prediction for y_t in period (a) $t = 14$; (b) $t = 17$; (c) $t = 20$.

$$y_t = 491 + 19t + 29\,Qtr1 - 18\,Qtr2 + 12\,Qtr3.$$

14.36 You want to invest \$1,000. Which growth curve would yield the largest principal 5 years from now? 10 years? 20 years? Explain. *Hint:* Show all the forecasts.

a. $y_t = 1000e^{0.021t}$
b. $y_t = 1000 + 25t$
c. $y_t = 1000 + 28t - 0.4t^2$

U.S. Manufactured General Aviation Shipments, 2012–2019 📁 **AirplanesQtr**

Year	Qtr 1	Qtr 2	Qtr 3	Qtr 4	Total
2012	305	358	339	514	1,516
2013	329	413	353	520	1,615
2014	345	380	379	527	1,631
2015	296	374	378	544	1,592
2016	280	357	385	509	1,531
2017	311	379	377	532	1,599
2018	315	428	416	587	1,746
2019	366	401	423	581	1,771

Note: Quarterly shipments may not add to annual total because some manufacturers report only annual totals.
Source: "U.S. Manufactured General Aviation Shipments," *Statistical Databook*, General Aviation Manufacturers Association, used with permission.

Coca-Cola Revenues (\$ millions), 2014–2019 📁 **CocaCola**

Quarter	2014	2015	2016	2017	2018	2019
Qtr1	10.58	10.71	10.28	9.12	7.63	8.02
Qtr2	12.57	12.16	11.54	9.70	9.42	10.00
Qtr3	11.98	11.43	10.63	9.08	8.78	9.51
Qtr4	10.87	10.00	9.41	7.51	5.36	9.07

Sources: 10-K reports of The Coca-Cola Company.

Student Pilot Certificates Issued by Month, 2013–2018 📁 **StudentPilots**

Month	2013	2014	2015	2016	2017	2018
Jan	4,480	3,882	3,805	3,714	2,173	3,202
Feb	3,921	3,154	3,327	3,700	2,180	3,462
Mar	4,662	3,451	3,833	5,287	3,250	4,110
Apr	3,693	3,881	3,918	1,753	2,495	3,441
May	4,029	4,159	3,882	2,948	2,828	3,958
Jun	4,336	4,614	4,856	3,001	3,128	3,611
Jul	4,789	4,833	4,659	3,096	3,141	4,460
Aug	5,492	5,104	4,867	3,670	4,536	3,998
Sep	4,025	4,195	4,188	3,921	2,588	4,242
Oct	3,926	3,963	3,863	2,815	5,534	4,635
Nov	3,293	3,133	3,061	1,302	3,945	3,140
Dec	2,920	3,038	3,122	938	2,603	3,095

Source: www.faa.gov/data_statistics/aviation_data_statistics.

Related Reading

Hanke, John E., and Dean W. Wichern. *Business Forecasting,* 9th ed. Pearson, 2014.

Ord, Keith, Robert Fildes, and Nikolaos Kourentzes. *Principles of Business Forecasting,* 2nd ed. Wessex Press, 2017.

Wilson, J. Holton, and Barry Keating. *Forecasting and Predictive Analytics with Forecast X*™, 7th ed. McGraw Hill, 2019.

CHAPTER 14 More Learning Resources

You can access these *LearningStats* demonstrations through Connect to help you understand time series analysis.

Topic	*LearningStats Demonstrations*
Trends and forecasting	Trend Forecasting Time Series Trend Analysis Using R Measures of Fit Exponential Trend Formula
Simulations	Time Series Components Trend Simulator Seasonal Time Series Generator
Exponential smoothing	Advanced Forecasting Methods Single Exponential Smoothing Brown's Double Smoothing Holt-Winters Seasonal Smoothing Exponential Smoothing Weights
ARIMA Models	ARIMA Terminology ARIMA Patterns ARIMA Calculations Seasonal ARIMA

Key: = PowerPoint = Excel = Adobe PDF

Software Supplement

Minitab Trend Analysis

Figure 14.28 shows Minitab's trend menu and time plot of fire loss claims. Minitab displays fit statistics *MAPE, MAD,* and *MSD* instead of R^2. An attractive feature is Minitab's separation between actual and forecasts, and using different colors for actual (blue) and forecasts (green).

Figure 14.28

Minitab's Trend Analysis
FireLosses

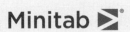

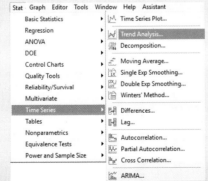

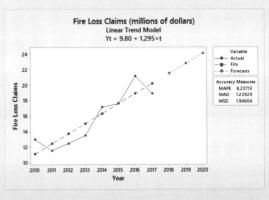

Source: Minitab.

Minitab Exponential Smoothing

Figure 14.29 shows Minitab's single exponential smoothing and four weeks' forecasts for the deck sealer data (Excel's exponential smoothing does not make forecasts). After week 18, the exponential smoothing method cannot be updated with actual data, so the forecasts are constant. The wide 95 percent confidence intervals (red triangles) reflect the erratic past sales pattern.

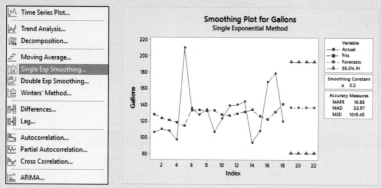

Source: Minitab.

Figure 14.29

Minitab's Exponential Smoothing 🗁 **DeckSealer**

Using Minitab to Deseasonalize

Minitab performs its deseasonalization by fitting a trend and then averaging the seasonal factors using *medians* instead of *means,* so the results are not exactly the same as with MegaStat. Minitab offers excellent graphical displays for decomposition, as well as forecasts, as shown in Figure 14.30. Minitab offers additive as well as multiplicative seasonality. In an additive model, the *CMA* is calculated in the same way, but the raw seasonals are *differences* (instead of ratios) and the seasonal indexes are forced to sum to *zero* (e.g., months with higher sales must exactly balance months with lower sales). Most analysts prefer multiplicative models for trended data.

Figure 14.30

Minitab's Graphs for Floor Covering Sales 🗁 **FloorSales**

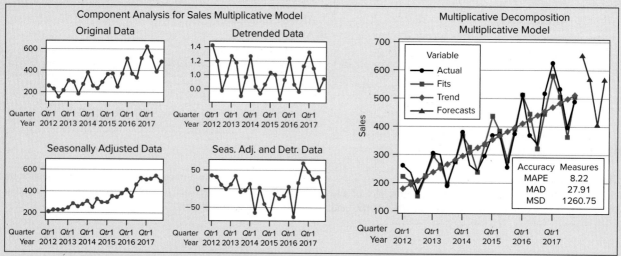

Source: Minitab.

Fitting Trends in R 🗁 FireLosses

We can fit and display several trend models (linear, quadratic, exponential) on the same graph using the 8 years of fire loss claims (Table 14.6). First, create a vector Claims containing the data and a vector Time as an index for the years 1, 2, . . . , 8. Put the variables in a data frame CData. Print the data to verify it:

```
> Claims = c(12.940, 11.510, 12.428, 13.457, 17.118, 17.586, 21.129, 18.874)
> Time=c(1:8)
> CData=data.frame(Claims,Time) # create a data frame for convenience
> CData                         # list the data
```

	Claims	Time
1	12.940	1
2	11.510	2
3	12.428	3
4	13.457	4
5	17.118	5
6	17.586	6
7	21.129	7
8	18.874	8

Fit the trend models. Save each result for later use. Display their fitted predictions so you can compare them. In this case, predictions are similar, which favors the simpler linear model:

```
> fit1=lm(Claims~Time)        # fit linear trend (1st deg poly)
> fit2=lm(Claims~poly(Time,2)) # fit quadratic trend (2nd deg poly)
> fit3=lm(log(Claims)~Time)    # fit exponential trend in logs

> fitted.values(fit1)
        1        2        3        4        5        6        7        8
 11.09825 12.39311 13.68796 14.98282 16.27768 17.57254 18.86739 20.16225

> fitted.values(fit2)
        1        2        3        4        5        6        7        8
 11.61575 12.46704 13.46618 14.61318 15.90804 17.35075 18.94132 20.67975

> fitted.values(fit3)
        1        2        3        4        5        6        7        8
 11.44177 12.43040 13.50446 14.67132 15.93901 17.31623 18.81245 20.43795
```

Save the predictions for each model. Plot the data and all 3 fitted curves on the same graph in colors with variable names as axis labels. Use exp() for exponential since its predictions are logs. Separately, plot the preferred linear model, using its fitted equation as a title (rounded to 3 or 4 decimals for brevity):

```
> Pred1=predict(fit1,data=CData)        # linear trend
> Pred2=predict(fit2,data=CData)        # quadratic trend
> Pred3=exp(predict(fit3,data=CData))   # exponential trend
> plot(Time,Claims,main=
    "Three Trend Models")               # plot x,y data
> lines(Time,Pred1,col="red")           # plot linear trend
> lines(Time,Pred2,col="blue")          # plot quadratic trend
> lines(Time,Pred3,col="green")         # plot exponential trend
> b0=round(fit1$coefficients[[1]],3)    # plot linear separately
> b1=round(fit1$coefficients[[2]],3)
> r2=round(summary(fit1)$r.squared,4)
> B1=as.character(abs(b1))
> B0=as.character(b0)
> R2=as.character(r2)
> if(b1<0) DispEqn=paste("y =",B0,"-",B1,"x ","Rsq =",R2) else
+  DispEqn=paste("y =",B0,"+",B1,"x ","Rsq =",R2)
> plot(Time,Claims,main=DispEqn)
> lines(Time,Pred1,col="red")
```

If you need more details (R^2, coefficients, standard error, t-values, p-values, residuals), you can use the summary() function for any of the fitted models. For example, here is a summary for the linear model:

```
> summary(fit1)
Residuals:
```

Min	1Q	Median	3Q	Max
−1.5258	−1.2670	−0.4348	1.0907	2.2616

Coefficients:

	Estimate Std.	Error	t value	Pr(>|t|)	
(Intercept)	9.8034	1.2551	7.811	0.000232	***
Time	1.2949	0.2486	5.210	0.001996	**

```
---
Signif. codes:   0 '***'  0.001 '**'  0.01 '*'  0.05 '.'  0.1 ' '  1
```

Residual standard error: 1.611 on 6 degrees of freedom
Multiple R-squared: 0.8189, Adjusted R-squared: 0.7888
F-statistic: 27.14 on 1 and 6 DF, p-value: 0.001996

Seasonal Decomposition in R   FloorSales

It is easy to calculate seasonal factors for monthly or quarterly time series data. For example, we have an Excel spreadsheet with quarterly floor sales data (24 quarters starting in year 2011) as shown in textbook Table 14.10:

Year	Quarter	Sales
2012	Qtr 1	259
	Qtr 2	236
	Qtr 3	164
	Qtr 4	222
	...	...
	...	...
	...	...
2017	Qtr 1	626
	Qtr 2	535
	Qtr 3	397
	Qtr 4	488

Import the Sales column (including the heading) into a data frame called FloorSales (we ignore the year and quarter labels):

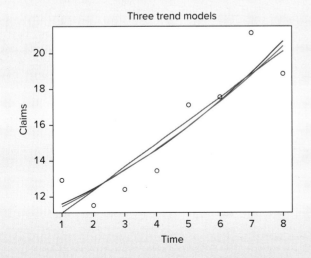

Three trend models

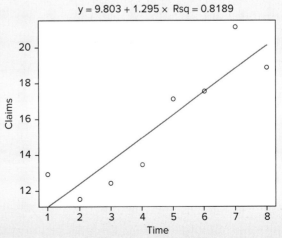

y = 9.803 + 1.295 × Rsq = 0.8189

```
> FloorSales=read.table(file="clipboard",sep="t",header=TRUE)
> FloorSales
      Sales
1      259
2      236
3      164
4      222
.       .
.       .
.       .
21     626
22     535
23     397
24     488
```

Use the ts() function to convert the data to a quarterly time series with frequency=4 and year labels beginning with 2012 using start=2011. Plot the data.

```
> ts.FloorSales=ts(FloorSales,frequency=4,start=2012)
> ts.FloorSales
```

	Qtr1	Qtr2	Qtr3	Qtr4
2012	259	236	164	222
2013	306	300	189	275
2014	379	262	242	296
2015	369	373	255	374
2016	515	373	339	519
2017	626	535	397	488

```
> plot(ts.FloorSales)
```

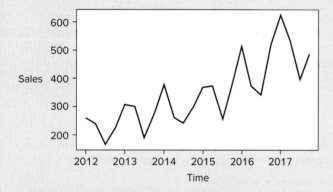

Use the decompose() function to obtain seasonal factors. The results match Table 14.11 in the textbook and show all the steps of decomposition. R also displays the deseasonalized time series and residuals around the fitted trend. The last line shows the four seasonal factors (they are adjusted so they sum to 1), which are identical to those in Table 14.12:

```
> decompose(ts.FloorSales,"multiplicative")
```

$x

	Qtr1	Qtr2	Qtr3	Qtr4
2012	259	236	164	222
2013	306	300	189	275
2014	379	262	242	296
2015	369	373	255	374
2016	515	373	339	519
2017	626	535	397	488

$seasonal

	Qtr1	Qtr2	Qtr3	Qtr4
2012	1.2520398	1.0210404	0.7396682	0.9872516
2013	1.2520398	1.0210404	0.7396682	0.9872516
2014	1.2520398	1.0210404	0.7396682	0.9872516
2015	1.2520398	1.0210404	0.7396682	0.9872516
2016	1.2520398	1.0210404	0.7396682	0.9872516
2017	1.2520398	1.0210404	0.7396682	0.9872516

$trend

	Qtr1	Qtr2	Qtr3	Qtr4
2012	NA	NA	226.125	240.000
2013	251.125	260.875	276.625	281.000
2014	282.875	292.125	293.500	306.125
2015	321.625	333.000	361.000	379.250
2016	389.750	418.375	450.375	484.500
2017	512.000	515.375	NA	NA

$random

	Qtr1	Qtr2	Qtr3	Qtr4
2012	NA	NA	0.9805242	0.9369445
2013	0.9732252	1.1262787	0.9237053	0.9912849
2014	1.0701053	0.8783946	1.1147316	0.9794112
2015	0.9163438	1.0970380	0.9549839	0.9988911
2016	1.0553657	0.8731728	1.0176267	1.0850399
2017	0.9765315	1.0166876	NA	NA

$figure
[1]   1.2520398   1.0210404   0.7396682   0.9872516

15 Chi-Square Tests

CHAPTER CONTENTS

CHAPTER LEARNING OBJECTIVES

When you finish this chapter, you should be able to

LO 15-1 Recognize a contingency table and understand how it is created.

LO 15-2 Find degrees of freedom and use the chi-square table of critical values.

LO 15-3 Perform a chi-square test for independence on a contingency table.

LO 15-4 Perform a goodness-of-fit (GOF) test for a multinomial distribution.

LO 15-5 Perform a GOF test for a uniform distribution.

LO 15-6 Explain the GOF test for a Poisson distribution.

LO 15-7 Explain the chi-square GOF test for normality.

LO 15-8 Interpret ECDF tests and know their advantages.

Ingram Publishing/AGE Fotostock

Not all information pertaining to business can be summarized numerically. We are often interested in answers to questions such as: Do employees in different age groups choose different types of health plans? Do consumers prefer red, yellow, or blue package lettering on our bread bags? Does the name of our new lawn mower influence how we perceive the quality? Answers to questions such as these are not measurements on a numerical scale. Rather, the variables that we are interested in learning about may be *categorical* or *ordinal*. Health plans are categorized by the way services are paid, so the variable *health plan* might have four different categories: Catastrophic, HMO (health maintenance organization), POS (point of service), and CDHP (consumer-driven health plan). The variable *package lettering color* would have categories red, yellow, and blue, and the variable *perceived quality* might have categories excellent, satisfactory, and poor.

We can collect observations on these variables to answer the types of questions posed either by surveying our customers and employees or by conducting carefully designed experiments. Once our data have been collected, we summarize by tallying response frequencies on a table that we call a *contingency table*. A **contingency table** is a cross-tabulation of *n* paired observations into categories. Each cell shows the count of observations that fall into the category defined by its row and column heading.

LO 15-1

Recognize a contingency table and understand how it is created.

Example 15.1

Webpages (4 × 3 Table)

As online shopping has grown, opportunity also has grown for personal data collection and invasion of privacy. Mainstream online retailers have policies known as "privacy disclaimers" that define the rules regarding their uses of information collected, the customer's right to refuse third-party promotional offers, and so on. You can access these policies through a web link, found either on the website's home page, on the order page (i.e., as you enter your credit card information), on a client webpage, or on some other webpage. Location of the privacy disclaimer is considered to be a measure of the degree of consumer protection (the further the link is from the home page, the less likely it is to be noticed). Marketing researchers did a survey of 291 websites in three nations (France, United Kingdom, United States) and obtained the *contingency table* shown here as Table 15.1. Is location of the privacy disclaimer *independent* of the website's nationality? This question can be answered by using a test based on the frequencies in this contingency table.

Table 15.1 Privacy Disclaimer Location and Website Nationality WebSites

Location of Disclaimer	Nationality of Website			Row Total
	France	U.K.	U.S.	
Home page	56	68	35	159
Order page	19	19	28	66
Client page	6	10	16	32
Other page	12	9	13	34
Col Total	93	106	92	291

Source: Calin Gurau, Ashok Ranchhod, and Claire Gauzente, "To Legislate or Not to Legislate: A Comparative Exploratory Study of Privacy/Personalisation Factors Affecting French, UK, and US Web Sites," *Journal of Consumer Marketing* 20, no. 7 (2003), p. 659.

Table 15.2 illustrates the terminology of a contingency table. Variable *A* has *r* levels (rows) and variable *B* has *c* levels (columns), so we call this an $r \times c$ contingency table. Each cell shows the observed frequency f_{jk} in row *j* and column *k*.

	Variable B				
Variable *A*	1	2	⋯	*c*	Row Total
1	f_{11}	f_{12}	⋯	f_{1c}	R_1
2	f_{21}	f_{22}	⋯	f_{2c}	R_2
⋮	⋮	⋮	⋮	⋮	⋮
r	f_{r1}	f_{r2}	⋯	f_{rc}	R_r
Col Total	C_1	C_2	⋯	C_c	*n*

Table 15.2

Table of Observed Frequencies

Chi-Square Test

In a test of independence for an $r \times c$ contingency table, the hypotheses are

H_0: Variable *A* is independent of variable *B*

H_1: Variable *A* is not independent of variable *B*

To test these hypotheses, we use the **chi-square test for independence**, developed by Karl Pearson (1857–1936). It is a test based on *frequencies*. It measures the association between the two variables *A* and *B* in the contingency table. The chi-square test for independence is called a *distribution-free test* because it requires no assumptions about the shape of the populations from which the samples are drawn. The only operation performed is classifying the *n* data pairs into *r* rows (variable *A*) and *c* columns (variable *B*) and then comparing the **observed frequency** f_{jk} in each cell of the contingency table with the **expected frequency** e_{jk} under the assumption of independence. The chi-square test statistic measures the *relative* difference between expected and observed frequencies:

$$\chi^2_{\text{calc}} = \sum_{j=1}^{r} \sum_{k=1}^{c} \frac{[f_{jk} - e_{jk}]^2}{e_{jk}} \tag{15.1}$$

If the two variables are **independent**, then f_{jk} should be close to e_{jk}, leading to a chi-square test statistic near zero. Conversely, large differences between f_{jk} and e_{jk} will lead to a large chi-square test statistic. The chi-square test statistic cannot be negative (due to squaring), so it is always a right-tailed test. If the test statistic is far enough in the right tail, we will reject the hypothesis of independence. Squaring each difference removes the sign, so it doesn't matter whether e_{jk} is greater than or less than f_{jk}. Each squared difference is expressed *relative* to e_{jk}.

Chi-Square Distribution

The test statistic is compared with a critical value from the **chi-square probability distribution**. It has one parameter called **degrees of freedom**. For the $r \times c$ contingency table, the degrees of freedom are

$$d.f. = \text{degress of freedom} = (r - 1)(c - 1) \tag{15.2}$$

where

r = the number of rows in the contingency table

c = the number of columns in the contingency table

The parameter *d.f.* is the number of nonredundant cells in the contingency table. There is a different chi-square distribution for each value of *d.f.* Appendix E contains critical values

LO 15-2

Find degrees of freedom and use the chi-square table of critical values.

for right-tail areas of the chi-square distribution. Its mean is *d.f.* and its variance is 2*d.f.* As illustrated in Figure 15.1, all chi-square distributions are skewed to the right but become more symmetric as *d.f.* increases. For *d.f.* = 1, the distribution is discontinuous at the origin. As *d.f.* increases, the shape begins to resemble a normal, bell-shaped curve. However, for any contingency table you are likely to encounter, degrees of freedom will not be large enough to assume normality.

Figure 15.1

Various Chi-Square Distributions

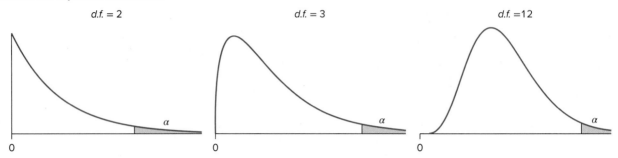

Expected Frequencies

Assuming that H_0 is true, the expected frequency of row j and column k is

(15.3) $e_{jk} = R_j C_k / n$ (expected frequency in now j and column k)

where

R_j = total for row j ($j = 1, 2, \ldots, r$)

C_k = total for column k ($k = 1, 2, \ldots, c$)

n = sample size (or number of responses)

This formula for expected frequencies stems from the definition of independent events (see Chapter 5). When two events are independent, their *joint* probability is the product of their marginal probabilities, so for a cell in row j and column k, the joint probability would be $(R_j/n)(C_k/n)$. To get the expected cell frequency, we multiply this joint probability by the sample size n to obtain $e_{jk} = R_j C_k / n$. The e_{jk} always sum to the same row and column frequencies as the observed frequencies. Expected frequencies will not, in general, be integers.

Illustration of the Chi-Square Calculations

LO 15-3

Perform a chi-square test for independence on a contingency table.

We will illustrate the chi-square test by using the webpage frequencies from the contingency table (Table 15.1). We follow the usual five-step hypothesis testing procedure:

Step 1: State the Hypotheses   For the webpage example, the hypotheses are

H_0: Privacy disclaimer location is independent of website nationality

H_1: Privacy disclaimer location is dependent on website nationality

Step 2: Specify the Decision Rule   For the webpage contingency table, we have $r = 4$ rows and $c = 3$ columns, so degrees of freedom are $d.f. = (r - 1)(c - 1) = (4 - 1)(3 - 1) = 6$. We will choose $\alpha = .05$ for the test. Figure 15.2 shows that the right-tail critical value from Appendix E with $d.f. = 6$ is $\chi^2_{.05} = 12.59$. This critical value also could be obtained from Excel using =CHISQ.INV.RT(.05,6) = 12.59159.

Figure 15.2

APPENDIX E

Critical Value of Chi-Square from Appendix E for *d.f.* = 6 and α = .05

CHI-SQUARE CRITICAL VALUES

This table shows the critical value of chi-square for each desired tail area and degrees of freedom (*d.f.*).

d.f.	.995	.990	.975	.95	.90	.10	.05	.025	.01	.005
						Area in Upper Tail				
1	0.000	0.000	0.001	0.004	0.016	2.706	3.841	5.024	6.635	7.879
2	0.010	0.020	0.051	0.103	0.211	4.605	5.991	7.378	9.210	10.60
3	0.072	0.115	0.216	0.352	0.584	6.251	7.815	9.348	11.34	12.84
4	0.207	0.297	0.484	0.711	1.064	7.779	9.488	11.14	13.28	14.86
5	0.412	0.554	0.831	1.145	1.610	9.236	11.07	12.83	15.09	16.75
6	0.676	0.872	1.237	1.635	2.204	10.64	12.59	14.45	16.81	18.55
7	0.989	1.239	1.690	2.167	2.833	12.02	14.07	16.01	18.48	20.28
8	1.344	1.646	2.180	2.733	3.490	13.36	15.51	17.53	20.09	21.95
9	1.735	2.088	2.700	3.325	4.168	14.68	16.92	19.02	21.67	23.59
10	2.156	2.558	3.247	3.940	4.865	15.99	18.31	20.48	23.21	25.19
100	67.33	70.06	74.22	77.93	82.36	118.5	124.3	129.6	135.8	140.2

For α = .05 in this right-tailed test, the decision rule is

Reject H_0 if $\chi^2_{calc} > 12.59$
Otherwise, do not reject H_0

The decision rule is illustrated in Figure 15.3.

Figure 15.3

Right-Tailed Chi-Square Test for *d.f.* = 6

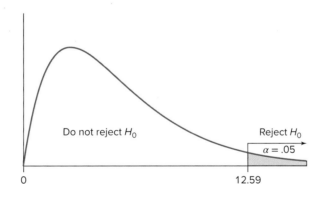

Do not reject H_0

Reject H_0

α = .05

0   12.59

Step 3: Calculate the Test Statistic   The expected frequency in row *j* and column *k* is $e_{jk} = R_j C_k / n$. The calculations are illustrated in Table 15.3. The expected frequencies (lower part of Table 15.3) must sum to the same row and column frequencies as the observed frequencies (upper part of Table 15.3).
The chi-square test statistic is

$$\chi^2_{calc} = \sum_{j=1}^{r} \sum_{k=1}^{c} \frac{[f_{jk} - e_{jk}]^2}{e_{jk}} = \frac{(56 - 50.81)^2}{50.81} + \cdots + \frac{(13 - 10.75)^2}{10.75}$$
$$= 0.53 + \cdots + 0.47 = 17.54$$

Table 15.3

Observed and Expected Frequencies Websites

Observed Frequencies

Location	France	U.K.	U.S.	Row Total
Home	56	68	35	159
Order	19	19	28	66
Client	6	10	16	32
Other	12	9	13	34
Col Total	93	106	92	291

Expected Frequencies (assuming independence)

Location	France	U.K.	U.S.	Row Total
Home	(159 × 93)/291 = 50.81	(159 × 106)/291 = 57.92	(159 × 92)/291 = 50.27	159
Order	(66 × 93)/291 = 21.09	(66 × 106)/291 = 24.04	(66 × 92)/291 = 20.87	66
Client	(32 × 93)/291 = 10.23	(32 × 106)/291 = 11.66	(32 × 92)/291 = 10.12	32
Other	(34 × 93)/291 = 10.87	(34 × 106)/291 = 12.38	(34 × 92)/291 = 10.75	34
Col Total	93	106	92	291

Even for this simple problem, the calculations are tedious without a spreadsheet. The Excel function =CHISQ.TEST(actual data range, expected data range) gives the p-value, although we still must calculate the e_{jk} ourselves. Fortunately, any statistical package will do a chi-square test. For example, Figure 15.4 shows MegaStat's setup and calculations arranged in a tabular form.

Figure 15.4

MegaStat's Chi-Square Test for Webpage Data

Source: MegaStat.

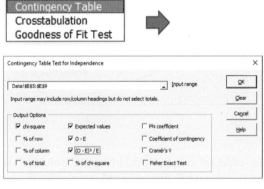

Step 4: Make the Decision Because the test statistic $\chi^2_{calc} = 17.54$ exceeds 12.59, we conclude that the observed differences between expected and observed frequencies differ significantly at $\alpha = .05$. The p-value (.0075) indicates that H_0 should be rejected at $\alpha = .05$. You can obtain this p-value using Excel's function =CHISQ.DIST.RT(χ^2, d.f.) or, in this case,

=CHISQ.DIST.RT(17.54,6), which gives a right-tail area of .0075. This *p*-value indicates that privacy disclaimer location is *not* independent of nationality at $\alpha = .05$, based on this sample of 291 websites.

Step 5: Take Action These results suggest that in the United States, the privacy disclaimer placement is more evenly distributed between various website pages when compared to France and the United Kingdom. This might have changed had the Commercial Privacy Bill of Rights Act of 2011 passed in the United States.

Discussion MegaStat rounds things off for display purposes, though it maintains full internal accuracy in the calculations (as you must, if you do these calculations by hand). Differences between observed and expected frequencies (O − E) must sum to zero across each row and down each column. If you are doing these calculations by hand, check these sums (if they are not zero, you have made an error). From Figure 15.4 we see that only three cells (column 3, rows 1, 2, and 3) contribute a majority (4.64, 2.44, 3.42) of the chi-square sum (17.54). The hypothesis of independence fails largely because of these three cells.

Using Excel Instead of MegaStat, we could have used Excel's =CHISQ.TEST(observed frequency range, expected frequency range) function. After using Excel to calculate expected frequencies for each cell using formula 15.3, we would enter the cell ranges for the observed frequencies and the cell ranges for the expected frequencies into the function. Excel then returns the *p*-value for our chi-square statistic. In this case, Excel would return the value .0075. To see the value of χ^2_{calc}, we could use Excel's function =CHISQ.INV.RT(probability, degrees of freedom), =CHI.INV. RT(.0075,6) = 17.54.

Example 15.2

Night Flying (2 × 2 Table)

For private pilots, past studies have shown that the percentage of nighttime flying accidents that are fatal is significantly higher than the percentage of daytime accidents that are fatal (www.avweb.com/flight-safety/technique/night-flying-safety/). A random telephone poll was taken in which 409 New Yorkers were asked, "Should private pilots be allowed to fly at night without an instrument rating?" The same question was posed to 70 aviation experts. Results are shown in Table 15.4. The totals exclude those who had "No Opinion" (1 expert and 25 general public).

Table 15.4

Should Noninstrument-Rated Pilots Fly at Night? Pilots

Opinion	Experienced Pilots	General Public	Row Total
Yes	40	61	101
No	29	323	352
Col Total	69	384	453

Source: Siena College Research Institute.

The hypotheses are

H_0: Opinion is independent of aviation expertise
H_1: Opinion is not independent of aviation expertise

The test results from MegaStat are shown in Figure 15.5. Degrees of freedom are $d.f. = (r - 1)(c - 1) = (2 - 1)(2 - 1) = 1$. Appendix E shows that the critical value of chi-square for $\alpha = .005$ is 7.879. Because the test statistic $\chi^2 = 59.80$ greatly exceeds 7.879, we firmly reject the hypothesis. The *p*-value (.0000) confirms that opinion is *not* independent of aviation experience.

Figure 15.5

Chi-Square Test for Night-Flying Data with *d.f.* = 1

Chi-square Contingency Table Test for Independence			
	Col 1	Col 2	Total
Row 1 Observed	**40**	**61**	101
Expected	15.38	85.62	101.00
O – E	24.62	–24.62	0.00
$(O - E)^2/E$	39.39	7.08	46.47
Row 2 Observed	**29**	**323**	352
Expected	53.62	298.38	352.00
O – E	–24.62	24.62	0.00
$(O - E)^2/E$	11.30	2.03	13.33
Total Observed	69	384	453
Expected	69.00	384.00	453.00
O – E	0.00	0.00	0.00
$(O - E)^2/E$	50.69	9.11	59.80

59.80	chi-square
1	df
1.05E-14	p-value

Test of Two Proportions

For a 2 × 2 contingency table, the chi-square test is equivalent to a two-tailed z test for two proportions, if the samples are large enough to ensure normality. The hypotheses are

$H_0: \pi_1 - \pi_2 = 0$

$H_1: \pi_1 - \pi_2 \neq 0$

In the aviation survey example, the proportion of aviation experts who said yes on the survey is $p_1 = x_1/n_1 = 40/69 = .57971$, or 58.0 percent, compared to the proportion of the general public $p_2 = x_2/n_2 = 61/384 = .15885$, or 15.9 percent. The pooled proportion is $p_c = (x_1 + x_2)/(n_1 + n_2) = 101/453 = .22296$. The z test statistic is then

$$z_{calc} = \frac{p_1 - p_2}{\sqrt{p_c(1 - p_c)\left(\frac{1}{n_1} + \frac{1}{n_2}\right)}} = \frac{.57971 - .15885}{\sqrt{.22296(1 - .22296)\left(\frac{1}{69} + \frac{1}{384}\right)}} = 7.7329$$

To find the two-tailed p-value, we use Excel's function =2*(1-NORM.S.DIST(7.7329,1)) = .0000. The square of the z test statistic for the two-tailed test of proportions is the same as the chi-square test statistic for the corresponding 2 × 2 contingency table. In the aviation example, $z^2 = 7.7329^2 = 59.80 = \chi^2$. Our conclusion is identical whether we used the chi-square test or the test for two proportions.

Small Expected Frequencies

The chi-square test is unreliable if the *expected* frequencies are too small. As you can see from the formula for the test statistic, when e_{jk} in the denominator is small, the chi-square statistic may be inflated. A commonly used rule of thumb known as **Cochran's Rule** requires that $e_{jk} > 5$ for all cells. Another rule of thumb says that up to 20 percent of the cells may have $e_{jk} < 5$. Statisticians generally become quite nervous when $e_{jk} < 2$, and there is agreement that a chi-square test is infeasible if $e_{jk} < 1$ in any cell. Computer packages may offer warnings or refuse to proceed when expected frequencies are too small. When this happens, it may be possible to salvage the test by combining adjacent rows or columns to enlarge the expected frequencies. In the webpage example, all the expected frequencies are safely greater than 5.

Cross-Tabulating Raw Data

Chi-square tests for independence are quite flexible. Although most often used with nominal data such as gender (male, female), we also can analyze quantitative variables (such as salary) by coding them into categories (e.g., under $25,000; $25,000 to $50,000; $50,000 and over). Open-ended classes are acceptable. We can mix data types as required (nominal, ordinal, interval, ratio) by defining the bins appropriately. Few statistical tests are so versatile. Continuous data may be classified into any categories that make sense. To tabulate a continuous variable into two classes, we would make the cut at the median. For three bins, we would use the 33rd and 67th percentiles as cutpoints. For four bins, we would use the 25th, 50th, and 75th percentiles as cutpoints. We prefer classes that yield approximately equal frequencies for each cell to help protect against small expected frequencies (recall that Cochran's Rule requires expected frequencies be at least 5). Our bin choices may be limited if we have integer data with a small range (e.g., a Likert scale with responses 1, 2, 3, 4, 5), but we can still define classes however we wish (e.g., 1 or 2, 3, 4 or 5).

Example 15.3

Doctors and Infant Mortality 📁 **Doctors**

Let X = doctors per 100,000 residents of a state and Y = infant deaths per 1,000 births in the state. We might reasonably hypothesize that states with more doctors relative to population would have lower infant mortality, but do they? We are reluctant to assume normality and equal variances, so we prefer to avoid a t test. Instead, we hypothesize:

H_0: Infant mortality rate is independent of doctors per 100,000 population
H_1: Infant mortality rate is not independent of doctors per 100,000 population

Figure 15.6 shows a 3×3 table of observed (bold font) and expected frequencies assuming the null hypothesis. From the p-value (.7173), we conclude that doctors and infant mortality are not strongly related. However, a *multivariate* regression might reveal other covariates (e.g., per capita income, per capita Medicaid spending, percent of college graduates) that might be related to infant mortality in a state.

Figure 15.6

Cross-Tabulation of Numeric Data

3 × 3 Table
Doctors per 100,000

Infant Deaths per 1,000 Births		Low	Med	High	Total
Low	Obs	**4**	**6**	**6**	16
	Exp	5.44	5.44	5.12	
Med	Obs	**5**	**6**	**6**	17
	Exp	5.78	5.78	5.44	
High	Obs	**8**	**5**	**4**	17
	Exp	5.78	5.78	5.44	
Total		17	17	16	50

2.10 chi-square (*d.f.* = 4)
.7173 *p*-value

Why Do a Chi-Square Test on Numerical Data?

Why would anyone convert numerical data (X, Y) into categorical data in order to make a contingency table and do a chi-square test? Why not use the (X, Y) data to calculate a correlation coefficient or fit a regression? Here are three reasons:

- The researcher may believe there is a relationship between X and Y but does not want to make an assumption about its form (linear, curvilinear, etc.) as required in a regression.

- There are outliers or other anomalies that prevent us from assuming that the data came from a normal population. Unlike correlation and regression, the chi-square test does not require any normality assumptions.
- The researcher has numerical data for one variable but not the other. A chi-square test can be used if we convert the numerical variable into categories.

3-Way Tables and Higher

There is no conceptual reason to limit ourselves to two-way contingency tables comparing two variables. However, such tables become rather hard to visualize, even when they are "sliced" into a series of 2-way tables. A table comparing three variables can be visualized as a cube or as a stack of tiled 2-way contingency tables. Major computer packages (SAS, SPSS, and others) permit 3-way contingency tables. For four or more variables, there is no physical analog to aid us, and their cumbersome nature would suggest analytical methods other than chi-square tests.

Analytics in Action

Confusion Matrix for Machine Learning

A special type of cross-tabulation table is used in machine learning to evaluate the performance of algorithms that visually classify images. This table is called a *confusion matrix*. To test such an algorithm, the algorithm might be asked to classify images of chickens and mice into their correct categories. A count of correct and incorrect classifications would then be shown in a table such as the one below.

		Actual Class	
		Chicken	Mouse
Predicted Class	Chicken	6	3
	Mouse	2	5

The confusion matrix is used to determine how accurately an algorithm can classify images. The confusion matrix (sometimes called a *table of confusion*) is then analyzed using language that you are familiar with. False positives, or Type I errors, are one measure used to determine the effectiveness of the algorithm. Other terms introduced in this textbook include *sensitivity, power,* and *specificity.* An new open source Python package called *Fairlearn* (www. fairlearn.org/) can help assess a system's fairness using a confusion matrix and other tools to detect algorithm bias arising from race, gender, or other factors. With the increased use of facial recognition software, it is important that these algorithms get it right!

Section Exercises

connect

Instructions: For each exercise, include software results (e.g., from Excel, MegaStat, or JMP) to support your chi-square calculations. (a) State the hypotheses. (b) Show how the degrees of freedom are calculated for the contingency table. (c) Using the level of significance specified in the exercise, find the critical value of chi-square from Appendix E or from Excel's function =CHISQ.INV.RT(alpha, deg_freedom). (d) Carry out the calculations for a chi-square test for independence and draw a conclusion. (e) Which cells of the contingency table contribute the most to the chi-square test statistic? (f) Are any of the expected frequencies too small? (g) Interpret the *p*-value. If necessary, you can calculate the *p*-value using Excel's function =CHISQ.DIST.RT(test statistic, deg_freedom). *(h) If it is a 2×2 table, perform a two-tailed, two-sample *z* test for $\pi_1 = \pi_2$ and verify that z^2 is the same as your chi-square statistic. *Note:* Exercises marked with an asterisk (*) are more difficult.

15.1 A random sample of California residents who had recently visited a car dealership were asked which type of vehicle they were most likely to purchase, with the results shown.

Research question: At $\alpha = .10$, is the choice of vehicle type independent of the buyer's age?

📁 **CarBuyers**

Vehicle Type	Buyer's Age			Row Total
	Under 30	30 Under 50	50 and Over	
Diesel	5	10	15	30
Gasoline	15	30	45	90
Hybrid	25	25	25	75
Electric	15	15	15	45
Col Total	60	80	100	240

15.2   Running shoe companies design marketing strategies based on the factors that influence purchasing decisions. If these factors differ across different age groups, marketing strategies must be designed to target individual age groups. *Research question:* At $\alpha = .10$, does this sample show that factors influencing running shoe purchase are independent of age group? 📁 **Running**

Age Group	Factors Influencing Running Shoe Purchase			Row Total
	Shoe Characteristics	Other People	Price	
10–25 years (Gen Z)	37	44	46	127
26–41 years (Millennials)	48	30	36	114
42–57 years (Gen X)	15	26	18	59
Col Total	100	100	100	300

15.3   Students applying for admission to an MBA program must submit scores from the GMAT test, which includes a verbal and a quantitative component. Shown here are raw scores for 100 randomly chosen MBA applicants at a Midwestern, public, AACSB-accredited business school. *Research question:* At $\alpha = .005$, is the quantitative score independent of the verbal score? 📁 **GMAT**

Verbal	Quantitative			Row Total
	Under 25	25 to 34	35 or More	
Under 25	25	9	1	35
25 to 34	4	28	18	50
35 or More	1	3	11	15
Col Total	30	40	30	100

15.4   Is HDTV ownership related to quantity of purchases of other electronics? A Best Buy retail outlet collected the following data for a random sample of its recent customers. *Research question:* At $\alpha = .10$, is the frequency of in-store purchases independent of the number of large-screen HDTVs owned (defined as 50 inches or more)? 📁 **Purchases**

HDTVs Owned	In-Store Purchases Last Month			Row Total
	None	One	More Than One	
None	13	14	13	40
One	18	31	31	80
Two or More	19	45	66	130
Col Total	50	90	110	250

15.5   Marketing researchers sent an advance e-mail notice announcing an upcoming Internet survey and describing the purpose of their research. Half the target customers received the prenotification, followed by the survey. The other half received only the survey. The survey completion frequencies are shown below. *Research question:* At $\alpha = .025$, is completion rate independent of prenotification? 📂 **Advance**

Cross-Tabulation of Completion by Notification			
Pre-Notified?	Completed	Not Completed	Row Total
Yes	39	155	194
No	22	170	192
Col Total	61	325	386

Mini Case 15.1

Student Work and Car Age

Do students work longer hours to pay for newer cars? This hypothesis was tested using data from a survey of introductory business statistics students at a large commuter university campus. The survey contained these two fill-in-the-blank questions:

About how many hours per week do you expect to work at an outside job this semester?

What is the age (in years) of the car you usually drive?

The contingency table shown in Table 15.5 summarizes the responses of 162 students. Very few students worked less than 15 hours. Most drove cars less than 3 years old, although a few drove cars 10 years old or more. Neither variable was normally distributed (and there were outliers), so a chi-square test was preferable to a correlation or regression model. The hypotheses to be tested are

H_0: Car age is independent of work hours

H_1: Car age is not independent of work hours

Table 15.5

Frequency Classification for Work Hours and Car Age 📂 **CarAge**

	Age of Car Usually Driven				
Hours of Outside Work per Week	Less than 3	3 to <6	6 to <10	10 or More	Row Total
Under 15	9	8	8	4	29
15 to <25	34	17	11	9	71
25 or More	28	20	8	6	62
Col Total	71	45	27	19	162

Figure 15.7 shows MegaStat's analysis of the 3×4 contingency table. Two expected frequencies (upper right) are below 5, so Cochran's Rule is not quite met. MegaStat has highlighted these cells to call attention to this concern. But the most striking feature of this table is that almost all of the actual frequencies are very close to the frequencies expected under the hypothesis of independence, leading to a very small chi-square test statistic (5.24). The test requires six degrees of freedom, i.e., $(r - 1)(c - 1) = (3 - 1)(4 - 1) = 6$. From Appendix E we obtain the right-tail critical value $\chi^2_{.10} = 10.64$ at $\alpha = .10$ and *d.f.* = 6. Even at this rather weak level of significance, we cannot reject H_0. MegaStat's p-value (.5132) says that a test statistic of this magnitude could arise by chance more than half the time in samples from a population in which the two variables really were independent. The data lend no support to the hypothesis that work hours are related to car age.

Figure 15.7

MegaStat's Analysis of Car Age Data

Chi-square Contingency Table Test for Independence

		Less than 3	3 to <6	6 to <10	10 or More	Total
Under 15	Observed	**9**	**8**	**8**	**4**	29
	Expected	12.71	8.06	4.83	3.40	29.00
	O − E	−3.71	−0.06	3.17	0.60	0.00
	$(O - E)^2/E$	1.08	0.00	2.07	0.11	3.26
15 to <25	Observed	**34**	**17**	**11**	**9**	71
	Expected	31.12	19.72	11.83	8.33	71.00
	O − E	2.88	−2.72	−0.83	0.67	0.00
	$(O - E)^2/E$	0.27	0.38	0.06	0.05	0.76
25 or More	Observed	**28**	**20**	**8**	**6**	62
	Expected	27.17	17.22	10.33	7.27	62.00
	O − E	0.83	2.78	−2.33	−1.27	0.00
	$(O - E)^2/E$	0.03	0.45	0.53	0.22	1.22
Total	Observed	71	45	27	19	162
	Expected	71.00	45.00	27.00	19.00	162.00
	O − E	0.00	0.00	0.00	0.00	0.00
	$(O - E)^2/E$	1.38	0.82	2.66	0.38	5.24

5.24 chi-square

6 df

.5132 p-value

15.2 CHI-SQUARE TESTS FOR GOODNESS OF FIT

Purpose of the Test

A **goodness-of-fit test** (or GOF test) is used to help you decide whether your sample resembles a particular kind of population. The chi-square test can be used to compare sample frequencies with any probability distribution. Tests for goodness of fit are easy to understand, but until spreadsheets came along, the calculations were tedious. Today, computers make it easy, and tests for departure from normality or any other distribution are routine. We will first illustrate the GOF test using a general type of distribution. A **multinomial distribution** is defined by any k probabilities $\pi_1, \pi_2, \ldots, \pi_k$ that sum to one. You can apply this same technique for the three familiar distributions we have already studied (uniform, Poisson, and normal). Although there are many tests for goodness-of-fit, the chi-square test is attractive because it is versatile and easy to understand.

LO 15-4

Perform a goodness-of-fit (GOF) test for a multinomial distribution.

Multinomial GOF Test: M&M Colors

According to the "official" M&M website, the distribution of M&M colors is

Brown (13%) Red (13%) Blue (24%)
Orange (20%) Yellow (16%) Green (14%)

But do bags of M&Ms shipped to retailers actually follow this distribution? We will use a sample of four bags of candy and conduct a chi-square GOF test. We will assume the distribution is the same as stated on the website *unless the sample shows us otherwise.*

The hypotheses are

H_0: $\pi_{brown} = .13$, $\pi_{red} = .13$, $\pi_{blue} = .24$, $\pi_{orange} = .20$, $\pi_{yellow} = .16$, $\pi_{green} = .14$

H_1: At least one of the π's differs from the hypothesized value

To test these hypotheses, statistics students opened four bags of M&Ms ($n = 220$ pieces) and counted the number of each color, with the results shown in Table 15.6. We assign an index to each of the six colors ($j = 1, 2, \ldots, 6$) and define

f_j = the actual frequency of M&Ms of color j

e_j = the expected frequency of M&Ms of color j assuming that H_0 is true

Each expected frequency (e_j) is calculated by multiplying the sample size (n) by the hypothesized proportion (π_j). We can now calculate a chi-square test statistic that compares the actual and expected frequencies:

$$(15.4) \qquad \chi^2_{calc} = \sum_{j=1}^{c} \frac{[f_i - e_j]^2}{e_j}$$

Table 15.6

Hypothesis Test of M&M Proportions 📁 **MM**

Source: Official proportions (in 2006) are from https://www.mms.com/.

Color	Official π_j	Observed f_j	Expected $e_j = n \times \pi_j$	$f_j - e_j$	$(f_j - e_j)^2/e_j$
Brown	0.13	38	28.6	+9.4	3.0895
Red	0.13	30	28.6	+1.4	0.0685
Blue	0.24	44	52.8	−8.8	1.4667
Orange	0.20	52	44.0	+8.0	1.4545
Yellow	0.16	30	35.2	−5.2	0.7682
Green	0.14	26	30.8	−4.8	0.7481
Sum	1.00	220	220.0	0.0	$\chi^2_{calc} = 7.5955$

If the proposed distribution gives a good fit to the sample, the chi-square statistic will be near zero because f_j and e_j will be approximately equal. Conversely, if f_j and e_j differ greatly, the chi-square statistic will be large. It is always a right-tail test. We will reject H_0 if the test statistic exceeds the chi-square critical value chosen from Appendix E. For any GOF test, the rule for degrees of freedom is

$$(15.5) \qquad d.f. = c - m - 1$$

where c is the number of classes used in the test and m is the number of parameters estimated. Table 15.6 summarizes the calculations in a worksheet. No parameters were estimated ($m = 0$) and we have six classes ($c = 6$), so degrees of freedom are

$$d.f. = c - m - 1 = 6 - 0 - 1 = 5$$

From Appendix E, the critical value of chi-square for $\alpha = .01$ and $d.f. = 5$ is $\chi^2_{.01} = 15.09$. Because the test statistic $\chi^2_{calc} = 7.5955$ (from Table 15.6) is smaller than the critical value, we cannot reject the hypothesis that the M&M's color distribution is as stated on the M&M website. Notice that the f_j and e_j always sum exactly to the sample size ($n = 220$ in this example) and the differences $f_j - e_j$ must sum to zero. If not, you have made a mistake in your calculations—a useful way to check your work.

Excel also can be used to conduct a chi-square goodness-of-fit test. Enter the observed frequencies on an Excel spreadsheet and then calculate the expected frequencies. Insert the function =CHISQ.TEST(observed frequency range, expected frequency range) into a cell. Excel will return the p-value for your test. In our M&M example, Excel's =CHISQ.TEST returns the value .17998. Because .17998 > .01, our chosen value of α, we would fail to reject the null hypothesis. Using Excel's function =CHISQ.INV.RT(probability, degrees of freedom), we can determine the value of χ^2_{calc}. In our example, χ^2_{calc} =CHISQ.INV.RT(.17998,5) = 7.5995.

Small Expected Frequencies

Goodness-of-fit tests may lack power in small samples. Further, small expected frequencies tend to inflate the χ^2 test statistic because e_j is in the denominator of formula 15.4. The minimum necessary sample size depends on the type of test being employed. As a guideline, a chi-square goodness-of-fit test should be avoided if $n < 25$ (some experts would suggest a higher number). Cochran's Rule that expected frequencies should be at least 5 (i.e., all $e_j \geq 5$) also provides a guideline, although some experts would weaken the rule to require only $e_j \geq 2$. In the M&M example, the expected frequencies are all large, so there is no reason to doubt the test.

GOF Tests for Other Distributions

We also can use the chi-square GOF test to compare a sample of data with a familiar distribution such as the uniform, Poisson, or normal. We would state the hypotheses as below:

H_0: The population follows a _____ distribution.

H_1: The population doesn't follow a _____ distribution.

The blank may contain the name of any theoretical distribution. Assuming that we have n observations, we group the observations into c classes and then find the *chi-square test statistic* using formula 15.4. In a GOF test, if we use sample data to *estimate* the distribution's parameters, then our degrees of freedom would be as follows:

Uniform:	$d.f. = c - m - 1 = c - 0 - 1 = c - 1$ (no parameters are estimated)	**(15.6)**
Poisson:	$d.f. = c - m - 1 = c - 1 - 1 = c - 2$ (if λ is estimated)	**(15.7)**
Normal:	$d.f. = c - m - 1 = c - 2 - 1 = c - 3$ (if μ and σ are estimated)	**(15.8)**

Data-Generating Situations

"Fishing" for a good-fitting model is inappropriate. Instead, we visualize *a priori* the characteristics of the underlying *data-generating process*. Obtaining a good fit is not sufficient justification for assuming a particular model. Each probability distribution has its own logic about the nature of the underlying process. The proposed model should be both logical and empirically apt. The most common GOF test is for the normal distribution because so many parametric tests assume normality, and that assumption must be tested. Also, the normal distribution may be used as a benchmark for any mound-shaped data that have centrality and tapering tails, as long as you have reason to believe that a constant mean and variance would be reasonable (e.g., weights of circulated dimes). You would not consider a Poisson distribution for continuous data (e.g., gasoline price per liter) because a Poisson model only applies to integer data on arrivals or rare, independent events (e.g., number of paint defects per square meter). We remind you of this because software makes it possible to fit inappropriate distributions all too easily.

Mixtures: A Problem

Your sample may not resemble any known distribution. One common problem is *mixtures*. A sample may have been created by more than one data-generating process superimposed on top of another. For example, adult heights of either sex would follow a normal distribution, but a combined sample of both genders will be bimodal, and its mean and standard deviation may be unrepresentative of either sex.

Eyeball Tests

A simple "eyeball" inspection of the histogram or dot plot may suffice to rule out a hypothesized population. For example, if the sample is strongly bimodal or skewed or if outliers are present, we would anticipate a poor fit to a normal distribution. The shape of the histogram can give you a rough idea whether a normal distribution is a likely candidate for a good fit. You can be fairly sure that a formal test will agree with what your common sense tells you, as long as the sample size is not too small. A limitation of eyeball tests is that we are sometimes unduly impressed by a small departure from the hypothesized distribution, when actually it is within chance.

15.6 In the second quarter of 2022, the U.S. smartphone market was dominated by Samsung at 48 percent and Apple at 30 percent. Other brands (such as Lenovo and Google's OnePlus) made up 22 percent of the market. Does smartphone brand ownership by college students follow this national pattern? A sample of 148 student smartphone owners showed that 63 owned a Samsung smartphone, 56 owned an Apple smartphone, and 29 owned a smartphone that was neither a Samsung nor an Apple. Conduct a chi-square GOF test to answer this research question.

15.7 Market research has shown that Americans will continue to eat out even in a depressed economy. A market analysis during the most recent recession showed the following distribution of visits to the various types of restaurants:

Fast food (36%) Quick casual (29%) Casual dining (15%)

Family style (15%) Fine dining (5%)

A new survey of 250 Americans reported the following types of visits on their last restaurant visit: 95 visited a fast-food establishment, 68 visited a quick-casual restaurant, 38 visited a casual-dining restaurant, 30 visited a family-style restaurant, and 19 visited a fine-dining establishment. Does this sample indicate that the distribution of type of restaurant visited has changed since the market analysis shown above was conducted? Conduct a chi-square GOF test to answer this question.

15.3 UNIFORM GOODNESS-OF-FIT TEST

LO 15-5

Perform a GOF test for a uniform distribution.

The uniform goodness-of-fit test is a special case of the multinomial in which every value has the same chance of occurrence. Uniform data-generating situations are rare, but some data must be from a **uniform distribution**, such as winning lottery numbers or random digits generated by a computer for random sampling. Another use of the uniform distribution is as a worst case scenario for an unknown distribution whose range is specified in a what-if analysis.

The chi-square test for a uniform distribution is a generalization of the test for equality of two proportions. The hypotheses are

$H_0: \pi_1 = \pi_2 = \cdots = \pi_c = 1/c$

$H_1:$ Not all the π_j are equal

The chi-square test compares all c groups *simultaneously*. Each discrete outcome should have probability $1/c$, so the test is very easy to perform. Evidence against H_0 would consist of sample frequencies that were not the same for all categories.

Classes need not represent numerical values. For example, we might compare the total number of items scanned per hour by four supermarket checkers (Bob, Frieda, Sam, and Wanda). The uniform test is quite versatile. For numerical variables, bins do not have to be of equal width and can be open-ended. For example, we might be interested in the ages of x-ray machines in a hospital (under 2 years, 2 to <5 years, 5 to <10 years, 10 years and over). In a uniform population, each category would be expected to have $e_j = n/c$ observations, so the calculation of expected frequencies is simple.

Uniform GOF Test: Grouped Data

The test is easiest if data are already tabulated into groups, which saves us the effort of defining the groups. For example, one year, a certain state had 756 traffic fatalities. Table 15.7 suggests that fatalities are not uniformly distributed by day of week, being higher on weekends. Can we reject the hypothesis of a uniform distribution, say, at $\alpha = .005$? The hypotheses are

$H_0:$ Traffic fatalities are uniformly distributed by day of the week

$H_1:$ Traffic fatalities are not uniformly distributed by day of the week

Day	f_j	e_j	$f_j - e_j$	$(f_j - e_j)^2$	$(f_j - e_j)^2/e_j$
Sun	121	108	13	169	1.565
Mon	96	108	−12	144	1.333
Tue	91	108	−17	289	2.676
Wed	92	108	−16	256	2.370
Thu	96	108	−12	144	1.333
Fri	122	108	14	196	1.815
Sat	138	108	30	900	8.333
Total	756	756	0		$\chi^2_{calc} = 19.426$

Under H_0 the expected frequency for each weekday is $e_j = n/c = 756/7 = 108$. The expected frequencies happen to be integers, although this is not true in general. Because no parameters were estimated ($m = 0$) to form the seven classes ($c = 7$), the chi-square test will have *d.f.* = $c - m - 1 = 7 - 0 - 1 = 6$ degrees of freedom. From Appendix E the critical value of chi-square for the 1 percent level of significance is $\chi^2_{.01} = 16.81$, so the hypothesis of a rectangular or uniform population can be rejected. The *p*-value (.0035) can be obtained from the Excel function =CHISQ.DIST.RT(19.426,6). The *p*-value tells us that such a sample result would occur by chance only about 35 times in 10,000 samples. There is a believable underlying causal mechanism at work (e.g., people may drink and drive more often on weekends).

Uniform GOF Test: Raw Data

When we are using raw data, we must form *c* bins of equal width and create our own frequency distribution. For example, suppose an auditor is checking the fairness of a state's "Daily 3" lottery. Table 15.8 shows winning three-digit lottery numbers for 100 consecutive days. All numbers from 000 to 999 are supposed to be equally likely, so the auditor is testing these hypotheses:

H_0: Lottery numbers are uniformly distributed

H_1: Lottery numbers are not uniformly distributed

367	865	438	437	596	567	121	244	036	337
152	260	470	821	452	606	417	674	786	311
739	611	359	739	184	229	418	565	547	403
103	344	303	531	054	496	167	550	403	785
341	237	913	991	656	661	178	983	431	472
315	792	676	299	738	080	450	991	673	846
500	001	016	581	154	677	457	617	261	807
452	048	052	018	037	517	760	522	711	898
294	605	135	333	886	257	533	119	882	899
814	490	490	885	329	033	033	707	551	651

We know that three-digit lottery numbers must lie in the range 000 to 999, so there are many ways we could define our classes (e.g., 5 bins of width 200, 10 bins of width 100, 20 bins of width 50). We will use 10 bins, with the realization that we might get a different result if we chose different bins. The steps are

- Step 1: Divide the range into 10 bins of equal width.
- Step 2: Calculate the observed frequency f_j for each bin.
- Step 3: Define $e_j = n/c = 100/10 = 10$.
- Step 4: Perform the chi-square calculations (see Table 15.9).
- Step 5: Make the decision.

Table 15.9

Uniform GOF Test for Lottery Numbers

Bin	f_j	e_j	$f_j - e_j$	$(f_j - e_j)^2$	$(f_j - e_j)^2/e_j$
0 < 100	11	10	1	1	0.100
100 < 200	9	10	−1	1	0.100
200 < 300	8	10	−2	4	0.400
300 < 400	10	10	0	0	0.000
400 < 500	16	10	6	36	3.600
500 < 600	12	10	2	4	0.400
600 < 700	11	10	1	1	0.100
700 < 800	9	10	−1	1	0.100
800 < 900	10	10	0	0	0.000
900 < 1,000	4	10	−6	36	3.600
Total	100	100	0		$\chi^2_{calc} = 8.400$

Because no parameters were estimated ($m = 0$) to form the 10 classes ($c = 10$), we have $d.f. = c - m - 1 = 10 - 0 - 1 = 9$ degrees of freedom. From Appendix E the critical value of chi-square for the 10 percent level of significance is $\chi^2_{.10} = 14.684$. Most of the chi-square sum arises from two bins: 400 < 500 (more than expected) and 900 < 1,000 (fewer than expected). Because the test statistic is 8.400, the hypothesis of a uniform distribution cannot be rejected.

If the bin limits cannot be set using *a priori* knowledge (as was possible in the lottery example), we maximize the test's power by defining bin width as the range divided by the number of classes:

(15.9) $\qquad \text{Bin Width} = \dfrac{x_{max} - x_{min}}{c} \qquad$ (setting bin width from sample data)

The resulting bin limits may not be aesthetically pleasing, but the expected frequencies will be as large as possible. (You might be able to round the bin limits to a "nice" number without affecting the calculations very much.) If the sample size is small, small expected frequencies could be a problem. Because all expected frequencies are the same in a uniform model, this problem will exist in all classes simultaneously. For example, we could not classify 25 observations into 10 classes without violating Cochran's Rule.

The histogram in Figure 15.8 suggests too many winning lottery numbers in the middle and too few at the top. But histogram appearance is affected by the way we define our bins and the number of classes, so the chi-square test is a more reliable guide. Humans are adept at finding patterns in sample distributions that actually are within the realm of chance.

Figure 15.8

Ten-Bin Histogram

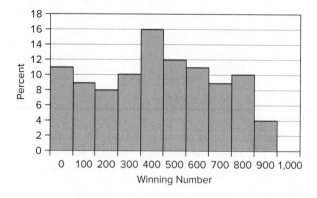

As you learned in Chapter 6, a discrete uniform distribution $U(a,b)$ is symmetric with mean $\mu = (a + b)/2$ and $\sigma = \sqrt{[(b - a + 1)^2 - 1]/12}$. For the lottery, we have $a = 000$ and $b = 999$, so we expect the mean to be $\mu = (0 + 999)/2 = 499.5$ and $\sigma = \sqrt{[(999 - 0 + 1)^2 - 1]/12} = 288.7$. For the sample, Table 15.10 shows that the low (001) and high (991) are near their theoretical values, as are the sample mean (472.2), standard deviation (271.0), and skewness coefficient (.01). The first quartile (260.5) is close to its expected value (.25 × 999 = 249.8), while the third quartile (675.5) is smaller than expected (.75 × 999 = 749.3).

Table 15.10

Descriptive Statistics
📁 **Lottery-A**

Statistic	Sample	If Uniform
Minimum	001	000
Maximum	991	999
Mean	472.2	499.5
Median	471.0	499.5
Standard Deviation	271.0	288.7
Quartile 1	260.5	249.8
Quartile 3	675.5	749.3
Skewness	0.01	0.00

Because the data are not skewed (mean $\approx$ median) and the sample size is large ($n \geq 30$), the mean is approximately normally distributed, so we can use the normal distribution to test the sample mean for a significant difference from the hypothesized uniform mean, assuming that $\sigma = 288.7$ as would be true if the data were uniform:

$$z = \frac{\bar{x} - \mu}{\frac{\sigma}{\sqrt{n}}} = \frac{472.2 - 499.5}{\frac{288.7}{\sqrt{100}}} = -0.95 \qquad \text{(two-tail } p\text{-value} = .34)$$

The difference is not significant at any common level of α. Overall, these statistics show no convincing evidence of departure from a uniform distribution, thereby confirming the chi-square test's conclusion, that is, the lottery is fair.

15.8 Online retailers often offer discounts to entice first-time visitors to their website to make a purchase. In addition to knowing if the discounts are effective, it is also important for the online retailers to know if the decision to make a purchase for the first time differs by age. One online retailer collected data from 120 first-time purchasers. Their ages are categorized below. (a) Make a bar chart and describe it. (b) Calculate expected frequencies for each class. (c) Perform the chi-square test for a uniform distribution. At $\alpha = .01$, does this sample contradict the assumption that first-time purchases are uniformly distributed among these six age groups? 📁 **FirstTimePurchases**

Purchaser Age	Number of Purchasers
18–24	38
25–34	28
35–44	19
45–54	16
55–64	10
65+	9
Total	120

15.9 One-year sales volume of four similar 20-oz. beverages on a college campus is shown. (a) Make a bar chart and describe it. (b) Calculate expected frequencies for each class. (c) Perform the chi-square test for a uniform distribution. At $\alpha = .05$, does this sample contradict the assumption that sales are the same for each beverage? 📁 **Frapp**

Beverage	Sales (Cases)
Frappuccino Coffee	18
Frappuccino Mocha	23
Frappuccino Vanilla	23
Frappuccino Caramel	20
Total	84

15.10 In a three-digit lottery, each of the three digits is supposed to have the same probability of occurrence (counting initial blanks as zeros, e.g., 32 is treated as 032). The table shows the frequency of occurrence of each digit for 90 consecutive daily three-digit drawings. (a) Make a bar chart and describe it. (b) Calculate expected frequencies for each class. (c) Perform the chi-square test for a uniform distribution. At $\alpha = .05$, can you reject the hypothesis that the digits are from a uniform population? **Lottery3**

Digit	Frequency
0	33
1	17
2	25
3	30
4	31
5	28
6	24
7	25
8	32
9	25
Total	270

15.11 Ages of 56 attendees of a recent Avengers movie are shown. (a) Form seven age classes (10 to 20, 20 to 30, etc.). Tabulate the frequency of attendees in each class. (b) Calculate expected frequencies for each class. (c) Perform a chi-square GOF test for a uniform distribution, using the 5 percent level of significance. **Avengers**

10	22	58	11	73	22	57
35	33	33	59	54	55	75
79	24	13	73	52	69	30
71	64	17	50	72	67	50
72	35	26	59	47	65	35
64	34	39	66	37	41	58
51	43	29	74	73	50	62
58	34	50	27	13	67	67

15.4 POISSON GOODNESS-OF-FIT TEST

Poisson Data-Generating Situations

LO 15-6

Explain the GOF test for a Poisson distribution.

In a **Poisson distribution model**, X represents the number of events per unit of time or space. By definition, X is a discrete nonnegative random variable with integer values $(0, 1, 2, \ldots)$. Event arrivals must be independent of one another. Events that tend to fit this definition might include customer arrivals per minute at an ATM, calls per minute at Ticketmaster, or alarms per hour at a fire station. In such cases, the mean arrival rate would vary by time of day, day of the week, and so on. The Poisson has been demonstrated to apply to scores in some sports events (goals scored per soccer game, goals in hockey games) and to defects in manufactured components such as LCDs, printed circuits, and automobile paint jobs. Typically X has a fairly small mean, which is why the Poisson is sometimes called a model of *rare events*. If the mean is large, we might fit a normal distribution instead. Poisson random number generators are used by researchers who model queues, an important application in dense urban cultures. The Poisson distribution is inappropriate for noninteger data or financial data such as you would find in company annual reports. Remembering these facts can spare you from wasted time trying to fit a Poisson model when it is inappropriate.

Poisson Goodness-of-Fit Test

A Poisson model is completely described by its one parameter, the mean λ. Assuming that λ is unknown and must be estimated from the sample, the initial steps are

- Step 1: Tally the observed frequency f_j of each x-value.
- Step 2: Estimate the mean λ from the sample.

- Step 3: Use the estimated λ to find the Poisson probability $P(X = x)$ for each x-value.
- Step 4: Multiply $P(X = x)$ by the sample size n to get expected Poisson frequencies e_j.
- Step 5: Perform the chi-square calculations.
- Step 6: Make the decision.

If the data are already tabulated, we can skip the first step. A Poisson test always has an open-ended class on the high end because technically X has no upper limit. Unfortunately, Poisson tail probabilities are very small, and so will be the corresponding expected frequencies. But classes can be combined from each end inward until expected frequencies become large enough for the test (at least until $e_j \geq 2$). Combining classes implies using fewer classes than you would wish, but a more detailed breakdown isn't justified unless the sample is very large.

Poisson GOF Test: Tabulated Data

The number of U.S. Supreme Court appointments in a given year might be hypothesized to be a Poisson variable because rare events that occur independently over time are often well approximated by the Poisson model. We formulate these hypotheses:

H_0: Supreme Court appointments follow a Poisson distribution

H_1: Supreme Court appointments do not follow a Poisson distribution

The frequency of U.S. Supreme Court appointments for the period 1900 through 1999 is summarized in Table 15.11. This sample of 100 years should be large enough to obtain a valid hypothesis test. In a typical year, there are no appointments, and only twice have there been three or four appointments (1910 and 1941).

x	f_j	$x_j f_j$
0	59	0
1	31	31
2	8	16
3	1	3
4	1	4
Total	100	54

Table 15.11

Number of Annual U.S. Supreme Court Appointments, 1900–1999 📁 **Supreme**

Source: www.wikipedia.org.

The total number of appointments is

$$\sum_{j=1}^{c} x_j f_j = (0)(59) + (1)(31) + (2)(8) + (3)(1) + 4(1) = 54$$

so the sample mean is

$$\hat{\lambda} = \frac{54}{100} = 0.54 \text{ appointment per year}$$

Using the estimated mean $\hat{\lambda} = 0.54$, we can calculate the Poisson probabilities, either by using the Poisson formula $P(X = x) = \lambda^x e^{-\lambda}/x!$ or Excel's function =POISSON.DIST(x, mean, 0). We multiply $P(X = x)$ by n to get the expected frequencies, with $n = 100$ years, as shown in Table 15.12.

x	P(X = x)	$e_j = nP(X = x)$
0	0.58275	100 × 0.58275 = 58.275
1	0.31468	100 × 0.31468 = 31.468
2	0.08496	100 × 0.08496 = 8.496
3	0.01529	100 × 0.01529 = 1.529
4	0.00206	100 × 0.00206 = 0.206
5	0.00022	100 × 0.00022 = 0.022
6 or more	0.00004	100 × 0.00004 = 0.004
Sum	1.00000	100.00

Table 15.12

Fitted Poisson Probabilities 📁 **Supreme**

The probabilities rapidly become small as X increases. To ensure that $e_j \geq 2$, it is necessary to combine the top classes to end up with only three classes, the top class being "2 or more," before doing the chi-square calculations shown in Table 15.13. Because f_j and e_j are almost identical, the Poisson distribution obviously gives an excellent fit, so we are not surprised that the test statistic (0.022) is very near zero.

Table 15.13

Chi-Square Test for Supreme Court Data

x	f_j	e_j	$f_j - e_j$	$(f_j - e_j)^2$	$(f_j - e_j)^2/e_j$
0	59	58.275	0.725	0.525625	0.009
1	31	31.468	−0.468	0.219024	0.007
2 or more	10	10.257	−0.257	0.066049	0.006
Total	100	100.000	0.000		$\chi^2_{calc} = 0.022$

Using $c = 3$ classes in the test and with $m = 1$ parameter estimated, the degrees of freedom are $c - m - 1 = 3 - 1 - 1 = 1$. From Appendix E, we see that the critical value for $\alpha = .10$ is $\chi^2_{.10} = 2.706$, so we clearly cannot reject the hypothesis of a Poisson distribution, even at a modest level of significance. Excel's function =CHISQ.DIST.RT(.022,1) gives the p-value .882, which indicates an excellent fit. Although we can never *prove* that the annual U.S. Supreme Court appointments follow a Poisson distribution, the Poisson distribution fits the sample well, as shown in the graph in Figure 15.9.

Figure 15.9

Supreme Court Appointments Poisson GOF Test

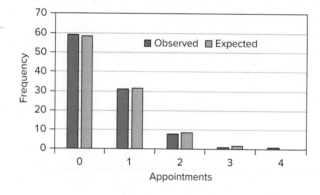

Generating Random Poisson Data

Figure 15.10 shows the menu from Excel's Data Analysis > Random Number Generation, which was used to create 100 Poisson random numbers with a mean of $\lambda = 4.0$. Generating random numbers from known distributions is useful when simulating business processes. Simulation will be discussed further in Chapter 18, found online at this textbook's website.

Figure 15.10

Excel's Poisson Random Number Generator

Source: Microsoft Excel.

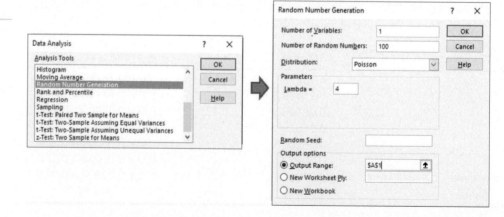

15.12 Excel was asked to generate 50 Poisson random numbers with mean $\lambda = 5$. (a) Calculate the sample mean. How close is it to the desired value? (b) Calculate the expected frequencies assuming a Poisson distribution with $\lambda = 5$. Show your calculations in a spreadsheet format. (c) Carry out the chi-square test at $\alpha = .05$, combining end categories as needed to ensure that all expected frequencies are at least five. Show your degrees of freedom calculation. (d) Do you think your calculations would have been materially different if you had used the sample mean instead of $\lambda = 5.0$? Explain.   **RandPois**

x	Frequency
0	1
1	1
2	4
3	7
4	7
5	8
6	10
7	3
8	3
9	4
10	1
11	1

15.13 During the 2021–22 hockey season, the Boston Bruins played 42 home games and scored 134 points, as shown below. (a) Estimate the mean from the sample. (b) Calculate the expected frequencies assuming a Poisson distribution. Show your calculations in a spreadsheet format. (c) Carry out the chi-square test, combining end categories as needed to ensure that all expected frequencies are at least five. Show your degrees of freedom calculation. (d) At $\alpha = .05$, can you reject the hypothesis that goals per game follow a Poisson process?   **Boston**

Number of Goals Scored (per game) by Boston Bruins, 2021–22

	0	1	2	3	4	5	6	7	8	Total
Frequency	2	4	9	10	7	8	1	0	1	42

15.14 At a local supermarket receiving dock, the number of truck arrivals per day is recorded for 100 days. (a) Estimate the mean from the sample. (b) Calculate the expected frequencies assuming a Poisson distribution. Show your calculations in a spreadsheet format. (c) Carry out the chi-square test, combining end categories as needed to ensure that all expected frequencies are at least five. Show your degrees of freedom calculation. (d) At $\alpha = .05$, can you reject the hypothesis that arrivals per day follow a Poisson process?   **Trucks**

Arrivals per Day at a Loading Dock

	0	1	2	3	4	5	6	7	Total
Frequency	4	23	28	22	8	9	4	2	100

15.5 NORMAL CHI-SQUARE GOODNESS-OF-FIT TEST

LO 15-7

Explain the chi-square GOF test for normality.

Normal Data-Generating Situations

Any normal population is fully described by the two parameters μ and σ. Many data-generating situations could be compatible with a **normal distribution** if the data possess a reasonable degree of central tendency and are not badly skewed. For example, measurements of continuous variables such as physical attributes (e.g., weight, size, travel time) would tend to have a constant mean and variance if the underlying process is stable and the population is homogeneous. The normal model might apply to discrete or integer data if the range is relatively large, such as the number of successes in a large binomial sample or Poisson occurrences if the mean is large. Unless μ and σ parameters are known *a priori* (a rare circumstance), they must be estimated from a sample by using $\bar{x}$ and s.

Normal Goodness-of-Fit Test

There are several ways we could perform a formal chi-square goodness-of-fit test for normality. The chi-square test requires c bins. The most efficient method is to define histogram bins in such a way that an equal number of observations would be *expected* within each bin under the null hypothesis. We define bin limits so that

(15.10) $\qquad e_j = n/c \quad$ (define bins to get equal expected frequencies)

We divide the area under the normal curve to put a probability of $1/c$ in each of the c bins. The first and last classes must be open-ended for a normal distribution, so to define c bins, we need $c - 1$ cutpoints. The upper limit of bin j can be found directly by using Excel's function =NORM.INV(j/c, $\bar{x}$, s). Alternatively, we can find z_j for bin j with Excel's normal function =NORM.INV(j/c, 0, 1) and then calculate the upper limit for bin j as $\bar{x} + z_j s$. Table 15.14 shows some typical z-values to put an area of $1/c$ in each bin.

Table 15.14

Standard Normal Cutpoints for Equal Area Bins

Bin	3 Bins	4 Bins	5 Bins	6 Bins	7 Bins	8 Bins
1	−0.431	−0.675	−0.842	−0.967	−1.068	−1.150
2	0.431	0.000	−0.253	−0.431	−0.566	−0.675
3		0.675	0.253	0.000	−0.180	−0.319
4			0.842	0.431	0.180	0.000
5				0.967	0.566	0.319
6					1.068	0.675
						1.150

Once the bins are defined, we count the observations f_j within each bin and compare them with the expected frequencies $e_j = n/c$. Although the bin limits will not be "nice," the compelling advantage of this method is that it guarantees the largest possible expected frequencies, and hence the most powerful test for c bins. MegaStat uses this method (Descriptive Statistics > Normal Curve Goodness of Fit) using the number of bins suggested by Sturges' Rule, $k = 1 + 3.3\log_{10}(n)$.

Application: Quality Management

A sample of 35 Hershey's Milk Chocolate Kisses was taken from a bag containing 84 Kisses. The population is assumed infinite. After removing the foil wrapper, each Kiss was weighed. The weights are shown in Table 15.15. Are these weights from a normal population?

4.666	4.854	4.868	4.849	4.700	4.683	5.064
4.800	4.694	4.760	5.075	4.780	4.781	5.103
4.568	4.983	5.076	4.808	5.084	4.749	5.092
4.783	4.520	4.698	5.084	4.880	4.883	4.880
4.928	4.651	4.797	4.682	4.756	5.041	4.906

Table 15.15

Weights of 35 Hershey's Milk Chocolate Kisses (in grams) 📁 **Kisses**

Source: An independent project by MBA student Frances Williams. Kisses were weighed on an American Scientific Model S/P 120 analytical balance accurate to 0.0001 gm.

It might be supposed *a priori* that Kiss weights would be normally distributed because the manufacturing process should have a single, constant mean and standard deviation. Variation is inevitable in any manufacturing process. Chocolate is especially difficult to handle because liquid chocolate must be dropped in precisely measured amounts, solidified, wrapped, and bagged. Because chocolate is soft and crumbles easily, even the process of weighing the Kisses may abrade some chocolate and introduce measurement error. We will test the following hypotheses:

H_0: Kisses' weights are from a normal distribution

H_1: Kisses' weights are not from a normal distribution

Before undertaking a GOF test, consider the histograms in Figure 15.11. The graphs show fitted normal distributions based on the estimated mean and standard deviation from the data. Although it is only a visual aid, the fitted normal gives you a clue as to the likely outcome of the test. The histograms reveal no apparent outliers, and nothing in conflict with the idea of a normal population except a second mode toward the high end of the scale and perhaps a flatter appearance than normal. Because histogram appearance can vary, depending on the number of classes and the way the bin limits are specified, further tests are needed.

Figure 15.11

Histograms of 35 Hershey's Kiss Weights

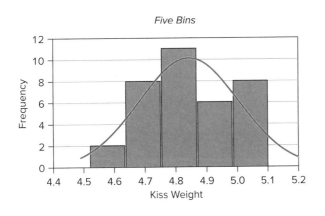
Five Bins

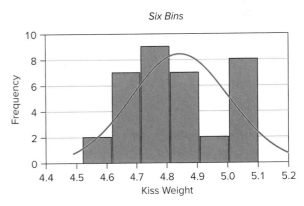
Six Bins

Table 15.16 compares the sample statistics with parameters expected for a normal distribution. The sample median (4.808) is near the sample mean (4.844), and the skewness coefficient (0.14) is fairly close to the value (0.00) that would be expected in a symmetric normal distribution. The sample quartiles (4.700 and 4.983) are nearly what we expect for a normal distribution using the 25th and 75th percentiles ($\bar{x} \pm 0.675s$). There are no outliers, as the smallest Kiss (4.520 grams) is 2.01 standard deviations below the mean, while the largest Kiss (5.103 grams) is 1.61 standard deviations above the mean.

Table 15.16

Sample versus Normal
📁 **Kisses**

Statistic	Kiss Weight	If Normal
Mean	4.844	4.844
Standard Deviation	0.161	0.161
Quartile 1	4.700	4.735
Median	4.808	4.844
Quartile 3	4.983	4.952
Skewness	0.14	0.00

For a chi-square GOF test, degrees of freedom are $d.f. = c - m - 1$, where c is the number of classes used in the test and m is the number of parameters estimated. Because two parameters, μ and σ, are estimated from the sample of Kiss weights, we use $m = 2$. We need at least four bins to ensure at least 1 degree of freedom, while Cochran's Rule (at least 5 *expected* observations per bin) suggests a maximum of 7 bins for $n = 35$ data points (because $35/7 = 5$). It therefore seems reasonable to use 6 bins for our GOF test with $d.f. = c - m - 1 = 6 - 2 - 1 = 3$.

The GOF test in Table 15.17 shows that the chi-square test statistic (3.571) is not significant at $\alpha = .10$ ($\chi^2_{.10} = 6.251$). The p-value =CHISQ.DIST.RT(3.571,3) = .312 indicates that such a result would be expected about 312 times in 1,000 samples if the population were normal. Bin five (highlighted) contributes heavily to the chi-square statistic.

Table 15.17

Chi-Square Test for Normality of 35 Kiss Weights

x	f_j	e_j	$f_j - e_j$	$(f_j - e_j)^2$	$(f_j - e_j)^2/e_j$
Under 4.688	6	5.8333	0.1667	0.0278	0.0048
4.688 < 4.774	6	5.8333	0.1667	0.0278	0.0048
4.774 < 4.844	6	5.8333	0.1667	0.0278	0.0048
4.884 < 4.913	7	5.8333	1.1667	1.3611	0.2333
4.913 < 4.999	2	5.8333	−3.8333	14.6944	2.5190
4.999 or more	8	5.8333	2.1667	4.6944	0.8048
Total	35	35.0000	0.0000		$\chi^2_{calc} = 3.571$

The chi-square test fails to reject the hypothesis of normality at $\alpha = .10$. However, the histograms do hint at a bimodal shape. This could occur if the Kisses were molded by two or more different machines. If each machine has a different μ and σ, this could lead to the "mixture of distributions" problem mentioned earlier. This issue bears further investigation. A quality control analyst would probably take a larger sample and study the manufacturing methods to see what could be learned.

Section Exercises

McGraw Hill **connect**

Hint: Check your work using MegaStat's Descriptive Statistics > Normal curve goodness of fit test or a computer package such as Minitab, JMP, or SPSS.

15.15 Exam scores of 40 students in a statistics class are shown. (a) Estimate the mean and standard deviation from the sample. (b) Assuming that the data are from a normal distribution, define bins by using method 3 (equal expected frequencies). Use 8 bins. (c) Set up an Excel worksheet for your chi-square calculations, with a column showing the expected frequency for each bin (they must add to 40). (d) Tabulate the observed frequency for each bin and record it in the next column. (e) Carry out the chi-square test, using $\alpha = .05$. Can you reject the hypothesis that the exam scores came from a normal population? 📁 **ExamScores**

79	75	77	57	81	70	83	66
81	89	59	83	75	60	96	86
78	76	71	78	78	70	54	60
71	81	79	88	77	82	75	68
77	69	83	79	79	76	78	71

15.16 One Friday night, there were 42 carry-out orders at Ashoka Curry Express. (a) Estimate the mean and standard deviation from the sample. (b) Assuming that the data are from a normal distribution, define bins by using method 3 (equal expected frequencies). Use 8 bins. (c) Set up an Excel worksheet for your chi-square calculations, with a column showing the expected frequency for each bin (they must add to 42). (d) Tabulate the observed frequency for each bin and record it in the next column. (e) Do the chi-square test at $\alpha = .025$. Can you reject the hypothesis that carry-out orders follow a normal population? TakeOut

18.74	21.05	31.19	23.06	20.17	25.12	24.30
46.04	33.96	45.04	34.63	35.24	30.13	29.93
52.33	26.52	19.68	19.62	32.96	42.07	47.82
38.62	31.88	44.97	36.35	21.50	41.42	33.87
26.43	35.28	21.88	24.80	27.49	18.30	44.47
28.40	36.72	26.30	47.08	34.33	13.15	15.51

15.6 ECDF TESTS (OPTIONAL)

There are alternatives to the chi-square test for goodness of fit. These alternatives are based on the **empirical cumulative distribution function (ECDF)**. The **Anderson-Darling test** is perhaps the most widely used test for non-normality because of its power. It is done on a computer because it requires the inverse CDF for the hypothesized distribution. The A-D test is based on a **probability plot**. When the data fit the hypothesized distribution closely, the probability plot will be close to a straight line. The A-D test statistic measures the overall distance between the actual and the hypothesized distributions using a weighted squared distance. It provides a *p*-value to complement the visual plot. The A-D statistic is not difficult to calculate, but its formula is rather complex, so it is omitted.

Figure 15.12 shows a graph displaying the probability plot and A-D statistic for the Hershey's Kiss data using Minitab's Stats > Basic Statistics > Normality Test. The *p*-value (.091) suggests a departure from normality at the 10 percent level of significance, but not at the 5 percent level. This result is consistent with our previous findings. The A-D test is more powerful than a chi-square test if raw data are available because it treats the observations individually. The probability plot has the attraction of visually revealing discrepancies between the sample and the hypothesized distribution, and it is usually easy to spot outliers.

LO 15-8

Interpret ECDF tests and know their advantages.

Figure 15.12

ECDF Tests for Normality of Kiss Weights   Kisses

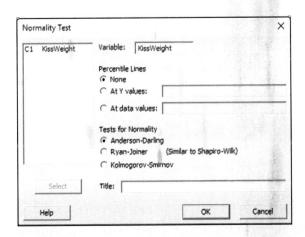

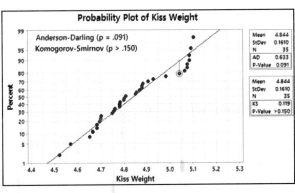

Another useful ECDF test is the **Kolmogorov-Smirnov test**. The K-S test statistic D is the largest absolute difference between the actual and expected cumulative relative frequency of the n data values:

$$(15.11) \qquad\qquad D = \text{Max}|F_a - F_e|$$

F_a is the actual cumulative frequency at observation i and F_e is the expected cumulative frequency at observation i under the assumption that the data came from the hypothesized distribution. The K-S test can be illustrated in the same probability plot as the A-D test (Figure 15.12). The largest difference ($D = .119$, shown by the red vertical line) occurs at observation 28 (circled). The K-S test result ($p > .150$) does not warrant rejection of the hypothesis of normality at $\alpha = .05$. We rely on software for the K-S test to calculate the inverse CDF for the hypothesized distribution as well as special tables for the p-value.

The K-S test is not recommended for grouped data, as it may be less powerful than the chi-square test. The K-S test assumes that no parameters are estimated. If the mean and variance were estimated, we would use a **Lilliefors test**, whose test statistic is the same but with a different table of critical values. Because these tests are done by computer, we will omit further details.

Section Exercises

Mc Graw Hill **connect**

*15.17 Use Minitab's Stat > Basic Statistics > Normality Test or other software to obtain a probability plot for the exam score data (see Exercise 15.15). Interpret the probability plot and Anderson-Darling statistic. 🖼 **ExamScores**

*15.18 Use Minitab's Stat > Basic Statistics > Normality Test or other software to obtain a probability plot for the Ashoka Curry House carry-out order data (see Exercise 15.16). Interpret the probability plot and Anderson-Darling statistic. 🖼 **TakeOut**

Chapter Summary

A **chi-square test for independence** requires an $r \times c$ **contingency table** that has r rows and c columns. Degrees of freedom for the chi-square test will be $(r - 1)(c - 1)$. In this test, the **observed frequencies** are compared with the **expected frequencies** under the hypothesis of independence. The test assumes categorical data (attribute data) but also can be used with numerical data grouped into classes. **Cochran's Rule** requires that expected frequencies be at least 5 in each cell, although this rule is often relaxed. A test for **goodness of fit (GOF)** uses the chi-square statistic to decide whether a sample is from a specified distribution (e.g., multinomial,

uniform, Poisson, normal). The **parameters** of the fitted distribution (e.g., the mean) may be specified *a priori,* but more often are estimated from the sample. Degrees of freedom for the GOF test are $c - m - 1$, where c is the number of categories and m is the number of parameters estimated. The **Kolmogorov-Smirnov** and **Lilliefors** tests are **ECDF-based tests** that look at differences between the sample's empirical cumulative distribution function (ECDF) and the hypothesized distribution. They are best used with n individual observations. The **Anderson-Darling** test and the **probability plot** are the most common ECDF tests, most often used to test for normality.

Key Terms

Anderson-Darling test	contingency table	goodness-of-fit test	observed frequency
chi-square probability distribution	degrees of freedom	independent	Poisson distribution model
	empirical cumulative	Kolmogorov-Smirnov test	probability plot
chi-square test for independence	distribution function (ECDF)	Lilliefors test	uniform distribution
Cochran's Rule	expected frequency	multinomial distribution	
		normal distribution	

Commonly Used Formulas

Chi-Square Test for Independence

Test statistic for independence in a contingency table with r rows and c columns: $\chi^2_{\text{calc}} = \sum_{j=1}^{r} \sum_{k=1}^{c} \dfrac{[f_{jk} - e_{jk}]^2}{e_{jk}}$

Degrees of freedom for a contingency table with r rows and c columns: $d.f. = (r - 1)(c - 1)$

Expected frequency in row j and column k: $e_{jk} = R_j C_k / n$

Chi-Square Test for Goodness of Fit

Test statistic for observed frequencies in c classes under a hypothesized distribution

H_0 (e.g., uniform, Poisson, normal): $\chi^2_{calc} = \sum_{j=1}^{c} \frac{[f_j - e_j]^2}{e_j}$

where

f_j = the observed frequency in class j

e_j = the expected frequency in class j

Degrees of freedom for the chi-square GOF test: $d.f. = c - m - 1$

where

c = the number of classes used in the test

m = the number of parameters estimated

Estimated mean of Poisson distribution with c classes: $\lambda = \sum_{j=1}^{c} x_j f_j$

where

x_j = the value of X in class j

f_j = the observed frequency in class j

Expected frequency in class j assuming a uniform distribution with c classes: $e_j = n/c$

Chapter Review

Note: Questions labeled * are based on optional material from this chapter.

1. (a) What are the hypotheses in a chi-square test for independence? (b) Why do we call it a test of frequencies? (c) What distribution is used in this test? (d) How do we calculate the degrees of freedom for an $r \times c$ contingency table?

2. How do we calculate the expected frequencies for each cell of the contingency table?

3. What is Cochran's Rule, and why is it needed? Why do we call it a "rule of thumb"?

4. (a) Explain why the 2×2 table is analogous to a z test for two proportions. (b) What is the relationship between z and χ^2 in the 2×2 table?

5. (a) What are the hypotheses for a GOF test? (b) Explain how a chi-square GOF test is carried out in general.

6. What is the general formula for degrees of freedom in a chi-square GOF test?

7. (a) In a uniform GOF test, how do we calculate the expected frequencies? (b) Why is the test easier if the data are already grouped?

8. (a) In a Poisson GOF test, how do we calculate the expected frequencies? (b) Why do we need the mean λ before carrying out the chi-square test?

9. (a) Very briefly describe three ways of calculating expected frequencies for a normal GOF test. (b) Name advantages and disadvantages of each way. (c) Why is a normal GOF test almost always done on a computer?

*10. What is an ECDF test? Give an example.

*11. (a) Name potential advantages of the Kolmogorov-Smirnov or Lilliefors tests. (b) Why would this type of test almost always be done on a computer?

*12. (a) What does a probability plot show? (b) If the hypothesized distribution is a good fit to the data, what would be the appearance of the probability plot? (c) What are the advantages and disadvantages of a probability plot?

*13. (a) Name two advantages of the Anderson-Darling test. (b) Why is it almost always done on a computer?

Chapter Exercises

Instructions: In all exercises, include software results (e.g., from Excel, MegaStat, or Minitab) to support your calculations. State the hypotheses, show how the degrees of freedom are calculated, find the critical value of chi-square from Appendix E or from Excel's function =CHISQ.INV.RT(alpha, deg_freedom), calculate the chi-square test statistic, and interpret the *p*-value. Tell whether the conclusion is sensitive to the level of significance chosen, identify cells that contribute the most to the chi-square test statistic, and check for small

expected frequencies. If necessary, you can calculate the *p*-value by using Excel's function =CHISQ.DIST.RT(test statistic, deg_freedom).

Note: Exercises marked * are harder or require optional material.

15.19 Employees of Axolotl Corporation were sampled at random from pay records and asked to complete an anonymous job satisfaction survey, yielding the tabulation shown. *Research question:* At $\alpha = .05$, is job satisfaction independent of pay category? 🖾 **Employees**

Pay Type	Satisfied	Neutral	Dissatisfied	Total
Salaried	24	10	2	36
Hourly	131	135	58	324
Total	155	145	60	360

15.20 Sixty-four students in an introductory college economics class were asked how many credits they had earned in college, and how certain they were about their choice of major. *Research question:* At $\alpha = .01$, is the degree of certainty independent of credits earned? Certainty

Credits Earned	Very Uncertain	Somewhat Certain	Very Certain	Row Total
0–9	12	8	4	24
10–59	8	2	10	20
60 or more	4	6	10	20
Col Total	24	16	24	64

15.21 To see whether students who finish an exam first get the same grades as those who finish later, a professor kept track of the order in which papers were handed in. Of the first 25 papers, 10 received a "B" or better, compared with 8 of the last 24 papers handed in. *Research question:* At $\alpha = .10$, is the grade independent of the order handed in? Because it is a 2 × 2 table, try also a two-tailed, two-sample z test for $\pi_1 = \pi_2$ (see Chapter 10) and verify that z^2 is the same as your chi-square statistic. Which test do you prefer? Why? Grades

Grade	Earlier Hand-In	Later Hand-In	Row Total
"B" or better	11	8	19
"C" or worse	14	17	31
Col Total	25	25	50

15.22 From 74 of its restaurants, Noodles & Company managers collected data on per person sales and the percent of sales due to "potstickers" (a popular food item). Both numerical variables failed tests for normality, so they tried a chi-square test. Each variable was converted into ordinal categories (low, medium, high) using cutoff points that produced roughly equal group sizes. *Research question:* At $\alpha = .05$, is per person spending independent of percent of sales from potstickers? Noodles

Per Person Spending	Potsticker % of Sales			Row Total
	Low	Medium	High	
Low	14	6	3	23
Medium	7	16	5	28
High	3	4	16	23
Col Total	24	26	24	74

15.23 A web-based anonymous survey of students asked for a self-rating on proficiency in a language other than English and the student's frequency of newspaper reading. *Research question:* At $\alpha = .10$, is frequency of newspaper reading independent of foreign language proficiency? WebSurvey

Non-English Proficiency	Daily Newspaper Reading			Row Total
	Never	Occasionally	Regularly	
None	4	13	5	22
Slight	11	45	9	65
Moderate	6	33	7	46
Fluent	5	19	1	25
Col Total	26	110	22	158

15.24 A student team examined parked cars in four different suburban shopping malls. One hundred vehicles were examined in each location. *Research question:* At $\alpha = .05$, does vehicle type vary by mall location? 📁 **Vehicles**

Vehicle Type	Somerset	Oakland	Great Lakes	Jamestown	Row Total
Car	44	49	36	64	193
Minivan	21	15	18	13	67
Full-sized van	2	3	3	2	10
SUV	19	27	26	12	84
Truck	14	6	17	9	46
Col Total	100	100	100	100	400

15.25 A recent survey of 400 adults provided information on social media platform use by different age groups: Silver Surfers (born before 1980), Digital Natives (born between 1980 and 1996), and Generation Z (born after 1996). Their first choice for social media platform is shown in the 3 × 3 contingency table below. Using $\alpha = .05$, is social media platform preference independent of age group? 📁 **SocialMedia**

Age and Social Media Platform Preference			
	Facebook	Instagram	Snapchat	Row Total
---	---	---	---	---
Generation Z	11	26	75	112
Digital Natives	90	60	30	180
Silver Surfers	85	15	8	108
Col Total	186	101	113	400

15.26 High levels of cockpit noise in an aircraft can damage the hearing of pilots who are exposed to this hazard for many hours. An airline co-pilot collected 61 noise observations in the cockpit of a particular type of aircraft using a handheld sound meter. Noise level is defined as "Low" (under 88 decibels), "Medium" (88 to 91 decibels), or "High" (92 decibels or more). There are three flight phases (Climb, Cruise, Descent). *Research question:* At $\alpha = .05$, is the cockpit noise level independent of flight phase? 📁 **Noise**

Noise Level	Climb	Cruise	Descent	Row Total
Low	6	2	6	14
Medium	18	3	8	29
High	1	3	14	18
Col Total	25	8	28	61

15.27 Forecasters' interest rate predictions over an eight-year period were studied to see whether the predictions corresponded to what actually happened. The 2 × 2 contingency table below shows the frequencies of actual and predicted interest rate movements. *Research question:* At $\alpha = .10$, is the actual change independent of the predicted change? 📁 **Forecasts**

Forecasted Change	Rates Fell	Rates Rose	Row Total
Rates would fall	7	12	19
Rates would rise	9	6	15
Col Total	16	18	34

15.28 In a research study, 4,317 observations were collected on education and smoking during pregnancy, shown in the 4×3 contingency table below. *Research question:* At $\alpha = .005$, is smoking during pregnancy independent of education level? Pregnancy

Education	No Smoking	<½ Pack	≥½ Pack	Row Total
< High School	641	196	196	1,033
High School	1,370	290	270	1,930
Some College	635	68	53	756
College	550	30	18	598
Col Total	3,196	584	537	4,317

15.29 The contingency table below shows return on investment (ROI) and percent of sales growth over the previous five years for 85 U.S. firms. ROI is defined as percentage of return on a combination of stockholders' equity (both common and preferred) plus capital from long-term debt including current maturities, minority stockholders' equity in consolidated subsidiaries, and accumulated deferred taxes and investment tax credits. *Research question:* At $\alpha = .05$, is ROI independent of sales growth? Would you expect it to be? Are small expected frequencies a problem? ROI

2 × 2 Cross-Tabulation of Companies

ROI	Low Growth	High Growth	Row Total
Low ROI	24	16	40
High ROI	14	31	45
Col Total	38	47	85

15.30 Can people really identify their favorite brand of cola? Volunteers tasted Coke, Pepsi, Diet Coke, and Diet Pepsi, with the results shown below. *Research question:* At $\alpha = .05$, is the correctness of the prediction different for regular cola and diet cola drinkers? Because it is a 2×2 table, try also a two-tailed, two-sample z test for $\pi_1 = \pi_2$ (see Chapter 10) and verify that z^2 is the same as your chi-square statistic. Cola

Correct?	Regular Cola	Diet Cola	Row Total
Yes, got it right	6	10	16
No, got it wrong	18	14	32
Col Total	24	24	48

15.31 A survey of randomly chosen new students at a certain university revealed the data below concerning the main reason for choosing this university instead of another. *Research question:* At $\alpha = .01$, is the main reason for choosing the university independent of student type? Students

New Student	Tuition	Location	Reputation	Row Total
Freshmen	51	32	36	119
Transfers	16	31	21	68
MBAs	3	17	65	85
Col Total	70	80	122	272

15.32 A survey of 189 statistics students asked the age of car usually driven and the student's political orientation. The car age was a numerical variable, which was converted into ordinal categories. *Research question:* At $\alpha = .10$, are students' political views independent of the age of car they usually drive? Politics

Age of Car Usually Driven

Politics	Under 3	3–6	7 or More	Row Total
Liberal	19	12	13	44
Middle-of-Road	33	31	28	92
Conservative	16	24	13	53
Col Total	68	67	54	189

15.33 The actual distribution of car colors for 2006 model car buyers is shown below. Based on the sample of 200 car buyers for 2016 model vehicles, use the multinomial chi-square GOF test at $\alpha = .05$ to test whether car buyers' color preferences have changed. 📄 **CarColor**

Car Color	2006 Percent	2016 Sample Frequencies
Silver/gray	22	70
Blue	12	16
White	16	32
Tan/brown	11	12
Black	15	42
Green	5	6
Red	13	20
Other	6	2
Total	100	200

15.34 Prof. Green's multiple-choice exam had 50 questions with the distribution of correct answers shown below. *Research question:* At $\alpha = .05$, can you reject the hypothesis that Green's exam answers came from a uniform population? 📄 **Correct**

Correct Answer	Frequency
A	8
B	8
C	9
D	11
E	14
Total	50

15.35 Oxnard Kortholt, Ltd., employs 50 workers. *Research question:* At $\alpha = .05$, do Oxnard employees differ significantly from the national percent distribution? 📄 **Oxnard**

Health Care Visits	National Percentage	Oxnard Employees Frequency
No visits	16.5	4
1–3 visits	45.8	20
4–9 visits	24.4	15
10 or more visits	13.3	11
Total	100.0	50

15.36 In a four-digit lottery, each of the four digits is supposed to have the same probability of occurrence. The table shows the frequency of occurrence of each digit for 89 consecutive daily four-digit drawings. *Research question:* At $\alpha = .01$, can you reject the hypothesis that the digits are from a uniform population? Why do the frequencies add to 356? 📄 **Lottery4**

Digit	Frequency
0	39
1	27
2	35
3	39
4	35
5	35
6	27
7	42
8	36
9	41
Total	356

15.37 A student rolled a supposedly fair die 60 times, resulting in the distribution of dots shown. *Research question:* At $\alpha = .10$, can you reject the hypothesis that the die is fair? 🗂 **Dice**

Number of Dots

	1	2	3	4	5	6	Total
Frequency	7	14	9	13	7	10	60

15.38 In the World Cup tournaments between 1990 and 2002, there were 232 games with the distribution of goals shown in this worksheet. *Research question:* At $\alpha = .025$, can you reject the hypothesis that goals per game follow a Poisson process? *Hint:* You must calculate the mean and look up the Poisson probabilities. 🗂 **WorldCup**

Goals	f_j	$P(X)$	e_j	$f_j - e_j$	$(f_j - e_j)^2$	$(f_j - e_j)^2/e_j$
0	19					
1	49					
2	60					
3	47					
4	32					
5	18					
6 or more	7					
Total games	232					
Total goals	575					
Mean goals/game						

Source: Singfat Chu, "Using Soccer Goals to Motivate the Poisson Process," *INFORMS Transactions on Education* 3, no. 2 (2003), pp. 62–68.

*15.39 The table below shows the number of ATM customer arrivals per minute in 60 randomly chosen minutes. *Research question:* At $\alpha = .025$, can you reject the hypothesis that the number of arrivals per minute follows a Poisson process? 🗂 **ATM**

0	0	0	1	3	0	0	0	2	5	2	0	1	1	1	2	1	1	0	2
3	0	0	3	0	1	0	1	1	1	1	2	0	2	0	3	0	2	0	1
1	0	0	0	0	1	3	2	1	0	0	0	4	1	0	1	0	3	3	1

15.40 Pick *one* Excel data set (A through F) and investigate whether the data could have come from a normal population using $\alpha = .01$. Use any test you wish, including a histogram, MegaStat's Descriptive Statistics > Normal curve goodness of fit test, or Minitab's Stats > Basic Statistics > Normality Test to obtain a probability plot with the Anderson-Darling statistic. Interpret the *p*-value from your tests. For larger data sets, only the first three and last three observations are shown.

DATA SET A   Kentucky Derby Winning Time (Seconds), 1950–2022 ($n = 73$)
🗂 **Derby**

Year	Derby Winner	Time
1950	Middleground	121.6
1951	Count Turf	122.6
1952	Hill Gail	121.6
⋮	⋮	⋮
2020	Authentic	120.61
2021	Mandaloun	121.02
2022	Rich Strike	122.61

Source: www.wikipedia.org.

DATA SET B National League Runs Scored Leader, 1900–2019 (*n* = 120) 📄 Runs

Year	Player	Runs
1900	Roy Thomas, Phil	131
1901	Jesse Burkett, StL	139
1902	Honus Wagner, Pitt	105
⋮	⋮	⋮
2017	Charlie Blackmon (COL)	137
2018	Charlie Blackmon (COL)	119
2019	Ronald Acuña Jr. (ATL)	127

Source: www.wikipedia.org.

DATA SET C Weight (in grams) of Pieces of Halloween Candy (*n* = 78) 📄 Candy

1.6931	1.8320	1.3167	0.5031	0.7097	1.4358
1.8851	1.6695	1.6101	1.6506	1.2105	1.4074
1.5836	1.1164	1.2953	1.4107	1.3212	1.6353
1.5435	1.7175	1.3489	1.1688	1.5543	1.3566
1.4844	1.4636	1.1701	1.5238	1.7346	1.1981
1.6601	1.8359	1.1334	1.7030	1.2481	1.4356
1.3756	1.3172	1.3700	1.0145	1.0002	0.9409
1.4942	1.2316	1.6505	1.7088	1.1850	1.3583
1.5188	1.3460	1.3928	1.6522	0.5303	1.6301
1.0474	1.4664	1.2902	1.9638	1.9687	1.2406
1.6759	1.6989	1.4959	1.4180	1.5218	2.1064
1.3213	1.1116	1.4535	1.4289	1.9156	1.8142
1.3676	1.7157	1.4493	1.4303	1.2912	1.7137

Source: Independent project by statistics student Frances Williams. Weighed on an American Scientific Model S/P 120 analytical balance, accurate to 0.0001 gram.

DATA SET D Price/Earnings Ratios for Specialty Retailers (*n* = 58) 📄 PERatios

Company	PE Ratio
Abercrombie and Fitch	19
Advance AutoParts	16
American Eagle Outfitters	30
⋮	⋮
United Auto Group	12
Williams-Sonoma	28
Zale	15

DATA SET E U.S. Presidents' Ages at Inauguration (*n* = 46) 📄 Presidents

President	Age
Washington	57
J. Adams	61
Jefferson	57
⋮	⋮
Obama	47
Trump	70
Biden	78

Source: www.wikipedia.org.

DATA SET F   Weights of 31 Randomly Chosen Circulated Nickels ($n = 31$)
 Nickels

5.043	4.980	4.967	5.043	4.956	4.999	4.917	4.927
4.893	5.003	4.951	5.040	5.043	5.004	5.014	5.035
4.883	5.022	4.932	4.998	5.032	4.948	5.001	4.983
4.912	4.796	4.970	4.956	5.036	5.045	4.801	

Note: Weighed by statistics student Dorothy Duffy as an independent project. Nickels were weighed on a Mettler PE 360 Delta Range scale, accurate to 0.001 gram.

INTEGRATIVE PROJECTS

*15.41 (a) Use Excel's function =NORM.INV(RAND(),0,1) or Excel's Data Analysis > Random Numbers to generate 100 normally distributed random numbers with a mean of 0 and a standard deviation of 1. (b) Make a histogram of your sample and assess its shape. Are there outliers? (c) Calculate descriptive statistics. Are the sample mean and standard deviation close to their intended values? (d) See if the first and third quartiles are approximately −0.675 and +0.675, as they should be. (e) Use a z test

$$z = \frac{\bar{x} - \mu}{\sigma/\sqrt{n}} = \frac{\bar{x} - 0}{1/\sqrt{100}} = 10\bar{x}$$

to compare the sample mean to the desired mean. *Note:* Use z instead of t because the hypothesized mean $\mu = 0$ and standard deviation $\sigma = 1$ are known. (f) What would happen if 100 statistics students performed similar experiments, assuming that the random number generator is working correctly?

*15.42 (a) Use Excel's function =RAND() or Excel's Data Analysis > Random Numbers to generate 100 uniformly distributed random numbers between 0 and 1. (b) Make a histogram of your sample and assess its shape. (c) Calculate descriptive statistics. Are the sample mean and standard deviation close to their intended values $\mu = (0 + 1)/2 = 0.5000$ and $\sigma = \sqrt{1/12} = 0.288675$? (d) See if the first and third

quartiles are approximately 0.25 and 0.75, as they should be. (e) Use a z test

$$z = \frac{\bar{x} - \mu}{\sigma/\sqrt{n}} = \frac{\bar{x} - 0.5000}{(0.288675)/\sqrt{100}}$$

to compare the sample mean to the desired mean. *Note:* Use z instead of t because the hypothesized mean $\mu = 0.5000$ and standard deviation $\sigma = 0.288675$ are known. (f) What would happen if 100 statistics students performed similar experiments, assuming that the random number generator is working correctly?

*15.43 (a) Use Excel's Data Analysis > Random Numbers to generate 100 Poisson-distributed random numbers with a mean of $\lambda = 4$. (b) Make a histogram of your sample and assess its shape. (c) Calculate descriptive statistics. Are the sample mean and standard deviation close to the intended values (mean $\mu = \lambda = 4$ and standard deviation $\sigma = \sqrt{\lambda} = \sqrt{4} = 2$)? (d) Perform a z test

$$z = \frac{\bar{x} - \mu}{\sigma/\sqrt{n}} = \frac{\bar{x} - 4}{2/\sqrt{100}} = 5\bar{x} - 20$$

to compare the sample mean to the desired mean. *Note:* Use z instead of t because the hypothesized mean (4) and standard deviation (2) are known. (e) What would happen if 100 statistics students performed similar experiments, assuming that Excel's random number generator is working correctly?

Related Readings

D'Agostino, Ralph B., and Michael A. Stephens. *Goodness-of-Fit Techniques.* Marcel Dekker, 1986.

Haber, Michael. "A Comparison of Some Continuity Corrections for the Chi-Squared Test on 2 × 2 Tables." *Journal of the American Statistical Association* 75, no. 371 (1980), pp. 510–15.

Rayner, J. C. W., O. Thas, and D. J. Best. *Smooth Tests of Goodness of Fit: Using R.* 2nd ed. Wiley, 2009.

CHAPTER 15   More Learning Resources

Mc Graw Hill **connect**

You can access these *LearningStats* demonstrations through Connect to help you understand chi-square tests.

Topic	LearningStats Demonstrations
Contingency tables	🅲 Testing for Independence 🅇 Contingency Tables: A Simulation 🅲 Chi Square Independence Tests Using R
Goodness-of-fit tests	🅇 Goodness-of-Fit Test 🅇 Normal and Uniform Tests 🅇 ECDF Plots Illustrated 🅇 Probability Plots: A Simulation 🅇 CDF Normality Test

Key: 🅇 = Excel   🅲 = PowerPoint

Software Supplement

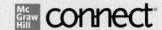

Chi-Square Tests in R

Independence Test on Contingency Table   CarAge

Functions in R (and Excel) generally expect each variable to be in a column. However, a *contingency table* is a specialized tabular arrangement designed for easy interpretation by humans but not by R. Its cells allow summation across rows and down columns. Headings are descriptive but not necessarily easy to import into R.

For example, R dislikes column headings that begin with a number (R will put an X in front of each) or that contain spaces (R will insert a period . in place of a space). R forbids unusual heading characters (e.g., ? or > or %). Table 15.5 shows an actual tabulation of survey responses by 162 statistics students. The table contains several of these no-nos. One way to accommodate R's preferences is simply to surrender and assign "legal" headings that retain the row-column terminology, as shown here.

Original Contingency Table					
	Age of Car Usually Driven (Years)				
Work Hours	Under 3	3 to 6	6 to 10	10+	Row Tot
Under 15	9	8	8	4	29
15 to 25	34	17	11	9	71
25 or More	28	20	8	6	62
Col Tot	71	45	27	19	162

Copied to R			
	Col1	Col2	Col3	Col4
Row1	9	8	8	4
Row2	34	17	11	9
Row3	28	20	8	6

We will import only the cell frequencies and their headings (*not* the totals). Despite the large sample, two cells (1,3 and 1,4) have expected frequencies less than 5 (see Mini Case 15.1), which we expect R to recognize and offer a warning in the chi-square test. Syntax of the R commands is precise, so be careful with brackets [] and parentheses (). Results agree with the textbook.

```
> CarAge=read.table(file="clipboard",sep="\t",header=TRUE)
> CarAge                        # list the data we imported
     X  Col1  Col2  Col3  Col4
1  Row1    9     8     8     4
2  Row2   34    17    11     9
3  Row3   28    20     8     6
> gof=chisq.test(CarAge[2:5])  # save the chi-square result in object gof
Warning message:
In chisq.test(CarAge[2:5]) : Chi-squared approximation may be incorrect
> CarAge[,2:5]                 # show the columns used in the test
   Col1  Col2  Col3  Col4
1     9     8     8     4
2    34    17    11     9
3    28    20     8     6
> gof                         # print result of the test
            Pearson's Chi-squared test
data: CarAge[2:5]
X-squared = 5.2417, df = 6, p-value = 0.5132
```

Test for Uniform Distribution in R   Traffic

The test for a *uniform* distribution is the default hypothesis in R, so instructions are simple. You can define the probabilities (also shown below), but this is unnecessary (they must sum to 1). For example, a sample of 756 traffic accidents suggests that they are not uniformly distributed (see textbook Table 15.7). The p-value shows that we should reject the hypothesis of a uniform distribution.

```
> ObsFatal=c(121,96,91,92,96,122,138)   # observed accidents by weekday
> chisq.test(ObsFatal)
            Chi-squared test for given probabilities
data:  ObsFatal
X-squared = 19.426, df = 6, p-value = 0.003502
> ExpFatal=c(1/7, 1/7, 1/7, 1/7, 1/7, 1/7, 1/7) # expected probabilities if
     uniform observed accidents by weekday

> chisq.test(ObsFatal,p=ExpFatal)
            Chi-squared test for given probabilities
data:  ObsFatal
X-squared = 19.426, df = 6, p-value = 0.003502
```

Test for Multinomial Distribution in R   MM

A sample of 220 M&Ms is tested to see whether its distribution of colors conforms to the published proportions (see textbook Table 15.6). The chi-square test is like a test for a uniform distribution except that you must specify the expected probabilities (they must sum to 1). In this case, the p-value suggests that we cannot reject the hypothesis that the sample colors follow the official proportions.

```
> ObsColors=c(38,30,44,52,30,26)
> ExpColors=c(.13,.13,.24,.20,.16,.14)
> chisq.test(ObsColors,p=ExpColors)
            Chi-squared test for given probabilities
data:  ObsColors
X-squared = 7.5955, df = 5, p-value = 0.18
```

16 Nonparametric Tests

CHAPTER CONTENTS

CHAPTER LEARNING OBJECTIVES

When you finish this chapter, you should be able to

LO 16-1 Define nonparametric tests and explain when they may be desirable.

LO 16-2 Use the one-sample runs test.

LO 16-3 Use the Wilcoxon signed-rank test.

LO 16-4 Use the Wilcoxon rank sum test for two samples.

LO 16-5 Use the Kruskal-Wallis test for c independent samples.

LO 16-6 Use the Friedman test for related samples.

LO 16-7 Use the Spearman rank correlation test.

WHY USE NONPARAMETRIC TESTS?

The hypothesis tests in previous chapters require the estimation of one or more unknown parameters (for example, the population mean or variance). These tests often make unrealistic assumptions about the normality of the underlying population or require large samples to invoke the Central Limit Theorem. In contrast, **nonparametric tests** or distribution-free tests usually focus on the sign or rank of the data rather than the exact numerical value of the variable, do not specify the shape of the parent population, can often be used in smaller samples, and can be used for ordinal data (when the measurement scale is not interval or ratio). Table 16.1 highlights the advantages and disadvantages of nonparametric tests.

Rejection of a hypothesis using a nonparametric test is especially convincing because nonparametric tests generally make fewer assumptions about the population. If two methods are justified and have similar **power**, the principle of Occam's Razor favors the simpler method. For this reason, statisticians are attracted to nonparametric tests, particularly in applications where data are likely to be ill-behaved and when samples are small.

insta_photos/Shutterstock

Table 16.1

Advantages and Disadvantages of Nonparametric Tests

LO 16-1

Define nonparametric tests and explain when they may be desirable.

Advantages	Disadvantages
1. Can often be used in small samples.	1. Require special tables for small samples.
2. Generally more powerful than parametric tests when normality cannot be assumed.	2. If normality *can* be assumed, parametric tests are generally more powerful.
3. Can be used for ordinal data.	

You might expect that nonparametric tests would primarily be used in areas of business where nominal or ordinal data are common (e.g., human resources, marketing). Yet business analysts who mostly use ratio data (e.g., accounting, finance) may encounter skewed populations that render parametric tests unreliable. These analysts might use nonparametric tests as a *complement* to their customary **parametric tests**. Figure 16.1 shows common nonparametric tests and their parametric counterparts, which you have seen in earlier chapters.

Figure 16.1

Some Common Nonparametric Tests

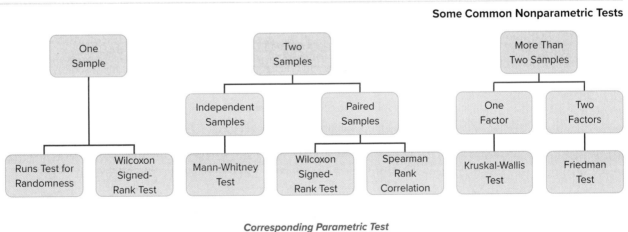

No equivalent	One-sample *t* test (Ch 9)	Two-sample *t* test (Ch 10)	Paired *t* test (Ch 10)	Pearson correlation (Ch 12)	One-factor ANOVA (Ch 11)	Randomized block ANOVA (Ch 11)

This chapter illustrates only a few of the many nonparametric techniques that are available. The selections are those you are most likely to encounter. Small-sample nonparametric tests are omitted, but references are shown at the end of the chapter for those who need them. Excel's Data Analysis Tools do not offer nonparametric tests, so you must perform the calculations manually or use a software package such as Minitab or MegaStat.

16.2 ONE-SAMPLE RUNS TEST

LO 16-2

Use the one-sample runs test.

The one-sample **runs test** is also called the **Wald-Wolfowitz test** after its inventor, Abraham Wald (1902–1950), and his student Jacob Wolfowitz. Its purpose is to detect nonrandomness. A nonrandom pattern suggests that the observations are not *independent*—a fundamental assumption of many statistical tests. We are asking whether each observation in a sequence is independent of its predecessor. In a time series, a nonrandom pattern of residuals indicates *autocorrelation* (as in Chapters 12 and 13). In quality control, a nonrandom pattern of deviations from the design specification may indicate an *out-of-control* process. We will illustrate only the large-sample version of this test (defined as samples of 10 or more).

Runs Test

This test is to determine whether a sequence of binary events follows a random pattern. A nonrandom sequence suggests nonindependent observations.

The hypotheses are

H_0: Events follow a random pattern

H_1: Events do not follow a random pattern

To test the hypothesis of randomness, we first count the number of outcomes of each type:

n_1 = number of outcomes of the first type

n_2 = number of outcomes of the second type

n = total sample size = $n_1 + n_2$

Application: Quality Inspection 📝 Defects

Inspection of 44 computer chips reveals the following sequence of defective (D) or acceptable (A) chips:

DAAAAAAADDDDAAAAAAAADDAAAAAAAADDDDAAAAAAAAAA

Do defective chips appear at random? A pattern could indicate that the assembly process has a cyclic problem due to unknown causes. The hypotheses are

H_0: Defects follow a random sequence

H_1: Defects follow a nonrandom sequence

A *run* is a series of consecutive outcomes of the same type, surrounded by a sequence of outcomes of the other type. We group sequences of similar outcomes and count the runs:

D	AAAAAAA	DDDD	AAAAAAAA	DD	AAAAAAAA	DDDD	AAAAAAAAAA
1	2	3	4	5	6	7	8

A run can be a single outcome if it is preceded and followed by outcomes of the other type. There are 8 runs in our sample ($R = 8$). The number of outcomes of each type is

n_1 = number of defective chips (D) = 11

n_2 = number of acceptable chips (A) = 33

n = total sample size = $n_1 + n_2 = 11 + 33 = 44$

In a large-sample situation (when $n_1 \geq 10$ and $n_2 \geq 10$), the number of runs R may be assumed to be normally distributed with mean μ_R and standard deviation σ_R.

$$z_{calc} = \frac{R - \mu_R}{\sigma_R} \quad \text{(test statistic comparing } R \text{ with its expected value } \mu_R) \qquad \textbf{(16.1)}$$

$$\mu_R = \frac{2n_1 n_2}{n} + 1 \quad \text{(expected value of } R \text{ if } H_0 \text{ is true)} \qquad \textbf{(16.2)}$$

$$\sigma_R = \sqrt{\frac{2n_1 n_2 (2n_1 n_2 - n)}{n^2 (n-1)}} \quad \text{(standard error of } R \text{ if } H_0 \text{ is true)} \qquad \textbf{(16.3)}$$

For our data, the expected number of runs would be

$$\mu_R = \frac{2n_1 n_2}{n} + 1 = \frac{2(11)(33)}{44} + 1 = 17.5$$

Because the actual number of runs ($R = 8$) is far less than expected ($\mu_R = 17.5$), our sample suggests that the null hypothesis may be false, depending on the standard deviation. For our data, the standard deviation is

$$\sigma_R = \sqrt{\frac{2n_1 n_2 (2n_1 n_2 - n)}{n^2 (n-1)}} = \sqrt{\frac{2(11)(33)[2(11)(33) - 44]}{44^2 (44 - 1)}} = 2.438785$$

The actual number of runs is $R = 8$, so the test statistic is

$$z_{calc} = \frac{R - \mu_R}{\sigma_R} = \frac{8 - 17.5}{2.438785} = -3.90$$

Because either too many runs or too few runs would be nonrandom, we choose a two-tailed test. The critical value $z_{.005}$ for a two-tailed test at $\alpha = .01$ is ± 2.576, so the decision rule is

Reject the hypothesis of a random pattern if $z < -2.576$ or $z > +2.576$

Otherwise, the observed difference is attributable to chance

The test statistic $z = -3.90$ is well below the lower critical limit, as shown in Figure 16.2, so we can easily reject the hypothesis of randomness. The difference between the observed number of runs and the expected number of runs is too great to be due to chance ($p = .0001$).

Figure 16.2

Decision Rule for Large-Sample Runs Test

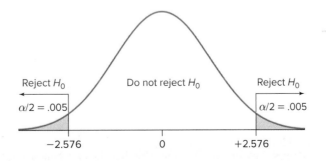

Reject H_0 Do not reject H_0 Reject H_0

$\alpha/2 = .005$ $\alpha/2 = .005$

-2.576 0 $+2.576$

Figure 16.3

MegaStat's Runs Test

Runs Test for Random Sequence

Number of Defects

n	runs	
11	4	D
33	4	A
44	8	total

17.50 expected value
 2.44 standard deviation
-3.895 z
.0001 p-value (two-tailed)

Figure 16.3 shows the MegaStat output for this problem. As with any hypothesis test, the smaller the p-value, the stronger the evidence against H_0. This small p-value provides very strong evidence that H_0 is false (i.e., that the sequence is not random).

Small Samples

See end-of-chapter Related Reading for small-sample procedures and tables of critical values that extend the test to small samples ($n < 10$). The problem with small samples is that they lack power. That is, in a small sample, it would take an extremely small or large number of runs to convince us that the sequence is nonrandom. While some researchers must deal with small samples, business analysts (e.g., quality control) often have hundreds of observations, so small samples rarely pose a problem.

Section Exercises

Mc Graw Hill connect

16.1 Using $\alpha = .05$, perform a runs test for randomness on the sample data ($n = 27$).

  A A B B A A B B A B A B A A B B B A A B B B B A A B A B B

16.2 Using $\alpha = .10$, perform a runs test for randomness on the sample data ($n = 24$).

  X O X X X X O O O O X O O O X O O O X O O X X O

16.3 On a professional certifying exam, there are 25 true-false questions. The correct answers are *T F T T F F F T T F T F T T T F F T T F F T T F T. Research question:* At $\alpha = .05$, is the *T/F* pattern random? 📄 **TrueFalse**

16.4 A baseball player was at bat 33 times during preseason exhibition games. His pattern of hits (*H*) and nonhits (*N*) is shown (a nonhit is a walk or a strikeout). *Research question:* At $\alpha = .01$, is the pattern of hits random? 📄 **Hits**

  N N N H N H N N H N N H H N N N N H N H N N N H N N H N H N N H H

16.3 WILCOXON SIGNED-RANK TEST

LO 16-3

Use the Wilcoxon signed-rank test.

The **Wilcoxon signed-rank test** was developed by Frank Wilcoxon (1892–1965) to compare a single sample with a benchmark using only **ranks** of the data instead of the original observations, as in a one-sample t test. It is more often used to compare *paired* observations, as an alternative to the paired-sample t test, which is a special case of the one-sample t test. The advantages of the Wilcoxon test are its freedom from the normality assumption, its robustness to outliers, and its applicability to ordinal data. Although the test does require the population to be roughly symmetric, it has fairly good power over a range of possible non-normal population shapes. It is slightly less powerful than the one-sample t test when the population is normal.

Wilcoxon Signed-Rank Test

The Wilcoxon signed-rank test is a nonparametric test to compare a sample median with a benchmark or to test the median difference in paired samples. It does not require normality but does assume symmetric populations. It corresponds to the parametric *t* test for one mean.

If we denote the hypothesized benchmark median as M_0, the hypotheses about the population median M are

Left-Tailed Test	*Two-Tailed Test*	*Right-Tailed Test*
$H_0: M \geq M_0$	$H_0: M = M_0$	$H_0: M \leq M_0$
$H_1: M < M_0$	$H_1: M \neq M_0$	$H_1: M > M_0$

When the variable of interest is the median difference between paired observations, the test is the same, but we use the symbol M_d for the population median *difference* and (generally) use zero as the benchmark:

Left-Tailed Test	*Two-Tailed Test*	*Right-Tailed Test*
$H_0: M_d \geq 0$	$H_0: M_d = 0$	$H_0: M_d \leq 0$
$H_1: M_d < 0$	$H_1: M_d \neq 0$	$H_1: M_d > 0$

We calculate the difference between each observation and the hypothesized median (or the differences between the paired observations), rank them from smallest to largest by absolute value, and add the ranks of the differences to obtain the Wilcoxon signed-rank test statistic *W*. Its expected value and variance depend only on the sample size *n*.

$$W = \min (\text{sum of } R^+, \text{ sum of } R^-) \tag{16.4}$$

$$\mu_W = \frac{n(n + 1)}{4} \qquad \text{(expected value of the } W \text{ statistic)} \tag{16.5}$$

$$\sigma_W = \sqrt{\frac{n(n + 1)(2n + 1)}{24}} \qquad \text{(standard deviation of the } W \text{ statistic)} \tag{16.6}$$

For large samples ($n \geq 20$), the test statistic is approximately normal:

$$z_{\text{calc}} = \frac{W - \dfrac{n(n + 1)}{4}}{\sqrt{\dfrac{n(n + 1)(2n + 1)}{24}}} \qquad \text{(Wilcoxon test statistic for large } n) \tag{16.7}$$

Application: Median versus Benchmark

Are price-earnings (P/E) ratios of stocks in *specialty* retail stores (e.g., Abercrombie & Fitch) the same as P/E ratios for stocks of *multiline* retail stores (e.g., Target)? Table 16.2 shows P/E ratios for a random sample of 21 specialty stores. The median P/E ratio for all multiline retail stores for the same date was $M_0 = 20.2$ (our benchmark). Our hypotheses are

$H_0: M = 20.2$ (the median P/E ratio for specialty stores is 20.2)

$H_1: M \neq 20.2$ (the median P/E ratio for specialty stores is not 20.2)

To perform the test, we subtract 20.2 (the benchmark) from each specialty store's P/E ratio, take absolute values, convert to ranks, and sum the ranks. Negative ranks are shown but are

Table 16.2

Wilcoxon Signed-Rank Test of P/E Ratios ($n = 21$ firms) 🖾 WilcoxonA

Company	X	X − 20.2	|X − 20.2|	Rank	R⁺	R⁻
Bebe Stores Inc	19.8	−0.4	0.4	2		2
Barnes & Noble Inc	19.8	−0.4	0.4	2		2
Aeropostale Inc	20.6	0.4	0.4	2	2	
Deb Shops	18.7	−1.5	1.5	4		4
Gap Inc	22.0	1.8	1.8	5	5	
PetSmart Inc	17.9	−2.3	2.3	6		6
Payless ShoeSource	17.0	−3.2	3.2	7		7
Abercrombie & Fitch Co	16.8	−3.4	3.4	8.5		8.5
AutoZone Inc	16.8	−3.4	3.4	8.5		8.5
Lithia Motors Inc A	16.3	−3.9	3.9	10		10
Genesco Inc	24.3	4.1	4.1	11	11	
Sherwin-Williams Co	16.0	−4.2	4.2	12		12
CSK Auto Corp	14.5	−5.7	5.7	13		13
Tiffany & Co	26.2	6.0	6.0	14	14	
Rex Stores	14.0	−6.2	6.2	15		15
Casual Male Retail Group	12.6	−7.6	7.6	16		16
Sally Beauty Co Inc	28.9	8.7	8.7	17	17	
Syms Corp	32.1	11.9	11.9	18	18	
Zale Corp	40.4	20.2	20.2	19	19	
Coldwater Creek Inc	41.0	20.8	20.8	20	20	
Talbots Inc	124.7	104.5	104.5	21	21	
			Sum	231.0	127.0	104.0

Source: Sample of 21 firms is from http://investing.businessweek.com, accessed June 19, 2007. Companies are sorted by rank of absolute differences.

not used. We assign tie ranks so that the sum of the tied values is the same as if they were not tied. For example, 3.4 occurs twice (Abercrombie & Fitch and AutoZone). If not tied, these data values would have ranks 8 and 9, so we assign a "tie" rank of 8.5 to each. Companies are shown in rank order of absolute differences. The test statistic is:

$$z_{calc} = \frac{W - \frac{n(n+1)}{4}}{\sqrt{\frac{n(n+1)(2n+1)}{24}}} = \frac{104.0 - \frac{21(21+1)}{4}}{\sqrt{\frac{21(21+1)(42+1)}{24}}} = \frac{104.0 - 115.5}{28.770645} = -0.3997$$

Using Excel, the two-tailed p-value is $p = 2*\text{NORM.S.DIST}(-0.3997,1) = .6894$. The R function wilcox.test() gives $p = .6766$ (see details in Software Supplement at the end of the chapter). At any customary level of significance, we cannot reject the hypothesis that specialty retail stores have the same median P/E as multiline retail stores. Because the P/E *data* do not seem to be from a normal population (for example, look at Talbot's extreme P/E ratio of 124.7), the Wilcoxon nonparametric test of the *median* is preferred to a one-sample t test of the *mean* (Chapter 9).

Application: Paired Data

Did P/E ratios decline between 2003 and 2007? We will perform a Wilcoxon test for *paired data* using a random sample of 23 common stocks. The parameter of interest is the median difference (M_d). Because the P/E ratios do not appear to be normally distributed (e.g., Rohm & Haas Co.'s

2003 P/E ratio), the Wilcoxon test of *medians* is attractive (instead of the paired *t* test for *means* in Chapter 10). Using $d = X_{2007} - X_{2003}$, a left-tailed test is appropriate:

H_0: $M_d \geq 0$ (the median difference is zero or positive)

H_1: $M_d < 0$ (the median difference is negative, i.e., 2007 P/E is less than 2003 P/E)

Table 16.3 shows the calculations for the Wilcoxon signed-rank statistic, with the companies in rank order of absolute differences. Because the first three have zero difference (neither positive nor negative), these three observations (highlighted) are *excluded* from the analysis.

Table 16.3

Wilcoxon Signed-Rank Paired Test (n = 23 firms) 📁 WilcoxonB

Company (Ticker Symbol)	2007 P/E	2003 P/E	d	|d|	Rank	R⁺	R⁻
FirstEnergy Corp (FE)	14	14	0	0	—		
Whirlpool Corp (WHR)	18	18	0	0	—		
Burlington Northern/Santa Fe (BNI)	14	14	0	0	—		
Constellation Energy (CEG)	16	15	1	1	1	1	
Mellon Financial (MEL)	17	19	−2	2	2.5		2.5
Yum! Brands Inc (YUM)	18	16	2	2	2.5	2.5	
Baxter International (BAX)	20	23	−3	3	5		5
Fluor Corp (FLR)	22	19	3	3	5	5	
Allied Waste Ind (AW)	21	18	3	3	5	5	
Ingersoll-Rand-A (IR)	12	16	−4	4	7.5		7.5
Lexmark Intl A (LXK)	17	21	−4	4	7.5		7.5
Moody's Corp (MCO)	29	24	5	5	9	9	
Electronic Data (EDS)	17	23	−6	6	10		10
Freeport-Mcmor-B (FCX)	11	18	−7	7	11		11
Family Dollar Stores (FDO)	19	27	−8	8	12.5		12.5
Leggett & Platt (LEG)	13	21	−8	8	12.5		12.5
Wendy's Intl Inc (WEN)	24	15	9	9	14	14	
Sara Lee Corp (SLE)	25	13	12	12	15	15	
Bed Bath & Beyond (BBBY)	20	37	−17	17	16		16
Ace Ltd (ACE)	8	26	−18	18	17.5		17.5
ConocoPhillips (COP)	8	26	−18	18	17.5		17.5
Baker Hughes Inc (BHI)	13	55	−42	42	19		19
Rohm & Haas Co (ROH)	15	68	−53	53	20		20
					Sum	51.5	158.5

Source: Sampled P/E values from *The Wall Street Journal*, July 31, 2003, and Standard & Poor's, *Security Owner's Stock Guide*, February 2007. Companies are sorted by rank of absolute differences.

Despite losing three observations due to zero differences, we still have $n \geq 20$, so we can use the large-sample test statistic:

$$z_{calc} = \frac{W - \dfrac{n(n+1)}{4}}{\sqrt{\dfrac{n(n+1)(2n+1)}{24}}} = \frac{51.5 - \dfrac{20(20+1)}{4}}{\sqrt{\dfrac{20(20+1)(40+1)}{24}}} = \frac{51.5 - 105.0}{26.7862} = -1.9973$$

Using Excel, we find the left-tailed *p*-value =NORM.S.DIST(−1.9973, 1) =.0229. so at $\alpha = .05$ we conclude that P/E ratios did decline between 2003 and 2007. Figure 16.4 shows that MegaStat confirms our calculations. The R function wilcox.test() gives a similar result (*p* = .0239).

Figure 16.4

MegaStat Signed-Rank Test for Paired Data

Wilcoxon Signed-Rank Test

variables: 2007 P/E-2003 P/E
   51.5 sum of positive ranks
   158.5 sum of negative ranks

   20 n
105.00 expected value
  26.79 standard deviation
  −1.997 z
  .0229 p-value (one-tailed, lower)

Section Exercises

Mc Graw Hill **connect**

16.5 A sample of 28 student scores on the chemistry midterm exam is shown. (a) At $\alpha = .10$, does the population median differ from 50? Make a worksheet in Excel for your calculations. (b) Make a histogram of the data. Would you be justified in using a parametric t test that assumes normality? Explain. 📄 **Chemistry**

74	60	7	97	62	2	100
5	99	78	93	32	43	64
87	37	70	54	60	62	17
26	45	84	24	66	7	48

16.6 Final exam scores for a sample of 20 students in a managerial accounting class are shown. (a) At $\alpha = .05$, is there a difference in the population median scores on the two exams? Make an Excel worksheet for your Wilcoxon signed-rank test calculations and check your work by using MegaStat or a similar computer package. (b) Perform a two-tailed parametric t test for paired two-sample means by using Excel or MegaStat. Do you get the same decision? 📄 **Accounting**

Student	Exam 1	Exam 2	Student	Exam 1	Exam 2	Student	Exam 1	Exam 2
1	70	81	8	71	69	15	59	68
2	74	89	9	52	53	16	54	47
3	65	59	10	79	84	17	75	84
4	60	68	11	84	96	18	92	100
5	63	75	12	95	96	19	70	81
6	58	77	13	83	99	20	54	58
7	72	82	14	81	76			

16.4 WILCOXON RANK SUM TEST

LO 16-4

Use the Wilcoxon rank sum test for two samples.

The **Wilcoxon rank sum test** (also known as the **Mann-Whitney test**) is named after statisticians Frank Wilcoxon (1892–1965), Henry B. Mann (1905–2000), and D. Ransom Whitney (1915–2007). It is a nonparametric test that compares two populations whose distributions are assumed to be the same except for a shift in location (e.g., all X values shifted by a given amount). It does not assume normality. Assuming that the populations differ only in centrality (i.e., location), it is a test for equality of *medians*. It is analogous to the t test for two independent sample means.

Wilcoxon Rank Sum Test

The Wilcoxon rank sum test is a nonparametric test to compare two populations, utilizing only the ranks of the data from two independent samples. If the populations differ only in location (center), it is a test for equality of medians, corresponding to the parametric t test for two means.

Studies suggest that the Wilcoxon test has only slightly less power in distinguishing between centrality of two populations than the t test for two independent sample means, which you studied in Chapter 10. The Wilcoxon test requires independent samples from populations with equal variances, but the populations need not be normal. To avoid the use of special tables, we will illustrate only a large-sample version of this test (defined as samples of 10 or more). We will illustrate two versions of the test.

Assuming that the only difference in the populations is in location, the hypotheses for a two-tailed test of the population medians would be

H_0: $M_1 - M_2 = 0$ (no difference in medians)

H_1: $M_1 - M_2 \neq 0$ (medians differ for the two groups)

Application: Restaurant Quality 🗁 Restaurants

Does spending more at a restaurant lead to greater customer satisfaction? Frequent diners were asked to rate 29 chain restaurants on a scale of 0 to 100, based mainly on the taste of the food. Results are shown in Table 16.4, sorted by satisfaction rating (note that, in this test, the lowest data value is assigned a rank of 1, which is rather counterintuitive for restaurant ratings). Each restaurant is assigned to one of two price groups: *Low* (under $15 per person) and *High* ($15 or more per person). Is there a significant difference in satisfaction between the higher-priced restaurants and the lower-priced ones? The parametric t test for two means would require that the variable be measured on a ratio or interval level. Because the satisfaction ratings are solely based on human perception, we are unwilling to assume the strong measurement properties associated with ratio or interval data. Instead, we treat these measurements as ordinal data (i.e., ranked data).

Table 16.4

Satisfaction and Ranks for 29 Chain Restaurants
🗁 **Restaurants**

Obs	Satisfaction	Rank	Price	Obs	Satisfaction	Rank	Price
1	74	1	Low	16	83	16.5	Low
2	78	2.5	Low	17	83	16.5	Low
3	78	2.5	High	18	84	18.5	High
4	79	4.5	Low	19	84	18.5	High
5	79	4.5	Low	20	85	20.5	Low
6	80	7	Low	21	85	20.5	Low
7	80	7	High	22	86	22.5	High
8	80	7	High	23	86	22.5	High
9	81	10	Low	24	87	25	High
10	81	10	Low	25	87	25	High
11	81	10	Low	26	87	25	High
12	82	13.5	Low	27	88	28	High
13	82	13.5	Low	28	88	28	High
14	82	13.5	Low	29	88	28	High
15	82	13.5	High				

In Table 16.4, we convert the customer satisfaction ratings into ranks by sorting the *combined* samples from lowest to highest satisfaction, and then assigning a rank to each satisfaction score. If values are tied, the average of the ranks is assigned to each. Restaurants are then separated into two groups based on the price category (*High*, *Low*), as displayed in Table 16.5. We will demonstrate two equivalent methods for the Wilcoxon test.

Method A T_1 is the sum of ranks for the *smaller* sample. The ranks are summed for each column to get $T_1 = 271$ and $T_2 = 164$. The sum $T_1 + T_2$ must be $n(n + 1)/2$, where $n = n_1 + n_2 = 14 + 15 = 29$. Because $n(n + 1)/2 = (29)(30)/2 = 435$ and the sample sums are $T_1 + T_2 = 271 + 164 = 435$, our calculations check.[1] Next, we calculate the mean rank sums $\overline{T}_1$ and $\overline{T}_2$. If there is no difference between groups, we would expect $\overline{T}_1 - \overline{T}_2$ to be near zero.

---

[1] If the sum $T_1 + T_2$ does not check, you have made an error in calculating the ranks. Avoid Excel's functions =RANK() and =RANK.EQ() because they do not adjust for ties. Instead, use Excel's =RANK.AVG() function.

Table 16.5

Chain Restaurant
Customer Satisfaction
Score

High-Priced Restaurants ($n_1 = 14$)		Low-Priced Restaurants ($n_2 = 15$)	
Satisfaction	**Rank**	**Satisfaction**	**Rank**
78	2.5	74	1
80	7	78	2.5
80	7	79	4.5
82	13.5	79	4.5
84	18.5	80	7
84	18.5	81	10
86	22.5	81	10
86	22.5	81	10
87	25	82	13.5
87	25	82	13.5
87	25	82	13.5
88	28	83	16.5
88	28	83	16.5
88	28	85	20.5
		85	20.5

Rank sum:	$T_1 = 271$	Rank sum:	$T_2 = 164$
Sample size	$n_1 = 14$	Sample size	$n_2 = 15$
Mean rank:	$\overline{T}_1 = 271/14 = 19.35714$	Mean rank:	$\overline{T}_2 = 164/15 = 10.93333$

For large samples, ($n_1 \geq 10$, $n_2 \geq 10$), we can use a z test. The test statistic is the difference in mean ranks, divided by its standard error:

(16.8) $$z_{calc} = \frac{\overline{T}_1 - \overline{T}_2}{(n_1 + n_2)\sqrt{\dfrac{n_1 + n_2 + 1}{12 n_1 n_2}}}$$

⟵ difference in mean ranks

⟵ standard error of difference

For our restaurant data:

$$z_{calc} = \frac{19.35714 - 10.93333}{(14 + 15)\sqrt{\dfrac{14 + 15 + 1}{(12)(14)(15)}}} = +2.662$$

At $\alpha = .01$, rejection in a two-tailed test requires $z > +2.576$ or $z < -2.576$, so we reject the hypothesis that the population medians are the same. The two-tail p-value from Excel is $=2*(1-\text{NORM.S.DIST}(2.662, 1)) = .0078$, which says that a sample difference of this magnitude would be expected only about 8 times in 1,000 samples if the populations were the same. The R function wilcox.test() gives a similar result ($p = .0080$).

Method B An alternative way to perform the large-sample Wilcoxon rank sum z-test is based on the expected value and variance of the sum of the ranks in the smaller sample T_1. Using this alternate form, the test statistic is

(16.9) $$z_{calc} = \frac{T_1 - E(T_1)}{\sqrt{\text{Var}(T_1)}} = \frac{T_1 - \dfrac{n_1(n_1 + n_2 + 1)}{2}}{\sqrt{\dfrac{n_1 n_2}{12}(n_1 + n_2 + 1)}}$$

For our restaurant example, $T_1 = 271$, so the test statistic is

$$z_{calc} = \frac{T_1 - \dfrac{n_1(n_1 + n_2 + 1)}{2}}{\sqrt{\dfrac{n_1 n_2}{12}(n_1 + n_2 + 1)}} = z_{calc} = \frac{271 - 14(14 + 15 + 1)}{\sqrt{\dfrac{(14)(15)}{12}(14 + 15 + 1)}}$$

$$= \frac{271 - 210}{\sqrt{525}} = \frac{61}{22.91287847} = +2.662$$

As you can see, either formula for z_{calc} will give the same result. MegaStat uses a different version of this test but obtains a similar result, as shown in Figure 16.5.

Wilcoxon Mann/Whitney Test

n	sum of ranks	
15	164	Group 1
14	271	Group 2
29	435	total

225.00 expected value
22.91 standard deviation
−2.662 z
.0078 p-value (two-tailed)

Figure 16.5

MegaStat's Wilcoxon Mann-Whitney Test

Section Exercises

connect

16.7   Bob and Tom are "paper investors." They each "buy" stocks they think will rise in value and "hold" them for a year. At the end of the year, they compare their stocks' appreciation (percent). (a) At $\alpha = .05$, is there a difference in the medians (assume these are samples of Bob's and Tom's stock-picking skills)? Use software (e.g., MegaStat or Minitab) for the Wilcoxon rank sum test (Mann-Whitney test) calculations. (b) Use Excel to perform a two-tailed parametric t test for two independent sample means. Do you get the same decision? **Investors**

Bob's Portfolio (10 stocks)	7.0, 2.5, 6.2, 4.4, 4.2, 8.5, 10.0, 6.4, 3.6, 7.6
Tom's Portfolio (12 stocks)	5.2, 0.4, 2.6, −0.2, 4.0, 5.2, 8.6, 4.3, 3.0, 0.0, 8.6, 7.5

16.8   An experimental bumper was designed to reduce damage in low-speed collisions. This bumper was installed on an experimental group of vans in a large fleet, but not on a control group. At the end of a trial period, there were 12 repair incidents (a "repair incident" is an accident that resulted in a repair invoice) for the experimental group and 9 repair incidents for the control group. The dollar cost per repair incident is shown below. (a) Use software (e.g., MegaStat or Minitab) to perform a two-tailed Wilcoxon rank sum test (Mann-Whitney test) at $\alpha = .05$. (b) Use Excel to perform a two-tailed parametric t test for two independent sample means. Do you get the same decision? (Data are from Floyd G. Willoughby and Thomas W. Lauer, confidential case study.) **Damage**

Old bumper: 1,185; 885; 2,955; 815; 2,852; 1,217; 1,762; 2,592; 1,632
New bumper: 1,973; 403; 509; 2,103; 1,153; 292; 1,916; 1,602; 1,559; 547; 801; 359

16.5   KRUSKAL-WALLIS TEST FOR INDEPENDENT SAMPLES

LO 16-5

Use the Kruskal-Wallis test for c independent samples.

William H. Kruskal and W. Allen Wallis proposed a test to compare c independent samples. It may be viewed as a generalization of the Wilcoxon (Mann-Whitney) rank sum test, which compares two independent samples. Groups can be of different sizes if each has five or more observations. If we assume that the populations differ only in centrality (i.e., location), the Kruskal-Wallis test (K-W test) compares the medians of c independent samples. It is analogous to one-factor ANOVA (completely randomized model). The K-W test requires that the populations be of similar shape but does not require normal populations as in ANOVA, making it an attractive alternative for applications in finance, engineering, and marketing.

Kruskal-Wallis Test

The K-W test compares the medians of c independent samples. It may be viewed as a generalization of the Mann-Whitney test and is a nonparametric alternative to one-factor ANOVA.

Assuming that the populations are otherwise similar, the hypotheses to be tested are

H_0: All c population medians are the same

H_1: Not all the population medians are the same

In testing for equality of location, the K-W test may be almost as powerful as one-factor ANOVA. It can be useful for ratio or interval data when there are outliers or if the population is thought to be non-normal. For a completely randomized design with c groups, the test statistic is

(16.10) $$H_{calc} = \frac{12}{n(n+1)} \sum_{j=1}^{c} \frac{T_j^2}{n_j} - 3(n+1) \qquad \text{(Kruskal-Wallis test statistic)}$$

where

$n = n_1 + n_2 + \cdots + n_c$

n_j = number of observations in group j

T_j = sum of ranks for group j

$d.f. = c - 1$

Application: Employee Absenteeism

The XYZ Corporation is interested in possible differences in days worked by salaried employees in three departments in the financial area. Table 16.6 shows annual days worked by 23 randomly chosen employees from these departments. Because the sampling methodology reflects the department sizes, the sample sizes are unequal.

Table 16.6

Annual Days Worked by Department 📄 **Days**

Department	Days Worked									
Budgets	278	260	265	245	258					
Payables	205	270	220	240	255	217	266	239	240	228
Pricing	240	258	233	256	233	242	244	249		

To get the test statistic, we combine the samples and assign a rank to each observation in each group, as shown in Table 16.7. We use a column worksheet so the calculations are easier to follow. When a tie occurs, each observation is assigned the average of the ranks.

Next, the data are arranged by groups, as shown in Table 16.8, and the ranks are summed to give T_1, T_2, and T_3. As a check on our work, the sum of the ranks must be $n(n+1)/2 = (23)$ $(23+1)/2 = 276$. This is easily verified because $T_1 + T_2 + T_3 = 92.5 + 93.0 + 90.5 = 276$.

The value of the test statistic is

$$H_{calc} = \frac{12}{n(n+1)} \sum_{j=1}^{c} \frac{T_j^2}{n_j} - 3(n+1)$$

$$= \frac{12}{(23)(23+1)} \left[\frac{92.5^2}{5} + \frac{93^2}{10} + \frac{90.5^2}{8} \right] - 3(23+1) = 6.259$$

The H test statistic follows a chi-square distribution with degrees of freedom $d.f. = c - 1 = 3 - 1 = 2$. This is a right-tailed test (i.e., we will reject the null hypothesis of equal medians if H exceeds its critical value). Using $d.f. = 2$, from Appendix E we obtain critical values for various levels of significance:

α	χ^2_{crit}	Interpretation
.10	4.605	Reject H_0—conclude that the medians differ
.05	5.991	Reject H_0—conclude that the medians differ
.025	7.378	Do not reject H_0—conclude that the medians are not different

Table 16.7

Merged Data Converted to Ranks

Obs	Rank	Days	Dept
1	1	205	Payables
2	2	217	Payables
3	3	220	Payables
4	4	228	Payables
5	5.5	233	Pricing
6	5.5	233	Pricing
7	7	239	Payables
8	9	240	Payables
9	9	240	Payables
10	9	240	Pricing
11	11	242	Pricing
12	12	244	Pricing
13	13	245	Budgets
14	14	249	Pricing
15	15	255	Payables
16	16	256	Pricing
17	17.5	258	Budgets
18	17.5	258	Pricing
19	19	260	Budgets
20	20	265	Budgets
21	21	266	Payables
22	22	270	Payables
23	23	278	Budgets

Table 16.8

Worksheet for Rank Sums

Budgets	Rank	Payables	Rank	Pricing	Rank
245	13	205	1	233	5.5
258	17.5	217	2	233	5.5
260	19	220	3	240	9
265	20	228	4	242	11
278	23	239	7	244	12
		240	9	249	14
		240	9	256	16
		255	15	258	17.5
		266	21		
		270	22		
Sum of ranks	92.5	Sum of ranks	93	Sum of ranks	90.5
Sample size	$n_1 = 5$	Sample size	$n_2 = 10$	Sample size	$n_3 = 8$

In this instance, our decision is sensitive to the level of significance chosen. The *p*-value from Excel is =CHISQ.DIST.RT(6.259,2) = .0437, which agrees with MegaStat (Figure 16.6). The R function kruskal.test() gives $p = .0433$. Because the *p*-value is between .05 and .025, we conclude that the difference among the three groups is not overwhelming. The stacked dot plots in Figure 16.7 reveal that the three distributions overlap quite a bit, so there may be little practical difference in the distributions. Minitab's K-W test is similar but requires unstacked data (one column for the data, one column for the group name). Minitab warns you if the sample size is too small.

Figure 16.6

MegaStat's Kruskal-Wallis Test

Kruskal-Wallis Test

Median	n	Avg. Rank	
260.00	5	18.50	Budgets
239.50	10	9.30	Payables
243.00	8	11.31	Pricing
244.00	23		Total
		6.259	H
		2	d.f.
		.0437	p-value

Figure 16.7

Stacked Dot Plots for Days Worked

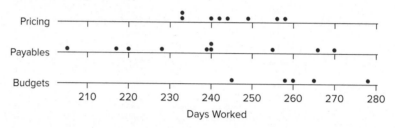

Days Worked

16.9 Samples are shown of volatility (coefficient of variation) for sector stocks over a certain period of time. (a) At $\alpha = .05$, is there a difference in median volatility in these four portfolios? Use MegaStat, Minitab, or a similar computer package for the calculations. (b) Use one-factor ANOVA to compare the means. Do you reach the same conclusion? (c) Make a histogram or other display of each sample. Would you be willing to assume normality? 📁 **Volatile**

Health	Energy	Retail	Leisure
14.5	23.0	19.4	17.6
18.4	19.9	20.7	18.1
13.7	24.5	18.5	16.1
16.9	24.2	15.5	23.2
16.2	19.4	17.7	17.6
21.6	22.1	21.4	25.5
25.6	31.6	26.5	24.1
21.4	22.4	21.5	25.9
26.6	31.3	22.8	25.5
19.0	32.5	27.4	26.3
12.6	12.8	22.0	12.9
13.5	14.4	17.1	11.1
13.5		24.8	4.9
13.0		13.4	
13.6			

16.10 The results shown below are mean productivity measurements (average number of assemblies completed per hour) for a random sample of workers at each of three work stations. (a) At $\alpha = .05$, is there a difference in median productivity? Use MegaStat, Minitab, or a similar computer package for the calculations. (b) Use one-factor ANOVA to compare the means. Do you reach the same conclusion? (c) Make a histogram or other display of the pooled data. Does the assumption of normality seem justified? 📁 **Workers**

Hourly Productivity of Assemblers in Plants

Work Station	Finished Units Produced per Hour								
A (9 workers)	3.6	5.1	2.8	4.6	4.7	4.1	3.4	2.9	4.5
B (6 workers)	2.7	3.1	5.0	1.9	2.2	3.2			
C (10 workers)	6.8	2.5	5.4	6.7	4.6	3.9	5.4	4.9	7.1	8.4

Mini Case 16.1

Price/Earnings Ratios

Based on the sample in Table 16.9, can we conclude that price-earnings ratios differ for firms in the five sectors shown? Because the data are interval, we could try either one-factor ANOVA or a Kruskal-Wallis test.

Combining the samples, the histogram in Figure 16.8 and the probability plot in Figure 16.9 suggest non-normality, so instead of one-factor ANOVA we would prefer the nonparametric Kruskal-Wallis test with $d.f. = c - 1 = 5 - 1 = 4$ degrees of freedom. Minitab's output in Figure 16.10 shows that the medians differ ($p = .000$). The test statistic ($H = 25.32$) exceeds the chi-square critical value for $\alpha = .01$ (13.28). We conclude that the P/E ratios are *not* the same for these five sectors.

Table 16.9   **Common Stock P/E Ratios of Selected Companies**   📁 **PERatios**

Automotive and Components (n = 17)								
9	13	14	29	10	32	16	14	9
21	17	21	10	7	20	13	17	
Energy Equipment and Services (n = 12)								
31	22	39	25	46	7	29	36	42
36	49	35						
Food and Staples Retailing (n = 22)								
25	22	18	24	27	21	66	30	24
22	21	9	11	16	13	32	15	25
36	29	25	18					
Hotels, Restaurants, and Leisure (n = 18)								
34	26	74	24	17	19	22	34	30
22	24	19	23	19	21	31	16	19
Multiline Retail Firms (n = 18)								
16	29	22	19	20	14	22	18	28
13	16	20	21	23	20	3	14	27

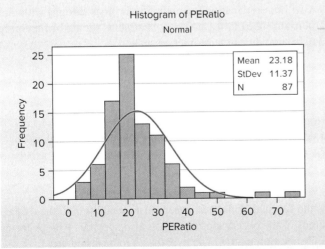

Histogram of PERatio
Normal

Mean	23.18
StDev	11.37
N	87

PERatio

FIGURE 16.8

Histogram of Combined Samples (n = 87)

Figure 16.9 **Probability Plot of Combined Samples ($n = 87$)**

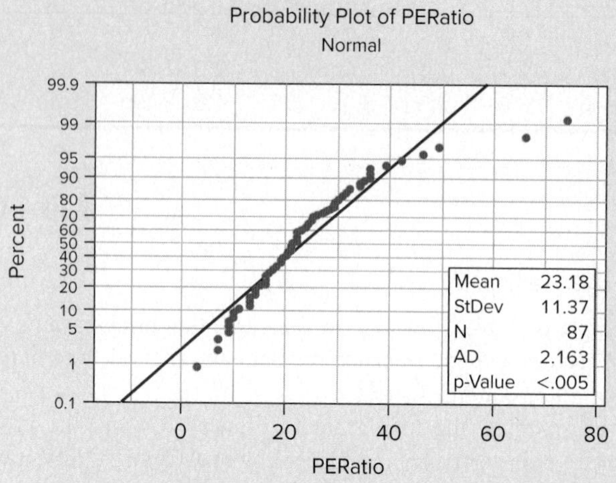

Probability Plot of PERatio
Normal

	Mean	23.18
	StDev	11.37
	N	87
	AD	2.163
	p-Value	<.005

Figure 16.10 **Minitab's Kruskal-Wallis Test**

Kruskal-Wallis Test: PERatio versus Sector

Sector	N	Median	Ave Rank	z
Auto	17	14.00	24.4	−3.58
EnergyEq	12	35.50	68.5	3.62
FoodDrug	22	23.00	46.9	0.62
Leisure	18	22.50	51.1	1.34
Retail	18	20.00	35.6	−1.58
Overall	87		44.0	

H = 25.27   DF = 4   P = 0.000
H = 25.32   DF = 4   P = 0.000   (adjusted for ties)

LO 16-6

Use the Friedman test for related samples.

16.6 FRIEDMAN TEST FOR RELATED SAMPLES

The **Friedman test** is a nonparametric test that will reveal whether c treatments have the same central tendency when there is a second factor with r levels. If the populations are assumed the same except for centrality (location), the test is a comparison of medians. The test is analogous to two-factor ANOVA without replication (or randomized block design) with one observation for each cell. The groups must be of the same size, treatments should be randomly assigned within the blocks, and data should be at least interval scale.

Friedman Test

The Friedman test is a nonparametric procedure to discover whether c population medians are the same or different when classification is based on two factors. It is analogous to randomized block ANOVA (two-factor without replication) but without the normality assumption.

The Friedman test resembles the Kruskal-Wallis test except that, in addition to the c treatment levels that define the columns of the observation matrix, it also specifies r block factor levels to define each row of the observation matrix. The hypotheses to be tested are

H_0: All c populations have the same median

H_1: Not all the populations have the same median

The Friedman test may be almost as powerful as two-way ANOVA without replication (randomized block design) and may be used with ratio or interval data when there is concern for outliers or non-normality of the underlying populations. It is a rare population that meets the normality requirement, so Friedman's test is quite useful.

Test Statistic

The test statistic is

$$F_{calc} = \frac{12}{rc(c+1)} \sum_{j=1}^{c} T_j^2 - 3r(c+1) \quad \text{(Friedman test statistic)} \quad \textbf{(16.11)}$$

where

r = the number of blocks (rows)

c = the number of treatments (columns)

T_j = the sum of ranks for treatment j

Although the Friedman formula resembles the Kruskal-Wallis formula, there is a difference: The ranks are computed *within each block* rather than within a pooled sample.

Application: Braking Effectiveness

Experiments are being conducted to test the effect of brake pad composition on stopping distance. Five prototype brake pads are prepared. Each pad is installed on the same automobile, which is accelerated to 100 kph and then braked to the shortest possible stop without loss of control. This test is repeated four times in rapid succession to reveal brake fade due to heating and lining abrasion. Car weight and balance are identical in all tests, and the same expert driver performs all tests. The pavement is dry and the outside air temperature is the same for all tests. To eliminate potential bias, the driver has no information about which pad is installed for a given test. The results are shown in Table 16.10.

Table 16.10

Stopping Distance from 100 kph 🖼 **Braking**

	Pad 1		Pad 2		Pad 3		Pad 4		Pad 5	
	Feet	Rank	Feet	Rank	Feet	Rank	Feet	Rank	Feet	Rank
Trial 1	166	3	176	4	152	2	198	5	148	1
Trial 2	174	4	170	3	148	1	206	5	152	2
Trial 3	184	3	186	4	160	1	212	5	168	2
Trial 4	220	5	204	3	184	1	216	4	196	2
Rank sum	$T_1 = 15$		$T_2 = 14$		$T_3 = 5$		$T_4 = 19$		$T_5 = 7$	

The Friedman test requires that either the number of blocks or the number of treatments be at least 5. Our matrix meets this requirement because $r = 4$ and $c = 5$. Ranks are computed *within each row.* As a check on our arithmetic, we may utilize the fact that the ranks must sum to $rc(c+1)/2 = (4)(5)(5+1)/2 = 60$. We see that our sums are correct because $T_1 + T_2 + T_3 + T_4 + T_5 = 15 + 14 + 5 + 19 + 7 = 60$.

We now compute the test statistic:

$$F_{calc} = \frac{12}{rc(c+1)} \sum_{j=1}^{c} T_j^2 - 3r(c+1)$$

$$= \frac{12}{(4)(5)(5+1)} [15^2 + 14^2 + 5^2 + 19^2 + 7^2] - 3(4)(5+1) = 13.6$$

The Friedman test statistic follows a chi-square distribution with degrees of freedom $d.f. = c - 1 = 5 - 1 = 4$. Using $d.f. = 4$, from Appendix E we obtain the critical values for various α levels:

α	χ_α^2	Interpretation
.025	11.143	Reject H_0—conclude that the medians differ
.01	13.277	Reject H_0—conclude that the medians differ
.005	14.861	Do not reject H_0—the medians do not differ

The p-value using Excel is =CHISQ.DIST.RT(13.6, 4) = .0087. The R function friedman.test() and MegaStat (Figure 16.11) give the same p-value (.0087). Because the p-value is between .01 and .005, we conclude that there is a significant difference in brake pads except at very strict Type I error levels.

Figure 16.11

MegaStat's Friedman Test for Brake Pads

Friedman Test

Sum of Ranks	Avg. Rank
15.00	3.75
14.00	3.50
5.00	1.25
19.00	4.75
7.00	1.75
60.00	3.00

n = 4
13.600 chi-square
4 d.f.
.0087 p-value

Section Exercises

connect

16.11 Consumers are asked to rate the attractiveness of four potential dashboard surface textures on an interval scale (1 = least attractive, 10 = most attractive). Use MegaStat or another software package to perform a Friedman test to see whether the median ratings of surfaces differ at $\alpha = .05$, using age as the blocking factor. **Texture**

	Shiny	Satin	Pebbled	Pattern	Embossed
Youth (Under 21)	6.7	6.6	5.5	4.3	4.4
Adult (21 to 39)	5.5	5.3	6.2	5.9	6.2
Middle Age (40 to 61)	4.5	5.1	6.7	5.5	5.4
Senior (62 and over)	3.9	4.5	6.1	4.1	4.9

16.12 JavaMax is a neighborhood take-out coffee shop that offers three sizes. Yesterday's sales are shown. Use MegaStat or another software package to perform a Friedman test to see whether the median sales of coffee sizes differ at $\alpha = .05$, using time of day as the blocking factor. **Coffee**

	Small	Medium	Large
6 a.m. < 8 a.m.	60	77	85
8 a.m. < 11 a.m.	65	74	76
11 a.m. < 3 p.m.	70	70	70
3 p.m. < 7 p.m.	61	60	55
7 p.m. < 11 p.m.	55	50	48

16.7 SPEARMAN RANK CORRELATION TEST

An overall nonparametric test of association between two variables can be performed by using Spearman's **rank correlation coefficient** (sometimes called **Spearman's rho**). This statistic is useful when it is inappropriate to assume an interval scale (a requirement of the Pearson correlation coefficient you learned in Chapter 12). The statistic is named for Charles E. Spearman (1863–1945), a British behavioral psychologist who was interested in assessment of human intelligence. The research question was the extent of agreement between different I.Q. tests (e.g., Stanford-Binet and Wechsler's WAIS). However, ordinal data are also common in business. For example, Moody's bond ratings (e.g., Aaa, Aa, A, Baa, Ba, B, etc.), bank safety ratings (e.g., by Veribanc or BankRate.com), or Morningstar's mutual fund ratings (e.g., 5/5, 5/4, 4/4, etc.) are ordinal (not interval) measurements. We could use Spearman's rank correlation to answer questions like these:

LO 16-7

Use the Spearman rank correlation test.

- When *n* corporate bonds are assigned a quality rating by two different agencies (e.g., Moody and Dominion), to what extent do the ratings agree?
- When creditworthiness scores are assigned to *n* individuals by different credit-rating agencies (e.g., Equifax and TransUnion), to what extent do the scores agree?

In cases like these, we would expect strong agreement because presumably the rating agencies are trying to measure the same thing. In other cases, we may have ratio or interval data, but prefer to rely on rank-based tests because of serious non-normality or outliers. For example:

- To what extent do rankings of *n* companies based on revenues agree with their rankings based on profits?
- To what extent do rankings of *n* mutual funds based on 1-year rates of return agree with their rankings based on 5-year rates of return?

Spearman Rank Correlation

Spearman rank correlation is a nonparametric test that measures the strength of the association, if any, between two variables using only ranks. It does not assume interval measurement.

The formula for Spearman's rank correlation coefficient for a sample is

$$r_s = 1 - \frac{6 \sum_{i=1}^{n} d_i^2}{n(n^2 - 1)} \quad \text{(Spearman rank correlation)} \quad \textbf{(16.12)}$$

where

d_i = difference in ranks for case *i*

n = sample size

The sample rank correlation coefficient r_s must fall in the range $-1 \leq r_s \leq +1$. Its sign tells whether the relationship is direct (ranks tend to vary in the same direction) or inverse (ranks tend to vary in opposite directions). If r_s is near zero, there is little or no agreement between the rankings. If r_s is near $+1$, there is strong agreement between the ranks, while if r_s is near -1, there is strong *inverse* agreement between the ranks.

Application: Calories and Fat

Calories come from fat but also from carbohydrates. How closely related are fat calories and total calories? As an experiment, a student team examined a sample of 20 brands of pasta sauce, obtaining the data shown in Table 16.11. The serving sizes (in grams) varied, so we divided each product's total calories and fat calories by serving size to obtain a per-gram measurement. Ranks were then calculated for each measure of calories. If more than one value was the same, they were assigned the average of the ranks. As a check, the sums of ranks within each

Table 16.11

Calories per Gram for 20 Pasta Sauces

📁 **Pasta**

Source: Data are from a project by statistics students Donna Bennett, Nicole Cook, Latrice Haywood, and Robert Malcolm. Data are for learning purposes and should not be viewed as a current nutrition guide due to changes in products since the sample was taken.

Product	Total Calories Per Gram	Total Calories Rank	Fat Calories Per Gram	Fat Calories Rank	d_i	d_i^2
Barilla Roasted Garlic & Onion	0.64	10	0.20	8	2	4
Barilla Tomato & Basil	0.56	13	0.12	13.5	−0.5	0.25
Classico Tomato & Basil	0.40	19.5	0.08	17	2.5	6.25
Del Monte Mushroom	0.48	17	0.04	19	−2	4
Five Bros. Tomato & Basil	0.64	10	0.12	13.5	−3.5	12.25
Healthy Choice Traditional	0.40	19.5	0.00	20	−0.5	0.25
Master Choice Chunky Garden Veg.	0.56	13	0.08	17	−4	16
Meijer All Natural Meatless	0.55	15	0.08	17	−2	4
Newman's Own Traditional	0.48	17	0.12	13.5	3.5	12.25
Paul Newman Venetian	0.48	17	0.12	13.5	3.5	12.25
Prego Fresh Mushrooms	1.25	1	0.38	1	0	0
Prego Hearty Meat—Hamburger	1.00	3.5	0.29	4	−0.5	0.25
Prego Hearty Meat—Pepperoni	1.00	3.5	0.33	2.5	1	1
Prego Roasted Red Pepper & Garlic	0.92	5	0.25	5.5	−0.5	0.25
Prego Traditional	1.17	2	0.33	2.5	−0.5	0.25
Ragu Old World Style w/meat	0.67	8	0.25	5.5	2.5	6.25
Ragu Roasted Garlic	0.70	7	0.19	10	−3	9
Ragu Roasted Red Pepper & Onion	0.86	6	0.20	8	−2	4
Ragu Traditional	0.56	13	0.20	8	5	25
Sutter Home Tomato & Garlic	0.64	10	0.16	11	−1	1
Column Sum		210		210	0	118.5

column must always be $n(n + 1)/2$, which in our case is $(20)(20 + 1)/2 = 210$. After checking the ranks, the difference in ranks d_i is computed for each observation. As a further check on our calculations, we verify that the rank differences sum to zero (if not, we have made an error somewhere). The sample rank correlation coefficient $r_s = .9109$ indicates positive agreement:

$$r_s = 1 - \frac{6 \sum_{i=1}^{n} d_i^2}{n(n^2 - 1)} = 1 - \frac{(6)(118.5)}{(20)(20^2 - 1)} = .9109$$

Our sample correlation r_s is 0.9109. For a right-tailed test, the hypotheses are

H_0: True rank correlation is zero ($\rho_s \leq 0$)

H_1: True rank correlation is positive ($\rho_s > 0$)

In this case, we choose a right-tailed test because *a priori* we would expect positive agreement. That is, a pasta sauce that ranks high in fat calories would be expected also to rank high in total calories. If the sample size is small, a special table is required (see end-of-chapter Related Reading). If n is large (usually defined as at least 20 observations), then r_s may be assumed to follow the normal distribution using the test statistic

(16.13) $$z_{calc} = r_s \sqrt{n - 1}$$

To illustrate this formula, we will plug in our previous sample result:

$$z_{calc} = (.9109) \sqrt{20 - 1} = 3.971$$

Using Appendix C, we obtain one-tail critical values of z for various levels of significance:

α	z_α	Interpretation
.025	1.960	Reject H_0
.01	2.326	Reject H_0
.005	2.576	Reject H_0

Clearly, we can reject the hypothesis of no correlation at any of the customary α levels. Using MegaStat, we can obtain equivalent results, as shown in Figure 16.12, except that the critical value of r_s is shown instead of the *t* statistic.

Figure 16.12

MegaStat's Rank Correlation Test

Spearman Coefficient of Rank Correlation

	Total Calories/gram	Fat Calories/gram
Total Calories/gram	1.000	
Fat Calories/gram	.911	1.000

20 sample size

±.444 critical value .05 (two-tail)
±.561 critical value .01 (two-tail)

Correlation versus Causation

One final word of caution: You should remember that correlation does not imply causation. Countless examples can be found of correlations that are "significant" even when there is no causal relation between the two variables. On the other hand, causation is not ruled out. More than one scientific discovery has occurred because of an unexpected correlation. Just bear in mind that if you look at 1,000 correlation coefficients in samples drawn from uncorrelated populations, approximately 50 will be "significant" at $\alpha = .05$, approximately 10 will be "significant" at $\alpha = .01$, and so on. Testing for significance is just one step in the scientific process.

Bear in mind also that multiple causes may be present. Correlation between *X* and *Y* could be caused by an unspecified third variable *Z*. Even more complex systems of causation may exist. Bivariate correlations of any kind must be regarded as potentially out of context if the true relationship is *multivariate* rather than *bivariate*.

16.13 Profits of 20 consumer food companies are shown. (a) Convert the data to ranks. Check the column sums. (b) Calculate Spearman's rank correlation coefficient. Show your calculations. (c) At $\alpha = .01$, can you reject the hypothesis of zero rank correlation? (d) Check your work by using MegaStat. (e) Calculate the Pearson correlation coefficient (using Excel). (f) Why might the rank correlation be preferred? **Food-B**

Section Exercises

Mc Graw Hill **connect**

Profit of 20 Food Consumer Products Firms ($ millions)		
Company	**2004**	**2005**
Campbell Soup	595	647
ConAgra Foods	775	880
Dean Foods	356	285
Del Monte Foods	134	165
Dole Food	105	134
Flowers Foods	15	51
General Mills	917	1,055
H. J. Heinz	566	804
Hershey Foods	458	591
Hormel Foods	186	232
Interstate Bakeries	27	−26
J. M. Smucker	96	111
Kellogg	787	891
Land O'Lakes	107	21
McCormick	211	215
PepsiCo	3,568	4,212
Ralcorp Holdings	7	65
Sara Lee	1,221	1,272
Smithfield Foods	26	227
Wm. Wrigley, Jr.	446	493

Source: Sampled firms and data are from *Fortune* 151, no. 8 (April 18, 2005), p. F-52.

16.14 Rates of return on 24 mutual funds are shown. (a) Convert the data to ranks. Check the column sums. (b) Calculate Spearman's rank correlation coefficient. Show your calculations. (c) At $\alpha = .01$, can you reject the hypothesis of zero rank correlation? (d) Check your work by using MegaStat. (e) Calculate the Pearson correlation coefficient (using Excel). (f) In this case, why might either test be used? 📷 **Funds**

Rates of Return on 24 Selected Mutual Funds (percent)

Fund	12-Mo.	5-Year	Fund	12-Mo.	5-Year
1	11.2	10.5	13	14.0	9.7
2	−2.4	5.0	14	11.6	14.7
3	8.6	8.6	15	13.2	11.8
4	3.4	3.7	16	−1.0	2.3
5	3.9	−2.9	17	6.2	10.5
6	10.3	9.6	18	21.1	9.0
7	16.1	14.1	19	−1.2	3.0
8	6.7	6.2	20	8.7	7.1
9	6.5	7.4	21	9.7	10.2
10	11.1	14.0	22	0.4	9.3
11	8.0	7.3	23	0.9	6.0
12	11.2	14.2	24	12.7	10.0

Chapter Summary

Statisticians are attracted to **nonparametric tests** because they avoid the restrictive assumption of normality, although often there are still assumptions to be met (e.g., similar population shape). Many nonparametric tests have similar power to their parametric counterparts (and superior power when samples are small). The **runs test** (or **Wald-Wolfowitz** test) checks for random order in binary data. The **Wilcoxon signed-rank test** resembles a parametric one-sample t test, most often being used as a substitute for the parametric paired-difference t test. The **Wilcoxon rank sum test** (also called the **Mann-Whitney test**) compares medians in independent samples, resembling a parametric two-sample t test. The **Kruskal-Wallis test** is a c-sample comparison of medians (similar to one-factor ANOVA). The **Friedman test** resembles a randomized block ANOVA except that it compares medians instead of means. **Spearman's rank correlation coefficient** is like the usual Pearson correlation coefficient except the data are ranks. Calculations of these tests are usually done by computer. Special tables are required when samples are small (see Related Reading or check the Internet for small-sample procedures and tables of critical values, e.g., http://en.wikipedia.org/wiki/Mann-Whitney_U_test).

Key Terms

Friedman test	parametric tests	Spearman's rank correlation	Wilcoxon rank sum test
Kruskal-Wallis test	power	coefficient	Wilcoxon signed-rank test
Mann-Whitney test	ranks	Spearman's rho	
nonparametric tests	runs test	Wald-Wolfowitz test	

Commonly Used Formulas

Wald-Wolfowitz one-sample runs test for randomness (for $n_1 \geq 10$, $n_2 \geq 10$):

$$z_{calc} = \frac{R - \dfrac{2 n_1 n_2}{n} + 1}{\sqrt{\dfrac{2 n_1 n_2 (2 n_1 n_2 - n)}{n^2 (n - 1)}}}$$

where

R = number of runs

n = total sample size = $n_1 + n_2$

Wilcoxon signed-rank test for one sample median (for $n \geq 20$):

$$z_{\text{calc}} = \frac{W - \frac{n(n+1)}{4}}{\sqrt{\frac{n(n+1)(2n+1)}{24}}}$$

where

$$W = \min(\text{sum of } R^+, \text{ sum of } R^-)$$

Wilcoxon rank sum (Mann-Whitney) test for equality of two medians (for $n_1 \geq 10, n_2 \geq 10$):

$$z_{\text{calc}} = \frac{\overline{T}_1 - \overline{T}_2}{(n_1 + n_2)\sqrt{\frac{n_1 + n_2 + 1}{12 n_1 n_2}}}$$

where

$\overline{T}_1$ = mean rank for smaller sample

$\overline{T}_2$ = mean rank for larger sample

Alternate form of Wilcoxon rank sum (Mann-Whitney) test for two medians ($n_1 \geq 10, n_2 \geq 10$):

$$z_{\text{calc}} = \frac{T_1 - E(T_1)}{\sqrt{\text{Var}(T_1)}} = \frac{T_1 - \frac{n_1(n_1 + n_2 + 1)}{2}}{\sqrt{\frac{n_1 n_2}{12}(n_1 + n_2 + 1)}}$$

Kruskal-Wallis test for equality of c medians:

$$H_{\text{calc}} = \frac{12}{n(n+1)} \sum_{j=1}^{c} \frac{T_j^2}{n_j} - 3(n+1) \quad \text{with } d.f. = c - 1$$

where

n_j = number of observations in group j

T_j = sum of ranks for group j

$n = n_1 + n_2 + \cdots + n_c$

Friedman test for equality of medians in an array with r rows and c columns:

$$F_{\text{calc}} = \frac{12}{rc(c+1)} \sum_{j=1}^{c} T_j^2 - 3r(c+1) \quad \text{with } d.f. = c - 1$$

where

r = the number of blocks (rows)

c = the number of treatments (columns)

T_j = the sum of ranks for treatment j

Spearman's rank correlation coefficient for n paired observations (for $n \geq 20$):

$$r_s = 1 - \frac{6 \sum_{i=1}^{n} d_i^2}{n(n^2 - 1)} \quad \text{and} \quad z_{\text{calc}} = r_s \sqrt{n - 1}$$

where

d_i = difference in ranks for case i

n = sample size

Chapter Review

1. (a) Name three advantages of nonparametric tests. (b) Name two deficiencies in data that might cause us to prefer a nonparametric test. (c) Why is significance in a nonparametric test especially convincing?

2. (a) What is the purpose of a runs test? (b) How many runs of each type are needed for a large-sample runs test? (c) Give an example of a sequence containing runs and count the runs. (d) What distribution do we use for the large-sample runs test?

3. (a) What is the purpose of a Wilcoxon signed-rank test? (b) How large a sample is needed to use a normal table for the test statistic? (c) The Wilcoxon signed-rank test resembles which parametric test(s)?

4. (a) What is the purpose of a Wilcoxon rank sum (Mann-Whitney) test? (b) The Wilcoxon rank sum (Mann-Whitney) test is a test of two medians under what assumption? (c) What sample sizes are needed for the large-sample Wilcoxon rank sum (Mann-Whitney) test? (d) The Wilcoxon rank sum (Mann-Whitney) test is analogous to which parametric test?

5. (a) In the Wilcoxon rank sum (Mann-Whitney) test, how are ranks assigned when there is a tie? (b) What distribution do we use for the large-sample version of this test?

6. (a) What is the purpose of a Kruskal-Wallis test? (b) The K-W test is a test of c medians under what assumption? (c) The K-W test is analogous to which parametric test?

7. (a) In the Kruskal-Wallis test, explain the procedure for assigning ranks to observations in each group. (b) What distribution do we use for the K-W test? (c) What are the degrees of freedom for the K-W test?

8. (a) What is the purpose of a Friedman test? (b) The Friedman test is analogous to what parametric test? (c) How does the Friedman test differ from the ANOVA test in the way it handles the blocking factor?

9. (a) Describe the assignment of ranks in the Friedman test. (b) What distribution do we use for the Friedman test? (c) What are the degrees of freedom for the Friedman test?

10. (a) What is the purpose of the Spearman rank correlation test? (b) Describe the way in which ranks are assigned in calculating the Spearman rank correlation test.

11. (a) Why is a significant correlation not proof of causation? (b) When might a bivariate correlation be misleading?

Chapter Exercises

Instructions: In all exercises, you may use a computer package (e.g., MegaStat, Minitab) or show Excel calculations in a worksheet, depending on your instructor's wishes. Include relevant output or screen shots to support your answers. If you do the calculations yourself, show your work. State the hypotheses and give the test statistic and its two-tailed p-value. Make the decision. If the decision is close, say so. Are there issues of sample size? Is non-normality a concern?

16.15 A supplier of laptop PC power supplies uses a control chart to track the output (in watts) of each unit produced. The pattern below shows whether each unit's output was above (A) or below (B) the desired specification. *Research question:* At $\alpha = .05$, do the deviations follow a random pattern? 📁 **Watts**

B A A B B B A B A B A B A A B A A B B A B B B A A B A B A
A A B B A A A A B B A A B A A A A B B A A B A A

16.16 A basketball player took 35 free throws during the season. Her sequence of hits (H) and misses (M) is shown. *Research question:* At $\alpha = .0$, is her hit/miss sequence random? 📁 **FreeThrows**

H M M H H M H M M H H H H H H M M H H M M H M H M H H H M H H
H H M M M H H

16.17 Thirty-four customers at Starbucks either ordered coffee (C) or did not order coffee (X). *Research question:* At $\alpha = .05$, is the sequence random? 📁 **Starbucks**

C X C X C C C C X X X X C X C X C X C C C X C X C C X C X X X
C C X

16.18 The price of a particular stock over a period of 60 days rises (+) or declines (−) in the following pattern. *Research question:* At $\alpha = .05$, is the pattern random? 📁 **Stock**

+ + − − − + + + + + + + − − − − + + − + − + − +
− − − − + + + + − + + + + − + + + − + − + − + +
+ − − − − − − + + + + − −

16.19 A forecasting model is fitted to sales data over 24 months. Forecasting errors are tabulated to reveal whether the model provides an overestimate (+) or an underestimate (−) for each month's sales. The results are − − + + + − + − − + + − − − − − − − + + + + + − −. *Research question:* At $\alpha = .05$, is the pattern random? 📁 **Forecast**

16.20 A cognitive retraining clinic assists outpatient victims of head injury, anoxia, or other conditions that result in cognitive impairment. Each incoming patient is evaluated to establish an appropriate treatment program and estimated length of stay (ELOS is always a multiple of 4 weeks because treatment sessions are scheduled on a monthly basis). To see if there is any difference in ELOS between the two clinics, a sample is taken, consisting of all patients evaluated at each clinic during October, with the results shown. *Research question:* At $\alpha = .10$, do the medians differ? 📁 **Cognitive**

Clinic A (10 patients):   24, 24, 52, 30, 40, 40, 18, 30, 18, 40

Clinic B (12 patients):   20, 20, 52, 36, 36, 36, 24, 32, 16, 40, 24, 16

16.21 Two manufacturing facilities produce 1280 × 1024 light-emitting diode (LED) displays. Twelve shipments are tested at random from each lab, and the number of bad pixels per billion is noted for each shipment. *Research question:* At $\alpha = .05$, do the medians differ? 📁 **LED**

Lab *A*   422, 319, 326, 410, 393, 368 497, 381, 515, 472, 423, 355

Lab *B*   497, 421, 408, 375, 410, 489 389, 418, 447, 429, 404, 477

16.22 A salary equity study compared two industries. Salaries of 30 randomly chosen individuals in the same occupation were selected from each industry. (Only the first 3 and last 3 observations are shown.) *Research question:* Without assuming normality of the populations, is there a difference in the medians at $\alpha = .01$? 📁 **Salaries**

Industry A	Industry B
53,599	62,092
56,107	64,928
69,957	64,993
.	.
.	.
.	.
99,180	122,001
92,863	123,189
108,101	120,571

16.23 Does a class break stimulate the pulse? Here are heart rates for a sample of 30 students before and after a class break. *Research question:* At $\alpha = .05$, do the medians differ? *Note:* Only the first 3 and last 3 observations are shown. 📁 **HeartRate**

Heart Rate before and after Class Break		
Student	**Before**	**After**
1	60	62
2	70	76
3	77	78
.	.	.
.	.	.
.	.	.
28	66	67
29	59	63
30	98	82

Thanks to colleague Gene Fliedner for having his evening students take their own pulses before and after the class break.

16.24 An experimental bumper was designed to reduce damage in low-speed collisions. This bumper was installed on an experimental group of vans in a large fleet, but not on a control group. At the end of a trial period, accident data showed 12 repair incidents for the experimental group and 9 repair incidents for the control group. The vehicle downtime (in days) per repair incident is shown. *Research question:* At $\alpha = .05$, do the medians differ? 📁 **Downtime**

New bumper: 9, 2, 5, 12, 5, 4, 7, 5, 11, 3, 7, 1
Old bumper: 7, 5, 7, 4, 18, 4, 8, 14, 13

16.25 The square footage of each of the last 11 homes sold in each of two suburban neighborhoods is noted. *Research question:* At $\alpha = .01$, do the medians differ? 📁 **SqFt**

Square Footage of Homes Sold

Grosse Hills (Built in 1985)	Haut Nez Estates (Built in 2003)	Grosse Hills (Built in 1985)	Haut Nez Estates (Built in 2003)
3,220	3,850	2,800	3,400
3,450	3,560	3,050	3,550
3,270	4,300	2,950	3,750
3,200	4,100	3,430	4,150
4,850	3,750	3,220	3,850
3,150	3,450		

16.26 Below are grade point averages for 25 randomly chosen university business students during a recent semester. *Research question:* At $\alpha = .01$, are the median grade point averages the same for students in these four class levels? 📁 **GPA**

Grade Point Averages of 25 Business Students

Freshman (5 students)	Sophomore (7 students)	Junior (7 students)	Senior (6 students)
1.91	3.89	3.01	3.32
2.14	2.02	2.89	2.45
3.47	2.96	3.45	3.81
2.19	3.32	3.67	3.02
2.71	2.29	3.33	3.01
	2.82	2.98	3.17
	3.11	3.26	

16.27 In a bumper test, three types of autos were deliberately crashed into a barrier at 5 mph, and the resulting damage (in dollars) was estimated. Five test vehicles of each type were crashed, with the results shown below. *Research question:* At $\alpha = .01$, are the median crash damages the same for these three vehicles? 📁 **Crash**

Crash Damage in Dollars

Goliath	Varmint	Weasel
1,600	1,290	1,090
760	1,400	2,100
880	1,390	1,830
1,950	1,850	1,250
1,220	950	1,920

16.28 The waiting time (in minutes) for emergency room patients with non-life-threatening injuries was measured at four hospitals for all patients who arrived between 6:00 and 6:30 p.m. on a certain Wednesday. The results are shown at the end of this exercise. *Research question:* At $\alpha = .05$, are the median waiting times the same for emergency patients in these four hospitals? 📁 **Emergency**

Emergency Room Waiting Time (minutes)

Hospital A (5 patients)	Hospital B (4 patients)	Hospital C (7 patients)	Hospital D (6 patients)
10	8	5	0
19	25	11	20
5	17	24	9
26	36	16	5
11		18	10
		29	12
		15	

16.29 Mean output of arrays of solar cells of three types are measured six times under random light intensity over a period of 5 minutes, yielding the results shown below. *Research question:* At $\alpha = .05$, is the median solar cell output the same for all three types? 📁 **Solar**

Solar Cell Output (Watts)

Cell Type	Output (Watts)					
A	123	121	123	124	125	127
B	125	122	122	121	122	126
C	126	128	125	129	131	128

16.30 Below are results of braking tests of a certain SUV on glare ice, packed snow, and split traction (one set of wheels on ice, the other on dry pavement), using three braking methods. *Research question:* At $\alpha = .01$, is braking method related to stopping distance? 📁 **Stopping**

Stopping Distance from 40 mph to Zero

Road Condition	Pumping	Locked	ABS
Glare Ice	441	455	460
Split Traction	223	148	183
Packed Snow	149	146	167

16.31 In a call center, the average waiting time for an answer (in seconds) is shown below by time of day. *Research question:* At $\alpha = .01$, does the waiting time differ by day of the week? *Note:* Only the first 3 and last 3 observations are shown. 📁 **Wait**

Average Waiting Time (in Seconds) for Answer (*n* = 26)

Time	Mon	Tue	Wed	Thu	Fri
06:00	34	71	33	39	39
06:30	52	70	88	53	49
07:00	36	103	47	32	91
.	.	.	.	.	.
.	.	.	.	.	.
.	.	.	.	.	.
17:30	28	31	27	22	26
18:00	35	14	115	26	22
18:30	25	34	9	5	47

16.32 The table below shows annual financial data for a sample of 20 companies in the food consumer products sector. *Research question:* At $\alpha = .01$, is there a significant correlation between revenue and profit? Why is a rank correlation preferred? What factors might result in a less-than-perfect correlation? *Note:* Only the first 3 and last 3 companies are shown. 🖫 **Food-A**

Food Consumer Products Companies' Revenue and Profit ($ millions)

Obs	Company	Revenue	Profit
1	Campbell Soup	7,109	647
2	ConAgra Foods	18,179	880
3	Dean Foods	10,822	285
.	.	.	.
.	.	.	.
.	.	.	.
18	Sara Lee	19,556	1,272
19	Smithfield Foods	10,107	227
20	Wm. Wrigley, Jr.	3,649	493

Source: Sample of firms was taken from *Fortune* 151, no. 8 (April 18, 2005), p. F-52.

16.33 Fertility rates (children born per woman) are shown for 27 EU member nations in 2 years. *Research question:* At $\alpha = .05$, is there a significant rank correlation? *Note:* Only the first 3 and last 3 nations are shown. 🖫 **Fertility**

Fertility Rates in EU Member Nations ($n = 27$)

Nation	2000	2009
Austria	1.364	1.390
Belgium	1.667	1.840
Bulgaria	1.261	1.568
.	.	.
.	.	.
Slovenia	1.259	1.533
Sweden	1.544	1.935
United Kingdom	1.641	1.938

Note: For recent data, see http://ec.europa.eu/eurostat/data/database.

16.34 A newspaper article listed nutritional facts for 56 frozen dinners. From that list, 16 frozen dinners were randomly selected by using the random number method. *Research question:* Choose any two variables. At $\alpha = .01$, based on this sample shown here, is there a significant rank correlation between the two variables? *Note:* Only the first 3 and last 3 observations are shown.

Frozen Dinner Nutritional Information ($n = 16$)
🖫 **Dinners**

Dinner/Entree	Fat (g)	Calories	Sodium (mg)
French Recipe Chicken	9	240	1,000
Chicken au Gratin	11	250	870
Stuffed Turkey Breast	6	230	520
.	.	.	.
.	.	.	.
.	.	.	.
Filet of Fish au Gratin	6	200	700
Beef Sirloin Tips	7	220	540
Lasagna with Meat Sauce	10	320	630

16.35 The table below shows ratings of 18 movies by two reviewers (on a 0 to 5 ★ scale using half ★ increments). *Research question:* At $\alpha = .05$, based on these data, is the rank correlation between reviewers significantly greater than zero (i.e., a right-tail test)? 🖫 **Reviews**

Movie	Reviewer A	B	Movie	Reviewer A	B
Escape Room	3.5	2.0	Pet Sematary	2.5	3.0
A Dog's Way Home	4.0	4.5	Shazam!	1.5	1.5
Glass	4.0	4.5	Hellboy	2.5	2.0
What Men Want	3.0	2.5	Her Smell	0.5	1.0
Captain Marvel	3.5	3.0	Long Shot	2.0	2.5
The Kid	2.5	1.5	The Sun Is Also a Star	3.0	2.5
Never Grow Old	2.5	3.5	Booksmart	4.0	3.5
Us	3.5	3.0	Shaft	3.0	2.0
Dumbo	2.5	2.0	Good Boys	3.0	3.0

16.36 Are gasoline prices a potential policy tool in controlling carbon emissions? The table below shows 2001 gasoline prices (dollars per liter) and carbon dioxide emissions per dollar of GDP. *Research question:* At $\alpha = .05$, based on these data, is there a significant rank correlation between these two variables? *Note:* Only the first 3 and last 3 observations are shown. 🖫 **Emissions**

Gasoline Prices and Carbon Emissions for Selected Nations (*n* = 31)

Nation	Gas Price ($/L)	CO$_2$/GDP (kg/$)
Australia	0.489	0.79
Austria	0.888	0.25
Belgium	0.984	0.37
.	.	.
.	.	.
.	.	.
Turkey	1.003	0.99
United Kingdom	1.165	0.41
United States	0.381	0.63

Source: International Energy Agency, www.iea.org.

16.37 The top 20 U.S. football teams in the seventh and eighth weeks of a recent season are shown here, along with the points awarded to each team by a poll of coaches. *Research question:* At α = .01, based on these data, does the true rank correlation differ from zero for these two ratings?

Football Ratings in Coaches Poll (*n* = 20)
(table is continued in next column)   Teams

Team	This Week	Last Week
Oklahoma	1575	1622
Southern Cal	1502	1470

Team	This Week	Last Week
Florida State	1412	1320
LSU	1337	1241
Virginia Tech	1281	1026
Miami	1263	1563
Ohio State	1208	1226
Michigan	1135	938
Georgia	951	1378
Iowa	932	762
Texas	881	605
TCU	875	727
Wash State	827	1260
Purdue	667	487
Michigan State	645	1041
Nebraska	558	924
Tennessee	544	449
Minnesota	490	149
Florida	480	246
Bowling Green	369	577

Related Reading

Corder, Gregory W., and Dale I. Foreman. *Nonparametric Statistics for Non-Statisticians: A Step-by-Step Approach.* 2nd ed. Wiley, 2014.

Higgins, James J. *Introduction to Modern Nonparametric Statistics.* Cengage, 2004.

Hollander, Myles, Douglas A. Wolfe, and Eric Chicken. *Nonparametric Statistical Methods.* 3rd ed. Wiley, 2014.

Huber, Peter J. *Robust Statistics.* 2nd ed. Wiley, 2009.

CHAPTER 16   More Learning Resources

You can access these *LearningStats* demonstrations through Connect to help you understand nonparametric tests.

Topic	LearningStats Demonstrations
Overview	What Are Nonparametric Tests? Nonparametric Tests Using R
Case studies	Runs Test: Baseball Streaks Wilcoxon Signed-Rank: Exam Scores Mann-Whitney: ATM Withdrawals Kruskal-Wallis: DVD Prices Friedman Test: Freeway Pollution Spearman's Rho: EU Nations Fertility
Tables	Chi-Square Critical Values
Supplemental Topics	Sign Test Mann-Whitney Small Sample Test

Key: = PowerPoint = Excel = Adobe PDF

Software Supplement

Nonparametric Tests in R

Excel has no nonparametric tests, so R offers a distinct advantage. Further, the syntax of R nonparametric functions is simple, there are many options, and the results are compact.

Kruskal-Wallis Test 🖅 **Days**

The Kruskal-Wallis test in R accepts data columns as they might appear in an Excel spreadsheet. We import three columns (with headings) showing annual days worked (Days) by employees in three departments. Unequal sample sizes are acceptable.

```
> Days=read.table(file="clipboard",sep="\t",header=TRUE)
> Days
```

	Budgets	Payables	Pricing
1	278	205	240
2	260	270	258
3	265	220	233
4	245	240	256
5	258	255	233
6	NA	217	242
7	NA	266	244
8	NA	239	249
9	NA	240	NA
10	NA	228	NA

```
< kruskal.test(Days)
   Kruskal-Wallis rank sum test
data: Days
Kruskal-Wallis chi-squared = 6.278, df = 2, p-value = 0.04333
```

Or we could use stacked data with Dept as a factor (abbreviated list shown):

```
> Days=read.table(file="clipboard",sep="\t",header=TRUE)
> Days
```

	Days	Dept
1	278	Budgets
2	260	Budgets
3	265	Budgets
.	.	.
.	.	.
21	242	Pricing
22	244	Pricing
23	249	Pricing

```
> kruskal.test(Days$Days,Days$Dept)
   Kruskal-Wallis rank sum test
data: Days$Days and Days$Dept
Kruskal-Wallis chi-squared = 6.278, df = 2, p-value = 0.04333
```

Wilcoxon Signed Rank Test One Sample 🖅 **WilcoxonA**

Does the median PE ratio from a sample of 21 *specialty* stores differ from the median of 20.2 for all *multiline* stores (e.g., Target)?

We have one data column (abbreviated below).

```
> PE=read.table(file="clipboard",sep="\t",header=TRUE)
> PE
```

1	19.8
2	19.8
3	20.6
.	.
.	.
.	.
19	40.4
20	41.0
21	124.7

```
> wilcox.test(PE$PE,mu=20.2,paired=FALSE,alternative="two.sided")
   Wilcoxon signed rank test with continuity correction
data: PE$PE
V = 128, p-value = 0.6766
alternative hypothesis: true location is not equal to 20.2
Warning message: cannot compute exact p-value with ties
```

Wilcoxon Signed Rank Test Two Sample Paired 🖅 **WilcoxonB**

Based on a sample of PE ratios from 23 firms, did the PE ratios decline between year 2 (PEYr2) and year 1 (PEYr1)? Paired columns are shown below (abbreviated).

```
> PEData=read.table(file="clipboard",sep="`\t",header=TRUE)
> PEData
```

	PEYr2	PEYr1
1	14	14
2	18	18
3	14	14
.	.	.
.	.	.
.	.	.
21	8	26
22	13	55
23	15	68

```
> wilcox.test(PEData$PEYr2,PEData$PEYr1,paired=TRUE,
alternative="less")
   Wilcoxon signed rank test with continuity correction
data: PEData$PEYr2 and PEData$PEYr1
V = 51.5, p-value = 0.02385
alternative hypothesis: true location shift is less than 0
Warning messages:
   cannot compute exact p-value with ties
   cannot compute exact p-value with zeroes
```

Wilcoxon Rank Sum (Mann-Whitney) Test 🖅 **Restaurants**

From a sample of 29 restaurants, can we conclude that customer satisfaction (0 to 100 scale) depends on price (low = 1, high = 2)? We import paired columns Satisfaction and Price. The Mann-Whitney test in R requires the factor (Price) to be numeric and uses the ~ to specify the model (Satisfaction ~ Price).

```
> RestData=read.table(file="clipboard",sep="\t",header=TRUE)
> RestData
       Satisfaction   Price
  1        74           1
  2        78           1
  3        78           2
  .         .           .
  .         .           .
  .         .           .
  27       88           2
  28       88           2
  29       88           2
> wilcox.test(RestData$Satisfaction~RestData$Price,alternative="two.
sided")
    Wilcoxon rank sum test with continuity correction
data: RestData$Satisfaction by RestData$Price
W = 44, p-value = 0.008028
alternative hypothesis: true location shift is not equal to 0
Warning message: cannot compute exact p-value with ties
```

Friedman Test Braking

Braking distance in feet (Dist) is measured for each of 5 types of brake pads (Pad) in 4 trials (Trial). We coded factors numerically, although this is not necessary (abbreviated list below).

```
> Dist=read.table(file="clipboard",sep="\t",header=TRUE)
> Dist Pad Trial
  1       166       1        1
  2       174       1        2
  3       184       1        3
  .        .        .        .
  .        .        .        .
  18      152       5        2
  19      168       5        3
  20      196       5        4
```

```
> friedman.test(Dist$Dist,Dist$Pad,Dist$Trial)
    Friedman rank sum test
data: Dist$Dist, Dist$Pad and Dist$Trial
Friedman chi-squared =13.6, df = 4, p-value = 0.008687
```

Spearman Test Pasta

Find the rank correlation between total calories (Calories) and fat calories (Fat) in 20 brands of pasta (abbreviated list below). Import the named data columns to an object named FatData.

```
> FatData=read.table(file="clipboard",sep="\t",header=TRUE)
> FatData
       Calories      Fat
  1      0.64        0.20
  2      0.56        0.12
  3      0.40        0.08
  .       .           .
  .       .           .
  18     0.70        0.19
  19     0.56        0.20
  20     0.64        0.16
> cor(FatData$Calories,FatData$Fat,method="spearman")
[1] 0.909751
```

CHAPTER CONTENTS

CHAPTER LEARNING OBJECTIVES

When you finish this chapter, you should be able to

LO 17-1   Define quality and explain how it may be measured.

LO 17-2   Name key individuals and their contributions to the quality movement.

LO 17-3   List steps and common analytical tools for quality improvement.

LO 17-4   Define a control chart and the types of variables displayed.

LO 17-5   Make and interpret control charts for a mean.

LO 17-6   Make and interpret control charts for a range.

LO 17-7   Make and interpret control charts for attribute data.

LO 17-8   Recognize abnormal patterns in control charts and their potential causes.

LO 17-9   Assess the capability of a process.

LO 17-10 Identify topics commonly associated with quality management (optional).

What Is Quality?

Quality can be measured in many ways. Quality may be a *physical* metric, such as the number of bad sectors on a computer hard disk or the quietness of an air-conditioning fan. Quality may be an *aesthetic* attribute such as the ripeness of a banana or cleanliness of a clinic waiting room. (Does the fig bar turned in the wrong direction in Figure 17.1 affect the aesthetic quality of the product?) Quality may be a *functional* characteristic such as ergonomic accessibility of car radio controls or convenience of hours that a bank is open. It may be a *personal* attribute such as friendliness of service at a restaurant or diligence of follow-up by a veterinary clinic. It may be an *efficiency* attribute such as promptness in delivery of an order or the waiting time at a dentist's. Quality is generally understood to include these attributes:

- Conformance to specifications.
- Performance in the intended use.
- As near to zero defects as possible.
- Reliability and durability.
- Serviceability when needed.
- Favorable customer perceptions.

Ariel Skelley/Blend Images/Getty Images

Measurement of quality is specific to the organization and its products. To improve quality, we must undertake systematic data collection and careful measurement of key metrics that describe the product or service that is valued by customers. In manufacturing, the focus is likely to be on physical characteristics (e.g., defects, reliability, consistency), while in services, the focus is likely to be on customer perceptions (e.g., courtesy, responsiveness, competence). Table 17.1 lists some typical quality indicators that might be important to firms engaged in manufacturing as compared to firms that deliver services.

LO 17-1

Define quality and explain how it may be measured.

David Doane

Figure 17.1

Unopened Fig Bars: Quality as an Aesthetic Attribute

Manufacturing	Services
Proportion of nonconforming output	Proportion of satisfied customers
Warranty claim costs	Average customer waiting time
Repeat purchase rate (loyalty)	Repeat client base (loyalty)

Table 17.1

Typical Quality Indicators

Productivity, Processes, and Quality

Productivity (ratio of output to input) is a measure of efficiency. High productivity lowers cost per unit, increases profit, and supports higher wages and salaries. Productivity, like quality, can be measured in various ways. There are single-factor ratios that compare process output to a single input such as labor. And there are multifactor ratios that compare process output to the sum of multiple inputs such as labor, energy, and materials. Measuring productivity in a manufacturing business is typically straightforward because the output and inputs are easy to quantify. Productivity in a service business is less straightforward and will depend on the type of service. For example, a professional service firm might compare number of clients serviced to billable hours. On the other hand, a restaurant might compare the number of tables serviced to operating costs for an evening.

In the past, many companies assumed an inverse relationship between quality and productivity. This view was based on a short-term perspective. The belief was that the only way to improve quality was to slow down and put more time into each product. While slowing or stopping an assembly line does imply less output, defective products lead to waste, rework, and lost customers. In the modern view, quality and productivity move in the same direction because doing it right the first time saves time and money in the long run. Similarly, in services (e.g., health care), reducing delays and avoiding errors will make customers happier and will reduce the burden of follow-up to fix problems. Our focus in the 21st century is on designing and operating effective business processes to meet customer requirements consistently.

A **process** is a sequence of interconnected tasks that result in the creation of a product or in the delivery of a service. Manufacturing, assembly, or packaging operations usually come to mind when we hear the term "process." Yet service operations such as filling mail orders, providing customer support, handling loan applications, delivering health services, and meeting payrolls are also processes. Because a majority of workers are in the service sector, it is reasonable to say that nonmanufacturing processes are predominant in today's economy. **Quality control** refers to methods used by organizations to ensure that their products and services meet customer expectations and to ensure that there is improvement over time. **Process control** refers to methods used by organizations to ensure that their processes are predictable and produce products and services that meet customers' expectations in an efficient manner.

Common Cause versus Special Cause Variation

Where does statistics enter the quality picture? Statistics focuses on the phenomenon of *variation.* Processes that produce, package, and deliver supposedly identical products and services will always contain some degree of variation. Although variation is normal and expected, firms still strive for consistency in their products and services. Excessive variation is often a sign of poor quality because this affects real or perceived performance of the product or service. The quest for *reduced variation* is a never-ending activity for any firm or not-for-profit organization.

Statisticians define two categories of variation. **Common cause variation** (random "noise") comes from within the process and is normal and expected. Processes that vary only due to common cause variability are considered stable and predictable. **Special cause variation** is due to factors that are outside of a process, producing a process that is unpredictable. Until special cause variation is identified and eliminated, a process is considered out of control. For example, waiting time at a ski lift is a random variable that follows a predictable pattern at different times of day (common cause variation, normal and expected). But when there is an equipment malfunction, waiting times may change dramatically (special cause variation).

Sources of variation in processes include human abilities, training, motivation, technology, materials, management, and organization. Some of these factors are under the control of the organization, while others cannot easily be changed. Most factors are fixed in the short run but may be changed in the long run. For example, technology can be changed through research and development and capital spending on new equipment, but such changes may take years. Human performance can change over time through education and training, but usually not in hours or days.

Role of Statisticians

Statisticians can help a company define appropriate metrics, set up a system to collect valid data, and track variation in the chosen metric(s). Trained statisticians know how to collect and analyze data to determine if processes are in control or out of control (i.e., contain only common cause variability or both common cause and special cause variability). We use statistics to measure variation, set attainable goals for variance reduction, and establish rules to decide whether processes are in control. The degree to which variation can be reduced depends on equipment, technology, and worker training.

Managers make the decisions necessary to invest in new equipment or technology and to train nonstatisticians, who comprise the majority of the workforce. A manufacturing firm may require broad-based statistical training for engineers, plant managers, supervisors, and even assembly workers. But financial, purchasing, marketing, and sales managers also must understand statistics because they interact with technical experts on cost control, waste management, and quality improvement. Even in banks or health care, broad-based training in statistical methods can be helpful in increasing efficiency and improving quality.

17.1 Define (a) productivity, (b) quality control, and (c) process control.

17.2 Explain the relationship between productivity and quality from a modern perspective. How does this differ from the past perspective?

17.3 Explain the difference between common cause variation and special cause variation.

17.4 Can zero variation be achieved? Explain.

17.5 Explain the role of statisticians in quality improvement.

Section Exercises

17.2 PIONEERS IN QUALITY MANAGEMENT

Brief History of Quality Control

LO 17-2

Name key individuals and their contributions to the quality movement.

During the early 1900s, quality control took the form of improved inspection and improvement in the methods of mass production, under the leadership of American experts. From about 1920 to just after World War II, techniques such as process control charts (Walter A. Shewhart) and acceptance sampling from lots (Harold F. Dodge and Harry G. Romig) were perfected and were widely applied in North America. But during the 1950s and 1960s, Japanese manufacturers (particularly automotive) shed their previous image as low-quality producers and began to apply American quality control techniques. The Japanese based their efforts largely on ideas and training from American statisticians W. Edwards Deming and Joseph M. Juran as well as Japanese statisticians Genichi Taguchi and Kaoru Ishikawa. They developed new approaches that focused on customer satisfaction and costs of quality. By the 1970s, despite exhortation by Deming and others, North American firms had lost their leadership in quality control, while the Japanese perfected new quality improvement methods, soon adopted by the Europeans.

During the 1980s, North American firms began a process of recommitment to quality improvement and Japanese production methods. In their quest for quality improvement, these firms sought training and advice by experts such as Deming, Juran, and Armand Feigenbaum. The Japanese, however, continued to push the quality frontier forward, under the teachings of Taguchi and the perfection of the Kaizen philosophy of continuous improvement. The Europeans articulated the ISO 9000 standards, now adopted by most world-class firms. North American firms now seek to build *quality* into their products and services, all the way down the supply chain. Quality is best viewed as a *management system* rather than purely as an application of statistics.

W. Edwards Deming

The late **W. Edwards Deming** (1900–1993) deserves special mention as an influential thinker. He was widely honored in his lifetime. Many know him primarily for his contributions to improving productivity and quality in Japan. In 1950, at the invitation of the Union of Japanese Scientists and Engineers, Deming gave a series of lectures to 230 leading Japanese industrialists who together controlled 80 percent of Japan's capital. His message was the same as to Americans he had taught during the previous decades. The Japanese listened carefully to his message, and their success in implementing Deming's ideas is a matter of historical record.

Deming said that *profound knowledge* of a system is needed for an individual to become a good listener who can teach others. He emphasized that all people are different, that management is not about ranking people, and that anyone's performance is governed largely by the system in which he/she works. He said that fear invites presentation of bad data. If bearers of bad news fare badly, the boss will hear only good news—guaranteeing bad management decisions.

Deming believed that most employees want to do a good job. He found that most quality problems do not stem from willful disregard of quality, but from flaws in the process or system, such as

- Inadequate equipment.
- Inadequate maintenance.
- Inadequate training.
- Inadequate supervision.
- Inadequate support systems.
- Inadequate task design.

It is difficult to encapsulate Deming's many ideas succinctly, but most observers would agree that his philosophy is reflected in his widely reproduced *14 Points,* which can be found in full on the web. The 14 Points are primarily statements about management, not statistics. They ask that management take responsibility for improving quality and avoid blaming workers. Deming spent much of his long life explaining his ideas, through a series of seminars aimed initially at management, an activity that continues today through the work of his followers at The W. Edwards Deming Institute (www.deming.org).

Other Influential Thinkers

Walter A. Shewhart (1891–1967) invented the control chart and the concepts of special cause and common cause variation. Shewhart's charts were adopted by the American Society for Testing Materials (ASTM) in 1933 and were used to improve production during World War II. *Joseph M. Juran* (1904–2008) also taught quality education in Japan, contemporaneously with Deming. Like Deming, he became more influential with North American management in the 1980s. He felt that most quality defects arise from management actions and therefore that quality control was management's responsibility. Juran articulated the idea of the *vital few*—a handful of causes that account for a vast majority of quality problems (the principle behind the **Pareto chart**). Effort, he said, should be concentrated on key problems rather than diffused over many less-important problems.

Kaoru Ishikawa is a Japanese quality expert who is associated with the idea of *quality circles,* which characterize the Japanese approach. He also pioneered the idea of company-wide quality control and was influential in popularizing statistical tools for quality control. He taught that elementary statistical tools (Pareto charts, histograms, scatter diagrams, and control charts) should be understood by everyone, while advanced tools (experimental design, regression) might best be left to specialists.

Armand V. Feigenbaum first used the term *total quality control* in 1951. He favored broad sharing of responsibility for quality assurance. This was at a time when many companies assumed that quality was the responsibility of the Quality Assurance Department alone. He felt that quality is an essential element of modern management, like marketing or finance. *Philip B. Crosby* was among the first to popularize the catch-phrase "zero defects."

Other quality gurus include *Claus Moller,* whose European company specializes in management training. Moller believes that people can be inspired to do their best through development of the individual's self-esteem. Moller is known for his 12 Golden Rules and 17 hallmarks of a quality company. *Shigeo Shingo* has had great impact on Japanese industry.

His basic idea is to stop a process whenever a defect occurs, define the cause, and prevent future occurrences. If source inspections are used, statistical sampling becomes unnecessary because the worker is prevented from making errors in the first place.

17.6 On the Internet, look up *two* influential thinkers in quality control, and briefly state their contributions.

17.7 Look up *two* of Deming's *14 Points* on the Internet and explain their meaning.

17.3 QUALITY IMPROVEMENT

Measuring Quality

Quality improvement begins with measurement of a *variable* (e.g., dimensions of an automobile door panel) or an *attribute* (e.g., number of emergency patients who wait more than 30 minutes). For a variable, quality improvement means reducing variation from the target specification. For an attribute, quality improvement means decreasing the rate of nonconformance. Statistical methods are used to ensure that the process is stable and in control by eliminating sources of *special cause* (nonrandom) variation, as opposed to *common cause* (random) variation that is normal and inherent in the process. We change the process whenever a way is discovered to reduce variation or to decrease nonconformance (especially if the process is incapable of meeting the target specifications).

> **LO 17-3**
>
> List steps and common analytical tools for quality improvement.

Data collection is essential for monitoring processes and improving quality. Employees and customers may not agree on the interpretation of quality measures, so customer input, planning, and training are essential to ensure that employees collect meaningful data that lead to improved quality and customer satisfaction. We may have to measure several aspects of the product or service to assess quality, while sometimes a single quality measurement may be sufficient. Examples of different quality measures are shown below.

Payroll department in a large hospital

- Percent of employees paid incorrectly or late.
- Percent of employees with insufficient taxes withheld.
- Number of weekly creditor telephone complaints.

Aluminum beverage container production
- Monthly hours of downtime.
- Number of defective cans per 100,000.
- Number of worker injuries per month.

Retail pharmacy
- Percent of prescriptions filled within 15 minutes.
- Average wait for phone to be answered.
- Time (in minutes) a cashier must wait for pharmacist.

Statistical quality control or **SQC** refers to a subset of quality improvement techniques that rely on statistics. A few of the descriptive tools (see Figure 17.2) have already been covered in earlier chapters, while others (e.g., control charts) will be discussed in this chapter.

Descriptive Tools	*Analytic Methods*
• Pareto diagrams	• Control charts
• Scatter plots	• Lot and batch inspection plans
• Box plots	• Acceptance sampling
• Fishbone diagrams	• Experimental design
• Check sheets	• Taguchi robust design

Figure 17.2

Descriptive SQC Tools

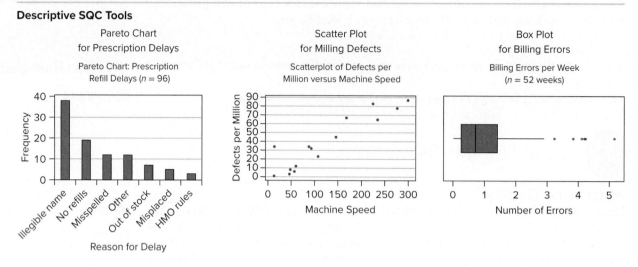

Pareto Chart
for Prescription Delays

Pareto Chart: Prescription
Refill Delays (n = 96)

Scatter Plot
for Milling Defects

Scatterplot of Defects per
Million versus Machine Speed

Box Plot
for Billing Errors

Billing Errors per Week
(n = 52 weeks)

A *check sheet* is a form for counting the frequency of sources of nonconformance. The **fishbone chart** (also called a *cause-and-effect diagram*) is a visual display that summarizes the factors that increase process variation or adversely affect achievement of the target. For example, Figure 17.3 shows a fishbone chart for factors affecting patient length of stay in a hospital. The six main categories (materials, methods, people, management, measurement, technology) are general and may apply to almost any process.

Figure 17.3

Fishbone (Cause-and-Effect) Chart for Patient Length of Stay

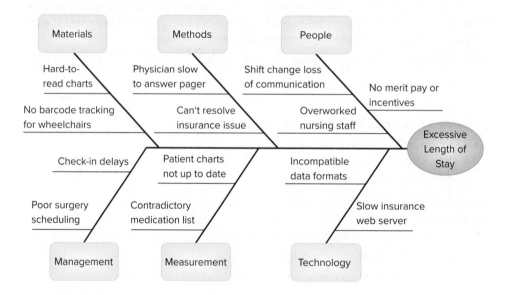

You can insert as many verbal descriptions ("fishbones") as you need to identify the causes of variation. The fishbone chart is not, strictly speaking, a statistical tool, but it is helpful in thinking about root causes. Minitab will make fishbone charts.

Statistical process control or **SPC** refers to the monitoring of ongoing repetitive processes to ensure conformance to standards by using methods of statistics. Its main tools are *capability analysis* and *control charts*. Control charts are used to monitor both quality characteristics and process characteristics. We will discuss how to create and use control charts in the next section and process capability measures later in this chapter.

Quality Improvement Programs

Businesses typically follow a disciplined approach for quality improvement using the tools we've identified in this section. Quality and process improvement programs are ubiquitous in the business world. A quick Google search using the terms *quality improvement* or *process improvement* will turn up hundreds of links to consulting firms advertising their programs to help improve your business. Each of these programs is based on some combination of the philosophies taught by the quality gurus described in Section 17.2. The following are a few that you should know before you enter the workforce.

Analytics in Action

Big Data Tracks a Virus

The Internet of things (IoT) refers to an evolving system of devices and technologies that automatically upload real-time data over networks. Examples include the "smart home" (thermostats, lighting, security systems, Amazon Echo) and personal devices (iPhones, Apple Watch, Fitbit, Siri). Applications include fields as diverse as manufacturing process control, agricultural crop monitoring, balancing electrical grid loads, and improving elder care. Digital control and predictive maintenance can save lives as well as money, although the value of data collected and benefits of remote control of homes must be weighed against privacy concerns.

During the coronavirus outbreak, a tech company called Kinsa found an unanticipated use for its "smart" thermometers. Already used by millions of customers, the devices electronically report real-time readings of body temperatures. Because customer locations are known, Kinsa's app automatically collects millions of data points that can be displayed anonymously on a map by zip code (www.healthweather.us). Statistical analysis can then be used to identify unusual spikes in fever that might indicate "hot spots" of virus activity (either seasonal flu or a new virus). Governments and hospitals are exploring whether this system or its future offshoots, linked to mobile phones, Fitbit, and other electronic devices, can become an "early warning" system of future pandemics to guide decisions about resource allocation for diagnostic testing and treatment.

Source: www.kinsahealth.co.

Total quality management or **TQM** requires that all business activities should be oriented toward meeting and exceeding customer needs, empowering employees, eliminating waste or rework, and ensuring the long-run viability of the enterprise through continuous quality improvement. TQM encompasses a broad spectrum of behavioral, managerial, and technical approaches. It includes diverse but complementary elements such as statistics, benchmarking, process redesign, team building, group communications, quality function deployment, and crossfunctional management.

Like TQM, **business process redesign** or **BPR** has a cross-functional orientation. But instead of focusing on incremental change and gradual improvement of processes, BPR seeks radical redesign of processes to achieve breakthrough improvement in performance measures—a lofty goal that is easier to state than to achieve. Business schools typically incorporate TQM and/or BPR concepts into a variety of nonstatistics core classes.

The concept of **continuous quality improvement** or **CQI** arises from the idea that we continue to seek ways to reduce variation and/or nonconformance to even lower levels. Many improvement programs have defined this never-ending cycle by stating a set of iterative steps. For example, Deming used an improvement cycle he called **PDCA**, or Plan-Do-Check-Act. In the **Six Sigma** school of thought (see Section 17.10), the steps to quality improvement are abbreviated as **DMAIC**, or define, measure, analyze, improve, and control.

Steps to Continuous Quality Improvement

- Step 1: Define a relevant, measurable parameter of the product or service.
- Step 2: Establish targets or desired specifications for the product or service.
- Step 3: Monitor the process to be sure it is stable and in control.
- Step 4: Is the process capable of meeting the desired specifications?
- Step 5: Identify sources of variation or nonconformance.
- Step 6: Change the process (technology, training, management, materials).
- Step 7: Repeat steps 3–6 indefinitely.

The Japanese are credited with perfecting and implementing the philosophy of continuous improvement, along with the related concepts of quality circles, just-in-time inventory, and robust design of products and processes (the **Taguchi method**). Different social, economic, and geographic factors prevent adoption of some Japanese approaches by North American firms, but there is general agreement on their main points. Continuous improvement now is a guiding principle for automobile manufacturers, health care providers, insurance companies, computer software designers, fast-food restaurants, and even universities, churches, entertainment, and sports teams. Permanent change and a continuous search for better ways of doing things cascade down the organizational chart and across departmental lines.

Service Quality

Service quality measures have been elusive because it is difficult to quantify intangible services. However, in 1988 a group of researchers developed a survey called **SERVQUAL** to assess customer satisfaction along five service quality dimensions. The dimensions are reliability, responsiveness, assurance, empathy, and tangibles. A *reliable* service is one that is dependable and consistent. A *responsive* service is one that meets the customers' needs. *Assurance* means that the customer can trust the service provider. *Empathy* refers to a caring service provider. *Tangibles* refer to the physical environment in which the service is delivered.

The SERVQUAL instrument measures both customer expectations and customer perceptions of the service in each of these five areas. A business uses the results of the survey to quantify the gap between customer expectations and perceptions. By identifying the largest gap, a business knows where to focus its quality improvement efforts. Quality improvement tools specific to service businesses include **service blueprints** and **service transaction analysis**. A service blueprint is a map of the service process used by a business to identify points of possible service failure. Service transaction analysis is a method of analyzing a service from the customer's perspective to determine how to close the gap between expectations and perceptions.

Section Exercises

17.8 Define a measurable aspect of quality for (a) the car dealership where you bought your car, (b) the bank or credit union where you usually make personal transactions, and (c) the movie theater where you usually go.

17.9 Explain the difference between SQC and SPC.

17.10 Identify three common quality improvement programs and give their acronyms.

17.11 Why is the quality improvement process never-ending? Identify the steps of one improvement cycle identified in the textbook.

17.12 Name the survey instrument used to measure service quality. What are the five service quality dimensions?

17.13 Describe two quality improvement tools unique to the service industry.

17.4 CONTROL CHARTS: OVERVIEW

What Is a Control Chart?

A **control chart** is a visual display used to study how a process changes over time. Data are plotted in time order. It compares the statistic with limits showing the range of expected common cause variation in the data. Control charts are tools for monitoring process stability and for alerting managers if the process changes. In some processes, inspection of every item may be possible. But random sampling is needed when measurements are costly, time-consuming, or destructive. For example, we can't test every cell phone battery's useful life, trigger every airbag to see whether it will deploy correctly, or cut open every watermelon to test for pesticides. Sample size and sampling frequency vary with the problem. In SPC the sample size n is referred to as the *subgroup size*.

For *numerical* data (sometimes called *variable* data), the control chart displays a measure of central tendency (e.g., the sample mean) and/or a measure of variation (e.g., the sample range or standard deviation). A *variable control chart* is used for measurable quantities like weight, diameter, or time. Typically, such data are found in manufacturing (e.g., dimensions of a metal fastener) but sometimes also in services (e.g., client waiting time). The subgroup size for numerical data may be quite small (e.g., under 10) or even a single item.

For *attribute* data (sometimes called *qualitative* data), the focus is on counting nonconforming items (those that do not meet the target specification). An *attribute control chart* may show the proportion nonconforming (binomial process) or the total number nonconforming (binomial or Poisson process). Attribute control charts are important in service or manufacturing environments when physical measurements are not appropriate. Subgroup size for attribute data may be large (e.g., over 100) depending on the rate of nonconformance. Because modern manufacturing processes may have very low nonconformance rates (e.g., .00001 or .0000001), even larger samples are needed.

LO **17-4**

Define a control chart and the types of variables displayed.

Three Common Control Charts

For a sample mean, the control chart is called an $\bar{x}$ **chart**. For a sample range, it is called an **R chart**. For a sample proportion, it is called a **p chart**. These three charts are illustrated in Figure 17.4. Each chart plots a sample statistic over time, as well as upper and lower *control limits* that define the expected range of the sample statistic. In these illustrations, all samples fall within the control limits. Control limits are based on the sampling distribution of the statistic. While there are many others, the basic concepts of SPC and setting control limits can be illustrated with these three chart types.

Figure 17.4

Three Common Control Charts (from Minitab)

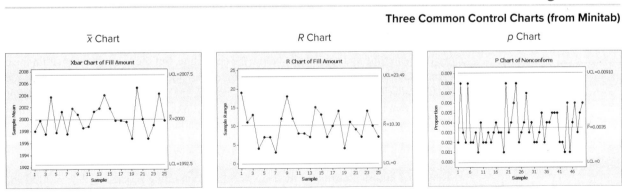

17.14 What is the difference between an attribute control chart and a variable control chart?

17.15 (a) What determines sampling frequency? (b) Why are variable samples often small? (c) Why are attribute samples often large?

Section Exercises

 connect

$\bar{x}$ Charts: Bottle-Filling Example

A bottling plant is filling 2-liter (2,000 ml) soft drink bottles. It is important that the equipment neither overfill nor underfill the bottle. The filling process is stable and in control with mean fill μ and standard deviation σ. The degree of variation depends on the process technology used in the plant. Every 10 minutes, n bottles are chosen at random and their fill is measured. The unit of measurement is milliliters. For a subgroup of size $n = 5$, the sample might look like this:

$x_1 = 2{,}001$ $x_2 = 1{,}998$ $x_3 = 2{,}001$ $x_4 = 2{,}001$ $x_5 = 1{,}997$

The mean fill for these five bottles is $\bar{x} = 1999.6$. Each time we take a sample of five bottles, we expect a different value of the sample mean due to random variation inherent in the process. From previous chapters, we know that the sample mean is an unbiased estimator of the true process mean (i.e., its expected value is μ):

(17.1) $\qquad\qquad E(\overline{X}) = \mu \qquad$ ($\overline{X}$ tends toward the true process mean)

We also know that the standard error of the sample mean is

(17.2) $\qquad\qquad \sigma_{\overline{X}} = \dfrac{\sigma}{\sqrt{n}} \qquad$ (larger n implies smaller variance of $\overline{X}$)

The sample mean follows a normal distribution if the population is normal or if the sample is large enough to assure normality by the Central Limit Theorem.

Control Limits: Known μ and σ

Sample means from a process that is in control should be near the process mean μ, which is the *centerline* of the control chart. The **upper control limit (UCL)** and **lower control limit (LCL)** are set at ± 3 standard errors from the centerline, using the Empirical Rule, which says that almost all the sample means (actually 99.73 percent) will fall within "3-sigma" limits:

(17.3) $\qquad\qquad$ UCL $= \mu + 3\dfrac{\sigma}{\sqrt{n}} \qquad$ (upper control limit for $\overline{X}$, known μ and σ)

(17.4) $\qquad\qquad$ LCL $= \mu - 3\dfrac{\sigma}{\sqrt{n}} \qquad$ (lower control limit for $\overline{X}$, known μ and σ)

The $\bar{x}$ chart provides a kind of visual hypothesis test. Sample means will vary, sometimes above the centerline and sometimes below the centerline, but they should stay within the control limits and be symmetrically distributed on either side of the centerline. You will recognize the similarity between *control limits* and *confidence limits* covered in previous chapters. The idea is that if a sample mean falls outside these limits, we suspect that the sample may be from a different population from the one we have specified.

EXAMPLE 17.1

Bottle Filling

Table 17.2 shows 25 samples of size $n = 5$ from a bottling process with $\mu = 2000$ and $\sigma = 4$. For each sample, the mean and range are calculated. The control limits are

$$\text{UCL} = \mu + 3\frac{\sigma}{\sqrt{n}} = 2000 + 3\frac{4}{\sqrt{5}} = 2000 + 5.367 = 2005.37$$

$$\text{LCL} = \mu - 3\frac{\sigma}{\sqrt{n}} = 2000 - 3\frac{4}{\sqrt{5}} = 2000 - 5.367 = 1994.63$$

Table 17.2 Twenty-Five Samples of Bottle Fill with $n = 5$ 📂 BottleFill

Sample	Bottle 1	Bottle 2	Bottle 3	Bottle 4	Bottle 5	Mean	Range
1	2001	1998	2001	2001	1997	1999.6	4
2	1997	2004	2001	2000	2002	2000.8	7
3	2001	2000	2003	1995	1994	1998.6	9
4	2007	2007	2001	2000	1997	2002.4	10
5	1999	2001	1998	2001	1996	1999.0	5
6	2002	2002	1988	1995	2004	1998.2	16
7	2003	1998	1998	1996	2001	1999.2	7
8	2005	2000	1991	1996	1996	1997.6	14
9	1999	1997	2006	1999	1999	2000.0	9
10	2005	1999	1998	2002	2000	2000.8	7
11	2001	1997	2002	2004	2007	2002.2	10
12	2002	1995	1995	1997	2000	1997.8	7
13	2006	2006	1997	1998	1994	2000.2	12
14	2003	1997	2000	2003	2004	2001.4	7
15	2003	2008	1994	1998	1999	2000.4	14
16	1998	1997	1999	2001	1994	1997.8	7
17	1988	1996	2001	2002	2002	1997.8	14
18	2003	2003	1997	1995	2001	1999.8	8
19	2003	2004	1998	1998	2006	2001.8	8
20	2005	2001	2005	2000	2004	2003.0	5
21	2004	1996	2003	2002	1993	1999.6	11
22	1998	1996	2005	1997	1999	1999.0	9
23	2002	2001	1995	2004	2007	2001.8	12
24	2002	2002	1997	1995	2002	1999.6	7
25	2002	1999	2001	1992	1993	1997.4	10
			Average over 25 samples of 5 bottles:			1999.832	9.160

Figure 17.5 shows a Minitab $\bar{x}$ chart for these 25 samples, with Minitab's option to specify μ and σ instead of estimating them from the data. Because all the sample means lie within the control limits, this chart shows a process that is *in control*. If a sample mean exceeds UCL or is below LCL, we suspect that the process may be *out of control*. More rules for detecting an out-of-control process will be explained shortly.

Figure 17.5

Minitab's $\bar{x}$ Chart with Known Control Limits

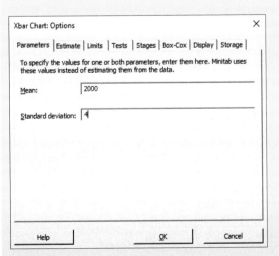

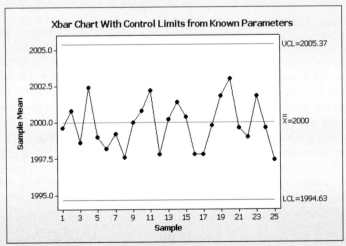

Minitab

Empirical Control Limits

When the process mean μ and standard deviation σ are unknown (as they often are), we can estimate them from sample data, replacing μ with $\bar{\bar{x}}$ (the average of the means of all samples) and replacing σ with the standard deviation s from a pooled sample of individual X-values. Generally, the centerline and control limits are based on *past* data but are to be used on *future* data to monitor the process. It is desirable to set the control limits from samples taken independently, rather than using the same data to create the control limits and to plot the control chart. However, this is not always possible.

(17.5) $\text{UCL} = \bar{\bar{x}} + 3\dfrac{s}{\sqrt{n}}$ (upper control limit for $\bar{X}$, unknown μ and σ)

(17.6) $\text{LCL} = \bar{\bar{x}} - 3\dfrac{s}{\sqrt{n}}$ (lower control limit for $\bar{X}$, unknown μ and σ)

There are other ways to estimate the process standard deviation σ. For example, we could use $\bar{s}$, the mean of the standard deviations over many subgroups of size n, with an adjustment for bias. Or we could replace σ with an estimate $\bar{R}/d_2$, where $\bar{R}$ is the average range for many samples and d_2 is a control chart factor that depends on the subgroup size (see Table 17.3). If the number of samples is large enough, any of these methods should give reliable control limits. The $\bar{R}$ method is still common for historical reasons (easier to use prior to the advent of computers). If the $\bar{R}$ method is used, the formulas become

(17.7) $\text{UCL} = \bar{\bar{x}} + 3\dfrac{\bar{R}}{d_2\sqrt{n}}$ (upper control limit for $\bar{X}$, unknown μ and σ)

(17.8) $\text{LCL} = \bar{\bar{x}} - 3\dfrac{\bar{R}}{d_2\sqrt{n}}$ (lower control limit for $\bar{X}$, unknown μ and σ)

Table 17.3

Control Chart Factors

Subgroup Size	d_2	D_3	D_4
2	1.128	0	3.267
3	1.693	0	2.574
4	2.059	0	2.282
5	2.326	0	2.114
6	2.534	0	2.004
7	2.704	0.076	1.924
8	2.847	0.136	1.864
9	2.970	0.184	1.816

See Laythe C. Alwan, *Statistical Process Analysis* (Irwin/McGraw-Hill, 2000), p. 740, for details of how these factors are derived.

Figure 17.6 shows Minitab's menu options for estimating control limits from a sample. By default, Minitab uses the pooled standard deviation, an attractive choice because it directly estimates σ. In Figure 17.6, using the $\bar{R}$ method, the $\bar{x}$ chart is similar to the chart in Figure 17.5, where σ was known, except that the LCL and UCL values are slightly different.

Control Chart Factors

To calculate $\bar{\bar{x}}$ and $\bar{R}$, we use averages of 25 sample means and ranges (see Table 17.2):

(17.9) $\bar{\bar{x}} = \dfrac{\bar{x}_1 + \bar{x}_2 + \cdots + \bar{x}_{25}}{25} = \dfrac{1999.6 + 2000.8 + \cdots + 1997.4}{25} = 1999.832$

(17.10) $\bar{R} = \dfrac{R_1 + R_2 + \cdots + R_{25}}{25} = \dfrac{4 + 7 + \cdots + 10}{25} = 9.160$

Figure 17.6

Minitab's $\bar{x}$ Chart with Estimated Control Limits

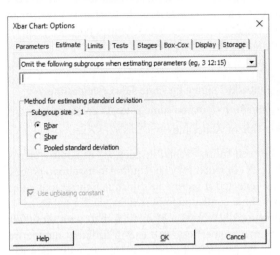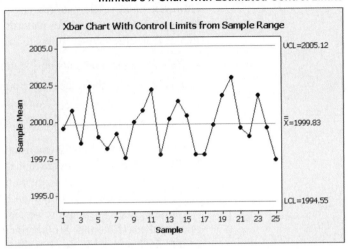

Source: Minitab.

Using the sample estimates $\bar{\bar{x}} = 1999.832$ and $\bar{R} = 9.160$, along with $d_2 = 2.326$ for $n = 5$ from Table 17.3, the estimated empirical control limits are

$$\text{UCL} = \bar{\bar{x}} + 3\frac{\bar{R}}{d_2\sqrt{n}} = 1999.832 + 3\frac{9.160}{2.326\sqrt{5}} = 2005.12$$

$$\text{LCL} = \bar{\bar{x}} - 3\frac{\bar{R}}{d_2\sqrt{n}} = 1999.832 - 3\frac{9.160}{2.326\sqrt{5}} = 1994.55$$

Note that these *empirical* control limits (2005.12 and 1994.55) differ somewhat from the *theoretical* control limits (2005.37 and 1994.63) that we obtained using $\mu = 2000$ and $\sigma = 4$, and the *empirical* centerline ($\bar{x} = 1999.83$) differs from $\mu = 2000$. In practice, it would be necessary to take more than 25 samples to ensure a good estimate of the true process mean and standard deviation. Indeed, engineers may run a manufacturing process for days or weeks before its characteristics are well understood. Figure 17.7 shows MegaStat's $\bar{x}$ chart using the sample data to estimate the control limits.

Figure 17.7

MegaStat's $\bar{x}$ Chart with Estimated Control Limits

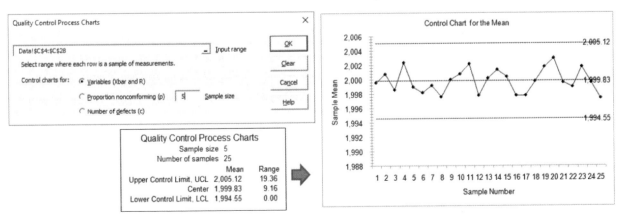

Source: MegaStat.

MegaStat *always* uses estimated control limits by the $\bar{R}$ method and does not permit you to specify known parameters. Also, MegaStat expects the observed data to be arranged as a rectangle, with each subgroup's observations comprising a *row* (like Table 17.2). MegaStat's $\bar{x}$ chart is similar to Minitab's except for details of scaling.

Detecting Abnormal Patterns

Sample means beyond the control limits are strong indicators of an out-of-control process. However, more subtle patterns also can indicate problems. Experts have developed many "rules of thumb" to check for patterns that might indicate an out-of-control process. Here are four of them (the "sigma" refers to the *standard error of the mean*):

- *Rule 1:* Single point outside 3 sigma.
- *Rule 2:* Two of three successive points outside 2 sigma on same side of centerline.
- *Rule 3:* Four of five successive points outside 1 sigma on same side of centerline.
- *Rule 4:* Nine successive points on same side of centerline.

Violations of Rules 1 and 2 can usually be seen from "eyeball inspection" of control charts. Violations of the other rules are more subtle. A computer may be required to monitor a process to be sure that control chart violations are detected. Figure 17.8 illustrates these four rules, applied to a service organization (an HMO clinic conducting physical exams for babies).

Multiple rule violations are possible. Figure 17.9 shows a Minitab $\bar{x}$ chart with these four rules applied. (Note that Minitab includes many other tests and uses a different numbering system for its rules.) In this illustration, an out-of-control process is shown, with eight rule violations. (Each violation is numbered and highlighted in red.)

Figure 17.8

Red Dots Indicate Rule Violations (from *Visual Statistics*)

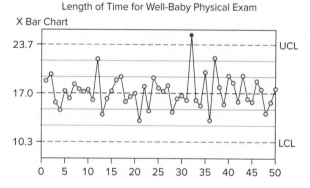

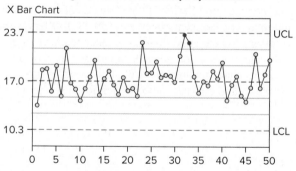

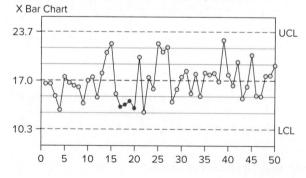

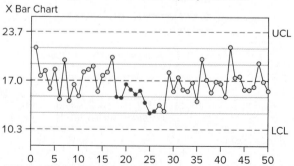

Figure 17.9

Violations of Rules of Thumb (from Minitab)

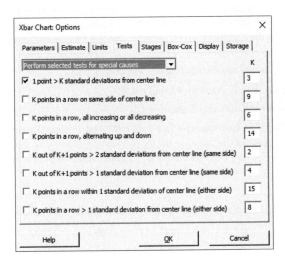

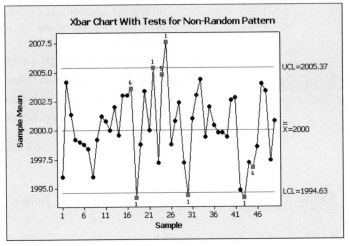

Minitab

Histograms

The normal curve is the reference point for variation inherent in the process or due to random sampling. UCL and LCL are set at ± 3 standard errors from the mean, but we could (and should) also examine ± 2 and ± 1 standard error ranges to see whether the percentage of sample means follows the normal distribution. Recall that the expected percent of samples within various distances from the centerline can be stated as normal areas or percentages:

- Within ± 1 standard deviation or 68.26 percent of the time.
- Within ± 2 standard deviations or 95.44 percent of the time.
- Within ± 3 standard deviations or 99.73 percent of the time.

The distribution of sample means can be scrutinized for symmetry and/or deviations from the expected normal percentages. Figure 17.10 shows an $\bar{x}$ chart and histogram for 100 samples of acidity for a commercial cleaning product. The histogram is roughly symmetric, with 60 sample means between -1 and $+1$ and 97 sample means between -2 and $+2$, while the normal distribution would predict 68 and 95, respectively.

Figure 17.10

$\bar{x}$ Chart and Histogram (from *Visual Statistics*)

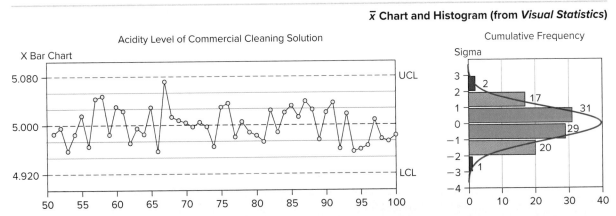

Section Exercises

17.16 (a) To construct control limits for an $\bar{x}$ chart, name three ways to estimate σ empirically. (b) Why is the $\bar{R}$ method often used? (c) Why is the s method the default in Minitab?

17.17 For an $\bar{x}$ chart, what percent of sample means should be (a) within 1 sigma of the centerline; (b) within 2 sigmas of the centerline; (c) within 3 sigmas of the centerline; (d) outside 2 sigmas of the centerline; (e) outside 3 sigmas of the centerline? *Note:* "Sigma" denotes the standard error of the mean.

17.18 List four rules for detecting abnormal (special cause) observations in a control chart.

17.19 Set up control limits for an $\bar{x}$ chart, given $\bar{\bar{x}} = 12.50$, $\bar{R} = .42$, and $n = 5$.

17.20 Set up control limits for an $\bar{x}$ chart, given $\mu = 400$, $\sigma = 2$, and $n = 4$.

17.21 Time (in seconds) to serve an early-morning customer at a fast-food restaurant is normally distributed. Set up a control chart for the mean serving time, assuming that serving times were sampled in random subgroups of 4 customers. *Note:* Use this sample of 36 observations to estimate μ and σ. 🖉 **ServeTime**

Sample 1	Sample 2	Sample 3	Sample 4	Sample 5	Sample 6	Sample 7	Sample 8	Sample 9
65	56	84	69	75	87	87	99	102
51	87	67	81	80	84	90	61	61
94	84	71	59	76	80	65	84	88
79	70	85	75	88	52	61	79	78

17.22 To print 8.5 × 5.5 note pads, a copy shop uses standard 8.5 × 11 paper, glues the long edge, and then cuts the pads in half so that the pad width is 5.5 inches. However, there is variation in the cutting process. Set up a control chart for the mean width of a note pad, assuming that, in the future, pads will be sampled in random subgroups of 5 pads. Use this sample of 40 observations (widths in inches) to estimate μ and σ. 🖉 **NotePads**

5.52	5.57	5.44	5.47	5.52	5.46	5.43	5.45
5.49	5.47	5.48	5.51	5.53	5.53	5.48	5.47
5.59	5.51	5.43	5.48	5.53	5.50	5.49	5.52
5.46	5.46	5.56	5.54	5.47	5.44	5.53	5.58
5.55	5.56	5.47	5.44	5.55	5.42	5.45	5.54

Mini Case 17.1

Control Limits for Jelly Beans 🖉 JellyBeans

The manufacture of jelly beans is a high-volume operation that is tricky to manage, with strict standards for food purity, worker safety, and environmental controls. Each bean's jelly core is soft and sticky and must be coated with a harder sugar shell of the appropriate color. Hundreds of thousands of beans must be cooled and bagged, with approximately the desired color proportions. To meet consumer expectations, the surface finish of each bean and its weight must be as uniform as possible. Jelly beans are a low-priced item, and the market is highly competitive (i.e., there are many substitutes and many producers), so it is not cost-effective to spend millions to achieve the same level of precision that might be used, say, in manufacturing a prescription drug.

So, how do we measure jelly bean quality? One obvious metric is weight. To set control limits, we need estimates of μ and σ. From a local grocery, a bag of Brach's jelly beans was purchased (see Figure 17.11). Each bean was weighed on a precise scale. The resulting sample of 182 jelly bean weights showed a bell-shaped distribution, except for three high outliers, easily visible in Figure 17.12. Once the outliers are removed, the sample presents a satisfactory normal probability plot, shown in Figure 17.13. The sample mean and standard deviation ($\bar{x} = 3.352$ grams and $s = .3622$ gram) can now be used to set control limits.

Referring to Figure 17.11, some differences in size are visible. Can you spot the three oversized beans? Some consumers may regard "double" beans as a treat rather than as a product defect. But manufacturers always strive for the most consistent product possible, subject to constraints of time, technology, and budget.

Figure 17.11

The Data Set (*n* = 182)

David Doane

David Doane

Figure 17.12

Dot Plot for All Data (*n* = 182)

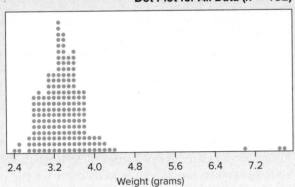

Figure 17.13

Minitab Normal Plot for Trimmed Data (*n* = 179)

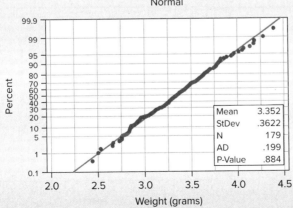

17.6 CONTROL CHARTS FOR A RANGE

LO 17-6

Make and interpret control charts for a range.

The $\bar{x}$ chart of sample means by itself is insufficient to tell whether a process is in control because it reveals only *centrality*. We also should examine a chart showing *variation* around the mean. We could track the sample standard deviations (using an *s chart*), but it is more traditional to track the *sample range* (the difference between the largest and smallest items in each sample) using the *R chart*. The sample range is sensitive to extreme values. Nonetheless, its behavior can be predicted statistically, and control limits can be established. The *R* chart has asymmetric control limits because the sample range is not a normally distributed statistic.

Control Limits for the Range

The centerline is obtained by calculating the average range $\bar{R}$ over many samples taken from the process. Estimation of $\bar{R}$ ideally would precede construction of the control chart, using a large number of independent samples, though this is not always possible in practice. Control limits based on samples may not be a good representation of the true process. It depends on the number of samples and the "luck of the draw." The control limits for the *R* chart can be set using either the average sample range $\bar{R}$ or an estimate $\hat{\sigma}$ of the process standard deviation:

(17.11) $\quad \text{UCL} = D_4 \bar{R} \quad \text{or} \quad \text{UCL} = D_4 d_2 \hat{\sigma} \quad$ (upper control limit of sample range)

(17.12) $\quad \text{LCL} = D_3 \bar{R} \quad \text{or} \quad \text{LCL} = D_3 d_2 \hat{\sigma} \quad$ (lower control limit of sample range)

Example 17.2

Bottle Filling: R Chart

The control limits depend upon factors that must be obtained from a table. For the bottle fill data with $n = 5$, we have $D_4 = 2.114$ and $D_3 = 0$ (from Table 17.3). Using $\bar{R} = 9.16$ from the 25 samples (from Table 17.2), the control limits are

$\bar{R} = 9.16 \quad$ (centerline for *R* chart)

$\text{UCL} = D_4 \bar{R} = (2.114)(9.160) = 19.36 \quad$ (upper control limit)

$\text{LCL} = D_3 \bar{R} = (0)(9.160) = 0 \quad$ (lower control limit)

Figure 17.14 shows Minitab's *R* chart for the data in Table 17.2. Note that the *R* chart control limits also could be based on a pooled standard deviation. Using the Parameters tab, Minitab also offers an option (not shown) to specify σ yourself (e.g., from historical experience). In this illustration, the process variation remains within the control limits.

Figure 17.14

Minitab's *R* Chart with Control Limits from Sample Data

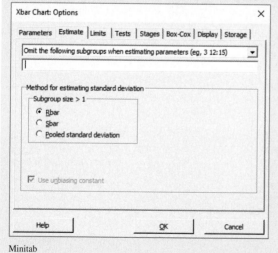

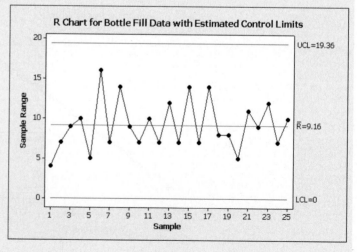

Figure 17.15 shows MegaStat's *R* chart using the same data to estimate the control limits. The Minitab and MegaStat charts are similar except for scaling. MegaStat always uses estimated control limits, whereas Minitab gives you the option.

Figure 17.15

MegaStat *R* Chart with Estimated Control Limits

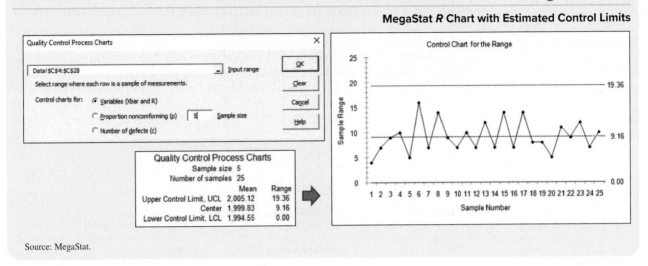

Source: MegaStat.

Section Exercises

McGraw Hill **connect**

17.23 Set up limits for the *R* chart, given $\overline{R} = 0.82$ and $n = 6$.

17.24 Set up limits for the *R* chart, given $\overline{R} = 12$ and $n = 3$.

 17.7 OTHER CONTROL CHARTS

Attribute Data: *p* Charts

The *p chart* for attribute data plots the *proportion* of nonconforming items using the familiar sample proportion *p*:

$$p = \frac{\text{number of nonconforming items}}{\text{sample size}} = \frac{x}{n} \qquad (17.13)$$

LO 17-7

Make and interpret control charts for attribute data.

In manufacturing, *p* used to be referred to as a "defect rate," but the term "nonconforming items" is preferred because it is more neutral and better adapted to applications outside manufacturing, such as service environments. For example, for a retailer, *p* might refer to the proportion of customers who return their purchases for a refund. For a bank, *p* might refer to the proportion of checking account customers who have insufficient funds to cover one or more checks. For Ticketmaster, *p* might refer to the proportion of customers who have to wait "on hold" more than 5 minutes to obtain concert tickets.

The number of nonconforming items in a sample of *n* items is a binomial random variable, so the control limits are constructed as a confidence interval for a population proportion using one of several methods to state the *population* nonconformance rate π:

- Use an assumed value of π (e.g., a target rate of nonconformance).
- Use an empirical estimate of π based on a large number of trials.
- Use an estimate *p* from the samples being tested (if no other choice).

If *n* is large enough to assume normality,[1] the control limits would be

$$\text{UCL} = \pi + 3\sqrt{\frac{\pi(1-\pi)}{n}} \quad (\pi \text{ is the process centerline}) \qquad (17.14)$$

$$\text{LCL} = \pi - 3\sqrt{\frac{\pi(1-\pi)}{n}} \quad (\pi \text{ is the process centerline}) \qquad (17.15)$$

---

[1]To assume normality, we want $n\pi \geq 10$ and $n(1-\pi) \geq 10$. If not, the binomial distribution may be used to set up control limits. Minitab will handle this, although the resulting control limits may be quite wide.

The logic is similar to a two-tailed hypothesis test of a proportion. Approximately 99.73 percent of the time, we expect the sample proportion p to fall within 3 standard deviations of the assumed centerline (π). If the LCL is negative, it is assumed to be zero (because a proportion cannot be negative). In manufacturing, the rate of nonconformance is likely to be a very small fraction (e.g., .02 or even smaller), so it is quite likely that LCL will be zero.

Example 17.3

Cell Phone Manufacture

A manufacturer of cell phones has a .002 historical rate of nonconformance to specifications (i.e., 2 nonconforming phones per 1,000). All phones are tested, and the nonconformance rates are plotted on a p chart, using an assumed value $\pi = .002$. Thus, the control limits are

$$\text{UCL} = .002 + 3\sqrt{\frac{(.002)(.998)}{n}} \quad \text{and} \quad \text{LCL} = .002 - 3\sqrt{\frac{(.002)(.998)}{n}}$$

Table 17.4 shows inspection data for 100 days of production. Each production run (n) is around 2,000 phones per day, but it does vary. Hence, the control limits are not constant, as shown in the p chart in Figure 17.16.

Table 17.4   Nonconforming Cell Phones   CellPhones

Day	Nonconforming (x)	Production (n)	x/n
1	3	2,056	0.00146
2	1	1,939	0.00052
3	4	2,079	0.00192
4	5	2,079	0.00241
5	4	1,955	0.00205
⋮	⋮	⋮	⋮
96	4	1,967	0.00203
97	6	2,077	0.00289
98	3	2,075	0.00145
99	5	1,908	0.00262
100	2	2,045	0.00098

Figure 17.16

Minitab *P* Chart for Cell Phones

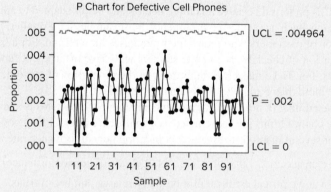

P Chart for Defective Cell Phones

Tests performed with unequal sample sizes

Notice that p stays within the control limits, although it touches the LCL twice (not a problem because zero defects is ideal). Although n varies, we can illustrate the control limit calculation by using $\pi = .002$ and $n = 2,000$:

$$\text{UCL} = .002 + 3\sqrt{\frac{(.002)(.998)}{2,000}} = .004997$$

$$\text{LCL} = .002 - 3\sqrt{\frac{(.002)(.998)}{2,000}} = -0.00997$$

Because a negative proportion is impossible, we just set LCL = 0. You will notice that Minitab's UCL is not quite the same as the calculation above because Minitab uses a binomial calculation rather than the normal approximation. The difference may be noticeable when $n\pi < 10$ (the criterion for a normal approximation to the binomial). In this example, $n\pi = (.002)(2,000) = 4$, so the binomial method is preferred.

Application: Emergency Patients

Instead of being a rate of *nonconformance* to specifications, p could be a rate of *conformance* to specifications. Then

$$p = \frac{\text{number of conforming items}}{\text{sample size}} = \frac{x}{n} \qquad (17.16)$$

Ardmore Hospital's emergency facility advertises that its goal is to ensure that, on average, 90 percent of patients receive treatment within 30 minutes of arrival. Table 17.5 shows data from 100 days of emergency department records.

Table 17.5

Emergency Patients Seen within 30 Minutes
ERPatients

Day	Seen in 30 Minutes (x)	Patient Volume (n)	x/n
1	87	97	0.900
2	113	122	0.924
3	106	115	0.920
4	84	90	0.928
5	82	92	0.896
⋮	⋮	⋮	⋮
96	128	142	0.900
97	101	112	0.900
98	123	135	0.908
99	128	141	0.908
100	141	149	0.944

The average number of patient arrivals per day is around 120, but there is considerable variation. Hence, the control limits are not constant, as shown in the p chart in Figure 17.17. Because the sample sizes are smaller than in the cell phone example, the LCL and UCL are more sensitive to the varying sample size, and hence appear more jagged. This process is in control.

Figure 17.17

***p* Chart for ER Patients**

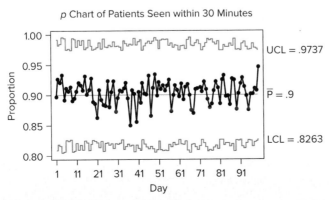

p Chart of Patients Seen within 30 Minutes

UCL = .9737
$\bar{P}$ = .9
LCL = .8263

Tests performed with unequal sample sizes

The *p* chart is likely to be used in service operations (e.g., for benchmarking in health care delivery). Rules of thumb for detecting outliers, runs, and patterns apply to the *p* chart, just as for the $\bar{x}$ chart. However, tests for patterns are rarely seen outside manufacturing, except when service sector tasks are ongoing, repeatable, and easily sampled.

Other Control Charts (*s, c, np, I, MR*)

There are many other types of control charts. For example, here are some control chart menus offered by Minitab:

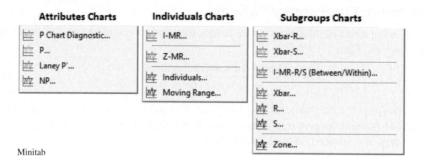

Minitab

Common types of control charts include

- *I charts* (for individual numerical observations).
- *MR charts* (moving range for individual observations).
- *s* charts (for standard deviations).
- *c* charts (for Poisson events with mean *c*).
- *np* charts (for binomial totals).
- zone charts (using six regions based on σ).

The first two are used when *continuous inspection* is possible. When $n = 1$, there is no range, so a *moving range* is used. *I* chart control limits simply are $\mu \pm 3\sigma$ when $n = 1$. Mini Case 17.2 gives an illustration. Interpretation is the same as for any other control chart.

Mini Case 17.2

I-MR Charts for Jelly Beans 🖾 JellyBeans2

Table 17.6 shows a sample of weights for 44 Brach's jelly beans (all black) from a randomly chosen bag of jelly beans. Is the weight of the jelly beans in control? To construct the control limits, we use the sample mean and standard deviation from the large trimmed sample in Mini Case 17.1 ($\bar{x} = 3.352$ grams and $s = 0.3622$ gram) with Minitab's I-MR chart option with assumed parameters $\mu = 3.352$ and $\sigma = 0.3622$.

In Figure 17.18, the *I* chart (upper one) reveals that two jelly beans (the 4th and 43rd observations) are not within the control limits. There is also evidence of a problem in sample-to-sample variation in the *MR* chart (lower one). The explanation turned out to be rather clear. The 4th jelly bean was a "double-bean" (where two jelly beans got stuck together), and the 43rd jelly bean was a "mini-bean" (where the jelly bean was only partially formed). Figure 17.19 shows that, if we remove these outliers, the trimmed sample means stay within the control limits on the *I* chart (upper one), although the *MR* chart (lower one) still has one odd point. Improved quality control for a high-volume, low-cost item like jelly beans is cost-effective only up to a point. A business case would have to be made before spending money on better technology, taking into account consumer preferences and competitors' quality levels.

Table 17.6   Weights of 44 Black Brach's Jelly Beans

Obs	Weight	Obs	Weight	Obs	Weight	Obs	Weight
1	3.498	12	3.181	23	3.976	34	3.168
2	3.603	13	3.545	24	3.321	35	2.656
3	4.223	14	3.925	25	3.609	36	2.624
4	7.250	15	3.686	26	3.604	37	3.254
5	3.830	16	3.938	27	3.668	38	3.411
6	3.563	17	3.667	28	3.433	39	2.553
7	2.505	18	3.152	29	3.678	40	4.217
8	3.034	19	3.325	30	3.264	41	3.417
9	3.408	20	3.905	31	3.743	42	3.615
10	3.564	21	3.714	32	3.446	43	1.218
11	3.042	22	3.359	33	3.036	44	3.612

Note: Measurements taken using a Mettler DF360 Delta Range Scale.

Figure 17.18

Before Outliers Removed

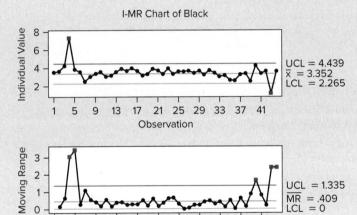

Figure 17.19

After Outliers Removed

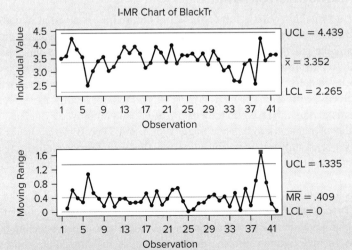

Ad Hoc Charts

We said earlier that any display of a quality metric over time is a kind of control chart. If we set aside the formalities of control chart theory, anyone can create a "control chart" to monitor something of importance. For example, Figure 17.20 is not a "classic" control chart, but it shows a quality metric (patient waiting time in an emergency department) plotted over time. A box plot showing the range and quartiles over time is an *ad hoc* chart, yet it's a useful one. Organizations must develop their own approaches to quality improvement. As long as they begin with measurement, charting, and analysis, they are heading in the right direction.

Figure 17.20

Box Plots over Time
AdHoc

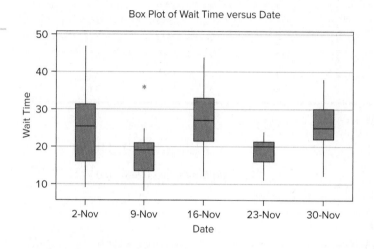

Box Plot of Wait Time versus Date

Section Exercises

connect

17.25 Why are *p* charts widely used in service applications like health care?

17.26 Create control limits for a *p* chart for a process with $\pi = .02$ and subgroup size $n = 500$. Is it safe to assume normality? Explain.

17.27 Create control limits for a *p* chart for a process with $\pi = .50$ and subgroup size $n = 20$. Is it safe to assume normality? Explain.

17.28 Create control limits for a *p* chart for a process with $\pi = .90$ and subgroup size $n = 40$. Is it safe to assume normality? Explain.

17.8 PATTERNS IN CONTROL CHARTS

LO 17-8

Recognize abnormal patterns in control charts and their potential causes.

The Overadjustment Problem

The $\bar{x}$ chart is a visual hypothesis test for μ, while the R chart is a visual hypothesis test for σ. In manufacturing, a control chart is used to guide decisions to continue the process or halt the process to make adjustments. *Overadjustment* or stopping to make unnecessary process corrections (Type I error) can lead to loss of production, downtime, unnecessary expense, forgone profit, delayed deliveries, stockout, or employee frustration. On the other hand, failing to make timely process corrections (Type II error) can lead to poor quality, excess scrap, rework, customer dissatisfaction, adverse publicity or litigation, and employee cynicism.

Statistics allows managers to balance these Type I and Type II errors. It has been shown that, in the absence of statistical decision rules, manufacturing process operators tend toward overadjustment, which will actually *increase* variation above the level the process is capable of attaining.

The actions to be taken when a control chart violation is detected will depend on the consequences of Type I and Type II errors. For example, if a health insurer notices that processing times for claim payments are out of control (i.e., relative to target benchmarks), the only action may be an investigation into the problem because the immediate consequences are not severe. But in car manufacturing, an out-of-control metal-forming process could require immediate shutdown of the assembly process to prevent costly rework or product liability.

Abnormal Patterns

Quality experts have given names to some of the more common abnormal control chart patterns, that is, patterns that indicate assignable causes:

- **Cycle**   Samples tend to follow a cyclic pattern.
- **Oscillation** Samples tend to alternate (high-low-high-low) in "sawtooth" fashion.
- **Instability** Samples vary more than expected.
- **Level shift** Samples shift abruptly either above or below the centerline.
- **Trend**   Samples drift slowly either upward or downward.
- **Mixture**   Samples come from two different populations (increased variation).

These names are intended to help you recognize symptoms that may be associated with known causes. These concepts extend to any time-series pattern (not just control charts).

Symptoms and Assignable Causes

Each $\bar{x}$ chart in Figure 17.21 displays 100 samples, which is a long enough run to show the patterns clearly. However, the $\bar{x}$ charts shown are exaggerated to emphasize the essential features of each pattern. Abnormal patterns like these would generate violations of Rules 1, 2, 3, or 4

Figure 17.21

Common Abnormal Patterns

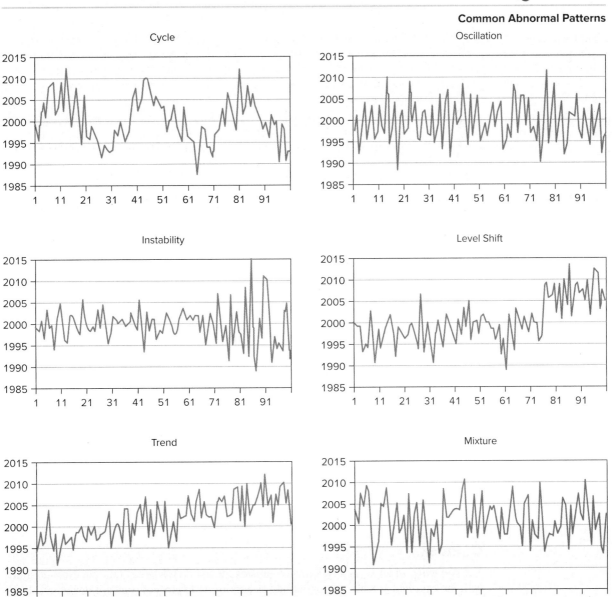

Table 17.7 Pattern Descriptions and Assignable Causes

Pattern Description	Likely Assignable Causes	Detected How?
Cycle is a repeated series of high measurements followed by a series of low measurements $(+ + + + - - - + + + - - - - + + + +$, etc.) relative to the centerline. Equivalent to positive autocorrelation in regression residuals.	*Industry:* worn threads or gears, humidity or temperature fluctuations, operator fatigue, voltage changes, overadjustment. *Services:* duty rotations, employee fatigue, poor scheduling, periodic distractions.	May be detected visually (fewer than $m/2$ centerline crossings in m samples) or by a runs test (see Chapter 16) or higher-than-expected tail frequencies in a histogram. Look for violations of Rule 4.
Oscillation is a pattern of alternating high and low measurements $(+ - + - + - + - +$, etc.) relative to the centerline (a zigzag or sawtooth pattern). Equivalent to negative autocorrelation in regression residuals.	*Industry:* alternating sampling of two machines, two settings, two inspectors, or two gauges. *Services:* attempts to compensate for performance variation on the last task, alternating task between two workers.	May be detected visually (more than $m/2$ centerline crossings in m samples) or by a runs test (see Chapter 16). Process mean stays near the centerline, though process variance may increase. May not violate any rules.
Instability is a larger-than-normal amount of variation preceded by a period of normal, stable variation.	*Industry:* untrained operators, overadjustment, tool wear, defective material. *Services:* distractions, poor job design, untrained employees, flawed sampling process.	May be detectable on the $\bar{x}$ chart, but shows up most clearly on the R chart and in higher-than-expected frequencies in the tails of the histogram. Violations of Rules 1, 2, and 3 are likely.
Level shift is a sudden change in measurements either above or below the centerline. It is a change in the actual process mean. Easily confused with trend.	*Industry:* new workers, change in equipment, new inspector, new machine setting, new lot of material. *Services:* changed environment, new supervisor, new work rules.	Center of the histogram shifts but with no change in variation. Violation of Rule 4 is likely, and perhaps others. May be too few centerline crossings (fewer than $m/2$).
Trend is a slow, continuous drifting of measurements either up or down from the chart centerline. Detectable visually if enough measurements are taken. Easily confused with level shift.	*Industry:* tool wear, inadequate maintenance, worker fatigue, gradual clogging (dirt, shavings, etc.), drying of lubricants. *Services:* inattention, rising workload, bottlenecks.	Process variance may be unchanged, but the histogram grows skewed in one tail. May be too few centerline crossings (fewer than $m/2$). Violation of Rule 4 is likely, and perhaps others.
Mixture is merged output from two or more separate processes. Both may be in control, but with different means, so the overall process variance is increased.	*Industry:* two machines, two gauges, two shifts (day, night), two inspectors, different lots of material. *Services:* different supervisors, two work teams, two shifts.	Difficult to detect, either visually or statistically, especially if more than two processes are mixed. Histogram may be bimodal. Use same tests as for instability.

(or multiple rule violations) so the process would actually have been stopped *before* the pattern developed to the degree shown in Figure 17.21. Although many patterns are discussed in terms of the $\bar{x}$ chart, the R chart and histogram of sample means also may reveal abnormal patterns. It may be impossible to identify a pattern or its assignable cause(s) if the period of observation is short. Table 17.7 summarizes the symptoms and likely underlying causes of abnormal patterns.

17.9 PROCESS CAPABILITY

LO 17-9

Assess the capability of a process.

A business must translate *customer requirements* into an **upper specification limit (USL)** and **lower specification limit (LSL)** of a quality metric. These limits do *not* depend on the process. Whether the process is *capable* of meeting these requirements depends on the magnitude of the process variation (σ) and whether the process is correctly centered (μ). We now will define ways to measure the **capability** of a process.

C_p Index

The *capability index C_p* is a ratio that compares the interval between the specification limits with the expected process range (defined as six times the process standard deviation). If the process range is small relative to the specification range, the capability index will be high, and conversely. A higher C_p index (a *more capable* process) is always better.

$$C_p = \frac{\text{USL} - \text{LSL}}{6\sigma} \qquad \text{(process capability index } C_p) \qquad \textbf{(17.17)}$$

A C_p value of 1.00 indicates a process that is barely capable of staying within the specifications *if* precisely centered. Managers typically require $C_p > 1.33$ to allow flexibility in case the process drifts off center. A much higher capability index may be required in some applications (e.g., manufacturing satellite components where repair is impossible).

$$\text{Example 1: If USL} - \text{LSL} = 6\sigma, \text{ then } C_p = \frac{6\sigma}{6\sigma} = 1.00$$

$$\text{Example 2: If USL} - \text{LSL} = 8\sigma, \text{ then } C_p = \frac{8\sigma}{6\sigma} = 1.33$$

$$\text{Example 3: If USL} - \text{LSL} = 10\sigma, \text{ then } C_p = \frac{10\sigma}{6\sigma} = 1.67$$

$$\text{Example 4: If USL} - \text{LSL} = 12\sigma, \text{ then } C_p = \frac{12\sigma}{6\sigma} = 2.00$$

Assuming that customer requirements (LSL and USL) are fixed, the only way to increase the capability index is to find ways to reduce σ. Figure 17.22 illustrates how a reduction in σ (e.g., through improved manufacturing technology) can improve C_p even though the specification limits remain the same. A smaller σ will narrow the $\pm 6\sigma$ range of expected process variation, widening the "safety margin" on either side (green area). Even if the process drifts outside the $\pm 6\sigma$ range, we will still be within the desired limits LSL and USL.

Figure 17.22

Reducing σ Improves Process Capability

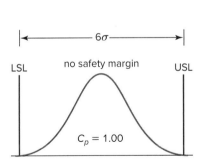

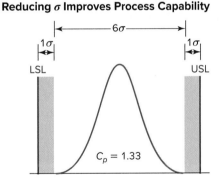

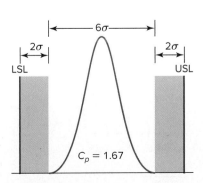

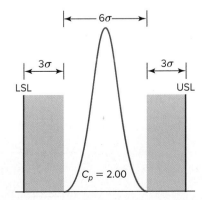

C_{pk} Index

The index C_p is easy to understand but fails to indicate if the process is off-centered. A process with acceptable variation could be off-centerline and yet have a high C_p capability index. To remedy this weakness, we define another process capability index called C_{pk} that compares the process distribution *individually* to each specification limit (LSL and USL) by using the distance between each specification limit and the process centerline.

The process capability index C_{pk} simply is the smaller of these two distances (USL $- \mu$ and $\mu -$ LSL) expressed as a fraction of the 3σ *distance above or below μ*:

(17.18) $$C_{pk} = \frac{\min(\mu - \text{LSL}, \text{USL} - \mu)}{3\sigma} \quad \text{(process capability index } C_{pk})$$

We are assuming that LSL lies below μ (the centerline) and USL lies above μ (the centerline) so that both distances are positive. If both $\mu -$ LSL and USL $- \mu$ are exactly 3σ, then $C_{pk} = 1.00$. A C_{pk} index of 1.00 is the minimum capability, but much higher values are preferred. If $\mu -$ LSL and USL $- \mu$ are the same, then C_{pk} will be identical to C_p. In contrast to the C_p index, the C_{pk} index imposes a penalty when the process is off-center. Thus, the C_{pk} index is more sensitive in detecting nonconformance. An animated illustration of both the C_p index and the C_{pk} index can be found at the website https://elsmar.com/Cp_vs_Cpk.html.

Example 17.4

Cookie Baking

A bakery wants to produce cookies whose average weight, after baking, is 31 grams. To meet quality requirements, management specifies USL = 35.0 grams and LSL = 28.0 grams. The process standard deviation is 0.8 gram and the process centerline is set at $\mu = 31$ grams. The company requires a capability index of at least 1.33. Is the process capable?

C_p *index:*

$$C_p = \frac{\text{USL} - \text{LSL}}{6\sigma} = \frac{35.0 - 28.0}{(6)(.8)} = 1.46$$

C_{pk} *index:*

$$C_{pk} = \frac{\min(\mu - \text{LSL}, \text{USL} - \mu)}{3\sigma} = \frac{\min(31.0 - 28.0, 35.0 - 31.0)}{(3)(0.8)} = \frac{3.0}{2.4} = 1.25$$

According to the C_p index, the process capability is barely acceptable ($C_p = 1.46$), but using the C_{pk} index ($C_{pk} = 1.25$), the process capability is unacceptable in view of the company's chosen minimum (1.33). The situation is illustrated in Figure 17.23. In cookie making, management is less concerned about oversized cookies than undersized ones (customers will rarely complain if a cookie is too big), so the limits are not symmetric, as can be seen in Figure 17.23. Note that their process is correctly centered at $\mu = 31$, even though the specification limits are asymmetric.

Figure 17.23

Process Capability for Cookie Making

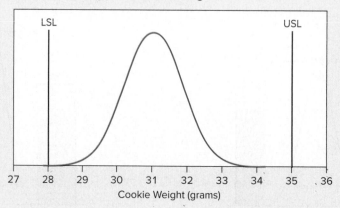

Cookie Weight (grams)

Bottle Filling Revisited

Figure 17.24 illustrates several possible situations, using the bottle-filling scenario with symmetric specification limits LSL = 1994 and USL = 2006 and a target $\mu = 2000$. Specification limits are based on customer demands (or engineering requirements) and not on the process itself. *Even if a process is in control, the process may not be capable of meeting the requirements.* If it is not, there is no choice but to find ways to improve the process (by reducing σ) through improved technology, worker training, or capital investment.

Figure 17.24

Small σ Can Prevent Bottle Overfill or Underfill

Highly Capable	**Barely Capable**	**Not Capable**
Process is capable of meeting specifications even if poorly centered	Process is capable of meeting specifications only if well centered	Process is incapable of meeting specifications even if well centered

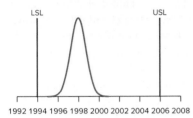
1992 1994 1996 1998 2000 2002 2004 2006 2008

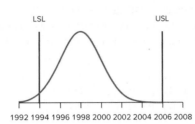
1992 1994 1996 1998 2000 2002 2004 2006 2008

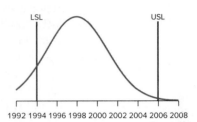
1992 1994 1996 1998 2000 2002 2004 2006 2008

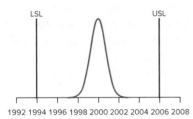
1992 1994 1996 1998 2000 2002 2004 2006 2008

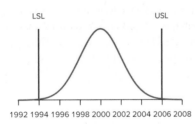
1992 1994 1996 1998 2000 2002 2004 2006 2008

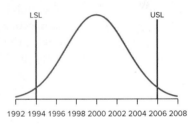
1992 1994 1996 1998 2000 2002 2004 2006 2008

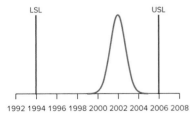
1992 1994 1996 1998 2000 2002 2004 2006 2008

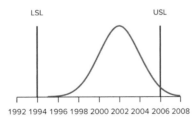
1992 1994 1996 1998 2000 2002 2004 2006 2008

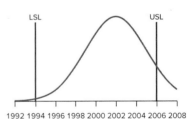
1992 1994 1996 1998 2000 2002 2004 2006 2008

Section Exercises

 connect

17.29 Find the C_p and C_{pk} indexes for a process with $\mu = 720$, $\sigma = 1.0$, LSL = 715, USL = 725. How would you rate the capability of this process? Explain.

17.30 Find the C_p and C_{pk} indexes for a process with $\mu = 0.426$, $\sigma = 0.001$, LSL = 0.423, USL = 0.432. How would you rate the capability of this process? Explain.

17.31 Find the C_p and C_{pk} indexes for a process with $\mu = 55.4$, $\sigma = 0.1$, LSL = 55.2, USL = 55.9. How would you rate the capability of this process? Explain.

17.10 ADDITIONAL QUALITY TOPICS (OPTIONAL)

LO 17-10

Identify topics commonly associated with quality management (optional).

Acceptance Sampling

The end quality of products and services is strongly affected by the quality of materials purchased from suppliers. Manufacturing firms and service businesses rely on random sampling inspection of shipments of incoming materials and supplies when trying out a new vendor or verifying the capability of a new business process. This inspection activity is called **acceptance sampling**.

In acceptance sampling, *the producer's risk* (α error) is the probability of rejecting material of desirable quality level, while the *consumer's risk* (β error) is the probability of accepting material of undesirable quality level. These two risks must be balanced because there is a trade-off between α and β for a given sample size. In its simplest form, lot sampling is based on the hypergeometric distribution, in which samples of n items are taken from a lot of size N containing s nonconforming items. Power curves and operating characteristic curves can be developed to guide decisions about acceptance or rejection of shipments, based on the attribute of interest (usually the proportion of nonconforming items).

Statisticians Walter Shewhart, Harry Romig, and Harold Dodge developed plans for conducting acceptance sampling, which were published by the U.S. Department of Defense in 1963 as MIL-STD-105. These plans spelled out details related to sample size (n) for various lot sizes (N), acceptance criteria (i.e., what is the maximum number of nonconforming sample items allowed for lot acceptance), and rejection criteria (i.e., how many nonconforming sample items lead to lot rejection). These parameters were based on the trade-off between the producer's risk and consumer's risk. In 1996 this standard was replaced by MIL-STD-1916. The primary change was to mandate *accept on zero* (AOZ) acceptance criteria. An AOZ plan states that acceptance of a lot is allowed only if zero nonconforming items are observed in the sample. Today's businesses recognize that ensuring quality through acceptance sampling is costly and not effective in the long run. The preferred approach for ensuring quality goods and services is to use process control techniques such as control charts and capability measures.

Supply-Chain Management

The problem with acceptance sampling is that it places the firm in the awkward position of rejecting shipments of purchased material that may be needed for production in the near future. This forces the firm to increase lead times and hold larger inventory to provide a buffer against defective material. It also strains relations with suppliers and creates incentives to cut corners on quality by accepting questionable shipments. Worst of all, it gives the firm no direct control over its suppliers, except the negative control of saying no to shipments.

Most firms believe that a more constructive approach is to reduce reliance on acceptance sampling and instead to engage in direct dialogue with suppliers to ensure that their quality control is adequate to meet the buyer's expectations. The idea is to prevent problems rather than merely spot them after they have occurred. If suppliers implement the TQM philosophy and utilize SPC to control and improve their processes, there is harmony of purpose between vendor and buyer. This is one principle behind ISO 9000, which will be discussed shortly.

But new problems arise from this supply-chain management approach. Suppliers may be smaller companies that lack the experience and resources needed to invest in training, research, and development, and there may be coordination problems between seller and buyer. Buyers may have to subsidize the process of implementing quality control at the supplier level, for example, by sponsoring training seminars, sharing their managerial experience, and working toward common database and decision support systems. Deming felt that suppliers should not be chosen solely on the basis of lowest cost. Rather, he thought buyers should develop long-term relationships with a small group of suppliers and then nurture the links with those suppliers. Many firms have done this. But changing the supply-chain relationships can be difficult. Overseas outsourcing makes quality control even more complex. What does a U.S., Canadian, or European original equipment manufacturer do if its low-cost Chinese supplier delivers nonconforming or defective raw materials or parts? How does it work with a Chinese supplier to resolve the problem across thousands of miles and language and cultural barriers?

These nonstatistical problems illustrate why quality management in a global environment requires an understanding of international business as well as behavioral, financial, and supply-chain management. Engineers and technical specialists often find it helpful to study business management (and maybe Chinese language). If you require a more detailed understanding of quality management, you will need further training (you can start with the Related Reading list).

Quality and Design

Quality is closely tied to design. A well-designed process, product, or service is more likely to yield better quality and more customer satisfaction, with less effort and for a longer time. A poorly designed process, product, or service is more likely to yield undesired outcomes, awkward or inconvenient working arrangements, employee frustration in trying to maintain quality, and more frequent problems, breakdowns, and dissatisfied customers.

Firms may know that their products and services could be designed in a better way, but it would take time and cost money. Customer needs must be met today, so they say, "We will nurse along the old design and do the best we can with it." The problem is that, in the longer run, the customers may not be there, as more dynamic competitors capture the market. One lesson of our time is that there is no such thing as a "safe job," even in a large organization. When we can see a better way to do it, change becomes an ally and inertia an enemy. The search for design improvement is an ongoing process, not something done once. If we improve the design tomorrow, even better solutions are likely to be found later on. Successful organizations try to create a climate in which employees are encouraged to suggest new ways of doing things.

Taguchi's Robust Design

The prominence of Japanese quality expert Genichi Taguchi is mainly due to his contributions in the field of *robust design,* which uses statistically planned experiments to identify process control parameter settings that reduce a process's sensitivity to manufacturing variation. In Taguchi's taxonomy, we identify the functional characteristics that measure the final product's performance, the control parameters that can be specified by process engineers, and the sources of noise that are expensive or impossible to control. By varying control parameters in a planned experiment, we can predict control parameter settings that would make the product's performance less sensitive to random variation. Parameter settings are first varied in a few experimental runs. Then a fractional factorial experimental design is selected (see Chapter 11), using a balancing property to choose pairs of parameter settings. Finally, predictions of improved parameter settings are made and verified in a confirming experiment.

Taguchi's methods are especially useful in manufacturing situations with many process control parameters, which imply complex experimental designs. Once the problem is defined, we rely on well-known experimental design methods. In addition, Taguchi is known for explicitly including in quality measures the total loss incurred from the time the product is shipped, using a quadratic loss penalty based on the squared difference between actual and target quality. His inclusion of customers in the model is considered a major innovation.

Six Sigma and Lean Six Sigma

Six Sigma is a broad philosophy to reduce cost, eliminate variability, and improve customer satisfaction through improved design and better management strategy. *Lean Six Sigma* integrates Six Sigma with supply-chain management to optimize resource flows while also lowering cost and raising quality. Most of us have heard of the Six Sigma goal of 3.4 defects per million through reduced process variation (i.e., extremely high C_p and C_{pk} indexes), essentially using the tools outlined in this chapter and the DMAIC steps for process improvement. However, there is more to it than statistics, and Six Sigma experts must be certified (Green Belts, Black Belts, Master Black Belts) through advanced training. Six Sigma implementation varies according to the organization, with health care being perhaps the latest major application. Six Sigma knowledge goes beyond the bounds of an introductory statistics class, but if you take a job that requires it, your company will give you advanced training.

ISO 9000

Since 1992, firms wishing to sell their products globally have had to comply with a series of ISO standards, first articulated in 1987 in Europe. These standards have continued to evolve. Now, ISO 9000 and ISO audits (both internal and of suppliers) have become a de facto quality system standard for any company wanting to be a world-class competitor. ISO 9001 and ISO/TS 16949 include customer service as well as design of products and services (not just manufacturing). Several other ISO standards have been developed that address environmental management, food safety management, social responsibility, and risk management. The broad scope of the standards requires special training that is not normally part of an introductory statistics class.

Malcolm Baldrige Award

To recognize the importance of achievement in attaining superior quality, in 1988 the United States initiated the *Malcolm Baldrige National Quality Award,* based on seven categories of quality: leadership, information/analysis, strategic planning, human resource development, process management, operational results, and customer satisfaction. The **Baldrige Award** is given by the president of the United States to firms (large or small, manufacturing or services) that have made notable achievements in design, manufacture, installation, sales, and service.

Software

A glance at Minitab's extensive menus will tell you that quality tools are one of its strengths. In addition to all types of control charts and cause-and-effect diagrams (fishbone or Ishikawa diagrams), Minitab offers capability analysis, variable transformations to achieve normality, alternative distributions where the assumption of normality is inappropriate, and gage study for variables and attributes. If you want further study of statistical quality tools, you could do worse than to explore Minitab's menus, help system, and data sets. Many other general-purpose software packages (e.g., SAS, SPSS) offer similar capabilities. Although basic R has no comprehensive quality tools, you can download supplementary CRAN packages such as qcc, ggQC, and ggplot to make control charts, cause-and-effect diagrams, and Pareto charts as well as to perform capability analysis.

Future of Statistical Process Control

Automation, numerical control, and continuous process monitoring have changed the meaning of SPC in manufacturing. The integration of manufacturing and factory floor quality monitoring systems in manufacturing planning and control; materials requirements planning (MRP); computer-aided design and manufacturing (CAD/CAM); order entry; and financial, customer service, and support systems have continued to redefine the role of SPC. Automation has made 100 percent testing and inspection attainable in some applications where it was previously thought to be either impossible or uneconomical.

It may be that SPC itself will become part of the background that is built in to every manufacturing organization, allowing managers to focus on higher-level issues. As an analogy, consider that only a few decades ago, chart-making required specialists who were skilled in drafting. Now, anyone with access to a computer can make excellent charts. In the service sector of the economy, quality improvement is still at an early level of implementation. In health care, financial services, and retailing, processes are harder to define and tasks are often not as repetitive or as standardized as in manufacturing. Thus, the role of SPC is still unfolding, and every business student needs to know its basic principles.

Mini Case 17.3

ISO 9001/14001 Certification

Firms wishing to compete globally have to comply with a series of standards developed and maintained by the International Organization for Standardization (ISO). ISO 9000 standards address Quality Management Systems (QMS) within an organization. ISO 9001 certification means that an organization has a QMS in place to measure, achieve, and continually improve its customers' quality requirements, whether the organization provides products, services, or a combination of both. The ISO 14000 standards address Environmental Management Systems (EMS). ISO 14001 certification means that an organization is setting environmental objectives, identifying and controlling its environmental impact, and continually improving its environmental performance.

Vail Resorts Hospitality manages contracts within the Grand Teton National Park near Jackson, Wyoming, through the Grand Teton Lodge Company (GTLC). Grand Teton Lodge Company was one of the first Wyoming tourism entities to achieve ISO 14001 certification, a designation that also places it among an elite group of national park concessionaires that have received such certification. GTLC is also ISO 9001 certified, the only hospitality company in the United States to certify its quality management system.

Each year its commitment is verified through independent, third-party audits. Vail Resorts reports that the environmental management system has been successfully recertified each year, including several years with no adverse findings in this process. The GTLC EMS has achieved goals related to the environment such as annually diverting 300 tons of waste from landfills through recycling and reuse and working with its food vendors to develop a line of eco-friendly disposable food containers.

You can read more about the ISO certification process and Vail Resorts' quality and environmental efforts at the following websites: www.iso.org and www.vailresorts.com.

Chapter Summary

Quality is measured by a set of attributes that affect customer satisfaction. Quality improvement is aimed at variance reduction. **Common cause variation** is normal and expected, while **special cause variation** is abnormal and requires action, such as adjusting the **process** for producing a good or service. Quality is affected by management, resources, technology, and human factors (e.g., training, employee involvement). **Statistical process control** (SPC) involves using **control charts** of key quality metrics to make sure that the processes are in control. The **upper control limit** (UCL) and **lower control limit** (LCL) define the range of allowable variation. These limits are usually set empirically by observing a process over time. Control charts are used to track the **mean** ($\bar{x}$ chart), **range** (*R* chart), **proportion** (*p* chart), and other statistics.

Samples may be taken by subgroups of *n* items, or by continuous monitoring with **individual charts** (*I* charts) and **moving range charts** (*MR* charts). There are rules of thumb to identify out-of-control patterns (**instability, trend, level shift, cycle, oscillation, mixture**) and their likely causes. The **capability** of a process is improved when variability (σ) is small in relation to the upper and lower **specification limits** (USL and LSL) as reflected in the C_p and C_{pk} capability indexes. Process control concepts were first applied to manufacturing but have been adapted to service environments such as finance, health care, and retailing. International standards such as **ISO 9000** now guide companies selling in world markets, and **Six Sigma** techniques are widely used to improve quality in service organizations, as well as in manufacturing.

Key Terms

acceptance sampling	DMAIC	Pareto chart	statistical process control
Baldrige Award	fishbone chart	PDCA	(SPC)
business process	*I* charts	process	statistical quality control
redesign (BPR)	instability	process control	(SQC)
C_p	ISO 9000	productivity	Taguchi method
C_{pk}	level shift	quality	total quality management
capability	lower control limit (LCL)	quality control	(TQM)
common cause variation	lower specification	*R* chart	trend
continuous quality	limit (LSL)	service blueprint	upper control limit
improvement (CQI)	mixture	service transaction analysis	(UCL)
control chart	*MR* charts	SERVQUAL	upper specification
cycle	oscillation	Six Sigma	limit (USL)
Deming, W. Edwards	*p* chart	special cause variation	$\bar{x}$ chart

Commonly Used Formulas

Control limits for $\bar{x}$ chart (known or historical σ): $\mu \pm 3\dfrac{\sigma}{\sqrt{n}}$

Control limits for $\bar{x}$ chart (sample estimate of σ): $\bar{\bar{x}} \pm 3\dfrac{s}{\sqrt{n}}$

Control limits for $\bar{x}$ chart (using average range): $\bar{\bar{x}} \pm 3\dfrac{\bar{R}}{d_2\sqrt{n}}$

Control limits for R chart (using average range or sample standard deviation with control chart factors from a table):

$$\text{UCL} = D_4\bar{R} \quad \text{or} \quad \text{UCL} = D_4 d_2 s$$
$$\text{LCL} = D_3\bar{R} \quad \text{or} \quad \text{LCL} = D_3 d_2 s$$

Control limits for p chart: $\pi \pm 3\sqrt{\dfrac{\pi(1-\pi)}{n}}$

Capability index (ignores centering): $C_p = \dfrac{\text{USL} - \text{LSL}}{6\sigma}$

Capability index (tests for centering): $C_{pk} = \dfrac{\min(\mu - \text{LSL}, \text{USL} - \mu)}{3\sigma}$

Chapter Review

Note: Questions with * are based on optional material.

1. Define (a) quality, (b) process, and (c) productivity. Why are they hard to define?

2. List six general attributes of quality.

3. Distinguish between common cause and special cause variation.

4. In quality improvement, list three roles played by statisticians.

5. Describe the five service quality dimensions.

6. In chronological order, list important phases in the evolution of the quality movement in North America. What is the main change in emphasis over the last 100 years?

7. (a) Who was W. Edwards Deming and why is he remembered? (b) List three of Deming's major ideas and explain them in your own terms.

8. List three influential thinkers other than Deming who made contributions to the quality movement and state their contributions.

9. (a) Briefly explain each acronym: TQM, BPR, SQC, SPC, CQI, DMAIC. (b) List the steps in the continuous quality improvement model.

10. (a) What is shown on the $\bar{x}$ chart? (b) Name three ways to set the control limits on the $\bar{x}$ chart. (c) How can we obtain good empirical control limits for the $\bar{x}$ chart? (d) Why are quality control samples sometimes small?

11. Explain the four rules of thumb for identifying an out-of-control process.

12. (a) What is shown on the R chart? (b) How do we set control limits for the R chart?

13. Name the six abnormal control chart patterns and tell (a) how they may be recognized and (b) what their likely causes might be.

14. (a) State the formulas for the two capability indexes C_p and C_{pk}. (b) Why isn't C_p alone sufficient? (c) What is considered an acceptable value for these indexes? (d) Why is an *in-control* process not necessarily *capable?*

15. (a) What is shown on the p chart? (b) How do we set control limits for the p chart? (c) Why might the p chart control limits vary from sample to sample?

*16. Briefly explain (a) the overadjustment problem, (b) *ad hoc* control charts, (c) acceptance sampling, (d) supply-chain management, (e) Taguchi's robust design, (f) the Six Sigma philosophy, (g) ISO 9000, and (h) the Malcolm Baldrige Award.

Chapter Exercises

Instructions: You may use Minitab, MegaStat, JMP, or similar software to assist you in the control chart questions. You may download the data files from McGraw Hill's Connect®.

17.32 Explain each chart's purpose and the parameters that must be known or estimated to establish its control limits.

   a. $\bar{x}$ chart
   b. R chart
   c. p chart
   d. I chart

17.33 Define three possible quality metrics (not necessarily the ones actually used) to describe and monitor (a) your performance in your college classes; (b) effectiveness of the professors in your college classes; (c) your effectiveness in managing your personal finances; (d) your textbook's effectiveness in helping you learn in a college statistics class.

17.34 Define three quality metrics that might be used to describe quality and performance for the following services: (a) your cellular phone service (e.g., Verizon); (b) your Internet service provider (e.g., AOL); (c) your dry cleaning and laundry service; (d) your physician's office; (e) your hairdresser or barber; (f) your favorite fast-food restaurant.

17.35 Define three quality metrics that might be used to describe quality and performance in the following consumer products: (a) your personal vehicle (e.g., car, SUV, truck, bicycle, motorcycle); (b) the printer for your computer; (c) the toilet in your bathroom; (d) your iPad; (e) an HDTV display screen; (f) a light bulb.

17.36 Based on the cost of sampling and the presumed accuracy required, would sampling or 100 percent inspection be used to collect data on (a) the horsepower of each engine being installed in new cars; (b) the fuel consumption per seat mile of each Southwest Airlines flight; (c) the daily percent of customers who order low-carb menu items for each McDonald's restaurant; (d) the life in hours of each lithium ion battery installed in new laptop computers; (e) the number of medication errors per month in a large hospital?

17.37 Why are the control limits for an *R* chart asymmetric, while those of an $\bar{x}$ chart are symmetric?

17.38 Bob said, "We use the normal distribution to set the control limits for the $\bar{x}$ chart because samples from processes follow a normal distribution." Is Bob right? Explain.

17.39 Bob said, "They must not be using quality control in automobile manufacturing. Just look at the J.D. Power data showing that new cars all seem to have defects." (a) Discuss Bob's assertion, focusing on the concept of variation. (b) Can you think of processes where zero defects *could* be attained on a regular basis? Explain. (c) Can you think of processes where zero defects *cannot* be attained on a regular basis? *Hint:* Consider activities like pass completion by a football quarterback or three-point shots by a college basketball player.

17.40 Use your favorite Internet search engine to look up any four of the following quality experts. Write a one-paragraph biographical sketch *in your own words* that lists his contributions to quality improvement.

   a. Walter A. Shewhart
   b. Harold F. Dodge
   c. Harry G. Romig
   d. Joseph M. Juran
   e. Genichi Taguchi
   f. Kaoru Ishikawa
   g. Armand V. Feigenbaum

17.41 Make a fishbone chart (cause-and-effect diagram) like the following for the reasons you have ever been (or could be) late to class. Use as many branches as necessary. Which factors are most important? Which are most easily controlled?

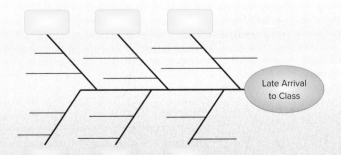

17.42 Make a fishbone chart (cause-and-effect diagram) for the reasons your end-of-month checkbook balance may not match your bank statement. Use as many branches as necessary. Which factors are most important? Which are most easily controlled?

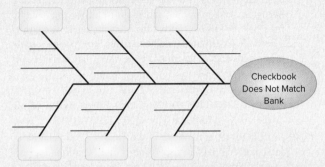

17.43 Make a fishbone chart (cause-and-effect diagram) for the reasons an airline flight might be late to arrive. Use as many branches as necessary. Which factors are most important? Which are most easily controlled?

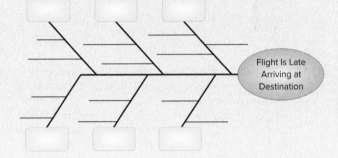

CAPABILITY

17.44 A process has specification limits of LSL = 540 and USL = 550. The process standard deviation is $\sigma = 1.25$. Find the C_p and C_{pk} capability indexes if (a) the process mean is 545; (b) the process mean is 543.

17.45 In painting an automobile, the thickness of the color coat has a lower specification limit of 0.80 mil and an upper specification limit of 1.20 mils. Find the C_p and C_{pk} capability indexes if (a) the process mean is 1.00 mil and the process standard deviation is 0.05 mil; (b) the process mean is .90 mil and the process standard deviation is 0.05 mil.

17.46 Moisture content per gram of a certain baked product has specification limits of 120 mg and 160 mg. Find the C_p and C_{pk} capability indexes if (a) the process mean is 140 mg and the process standard deviation is 5 mg; (b) the process mean is 140 mg and the process standard deviation is 3 mg.

$\bar{X}$ CHARTS

17.47 The yield strength of a metal bolt has a mean of 6,050 pounds with a standard deviation of 100 pounds. Twenty samples of three bolts were tested, resulting in the means shown below. (a) Construct upper and lower control limits for the $\bar{x}$ chart, using the given product parameters. (b) Plot the data on the control chart. (c) Is this process in control? Explain your reasoning. **Bolts-M**

6,107 6,031 6,075 6,115 6,039 6,079 5,995 6,097 6,114 6,039
6,154 6,054 6,028 6,002 6,062 6,094 6,051 6,031 5,965 6,082

17.48 Refer to the bolt strength problem 17.47. Assume $\mu = 6,050$ and $\sigma = 100$. Use the following 24 *individual* bolt strength observations to answer the questions posed. (a) Prepare a histogram and/or normal probability plot for the sample. (b) Does the sample support the view that yield strength is a normally distributed random variable? (c) Are the sample mean and standard deviation about where they are expected to be? 📷 **Bolts-I**

6,121 6,100 6,007 6,166 6,164 6,032 6,276 6,151
6,054 5,836 6,024 6,105 6,033 6,066 6,079 6,192
6,028 6,087 5,983 6,040 6,062 6,054 6,100 5,983

17.49 In painting an automobile at the factory, the thickness of the color coat has a process mean of 1.00 mil and a process standard deviation of 0.07 mil. Twenty samples of five cars were tested, resulting in the mean paint thicknesses shown below. (a) Construct upper and lower control limits for the $\bar{x}$ chart, using the given process parameters. (b) Plot the data on the control chart. (c) Is this process in control? Explain your reasoning. 📷 **Paint-M**

0.996 0.960 1.016 1.017 1.001 0.988 1.006
1.073 1.032 1.021 0.984 1.019 0.997 1.024
1.033 1.030 0.994 0.980 0.977 1.037

17.50 Refer to the paint thickness problem 17.49. Assume $\mu = 1.00$ and $\sigma = 0.07$. Use the following 35 *individual* observations on paint thickness to answer the questions posed. (a) Prepare a histogram and/or normal probability plot for the sample. (b) Does the sample support the view that paint thickness is a normally distributed random variable? (c) Are the mean and standard deviation about as expected? 📷 **Paint-I**

1.026 0.949 1.069 1.105 0.995 0.955 1.080
0.932 1.014 0.899 1.031 1.042 1.022 1.082
1.111 0.995 1.005 1.004 0.964 1.065 0.909
0.912 0.978 1.037 0.992 1.010 0.974 0.977
0.905 1.008 0.971 0.951 1.200 1.065 0.972

17.51 The temperature control unit on a commercial freezer in a 24-hour grocery store is set to maintain a mean temperature of 23 degrees Fahrenheit. The temperature varies because people are constantly opening the freezer door to remove items, but the thermostat is capable of maintaining temperature with a standard deviation of 2 degrees Fahrenheit. The desired range is 18 to 30 degrees Fahrenheit. (a) Find the C_p and C_{pk} capability indexes. (b) In words, how would you describe the process capability? (c) If improvement is desired, what might be some obstacles to increasing the capability?

17.52 Refer to the freezer problem 17.51 with $\mu = 23$ and $\sigma = 2$. Temperature measurements are recorded four times a day (at midnight, 0600, 1200, and 1800). Twenty samples of four observations are shown below. (a) Construct upper and lower control limits for the $\bar{x}$ chart, using the given process parameters. (b) Plot the data on the control chart. (c) Is this process in control? Explain your reasoning. 📷 **Freezer**

Sample	Midnight	At 0600	At 1200	At 1800	Mean
1	25	26	23	23	24.25
2	22	23	28	22	23.75
3	20	24	25	21	22.50
4	21	25	22	23	22.75
5	21	23	21	23	22.00
6	26	25	27	26	26.00
7	21	23	25	20	22.25
8	25	23	22	25	23.75
9	22	24	24	22	23.00
10	27	23	26	25	25.25
11	24	23	20	21	22.00
12	25	21	23	20	22.25
13	26	21	21	23	22.75
14	26	22	26	22	24.00
15	21	24	20	19	21.00
16	23	26	23	23	23.75
17	23	21	24	21	22.25
18	25	22	22	23	23.00
19	24	20	21	22	21.75
20	24	21	23	21	22.25

17.53 Refer to the freezer data's 80 *individual* temperature observations in problem 17.52. (a) Prepare a histogram and/or normal probability plot for the sample. (b) Does the sample support the view that freezer temperature is a normally distributed random variable? (c) Are the sample mean and standard deviation about where they are expected to be? 📷 **Freezer**

17.54 A Nabisco Fig Newton has a process mean weight of 14.00 g with a standard deviation of 0.10 g. The lower specification limit is 13.40 g and the upper specification limit is 14.60 g. (a) Describe the capability of this process, using the techniques you have learned. (b) Would you think that further variance reduction efforts would be a good idea? Explain the pros and cons of such an effort.

17.55 A new type of smoke detector battery is developed. From laboratory tests under standard conditions, the half-lives (defined as less than 50 percent of full charge) of 20 batteries are shown below. (a) Make a histogram of the data and/or a probability plot. Do you think that battery half-life can be assumed normal? (b) The engineers say that the mean battery half-life will be 8,760 hours with a standard deviation of 200 hours. Using these parameters (not the sample), set up the centerline and control limits for the $\bar{x}$ chart for a subgroup size of $n = 5$ batteries to be sampled in future production runs. (c) Repeat the previous exercise, but this time, use the sample mean and standard deviation. (d) Do you think that the control limits from this sample would be reliable? Explain, and suggest alternatives. 📷 **Battery**

8,502 8,660 8,785 8,778 8,804 9,069 8,516 9,048 8,628 9,213
8,511 8,965 8,688 8,892 8,638 8,440 8,900 8,993 8,958 8,707

17.56 A box of Wheat Chex cereal is to be filled to a mean weight of 466 grams. The lower specification limit is 453 grams (the labeled weight is 453 grams) and the upper specification limit is 477 grams (so as not to overfill the box). The process standard deviation is 2 grams. (a) Find the C_p and C_{pk} capability indexes. (b) Assess the process capability. (c) Why might it be difficult to reduce the variance in this process to raise the capability indices? *Hint:* A single Wheat Chex weighs .3 g (30 mg).

17.57 Refer to the Wheat Chex problem 17.56 with $\mu = 466$ and $\sigma = 3$. During production, samples of three boxes are weighed every 5 minutes. (a) Find the upper and lower control limit for the $\bar{x}$ chart. (b) Plot the following 20 sample means on the chart. Is the process in control? 📷 **Chex-M**

465.7 463.7 466.0 466.3 463.0 468.3 465.0 463.3 462.0 463.0
465.7 467.0 463.3 466.0 465.3 465.3 463.0 466.7 466.3 466.3

17.58 Refer to the Wheat Chex box fill problem 17.56 with $\mu = 466$ and $\sigma = 3$. Below are 30 *individual* observations on box fill. (a) Prepare a histogram and/or normal probability plot for the sample. Does the sample support the view that box fill is a normally distributed random variable? Explain. (b) Is the mean of these 30 *individual* observations where it should be? 📷 **Chex-I**

461 465 462 469 463 465 462 465 467 467
460 467 466 466 465 465 462 458 470 460
465 466 464 460 465 465 466 464 465 461

17.59 Each gum drop in two bags of Sathers Gum Drops was weighed (to the nearest .001 g) on a sensitive Mettler PE 360 Delta Range scale. After removing one outlier (to improve normality), there were 84 gum drops in the sample, yielding an overall mean $\bar{x} = 11.988$ g and a pooled standard deviation $s = .2208$ g. (a) Use these sample statistics to construct control limits for an $\bar{x}$ chart, using a subgroup size $n = 6$. (b) Plot the means shown in Exhibit "A"

on your control chart. Is the process in control? (c) Prepare a histogram and/or normal probability plot for the pooled sample. Does the sample support the view that gum drop weight is a normally distributed random variable? Explain. 📷 **GumDrops**

P CHARTS

17.60 Past experience indicates that the probability of a postsurgical complication in a certain procedure is 6 percent. A hospital typically performs 200 such surgeries per month. (a) Find the control limits for the monthly p chart. (b) Would it be reasonably safe to assume that the sample proportion x/n is normally distributed? Explain.

17.61 A large retail toy store finds that, on average, a certain cheap (under $20) electronic toy has a 5 percent damage rate during shipping. From each incoming shipment, a sample of 100 is inspected. (a) Find the control limits for a p chart. (b) Plot the 10 samples below on the p chart. Is the process in control? (c) Is the sample size large enough to assume normality of the sample proportion? Explain. 📷 **Toys**

Sample	X	n	X/n
1	3	100	0.03
2	5	100	0.05
3	4	100	0.04
4	7	100	0.07
5	2	100	0.02
6	2	100	0.02
7	0	100	0.00
8	2	100	0.02
9	7	100	0.07
10	6	100	0.06

EXHIBIT A
📷 **GumDrops**

Sample	x_1	x_2	x_3	x_4	x_5	x_6	Mean
1	11.741	11.975	11.985	12.163	12.317	12.032	12.036
2	12.206	11.970	12.179	12.182	11.756	11.975	12.045
3	12.041	12.120	11.855	12.036	11.750	11.870	11.945
4	12.002	11.800	12.092	12.017	12.340	12.488	12.123
5	12.305	12.134	11.949	12.050	12.246	11.839	12.087
6	11.862	12.049	12.105	11.894	11.995	11.722	11.938
7	11.979	12.124	12.171	12.093	12.224	11.965	12.093
8	11.941	11.855	11.587	11.574	11.752	12.345	11.842
9	12.297	12.078	12.137	11.869	11.609	11.732	11.954
10	11.677	11.879	11.926	11.852	11.781	11.932	11.841
11	12.113	12.129	12.156	12.284	12.207	12.247	12.189
12	12.510	11.904	11.675	11.880	12.086	12.458	12.086
13	12.193	11.975	12.173	11.635	11.549	11.744	11.878
14	11.880	11.784	11.696	11.804	11.823	11.693	11.780

PATTERNS IN CONTROL CHARTS

17.62 Referring to Charts A–F, which abnormal pattern (cycle, instability, level shift, oscillation, trend, mixture), if any, exists in each of the $\bar{x}$ charts shown earlier? If you see none, say so. If you see more than one possibility, say so. Explain your reasoning.

17.63 Referring to Charts A–F, which Rules (1, 2, 3, 4) are violated in each chart? Make a photocopy and circle the points that violate each rule.

Chart A

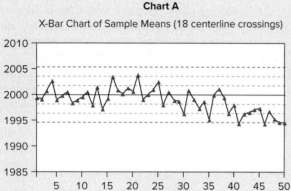

Chart D

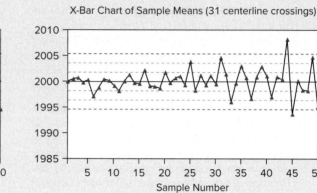

Chart B

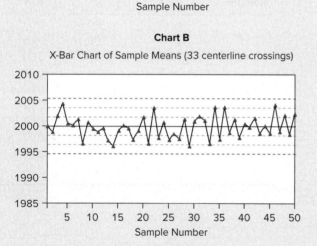

Chart E

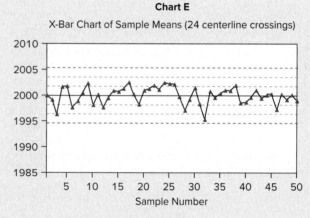

Chart C

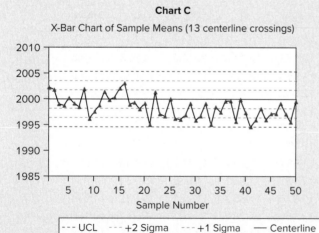

Chart F

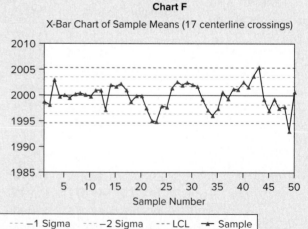

UCL   +2 Sigma   +1 Sigma   — Centerline   −1 Sigma   −2 Sigma   LCL   ▲ Sample

17.64 Refer to the bolt strength problem 17.47. Assuming $\mu = 6,050$ and $\sigma = 100$ with $n = 3$, then LCL = 5,876.8 and UCL = 6,223.2. Below are five sets of 20 sample means using $n = 3$. Test each set of means for the pattern suggested in the column heading. This is a visual judgment question, though you can apply Rules 1–4 if you wish. Bolts-P

Up Trend?	Down Trend?	Unstable?	Cycle?	Oscillate?
5,907	6,100	6,048	6,079	6,122
6,060	6,009	5,975	6,029	5,983
5,987	6,145	6,092	6,006	6,105
5,919	6,049	5,894	6,012	6,024
6,029	6,039	6,083	6,098	6,123
6,114	5,956	6,069	6,124	6,022
6,063	6,103	6,073	6,092	6,082
6,084	6,140	5,972	6,114	6,018
5,980	6,054	6,112	6,071	6,031
6,056	6,062	5,988	6,097	6,107
6,078	6,042	6,006	6,038	6,031
6,118	6,152	6,226	6,099	6,047
6,051	5,961	5,989	6,000	6,055
6,021	5,926	6,111	6,004	6,041
6,068	6,109	6,026	6,054	5,972
6,157	5,904	6,057	6,083	5,987
6,041	6,049	6,098	6,148	6,043
6,129	6,042	6,082	6,071	6,137
6,026	5,847	6,050	6,095	5,930
6,174	6,033	6,084	6,092	6,057

17.65 Refer to the paint problem 17.49 with $\mu = 1.00$ and $\sigma = .07$. With $n = 5$, LCL = .906 and UCL = 1.094. Below are five sets of 20 sample means using $n = 5$. Test each set of means for the pattern suggested in the column heading. This is a visual judgment question, though you can apply Rules 1–4 if you wish. Paint-P

No Pattern?	Up Trend?	Down Trend?	Unstable?	Cycle?
0.996	0.995	1.007	0.999	0.964
0.960	0.942	1.000	0.986	1.025
1.016	0.947	1.011	0.950	0.988
1.017	1.011	0.989	0.982	1.000
1.001	0.983	0.999	0.967	1.023
0.988	0.989	1.000	0.972	1.019
1.006	0.978	1.025	0.977	1.035
1.073	0.958	0.963	1.015	1.043
1.032	1.034	1.060	0.970	1.044
1.021	1.058	1.020	1.016	0.993
0.984	1.058	0.977	0.979	0.994
1.019	0.958	0.985	0.934	0.988
0.997	1.030	1.033	0.975	0.991
1.024	1.022	0.975	1.100	1.001
1.033	0.976	0.939	0.976	1.011
1.030	1.024	1.007	0.976	1.015
0.994	1.032	0.994	1.029	1.000
0.980	0.994	0.990	0.987	1.010
0.977	1.016	0.925	0.954	1.061
1.037	1.039	0.907	1.011	1.001

DO IT YOURSELF

17.66 Buy a bag of M&Ms. (a) As a measure of quality, take a sample of 100 M&Ms and count the number with an incomplete or illegible "M" printed on them. (b) Calculate the sample proportion with defects. (c) What ambiguity (if any) did you encounter in this task? (d) Do you feel that your sample was large enough? Explain.

17.67 Examine a square meter (or another convenient unit) of paint on your car's driver door. Be sure the area is clean. (a) Tally the number of paint defects (scratch, abrasion, embedded dirt, chip, dent, rust, other). You may add your own defect categories. (b) Repeat, using a friend's car that is either older or newer than yours. (c) State your findings succinctly.

17.68 Buy a box of Cheerios. (a) As a measure of quality, take a sample of 100 Cheerios and count the number of Cheerios that are broken. (b) Calculate the sample proportion with defects. (c) What ambiguity (if any) did you encounter in this task?

Web Data Sources

Source	Website
American Society for Quality (books, training, videos)	www.qualitypress.asq.org
Deming Institute (philosophy, references)	www.deming.org
Juran Institute	www.juran.com
Quality Digest Magazine (current issues)	www.qualitydigest.com
Quality University (training videos)	www.qualityuniversity.com

Related Reading

Besterfield, Dale H. *Quality Control.* 9th ed. Pearson, 2013.

Brussee, Warren. *Statistics for Six Sigma Made Easy.* 2nd ed. McGraw Hill, 2012.

Cano, Emilio L., Javier M. Moguerza, and Mariano Prieto Corcoba. *Quality Control with R: An ISO Standards Approach.* Springer, 2015.

Evans, James R., and William M. Lindsay. *Management and the Control of Quality.* 10th ed. Cengage, 2017.

Foster, S. Thomas. *Managing Quality.* 6th ed. Pearson. 2020.

Montgomery, Douglas C. *Introduction to Statistical Quality Control.* 8th ed. Wiley, 2019.

Pyzdek, Thomas. *The Six Sigma Handbook.* 5th ed. McGraw Hill, 2019.

CHAPTER 17   More Learning Resources

You can access these LearningStats demonstrations through Connect to help you understand quality management.

Topic	LearningStats Demonstrations
Quality overview	Quality Overview Process Control Overview
Capability	Capability Explained Capability Indexes Moving Range
Control chart patterns and rules	Control Chart Patterns
Other	What Is Six Sigma?

Key: = PowerPoint   = Excel

BINOMIAL PROBABILITIES

Example: $P(X = 3 \mid n = 8, \pi = .50) = .2188$

This table shows $P(X = x)$.

									π									
n	x	.01	.02	.05	.10	.15	.20	.30	.40	.50	.60	.70	.80	.85	.90	.95	.98	.99
2	0	.9801	.9604	.9025	.8100	.7225	.6400	.4900	.3600	.2500	.1600	.0900	.0400	.0225	.0100	.0025	.0004	.0001
	1	.0198	.0392	.0950	.1800	.2550	.3200	.4200	.4800	.5000	.4800	.4200	.3200	.2550	.1800	.0950	.0392	.0198
	2	.0001	.0004	.0025	.0100	.0225	.0400	.0900	.1600	.2500	.3600	.4900	.6400	.7225	.8100	.9025	.9604	.9801
3	0	.9703	.9412	.8574	.7290	.6141	.5120	.3430	.2160	.1250	.0640	.0270	.0080	.0034	.0010	.0001	—	—
	1	.0294	.0576	.1354	.2430	.3251	.3840	.4410	.4320	.3750	.2880	.1890	.0960	.0574	.0270	.0071	.0012	.0003
	2	.0003	.0012	.0071	.0270	.0574	.0960	.1890	.2880	.3750	.4320	.4410	.3840	.3251	.2430	.1354	.0576	.0294
	3	—	—	.0001	.0010	.0034	.0080	.0270	.0640	.1250	.2160	.3430	.5120	.6141	.7290	.8574	.9412	.9703
4	0	.9606	.9224	.8145	.6561	.5220	.4096	.2401	.1296	.0625	.0256	.0081	.0016	.0005	.0001	—	—	—
	1	.0388	.0753	.1715	.2916	.3685	.4096	.4116	.3456	.2500	.1536	.0756	.0256	.0115	.0036	.0005	—	—
	2	.0006	.0023	.0135	.0486	.0975	.1536	.2646	.3456	.3750	.3456	.2646	.1536	.0975	.0486	.0135	.0023	.0006
	3	—	—	.0005	.0036	.0115	.0256	.0756	.1536	.2500	.3456	.4116	.4096	.3685	.2916	.1715	.0753	.0388
	4	—	—	—	.0001	.0005	.0016	.0081	.0256	.0625	.1296	.2401	.4096	.5220	.6561	.8145	.9224	.9606
5	0	.9510	.9039	.7738	.5905	.4437	.3277	.1681	.0778	.0313	.0102	.0024	.0003	.0001	—	—	—	—
	1	.0480	.0922	.2036	.3281	.3915	.4096	.3602	.2592	.1563	.0768	.0284	.0064	.0022	.0005	—	—	—
	2	.0010	.0038	.0214	.0729	.1382	.2048	.3087	.3456	.3125	.2304	.1323	.0512	.0244	.0081	.0011	.0001	—
	3	—	.0001	.0011	.0081	.0244	.0512	.1323	.2304	.3125	.3456	.3087	.2048	.1382	.0729	.0214	.0038	.0010
	4	—	—	.0005	.0022	.0064	.0284	.0768	.1563	.2592	.3602	.4096	.3915	.3281	.2036	.0922	.0480	
	5	—	—	—	—	.0001	.0003	.0024	.0102	.0313	.0778	.1681	.3277	.4437	.5905	.7738	.9039	.9510
6	0	.9415	.8858	.7351	.5314	.3771	.2621	.1176	.0467	.0156	.0041	.0007	.0001	—	—	—	—	—
	1	.0571	.1085	.2321	.3543	.3993	.3932	.3025	.1866	.0938	.0369	.0102	.0015	.0004	.0001	—	—	—
	2	.0014	.0055	.0305	.0984	.1762	.2458	.3241	.3110	.2344	.1382	.0595	.0154	.0055	.0012	.0001	—	—
	3	—	.0002	.0021	.0146	.0415	.0819	.1852	.2765	.3125	.2765	.1852	.0819	.0415	.0146	.0021	.0002	—
	4	—	—	.0001	.0012	.0055	.0154	.0595	.1382	.2344	.3110	.3241	.2458	.1762	.0984	.0305	.0055	.0014
	5	—	—	—	.0001	.0004	.0015	.0102	.0369	.0938	.1866	.3025	.3932	.3993	.3543	.2321	.1085	.0571
	6	—	—	—	—	—	.0001	.0007	.0041	.0156	.0467	.1176	.2621	.3771	.5314	.7351	.8858	.9415
7	0	.9321	.8681	.6983	.4783	.3206	.2097	.0824	.0280	.0078	.0016	.0002	—	—	—	—	—	—
	1	.0659	.1240	.2573	.3720	.3960	.3670	.2471	.1306	.0547	.0172	.0036	.0004	.0001	—	—	—	—
	2	.0020	.0076	.0406	.1240	.2097	.2753	.3177	.2613	.1641	.0774	.0250	.0043	.0012	.0002	—	—	—
	3	—	.0003	.0036	.0230	.0617	.1147	.2269	.2903	.2734	.1935	.0972	.0287	.0109	.0026	.0002	—	—
	4	—	—	.0002	.0026	.0109	.0287	.0972	.1935	.2734	.2903	.2269	.1147	.0617	.0230	.0036	.0003	—
	5	—	—	—	.0002	.0012	.0043	.0250	.0774	.1641	.2613	.3177	.2753	.2097	.1240	.0406	.0076	.0020
	6	—	—	—	—	.0001	.0004	.0036	.0172	.0547	.1306	.2471	.3670	.3960	.3720	.2573	.1240	.0659
	7	—	—	—	—	—	—	.0002	.0016	.0078	.0280	.0824	.2097	.3206	.4783	.6983	.8681	.9321
8	0	.9227	.8508	.6634	.4305	.2725	.1678	.0576	.0168	.0039	.0007	.0001	—	—	—	—	—	—
	1	.0746	.1389	.2793	.3826	.3847	.3355	.1977	.0896	.0313	.0079	.0012	.0001	—	—	—	—	—
	2	.0026	.0099	.0515	.1488	.2376	.2936	.2965	.2090	.1094	.0413	.0100	.0011	.0002	—	—	—	—
	3	.0001	.0004	.0054	.0331	.0839	.1468	.2541	.2787	.2188	.1239	.0467	.0092	.0026	.0004	—	—	—
	4	—	—	.0004	.0046	.0185	.0459	.1361	.2322	.2734	.2322	.1361	.0459	.0185	.0046	.0004	—	—
	5	—	—	—	.0004	.0026	.0092	.0467	.1239	.2188	.2787	.2541	.1468	.0839	.0331	.0054	.0004	.0001
	6	—	—	—	—	.0002	.0011	.0100	.0413	.1094	.2090	.2965	.2936	.2376	.1488	.0515	.0099	.0026
	7	—	—	—	—	—	.0001	.0012	.0079	.0313	.0896	.1977	.3355	.3847	.3826	.2793	.1389	.0746
	8	—	—	—	—	—	—	.0001	.0007	.0039	.0168	.0576	.1678	.2725	.4305	.6634	.8508	.9227
9	0	.9135	.8337	.6302	.3874	.2316	.1342	.0404	.0101	.0020	.0003	—	—	—	—	—	—	—
	1	.0830	.1531	.2985	.3874	.3679	.3020	.1556	.0605	.0176	.0035	.0004	—	—	—	—	—	—
	2	.0034	.0125	.0629	.1722	.2597	.3020	.2668	.1612	.0703	.0212	.0039	.0003	—	—	—	—	—
	3	.0001	.0006	.0077	.0446	.1069	.1762	.2668	.2508	.1641	.0743	.0210	.0028	.0006	.0001	—	—	—
	4	—	—	.0006	.0074	.0283	.0661	.1715	.2508	.2461	.1672	.0735	.0165	.0050	.0008	—	—	—
	5	—	—	—	.0008	.0050	.0165	.0735	.1672	.2461	.2508	.1715	.0661	.0283	.0074	.0006	—	—

n	x	.01	.02	.05	.10	.15	.20	.30	.40	.50	.60	.70	.80	.85	.90	.95	.98	.99
	6	—	—	—	.0001	.0006	.0028	.0210	.0743	.1641	.2508	.2668	.1762	.1069	.0446	.0077	.0006	.0001
	7	—	—	—	—	—	.0003	.0039	.0212	.0703	.1612	.2668	.3020	.2597	.1722	.0629	.0125	.0034
	8	—	—	—	—	—	—	.0004	.0035	.0176	.0605	.1556	.3020	.3679	.3874	.2985	.1531	.0830
	9	—	—	—	—	—	—	—	.0003	.0020	.0101	.0404	.1342	.2316	.3874	.6302	.8337	.9135
10	0	.9044	.8171	.5987	.3487	.1969	.1074	.0282	.0060	.0010	.0001	—	—	—	—	—	—	—
	1	.0914	.1667	.3151	.3874	.3474	.2684	.1211	.0403	.0098	.0016	.0001	—	—	—	—	—	—
	2	.0042	.0153	.0746	.1937	.2759	.3020	.2335	.1209	.0439	.0106	.0014	.0001	—	—	—	—	—
	3	.0001	.0008	.0105	.0574	.1298	.2013	.2668	.2150	.1172	.0425	.0090	.0008	.0001	—	—	—	—
	4	—	—	.0010	.0112	.0401	.0881	.2001	.2508	.2051	.1115	.0368	.0055	.0012	.0001	—	—	—
	5	—	—	.0001	.0015	.0085	.0264	.1029	.2007	.2461	.2007	.1029	.0264	.0085	.0015	.0001	—	—
	6	—	—	—	.0001	.0012	.0055	.0368	.1115	.2051	.2508	.2001	.0881	.0401	.0112	.0010	—	—
	7	—	—	—	—	.0001	.0008	.0090	.0425	.1172	.2150	.2668	.2013	.1298	.0574	.0105	.0008	.0001
	8	—	—	—	—	—	.0001	.0014	.0106	.0439	.1209	.2335	.3020	.2759	.1937	.0746	.0153	.0042
	9	—	—	—	—	—	—	.0001	.0016	.0098	.0403	.1211	.2684	.3474	.3874	.3151	.1667	.0914
	10	—	—	—	—	—	—	—	.0001	.0010	.0060	.0282	.1074	.1969	.3487	.5987	.8171	.9044
12	0	.8864	.7847	.5404	.2824	.1422	.0687	.0138	.0022	.0002	—	—	—	—	—	—	—	—
	1	.1074	.1922	.3413	.3766	.3012	.2062	.0712	.0174	.0029	.0003	—	—	—	—	—	—	—
	2	.0060	.0216	.0988	.2301	.2924	.2835	.1678	.0639	.0161	.0025	.0002	—	—	—	—	—	—
	3	.0002	.0015	.0173	.0852	.1720	.2362	.2397	.1419	.0537	.0125	.0015	.0001	—	—	—	—	—
	4	—	.0001	.0021	.0213	.0683	.1329	.2311	.2128	.1208	.0420	.0078	.0005	.0001	—	—	—	—
	5	—	—	.0002	.0038	.0193	.0532	.1585	.2270	.1934	.1009	.0291	.0033	.0006	—	—	—	—
	6	—	—	—	.0005	.0040	.0155	.0792	.1766	.2256	.1766	.0792	.0155	.0040	.0005	—	—	—
	7	—	—	—	—	.0006	.0033	.0291	.1009	.1934	.2270	.1585	.0532	.0193	.0038	.0002	—	—
	8	—	—	—	—	.0001	.0005	.0078	.0420	.1208	.2128	.2311	.1329	.0683	.0213	.0021	.0001	—
	9	—	—	—	—	—	.0001	.0015	.0125	.0537	.1419	.2397	.2362	.1720	.0852	.0173	.0015	.0002
	10	—	—	—	—	—	—	.0002	.0025	.0161	.0639	.1678	.2835	.2924	.2301	.0988	.0216	.0060
	11	—	—	—	—	—	—	—	.0003	.0029	.0174	.0712	.2062	.3012	.3766	.3413	.1922	.1074
	12	—	—	—	—	—	—	—	—	.0002	.0022	.0138	.0687	.1422	.2824	.5404	.7847	.8864
14	0	.8687	.7536	.4877	.2288	.1028	.0440	.0068	.0008	.0001	—	—	—	—	—	—	—	—
	1	.1229	.2153	.3593	.3559	.2539	.1539	.0407	.0073	.0009	.0001	—	—	—	—	—	—	—
	2	.0081	.0286	.1229	.2570	.2912	.2501	.1134	.0317	.0056	.0005	—	—	—	—	—	—	—
	3	.0003	.0023	.0259	.1142	.2056	.2501	.1943	.0845	.0222	.0033	.0002	—	—	—	—	—	—
	4	—	.0001	.0037	.0349	.0998	.1720	.2290	.1549	.0611	.0136	.0014	—	—	—	—	—	—
	5	—	—	.0004	.0078	.0352	.0860	.1963	.2066	.1222	.0408	.0066	.0003	—	—	—	—	—
	6	—	—	—	.0013	.0093	.0322	.1262	.2066	.1833	.0918	.0232	.0020	.0003	—	—	—	—
	7	—	—	—	.0002	.0019	.0092	.0618	.1574	.2095	.1574	.0618	.0092	.0019	.0002	—	—	—
	8	—	—	—	—	.0003	.0020	.0232	.0918	.1833	.2066	.1262	.0322	.0093	.0013	—	—	—
	9	—	—	—	—	—	.0003	.0066	.0408	.1222	.2066	.1963	.0860	.0352	.0078	.0004	—	—
	10	—	—	—	—	—	—	.0014	.0136	.0611	.1549	.2290	.1720	.0998	.0349	.0037	.0001	—
	11	—	—	—	—	—	—	.0002	.0033	.0222	.0845	.1943	.2501	.2056	.1142	.0259	.0023	.0003
	12	—	—	—	—	—	—	—	.0005	.0056	.0317	.1134	.2501	.2912	.2570	.1229	.0286	.0081
	13	—	—	—	—	—	—	—	.0001	.0009	.0073	.0407	.1539	.2539	.3559	.3593	.2153	.1229
	14	—	—	—	—	—	—	—	—	.0001	.0008	.0068	.0440	.1028	.2288	.4877	.7536	.8687
16	0	.8515	.7238	.4401	.1853	.0743	.0281	.0033	.0003	—	—	—	—	—	—	—	—	—
	1	.1376	.2363	.3706	.3294	.2097	.1126	.0228	.0030	.0002	—	—	—	—	—	—	—	—
	2	.0104	.0362	.1463	.2745	.2775	.2111	.0732	.0150	.0018	.0001	—	—	—	—	—	—	—
	3	.0005	.0034	.0359	.1423	.2285	.2463	.1465	.0468	.0085	.0008	—	—	—	—	—	—	—
	4	—	.0002	.0061	.0514	.1311	.2001	.2040	.1014	.0278	.0040	.0002	—	—	—	—	—	—
	5	—	—	.0008	.0137	.0555	.1201	.2099	.1623	.0667	.0142	.0013	—	—	—	—	—	—
	6	—	—	.0001	.0028	.0180	.0550	.1649	.1983	.1222	.0392	.0056	.0002	—	—	—	—	—
	7	—	—	—	.0004	.0045	.0197	.1010	.1889	.1746	.0840	.0185	.0012	.0001	—	—	—	—
	8	—	—	—	.0001	.0009	.0055	.0487	.1417	.1964	.1417	.0487	.0055	.0009	.0001	—	—	—
	9	—	—	—	—	.0001	.0012	.0185	.0840	.1746	.1889	.1010	.0197	.0045	.0004	—	—	—
	10	—	—	—	—	—	.0002	.0056	.0392	.1222	.1983	.1649	.0550	.0180	.0028	.0001	—	—
	11	—	—	—	—	—	—	.0013	.0142	.0667	.1623	.2099	.1201	.0555	.0137	.0008	—	—
	12	—	—	—	—	—	—	.0002	.0040	.0278	.1014	.2040	.2001	.1311	.0514	.0061	.0002	—
	13	—	—	—	—	—	—	—	.0008	.0085	.0468	.1465	.2463	.2285	.1423	.0359	.0034	.0005
	14	—	—	—	—	—	—	—	.0001	.0018	.0150	.0732	.2111	.2775	.2745	.1463	.0362	.0104
	15	—	—	—	—	—	—	—	—	.0002	.0030	.0228	.1126	.2097	.3294	.3706	.2363	.1376
	16	—	—	—	—	—	—	—	—	—	.0003	.0033	.0281	.0743	.1853	.4401	.7238	.8515

POISSON PROBABILITIES

Example: $P(X = 3 \mid \lambda = 2.3) = .2033$

This table shows $P(X = x)$.

| | | | | | | | | λ | | | | | | | |
x	0.1	0.2	0.3	0.4	0.5	0.6	0.7	0.8	0.9	1.0	1.1	1.2	1.3	1.4	1.5
0	.9048	.8187	.7408	.6703	.6065	.5488	.4966	.4493	.4066	.3679	.3329	.3012	.2725	.2466	.2231
1	.0905	.1637	.2222	.2681	.3033	.3293	.3476	.3595	.3659	.3679	.3662	.3614	.3543	.3452	.3347
2	.0045	.0164	.0333	.0536	.0758	.0988	.1217	.1438	.1647	.1839	.2014	.2169	.2303	.2417	.2510
3	.0002	.0011	.0033	.0072	.0126	.0198	.0284	.0383	.0494	.0613	.0738	.0867	.0998	.1128	.1255
4	—	.0001	.0003	.0007	.0016	.0030	.0050	.0077	.0111	.0153	.0203	.0260	.0324	.0395	.0471
5	—	—	—	.0001	.0002	.0004	.0007	.0012	.0020	.0031	.0045	.0062	.0084	.0111	.0141
6	—	—	—	—	—	—	.0001	.0002	.0003	.0005	.0008	.0012	.0018	.0026	.0035
7	—	—	—	—	—	—	—	—	—	.0001	.0001	.0002	.0003	.0005	.0008
8	—	—	—	—	—	—	—	—	—	—	—	—	.0001	.0001	.0001

| | | | | | | | | λ | | | | | | | |
x	1.6	1.7	1.8	1.9	2.0	2.1	2.2	2.3	2.4	2.5	2.6	2.7	2.8	2.9	3.0
0	.2019	.1827	.1653	.1496	.1353	.1225	.1108	.1003	.0907	.0821	.0743	.0672	.0608	.0550	.0498
1	.3230	.3106	.2975	.2842	.2707	.2572	.2438	.2306	.2177	.2052	.1931	.1815	.1703	.1596	.1494
2	.2584	.2640	.2678	.2700	.2707	.2700	.2681	.2652	.2613	.2565	.2510	.2450	.2384	.2314	.2240
3	.1378	.1496	.1607	.1710	.1804	.1890	.1966	.2033	.2090	.2138	.2176	.2205	.2225	.2237	.2240
4	.0551	.0636	.0723	.0812	.0902	.0992	.1082	.1169	.1254	.1336	.1414	.1488	.1557	.1622	.1680
5	.0176	.0216	.0260	.0309	.0361	.0417	.0476	.0538	.0602	.0668	.0735	.0804	.0872	.0940	.1008
6	.0047	.0061	.0078	.0098	.0120	.0146	.0174	.0206	.0241	.0278	.0319	.0362	.0407	.0455	.0504
7	.0011	.0015	.0020	.0027	.0034	.0044	.0055	.0068	.0083	.0099	.0118	.0139	.0163	.0188	.0216
8	.0002	.0003	.0005	.0006	.0009	.0011	.0015	.0019	.0025	.0031	.0038	.0047	.0057	.0068	.0081
9	—	.0001	.0001	.0001	.0002	.0003	.0004	.0005	.0007	.0009	.0011	.0014	.0018	.0022	.0027
10	—	—	—	—	—	.0001	.0001	.0001	.0002	.0002	.0003	.0004	.0005	.0006	.0008
11	—	—	—	—	—	—	—	—	—	—	.0001	.0001	.0001	.0002	.0002
12	—	—	—	—	—	—	—	—	—	—	—	—	—	—	.0001

| | | | | | | | | λ | | | | | | | |
x	3.1	3.2	3.3	3.4	3.5	3.6	3.7	3.8	3.9	4.0	4.1	4.2	4.3	4.4	4.5
0	.0450	.0408	.0369	.0334	.0302	.0273	.0247	.0224	.0202	.0183	.0166	.0150	.0136	.0123	.0111
1	.1397	.1304	.1217	.1135	.1057	.0984	.0915	.0850	.0789	.0733	.0679	.0630	.0583	.0540	.0500
2	.2165	.2087	.2008	.1929	.1850	.1771	.1692	.1615	.1539	.1465	.1393	.1323	.1254	.1188	.1125
3	.2237	.2226	.2209	.2186	.2158	.2125	.2087	.2046	.2001	.1954	.1904	.1852	.1798	.1743	.1687
4	.1733	.1781	.1823	.1858	.1888	.1912	.1931	.1944	.1951	.1954	.1951	.1944	.1933	.1917	.1898
5	.1075	.1140	.1203	.1264	.1322	.1377	.1429	.1477	.1522	.1563	.1600	.1633	.1662	.1687	.1708
6	.0555	.0608	.0662	.0716	.0771	.0826	.0881	.0936	.0989	.1042	.1093	.1143	.1191	.1237	.1281
7	.0246	.0278	.0312	.0348	.0385	.0425	.0466	.0508	.0551	.0595	.0640	.0686	.0732	.0778	.0824
8	.0095	.0111	.0129	.0148	.0169	.0191	.0215	.0241	.0269	.0298	.0328	.0360	.0393	.0428	.0463
9	.0033	.0040	.0047	.0056	.0066	.0076	.0089	.0102	.0116	.0132	.0150	.0168	.0188	.0209	.0232
10	.0010	.0013	.0016	.0019	.0023	.0028	.0033	.0039	.0045	.0053	.0061	.0071	.0081	.0092	.0104
11	.0003	.0004	.0005	.0006	.0007	.0009	.0011	.0013	.0016	.0019	.0023	.0027	.0032	.0037	.0043
12	.0001	.0001	.0001	.0002	.0002	.0003	.0003	.0004	.0005	.0006	.0008	.0009	.0011	.0013	.0016
13	—	—	—	—	.0001	.0001	.0001	.0001	.0002	.0002	.0002	.0003	.0004	.0005	.0006
14	—	—	—	—	—	—	—	—	—	.0001	.0001	.0001	.0001	.0001	.0002
15	—	—	—	—	—	—	—	—	—	—	—	—	—	—	.0001

(continued)

							λ								
x	**4.6**	**4.7**	**4.8**	**4.9**	**5.0**	**5.1**	**5.2**	**5.3**	**5.4**	**5.5**	**5.6**	**5.7**	**5.8**	**5.9**	**6.0**
0	.0101	.0091	.0082	.0074	.0067	.0061	.0055	.0050	.0045	.0041	.0037	.0033	.0030	.0027	.0025
1	.0462	.0427	.0395	.0365	.0337	.0311	.0287	.0265	.0244	.0225	.0207	.0191	.0176	.0162	.0149
2	.1063	.1005	.0948	.0894	.0842	.0793	.0746	.0701	.0659	.0618	.0580	.0544	.0509	.0477	.0446
3	.1631	.1574	.1517	.1460	.1404	.1348	.1293	.1239	.1185	.1133	.1082	.1033	.0985	.0938	.0892
4	.1875	.1849	.1820	.1789	.1755	.1719	.1681	.1641	.1600	.1558	.1515	.1472	.1428	.1383	.1339
5	.1725	.1738	.1747	.1753	.1755	.1753	.1748	.1740	.1728	.1714	.1697	.1678	.1656	.1632	.1606
6	.1323	.1362	.1398	.1432	.1462	.1490	.1515	.1537	.1555	.1571	.1584	.1594	.1601	.1605	.1606
7	.0869	.0914	.0959	.1002	.1044	.1086	.1125	.1163	.1200	.1234	.1267	.1298	.1326	.1353	.1377
8	.0500	.0537	.0575	.0614	.0653	.0692	.0731	.0771	.0810	.0849	.0887	.0925	.0962	.0998	.1033
9	.0255	.0281	.0307	.0334	.0363	.0392	.0423	.0454	.0486	.0519	.0552	.0586	.0620	.0654	.0688
10	.0118	.0132	.0147	.0164	.0181	.0200	.0220	.0241	.0262	.0285	.0309	.0334	.0359	.0386	.0413
11	.0049	.0056	.0064	.0073	.0082	.0093	.0104	.0116	.0129	.0143	.0157	.0173	.0190	.0207	.0225
12	.0019	.0022	.0026	.0030	.0034	.0039	.0045	.0051	.0058	.0065	.0073	.0082	.0092	.0102	.0113
13	.0007	.0008	.0009	.0011	.0013	.0015	.0018	.0021	.0024	.0028	.0032	.0036	.0041	.0046	.0052
14	.0002	.0003	.0003	.0004	.0005	.0006	.0007	.0008	.0009	.0011	.0013	.0015	.0017	.0019	.0022
15	.0001	.0001	.0001	.0001	.0002	.0002	.0002	.0003	.0003	.0004	.0005	.0006	.0007	.0008	.0009
16	—	—	—	—	—	.0001	.0001	.0001	.0001	.0001	.0002	.0002	.0002	.0003	.0003
17	—	—	—	—	—	—	—	—	—	—	.0001	.0001	.0001	.0001	.0001

							λ								
x	**6.1**	**6.2**	**6.3**	**6.4**	**6.5**	**6.6**	**6.7**	**6.8**	**6.9**	**7.0**	**7.1**	**7.2**	**7.3**	**7.4**	**7.5**
0	.0022	.0020	.0018	.0017	.0015	.0014	.0012	.0011	.0010	.0009	.0008	.0007	.0007	.0006	.0006
1	.0137	.0126	.0116	.0106	.0098	.0090	.0082	.0076	.0070	.0064	.0059	.0054	.0049	.0045	.0041
2	.0417	.0390	.0364	.0340	.0318	.0296	.0276	.0258	.0240	.0223	.0208	.0194	.0180	.0167	.0156
3	.0848	.0806	.0765	.0726	.0688	.0652	.0617	.0584	.0552	.0521	.0492	.0464	.0438	.0413	.0389
4	.1294	.1249	.1205	.1162	.1118	.1076	.1034	.0992	.0952	.0912	.0874	.0836	.0799	.0764	.0729
5	.1579	.1549	.1519	.1487	.1454	.1420	.1385	.1349	.1314	.1277	.1241	.1204	.1167	.1130	.1094
6	.1605	.1601	.1595	.1586	.1575	.1562	.1546	.1529	.1511	.1490	.1468	.1445	.1420	.1394	.1367
7	.1399	.1418	.1435	.1450	.1462	.1472	.1480	.1486	.1489	.1490	.1489	.1486	.1481	.1474	.1465
8	.1066	.1099	.1130	.1160	.1188	.1215	.1240	.1263	.1284	.1304	.1321	.1337	.1351	.1363	.1373
9	.0723	.0757	.0791	.0825	.0858	.0891	.0923	.0954	.0985	.1014	.1042	.1070	.1096	.1121	.1144
10	.0441	.0469	.0498	.0528	.0558	.0588	.0618	.0649	.0679	.0710	.0740	.0770	.0800	.0829	.0858
11	.0244	.0265	.0285	.0307	.0330	.0353	.0377	.0401	.0426	.0452	.0478	.0504	.0531	.0558	.0585
12	.0124	.0137	.0150	.0164	.0179	.0194	.0210	.0227	.0245	.0263	.0283	.0303	.0323	.0344	.0366
13	.0058	.0065	.0073	.0081	.0089	.0099	.0108	.0119	.0130	.0142	.0154	.0168	.0181	.0196	.0211
14	.0025	.0029	.0033	.0037	.0041	.0046	.0052	.0058	.0064	.0071	.0078	.0086	.0095	.0104	.0113
15	.0010	.0012	.0014	.0016	.0018	.0020	.0023	.0026	.0029	.0033	.0037	.0041	.0046	.0051	.0057
16	.0004	.0005	.0005	.0006	.0007	.0008	.0010	.0011	.0013	.0014	.0016	.0019	.0021	.0024	.0026
17	.0001	.0002	.0002	.0002	.0003	.0003	.0004	.0004	.0005	.0006	.0007	.0008	.0009	.0010	.0012
18	—	.0001	.0001	.0001	.0001	.0001	.0001	.0002	.0002	.0002	.0003	.0003	.0004	.0004	.0005
19	—	—	—	—	—	—	.0001	.0001	.0001	.0001	.0001	.0001	.0001	.0002	.0002
20	—	—	—	—	—	—	—	—	—	—	—	—	.0001	.0001	.0001

	λ														
x	8.0	8.5	9.0	9.5	10.0	11.0	12.0	13.0	14.0	15.0	16.0	17.0	18.0	19.0	20.0
0	.0003	.0002	.0001	.0001	—	—	—	—	—	—	—	—	—	—	—
1	.0027	.0017	.0011	.0007	.0005	.0002	.0001	—	—	—	—	—	—	—	—
2	.0107	.0074	.0050	.0034	.0023	.0010	.0004	.0002	.0001	—	—	—	—	—	—
3	.0286	.0208	.0150	.0107	.0076	.0037	.0018	.0008	.0004	.0002	.0001	—	—	—	—
4	.0573	.0443	.0337	.0254	.0189	.0102	.0053	.0027	.0013	.0006	.0003	.0001	.0001	—	—
5	.0916	.0752	.0607	.0483	.0378	.0224	.0127	.0070	.0037	.0019	.0010	.0005	.0002	.0001	.0001
6	.1221	.1066	.0911	.0764	.0631	.0411	.0255	.0152	.0087	.0048	.0026	.0014	.0007	.0004	.0002
7	.1396	.1294	.1171	.1037	.0901	.0646	.0437	.0281	.0174	.0104	.0060	.0034	.0019	.0010	.0005
8	.1396	.1375	.1318	.1232	.1126	.0888	.0655	.0457	.0304	.0194	.0120	.0072	.0042	.0024	.0013
9	.1241	.1299	.1318	.1300	.1251	.1085	.0874	.0661	.0473	.0324	.0213	.0135	.0083	.0050	.0029
10	.0993	.1104	.1186	.1235	.1251	.1194	.1048	.0859	.0663	.0486	.0341	.0230	.0150	.0095	.0058
11	.0722	.0853	.0970	.1067	.1137	.1194	.1144	.1015	.0844	.0663	.0496	.0355	.0245	.0164	.0106
12	.0481	.0604	.0728	.0844	.0948	.1094	.1144	.1099	.0984	.0829	.0661	.0504	.0368	.0259	.0176
13	.0296	.0395	.0504	.0617	.0729	.0926	.1056	.1099	.1060	.0956	.0814	.0658	.0509	.0378	.0271
14	.0169	.0240	.0324	.0419	.0521	.0728	.0905	.1021	.1060	.1024	.0930	.0800	.0655	.0514	.0387
15	.0090	.0136	.0194	.0265	.0347	.0534	.0724	.0885	.0989	.1024	.0992	.0906	.0786	.0650	.0516
16	.0045	.0072	.0109	.0157	.0217	.0367	.0543	.0719	.0866	.0960	.0992	.0963	.0884	.0772	.0646
17	.0021	.0036	.0058	.0088	.0128	.0237	.0383	.0550	.0713	.0847	.0934	.0963	.0936	.0863	.0760
18	.0009	.0017	.0029	.0046	.0071	.0145	.0255	.0397	.0554	.0706	.0830	.0909	.0936	.0911	.0844
19	.0004	.0008	.0014	.0023	.0037	.0084	.0161	.0272	.0409	.0557	.0699	.0814	.0887	.0911	.0888
20	.0002	.0003	.0006	.0011	.0019	.0046	.0097	.0177	.0286	.0418	.0559	.0692	.0798	.0866	.0888
21	.0001	.0001	.0003	.0005	.0009	.0024	.0055	.0109	.0191	.0299	.0426	.0560	.0684	.0783	.0846
22	—	.0001	.0001	.0002	.0004	.0012	.0030	.0065	.0121	.0204	.0310	.0433	.0560	.0676	.0769
23	—	—	—	.0001	.0002	.0006	.0016	.0037	.0074	.0133	.0216	.0320	.0438	.0559	.0669
24	—	—	—	—	.0001	.0003	.0008	.0020	.0043	.0083	.0144	.0226	.0328	.0442	.0557
25	—	—	—	—	—	.0001	.0004	.0010	.0024	.0050	.0092	.0154	.0237	.0336	.0446
26	—	—	—	—	—	—	.0002	.0005	.0013	.0029	.0057	.0101	.0164	.0246	.0343
27	—	—	—	—	—	—	.0001	.0002	.0007	.0016	.0034	.0063	.0109	.0173	.0254
28	—	—	—	—	—	—	—	.0001	.0003	.0009	.0019	.0038	.0070	.0117	.0181
29	—	—	—	—	—	—	—	.0001	.0002	.0004	.0011	.0023	.0044	.0077	.0125
30	—	—	—	—	—	—	—	—	.0001	.0002	.0006	.0013	.0026	.0049	.0083
31	—	—	—	—	—	—	—	—	—	.0001	.0003	.0007	.0015	.0030	.0054
32	—	—	—	—	—	—	—	—	—	.0001	.0001	.0004	.0009	.0018	.0034
33	—	—	—	—	—	—	—	—	—	—	.0001	.0002	.0005	.0010	.0020
34	—	—	—	—	—	—	—	—	—	—	—	.0001	.0002	.0006	.0012
35	—	—	—	—	—	—	—	—	—	—	—	—	.0001	.0003	.0007
36	—	—	—	—	—	—	—	—	—	—	—	—	.0001	.0002	.0004
37	—	—	—	—	—	—	—	—	—	—	—	—	—	.0001	.0002
38	—	—	—	—	—	—	—	—	—	—	—	—	—	—	.0001
39	—	—	—	—	—	—	—	—	—	—	—	—	—	—	.0001

APPENDIX

 **C-1**

STANDARD NORMAL AREAS

Example: $P(0 \leq z \leq 1.96) = .4750$

This table shows the normal area between 0 and z.

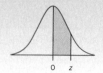

z	.00	.01	.02	.03	.04	.05	.06	.07	.08	.09
0.0	.0000	.0040	.0080	.0120	.0160	.0199	.0239	.0279	.0319	.0359
0.1	.0398	.0438	.0478	.0517	.0557	.0596	.0636	.0675	.0714	.0753
0.2	.0793	.0832	.0871	.0910	.0948	.0987	.1026	.1064	.1103	.1141
0.3	.1179	.1217	.1255	.1293	.1331	.1368	.1406	.1443	.1480	.1517
0.4	.1554	.1591	.1628	.1664	.1700	.1736	.1772	.1808	.1844	.1879
0.5	.1915	.1950	.1985	.2019	.2054	.2088	.2123	.2157	.2190	.2224
0.6	.2257	.2291	.2324	.2357	.2389	.2422	.2454	.2486	.2517	.2549
0.7	.2580	.2611	.2642	.2673	.2704	.2734	.2764	.2794	.2823	.2852
0.8	.2881	.2910	.2939	.2967	.2995	.3023	.3051	.3078	.3106	.3133
0.9	.3159	.3186	.3212	.3238	.3264	.3289	.3315	.3340	.3365	.3389
1.0	.3413	.3438	.3461	.3485	.3508	.3531	.3554	.3577	.3599	.3621
1.1	.3643	.3665	.3686	.3708	.3729	.3749	.3770	.3790	.3810	.3830
1.2	.3849	.3869	.3888	.3907	.3925	.3944	.3962	.3980	.3997	.4015
1.3	.4032	.4049	.4066	.4082	.4099	.4115	.4131	.4147	.4162	.4177
1.4	.4192	.4207	.4222	.4236	.4251	.4265	.4279	.4292	.4306	.4319
1.5	.4332	.4345	.4357	.4370	.4382	.4394	.4406	.4418	.4429	.4441
1.6	.4452	.4463	.4474	.4484	.4495	.4505	.4515	.4525	.4535	.4545
1.7	.4554	.4564	.4573	.4582	.4591	.4599	.4608	.4616	.4625	.4633
1.8	.4641	.4649	.4656	.4664	.4671	.4678	.4686	.4693	.4699	.4706
1.9	.4713	.4719	.4726	.4732	.4738	.4744	.4750	.4756	.4761	.4767
2.0	.4772	.4778	.4783	.4788	.4793	.4798	.4803	.4808	.4812	.4817
2.1	.4821	.4826	.4830	.4834	.4838	.4842	.4846	.4850	.4854	.4857
2.2	.4861	.4864	.4868	.4871	.4875	.4878	.4881	.4884	.4887	.4890
2.3	.4893	.4896	.4898	.4901	.4904	.4906	.4909	.4911	.4913	.4916
2.4	.4918	.4920	.4922	.4925	.4927	.4929	.4931	.4932	.4934	.4936
2.5	.4938	.4940	.4941	.4943	.4945	.4946	.4948	.4949	.4951	.4952
2.6	.4953	.4955	.4956	.4957	.4959	.4960	.4961	.4962	.4963	.4964
2.7	.4965	.4966	.4967	.4968	.4969	.4970	.4971	.4972	.4973	.4974
2.8	.4974	.4975	.4976	.4977	.4977	.4978	.4979	.4979	.4980	.4981
2.9	.4981	.4982	.4982	.4983	.4984	.4984	.4985	.4985	.4986	.4986
3.0	.49865	.49869	.49874	.49878	.49882	.49886	.49889	.49893	.49896	.49900
3.1	.49903	.49906	.49910	.49913	.49916	.49918	.49921	.49924	.49926	.49929
3.2	.49931	.49934	.49936	.49938	.49940	.49942	.49944	.49946	.49948	.49950
3.3	.49952	.49953	.49955	.49957	.49958	.49960	.49961	.49962	.49964	.49965
3.4	.49966	.49968	.49969	.49970	.49971	.49972	.49973	.49974	.49975	.49976
3.5	.49977	.49978	.49978	.49979	.49980	.49981	.49981	.49982	.49983	.49983
3.6	.49984	.49985	.49985	.49986	.49986	.49987	.49987	.49988	.49988	.49989
3.7	.49989	.49990	.49990	.49990	.49991	.49991	.49992	.49992	.49992	.49992

CUMULATIVE STANDARD
NORMAL DISTRIBUTION

Example: $P(z \le -1.96) = .0250$

This table shows the normal area less than z.

C-2

z	.00	.01	.02	.03	.04	.05	.06	.07	.08	.09
−3.7	.00011	.00010	.00010	.00010	.00009	.00009	.00008	.00008	.00008	.00008
−3.6	.00016	.00015	.00015	.00014	.00014	.00013	.00013	.00012	.00012	.00011
−3.5	.00023	.00022	.00022	.00021	.00020	.00019	.00019	.00018	.00017	.00017
−3.4	.00034	.00032	.00031	.00030	.00029	.00028	.00027	.00026	.00025	.00024
−3.3	.00048	.00047	.00045	.00043	.00042	.00040	.00039	.00038	.00036	.00035
−3.2	.00069	.00066	.00064	.00062	.00060	.00058	.00056	.00054	.00052	.00050
−3.1	.00097	.00094	.00090	.00087	.00084	.00082	.00079	.00076	.00074	.00071
−3.0	.00135	.00131	.00126	.00122	.00118	.00114	.00111	.00107	.00104	.00100
−2.9	.0019	.0018	.0018	.0017	.0016	.0016	.0015	.0015	.0014	.0014
−2.8	.0026	.0025	.0024	.0023	.0023	.0022	.0021	.0021	.0020	.0019
−2.7	.0035	.0034	.0033	.0032	.0031	.0030	.0029	.0028	.0027	.0026
−2.6	.0047	.0045	.0044	.0043	.0041	.0040	.0039	.0038	.0037	.0036
−2.5	.0062	.0060	.0059	.0057	.0055	.0054	.0052	.0051	.0049	.0048
−2.4	.0082	.0080	.0078	.0075	.0073	.0071	.0069	.0068	.0066	.0064
−2.3	.0107	.0104	.0102	.0099	.0096	.0094	.0091	.0089	.0087	.0084
−2.2	.0139	.0136	.0132	.0129	.0125	.0122	.0119	.0116	.0113	.0110
−2.1	.0179	.0174	.0170	.0166	.0162	.0158	.0154	.0150	.0146	.0143
−2.0	.0228	.0222	.0217	.0212	.0207	.0202	.0197	.0192	.0188	.0183
−1.9	.0287	.0281	.0274	.0268	.0262	.0256	.0250	.0244	.0239	.0233
−1.8	.0359	.0351	.0344	.0336	.0329	.0322	.0314	.0307	.0301	.0294
−1.7	.0446	.0436	.0427	.0418	.0409	.0401	.0392	.0384	.0375	.0367
−1.6	.0548	.0537	.0526	.0516	.0505	.0495	.0485	.0475	.0465	.0455
−1.5	.0668	.0655	.0643	.0630	.0618	.0606	.0594	.0582	.0571	.0559
−1.4	.0808	.0793	.0778	.0764	.0749	.0735	.0721	.0708	.0694	.0681
−1.3	.0968	.0951	.0934	.0918	.0901	.0885	.0869	.0853	.0838	.0823
−1.2	.1151	.1131	.1112	.1093	.1075	.1056	.1038	.1020	.1003	.0985
−1.1	.1357	.1335	.1314	.1292	.1271	.1251	.1230	.1210	.1190	.1170
−1.0	.1587	.1562	.1539	.1515	.1492	.1469	.1446	.1423	.1401	.1379
−0.9	.1841	.1814	.1788	.1762	.1736	.1711	.1685	.1660	.1635	.1611
−0.8	.2119	.2090	.2061	.2033	.2005	.1977	.1949	.1922	.1894	.1867
−0.7	.2420	.2389	.2358	.2327	.2296	.2266	.2236	.2206	.2177	.2148
−0.6	.2743	.2709	.2676	.2643	.2611	.2578	.2546	.2514	.2483	.2451
−0.5	.3085	.3050	.3015	.2981	.2946	.2912	.2877	.2843	.2810	.2776
−0.4	.3446	.3409	.3372	.3336	.3300	.3264	.3228	.3192	.3156	.3121
−0.3	.3821	.3783	.3745	.3707	.3669	.3632	.3594	.3557	.3520	.3483
−0.2	.4207	.4168	.4129	.4090	.4052	.4013	.3974	.3936	.3897	.3859
−0.1	.4602	.4562	.4522	.4483	.4443	.4404	.4364	.4325	.4286	.4247
−0.0	.5000	.4960	.4920	.4880	.4841	.4801	.4761	.4721	.4681	.4641

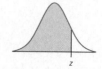

This table shows the normal area less than *z*.

z	.00	.01	.02	.03	.04	.05	.06	.07	.08	.09
0.0	.5000	.5040	.5080	.5120	.5160	.5199	.5239	.5279	.5319	.5359
0.1	.5398	.5438	.5478	.5517	.5557	.5596	.5636	.5675	.5714	.5753
0.2	.5793	.5832	.5871	.5910	.5948	.5987	.6026	.6064	.6103	.6141
0.3	.6179	.6217	.6255	.6293	.6331	.6368	.6406	.6443	.6480	.6517
0.4	.6554	.6591	.6628	.6664	.6700	.6736	.6772	.6808	.6844	.6879
0.5	.6915	.6950	.6985	.7019	.7054	.7088	.7123	.7157	.7190	.7224
0.6	.7257	.7291	.7324	.7357	.7389	.7422	.7454	.7486	.7517	.7549
0.7	.7580	.7611	.7642	.7673	.7704	.7734	.7764	.7794	.7823	.7852
0.8	.7881	.7910	.7939	.7967	.7995	.8023	.8051	.8078	.8106	.8133
0.9	.8159	.8186	.8212	.8238	.8264	.8289	.8315	.8340	.8365	.8389
1.0	.8413	.8438	.8461	.8485	.8508	.8531	.8554	.8577	.8599	.8621
1.1	.8643	.8665	.8686	.8708	.8729	.8749	.8770	.8790	.8810	.8830
1.2	.8849	.8869	.8888	.8907	.8925	.8944	.8962	.8980	.8997	.9015
1.3	.9032	.9049	.9066	.9082	.9099	.9115	.9131	.9147	.9162	.9177
1.4	.9192	.9207	.9222	.9236	.9251	.9265	.9279	.9292	.9306	.9319
1.5	.9332	.9345	.9357	.9370	.9382	.9394	.9406	.9418	.9429	.9441
1.6	.9452	.9463	.9474	.9484	.9495	.9505	.9515	.9525	.9535	.9545
1.7	.9554	.9564	.9573	.9582	.9591	.9599	.9608	.9616	.9625	.9633
1.8	.9641	.9649	.9656	.9664	.9671	.9678	.9686	.9693	.9699	.9706
1.9	.9713	.9719	.9726	.9732	.9738	.9744	.9750	.9756	.9761	.9767
2.0	.9772	.9778	.9783	.9788	.9793	.9798	.9803	.9808	.9812	.9817
2.1	.9821	.9826	.9830	.9834	.9838	.9842	.9846	.9850	.9854	.9857
2.2	.9861	.9864	.9868	.9871	.9875	.9878	.9881	.9884	.9887	.9890
2.3	.9893	.9896	.9898	.9901	.9904	.9906	.9909	.9911	.9913	.9916
2.4	.9918	.9920	.9922	.9925	.9927	.9929	.9931	.9932	.9934	.9936
2.5	.9938	.9940	.9941	.9943	.9945	.9946	.9948	.9949	.9951	.9952
2.6	.9953	.9955	.9956	.9957	.9959	.9960	.9961	.9962	.9963	.9964
2.7	.9965	.9966	.9967	.9968	.9969	.9970	.9971	.9972	.9973	.9974
2.8	.9974	.9975	.9976	.9977	.9977	.9978	.9979	.9979	.9980	.9981
2.9	.9981	.9982	.9982	.9983	.9984	.9984	.9985	.9985	.9986	.9986
3.0	.99865	.99869	.99874	.99878	.99882	.99886	.99889	.99893	.99896	.99900
3.1	.99903	.99906	.99910	.99913	.99916	.99918	.99921	.99924	.99926	.99929
3.2	.99931	.99934	.99936	.99938	.99940	.99942	.99944	.99946	.99948	.99950
3.3	.99952	.99953	.99955	.99957	.99958	.99960	.99961	.99962	.99964	.99965
3.4	.99966	.99968	.99969	.99970	.99971	.99972	.99973	.99974	.99975	.99976
3.5	.99977	.99978	.99978	.99979	.99980	.99981	.99981	.99982	.99983	.99983
3.6	.99984	.99985	.99985	.99986	.99986	.99987	.99987	.99988	.99988	.99989
3.7	.99989	.99990	.99990	.99990	.99991	.99991	.99992	.99992	.99992	.99992

STUDENT'S *t* CRITICAL VALUES

This table shows the *t*-value that defines the area for the stated degrees of freedom (*d.f.*).

	Confidence Level						**Confidence Level**				
	.80	.90	.95	.98	.99		.80	.90	.95	.98	.99
	Significance Level for Two-Tailed Test						**Significance Level for Two-Tailed Test**				
	.20	.10	.05	.02	.01		.20	.10	.05	.02	.01
	Significance Level for One-Tailed Test						**Significance Level for One-Tailed Test**				
d.f.	.10	.05	.025	.01	.005	*d.f.*	.10	.05	.025	.01	.005
1	3.078	6.314	12.706	31.821	63.657	37	1.305	1.687	2.026	2.431	2.715
2	1.886	2.920	4.303	6.965	9.925	38	1.304	1.686	2.024	2.429	2.712
3	1.638	2.353	3.182	4.541	5.841	39	1.304	1.685	2.023	2.426	2.708
4	1.533	2.132	2.776	3.747	4.604	40	1.303	1.684	2.021	2.423	2.704
5	1.476	2.015	2.571	3.365	4.032	41	1.303	1.683	2.020	2.421	2.701
6	1.440	1.943	2.447	3.143	3.707	42	1.302	1.682	2.018	2.418	2.698
7	1.415	1.895	2.365	2.998	3.499	43	1.302	1.681	2.017	2.416	2.695
8	1.397	1.860	2.306	2.896	3.355	44	1.301	1.680	2.015	2.414	2.692
9	1.383	1.833	2.262	2.821	3.250	45	1.301	1.679	2.014	2.412	2.690
10	1.372	1.812	2.228	2.764	3.169	46	1.300	1.679	2.013	2.410	2.687
11	1.363	1.796	2.201	2.718	3.106	47	1.300	1.678	2.012	2.408	2.685
12	1.356	1.782	2.179	2.681	3.055	48	1.299	1.677	2.011	2.407	2.682
13	1.350	1.771	2.160	2.650	3.012	49	1.299	1.677	2.010	2.405	2.680
14	1.345	1.761	2.145	2.624	2.977	50	1.299	1.676	2.009	2.403	2.678
15	1.341	1.753	2.131	2.602	2.947	55	1.297	1.673	2.004	2.396	2.668
16	1.337	1.746	2.120	2.583	2.921	60	1.296	1.671	2.000	2.390	2.660
17	1.333	1.740	2.110	2.567	2.898	65	1.295	1.669	1.997	2.385	2.654
18	1.330	1.734	2.101	2.552	2.878	70	1.294	1.667	1.994	2.381	2.648
19	1.328	1.729	2.093	2.539	2.861	75	1.293	1.665	1.992	2.377	2.643
20	1.325	1.725	2.086	2.528	2.845	80	1.292	1.664	1.990	2.374	2.639
21	1.323	1.721	2.080	2.518	2.831	85	1.292	1.663	1.988	2.371	2.635
22	1.321	1.717	2.074	2.508	2.819	90	1.291	1.662	1.987	2.368	2.632
23	1.319	1.714	2.069	2.500	2.807	95	1.291	1.661	1.985	2.366	2.629
24	1.318	1.711	2.064	2.492	2.797	100	1.290	1.660	1.984	2.364	2.626
25	1.316	1.708	2.060	2.485	2.787	110	1.289	1.659	1.982	2.361	2.621
26	1.315	1.706	2.056	2.479	2.779	120	1.289	1.658	1.980	2.358	2.617
27	1.314	1.703	2.052	2.473	2.771	130	1.288	1.657	1.978	2.355	2.614
28	1.313	1.701	2.048	2.467	2.763	140	1.288	1.656	1.977	2.353	2.611
29	1.311	1.699	2.045	2.462	2.756	150	1.287	1.655	1.976	2.351	2.609
30	1.310	1.697	2.042	2.457	2.750	∞	1.282	1.645	1.960	2.326	2.576
31	1.309	1.696	2.040	2.453	2.744						
32	1.309	1.694	2.037	2.449	2.738						
33	1.308	1.692	2.035	2.445	2.733						
34	1.307	1.691	2.032	2.441	2.728						
35	1.306	1.690	2.030	2.438	2.724						
36	1.306	1.688	2.028	2.434	2.719						

Note: As *n* increases, critical values of Student's *t* approach the *z*-values in the last line of this table. A common rule of thumb is to use *z* when *n* > 30, but that is *not* conservative.

APPENDIX

E

CHI-SQUARE CRITICAL VALUES

This table shows the critical value of chi-square for each desired right-tail area and degrees of freedom (*d.f.*)

Example for *d.f.* = 4

.05

0   9.488

d.f.	.995	.990	.975	.95	.90	.10	.05	.025	.01	.005
					Area in Upper Tail					
1	0.000	0.000	0.001	0.004	0.016	2.706	3.841	5.024	6.635	7.879
2	0.010	0.020	0.051	0.103	0.211	4.605	5.991	7.378	9.210	10.60
3	0.072	0.115	0.216	0.352	0.584	6.251	7.815	9.348	11.34	12.84
4	0.207	0.297	0.484	0.711	1.064	7.779	9.488	11.14	13.28	14.86
5	0.412	0.554	0.831	1.145	1.610	9.236	11.07	12.83	15.09	16.75
6	0.676	0.872	1.237	1.635	2.204	10.64	12.59	14.45	16.81	18.55
7	0.989	1.239	1.690	2.167	2.833	12.02	14.07	16.01	18.48	20.28
8	1.344	1.646	2.180	2.733	3.490	13.36	15.51	17.53	20.09	21.95
9	1.735	2.088	2.700	3.325	4.168	14.68	16.92	19.02	21.67	23.59
10	2.156	2.558	3.247	3.940	4.865	15.99	18.31	20.48	23.21	25.19
11	2.603	3.053	3.816	4.575	5.578	17.28	19.68	21.92	24.72	26.76
12	3.074	3.571	4.404	5.226	6.304	18.55	21.03	23.34	26.22	28.30
13	3.565	4.107	5.009	5.892	7.042	19.81	22.36	24.74	27.69	29.82
14	4.075	4.660	5.629	6.571	7.790	21.06	23.68	26.12	29.14	31.32
15	4.601	5.229	6.262	7.261	8.547	22.31	25.00	27.49	30.58	32.80
16	5.142	5.812	6.908	7.962	9.312	23.54	26.30	28.85	32.00	34.27
17	5.697	6.408	7.564	8.672	10.09	24.77	27.59	30.19	33.41	35.72
18	6.265	7.015	8.231	9.390	10.86	25.99	28.87	31.53	34.81	37.16
19	6.844	7.633	8.907	10.12	11.65	27.20	30.14	32.85	36.19	38.58
20	7.434	8.260	9.591	10.85	12.44	28.41	31.41	34.17	37.57	40.00
21	8.034	8.897	10.28	11.59	13.24	29.62	32.67	35.48	38.93	41.40
22	8.643	9.542	10.98	12.34	14.04	30.81	33.92	36.78	40.29	42.80
23	9.260	10.20	11.69	13.09	14.85	32.01	35.17	38.08	41.64	44.18
24	9.886	10.86	12.40	13.85	15.66	33.20	36.42	39.36	42.98	45.56
25	10.52	11.52	13.12	14.61	16.47	34.38	37.65	40.65	44.31	46.93
26	11.16	12.20	13.84	15.38	17.29	35.56	38.89	41.92	45.64	48.29
27	11.81	12.88	14.57	16.15	18.11	36.74	40.11	43.19	46.96	49.64
28	12.46	13.56	15.31	16.93	18.94	37.92	41.34	44.46	48.28	50.99
29	13.12	14.26	16.05	17.71	19.77	39.09	42.56	45.72	49.59	52.34
30	13.79	14.95	16.79	18.49	20.60	40.26	43.77	46.98	50.89	53.67
31	14.46	15.66	17.54	19.28	21.43	41.42	44.99	48.23	52.19	55.00
32	15.13	16.36	18.29	20.07	22.27	42.58	46.19	49.48	53.49	56.33
33	15.82	17.07	19.05	20.87	23.11	43.75	47.40	50.73	54.78	57.65
34	16.50	17.79	19.81	21.66	23.95	44.90	48.60	51.97	56.06	58.96
35	17.19	18.51	20.57	22.47	24.80	46.06	49.80	53.20	57.34	60.27
36	17.89	19.23	21.34	23.27	25.64	47.21	51.00	54.44	58.62	61.58
37	18.59	19.96	22.11	24.07	26.49	48.36	52.19	55.67	59.89	62.88
38	19.29	20.69	22.88	24.88	27.34	49.51	53.38	56.90	61.16	64.18
39	20.00	21.43	23.65	25.70	28.20	50.66	54.57	58.12	62.43	65.48
40	20.71	22.16	24.43	26.51	29.05	51.81	55.76	59.34	63.69	66.77
50	27.99	29.71	32.36	34.76	37.69	63.17	67.50	71.42	76.15	79.49
60	35.53	37.48	40.48	43.19	46.46	74.40	79.08	83.30	88.38	91.95
70	43.28	45.44	48.76	51.74	55.33	85.53	90.53	95.02	100.4	104.2
80	51.17	53.54	57.15	60.39	64.28	96.58	101.9	106.6	112.3	116.3
90	59.20	61.75	65.65	69.13	73.29	107.6	113.1	118.1	124.1	128.3
100	67.33	70.06	74.22	77.93	82.36	118.5	124.3	129.6	135.8	140.2

Note: For *d.f.* > 100, use the Excel function = CHISQ.INV.RT(*a*, degrees of freedom).

CRITICAL VALUES OF $F_{.10}$

This table shows the 10 percent right-tail critical values of F for the stated degrees of freedom (d.f.).

Denominator Degrees of Freedom (df_2)	Numerator Degrees of Freedom (df_1)										
	1	**2**	**3**	**4**	**5**	**6**	**7**	**8**	**9**	**10**	**12**
1	39.86	49.50	53.59	55.83	57.24	58.20	58.91	59.44	59.86	60.19	60.71
2	8.53	9.00	9.16	9.24	9.29	9.33	9.35	9.37	9.38	9.39	9.41
3	5.54	5.46	5.39	5.34	5.31	5.28	5.27	5.25	5.24	5.23	5.22
4	4.54	4.32	4.19	4.11	4.05	4.01	3.98	3.95	3.94	3.92	3.90
5	4.06	3.78	3.62	3.52	3.45	3.40	3.37	3.34	3.32	3.30	3.27
6	3.78	3.46	3.29	3.18	3.11	3.05	3.01	2.98	2.96	2.94	2.90
7	3.59	3.26	3.07	2.96	2.88	2.83	2.78	2.75	2.72	2.70	2.67
8	3.46	3.11	2.92	2.81	2.73	2.67	2.62	2.59	2.56	2.54	2.50
9	3.36	3.01	2.81	2.69	2.61	2.55	2.51	2.47	2.44	2.42	2.38
10	3.29	2.92	2.73	2.61	2.52	2.46	2.41	2.38	2.35	2.32	2.28
11	3.23	2.86	2.66	2.54	2.45	2.39	2.34	2.30	2.27	2.25	2.21
12	3.18	2.81	2.61	2.48	2.39	2.33	2.28	2.24	2.21	2.19	2.15
13	3.14	2.76	2.56	2.43	2.35	2.28	2.23	2.20	2.16	2.14	2.10
14	3.10	2.73	2.52	2.39	2.31	2.24	2.19	2.15	2.12	2.10	2.05
15	3.07	2.70	2.49	2.36	2.27	2.21	2.16	2.12	2.09	2.06	2.02
16	3.05	2.67	2.46	2.33	2.24	2.18	2.13	2.09	2.06	2.03	1.99
17	3.03	2.64	2.44	2.31	2.22	2.15	2.10	2.06	2.03	2.00	1.96
18	3.01	2.62	2.42	2.29	2.20	2.13	2.08	2.04	2.00	1.98	1.93
19	2.99	2.61	2.40	2.27	2.18	2.11	2.06	2.02	1.98	1.96	1.91
20	2.97	2.59	2.38	2.25	2.16	2.09	2.04	2.00	1.96	1.94	1.89
21	2.96	2.57	2.36	2.23	2.14	2.08	2.02	1.98	1.95	1.92	1.87
22	2.95	2.56	2.35	2.22	2.13	2.06	2.01	1.97	1.93	1.90	1.86
23	2.94	2.55	2.34	2.21	2.11	2.05	1.99	1.95	1.92	1.89	1.84
24	2.93	2.54	2.33	2.19	2.10	2.04	1.98	1.94	1.91	1.88	1.83
25	2.92	2.53	2.32	2.18	2.09	2.02	1.97	1.93	1.89	1.87	1.82
26	2.91	2.52	2.31	2.17	2.08	2.01	1.96	1.92	1.88	1.86	1.81
27	2.90	2.51	2.30	2.17	2.07	2.00	1.95	1.91	1.87	1.85	1.80
28	2.89	2.50	2.29	2.16	2.06	2.00	1.94	1.90	1.87	1.84	1.79
29	2.89	2.50	2.28	2.15	2.06	1.99	1.93	1.89	1.86	1.83	1.78
30	2.88	2.49	2.28	2.14	2.05	1.98	1.93	1.88	1.85	1.82	1.77
40	2.84	2.44	2.23	2.09	2.00	1.93	1.87	1.83	1.79	1.76	1.71
50	2.81	2.41	2.20	2.06	1.97	1.90	1.84	1.80	1.76	1.73	1.68
60	2.79	2.39	2.18	2.04	1.95	1.87	1.82	1.77	1.74	1.71	1.66
120	2.75	2.35	2.13	1.99	1.90	1.82	1.77	1.72	1.68	1.65	1.60
200	2.73	2.33	2.11	1.97	1.88	1.80	1.75	1.70	1.66	1.63	1.58
∞	2.71	2.30	2.08	1.94	1.85	1.77	1.72	1.67	1.63	1.60	1.55

Denominator Degrees of Freedom (df₂)	Numerator Degrees of Freedom (df₁)										
	15	**20**	**25**	**30**	**35**	**40**	**50**	**60**	**120**	**200**	**∞**
1	61.22	61.74	62.05	62.26	62.42	62.53	62.69	62.79	63.06	63.17	63.32
2	9.42	9.44	9.45	9.46	9.46	9.47	9.47	9.47	9.48	9.49	9.49
3	5.20	5.18	5.17	5.17	5.16	5.16	5.15	5.15	5.14	5.14	5.13
4	3.87	3.84	3.83	3.82	3.81	3.80	3.80	3.79	3.78	3.77	3.76
5	3.24	3.21	3.19	3.17	3.16	3.16	3.15	3.14	3.12	3.12	3.11
6	2.87	2.84	2.81	2.80	2.79	2.78	2.77	2.76	2.74	2.73	2.72
7	2.63	2.59	2.57	2.56	2.54	2.54	2.52	2.51	2.49	2.48	2.47
8	2.46	2.42	2.40	2.38	2.37	2.36	2.35	2.34	2.32	2.31	2.29
9	2.34	2.30	2.27	2.25	2.24	2.23	2.22	2.21	2.18	2.17	2.16
10	2.24	2.20	2.17	2.16	2.14	2.13	2.12	2.11	2.08	2.07	2.06
11	2.17	2.12	2.10	2.08	2.06	2.05	2.04	2.03	2.00	1.99	1.97
12	2.10	2.06	2.03	2.01	2.00	1.99	1.97	1.96	1.93	1.92	1.90
13	2.05	2.01	1.98	1.96	1.94	1.93	1.92	1.90	1.88	1.86	1.85
14	2.01	1.96	1.93	1.91	1.90	1.89	1.87	1.86	1.83	1.82	1.80
15	1.97	1.92	1.89	1.87	1.86	1.85	1.83	1.82	1.79	1.77	1.76
16	1.94	1.89	1.86	1.84	1.82	1.81	1.79	1.78	1.75	1.74	1.72
17	1.91	1.86	1.83	1.81	1.79	1.78	1.76	1.75	1.72	1.71	1.69
18	1.89	1.84	1.80	1.78	1.77	1.75	1.74	1.72	1.69	1.68	1.66
19	1.86	1.81	1.78	1.76	1.74	1.73	1.71	1.70	1.67	1.65	1.63
20	1.84	1.79	1.76	1.74	1.72	1.71	1.69	1.68	1.64	1.63	1.61
21	1.83	1.78	1.74	1.72	1.70	1.69	1.67	1.66	1.62	1.61	1.59
22	1.81	1.76	1.73	1.70	1.68	1.67	1.65	1.64	1.60	1.59	1.57
23	1.80	1.74	1.71	1.69	1.67	1.66	1.64	1.62	1.59	1.57	1.55
24	1.78	1.73	1.70	1.67	1.65	1.64	1.62	1.61	1.57	1.56	1.53
25	1.77	1.72	1.68	1.66	1.64	1.63	1.61	1.59	1.56	1.54	1.52
26	1.76	1.71	1.67	1.65	1.63	1.61	1.59	1.58	1.54	1.53	1.50
27	1.75	1.70	1.66	1.64	1.62	1.60	1.58	1.57	1.53	1.52	1.49
28	1.74	1.69	1.65	1.63	1.61	1.59	1.57	1.56	1.52	1.50	1.48
29	1.73	1.68	1.64	1.62	1.60	1.58	1.56	1.55	1.51	1.49	1.47
30	1.72	1.67	1.63	1.61	1.59	1.57	1.55	1.54	1.50	1.48	1.46
40	1.66	1.61	1.57	1.54	1.52	1.51	1.48	1.47	1.42	1.41	1.38
50	1.63	1.57	1.53	1.50	1.48	1.46	1.44	1.42	1.38	1.36	1.33
60	1.60	1.54	1.50	1.48	1.45	1.44	1.41	1.40	1.35	1.33	1.29
120	1.55	1.48	1.44	1.41	1.39	1.37	1.34	1.32	1.26	1.24	1.19
200	1.52	1.46	1.41	1.38	1.36	1.34	1.31	1.29	1.23	1.20	1.15
∞	1.49	1.42	1.38	1.34	1.32	1.30	1.26	1.24	1.17	1.13	1.00

CRITICAL VALUES OF $F_{.05}$

This table shows the 5 percent right-tail critical values of F for the stated degrees of freedom (*d.f.*).

Denominator Degrees of Freedom (df_2)	Numerator Degrees of Freedom (df_1)										
	1	**2**	**3**	**4**	**5**	**6**	**7**	**8**	**9**	**10**	**12**
1	161.4	199.5	215.7	224.6	230.2	234.0	236.8	238.9	240.5	241.9	243.9
2	18.51	19.00	19.16	19.25	19.30	19.33	19.35	19.37	19.38	19.40	19.41
3	10.13	9.55	9.28	9.12	9.01	8.94	8.89	8.85	8.81	8.79	8.74
4	7.71	6.94	6.59	6.39	6.26	6.16	6.09	6.04	6.00	5.96	5.91
5	6.61	5.79	5.41	5.19	5.05	4.95	4.88	4.82	4.77	4.74	4.68
6	5.99	5.14	4.76	4.53	4.39	4.28	4.21	4.15	4.10	4.06	4.00
7	5.59	4.74	4.35	4.12	3.97	3.87	3.79	3.73	3.68	3.64	3.57
8	5.32	4.46	4.07	3.84	3.69	3.58	3.50	3.44	3.39	3.35	3.28
9	5.12	4.26	3.86	3.63	3.48	3.37	3.29	3.23	3.18	3.14	3.07
10	4.96	4.10	3.71	3.48	3.33	3.22	3.14	3.07	3.02	2.98	2.91
11	4.84	3.98	3.59	3.36	3.20	3.09	3.01	2.95	2.90	2.85	2.79
12	4.75	3.89	3.49	3.26	3.11	3.00	2.91	2.85	2.80	2.75	2.69
13	4.67	3.81	3.41	3.18	3.03	2.92	2.83	2.77	2.71	2.67	2.60
14	4.60	3.74	3.34	3.11	2.96	2.85	2.76	2.70	2.65	2.60	2.53
15	4.54	3.68	3.29	3.06	2.90	2.79	2.71	2.64	2.59	2.54	2.48
16	4.49	3.63	3.24	3.01	2.85	2.74	2.66	2.59	2.54	2.49	2.42
17	4.45	3.59	3.20	2.96	2.81	2.70	2.61	2.55	2.49	2.45	2.38
18	4.41	3.55	3.16	2.93	2.77	2.66	2.58	2.51	2.46	2.41	2.34
19	4.38	3.52	3.13	2.90	2.74	2.63	2.54	2.48	2.42	2.38	2.31
20	4.35	3.49	3.10	2.87	2.71	2.60	2.51	2.45	2.39	2.35	2.28
21	4.32	3.47	3.07	2.84	2.68	2.57	2.49	2.42	2.37	2.32	2.25
22	4.30	3.44	3.05	2.82	2.66	2.55	2.46	2.40	2.34	2.30	2.23
23	4.28	3.42	3.03	2.80	2.64	2.53	2.44	2.37	2.32	2.27	2.20
24	4.26	3.40	3.01	2.78	2.62	2.51	2.42	2.36	2.30	2.25	2.18
25	4.24	3.39	2.99	2.76	2.60	2.49	2.40	2.34	2.28	2.24	2.16
26	4.23	3.37	2.98	2.74	2.59	2.47	2.39	2.32	2.27	2.22	2.15
27	4.21	3.35	2.96	2.73	2.57	2.46	2.37	2.31	2.25	2.20	2.13
28	4.20	3.34	2.95	2.71	2.56	2.45	2.36	2.29	2.24	2.19	2.12
29	4.18	3.33	2.93	2.70	2.55	2.43	2.35	2.28	2.22	2.18	2.10
30	4.17	3.32	2.92	2.69	2.53	2.42	2.33	2.27	2.21	2.16	2.09
40	4.08	3.23	2.84	2.61	2.45	2.34	2.25	2.18	2.12	2.08	2.00
50	4.03	3.18	2.79	2.56	2.40	2.29	2.20	2.13	2.07	2.03	1.95
60	4.00	3.15	2.76	2.53	2.37	2.25	2.17	2.10	2.04	1.99	1.92
120	3.92	3.07	2.68	2.45	2.29	2.18	2.09	2.02	1.96	1.91	1.83
200	3.89	3.04	2.65	2.42	2.26	2.14	2.06	1.98	1.93	1.88	1.80
∞	3.84	3.00	2.60	2.37	2.21	2.10	2.01	1.94	1.88	1.83	1.75

Denominator Degrees of Freedom (df₂)	Numerator Degrees of Freedom (df₁)										
	15	**20**	**25**	**30**	**35**	**40**	**50**	**60**	**120**	**200**	**∞**
1	245.9	248.0	249.3	250.1	250.7	251.1	251.8	252.2	253.3	253.7	254.3
2	19.43	19.45	19.46	19.46	19.47	19.47	19.48	19.48	19.49	19.49	19.50
3	8.70	8.66	8.63	8.62	8.60	8.59	8.58	8.57	8.55	8.54	8.53
4	5.86	5.80	5.77	5.75	5.73	5.72	5.70	5.69	5.66	5.65	5.63
5	4.62	4.56	4.52	4.50	4.48	4.46	4.44	4.43	4.40	4.39	4.37
6	3.94	3.87	3.83	3.81	3.79	3.77	3.75	3.74	3.70	3.69	3.67
7	3.51	3.44	3.40	3.38	3.36	3.34	3.32	3.30	3.27	3.25	3.23
8	3.22	3.15	3.11	3.08	3.06	3.04	3.02	3.01	2.97	2.95	2.93
9	3.01	2.94	2.89	2.86	2.84	2.83	2.80	2.79	2.75	2.73	2.71
10	2.85	2.77	2.73	2.70	2.68	2.66	2.64	2.62	2.58	2.56	2.54
11	2.72	2.65	2.60	2.57	2.55	2.53	2.51	2.49	2.45	2.43	2.41
12	2.62	2.54	2.50	2.47	2.44	2.43	2.40	2.38	2.34	2.32	2.30
13	2.53	2.46	2.41	2.38	2.36	2.34	2.31	2.30	2.25	2.23	2.21
14	2.46	2.39	2.34	2.31	2.28	2.27	2.24	2.22	2.18	2.16	2.13
15	2.40	2.33	2.28	2.25	2.22	2.20	2.18	2.16	2.11	2.10	2.07
16	2.35	2.28	2.23	2.19	2.17	2.15	2.12	2.11	2.06	2.04	2.01
17	2.31	2.23	2.18	2.15	2.12	2.10	2.08	2.06	2.01	1.99	1.96
18	2.27	2.19	2.14	2.11	2.08	2.06	2.04	2.02	1.97	1.95	1.92
19	2.23	2.16	2.11	2.07	2.05	2.03	2.00	1.98	1.93	1.91	1.88
20	2.20	2.12	2.07	2.04	2.01	1.99	1.97	1.95	1.90	1.88	1.84
21	2.18	2.10	2.05	2.01	1.98	1.96	1.94	1.92	1.87	1.84	1.81
22	2.15	2.07	2.02	1.98	1.96	1.94	1.91	1.89	1.84	1.82	1.78
23	2.13	2.05	2.00	1.96	1.93	1.91	1.88	1.86	1.81	1.79	1.76
24	2.11	2.03	1.97	1.94	1.91	1.89	1.86	1.84	1.79	1.77	1.73
25	2.09	2.01	1.96	1.92	1.89	1.87	1.84	1.82	1.77	1.75	1.71
26	2.07	1.99	1.94	1.90	1.87	1.85	1.82	1.80	1.75	1.73	1.69
27	2.06	1.97	1.92	1.88	1.86	1.84	1.81	1.79	1.73	1.71	1.67
28	2.04	1.96	1.91	1.87	1.84	1.82	1.79	1.77	1.71	1.69	1.66
29	2.03	1.94	1.89	1.85	1.83	1.81	1.77	1.75	1.70	1.67	1.64
30	2.01	1.93	1.88	1.84	1.81	1.79	1.76	1.74	1.68	1.66	1.62
40	1.92	1.84	1.78	1.74	1.72	1.69	1.66	1.64	1.58	1.55	1.51
50	1.87	1.78	1.73	1.69	1.66	1.63	1.60	1.58	1.51	1.48	1.44
60	1.84	1.75	1.69	1.65	1.62	1.59	1.56	1.53	1.47	1.44	1.39
120	1.75	1.66	1.60	1.55	1.52	1.50	1.46	1.43	1.35	1.32	1.26
200	1.72	1.62	1.56	1.52	1.48	1.46	1.41	1.39	1.30	1.26	1.19
∞	1.67	1.57	1.51	1.46	1.42	1.39	1.35	1.32	1.22	1.17	1.00

CRITICAL VALUES OF $F_{.025}$

This table shows the 2.5 percent right-tail critical values of F for the stated degrees of freedom (d.f.).

Denominator Degrees of Freedom (df_2)	Numerator Degrees of Freedom (df_1)										
	1	2	3	4	5	6	7	8	9	10	12
1	647.8	799.5	864.2	899.6	921.8	937.1	948.2	956.6	963.3	968.6	976.7
2	38.51	39.00	39.17	39.25	39.30	39.33	39.36	39.37	39.39	39.40	39.41
3	17.44	16.04	15.44	15.10	14.88	14.73	14.62	14.54	14.47	14.42	14.34
4	12.22	10.65	9.98	9.60	9.36	9.20	9.07	8.98	8.90	8.84	8.75
5	10.01	8.43	7.76	7.39	7.15	6.98	6.85	6.76	6.68	6.62	6.52
6	8.81	7.26	6.60	6.23	5.99	5.82	5.70	5.60	5.52	5.46	5.37
7	8.07	6.54	5.89	5.52	5.29	5.12	4.99	4.90	4.82	4.76	4.67
8	7.57	6.06	5.42	5.05	4.82	4.65	4.53	4.43	4.36	4.30	4.20
9	7.21	5.71	5.08	4.72	4.48	4.32	4.20	4.10	4.03	3.96	3.87
10	6.94	5.46	4.83	4.47	4.24	4.07	3.95	3.85	3.78	3.72	3.62
11	6.72	5.26	4.63	4.28	4.04	3.88	3.76	3.66	3.59	3.53	3.43
12	6.55	5.10	4.47	4.12	3.89	3.73	3.61	3.51	3.44	3.37	3.28
13	6.41	4.97	4.35	4.00	3.77	3.60	3.48	3.39	3.31	3.25	3.15
14	6.30	4.86	4.24	3.89	3.66	3.50	3.38	3.29	3.21	3.15	3.05
15	6.20	4.77	4.15	3.80	3.58	3.41	3.29	3.20	3.12	3.06	2.96
16	6.12	4.69	4.08	3.73	3.50	3.34	3.22	3.12	3.05	2.99	2.89
17	6.04	4.62	4.01	3.66	3.44	3.28	3.16	3.06	2.98	2.92	2.82
18	5.98	4.56	3.95	3.61	3.38	3.22	3.10	3.01	2.93	2.87	2.77
19	5.92	4.51	3.90	3.56	3.33	3.17	3.05	2.96	2.88	2.82	2.72
20	5.87	4.46	3.86	3.51	3.29	3.13	3.01	2.91	2.84	2.77	2.68
21	5.83	4.42	3.82	3.48	3.25	3.09	2.97	2.87	2.80	2.73	2.64
22	5.79	4.38	3.78	3.44	3.22	3.05	2.93	2.84	2.76	2.70	2.60
23	5.75	4.35	3.75	3.41	3.18	3.02	2.90	2.81	2.73	2.67	2.57
24	5.72	4.32	3.72	3.38	3.15	2.99	2.87	2.78	2.70	2.64	2.54
25	5.69	4.29	3.69	3.35	3.13	2.97	2.85	2.75	2.68	2.61	2.51
26	5.66	4.27	3.67	3.33	3.10	2.94	2.82	2.73	2.65	2.59	2.49
27	5.63	4.24	3.65	3.31	3.08	2.92	2.80	2.71	2.63	2.57	2.47
28	5.61	4.22	3.63	3.29	3.06	2.90	2.78	2.69	2.61	2.55	2.45
29	5.59	4.20	3.61	3.27	3.04	2.88	2.76	2.67	2.59	2.53	2.43
30	5.57	4.18	3.59	3.25	3.03	2.87	2.75	2.65	2.57	2.51	2.41
40	5.42	4.05	3.46	3.13	2.90	2.74	2.62	2.53	2.45	2.39	2.29
50	5.34	3.97	3.39	3.05	2.83	2.67	2.55	2.46	2.38	2.32	2.22
60	5.29	3.93	3.34	3.01	2.79	2.63	2.51	2.41	2.33	2.27	2.17
120	5.15	3.80	3.23	2.89	2.67	2.52	2.39	2.30	2.22	2.16	2.05
200	5.10	3.76	3.18	2.85	2.63	2.47	2.35	2.26	2.18	2.11	2.01
∞	5.02	3.69	3.12	2.79	2.57	2.41	2.29	2.19	2.11	2.05	1.94

Denominator Degrees of Freedom (df$_2$)	Numerator Degrees of Freedom (df$_1$)										
	15	20	25	30	35	40	50	60	120	200	∞
1	984.9	993.1	998.1	1001	1004	1006	1008	1010	1014	1016	1018
2	39.43	39.45	39.46	39.46	39.47	39.47	39.48	39.48	39.49	39.49	39.50
3	14.25	14.17	14.12	14.08	14.06	14.04	14.01	13.99	13.95	13.93	13.90
4	8.66	8.56	8.50	8.46	8.43	8.41	8.38	8.36	8.31	8.29	8.26
5	6.43	6.33	6.27	6.23	6.20	6.18	6.14	6.12	6.07	6.05	6.02
6	5.27	5.17	5.11	5.07	5.04	5.01	4.98	4.96	4.90	4.88	4.85
7	4.57	4.47	4.40	4.36	4.33	4.31	4.28	4.25	4.20	4.18	4.14
8	4.10	4.00	3.94	3.89	3.86	3.84	3.81	3.78	3.73	3.70	3.67
9	3.77	3.67	3.60	3.56	3.53	3.51	3.47	3.45	3.39	3.37	3.33
10	3.52	3.42	3.35	3.31	3.28	3.26	3.22	3.20	3.14	3.12	3.08
11	3.33	3.23	3.16	3.12	3.09	3.06	3.03	3.00	2.94	2.92	2.88
12	3.18	3.07	3.01	2.96	2.93	2.91	2.87	2.85	2.79	2.76	2.73
13	3.05	2.95	2.88	2.84	2.80	2.78	2.74	2.72	2.66	2.63	2.60
14	2.95	2.84	2.78	2.73	2.70	2.67	2.64	2.61	2.55	2.53	2.49
15	2.86	2.76	2.69	2.64	2.61	2.59	2.55	2.52	2.46	2.44	2.40
16	2.79	2.68	2.61	2.57	2.53	2.51	2.47	2.45	2.38	2.36	2.32
17	2.72	2.62	2.55	2.50	2.47	2.44	2.41	2.38	2.32	2.29	2.25
18	2.67	2.56	2.49	2.44	2.41	2.38	2.35	2.32	2.26	2.23	2.19
19	2.62	2.51	2.44	2.39	2.36	2.33	2.30	2.27	2.20	2.18	2.13
20	2.57	2.46	2.40	2.35	2.31	2.29	2.25	2.22	2.16	2.13	2.09
21	2.53	2.42	2.36	2.31	2.27	2.25	2.21	2.18	2.11	2.09	2.04
22	2.50	2.39	2.32	2.27	2.24	2.21	2.17	2.14	2.08	2.05	2.01
23	2.47	2.36	2.29	2.24	2.20	2.18	2.14	2.11	2.04	2.01	1.97
24	2.44	2.33	2.26	2.21	2.17	2.15	2.11	2.08	2.01	1.98	1.94
25	2.41	2.30	2.23	2.18	2.15	2.12	2.08	2.05	1.98	1.95	1.91
26	2.39	2.28	2.21	2.16	2.12	2.09	2.05	2.03	1.95	1.92	1.88
27	2.36	2.25	2.18	2.13	2.10	2.07	2.03	2.00	1.93	1.90	1.85
28	2.34	2.23	2.16	2.11	2.08	2.05	2.01	1.98	1.91	1.88	1.83
29	2.32	2.21	2.14	2.09	2.06	2.03	1.99	1.96	1.89	1.86	1.81
30	2.31	2.20	2.12	2.07	2.04	2.01	1.97	1.94	1.87	1.84	1.79
40	2.18	2.07	1.99	1.94	1.90	1.88	1.83	1.80	1.72	1.69	1.64
50	2.11	1.99	1.92	1.87	1.83	1.80	1.75	1.72	1.64	1.60	1.55
60	2.06	1.94	1.87	1.82	1.78	1.74	1.70	1.67	1.58	1.54	1.48
120	1.94	1.82	1.75	1.69	1.65	1.61	1.56	1.53	1.43	1.39	1.31
200	1.90	1.78	1.70	1.64	1.60	1.56	1.51	1.47	1.37	1.32	1.23
∞	1.83	1.71	1.63	1.57	1.52	1.48	1.43	1.39	1.27	1.21	1.00

CRITICAL VALUES OF $F_{.01}$

This table shows the 1 percent right-tail critical values of F for the stated degrees of freedom ($d.f.$).

Denominator Degrees of Freedom (df_2)	Numerator Degrees of Freedom (df_1)										
	1	**2**	**3**	**4**	**5**	**6**	**7**	**8**	**9**	**10**	**12**
1	4052	4999	5404	5624	5764	5859	5928	5981	6022	6056	6107
2	98.50	99.00	99.16	99.25	99.30	99.33	99.36	99.38	99.39	99.40	99.42
3	34.12	30.82	29.46	28.71	28.24	27.91	27.67	27.49	27.34	27.23	27.05
4	21.20	18.00	16.69	15.98	15.52	15.21	14.98	14.80	14.66	14.55	14.37
5	16.26	13.27	12.06	11.39	10.97	10.67	10.46	10.29	10.16	10.05	9.89
6	13.75	10.92	9.78	9.15	8.75	8.47	8.26	8.10	7.98	7.87	7.72
7	12.25	9.55	8.45	7.85	7.46	7.19	6.99	6.84	6.72	6.62	6.47
8	11.26	8.65	7.59	7.01	6.63	6.37	6.18	6.03	5.91	5.81	5.67
9	10.56	8.02	6.99	6.42	6.06	5.80	5.61	5.47	5.35	5.26	5.11
10	10.04	7.56	6.55	5.99	5.64	5.39	5.20	5.06	4.94	4.85	4.71
11	9.65	7.21	6.22	5.67	5.32	5.07	4.89	4.74	4.63	4.54	4.40
12	9.33	6.93	5.95	5.41	5.06	4.82	4.64	4.50	4.39	4.30	4.16
13	9.07	6.70	5.74	5.21	4.86	4.62	4.44	4.30	4.19	4.10	3.96
14	8.86	6.51	5.56	5.04	4.69	4.46	4.28	4.14	4.03	3.94	3.80
15	8.68	6.36	5.42	4.89	4.56	4.32	4.14	4.00	3.89	3.80	3.67
16	8.53	6.23	5.29	4.77	4.44	4.20	4.03	3.89	3.78	3.69	3.55
17	8.40	6.11	5.19	4.67	4.34	4.10	3.93	3.79	3.68	3.59	3.46
18	8.29	6.01	5.09	4.58	4.25	4.01	3.84	3.71	3.60	3.51	3.37
19	8.18	5.93	5.01	4.50	4.17	3.94	3.77	3.63	3.52	3.43	3.30
20	8.10	5.85	4.94	4.43	4.10	3.87	3.70	3.56	3.46	3.37	3.23
21	8.02	5.78	4.87	4.37	4.04	3.81	3.64	3.51	3.40	3.31	3.17
22	7.95	5.72	4.82	4.31	3.99	3.76	3.59	3.45	3.35	3.26	3.12
23	7.88	5.66	4.76	4.26	3.94	3.71	3.54	3.41	3.30	3.21	3.07
24	7.82	5.61	4.72	4.22	3.90	3.67	3.50	3.36	3.26	3.17	3.03
25	7.77	5.57	4.68	4.18	3.85	3.63	3.46	3.32	3.22	3.13	2.99
26	7.72	5.53	4.64	4.14	3.82	3.59	3.42	3.29	3.18	3.09	2.96
27	7.68	5.49	4.60	4.11	3.78	3.56	3.39	3.26	3.15	3.06	2.93
28	7.64	5.45	4.57	4.07	3.75	3.53	3.36	3.23	3.12	3.03	2.90
29	7.60	5.42	4.54	4.04	3.73	3.50	3.33	3.20	3.09	3.00	2.87
30	7.56	5.39	4.51	4.02	3.70	3.47	3.30	3.17	3.07	2.98	2.84
40	7.31	5.18	4.31	3.83	3.51	3.29	3.12	2.99	2.89	2.80	2.66
50	7.17	5.06	4.20	3.72	3.41	3.19	3.02	2.89	2.78	2.70	2.56
60	7.08	4.98	4.13	3.65	3.34	3.12	2.95	2.82	2.72	2.63	2.50
120	6.85	4.79	3.95	3.48	3.17	2.96	2.79	2.66	2.56	2.47	2.34
200	6.76	4.71	3.88	3.41	3.11	2.89	2.73	2.60	2.50	2.41	2.27
∞	6.63	4.61	3.78	3.32	3.02	2.80	2.64	2.51	2.41	2.32	2.18

Denominator Degrees of Freedom (df_2)	Numerator Degrees of Freedom (df_1)										
	15	20	25	30	35	40	50	60	120	200	∞
1	6157	6209	6240	6260	6275	6286	6302	6313	6340	6350	6366
2	99.43	99.45	99.46	99.47	99.47	99.48	99.48	99.48	99.49	99.49	99.50
3	26.87	26.69	26.58	26.50	26.45	26.41	26.35	26.32	26.22	26.18	26.13
4	14.20	14.02	13.91	13.84	13.79	13.75	13.69	13.65	13.56	13.52	13.47
5	9.72	9.55	9.45	9.38	9.33	9.29	9.24	9.20	9.11	9.08	9.02
6	7.56	7.40	7.30	7.23	7.18	7.14	7.09	7.06	6.97	6.93	6.88
7	6.31	6.16	6.06	5.99	5.94	5.91	5.86	5.82	5.74	5.70	5.65
8	5.52	5.36	5.26	5.20	5.15	5.12	5.07	5.03	4.95	4.91	4.86
9	4.96	4.81	4.71	4.65	4.60	4.57	4.52	4.48	4.40	4.36	4.31
10	4.56	4.41	4.31	4.25	4.20	4.17	4.12	4.08	4.00	3.96	3.91
11	4.25	4.10	4.01	3.94	3.89	3.86	3.81	3.78	3.69	3.66	3.60
12	4.01	3.86	3.76	3.70	3.65	3.62	3.57	3.54	3.45	3.41	3.36
13	3.82	3.66	3.57	3.51	3.46	3.43	3.38	3.34	3.25	3.22	3.17
14	3.66	3.51	3.41	3.35	3.30	3.27	3.22	3.18	3.09	3.06	3.01
15	3.52	3.37	3.28	3.21	3.17	3.13	3.08	3.05	2.96	2.92	2.87
16	3.41	3.26	3.16	3.10	3.05	3.02	2.97	2.93	2.84	2.81	2.76
17	3.31	3.16	3.07	3.00	2.96	2.92	2.87	2.83	2.75	2.71	2.66
18	3.23	3.08	2.98	2.92	2.87	2.84	2.78	2.75	2.66	2.62	2.57
19	3.15	3.00	2.91	2.84	2.80	2.76	2.71	2.67	2.58	2.55	2.49
20	3.09	2.94	2.84	2.78	2.73	2.69	2.64	2.61	2.52	2.48	2.42
21	3.03	2.88	2.79	2.72	2.67	2.64	2.58	2.55	2.46	2.42	2.36
22	2.98	2.83	2.73	2.67	2.62	2.58	2.53	2.50	2.40	2.36	2.31
23	2.93	2.78	2.69	2.62	2.57	2.54	2.48	2.45	2.35	2.32	2.26
24	2.89	2.74	2.64	2.58	2.53	2.49	2.44	2.40	2.31	2.27	2.21
25	2.85	2.70	2.60	2.54	2.49	2.45	2.40	2.36	2.27	2.23	2.17
26	2.81	2.66	2.57	2.50	2.45	2.42	2.36	2.33	2.23	2.19	2.13
27	2.78	2.63	2.54	2.47	2.42	2.38	2.33	2.29	2.20	2.16	2.10
28	2.75	2.60	2.51	2.44	2.39	2.35	2.30	2.26	2.17	2.13	2.07
29	2.73	2.57	2.48	2.41	2.36	2.33	2.27	2.23	2.14	2.10	2.04
30	2.70	2.55	2.45	2.39	2.34	2.30	2.25	2.21	2.11	2.07	2.01
40	2.52	2.37	2.27	2.20	2.15	2.11	2.06	2.02	1.92	1.87	1.81
50	2.42	2.27	2.17	2.10	2.05	2.01	1.95	1.91	1.80	1.76	1.69
60	2.35	2.20	2.10	2.03	1.98	1.94	1.88	1.84	1.73	1.68	1.60
120	2.19	2.03	1.93	1.86	1.81	1.76	1.70	1.66	1.53	1.48	1.38
200	2.13	1.97	1.87	1.79	1.74	1.69	1.63	1.58	1.45	1.39	1.28
∞	2.04	1.88	1.77	1.70	1.64	1.59	1.52	1.47	1.32	1.25	1.00

Solutions to Odd-Numbered Exercises

CHAPTER 1

1.9 No, association does not imply causation. See Pitfall 5.

1.11 a. All combinations have same chance of winning so method did not "work."

b. No, same as any other six numbers.

1.13 A reduction of 0.2% may not seem important to the individual customer, but from the company's perspective it could be significant depending on how many customers it has.

1.15 Disagree. The difference is practically important. 0.9% of 231,164 is 2,080 patients.

1.19 a. Analyze the 80 responses but make no conclusions about nonrespondents.

b. No, study seems too flawed.

1.21 Agree. Be wary of Pitfall 2.

1.23 Disagree. Tom fell for Pitfall 2.

1.25 a. Attendance, study time, ability level, interest level, instructor's ability, prerequisites.

b. Reverse causation? Good students make better decisions about their health.

c. No, causation is not shown.

1.27 A major problem is that we don't know the number of students in each major.

a. Likely more engineers want an MBA, so they take it.

b. Causation not shown. Physics may differ from marketing majors (e.g., math skills).

c. The GMAT is just an indicator of academic skills.

1.29 a. Most would prefer the graph, but both are clear.

b. The number of salads sold reached a maximum in May and decreased steadily toward the end of the year.

CHAPTER 2

2.1 a. Categorical. b. Categorical. c. Discrete numerical.

2.3 a. Continuous numerical.

b. Continuous numerical (often reported as an integer).

c. Categorical. d. Categorical.

2.5 a. Cross-sectional. b. Time series.

c. Time series. d. Cross-sectional.

2.7 a. Time series. b. Cross-sectional. c. Time series.

d. Cross-sectional.

2.9 a. Ratio. b. Nominal. c. Interval.

d. Ratio. e. Ordinal.

2.11 a. Ratio. b. Ratio. c. Nominal. d. Ordinal.

e. Interval.

Likert scales are typically assumed to be interval.

2.13 a. Interval, assuming intervals are equal.

b. Yes (assuming interval data).

c. 10-point scale might give too many points and make it hard for guests to choose between.

2.15 a. Census. b. Sample or census.

c. Sample. d. Census.

2.17 a. Parameter. b. Parameter.

c. Statistic. d. Statistic.

2.19 a. Convenience. b. Systematic.

c. Judgment or biased.

2.25 a. Telephone or web. b. Direct observation.

c. Interview, web, or mail.

2.27 Version 1: Most would say yes. Version 2: More varied responses.

2.29 a. Continuous numerical. b. Categorical.

c. Discrete numerical.

2.31 a. Ordinal. b. Interval. c. Ratio.

2.33 Q1 Categorical, nominal. Q2 Continuous, ratio.

Q3 Continuous, ratio. Q4 Discrete, ratio.

Q5 Categorical, ordinal.

2.35 a. =RANDBETWEEN(1,100)

b. =RAND()

c. =RAND()*100

2.37 a. Time series. b. Cross-sectional. c. Cross-sectional.

d. Time series.

2.39 a. Census. b. Sample.

c. Census.

2.41 a. Statistic. b. Parameter. c. Parameter.

2.43 a. Complaint. b. Patient visits, discrete. Waiting time, continuous.

2.45 a. No, census costly, perhaps impossible.

2.47 a. Cluster sampling.

b. No, population effectively infinite.

2.49 a. Census. b. Sample.

c. Sample. d. Census.

2.51 a. Cluster sample. b. Cluster sample.

c. Random sample of tax returns. d. Statistic based on sales, not a sample.

2.53 a. Cluster sampling; neighborhoods are natural clusters.

b. Picking a day near a holiday with light trash.

2.55 a. Yes, 11,000/18 > 20. b. 1/39

2.57 Education and income could affect who uses the no-call list.

a. They won't reach those who purchase such services. Same response for b and c.

Surveys and Scales

2.59 a. Rate the effectiveness of this professor. 1—Excellent to 5—Poor.

b. Rate your satisfaction with the president's economic policy. 1—Very satisfied to 5—Very dissatisfied.

c. How long did you wait to see your doctor? Less than 15 minutes, between 15 and 30 minutes, between 30 minutes and 1 hour, more than 1 hour.

2.61 a. Ordinal.

b. Intervals are equal.

CHAPTER 3

3.1 a.

Frequency	Stem	Leaf
1	0	9
5	1	56889
7	2	1145667
10	3	1222334559
1	4	2
24		

b.

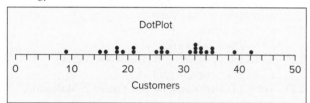

c. Skewed left, central tendency approximately 30, range from 9 to 42.

3.3 a.

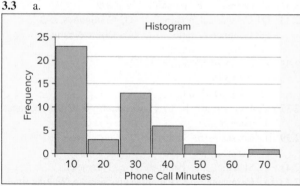

b. Sarah's phone call minutes are right skewed with one lengthy phone call lasting between 60 and 70 minutes.

c. Yes, this follows Sturges' Rule, which says we should have 6 or 7 bins.

3.5 a.

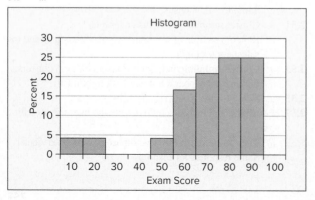

b. Skewed left, central tendency approximately 80, majority of data from 50 to 100, two outliers at 18 and 27.

3.7 Sturges' Rule suggests about 6 bins. Slight right skew.

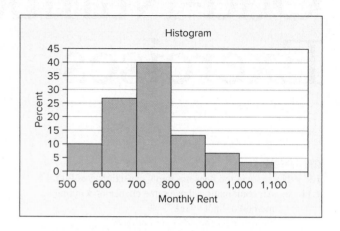

3.9 a. 7 bins, width = 5, Sturges' Rule = 6 bins.

b. 8 bins, width = 10, Sturges' Rule = 6 or 7 bins.

c. 10 bins, width = 0.15, Sturges' Rule = 9 bins.

d. 8 bins, width = 0.01, Sturges' Rule = 8 bins.

3.11 a.

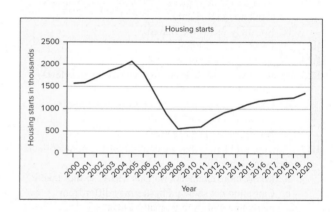

b. Decreasing trend from 2005 to 2010, then increasing at slower rate up to 2020.

3.13 Decreasing at a decreasing rate.

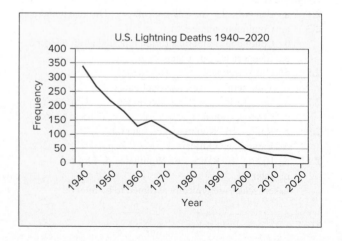

3.15 a. To show more detail, you could start the graph at (.80,100).

  b. There is a moderate negative linear relationship.

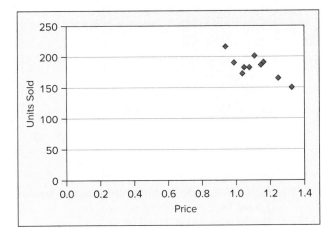

3.17 a. See graph below.

  b. There is a moderate positive linear relationship.

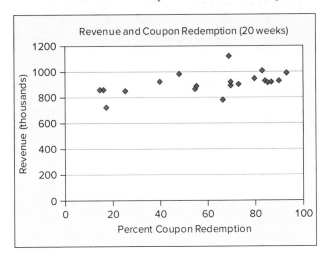

3.19 a.

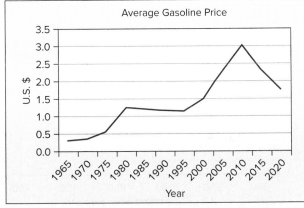

Source: www.fueleconomy.gov.

b.

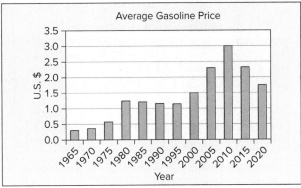

Source: www.fueleconomy.gov.

  c. Line chart is preferred by most analysts.

3.21 a.

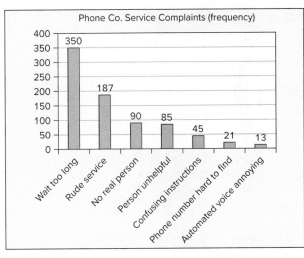

  b. Wait too long, Rude service, No real person on line.

  c. Wait too long.

3.23 a.

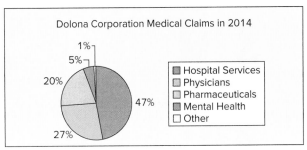

b.

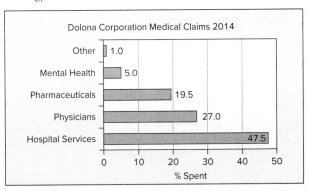

c.

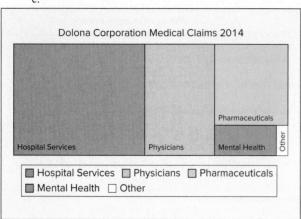

Dolona Corporation Medical Claims 2014

Hospital Services   Physicians   Pharmaceuticals
Mental Health   Other

3.25 a.

Frequency	Stem	Leaf
3	0	249
6	1	227778
5	2	14557
4	3	0025
1	4	4
3	5	013
2	6	26
0	7	
1	8	4
1	9	9

b.

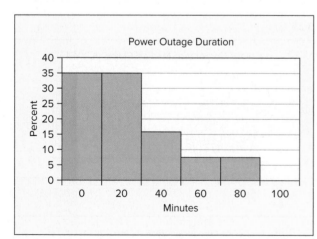

Power Outage Duration

c. Skewed right.

3.27 a.

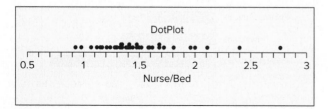

DotPlot

b.

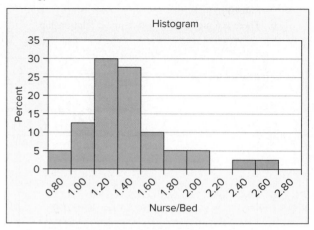

Histogram

c. Skewed to the right. Half of the data values are between 1.2 and 1.6.

3.29 a. Dotplot created using MegaStat

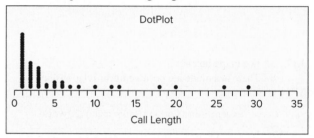

DotPlot

b. Histogram with 29 bins

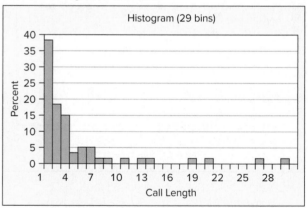

Histogram (29 bins)

c. Heavily skewed to the right. Central tendency approximately 3 minutes.

3.31 b. Scatter plot shows strong positive linear relationship.

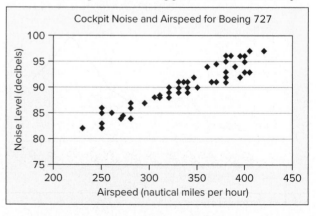

Cockpit Noise and Airspeed for Boeing 727

3.33 a.

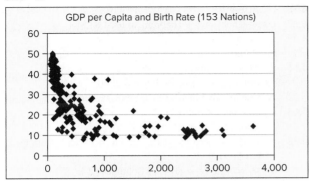

b. Negative, Nonlinear, Moderate relationship.

3.35 a. Horizontal bar chart with 3D visual effect.

b. Strengths: Good proportions and no distracting pictures. Weaknesses: No labels on *X* and *Y* axes, title unclear, 3D effect does not add to presentation.

c. Vertical bar chart without visual effect and label on *X* axis.

3.37 a. Exploded pie chart.

b. Strengths: Information complete, colorful. Weaknesses: Hard to assess differences in size of pie slices.

c. Sorted column chart with OPEC and non-OPEC countries color coded.

3.39 a. Line chart with pictures.

b. Pictures distracting, implies irresponsibility, does show source of data.

c. Take out pictures, show a simple line chart.

3.41 a.

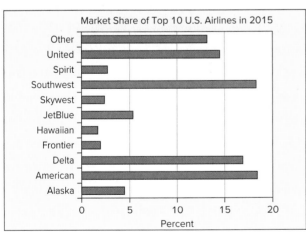

b. Yes, a pie chart could be used.

3.43 a.

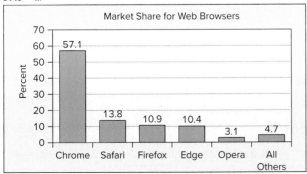

b. Yes, a pie chart would work.

3.45 a.

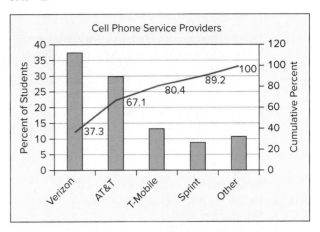

b. Verizon, Cingular, T-Mobile.

CHAPTER 4

4.1 a. mean = 2.83, median = 1.5, mode = 0.

b. mean = 68.33, median = 72, mode = 40.

c. mean = 304, median = 303, no mode.

4.3 a. No, all the data values are unique.

b. Yes, these are categorical data with repeated values. (Mode = C)

c. No, the data are continuous numerical.

4.5 a. Continuous data, skewed right, no mode. Median best choice.

b. Mostly one rider, mode best choice.

c. Symmetric distribution, mean and median the same, and two modes. Mean best choice.

4.7 a. Mean = 75.5, median = 80.5, mode = 93.

b. Skewed left.

c. Mode not a useful measure. The value 93 appears only 3 times out of 24 observations.

4.9 a. $\bar{x}$ = 27.34, median = 26, mode = 26.

b. No, $\bar{x}$ is greater than the median and mode.

c. Slightly skewed right.

4.11 a. $\bar{x}$ = 4.27, median = 2, mode = 1.

b. No, $\bar{x}$ > median > mode.

c. Skewed right.

4.13 a. =TRIMMEAN(A1:A50,.2)   b. 5.   c. 10.

4.15 a. Mean = 100, median = 0, mode = 0, midrange = 325. Geometric mean is undefined because there are values equal to 0.

b. Choose the mean. Midrange would overestimate total expected expenses.

4.17 a. $\bar{x}$ = 27.34, midrange = 25.50, geometric mean = 26.08, 10% trimmed mean = 27.46.

b. The measures are all close, especially the mean and trimmed mean.

4.19 a. $\bar{x}$ = 4.27, midrange = 15.0, geometric mean = 2.48, 10% trimmed mean = 2.75.

b. The data are skewed right.

4.21 a. Sample A: $\bar{x}$ = 7, *s* = 1. Sample B: $\bar{x}$ = 62, *s* = 1. Sample C: $\bar{x}$ = 1001, *s* = 1.

b. The standard deviation is not a function of the mean.

4.23 Hybrid: 5.1%. Gas: 7%. The hybrid had more consistent gas mileage relative to the mean than the gasoline vehicle.

4.25 =AVEDEV(1,2,3,4,5,6,7,8,9,10) = 2.5.

4.27 a. Stock A: CV = 21.43%. Stock B: CV = 8.32%. Stock C: CV = 36.17%.

b. Stock C.

c. Directly comparing standard deviation would not be helpful in this case because the means have different magnitudes.

4.29 a. $\bar{x}_A$ = 6.857, s_A = 1.497, $\bar{x}_B$ = 7.243, s_B = 1.209.

b. CV_A = 0.218, CV_B = 0.167.

c. Consumers preferred sauce B.

4.31 At least 75% of the data will fall within ±2 standard deviations, so .75 × 400 = 300.

4.33 a. z = 2.396.

b. Unusual. 2 < z < 3.

c. Outlier would be 33 days or longer.

4.35 a. z = 2.4. b. z = 1. c. z = 0.6.

4.37 a. Bob's GPA = 3.596.

b. Sarah's weekly work hours = 29.978.

c. Dave's bowling score = 96.

4.39 b. 18 (z = 2.34) and 20 (z = 2.84) are unusual observations. 26 (z = 3.6) and 29 (z = 4.21) are outliers.

c. 90% lie within 1 standard deviation and 93.3% lie within 2 standard deviations. 90% is much greater than the 68% specified by the empirical rule. The distribution does not appear normal.

4.41 b. Strongly skewed right.

4.43 Inner fences: [145, 185]. Not an outlier: 145 < 149 < 185.

4.45 a. $Q_1 \approx 3300$, $Q_2 \approx 3900$, and $Q_3 \approx 4300$.

b. $x_{min} \approx 2400$ and $x_{max} \approx 4800$.

c. Left skewed.

4.47 a. Q_1 = 1, Q_3 = 4.75. The middle 50% of the calls last between 1 and 4.25 minutes.

b. Midhinge = 2.875. Calls typically last a little over 2.5 minutes.

c. The data are heavily skewed to the right.

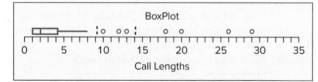

4.49 a. −.8571.

b. Strong inverse (negative) correlation.

4.51 a.

b. r = .8338.

c. Yes, there is a strong positive linear relationship.

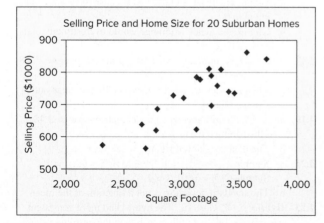

4.53 a. Mean = 8.48.

4.55 [1.58, 2.46].

4.57 Using Chebyshev's Theorem, at least 88.9%.

4.59 a. z = 3.128.

b. Outlier, z > 3.

4.61 a. Allison's final exam = 90.1.

b. Jim's weekly grocery bill = $35.60.

c. Eric's daily video game time = 3.09 hours.

4.63 a. Standard deviation = 2.

b. Assumed a normal distribution.

4.65 a. $\bar{x}$ = 26.71, median = 14.5, mode = 11, and midrange = 124.5.

b. Q_1 = 7.25, Q_3 = 20.75, Midhinge = 14.

c. The geometric mean is only valid for data greater than zero.

4.67 a. Mean = 66.85, median = 69.5, and mode = 86.

b. Mean and median fairly close.

c. No, mode not typical; continuous numerical data with few repeats.

d. Difficult to describe shape based only on mean and median. Might conclude somewhat symmetric.

4.69 a. Stock funds: $\bar{x}$ = 1.329, median = 1.22. Bond funds: $\bar{x}$ = 0.875, median = 0.85.

b. Stock funds: s = 0.5933, CV = 44.65%. Bond funds: s = 0.4489, CV = 51.32%.

c. The stock funds have less variability relative to the mean.

4.71 z = −3.33 is an outlier.

4.73 a.

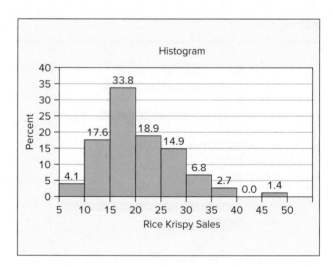

b. Skewed right.

c. Mean = 20.12, standard deviation = 7.64.

d. One possible outlier at 49 (store 22).

4.75 a. Tuition plans: CV = 42.86%, S&P 500: CV = 122.48%.

b. The CV shows *relative* risk for each investment.

4.77 a. Midrange = 0.855.

b. Standard deviation = .0217.

4.79 The distribution is skewed to the left (negative skew less than lower 5% range) and has a sharper peak (positive kurtosis greater than upper 5% range).

4.81 a. The distribution is skewed to the right.

b. This makes sense; most patrons would keep books about 10 days, with a few keeping them much longer.

4.83 a. Would expect mean to be close in value to the median, or slightly higher.
  b. Life span would have normal distribution. If skewed, more likely skewed right than left. Life span is bounded below by zero, but is unbounded in the positive direction.

4.85 a. It is the midrange, not the median.
  b. The midrange is influenced by outliers. Salaries tend to be skewed to the right. Community should use the median.

4.87 a. and c.

	Week 1	Week 2	Week 3	Week 4
Mean	50.00	50.00	50.00	50.00
Sample standard deviation	10.61	10.61	10.61	10.61
Median	50.00	52.00	56.00	47.00

  b. Based on the mean and standard deviation, it appears that the distributions are the same.
  d.

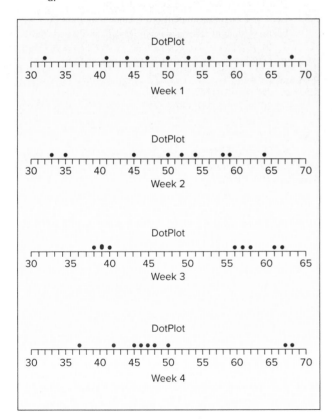

  e. Based on the medians and dotplots, distributions are quite different.

4.89 a. $\bar{x} = 9.458$.
  b. $s = 10.855$.
  c. No, the distribution is skewed right.
  d. To prevent bins with zero frequencies.

4.91 a. $\bar{x} = 60.2$, $s = 8.54$, $CV = 14.2\%$.
  b. No. The class widths increase as the data values get more spread out.

4.95 a.

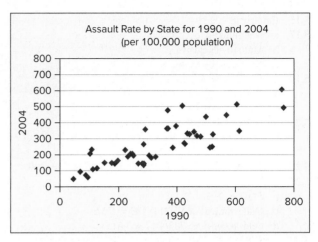

  b. $r = .8332$.
  c. The rates are positively correlated.
  d. 1990: mean = 331.92, median = 3.7, Std Dev = 172.914. 2004: mean = 256.6, median = 232, Std Dev = 131.259.

4.97 a.

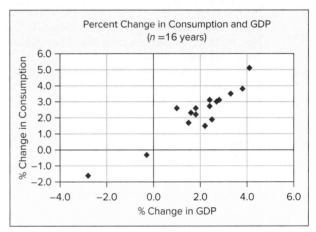

  b. Strong positive linear relationship.
  c. As GDP increases, people can afford to purchase more consumer goods.
  d. $r = .9314$.

CHAPTER 5

5.1 a. $S = \{(V,B), (V,E), (V,O), (M,B), (M,E), (M,O), (A,B), (A,E), (A,O)\}$.
  b. Events are not equally likely. Barnes and Noble probably carries more books than other merchandise.

5.3 a. $S = \{(L,B), (L,B'), (R,B), (R,B')\}$.
  b. Events are not equally likely. More right-handed people than left-handed people.

5.5 a. There are 20 events. $S = \{(A, F), (A, S), (A, J), (A, R), (B, F), \ldots (M, J), (M, R)\}$.
  b. Events are not equally likely. Arts and sciences enrolls more students than music for example. Also, there are more freshmen than seniors possibly.

5.7 Subjective

5.9 Empirical.

5.11 Empirical.

5.13 Classical.

5.15 Empirical.

5.17 a. Not mutually exclusive.

b. Mutually exclusive.

c. Not mutually exclusive.

5.19 a. $P(A \cup B) = .4 + .5 - .05 = .85$.

b. $P(A \mid B) = .05/.50 = .10$.

c. $P(B \mid A) = .05/.4 = .125$.

5.21 a. $P(A$ and $B) = .40 \times .70 = .28$.

b. $P(A \cup B) = .70 + .30 - .28 = .72$.

c. $P(A \mid B) = .28/.3 = .9333$.

5.23 $P(F) = .62$. $P(S \mid F) = .8$, $P(S$ and $F) = .62 \times .8 = .496$.

5.25 a. $P(S) = .217$.

b. $P(S') = .783$.

c. Odds in favor of S: $.217/.783 = .277$.

d. Odds against S: $.783/.217 = 3.61$.

5.27 a. $P(S') = 1 - .246$. There is a 75.4% chance that a female aged 18–24 is a nonsmoker.

b. $P(S \cup C) = .246 + .830 - .232 = .844$. There is an 84.4% chance that a female aged 18–24 is a smoker or is Caucasian.

c. $P(S \mid C) = .232/.830 = .2795$. Given that the female aged 18–24 is a Caucasian, there is a 27.95% chance that she is a smoker.

d. $P(S \cap C') = P(S) - P(S \cap C) = .246 - .232 = .014$. $P(S \mid C') = .014/.17 = .0824$. Given that the female aged 18–24 is *not* Caucasian, there is an 8.24% chance that she smokes.

5.29 $P(J \cap K) = P(J) \times P(K) = .26 \times .48 = .1248$, so $P(J \cup K) = P(J) + P(K) - P(J \cap K) = .26 + .48 - .1248 = .6152$.

5.31 a. $P(A \mid B) = P(A \cap B)/P(B) = .05/.50 = .10$.

b. No, A and B are not independent because $P(A \mid B) \neq P(A)$.

5.33 a. $P(J \mid K) = P(J \cap K)/P(K) = .15/.4 = .375$.

b. No, J and K are not independent because $P(J \cap K) \neq P(J)$.

5.35 a. $P(V \cup M) = .73 + .18 - .03 = .88$.

b. $P(V \cap M) \neq P(V)P(M)$; therefore, V and M are not independent.

5.37 "Five nines" reliability means P(not failing) $= .99999$. P(power system failure) $= 1 - (.05)^3 = .999875$. The system does not meet the test.

5.39 Ordering a soft drink is independent of ordering a square pizza. P(ordering a soft drink) $\times P$(ordering a square pizza) $= .5(.8) = .4$. This is equal to P(ordering both a soft drink and a square pizza).

5.41 a. P(Recycles) $= .34$.

b. P(Don't Recycle | Lives in Deposit Law State) $= .30$.

c. P(Recycle and Live in Deposit Law State) $= .154$.

d. P(Recycle | Lives in Deposit Law State) $= .70$.

5.43 a. $P(D) = .5064$.

b. $P(R) = .1410$.

c. $P(D \cap R) = .0513$.

d. $P(D \cup R) = .5962$.

e. $P(R \mid D) = .1013$.

f. $P(R \mid P) = .1628$.

5.45 *Gender* and *Major* are not independent. For example, $P(A \cap F) = .22$. $P(A)P(F) = .245$. Because the values are not equal, the events are not independent.

5.47

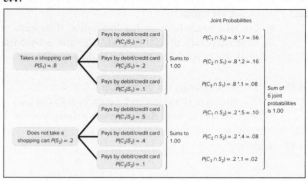

5.49 Let A = using the drug. $P(A) = .04$. $P(A') = .96$. Let T be a positive result. False positive: $P(T \mid A') = .05$. False negative: $P(T' \mid A) = .10$. $P(T \mid A) = 1 - .10 = .90$. $P(T) = (.04)(.90) + (.05)(.96) = .084$. $P(A \mid T) = (.9)(.04)/.084 = .4286$.

5.51 Let W = suitcase contains a weapon. $P(W) = .001$. $P(W') = .999$. Let A be the alarm trigger. False positive: $P(A \mid W') = .02$. False negative: $P(A' \mid W) = .02$. $P(A \mid W) = 1 - .02 = .98$. $P(A) = (.001)(.98) + (.02)(.999) = .02096$. $P(W \mid A) = (.98)(.001)/.02096 = .04676$.

5.53 Let F = a business student is a finance major and R = a business student takes a class in R. $P(F) = .20$ and $P(F') = .80$. $P(R \mid F) = .30$. $P(R \mid F') = .15$. $P(F \mid R) = P(F$ and $R)/P(R) = (.2 \times .3)/(.2 \times .3 + .15 \times .8) = .3333$.

5.55 $_{20}C_5 = 15,504$.

5.57 a. $26^6 = 308,915,776$.

b. $36^6 = 2,176,782,336$.

c. $32^6 = 1,073,741,824$.

5.59 a. $10^6 = 1,000,000$.

b. $10^5 = 100,000$.

c. $10^6 = 1,000,000$.

5.61 a. $7! = 5,040$ ways.

b. No, too many!

5.63 a. $_8C_3 = 56$.

b. $_8C_5 = 56$.

c. $_8C_1 = 8$.

d. $_8C_8 = 1$.

5.69 Empirical.

5.71 a. An empirical probability using response frequencies from the survey.

5.73 a. Classical.

5.75 a. Subjective.

5.77 a. Empirical or subjective.

5.81 P(vehicle is recovered) $= 1 - .436 = .564$. Odds in favor of a vehicle being recovered are $.564/.436$ or 1.3 to 1.

5.83 P(Butler making it to the NCAA finals) $= 1/(125 + 1) = .008$.

5.85 a. $26^3 10^3 = 17,576,000$.

b. $36^6 = 2,176,782,336$.

c. 0 and 1 might be disallowed because they are similar in appearance to letters like O and I.

d. Yes, 2.2 billion unique plates should be enough.

e. $34^6 = 1,544,804,416$.

5.87 Order does not matter. $_7C_3 = 35$.

5.89 a. $P(D \mid P) = .12/.57 = .2105$.

b. No, $P(D) = .12$ and $P(D \mid P) = .2105$, which means the two events are dependent.

5.91 No, $P(A)P(B) \neq .05$.

5.93 P(at least one gyro will operate) = .99999936. Yes, they've achieved "five-nines" reliability.

5.95 Assuming independence, P(3 cases won out of next 3) = $.7^3 = .343$.

5.97* Assuming independence, P(4 adults say yes) = $.56^4 = 0.0983$.

5.99* See the Excel Spreadsheet in *Learning Stats:* 05-13 Birthday Problem.xls.
For 2 riders: P(no match) = .9973.
For 10 riders: P(no match) = 0.8831.
For 20 riders: P(no match) = 0.5886.
For 50 riders: P(no match) = 0.0296.

5.101 a. i. .4825 ii. .25 iii. .19 iv. .64 v. .09 vi. .015
b. Yes, the vehicle type and mall location are dependent.

5.103 a. i. .5588 ii. .5294 iii. .19 iv. .64 v. .09 vi. .015

5.105*

	Cancer	No Cancer	Totals
Positive Test	4	500	504
Negative Test	0	9496	9496
Totals	4	9996	10000

P(Cancer | Positive Test) = 4/504 = 0.00794.

5.107* Let D = applicant uses drugs and T = applicant has positive test result.

$$P(D|T) = \frac{P(T|D)P(D)}{P(T|D)P(D) + P(T|D')P(D')}$$

$$= \frac{.036}{.036 + .144} = .20.$$

CHAPTER 6

6.1 Only A is a PDF because $P(x)$ sum to 1.

6.3 a. $P(X \geq 3) = .2 + .15 + .05 = .4$.
b. $P(X \leq 2) = .05 + .3 + .25 = .6$.
c. $P(X > 4) = .05 + .3 + .25 + .2 = .8$.
d. $P(X = 1) = .3$.
e. (b) is a CDF.

6.5 $E(X) = 2.25, V(X) = 1.6875, \sigma = 1.299$, right-skewed.

6.7 $E(X) = 5000(.01) + (0)(.999) = \$50$, add \$25, charges \$75.

6.9 $E(X) = 250(.3) + 950(.3) + 0(.4) = \$360$ million.

6.11 a. $\mu = (20 + 60)/2 = 40$,
$\sigma = \sqrt{[(160 - 20 + 1)^2 - 1]/12} = 11.83$.
b. $P(X \geq 40) = .5122, P(X \geq 30) = .7561$.

6.13 a. $X = 0, 1$, or 2.
b. $X = 4, 5, 6$, or 7.
c. $X = 4, 5, 6, 7, 8, 9$, or 10.

6.15 a. $\mu = 0.8, \sigma = 0.8485$.
b. $\mu = 4, \sigma = 1.5492$.
c. $\mu = 6, \sigma = 1.7321$.

6.17 a. $P(X = 5) = .0074$.
b. $P(X = 0) = .2621$.
c. $P(X = 9) = .1342$.

6.19 a. $P(X \leq 3) = .9437$.
b. $P(X > 7) = 1 - P(X \leq 7) = .0547$.
c. $P(X < 3) = P(X \leq 2) = .0705$.

6.21 a. $P(X > 10) = .9183$. b. $P(X \geq 4) = .4059$.
c. $P(X \leq 2) = .9011$.

6.23 a. $P(X = 0) = .10737$.
b. $P(X \geq 2) = .62419$.
c. $P(X < 3) = .67780$.
d. $\mu = n\pi = (10)(.2) = 2$.
e. $\sigma = \sqrt{(10)(.2)(1 - .2)} = 1.2649$.
g. Skewed right.

6.25 a. $P(X = 0) = .0916$.
b. $P(X \geq 2) = .6276$.
c. $P(X < 4) = .9274$.
d. $P(X = 5) = .0079$.

6.27 a. $P(X = 16) = .0033$. c. $P(X \geq 10) = .8247$.
b. $P(X < 10) = .1753$.

6.29 a. $\lambda = 1, \mu = 1.0, \sigma = 1$.
b. $\lambda = 2, \mu = 2.0, \sigma = 1.414$.
c. $\lambda = 4, \mu = 4.0, \sigma = 2.0$.

6.31 a. $P(X = 6) = .1042$. c. $P(X = 4) = .0912$.
b. $P(X = 10) = .1048$.

6.33 a. $\lambda = 4.3, P(X \leq 3) = .37715$.
b. $\lambda = 5.2, P(X > 7) = .15508$.
c. $\lambda = 2.7, P(X < 3) = .49362$.

6.35 a. $P(X > 10) = .1841$.
b. $P(X \leq 5) = .7851$.
c. $P(X \geq 2) = .9596$.

6.37 a. $P(X \geq 1) = .8173$. c. $P(X > 3) = .0932$.
b. $P(X = 0) = .1827$. d. Skewed right.

6.39 a. Add-ons are ordered independently.
b. $P(X \geq 2) = .4082$.
c. $P(X = 0) = .2466$.
d. Skewed right.

6.41* Let $\lambda = n\pi = (500)(.003) = 1.5$.
a. $P(X \geq 2) = 1 - .55783 = .44217$.
b. $P(X \leq 4) = .93436$.
c. Yes, $n \geq 20$ and $\pi \leq .05$.

6.43* a. Set $\lambda = \mu = (200)(.03) = 6$.
b. $\sigma = \sqrt{(200)(.03)(1 - .03)} = 2.412$.
c. $P(X \geq 10) = 1 - .91608 = .08392$.
d. $P(X \leq 4) = .28506$.
e. Yes, $n \geq 20$ and $\pi \leq .05$.

6.45* a. $E(X) = 2.3$.
b. $P(X = 0) \approx .1003, P(X > 2) = .4040$.
c. Yes, $n \geq 20$ and $\pi \leq .05$.

6.47 Distribution is symmetric with small range.

6.49 a. X = number of incorrect vouchers in sample.
b. $P(X = 0) = .06726$.
c. $P(X = 1) = .25869$.
d. $P(X \geq 3) = 1 - .69003 = .30997$.
e. Fairly symmetric.

6.51* a. $3/100 < .05$, OK.
b. $10/200 = .05$, OK.
c. $12/160 > .05$, not OK.
d. $7/500 < .05$, OK.

6.53* a. $P(X = 0) = 0.34868$ (B) or .34516 (H).
b. $P(X \geq 2) = .26390$ (B) or .26350 (H).
c. $P(X < 4) = .98720$ (B) or .98814 (H).
d. $n/N = 10/500 = .02$, so set $\pi = s/N = 50/500 = .1$.

6.55* a. $P(X = 5) = .03125$ when $\pi = .50$.
b. $P(X = 3) = .14063$ when $\pi = .25$.
c. $P(X = 4) = .03840$ when $\pi = .60$.

6.57* a. $\mu = 1/\pi = 1/(.50) = 2$.
b. $P(X > 10) = (.50)^{10} = .00098$.

6.59* a. $\mu = 9500 + 7400 + 8600 = \$25,500$ (Rule 3), $\sigma^2 = 1250 + 1425 + 1610 = 4285$ (Rule 4), $\sigma = 65.4599$.
b. Rule 4 assumes independent monthly sales (unlikely).

6.61* a. If $Y = $ Bob's point total, then $\mu_Y = 400$, $\sigma_Y = 11.18$.
b. No, 450 is more than 3 standard deviations from the mean.

6.63 $E(\text{loss}) = \$0(.98) + \$250(.02)$
$= \$5$.
$E(\text{loss}) > $ Insurance, so purchase insurance.

6.65 a. $\pi = .80$ (answers will vary).
b. $\pi = .300$ (answers will vary).
c. $\pi = .50$ (answers will vary).
d. $\pi = .80$ (answers will vary).
e. One trial may influence the next.

6.67 a. $P(X = 5) = .59049$.
b. $P(X = 4) = .32805$.
c. Strongly right-skewed.

6.69 a. $P(X = 0) = .06250$.
b. $P(X \geq 2) = 1 - .31250 = .68750$.
c. $P(X \leq 2) = .68750$.
d. Symmetric.

6.71 a. =BINOM.DIST(3,20,0.3,0)
b. =BINOM.DIST(7,50,0.1,0)
c. =BINOM.DIST(6,80,0.05,1)
d. =1−BINOM.DIST(29,120,0.2,1)

6.73 a. $P(X = 0) = .48398$.
b. $P(X \geq 3) = 1 - P(X \leq 2) = 1 - .97166 = .02834$.
c. $\mu = n\pi = (10)(.07) = 0.7$ default.

6.75 Binomial with $n = 16$, $\pi = .8$:
a. $P(X \geq 10) = 1 - P(X \leq 9) = 1 - .02666 = .97334$.
b. $P(X < 8) = P(X \leq 7) = .00148$.

6.77 Let $X = $ number of no shows. Then:
a. If $n = 10$ and $\pi = .10$, then $P(X = 0) = .34868$.
b. If $n = 11$ and $\pi = .10$, then $P(X \geq 1) = 1 - P(X = 0) = 1 - .31381 = .68619$.
c. If they sell 11 seats, not more than 1 will be bumped.
d. If $X = $ number who show ($\pi = .90$), using =1−BINOM.DIST(9,n,.9,1), we find that $n = 13$ will ensure that $P(X \geq 10) \geq .95$.

6.79 a. Because calls to a fire station within a minute are most likely all about the same fire, the calls are not independent.
b. Answers will vary.

6.81 a. $P(X = 5) = .0872$.
b. $P(X \leq 5) = .9349$.
c. $\lambda = 14$ arrivals/5 min interval.
d. Independence.

6.83 a. $P(X = 0) = .7408$.
b. $P(X \geq 2) = .0369$.

6.85 a. Assume independent cancellations.
b. $P(X = 0) = .22313$.
c. $P(X = 1) = .33470$.
d. $P(X > 2) = 1 - .80885 = .19115$.
e. $P(X \geq 5) = 1 - .98142 = .01858$.

6.87 a. Assume independent defects with $\lambda = 2.4$.
b. $P(X = 0) = .09072$.
c. $P(X = 1) = .21772$.
d. $P(X \leq 1) = .30844$.

6.89* $P(\text{at least one rogue wave in 5 days}) = 1 - P(X = 0) = .9892$.

6.91 a. Assume independent crashes.
b. $P(X \geq 1) = 1 - .13534 = .86466$.
c. $P(X < 5) = P(X \leq 4) = .94735$.
d. Skewed right.

6.93* a. $E(X) = (200)(.02) = 4$.
b. Poisson approximation: $P(X = 0) = .0183$, binomial function: $P(X = 0) = .0176$.
c. Poisson approximation: $P(X = 1) = .0733$, binomial function: $P(X = 1) = .0718$.
e. Yes, $n \geq 20$ and $\pi \leq .05$.

6.95* a. $E(X) = n\pi = (5708)(.00128) = 7.31$.
b. Poisson approximation: $P(X < 10) = .7977$, $P(X > 5) = .7371$.
c. Yes, $n \geq 20$ and $\pi \leq .05$.

6.97* a. $P(X \geq 4) = 1 - P(X \leq 3) = .5182$.
b. Assume calls are independent.

6.99* a. $\mu = 1/.25 = 4$.
b. $P(X \leq 6) = .8220$.

6.101* a. $\mu = 1/\pi = 1/(.08) = 12.5$ cars.
b. $P(X \leq 5) = 1 - (1 - .08)^5 = .3409$.

6.103* a. $\mu = 1/\pi = 1/(.05) = 20$.
b. $P(X \geq 30) = 1 - P(X \leq 29) = .2259$.

6.105* a. $(233.1)(0.4536) = 105.734$ kg.
b. $(34.95)(0.4536) = 15.8533$ kg.

6.107 $\mu_{X_1+X_2} = \mu_1 + \mu_2 = 3420 + 390 = 3810$ ml,
$\sigma_{X_1+X_2} = \sqrt{\sigma_1^2 + \sigma_2^2} = \sqrt{10^2 + 2^2} = 10.2$ ml.

6.109* Rule 1, $\mu_{vQ+F} = v\mu_Q + F = (2225)(7) + 500 = \$16,075$.
Rule 2, $\sigma_{vQ+F} = v\sigma_Q = (2225)(2) = \$4,450$.
Rule 1, $E(PQ) = P\mu_Q = (2850)(7) = \$19,950$.
$E(TR) - E(TC) = 19,950 - 16,075 = \$3,875$.

6.111* a. $\mu_{X+Y} = \$70 + \$200 = \$270$.
b. $\sigma_{X+Y} = \sqrt{10^2 + 30^2 + 2 \times 400} = \$42.43$.
c. The variance of the total is greater than either of the individual variances.

CHAPTER 7

Note: Using Appendix C or Excel will lead to somewhat different answers.

7.1 a. D. b. C. c. C.

7.3 a. Area $= bh = (1)(.25) = .25$, so not a PDF (area is not 1).
b. Area $= bh = (4)(.25) = 1$, so could be a PDF (area is 1).
c. Area $= bh = (2)(2) = 2$, so not a PDF (area is not 1).

7.5 a. $\mu = (0 + 10)/2 = 5$, $\sigma = \sqrt{\dfrac{(10 - 0)^2}{12}} = 2.886751$.
b. $\mu = (200 + 100)/2 = 150$, $\sigma = \sqrt{\dfrac{(200 - 100)^2}{12}} = 28.86751$.
c. $\mu = (1 + 99)/2 = 50$, $\sigma = \sqrt{\dfrac{(99 - 1)^2}{12}} = 28.29016$.

7.7 A point has no area in a continuous distribution, so < or $\leq$ yields the same result.

7.9 a. 10:20:30.
b. The arrival window is a five-minute range (0,5). $\sigma = 1.4434$ min.
c. $P(\text{Bus is early}) = P(\text{arrival time} < 2 \text{ min}) = .4$.
d. $P(\text{Bus arrives between 10:19 and 10:21}) = P(1 \text{ min} \leq \text{arrival time} \leq 3 \text{ min}) = .4$.

7.11 Means and standard deviations differ (X axis scales are different) and so do $f(x)$ heights.

7.13 For samples from a *normal distribution,* we expect about 68.26% within $\mu \pm 1\sigma$, about 95.44% within $\mu \pm 2\sigma$, and about 99.73% within $\mu \pm 3\sigma$.

7.15 Using Appendix C-1:
 a. $P(0 < Z < 0.50) = .1915$.
 b. $P(-0.50 < Z < 0) = P(0 < Z < 0.50) = .1915$.
 c. $P(Z > 0) = .5000$.
 d. Probability of any point is 0.

7.17 Using Appendix C-2:
 a. $P(Z < 2.15) - P(Z < -1.22) = .9842 - .1112 = .8730$.
 b. $P(Z < 2.00) - P(Z < -3.00) = .9772 - .00135 = .97585$.
 c. $P(Z < 2.00) = .9772$.
 d. Probability of any point is 0.

7.19 Using Appendix C-2:
 a. $P(Z < -1.28) = .1003$.
 b. $P(Z > 1.28) = .1003$.
 c. $P(Z < 1.96) - P(Z < -1.96) = .975 - .025 = .95$.
 d. $P(Z < 1.65) - P(Z < -1.65) = .9505 - .0485 = .902$.

7.21 $P(Z > 1.75) = .0401$, $.0401 \times 405 = 16.24$. About 16 women ran faster than Joan.

7.23 Using Appendix C-2:
 a. $z = -1.555$ (the area .06 is halfway between .0606 and .0594).
 b. $z = .25$ (the closest area to .6 is .5987).
 c. $z = -1.48$ (closest area is .0694).

7.25 Using Appendix C-2:
 a. $-0.84 < Z < 0.84$ (the closest area is .2995).
 b. $z = 2.05$ (closest area is .9798).
 c. $-1.96 < Z < 1.96$.

7.27 =NORM.S.INV(.10) $= -1.28$. The runners must finish 1.28 standard deviations below the mean.

7.29 a. =1−NORM.DIST(592,600,5,1)
 b. =NORM.DIST(603,600,5,1)
 c. =NORM.DIST(603,600,5,1) − NORM.DIST(592,600,5,1)

7.31 a. $P(X < 300) = P(Z < 0.7143) = .7625$
 b. $P(X > 250) = 1 - P(Z < -2.86) = .9979$
 c. $P(275 < X < 310) = P(Z < 1.43) - P(Z < -1.07) = .9236 - .1423 = .7813$

7.33 $P(X \geq 24) = P(Z \geq 1.92) = 1 - .9726 = .0274$.

7.35 $P(X > .5) = .0062$ or 0.62%.

7.37 a. =NORM.INV(.8,56,4)
 b. X_{upper} =NORM.INV(.8,56,4), X_{lower} =NORM.INV(.2,56,4)
 c. =NORM.INV(.3,56,4)

7.39 a. $Z = 1.282$, $X = 13.85$.
 b. $X_L = 7.98$, $X_U = 12.02$.
 c. $Z = -0.842$, $X = 7.47$.
 d. $Z = -1.282$, $X = 6.15$.

7.41 a. $P(X > x) = .05$, $z = 1.645$, $x = 124.52$ oz.
 b. $P(X < x) = .50$, $z = 0$, $x = 114$ oz.
 c. $P(x_L < X < x_U) = .95$, $z = \pm 1.96$, $x_L = 100.28$ oz., $x_U = 127.72$ oz.
 d. $P(X < x) = .80$, $z = .842$, $x = 119.89$.

7.43 a. $Z = (8.0 - 6.9)/1.2 = 0.92$, so $P(Z < 0.92) = .8212$ (82nd percentile).
 b. $Z = 1.282$, $X = 8.44$ lbs.
 c. $Z = \pm 1.960$, $X = 4.55$ lbs to 9.25 lbs.

7.45 $P(X \leq X_L) = .25$ and $P(X \geq X_U) = .25$. Solve for X_L and X_U using $z = \pm 0.67$. $X_L = 18$ and $X_U = 21$. The middle 50% of occupied beds falls between 18 and 21.

7.47 Using Appendix C-2: Use $z = 0.52$ (closest area is .6985). Solve the following for σ: $0.52 = (\$171 - \$157)/\sigma$, $\sigma = \$26.92$.

7.49 a. =NORM.DIST(110,100,15,TRUE) − NORM.DIST(80,100,15,TRUE) = .6563.
 b. =NORM.DIST(2,0,1,TRUE) − NORM.DIST(1.5,0,1,TRUE) = .0441.
 c. =NORM.DIST(7000,6000,1000,TRUE) − NORM.DIST(4500,6000,1000,TRUE) = .7745.
 d. =NORM.DIST(450,600,100,TRUE) − NORM.DIST(225,600,100,TRUE) = .0667.

7.51 a. =1−NORM.DIST(60,40,28,TRUE) = 0.2375.
 b. =NORM.DIST(20,40,28,TRUE) = 0.2375.
 c. =1−NORM.DIST(10,40,28,TRUE) = 0.8580.

7.53 Normality OK because $n\pi = (1000)(.07) = 70 \geq 10$, $n(1 - \pi) = (1000)(.93) = 930 \geq 10$. Set $\mu = n\pi = 70$ and $\sigma = \sqrt{n\pi(1 - \pi)} = 8.0684571$.
 a. $P(X < 50) = P(Z < -2.54) = .0055$ (using $X = 49.5$).
 b. $P(X > 100) = P(Z > 3.78) = 1 - P(Z \leq 3.78) = 1 - .99992 = .00008$ (using $X = 100.5$).

7.55 Normality OK. Set $\mu = 180$, $\sigma = 4.242641$.
 a. $P(X \geq 175) = P(Z \geq -1.30) = 1 - P(Z \leq 1.30) = 1 - .0968 = .9032$ (using $X = 174.5$).
 b. $P(X < 190) = P(Z < 2.24) = .9875$ (using $X = 189.5$).

7.57 Set $\mu = \lambda = 50$ and $\sigma = \sqrt{50} = 7.071$.
 a. $P(X \geq 60) = 1 - P(Z \leq 1.34) = 1 - .9099 = .0901$ (using $X = 59.5$). Excel: =1−NORM.DIST(59.5, 50, 7.071, 1) = .0896.
 b. $P(X < 35) = P(Z \leq -2.19) = .0143$ (using $X = 34.5$). Excel: =NORM.DIST(34.5, 50, 7.071,1) = .0142.
 c. .0923 and .0108. Estimates are off in the third decimal.

7.59 Set $\mu = \lambda = 28$ and $\sigma = \sqrt{28} = 5.2915$.
 a. $P(X > 35) = 1 - P(Z \leq 1.42) = 1 - .9222 = .0788$ (using $X = 35.5$). Excel: =1−NORM.DIST(35.5, 28, 5.29, 1) = .0781.
 b. $P(X < 25) = P(Z \leq -0.66) = .2546$ (using $X = 24.5$). Excel: =NORM.DIST(24.5, 28, 5.29,1) = .2541.
 c. $\lambda = 28 \geq 10$, so OK to use normal.
 d. .0823 and .2599. Fairly close.

7.61 a. $P(X > 7) = e^{-\lambda x} = e^{-(0.3)(7)} = e^{-2.1} = .1225$.
 b. $P(X < 2) = 1 - e^{-\lambda x} = 1 - e^{-(0.3)(2)} = 1 - e^{-0.6} = 1 - .5488 = .4512$.

7.63 $\lambda = 2.1$ alarms/minute or $\lambda = .035$ alarm/second.
 a. $P(X < 60 \text{ seconds}) = 1 - e^{-\lambda x} = 1 - e^{(-0.035)(60)} = 1 - .1225 = .8775$.
 b. $P(X > 30 \text{ seconds}) = e^{-\lambda x} = e^{(-0.035)(30)} = .3499$.
 c. $P(X > 45 \text{ seconds}) = e^{-\lambda x} = e^{(-0.035)(45)} = .2070$.

7.65 a. $P(X > 30 \text{ sec}) = .2466$.
 b. $P(X \leq 15 \text{ sec}) = .5034$.
 c. $P(X > 1 \text{ min}) = .0608$.

7.67 $\lambda = 4.2$ orders/hour or $\lambda = .07$ order/minute.
 a. Set $e^{-\lambda x} = .50$, take logs, $x = 0.165035$ hr (9.9 min).
 b. Set $e^{-\lambda x} = .25$, take logs, $x = 0.33007$ hr (19.8 min).
 c. Set $e^{-\lambda x} = .10$, take logs, $x = 0.548235$ hr (32.89 min).

7.69 MTBE = 20 min/order, so $\lambda = 1/\text{MTBE} = 1/20$ order/min.
 a. Set $e^{-\lambda x} = .50$, take logs, $x = 13.86$ min.
 b. Distribution is very right-skewed.
 c. Set $e^{-\lambda x} = .25$, take logs, $x = 27.7$ min.

7.71 a. $\mu = (0 + 25 + 75)/3 = 33.3333$

b. $\sigma = \sqrt{\dfrac{0^2 + 75^2 + 25^2 - (0)(75) - (0)(25) - (75)(25)}{18}}$

$= 15.5902$

c. $P(X < 25) = (25 - 0)^2/((75 - 0)*(25 - 0)) = .3333$

d. Shaded area represents the probability.

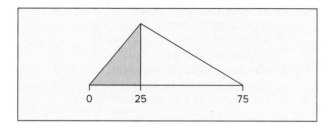

7.73 a. Discrete.

b. Continuous.

c. Continuous.

7.75 a. $\mu = (25 + 65)/2 = 45$.

b. $\sigma = 11.54701$.

c. $P(X > 45) = (65 - 45)/(65 - 25) = 0.5$.

d. $P(X > 55) = (65 - 55)/(65 - 25) = 0.25$.

e. $P(30 < X < 60) = (60 - 30)/(65 - 25) = 0.75$.

7.77 a. Right-skewed (zero low bound, high outliers likely).

b. Right-skewed (zero low bound, high outliers likely).

c. Normal.   d. Normal.

7.79 a. =NORM.S.DIST(1,1) = .8413, 84th percentile.

b. =NORM.S.DIST(2.571) = .9949, 99th percentile.

c. =NORM.S.DIST(-1.7141) = .0433, 4th percentile.

7.81 a. =NORM.INV(0.5,450,80) = 450.

b. =NORM.INV(0.25,450,80) = 396.04.

c. =NORM.INV(.1,450,80) = 347.48, =NORM.INV(.9,450,80) = 552.52.

d. =NORM.INV(0.05,450,80) = 318.42.

7.83 a. $Q_1 = \$6.77$.

b. $Q_2 = \$7.00$.

c. 90th percentile = $7.45.

7.85 a. =1−NORM.DIST(130,115,20,TRUE) = .2266.

b. =NORM.DIST(100,115,20,TRUE) = .2266.

c. =NORM.DIST(91,115,20,TRUE) = .1151.

7.87 a. 75% of 30 is 22.5 psi.

b. $P(X < 22.5) = P(Z < -3.75) = .00009$.

c. $P(28 < X < 32) = P(X < 32) - P(X < 28) = .8413 - .1587 = .6826$.

7.89 $P(1.975 < X < 2.095) = P(-2.00 < Z < +2.00) = .9544$, so 4.56% will not meet specs.

7.91 Using Appendix C-2: $P(X \geq 230) = P(Z \geq 1.64) = .0505$.

7.93 $P(X \leq 90)$ =NORM.DIST(90,84,10,TRUE) = .7257.

7.95 $P(X > 5200)$ =1−NORM.DIST(5200,4905,355,TRUE) = 0.2030.

7.97 a. 5.3% below John.

b. 69.2% below Mary.

c. 96.3% below Zak.

d. 99.3% below Frieda.

7.99* a. $P(X < 54)$ *Route A:* =NORM.DIST(54,54,6,TRUE) = .5000.

Route B: =NORM.DIST(54,60,3,TRUE) =.0228.

Probability of making it to the airport in 54 minutes or less is .5000 for *A* and .0228 for *B,* so use route *A*.

b. $P(X < 60)$ *Route A:* =NORM.DIST(60,54,6,TRUE) = .8413.

Route B: =NORM.DIST(60,60,3,TRUE) = .5000.

Probability of making it to the airport in 60 minutes or less is .8413 for *A* and .5000 for *B,* so use route *A*.

c. $P(X < 66)$ *Route A:* =NORM.DIST(66,54,6,TRUE) = .9772.

Route B: =NORM.DIST(66,60,3,TRUE) = .9772.

Probability of making it to the airport in 66 minutes or less is the same for routes *A* and *B*.

7.101 =NORM.S.INV(.20) = −0.842, so $x = \mu + z\sigma = 12.5 + (-0.842)(1.2) = 11.49$ inches.

7.103 For any normal distribution, $P(X > \mu) = .5$ or $P(X < \mu) = .5$. Assuming independent events:

a. Probability that both exceed the mean is $(.5)(.5) = .25$.

b. Probability that both are less than the mean is $(.5)(.5) = .25$.

c. Probability that one is above and one is less than the mean is $(.5)(.5) = .25$, but there are two combinations that yield this, so the likelihood is $.25 + .25 = .50$.

d. $P(X = \mu) = 0$ for any continuous random variable.

7.105 Normality OK because $n\pi \geq 10$ and $n(1 - \pi) \geq 10$. Set $\mu = n\pi = (.25)(100) = 25$ and $\sigma = $ SQRT(.25*100*(1 − .25)) = 4.3301. Then $P(X < 19.5)$ =NORM.DIST(19.5,25,4.3301,1) = .1020.

7.107 Set $\mu = n\pi = (.25)(100) = 25$ and $\sigma = $ SQRT(.25*100*(1 − .25)) = 4.3301.

a. z =NORM.S.INV(.95) = 1.645, so $x = \mu + z\sigma = 25 + 1.645(4.3301) = 32.12$.

b. z =NORM.S.INV(.99) = 2.326, so $x = \mu + z\sigma = 25 + 2.326(4.3301) = 35.07$.

c. Q_1 =NORM.INV(0.25,25,4.3301) = 22.08.

Q_2 =NORM.INV(0.5,25,4.3301) = 25.00.

Q_3 =NORM.INV(0.75,25,4.3301) = 27.92.

7.109 Set $\mu = n\pi = (.02)(1500) = 30$ and $\sigma = $ SQRT(.02*1500*(1 − .02)) = 5.4222. Then

a. $P(X > 24.5) = 1 - P(X < 24.5)$ =1−NORM.DIST(24.5,30,5.4222,1) = .8448.

b. $P(X > 40.5) = 1 - P(X < 40.5)$ =1−NORM.DIST(40.5,30,5.4222,1) = .0264.

7.111 a. $P(X > 100,000$ hrs.$) = .2397$.

b. $P(X \leq 50,000$ hrs.$) = .5105$.

c. $P(50,000 \leq X \leq 80,000) = .6811 - .5105 = .1706$.

7.113 a. $P(X \leq 3$ min$) = .6321$.

b. The distribution is skewed right, so the mean is greater than the median.

7.115* a. $\mu = (300 + 350 + 490)/3 = 380$.

b.

$$\sigma = \sqrt{\dfrac{300^2 + 350^2 + 490^2 - (300)(350) - (300)(490) - (350)(490)}{18}}$$

$= 40.21$.

c. $P(X > 400) = (490 - 400)^2/((490 - 300)(490 - 350)) = .3045$.

7.117* a. $\mu = (500 + 700 + 2100)/3 = 1100$.

b.

$$\sigma = \sqrt{\dfrac{500^2 + 700^2 + 2100^2 - (500)(700) - (500)(2100) - (700)(2100)}{18}}$$

$= 355.90$.

c. $P(X > 750) = (2100 - 750)^2/((2100 - 500)(2100 - 700)) = .8136$.

d. Shaded area represents the probability.

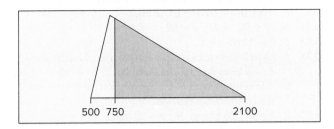

500 750   2100

CHAPTER 8

8.1 a. (96.71, 103.29).   b. (1917.75, 2082.25).
c. (496.71, 503.29).

8.3 a. (4.0252, 4.0448).   b. (4.0330, 4.0370).
c. Both are outside expected range.

8.5 a. 16.   b. 8.   c. 4.

8.7 a. $\sigma_{\bar{x}} = \sigma/\sqrt{n} = 0.25/\sqrt{10} = 0.0791$.
b. (3.345, 3.655).

8.9 a. (11.06, 16.94).
b. (33.68, 40.33).
c. (115.12, 126.88).

8.11 a. (2.3177, 2.4823).
b. (2.302, 2.498).
c. (2.2712, 2.5288).

8.13 a. (254.32, 285.68).
b. (258.91, 281.09).
c. (262.16, 277.84).
d. Width decreases as n increases.

8.15 (41.295, 41.705).

8.17 (0.27278, 0.27342).

8.19 a. 2.262, 2.2622.   b. 2.602, 2.6025.
c. 1.678, 1.6779.

8.21 a. $d.f. = 9$, $t = 2.262$, (255.6939, 284.3061).
b. $d.f. = 19$, $t = 2.093$, (260.6398, 279.3602).
c. $d.f. = 39$, $t = 2.023$, (263.6027, 276.3973).
d. Width decreases.

8.23 a. (33.01, 58.31).
b. Increase n or decrease 95%.

8.25 (1742.20, 1882.80).

8.27 a. (81.87, 88.13) for Exam 1, (82.79, 94.41) for Exam 2, (73.34, 78.66) for Exam 3.
b. Exams 1 and 2 overlap.
c. Unknown σ.

8.29 a. .0725.   b. 0212.   c. 0894.   d. 0085.

8.31 a. Yes.   b. Yes.   c. No, .08 × 100 = 8 < 10.

8.33 a. .062.   b. .0877.   c. .1216.

8.35 a. (.2556, .4752).   b. Yes, normal.

8.37 a. (.013, .083).   b. Yes, normal.

8.39 a. (47.518, 52.482).   b. (47.042, 52.958).
c. (46.113, 53.887).

8.41 (.4575, .5225).

8.43 25.

8.45 a. $\sigma = R/6 = 16.67$.   b. 43.

8.47 385.

8.49 a. $\sigma = R/6 = (22 - 17)/6 = 0.833$.   b. 31.

8.51 a. 1692.

8.53 a. 2401.

8.55 a. 125.
b. Conduct inspection of a sample of 125. AA fleet size is much larger.

8.57 (1.01338, 1.94627).

8.59 (5.88, 10.91).

8.61 a. Uneven wear.   b. (0.8332, 0.8355).
c. Normality.   d. 95.

8.63 a. (29.443, 39.634).   b. Varying methods.
c. Raisin clumps.

8.65 a. FCPF = .9333.   b. No, $n*20 = 500 > 187$.

8.67 a. (19.245, 20.69).   b. Small n.

8.69 a. ($133.013, $158.315). Using the FPCF ($134.0690, $157.2590).
b. 119.
c.* (21.26, 40.14).

8.71 a. (29.078, 29.982).
b. 116.

8.73 (4.38, 6.80), $\chi^2_L = 17.71$, $\chi^2_U = 42.56$.

8.75 a. (3.443, 3.641).
b. Outliers.
c. 6 seconds is 0.1 min. $n = 37.2$, round up to 38.

8.77 a. (.125, .255).
b. Yes.
c. 463.

8.79 136.

8.81 a. (.258, .322).
b. Yes.

8.83 a. (.595, .733).

8.85 a. Margin of error = .035 for 95% CI.
b. Greater.

8.87 a. (.393, .527).
b. Yes.

8.89 a. No.
b. (.0044, .0405).

8.91 .04 for 95% CI.

8.93 a. 385.   b. n would increase to 1537.

CHAPTER 9

9.1 a. 50.   b. 10.   c. 1.

9.3 a. H_0: Employee not using drugs.
H_1: Employee is using drugs.
b. Type I error: Test positive for drugs when not using.
Type II error: Test negative for drugs when using.
c. Employees fear Type I, while employers fear both for legal reasons.

9.5 False negative: Brakes are OK when they are not.
Consequences can be dangerous, such as a car wreck.

9.7 H_0: $\mu = 2.5$ mg vs. H_1: $\mu \neq 2.5$ mg.

9.9 H_0: $\mu \leq 4$ min vs. H_1: $\mu > 4$ min.

9.11 a. Reject in lower tail.
b. Reject in both tails.
c. Reject in upper tail.

9.13 a. $z_{calc} = 2.98$.   b. $z_{calc} = -1.58$.
c. $z_{calc} = 2.22$.

9.15 a. $z_{.025} = \pm 1.96$.   b. $z_{.10} = 1.2816$.
c. $z_{.01} = -2.3264$.

9.17 a. H_0: $\mu \leq 3.5$ mg vs. H_1: $\mu > 3.5$ mg.
b. $z_{calc} = 2.50$.
c. Yes, $z_{crit} = 2.33$ and $2.50 > 2.33$.
d. p-value = .0062.

9.19 a. $z = 1.50$, p-value = .1336.
b. $z = -2.0$, p-value = .0228.
c. $z = 3.75$, p-value = .0001.

9.21 p-value = .0062. The mean weight is heavier than it should be.

9.23 a. Reject H_0 if $z > 1.96$ or $z < -1.96$.
 b. $z = 0.78$. Fail to reject H_0.

9.25 $z = 3.26$, p-value = .0006.

9.27 H_0: $\mu = 1.967$ vs. H_1: $\mu \neq 1.967$, $t_{calc} = -1.80$, p-value = .0719. Fail to reject H_0.

9.29 a. Using $d.f. = 20$, $t_{.05} = \pm 1.725$.
 b. Using $d.f. = 8$, $t_{.01} = 2.896$.
 c. Using $d.f. = 27$, $t_{.05} = -1.703$.

9.31 a. $t_{calc} = -3.33$.
 b. $t_{calc} = -1.67$.
 c. $t_{calc} = 2.02$.

9.33 a. p-value = .0836.
 b. p-value = .0316.
 c. p-value = .0391.

9.35 a. $t = 1.5$, p-value = .1544.
 b. $t = -2.0$, p-value = .0285.
 c. $t = 3.75$, p-value = .0003.

9.37 a. H_0: $\mu \geq 400$ vs. H_1: $\mu < 400$. Reject H_0 if $t_{calc} < -1.476$, $t_{calc} = -1.977$; therefore, reject H_0.
 b. Yes, decision close at $\alpha = .05$.
 c. p-value = .0525.
 d. Could be important to a contractor.

9.39 a. H_0: $\mu \leq \$818$ vs. H_1: $\mu > \$818$. Reject H_0 if $t_{calc} > 1.729$. $t_{calc} = 4.472$, so reject H_0.
 b. p-value = .00013.

9.41 H_0: $\mu \geq 1.6$ vs. H_1: $\mu < 1.6$, $t_{calc} = 1.14$, p-value = .1306. Fail to reject H_0.

9.43 a. p-value = .1079. Fail to reject H_0.
 b. (3.226, 3.474) includes 3.25.

9.45 a. $z = 2.0$, p-value = .046.
 b. $z = 1.90$, p-value = .029.
 c. $z = 1.14$, p-value = .127.

9.47 a. No. b. No.
 c. Yes.

9.49 a. H_0: $\pi \geq .997$ versus H_1: $\pi < .997$. Reject H_0 if the p-value is less than 0.05.
 b. Yes.
 c. Type I error: Throw away a good syringe. Type II error: Keep a bad syringe.
 d. p-value = .1401 ($z = -1.08$).

9.51 a. H_0: $\pi \geq .50$ versus H_1: $\pi < .50$.
 b. p-value = .0228.
 c. Because the p-value is less than .05, reject H_0.
 d. Yes, the cost could be high if call volume is large.

9.53 p-value ≈ 0. More than half support the ban.

9.55 a. p-value = .143. Standard is being met.
 b. Less than five defects observed; cannot assume normality.

9.57 a. .4622. b. .7974.
 c. .9459.

9.59 a. .0924. b. .3721.
 c. .7497.

9.61 No, p-value = .2233.

9.63 p-value = .0465. Reject H_0.

9.65 p-value = .6202. Fail to reject H_0.

9.67 a. P(Type I error) = 0.
 b. You increase the P(Type II error).

9.69 a. H_0: User is authorized.
 H_1: User is unauthorized.
 b. Type I error: Scanner fails to admit an authorized user. Type II error: Scanner admits an unauthorized user.
 c. Type II is feared by the public.

9.71 H_0: $\pi \leq .02$ vs. H_1: $\pi > .02$.

9.73 P(Type I error) = 0.

9.75 a. Type I: deny access to authorized user; Type II: allow access to an unauthorized user.
 b. The consequences of a false rejection are less serious than a false authorization.

9.77 a. H_0: $\mu \leq 20$ min vs. H_1: $\mu > 20$ min.
 b. $t_{calc} = 2.545$.
 c. Yes. Using $d.f. = 14$, $t_{.05} = 1.761$, $2.545 > 1.761$.
 d. p-value = .0117 < .05. Yes.

9.79 a. A two-tailed test.
 b. Overfill is unnecessary while underfill is illegal.
 c. Normal (known σ).
 d. Reject if $z > 2.576$ or if $z < -2.576$.

9.81 a. H_0: $\mu \geq 90$ vs. H_1: $\mu < 90$.
 b. t_{crit} =T.INV(.01,7) = −2.998.
 c. $t = -0.92$. Fail to reject H_0.
 d. At least a symmetric population.
 e. p-value = .1936.

9.83 a. H_0: $\mu \geq 2.268$ vs. H_1: $\mu < 2.268$. If the p-value is less than .05, reject H_0. p-value = .0478. Reject H_0.
 b. Usage wears them down.

9.85 p-value = .0228. Reject H_0.

9.87 p-value = .0258. Fail to reject (close decision) H_0.

9.89 a. H_0: $\pi \leq .1$ vs. H_1: $\pi > .1$, $z_{calc} = 0.8819$. $0.8819 < 1.645$ (z_{crit}); therefore, fail to reject H_0.
 b. p-value = .1889.

9.91 p-value = .0192. Reject H_0. Important to players and universities.

9.93 p-value = .0017. Reject H_0.

9.95 a. Yes, $z = 1.95$, p-value = .0253. b. Yes.

9.99 a. H_0: $\mu \geq 880$ vs. H_1: $\mu < 880$.
 b. $t_{crit} = -1.86$.
 c. Because $t_{calc} = -1.731 > -1.86$, do not reject H_0.

9.101 p-value = .0794. Fail to reject H_0.

9.103 H_0: $\pi = .50$ vs. H_1: $\pi > .50$. $P(X \geq 10 \mid n = 16, \pi = .5) = .2272$. Fail to reject H_0.

9.105 a. (0, .0125).
 b. $np < 10$.
 c. Goal is being achieved.

9.107 $\beta = 1 -$ power. The power values are:
 $n = 4$: .2085, .5087, .8038, .9543.
 $n = 16$: .5087, .9543, .9996, 1.0000.

9.109 a. H_0: $\mu \leq 106$ vs. H_1: $\mu > 106$. $t_{calc} = 130.95$; so reject the null hypothesis.
 b. H_0: $\sigma^2 \geq .0025$ vs. H_1: $\sigma^2 < .0025$. $\chi^2 = 12.77$; therefore, we would fail to reject the null hypothesis.

CHAPTER 10

Note: Results from Excel except as noted (may not agree with Appendix C, D, or E due to rounding or use of exact $d.f.$).

10.1 a. H_0: $\mu_1 - \mu_2 \geq 0$ vs. H_1: $\mu_1 - \mu_2 < 0$, $t = -2.148$, $d.f. = 28$, $t_{.025} = -2.048$, p-value = .0202, so reject H_0.

b. $H_0: \mu_1 - \mu_2 = 0$ vs. $H_1: \mu_1 - \mu_2 \neq 0$, $t = -1.595$, $d.f. = 39$, $t_{.05} = \pm 2.023$, p-value $= .1188$, so can't reject H_0.

c. $H_0: \mu_1 - \mu_2 \leq 0$ vs. $H_1: \mu_1 - \mu_2 > 0$, $t = 1.935$, $d.f. = 27$, $t_{.05} = 1.703$, p-value $= .0318$, so reject H_0.

10.3 a. $H_0: \mu_1 - \mu_2 = 0$ vs. $H_1: \mu_1 - \mu_2 \neq 0$. Using $d.f. = 190$, $t_{.005} = 2.602$, $t_{calc} = -6.184 < -2.602$. Reject H_0.

b. Using $d.f. = 190$, p-value $= 3.713\text{E}-09$.

10.5 a. $H_0: \mu_1 - \mu_2 \leq 0$ vs. $H_1: \mu_1 - \mu_2 > 0$, $t = 1.902$, $d.f. = 29$, $t_{.01} = 2.462$, can't reject H_0.

b. p-value $= .0336$. Would be significant at $\alpha = .05$.

10.7 $H_0: \mu_1 - \mu_2 = 0$ vs. $H_1: \mu_1 - \mu_2 \neq 0$. $t = -3.55$, $d.f. = 11$, p-value $= .0045$. Reject H_0.

10.9 $H_0: \mu_S - \mu_F \geq 0$ vs. $H_1: \mu_S - \mu_F < 0$, $t_{calc} = -1.813$, p-value $= .0424 < .05$. Reject H_0.

10.11 a. $(-1.163, 0.830)$.

b. $(-1.151, 0.791)$.

c. Assumptions did not change conclusion.

d. $\mu_1: (7.95, 9.33)$, $\mu_2: (8.10, 9.54)$.

10.13 $H_0: \mu_d \leq 0$ vs. $H_1: \mu_d > 0$, $t = 1.93$, $d.f. = 6$, and p-value $= .0509$, so can't quite reject H_0 at $\alpha = .05$.

10.15 $H_0: \mu_d = 0$ vs. $H_1: \mu_d \neq 0$, $t = -1.112$, $d.f. = 9$, and p-value $= .1475$, so do not reject H_0 at $\alpha = .05$.

10.17 $H_0: \mu_d \leq 0.5$ vs. $H_1: \mu_d > 0.5$, $t_{calc} = 1.82$, p-value $= .0404 < .05$. Reject H_0.

10.19 $H_0: \mu_d = 0$ vs. $H_1: \mu_d \neq 0$, $t = -1.71$, $d.f. = 7$, and p-value $= .1307$, so can't reject H_0 at $\alpha = .01$.

10.21 a. $z_{calc} = -1.310$, p-value $= .0951$.

b. $z_{calc} = 1.787$, p-value $= .0370$.

c. $z_{calc} = 1.577$, p-value $= .1149$.

10.23 a. $H_0: \pi_1 - \pi_2 \geq 0$ vs. $H_1: \pi_1 - \pi_2 < 0$, $\overline{p} = .4200$, $z = -2.431$, $z_{.01} = -2.326$, p-value $= .0075$, so reject at $\alpha = .01$.

b. $H_0: \pi_1 - \pi_2 = 0$ vs. $H_1: \pi_1 - \pi_2 \neq 0$, $\overline{p} = .37500$, $z = 2.263$, $z_{.05} = \pm 1.645$, p-value $= .0237$, so reject at $\alpha = .10$.

c. $H_0: \pi_1 - \pi_2 \geq 0$ vs. $H_1: \pi_1 - \pi_2 < 0$, $\overline{p} = .25806$, $z = -1.706$, $z_{.05} = -1.645$, p-value $= .0440$, so reject at $\alpha = .05$.

10.25 a. $H_0: \pi_1 - \pi_2 \geq 0$ vs. $H_1: \pi_1 - \pi_2 < 0$, $p_c = .38$, $z_{calc} = -2.060$.

b. $z_{.01} = -2.326$, can't reject at $\alpha = .01$ (close decision).

c. p-value $= .0197$

d. Normality OK because $n_1 p_1 = 66$, $n_2 p_2 = 86$; both exceed 10.

10.27 $H_0: \pi_1 - \pi_2 = 0$ vs. $H_1: \pi_1 - \pi_2 \neq 0$, $\overline{p} = .11$, $z = 2.021$, $z_{.025} = \pm 1.960$ ($p = .0432$), so reject at $\alpha = .05$ (close decision).

10.29 a. $H_0: \pi_1 - \pi_2 = 0$ vs. $H_1: \pi_1 - \pi_2 \neq 0$, $p_1 = .07778$, $p_2 = .10448$, $\overline{p} = .08502$, $z = -0.669$, critical value is $z_{.025} = 1.960$, and p-value $= .5036$, so cannot reject at $\alpha = .05$.

b. Normality not OK because $n_1 p_1 = 14$, but $n_2 p_2 = 7$.

10.31 a. $H_0: \pi_1 - \pi_2 \leq .10$ vs. $H_1: \pi_1 - \pi_2 > .10$, $p_1 = .28125$, $p_2 = .14583$, $\overline{p} = .22321$, $z = 0.66$, and critical value is $z_{.05} = 1.645$.

b. p-value $= .2546$, so cannot reject at $\alpha = .05$

10.33 $(-.1584, .1184)$.

10.35 $(.0063, .1937)$.

10.37 a. $H_0: \sigma_1^2 = \sigma_2^2$ vs. $H_1: \sigma_1^2 \neq \sigma_2^2$. Reject H_0 if $F > 4.76$ or $F < .253$ ($df_1 = 10$, $df_2 = 7$). $F = 2.54$, so we fail to reject the null hypothesis.

b. $H_0: \sigma_1^2 = \sigma_2^2$ vs. $H_1: \sigma_1^2 < \sigma_2^2$. Reject H_0 if $F < .264$ ($df_1 = 7$, $df_2 = 7$). $F = .247$, so we reject the null hypothesis.

c. $H_0: \sigma_1^2 = \sigma_2^2$ vs. $H_1: \sigma_1^2 > \sigma_2^2$. Reject H_0 if $F > 2.80$ ($df_1 = 9$, $df_2 = 12$). $F = 19.95$, so we reject the null hypothesis.

10.39 $H_0: \sigma_1^2 = \sigma_2^2$ vs. $H_1: \sigma_1^2 < \sigma_2^2$. Reject H_0 if $F < .355$ ($df_1 = 11$, $df_2 = 11$). $F = .103$, so we reject the null hypothesis. The new drill has a reduced variance.

10.41 a. $H_0: \pi_M - \pi_W = 0$ vs. $H_1: \pi_M - \pi_W \neq 0$. Reject the null hypothesis if $z < -1.645$ or $z > 1.645$.

b. $p_M = .60$ and $p_W = .6875$.

c. $z = -.69$, p-value $= .492$. The sample does not show a significant difference in proportions.

d. Normality can be assumed because both $n_1 p_1 \geq 10$ and $n_2 p_2 \geq 10$.

10.43 a. $H_0: \pi_1 - \pi_2 \leq 0$ vs. $H_1: \pi_1 - \pi_2 > 0$.

b. Reject if $z > z_{.05} = 1.645$.

c. $p_1 = .98000$, $p_2 = .93514$, $\overline{p} = .95912$, $z = 4.507$.

d. Reject at $\alpha = .05$.

e. p-value $= .0000$.

f. Normality is OK because $n_1(1 - \pi_1) = 17$ and $n_2(1 - \pi_2) = 48$, both > 10.

10.45 a. $H_0: \pi_1 - \pi_2 = 0$ vs. $H_1: \pi_1 - \pi_2 \neq 0$.

b. $p_1 = .17822$, $p_2 = .14300$, $\overline{p} = .14895$, $z = 1.282$, p-value $= .2000$. Because z is within ± 1.960 for a two-tail test at $\alpha = .05$ and p-value exceeds .05, we fail to reject H_0.

10.47 a. $H_0: \pi_1 - \pi_2 = 0$ vs. $H_1: \pi_1 - \pi_2 \neq 0$, $p_1 = .38492$, $p_2 = .48830$, $\overline{p} = .44444$, $z = -2.506$. Because z does not exceed ± 2.576, we cannot reject H_0.

b. Two-tailed p-value $= .0122$.

c. Normality OK because $n_1 p_1 = 97$, $n_2 p_2 = 167$; both exceed 10.

d. Gender interests may imply different marketing strategies.

10.49 a. $H_0: \pi_1 - \pi_2 \geq 0$ vs. $H_1: \pi_1 - \pi_2 < 0$, $p_1 = .14914$, $p_2 = .57143$, $\overline{p} = .21086$, $z = -8.003$. Because $z < -2.326$, we conclude that pilots are more likely to approve of night-flying without noninstrument rating.

b. Left-tailed p-value $= .0000$.

c. Normality assumption OK because $n_1 p_1 = 61$, $n_2(1 - p_2) = 30$; both exceed 10.

10.51 $H_0: \mu_B - \mu_A \geq 0$ vs. $H_1: \mu_B - \mu_A < 0$, $t_{calc} = -1.581$, p-value $= .0607 > .05$. Fail to reject H_0.

10.53 a. $H_0: \pi_1 - \pi_2 \geq 0$ vs. $H_1: \pi_1 - \pi_2 < 0$. Reject the null hypothesis if $z < -1.645$ or p-value $< .05$. p-value $= .0914$, so fail to reject H_0.

b. Yes, normality is met.

10.55 $H_0: \mu_1 - \mu_2 \leq 0$ vs. $H_1: \mu_1 - \mu_2 > 0$. Assuming equal variances, $t = 4.089$ with $d.f. = 84$. Because the p-value is .0000, reject H_0 at $\alpha = .01$.

10.57 a. $H_0: \pi_1 - \pi_2 \geq 0$ vs. $H_1: \pi_1 - \pi_2 < 0$, $p_1 = .1402$, $p_2 = .2000$, $\overline{p} = .16396$.

b. $z = -2.777$ and left-tailed p-value $= .0027$. Because $z < -2.326$, reject H_0.

c. Normality OK because $n_1 p_1 = 104$, $n_2 p_2 = 98$; both exceed 10.

d. Many people can't afford them or lack insurance to pay for them.

10.59 a. $H_0: \mu_1 - \mu_2 \le 0$ vs. $H_1: \mu_1 - \mu_2 > 0$. Assuming unequal variances, $t = 1.718$ with $d.f. = 16$ (using Welch's adjustment). Because the p-value is .0525, we fail to reject H_0 at $\alpha = .05$.

b. If we had looked at the same firm in each year, the test would have more power.

10.61 a. Dot plots suggest that the new bumper has less downtime, but variation is similar.

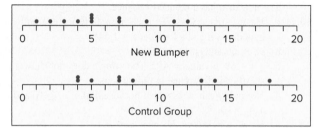

b. $H_0: \mu_1 - \mu_2 \le 0$ vs. $H_1: \mu_1 - \mu_2 > 0$.

c. Assuming equal variances, reject H_0 if $t < -1.729$ with $d.f. = 19$.

d. $\bar{x}_1 = 5.917$, $s_1 = 3.423$, $\bar{x}_2 = 8.889$, $s_2 = 4.961$, $s_p^2 = 17.148$, $t = -1.63$, p-value $= .0600$, so fail to reject H_0 at $\alpha = .05$.

10.63 a. $H_0: \mu_1 - \mu_2 \le 0$ vs. $H_1: \mu_1 - \mu_2 > 0$, $t = 7.08$, $d.f. = 28$, p-value ≈ 0. Reject H_0.

b. $H_0: \sigma_1^2 = \sigma_2^2$ vs. $H_1: \sigma_1^2 \ne \sigma_2^2$. Reject H_0 if $F < .3357$ or $F > 2.9786$ ($df_1 = 14$, $df_2 = 14$). $F = 2.778$, so we fail to reject the null hypothesis.

10.65 a. $H_0: \mu_1 - \mu_2 \le 0$ vs. $H_1: \mu_1 - \mu_2 > 0$.

b. Reject H_0 if $t > 2.438$ with $d.f. = 35$.

c. $\bar{x}_1 = 117{,}853$, $s_1 = 10{,}115$, $\bar{x}_2 = 98{,}554$, $s_2 = 14{,}541$, $s_p^2 = 152{,}192{,}286$, $t = 4.742$.

d. Reject H_0 at $\alpha = .01$. Men are paid significantly more.

e. p-value $= .0000$. Unlikely result if H_0 is true.

10.67 a. Dot plots show strong skewness, but means could be similar.

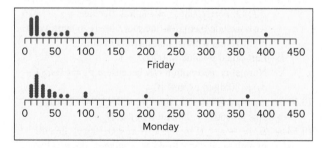

b. $H_0: \mu_1 - \mu_2 = 0$ vs. $H_1: \mu_1 - \mu_2 \ne 0$. Assume equal variances.

c. Reject H_0 if $t > 2.663$ or if $t < -2.663$ with $d.f. = 58$.

d. $\bar{x}_1 = 50.333$, $s_1 = 81.684$, $\bar{x}_2 = 50.000$, $s_2 = 71.631$, $s_p^2 = 5{,}901.667$. Because $t = .017$, we cannot reject H_0.

e. p-value $= .9866$. Sample result well within chance range.

10.69 a. $H_0: \mu_1 - \mu_2 = 0$ vs. $H_1: \mu_1 - \mu_2 \ne 0$.

b. For equal variances and $d.f. = 55$, reject H_0 if $t < -2.004$.

c. Because $t = -3.162$ ($p = .0025$), we reject H_0 at $\alpha = .05$. Mean sales are lower on the east side.

10.71 $H_0: \sigma_1^2 = \sigma_2^2$ vs. $H_1: \sigma_1^2 \ne \sigma_2^2$, $df_1 = 30$, $df_2 = 29$. For $\alpha/2 = .025$, $F_R = F_{30,29} = 2.09$, and $F_L = 1/F_{29,30} \cong 1/F_{25,30} = 1/2.12 = .48$. Test statistic is $F = (13.482)^2/(15.427)^2 = 0.76$, so we can't reject H_0.

10.73 $H_0: \mu_d = 0$ vs. $H_1: \mu_d \ne 0$, $t = -0.87$, p-value $= .4154$ (from MegaStat). Fail to reject the null hypothesis.

10.75 a. $H_0: \mu_1 - \mu_2 = 0$ vs. $H_1: \mu_1 - \mu_2 \ne 0$, $t = -1.10$, $d.f. = 22$, p-value $= .2839$. Fail to reject H_0.

b. $H_0: \sigma_1^2 = \sigma_2^2$ versus $H_1: \sigma_1^2 \ne \sigma_2^2$. Reject H_0 if $F < 0.188$ or $F > 5.3197$ ($df_1 = 11$, $df_2 = 11$). $F = 2.59$, so we fail to reject the null hypothesis.

10.77 a. $H_0: \sigma_1^2 \le \sigma_2^2$, $H_1: \sigma_1^2 > \sigma_2^2$, $df_1 = 11$, $df_2 = 11$. For a right-tail test at $\alpha = .025$, Appendix F gives $F_R = F_{11,11} \cong F_{10,11} = 3.53$. The test statistic is $F = (2.9386)^2/(0.9359)^2 = 9.86$, so conclude that Portfolio A has a greater variance than Portfolio B.

b. These are independent samples. $H_0: \mu_1 - \mu_2 = 0$ vs. $H_1: \mu_1 - \mu_2 \ne 0$. $\bar{x}_1 = 8.5358$, $s_1 = 2.9386$, $\bar{x}_2 = 8.1000$, $s_2 = .9359$. Assuming unequal variances with $d.f. = 13$ (with Welch's adjustment), we get $t = 0.49$ with p-value $= .6326$, so we cannot reject H_0 at $\alpha = .025$.

10.79 $(-.0153, .2553)$. Yes, the interval includes zero.

10.81 a. $(-2.51, 0.11)$.

b. No.

10.83 a. $(-.4431, -.0569)$.

b. Normality is not met.

10.85 a. $H_0: \mu_M - \mu_F \le 0$ vs. $H_1: \mu_M - \mu_F > 0$.

b. $t_{calc} = 2.616$.

c. Reject H_0 if $t_{calc} > 2.5835$.

d. p-value $= .0094 < .01$. Reject H_0.

e. $F_{calc} = 7.169$, p-value $= .0047$. Assume unequal variances.

10.87 $H_0: \pi_A - \pi_B \le 0$ vs. $H_1: \pi_A - \pi_B > 0$. Reject the null hypothesis if $z_{calc} > 2.326$. $z_{calc} = 2.294 < 2.326$. Fail to reject the null hypothesis.

10.89 $H_0: \mu_d \le 0$ vs. $H_1: \mu_d > 0$. Reject the null hypothesis if $t_{calc} > 2.132$. $t_{calc} = 1.851 < 2.132$. Fail to reject the null hypothesis.

CHAPTER 11

11.1 a. 40.

b. 5.

c. $H_0: \mu_1 = \mu_2 = \mu_3 = \mu_4 = \mu_5$ vs. H_1: At least one mean is different.

d. $F_{crit} = 2.6415$.

e. $F_{calc} = 1.80$.

f. Fail to reject H_0.

11.3 a. $H_0: \mu_1 = \mu_2 = \mu_3 = \mu_4 = \mu_5$ vs. H_1: At least one mean is different.

b. $df_1 = 4$, $df_2 = 25$.

c. $F_{crit} = 2.18$.

d. Reject H_0.

e. p-value =F.DIST.RT(2.447,4,25) = .0726

11.5 a. $H_0: \mu_A = \mu_B = \mu_C$ vs. H_1: Not all means are equal.

b. $df_1 = 2$, $df_2 = 12$.

c. For $\alpha = .05$, $F_{2,12} = 3.89$.

d. $F_{calc} = 5.31$; reject H_0 at $\alpha = .05$.

e. p-value $= .0223$.

f. Plant B mean likely higher, Plant C lower.

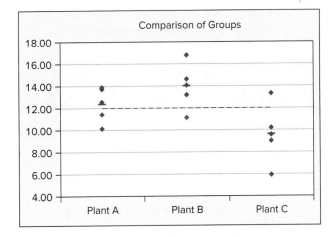

ANOVA table: Two factor without replication

Source	SS	df	MS	F	p-value
Treatments (plant)	216.25	3	72.083	41.19	.0002
Blocks (date)	30.17	2	15.083	8.62	.0172
Error	10.50	6	1.750		
Total	256.92	11			

Mean	n	Std. Dev	Factor Level
20.333	3	1.528	Plant 1
18.000	3	2.000	Plant 2
29.000	3	2.646	Plant 3
25.000	3	2.646	Plant 4
21.500	4	4.041	04-Mar
25.250	4	5.377	11-Mar
22.50	4	5.508	18-Mar

11.7 a. $H_0: \mu_1 = \mu_2 = \mu_3 = \mu_4$ vs. H_1: Not all equal.
b. $df_1 = 3$, $df_2 = 24$.
c. For $\alpha = .05$, $F_{3,24} = 3.01$.
d. $F_{calc} = 3.52$; reject H_0 at $\alpha = .05$. GPAs not the same.
e. p-value = .0304.
f. Marketing and HR likely higher, accounting and finance lower.

11.9 a. 6.
b. $c = 4$, $n - c = 30$.
c. $T_{4,30} = 2.72$.

11.11 a. 3.
b. $c = 3$, $n - c = 12$.
c. $T_{3,12} = 2.74$. (Table 11.4 used.)
d. Only Plant B and Plant C differ at $\alpha = .05$ ($t = 3.23$) using MegaStat Tukey test.

11.13 a. 6.
b. $c = 4$, $n - c = 24$.
c. $T_{4,24} = 2.80$. (Table 11.4 used.)
d. Only marketing and accounting differ at $\alpha = .05$ (Tukey $t = 3.00$).

11.15 a. $H_0: \sigma_1^2 = \sigma_2^2 = \sigma_3^2$ vs. H_1: At least one variance is different.
b. $df_1 = 3$, $df_2 = 4$.
c. $H_{3,4} = 15.5$.
d. $H_{calc} = 7.68$.
e. Fail to reject equal variances.

11.17 a. $df_1 = c = 3$, $df_2 = n/c - 1 = 4$.
b. $H_{critical} = 15.5$.
c. $H_{calc} = 7.027/2.475 = 2.839$.
d. Fail to reject equal variances.

11.19 a. $df_1 = c = 4$, $df_2 = n/c - 1 = 6$.
b. $H_{critical} = 10.4$.
c. $H_{calc} = 8.097$.
d. Fail to reject equal variances.

11.21 a. H_0: Mean absenteeism same in all four plants, H_1: Mean absenteeism not the same in all four plants. (Date is the blocking factor.)
b. Plant means differ: Plants 1, 2 below overall mean; Plants 3, 4 above.
c. $F = 41.19$ ($p = .0002$). Blocking factor $F = 8.62$ ($p = .0172$) also significant.

11.23 a. H_0: Mean driving times are the same for all five drivers vs. H_1: Mean times are not the same. (Hotel is the blocking factor.)
b. The driver is barely significant and hotel is not significant. Driver C appears to have a slightly higher average than the other drivers.
c. Driver ($F = 3.26$, $p = .05$), Hotel ($F = 1.11$, $p = .3824$).

11.25 a. *Rows:* H_0: Year means the same vs. H_1: Year means differ. *Columns:* H_0: Portfolio type means the same vs. H_1: Portfolio type means differ. *Interaction(Year $\times$ Type)*: H_0: No interaction vs. H_1: There is an interaction effect.
b. ANOVA table: Two factor with replication (5 observations per cell).
c. Year ($F = 66.82$, $p < .0001$) is highly significant. Portfolio ($F = 5.48$, $p = .0026$) differs significantly, and significant interaction ($F = 4.96$, $p = .0005$).
d. p-values are very small, indicating significant effects at $\alpha = .05$. Year is strongest result.
e. Interaction plot lines do cross and support the interaction found and reported above.

ANOVA table: Two factor with replication (5 observations per cell)

Source	SS	df	MS	F	p-value
Factor 1 (portfolio)	1,191.584	2	595.7922	66.82	<.0001
Factor 2 (year)	146.553	3	48.8511	5.48	.0026
Interaction	265.192	6	44.1986	4.96	.0005
Error	427.980	48	8.9162		
Total	2,031.309	59			

11.27 a. *Rows:* H_0: Age group means same vs. H_1: Age group means differ. *Columns:* H_0: Region means same vs. H_1: Region means differ. *Interaction (Age $\times$ Region)*: H_0: No interaction vs. H_1: Interaction.
b. ANOVA table: Two factor with replication (5 observations per cell).

c. Age group means ($F = 36.96$, $p < .0001$) differ dramatically. Region means ($F = 0.5$, $p = .6493$) don't differ significantly. Significant interaction ($F = 3.66$, $p = .0010$).

d. Age group p-value indicates very strong result; interaction is significant at $\alpha = .05$.

e. Interaction plot lines do cross and support the interaction found and reported above.

11.29 a. H_0: $\mu_1 = \mu_2 = \mu_3 = \mu_4$ vs. H_1: Not all the means are equal.

b. Graph shows mean Freshmen GPA is lower than overall mean.

c. $F = 2.36$ ($p = .1000$); fail to reject H_0 at $\alpha = .05$, no significant difference among GPAs.

d. Reject H_0 if $F > F_{3,21} = 3.07$.

e. Differences in mean grades large enough (.4 to .7) to matter, but not significant, so cannot be considered important.

f. Large variances within groups and small samples rob the test of power, suggests larger sample within each group.

g. Tukey confirms no significant difference in any pairs of means.

h. $H_{calc} = (0.6265)^2/(0.2826)^2 = 4.91$, which is less than Hartley's critical value 13.7 with $df_1 = c = 4$ and $df_2 = n/c - 1 = 5$, so conclude equal variances.

11.31 a. H_0: $\mu_1 = \mu_2 = \mu_3$ vs. H_1: Not all means are equal.

b. Graph suggests Type B a bit lower, C higher than the overall mean.

c. $F = 9.44$ ($p = .0022$), so there is a significant difference in mean cell outputs.

d. Reject H_0 if $F > F_{2,15} = 3.68$.

e. Small differences in means but could be important in a large solar cell array.

f. Sounds like a controlled experiment and variances are small, so a small sample suffices.

g. Tukey test shows that C differs from B at $\alpha = .01$ and from A at $\alpha = .05$.

h. $H_{calc} = (4.57)/(4.00) = 1.14$, less than Hartley's 10.8 with $df_1 = c = 3$ and $df_2 = n/c - 1 = 5$; conclude equal variances.

11.33 a. H_0: $\mu_1 = \mu_2 = \mu_3 = \mu_4$ vs. H_1: Not all means equal.

b. Graph shows B higher, D lower.

c. $F = 1.79$ ($p = .1857$), so at $\alpha = .05$, no significant difference in mean waiting times.

d. Reject H_0 if $F > F_{3,18} = 3.16$.

e. Differences in means might matter to patient, but not significant so can't be considered important.

f. Variances large, samples small, so test has low power.

g. Tukey test shows no significant differences in pairs of means.

h. $H_{calc} = (11.90)^2/(6.74)^2 = 3.121$, less than the Hartley's 20.6 with $df_1 = c = 4$ and $df_2 = n/c - 1 = 4$, so conclude equal variances.

11.35 a. *Columns:* H_0: Surface has no effect on mean braking distance vs. H_1: Surface does affect distance. *Rows:* H_0: Pumping method has no effect on mean braking distance vs. H_1: Pumping method does affect distance.

b. Graph suggests differences (ice is greater than the other two).

c. Surface: $F = 134.39$ ($p = .0002$); reject H_0. Surface has a significant effect on mean stopping distance.

Braking method: $F = 0.72$ ($p = .5387$); cannot reject H_0. Braking method has no significant effect on stopping distance.

d. Reject H_0 if $F > F_{2,4} = 6.94$.

e. For surface, differences large enough to be very important in preventing accidents.

f. Replication would be desirable, if tests are not too costly.

g. Tukey test shows a difference between ice and the other two surfaces.

h. $H_{calc} = 1.37$ (for Method) and $H_{calc} = 14.5$ (for Surface). Cannot reject the hypothesis of equal variances.

11.37 a. H_0: $\mu_1 = \mu_2 = \mu_3 = \mu_4 = \mu_5$ vs. H_1: Not all means equal.

b. Graph suggests that Chalmers is higher and Ulysses is lower.

c. $F_{calc} = 6.19$ ($p = .0019$), so reject H_0. There are significant differences in means.

d. Reject H_0 if $F_{calc} > F_{4,21} = 2.84$ at $\alpha = .05$.

e. Significant and probably important to clients.

f. Sample may be limited by number of clinics in each town.

g. Chalmers differs from all except Villa Nueve, while other means do not differ significantly.

h. $H_{calc} = (11.171)^2/(6.850)^2 = 2.659$; does not exceed Hartley's $H_{crit} = 25.2$ with $df_1 = c = 5$ and $df_2 = n/c - 1 = 26/5 - 1 = 4$, so conclude equal variances.

11.39 a. H_0: $\mu_1 = \mu_2 = \mu_3 = \mu_4 = \mu_5$ vs. H_1: Not all means equal.

b. Graph shows no differences in means.

c. $F = 0.39$ ($p = .8166$); cannot reject H_0. No significant difference in the mean dropout rates.

d. Reject H_0 if $F > F_{4,45} = 2.61$.

e. Differences not significant, not important.

f. Could look at a different year, but sample is already fairly large.

g. Tukey shows no significant differences in pairs.

h. $H_{calc} = (10.585)^2/(3.759)^2 = 7.93$; exceeds Hartley's 7.11 with $df_1 = c = 5$ and $df_2 = n/c - 1 = 9$, so conclude unequal variances.

11.41 In this replicated two-factor ANOVA, the response (days until expiration) is significantly related to *Brand* (row factor, $F_{calc} = 3.39$, $p = .0284$) and strongly related to *Store* (column factor, $F_{calc} = 7.36$, $p = .0021$). Freshness is important to customers. Sample sizes could be increased because bag inspection is not difficult.

11.43 In this two-factor ANOVA without replication (random-ized block), the response (trucks produced per shift) is weakly related to *Plant* (row factor, $F_{calc} = 2.72$, $p = .0912$) and strongly related to *Day* (column factor, $F_{calc} = 9.18$, $p = .0012$). Productivity is important to car companies. Sample sizes could be increased because daily production is routinely recorded for each shift.

11.45 a. Two-factor ANOVA.

b. Instructor gender p-value ($p = .43$) exceeds $\alpha = .10$; instructor gender means do not differ. Student gender p-value ($p = .24$) exceeds $\alpha = .10$; student gender means do not differ (p-value $> \alpha = .10$). For interaction, the p-value ($p = .03$) suggests significant interaction effect (at $\alpha = .05$).

c. Unlikely that a gender effect was overlooked due to sample size; test should have very good power.

11.47 a. $F_{calc} = (1069.17)/(12270.28) = 0.0871$.

b. p-value $= .9666$.

c. $F_{3,36} = 2.87$.

d. No significant difference in means.

11.49 a. Two-factor ANOVA without replication (randomized block).

b. 4 plant locations, 4 noise levels, 1 observation per cell.

c. At $\alpha = .05$, plant location is not significant ($p = .1200$), while noise level is quite significant ($p = .0093$).

11.51 a. Two-factor, either factor could be of research interest.

b. Pollution affected freeway ($F = 24.90$, $p = .0000$) and by time of day ($F = 21.51$, $p = .0000$).

c. Variances for freeway 2926.7 to 14333.7, for time of day 872.9 to 14333.6. Ratios are large (4.90 and 16.42), suggesting possibly unequal variances.

d. For freeway, $df_1 = c = 4$ and $df_2 = n/c - 1 = 20/4 - 1 = 4$, Hartley's critical value is 20.6, so conclude equal variances. Time of day, $df_1 = c = 5$ and $df_2 = n/c - 1 = 20/5 - 1 = 3$, Hartley's critical value is 50.7, so conclude equal variances.

11.53 a. One factor.

b. Between groups $df_1 = 4$ and $df_1 = c - 1$, so $c = 5$ bowlers.

c. p-value 0.000; reject null, conclude at least two samples are significantly different.

d. Sample variances 83.66 to 200.797, $F_{max} = 200.797/77.067 = 2.61$. Hartley's test, $df_1 = c = 5$ and $df_2 = n/c - 1 = 67/5 - 1 = 12$; critical value is 5.30. Not enough variation to reject null hypothesis of homogeneity.

CHAPTER 12

12.1 For each sample: H_0: $\rho = 0$ vs. H_1: $\rho \neq 0$

Sample	df	r	t	t_α	Decision
a	18	.45	2.138	2.101	Reject
b	28	−.35	−1.977	1.701	Reject
c	5	.6	1.677	2.015	Fail to reject
d	59	−.3	−2.416	2.39	Reject

12.3 b. −.7328.

c. $t_{.025} = 3.182$.

d. $t = −1.865$, fail to reject.

e. p-value $= .159$.

12.5 b. .531.

c. 2.131.

d. 2.429.

e. Yes, reject.

12.7 a. Each additional *SquareFeet* increases price $150.

b. $425,000.

c. No, *SquareFeet* cannot be zero.

12.9 a. For each additional year in median age, there is an average of 35.3 fewer cars stolen per 100,000 people.

b. 255 cars per 100,000 people.

c. No, median age cannot be zero.

12.11 a. Each one-unit increase in *PowerDistanceIndex* means an increase of 1.75 international franchises.

b. 101.25.

c. No, cannot have negative number of franchises.

12.13 a. Earning an extra $1,000 raises home price by $2,610.

b. No.

c. $181,800, $312,300.

12.15 a. Blazer: Each year reduces price by $1,050. Silverado: Each year reduces price by $1,339.

b. Intercept could indicate price of new car.

c. $10,939.

d. $15,896.

12.17 a. $\hat{y} = 14.42$, $e = 3.58$, underestimate.

b. $\hat{y} = 13.3$, $e = −7.3$, overestimate.

12.19 b. *Wait Time* $= 458 − 18.5$ *ClerkWindowsOpen*.

d. $R^2 = .5369$.

12.21 $\hat{y} = 0.458x + 11.155$, $R^2 = .2823$.

12.23 b. H_0: $\beta_1 = 0$ vs. H_0: $\beta_1 \neq 0$.

c. p-value $= .0269$, (1.3192, 10.8522).

d. Slope is significantly different from zero because the p-value is less than .05.

12.25 b. $\hat{y} = 557.45 + 3.00x$.

c. (1.2034, 4.806).

d. H_0: $\beta_1 \leq 0$ vs. H_1: $\beta_1 > 0$, p-value $= .0009$, reject H_0.

12.27 a. $\hat{y} = 1.8064 + .0039x$.

b. Intercept: $1.8064/.6116 = 2.954$, slope: $.0039/.0014 = 2.786$ (may be off due to rounding).

c. $d.f. = 10$, $t_{.025} = \pm2.228$.

12.29 a. $\hat{y} = 10.960 − 0.053x$.

b. (−0.1946, 0.0886). Interval does contain zero; slope is not significantly different from zero.

c. t test p-value $= .4133$. Conclusion: Slope is not significantly different from zero.

d. F statistic p-value $= .4133$. Conclusion: No significant relationship between variables.

e. $0.74 = (−0.863)^2$.

12.31 a. $\hat{y} = −31.1895 − 4.9322x$.

b. (2.502, 7.362). Interval does not contain zero; the slope is greater than zero.

c. t test p-value is 0.0011. Conclusion: The slope is positive.

d. F statistic p-value $= .0011$. Conclusion: Significant relationship between variables.

e. $(4.523)^2 = 20.46$.

12.33 a. 95% confidence interval: (0.3671, 1.1477); 95% prediction interval: (−0.4662, 1.9810).

b. 95% confidence interval for μ_Y: (−0.205, 1.4603).

c. The second interval is much wider.

12.35 Normality assumption is reasonable. Residual plot shows signs of nonconstant variance.

12.37 a. $e_i = 28$. b. $e_i = −21$. c. $e_i = 7$.

12.39 a. 33.02 mpg. b. 5.13. c. 2.527.

d. This is an unusual observation, not an outlier.

12.41 a. .16. b. 0.10. c. 0.02.

12.43 a. $h = .1850 > .0541$, high leverage.

b. $h = .0304 < .0541$, not high leverage.

c. $h = .0952 > .0541$, high leverage.

12.45 a. .0028. b. .0621. c. .6106.

12.63 $t_{critical} = 2.3069$ (from Excel). From sample: $t = 2.3256$. Reject H_0.

12.65 a. 1515.2. b. No.

c. (1406.03, 1624.37).

12.67 a. $y = 1743.57 − 1.2163x$.

b. $d.f. = 13$. $t_{critical} = 2.160$.

c. Slope is significantly different from zero.

d. $(-2.1671, -0.2656)$. Interval indicates slope is significantly less than zero.

e. $7.64 = (-2.764)^2$.

12.69 a. $r = +.9466$.

b. From sample: $t = 18.566$. For a two-tailed test, $t_{.005} = 2.704$. Reject H_0.

12.71 a. $r = .749$.

b. $d.f. = 13$. For a two-tailed test, $t_{.025} = 2.160$, $t_{calc} = 4.076$. Reject H_0.

12.75 b. $y = -4.2896 + 0.171x$, $R^2 = .2474$. Fit is poor.

12.77 a. State participation in the SAT is strongly related to the state's average SAT score ($p \approx 0$). As a state's participation rate increases, the average SAT score decreases.

b. Yes, this relationship is strong ($R = .9178$, $R^2 = .8419$).

c. When more students take the SAT, there will be a wider range of abilities, which will bring the average down.

12.79 a. The negative slope means that as age increases, price decreases.

b. Intercepts could be asking price of a new car.

c. The fit is good for the Explorer, Pickup, and Taurus.

d. Additional predictors: condition of car, mileage.

12.81 a. .0297. b. .2563. c. .7950.

CHAPTER 13

13.1 a. $Net\ Revenue = 4.31 - 0.082\ ShipCost + 2.265\ PrintAds + 2.498\ WebAds + 16.7\ Rebate\%$.

b. Positive coefficients indicate an increase in net revenue; negative coefficients indicate a decrease in net revenue.

c. The intercept is meaningless.

d. \$467,111.

13.3 a. $Ovalue = 2.8931 + 0.1542\ LiftWait + 0.2495\ AmountGroomed + 0.0539\ SkiPatrolVisibility - 0.1196\ FriendlinessHosts$.

b. Overall satisfaction increases with an increase in satisfaction for each coefficient except for friendliness of hosts. This counterintuitive result could be due to an interaction effect.

c. No.

d. 4.5831.

13.5 a. $df_1 = 4$ (numerator) and $df_2 = 45$ (denominator).

b. $F_{.05} = 2.61$ using $df_1 = 4$ and $df_2 = 40$.

c. $F = 12.997$. Yes, overall regression is significant.

d. $R^2 = .536$. $R^2_{adj} = .495$.

13.7 a. $df_1 = 4$, $df_2 = 497$.

b. Using Appendix F, $df_1 = 4$ and $df_2 = 200$, $F_{.05} = 2.42$.

c. $F_{calc} = 12.923$. Yes, overall regression is significant.

d. $R^2 = .0942$, $R^2_{adj} = .0869$.

13.9 a. and c. See Table.

Predictor	Coef	std error coef	t-value	p-value
Intercept	4.31	70.82	0.0608585	.9517414
ShipCost	−0.082	4.678	−0.0175289	.9860922
PrintAds	2.265	1.05	2.1571429	.0363725
WebAds	2.498	0.8457	2.9537661	.0049772
Rebate%	16.697	3.57	4.6770308	.0003

b. $t_{critical} = 2.69$. *WebAds* and *Rebate%* differ significantly from zero.

13.11 a. and c.

b. $t_{.005} = 2.586$. *LiftWait* and *AmountGroomed* differ significantly from zero.

Predictor	Coef	std error coef	t-value	p-value
Intercept	2.8931	0.3680	7.8617	2.37E-14
LiftWait	0.1542	0.0440	3.5045	.0005
AmountGroomed	0.2495	0.0529	4.7164	3.07E-06
SkiPatrolVisibility	0.0539	0.0443	1.2167	.2245
FriendlinessHosts	−0.1196	0.0623	−1.9197	.0557

13.13 $\hat{y}_i \pm 2.032(3620)$: $\hat{y}_i \pm 7355.84$.

13.15 a. Number of nights needed and number of bedrooms.

b. Two: *SwimPool* $= 1$ if there is a swimming pool and *ParkGarage* $= 1$ if there is a parking garage.

c. $CondoPrice = \beta_0 + \beta_1 NumNights + \beta_2 NumBedrooms + \beta_3 SwimPool + \beta_4 ParkGarage$.

13.17 a. $\ln(Price) = 5.4841 - 0.0733\ SalePrice + 1.1196\ Sub\text{-}Zero + 0.0696\ Capacity + 0.0466\ 2DoorFzBot - 0.3432\ 2DoorFzTop - 0.7096\ 1DoorFz - 0.8820\ 1DoorNoFz$.

b. *SalePrice:* p-value $= .0019$, *Sub-Zero:* p-value $= 2.24$E-13, *Capacity:* p-value $= 2.69$E-31, *2DoorFzBot:* p-value $= .5623$, *2DoorFzTop:* p-value $= 3.57$E-19, *1DoorFz:* p-value $= 1.18$E-07, *1DoorNoFz:* p-value $= 8.32$E-09.

c. 0.3432.

d. Side freezer.

13.19 a. Positive nonlinear relationship. Yes, nonlinear model appropriate.

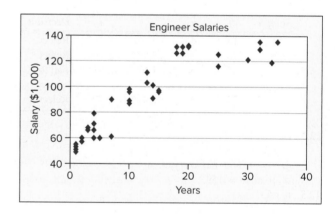

b. $R^2 = .915$, $F_{calc} = 194.99$, p-value $= 4.84 \times 10^{-20}$. Yes, model significant.

c. *Years:* p-value $= 5.5 \times 10^{-14}$, *YearsSq:* p-value $= 6.21 \times 10^{-8}$. Yes, the predictors are both significant.

13.21 a. *LiftOps* and *Scanners* ($r = .635$), *Crowds* and *LiftWait* ($r = .577$), *AmountGroomed* and *TrailGr* ($r = .531$), *SkiSafe* and *SkiPatrolVisibility* ($r = .488$).

b. All VIFs are less than 2. No cause for concern.

13.23 a. Not high.

b. High.

c. High.

13.43 The sample size is too small relative to number of predictors.

13.45 a. *CostPerLoad* = 26 – 6.3 *Top-Load* – 0.2714 *Powder*.

    b. Overall regression not significant (*F* statistic *p*-value = .371).

    c. No apparent relationship between cost per load and predictors for type of washer and type of detergent used.

13.47 a. Coefficients make sense, except for *TrnOvr*, which would be expected to be negative.

    b. No.

    c. With 6 predictors, should have minimum of 30 observations. We have only 23, so sample is small.

    d. Rebounds and points highly correlated.

13.49 a. Experience lowers the predicted finish time.

    b. No.

    c. If the relationship is not strictly linear, it can make sense to include a squared predictor, such as seen here.

13.51 The first three predictors (*Income, Unem, Pupil/Tea*) are significant at $\alpha = .05$, but adding the fourth predictor (*Divorce*) yields a weak *p*-value (.1987) and R^2 and R^2_{adj} barely improve when *Divorce* is added.

CHAPTER 14

14.1 a.

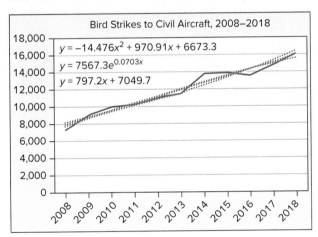

Bird Strikes to Civil Aircraft, 2008–2018

$y = -14.476x^2 + 970.91x + 6673.3$

$y = 7567.3e^{0.0703x}$

$y = 797.2x + 7049.7$

    b. More planes in the air.

    c. See graph.

    d.

t	*Linear*	*Quadratic*	*Exponential*
2019	16,616	16,240	17,592
2020	17,413	16,849	18,873
2021	18,211	17,429	20,248

14.3 a. No obvious increasing or decreasing trend.

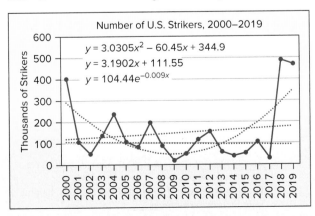

Number of U.S. Strikers, 2000–2019

$y = 3.0305x^2 – 60.45x + 344.9$

$y = 3.1902x + 111.55$

$y = 104.44e^{-0.009x}$

    b. Erosion of unskilled jobs, globalization, tougher bargaining.

    c. See graph.

    d. None of the trend models considers cyclical effects.

    e. The COVID-19 pandemic beginning March 2020 created a disruption in the workforce that couldn't be predicted by past data.

14.5 a. 627.   b. 2,097.   c. 1,587.

14.7 Graph shows negative trend and cyclical pattern. Fit improves as α increases (i.e., as we give more weight to recent data). Forecasts are similar for each value of α.

Alpha (Using Method A)	.10	.20	.30
Mean Squared Error	0.062	0.029	0.020
Mean Absolute Percent Error	9.8%	6.7%	5.6%
Percent Positive Errors	19.6%	27.5%	35.3%
Forecast for Period 52	1.84	1.84	1.85

14.9 a. A trend is not obvious in the time plot.

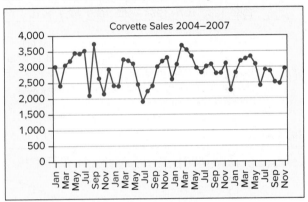

Corvette Sales 2004–2007

    b. Seasonality is not obvious, although one might conclude that spring months see higher sales.

    c. Slight seasonality based on index values: Jan (0.836), Feb (0.951), Mar (1.167), Apr (1.162), May (1.134), Jun (0.979), Jul (0.934), Aug (0.835), Sep (1.049), Oct (0.961), Nov (0.926), Dec (1.066).

    d. Performance sports cars more desirable in warm weather.

    e. Fit of the multiple regression model is not very good ($R^2 = .416$, $R^2adj = .216$), and no predictors are significant at = .05.

14.11 a.

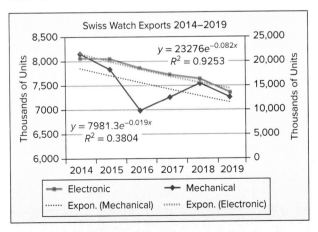

Swiss Watch Exports 2014–2019

$y = 23276e^{-0.082x}$

$R^2 = 0.9253$

$y = 7981.3e^{-0.019x}$

$R^2 = 0.3804$

Electronic   Mechanical

Expon. (Mechanical)   Expon. (Electronic)

    b. Overall trends are both negative with mechanical showing a sharp dip in 2016.

c. See graph. Fit for electronic sales much better than mechanical sales ($R^2 = .9253 >> R^2 = .3804$).

d. Fitbit and Apple watches have hurt sales of Swiss watches. A strong marketing campaign in 2017 helped recover some lost sales in mechanical Swiss watches.

14.13 a.

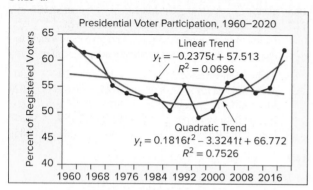

Presidential Voter Participation, 1960–2020

Linear Trend
$y_t = -0.2375t + 57.513$
$R^2 = 0.0696$

Quadratic Trend
$y_t = 0.1816t^2 - 3.3241t + 66.772$
$R^2 = 0.7526$

b. The overall linear trend is decreasing, but the last four elections have had a higher turnout than the low in 1996.

c. See graph for trend models.

d. Choose the quadratic, which shows an increase after 1996 and has a better fit.

e. Using the quadratic model: 2020 forecast is 66.74%.

14.15 a.

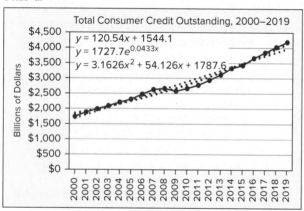

Total Consumer Credit Outstanding, 2000–2019

$y = 120.54x + 1544.1$
$y = 1727.7e^{0.0433x}$
$y = 3.1626x^2 + 54.126x + 1787.6$

b. Increasing trend.

c. See graph in part a.

d. The trends are positive with nonrevolving credit increasing faster than revolving.

14.17 a. 227.   b. 839.   c. 1,695.

14.19 a.

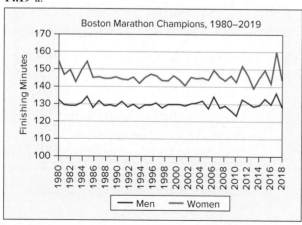

Boston Marathon Champions, 1980–2019

— Men — Women

b. It is unlikely.

d. Yes, a moving average would make sense. There is not much of a decreasing trend.

14.21 a.

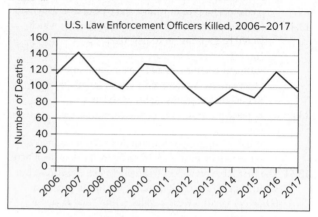

U.S. Law Enforcement Officers Killed, 2006–2017

b. No overall trend.

c. No, a trend would not be helpful.

14.23 a.

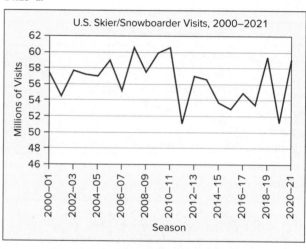

U.S. Skier/Snowboarder Visits, 2000–2021

b. No consistent increasing or decreasing trend. Smoothing average method preferred.

14.25

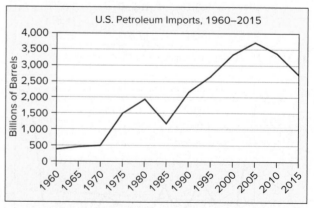

U.S. Petroleum Imports, 1960–2015

a. Steady upward trend until 2005. Last 10 years have shown a decrease.

b. Greater awareness of fossil fuel use and U.S. tapping in to their own resources.

c. Linear: $y_t = 16.242 + 302.05t$; quadratic: $y_t = -15.649t^2 + 505.48t - 458.43$; exponential: $y_t = 396.42e^{0.2084t}$

d. Forecast for 2020: Quadratic: $y_{13} = 3{,}468$.

14.27 a. Rate increases then levels off in later years.

b. The following graph shows $\alpha = .20$.

c. The degree of smoothing is most apparent for smaller values of α because less weight is given to the most recent values. Experts might choose $\alpha = .20$ as providing enough smoothing so that forecasts aren't overreacting to noise in the data but still capture short shifts in exchange rate.

d. Yes, exponential smoothing works well for forecasting short time periods ahead with data that does not have an obvious trend.

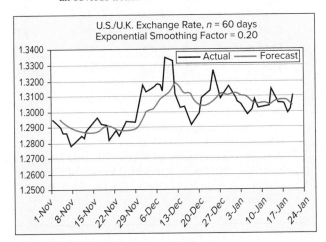

U.S./U.K. Exchange Rate, $n = 60$ days
Exponential Smoothing Factor = 0.20

14.29 a.

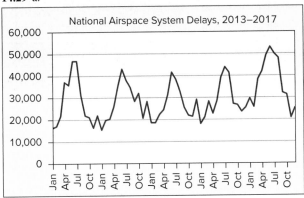

National Airspace System Delays, 2013–2017

b. Seasonality but no overall trend.

c. Megastat indexes are adjusted so they sum to 12.000.

Month	Index	Month	Index
Jan	0.684	Jul	1.485
Feb	0.721	Aug	1.229
Mar	0.902	Sep	0.897
Apr	0.944	Oct	0.880
May	1.184	Nov	0.705
Jun	1.465	Dec	0.904

d. Highest: summer (June, July, August); Lowest: winter (January, February).

14.31 Fund A: $[\ln(2509) - \ln(1000)]/(10 - 1) = 0.1022$
Fund B: $[\ln(2096) - \ln(1000)]/(10 - 1) = 0.0822$
Fund C: $[\ln(3034) - \ln(1000)]/(10 - 1) = 0.1233$

14.33

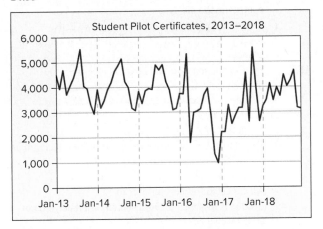

Student Pilot Certificates, 2013–2018

a.

Month	Index	Month	Index
Jan	0.931	Jul	1.122
Feb	0.888	Aug	1.312
Mar	1.139	Sep	1.070
Apr	0.863	Oct	1.127
May	0.993	Nov	0.804
Jun	1.069	Dec	0.682

b. There is a seasonal component, but the trend is slight. Summer and early fall months see higher activity. Nicer weather would be conducive to learning how to fly.

c. The model is significant and shows a significant negative trend (p-value = .007). Months August (p-value = .0087) and December (p-value = .0670) are significantly different from January, which was the base month. August has higher average students than January, and December has lower average students than January.

14.35 a. 739.

b. 843.

c. 871.

CHAPTER 15

15.1 a. H_0: *Choice of vehicle* and *Buyer's age* are independent.

b. Degrees of freedom $= (r - 1)(c - 1) = (4 - 1)(3 - 1) = 6$.

c. =CHISQ.INV.RT(.01,6) = 16.81 and test statistic = 10.667.

d. Test statistic is 10.667 (p-value = .0992), so reject the null at $\alpha = .10$.

e. *Gasoline* and *Under 30* contribute the most.

f. All expected frequencies exceed 5.

g. p-value is close to level of significance but less, so we would reject the null hypothesis.

15.3 a. H_0: *Verbal* and *Quantitative* are independent.

b. Degrees of freedom $= (r - 1)(c - 1) = (3 - 1)(3 - 1) = 4$.

c. =CHISQ.INV(.005,4) = 14.86.

d. Test statistic is 55.88 (p-value = .0000), so reject null at $\alpha = .005$.

e. *Under 25* (Quantitative) and *Under 25* (Verbal) contribute the most.

f. Expected frequency is less than 5 in two cells.

g. *p*-value is nearly zero (observed difference not due to chance).

15.5 a. H_0: *Completion Rate* and *Email Notice* are independent.

b. Degrees of freedom = $(r - 1)(c - 1) = (2 - 1)$ $(2 - 1) = 1$.

c. =CHISQ.INV.RT(.025,1) = 5.024.

d. Test statistic is 5.42 (*p*-value = .0199), so reject null at $\alpha = .025$.

e. *Completed* and *No* contribute the most.

f. All expected frequencies exceed 5.

g. *p*-value is less than .025 (observed difference did not arise by chance).

h. $z = 2.33$ (*p*-value = .0199 for two-tailed test).

15.7 $\chi^2_{calc} = 5.44$, *p*-value = .2447, fail to reject H_0. The sample shows the distribution of visits has stayed the same.

15.9 a. Bars are similar in length. Vanilla and Mocha are the leading flavors.

b. If uniform, $e_j = 84/4 = 21$ for each flavor.

c. Test statistic is 0.86 with *d.f.* = 4 − 1 = 3 (*p*-value = .8358). Chi-square critical value for $\alpha = .05$ is 7.815, so sample does not contradict the hypothesis that sales are the same for each beverage.

15.11 Expected frequency is 56/7 = 8 for each age group. Test statistic is 10.000 (*p*-value = .1247). At $\alpha = .05$, critical value for *d.f.* = 7 − 1 = 6 is 12.59. Cannot reject the hypothesis that moviegoers are from a uniform population.

Age Class	Obs	Exp	O − E	$(O - E)^2/E$
10 < 20	5	8.000	−3.000	1.125
20 < 30	6	8.000	−2.000	0.500
30 < 40	10	8.000	2.000	0.500
40 < 50	3	8.000	−5.000	3.125
50 < 60	14	8.000	6.000	4.500
60 < 70	9	8.000	1.000	0.125
70 < 80	9	8.000	1.000	0.125
Total	56	56.000	0.000	10.000

15.13 Sample mean $\lambda = 3.1905$, test statistic 0.371 (*p*-value = .9461) with *d.f.* = 5 − 1 − 1 = 3. The critical value for $\alpha = .05$ is 7.8147; cannot reject the hypothesis of a Poisson distribution.

X	P(X)	Obs	Exp	O − E	$(O - E)^2/E$
1 or less	0.172	6.000	7.243	−1.243	0.213
2	0.209	9.000	8.797	0.203	0.005
3	0.223	10.000	9.355	0.645	0.044
4	0.178	7.000	7.462	−0.462	0.029
5 or more	0.218	10.000	9.143	0.847	0.080
	1.000	42.000	42.000	0.000	0.371

15.15 From sample, $\bar{x} = 75.375$, $s = 8.943376$. Set $e_j = 40/8 = 5$. Test statistic is 6.000 (*p*-value = .306) using *d.f.* = 8 − 2 − 1 = 5. Critical value for $\alpha = .05$ is 11.07; cannot reject the hypothesis of a normal distribution.

Score	Obs	Exp	Obs − Exp	Chi-Square
Under 65.09	5	5.000	0.000	0.000
65.09 < 69.34	3	5.000	−2.000	0.800
69.34 < 72.53	5	5.000	0.000	0.000
72.53 < 75.38	3	5.000	−2.000	0.800
75.38 < 78.22	9	5.000	4.000	3.200
78.22 < 81.41	7	5.000	2.000	0.800
81.41 < 85.66	4	5.000	−1.000	0.200
85.66 or more	4	5.000	−1.000	0.200
Total	40	40.000	0.000	6.000

15.17*The probability plot looks linear, but *p*-value (.033) for the Anderson-Darling test is less than $\alpha = .05$. This tends to contradict the chi-square test used in Exercise 15.15. However, the Kolmogorov-Smirnov test ($D_{Max} = .158$) does not reject normality (*p*-value > .20). Data are a borderline case, having some characteristics of a normal distribution. If we have to choose one test, the A-D is the most powerful.

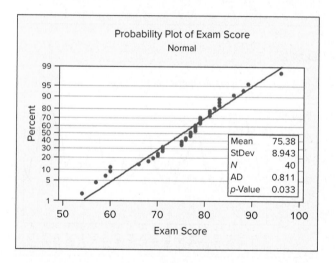

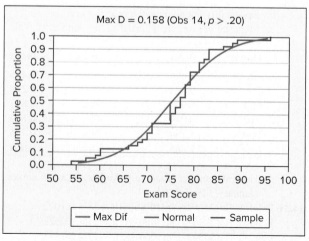

15.19 Is *Satisfaction* independent of *Pay Category?* For $d.f. = (r-1)(c-1) = (3-1)(2-1) = 2$, critical value is =CHISQ.INV.RT(.05,2) = 5.991. Test statistic is 9.69 (*p*-value = .0079); reject the null at $\alpha = .05$. *Salaried* and *Satisfied* contribute the most. All expected frequencies exceed 5. The *p*-value suggests that observed difference would arise by chance 7.9 times in 1,000 samples if the two variables really were independent.

15.21 Is *Grade* independent of *Hand-In Order?* For $d.f. = (r-1)(c-1) = (2-1)(2-1) = 1$, critical value =CHISQ.INV.RT(.10,1) = 2.706. Test statistic is 0.76 (*p*-value = .3821), so cannot reject the null at $\alpha = .10$. *"B" or Better* and *Later Hand-In* contribute the most. All expected frequencies exceed 5. For a two-tailed test of proportions, $p_1 = .44$, $p_2 = .32$, $p_c = .38$, $z = 0.87$ (*p*-value = .3821), which agrees with the chi-square test.

15.23 Is *Reading* independent of *Language?* For $d.f. = (r-1)(c-1) = (3-1)(4-1) = 6$, the critical value is =CHISQ.INV.RT(.10,6) = 10.64. Test statistic is 4.14 (*p*-value = .6577), so we cannot reject the null at $\alpha = .10$. Four cells (each corner) have expected frequencies below 5.

15.25 Are *Age* and *Social Media Platform Preference* independent? For $d.f. = (r-1)(c-1) = 4$, critical value is =CHISQ.INV.RT(.05,4) = 9.488. The calculated test statistic is 151.91 (*p*-value $\approx$ 0); reject the null hypothesis at $\alpha = .05$. *Age* and *Social Media Platform Preference* are not independent. All expected frequencies exceed 5.

15.27 For $d.f. = (r-1)(c-1) = (2-1)(2-1) = 1$, critical value =CHIINV(.10,1) = 2.706, test statistic is 1.80 (*p*-value = .1792), so fail to reject the null at $\alpha = .10$. The lower left cell contributes most. All expected frequencies exceed 5. The two-tailed test of proportions ($z = 1.342$) agrees with the chi-square test. Interestingly, the relationship seems to be inverse (i.e., rates tend to rise when they are predicted to fall).

15.29 For the 2×2 table, $d.f. = 1$, critical value is =CHISQ.INV.RT(.05,1) = 3.841, test statistic is 7.15 (*p*-value = .0075), so reject null at $\alpha = .05$. All expected frequencies exceed 5.

15.31 With $d.f. = (r-1)(c-1) = (3-1)(3-1) = 4$, the critical value is =CHISQ.INV.RT(.01,4) = 13.28. The test statistic is 66.40 (*p*-value = .0000), so we reject the null at $\alpha = .01$. All expected frequencies exceed 5.

15.33 $\chi_{calc}^2 = 38.69$, *p*-value = 2.24×10^{-6}; therefore, reject H_0. The sample shows the distribution of car colors has changed.

15.35 For $d.f. = 4-1 = 3$, critical value is =CHISQ.INV.RT(.05,3) = 7.815, test statistic is 6.045 (*p*-value = .1095), so we cannot reject hypothesis that Oxnard follows U.S. distribution.

15.37 For $d.f. = 6-1 = 5$, critical value is =CHISQ.INV.RT(.10,5) = 9.236, test statistic is 4.40 (*p*-value = .4934), so we can't reject the hypothesis that the die is fair.

15.39 Estimated mean is $\lambda = 1.06666667$. For $d.f. = 4-1-1 = 2$, critical value is =CHISQ.INV.RT(.025,2) = 7.378, test statistic is 4.947 (*p*-value = .0843), so can't reject the hypothesis of a Poisson distribution.

X	f_j	P(X)	e_j	$f_j - e_j$	$(f_j - e_j)^2/e_j$
0	25	0.344154	20.64923	4.35077	0.917
1	18	0.367097	22.02584	−4.02584	0.736
2	8	0.195785	11.74712	−3.74712	1.195
3 or more	9	0.092964	5.57781	3.42219	2.100
Total	60	1.000000	60.00000	0.00000	4.947

15.41* Answers will vary, but most should confirm the normal distribution and intended μ and σ.

15.43* Answers will vary, but most should confirm the Poisson distribution and intended λ.

CHAPTER 16

16.1 $R = 14$, $z = -0.133$ (*p*-value = .8942). Fail to reject H_0.

16.3 $R = 15$, $z = 0.697$ (*p*-value = .4858). Results are random.

16.5 a. Sample median = 53.75. $W = 234.5$, $z = 0.7174$, *p*-value = .48. Median is not significantly different from 50.

  b. Close to a normal distribution. Parametric *t* test could be justified.

16.7 a. $z = 1.319$, *p*-value = .1872. No difference in medians.

  b. $t = 1.62$, *p*-value = .0606 (assuming equal variances). Same decision but *p*-value closer to .05.

16.9 a. $H = 5.724$, *p*-value = .1258. No difference in medians.

  b. Yes, $F = 2.71$, *p*-value = .055.

  c. Can assume normality for Energy and Retail. Health and Leisure are less obvious.

16.11 $\chi^2 = 4.950$, *p*-value = .2925. No difference in median ratings.

16.13 a.

2004	2005	2004	2005
6	7	17	20
5	5	16	16
10	10	4	4
13	14	14	19
15	15	11	13
19	18	1	1
3	3	20	17
7	6	2	2
8	8	18	12
12	11	9	9

  b. $r_s = .9338$.

  c. Yes, $z_{calc} = 4.07$, *p*-value = .0000.

  e. Pearson: $r = .996$.

  f. Nonnormal data justify use of Spearman rank correlation.

16.15 $R = 28$, $z = 0.775$ (*p*-value = .4383). Results are random.

16.17 $R = 22$, $z = 1.419$ (*p*-value = .1559). Results are random.

16.19 $R = 9$, $z = -1.647$ (*p*-value = .0996). Results are random.

16.21 $z = -1.039$, *p*-value = .2988. No difference in medians.

16.23 From MegaStat Wilcoxon Signed-Rank Paired Data Test: $z = -1.481$, *p*-value = .1386. Medians do not differ.

16.25 $z = -3.086$, *p*-value = .0020. The medians differ.

16.27 $H = 1.46$, *p*-value = .4819. No difference in medians.

16.29 $H = 9.026$, *p*-value = .0110. The medians differ.

16.31 $\chi_{calc}^2 = 2.731$, *p*-value = .6038. No difference in median waiting times by day of week.

16.33 $r_s = .67$, $r_{.05} = .374$. Significant rank correlation.

16.35 $r_s = .696$, $r_{.05} = .468$. Significant rank correlation.

16.37 $r_s = .812$, $r_{.05} = .444$. Significant rank correlation.

CHAPTER 17

17.1 a. *Productivity:* ratio of output to input, measures efficiency.

  b. *Quality control:* used to ensure product/service quality.

  c. *Process control:* used to ensure process conformance to specifications.

17.3 Common cause comes from within the process. Special cause originates outside the process.

17.5 Define metrics, collect data, track variation.

17.7 See www.deming.org.

17.9 SQC applies statistical controls to the end product. SPC applies statistical controls to the process.

17.11 Quality improvement is a continuous cycle that repeats. See Section 17.3 for the steps.

17.13 Service blueprints and service transaction analysis.

17.15 a. Sampling frequency depends on cost and physical possibility of sampling.

b. For normal data, small samples may suffice for a mean (Central Limit Theorem).

c. Large samples may be needed for a proportion to get sufficient precision.

17.17 Expect 68.26 percent, 95.44 percent, 99.73 percent, respectively.

17.19 $UCL = \overline{\overline{x}} + 3\dfrac{\overline{R}}{d_2\sqrt{n}} = 12.5 + 3\dfrac{.42}{2.326\sqrt{5}} = 12.742$

$LCL = \overline{\overline{x}} - 3\dfrac{\overline{R}}{d_2\sqrt{n}} = 12.5 - 3\dfrac{.42}{2.326\sqrt{5}} = 12.258$

17.21 Estimated σ is $\overline{R}/d_2 = 30/2.059 = 14.572$, UCL = 98.37, LCL = 54.63.

$\overline{\overline{x}} = \dfrac{\overline{x}_1 + \overline{x}_2 + \cdots + \overline{x}_9}{9}$

$= \dfrac{72.25 + 74.25 + \cdots + 82.25}{9} = 76.5$

$\overline{R} = \dfrac{R_1 + R_2 + \cdots + R_9}{9}$

$= \dfrac{43 + 31 + \cdots + 41}{9} = 30$

17.23 $\overline{R} = .82$ (centerline)

$UCL = D_4\overline{R} = (2.004)(.82) = 1.64328$

$LCL = D_3\overline{R} = (0)(.82) = 0$

17.25 Services are often assessed using percent conforming or acceptable quality, so we use p charts.

17.27 Yes, safe to assume normality. UCL = .8354, LCL = .1646.

17.29 By either criterion, process is within acceptable standard ($C_p = 1.67$, $C_{pk} = 1.67$).

17.31 Fails both criteria, especially C_{pk} due to bad centering ($C_p = 1.17$, $C_{pk} = 0.67$).

17.33 Answers will vary. Examples:

a. GPA, number of classes retaken, faculty recommendation letters (Likert).

b. Knowledge of material, enthusiasm, organization, fairness (Likert scales for all).

c. Number of bounced checks, size of monthly bank balance errors, unpaid Visa balance.

d. Number of print errors, clarity of graphs, useful case studies (Likert scales for last two).

17.35 Answers will vary. Examples:

a. MPG, repair cost.

b. Frequency of jams, ink cost.

c. Frequency of re-flushes, water consumption.

d. Battery life, ease of use (Likert scale).

e. Cost, useful life, image sharpness (Likert scale).

f. Cost, useful life, watts per lumen.

17.37 $\overline{x}$ is normally distributed from the Central Limit Theorem for sufficiently large values of n (i.e., symmetric distribution) while the range and standard deviation are not.

17.39 a. Variation and chance defects are inevitable in all human endeavors.

b. Some processes have very *few* defects (maybe zero in short run, but not in long run).

c. Quarterbacks cannot complete all their passes, etc.

17.41 Answers will vary (e.g., forgot to set clock, clock set incorrectly, couldn't find backpack, stopped to charge cell phone, had to shovel snow, clock didn't go off, traffic, car won't start, can't find parking).

17.43 Answers will vary (e.g., weather, union slowdown, pilot arrived late, crew change required, deicing planes, traffic congestion at takeoff, no arrival gate available).

17.45 a. $C_p = 1.33$, $C_{pk} = 1.33$. b. $C_p = 1.33$, $C_{pk} = 0.67$.

17.47 a. UCL = 6223, LCL = 5877.

b. Chart violates no rules.

c. Process is in control.

17.49 a. UCL = 1.0939, LCL = 0.9061.

b. Chart violates no rules.

c. Process is in control.

17.51 a. $C_p = 1.00$, $C_{pk} = 0.83$.

b. Process well below capability standards (both indices less than 1.33).

c. Technology, cost, door not closed tightly, frequency of door opening.

17.53 Sample mean of 23.025 and standard deviation of 2.006 are very close to the process values ($\mu = 23$, $\sigma = 2$). Histogram is symmetric, but perhaps platykurtic (chi-square or Anderson-Darling test needed).

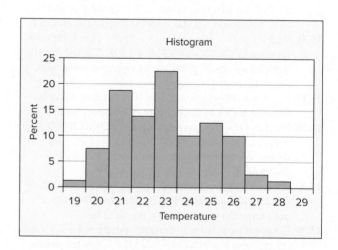

17.55 a. Histogram is arguably normal, but somewhat bimodal (chi-square or Anderson-Darling test needed).

b. $\mu = 8760$, $\sigma = 200$, UCL = 9028, LCL = 8492.

c. $\bar{x} = 8785$, $s = 216.14$, UCL = 9075, LCL = 8495,

d. Sample is small; may have unreliable estimates of μ and σ.

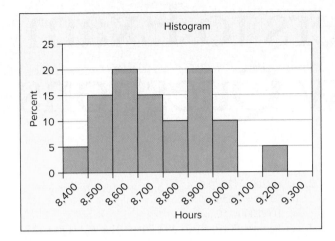

17.57 a. UCL = 470.2, LCL = 459.8.

b. No rules violated. Process in control.

17.59 a. UCL = 12.22095, LCL = 11.75569, centerline = 11.98832.

b. Process appears to be in control.

c. Histogram approximates normal distribution.

17.61 a. UCL = .1154, LCL = 0.

b. Sample 7 hits the LCL, otherwise in control.

c. Samples are too small to assume normality ($n\pi = 5$). (Better to use Minitab's binomial option.)

17.63 Chart A: Rule 4.

Chart B: No rules violated.

Chart C: Rule 4.

Chart D: Rules 1, 4.

Chart E: No rules violated.

Chart F: Rules 1, 2.

17.65 Each pattern is clearly evident, except possibly instability in third series.

CHAPTERS 1–4

1. a. inferential; b. descriptive; c. inferential
2. c. independent judgment is needed
3. b. anecdotal data ($n = 1$)
4. a. numerical;   b. categorical;   c. numerical
5. a. ratio (true zero);
   b. ordinal;
   c. nominal
6. a. continuous;   b. continuous;   c. discrete
7. a. convenience;
   b. simple random;
   c. systematic
8. c. Computer software makes it easy, and inexpensive, to generate random numbers.
9. a. Likert only if distances have meaning
10. a. sampling error cannot be eliminated
11. skewed right, no outliers, Sturges $k \cong 6$
12. c. range is $-1 \le r \le +1$
13. a. small n and sum to 100%
14. $\bar{x} = 12$, $s = 5.701$, $CV = 47.5\%$
15. a. $\bar{x} = 59.3$, median $= 58.5$, modes 55, 58, 62 (not unique)
    b. mean or median best
    c. $Q_1 = 55$, $Q_2 = 62.25$
16. a. slight positive correlation;
    b. $r = 0.5656$ (not very linear)

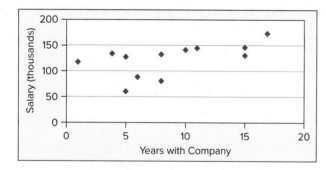

17. b. GEOMEAN(Data) requires all $x_i > 0$
18. a. $z = (x - \mu)/\sigma = (81 - 52)/15 = 1.93$ (not an outlier)
19. b. log scales are less familiar to most

CHAPTERS 5–7

1. a. empirical;
   b. subjective;
   c. classical

2. a. $40/200 = .20$;
   b. $50/90 = .5556$;
   c. $100/200 = .50$
3. no, because $P(A)P(B) = (.30)(.70) = .21 \ne P(A \cap B) = .25$
4. b. would be true if $P(A \cap B) = 0$
5. c. $U(a, b)$ has two parameters
6. a. $.60(1000) + .30(2000) + .10(5000) = 1700$
7. a. .2565;
   b. .4562;
   c. .7576;
   d. Poisson, $\lambda = 2.5$
8. a. .1468;
   b. .0563;
   c. .7969;
   d. binomial, $n = 8$, $\pi = .20$
9. a. $\mu = n\pi = (50)(.30) = 15$;
   b. $\sigma = \sqrt{n\pi(1 - \pi)} = \sqrt{(50)(.30)(.70)} = 3.24$
10. a. binomial ($n = 8$ trials, π unknown);
    b. Poisson (arrivals, λ unknown);
    c. discrete uniform ($a = 0$, $b = 9$)
11. c. $X = $ trials until first success in geometric
12. a. points have no area, hence no probability
13. a. Normal PDF is always *symmetric* about the mean
14. Using Appendix C:
    a. $P(Z > 1.14) = .1271$;
    b. $P(-.71 < Z < +0.71) = .5222$;
    c. $P(Z < 0) = .5000$
    Using Excel:
    a. .1265;   b. .5249;   c. .5000
15. Using Table 7.9:
    a. $\mu + 1.645\sigma = 70 + 1.645(7) = 81.52$;
    b. $\mu - 1.282\sigma = 70 - 1.282(7) = 61.04$;
    c. $\mu + 0.675\sigma = 70 + 0.675(7) = 74.73$
16. a. gives cumulative left tail area
17. Using $\lambda = 1.2$:
    a. $P(X < 1.5) = 1 - e^{-\lambda x} = 1 - e^{(-1.2)(1.5)} = .8347$;
    b. $P(X > 0.5) = e^{-\lambda x} = e^{(-1.2)(0.5)} = .5488$;
    c. $P(X > 1) - P(X > 2) = e^{(-1.2)(1)} - e^{(-1.2)(2)} = .3012 - .0907 = .2105$ (if X is expressed in minutes)
18. Using $\lambda = 1.2$:
    a. Solve $e^{-\lambda x} = .05$ to get $x = 2.496$ min (149.8 sec);
    b. Solve $e^{-\lambda x} = .75$ to get $x = 0.2397$ min (14.38 sec);
    c. MTBE $= 1/\lambda = (1/1.2) = 0.83$ min (50 sec)
19. a. This is a correct rule of thumb (set $\mu = \lambda$ and $\sigma = \sqrt{\lambda}$)
20. c. Triangular is skewed unless b is halfway between a and c.

CHAPTERS 8–10

1. a. CLT applies to $\overline{X}$. Sample *data* may not be normal.

2. a. consistent;
b. efficient;
c. unbiased

3. b. It is conservative to use t *whenever* σ is unknown, regardless of n.

4. a. $d.f. = n - 1 = 8$, $t_{.025} = 2.306$, so $\overline{x} \pm t \dfrac{s}{\sqrt{n}}$ gives $13.14 < \mu < 16.36$;
b. Unknown σ

5. a. $n = 200$, $z = 1.96$, $p = 28/200 = .14$, so $p \pm z \sqrt{\dfrac{p(1-p)}{n}}$ gives $.092 < \pi < .188$;
b. $np = 28 > 10$ and $n(1 - p) = 172 > 10$;
c. Using $z = 1.645$ and $E = \pm.03$, the formula $n = \left(\dfrac{z}{E}\right)^2 \pi(1 - \pi)$ gives $n = 363$ (using $p = .14$ for π from preliminary sample) or $n = 752$ (using $\pi = .50$ if we want to be very conservative)

6. c. Normality OK because $np = 17.5 > 10$.

7. b. Type I error is rejecting a true H_0.

8. b. $z_{.025} = \pm1.960$

9. a. $H_0: \mu \geq 56$, $H_1: \mu < 56$;
b. Using $\overline{x} = 55.82$, $\sigma = 0.75$ (known), and $n = 49$, we get $z_{calc} = \dfrac{\overline{x} - \mu_o}{\sigma/\sqrt{n}} = -1.636$;
c. $z_{.05} = -1.645$;
d. fail to reject (but a very close decision)

10. a. $H_0: \mu \leq 60$, $H_1: \mu > 60$;
b. Using $\overline{x} = 67$, $s = 12$, and $n = 16$, we get $t_{calc} = \dfrac{\overline{x} - \mu_0}{\sigma/\sqrt{\pi}} = 2.333$;
c. For $d.f. = n - 1 = 15$, $t_{.025} = 2.131$;
d. reject

11. a. $\alpha = P(\text{reject } H_0 \mid H_0 \text{ is true})$
b. True. As the sample size increases, critical values of $t_{.05}$ increase because the $d.f.$ increase, gradually approaching $z_{.05}$.

12. a. $H_0: \pi \leq .85$, $H_1: \pi > .85$, $p = 435/500 = .87$, $z_{calc} = \dfrac{p - \pi_0}{\sqrt{\dfrac{\pi_0(1 - \pi_0)}{n}}} = 1.252$, $z_{.05} = 1.645$, not a significant increase;
b. $n\pi_0 = (500)(.85) = 425 > 10$ and $n(1 - \pi_0) = (500)(.15) = 75 > 10$

13. a. independent samples, unknown variances, $t_{calc} = -2.034$ (regardless whether equal or unequal variances assumed);
b. two-tailed test, $t_{.025} = \pm2.0739$ (if equal variances assumed, $d.f. = 22$) or $t_{.025} = \pm2.0796$ (if unequal variances assumed, $d.f. = 21$);
c. reject $H_0: \mu_1 = \mu_2$ in favor of $H_1: \mu_1 \neq \mu_2$

14. a. $H_0: \pi_1 \leq \pi_2$, $H_1: \pi_1 > \pi_2$, $p_1 = 150/200 = .75$, $p_2 = 140/200 = .70$, $\overline{p} = .725$, $z_{calc} = 1.120$, $z_{.025} = 1.96$. Colorado not significantly greater

15. a. paired t test;
b. $d.f. = n - 1 = 5 - 1 = 4$, left-tailed test, $t_{.10} = -1.533$;
c. $t_{calc} = -1.251$, fail to reject, second exam not significantly greater

16. a. Reject if *small p-value*

17. a. $F_{calc} = s_1^2/s_2^2 = (14^2)/(7^2) = 4.00$;
b. $\alpha/2 = .05/2 = .025$, $F_L = 0.2123$ ($d.f. = 7, 11$) and $F_R = 3.7586$ ($d.f. = 7, 11$), reject $H_0: \sigma_1^2 = \sigma_2^2$.

CHAPTERS 11–13

1. c. In ANOVA, each population is assumed normal.

2. b. Hartley's F_{max} test compares variances (not means).

3. $F_{calc} = (744/4)/(751.5/15) = 3.71$, $F_{4,15} = 3.06$

4. a. 3;
b. 210;
c. No, *p*-value $= .9055 > .05$;
d. No, *p*-value $= .3740 > .05$

5. Two-tailed test, $t_{calc} = 2.127$, $d.f. = 28$, $t_{.005} = 2.763$, fail to reject.

6. b. In correlation analysis, neither variable is assumed dependent.

7. a. $R^2 = SSR/SST = (158.3268)/(317.4074) = .4988$.

8. b. $d.f. = n - 2 = 25$, $t_{.025} = 2.060$

9. a., c. Both formulas give the same t_{calc}.

10. a. false (residual is within $\pm 1 s_{yx}$);
b. true;
c. true

11. a. Evans' Rule suggests $n/k \geq 10$.

12. b. $R_{adj}^2 \leq R^2$ always; big difference would suggest weak predictors

13. a. because their 95% CIs do not include zero

14. c. *p*-value $< .05$ for $X3$ (clearly) and $X4$ (barely)

15. b. $d.f. = 38$, $t_{.005} = \pm2.712$, so only $X3$ is significant ($t_{calc} = -5.378$)

Writing and Presenting Reports

Business recruiters say that written and oral communication skills are critical for success in business. Susan R. Meisinger, president and CEO of the Society for Human Resource Management, says that "[i]n a knowledge-based economy a talented workforce with communication and critical thinking skills is necessary for organizations and the United States to be successful." Yet a survey of 431 human-resource officials in corporate America found a need for improvement in writing (www.conference-board.org). Table I.1 lists the key business skills needed for *initial* and *long-range* success as well as some common *weaknesses*.

Table I.1

Skills Needed for Success in Business

For Initial Job Success	For Long-Range Job Success	Common Weaknesses
Report writing	Managerial accounting	Communication skills
Accounting principles	Managerial economics	Writing skills
Mathematics	Managerial finance	Immaturity
Statistics	Oral communication	Unrealistic expectations

Mini Case I.1

Can You Read a Company Annual Report?

Many people say that company annual reports are hard to read. To investigate this claim, Prof. Feng Li of the University of Michigan's Ross School of Business analyzed the readability of more than 50,000 annual reports. One of his readability measures was the Gunning-Fog Index (GFI), which estimates how many years of formal education would be needed in order to read and understand a block of text. For company annual reports, the average GFI was 19.4. Because a college graduate will have 16 years of education, almost a PhD. level of education is apparently required to read a typical firm's annual report. Li also found that annual reports of firms with lower earnings were harder to read.

(Sources: http://accounting.smartpros.com/x53453.xml; and *Detroit Free Press,* June 7, 2006, p. E1.)

Rules for "Power" Writing

Why is writing so important? Because someone may mention your report on warranty repairs during a meeting of department heads, and your boss may say, "OK, make copies of that report so we can all see it." Next thing you know, the CEO is looking at it! Wish you'd taken more care in writing it? To avoid this awkward situation, set aside 25 percent of your allotted project time to *write* the report. You should always outline the report *before* you begin. Then complete the report in sections. Finally, ask trusted peers to review the report, and make revisions as necessary. Keep in mind that you may need to revise more than once. If you have trouble getting started, consult a good reference on technical report writing.

While you may have creative latitude in how to organize the flow of ideas in the report, it is essential to answer the assigned question succinctly. Describe what you did and what conclusions you reached, listing the most important results first.

Use section headings to group related material and avoid lengthy paragraphs. Your report is your legacy to others who may rely on it. They will find it instructive to know about difficulties you encountered. Provide clear data so others will not need to waste time checking your data and sources. Consider placing technical details in an appendix to keep the main report simple.

If you are writing the report as part of a team, an "editor-in-chief" must be empowered to edit the material so that it is stylistically consistent, has a common voice, and flows together. Allow enough lead time so that all team members can read the final report and give their comments and corrections to the editor-in-chief.

Avoid Jargon   Experts use jargon to talk to one another, but outsiders may find it obscure or even annoying. Technical concepts must be presented so that others can understand them. If you can't communicate the importance of your work, your potential for advancement will be limited. Even if your ideas are good and hundreds of hours went into your analysis, readers up the food chain will toss your report aside if it contains too many cryptic references like SSE, MAPE, or 3-Sigma Limits.

Make It Attractive   Reports should have a title page, descriptive title, date, and author names. It's a good idea to use footers with page numbers and dates (e.g., Page 7 of 23—Draft of 10/8/2021) to distinguish revised drafts.

Use wide margins so readers can take notes or write comments. Select an appropriate typeface and point size. Times Roman, Garamond, and Arial are widely accepted.

Call attention to your main points by using subheadings, bullets, **boldfaced type,** *italics,* large fonts, or color, but use special effects sparingly.

Watch Your Spelling and Grammar   To an educated reader, incorrect grammar or spelling errors are conspicuous signs of sloppy work. You don't recognize your errors—that's why you make them. Get someone you trust to red-pencil your work. Study your errors until you're sure you won't repeat them. Your best bet? Keep a dictionary handy! You can refer to it for both proper spelling and grammatical usage. Remember that Microsoft specializes in software, not English, so don't rely on spelling and grammar checkers. Here are some examples from student papers that passed the spell-checker, but each contains two errors. Can you spot them quickly?

Original	*Correction*
• "It's effects will transcend our nation's boarders."	(its, borders)
• "We cannot except this shipment on principal."	(accept, principle)
• "They seceded despite there faults."	(succeeded, their)
• "This plan won't fair well because it's to rigid."	(fare, too)
• "The amount of unhappy employees is raising."	(number, rising)

Organizing a Technical Report

Report formats vary, but a business report usually begins with an *executive summary* limited to a *single page* (or less). Attach the full report containing discussion, explanations, tables, graphs, interpretations, and (if needed) footnotes and appendices. Use appendices for backup material. Paste your key graphs and tables *into the main report* where you refer to them, and format them nicely. Each table or graph needs a title and a number. A common beginner's error is to attach a bunch of Excel printouts and graphs at the end of a technical report; most readers won't take the time to flip pages to look at them. Worse, if you put all your tables and graphs at the end, you may be tempted not to spend time formatting them nicely. There is no single acceptable style for a business report, but the following would be typical:

- Executive Summary (1 page maximum)
- Introduction (1 to 3 paragraphs)
  - Statement of the problem
  - Data sources and definitions
  - Methods utilized
- Body of the Report (as long as necessary)
  - Break it into sections
  - Each section has a descriptive heading
  - Use subsection headings as necessary
  - Tables and graphs, as needed
- Conclusions (1 to 3 paragraphs)
  - Restatement of findings
  - Limitations of your analysis
  - Future research suggestions
- Bibliography and Sources
- Appendices (if needed for lengthy or technical material)

General tips:

- Avoid huge paragraphs (break them up).
- Include page numbers to help the reader take notes.
- Check spelling and grammar. Ask others to proofread the report.

Writing an Executive Summary

The goal of an **executive summary** is to permit a busy decision maker to understand what you did and what you found out *without reading the rest of the report.* In a statistical report, the executive summary *briefly* describes the task and goals, data and data sources, methods that were used, main findings of the analysis, and (if necessary) any limitations of the analysis. The executive summary is limited to a single page (maybe only two or three paragraphs) and should avoid technical language.

An excellent way to evaluate your executive summary is to hand it to a peer. Ask the peer to read it and then tell you what you did and what you found out. If the peer cannot answer precisely, then your summary is deficient. The executive summary must make it *impossible to miss your main findings.* Your boss may judge you and your team by the executive summary alone and may merely leaf through the report to examine key tables or graphs.

Rules for Presenting Oral Reports

The goals of an oral report are *not the same* as those of a written report. Your oral presentation must only *highlight* the main points. If your presentation does not provide the answer to an audience question, you can say, "Good question. We don't have time to discuss that further here, but it's covered in the full report. I'll be happy to talk to you about it at the end of the presentation." Or give a brief answer so they know you did consider the matter. Keep these tips in mind while preparing your oral presentation:

- Select just a few key points you most want to convey.
- Use simple charts and diagrams to get the point across.
- Use **color** and *fonts* creatively to **emphasize a point.**
- Levity is nice on occasion, but avoid gratuitous jokes.
- Have backup slides or transparencies just in case.

- Rehearse to get the timing right (don't go too long).
- Refer the audience to the written report for details.
- Imagine yourself in the audience. Don't bore yourself!

The Three Ps

Pace   Many presenters speak too rapidly—partly because they are nervous and partly because they think it makes them look smarter.

Federal Revenue Sources

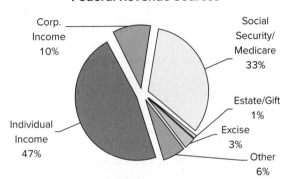

Figure I.1

Pictures Help Make the Point

Source: Based on 2016 data from United States Bureau of the Budget.

Slow down! Take a little time to introduce yourself, introduce your data, and explain what you are trying to do. If you skip the basic background and definitions, many members of the audience will not be able to follow the presentation and will have only a vague idea what you are talking about.

Planning   Create an outline to organize the ideas you want to present. Remember to keep it simple! You'll also need to prepare a verbal "executive summary" to tell your audience what your talk is about. Ask yourself: How much does my audience already know about this problem, these data, and this terminology? You don't want to bore them, but it's often better to err on the side of providing a little extra background. Don't bury them in detail, but make the first minute count. If you ran into problems or made errors in your analysis, it's OK to say so. The audience will sympathize.

Check the raw data carefully—you may be called on to answer questions. It's hard to defend yourself when you failed to catch serious errors or didn't understand a key definition.

Practice   Rehearse the oral presentation to get the timing right. Maybe your employer will send you to training classes to bolster your presentation skills. Otherwise, consider videotaping yourself or practicing in front of a few peers for valuable feedback. Technical presentations may demand skills different from the ones you used in English class, so don't panic if you have a few problems.

FOR MORE INFORMATION

Use your favorite search engine (e.g., Google) to search for "effective business writing," "technical writing," "scientific writing," and "presentation tips." Here are a few websites to get you started

- https://owl.english.purdue.edu/owl/resource/672/1/
- www.cs.columbia.edu/~hgs/etc/writing-style.html
- http://en.wikipedia.org/wiki/Business_communication

J Statistics in Excel and R

Descriptive Statistics*	Excel Function	R Function
Number of data items	COUNT(x)	length(x)
Largest data value	MAX(x)	max(x)
Smallest data value	MIN(x)	min(x)
Mean	AVERAGE(x)	mean(x)
Median	MEDIAN(x)	median(x)
Mode (returns first mode only)	MODE.SNGL(x)	----------------
Mode (array function for multiple modes; highlight output range and use Ctrl-Shift-Enter)	{MODE.MULT(x)}	----------------
Geometric mean (positive data values)	GEOMEAN(x)	exp(mean(log(x)))
Average deviation around the mean (R uses the *median* not the *mean*)	AVDEV(x)	mad(x)
Quartile k where k = 1, 2, 3 for 25%, 50%, 75%	QUARTILE.EXC(x, k)	quantile(x, p, type=6) where $p = .25, .50,$ or .75
Percentile p (e.g., p = .25 for 25%)	PERCENTILE.EXC(x, p)	quantile(x, p, type=6)
Sample variance for (X, Y) data pairs	VAR.S(x, y)	var(x, y)
Sample standard deviation	STDEV.S(x)	sd(x)
Sample covariance for (X, Y) data pairs	COVARIANCE.S(x, y)	cov(x, y)
Population standard deviation	STDEV.P(x)	sqrt(var(x))
Population covariance for (X, Y) data pairs	COVARIANCE.P(x, y)	cov(x, y)*(length(x)-1)/length(x)
Standardize an X value (in Excel you must first calculate the sample mean and standard deviation if μ and σ unknown)	STANDARDIZE(x, μ, σ)	(x-mean(x))/sd(x)
Sample correlation for (X, Y) data pairs	CORREL(x, y)	cor(x, y)
Slope of simple X-Y regression	SLOPE(y, x)	lm(y~x)
Intercept of simple X-Y regression	INTERCEPT(y, x)	lm(y~x)
R-squared for simple X-Y regression	RSQ(y, x)	cor(x, y)^2

*In these formulas, x and y denote *vectors* of data values (in Excel x and y would typically be data *columns*). We use the same notation as this textbook for parameters. For simplicity, R functions shown here often omit optional arguments. To see its full syntax, enter ? before the R function name (with no spaces). When built-in functions lack equivalents, a formula is shown if it's not complex (otherwise omitted). For R quantiles, we specify type=6 for compatibility with Minitab and SPSS (see textbook section 4.5).

Discrete Probability Distributions*	Excel Function	R Function*
Binomial distribution (n trials, prob π)		
PDF: Returns probability $P(X = x)$	BINOM.DIST(x, n, π, 0)	dbinom(x, n, π)
CDF: Returns probability $P(X \leq x)$	BINOM.DIST(x, n, π, 1)	pbinom(x, n, π)
Inverse CDF: Returns x for $P(X \leq x) = \alpha$	BINOM.INV(n, π, α)	qbinom(α, n, π)
…Random data (makes k samples in R)	BINOM.INV(n, π, RAND())	rbinom(k, n, π)
Poisson distribution (with mean λ)		
PDF: Returns probability $P(X = x)$	POISSON.DIST(x, λ, 0)	dpois(x, λ)
CDF: Returns probability $P(X \leq x)$	POISSON.DIST(x, λ, 1)	ppois(x, λ)
Inverse CDF: Returns x for $P(X \leq x) = \alpha$	----------------	qpois(α, λ)
…Random data (makes k samples in R)	----------------	rpois(k, λ)
Hypergeometric distribution (N pop size)		
PDF: Returns probability $P(X = x)$	HYPGEOM.DIST(x, n, s, N, 0)	dhyper(x, s, N-s, n)
CDF: Returns probability $P(X \leq x)$	HYPGEOM.DIST(x, n, s, N, 1)	phyper(x, s, N-s, n)
Inverse CDF: Returns x for $P(X \leq x) = \alpha$	----------------	qhyper(α, s, N-s, n)
…Random data (makes k samples in R)	----------------	rhyper(k, s, N-s, n)
Geometric distribution (success prob p)	x = num *trials* before success	x = num *failures* before success
PDF: Returns probability $P(X = x)$	π*$(1-\pi)$^$(x-1)$	dgeom(x, π)
CDF: Returns probability $P(X \leq x)$	$1 - (1-\pi)$^x	pgeom(x, π)
Inverse CDF: Returns x for $P(X \leq x) = \alpha$	----------------	qgeom(α, π)
…Random data (makes k samples in R)	1 + INT(LN(1-RAND())/LN($1-\pi$))	rgeom(k, π)

*In probability distributions, we use x to denote a *single* data value.

Continuous Probability Distributions*	Excel Function	R Function*
Normal distribution		
PDF: Returns height of $f(x)$	NORM.DIST(x, μ, σ, 0)	dnorm(x, μ, σ)
CDF: Returns probability $P(X \leq x)$	NORM.DIST(x, μ, σ, 1)	pnorm(x, μ, σ)
Inverse CDF: Returns x for $P(X \leq x) = \alpha$	NORM.INV(α, μ, σ)	qnorm(α, μ, σ)
…Random data (makes k samples in R)	NORM.INV(RAND(), μ, σ)	rnorm(k, μ, σ)
Standard normal distribution		
PDF: Returns height of $f(z)$	NORM.S.DIST(z, 0)	dnorm(z)
CDF: Returns probability $P(Z \leq z)$	NORM.S.DIST(z, 1)	pnorm(z)
Inverse CDF: Returns z for $P(Z \leq z) = \alpha$	NORM.S.INV(α)	qnorm(α)
…Random data (makes k samples in R)	NORM.S.INV(RAND())	rnorm(k)
Uniform distribution $a < x < b$		
PDF: Returns height of $f(x)$	$1/(b-a)$	dunif(x, a, b)
CDF: Returns probability $P(X \leq x)$	$(x-a)/(b-a)$	punif(x, a, b)
Inverse CDF: Returns x for $P(X \leq x) = \alpha$	$a + \alpha$*$(b-a)$	qunif(α, a, b)
…Random data (makes k samples in R)	$a + (b-a)$*RAND()	runif(k, a, b)
Exponential distribution		
PDF: Returns height of $f(x)$	EXPON.DIST(x, λ, 0)	dexp(x, λ)
CDF: Returns probability $P(X \leq x)$	EXPON.DIST(x, λ, 1)	pexp(x, λ)
Inverse CDF: Returns x for $P(X \leq x) = \alpha$	----------------	qexp(α, λ)
…Random data (makes k samples in R)	----------------	rexp(k, λ)
Student's t distribution		
PDF: Returns height of $f(t)$	T.DIST(t, df, 0)	dt(t, df)
CDF: Returns probability $P(t \leq t_0)$	T.DIST(t_0, df, 1)	pt(t_0, df)
Inverse CDF: Returns t_0 for $P(t \leq t_0) = \alpha$	T.INV(α, df)	qt(α, df)
…Random data (makes k samples in R)	T.INV(RAND(), df)	rt(k, df)
F distribution		
PDF: Returns height of $f(x)$	F.DIST(x, df_1, df_2, 0)	df(x, df_1, df_2)
CDF: Returns probability $P(X \leq x)$	F.DIST(x, df_1, df_2, 1)	pf(x, df_1, df_2)
Inverse CDF: Returns F_0 for $P(F \leq F_0) = \alpha$	F.INV(α, df_1, df_2)	qf(α, df_1, df_2)
…Random data (makes k samples in R)	F.INV(RAND(), df_1, df_2)	rf(k, df_1, df_2)

*In probability distributions, we use x to denote a *single* data value.

Common Hypothesis Tests	*Excel Function*	*R Function*
Normal distribution[1]		
Left-tailed p-value for test statistic z_{calc}	NORM.S.DIST(z_{calc}, 1)	pnorm(z_{calc} ,,, 1)
Right-tailed p-value for test statistic z_{calc}	1-NORM.S.DIST(z_{calc}, 1)	pnorm(z_{calc} ,,, 0)
Two-tailed p-value for test statistic z_{calc}	2*(1-NORM.S.DIST($|z_{calc}|$, 1))	2*pnorm($|z_{calc}|$,,, 0)
Critical z value for left-tailed test at α	NORM.S.INV(α)	qnorm(α)
Critical z value for right-tailed test at α	NORM.S.INV($1-\alpha$)	qnorm($1-\alpha$)
Critical z values for two-tailed test at α	$\pm$NORM.S.INV($\alpha/2$)	$\pm$qnorm($\alpha/2$)
Student's t distribution[1]		
Left-tailed p-value for test statistic t_{calc}	T.DIST(t_{calc}, df, 1)	pt(t_{calc}, df ,, 1)
Right-tailed p-value for test statistic t_{calc}	T.DIST.RT(t_{calc}, df)	pt(t_{calc}, df ,, 0)
Two-tailed p-value for test statistic t_{calc}	T.DIST.2T($|t_{calc}|$, df)	2*pt($|t_{calc}|$, df ,, 0)
Critical value of t_α for left-tailed test at α	T.INV(α, df)	qt(α, df)
Critical value of t_α for right-tailed test at α	T.INV($1-\alpha$, df)	qt($1-\alpha$, df)
Critical values of $t_{\alpha/2}$ for two-tailed test at α	$\pm$T.INV.2T(α, df)	$\pm$qt($\alpha/2$)
F distribution		
Left-tailed p-value for test statistic F_{calc}	F.DIST(F_{calc}, df_1, df_2, 1)	pf(F_{calc}, df_1, df_2 ,, 1)
Right-tailed p-value for test statistic F_{calc}	F.DIST.RT(F_{calc}, df_1, df_2)	pf(F_{calc}, df_1, df_2 ,, 0)
Two-tailed p-value for folded[2] F_{max} test	2*F.DIST.RT(F_{max}, df_1, df_2)	2*pf(F_{max}, df_1, df_2 ,, 0)
Critical value for left-tailed test at α	F.INV(α, df_1, df_2)	qf(α, df_1, df_2 ,, 1)
Critical value for right-tailed test at α	F.INV($1-\alpha$, df_1, df_2)	qf(α, df_1, df_2 ,, 0)
Critical value for folded[2] F test at α	F.INV.RT($\alpha/2$, df_1, df_2)	$\pm$qf($\alpha/2$, df_1, df_2 ,, 0)
Chi-square distribution		
Left-tailed p-value for test statistic χ^2_{calc}	CHISQ.DIST(χ^2_{calc}, df, 1)	pchisq(χ^2_{calc}, df ,, 1)
Right-tailed p-value for test statistic χ^2_{calc}	CHISQ.DIST.RT(χ^2_{calc}, df)	pchisq(χ^2_{calc}, df ,, 0)
Two-tailed p-value for test statistic χ^2_{calc}	2*CHISQ.DIST.RT(χ^2_{calc}, df)	2*pchisq(χ^2_{calc}, df ,, 0)
Critical value for left-tailed test at α	CHISQ.INV(α, df)	qchisq(α, df ,, 1)
Critical value for right-tailed test at α	CHISQ.INV.RT(α, df)	qchisq(α, df ,, 0)
Critical value for two-tailed test at α	CHISQ.INV.2T(α, df)	$\pm$qchisq($\alpha/2$, df ,, 0)

[1]In two-tailed z and t tests, we use absolute values $|z_{calc}|$ and $|t_{calc}|$ of the test statistic and $\pm$ to indicate that the left-tail and right-tail critical values are the same except for sign. For simplicity, some optional arguments have been omitted for R functions (check R help to see all available options).

[2]The folded F test forces a right-tailed test using $F_{max} = s^2_{max}/s^2_{min}$ with df_1 and df_2 reversed if necessary.

*Hypothesis Test Calculations**	*Excel Function*	*R Function**
t-test of two means for data arrays x and y. Excel returns a p-value to test for zero difference of means. R returns the test statistic and p-value, allows testing for a difference other than zero (default is zero), states the alternative hypothesis, and returns a confidence interval for the difference of means. For a right-tailed test in Excel, the larger mean must come first to agree with R's "greater" tail option.	T.TEST(x, y, Tails, Type) where Tails = 1 or 2 Type = 1, 2, or 3 where 1 = paired (must have $n_1 = n_2$) 2 = equal variances assumed 3 = unequal variances assumed Example: left tail, not paired, equal variances assumed T.TEST(x, y, 1, 2)	t.test(x, y, Tails, Mu, Paired, VarEq, Conf) where Tails = "two.sided," "less," or "greater" Mu = hypothesized mean difference (default is 0) VarEq = 0 (unequal variances assumed) or 1 (variances assumed equal) Conf = confidence (.95 default) Example: left tail, equal variances, 90% confidence t.test(x, y, "less", 0, 0, 1, .90)
F-test of two variances for two data arrays x and y. Excel returns a two-tailed p-value for equality of variances (ratio of 1). R allows testing for a ratio other than 1 (default is 1), returns a p-value, states the alternative hypothesis, gives a confidence interval for the ratio of variances, and offers choice of confidence level (default is 95%).	F.TEST(x, y) Excel simply returns the p-value for a two-tailed test.	var.test(x, y, Ratio, Tails, Conf) where Ratio = hypothesized ratio of variances (default is 1) Tails = "two.sided," "less," or "greater" Conf = confidence (.95 default) Example: right-tailed test for ratio of 2 at 90% confidence var.test(x, y, "greater", 2, .90)

Hypothesis Test Calculations*	Excel Function	R Function*
Chi-square test on frequencies. Both Excel and R return right-tailed p-values for a test on observed frequencies (x) and expected frequencies (y). Excel expects you to provide expected frequencies, while R will calculate them.	CHISQ.TEST(x,y) performs a goodness-of-fit test with $df = k - 1$ if x and y are vectors of length k or $df = (r - 1)(c - 1)$ if x and y are $r \times c$ arrays in contingency table format.	chisq.test(x) performs a uniform goodness-of-fit test with $df = k - 1$ if x is a vector of length k. chisq.test($x, p = y$) performs a multinomial goodness-of-fit test if you supply the k probabilities in y (must sum to 1). chisq.test(M) performs a test for independence with $df = (r - 1)(c - 1)$ for a matrix M of observed frequencies.

*In most formulas, x and y denote vectors of data values (in Excel x and y would typically be data columns).

Other Useful Stats Functions	Excel Function*	R Function
Rank (average for ties)	RANK.AVG(x, Data, k) where x is a cell in array Data, $k = 0$ (descending), $k = 1$ (ascending)	rank(x,,"average") where x is a vector of data (returns entire ranked data array)
Rank (no correction for ties)	RANK(x, Data, k) where x is a cell reference in array Data, $k = 0$ (descending), $k = 1$ (ascending)	rank(x,,"first") where x is a vector of data (returns entire ranked data array)
Random uniform ($0 < x < 1$)	RAND()	runif(n) to create n values
Random integer ($a \leq x \leq b$)	RANDBETWEEN(a, b)	sample($a{:}b$, 1, 1)
Sample of n random integers ($a \leq x \leq b$) with replacement	----------------	sample($a{:}b, n$, 1)
Sample of n random integers ($a \leq x \leq b$) without replacement	----------------	sample($a{:}b, n$, 0)
Confidence interval half-width $\pm z\, \sigma/\sqrt{(n)}$ using normal distribution with known standard deviation σ and confidence $1 - a$	CONFIDENCE.NORM(a, σ, n)	----------------
Sum of squares of an array of data values around their mean	DEVSQ(x)	sum((x-mean(x))^2)
Frequency of items in a data array x	{FREQUENCY(x, Bins)} where Bins is an array of upper limits for classes. This is an *array function*. Highlight output range, type function, and then use Ctrl-Shift-Enter to execute.	Default bins (Sturges' Rule): r = hist(x) or choose bin breaks (cutpoints) in vector y: r = hist(x, breaks=y) Then print results from r: r$breaks r$count

*For more information, see www.excelfunctions.net/excel-functions-list.html.

K Using R and RStudio

Introduction to R and RStudio

R is a free software environment for statistical computing and graphics created by the R Foundation (a not-for-profit organization) that runs on UNIX, Windows, and MacOS. RStudio is an integrated development environment (IDE) for R. It includes a console, an editor for direct code execution, and tools for plotting, history, debugging, and data management. RStudio is available in open source and commercial editions (students would likely choose RStudio Desktop). R and RStudio together provide a versatile programming language and an integrated suite of well-documented (www.r-project.org/other-docs.html) functions and procedures for data manipulation, calculation, and graphical display. **Appendix J** lists useful R functions and procedures alongside the corresponding Excel functions and procedures, using the notation in this textbook.

Installing R

You must download and install R before installing RStudio. Follow these steps:

- Step 1: Visit the R project website (www.r-project.org).
- Step 2: Choose a CRAN mirror site near you (https://cran.r-project.org/mirrors.html).
- Step 3: Download the version of R for your system (Linux, MacOS X, or Windows).
- Step 4: Select Install R for the first time.
- Step 5: Follow the setup instructions to install. Look for the .EXE file in your Downloads (Windows) or .PKG file (Mac).

Installing RStudio

After R has been successfully installed, follow these steps to download and install RStudio:

- Step 1: Install R.
- Step 2: Visit the Rstudio website (https://rstudio.com/products/rstudio/).
- Step 3: Choose Rstudio Desktop Free.
- Step 4: Click Download R Studio Desktop for your system (Linux, Mac OS X, or Windows).
- Step 5: Select Run or Open and follow the instructions to install. Look for the .EXE file in your Downloads (Windows) or .PKG file (Mac).

RStudio Interface

When you first run RStudio, you will see three panes. In the Console pane on the left, you can type R code and execute it by pressing Enter. The Environment pane on the upper right keeps track of objects that you create. In the lower right pane, the Files and Plots tabs will show the files and graphs that you have created, the Packages tab will list available R procedures, and the Help tab will let you search for help on any R topic (e.g., a particular function or procedure) to see its syntax, arguments, and examples.

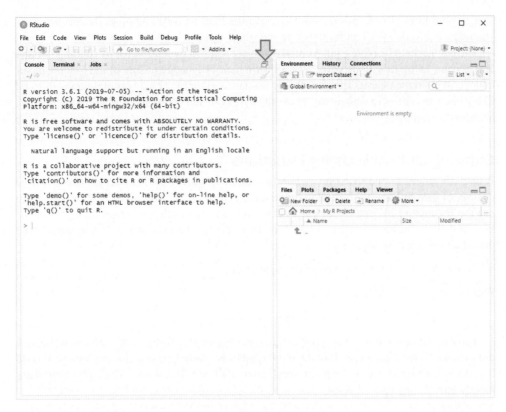

Click the "expand" symbol in the Console window (see red arrow) to add a fourth pane called the Source pane that will appear in the upper left.

Unlike code entered in the Console, code entered in the Source pane will not be executed until you press the Run arrow →Run (circled above). The result will appear in the Console pane (lower left) or in the lower right pane if it is a Plot (i.e., a graph). This can be useful when you have several lines of code to execute. Highlight the lines of code and press the Run arrow. If you make a mistake, you can fix the code and run it again.

You can get help by typing ? in front of an object name in the Console (or by running it in the Source pane, as we have done with ?sample shown here). If you are unsure of the exact name of a function but remember its first few characters and start typing, R may prompt you with some likely choices. The help screen will show the syntax of the command and all its arguments, along with examples.

Entering Data and Using Functions

R data are stored in one or more *vectors* (a vector is an array of data values, like a row or column of data in Excel). You can assign data to a vector in several ways. The simplest is to use the R function c(). Enter your data values inside the parentheses, separated by commas. For example, to make a vector named x containing the three numbers 5, 6, 7, you would type the following after the > prompt:

```
> x = c(5, 6, 7)     #to create a vector with three data values
```

Equivalently, you could have typed:

```
> x <− c(5, 6, 7)    #to create a vector with three data values
```

You can use either = or <− to assign values to a vector. It is a matter of preference (we will use = in our examples). Text following the # symbol is ignored, so you can annotate your code if you wish. This is useful to help you recall later what you were doing. To display your data, merely type the name of a vector:

```
> x
[1] 5 6 7
```

The square bracket [1] is the index of the first element of the vector that follows on that line. In the preceding example, x is a 3-element vector, and the [1] indicates that 5 is the first index. If the vector was much longer—say, 40 elements—R would need multiple lines to display it. The square bracket on the second line would indicate the index of the first element on the second line. For example, here are the integers from 500 to 539:

```
> c(500:539)
 [1] 500 501 502 503 504 505 506 507 508 509 510  511 512 513 514 515 516 517 518 519
[21] 520 521 522 523 524 525 526 527 528 529 530 531 532 533 534 535 536 537 538 539
```

You can perform simple statistical functions on a vector. For example,

```
> sum(x)        #calculate the sum of our data values
[1] 18
> mean(x)       #calculate an average of our data values
[1] 6
```

You can name a vector using any combination of letters, digits, period (.), and underscore (_). It must start with a letter or a period. If it starts with a period, it cannot be followed by a digit. Reserved words in R cannot be used as identifiers. Case is important (hence BodyStyle is not the same object as bodystyle). A vector can contain attribute data (nonnumerical) by enclosing each data value in quotation marks. For example:

```
> BodyStyle = c("Sedan", "SUV", "Truck", "Other")
> BodyStyle
[1] "Sedan" "SUV" "Truck" "Other"
> bodystyle
Error: object 'bodystyle' not found
```

R may not recognize a named vector if we are careless about uppercase and lowercase letters (as we deliberately did here to illustrate the point).

You can create data values in other ways, using R functions or rules. For example, let's make a vector that we call MyIntegers containing the integers from 1 to 10:

```
> MyIntegers = c(1:10)
> MyIntegers
[1] 1 2 3 4 5 6 7 8 9 10
> mean(MyIntegers)
[1] 5.5
```

Another way to make some data is using a function. For example, to create a vector (we'll call it MySample) containing a sample of 5 random integers in the range 0 to 1,000, we can use the R function sample():

```
> MySample = sample(0:1000,5)
> MySample
[1] 521 989 766 300 922
```

R functions have arguments that allow various options. Some arguments can be omitted if you are willing to accept the default values. We have used a "stripped-down" version of the sample() function here for clarity. You can type ? followed by a function name to see its syntax and arguments in the lower right pane under the Help tab (look at the R screen above). As you start typing a function, R may prompt you with suggestions (it tries to anticipate what you are looking for).

Importing Data from a Spreadsheet

How can we import data from Excel? We will illustrate three methods. The first is simple— just use the clipboard. To illustrate, here is an Excel spreadsheet (50 rows, 13 columns) that contains data for a sample of vehicles for the 2020 model year. For brevity, only the first three and last three rows are displayed:

	A	B	C	D	E	F	G	H	I	J	K	L	M
1	Obs	Brand and Model*	Style	Drive	Doors	HP	Engine	CityMPG	HwyMPG	Weight	Length	Width	Height
2	1	Acura RDX SH SWD 4dr SUV	SUV	AWD	4	272	2000	21	27	4026	186.7	74.8	65.7
3	2	Acura TLX sedan	Sedan	FWD	4	206	2400	23	33	3505	190.7	73.0	57.0
4	3	Audi V10 2dr quattro convertible	Other	AWD	2	602	5200	13	20	3572	174.3	76.4	49.0
	...	...		...	...	...	...	...	...	...	...	...	...
49	48	Toyota Yaris L 4dr sedan	Sedan	FWD	4	106	1500	30	39	2385	171.2	66.7	58.5
50	49	Volvo XC40 T5 4dr SUV	SUV	AWD	4	248	2000	22	30	3756	174.2	75.2	65.3
51	50	VW Tiguan SE R-Line 4dr SUV	SUV	FWD	4	184	2000	22	29	3757	185.1	72.4	66.3

Suppose we want to import only data columns C through M into a data frame in R (we choose to skip columns A and B). A data frame is similar to a matrix except that its columns can have names. That is appropriate in our illustration because each of our columns already has a caption. Let's call our R data frame VehicleData. An easy way to import Excel data to R is to highlight the data block in Excel (including the column headings) and copy it to the clipboard (Ctrl+C in Windows, command+C in MacOS). Then enter the following R instruction (for Windows):

```
> VehicleData = read.table(file="clipboard", sep="\t", header=TRUE)
```

For MacOS, you would enter

```
> VehicleData = read.table(pipe("pbpaste", sep="\t", header=TRUE)
```

The argument sep="\t" tells R that the data columns are tab-separated. The argument header=TRUE tells R to expect text headings on each column. Once the data are imported to data frame VehicleData, we can view the data, calculate statistics, and so on. For example, to calculate average city miles per gallon for our 50 vehicles (the column named CityMPG in the imported data), we use the function mean() and name the column in the data frame followed by $ (the dollar symbol) as follows:

```
> mean(VehicleData$CityMPG)
[1] 21.98
```

If our data frame columns have names (as they do here), we can get summary statistics for variables of interest using the summary() command. For example:

```
> summary(VehicleData[c("Weight","Length")])
```

Weight		Length	
Min.	:2385	Min.	:151.1
1st Qu.	:3356	1st Qu.	:181.9
Median	:3662	Median	:192.2
Mean	:3954	Mean	:190.9
3rd Qu.	:4661	3rd Qu.	:198.7
Max.	:5917	Max.	:231.9

Be careful of the syntax (e.g., the square brackets [ ]). You can also refer to columns in your data frame by number. You can use a colon : if your columns are contiguous (e.g., 6 through 8):

```
> summary(VehicleData[c(6:8)])
```

HP		Engine		CityMPG	
Min.	:106.0	Min.	:1400	Min.	:13.00
1st Qu.	:182.5	1st Qu.	:2000	1st Qu.	:18.00
Median	:249.0	Median	:2500	Median	:22.00
Mean	:262.3	Mean	:2898	Mean	:21.98
3rd Qu.	:308.0	3rd Qu.	:3500	3rd Qu.	:26.00
Max.	:602.0	Max.	:6200	Max.	:31.00

If the columns are not contiguous (e.g., 6, 8, and 11), use commas as separators:

```
> summary(VehicleData[c(6,8,11)])
```

HP		CityMPG		Length	
Min.	:106.0	Min.	:13.00	Min.	:151.1
1st Qu.	:182.5	1st Qu.	:18.00	1st Qu.	:181.9
Median	:249.0	Median	:22.00	Median	:192.2
Mean	:262.3	Mean	:21.98	Mean	:190.9
3rd Qu.	:308.0	3rd Qu.	:26.00	3rd Qu.	:198.7
Max.	:602.0	Max.	:31.00	Max.	:231.9

Importing an Excel File

To import an *entire* Excel spreadsheet, we choose File > Import Dataset > From Excel in the upper left corner of the Scripts pane:

Depending on your R version, you may be asked to download an update:

Our Excel file is named MPGData.xlsx. When prompted, click the Browse button to locate MPGData.xlsx and then click the Import button. After importing MPGData.xlsx, we see a new tab named MPGData (same as our Excel file name) in the Scripts window. In effect, R has created a new object (like a data frame) called a tibble.

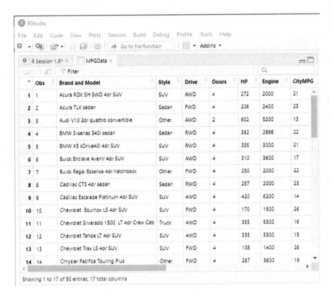

Here is what the CRAN project says:

> **Tibbles** are a modern take on data frames. They keep the features that have stood the test of time, and drop the features that used to be convenient but are now frustrating. . . .

Actions on the object MPGData behave similarly to those on a data frame. For example:

```
> summary(MPGData[c("Weight","Length")])
    Weight              Length
Min.    :2385      Min.    :151.1
1st Qu. :3356      1st Qu. :181.9
Median  :3662      Median  :192.2
Mean    :3954      Mean    :190.9
3rd Qu. :4661      3rd Qu. :198.7
Max.    :5917      Max.    :231.9
```

Importing an Excel file works fine if your spreadsheet contains *only* your desired data in contiguous columns. However, it's unpredictable if your Excel spreadsheet has comments, pictures, or blank lines for spacing. We provide this textbook's Excel files in a "clean" format for import into R.

Importing a CSV File

A third way to import an Excel spreadsheet is to save it in .csv format (an abbreviation for "comma-separated format," which Excel calls "comma delimited"). For example, saving our MPGData would look like this:

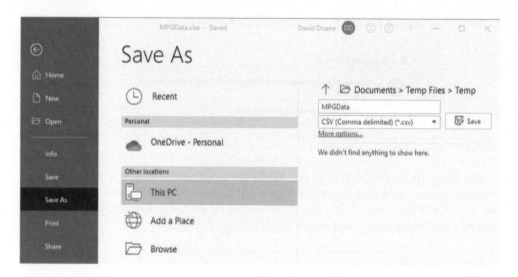

After saving your file in a convenient folder, use the read.csv() command in R to put the data into a data frame (we decided to name it VehicleData2). This can be tricky because R expects a forward slash / where Windows would use a back slash \ in the folder path. Here is how it looks for our example:

> VehicleData2=read.csv("C:/Users/David Doane/Documents/Temp/MPGData.csv")

Finding the exact folder name in Windows can be difficult. When you click on the file name in Windows Explorer, you could click in the search box above to get its full path name, including the working directory. Here is how it might look:

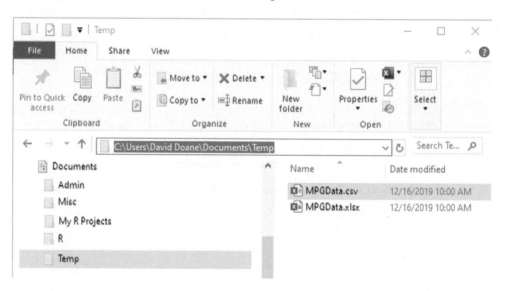

A simpler way to find your working directory is to enter getwd() in R. In our example, this is how it would look:

> getwd()
[1] "C:/Users/David Doane/Documents"

Notice that R supplies the forward slash / that it prefers in the working directory. Then you can just append the file name MPGData.csv as shown previously.

As with .xlsx files, importing a .csv file works well if your spreadsheet contains *only* your desired data in contiguous columns. But it's unpredictable if your Excel spreadsheet has comments, pictures, or blank lines for spacing. Finding the file and folder path name and entering it correctly are also a challenges if your computer has many folders (as yours probably does). The two options explained earlier (cut and paste to clipboard or .xlsx import) may seem easier.

Caution

To avoid import problems, avoid Excel column headings that begin with a number (e.g., "5 to 10") or symbols (e.g., comma or $ or %). R will replace spaces in imported column headings with a period. This is important because business spreadsheets often use the $ symbol and comma format (e.g., $24,216). For this reason, our textbook examples and exercise data sets for R are available in Connect in an R-compatible format.

Keeping Track of Files

The History pane (lower right) has a tab named Environment. Click on this tab to see all the variables you have created along with their characteristics. For example:

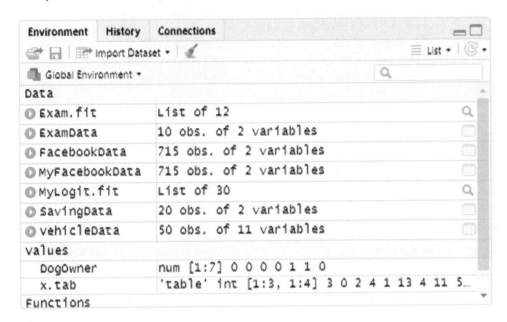

Useful R Procedures

Here are a few common procedures that you are likely to encounter in an introductory statistics class. Explore them (syntax, arguments) on your own. Use **Appendix J** as a guide to other procedures.

```
> summary(x)    # basic statistics for one or more variables
> plot(x,y)     # make a scatter plot of variable y against variable x
> hist(x)       # make a histogram of a variable x
> boxplot(x)    # make a box plot of one or more variables
> t.test(x,y)   # Student's t test for variables x and y
> lm(y~x)       # regress y on x
> lm(y,x1+x2)   # regress y on x1 and x2 (or more x's)
```

Useful R Tips

To save a script that you have developed, click File > Save As in the Script pane to name a folder for your own files (e.g., MyRScripts). Instructors may suggest downloading specific CRAN packages that can make some R tasks a bit easier. To see what packages are already available, click the Packages tab in lower right pane, or search the CRAN website. Specific R topics will be illustrated in the context of chapters in this textbook. Use **Appendix J** as a reference source for common functions and procedures. Here are a few miscellaneous useful commands:

```
rm()        #to remove a variable
ls()        #to see all the variables you have created
Ctrl-L      #to clear the Console (in Windows)
Option-L    #to clear the Console (in MacOS)
```

Index

STANDARD NORMAL AREAS

This table shows the normal area between 0 and z. Example: $P(0 \le z \le 1.96) = .4750$

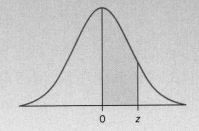

z	.00	.01	.02	.03	.04	.05	.06	.07	.08	.09
0.0	.0000	.0040	.0080	.0120	.0160	.0199	.0239	.0279	.0319	.0359
0.1	.0398	.0438	.0478	.0517	.0557	.0596	.0636	.0675	.0714	.0753
0.2	.0793	.0832	.0871	.0910	.0948	.0987	.1026	.1064	.1103	.1141
0.3	.1179	.1217	.1255	.1293	.1331	.1368	.1406	.1443	.1480	.1517
0.4	.1554	.1591	.1628	.1664	.1700	.1736	.1772	.1808	.1844	.1879
0.5	.1915	.1950	.1985	.2019	.2054	.2088	.2123	.2157	.2190	.2224
0.6	.2257	.2291	.2324	.2357	.2389	.2422	.2454	.2486	.2517	.2549
0.7	.2580	.2611	.2642	.2673	.2704	.2734	.2764	.2794	.2823	.2852
0.8	.2881	.2910	.2939	.2967	.2995	.3023	.3051	.3078	.3106	.3133
0.9	.3159	.3186	.3212	.3238	.3264	.3289	.3315	.3340	.3365	.3389
1.0	.3413	.3438	.3461	.3485	.3508	.3531	.3554	.3577	.3599	.3621
1.1	.3643	.3665	.3686	.3708	.3729	.3749	.3770	.3790	.3810	.3830
1.2	.3849	.3869	.3888	.3907	.3925	.3944	.3962	.3980	.3997	.4015
1.3	.4032	.4049	.4066	.4082	.4099	.4115	.4131	.4147	.4162	.4177
1.4	.4192	.4207	.4222	.4236	.4251	.4265	.4279	.4292	.4306	.4319
1.5	.4332	.4345	.4357	.4370	.4382	.4394	.4406	.4418	.4429	.4441
1.6	.4452	.4463	.4474	.4484	.4495	.4505	.4515	.4525	.4535	.4545
1.7	.4554	.4564	.4573	.4582	.4591	.4599	.4608	.4616	.4625	.4633
1.8	.4641	.4649	.4656	.4664	.4671	.4678	.4686	.4693	.4699	.4706
1.9	.4713	.4719	.4726	.4732	.4738	.4744	.4750	.4756	.4761	.4767
2.0	.4772	.4778	.4783	.4788	.4793	.4798	.4803	.4808	.4812	.4817
2.1	.4821	.4826	.4830	.4834	.4838	.4842	.4846	.4850	.4854	.4857
2.2	.4861	.4864	.4868	.4871	.4875	.4878	.4881	.4884	.4887	.4890
2.3	.4893	.4896	.4898	.4901	.4904	.4906	.4909	.4911	.4913	.4916
2.4	.4918	.4920	.4922	.4925	.4927	.4929	.4931	.4932	.4934	.4936
2.5	.4938	.4940	.4941	.4943	.4945	.4946	.4948	.4949	.4951	.4952
2.6	.4953	.4955	.4956	.4957	.4959	.4960	.4961	.4962	.4963	.4964
2.7	.4965	.4966	.4967	.4968	.4969	.4970	.4971	.4972	.4973	.4974
2.8	.4974	.4975	.4976	.4977	.4977	.4978	.4979	.4979	.4980	.4981
2.9	.4981	.4982	.4982	.4983	.4984	.4984	.4985	.4985	.4986	.4986
3.0	.49865	.49869	.49874	.49878	.49882	.49886	.49889	.49893	.49896	.49900
3.1	.49903	.49906	.49910	.49913	.49916	.49918	.49921	.49924	.49926	.49929
3.2	.49931	.49934	.49936	.49938	.49940	.49942	.49944	.49946	.49948	.49950
3.3	.49952	.49953	.49955	.49957	.49958	.49960	.49961	.49962	.49964	.49965
3.4	.49966	.49968	.49969	.49970	.49971	.49972	.49973	.49974	.49975	.49976
3.5	.49977	.49978	.49978	.49979	.49980	.49981	.49981	.49982	.49983	.49983
3.6	.49984	.49985	.49985	.49986	.49986	.49987	.49987	.49988	.49988	.49989
3.7	.49989	.49990	.49990	.49990	.49991	.49991	.49992	.49992	.49992	.49992

CUMULATIVE STANDARD NORMAL DISTRIBUTION

This table shows the normal area less than z. Example: $P(z \leq -1.96) = .0250$

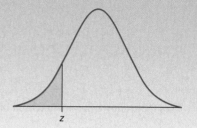

z	.00	.01	.02	.03	.04	.05	.06	.07	.08	.09
−3.7	.00011	.00010	.00010	.00010	.00009	.00009	.00008	.00008	.00008	.00008
−3.6	.00016	.00015	.00015	.00014	.00014	.00013	.00013	.00012	.00012	.00011
−3.5	.00023	.00022	.00022	.00021	.00020	.00019	.00019	.00018	.00017	.00017
−3.4	.00034	.00032	.00031	.00030	.00029	.00028	.00027	.00026	.00025	.00024
−3.3	.00048	.00047	.00045	.00043	.00042	.00040	.00039	.00038	.00036	.00035
−3.2	.00069	.00066	.00064	.00062	.00060	.00058	.00056	.00054	.00052	.00050
−3.1	.00097	.00094	.00090	.00087	.00084	.00082	.00079	.00076	.00074	.00071
−3.0	.00135	.00131	.00126	.00122	.00118	.00114	.00111	.00107	.00104	.00100
−2.9	.0019	.0018	.0018	.0017	.0016	.0016	.0015	.0015	.0014	.0014
−2.8	.0026	.0025	.0024	.0023	.0023	.0022	.0021	.0021	.0020	.0019
−2.7	.0035	.0034	.0033	.0032	.0031	.0030	.0029	.0028	.0027	.0026
−2.6	.0047	.0045	.0044	.0043	.0041	.0040	.0039	.0038	.0037	.0036
−2.5	.0062	.0060	.0059	.0057	.0055	.0054	.0052	.0051	.0049	.0048
−2.4	.0082	.0080	.0078	.0075	.0073	.0071	.0069	.0068	.0066	.0064
−2.3	.0107	.0104	.0102	.0099	.0096	.0094	.0091	.0089	.0087	.0084
−2.2	.0139	.0136	.0132	.0129	.0125	.0122	.0119	.0116	.0113	.0110
−2.1	.0179	.0174	.0170	.0166	.0162	.0158	.0154	.0150	.0146	.0143
−2.0	.0228	.0222	.0217	.0212	.0207	.0202	.0197	.0192	.0188	.0183
−1.9	.0287	.0281	.0274	.0268	.0262	.0256	.0250	.0244	.0239	.0233
−1.8	.0359	.0351	.0344	.0336	.0329	.0322	.0314	.0307	.0301	.0294
−1.7	.0446	.0436	.0427	.0418	.0409	.0401	.0392	.0384	.0375	.0367
−1.6	.0548	.0537	.0526	.0516	.0505	.0495	.0485	.0475	.0465	.0455
−1.5	.0668	.0655	.0643	.0630	.0618	.0606	.0594	.0582	.0571	.0559
−1.4	.0808	.0793	.0778	.0764	.0749	.0735	.0721	.0708	.0694	.0681
−1.3	.0968	.0951	.0934	.0918	.0901	.0885	.0869	.0853	.0838	.0823
−1.2	.1151	.1131	.1112	.1093	.1075	.1056	.1038	.1020	.1003	.0985
−1.1	.1357	.1335	.1314	.1292	.1271	.1251	.1230	.1210	.1190	.1170
−1.0	.1587	.1562	.1539	.1515	.1492	.1469	.1446	.1423	.1401	.1379
−0.9	.1841	.1814	.1788	.1762	.1736	.1711	.1685	.1660	.1635	.1611
−0.8	.2119	.2090	.2061	.2033	.2005	.1977	.1949	.1922	.1894	.1867
−0.7	.2420	.2389	.2358	.2327	.2296	.2266	.2236	.2206	.2177	.2148
−0.6	.2743	.2709	.2676	.2643	.2611	.2578	.2546	.2514	.2483	.2451
−0.5	.3085	.3050	.3015	.2981	.2946	.2912	.2877	.2843	.2810	.2776
−0.4	.3446	.3409	.3372	.3336	.3300	.3264	.3228	.3192	.3156	.3121
−0.3	.3821	.3783	.3745	.3707	.3669	.3632	.3594	.3557	.3520	.3483
−0.2	.4207	.4168	.4129	.4090	.4052	.4013	.3974	.3936	.3897	.3859
−0.1	.4602	.4562	.4522	.4483	.4443	.4404	.4364	.4325	.4286	.4247
−0.0	.5000	.4960	.4920	.4880	.4841	.4801	.4761	.4721	.4681	.4641

This table shows the normal area less than z. Example: $P(z \le 1.96) = \mathbf{.9750}$

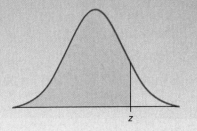

z	.00	.01	.02	.03	.04	.05	.06	.07	.08	.09
0.0	.5000	.5040	.5080	.5120	.5160	.5199	.5239	.5279	.5319	.5359
0.1	.5398	.5438	.5478	.5517	.5557	.5596	.5636	.5675	.5714	.5753
0.2	.5793	.5832	.5871	.5910	.5948	.5987	.6026	.6064	.6103	.6141
0.3	.6179	.6217	.6255	.6293	.6331	.6368	.6406	.6443	.6480	.6517
0.4	.6554	.6591	.6628	.6664	.6700	.6736	.6772	.6808	.6844	.6879
0.5	.6915	.6950	.6985	.7019	.7054	.7088	.7123	.7157	.7190	.7224
0.6	.7257	.7291	.7324	.7357	.7389	.7422	.7454	.7486	.7517	.7549
0.7	.7580	.7611	.7642	.7673	.7704	.7734	.7764	.7794	.7823	.7852
0.8	.7881	.7910	.7939	.7967	.7995	.8023	.8051	.8078	.8106	.8133
0.9	.8159	.8186	.8212	.8238	.8264	.8289	.8315	.8340	.8365	.8389
1.0	.8413	.8438	.8461	.8485	.8508	.8531	.8554	.8577	.8599	.8621
1.1	.8643	.8665	.8686	.8708	.8729	.8749	.8770	.8790	.8810	.8830
1.2	.8849	.8869	.8888	.8907	.8925	.8944	.8962	.8980	.8997	.9015
1.3	.9032	.9049	.9066	.9082	.9099	.9115	.9131	.9147	.9162	.9177
1.4	.9192	.9207	.9222	.9236	.9251	.9265	.9279	.9292	.9306	.9319
1.5	.9332	.9345	.9357	.9370	.9382	.9394	.9406	.9418	.9429	.9441
1.6	.9452	.9463	.9474	.9484	.9495	.9505	.9515	.9525	.9535	.9545
1.7	.9554	.9564	.9573	.9582	.9591	.9599	.9608	.9616	.9625	.9633
1.8	.9641	.9649	.9656	.9664	.9671	.9678	.9686	.9693	.9699	.9706
1.9	.9713	.9719	.9726	.9732	.9738	.9744	**.9750**	.9756	.9761	.9767
2.0	.9772	.9778	.9783	.9788	.9793	.9798	.9803	.9808	.9812	.9817
2.1	.9821	.9826	.9830	.9834	.9838	.9842	.9846	.9850	.9854	.9857
2.2	.9861	.9864	.9868	.9871	.9875	.9878	.9881	.9884	.9887	.9890
2.3	.9893	.9896	.9898	.9901	.9904	.9906	.9909	.9911	.9913	.9916
2.4	.9918	.9920	.9922	.9925	.9927	.9929	.9931	.9932	.9934	.9936
2.5	.9938	.9940	.9941	.9943	.9945	.9946	.9948	.9949	.9951	.9952
2.6	.9953	.9955	.9956	.9957	.9959	.9960	.9961	.9962	.9963	.9964
2.7	.9965	.9966	.9967	.9968	.9969	.9970	.9971	.9972	.9973	.9974
2.8	.9974	.9975	.9976	.9977	.9977	.9978	.9979	.9979	.9980	.9981
2.9	.9981	.9982	.9982	.9983	.9984	.9984	.9985	.9985	.9986	.9986
3.0	.99865	.99869	.99874	.99878	.99882	.99886	.99889	.99893	.99896	.99900
3.1	.99903	.99906	.99910	.99913	.99916	.99918	.99921	.99924	.99926	.99929
3.2	.99931	.99934	.99936	.99938	.99940	.99942	.99944	.99946	.99948	.99950
3.3	.99952	.99953	.99955	.99957	.99958	.99960	.99961	.99962	.99964	.99965
3.4	.99966	.99968	.99969	.99970	.99971	.99972	.99973	.99974	.99975	.99976
3.5	.99977	.99978	.99978	.99979	.99980	.99981	.99981	.99982	.99983	.99983
3.6	.99984	.99985	.99985	.99986	.99986	.99987	.99987	.99988	.99988	.99989
3.7	.99989	.99990	.99990	.99990	.99991	.99991	.99992	.99992	.99992	.99992

STUDENT'S *t* CRITICAL VALUES

This table shows the *t*-value that defines the area for the stated degrees of freedom (*d.f.*).

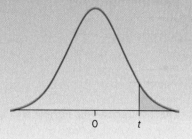

	Confidence Level						Confidence Level				
	.80	.90	.95	.98	.99		.80	.90	.95	.98	.99
	Significance Level for Two-Tailed Test						Significance Level for Two-Tailed Test				
	.20	.10	.05	.02	.01		.20	.10	.05	.02	.01
	Significance Level for One-Tailed Test						Significance Level for One-Tailed Test				
d.f.	.10	.05	.025	.01	.005	d.f.	.10	.05	.025	.01	.005
1	3.078	6.314	12.706	31.821	63.657	36	1.306	1.688	2.028	2.434	2.719
2	1.886	2.920	4.303	6.965	9.925	37	1.305	1.687	2.026	2.431	2.715
3	1.638	2.353	3.182	4.541	5.841	38	1.304	1.686	2.024	2.429	2.712
4	1.533	2.132	2.776	3.747	4.604	39	1.304	1.685	2.023	2.426	2.708
5	1.476	2.015	2.571	3.365	4.032	40	1.303	1.684	2.021	2.423	2.704
6	1.440	1.943	2.447	3.143	3.707	41	1.303	1.683	2.020	2.421	2.701
7	1.415	1.895	2.365	2.998	3.499	42	1.302	1.682	2.018	2.418	2.698
8	1.397	1.860	2.306	2.896	3.355	43	1.302	1.681	2.017	2.416	2.695
9	1.383	1.833	2.262	2.821	3.250	44	1.301	1.680	2.015	2.414	2.692
10	1.372	1.812	2.228	2.764	3.169	45	1.301	1.679	2.014	2.412	2.690
11	1.363	1.796	2.201	2.718	3.106	46	1.300	1.679	2.013	2.410	2.687
12	1.356	1.782	2.179	2.681	3.055	47	1.300	1.678	2.012	2.408	2.685
13	1.350	1.771	2.160	2.650	3.012	48	1.299	1.677	2.011	2.407	2.682
14	1.345	1.761	2.145	2.624	2.977	49	1.299	1.677	2.010	2.405	2.680
15	1.341	1.753	2.131	2.602	2.947	50	1.299	1.676	2.009	2.403	2.678
16	1.337	1.746	2.120	2.583	2.921	55	1.297	1.673	2.004	2.396	2.668
17	1.333	1.740	2.110	2.567	2.898	60	1.296	1.671	2.000	2.390	2.660
18	1.330	1.734	2.101	2.552	2.878	65	1.295	1.669	1.997	2.385	2.654
19	1.328	1.729	2.093	2.539	2.861	70	1.294	1.667	1.994	2.381	2.648
20	1.325	1.725	2.086	2.528	2.845	75	1.293	1.665	1.992	2.377	2.643
21	1.323	1.721	2.080	2.518	2.831	80	1.292	1.664	1.990	2.374	2.639
22	1.321	1.717	2.074	2.508	2.819	85	1.292	1.663	1.988	2.371	2.635
23	1.319	1.714	2.069	2.500	2.807	90	1.291	1.662	1.987	2.368	2.632
24	1.318	1.711	2.064	2.492	2.797	95	1.291	1.661	1.985	2.366	2.629
25	1.316	1.708	2.060	2.485	2.787	100	1.290	1.660	1.984	2.364	2.626
26	1.315	1.706	2.056	2.479	2.779	110	1.289	1.659	1.982	2.361	2.621
27	1.314	1.703	2.052	2.473	2.771	120	1.289	1.658	1.980	2.358	2.617
28	1.313	1.701	2.048	2.467	2.763	130	1.288	1.657	1.978	2.355	2.614
29	1.311	1.699	2.045	2.462	2.756	140	1.288	1.656	1.977	2.353	2.611
30	1.310	1.697	2.042	2.457	2.750	150	1.287	1.655	1.976	2.351	2.609
31	1.309	1.696	2.040	2.453	2.744	∞	1.282	1.645	1.960	2.326	2.576
32	1.309	1.694	2.037	2.449	2.738						
33	1.308	1.692	2.035	2.445	2.733						
34	1.307	1.691	2.032	2.441	2.728						
35	1.306	1.690	2.030	2.438	2.724						

Note: As *n* increases, critical values of Student's *t* approach the *z*-values in the last line of this table. A common rule of thumb is to use *z* when *n* > 30, but that is *not* conservative.